Contents

ASM Handbook®

Volume 3
Alloy Phase Diagrams

Prepared under the direction of the ASM International
Alloy Phase Diagram and Handbook Committees

Hugh Baker, Editor
Hiroaki Okamoto, Senior Technical Editor
Scott D. Henry, Manager of Handbook Development
Grace M. Davidson, Manager, Production Systems
Mary Anne Fleming, Manager, APD Publications
Linda Kacprzak, Manager of Production
Heather F. Lampman, Editorial/Production Assistant

William W. Scott, Jr., Technical Director
Robert C. Uhl, Director of Reference Publications

Editorial Assistance
Nikki D. Wheaton
Kathleen Mills

Production Assistance
Donna Sue Plickert
Steve Starr
Karen Skiba
Patricia Eland
Jeff Fenstermaker

**The Materials
Information Society**

First printing, December 1992

ASM Handbook is a collective effort involving thousands of technical specialists. It brings together in one book a wealth of information from world-wide sources to help scientists, engineers, and technicians solve current and long-range problems.

Great care is taken in the compilation and production of this Volume, but it should be made clear that NO WARRANTIES, EXPRESS OR IMPLIED, INCLUDING, WITHOUT LIMITATION, WARRANTIES OF MERCHANTABILITY OR FITNESS FOR A PARTICULAR PURPOSE, ARE GIVEN IN CONNECTION WITH THIS PUBLICATION. Although this information is believed to be accurate by ASM, ASM cannot guarantee that favorable results will be obtained from the use of this publication alone. This publication is intended for use by persons having technical skill, at their sole discretion and risk. Since the conditions of product or material use are outside of ASM's control, ASM assumes no liability or obligation in connection with any use of this information. No claim of any kind, whether as to products or information in this publication, and whether or not based on negligence, shall be greater in amount than the purchase price of this product or publication in respect of which damages are claimed. THE REMEDY HEREBY PROVIDED SHALL BE THE EXCLUSIVE AND SOLE REMEDY OF BUYER, AND IN NO EVENT SHALL EITHER PARTY BE LIABLE FOR SPECIAL, INDIRECT OR CONSEQUENTIAL DAMAGES WHETHER OR NOT CAUSED BY OR RESULTING FROM THE NEGLIGENCE OF SUCH PARTY. As with any material, evaluation of the material under end-use conditions prior to specification is essential. Therefore, specific testing under actual conditions is recommended.

Nothing contained in this book shall be construed as a grant of any right of manufacture, sale, use, or reproduction, in connection with any method, process, apparatus, product, composition, or system, whether or not covered by letters patent, copyright, or trademark, and nothing contained in this book shall be construed as a defense against any alleged infringement of letters patent, copyright, or trademark, or as a defense against liability for such infringement.

Comments, criticisms, and suggestions are invited, and should be forwarded to ASM International.

Library of Congress Cataloging-in-Publication Data

ASM International

ASM handbook.

(Revised for vol. 3)
Vols. 1-2 have title: Metals handbook.
Includes biographical references and indexes
Contents: v. 1. Properties and selection—irons, steels, and high-performance alloys—
v. 2. Properties and selection—nonferrous alloys and special-purpose—
v. 3. Alloy phase diagrams
1. Metals—Handbooks, manuals, etc.
I. ASM International. Handbook Committee
II. Metals handbook.
TA459.M43 1990 620.1'6 90-115
ISBN: 0-87170-377-7 (v.1)
0-87170-381-5 (v.3)
SAN: 204-7586

ASM International®
Materials Park, Ohio 44073-0002

Printed in the United States of America

Foreword

Phase diagrams, thermodynamic data in graphical form, are one of the basic tools of the metallurgist, materials scientist, and materials engineer. They can be used for alloy design, selection of hot-working and fabricating parameters, prediction of performance, guidance in selection of hot-working and fabricating parameters, prediction of performance, guidance in selection of heat-treating process parameters, solving performance problems, including failure analysis, and for many other purposes.

The formation of The American Society of Steel Treating, the forerunner of ASM International, was based on better understanding of heat-treating technology; this understanding was, of course, rooted in part in the proper utilization of phase diagrams. Experimental tools such as metallography were used in those early days, both to determine phase diagrams and to link the heat-treating process with the desired microstructure.

In 1978 ASM International joined with the National Bureau of Standards (now the National Institute of Standards of Technology, or NIST) in an effort to improve the reliability of phase diagrams by evaluating the existing data on a system-by-system basis. ASM raised $4 million from industry and government sources and NIST provided a similar amount of financial and in-kind support for this historic undertaking. An international effort was mounted simultaneously with similar objectives. As a result, all of the important binary systems have been evaluated, and international partners have evaluated more than 2000 ternary systems.

ASM actively participates in the Alloy Phase Diagram International Commission (APDIC), which comprises cooperative national or regional committees in 13 countries. APDIC was formed "to set overall objectives, determine priorities for alloy systems to be assessed, coordinate the assessment programs of APDIC members and associate members, establish scope and quality standards for assessment programs in other countries, and assist in the timely dissemination of the resultant phase diagram data."

The complete results of the international effort are recorded in various periodical and reference publications. However, we have continued to hear from ASM members that a summary version consisting primarily of phase diagrams should be published as an ASM Handbook for the practicing engineer. While such a Handbook could not contain all the diagrams and data, careful selection would ensure the inclusion of the most important systems, with references to other more complete sources. The present Handbook is the result of our attempts to meet these criteria and the stated need.

No reference book of this nature could be published without the contributions of literally hundreds of technical and staff workers. On behalf of ASM International, we extend our sincere thanks and appreciation to the category editors, contributors, reviewers, and staff who worked in this international effort. Thanks are also due to the ASM Alloy Phase Diagram and Handbook Committees for their guidance and support of the project.

Edward H. Kottcamp, Jr.
President
ASM International

Edward L. Langer
Managing Director
ASM International

iv

Preface

Alloy phase diagrams have long been used successfully by the scientific, engineering, and industrial communities as "road maps" to solve a variety of practical problems. It is, thus, not surprising that such diagrams have always been an important part of ASM Handbooks. The previous ASM compilation of commercially important diagrams appeared in Volume 8 of the 8th Edition of *Metals Handbook*.

Shortly after publication of the earlier volume in 1973, recognition of the universal importance of alloy phase diagrams led to the formation of several national phase diagram programs, as well as the International Programme for Alloy Phase Diagrams to act as the coordinating body for these activities. In the U.S., the national program has been spearheaded jointly by ASM International and the National Institute of Standards and Technology.

To meet the pressing need for diagrams, the national programs and the entire International Programme had two main goals: to increase the availability of phase diagrams and to ensure that the diagrams made available were of the highest possible quality. The specific tasks that were undertaken to accomplish these goals included assembling *all* existing data related to alloy phase diagrams, critically evaluating these data, using the data to construct the most up-to-date and accurate diagrams possible, and making the resulting diagrams readily available for use.

With the publication of the three-volume set of *Binary Alloy Phase Diagrams*, Second Edition, by ASM in 1991, the binary alloy portion of this monumental task is virtually complete. In addition, the first-ever truly comprehensive collection of ternary diagrams, the multivolume *Handbook of Ternary Alloy Phase Diagrams*, is scheduled for publication by ASM in 1994. Information from these two extensive and current diagram sources have been used as the basis of this updated engineering reference book, which reproduces the diagrams of the most commercially important systems (1046 binaries plus 80 ternaries) in a single, convenient volume. These alloy systems are represented by more than 1100 binary diagrams and 313 ternary diagrams, all plotted in weight percent as the primary scale.

The binary diagrams reproduced in this Handbook were selected from the 2965 systems covered in *Binary Alloy Phase Diagrams*, with updated diagrams from literature published since January 1991. Included with the binary diagrams is a complete index of all known alloy phase diagrams from *all* sources, listing where each can be found should a problem arise concerning a binary system not covered in this Handbook. Although many of the diagrams listed in this index (and a few of those reproduced in this volume) have not been evaluated under the Programme, they were selected to represent the best available. Updated binary diagrams from the phase diagram update section of the *Journal of Phase Equilibria* and abstracts of new full-length evaluation from the *Journal of Phase Equilibria* and the Monograph Series on Alloy Phase Diagrams are available from ASM International on a continuing basis through the Binary Alloy Phase Diagrams Updating Service.

The ternary diagrams reproduced here were selected from more than 12,000 diagrams being assembled for the ternary handbook. Where available, diagrams from recently published evaluated compilations were selected. The remainder were selected to represent the best available.

To aid in the full and effective use of these diagrams to solve practical problems, we have included an Introduction to Alloy Phase Diagrams, which contains sections on the theory and use of phase diagrams, and an Appendix listing the relevant properties of the elements and their crystal structures.

While the work of developing additional data, expanding alloy system coverage, and refining existing diagrams must and will continue, the quality checks built into the programme ensure that the diagrams reproduced here are as accurate and reliable as possible. Credit for this belongs to the conscientious work of all the experts involved in the worldwide Programme, especially Prof. Thaddeus B. Massalski and Dr. Alan A. Prince, who coordinated the evaluation efforts during the period of greatest activity.

The Editors

Contents

(continued)

Section 1
Introduction to Alloy Phase Diagrams

Hugh Baker, Editor

ALLOY PHASE DIAGRAMS are useful to metallurgists, materials engineers, and materials scientists in four major areas: (1) development of new alloys for specific applications, (2) fabrication of these alloys into useful configurations, (3) design and control of heat treatment procedures for specific alloys that will produce the required mechanical, physical, and chemical properties, and (4) solving problems that arise with specific alloys in their performance in commercial applications, thus improving product predictability. In all these areas, the use of phase diagrams allows research, development, and production to be done more efficiently and cost effectively.

In the area of alloy development, phase diagrams have proved invaluable for tailoring existing alloys to avoid overdesign in current applications, designing improved alloys for existing and new applications, designing special alloys for special applications, and developing alternative alloys or alloys with substitute alloying elements to replace those containing scarce, expensive, hazardous, or "critical" alloying elements. Application of alloy phase diagrams in processing includes their use to select proper parameters for working ingots, blooms, and billets, finding causes and cures for microporosity and cracks in castings and welds, controlling solution heat treating to prevent damage caused by incipient melting, and developing new processing technology.

In the area of performance, phase diagrams give an indication of which phases are thermodynamically stable in an alloy and can be expected to be present over a long time when the part is subjected to a particular temperature (e.g., in an automotive exhaust system). Phase diagrams also are consulted when attacking service problems such as pitting and intergranular corrosion, hydrogen damage, and hot corrosion.

In a majority of the more widely used commercial alloys, the allowable composition range encompasses only a small portion of the relevant phase diagram. The nonequilibrium conditions that are usually encountered in practice, however, necessitate the knowledge of a much greater portion of the diagram. Therefore, a thorough understanding of alloy phase diagrams in general and their practical use will prove to be of great help to a metallurgist expected to solve problems in any of the areas mentioned above.

Common Terms

Before the subject of alloy phase diagrams is discussed in detail, several of the commonly used terms will be discussed.

Phases. All materials exist in gaseous, liquid, or solid form (usually referred to as a *phase*), depending on the conditions of state. *State variables* include composition, temperature, pressure, magnetic field, electrostatic field, gravitational field, and so on. The term "phase" refers to that region of space occupied by a physically homogeneous material. However, there are two uses of the term: the strict sense normally used by physical scientists and the somewhat looser sense normally used by materials engineers.

In the strictest sense, homogeneous means that the physical properties throughout the region of space occupied by the phase are absolutely identical, and any change in condition of state, no matter how small, will result in a different phase. For example, a sample of solid metal with an apparently homogeneous appearance is not truly a single-phase material, because the pressure condition varies in the sample due to its own weight in the gravitational field.

In a phase diagram, however, each single-phase field (phase fields are discussed in a following section) is usually given a single label, and engineers often find it convenient to use this label to refer to all the materials lying within the field, regardless of how much the physical properties of the materials continuously change from one part of the field to another. This means that in engineering practice, the distinction between the terms "phase" and "phase field" is seldom made, and all materials having the same phase name are referred to as the same phase.

Equilibrium. There are three types of equilibria: stable, metastable, and unstable. These three conditions are illustrated in a mechanical sense in Fig. 1. Stable equilibrium exists when the object is in its lowest energy condition; metastable equilibrium exists when additional energy must be introduced before the object can reach true stability; unstable equilibrium exists when no additional energy is needed before reaching metastability or stability. Although true stable equilibrium conditions seldom exist in metal objects, the study of equilibrium systems is extremely valuable, because it constitutes a limiting condition from which actual conditions can be estimated.

Polymorphism. The structure of solid elements and compounds under stable equilibrium conditions is crystalline, and the crystal structure of each is unique. Some elements and compounds, however, are *polymorphic* (multishaped); that is, their structure transforms from one crystal structure to another with changes in temperature and pressure, each unique structure constituting a distinctively separate phase. The term *allotropy* (existing in another form) is usually used to describe polymorphic changes in chemical elements. Crystal structure of metals and alloys is discussed in a later section of this Introduction; the allotropic transformations of the elements are listed in the Appendix to this Volume.

Metastable Phases. Under some conditions, metastable crystal structures can form instead of stable structures. Rapid freezing is a common method of producing metastable structures, but some (such as Fe_3C, or "cementite") are produced at moderately slow cooling rates. With extremely rapid freezing, even thermodynamically unstable structures (such as amorphous metal "glasses") can be produced.

Systems. A physical *system* consists of a substance (or a group of substances) that is isolated from its surroundings, a concept used to facilitate study of the effects of conditions of state. "Isolated" means that there is no interchange of mass between the substance and its surroundings. The substances in alloy systems, for example, might be two metals, such as copper and zinc; a metal and a nonmetal, such as iron and carbon; a metal and an intermetallic compound, such as iron and cementite; or several metals, such as aluminum,

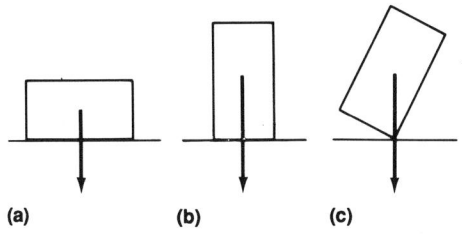

Fig. 1 Mechanical equilibria: (a) Stable. (b) Metastable. (c) Unstable

magnesium, and manganese. These substances constitute the *components* comprising the system and should not be confused with the various phases found within the system. A system, however, also can consist of a single component, such as an element or compound.

Phase Diagrams. In order to record and visualize the results of studying the effects of state variables on a system, diagrams were devised to show the relationships between the various phases that appear within the system under equilibrium conditions. As such, the diagrams are variously called *constitutional diagrams, equilibrium diagrams,* or *phase diagrams.* A single-component phase diagram can be simply a one- or two-dimensional plot showing the phase changes in the substance as temperature and/or pressure change. Most diagrams, however, are two- or three-dimensional plots describing the phase relationships in systems made up of two or more components, and these usually contain fields (areas) consisting of mixed-phase fields, as well as single-phase fields. The plotting schemes in common use are described in greater detail in subsequent sections of this Introduction.

System Components. Phase diagrams and the systems they describe are often classified and named for the number (in Latin) of components in the system:

Number of components	Name of system or diagram
One	Unary
Two	Binary
Three	Ternary
Four	Quaternary
Five	Quinary
Six	Sexinary
Seven	Septenary
Eight	Octanary
Nine	Nonary
Ten	Decinary

Phase Rule. The *phase rule,* first announced by J. Willard Gibbs in 1876, relates the physical state of a mixture to the number of constituents in the system and to its conditions. It was also Gibbs who first called each homogeneous region in a system by the term "phase." When pressure and temperature are the state variables, the rule can be written as follows:

$$f = c - p + 2$$

where f is the number of independent variables (called *degrees of freedom*), c is the number of components, and p is the number of stable phases in the system.

Unary Diagrams

Invariant Equilibrium. According to the phase rule, three phases can exist in stable equilibrium only at a single point on a unary diagram ($f = 1 - 3 + 2 = 0$). This limitation is illustrated as point O in the hypothetical unary pressure-temperature (PT) diagram shown in Fig. 2. In this diagram, the three states (or phases)—solid, liquid, and gas—are represented by the three correspondingly la-

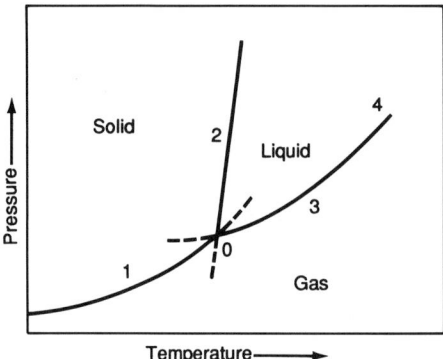

Fig. 2 Schematic pressure-temperature phase diagram

beled fields. Stable equilibrium between any two phases occurs along their mutual boundary, and *invariant equilibrium* among all three phases occurs at the so-called *triple point, O,* where the three boundaries intersect. This point also is called an *invariant point* because, at that location on the diagram, all externally controllable factors are fixed (no degrees of freedom). At this point, all three states (phases) are in equilibrium, but any changes in pressure and/or temperature will cause one or two of the states (phases) to disappear.

Univariant Equilibrium. The phase rule says that stable equilibrium between two phases in a unary system allows one degree of freedom ($f = 1 - 2 + 2$). This condition, called *univariant equilibrium* or *monovariant equilibrium,* is illustrated as lines 1, 2, and 3 separating the single-phase fields in Fig. 2. Either pressure or temperature may be freely selected, but not both. Once a pressure is selected, there is only one temperature that will satisfy equilibrium conditions, and conversely. The three curves that issue from the triple point are called *triple curves*: line 1, representing the reaction between the solid and the gas phases, is the *sublimation curve*; line 2 is the *melting curve*; and line 3 is the *vaporization curve*. The vaporization curve ends at point 4, called a *critical point*, where the physical distinction between the liquid and gas phases disappears.

Bivariant Equilibrium. If both the pressure and temperature in a unary system are freely and arbitrarily selected, the situation corresponds to having two degrees of freedom, and the phase rule says that only one phase can exit in stable equilibrium ($p = 1 - 2 + 2$). This situation is called *bivariant equilibrium.*

Binary Diagrams

If the system being considered comprises two components, a composition axis must be added to the PT plot, requiring construction of a three-dimensional graph. Most metallurgical problems, however, are concerned only with a fixed pressure of one atmosphere, and the graph reduces to a two-dimensional plot of temperature and composition (TX diagram).

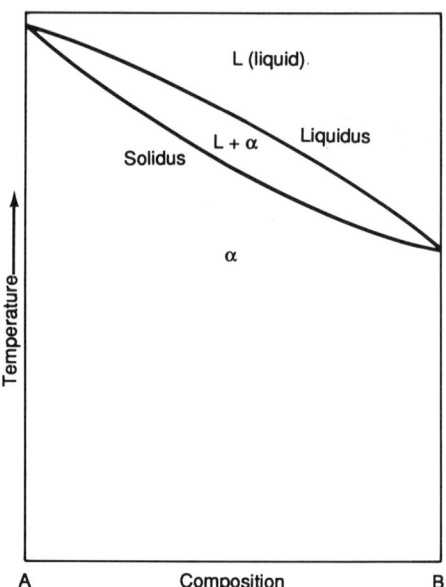

Fig. 3 Schematic binary phase diagram showing miscibility in both the liquid and solid states

The Gibbs phase rule applies to all states of matter (solid, liquid, and gaseous), but when the effect of pressure is constant, the rule reduces to:

$$f = c - p + 1$$

The stable equilibria for binary systems are summarized as follows:

Number of components	Number of phases	Degrees of freedom	Equilibrium
2	3	0	Invariant
2	2	1	Univariant
2	1	2	Bivariant

Miscible Solids. Many systems are comprised of components having the same crystal structure, and the components of some of these systems are completely miscible (completely soluble in each other) in the solid form, thus forming a *continuous solid solution.* When this occurs in a binary system, the phase diagram usually has the general appearance of that shown in Fig. 3. The diagram consists of two single-phase fields separated by a two-phase field. The boundary between the liquid field and the two-phase field in Fig. 3 is called the *liquidus*; that between the two-phase field and the solid field is the *solidus*. In general, a liquidus is the locus of points in a phase diagram representing the temperatures at which alloys of the various compositions of the system begin to freeze on cooling or finish melting on heating; a solidus is the locus of points representing the temperatures at which the various alloys finish freezing on cooling or begin melting on heating. The phases in equilibrium across the two-phase field in Fig. 3 (the liquid and solid solutions) are called *conjugate phases.*

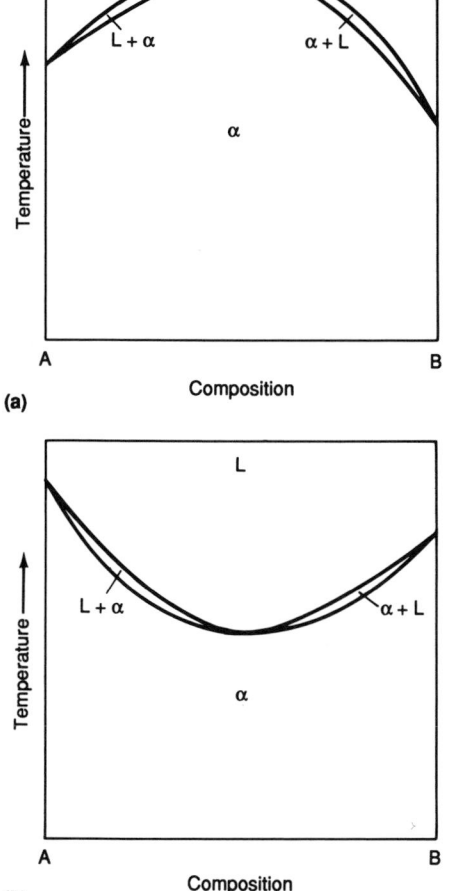

(a)

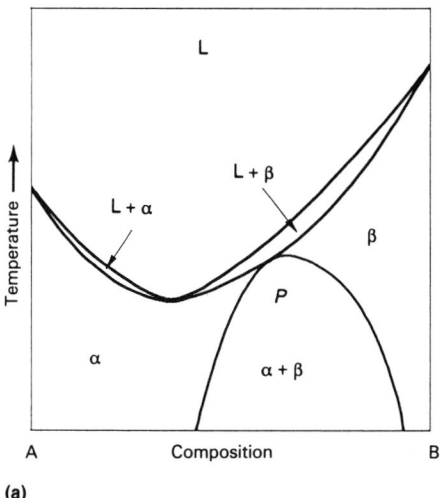

(b)

Fig. 4 Schematic binary phase diagrams with solid-state miscibility where the liquidus shows a maximum (a) and a minimum (b)

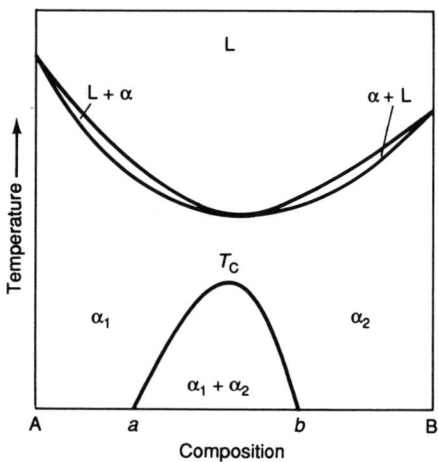

Fig. 5 Schematic binary phase diagram with a minimum in the liquidus and a miscibility gap in the solid state

If the solidus and liquidus meet tangentially at some point, a maximum or minimum is produced in the two-phase field, splitting it into two portions as shown in Fig. 4. It also is possible to have a gap in miscibility in a single-phase field; this is shown in Fig. 5. Point T_C, above which phases α_1 and α_2 become indistinguishable, is a critical point similar to point 4 in Fig. 2. Lines a-T_C and b-T_C, called *solvus* lines, indicate the limits of solubility of component B in A and A in B, respectively. The configurations of these and all other phase diagrams depend on the thermodynamics of the system, as discussed later in this Introduction.

Eutectic Reactions. If the two-phase field in the solid region of Fig. 5 is expanded so that it touches the solidus at some point, as shown in Fig. 6(a), complete miscibility of the components is lost. Instead of a single solid phase, the diagram now shows two separate solid *terminal phases*, which are in three-phase equilibrium with the liquid at

point P, an invariant point that occurred by coincidence. (Three-phase equilibrium is discussed in the following section.) Then, if this two-phase field in the solid region is even further widened so that the solvus lines no longer touch at the invariant point, the diagram passes through a series of configurations, finally taking on the more familiar shape shown in Fig. 6(b). The three-phase reaction that takes place at the invariant point E, where a liquid phase freezes into a mixture of two solid phases, is called a *eutectic* reaction (from the Greek word for "easily melted"). The alloy that corresponds to the eutectic composition is called a *eutectic alloy*. An alloy having a composition to the left of the eutectic point is called a *hypoeutectic alloy* (from the Greek word for "less than"); an alloy to the right is a *hypereutectic alloy* (meaning "greater than").

In the eutectic system described above, the two components of the system have the same crystal structure. This, and other factors, allows complete miscibility between them. Eutectic systems, however, also can be formed by two components having different crystal structures. When this occurs, the liquidus and solidus curves (and their extensions into the two-phase field) for each of the terminal phases (see Fig. 6c) resemble those for the situation of complete miscibility between system components shown in Fig. 3.

Three-Phase Equilibrium. Reactions involving three conjugate phases are not limited to the eutectic reaction. For example, upon cooling, a single solid phase can change into a mixture of two new solid phases or, conversely, two solid phases can react to form a single new phase. These and the other various types of invariant reactions observed in binary systems are listed in Table 1 and illustrated in Fig. 7 and 8.

Intermediate Phases. In addition to the three solid terminal-phase fields, α, β, and ε, the diagram in Fig. 7 displays five other solid-phase fields, γ, δ, δ', η, and σ, at intermediate compositions. Such phases are called *intermediate phases*. Many intermediate phases, such as those

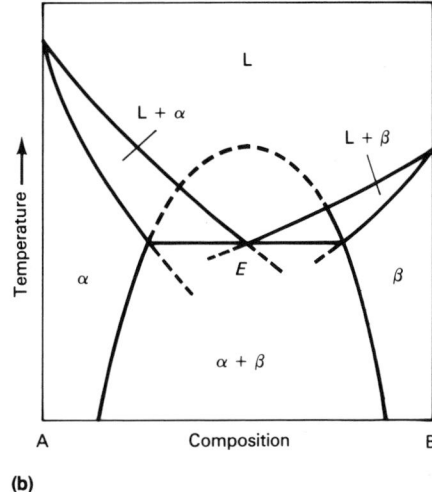

(a)

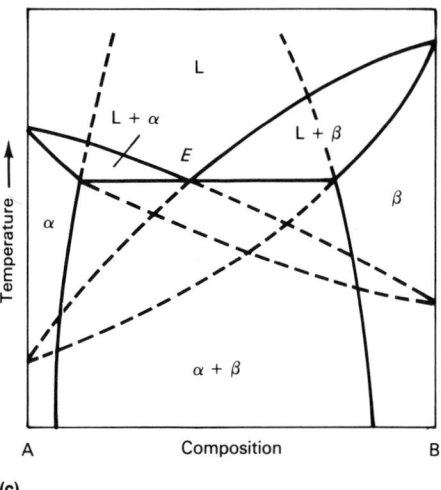

(b)

(c)

Fig. 6 Schematic binary phase diagrams with invariant points. (a) Hypothetical diagram of the type shown in Fig. 5, except that the miscibility gap in the solid touches the solidus curve at invariant point P; an actual diagram of this type probably does not exist. (b) and (c) Typical eutectic diagrams for components having the same crystal structure (b) and components having different crystal structures (c); the eutectic (invariant) points are labeled E. The dashed lines in (b) and (c) are metastable extensions of the stable-equilibria lines.

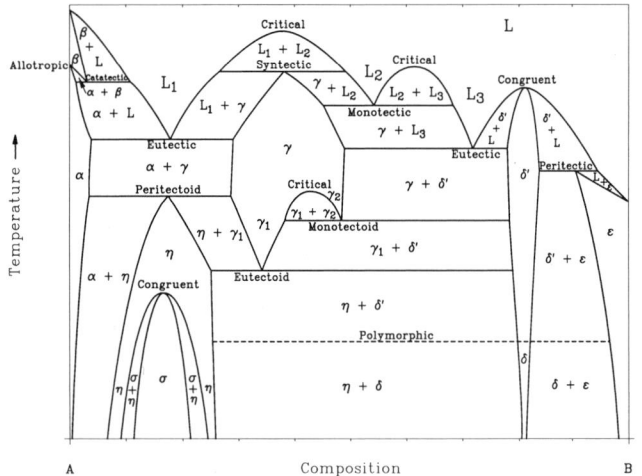

Fig. 7 Hypothetical binary phase diagram showing intermediate phases formed by various invariant reactions and a polymorphic transformation

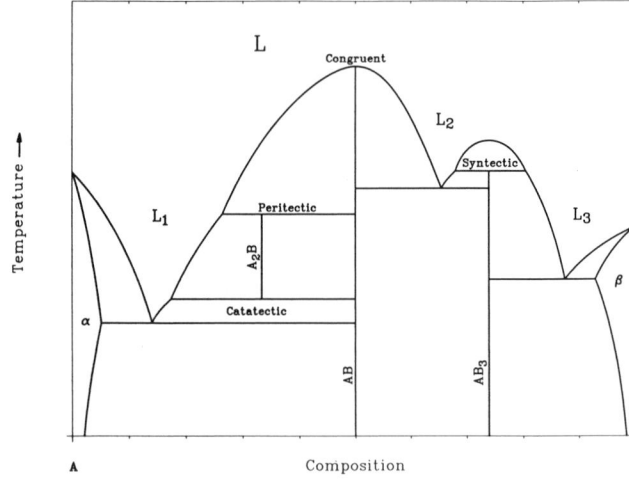

Fig. 8 Hypothetical binary phase diagram showing three intermetallic line compounds and four melting reactions

illustrated in Fig. 7, have fairly wide ranges of homogeneity. However, many others have very limited or no significant homogeneity range.

When an intermediate phase of limited (or no) homogeneity range is located at or near a specific ratio of component elements that reflects the normal positioning of the component atoms in the crystal structure of the phase, it is often called a compound (or *line compound*). When the components of the system are metallic, such an intermediate phase is often called an *intermetallic compound*. (Intermetallic compounds should not be confused with chemical compounds, where the type of bonding is different from that in crystals and where the ratio has chemical significance.) Three intermetallic compounds (with four types of melting reactions) are shown in Fig. 8.

In the hypothetical diagram shown in Fig. 8, an alloy of composition AB will freeze and melt isothermally, without the liquid or solid phases undergoing changes in composition; such a phase change is alled *congruent*. All other reactions are *incongruent*; that is, two phases are formed from one phase on melting. Congruent and incongruent phase changes, however, are not limited to line compounds: the terminal component B (pure phase ε) and the highest-melting composition of intermediate phase δ′ in Fig. 7, for example, freeze and melt congruently, while δ′ and ε freeze and melt incongruently at other compositions.

Metastable Equilibrium. In Fig. 6(c), dashed lines indicate the portions of the liquidus and solidus lines that disappear into the two-phase solid region. These dashed lines represent valuable information, as they indicate conditions that would exist under metastable equilibrium, such as might theoretically occur during extremely rapid cooling. Metastable extensions of some stable-equilibria lines also appear in Fig. 2 and 6(b).

Ternary Diagrams

When a third component is added to a binary system, illustrating equilibrium conditions in two

dimensions becomes more complicated. One option is to add a third composition dimension to the base, forming a solid diagram having binary diagrams as its vertical sides. This can be represented as a modified isometric projection, such as shown in Fig. 9. Here, boundaries of single-phase fields (liquidus, solidus, and solvus lines in the binary diagrams) become surfaces; single- and two-phase areas become volumes; three-phase lines become volumes; and four-phase points, while not shown in Fig. 9, can exist as an invariant plane. The composition of a binary eutectic liquid, which is a point in a two-component system, becomes a line in a ternary diagram, as shown in Fig. 9.

Although three-dimensional projections can be helpful in understanding the relationships in a

diagram, reading values from them is difficult. Therefore, ternary systems are often represented by views of the binary diagrams that comprise the faces and two-dimensional projections of the liquidus and solidus surfaces, along with a series of two-dimensional horizontal sections (*isotherms*) and vertical sections (*isopleths*) through the solid diagram.

Vertical sections are often taken through one corner (one component) and a congruently melting binary compound that appears on the opposite face; when such a plot can be read like any other true binary diagram, it is called a *quasibinary* section. One possibility is illustrated by line 1-2 in the isothermal section shown in Fig. 10. A vertical section between a congruently melting binary compound on one face and one on a dif-

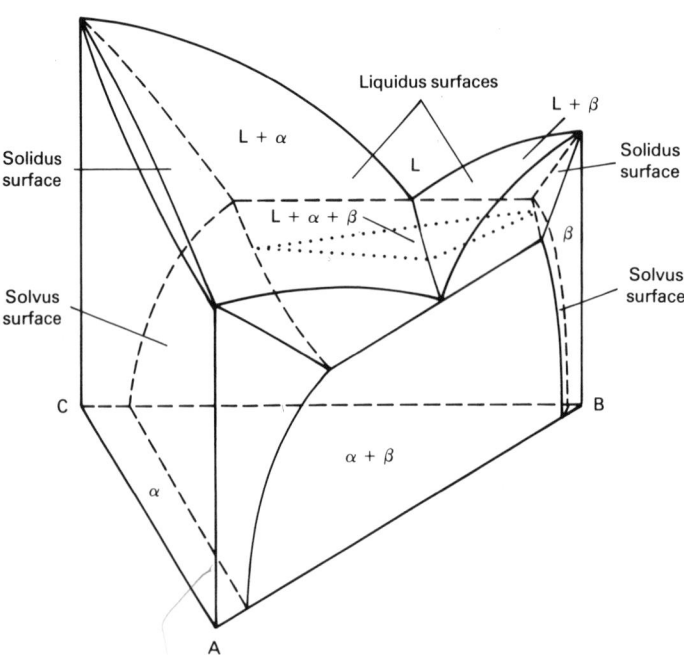

Fig. 9 Ternary phase diagram showing three-phase equilibrium. Source: 56Rhi

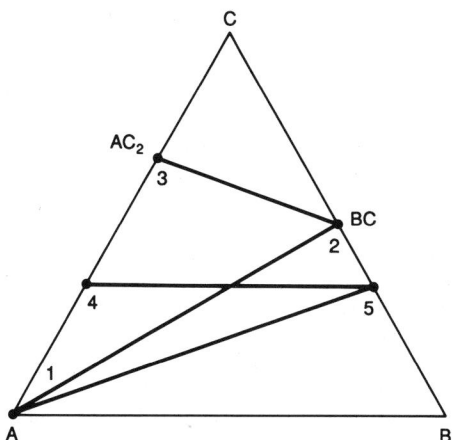

Fig. 10 Isothermal section of a ternary diagram with phase boundaries deleted for simplification

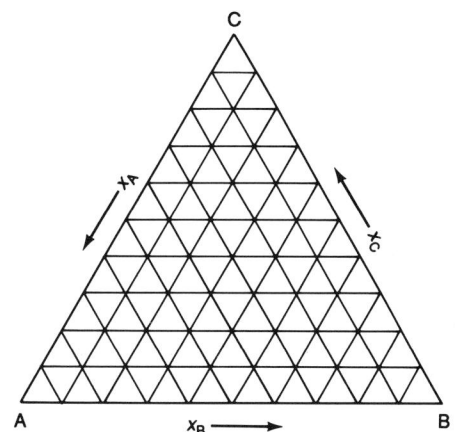

Fig. 11 Triangular composition grid for isothermal sections; x is the composition of each constituent in mole fraction or percent

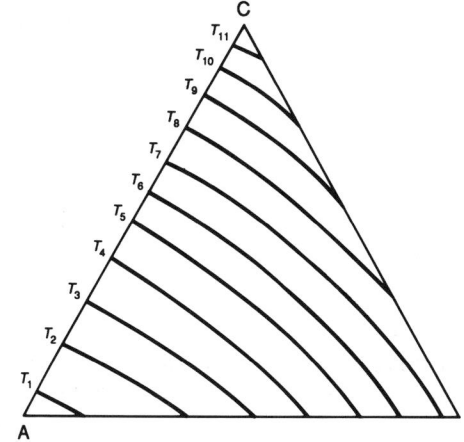

Fig. 12 Liquidus projection of a ternary phase diagram showing isothermal contour lines. Source: Adapted from 56Rhi

ferent face might also form a quasibinary section (see line 2-3).

All other vertical sections are not true binary diagrams, and the term *pseudobinary* is applied to them. A common pseudobinary section is one where the percentage of one of the components is held constant (the section is parallel to one of the faces), as shown by line 4-5 in Fig. 10. Another is one where the ratio of two constituents is held constant and the amount of the third is varied from 0 to 100% (line 1-5).

Isothermal Sections. Composition values in the triangular isothermal sections are read from a triangular grid consisting of three sets of lines parallel to the faces and placed at regular composition intervals (see Fig. 11). Normally, the point of the triangle is placed at the top of the illustra-

tion, component A is placed at the bottom left, B at the bottom right, and C at the top. The amount of component A is normally indicated from point C to point A, the amount of component B from point A to point B, and the amount of component C from point B to point C. This scale arrangement is often modified when only a corner area of the diagram is shown.

Projected Views. Liquidus, solidus, and solvus surfaces by their nature are not isothermal. Therefore, equal-temperature (isothermal) contour lines are often added to the projected views of these surfaces to indicate their shape (see Fig. 12). In addition to (or instead of) contour lines, views often show lines indicating the temperature troughs (also called "valleys" or "grooves")

formed at the intersections of two surfaces. Arrowheads are often added to these lines to indicate the direction of decreasing temperature in the trough.

Thermodynamic Principles

The reactions between components, the phases formed in a system, and the shape of the resulting phase diagram can be explained and understood through knowledge of the principles, laws, and terms of thermodynamics, and how they apply to the system.

Internal Energy. The sum of the kinetic energy (energy of motion) and potential energy (stored energy) of a system is called its *internal energy*, E. Internal energy is characterized solely by the state of the system.

Closed System. A thermodynamic system that undergoes no interchange of mass (material) with its surroundings is called a *closed system*. A closed system, however, can interchange energy with its surroundings.

First Law. The *First Law of Thermodynamics*, as stated by Julius von Mayer, James Joule, and Hermann von Helmholtz in the 1840s, states that *energy can be neither created nor destroyed*. Therefore, it is called the *Law of Conservation of Energy*. This law means that the total energy of an isolated system remains constant throughout any operations that are carried out on it; that is, for any quantity of energy in one form that disappears from the system, an equal quantity of another form (or other forms) will appear.

For example, consider a closed gaseous system to which a quantity of heat energy, δQ, is added and a quantity of work, δW, is extracted. The First Law describes the change in internal energy, dE, of the system as follows:

$$dE = \delta Q - \delta W$$

In the vast majority of industrial processes and material applications, the only work done by or on a system is limited to pressure/volume terms.

Table 1 Invariant reactions

Type	Reaction		
Eutectic (involves liquid and solid)	L_2 >——$\underset{L_1}{\vee}$——< S		Monotectic
	S_1 >——$\underset{L}{\vee}$——< S_2		Eutectic
	L >——$\underset{S_1}{\vee}$——< S_2		Catatectic (Metatectic)
Eutectoid (involves solid only)	S_1 >——$\underset{S_1}{\vee}$——< S_2		Monotectoid
	S_2 >——$\underset{S_1}{\vee}$——< S_3		Eutectoid
Peritectic (involves liquid and solid)	L_1 >——$\underset{S}{\wedge}$——< L_2		Syntectic
	L >——$\underset{S_2}{\wedge}$——< S_1		Peritectic
Peritectoid (involves solid only)	S_1 >——$\underset{S_3}{\wedge}$——< S_2		Peritectoid

Any energy contributions from electric, magnetic, or gravitational fields are neglected, except for electrowinning and electrorefining processes such as those used in the production of copper, aluminum, magnesium, the alkaline metals, and the alkaline earths. With the neglect of field effects, the work done by a system can be measured by summing the changes in volume, dV, times each pressure causing a change. Therefore, when field effects are neglected, the First Law can be written:

$$dE = \delta Q - PdV$$

Enthalpy. Thermal energy changes under constant pressure (again neglecting any field effects) are most conveniently expressed in terms of the *enthalpy, H,* of a system. Enthalpy, also called *heat content,* is defined by:

$$H = E + PV$$

Enthalpy, like internal energy, is a function of the state of the system, as is the product PV.

Heat Capacity. The *heat capacity, C,* of a substance is the amount of heat required to raise its temperature one degree; that is:

$$C = \frac{\delta Q}{\delta T}$$

However, if the substance is kept at constant volume ($dV = 0$):

$$\delta Q = dE$$

and

$$C_v = \left(\frac{\delta Q}{\delta T}\right)_v = \left(\frac{dE}{dT}\right)_v$$

If, instead, the substance is kept at constant pressure (as in many metallurgical systems),

$$C_p = \left(\frac{dE}{dT} + \frac{PdV}{dT}\right)_p$$

$$C_p = \left[\frac{d(E + PV)}{dT}\right]_p$$

and

$$C_p = \left(\frac{dH}{dT}\right)_p$$

Second Law. While the First Law establishes the relationship between the heat absorbed and the work performed by a system, it places no restriction on the source of the heat or its flow direction. This restriction, however, is set by the *Second Law of Thermodynamics,* which was advanced by Rudolf Clausius and William Thomson (Lord Kelvin). The Second Law states that *the spontaneous flow of heat always is from the higher temperature body to the lower temperature body.* In other words, *all naturally occurring processes tend to take place spontaneously in the direction that will lead to equilibrium.*

Entropy. The Second Law is most conveniently stated in terms of *entropy, S,* another property of state possessed by all systems. Entropy represents the energy (per degree of absolute temperature, T) in a system that is not available for work. In terms of entropy, the Second Law states that *all natural processes tend to occur only with an increase in entropy, and the direction of the process always is such as to lead to an increase in entropy.* For processes taking place in a system in equilibrium with its surroundings, the change in entropy is defined as follows:

$$dS \equiv \frac{\delta Q}{T} \equiv \frac{dE + PdV}{T}$$

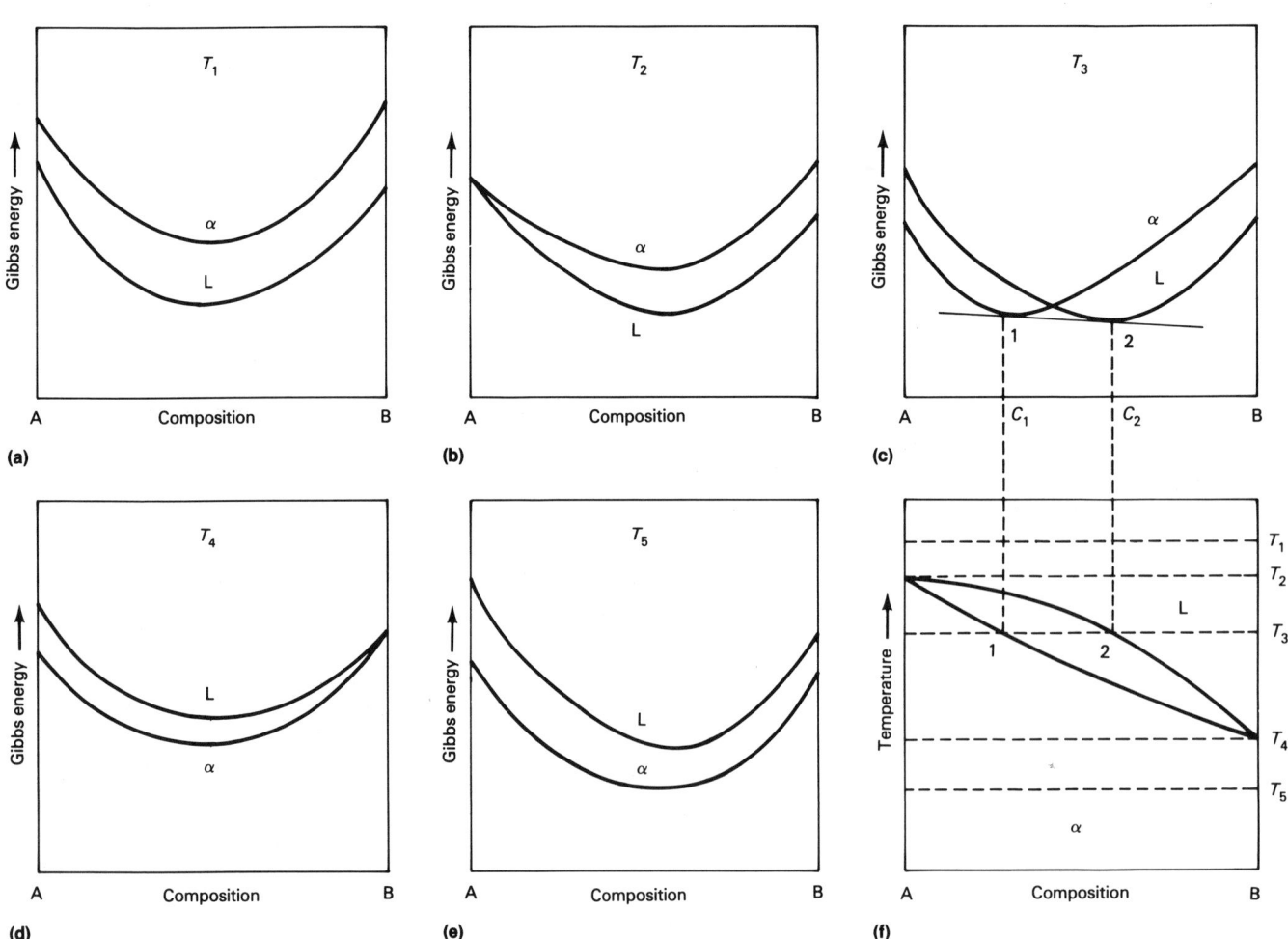

Fig. 13 Use of Gibbs energy curves to construct a binary phase diagram that shows miscibility in both the liquid and solid states. Source: Adapted from 66Pri

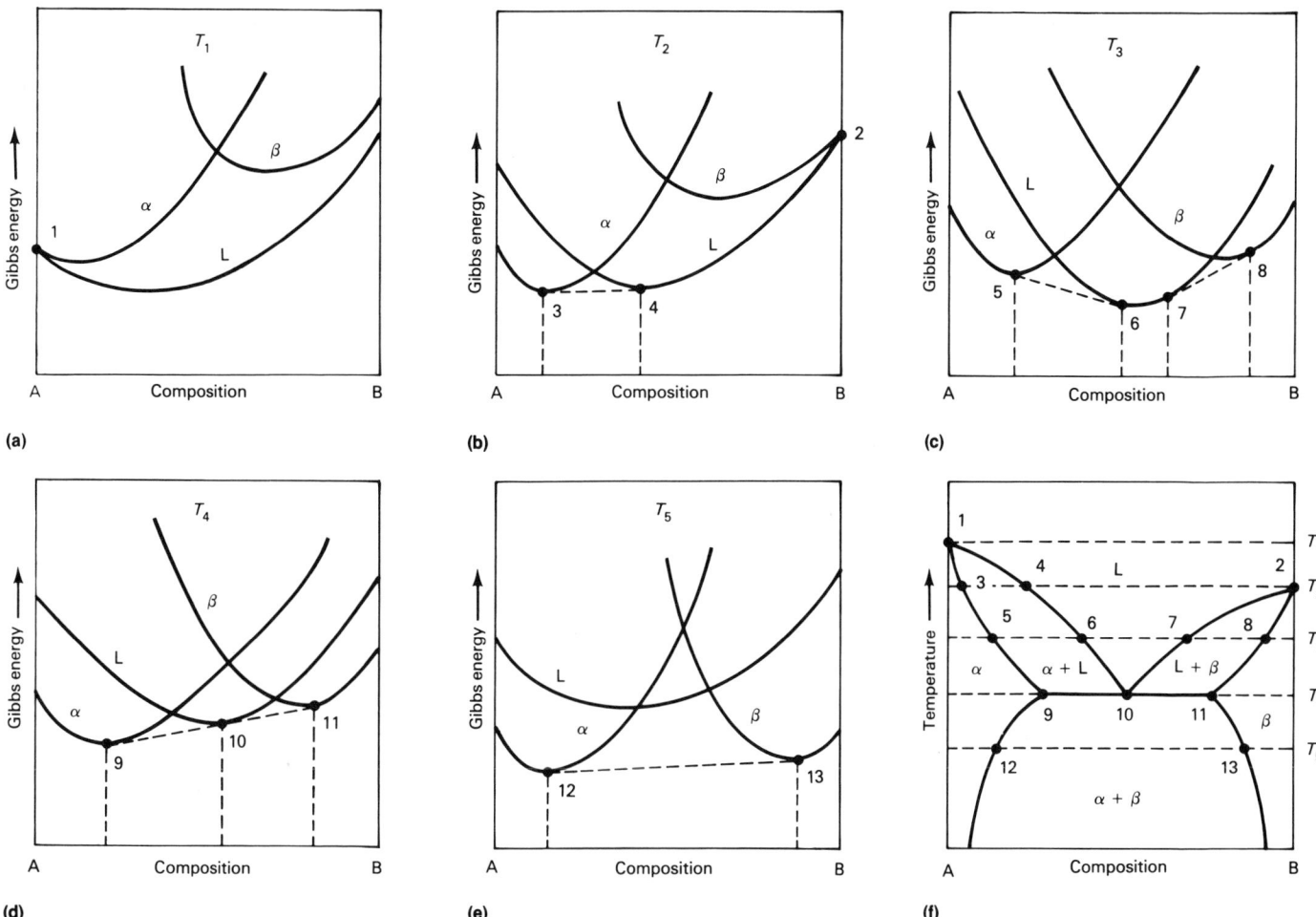

Fig. 14 Use of Gibbs energy curves to construct a binary phase diagram of the eutectic type. Source: Adapted from 68Gor

Third Law. A principle advanced by Theodore Richards, Walter Nernst, Max Planck, and others, often called the *Third Law of Thermodynamics*, states that *the entropy of all chemically homogeneous materials can be taken as zero at absolute zero temperature (0 K)*. This principle allows calculation of the absolute values of entropy of pure substances solely from heat capacity.

Gibbs Energy. Because both S and V are difficult to control experimentally, an additional term, *Gibbs energy, G*, is introduced, whereby:

$$G \equiv E + PV - TS \equiv H - TS$$

and

$$dG = dE + PdV + VdP - TdS - SdT$$

However,

$$dE = TdS - PdV$$

Therefore,

$$dG = VdP - SdT$$

Here, the change in Gibbs energy of a system undergoing a process is expressed in terms of two independent variables, pressure and absolute temperature, which are readily controlled experimentally. If the process is carried out under conditions of constant pressure and temperature, the change in Gibbs energy of a system at equilibrium with its surroundings (a reversible process) is zero. For a spontaneous (irreversible) process, the change in Gibbs energy is less than zero (negative); that is, the Gibbs energy decreases during the process, and it reaches a minimum at equilibrium.

Features of Phase Diagrams

The areas (fields) in a phase diagram, and the position and shapes of the points, lines, surfaces, and intersections in it, are controlled by thermodynamic principles and the thermodynamic properties of all of the phases that constitute the system.

Phase-field Rule. The *phase-field rule* specifies that at constant temperature and pressure, the number of phases in adjacent fields in a multicomponent diagram must differ by one.

Theorem of Le Châtelier. The *theorem of Henri Le Châtelier*, which is based on thermodynamic principles, states that *if a system in equilibrium is subjected to a constraint by which the equilibrium is altered, a reaction occurs that opposes the constraint, i.e., a reaction that partially nullifies the alteration*. The effect of this theorem on lines in a phase diagram can be seen in Fig. 2. The slopes of the sublimation line (1) and the vaporization line (3) show that the system reacts to increasing pressure by making the denser phases (solid and liquid) more stable at higher pressure. The slope of the melting line (2) indicates that this hypothetical substance contracts on freezing. (Note that the boundary between liquid water and ordinary ice, which expands on freezing, slopes toward the pressure axis.)

Clausius-Clapeyron Equation. The theorem of Le Châtelier was quantified by Benoit Clapeyron and Rudolf Clausius to give the following equation:

$$\frac{dP}{dT} = \frac{\Delta H}{T\Delta V}$$

where dP/dT is the slope of the univariant lines in a PT diagram such as those shown in Fig. 2, ΔV is the difference in molar volume of the two phases in the reaction, and ΔH is the difference in molar enthalpy of the two phases (the heat of the reaction).

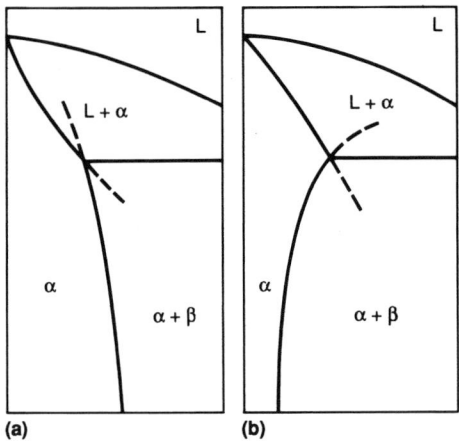

Fig. 15 Examples of acceptable intersection angles for boundaries of two-phase fields. Source: 56Rhi

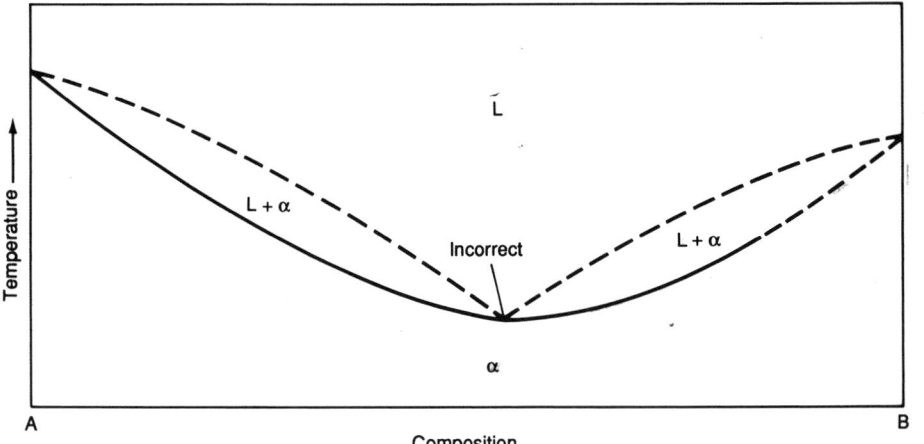

Fig. 16 An example of a binary phase diagram with a minimum in the liquidus that violates the Gibbs-Konovalov Rule. Source: 81Goo

Solutions. The shapes of liquidus, solidus, and solvus curves (or surfaces) in a phase diagram are determined by the Gibbs energies of the relevant phases. In this instance, the Gibbs energy must include not only the energy of the constituent components, but also the energy of mixing of these components in the phase.

Consider, for example, the situation of complete miscibility shown in Fig. 3. The two phases, liquid and solid α, are in stable equilibrium in the two-phase field between the liquidus and solidus lines. The Gibbs energies at various temperatures are calculated as a function of composition for ideal liquid solutions and for ideal solid solutions of the two components, A and B. The result is a series of plots similar to those shown in Fig. 13(a) to (e).

At temperature T_1, the liquid solution has the lower Gibbs energy and, therefore, is the more stable phase. At T_2, the melting temperature of A, the liquid and solid are equally stable only at a composition of pure A. At temperature T_3, between the melting temperatures of A and B, the Gibbs energy curves cross. Temperature T_4 is the melting temperature of B, while T_5 is below it.

Construction of the two-phase liquid-plus-solid field of the phase diagram in Fig. 13(f) is as follows. According to thermodynamic principles, the compositions of the two phases in equilibrium with each other at temperature T_3 can be determined by constructing a straight line that is tangential to both curves in Fig. 13(c). The points of tangency, 1 and 2, are then transferred to the phase diagram as points on the solidus and liquidus, respectively. This is repeated at sufficient temperatures to determine the curves accurately.

If, at some temperature, the Gibbs energy curves for the liquid and the solid tangentially touch at some point, the resulting phase diagram will be similar to those shown in Fig. 4(a) and (b), where a maximum or minimum appears in the liquidus and solidus curves.

Mixtures. The two-phase field in Fig. 13(f) consists of a mixture of liquid and solid phases. As stated above, the compositions of the two phases in equilibrium at temperature T_3 are C_1 and C_2. The horizontal isothermal line connecting points 1 and 2, where these compositions intersect temperature T_3, is called a *tie line*. Similar tie lines connect the coexisting phases throughout all two-phase fields (areas) in binary and (volumes) in ternary systems, while *tie triangles* connect the coexisting phases throughout all three-phase regions (volumes) in ternary systems.

Eutectic phase diagrams, a feature of which is a field where there is a mixture of two solid phases, also can be constructed from Gibbs energy curves. Consider the temperatures indicated on the phase diagram in Fig. 14(f) and the Gibbs energy curves for these temperatures (Fig. 14a-e). When the points of tangency on the energy curves are transferred to the diagram, the typical shape of a eutectic system results. The mixture of solid α and ß that forms upon cooling through the eutectic point k has a special microstructure, as discussed later.

Binary phase diagrams that have three-phase reactions other than the eutectic reaction, as well as diagrams with multiple three-phase reactions, also can be constructed from appropriate Gibbs energy curves. Likewise, Gibbs energy surfaces and tangential planes can be used to construct ternary phase diagrams.

Curves and Intersections. Thermodynamic principles also limit the shape of the various boundary curves (or surfaces) and their intersections. For example, see the *PT* diagram shown in Fig. 2. The Clausius-Clapeyron equation requires that at the intersection of the triple curves in such a diagram, the angle between adjacent curves should never exceed 180° or, alternatively, the extension of each triple curve between two phases must lie within the field of third phase.

The angle at which the boundaries of two-phase fields meet also is limited by thermodynamics. That is, the angle must be such that the extension of each beyond the point of intersection projects into a two-phase field, rather than a one-phase field. An example of correct intersections can be

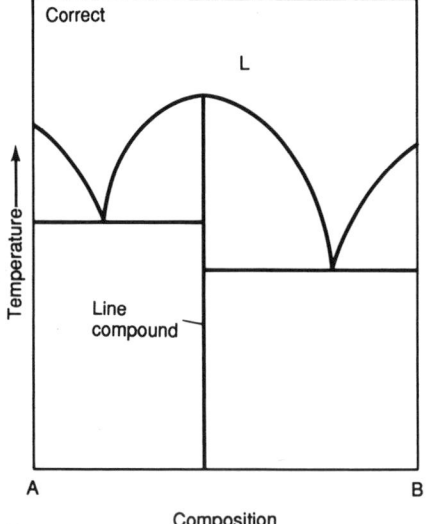

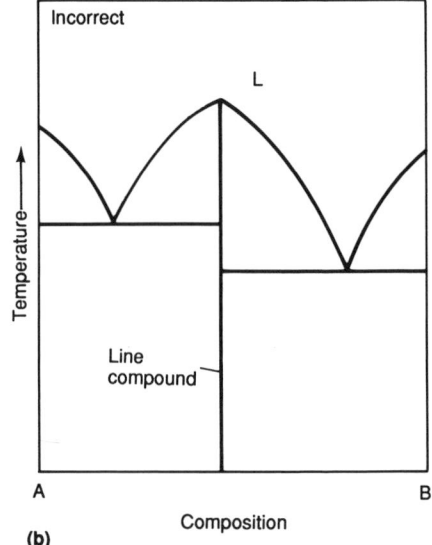

Fig. 17 Schematic diagrams of binary systems containing congruent-melting compounds but having no association of the component atoms in the melt common. The diagram in (a) is consistent with the Gibbs-Konovalov Rule, whereas that in (b) violates the rule. Source: 81Goo

Typical Phase-Rule Violations

(See Fig. 18)

1. A two-phase field cannot be extended to become part of a pure-element side of a phase diagram at zero solute. In example 1, the liquidus and the solidus must meet at the melting point of the pure element.
2. Two liquidus curves must meet at one composition at a eutectic temperature.
3. A tie line must terminate at a phase boundary.
4. Two solvus boundaries (or two liquidus, or two solidus, or a solidus and a solvus) of the same phase must meet (i.e., intersect) at one composition at an invariant temperature. (There should not be two solubility values for a phase boundary at one temperature.)
5. A phase boundary must extrapolate into a two-phase field after crossing an invariant point. The validity of this feature, and similar features related to invariant temperatures, is easily demonstrated by constructing hypothetical free-energy diagrams slightly below and slightly above the invariant temperature and by observing the relative positions of the relevant tangent points to the free energy curves. After intersection, such boundaries can also be extrapolated into metastable regions of the phase diagram. Such extrapolations are sometimes indicated by dashed or dotted lines.
6. Two single-phase fields (α and β) should not be in contact along a horizontal line. (An invariant-temperature line separates two-phase fields in contact.)
7. A single-phase field (α in this instance) should not be apportioned into subdivisions by a single line. Having created a horizontal (invariant) line at 6 (which is an error), there may be a temptation to extend this line into a single-phase field, α, creating an additional error.
8. In a binary system, an invariant-temperature line should involve equilibrium among three phases.
9. There should be a two-phase field between two single-phase fields (Two single phases cannot touch except at a point. However, second-order and higher-order transformations may be exceptions to this rule.)

10. When two phase boundaries touch at a point, they should touch at an extremity of temperature.
11. A touching liquidus and solidus (or any two touching boundaries) must have a horizontal common tangent at the congruent point. In this instance, the solidus at the melting point is too "sharp" and appears to be discontinuous.
12. A local minimum point in the lower part of a single-phase field (in this instance, the liquid) cannot be drawn without an additional boundary in contact with it. (In this instance, a horizontal monotectic line is most likely missing.)
13. A local maximum point in the lower part of a single-phase field cannot be drawn without a monotectic, monotectoid, syntectic, and sintectoid reaction occurring below it at a lower temperature. Alternatively, a solidus curve must be drawn to touch the liquidus at point 13.
14. A local maximum point in the upper part of a single-phase field cannot be drawn without the phase boundary touching a reversed monotectic, or a monotectoid, horizontal reaction line coinciding with the temperature of the maximum. When a 14 type of error is introduced, a minimum may be created on either side (or on one side) of 14. This introduces an additional error, which is the opposite of 13, but equivalent to 13 in kind.
15. A phase boundary cannot terminate within a phase field. (Termination due to lack of data is, of course, often shown in phase diagrams, but this is recognized to be artificial.)
16. The temperature of an invariant reaction in a binary system must be constant. (The reaction line must be horizontal.)
17. The liquidus should not have a discontinuous sharp peak at the melting point of a compound. (This rule is not applicable if the liquid retains the molecular state of the compound, i.e., in the situation of an ideal association.)
18. The compositions of all three phases at an invariant reaction must be different.
19. A four-phase equilibrium is not allowed in a binary system.
20. Two separate phase boundaries that create a two-phase field between two phases in equilibrium should not cross each other.

Problems Connected With Phase-Boundary Curvatures

Although phase rules are not violated, three additional unusual situations (21, 22, and 23) have also been included in Fig. 18. In each instance, a more subtle thermodynamic problem may exist related to these situations. Examples are discussed below where several thermodynamically unlikely diagrams are considered. The problems with each of these situations involve an indicated rapid change of slope of a phase boundary. If such situations are to be associated with realistic thermodynamics, the temperature (or the composition) dependence of the thermodynamic functions of the phase (or phases) involved would be expected to show corresponding abrupt and unrealistic variations in the phase diagram regions where such abrupt phase boundary changes are proposed, without any clear reason for them. Even the onset of ferromagnetism in a phase does not normally cause an abrupt change of slope of the related phase boundaries. The unusual changes of slope considered here are:

21. Two inflection points are located too closely to each other.
22. An abrupt reversal of the boundary direction (more abrupt than a typical smooth "retrograde"). This particular change can occur only if there is an accompanying abrupt change in the temperature dependence of the thermodynamic properties of either of the two phases involved (in this instance, δ or λ in relation to the boundary). The boundary turn at 22 is very unlikely to be explained by any realistic change in the composition dependence of the Gibbs energy functions.
23. An abrupt change in the slope of a single-phase boundary. This particular change can occur only by an abrupt change in the composition dependence of the thermodynamic properties of the single phase involved (in this instance, the δ phase). It cannot be explained by any possible abrupt change in the temperature dependence of the Gibbs energy function of the phase. (If the temperature dependence were involved, there would also be a change in the boundary of the ε phase.)

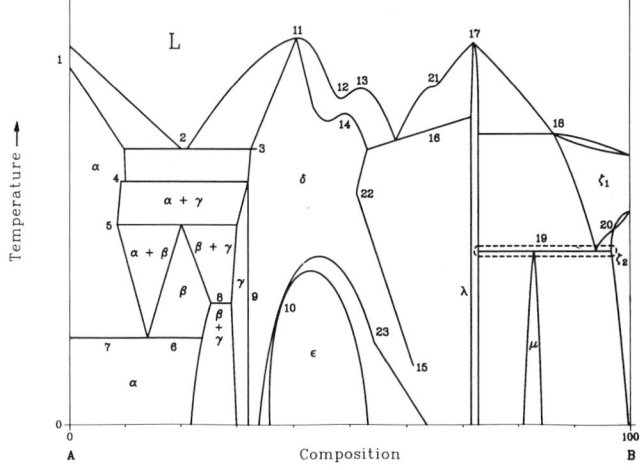

Fig. 18 Hypothetical binary phase diagram showing many typical errors of construction. See the accompanying text for discussion of the errors at points 1 to 23. Source: 91Oka1

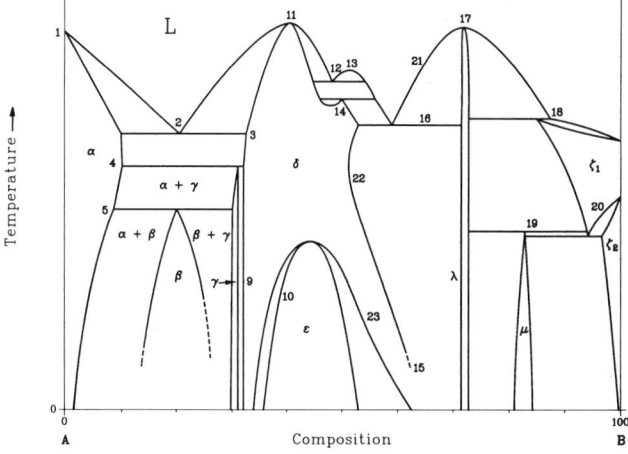

Fig. 19 Error-free version of the phase diagram shown in Fig. 18. Source: 91Oka1

seen in Fig. 6(b), where both the solidus and solvus lines are concave. However, the curvature of both boundaries need not be concave; Fig. 15 shows two equally acceptable (but unlikely) intersections where convex and concave lines are mixed.

Congruent Transformations. The *congruent point* on a phase diagram is where different phases of same composition are in equilibrium. The *Gibbs-Konovalov Rule* for congruent points, which was developed by Dmitry Konovalov from a thermodynamic expression given by J. Willard Gibbs, states that the slope of phase boundaries at congruent transformations must be zero (horizontal). Examples of correct slope at the maximum and minimum points on liquidus and solidus curves can be seen in Fig. 4. Often, the inner curve on a diagram such as that shown in Fig. 4 is erroneously drawn with a sharp inflection (see Fig. 16).

A similar common construction error is found in the diagrams of systems containing congruently melting compounds (such as the line compounds shown in Fig. 17) but having little or no association of the component atoms in the melt (as with most metallic systems). This type of error is especially common in partial diagrams, where one or more system components is a compound instead of an element. (The slope of liquidus and solidus curves, however, must *not* be zero when they terminate at an element, or at a compound having complete association in the melt.)

Common Construction Errors. Hiroaki Okamoto and Thaddeus Massalski have prepared the hypothetical binary phase shown in Fig. 18, which exhibits many typical errors of construction (marked as points 1 to 23). The explanation for each error is given in the accompanying text; one possible error-free version of the same diagram is shown in Fig. 19.

Higher-Order Transitions. The transitions considered in this Introduction up to this point have been limited to the common thermodynamic types called *first-order transitions*—that is, changes involving distinct phases having different lattice parameters, enthalpies, entropies, densities, and so on. Transitions not involving discontinuities in composition, enthalpy, entropy, or molar volume are called *higher-order transitions* and occur less frequently. The change in the magnetic quality of iron from ferromagnetic to paramagnetic as the temperature is raised above 771 °C (1420 °F) is an example of a second-order transition: no phase change is involved and the Gibbs phase rule does not come into play in the transition. Another example of a higher-order transition is the continuous change from a random arrangement of the various kinds of atoms in a multicomponent crystal structure (a *disordered structure*) to an arrangement where there is some degree of *crystal ordering* of the atoms (an *ordered structure*, or *superlattice*), or the reverse reaction.

Crystal Structure

A *crystal* is a solid consisting of atoms or molecules arranged in a pattern that is repetitive in three dimensions. The arrangement of the atoms

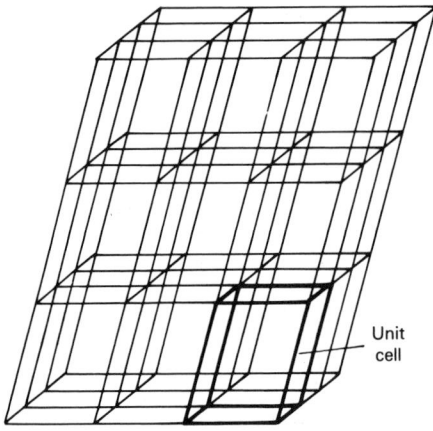

Fig. 20 A space lattice

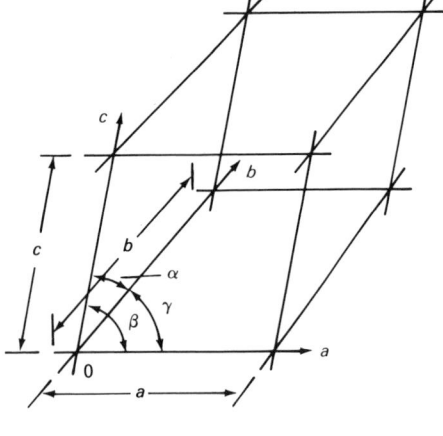

Fig. 21 Crystal axes and unit-cell edge lengths. Unit-cell faces are shown, but to avoid confusion they are not labeled.

or molecules in the interior of a crystal is called its *crystal structure*. The *unit cell* of a crystal is the smallest pattern of arrangement that can be contained in a parallelepiped, the edges of which form the a, b, and c axes of the crystal. The three-dimensional aggregation of unit cells in the crystal forms a *space lattice*, or *Bravais lattice* (see Fig. 20).

Crystal Systems. Seven different *crystal systems* are recognized in crystallography, each having a different set of axes, unit-cell edge lengths, and interaxial angles (see Table 2). Unit-cell *edge lengths* a, b, and c are measured along the corresponding a, b, and c axes (see Fig. 21). Unit-cell faces are identified by capital letters: face A contains axes b and c, face B contains c and a, and face C contains a and b. (Faces are not labeled in Fig. 21.) *Interaxial angle* α occurs in face A, angle β in face B, and angle γ in face C (see Fig. 21).

Lattice Dimensions. It should be noted that the unit-cell edge lengths and interaxial angles are unique for each crystalline substance. The unique edge lengths are called *lattice parameters*. The term *lattice constant* also has been used for the length of an edge, but the values of edge length are not constant, varying with composition within a phase field and also with temperature due to thermal expansion and contraction. (Reported lattice parameter values are assumed to be room-temperature values unless otherwise specified.) Interaxial angles other than 90° or 120° also can change slightly with changes in composition. When the edges of the unit cell are not equal in

all three directions, all unequal lengths must be stated to completely define the crystal. The same is true if all interaxial angles are not equal. When defining the unit-cell size of an alloy phase, the possibility of crystal ordering occurring over several unit cells should be considered. For example, in the copper-gold system, a superlattice forms that is made up of 10 cells of the disordered lattice, creating what is called *long-period ordering*.

Lattice Points. As shown in Fig. 20, a space lattice can be viewed as a three-dimensional network of straight lines. The intersections of the lines (called *lattice points*) represent locations in space for the same kind of atom or group of atoms of identical composition, arrangement, and orientation. There are five basic arrangements for lattice points within a unit cell. The first four are: primitive (simple), having lattice points solely at cell corners; base-face centered (end-centered), having lattice points centered on the C faces, or ends of the cell; all-face centered, having lattice points centered on all faces; and innercentered (body-centered), having lattice points at the center of the volume of the unit cell. The fifth arrangement, the primitive rhombohedral unit cell, is considered a separate basic arrangement, as shown in the following section on crystal structure nomenclature. These five basic arrangements are identified by capital letters as follows: P for the primitive cubic, C for the cubic cell with lattice points on the two C faces, F for all-face-centered cubic, I for innercentered (body-centered) cubic, and R for primitive rhombohedral.

Table 2 Relationships of edge lengths and of interaxial angles for the seven crystal systems

Crystal system	Edge lengths	Interaxial angles	Examples
Triclinic (anorthic)	$a \neq b \neq c$	$\alpha \neq \beta \neq \gamma \neq 90°$	HgK
Monoclinic	$a \neq b \neq c$	$\alpha = \gamma = 90° \neq \beta$	β-S; CoSb$_2$
Orthorhombic	$a \neq b \neq c$	$\alpha = \beta = \gamma = 90°$	α-S; Ga; Fe$_3$C (cementite)
Tetragonal	$a = b \neq c$	$\alpha = \beta = \gamma = 90°$	β-Sn (white); TiO$_2$
Hexagonal	$a = b \neq c$	$\alpha = \beta = 90°$; $\gamma = 120°$	Zn; Cd; NiAs
Rhombohedral(a)	$a = b = c$	$\alpha = \beta = \gamma \neq 90°$	As; Sb; Bi; calcite
Cubic	$a = b = c$	$\alpha = \beta = \gamma = 90°$	Cu; Ag; Au; Fe; NaCl

(a) Rhombohedral crystals (sometimes called trigonal) also can be described by using hexagonal axes (rhombohedral-hexagonal).

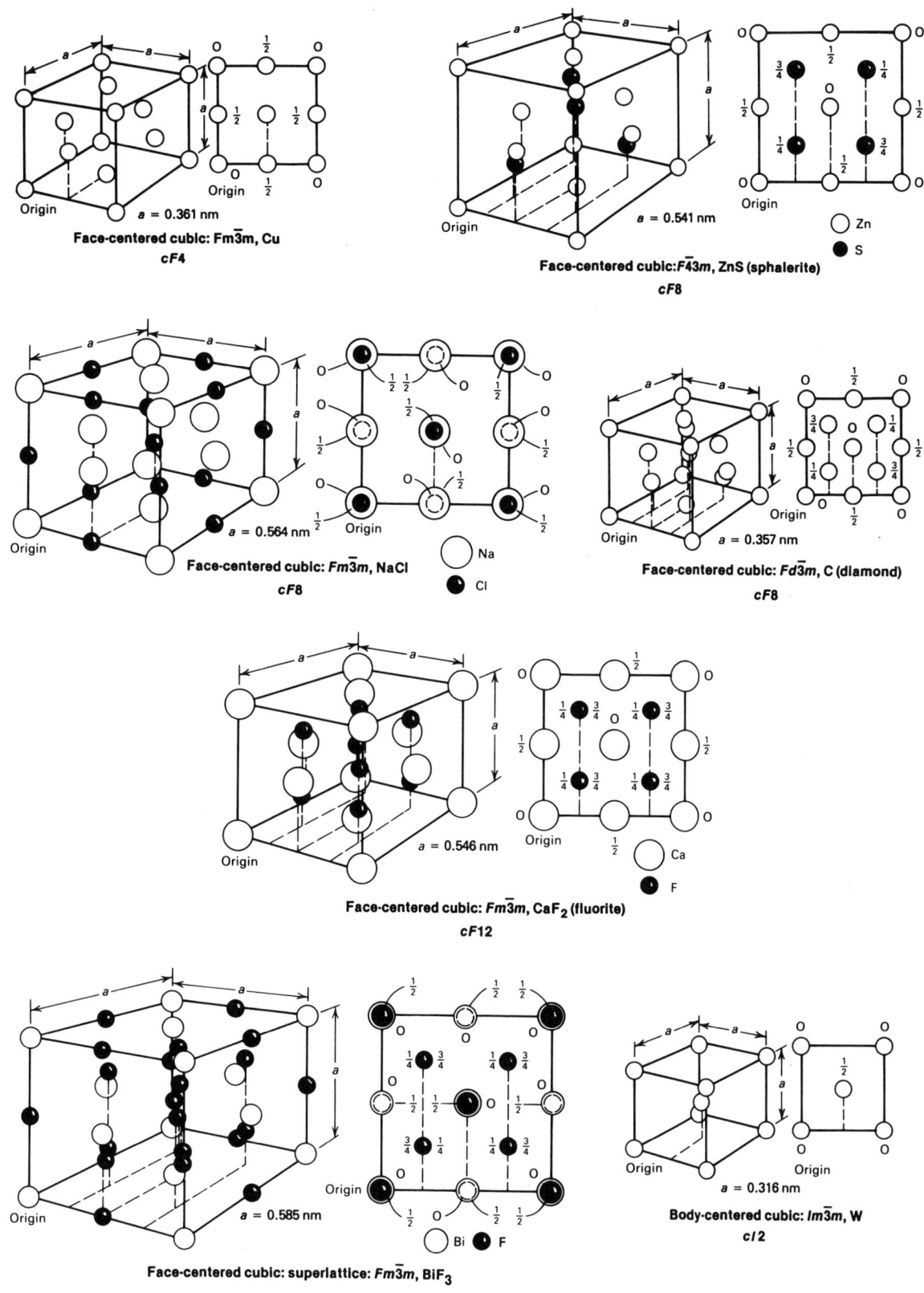

Face-centered cubic: Fm3̄m, Cu
cF4
$a = 0.361$ nm

Face-centered cubic: F4̄3m, ZnS (sphalerite)
cF8
$a = 0.541$ nm

○ Zn
● S

Face-centered cubic: Fm3̄m, NaCl
cF8
$a = 0.564$ nm

○ Na
● Cl

Face-centered cubic: Fd3̄m, C (diamond)
cF8
$a = 0.357$ nm

Face-centered cubic: Fm3̄m, CaF₂ (fluorite)
cF12
$a = 0.546$ nm

○ Ca
● F

Face-centered cubic: superlattice: Fm3̄m, BiF₃
cF16
$a = 0.585$ nm

○ Bi ● F

Body-centered cubic: Im3̄m, W
cI2
$a = 0.316$ nm

Fig. 22 Schematic drawings of the unit cells and ion positions for some simple metal crystals, arranged alphabetically according to Pearson symbol. Also listed are the space lattice and crystal system, space-group notation, and prototype for each crystal. Reported lattice parameters are for the prototype crystal. **(continued)**

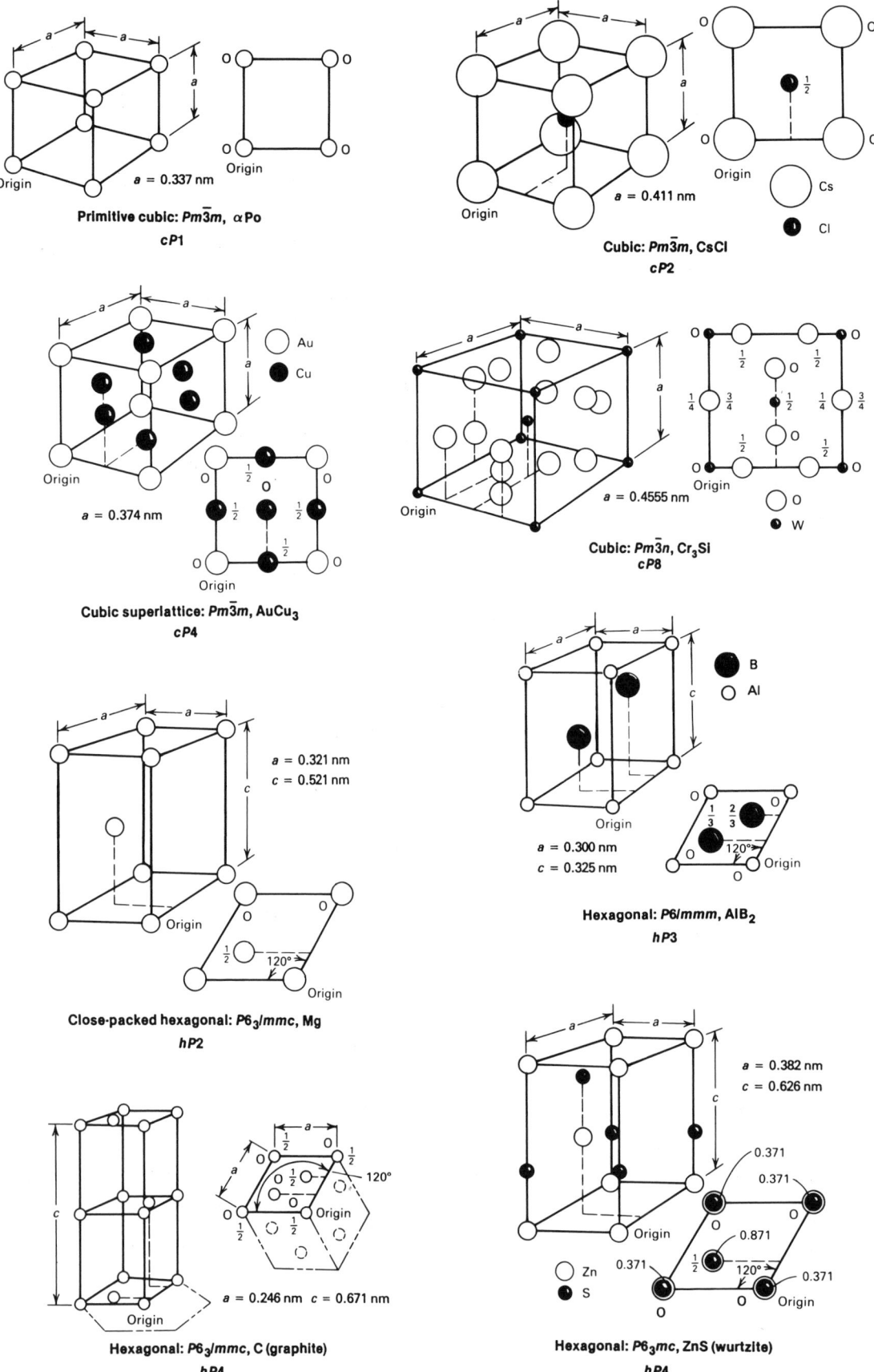

Primitive cubic: $Pm\bar{3}m$, αPo
*c*P1

$a = 0.337$ nm

Cubic: $Pm\bar{3}m$, CsCl
*c*P2

$a = 0.411$ nm

○ Cs
● Cl

Cubic superlattice: $Pm\bar{3}m$, AuCu₃
*c*P4

$a = 0.374$ nm

○ Au
● Cu

Cubic: $Pm\bar{3}n$, Cr₃Si
*c*P8

$a = 0.4555$ nm

○ O
● W

Close-packed hexagonal: $P6_3/mmc$, Mg
*h*P2

$a = 0.321$ nm
$c = 0.521$ nm

Hexagonal: $P6/mmm$, AlB₂
*h*P3

$a = 0.300$ nm
$c = 0.325$ nm

● B
○ Al

Hexagonal: $P6_3/mmc$, C (graphite)
*h*P4

$a = 0.246$ nm $c = 0.671$ nm

Hexagonal: $P6_3mc$, ZnS (wurtzite)
*h*P4

$a = 0.382$ nm
$c = 0.626$ nm

○ Zn
● S

Fig. 22 Schematic drawings of the unit cells and ion positions for some simple metal crystals, arranged alphabetically according to Pearson symbol. Also listed are the space lattice and crystal system, space-group notation, and prototype for each crystal. Reported lattice parameters are for the prototype crystal.　　**(continued)**

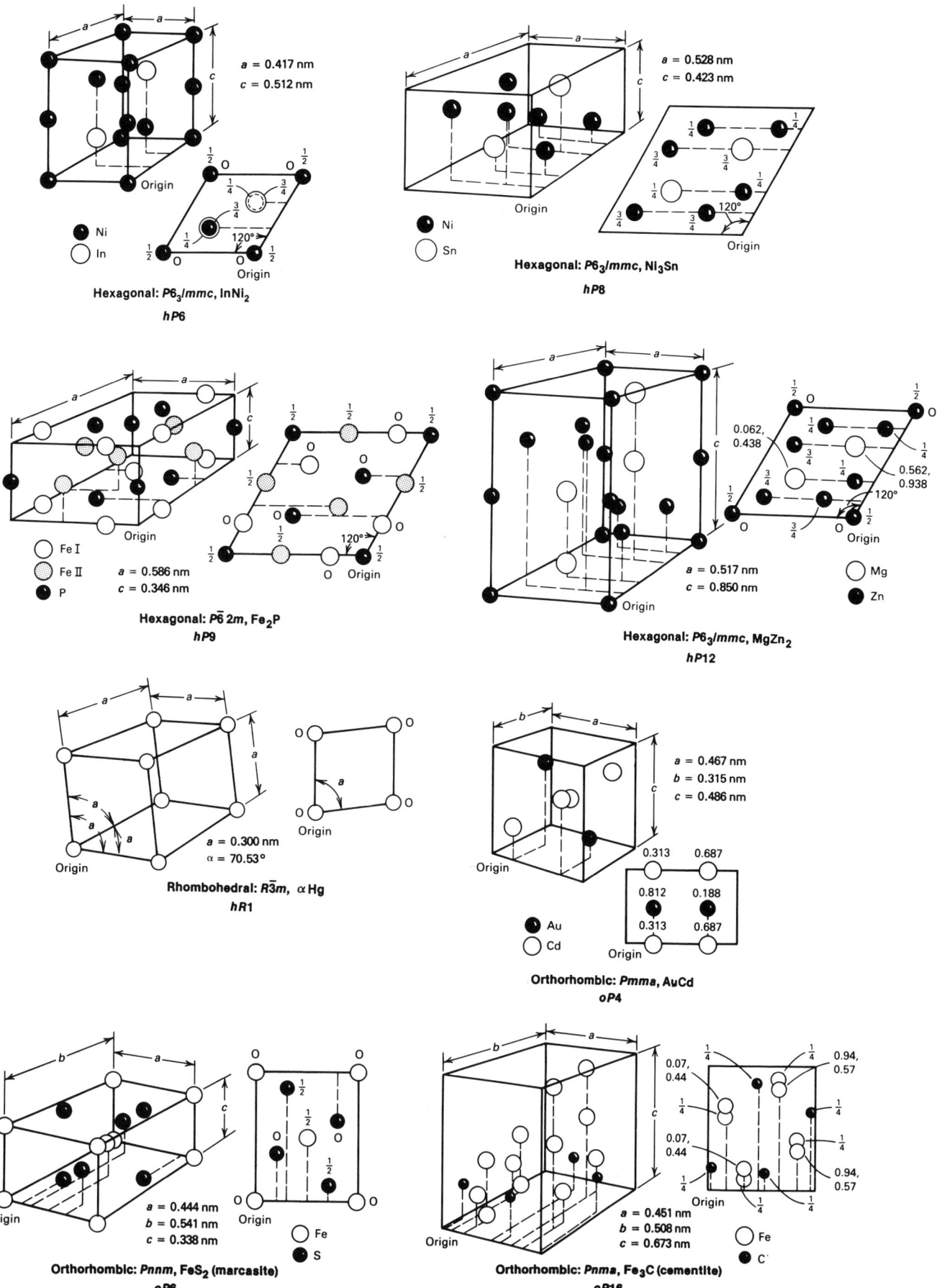

Fig. 22 Schematic drawings of the unit cells and ion positions for some simple metal crystals, arranged alphabetically according to Pearson symbol. Also listed are the space lattice and crystal system, space-group notation, and prototype for each crystal. Reported lattice parameters are for the prototype crystal. **(continued)**

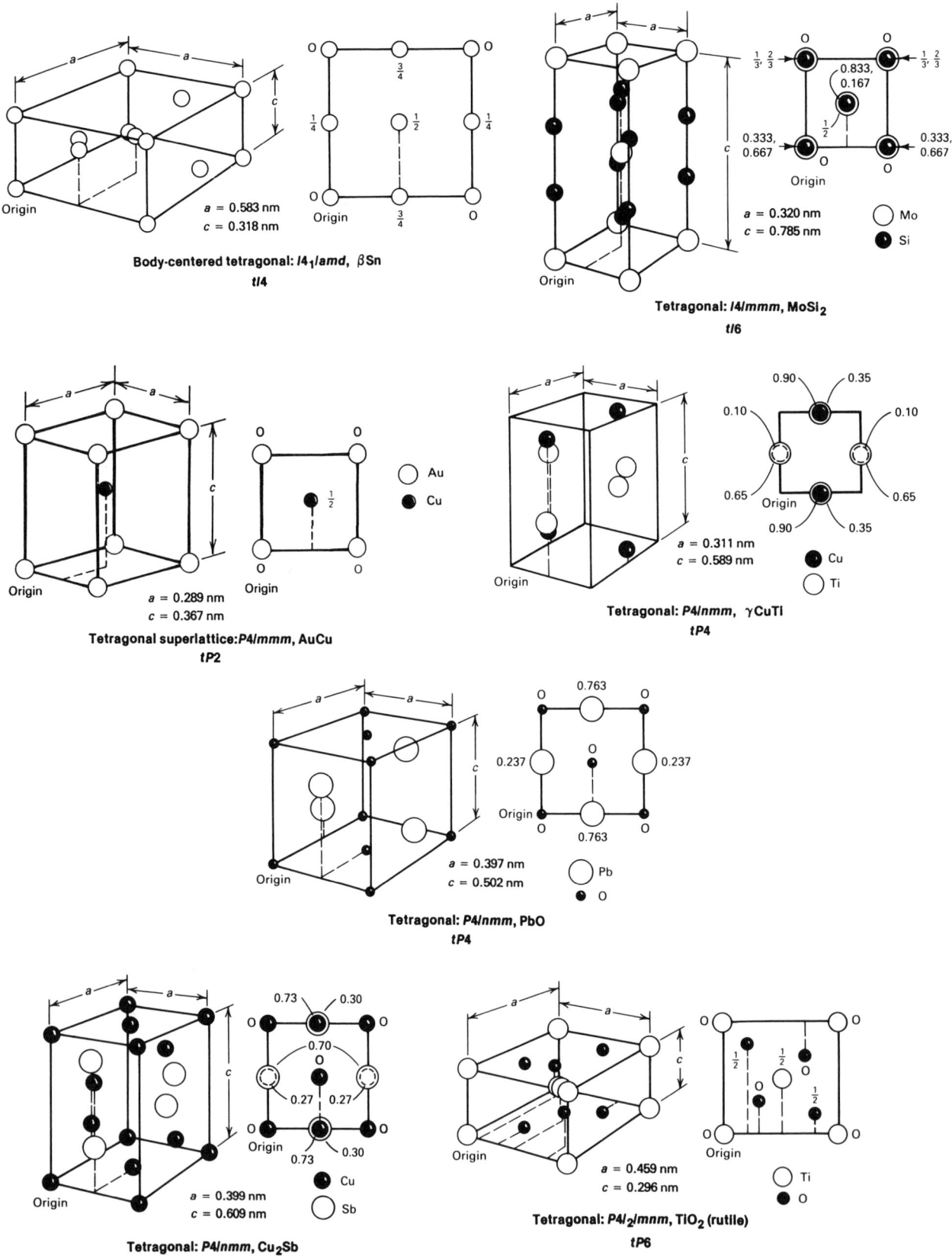

Fig. 22 Schematic drawings of the unit cells and ion positions for some simple metal crystals, arranged alphabetically according to Pearson symbol. Also listed are the space lattice and crystal system, space-group notation, and prototype for each crystal. Reported lattice parameters are for the prototype crystal.

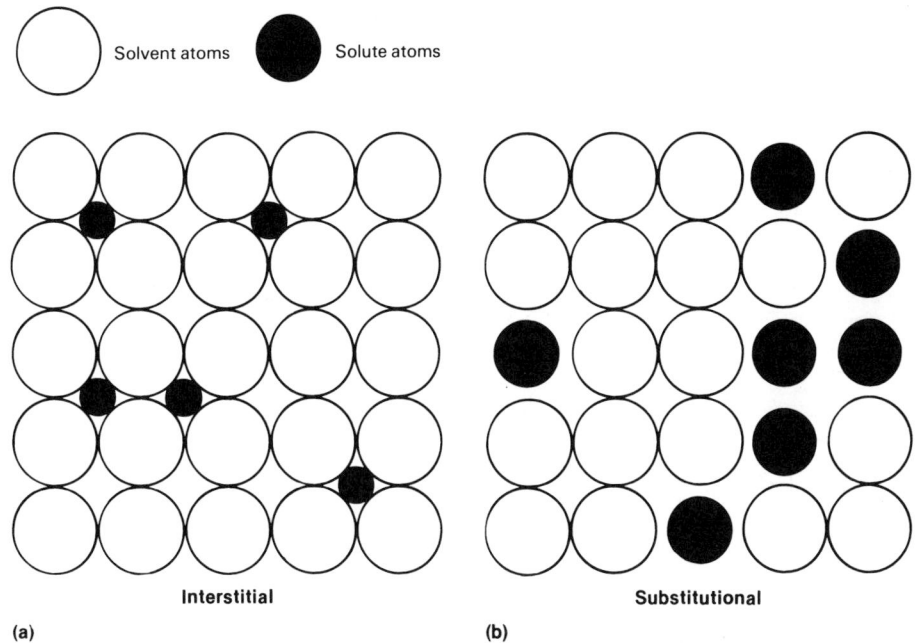

Fig. 23 Solid-solution mechanisms. (a) Interstitial. (b) Substitutional

Table 3 The 14 space (Bravais) lattices and their Pearson symbols

Crystal system	Space lattice	Pearson symbol
Triclinic (anorthic)	Primitive	*aP*
Monoclinic	Primitive	*mP*
	Base-centered(a)	*mC*
Orthorhombic	Primitive	*oP*
	Base-centered(a)	*oC*
	Face-centered	*oF*
	Body-centered	*oI*
Tetragonal	Primitive	*tP*
	Body-centered	*tI*
Hexagonal	Primitive	*hP*
Rhombohedral	Primitive	*hR*
Cubic	Primitive	*cP*
	Face-centered	*cF*
	Body-centered	*cI*

(a) The face that has a lattice point at its center may be chosen as the *c* face (the *xy* plane), denoted by the symbol *C*, or as the *a* or *b* face, denoted by *A* or *B*, because the choice of axes is arbitrary and does not alter the actual translations of the lattice.

Crystal Structure Nomenclature. When the seven crystal systems are considered together with the five space lattices, the combinations listed in Table 3 are obtained. These 14 combinations form the basis of the system of *Pearson symbols* developed by William B. Pearson, which are widely used to identify crystal types. As can be seen in Table 3, the Pearson symbol uses a small letter to identify the crystal system and a capital letter to identify the space lattice. To these is added a number equal to the number of atoms in the unit cell conventionally selected for the particular crystal type. When determining the number of atoms in the unit cell, it should be remembered that each atom that is shared with an adjacent cell (or cells) must be counted as only a fraction of an atom. The Pearson symbols for some simple metal crystals are shown in Fig. 22, along with schematic drawings illustrating the atom arrangements in the unit cell. It should be noted that in these schematic representations, the different kinds of atoms in the prototype crystal illustrated are drawn to represent their relative sizes, but in order to show the arrangements more clearly, all the atoms are shown much smaller than their true effective size in real crystals.

Several of the many possible crystal structures are so commonly found in metallic systems that they are often identified by three-letter abbreviations that combine the space lattice with the crystal system. For example, bcc is used for body-centered cubic (two atoms per unit cell), fcc for face-centered cubic (four atoms per unit cell), and cph for close-packed hexagonal (two atoms per unit cell).

Space-group notation is a symbolic description of the space lattice and symmetry of a crystal. It consists of the symbol for the space lattice followed by letters and numbers that designate the symmetry of the crystal. The space-group notation for each unit cell illustrated in Fig. 22 is identified next to it. For a more complete list of Pearson symbols and space-group notations, consult the Appendix.

To assist in classification and identification, each crystal structure type is assigned a representative substance (element or phase) having that structure. The substance selected is called the *structure prototype*. Generally accepted prototypes for some metal crystals are listed in Fig. 22.

An important source of information on crystal structures for many years was *Structure Reports* (*Strukturbericht* in German). In this publication, crystal structures were classified by a designation consisting of a capital letter (*A* for elements, *B* for AB-type phases, *C* for AB₂-type phases, *D* for other binary phases, *E* for ternary phases, and *L* for superlattices), followed by a number consecutively assigned (within each group) at the time the type was reported. To further distinguish among crystal types, inferior letters and numbers, as well as prime marks, were added to some designations. Because the Strukturbericht designation cannot be conveniently and systematically expanded to

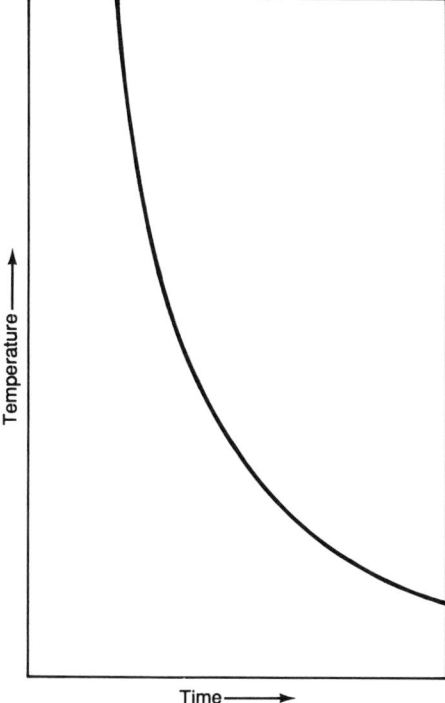

Fig. 24 Ideal cooling curve with no phase change

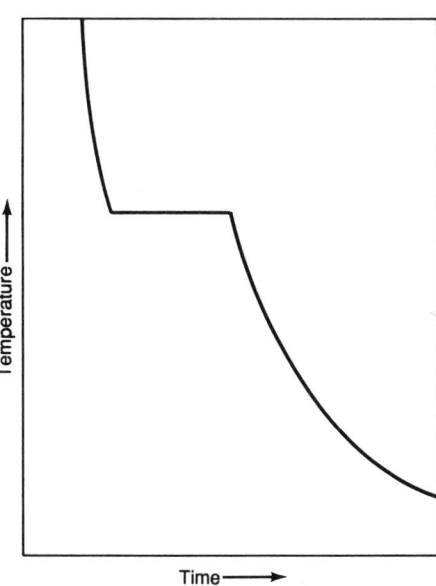

Fig. 25 Ideal freezing curve of a pure metal.

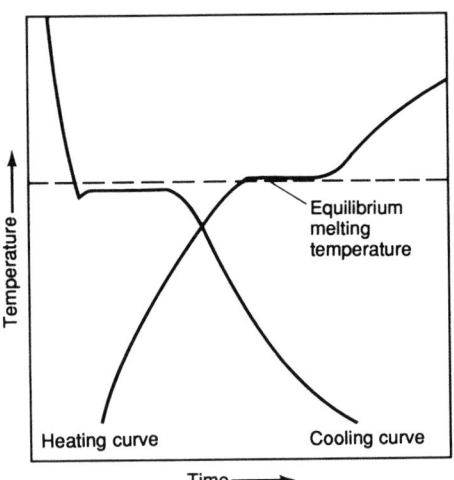

Fig. 26 Natural freezing and melting curves of a pure metal. Source: 56Rhi

cover the large variety of crystal structures currently being encountered, the system is falling into disuse.

The relations among common Pearson symbols, space groups, structure prototypes, and Strukturbericht designations for crystal systems are given in various tables in the Appendix. Crystallographic information for the metallic elements can be found in the table of allotropes in the Appendix; data for intermetallic phases of the systems included in this Volume are listed with the phase diagrams. Crystallographic data for an exhaustive list of intermediate phases are presented in 91Vil (see the Bibliography at the end of this Introduction).

Solid-Solution Mechanisms. There are only two mechanisms by which a crystal can dissolve atoms of a different element. If the atoms of the solute element are sufficiently smaller than the atoms comprising the solvent crystal, the solute atoms can fit into the spaces between the larger atoms to form an *interstitial solid solution* (see Fig. 23a). The only solute atoms small enough to fit into the interstices of metal crystals, however, are hydrogen, nitrogen, carbon, and boron. (The other small-diameter atoms, such as oxygen, tend to form compounds with metals rather than dissolve in them.) The rest of the elements dissolve in solid metals by replacing a solvent atom at a lattice point to form a *substitutional solid solution* (see Fig. 23b). When both small and large solute atoms are present, the solid solution can be both interstitial and substitutional. The addition of foreign atoms by either mechanism results in distortion of the crystal lattice and an increase in its internal energy. This distortion energy causes some hardening and strengthening of the alloy, called *solution hardening*. The solvent phase becomes saturated with the solute atoms and reaches its limit of homogeneity when the distortion energy reaches a critical value determined by the thermodynamics of the system.

Determination of Phase Diagrams

The data used to construct phase diagrams are obtained from a wide variety of measurements, many of which are conducted for reasons other than the determination of phase diagrams. No one research method will yield all of the information needed to construct an accurate diagram, and no diagram can be considered fully reliable without corroborating results obtained from the use of at least one other method.

Knowledge of the chemical composition of the sample and the individual phases is important in the construction of accurate phase diagrams. For example, the samples used should be prepared from high-purity constituents and accurately analyzed.

Chemical analysis is used in the determination of phase-field boundaries by measuring compositions of phases in a sample equilibrated at a fixed temperature by means of such methods as the diffusion-couple technique. The composition of individual phases can be measured by wet chemical methods, electron probe microanalysis, and so on.

Cooling Curves. One of the most widely used methods for the determination of phase boundaries is thermal analysis. The temperature of a sample is monitored while allowed to cool naturally from an elevated temperature (usually in the

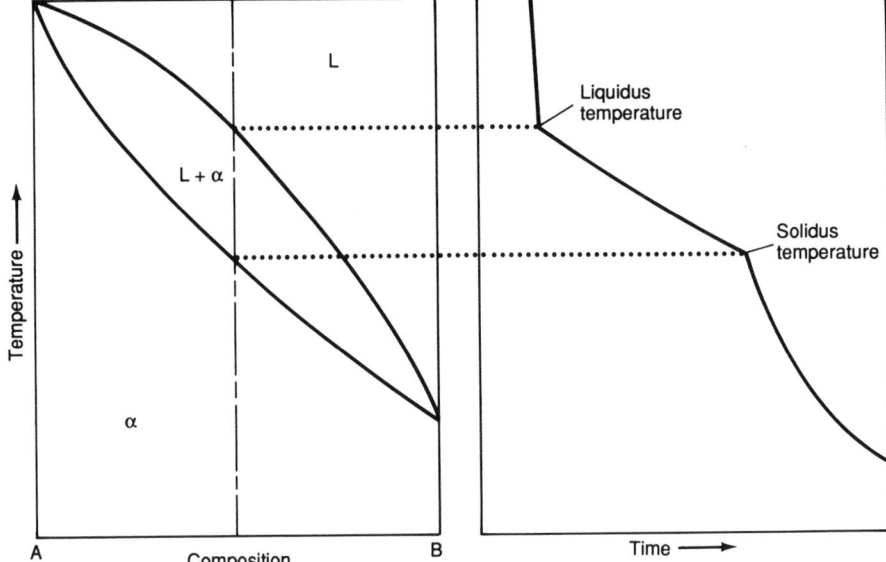

Fig. 27 Ideal freezing curve of a solid-solution alloy

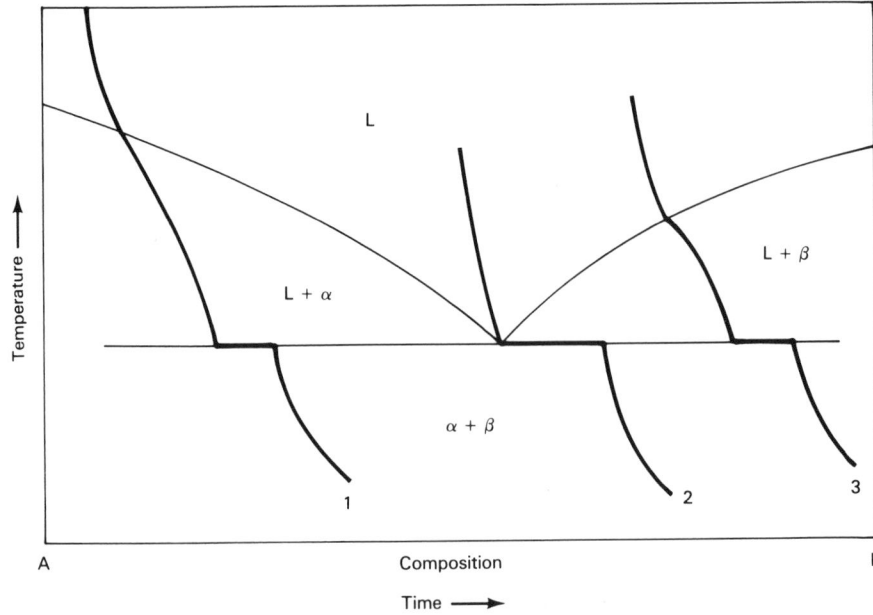

Fig. 28 Ideal freezing curves of (1) a hypoeutectic alloy, (2) a eutectic alloy, and (3) a hypereutectic alloy superimposed on a portion of a eutectic phase diagram. Source: Adapted from 66Pri

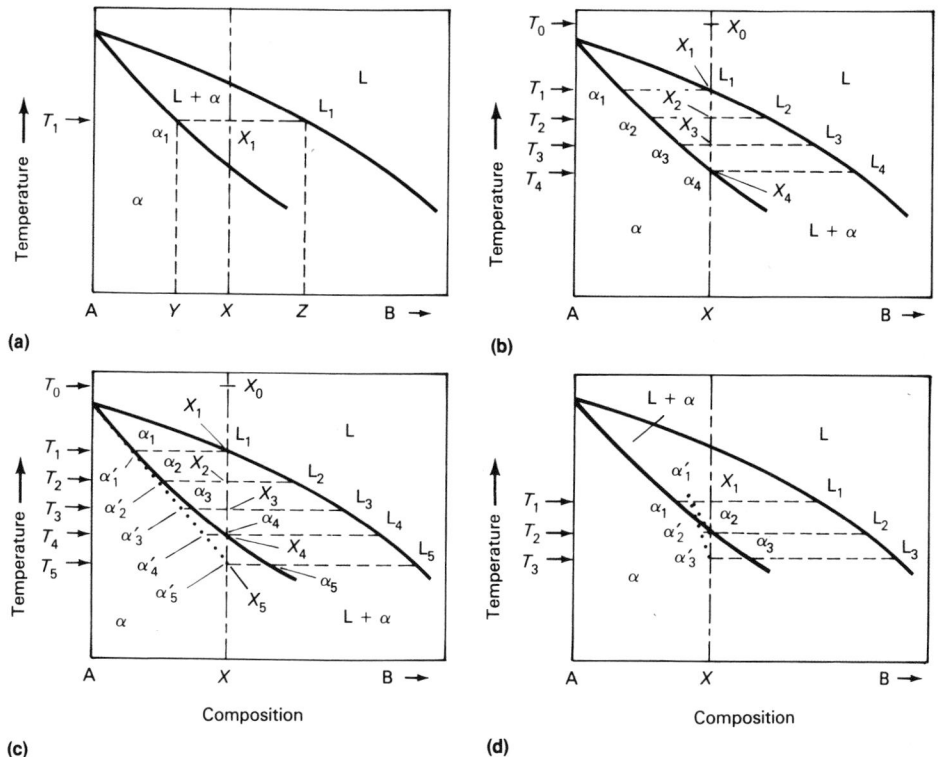

Fig. 29 Portion of a binary phase diagram containing a two-phase liquid-plus-solid field illustrating (a) the lever rule and its application to (b) equilibrium freezing, (c) nonequilibrium freezing and (d) heating of a homogenized sample. Source: 56Rhi

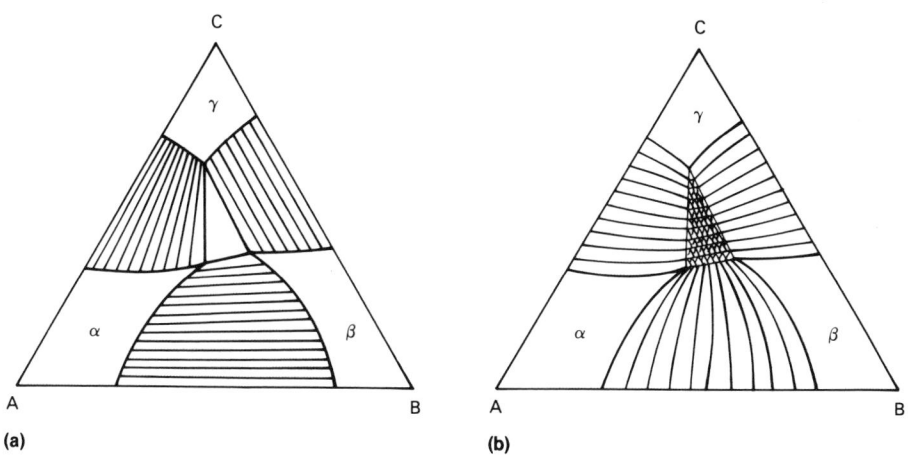

Fig. 30 Alternative systems for showing phase relationships in multiphase regions of ternary diagram isothermal sections. (a) Tie lines. (b) Phase-fraction lines. Source: 84Mor

liquid field). The shape of the resulting curves of temperature versus time are then analyzed for deviations from the smooth curve found for materials undergoing no phase changes (see Fig. 24).

When a pure element is cooled through its freezing temperature, its temperature is maintained near that temperature until freezing is complete (see Fig. 25). The true freezing/melting temperature, however, is difficult to determine from a cooling curve because of the nonequilibrium conditions inherent in such a dynamic test. This is illustrated in the cooling and heating curves shown in Fig. 26, where the effects of both supercooling and superheating can be seen. The dip in the cooling curve often found at the start of freezing is caused by a delay in the start of crystallization.

The continual freezing that occurs during cooling through a two-phase liquid-plus-solid field results in a reduced slope to the curve between the liquidus and solidus temperatures (see Fig. 27). By preparing several samples having composi-

tions across the diagram, the shape of the liquidus curves and the eutectic temperature of eutectic system can be determined (see Fig. 28). Cooling curves can be similarly used to investigate all other types of phase boundaries.

Differential thermal analysis is a technique used to increase test sensitivity by measuring the difference between the temperature of the sample and a reference material that does not undergo phase transformation in the temperature range being investigated.

Crystal Properties. X-ray diffraction methods are used to determine both crystal structure and lattice parameters of solid phases present in a system at various temperatures (phase identification). Lattice parameter scans across a phase field are useful in determining the limits of homogeneity of the phase; the parameters change with changing composition within the single-phase field, but they remain constant once the boundary is crossed into a two-phase field.

Physical Properties. Phase transformations within a sample are usually accompanied by changes in its physical properties (linear dimensions and specific volume, electrical properties, magnetic properties, hardness, etc.). Plots of these changes versus temperature or composition can be used in a manner similar to cooling curves to locate phase boundaries.

Metallographic Methods. Metallography can be used in many ways to aid in phase diagram determination. The most important problem with metallographic methods is that they usually rely on rapid quenching to preserve (or indicate) elevated-temperature microstructures for room-temperature observation. Hot-stage metallography, however, is an alternative. The application of metallographic techniques is discussed in the section on reading phase diagrams.

Thermodynamic Modeling. Because a phase diagram is a representation of the thermodynamic relationships between competing phases, it is theoretically possible to determine a diagram by considering the behavior of relevant Gibbs energy functions for each phase present in the system and physical models for the reactions in the system. How this can be accomplished is demonstrated for the simple problem of complete solid miscibility shown in Fig. 13. The models required to calculate the possible boundaries in the more complicated diagrams usually encountered are, of course, also more complicated, and involve the use of the equations governing solutions and solution interaction originally developed for physical chemistry. Although modeling alone cannot produce a reliable phase diagram, it is a powerful technique for validating those portions of a phase diagram already derived from experimental data. In addition, modeling can be used to estimate the relations in areas of diagrams where no experimental data exist, allowing much more efficient design of subsequent experiments.

Reading Phase Diagrams

Composition Scales. Phase diagrams to be used by scientists are usually plotted in atomic percentage (or mole fraction), while those to be used by

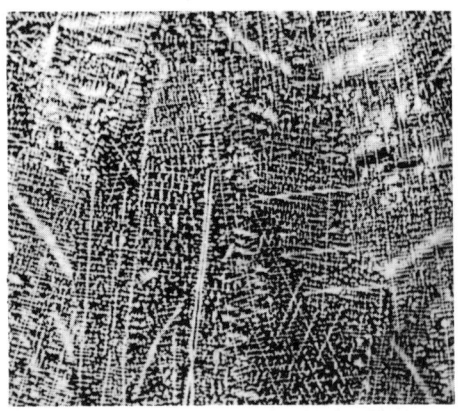

Fig. 31 Copper alloy C71500 (copper nickel, 30%) ingot. Dendritic structure shows coring: light areas are nickel rich; dark areas are low in nickel. 20×. Source: 85ASM

engineers are usually plotted in weight percentage. Conversions between weight and atomic composition also can be made using the equations given in the box below and standard atomic weights listed in the Appendix.

Lines and Labels. Magnetic transitions (Curie temperature and Néel temperature) and uncertain or speculative boundaries are usually shown in phase diagrams as nonsolid lines of various types. The components of metallic systems, which usually are pure elements, are identified in phase diagrams by their symbols. (The symbols used for chemical elements are listed in the Appendix.) Allotropes of polymorphic elements are distinguished by small (lower-case) Greek letter prefixes. (The Greek alphabet appears in the Appendix.)

Terminal solid phases are normally designated by the symbol (in parentheses) for the allotrope of the component element, such as (Cr) or (αTi). Continuous solid solutions are designated by the names of both elements, such as (Cu,Pd) or (βTi,βY).

Intermediate phases in phase diagrams are normally labeled with small (lower-case) Greek letters. However, certain Greek letters are conventionally used for certain phases, particularly disordered solutions: for example, β for disordered bcc, ζ or ϵ for disordered cph, γ for the γ-brass-type structure, and σ for the σCrFe-type structure.

For line compounds, a stoichiometric phase name is used in preference to a Greek letter (for example, A_2B_3 rather than δ). Greek letter prefixes are used to indicate high- and low-temperature forms of the compound (for example, αA_2B_3 for the low-temperature form and βA_2B_3 for the high-temperature form).

Lever Rule. As explained in the section on the features of phase diagrams, a tie line is an imaginary horizontal line drawn in a two-phase field connecting two points that represent two coexisting phases in equilibrium at the temperature indicated by the line. Tie lines can be used to determine the fractional amounts of the phases in

equilibrium by employing the *lever rule*. The lever rule is a mathematical expression derived by the principle of conservation of matter in which the phase amounts can be calculated from the bulk composition of the alloy and compositions of the conjugate phases, as shown in Fig. 29(a).

At the left end of the line between α_1 and L_1, the bulk composition is $Y\%$ component B and $100 - Y\%$ component A, and consists of 100% α solid solution. As the percentage of component B in the bulk composition moves to the right, some liquid appears along with the solid. With further increases in the amount of B in the alloy, more of the mixture consists of liquid, until the material becomes entirely liquid at the right end of the tie line. At bulk composition X, which is less than halfway to point L_1, there is more solid present than liquid. According to the lever rule, the percentages of the two phases present can be calculated as follows:

$$\% \text{ liquid} = \frac{\text{length of line } \alpha_1 X_1}{\text{length of line } \alpha_1 L_1} \times 100$$

$$\% \text{ solid } \alpha = \frac{\text{length of line } X_1 L_1}{\text{length of line } \alpha_1 L_1} \times 100$$

It should be remembered that the calculated amounts of the phases present are either in weight or atomic percentages and, as shown in the box on page 29, do not directly indicate the area or volume percentages of the phases observed in microstructures.

Phase-Fraction Lines. Reading the phase relationships in many ternary diagram sections (and other types of sections) often can be difficult because of the great many lines and areas present. *Phase-fraction lines* are used by some to simplify this task. In this approach, the sets of often non-

parallel tie lines in the two-phase fields of isothermal sections (see Fig. 30a) are replaced with sets of curving lines of equal phase fraction (Fig. 30b). Note that the phase-fraction lines extend through the three-phase region, where they appear as a triangular network. As with tie lines, the number of phase-fraction lines used is up to the individual using the diagram. Although this approach to reading diagrams may not seem helpful for such a simple diagram, it can be a useful aid in more complicated systems. For more information on this topic, see 84Mor and 91Mor.

Solidification. Tie lines and the lever rule can be used to understand the freezing of a solid-solution alloy. Consider the series of tie lines at different temperatures shown in Fig. 29(b), all of which intersect the bulk composition X. The first crystals to freeze have the composition α_1. As the temperature is reduced to T_2 and the solid crystals grow, more A atoms are removed from the liquid than B atoms, thus shifting the composition of the remaining liquid to L_2. Therefore, during freezing, the compositions of both the layer of solid freezing out on the crystals and the remaining liquid continuously shift to higher B contents and become leaner in A. Therefore, for equilibrium to be maintained, the solid crystals must absorb B atoms from the liquid and B atoms must migrate (diffuse) from the previously frozen material into subsequently deposited layers. When this happens, the average composition of the solid material follows the solidus line to temperature T_4, where it equals the bulk composition of the alloy.

Coring. If cooling takes place too rapidly for maintenance of equilibrium, the successive layers deposited on the crystals will have a range of local compositions from their centers to their edges (a condition known as *coring*). The development of

---

Composition Conversions

The following equations can be used to make conversions in binary systems:

$$\text{wt\% A} = \frac{\text{at.\% A} \times \text{at. wt of A}}{(\text{at.\% A} \times \text{at. wt of A}) + (\text{at.\% B} \times \text{at. wt of B})} \times 100$$

$$\text{at.\% A} = \frac{\text{wt\% A} / \text{at. wt of A}}{(\text{at.\% A} / \text{at. wt of A}) + (\text{wt\% B} / \text{at. wt of B})} \times 100$$

The equation for converting from atomic percentages to weight percentages in higher-order systems is similar to that for binary systems, except that an additional term is added to the denominator for each additional component. For ternary systems, for example:

$$\text{wt\% A} = \frac{\text{at.\% A} \times \text{at. wt of A}}{(\text{at.\% A} \times \text{at. wt of A}) + (\text{at.\% B} \times \text{at. wt of B}) + (\text{at.\% C} \times \text{at. wt of C})} \times 100$$

$$\text{at.\% A} = \frac{\text{wt\% A} / \text{at. wt of A}}{(\text{wt\% A} / \text{at. wt of A}) + (\text{wt\% B} / \text{at. wt of B}) + (\text{wt\% C} / \text{at. wt of C})} \times 100$$

The conversion from weight to atomic percentages for higher-order systems is easy to accomplish on a computer with a spreadsheet program.

this condition is illustrated in Fig. 29(c). Without diffusion of B atoms from the material that solidified at temperature T_1 into the material freezing at T_2, the average composition of the solid formed up to that point will not follow the solidus line. Instead it will remain to the left of the solidus, following compositions α'_1 through α'_5. Note that final freezing does not occur until temperature T_5, which means that nonequilibrium solidification takes place over a greater temperature range than equilibrium freezing. Because most metals freeze by the formation and growth of "treelike" crystals, called *dendrites*, coring is sometimes called *dendritic segregation*. An example of cored dendrites is shown in Fig. 31.

Liquation. Because the lowest freezing material in a cored microstructure is segregated to the edges of the solidifying crystals (the grain boundaries), this material can remelt when the alloy sample is heated to temperatures below the equilibrium solidus line. If grain-boundary melting (called *liquation*, or "burning") occurs while the sample also is under stress, such as during hot forming, the liquefied grain boundaries will rupture and the sample will lose its ductility and be characterized as *hot short*.

Liquation also can have a deleterious effect on the mechanical properties (and microstructure) of the sample after it returns to room temperature. This is illustrated in Fig. 29(d) for a homogenized sample. If homogenized alloy X is heated into the liquid-plus-solid region for some reason (inadvertently or during welding, etc.), it will begin to melt when it reaches temperature T_2; the first liquid to appear will have the composition L_2. When the sample is heated at normal rates to temperature T_1, the liquid formed so far will have a composition L_1, but the solid will not have time to reach the equilibrium composition α_1. The average composition will instead lie at some intermediate value, such as α'_1. According to the lever rule, this means that less than the equilibrium amount of liquid will form at this temperature. If the sample is then rapidly cooled from temperature T_1, solidification will occur in the normal manner, with a layer of material having composition α_1 deposited on existing solid grains. This is followed by layers of increasing B content up to composition α_3 at temperature T_3, where all of the liquid is converted to solid. This produces coring in the previously melted regions along the grain boundaries, and sometimes even

voids that decrease the strength of the sample. Homogenization heat treatment will eliminate the coring, but not the voids.

Eutectic Microstructures. When an alloy of eutectic composition (such as alloy 2 in Fig. 28) is cooled from the liquid state, the eutectic reaction occurs at the eutectic temperature, where the two distinct liquidus curves meet. At this temperature, both α and β solid phases must deposit on the grain nuclei until all of the liquid is converted to solid. This simultaneous deposition results in microstructures made up of distinctively shaped particles of one phase in a matrix of the other phase, or alternate layers of the two phases. Examples of characteristic eutectic microstructures include spheroidal, nodular, or globular; acicular (needles) or rod; and lamellar (platelets, Chinese script or dendritic, or filigreed). Each eutectic alloy has its own characteristic microstructure when slowly cooled (see Fig. 32). More rapid cooling, however, can affect the microstructure obtained (see Fig. 33). Care must be taken in characterizing eutectic structures, because elongated particles can appear nodular and flat platelets can appear elongated or needlelike when viewed in cross section.

If the alloy has a composition different from the eutectic composition (such as alloy 1 or 3 in Fig. 28), the alloy will begin to solidify before the eutectic temperature is reached. If the alloy is hypoeutectic (such as alloy 1), some dendrites of α will form in the liquid before the remaining liquid solidifies at the eutectic temperature. If the alloy is hypereutectic (such as alloy 3), the first (primary) material to solidify will be dendrites of β. The microstructure produced by slow cooling of a hypoeutectic and hypereutectic alloy will consist of relatively large particles of *primary constituent*, consisting of the phase that begins to freeze first surrounded by relatively fine eutectic structure. In many instances, the shape of the particles will show a relationship to their dendritic origin (see Fig. 34a). In other instances, the initial dendrites will have filled out somewhat into *idiomorphic particles* (particles having their own characteristic shape) that reflect the crystal structure of the phase (see Fig. 34b).

As stated earlier, cooling at a rate that does not allow sufficient time to reach equilibrium conditions will affect the resulting microstructure. For example, it is possible for an alloy in a eutectic system to obtain some eutectic structure in an alloy outside the normal composition range for such a structure. This is illustrated in Fig. 35. With relatively rapid cooling of alloy X, the composition of the solid material that forms will follow line α_1-α'_4 rather than the solidus line to α_4. As a result, the last liquid to solidify will have the eutectic composition L_4, rather than L_3, and will form some eutectic structure in the microstructure. The question of what takes place when the temperature reaches T_5 is discussed later.

Eutectoid Microstructures. Because the diffusion rates of atoms are so much lower in solids than in liquids, nonequilibrium transformation is even more important in solid/solid reactions (such as the eutectoid reaction) than in liquid/solid reactions (such as the eutectic reaction). With slow cooling through the eutectoid tempera-

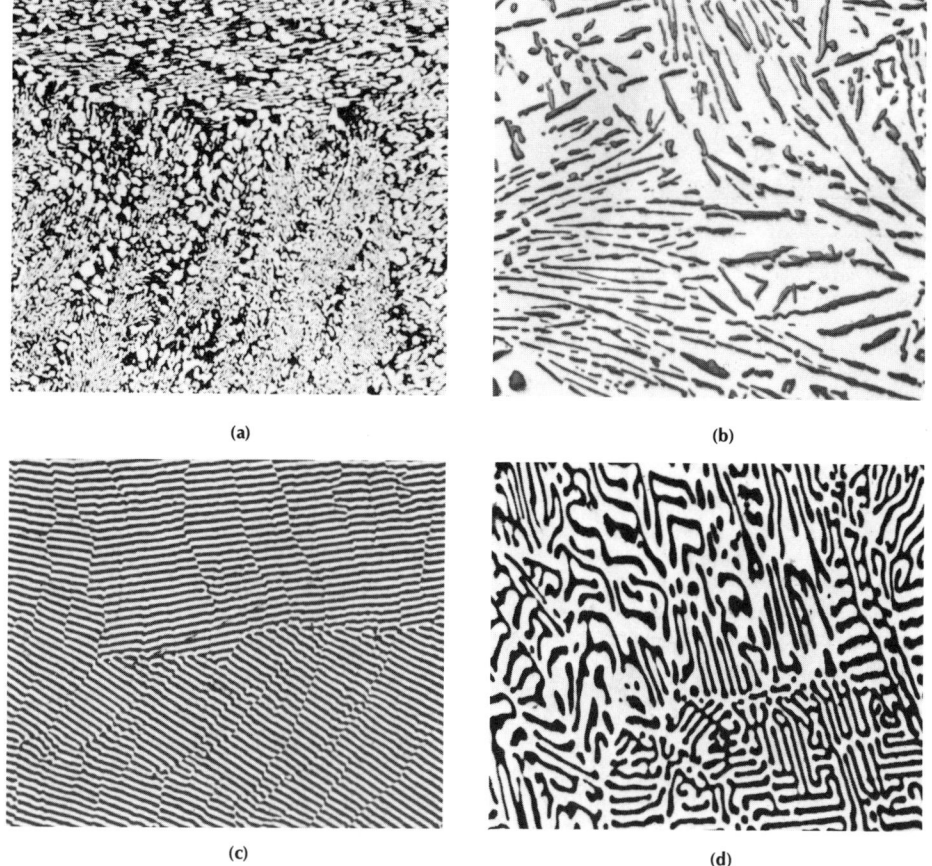

(a)

(b)

(c)

(d)

Fig. 32 Examples of characteristic eutectic microstructures in slowly cooled alloys. (a) 50Sn-50In alloy showing globules of tin-rich intermetallic phase (light) in a matrix of dark indium-rich intermetallic phase. 150×. (b) Al-13Si alloy showing an acicular structure consisting of short, angular particles of silicon (dark) in a matrix of aluminum. 200×. (c) Al-33Cu alloy showing a lamellar structure consisting of dark platelets of CuAl$_2$ and light platelets of aluminum solid solution. 180×. (d) Mg-37Sn alloy showing a lamellar structure consisting of Mg$_2$Sn "Chinese script" (dark) in a matrix of magnesium solid solution. 250×. Source: 85ASM

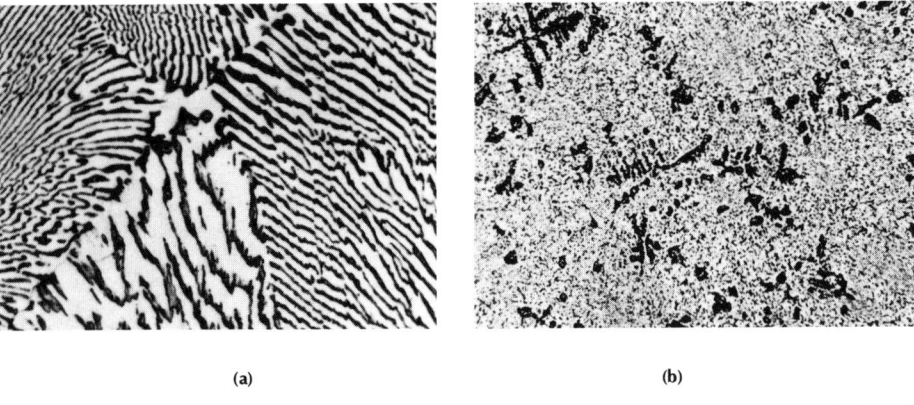

(a)   (b)

Fig. 33 Effect of cooling rate on the microstructure of Sn-37Pb alloy (eutectic soft solder). (a) Slowly cooled sample shows a lamellar structure consisting of dark platelets of lead-rich solid solution and light platelets of tin. 375×. (b) More rapidly cooled sample shows globules of lead-rich solid solution, some of which exhibit a slightly dendritic structure, in a matrix of tin. 375×. Source: 85ASM

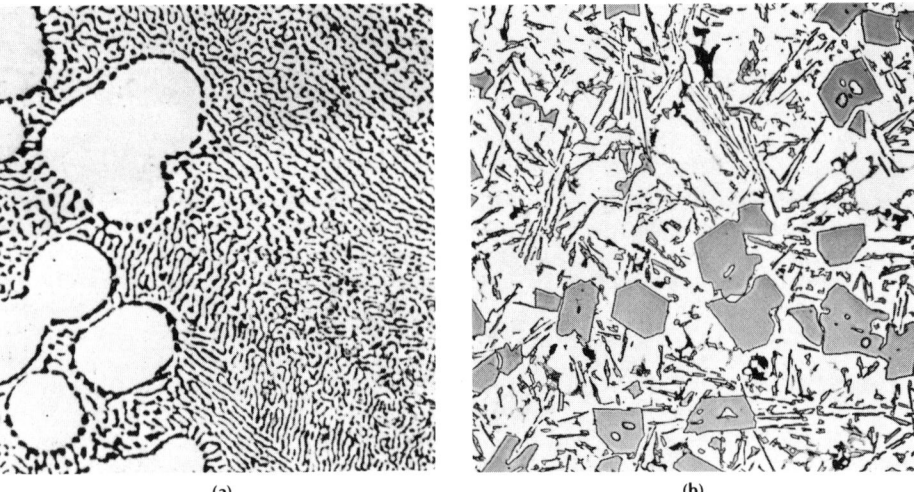

(a)   (b)

Fig. 34 Examples of primary particle shape. (a) Sn-30Pb hypoeutectic alloy showing dendritic particles of tin-rich solid solution in a matrix of tin-lead eutectic. 500×. (b) Al-19Si hypereutectic alloy, phosphorus-modified, showing idiomorphic particles of silicon in a matrix of aluminum-silicon eutectic. 100×. Source: 85ASM

ture, most alloys of eutectoid composition, such as alloy 2 in Fig. 36, transform from a single-phase microstructure to a lamellar structure consisting of alternate platelets of α and β arranged in groups (or "colonies"). The appearance of this structure is very similar to lamellar eutectic structure (see Fig. 37). When found in cast irons and steels, this structure is called "pearlite" because of its shiny mother-of-pearl appearance under the microscope (especially under oblique illumination); when similar eutectoid structure is found in nonferrous alloys, it often is called "pearlite-like" or "pearlitic."

The terms *hypoeutectoid* and *hypereutectoid* have the same relationship to the eutectoid composition as hypoeutectic and hypereutectic do in a eutectic system; alloy 1 in Fig. 36 is a hypoeutectoid alloy, whereas alloy 3 is hypereutectoid. The solid-state transformation of such alloys takes place in two steps, much like the freezing of hypoeutectic and hypereutectic alloys, except that the microconstituents that form before the eutectoid temperature is reached are referred to as *proeutectoid constituents* rather than "primary."

Microstructures of Other Invariant Reactions. Phase diagrams can be used in a manner

similar to that described in the discussion of eutectic and eutectoid reactions to determine the microstructures expected to result from cooling an alloy through any of the other six types of reactions listed in Table 1.

Solid-State Precipitation. If alloy X in Fig. 35 is homogenized at a temperature between T_3 and T_5, it will reach an equilibrium condition; that is, the β portion of the eutectic constituent will dissolve and the microstructure will consist solely of α grains. Upon cooling below temperature T_5, this microstructure will no longer represent equilibrium conditions, but instead will be supersaturated with B atoms. In order for the sample to return to equilibrium, some of the B atoms will tend to congregate in various regions of the sample to form colonies of new β material. The B atoms in some of these colonies, called *Guinier-Preston zones*, will drift apart, while other colonies will grow large enough to form incipient, but not distinct, particles. The difference in crystal structures and lattice parameters between the α and β phases causes lattice strain at the boundary between the two materials, thereby raising the total energy level of the sample and hardening and strengthening it. At this stage, the incipient parti-

cles are difficult to distinguish in the microstructure. Instead, there usually is only a general darkening of the structure. If sufficient time is allowed, the β regions will break away from their host grains of α and precipitate as distinct particles, thereby relieving the lattice strain and returning the hardness and strength to the former levels. This process is illustrated for a simple eutectic system, but it can occur wherever similar conditions exist in a phase diagram; that is, there is a range of alloy compositions in the system for which there is a transition on cooling from a single-solid region to a region that also contains a second solid phase, and where the boundary between the regions slopes away from the composition line as cooling continues. Several examples of such systems are shown schematically in Fig. 38.

Although this entire process is called *precipitation hardening*, the term normally refers only to the portion before much actual precipitation takes place. Because the process takes some time, the term *age hardening* is often used instead. The rate at which aging occurs depends on the level of supersaturation (how far from equilibrium), the amount of lattice strain originally developed (amount of lattice mismatch), the fraction left to be relieved (how far along the process has progressed), and the aging temperature (the mobility of the atoms to migrate). The β precipitate usually takes the form of small idiomorphic particles situated along the grain boundaries and within the grains of α phase. In most instances, the particles are more or less uniform in size and oriented in a systematic fashion. Examples of precipitation microstructures are shown in Fig. 39.

Examples of Phase Diagrams

The general principles of reading alloy phase diagrams are discussed in the preceding section. The application of these principles to actual diagrams for typical alloy systems is illustrated below.

The Copper-Zinc System. The metallurgy of brass alloys has long been of great commercial importance. The copper and zinc contents of five of the most common wrought brasses are:

| UNS No. | Common name | Zinc content, wt% | |
		Nominal	Range
C23000	Red brass, 85%	15	14.0-16.0
C24000	Low brass, 80%	20	18.5-21.5
C26000	Cartridge brass, 70%	30	28.5-31.5
C27000	Yellow brass, 65%	35	32.5-37.0
C28000	Muntz metal, 60%	40	37.0-41.0

As can be seen in Fig. 40, these alloys encompass a wide range of the copper-zinc phase diagram. The alloys on the high-copper end (red brass, low brass, and cartridge brass) lie within the copper solid-solution phase field and are called alpha brasses after the old designation for this field. As expected, the microstructure of these brasses consists solely of grains of copper solid solution (see

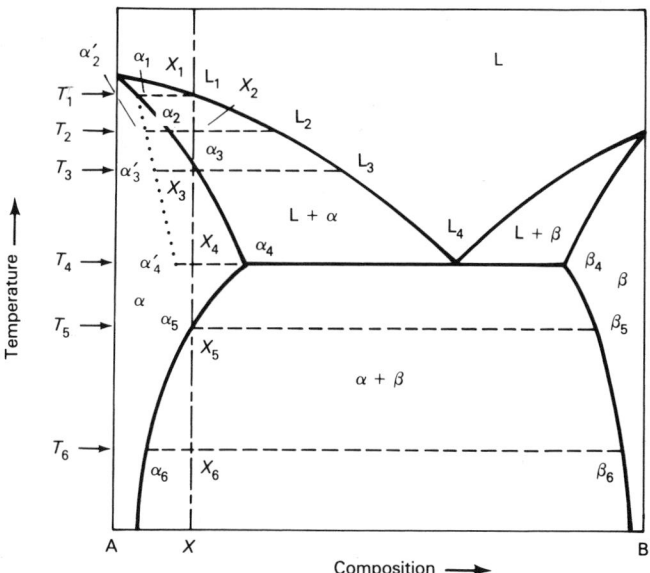

Fig. 35 Schematic binary phase diagram, illustrating the effect of cooling rate on an alloy lying outside the equilibrium eutectic transformation line. Rapid solidification into a terminal phase field can result in some eutectic structure being formed; homogenization at temperatures in the single-phase field will eliminate the eutectic structure; β phase will precipitate out of solution upon slow cooling into the α-plus-β field. Source: Adapted from 56Rhi

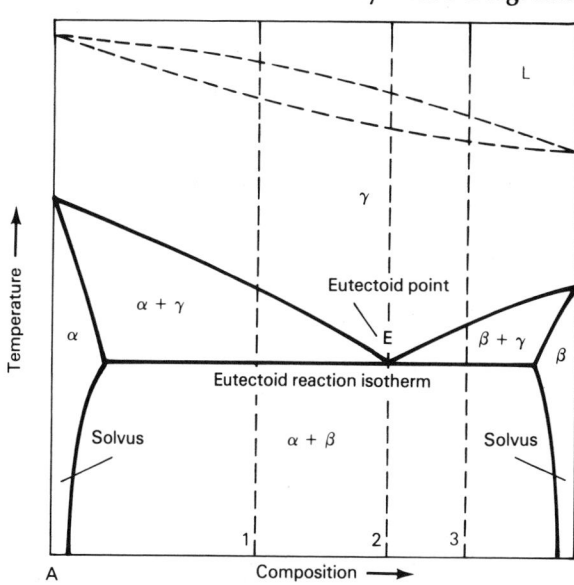

Fig. 36 Schematic binary phase diagram of a eutectoid system. Source: Adapted from 56Rhi

---

Fig. 41a). The strain on the copper crystals caused by the presence of the zinc atoms, however, produces solution hardening in the alloys. As a result, the strength of the brasses, in both the work-hardened and the annealed conditions, increases with increasing zinc content.

The composition range for those brasses containing higher amounts of zinc (yellow brass and Muntz metal), however, overlaps into the two-phase (Cu)-plus-β field. Therefore, the microstructure of these so-called alpha-beta alloys shows various amounts of β phase (see Fig. 41b and c), and their strengths are further increased over those of the alpha brasses.

The Aluminum-Copper System. Another alloy system of great commercial importance is aluminum-copper. Although the phase diagram of this system is fairly complicated (see Fig. 42), the alloys of concern in this discussion are limited to the region at the aluminum side of the diagram where a simple eutectic is formed between the aluminum solid solution and the θ (Al₂Cu) phase. This family of alloys (designated the 2xxx series) has nominal copper contents ranging from 2.3 to 6.3 wt%, making them hypoeutectic alloys.

A critical feature of this region of the diagram is the shape of the aluminum solvus line. At the eutectic temperature (548.2 °C, or 1018.8 °F), 5.65 wt% Cu will dissolve in aluminum. At lower temperatures, however, the amount of copper that can remain in the aluminum solid solution under equilibrium conditions drastically decreases, reaching less than 1% at room temperature. This is the typical shape of the solvus line for precipitation hardening; if any of these alloys are homogenized at temperatures in or near the solid-solution phase field, they can be strengthened by aging at a substantially lower temperature.

The Titanium-Aluminum, Titanium-Chromium, and Titanium-Vanadium Systems. The phase diagrams of titanium systems are domi-

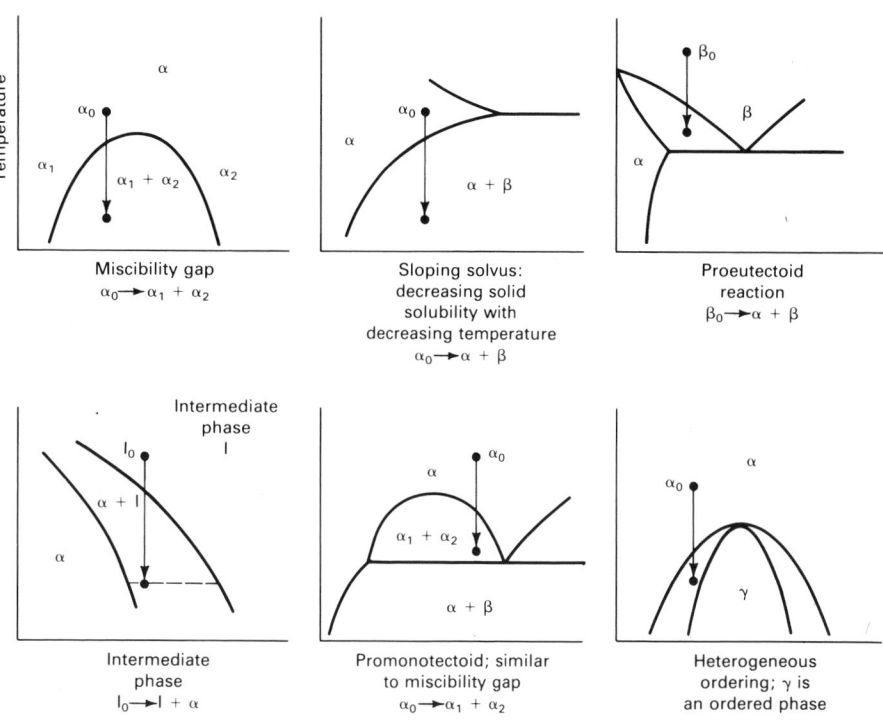

Fig. 38 Examples of binary phase diagrams that give rise to precipitation reactions. Source: 85ASM

Fig. 37 Fe-0.8C alloy showing a typical pearlite eutectoid structure of alternate layers of light ferrite and dark cementite. 500×. Source: 85ASM

nated by the fact that there are two allotropic forms of solid titanium: cph αTi is stable at room temperature and up to 882 °C (1620 °F); bcc βTi is stable from 882 °C (1620 °F) to the melting temperature. Most alloying elements used in commercial titanium alloys can be classified as alpha stabilizers (such as aluminum) or beta stabilizers (such as vanadium and chromium), depending on whether the allotropic transformation temperature is raised or lowered by the alloying addition (see Fig. 43). Beta stabilizers are further classified as those that are completely miscible with βTi (such as vanadium, molybdenum, tantalum, and niobium) and those that form eutectoid systems with titanium (such as chromium and iron). Tin and zirconium also are often alloyed in titanium, but instead of stabilizing either phase, they have extensive solubilities in both αTi and βTi. The microstructures of commercial titanium alloys are complicated, because most contain more than one of these four types of alloying elements.

The Iron-Carbon System. The iron-carbon diagram maps out the stable equilibrium conditions between iron and the graphitic form of carbon (see Fig. 44). Note that there are three allotropic forms of solid iron: the low-temperature phase, α; the medium-temperature phase, γ; and the high-temperature phase, δ. In addition, ferritic iron undergoes a magnetic phase transition at 771 °C (1420 °F) between the low-temperature ferromagnetic state and the higher-temperature paramagnetic state. The common name for bcc α-iron is "ferrite" (from *ferrum*, Latin for "iron"); the fcc γ phase is called "austenite" after William Roberts-Austen; bcc δ-iron is also commonly called ferrite, because (except for its temperature range) it is the same as α-iron. The main feature of the iron-carbon diagram is the presence of both a eutectic and a eutectoid reaction, along with the great difference between the solid solubilities of carbon in ferrite and austenite. It is these features that allow such a wide variety of microstructures and mechanical properties to be developed in iron-carbon alloys through proper heat treatment.

The Iron-Cementite System. In the solidification of steels, stable equilibrium conditions do not exist. Instead, any carbon not dissolved in the iron is tied up in the form of the metastable intermetallic compound, Fe₃C (also called cementite because of its hardness), rather than remaining as free graphite (see Fig. 45). It is, therefore, the iron-cementite phase diagram, rather than the iron-carbon diagram, that is important to industrial metallurgy. It should be remembered, however, that although cementite is an extremely enduring phase, given sufficient time, or the presence of a catalyzing substance, it will break down to iron and carbon. In cast irons, silicon is the catalyzing agent that allows free carbon (flakes, nodules, etc.) to appear in the microstructure (see Fig. 46).

The boundary lines on the iron-carbon and iron-cementite diagrams that are important to the heat treatment of steel and cast iron have been assigned special designations, which have been found useful in describing the treatments. These lines, where thermal arrest takes place during heating or cooling due to a solid-state reaction,

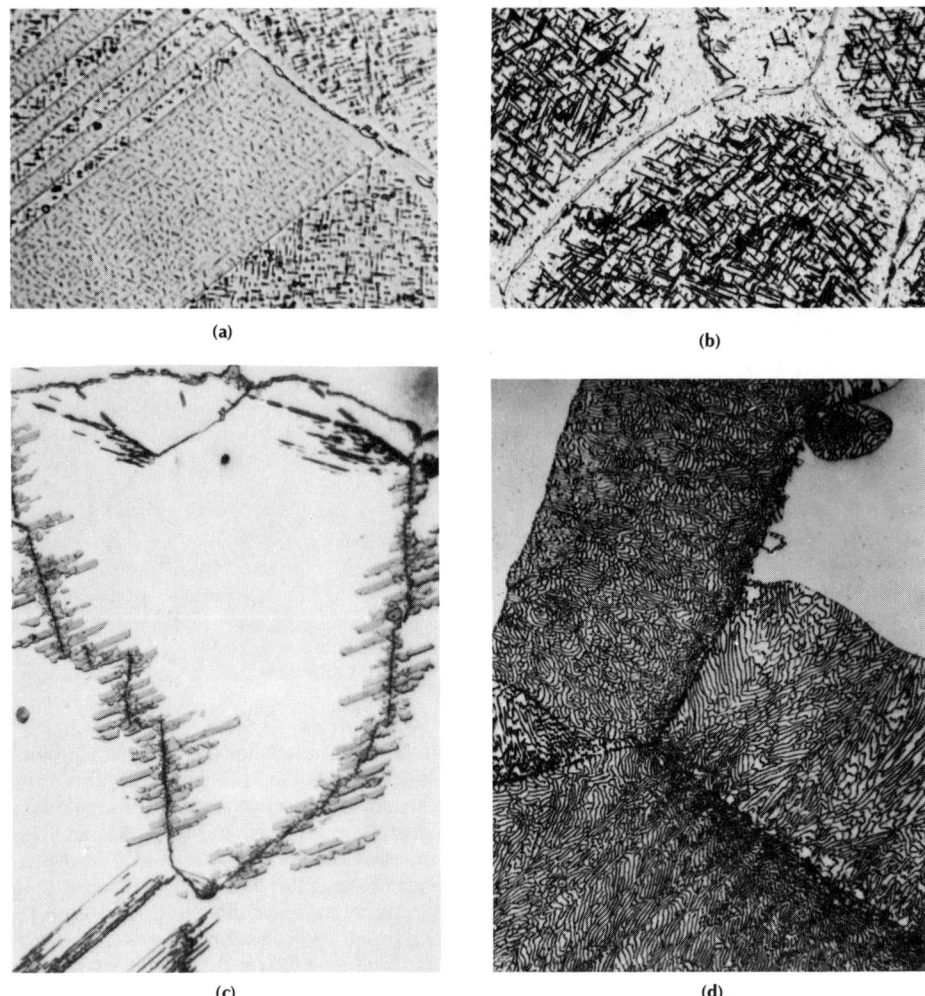

Fig. 39 Examples of characteristic precipitation microstructures. (a) General and grain-boundary precipitation of Co₃Ti (γ′ phase) in a Co-12Fe-6Ti alloy aged 3 × 10³ min at 800 °C (1470 °F). 1260×. (b) General precipitation (intragranular Widmanstätten) and localized grain-boundary precipitation in an Al-18Ag alloy aged 90 h at 375 °C (710 °F), with a distinct precipitation-free zone near the grain boundaries. 500×. (c) Preferential, or localized, precipitation along grain boundaries in a Ni-20Cr-1Al alloy. 500×. (d) Cellular, or discontinuous, precipitation growing out uniformly from the grain boundaries in an Fe-24.8Zn alloy aged 6 min at 600 °C (1110 °F). 1000×. Source: 85ASM

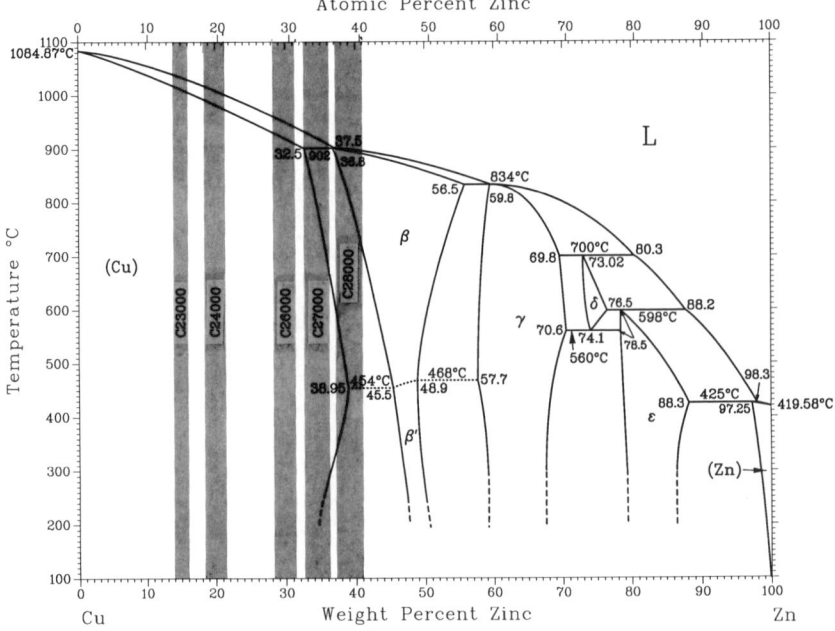

Fig. 40 The copper-zinc phase diagram, showing the composition range for five common brasses. Source: Adapted from 90Mas

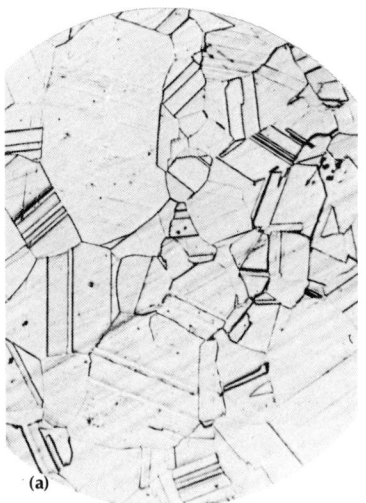

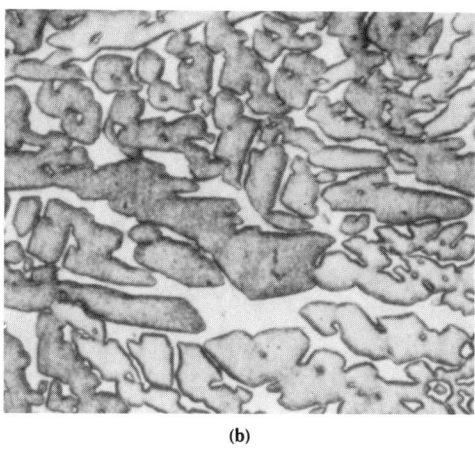

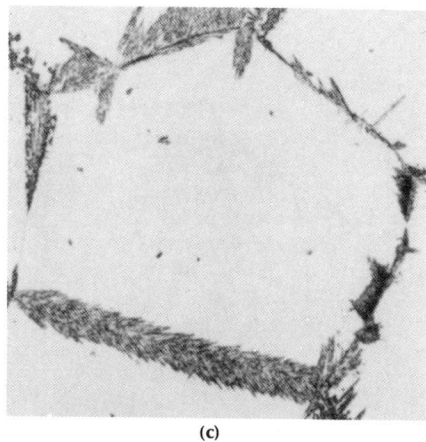

Fig. 41 The microstructures of two common brasses. (a) C26000 (cartridge brass, 70%), hot rolled, annealed, cold rolled 70%, and annealed at 638 °C (1180 °F), showing equiaxed grains of copper solid solution. Some grains are twinned. 75×. (b) C28000 (Muntz metal, 60%) ingot, showing dendrites of copper solid solution in a matrix of β. 200×. (c) C28000 (Muntz metal, 60%), showing feathers of copper solid solution that formed at β grain boundaries during quenching of the all-β structure. 100×. Source: 85ASM

are assigned the letter "A" for *arrêt* (French for "arrest"). These designations are shown in Fig. 45. To further differentiate the lines, an "e" is added to identify those indicating the changes occurring at equilibrium (to give Ae_1, Ae_3, Ae_4, and Ae_{cm}). Also, because the temperatures at which changes actually occur on heating or cooling are displaced somewhat from the equilibrium values, the "e" is replaced with "c" (for *chauffage*, French for "heating") when identifying the slightly higher temperatures associated with changes that occur on heating. Likewise, "e" is replaced with "r" (for *refroidissement*, French for "cooling") when identifying those slightly lower temperatures associated with changes occurring on cooling. These designations are convenient terms because they are used not only for binary alloys of iron and carbon, but also for commercial steels and cast irons, regardless of the other elements present in them. Alloying elements such as manganese, chromium, nickel, and molybdenum, however, do affect these temperatures (mainly A_3). For example, nickel lowers A_3, whereas chromium raises it.

The microstructures obtained in steels by slowly cooling are as follows. At carbon contents from 0.007 to 0.022%, the microstructure consists of ferrite grains with cementite precipitated in from ferrite, usually in too fine a form to be visible by light microscopy. (Because certain other metal atoms that may be present can substitute for some of the iron atoms in Fe_3C, the more general term, "carbide," is often used instead of "cementite" when describing microstructures.) In the hypoeutectoid range (from 0.022 to 0.76% C), ferrite and pearlite grains constitute the microstructure. In the hypereutectoid range (from 0.76 to 2.14% C), pearlite grains plus carbide precipitated from austenite are visible.

Slowly cooled hypoeutectic cast irons (from 2.14 to 4.3% C) have a microstructure consisting of dendritic pearlite grains (transformed from hypoeutectic primary austenite) and grains of iron-cementite eutectic (called "ledeburite") con-

sisting of carbide and transformed austenite, plus carbide precipitated from austenite and particles of free carbon. For slowly cooled hypereutectic cast iron (between 4.3 and 6.67% C), the microstructure shows primary particles of carbide and free carbon, plus grains of transformed austenite.

Cast irons and steels, of course, are not used in their slowly cooled as-cast condition. Instead, they are more rapidly cooled from the melt, then subjected to some type of heat treatment and, for wrought steels, some type of hot and/or cold work. The great variety of microconstituents and

microstructures that result from these treatments is beyond the scope of a discussion of stable and metastable equilibrium phase diagrams. Phase diagrams are invaluable, however, when designing heat treatments. For example, normalizing is usually accomplished by air cooling from about 55 °C (100 °F) above the upper transformation temperature (A_3 for hypoeutectoid alloys and A_{cm} for hypereutectoid alloys). Full annealing is done by controlled cooling from about 28 to 42 °C (50 to 75 °F) above A_3 for both hypoeutectoid and hypereutectoid alloys. All tempering and process

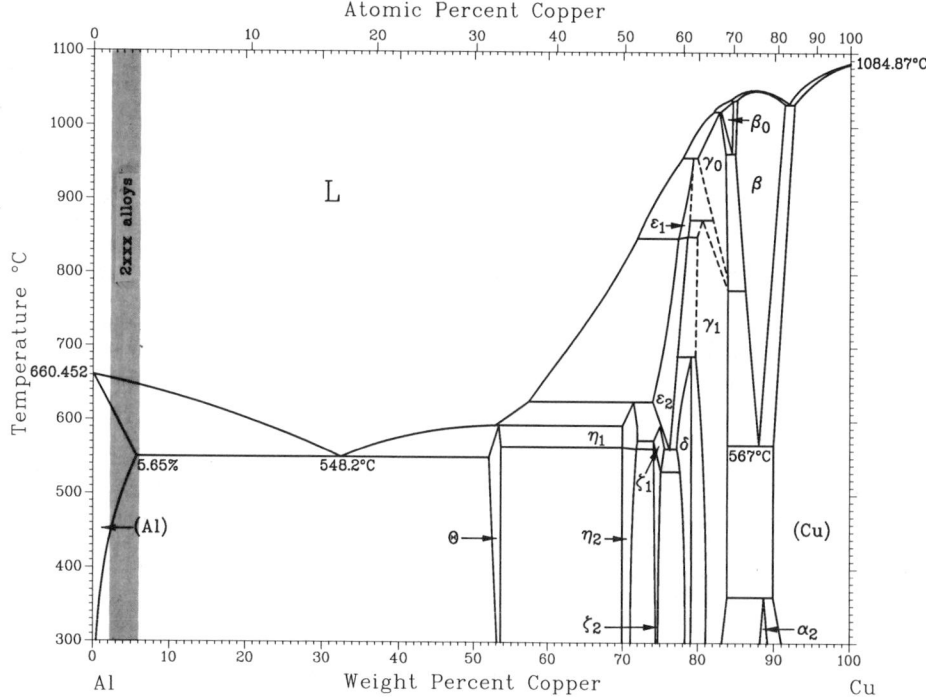

Fig. 42 The aluminum-copper phase diagram, showing the composition range for the 2xxx series of precipitation-hardenable aluminum alloys. Source: 90Mas

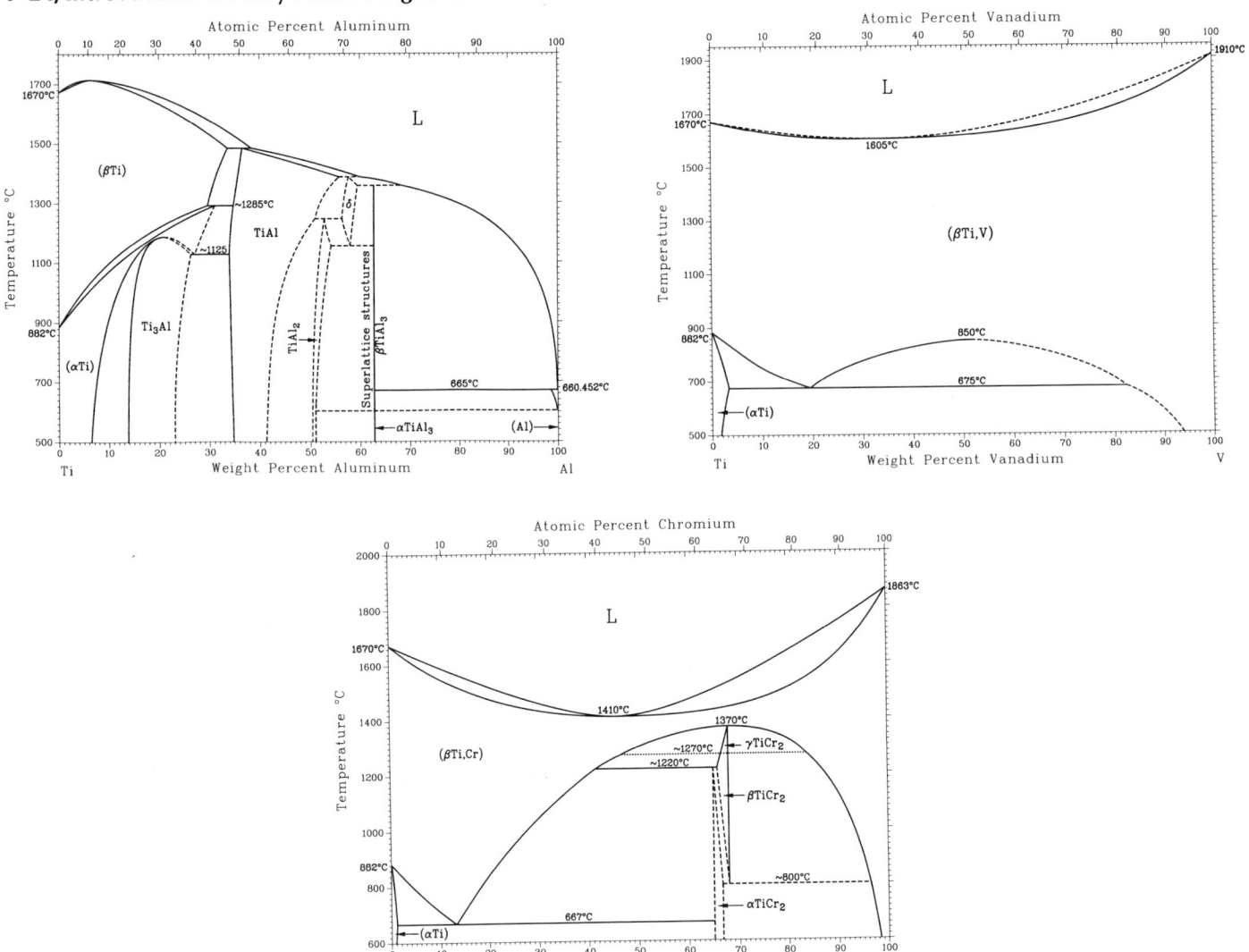

Fig. 43 Three representative binary titanium phase diagrams, showing alpha stabilization (Ti-Al), beta stabilization with complete miscibility (Ti-V), and beta stabilization with a eutectoid reaction (Ti-Cr). Source: 90Mas

annealing operations are done at temperatures below the lower transformation temperature (A₁). Austenitizing is done at a temperature sufficiently above A₃ and A_cm to ensure complete transformation to austenite, but low enough to prevent grain growth from being too rapid.

The Iron-Chromium-Nickel System. Many commercial cast irons and steels contain ferrite-stabilizing elements (such as silicon, chromium, molybdenum, and vanadium) and/or austenite stabilizers (such as manganese and nickel). The diagram for the binary iron-chromium system is representative of the effect of a ferrite stabilizer (see Fig. 47). At temperatures just below the solidus, bcc chromium forms a continuous solid solution with bcc (δ) ferrite. At lower temperatures, the γ-iron phase appears on the iron side of the diagram and forms a "loop" extending to about 11.2% Cr. Alloys containing up to 11.2% Cr, and sufficient carbon, are hardenable by quenching from temperatures within the loop.

At still lower temperatures, the bcc solid solution is again continuous bcc ferrite, but this time with αFe. This continuous bcc phase field confirms that δ-ferrite is the same as α-ferrite. The nonexistence of γ-iron in Fe-Cr alloys having more than about 13% Cr, in the absence of carbon, is an important factor in both the hardenable and nonhardenable grades of iron-chromium stainless steels. At these lower temperatures, a material known as sigma phase also appears in different amounts from about 14 to 90% Cr. Sigma is a hard, brittle phase and usually should be avoided in commercial stainless steels. Formation of sigma, however, is time dependent; long periods at elevated temperatures are usually required.

The diagram for the binary iron-nickel system is representative of the effect of an austenite stabilizer (see Fig. 47). The fcc nickel forms a continuous solid solution with fcc (γ) austenite that dominates the diagram, although the α-ferrite phase field extends to about 6% Ni. The diagram for the ternary iron-chromium-nickel system shows how the addition of ferrite-stabilizing chromium affects the iron-nickel system (see Fig. 48). As can be seen, the popular 18-8 stainless steel, which contains about 8% Ni, is an all-austenite alloy at 900 °C (1652 °F), even though it also contains about 18% Cr.

Practical Applications of Phase Diagrams

The following are but a few of the many instances where phase diagrams and phase relationships have proved invaluable in the efficient solving of practical metallurgical problems.

Alloy Design

Age Hardening Alloys. One of the earliest uses of phase diagrams in alloy development was in the suggestion in 1919 by the U.S. Bureau of Standards that precipitation of a second phase from solid solution would harden an alloy. The age hardening of certain aluminum-copper alloys (then called "Duralumin" alloys) had been accidentally discovered in 1904, but this process was

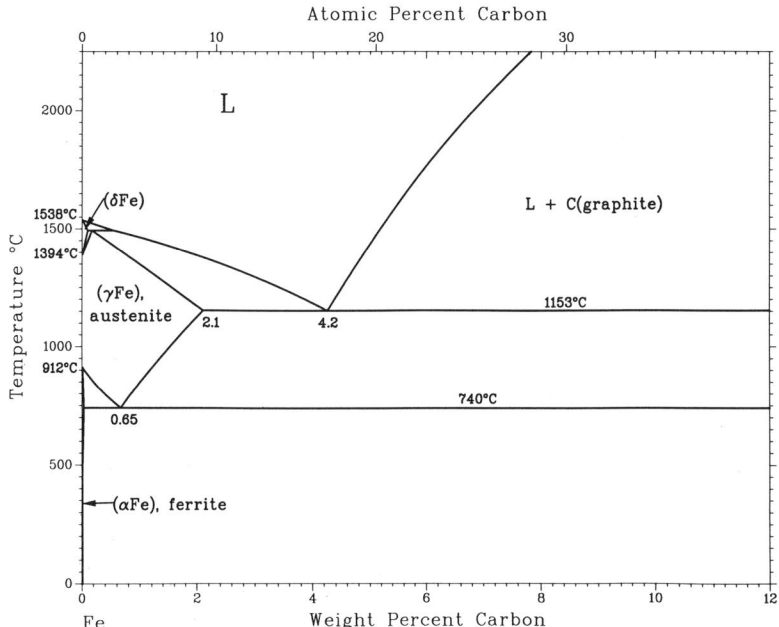

Fig. 44 The iron-carbon phase diagram. Source: Adapted from 90Mas

thought to be a unique and curious phenomenon. The work at the Bureau, however, showed the scientific basis of this process (which was discussed in previous sections of this Introduction). This work has now led to the development of several families of commercial "age hardening" alloys covering different base metals.

Austenitic Stainless Steel. In connection with a research project aimed at the conservation of always expensive, sometimes scarce, materials, the question arose: Can manganese and aluminum be substituted for nickel and chromium in stainless steels? (In other words, can standard chromium-nickel stainless steels be replaced with an austenitic alloy system?) The answer came in two stages—in both instances with the help of phase diagrams. It was first determined that manganese should be capable of replacing nickel because it stabilizes the γ-iron phase (austenite), and

aluminum may substitute for chromium because it stabilizes the α-iron phase (ferrite), leaving only a small γ loop (see Fig. 47 and 49). Aluminum is known to impart good high-temperature oxidation resistance to iron. Next, the literature on phase diagrams of the aluminum-iron-manganese system was reviewed, which suggested that a range of compositions exists where the alloy would be austenitic at room temperature. A nonmagnetic alloy with austenitic structure containing 44% Fe, 45% Mn, and 11% Al was prepared. However, it proved to be very brittle, presumably because of the precipitation of a phase based on β-Mn. By examining the phase diagram for carbon-iron-manganese (Fig. 50), as well as the diagram for aluminum-carbon-iron, the researcher determined that the problem could be solved through the addition of carbon to the aluminum-iron-manganese system, which would move the

composition away from the βMn phase field. The carbon addition also would further stabilize the austenite phase, permitting reduced manganese content. With this information, the composition of the alloy was modified to 7 to 10% Al, 30 to 35% Mn, and 0.75 to 1% C, with the balance iron. It had good mechanical properties, oxidation resistance, and moderate stainlessness.

Permanent Magnets. A problem with permanent magnets based on Fe-Nd-B is that they show high magnetization and coercivity at room temperature, but unfavorable properties at higher temperatures. Because hard magnetic properties are limited by nucleation of severed magnetic domains, the surface and interfaces of grains in the sintered and heat-treated material are the controlling factor. Therefore, the effects of alloying additives on the phase diagrams and microstructural development of the Fe-Nd-B alloy system plus additives were studied. These studies showed that the phase relationships and domain-nucleation difficulties were very unfavorable for the production of a magnet with good magnetic properties at elevated temperatures by the sintering method. However, such a magnet might be produced from Fe-Nd-C material by some other process, such as melt spinning or bonding (see 91Hay).

Processing

Hacksaw Blades. In the production of hacksaw blades, a strip of high-speed steel for the cutting edges is joined to a backing strip of low-alloy steel by laser or electron beam welding. As a result, a very hard martensitic structure forms in the weld area that must be softened by heat treatment before the composite strip can be further rolled or set. To avoid the cost of the heat treatment, an alternative technique was investigated. This technique involved alloy additions during welding to create a microstructure that would not require subsequent heat treatment. Instead of expensive experiments, several mathematical simulations were made based on additions of various steels or pure metals. In these simulations, the hardness of

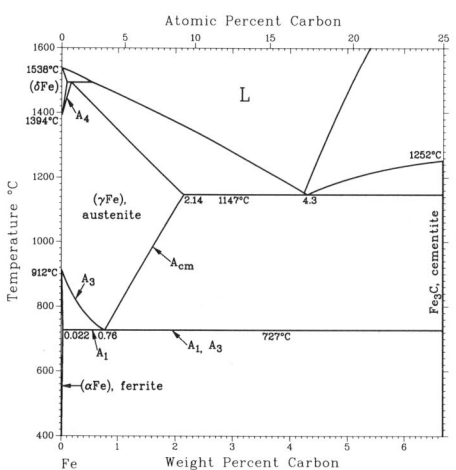
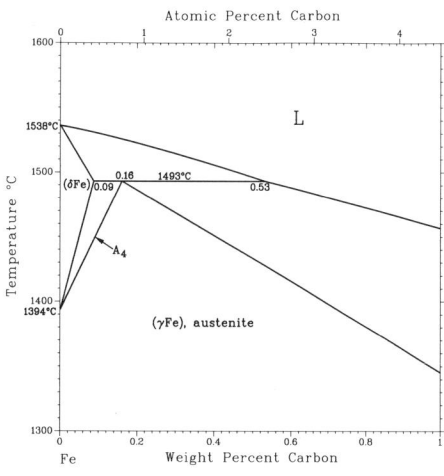
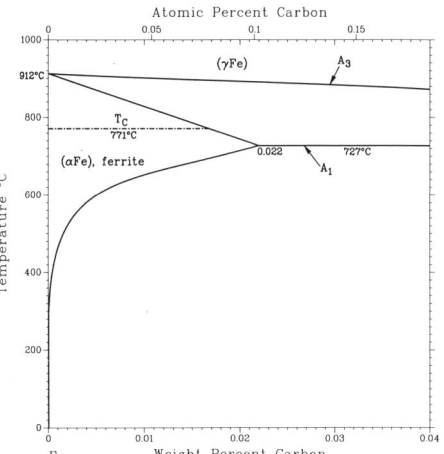

Fig. 45 The iron-cementite phase diagram and details of the (δFe) and (αFe) phase fields. Source: Adapted from 90Mas

the weld was determined by combining calculations of the equilibrium phase diagrams and available information to calculate (assuming the average composition of the weld) the martensite transformation temperatures and amounts of retained austenite, untransformed ferrite, and carbides formed in the postweld microstructure. Of those alloy additions considered, chromium was found to be the most efficient (see 91Hay).

Hardfacing. A phase diagram was used to design a nickel-base hardfacing alloy for corrosion and wear resistance. For corrosion resistance, a matrix of at least 15% Cr was desired; for abrasion resistance, a minimum amount of primary chromium-boride particles was desired. After consulting the B-Cr-Ni phase diagram, a series of samples having acceptable amounts of total chro-

mium borides and chromium matrix were made and tested. Subsequent fine tuning of the composition to ensure fabricability of welding rods, weldability, and the desired combination of corrosion, abrasion, and impact resistance led to a patented alloy.

Performance

Heating elements made of Nichrome (a nickel-chromium-iron alloy registered by Driver-Harris Company, Inc., Harrison, NJ) in a heat treating furnace were failing prematurely. Reference to nickel-base phase diagrams suggested that low-melting eutectics can be produced by very small quantities of the chalcogens (sulfur, selenium, or tellurium), and it was thought that one of these

eutectics could be causing the problem. Investigation of the furnace system resulted in the discovery that the tubes conveying protective atmosphere to the furnace were made of sulfur-cured rubber, which could result in liquid metal being formed at temperatures as low as 637 °C (1179 °F) (see Fig. 51). Armed with this information, a metallurgist solved the problem by substituting neoprene for the rubber.

Electric Motor Housings. At moderately high service temperatures, cracks developed in electric motor housings that had been extruded from aluminum produced from a combination of recycled and virgin metal. Extensive studies revealed that the cracking was caused by small amounts of lead and bismuth in the recycled metal reacting to form bismuth-lead eutectic at the grain boundaries at 327 and ~270 °C (621 and ~518 °F), respectively, much below the melting point of pure aluminum (660.45 °C, or 1220.81 °F) (see Fig. 52). The question became: How much lead and bismuth can be tolerated in this instance? The phase diagrams showed that aluminum alloys containing either lead or bismuth in amounts exceeding their respective solubility limits (<0.05% and ~0.2%) can lead to hot cracking of the aluminum.

Carbide Cutting Tools. A manufacturer of carbide cutting tools once experienced serious trouble with brittleness of the sintered carbide. No impurities were found. The range of compositions for cobalt-bonded sintered carbides is shown in the shaded area of Fig. 53, along the dashed line connecting pure tungsten carbide (marked "WC") on the right and pure cobalt at the lower left. At 1400 °C (2552 °F), materials with these compositions consist of particles of tungsten carbide suspended in liquid metal. However, when there is a deficiency of carbon, compositions drop into the region labeled WC + η + liquid, or the region labeled WC + η where tungsten carbide particles are surrounded by a matrix of η phase. The η

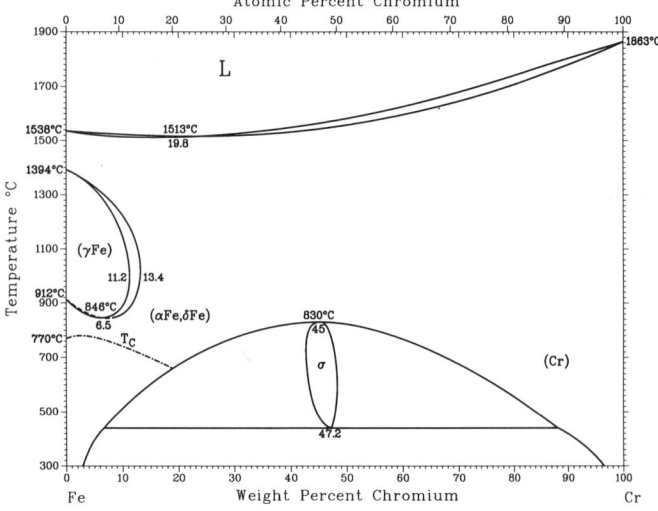

(a)

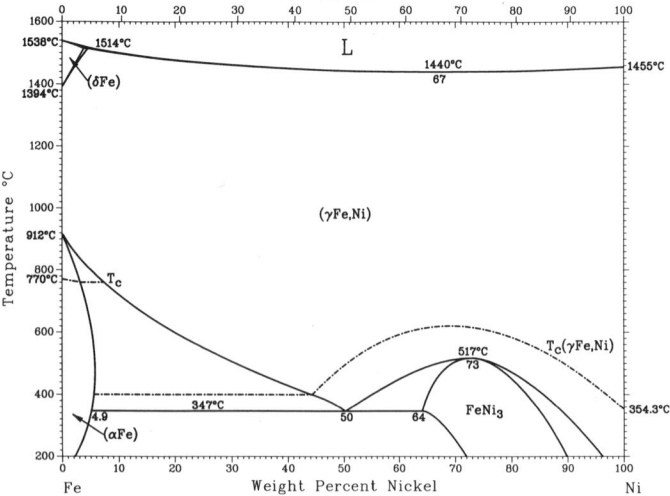

(b)

Fig. 46 The microstructures of two types of cast irons. (a) As-cast class 30 gray iron, showing type A graphite flakes in a matrix of pearlite. 500×. (b) As-cast grade 60-45-12 ductile iron, showing graphite nodules (produced by the addition of a calcium-silicon compound during pouring) in a ferrite matrix. 100×. Source: 85ASM

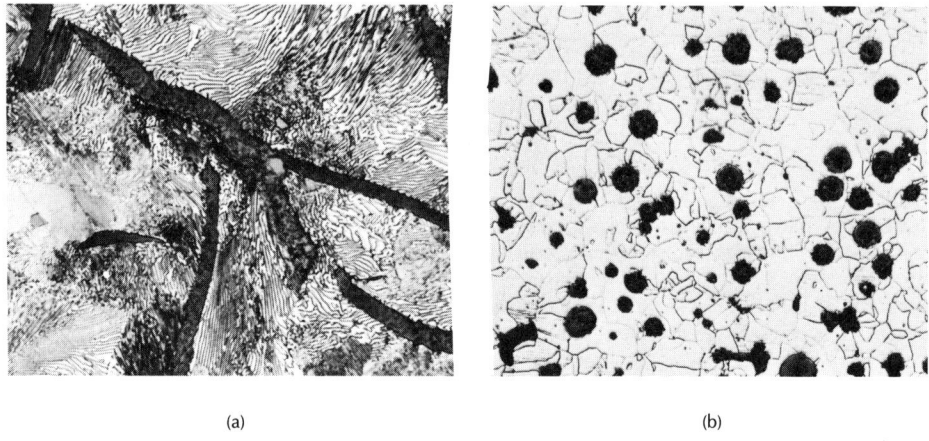

Fig. 47 Two representative binary iron phase diagrams, showing ferrite stabilization (Fe-Cr) and austenite stabilization (Fe-Ni). Source: 90Mas

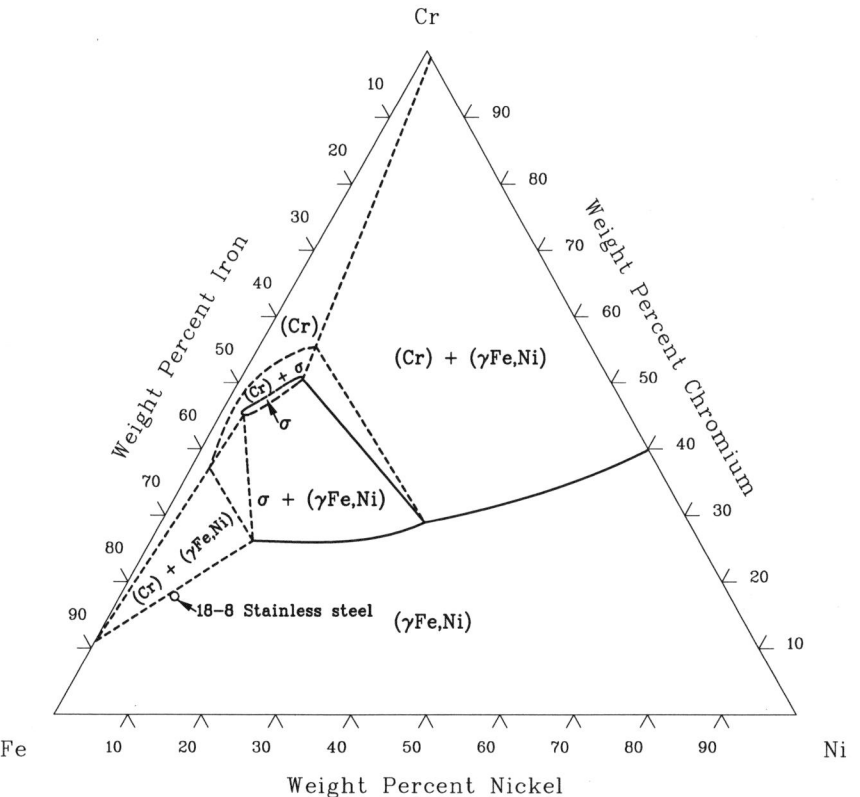

Fig. 48 The isothermal section at 900 °C (1652 °F) of the iron-chromium-nickel ternary phase diagram, showing the nominal composition of 18-8 stainless steel. Source: Adapted from Ref 1

problem and its solution, which could have been avoided had the proper phase diagram been examined (see Fig. 54).

A question concerning purple plague problems, however, has remained unresolved: whether or not the presence of silicon near the gold-aluminum interface has an influence on the stability and rate of formation of the damaging intermetallic phase. An examination of the phase relationships in the Al-Al₂Au-Si subternary system showed no stable ternary Al-Au-Si phases (see 91Hay). It was suggested instead that the reported effect of silicon may be due to a reaction between silicon and alumina (Al₂O₃) at the aluminum-gold interface that becomes thermodynamically feasible in the presence of gold.

BIBLIOGRAPHY

35Mar: J.S. Marsh, *Principles of Phase Diagrams*, McGraw-Hill, 1935. *This out-of-print book is an early thorough presentation of the principles of heterogeneous equilibrium in organic, inorganic salt, and metallic systems.*

44Mas: G. Masing (B.A. Rogers, transl.), *Ternary Alloys: Introduction to the Theory of Three Component Systems*, Reinhold, 1944; available from U·M·I, 300 North Zeeb Rd., Ann Arbor, MI 48106. *This out-of-print book, originally published in German in 1932, is one of the first to thoroughly discuss the theory underlying ternary alloy systems and their application to industrial alloys.*

56Rhi: F.N. Rhines, *Phase Diagrams in Metallurgy: Their Development and Application*, McGraw-Hill, 1956. *This out-of-print book is a basic text designed for undergraduate students in metallurgy.*

66Pri: A. Prince, *Alloy Phase Equilibria*, Elsevier, 1966. *This out-of-print book covers the thermodynamic approach to binary, ternary, and quaternary phase diagrams.*

68Gor: P. Gordon, *Principles of Phase Diagrams in Materials Systems*, McGraw-Hill, 1968;

phase is known to be brittle. The upward adjustment of the carbon content by only a few hundredths of a weight percent eliminated this problem.

Solid-State Electronics. In the early stages of the solid-state industry, a phenomenon known as the "purple plague" nearly destroyed the fledgling industry. Components were failing where the

gold lead wires were fused to aluminized transistor and integrated circuits. A purple residue was formed, which was thought to be a product of corrosion. Actually, what was happening was the formation of an intermetallic compound, an aluminum-gold precipitate (Al₂Au) that is purple in color and very brittle. Millions of actual and opportunity dollars were lost in identifying the

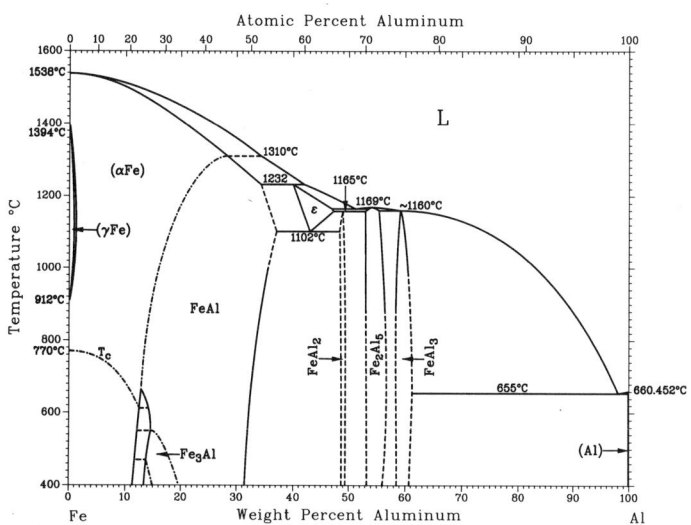

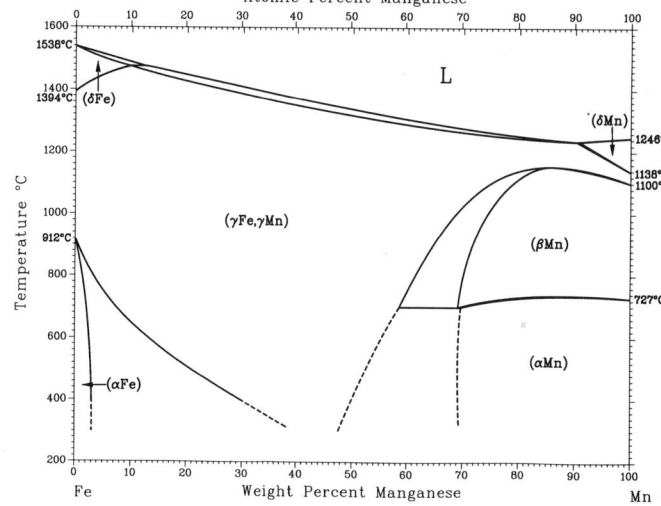

Fig. 49 The aluminum-iron and iron-manganese phase diagrams. Source: Ref 2

reprinted by Robert E. Krieger Publishing, 1983. *Covers the thermodynamic basis of phase diagrams; the presentation is aimed at materials engineers and scientists.*

70Kau: L. Kaufman and H. Bernstein, *Computer Calculations of Phase Diagrams*, Academic Press, 1970. *A comprehensive presentation of thermodynamic modeling with the aid of computers.*

75Gok: N.A. Gokcen, *Thermodynamics*, Techscience, 1975. *Chapter XV discusses the role of thermodynamics in phase diagrams and Gibbs energy diagrams.*

77Luk: H.L. Lukas, E.T. Henig, and B. Zimmerman, Optimization of Phase Diagrams by a Least Squares Method Using Simultaneously Different Types of Data, *Calphad*, Vol 1 (No. 3), 1977, p 225-236. *Presents the use of a computer-aided program for determining phase boundary lines that best fit scattered data points.*

81Goo: D.A. Goodman, J.W. Cahn, and L.H. Bennett, The Centennial of the Gibbs-Konovalov Rule for Congruent Points, *Bull. Alloy Phase Diagrams*, Vol 2 (No. 1), 1981, p 29-34. *Presents the theoretical basis for the rule and its application to phase diagram evaluation.*

81Hil: M. Hillert, A Discussion of Methods of Calculating Phase Diagrams, *Bull. Alloy Phase Diagrams*, Vol 2 (No. 3), 1981, p 265-268. *Presents a brief description of the various methods for thermodynamic modeling of phase diagrams.*

82Pel: A.D. Pelton, W.T. Thompson, and C.W. Bale, *F*A*C*T* (Facility for the Analysis of Chemical Thermodynamics)*, McGill University, 1982. *Describes a thermodynamic database and computer program for modeling phase diagrams.*

84Mor: J.E. Morral, Two-Dimensional Phase Fraction Charts, *Scr. Metall.*, Vol 18 (No. 4), 1984, p 407-410. *Gives a general description of phase-fraction charts.*

85ASM: *Metals Handbook*, 9th ed., Vol 9, *Metallography and Microstructures*, American Society for Metals, 1985. *A comprehensive reference covering terms and definitions, metallographic techniques, microstructures of industrial metals and alloys, and principles of microstructures and crystal structures.*

89Mas: T.B. Massalski, Phase Diagrams in Materials Science, *ASM News*, Vol 20 (No. 7), July 1989, p 8-9. *A concise presentation of the role of phase diagrams in materials science, and the worldwide efforts to make reliable diagrams readily available.*

90Mas: T.B. Massalski, Ed., *Binary Alloy Phase Diagrams*, 2nd ed., ASM International, 1990. *The most comprehensive collection of binary phase diagrams published to date: diagrams for 2965 systems, presented in both atomic and weight percent, with crystal data and discussion.*

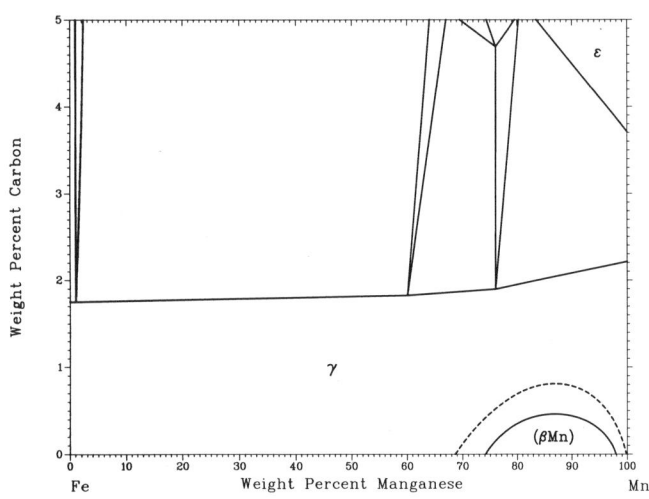
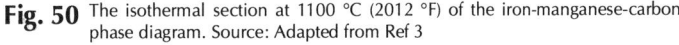

Fig. 50 The isothermal section at 1100 °C (2012 °F) of the iron-manganese-carbon phase diagram. Source: Adapted from Ref 3

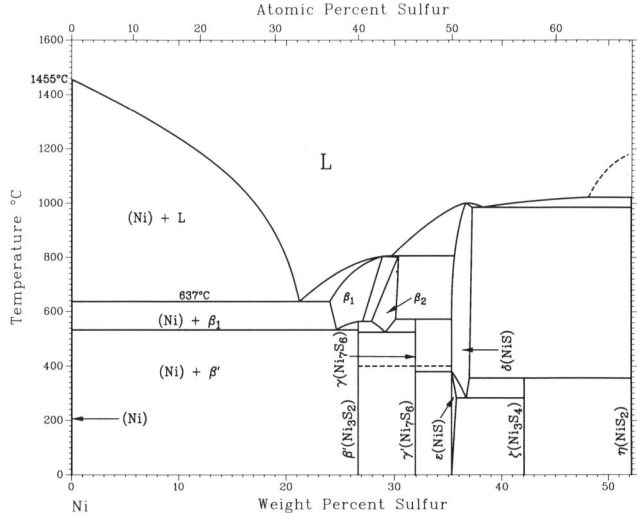

Fig. 51 The nickel-sulfur phase diagram. Source: Adapted from 90Mas

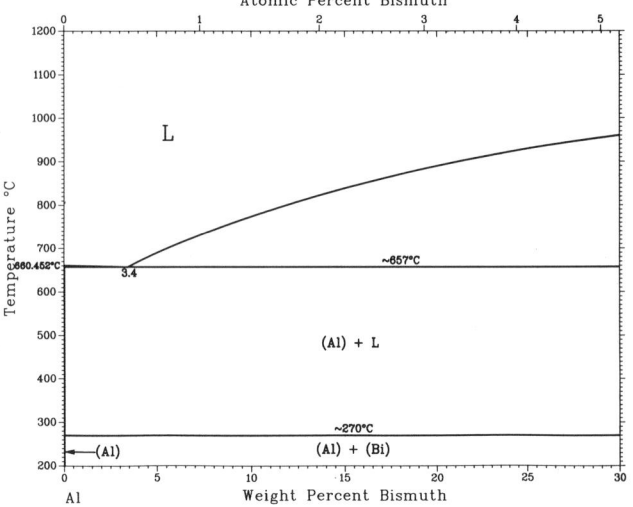
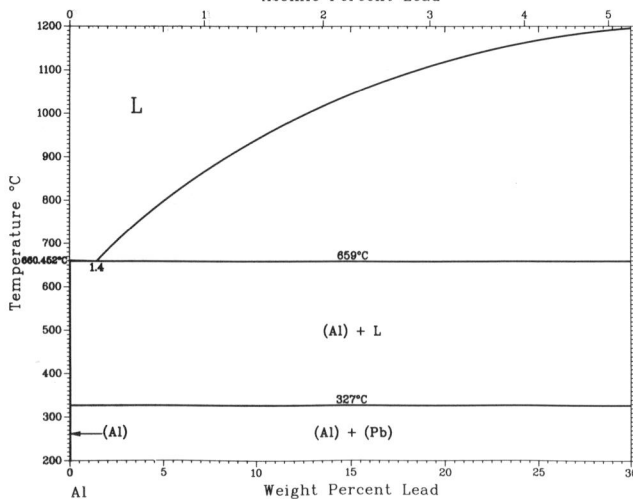

Fig. 52 The aluminum-bismuth and aluminum-lead phase diagrams. Source: Adapted from 90Mas

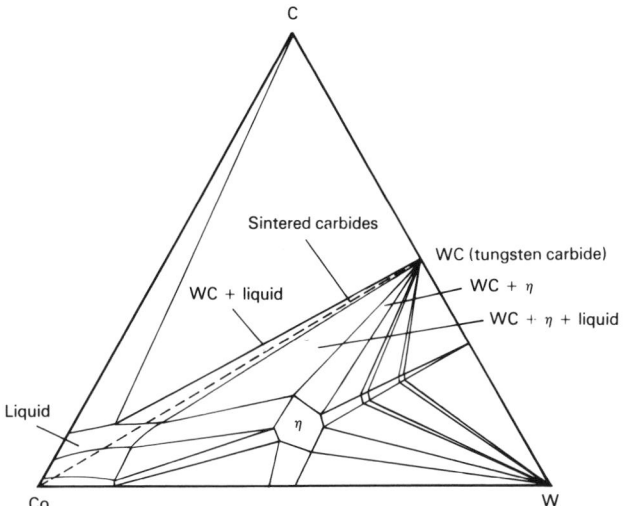

Fig. 53 The isothermal section at 1400 °C (2552 °F) of the cobalt-tungsten-carbon phase diagram. Source: Adapted from Ref 4

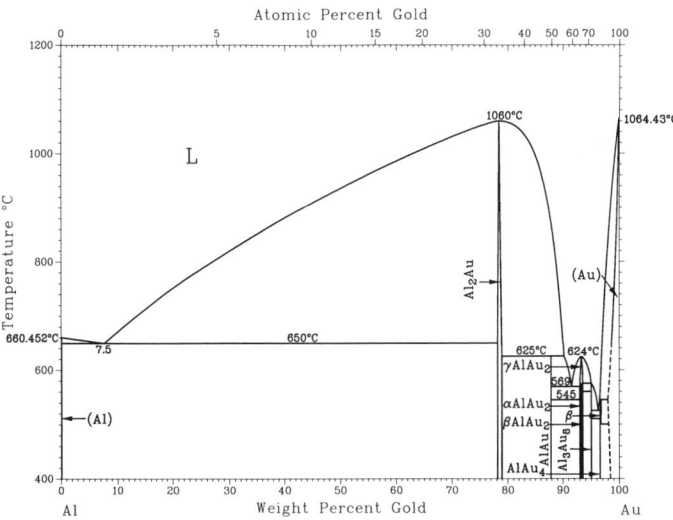

Fig. 54 The aluminum-gold phase diagram. Source: Ref 5

Volume Fraction

In order to relate the weight fraction of a phase present in an alloy specimen as determined from a phase diagram to its two-dimensional appearance as observed in a micrograph, it is necessary to be able to convert between weight-fraction values and areal-fraction values, both in decimal fractions. This conversion can be developed as follows:

The weight fraction of the phase is determined from the phase diagram, using the lever rule.

$$\text{Volume portion of the phase} = \frac{\text{weight fraction of the phase}}{\text{phase density}}$$

Total volume of all phases present = sum of the volume portions of each phase.

$$\text{Volume fraction of the phase} = \frac{\text{weight fraction of the phase}}{\text{phase density} \times \text{total volume}}$$

It has been shown by stereology and quantitative metallography that areal fraction is equal to volume fraction [85ASM]. (Areal fraction of a phase is the sum of areas of the phase intercepted by a microscopic traverse of the observed region of the specimen divided by the total area of the observed region.) Therefore:

$$\text{Areal fraction of the phase} = \frac{\text{weight fraction of the phase}}{\text{phase density} \times \text{total volume}}$$

The phase density value for the preceding equation can be obtained by measurement or calculation. The densities of chemical elements, and some line compounds, can be found in the literature. Alternatively, the density of a unit cell of a phase comprising one or more elements can be calculated from information about its crystal structure and the atomic weights of the elements comprising it as follows:

$$\text{Weight of each element} = \text{number of atoms} \times \frac{\text{atomic weight}}{\text{Avogadro's number}}$$

Total cell weight = sum of weights of each element

Density = total cell weight / cell volume

For example, the calculated density of pure copper, which has a fcc structure and a lattice parameter of 0.36146 nm, is:

$$\rho = \frac{4 \text{ atoms/cell} \times 63.546 \text{ g/mol}}{6.0227 \times 10^{23} \text{ atoms/mol} \times (0.36146 \times 10^{-9} \text{ m})^3} = 8.937 \text{Mg/m}^3$$

This compares favorably with the published value of 8.93.

91Hay: F.H. Hayes, Ed., *User Aspects of Phase Diagrams*, The Institute of Metals, London, 1991. *A collection of 35 papers and posters presented at a conference held June 1990 in Petten, The Netherlands.*

91Mor: J.E. Morral and H. Gupta, Phase Boundary, ZPF, and Topological Lines on Phase Diagrams, *Scr. Metall.*, Vol 25 (No. 6), 1991, p 1393-1396. *Reviews three different ways of considering the lines on a phase diagram.*

91Oka1: H. Okamoto and T.B. Massalski, Thermodynamically Improbable Phase Diagrams, *J. Phase Equilibria*, Vol 12 (No. 2), 1991, p 148-168. *Presents examples of phase-rule violations and problems with phase-boundary curvatures; also discusses unusual diagrams.*

91Oka2: H. Okamoto, Reevaluation of Thermodynamic Models for Phase Diagram Evaluation, *J. Phase Equilibria*, Vol 12 (No. 6), 1991, p 623-643. *Reviews the basic principles of thermodynamic calculation of phase diagrams, simplification of thermodynamic models, and reliability of thermodynamic data and parameters; also presents examples of unlikely calculated phase diagrams.*

91Vil: P. Villars and L.D. Calvert, *Pearson's Handbook of Crystallographic Data for Intermediate Phases*, ASM International, 1991. *This third edition of Pearson's comprehensive compilation includes data from all the international literature from 1913 to 1989.*

OTHER REFERENCES

1. G.V. Raynor and V.G. Rivlin, *Phase Equilibria in Iron Ternary Alloys*, Vol 4, The Institute of Metals, London, 1988
2. H. Okamoto, *Phase Diagrams of Binary Iron Alloys*, ASM International, 1992
3. R. Benz, J.F. Elliott, and J. Chipman, *Metall. Trans.*, Vol 4, 1973, p 1449
4. P. Rautala and J.T. Norton, *Trans. AIME*, Vol 194, 1952, p 1047
5. H. Okamoto, Ed., Binary Alloy Phase Diagrams Updating Service, ASM International, 1992

Index of Terms

Section 2
Binary Alloy Phase Diagrams

List of Binary Systems Included:

Introduction to Binary Alloy Phase Diagrams

THE 1046 BINARY SYSTEMS presented in this Section have been selected for their commercial importance from the almost 3000 systems covered in *Binary Alloy Phase Diagrams*, Second Edition. The diagrams used were reproduced from that compilation, from more recent evaluations, or, in some instances, updated evaluations based on the most recent literature. The source is indicated with each phase diagram. "Unpublished" indicates the source is a complete evaluation that has not yet been published in the *Journal of Phase Equilibria* or in a monograph. The crystal structure data shown with the diagrams have been updated in some instances with information from *Pearson's Handbook of Crystallographic Data for Intermetallic Phases*, Second Edition.

Except when the information for a system is from one of the General References listed in the following pages, the specific author of the information is listed as the source, along with the year the investigation was completed. To locate the author's complete investigation of a system, consult the Binary Alloy Phase Diagrams Index in this Section, which lists source information for all 2965 binary alloy systems for which data exist.

Because this Handbook is designed to be used mainly by engineers to solve industrial problems, the primary composition scale is plotted in weight percent. Atomic percentages are shown as a secondary scale at the top of the diagrams. Conversions between weight and atomic composition also can be made using the standard atomic weights listed in the Appendix. For the sake of clarity, grid lines are not superimposed on the phase diagrams. However, tick marks are provided along the composition scale as well as the temperature scale, which is shown in degrees Celsius. Celsius temperatures can be easily converted to degrees Fahrenheit using the table in the Appendix. Magnetic transitions (Curie temperature and Néel temperature) are shown as dot-dashed lines. Dashed lines are used to denote uncertain or speculative boundaries.

All diagrams presented in this Section of the Handbook are for stable equilibrium conditions, except where metastable conditions are indicated.

Binary Alloy Phase Diagrams Index

This index gives source information for all 2965 binary alloy systems. Column 2 designates all binary abstracts published in *Binary Alloy Phase Diagrams*, Second Edition (called "M2") and indicates if information for the system has been updated in the *Binary Alloy Phase Diagrams Updating Service* by listing the update year. Abstracts are a shortened version of the full evaluation giving concise descriptions of key features of the system, crystal structure data, primary references, and the equilibrium diagram, if any. Column 3 gives the source of the original abstract or the most recent full evaluation. Full evaluations include expanded information on the phase diagram, and any lattice parameter, thermodynamic, magnetism, and pressure information and ancillary figures available. A key to abbreviated titles of Alloy Phase Diagram Program source publications and General References used in column 3 precede the index. Systems marked "unpublished" have been submitted to the Alloy Phase Diagram Program, but have not yet been published. References to sources that are non-Alloy Phase Diagram publications follow the index. Column 4 indicates whether the evaluation includes a phase diagram (D) or is text only (T). Diagrams for systems marked by an asterisk are published in this handbook.

Binary General References

The following list of references has provided the foundation of much of the phase diagram data that is currently cited in the literature. To conserve space, these references will be cited by their general reference symbol in the index.

[Brandes]: E.A. Brandes and R.F. Flint, Ed., *Manganese Phase Diagrams,* The Manganese Centre, 17 Avenue Hoche, 75008 Paris, France (1980).

[Chiotti]: P. Chiotti, V.V. Akhachinskij, and I. Ansara, *The Chemical Thermodynamics of Actinide Elements and Compounds,* Part 5: The Actinide Binary Alloys, V. Medvedev, M.H. Rand, E.F. Westrum, Jr., and F.L. Oetting, Ed., International Atomic Energy Agency, Vienna (1981).

[Elliott]: R.P. Elliott, *Constitution of Binary Alloys, First Supplement,* McGraw-Hill, New York or General Electric Co., Business Growth Services, Schenectady, New York (1965).

[Hafnium]: P.J. Spencer, O. von Goldbeck, R. Ferro, R. Marazza, K. Girgis, and O. Kubaschewski, *Hafnium: Physico-Chemical Properties of Its Compounds and Alloys,* K.L. Komerek, Ed., Atomic Energy Review Special Issue No. 8, International Atomic Energy Agency, Vienna (1981).

[Hansen]: M. Hansen and K. Anderko, *Constitution of Binary Alloys,* McGraw-Hill, New York or General Electric Co., Business Growth Services, Schenectady, New York (1958).

[Hultgren, B]: R. Hultgren, P.D. Desai, D.T. Hawkins, M. Gleiser, and K.K. Kelley, *Selected Values of the Thermodynamic Properties of Binary Alloys,* American Society for Metals, Metals Park, Ohio (1973).

[Ivanov]: O.S. Ivanov, T.A. Badaeva, R.M. Sofronova, V.B. Kishenevskii, and N.P. Kushnir, *Phase Diagrams of Uranium Alloys,* Nauka, Moscow (1972).

[Kubaschewski]: O. Kubaschewski, *Iron-- Binary Phase Diagrams,* Springer-Verlag, New York (1982).

[Metals]: *Metals Handbook,* Metallography, Structures and Phase Diagrams, Vol. 8, 8th ed., American Society for Metals, Metals Park, OH (1973).

[Moffatt]: W.G. Moffatt, Ed., *Handbook of Binary Phase Diagrams,* Business Growth Services, General Electric Co., Schenectady, NY (1976).

[Molybdenum]: L. Brewer, *Molybdenum: Physico-Chemical Properties of Its Compounds and Alloys,* O. Kubaschewski, Ed., Atomic Energy Review Special Issue No. 7, International Atomic Energy Agency, Vienna (1980).

[Pearson3]: P. Villars and L.D. Calvert, *Pearson's Handbook of Crystallographic Data for Intermetallic Phases,* Vol. 1, 2, and 3, American Society for Metals, Metals Park, OH (1985).

[Pearson4]: P. Villars and L.D. Calvert, *Pearson's Handbook of Crystallographic Data for Intermetallic Phases,* 2nd ed., Vol. 1, 2, 3, and 4, ASM International, Materials Park, OH (1991).

[Plutonium]: M.H. Rand, D.T. Livey, P. Feschotte, H. Nowotny, K. Seifert, and R. Ferro, *Plutonium: Physico-Chemical Properties of Its Compounds and Alloys,* O. Kubaschewski, Ed., Atomic Energy Review Special Issues No. 1, International Atomic Energy Agency, Vienna (1966).

[Shunk]: F.A. Shunk, *Constitution of Binary Alloys, Second Supplement,* McGraw-Hill, New York or General Electric Co., Business Growth Services, Schenectady, New York (1969).

[Smith]: J.F. Smith, O.N. Carlson, D.T. Peterson, and T.E. Scott, *Thorium: Preparation and Properties,* Iowa State University Press, Ames, IA (1975).

[Smithells]: C.J. Smithells and E.A. Brandes, *Metals Reference Book,* 5th ed., Butterworths, Woburn, MA (1976).

[Thorium]: M.H. Rand, O. von Goldbeck, R. Ferro, K. Girgis, and A.L. Dragoo, *Thorium: Physico-Chemical Properties of Its Compounds and Alloys,* O. Kubaschewski, Ed., Atomic Energy Review Special Issue No. 5, International Atomic Energy Agency, Vienna (1975).

[Zirconium]: C.B. Alcock, K.T. Jacob, S. Zador, O. von Goldbeck, H. Nowotny, K. Seifert, and O. Kubaschewski, *Zirconium: Physico-Chemical Properties of Its Compounds and Alloys,* O. Kubaschewski, Ed., Atomic Energy Review Special Issue No. 6, International Atomic Energy Agency, Vienna (1976).

Key to Titles

Key to titles of Alloy Phase Diagram Publications abbreviated under "Published" and "Data Source":

BAPD
Bulletin of Alloy Phase Diagrams
ASM International

Binary Beryllium
Phase Diagrams of Binary Beryllium Alloys
ASM International, 1987

Binary Gold
Phase Diagrams of Binary Gold Alloys
ASM International, 1988

Binary Iron
Phase Diagrams of Binary Iron Alloys
ASM International, 1993

Binary Magnesium
Phase Diagrams of Binary Magnesium Alloys
ASM International, 1988

Binary Nickel
Phase Diagrams of Binary Nickel Alloys
ASM International, 1991

Binary Titanium
Phase Diagrams of Binary Titanium Alloys
ASM International, 1987

Binary Tungsten
Phase Diagrams of Binary Tungsten Alloys
The Indian Institute of Metals, 1991

Binary Vanadium
Phase Diagrams of Binary Vanadium Alloys
ASM International, 1989

Indium
Phase Diagrams of Indium Alloys and Their Engineering Applications
ASM International, 1992

JAPD
Journal of Alloy Phase Diagrams
The Indian Institute of Metals

JPE
Journal of Phase Equilibria
ASM International

M2
Binary Alloy Phase Diagrams, 2nd edition
ASM International, 1990

91
Binary Alloy Phase Diagrams Updating Service
ASM International, Dec. 1991

92
Binary Alloy Phase Diagrams Updating Service
ASM International, July and Dec. 1992

Binary Alloys Index

System	Published	Data source	Data type	System	Published	Data source	Data type	System	Published	Data source	Data type
Ac-Ag	M2	Unpublished	T	*Ag-Mo	M2	BAPD 11(6)	D	*Al-Bi	M2	BAPD 5(3)	D
Ac-Au	M2	Binary Gold	T	Ag-N	M2	BAPD 11(5)	T	Al-Br	No Data		
Ac-B	M2	M2	D	*Ag-Na	M2	BAPD 7(2)	D	Al-C	M2,91,92	M2	D
Ac-Cr	M2	BAPD 6(5)	D	Ag-Nb	M2	BAPD 10(6)	T	*Al-Ca	M2	BAPD 9(6)	D
Ac-Cu	M2	M2	T	*Ag-Nd	M2	BAPD 6(1)	D	*Al-Cd	M2	BAPD 3(2)	D
Ac-H	M2	M2	T	Ag-Ne	M2	Unpublished	T	*Al-Ce	M2	BAPD 9(6)	D
Ac-Mg	M2	Unpublished	T	*Ag-Ni	M2	Binary Nickel	D	Al-Cl	No Data		
Ac-Mo	M2	M2	D	Ag-Np	M2	M2	T	Al-Cm	No Data		
Ac-O	M2	M2	T	Ag-O	M2	JPE 13(2)	D	*Al-Co	M2	BAPD 10(6)	D
Ac-Pt	M2	BAPD 10(4a)	D	Ag-Os	M2	BAPD 7(4)	D	*Al-Cr	M2	Unpublished	D
Ac-S	M2	M2	T	*Ag-P	M2	BAPD 9(3)	D	Al-Cs	M2	Unpublished	T
Ac-W		Binary Tungsten	T	Ag-Pa	M2	M2	T	*Al-Cu	M2	[85Mur]	D
*Ag-Al	M2	BAPD 8(6)	D	*Ag-Pb	M2	BAPD 8(4)	D	Al-Dy	M2	M2	D
Ag-Am	No Data			*Ag-Pd	M2	BAPD 9(3)	D	*Al-Er	M2	BAPD 9(6)	D
Ag-Ar	M2	Unpublished	T	Ag-Pm	M2,91	M2	T	*Al-Eu	M2,91	M2	D
*Ag-As	M2	BAPD 11(2)	D	Ag-Po	M2	M2	T	Al-F	No Data		
Ag-At	M2	Unpublished	T	*Ag-Pr	M2	BAPD 6(1)	D	*Al-Fe	M2	Binary Iron	D
*Ag-Au	M2	Binary Gold	D	*Ag-Pt	M2	BAPD 8(4)	D	*Al-Ga	M2	BAPD 4(2)	D
Ag-B	M2,92	BAPD 11(6)	T	Ag-Pu	M2	[70Woo]	D	*Al-Gd	M2	BAPD 9(6)	D
Ag-Ba	M2,92	Unpublished	D	Ag-Ra	M2	Unpublished	T	*Al-Ge	M2	BAPD 5(4)	D
*Ag-Be	M2	Binary Beryllium	D	Ag-Rb	M2	BAPD 7(1)	T	*Al-H	M2	JPE 13(1)	D
*Ag-Bi	M2	BAPD 1(2)	D	Ag-Re	M2	BAPD 9(3)	D	Al-Hf	M2	Unpublished	D
Ag-Br	M2	M2	T	Ag-Rh	M2	BAPD 7(4)	D	*Al-Hg	M2	BAPD 6(3)	D
Ag-C	M2	BAPD 9(3)	D	Ag-Rn	M2	Unpublished	T	*Al-Ho	M2	BAPD 9(6)	D
*Ag-Ca	M2	BAPD 9(3)	D	Ag-Ru	M2	BAPD 7(4)	D	Al-I	No Data		
*Ag-Cd	M2	[Hansen]	D	*Ag-S	M2	BAPD 7(3)	D	*Al-In	M2	Indium	D
*Ag-Ce	M2	BAPD 6(5)	D	*Ag-Sb	M2	[Hansen]	D	Al-Ir	M2	M2	D
Ag-Cl	M2	M2	T	*Ag-Sc	M2	BAPD 4(4)	D	Al-K	M2	Unpublished	T
*Ag-Co	M2	BAPD 7(3)	D	*Ag-Se	M2	BAPD 11(3)	D	*Al-La	M2	BAPD 9(6)	D
Ag-Cr	M2	BAPD 11(3)	D	*Ag-Si	M2	BAPD 10(6)	D	*Al-Li	M2,91	BAPD 3(2)	D
Ag-Cs	M2	BAPD 7(3)	T	*Ag-Sm	M2,91	BAPD 6(2)	D	Al-Lu	M2	M2	D
*Ag-Cu	M2	Unpublished	D	*Ag-Sn	M2	BAPD 8(4)	D	*Al-Mg	M2	Binary Magnesium	D
*Ag-Dy	M2	BAPD 6(1)	D	*Ag-Sr	M2	BAPD 11(2)	D	*Al-Mn	M2	BAPD 8(5)	D
*Ag-Er	M2	BAPD 6(1)	D	Ag-Ta	M2	BAPD 9(3)	T	Al-Mo	M2,91	Unpublished	D
*Ag-Eu	M2	BAPD 6(1)	D	Ag-Tb	M2	BAPD 6(2)	D	Al-N	M2	BAPD 7(4)	D
Ag-F	M2	M2	T	Ag-Tc	M2	Unpublished	T	Al-Na	M2	BAPD 4(4)	D
*Ag-Fe	M2	Binary Iron	D	*Ag-Te	M2	JPE 12(1)	D	*Al-Nb	M2	Unpublished	D
Ag-Fr	M2	Unpublished	T	Ag-Th	M2,92	JPE 12(3)	D	*Al-Nd	M2,91	BAPD 10(1)	D
*Ag-Ga	M2,92	JPE 13(3)	D	*Ag-Ti	M2	Binary Titanium	D	*Al-Ni	M2	Binary Nickel	D
*Ag-Gd	M2	BAPD 6(2)	D	*Ag-Tl	M2	BAPD 10(6)	D	Al-Np	M2	BAPD 10(2)	T
*Ag-Ge	M2	BAPD 9(1)	D	*Ag-Tm	M2	M2	D	*Al-O	M2	BAPD 6(6)	D
Ag-H	M2,92	JPE 12(6)	T	Ag-U	M2	BAPD 10(6)	D	Al-Os	M2	Unpublished	T
Ag-He	M2	Unpublished	T	Ag-V	M2	Binary Vanadium	D	Al-P	M2	BAPD 6(3)	D
Ag-Hf	M2	BAPD 10(2)	T					*Al-Pb	M2	BAPD 5(1)	D
*Ag-Hg	M2	Unpublished	D	Ag-W	M2,92	Binary Tungsten	D	*Al-Pd	M2	BAPD 7(4)	D
*Ag-Ho	M2	BAPD 6(2)	D					Al-Pm	M2	M2	D
Ag-I	M2	M2	T	Ag-Xe	M2	Unpublished	T	*Al-Pr	M2	BAPD 10(1)	D
*Ag-In	M2	Indium	D	*Ag-Y	M2	BAPD 4(4)	D	*Al-Pt	M2	BAPD 7(1)	D
Ag-Ir	M2	BAPD 7(4)	D	*Ag-Yb	M2	BAPD 6(2)	D	Al-Pu	M2	BAPD 10(4a)	D
Ag-K	M2	BAPD 7(3)	T	*Ag-Zn	M2	[40And]	D	Al-Rb	M2	Unpublished	T
Ag-Kr	M2	Unpublished	T	*Ag-Zr	M2	JPE 13(2)	D	Al-Re	M2	Unpublished	T
*Ag-La	M2	BAPD 4(4)	D	Al-Am	M2	BAPD 10(3)	T	Al-Rh	M2	M2	D
*Ag-Li	M2	BAPD 7(3)	D	*Al-As	M2	BAPD 5(6)	D	Al-Ru	M2	M2	D
Ag-Lu	M2	BAPD 4(4)	D	*Al-Au	M2,91	BAPD 8(2)	D	*Al-S	M2,91	BAPD 8(2)	D
*Ag-Mg	M2	Binary Magnesium	D	Al-B	M2	BAPD 11(6)	D	*Al-Sb	M2	BAPD 5(5)	D
				*Al-Ba	M2,92	BAPD 2(3)	D	Al-Sc	M2,91	BAPD 10(1)	D
Ag-Mn	M2	BAPD 11(5)	D	*Al-Be	M2	Binary Beryllium	D	*Al-Se	M2	BAPD 10(6)	D
								*Al-Si	M2	BAPD 5(1)	D

(continued)

System	Published	Data source	Data type
Al-Sm	M2	BAPD 10(1)	D
*Al-Sn	M2	BAPD 4(4)	D
*Al-Sr	M2	BAPD 10(6)	D
*Al-Ta	M2	Unpublished	D
Al-Tb	M2	M2	D
Al-Tc	M2	M2	T
*Al-Te	M2	BAPD 11(2)	D
*Al-Th	M2	BAPD 10(4a)	D
*Al-Ti	M2	Binary Titanium	D
Al-Tl	M2,92	BAPD 10(2)	D
Al-Tm	M2	M2	D
*Al-U	M2,91	BAPD 11(1)	D
*Al-V	M2	Binary Vanadium	D
*Al-W	M2	Binary Tungsten	D
*Al-Y	M2	BAPD 10(1)	D
*Al-Yb	M2	BAPD 10(1)	D
*Al-Zn	M2	BAPD 4(1)	D
*Al-Zr	M2,92	JPE 13(3)	D
Am-As	M2	M2	T
Am-B	M2	M2	D
Am-Be	M2	Binary Beryllium	T
Am-Bi	M2	M2	T
Am-C	M2	M2	T
Am-Co	M2	M2	T
Am-Cr	M2	BAPD 6(5)	D
Am-Cu	M2	M2	T
Am-Fe	M2	M2	T
Am-H	M2	M2	T
Am-Ir	M2	M2	T
Am-Mo	M2	[Molybdenum]	D
Am-N	M2	M2	T
Am-Ni	M2	Binary Nickel	T
Am-O	M2,91	[Elliott]	T
Am-Os	M2	M2	T
Am-P	M2	M2	T
Am-Pd	M2	M2	T
Am-Pt	M2	BAPD 10(2)	D
Am-Pu	M2	[66Ell]	D
Am-Rh	M2	M2	T
Am-Ru	M2	M2	T
Am-S	M2	M2	T
Am-Sb	M2	M2	T
Am-Se	M2	M2	T
Am-Si	M2	M2	D
Am-Te	M2	M2	T
Am-W		Binary Tungsten	T
Ar-Au	M2	Binary Gold	T
Ar-Be	M2	Binary Beryllium	T
Ar-Cu	M2	Unpublished	T
Ar-Mg	M2	Binary Magnesium	T
Ar-Mo	M2	[Molybdenum]	D
Ar-W		Binary Tungsten	T
*As-Au	M2	Binary Gold	D
As-B	M2	M2	D
As-Ba	M2	M2	T
As-Be	M2	Binary Beryllium	T
*As-Bi	M2	[53Gea]	D
As-Bk	M2	M2	T
As-Br	No Data		
As-C	M2	M2	T
As-Ca	M2	M2	T
*As-Cd	M2	JPE 13(2)	D
As-Ce	M2	BAPD 7(3)	T
As-Cf	M2	M2	T

System	Published	Data source	Data type
As-Cl	No Data		
As-Cm	M2	M2	T
*As-Co	M2	BAPD 11(6)	D
As-Cr	M2	BAPD 11(5)	D
As-Cs	M2	M2	T
*As-Cu	M2	BAPD 9(5)	D
As-Dy	M2	M2	D
As-Er	M2	M2	D
As-Eu	M2	BAPD 7(3)	D
As-F	No Data		
*As-Fe	M2	Binary Iron	D
*As-Ga	M2	M2	D
As-Gd	M2	BAPD 7(4)	T
*As-Ge	M2,91	BAPD 6(3)	D
As-H	M2	M2	T
As-Hf	M2	M2	T
As-Hg	M2	M2	T
As-Ho	M2	M2	D
As-I	No Data		
*As-In	M2	Indium	D
*As-Ir	M2	M2	T
*As-K	M2	[61Dor1]	D
As-La	M2	BAPD 7(4)	T
As-Li	M2	M2	T
As-Lu	M2	M2	D
As-Mg	M2	Binary Magnesium	D
*As-Mn	M2	BAPD 10(5)	D
As-Mo	M2,91	[Molybdenum]	D
As-N	M2	M2	T
As-Na	M2	M2	T
As-Nb	M2	M2	T
*As-Nd	M2	BAPD 7(4)	D
*As-Ni	M2	Binary Nickel	D
As-Np	M2	M2	T
As-O	M2	M2	T
As-Os	M2	M2	D
*As-P	M2	JPE 12(3)	D
As-Pa	M2	M2	T
*As-Pb	M2	BAPD 11(2)	D
*As-Pd	M2,91,92	BAPD 11(5)	D
As-Pm	No Data		
As-Pr	M2	BAPD 7(4)	T
As-Pt	M2	BAPD 11(5)	D
As-Pu	M2	M2	T
As-Rb	M2	M2	T
As-Re	M2	M2	T
As-Rh	M2	M2	T
As-Ru	M2	M2	T
*As-S	M2	M2	D
*As-Sb	M2	M2	D
*As-Sc	M2	BAPD 7(4)	T
*As-Se	M2	M2	D
*As-Si	M2	BAPD 6(3)	D
As-Sm	M2	M2	D
*As-Sn	M2,91	BAPD 11(3)	D
As-Sr	M2	M2	T
As-Ta	M2	M2	T
As-Tb	M2	M2	D
As-Tc	M2	M2	T
*As-Te	M2	M2	D
As-Th	M2	[Smith]	D
As-Ti	M2	Binary Titanium	D
*As-Tl	M2	Unpublished	D
As-Tm	M2	M2	D
As-U	M2	M2	D
As-V	M2	JPE 12(4)	T
As-W	M2	Binary Tungsten	T
As-Y	M2	BAPD 7(4)	T
*As-Yb	M2	M2	D
*As-Zn	M2	JPE 13(2)	D

System	Published	Data source	Data type
As-Zr	M2,91	BAPD 11(6)	T
At-Au	M2	Binary Gold	T
At-Mo	M2	M2	D
At-W		Binary Tungsten	T
Au-B	M2	Binary Gold	D
Au-Ba	M2	Binary Gold	T
*Au-Be	M2	Binary Beryllium	D
*Au-Bi	M2	M2	D
Au-Br	M2	Binary Gold	T
Au-C	M2	Binary Gold	D
*Au-Ca	M2	Binary Gold	D
*Au-Cd	M2	Binary Gold	D
*Au-Ce	M2	Binary Gold	D
Au-Cl	M2	Binary Gold	D
Au-Cm	No Data		
*Au-Co	M2	Binary Gold	D
*Au-Cr	M2	Binary Gold	D
Au-Cs	M2	Binary Gold	D
*Au-Cu	M2	Binary Gold	D
*Au-Dy	M2	Binary Gold	D
Au-Er	M2	Binary Gold	D
*Au-Eu	M2	M2	D
Au-F	M2	Binary Gold	T
*Au-Fe	M2	Binary Iron	D
Au-Fr	M2	[68Gul1]	T
*Au-Ga	M2	Binary Gold	D
Au-Gd	M2	Binary Gold	D
*Au-Ge	M2	Binary Gold	D
Au-H	M2	Binary Gold	T
Au-He	M2	Binary Gold	T
Au-Hf	M2	Binary Gold	D
*Au-Hg	M2	BAPD 10(1)	D
*Au-Ho	M2	Binary Gold	D
Au-I	M2	Binary Gold	T
*Au-In	M2	Indium	D
Au-Ir	M2	Binary Gold	T
*Au-K	M2	Binary Gold	D
Au-Kr	M2	Binary Gold	T
*Au-La	M2	Binary Gold	D
*Au-Li	M2	Binary Gold	D
Au-Lu	M2	Binary Gold	D
*Au-Mg	M2	Binary Magnesium	D
*Au-Mn	M2	Binary Gold	D
Au-Mo	M2	Binary Gold	D
Au-N	M2	Binary Gold	T
*Au-Na	M2	Binary Gold	D
*Au-Nb	M2	Binary Gold	D
Au-Nd	M2	Binary Gold	D
Au-Ne	M2	Binary Gold	T
*Au-Ni	M2	Binary Gold	D
Au-Np	M2	Binary Gold	T
Au-O	M2	Binary Gold	T
Au-Os	M2	Binary Gold	T
Au-P	M2	Binary Gold	D
Au-Pa	M2	Binary Gold	T
*Au-Pb	M2	Binary Gold	D
*Au-Pd	M2	Binary Gold	D
Au-Pm	M2	Binary Gold	D
Au-Po	M2	Binary Gold	T
*Au-Pr	M2	Binary Gold	D
*Au-Pt	M2	Binary Gold	D
*Au-Pu	M2	Binary Gold	D
Au-Ra	M2	Binary Gold	T
*Au-Rb	M2	Binary Gold	D
Au-Re	M2	Binary Gold	T
Au-Rh	M2	Binary Gold	D
Au-Ru	M2	Binary Gold	D
Au-S	M2	Binary Gold	D
*Au-Sb	M2	Binary Gold	D

(continued)

System	Published	Data source	Data type	System	Published	Data source	Data type	System	Published	Data source	Data type
Au-Sc	M2	Binary Gold	D	B-Pu	M2	Unpublished	D	Ba-Np	No Data		
*Au-Se	M2	Binary Gold	D	B-Rb	No Data			Ba-O	M2	M2	D
*Au-Si	M2	Binary Gold	D	*B-Re	M2	[72Por]	D	Ba-Os	No Data		
Au-Sm	M2	Binary Gold	D	B-Rh	M2	[Moffatt]	D	*Ba-P	M2	M2	D
*Au-Sn	M2	Binary Gold	D	*B-Ru	M2	[63Obr]	D	*Ba-Pb	M2	[Hansen]	D
*Au-Sr	M2	Binary Gold	D	B-S	M2	[Moffatt]	D	Ba-Pd	M2,91	JPE 12(4)	D
Au-Ta	M2	Binary Gold	D	B-Sb	M2,91	M2	D	Ba-Pm	M2	M2	D
Au-Tb	M2	Binary Gold	D	*B-Sc	M2	BAPD 11(4)	D	Ba-Po	M2	M2	T
Au-Tc	M2	Binary Gold	T	B-Se	M2	[69Bor]	D	Ba-Pr	M2	BAPD 9(3)	D
*Au-Te	M2	Binary Gold	D	*B-Si	M2	BAPD 5(5)	D	Ba-Pt	M2,91	JPE 12(4)	D
*Au-Th	M2,91	Binary Gold	D	B-Sm	M2	Unpublished	D	Ba-Pu	M2	M2	T
*Au-Ti	M2	Binary Gold	D	B-Sn	M2	M2	D	Ba-Rb	M2	BAPD 5(5)	T
*Au-Tl	M2	Binary Gold	D	B-Sr	M2	M2	D	Ba-Re	No Data		
Au-Tm	M2	Binary Gold	D	*B-Ta	M2	M2	D	Ba-Rh	M2	M2	T
*Au-U	M2	M2	D	B-Tb	M2	BAPD 11(4)	D	Ba-Ru	No Data		
*Au-V	M2	Binary Vanadium	D	B-Tc	M2	M2	D	Ba-S	M2	M2	D
				B-Te	No Data			Ba-Sb	M2	M2	T
Au-W	M2	Binary Tungsten	D	B-Th	M2	[Moffatt]	D	Ba-Sc	M2	M2	D
				*B-Ti	M2	Binary Titanium	D	*Ba-Se	M2	JPE 12(4)	D
Au-Xe	M2	Binary Gold	T					*Ba-Si	M2	[64Obi2]	D
Au-Y	M2	Binary Gold	T	B-Tl	M2,91	M2	D	Ba-Sm	M2	BAPD 9(3)	D
*Au-Yb	M2	Binary Gold	D	B-Tm	M2	Unpublished	D	Ba-Sn	M2,91	M2	D
*Au-Zn	M2	BAPD 10(1)	D	B-U	M2	M2	D	Ba-Sr	M2,91	BAPD 8(6)	D
*Au-Zr	M2	Binary Gold	D	*B-V	M2,91	Binary Vanadium	D	Ba-Ta	No Data		
B-Ba	M2	M2	D					Ba-Tb	M2	M2	T
B-Be	M2	Binary Beryllium	D	*B-W	M2,92	Binary Tungsten	D	Ba-Tc	No Data		
								*Ba-Te	M2	Unpublished	D
B-Bi	M2,91	M2	D	*B-Y	M2	Unpublished	D	Ba-Th	No Data		
*B-C	M2,92	M2	D	B-Yb	M2	Unpublished	D	Ba-Ti	M2	Binary Titanium	D
B-Ca	M2	M2	D	B-Zn	M2,91	M2	D				
B-Cd	M2	Unpublished	D	*B-Zr	M2	[Zirconium]	D	*Ba-Tl	M2	[66Bru]	D
B-Ce	M2	Unpublished	D	Ba-Be	M2,91	Binary Beryllium	D	Ba-Tm	M2	M2	T
B-Cm	No Data							Ba-U	No Data		
*B-Co	M2	BAPD 9(4)	D	Ba-Bi	M2	[38Gru]	D	Ba-V	M2	Binary Vanadium	D
*B-Cr	M2	BAPD 7(3)	D	Ba-Br	M2	M2	D				
B-Cs	No Data			Ba-C	M2	M2	T	Ba-W	M2	Binary Tungsten	T
*B-Cu	M2	BAPD 3(1)	D	*Ba-Ca	M2	BAPD 7(4)	D				
B-Dy	M2	Unpublished	D	*Ba-Cd	M2	M2	D	Ba-Y	M2	M2	D
B-Er	M2	Unpublished	D	*Ba-Ce	M2	M2	T	Ba-Yb	M2,91	BAPD 9(3)	D
B-Eu	M2	Unpublished	D	Ba-Cl	M2	M2	D	*Ba-Zn	M2	JPE 12(4)	D
*B-Fe	M2	Binary Iron	D	Ba-Cm	No Data			Ba-Zr	No Data		
B-Ga	M2,91	M2	D	Ba-Co	M2	Unpublished	T	Be-Bi	M2	Binary Beryllium	D
B-Gd	M2	Unpublished	D	Ba-Cr	M2	BAPD 6(3)	T				
B-Ge	M2	BAPD 5(5)	D	Ba-Cs	M2	BAPD 5(5)	T	Be-Br	M2	Binary Beryllium	T
B-H	M2	M2	T	*Ba-Cu	M2	BAPD 5(6)	D				
B-Hf	M2	M2	D	Ba-Dy	M2	M2	T	Be-C	M2	Binary Beryllium	T
B-Hg	M2	Unpublished	D	Ba-Er	M2	M2	T				
B-Ho	M2	Unpublished	D	Ba-Eu	M2,91	BAPD 9(3)	D	Be-Ca	M2,91	Binary Beryllium	D
B-In	M2	Indium	D	Ba-F	M2	M2	D				
B-Ir	M2	M2	T	Ba-Fe	M2	M2	D	Be-Cd	M2	Binary Beryllium	T
B-K	M2	M2	T	*Ba-Ga	M2	JPE 12(5)	D				
B-La	M2,91	Unpublished	D	Ba-Gd	M2	M2	T	Be-Ce	M2	Binary Beryllium	D
B-Li	M2	BAPD 10(3)	T	*Ba-Ge	M2	M2	D				
B-Lu	M2	Unpublished	D	*Ba-H	M2	[60Pet1]	D	Be-Cl	M2	Binary Beryllium	D
B-Mg	M2	Binary Magnesium	D	Ba-Hf	No Data			Be-Cm	M2	Binary Beryllium	T
				*Ba-Hg	M2	M2	D				
*B-Mn	M2,91	BAPD 7(6)	D	Ba-Ho	M2	M2	T	*Be-Co	M2	BAPD 9(5)	D
*B-Mo	M2,91	BAPD 9(4)	D	Ba-I	M2	M2	D	*Be-Cr	M2	Binary Beryllium	D
B-N	M2	M2	D	*Ba-In	M2	Indium	D				
B-Na	M2	M2	T	Ba-Ir	No Data			Be-Cs	M2	Binary Beryllium	T
*B-Nb	M2	M2	D	Ba-K	M2	BAPD 5(5)	T				
B-Nd	M2	Unpublished	D	Ba-La	M2	M2	D	*Be-Cu	M2,92	BAPD 8(3)	D
*B-Ni	M2	Binary Nickel	D	*Ba-Li	M2	BAPD 5(5)	D	Be-Dy	M2	Binary Beryllium	T
B-Np	M2	M2	D	*Ba-Mg	M2	Binary Magnesium	D				
B-O	M2	M2	D					Be-Er	M2	Binary Beryllium	T
B-Os	M2	M2	D	Ba-Mn	M2	[64Obi1]	D				
B-P	M2	M2	T	Ba-Mo	M2	M2	D	Be-Eu	M2	Binary Beryllium	T
B-Pa	M2	M2	D	Ba-N	M2	M2	T				
B-Pb	M2	M2	D	*Ba-Na	M2	BAPD 6(1)	D	Be-F	M2	Binary Beryllium	T
*B-Pd	M2	Unpublished	D	Ba-Nb	No Data						
B-Pm	M2	Unpublished	D	Ba-Nd	M2	BAPD 9(3)	D	*Be-Fe	M2	Binary Iron	D
B-Pr	M2	Unpublished	D	Ba-Ni	M2	Binary Nickel	D				
*B-Pt	M2	M2	D								

System	Published	Data source	Data type
Be-Ga	M2	Binary Beryllium	D
Be-Gd	M2	Binary Beryllium	T
Be-Ge	M2	Binary Beryllium	D
Be-H	M2,91	Binary Beryllium	D
*Be-Hf	M2	Binary Beryllium	D
Be-Hg	M2	Binary Beryllium	T
Be-Ho	M2	Binary Beryllium	T
Be-I	M2	Binary Beryllium	T
Be-In	M2	Indium	D
Be-Ir	M2	Binary Beryllium	T
Be-K	M2	Binary Beryllium	T
Be-La	M2	Binary Beryllium	T
Be-Li	M2	Binary Beryllium	D
Be-Lu	M2	Binary Beryllium	T
Be-Mg	M2	Binary Magnesium	D
Be-Mn	M2	Binary Beryllium	T
Be-Mo	M2	Binary Beryllium	D
Be-N	M2	Binary Beryllium	T
Be-Na	M2	Binary Beryllium	D
*Be-Nb	M2,91	Binary Beryllium	D
Be-Nd	M2	Binary Beryllium	T
*Be-Ni	M2	Binary Nickel	D
Be-Np	M2	Binary Beryllium	T
Be-O	M2,91	Binary Beryllium	D
Be-Os	M2	Binary Beryllium	T
Be-P	M2	Binary Beryllium	T
Be-Pa	M2	Binary Beryllium	T
Be-Pb	M2	Binary Beryllium	T
*Be-Pd	M2	Binary Beryllium	D
Be-Pm	M2	M2	T
Be-Po	M2	Binary Beryllium	T
Be-Pr	M2	Binary Beryllium	T
Be-Pt	M2	Binary Beryllium	D
Be-Pu	M2	Binary Beryllium	D
Be-Rb	M2	Binary Beryllium	T
Be-Re	M2	Binary Beryllium	D
Be-Rh	M2	Binary Beryllium	T
Be-Ru	M2	Binary Beryllium	D

System	Published	Data source	Data type
Be-S	M2	Binary Beryllium	T
Be-Sb	M2	Binary Beryllium	D
Be-Sc	M2	Binary Beryllium	T
Be-Se	M2	Binary Beryllium	T
*Be-Si	M2	Binary Beryllium	D
Be-Sm	M2	Binary Beryllium	T
Be-Sn	M2	Binary Beryllium	D
Be-Sr	M2	Binary Beryllium	D
Be-Ta	M2	Binary Beryllium	D
Be-Tb	M2	Binary Beryllium	T
Be-Tc	M2	Binary Beryllium	T
Be-Te	M2	Binary Beryllium	T
*Be-Th	M2	Binary Beryllium	D
*Be-Ti	M2	Binary Beryllium	D
Be-Tl	No Data		
Be-Tm	M2	Binary Beryllium	T
Be-U	M2	Binary Beryllium	D
Be-V	M2	Binary Vanadium	D
*Be-W	M2	Binary Tungsten	D
Be-Y	M2	Binary Beryllium	D
Be-Yb	M2	Binary Beryllium	D
Be-Zn	M2	Binary Beryllium	D
*Be-Zr	M2	Binary Beryllium	D
Bi-Br	M2	M2	D
Bi-C	M2	Unpublished	T
*Bi-Ca	M2,91	M2	D
*Bi-Cd	M2	BAPD 9(4)	D
Bi-Ce	M2	BAPD 9(4)	D
Bi-Cm	M2	M2	T
Bi-Co	M2	JPE 12(3)	D
Bi-Cr	M2	BAPD 9(3)	D
*Bi-Cs	M2	JPE 12(4)	D
*Bi-Cu	M2	BAPD 5(2)	D
Bi-Dy	M2	BAPD 10(4a)	D
Bi-Er	M2	M2	D
Bi-Eu	M2	BAPD 10(4a)	T
*Bi-Ga	M2	M2	D
Bi-Gd	M2	BAPD 10(4a)	D
*Bi-Ge	M2	BAPD 7(6)	D
Bi-H	M2	M2	T
Bi-Hf	M2	M2	D
*Bi-Hg	M2	Unpublished	D
Bi-Ho	M2	M2	D
Bi-I	M2	M2	D
*Bi-In	M2	Indium	D
Bi-Ir	M2	M2	D
*Bi-K	M2	JPE 12 (1)	D
*Bi-La	M2	BAPD 10(4a)	D
*Bi-Li	M2	JPE 12(4)	D

System	Published	Data source	Data type
Bi-Lu	M2	M2	D
*Bi-Mg	M2	Binary Magnesium	D
*Bi-Mn	M2	M2	D
Bi-Mo	M2	M2	D
Bi-N	M2	M2	T
*Bi-Na	M2	JPE 12(4)	D
Bi-Nb	M2	[Moffatt]	D
*Bi-Nd	M2	BAPD 10(4a)	D
*Bi-Ni	M2	Binary Nickel	D
Bi-Np	M2	M2	T
Bi-O	M2	M2	T
Bi-Os	M2	Unpublished	D
Bi-P	M2	M2	T
Bi-Pa	M2	BAPD 2(4)	T
*Bi-Pb	M2,92	JPE 13(1)	D
*Bi-Pd	M2	Unpublished	D
Bi-Pm	No Data		
Bi-Po	M2	M2	D
Bi-Pr	M2	M2	D
*Bi-Pt	M2	JPE 12(2)	D
Bi-Pu	M2	[Chiotti]	D
*Bi-Rb	M2	Unpublished	D
Bi-Re	M2	M2	D
Bi-Rh	M2	[Elliott]	D
Bi-Ru	M2	[Moffatt]	D
*Bi-S	M2	Unpublished	D
*Bi-Sb	M2	Unpublished	D
Bi-Sc	M2	BAPD 10(4a)	T
*Bi-Se	M2	Unpublished	D
Bi-Si	M2	BAPD 6(4)	D
*Bi-Sm	M2	M2	D
*Bi-Sn	M2	M2	D
*Bi-Sr	M2	[Elliott]	D
Bi-Ta	M2,92	JPE 13(3)	T
Bi-Tb	M2	M2	D
Bi-Tc	No Data		
*Bi-Te	M2	Unpublished	D
Bi-Th	M2	M2	D
Bi-Ti	M2	Binary Titanium	D
*Bi-Tl	M2	Unpublished	D
Bi-Tm	M2	M2	D
*Bi-U	M2	[Chiotti]	D
Bi-V	M2	Binary Vanadium	D
Bi-W	M2	Binary Tungsten	T
Bi-Xe	M2	[Elliott]	T
*Bi-Y	M2	BAPD 10(4a)	D
*Bi-Yb	M2	M2	D
*Bi-Zn	M2,91	M2	D
*Bi-Zr	M2,91	BAPD 11(3)	D
Bk-Mo	M2	[Molybdenum]	D
Bk-N	M2	M2	T
Bk-O	M2	M2	T
Bk-P	M2	M2	T
Bk-S	M2	M2	T
Bk-Sb	M2	M2	T
Bk-W		Binary Tungsten	T
Br-Cu	M2	Unpublished	D
Br-In	M2	Indium	D
Br-K	M2	M2	D
Br-Mg	M2	Binary Magnesium	T
Br-Mo	M2	M2	D
Br-Na	M2	M2	D
Br-Ni	M2	Binary Nickel	T
Br-Rb	M2	M2	D
Br-Sc	M2	M2	D
Br-Sr	M2	M2	D
Br-Te	M2	M2	D

(continued)

System	Published	Data source	Data type	System	Published	Data source	Data type	System	Published	Data source	Data type
Br-W	M2	Binary Tungsten	T	Ca-Ce	M2	BAPD 8(6)	D	Cd-Ce	M2	BAPD 9(1)	D
C-Ca	M2	M2	T	Ca-Cl	M2	M2	D	Cd-Co	M2	M2	T
C-Cd	M2	M2	T	Ca-Cm	No Data			Cd-Cr	M2	JPE 13(2)	D
C-Ce	M2	M2	D	Ca-Co	M2	M2	D	Cd-Cs	M2	BAPD 8(6)	D
*C-Co	M2	JPE 12(4)	D	Ca-Cr	M2	BAPD 6(3)	T	*Cd-Cu	M2	BAPD 11(2)	D
*C-Cr	M2	BAPD 11(2)	D	Ca-Cs	M2	BAPD 6(2)	T	Cd-Dy	M2	BAPD 9(1)	T
C-Cs	M2	[87Gor]	T	*Ca-Cu	M2	BAPD 5(6)	D	Cd-Er	M2	BAPD 9(1)	T
*C-Cu	M2	Unpublished	D	Ca-Dy	M2	BAPD 8(6)	T	*Cd-Eu	M2	BAPD 9(1)	D
C-Dy	M2	BAPD 7(5)	T	Ca-Er	M2	BAPD 8(6)	T	Cd-Fe	M2	Binary Iron	D
C-Er	M2	BAPD 7(5)	T	Ca-Eu	M2	BAPD 8(6)	D	*Cd-Ga	M2	Unpublished	D
C-Eu	M2	BAPD 7(5)	T	Ca-F	M2	M2	D	*Cd-Gd	M2	BAPD 9(1)	D
*C-Fe	M2	Binary Iron	D	Ca-Fe	M2	Binary Iron	D	*Cd-Ge	M2	BAPD 7(2)	D
C-Ga	No Data			*Ca-Ga	M2,92	JPE 13(3)	D	Cd-H	M2	M2	T
C-Gd	M2	BAPD 7(5)	T	Ca-Gd	M2	BAPD 8(6)	T	Cd-Hf	M2	M2	T
C-Ge	M2	BAPD 5(5)	D	*Ca-Ge	M2	M2	D	*Cd-Hg	M2	JPE 13(4)	D
*C-Hf	M2	BAPD 11(4)	D	Ca-H	M2	M2	D	Cd-Ho	M2	BAPD 9(1)	T
C-Hg	M2	M2	T	*Ca-Hg	M2	M2	D	*Cd-In	M2	JPE 13(3)	D
C-Ho	M2	BAPD 7(5)	T	Ca-Ho	M2	M2	T	Cd-Ir	No Data		
C-In	M2	Indium	T	*Ca-In	M2	Indium	D	Cd-K	M2	BAPD 8(6)	D
C-Ir	M2	M2	D	Ca-Ir	M2	Unpublished	T	Cd-Kr	M2	M2	T
C-K	M2	M2	T	Ca-K	M2	BAPD 6(1)	T	*Cd-La	M2	BAPD 9(1)	D
*C-La	M2	BAPD 7(5)	D	Ca-La	M2	BAPD 8(6)	D	*Cd-Li	M2	BAPD 9(1)	D
C-Li	M2	BAPD 10(1)	D	*Ca-Li	M2	BAPD 8(2)	D	Cd-Lu	M2	BAPD 9(1)	T
C-Lu	M2	BAPD 7(6)	T	Ca-Lu	M2	M2	D	*Cd-Mg	M2	BAPD 5(1)	D
C-Mg	M2	Binary Magnesium	T	*Ca-Mg	M2	Binary Magnesium	D	Cd-Mn	M2	Unpublished	D
								Cd-Mo	M2	M2	D
*C-Mn	M2	M2	D	Ca-Mn	M2	[Shunk]	D	Cd-N	M2	BAPD 9(3)	T
*C-Mo	M2	M2	D	Ca-Mo	M2	M2	D	*Cd-Na	M2	BAPD 9(1)	D
C-Na	M2	M2	T	Ca-N	M2	BAPD 11(5)	D	Cd-Nb	M2	M2	T
C-Nb	M2	Unpublished	D	*Ca-Na	M2	BAPD 6(1)	D	Cd-Nd	M2	BAPD 9(2)	D
C-Nd	M2	BAPD 7(6)	T	Ca-Nb	M2	M2	T	*Cd-Ni	M2	Binary Nickel	D
*C-Ni	M2	Binary Nickel	D	*Ca-Nd	M2	BAPD 8(6)	D	Cd-Np	M2	BAPD 2(4)	D
C-Np	M2	M2	T	*Ca-Ni	M2,91	Binary Nickel	D	Cd-O	M2	BAPD 8(2)	D
C-Os	M2	[Moffatt]	D	Ca-Np	No Data			Cd-Os	M2	M2	T
C-Pa	M2	M2	T	*Ca-O	M2	BAPD 6(4)	D	*Cd-P	M2	M2	D
C-Pb	M2	M2	T	Ca-Os	No Data			*Cd-Pb	M2	BAPD 9(6)	D
C-Pd	M2	M2	D	Ca-P	M2	M2	T	Cd-Pd	M2	M2	D
C-Po	M2	M2	T	*Ca-Pb	M2	JPE 13(2)	D	Cd-Pm	M2	M2	D
*C-Pr	M2	M2	D	*Ca-Pd	M2,92	M2	D	Cd-Po	M2	M2	T
C-Pt	M2	M2	D	Ca-Pm	M2	M2	D	Cd-Pr	M2	BAPD 9(2)	D
C-Pu	M2	M2	D	Ca-Po	M2	M2	T	Cd-Pt	M2	[52Now]	D
C-Rb	M2	M2	T	Ca-Pr	M2	BAPD 8(6)	T	Cd-Pu	M2	[64Wit]	D
C-Re	M2	M2	D	*Ca-Pt	M2	M2	D	Cd-Rb	M2	BAPD 8(6)	D
C-Rh	M2	M2	D	Ca-Pu	M2	BAPD 10(4a)	D	Cd-Re	M2	M2	T
C-Ru	M2	M2	D	Ca-Rb	M2	BAPD 6(1)	T	Cd-Rh	M2	M2	T
C-Sb	M2	M2	T	Ca-Re	No Data			Cd-Ru	No Data		
*C-Sc	M2	M2	D	Ca-Rh	M2	M2	T	Cd-S	M2	Unpublished	D
C-Se	M2	M2	T	Ca-Ru	No Data			*Cd-Sb	M2	M2	D
*C-Si	M2	BAPD 5(5)	D	Ca-S	M2	M2	T	Cd-Sc	M2	BAPD 9(2)	T
C-Sm	M2	BAPD 7(6)	T	*Ca-Sb	M2	M2	D	*Cd-Se	M2	Unpublished	D
C-Sn	M2	M2	T	Ca-Sc	M2	M2	D	Cd-Si	M2	BAPD 6(6)	D
C-Sr	M2	M2	T	Ca-Se	M2	M2	T	*Cd-Sm	M2	BAPD 9(2)	D
*C-Ta	M2	[86Bar1]	D	*Ca-Si	M2	M2	D	*Cd-Sn	M2	BAPD 10(3)	D
C-Tb	M2	BAPD 7(6)	T	Ca-Sm	M2	BAPD 8(6)	T	*Cd-Sr	M2	M2	D
C-Tc	M2,91	M2	D	Ca-Sn	M2,91	M2	D	Cd-Ta	No Data		
C-Te	No Data			*Ca-Sr	M2	BAPD 7(5)	D	Cd-Tb	M2	BAPD 9(2)	T
*C-Th	M2	[69Ben1]	D	Ca-Ta	No Data			Cd-Tc	M2	M2	T
*C-Ti	M2	Binary Titanium	D	Ca-Tb	M2	M2	D	*Cd-Te	M2	BAPD 10(4)	D
				Ca-Te	M2	M2	T	*Cd-Th	M2	Unpublished	D
C-Tl	M2	M2	T	Ca-Th	No Data			Cd-Ti	M2	Binary Titanium	D
C-Tm	M2	BAPD 7(6)	T	Ca-Ti	M2	Binary Titanium	D				
*C-U	M2	[67Sto,69Ben2]	D					*Cd-Tl	M2	M2	D
				*Ca-Tl	M2	M2	D	Cd-Tm	M2	BAPD 9(2)	T
				Ca-Tm	M2	M2	D	Cd-U	M2	BAPD 1(2)	D
*C-V	M2,91	Binary Vanadium	D	Ca-U	M2	M2	T	Cd-V	M2	Binary Vanadium	D
				Ca-V	M2	Binary Vanadium	D				
*C-W	M2,91	Binary Tungsten	D	Ca-W	M2	Binary Tungsten	T	Cd-W	M2	Binary Tungsten	T
*C-Y	M2	BAPD 7(6)	D	Ca-Y	M2	BAPD 8(6)	D	*Cd-Y	M2	BAPD 9(2)	D
C-Yb	M2	BAPD 7(6)	T	*Ca-Yb	M2	BAPD 8(6)	D	*Cd-Yb	M2	BAPD 9(2)	D
C-Zn	M2	M2	T	*Ca-Zn	M2	BAPD 11(4)	D	*Cd-Zn	M2	BAPD 5(1)	D
*C-Zr	M2	M2	D	Ca-Zr	No Data			Cd-Zr	M2	[Zirconium]	D
*Ca-Cd	M2	M2	D					Ce-Cl	M2	M2	D

(continued)

System	Published	Data source	Data type	System	Published	Data source	Data type	System	Published	Data source	Data type
Ce-Cm	No Data			Cf-Pt	M2	M2	T	*Co-Mo	M2	[Molybdenum]	D
*Ce-Co	M2	[74Gsc1]	D	Cf-S	M2	M2	T	Co-N	M2	Unpublished	T
Ce-Cr	M2	BAPD 11(5)	D	Cf-Sb	M2	M2	T	Co-Na	M2	BAPD 11(5)	T
Ce-Cs	No Data			Cf-W		Binary	T	*Co-Nb	M2	[67Par]	D
*Ce-Cu	M2	BAPD 9(3a)	D			Tungsten		*Co-Nd	M2	[74Ray2]	D
Ce-Dy	M2	BAPD 3(1)	T	*Cl-Cs	M2	M2	D	*Co-Ni	M2	Binary Nickel	D
Ce-Er	M2	M2	D	Cl-Cu	M2	Unpublished	D	Co-Np	M2	M2	T
Ce-Eu	M2	BAPD 3(2)	D	Cl-Dy	M2	M2	D	Co-O	M2	M2	D
*Ce-Fe	M2	Binary Iron	D	Cl-Er	M2	M2	D	Co-Os	M2	[52Kos]	D
*Ce-Ga	M2	M2	D	*Cl-Ga	M2	M2	D	*Co-P	M2	BAPD 11(6)	D
Ce-Gd	M2	BAPD 3(2)	D	Cl-Gd	M2	M2	D	Co-Pb	M2	M2	D
*Ce-Ge	M2	BAPD 10(2)	D	*Cl-Hg	M2	M2	D	*Co-Pd	M2,91	JPE 12(1)	D
Ce-H	M2	M2	D	*Cl-In		Indium	D	Co-Pm	No Data		
Ce-Hf	M2	M2	D	Cl-K	M2	M2	D	*Co-Pr	M2	[74Ray1]	D
Ce-Hg	M2	Unpublished	D	Cl-La	M2	M2	D	*Co-Pt	M2	M2	D
Ce-Ho	M2	M2	D	Cl-Mg	M2	Binary	T	*Co-Pu	M2	[61Poo]	D
*Ce-In	M2	Indium	D			Magnesium		Co-Rb	M2	Unpublished	T
*Ce-Ir	M2	JPE 12(5)	D	Cl-Mo	M2	M2	D	*Co-Re	M2	M2	D
Ce-La	M2	BAPD 2(4)	D	*Cl-Na	M2	M2	D	Co-Rh	M2	[52Kos]	D
Ce-Li	No Data			Cl-Ni	M2	Binary Nickel	D	Co-Ru	M2	[52Kos]	D
Ce-Lu	M2	M2	D	Cl-Pd	M2	M2	D	*Co-S	M2	[08Fri]	D
*Ce-Mg	M2	Binary	D	Cl-Rb	M2	M2	D	*Co-Sb	M2,91	BAPD 11(3)	D
		Magnesium		Cl-Sc	M2	M2	D	Co-Sc	M2	[Moffatt]	D
*Ce-Mn	M2	Unpublished	D	Cl-Sn	M2	M2	D	*Co-Se	M2	M2	D
Ce-Mo	M2	Unpublished	D	Cl-Sr	M2	M2	D	*Co-Si	M2	JPE 12(5)	D
Ce-N	M2	[74Gsc2]	D	Cl-Te	M2	M2	D	*Co-Sm	M2	[Moffatt]	D
Ce-Na	No Data			Cl-Th	M2	M2	D	*Co-Sn	M2	JPE 12(1)	D
Ce-Nb	M2	M2	D	Cl-Tl	M2	M2	D	Co-Sr	M2	JPE 13(3)	T
Ce-Nd	M2	M2	D	Cl-Tm	M2	M2	D	*Co-Ta	M2,91	[86Bar2]	D
*Ce-Ni	M2	Binary Nickel	D	Cl-W	M2	Binary	T	*Co-Tb	M2	M2	D
Ce-Np	No Data					Tungsten		Co-Tc	M2	M2	T
*Ce-O	M2	M2	D	Cl-Y	M2	M2	D	*Co-Te	M2	Unpublished	D
Ce-Os	M2	M2	T	Cl-Yb	M2	M2	D	*Co-Th	M2	Unpublished	D
Ce-P	M2	M2	T	Cm-Cr	M2	BAPD 6(5)	D	Co-Ti	M2	Binary	D
Ce-Pb	M2	M2	D	Cm-Cu	M2	M2	T			Titanium	
*Ce-Pd	M2,91	M2	D	Cm-Ir	M2	M2	T	*Co-Tl	M2	M2	T
Ce-Pm	M2	M2	D	Cm-Mo	M2	[Molybdenum]	D	Co-Tm	M2	M2	T
Ce-Po	M2	[Shunk]	T	Cm-N	M2	M2	T	Co-U	M2	Unpublished	D
Ce-Pr	M2	BAPD 3(2)	D	Cm-O	M2	M2	T	*Co-V	M2	JPE 12(3)	D
Ce-Pt	M2	M2	D	Cm-P	M2	M2	T	*Co-W	M2	Binary	D
*Ce-Pu	M2	[Plutonium]	D	Cm-Pd	M2	M2	T			Tungsten	
Ce-Rb	No Data			Cm-Pt	M2	BAPD 10(2)	D	*Co-Y	M2,92	JPE 12(5)	D
Ce-Re	M2	M2	T	Cm-Rh	M2	M2	T	Co-Yb	M2	[76Ian]	D
Ce-Rh	M2	M2	D	Cm-S	M2	M2	T	*Co-Zn	M2	M2	D
Ce-Ru	M2,92	M2	D	Cm-Sb	M2	M2	T	Co-Zr	M2	[64Pec]	D
*Ce-S	M2	[74Gsc1]	D	Cm-Se	M2	M2	T	Cr-Cs	M2	BAPD 5(4)	D
Ce-Sb	M2	M2	D	Cm-Si	M2	M2	T	*Cr-Cu	M2	BAPD 5(4)	D
Ce-Sc	M2	BAPD 3(2)	D	Cm-Te	M2	M2	T	Cr-Dy	M2	M2	D
Ce-Se	M2	M2	D	Cm-W	M2	Binary	T	Cr-Er	M2	M2	D
*Ce-Si	M2	BAPD 10(1)	D			Tungsten		Cr-Eu	M2	M2	D
Ce-Sm	M2	BAPD 3(2)	D	*Co-Cr	M2	BAPD 11(4)	D	*Cr-Fe	M2	Binary Iron	D
*Ce-Sn	M2	M2	D	Co-Cs	No Data			*Cr-Ga	M2	[72Bor]	D
Ce-Sr	No Data			*Co-Cu	M2	BAPD 5(2)	D	Cr-Gd	M2	[Elliott]	D
Ce-Ta	M2	[66Den1]	D	*Co-Dy	M2	M2	D	*Cr-Ge	M2	BAPD 7(5)	D
Ce-Tb	M2	M2	D	*Co-Er	M2	M2	D	Cr-H	M2	JPE 12(6)	D
Ce-Tc	M2	M2	T	Co-Eu	No Data			*Cr-Hf	M2	BAPD 7(6)	D
*Ce-Te	M2	M2	D	*Co-Fe	M2	Binary Iron	D	Cr-Hg	M2	BAPD 10(2)	D
*Ce-Th	M2	M2	D	*Co-Ga	M2	M2	D	Cr-Ho	M2	[75Sve]	D
*Ce-Ti	M2	Binary	D	*Co-Gd	M2	M2	D	Cr-In	M2	Indium	D
		Titanium		*Co-Ge	M2	JPE 12(1)	D	*Cr-Ir	M2	BAPD 11(1)	D
*Ce-Tl	M2	Unpublished	D	Co-H	M2	M2	D	Cr-K	M2	BAPD 5(4)	D
Ce-Tm	M2	M2	D	*Co-Hf	M2	JPE 12(4)	D	Cr-La	M2	M2	D
Ce-U	M2	[Elliott]	D	Co-Hg	M2	M2	T	Cr-Li	M2	BAPD 5(4)	D
Ce-V	M2	Binary	D	*Co-Ho	M2	M2	D	*Cr-Lu	M2,92	[Moffatt]	D
		Vanadium		Co-In	M2	Indium	D	Cr-Mg	M2	Binary	T
Ce-W	M2	Binary	D	Co-Ir	M2	[52Kos]	D			Magnesium	
		Tungsten		Co-K	M2	Unpublished	T	*Cr-Mn	M2	BAPD 7(5)	D
Ce-Y	M2	BAPD 3(2)	D	Co-La	M2	[74Ray1]	D	*Cr-Mo	M2	BAPD 8(3)	D
Ce-Yb	M2	BAPD 3(1)	T	Co-Li	M2	BAPD 11(5)	T	Cr-N	M2	Unpublished	D
*Ce-Zn	M2	M2	D	Co-Lu	M2	M2	D	*Cr-Nb	M2	BAPD 7(5)	D
Ce-Zr	M2,91	JPE 12(1)	D	Co-Mg	M2	Binary	D	Cr-Nd	M2	[Moffatt]	D
Cf-Mo	M2	[Molybdenum]	D			Magnesium		*Cr-Ni	M2	Binary Nickel	D
Cf-O	M2	M2	T	*Co-Mn	M2	BAPD 11(2)	D				

(continued)

System	Published	Data source	Data type	System	Published	Data source	Data type	System	Published	Data source	Data type
Cr-Np	M2	BAPD 6(5)	D	Cs-Pt	M2	[81Loe]	T	Cu-Ra	M2	[68Gul1]	T
*Cr-O	M2	[80Ban]	D	Cs-Pu	No Data			Cu-Rb	M2	BAPD 7(1)	T
*Cr-Os	M2	BAPD 11(1)	D	*Cs-Rb	M2	BAPD 4(4)	D	Cu-Re	M2	Unpublished	D
Cr-P	M2	BAPD 11(5)	D	*Cs-Re	No Data			*Cu-Rh	M2	BAPD 2(4)	D
Cr-Pb	M2	BAPD 9(2)	D	Cs-Rh	M2	[81Loe]	T	Cu-Rn	M2	Unpublished	T
*Cr-Pd	M2	BAPD 11(1)	D	Cs-Ru	M2	[81Loe]	T	Cu-Ru	M2,92	Unpublished	D
Cr-Pm	No Data			*Cs-S	M2	[Smithells]	D	*Cu-S	M2	BAPD 4(3)	D
Cr-Po		BAPD 9(2)	T	*Cs-Sb	M2	[61Dor2]	D	*Cu-Sb	M2	M2	D
Cr-Pr	M2	M2	D	*Cs-Sc	No Data			Cu-Sc	M2	BAPD 9(3a)	D
*Cr-Pt	M2	BAPD 11(1)	D	*Cs-Se	M2	M2	D	*Cu-Se	M2	BAPD 2(3)	D
Cr-Pu	M2	BAPD 6(5)	D	Cs-Si	M2	M2	T	*Cu-Si	M2	BAPD 7(2)	D
Cr-Ra	M2	BAPD 6(4)	T	Cs-Sm	No Data			Cu-Sm	M2	BAPD 9(3a)	D
Cr-Rb	M2	BAPD 5(4)	D	*Cs-Sn	M2	[87Mel]	D	*Cu-Sn	M2	BAPD 11(3)	D
*Cr-Re	M2	BAPD 8(2)	D	*Cs-Sr	M2	BAPD 6(1)	T	*Cu-Sr	M2	BAPD 5(4)	D
*Cr-Rh	M2	BAPD 8(2)	D	Cs-Ta	M2,91	JAPD 6(2)	D	Cu-Ta	M2	BAPD 10(6)	D
*Cr-Ru	M2	BAPD 8(2)	D	*Cs-Te	M2	Unpublished	D	Cu-Tb	M2	BAPD 9(3a)	D
*Cr-S	M2,91	Unpublished	D	Cs-Th	No Data			Cu-Tc	M2	Unpublished	D
*Cr-Sb	M2,92	BAPD 11(5)	D	Cs-Ti	M2	BAPD 10(2)	D	*Cu-Te	M2	Unpublished	D
*Cr-Sc	M2	BAPD 6(5)	D	*Cs-Tl	M2	[81Bus]	D	*Cu-Th	M2	BAPD 7(1)	D
*Cr-Se		Unpublished	D	Cs-Tm	No Data			*Cu-Ti	M2	Binary Titanium	D
*Cr-Si	M2	BAPD 8(5)	D	Cs-U	No Data						
Cr-Sm	M2	[73Sve]	D	Cs-V	M2	Binary Vanadium	D	*Cu-Tl	M2	BAPD 5(2)	D
*Cr-Sn	M2	BAPD 9(2)	D					Cu-Tm	M2	BAPD 9(3a)	D
Cr-Sr	M2	BAPD 6(4)	T	Cs-W	M2	Binary Tungsten	T	Cu-U	M2	[Metals]	D
*Cr-Ta	M2	BAPD 8(2)	D					*Cu-V	M2	Binary Vanadium	D
Cr-Tb	M2	[71Sve]	D	Cs-Y	No Data						
Cr-Tc	M2	BAPD 7(6)	D	Cs-Yb	No Data			Cu-W	M2	Binary Tungsten	D
*Cr-Te	M2	Unpublished	D	Cs-Zn	M2	BAPD 8(5)	T				
Cr-Th	M2	BAPD 6(5)	D	Cs-Zr	M2	BAPD 8(1)	D	Cu-Xe	M2	Unpublished	T
*Cr-Ti	M2	Binary Titanium	D	*Cu-Dy	M2	BAPD 9(3a)	D	Cu-Y	M2,92	BAPD 2(3)	D
				*Cu-Er	M2	BAPD 9(3a)	D	*Cu-Yb	M2	BAPD 9(3a)	D
Cr-Tl	No Data			*Cu-Eu	M2	BAPD 9(3a)	D	*Cu-Zn	M2	Unpublished	D
Cr-Tm	M2	M2	D	Cu-F	M2	Unpublished	T	*Cu-Zr	M2	BAPD 11(5)	D
*Cr-U	M2	BAPD 6(5)	D	*Cu-Fe	M2	Binary Iron	D	D-Fe		Binary Iron	D
*Cr-V	M2	Binary Vanadium	D	Cu-Fr	M2	M2	T	D-Nb	M2	BAPD 4(1)	T
				*Cu-Ga	M2	Unpublished	D	D-Ta	92	[90Con]	D
*Cr-W	M2	Binary Tungsten	D	*Cu-Gd	M2	BAPD 9(3a)	D	D-V	M2	Binary Vanadium	D
				*Cu-Ge	M2	BAPD 7(1)	D				
Cr-Y	M2,92	BAPD 6(5)	D	*Cu-H	M2	[86Bar3]	D	Dy-Er	M2	BAPD 4(3)	D
Cr-Yb	M2	M2	D	Cu-He	M2	Unpublished	T	*Dy-Fe	M2	Binary Iron	D
Cr-Zn	M2	JPE 13(2)	D	*Cu-Hf	M2	BAPD 9(1)	D	*Dy-Ga	M2	[Moffatt]	D
*Cr-Zr	M2	BAPD 7(3)	D	*Cu-Hg	M2	BAPD 6(6)	D	Dy-Gd	M2	BAPD 4(3)	D
Cs-Cu	M2	BAPD 8(1)	T	Cu-Ho	M2	BAPD 9(3a)	D	*Dy-Ge	M2,91	[77Ere]	D
Cs-F	M2	M2	D	Cu-I	M2	M2	T	Dy-H	M2	[58Mul]	D
Cs-Fe		Binary Iron	T	*Cu-In	M2,91	Indium	D	Dy-Hf	No Data		
Cs-Ga	M2	BAPD 11(4)	D	*Cu-Ir	M2	BAPD 8(2)	D	Dy-Hg		Unpublished	D
*Cs-Ge	M2	M2	D	Cu-K	M2	BAPD 7(3)	T	Dy-Ho	M2	BAPD 4(3)	D
Cs-H	M2	M2	T	Cu-Kr	M2	Unpublished	T	Dy-I	M2	M2	D
Cs-Hf	M2	BAPD 8(1)	D	*Cu-La	M2,91	BAPD 2(3)	D	*Dy-In	M2	Indium	D
*Cs-Hg	M2	[Hansen]	D	*Cu-Li	M2	BAPD 7(2)	D	Dy-Ir	M2	JPE 13(2)	D
Cs-Ho	No Data			Cu-Lu	M2	BAPD 9(3a)	D	Dy-K	No Data		
Cs-I	M2	M2	D	*Cu-Mg	M2,92	Binary Magnesium	D	Dy-La	M2	M2	D
*Cs-In	M2	Indium	D					Dy-Lu	M2	M2	D
Cs-Ir	M2	M2	T	*Cu-Mn	M2	Unpublished	D	Dy-Mg	M2,92	Binary Magnesium	D
*Cs-K	M2	BAPD 4(4)	D	Cu-Mo	M2	BAPD 11(2)	D				
Cs-La	No Data			Cu-N	M2	M2	T	*Dy-Mn	M2	[67Kir1]	D
Cs-Li	M2	BAPD 10(3)	D	Cu-Na	M2	BAPD 7(2)	D	Dy-Mo	M2	M2	D
Cs-Lu	No Data			*Cu-Nb	M2,91	BAPD 2(4)	D	Dy-N	M2	M2	T
Cs-Mg	M2	Binary Magnesium	D	*Cu-Nd	M2	BAPD 9(3a)	D	Dy-Na	No Data		
				Cu-Ne	M2	Unpublished	T	Dy-Nb	No Data		
Cs-Mo	M2,91	M2	D	*Cu-Ni	M2	Binary Nickel	D	Dy-Nd	M2	BAPD 3(3)	D
Cs-N	M2	M2	T	Cu-Np	M2	M2	T	*Dy-Ni	M2	Binary Nickel	D
*Cs-Na	M2	BAPD 3(3)	D	*Cu-O	M2,91	BAPD 5(2)	D	Dy-Np	No Data		
Cs-Nb	M2	BAPD 9(1)	D	Cu-Os	M2	Unpublished	D	Dy-O	M2	M2	T
Cs-Nd	No Data			*Cu-P	M2	M2	D	Dy-Os	M2	[80Pal,59Boz]	T
Cs-Ni	No Data			*Cu-Pa	M2	M2	T	Dy-P	M2	M2	T
Cs-Np	No Data			*Cu-Pb	M2	BAPD 5(5)	D	*Dy-Pb	M2	[68Mcm]	D
*Cs-O	M2	M2	D	*Cu-Pd	M2	JPE 12(2)	D	*Dy-Pd	M2	M2	D
Cs-Os	M2	[81Loe]	T	Cu-Pm	M2	BAPD 9(3a)	D	Dy-Pm	M2	M2	D
Cs-P	M2	M2	T	Cu-Po	M2	Unpublished	T	Dy-Po	M2	M2	T
Cs-Pb	M2	M2	T	Cu-Pr	M2	BAPD 9(3a)	D	Dy-Pr	M2	M2	D
Cs-Pd	M2	[81Loe]	T	*Cu-Pt	M2	Unpublished	D	Dy-Pt	M2	M2	D
Cs-Pr	M2	M2	D	*Cu-Pu	M2	[67Kut1]	D	Dy-Pu	M2,92	M2	D

(continued)

System	Published	Data source	Data type
Dy-Re	M2	[65Ell]	T
Dy-Rh	M2	M2	D
Dy-Ru	M2	M2	D
*Dy-S	M2	M2	D
*Dy-Sb	M2	M2	D
Dy-Sc	No Data		
Dy-Se	M2	M2	T
Dy-Si	M2	M2	T
Dy-Sm	M2	M2	D
*Dy-Sn	M2	M2	D
Dy-Sr	No Data		
Dy-Ta	M2	[66Den1]	D
Dy-Tb	M2	M2	D
Dy-Tc	M2	M2	T
*Dy-Te	M2	M2	D
Dy-Th	M2	[69Bad]	D
Dy-Ti	M2	M2	D
*Dy-Tl	M2	Unpublished	D
Dy-Tm	M2	M2	D
Dy-U	M2	M2	T
Dy-V	M2	Binary Vanadium	D
Dy-W	M2	Binary Tungsten	D
Dy-Y	M2	BAPD 4(1)	D
Dy-Yb	M2	M2	D
Dy-Zn	M2	M2	D
*Dy-Zr	M2	[60Cro]	D
*Er-Fe	M2	Binary Iron	D
*Er-Ga	M2	M2	D
Er-Gd	M2	BAPD 4(3)	D
*Er-Ge	M2	M2	D
Er-H	M2	[58Mul]	D
Er-Hf	M2	[Hafnium]	D
Er-Hg	M2	Unpublished	D
Er-Ho	M2	BAPD 4(3)	D
Er-I	M2	M2	D
*Er-In	M2	Indium	D
Er-Ir	M2	JPE 13(2)	D
Er-K	No Data		
Er-La	M2	M2	D
Er-Li	No Data		
Er-Lu	M2	M2	D
Er-Mg	M2	Binary Magnesium	D
*Er-Mn	M2	[67Kir2]	D
Er-Mo	M2	M2	D
Er-N	M2	M2	T
Er-Na	No Data		
Er-Nb	M2	[61Lov]	T
Er-Nd	M2	BAPD 3(3)	D
*Er-Ni	M2	Binary Nickel	D
Er-Np	No Data		
Er-O	M2	[61Lov]	D
Er-Os	M2	M2	T
Er-P	M2	M2	T
Er-Pb	M2	M2	T
*Er-Pd	M2,91	[73Loe]	D
Er-Pm	M2	M2	D
Er-Po	M2	[Shunk]	T
Er-Pr	M2	M2	D
*Er-Pt	M2	M2	D
Er-Pu	M2	M2	D
Er-Re	M2	M2	D
Er-Rh	M2	[73Gha]	D
*Er-Ru	M2	M2	D
Er-S	M2	M2	T
Er-Sb	M2	M2	T
Er-Sc	M2	BAPD 4(1)	D
*Er-Se	M2	M2	D
Er-Si	M2	M2	D
Er-Sm	M2	M2	D
Er-Sn	M2	M2	D
Er-Ta	M2	[66Den1]	D
Er-Tb	M2	BAPD 4(3)	D
Er-Tc	M2	M2	T
*Er-Te	M2	M2	D
Er-Th	M2	M2	D
*Er-Ti	M2	Binary Titanium	D
*Er-Tl	M2	Unpublished	D
Er-Tm	M2	M2	D
Er-U	M2	M2	T
Er-V	M2	Binary Vanadium	D
Er-W	M2	Binary Tungsten	D
Er-Y	M2	BAPD 4(1)	D
Er-Yb	M2	M2	D
Er-Zn	M2	M2	D
Er-Zr	M2	[Zirconium]	D
Es-Mo	M2	[Molybdenum]	D
Es-O	M2	M2	T
Es-W		Binary Tungsten	T
Eu-Fe	M2	Binary Iron	D
*Eu-Ga	M2	[78Yat]	D
*Eu-Ge	M2	JPE 12(4)	D
Eu-H	M2	M2	T
Eu-Hf	M2	M2	D
Eu-Hg	M2	Unpublished	T
Eu-Ho	M2	BAPD 4(2)	T
*Eu-In	M2	Indium	D
Eu-Ir	M2	Unpublished	T
Eu-K	No Data		
Eu-La	M2,91		D
*Eu-Mg	M2,92	Binary Magnesium	D
Eu-Mn	M2	M2	D
Eu-Mo	M2	M2	D
Eu-N	M2	M2	T
Eu-Na	No Data		
Eu-Nb	M2	M2	D
Eu-Ni	M2,92	Binary Nickel	D
Eu-Np	No Data		
Eu-O	M2	M2	D
Eu-Os	No Data		
Eu-P	M2	M2	T
*Eu-Pb	M2	[67Mcm]	D
*Eu-Pd	M2	M2	D
Eu-Po	M2	M2	T
Eu-Pr	No Data		
*Eu-Pt	M2	[81Ian]	D
Eu-Pu	M2	M2	D
Eu-Re	M2	M2	T
Eu-Rh	No Data		
Eu-Ru	No Data		
Eu-S	M2	M2	D
Eu-Sb	M2	M2	T
Eu-Sc	M2	M2	D
Eu-Se	M2	M2	T
Eu-Si	M2	M2	T
Eu-Sm	M2	M2	T
Eu-Sn	M2	M2	T
Eu-Sr	No Data		
Eu-Ta	M2	M2	D
Eu-Tb	No Data		
*Eu-Te	M2	[70Sad]	D
Eu-Th	M2	M2	D
Eu-Ti	M2	Binary Titanium	T
Eu-Tl	M2	M2	T
Eu-U	M2	M2	D
Eu-V	M2	Binary Vanadium	D
Eu-W	M2	Binary Tungsten	D
Eu-Y	M2	M2	D
Eu-Yb	M2	M2	D
Eu-Zn	M2	M2	D
Eu-Zr	M2	M2	D
F-In	M2	Indium	T
F-K	M2	M2	D
F-Mg	M2	Binary Magnesium	T
F-Mo	M2	M2	D
F-Na	M2	M2	D
F-Ni	M2	Binary Nickel	T
F-Rb	M2	M2	D
F-Sm	M2	M2	D
F-Sn	M2	M2	D
F-W	92	Binary Tungsten	T
F-Yb	M2	M2	D
*Fe-Ga	M2,91	Binary Iron	D
*Fe-Gd	M2	Binary Iron	D
*Fe-Ge	M2	Binary Iron	D
*Fe-H	M2	Binary Iron	D
*Fe-Hf	M2	Binary Iron	D
Fe-Hg	M2	Binary Iron	D
*Fe-Ho	M2,91	Binary Iron	D
Fe-In	M2	Binary Iron	D
*Fe-Ir	M2	Binary Iron	D
Fe-K	M2	Binary Iron	D
*Fe-La	M2	Binary Iron	D
Fe-Li	M2	Binary Iron	D
*Fe-Lu	M2	Binary Iron	D
Fe-Mg	M2	Binary Iron	D
*Fe-Mn	M2	Binary Iron	D
*Fe-Mo	M2	Binary Iron	D
*Fe-N	M2	Binary Iron	D
Fe-Na	M2	Binary Iron	D
*Fe-Nb	M2	Binary Iron	D
*Fe-Nd	M2	Binary Iron	D
*Fe-Ni	M2	Binary Iron	D
Fe-Np	M2	Binary Iron	T
*Fe-O	M2	Binary Iron	D
Fe-Os	M2	Binary Iron	D
*Fe-P	M2	Binary Iron	D
Fe-Pb	M2	Binary Iron	D
Fe-Pd	M2	Binary Iron	D
Fe-Pm	M2	Binary Iron	D
Fe-Pr	M2	Binary Iron	D
Fe-Pt	M2	Binary Iron	D
*Fe-Pu	M2	Binary Iron	D
Fe-Rb	M2	Binary Iron	T
Fe-Re		Binary Iron	D
*Fe-Rh	M2	Binary Iron	D
Fe-Ru	M2	Binary Iron	D
*Fe-S	M2	Binary Iron	D
*Fe-Sb	M2	Binary Iron	D
*Fe-Sc	M2	Binary Iron	D
*Fe-Se	M2,91	Binary Iron	D
*Fe-Si	M2	Binary Iron	D
*Fe-Sm	M2	Binary Iron	D
*Fe-Sn	M2,92	Binary Iron	D
Fe-Sr	M2	Binary Iron	D
Fe-Ta	M2	Binary Iron	D
*Fe-Tb	M2	Binary Iron	D
Fe-Tc	M2	Binary Iron	D
*Fe-Te	M2	Binary Iron	D
*Fe-Th	M2,91	Binary Iron	D
*Fe-Ti	M2	Binary Iron	D
Fe-Tl	M2	Binary Iron	T
*Fe-Tm	M2	Binary Iron	D
*Fe-U	M2	Binary Iron	D
*Fe-V	M2	Binary Iron	D
*Fe-W	M2	Binary Iron	D

(continued)

System	Published	Data source	Data type	System	Published	Data source	Data type	System	Published	Data source	Data type
Fe-Y	M2	Binary Iron	D	Gd-Ho	M2	BAPD 4(3)	D	*Ge-Na	M2	M2	D
Fe-Yb	M2	Binary Iron	D	Gd-I	M2	M2	D	*Ge-Nb	M2	[Moffatt]	D
*Fe-Zn	M2	Binary Iron	D	*Gd-In	M2	Indium	D	*Ge-Nd	M2	BAPD 10(2)	D
*Fe-Zr	M2	Binary Iron	D	Gd-Ir	M2	Unpublished	D	*Ge-Ni	M2	Binary Nickel	D
Fm-Mo	M2	[Molybdenum]	D	Gd-K	No Data			Ge-Np	No Data		
Fr-Mg	M2	[68Gul2]	T	Gd-La	M2	BAPD 2(4)	D	Ge-O	M2	[56Tru]	D
Fr-Mo	M2	[Molybdenum]	D	Gd-Li	No Data			Ge-Os	M2	M2	T
Fr-W		Binary	T	Gd-Lu	M2	M2	D	*Ge-P	M2,91	BAPD 6(3)	D
		Tungsten		*Gd-Mg	M2	Binary	D	*Ge-Pb	M2	BAPD 5(4)	D
*Ga-Gd	M2	BAPD 11(1)	D			Magnesium		*Ge-Pd	M2	JPE 13(4)	D
Ga-Ge	M2	BAPD 6(3)	D	Gd-Mn	M2	M2	D	Ge-Pm	No Data		
Ga-H	No Data			Gd-Mo	M2	BAPD 1(2)	D	*Ge-Pr	M2,91	BAPD 10(3)	D
Ga-Hf	M2	M2	D	Gd-N	M2	M2	T	*Ge-Pt	M2	JPE 13(4)	D
Ga-Hg	M2	[60Pre]	D	Gd-Na	No Data			Ge-Pu	M2	M2	T
*Ga-Ho	M2	M2	D	Gd-Nb	M2	M2	T	Ge-Rb	M2	M2	D
Ga-I	M2	M2	D	Gd-Nd	M2	BAPD 3(3)	D	Ge-Re	M2	[Moffatt]	D
*Ga-In	M2	Indium	D	*Gd-Ni	M2	Binary Nickel	D	Ge-Rh	M2	M2	D
Ga-Ir	M2	M2	T	Gd-Np	No Data			Ge-Ru	M2	M2	D
Ga-K	M2	BAPD 11(4)	D	Gd-O	M2	M2	T	*Ge-S	M2	[63Liu]	D
*Ga-La	M2	BAPD 11(1)	D	Gd-Os	M2	[80Pal]	T	*Ge-Sb	M2	BAPD 7(3)	D
*Ga-Li	M2	JPE 12(1)	D	Gd-P	M2	M2	T	*Ge-Sc	M2	BAPD 7(6)	D
*Ga-Lu	M2	[79Yat]	D	*Gd-Pb	M2	JPE 12(6)	D	*Ge-Se	M2	BAPD 11(3)	D
*Ga-Mg	M2,91	Binary	D	*Gd-Pd	M2	M2	D	*Ge-Si	M2	BAPD 5(2)	D
		Magnesium		Gd-Pm	M2	M2	D	*Ge-Sm	M2	BDPD 9(5)	D
*Ga-Mn	M2	[80Lu]	D	Gd-Po	M2	M2	T	*Ge-Sn	M2	BAPD 5(3)	D
*Ga-Mo	M2	[Molybdenum]	D	Gd-Pr	M2	M2	D	*Ge-Sr	M2	M2	D
Ga-N	M2	M2	T	Gd-Pt	M2	M2	D	Ge-Ta	M2,92	JPE 12(6)	T
*Ga-Na	M2	BAPD 11(4)	D	Gd-Pu	M2	M2	D	*Ge-Tb	M2	M2	D
*Ga-Nb	M2	M2	D	Gd-Re	M2	M2	D	*Ge-Te	M2	M2	D
*Ga-Nd	M2	[Moffatt]	D	*Gd-Rh	M2	M2	D	Ge-Th	M2	[Thorium]	D
*Ga-Ni	M2	Binary Nickel	D	Gd-Ru	M2	[Moffatt]	D	*Ge-Ti	M2	Binary	D
Ga-Np	M2	M2	T	Gd-S	M2	M2	D			Titanium	
Ga-O	M2	M2	T	*Gd-Sb	M2	M2	D	*Ge-Tl	M2	BAPD 6(2)	D
Ga-Os	M2	M2	T	*Gd-Sc	M2	BAPD 4(2)	D	*Ge-Tm	M2	M2	D
Ga-P	M2	[Shunk]	D	*Gd-Se	M2	[82Pri]	D	*Ge-U	M2	[60Lya]	D
*Ga-Pb	M2	JPE 12(1)	D	Gd-Si	M2	BAPD 9(5)	D	Ge-V	M2	Binary	D
*Ga-Pd	M2	M2	D	Gd-Sm	M2	BAPD 4(2)	D			Vanadium	
Ga-Pm	M2,92	M2	D	*Gd-Sn	M2	JPE 12(6)	D	Ge-W	M2	Binary	D
*Ga-Pr	M2	M2	D	Gd-Sr	No Data					Tungsten	
*Ga-Pt	M2	M2	D	Gd-Ta	M2	[66Den1]	D	*Ge-Y	M2	BAPD 9(1)	D
*Ga-Pu	M2	BAPD 9(3)	D	Gd-Tb	M2	BAPD 4(3)	D	*Ge-Yb	M2	[83Ere]	D
Ga-Rb	M2,92	BAPD 11(4)	D	Gd-Tc	M2	M2	T	*Ge-Zn	M2	BAPD 6(6)	D
Ga-Re	M2	M2	D	*Gd-Te	M2	M2	D	Ge-Zr	M2	BAPD 7(1)	D
Ga-Rh	M2	M2	T	Gd-Th	M2	[69Bad]	D	H-Hf	M2,91	M2	D
Ga-Ru	M2	M2	T	*Gd-Ti	M2	Binary	D	H-Hg	M2	M2	T
*Ga-S	M2	[67Rus]	D			Titanium		H-Ho	M2	M2	T
*Ga-Sb	M2	BAPD 9(5)	D	*Gd-Tl	M2	Unpublished	D	H-In	M2	Indium	T
*Ga-Sc	M2	[79Yat]	D	Gd-Tm	M2	M2	D	H-Ir	M2,91	Unpublished	T
*Ga-Se	M2	[Moffatt]	D	Gd-U	M2	[Elliott]	T	H-K	M2	M2	T
Ga-Si	M2	BAPD 6(4)	D	Gd-V	M2	Binary	D	*H-La	M2	BAPD 11(1)	D
*Ga-Sm	M2	[Moffatt]	D			Vanadium		H-Li	M2	M2	D
*Ga-Sn	M2	JPE 13(2)	D	Gd-W	M2	Binary	D	H-Lu	M2	[82Sub]	D
*Ga-Sr	M2	JPE 13(2)	D			Tungsten		H-Mg	M2	BAPD 8(5)	D
Ga-Ta	M2	M2	D	Gd-Y	M2	BAPD 4(2)	D	H-Mn	M2	Unpublished	D
*Ga-Tb	M2	[Moffatt]	D	Gd-Yb	M2	BAPD 4(3)	D	H-Mo	M2	[Molybdenum]	D
*Ga-Te	M2	Unpublished	D	Gd-Zn	M2	M2	D	H-Na	M2	BAPD 11(3)	D
Ga-Th	M2	M2	T	Gd-Zr	M2	M2	D	*H-Nb	M2	BAPD 4(1)	D
Ga-Ti	M2	Binary	D	Ge-H	M2	[Elliott]	T	*H-Nd	M2	M2	D
		Titanium		Ge-Hf	M2	BAPD 11(3)	D	*H-Ni	M2	Binary Nickel	D
*Ga-Tl	M2	JPE 12(6)	D	Ge-Hg	M2	Unpublished	T	H-Np	M2	M2	T
*Ga-Tm	M2	[Moffatt]	D	*Ge-Ho	M2	[80Ere]	D	H-Os	M2	Unpublished	T
*Ga-U	M2	[73Bus]	D	Ge-I	M2	M2	D	H-Pa	M2	M2	T
*Ga-V	M2	Binary	D	*Ge-In	M2	Indium	D	H-Pb	M2	M2	T
		Vanadium		Ge-Ir	M2	M2	T	*H-Pd	M2	Unpublished	D
Ga-W	M2	Binary	T	*Ge-K	M2	M2	D	H-Po	M2	[Shunk]	T
		Tungsten		*Ge-La	M2	BAPD 10(4)	D	H-Pr	M2	M2	D
*Ga-Y	M2	[77Yat]	D	*Ge-Li	M2	M2	D	H-Pt	M2,91	Unpublished	T
*Ga-Yb	M2,92	JPE 13(1)	D	*Ge-Lu	M2	M2	D	H-Pu	M2	[56Mul]	D
*Ga-Zn	M2	BAPD 11(1)	D	*Ge-Mg	M2	Binary	D	H-Rb	M2	M2	T
*Ga-Zr	M2	[Shunk]	D			Magnesium		H-Re	M2	Unpublished	T
*Gd-Ge	M2	BAPD 10(2)	D	*Ge-Mn	M2	BAPD 11(5)	D	H-Rh	M2,91	Unpublished	T
Gd-H	M2	[60Bec]	D	*Ge-Mo	M2	BAPD 8(1)	D	H-Ru	M2,91	Unpublished	T
Gd-Hg	M2	Unpublished	D	Ge-N	M2	BAPD 11(6)	T	H-Sb	M2	M2	T

(continued)

System	Published	Data source	Data type
H-Sc	M2	M2	D
H-Se	M2	M2	T
H-Si	M2	Unpublished	D
H-Sm	M2	M2	D
H-Sn	M2	M2	T
*H-Sr	M2	[64Pet]	D
*H-Ta	M2	JPE 12(3)	D
H-Tb	M2	M2	T
H-Te	No Data		
H-Th	M2	[Smith]	D
*H-Ti	M2,92	Binary Titanium	D
H-Tl	M2	M2	T
H-Tm	M2	M2	D
*H-U	M2	Unpublished	D
*H-V	M2	Binary Vanadium	D
H-W	M2	Binary Tungsten	T
H-Y	M2	BAPD 9(3)	D
H-Yb	M2	M2	D
H-Zn	M2	BAPD 10(6)	D
*H-Zr	M2	BAPD 11(4)	D
He-Mo	M2	[Molybdenum]	D
He-W		Binary Tungsten	T
Hf-Hg	M2	M2	D
Hf-In	M2	Indium	T
*Hf-Ir	M2	M2	D
Hf-K	M2	BAPD 8(1)	D
Hf-La	No Data		
Hf-Li	M2	BAPD 10(3)	D
Hf-Lu	No Data		
Hf-Mg	M2	Binary Magnesium	D
*Hf-Mn	M2	Unpublished	D
*Hf-Mo	M2	[Molybdenum]	D
*Hf-N	M2	BAPD 11(2)	D
Hf-Na	M2	BAPD 8(1)	D
*Hf-Nb	M2,91	JPE 12(2)	D
*Hf-Ni	M2,91	Binary Nickel	D
Hf-Np	No Data		
*Hf-O	M2	[Hafnium]	D
*Hf-Os	M2	M2	D
Hf-P	M2	M2	T
Hf-Pd	M2	[72Shu]	D
Hf-Po	M2	M2	T
Hf-Pr	M2	[71Gri]	D
Hf-Pt	M2	M2	T
Hf-Pu	M2	M2	D
Hf-Rb	M2	BAPD 8(1)	D
Hf-Re	M2	[63Tay]	D
*Hf-Rh	M2	M2	D
Hf-Ru	M2	M2	D
Hf-Si	M2	M2	T
Hf-Sb	M2	M2	T
Hf-Sc	M2	M2	D
Hf-Se	M2	M2	T
*Hf-Si	M2	BAPD 10(4)	D
Hf-Sm	No Data		
Hf-Sn	M2	JPE 12(4)	D
Hf-Sr	No Data		
*Hf-Ta	M2	JAPD 5(2)	D
Hf-Tb	No Data		
Hf-Tc	M2	M2	T
Hf-Te	M2	M2	D
Hf-Th	M2	[58Gib]	D
Hf-Ti	M2	Binary Titanium	D
Hf-Tl	No Data		
Hf-Tm	No Data		
*Hf-U	M2	[60Pet2]	D

System	Published	Data source	Data type
*Hf-V	M2	Binary Vanadium	D
*Hf-W	M2,92	Binary Tungsten	D
Hf-Y	M2	[62Lun]	D
Hf-Yb	M2	[Moffatt]	D
Hf-Zn	M2	M2	T
*Hf-Zr	M2	BAPD 3(1)	D
Hg-Ho	M2	Unpublished	D
*Hg-In	M2	Indium	D
Hg-Ir	M2	M2	D
*Hg-K	M2	[79Vol]	D
*Hg-La	M2	Unpublished	D
*Hg-Li	M2	[Hansen]	D
Hg-Lu	M2	Unpublished	T
*Hg-Mg	M2	Binary Magnesium	D
Hg-Mn	M2	M2	D
Hg-Mo	M2	M2	D
Hg-N	M2	M2	T
*Hg-Na	M2	M2	D
Hg-Nb	M2	Unpublished	D
Hg-Nd	M2	Unpublished	D
Hg-Ni	M2,91	Binary Nickel	D
Hg-Np	No Data		
Hg-O	M2	M2	T
Hg-Os	M2	M2	D
Hg-P	No Data		
*Hg-Pb	M2	[Hansen]	D
Hg-Pd	M2	BAPD 11(1)	D
Hg-Po	M2	M2	T
Hg-Pr	M2	Unpublished	D
Hg-Pt	M2	BAPD 11(1)	D
Hg-Pu	M2	[59Sch]	D
*Hg-Rb	M2	[Hansen]	D
Hg-Re	M2	M2	D
Hg-Rh	M2	[67Jan]	D
Hg-Ru	M2	M2	D
*Hg-S	M2	JPE 13(5)	D
Hg-Sb	M2	BAPD 11(4)	D
Hg-Sc	M2	Unpublished	T
*Hg-Se	M2	JPE 13(5)	D
Hg-Si	M2	Unpublished	T
Hg-Sm	M2	Unpublished	D
*Hg-Sn	M2	M2	D
*Hg-Sr	M2	M2	D
Hg-Ta	M2	[05Bol]	T
Hg-Tb	M2	Unpublished	D
*Hg-Te	M2	Unpublished	D
Hg-Th	M2	[58Dom]	D
Hg-Ti	M2	Binary Titanium	D
*Hg-Tl	M2	Unpublished	D
Hg-Tm	M2	Unpublished	T
Hg-U	M2	M2	D
Hg-V	M2	Binary Vanadium	D
Hg-W	M2	Binary Tungsten	T
Hg-Y	M2	Unpublished	D
Hg-Yb	M2	Unpublished	D
*Hg-Zn	M2	Unpublished	D
Hg-Zr	M2	M2	D
Ho-I	M2	M2	D
*Ho-In	M2	Indium	D
Ho-Ir	M2	Unpublished	D
Ho-K	No Data		
Ho-La	M2	M2	D
Ho-Li	No Data		
Ho-Lu	M2	M2	D
Ho-Mg	M2	Binary Magnesium	D
*Ho-Mn	M2	[67Kir2]	D

System	Published	Data source	Data type
Ho-Mo	M2	M2	D
Ho-N	M2	M2	T
Ho-Na	No Data		
Ho-Nb	No Data		
Ho-Nd	M2	M2	D
Ho-Ni	M2,92	Binary Nickel	D
Ho-Np	No Data		
Ho-O	M2	M2	T
Ho-Os	M2	M2	T
Ho-P	M2	M2	T
Ho-Pb	M2	M2	T
*Ho-Pd	M2,91	M2	D
Ho-Pm	M2	M2	D
Ho-Po	M2	M2	T
Ho-Pr	M2	M2	D
Ho-Pt	M2	M2	D
Ho-Pu	M2,91	M2	D
Ho-Rb	No Data		
Ho-Re	M2	M2	T
Ho-Rh	M2	M2	D
Ho-Ru	M2	M2	D
Ho-S	M2	M2	T
*Ho-Sb	M2	M2	D
Ho-Sc	M2	M2	D
Ho-Se	M2,91	M2	T
Ho-Si	M2	[Pearson3]	T
Ho-Sm	M2	M2	D
Ho-Sn	M2	M2	T
Ho-Sr	No Data		
Ho-Ta	M2	[Moffatt]	D
Ho-Tb	M2	BAPD 4(3)	D
Ho-Tc	M2	M2	T
*Ho-Te	M2	[74Yar]	D
Ho-Th	M2	M2	D
Ho-Ti	No Data		
*Ho-Tl	M2	Unpublished	D
Ho-Tm	M2	M2	D
Ho-U	M2	M2	T
Ho-V	M2	Binary Vanadium	D
Ho-W	M2	Binary Tungsten	D
Ho-Y	M2	BAPD 4(1)	D
Ho-Yb	M2	M2	D
Ho-Zn	M2	M2	D
Ho-Zr	M2	M2	D
I-In	M2	Indium	D
I-K	M2	M2	D
I-Mg	M2	Binary Magnesium	T
I-Mo	M2	M2	D
I-Na	M2	M2	D
I-Ni	M2	Binary Nickel	T
I-Rb	M2,91	M2	D
I-Se	M2	M2	D
I-Sr	M2	M2	D
I-Tb	M2	M2	D
I-Te	M2	M2	D
I-Th	M2	[Smith]	D
I-Tl	M2	M2	D
I-W	M2	Binary Tungsten	T
I-Y	M2	M2	D
In-Ir	M2	Indium	T
*In-K	M2,92	Indium	D
In-Kr	M2	M2	T
*In-La	M2	Indium	D
*In-Li	M2	Indium	D
*In-Lu	M2	Indium	D
*In-Mg	M2	Indium	D
*In-Mn	M2,92	Indium	D
In-Mo	M2	Indium	D
In-N	M2	Indium	T

(continued)

System	Published	Data source	Data type	System	Published	Data source	Data type	System	Published	Data source	Data type
*In-Na	M2	Indium	D	Ir-Se	M2	M2	T	*La-Mg	M2	Binary Magnesium	D
*In-Nb	M2	Indium	D	Ir-Si	M2	M2	T				
*In-Nd	M2,91	Indium	D	Ir-Sm	M2	Unpublished	D	*La-Mn	M2	BAPD 11(5)	D
*In-Ni	M2	Indium	D	Ir-Sn	M2	M2	T	La-Mo	M2	M2	D
In-Np	No Data			Ir-Sr	M2	M2	T	La-N	M2	M2	T
In-O	M2,91	Indium	D	*Ir-Ta	M2	[Metals]	D	La-Na	No Data		
In-Os	M2	Indium	T	Ir-Tb	M2	Unpublished	D	La-Nb	M2	Unpublished	D
*In-P	M2	Indium	D	Ir-Tc	M2	M2	D	La-Nd	M2	BAPD 2(4)	D
*In-Pb	M2	Indium	D	Ir-Te	M2	M2	D	*La-Ni	M2,91	Binary Nickel	D
*In-Pd	M2	Indium	D	*Ir-Th	M2	JPE 12(5)	D	La-Np	No Data		
In-Pm	M2	Indium	D	*Ir-Ti	M2,92	Binary Titanium	D	La-O	M2	M2	T
*In-Pr	M2	Indium	D					La-Os	M2	M2	T
*In-Pt	M2,91	Indium	D	Ir-Tl	No Data			La-P	M2	M2	T
*In-Pu	M2	Indium	D	Ir-Tm	M2	Unpublished	D	*La-Pb	M2,92	JPE 13(1)	D
*In-Rb	M2	Indium	D	*Ir-U	M2	JPE 13(5)	D	La-Pd	M2	M2	T
In-Re	M2	Indium	T	*Ir-V	M2	Binary Vanadium	D	La-Pm	M2	M2	D
In-Rh	M2	Indium	T					La-Pr	M2	M2	D
In-Ru	M2	Indium	T	*Ir-W	M2,92	Binary Tungsten	D	La-Pt	M2	M2	D
*In-S	M2	Indium	D					La-Pu	M2	M2	D
*In-Sb	M2	Indium	D	Ir-Y	M2	Unpublished	D	La-Rb	No Data		
*In-Sc	M2	Indium	D	Ir-Yb	M2	JPE 13(2)	D	La-Re	M2	M2	D
*In-Se	M2,91	Indium	D	Ir-Zn	M2	[64Rhy]	T	La-Rh	M2	M2	D
*In-Si	M2	Indium	D	*Ir-Zr	M2	JPE 13(5)	D	La-Ru	M2,91	M2	D
*In-Sm	M2	Indium	D	K-La	No Data			*La-S	M2	Unpublished	D
*In-Sn	M2,91	Indium	D	K-Li	M2	BAPD 10(3)	D	*La-Sb	M2	[54Vog]	D
*In-Sr	M2	Indium	D	K-Mg	M2	Binary Magnesium	D	*La-Sc	M2	BAPD 3(1)	D
In-Ta	M2	Indium	D					*La-Se	M2	M2	D
*In-Tb	M2	Indium	D	K-Mo	M2	M2	D	La-Si	M2	M2	T
*In-Te	M2,91	Indium	D	K-N	M2	M2	T	La-Sm	M2	M2	D
*In-Th	M2	Indium	D	*K-Na	M2	BAPD 3(3)	D	*La-Sn	M2,92	JPE 13(1)	D
*In-Ti	M2	Indium	D	K-Nb	M2	BAPD 9(4)	D	La-Sr	No Data		
*In-Tl	M2	Indium	D	K-Nd	No Data			La-Ta	M2	[Moffatt]	D
*In-Tm	M2	Indium	D	K-Ni	M2	[65Swi]	T	La-Tb	M2	M2	D
In-U	M2	Indium	D	K-Np	No Data			La-Te	M2	[65Haa]	D
*In-V	M2	Indium	D	K-O	M2	M2	T	La-Th	M2	[69Bad]	D
In-W	M2	Binary Tungsten	T	K-Os	M2	M2	T	La-Ti	M2	Binary Titanium	D
				K-P	M2	M2	T				
*In-Y	M2	Indium	D	*K-Pb	M2	M2	D	*La-Tl	M2	Unpublished	D
*In-Yb	M2	Indium	D	K-Pd	M2	M2	T	La-Tm	M2	M2	D
*In-Zn	M2	Indium	D	K-Pr	No Data			La-U	M2	M2	T
In-Zr	M2	Indium	D	K-Pu	M2	[59Sch]	T	La-V	M2	Binary Vanadium	D
Ir-K	M2	[64Rhy]	T	*K-Rb	M2	BAPD 4(4)	D				
*Ir-La	M2	JPE 12(5)	D	K-Re	No Data			La-W	M2	Binary Tungsten	D
Ir-Li	M2	JPE 13(1)	D	K-Rh	M2	M2	T				
Ir-Lu	M2	Unpublished	D	K-Ru	M2	M2	T	La-Y	M2	BAPD 3(1)	D
Ir-Mg	M2	Binary Magnesium	D	*K-S	M2	M2	D	La-Yb	M2	M2	D
				*K-Sb	M2	[61Dor2]	D	*La-Zn	M2	[41Rol]	D
Ir-Mn	M2	Unpublished	D	*K-Se	M2	M2	D	La-Zr	M2	M2	T
*Ir-Mo	M2	[Molybdenum]	D	K-Si	M2	M2	T	*Li-Mg	M2	Binary Magnesium	D
Ir-N	M2	[05Emi]	T	K-Sm	No Data						
Ir-Na	M2	[64Rhy]	T	*K-Sn	M2	M2	D	Li-Mn	M2	[64Obi1]	D
*Ir-Nb	M2	Unpublished	D	K-Sr	M2	BAPD 6(2)	T	Li-Mo	M2	M2	D
Ir-Nd	M2	Unpublished	D	K-Ta	M2	JAPD 6(1)	D	Li-N	M2,92	JPE 13(3)	D
*Ir-Ni	M2	Binary Nickel	D	K-Tb	No Data			*Li-Na	M2	BAPD 10(3)	D
Ir-Np	M2	M2	T	*K-Te	M2	BAPD 11(5)	D	Li-Nb	M2	BAPD 9(4)	D
Ir-O	M2	M2	T	K-Th	M2	M2	T	Li-Ni	M2	Binary Nickel	D
Ir-Os	M2	Unpublished	D	K-Ti	M2	BAPD 10(2)	D	Li-Np	No Data		
Ir-P	M2	BAPD 11(4)	D	*K-Tl	M2	M2	T	Li-O	M2,92	JPE 13(3)	T
Ir-Pa	M2	M2	T	K-Tm	No Data			Li-Os	M2	JPE 13(1)	T
Ir-Pb	M2	M2	T	K-U	M2	M2	T	Li-P	M2	Unpublished	T
*Ir-Pd	M2,91	JPE 12(5)	D	K-V	M2	Binary Vanadium	D	*Li-Pb	M2	[Hansen]	D
Ir-Pm	M2	Unpublished	D					*Li-Pd	M2	JPE 13(1)	D
Ir-Pr	M2	Unpublished	D	K-W	M2	Binary Tungsten	T	Li-Pt	M2	JPE 12(6)	D
*Ir-Pt	M2	[30Mul, 56Rau]	D					Li-Pu	M2	M2	D
				K-Y	No Data			Li-Rb	M2	BAPD 10(3)	D
Ir-Pu	M2	M2	T	K-Yb	No Data			Li-Re	M2	JPE 12(6)	T
Ir-Rb	M2	[76Vol]	T	K-Zn	M2	BAPD 8(6)	D	Li-Rh	M2	JPE 12(6)	D
Ir-Re	M2	Unpublished	D	K-Zr	M2	BAPD 10(3)	D	Li-Ru	M2	JPE 12(6)	T
*Ir-Rh	M2,91	JPE 12(5)	D	Kr-Mo	M2	[Molybdenum]	D	*Li-S	M2	Unpublished	D
*Ir-Ru	M2	JPE 13(5)	D	Kr-W	M2	Binary Tungsten	T	Li-Sb	M2	M2	D
Ir-S	M2	M2	T					*Li-Se	M2	[71Cun]	D
Ir-Sb	M2	Unpublished	D	La-Li	No Data			*Li-Si	M2	BAPD 11(3)	D
Ir-Sc	M2	Unpublished	T	La-Lu	M2	M2	D	*Li-Sn	M2	[Moffatt]	D

(continued)

System	Published	Data source	Data type	System	Published	Data source	Data type	System	Published	Data source	Data type
*Li-Sr	M2	BAPD 10(3)	D	Mg-N	M2	Binary Magnesium	D	*Mg-Yb	M2	Binary Magnesium	D
Li-Ta	M2	JAPD 6(1)	D								
Li-Tb	No Data			Mg-Na	M2	Binary Magnesium	D	*Mg-Zn	M2	Binary Magnesium	D
Li-Tc	M2	Unpublished	T								
*Li-Te	M2	JPE 13(3)	D	Mg-Nb	M2	Binary Magnesium	D	*Mg-Zr	M2	Binary Magnesium	D
Li-Th	No Data										
Li-Ti	M2	BAPD 10(2)	D	Mg-Nd	M2,91	Binary Magnesium	D	*Mn-Mo	M2	[Molybdenum]	D
*Li-Tl	M2	[34Gru]	D	*Mg-Ni	M2	Binary Nickel	D	*Mn-N	M2	BAPD 11(1)	D
Li-Tm	No Data			Mg-Np	M2	[68Gul1]	T	Mn-Na	No Data		
Li-U	M2	M2	T	Mg-O	M2	Binary Magnesium	D	Mn-Nb	M2	M2	D
Li-V	M2	Binary Vanadium	D	Mg-Os	M2	[68Gul2]	T	*Mn-Nd	M2,92	[70Kir]	D
				Mg-P	M2	Binary Magnesium	T	*Mn-Ni	M2	JPE 12(3)	D
Li-W	M2	Binary Tungsten	T					Mn-Np	M2	M2	T
				Mg-Pa	M2	[68Gul1]	T	*Mn-O	M2	M2	D
Li-Y	No Data			*Mg-Pb	M2	Binary Magnesium	D	*Mn-P	M2	[50Ber]	D
Li-Yb	No Data							Mn-Pb	M2	[56Pel]	D
*Li-Zn	M2	JPE 12(1)	D	Mg-Pd	M2	Binary Magnesium	D	*Mn-Pd	M2	[Hansen]	D
Li-Zr	M2	BAPD 8(1)	D					Mn-Pm	91	[90Sac]	D
Lr-Mo	M2	[Molybdenum]	D	Mg-Pm	M2	[68Gul2]	T	*Mn-Pr	M2	M2	D
Lu-Mg	M2	Binary Magnesium	D	Mg-Po	M2	Binary Magnesium	T	Mn-Pt	M2	[55Rau]	D
								*Mn-Pu	M2	[55Kon]	D
Lu-Mn	M2	M2	D	Mg-Pr	M2	BAPD 10(1)	D	Mn-Rb	No Data		
Lu-Mo	M2	[Molybdenum]	D	Mg-Pt	M2	Binary Magnesium	T	Mn-Re	M2	[61Sav]	D
Lu-N	M2	M2	T					Mn-Rh	M2	[55Rau,59Hel]	D
Lu-Na	No Data			Mg-Pu	M2	M2	D	Mn-Ru	M2	M2	D
Lu-Nb	No Data			Mg-Ra	M2	[68Gul2]	T	Mn-S	M2	Unpublished	D
Lu-Nd	M2	M2	D	Mg-Rb	M2	Binary Magnesium	D	*Mn-Sb	M2	M2	D
Lu-Ni	M2	Binary Nickel	T					Mn-Sc	M2	M2	D
Lu-Np	No Data			Mg-Re	M2	[68Gul2]	T	Mn-Se	M2	Unpublished	D
Lu-O	M2	M2	T	Mg-Rh	M2	Binary Magnesium	T	*Mn-Si	M2,91	BAPD 11(5)	D
Lu-Os	M2	M2	T					*Mn-Sm	M2	[70Kir]	D
Lu-P	M2	M2	T	Mg-Ru	M2	Binary Magnesium	T	*Mn-Sn	M2	M2	D
*Lu-Pb	M2	[69Mcm]	D					Mn-Sr	M2	M2	D
Lu-Pd	M2	M2	T	Mg-S	M2	Binary Magnesium	D	Mn-Ta	M2	[60Sav]	D
Lu-Pm	M2	M2	D					Mn-Tb	M2	[70Kir]	D
Lu-Po	M2	M2	T	*Mg-Sb	M2	Binary Magnesium	D	Mn-Tc	M2	M2	T
Lu-Pr	M2	M2	D					Mn-Te	M2	Unpublished	D
Lu-Pt	M2	M2	D	*Mg-Sc	M2	Binary Magnesium	D	Mn-Th	M2	[Brandes]	D
Lu-Pu	M2,91	M2	D					*Mn-Ti	M2	Binary Titanium	D
Lu-Rb	No Data			Mg-Se	M2	Binary Magnesium	T				
Lu-Re	M2	M2	T					Mn-Tl	M2	M2	D
Lu-Rh	M2	M2	D	*Mg-Si	M2	Binary Magnesium	D	Mn-Tm	M2	M2	D
Lu-Ru	M2	M2	D					*Mn-U	M2	[Hansen]	D
Lu-S	M2	M2	T	*Mg-Sm	M2	M2	D	*Mn-V	M2,92	Binary Vanadium	D
Lu-Sb	M2,91	M2	T	*Mg-Sn	M2	Binary Magnesium	D				
Lu-Sc	No Data							Mn-W	M2	Binary Tungsten	T
Lu-Se	M2	M2	T	*Mg-Sr	M2	Binary Magnesium	D				
Lu-Si	M2	M2	D					*Mn-Y	M2,91	JPE 12(4)	D
Lu-Sm	M2	M2	D	Mg-Ta	M2	[68Gul2]	T	Mn-Yb	M2	M2	D
Lu-Sn	M2	M2	D	Mg-Tb	M2	Binary Magnesium	D	*Mn-Zn	M2	BAPD 11(4)	D
Lu-Sr	No Data							*Mn-Zr	M2	Unpublished	D
Lu-Ta	M2	[66Den1]	D	Mg-Tc	M2	[68Gul2]	T	*Mo-N	M2	M2	D
Lu-Tb	M2	M2	D	Mg-Te	M2	Binary Magnesium	T	Mo-Na	M2	M2	D
Lu-Tc	M2	M2	T					*Mo-Nb	M2,91	[Molybdenum]	D
Lu-Te	M2	M2	T	*Mg-Th	M2	Binary Magnesium	D	Mo-Nd	M2	M2	D
Lu-Th	M2	M2	D					Mo-Ne	M2	[Molybdenum]	D
Lu-Ti	M2	M2	D	Mg-Ti	M2	Binary Magnesium	D	*Mo-Ni	M2,91	Binary Nickel	D
*Lu-Tl	M2	M2	D					Mo-No	M2	[Molybdenum]	D
Lu-Tm	M2	M2	D	*Mg-Tl	M2	Binary Magnesium	D	Mo-Np	M2	[Molybdenum]	D
Lu-U	M2	M2	T					*Mo-O	M2	BAPD 1(2)	D
Lu-V	M2	Binary Vanadium	D	Mg-Tm	M2	Binary Magnesium	D	*Mo-Os	M2	[Molybdenum]	D
								*Mo-P	M2	[Molybdenum]	D
Lu-W	M2	Binary Tungsten	D	Mg-U	M2	Binary Magnesium	D	Mo-Pa	M2	[Molybdenum]	D
								Mo-Pb	M2	M2	D
Lu-Y	M2	M2	D	Mg-V	M2	Binary Magnesium	D	*Mo-Pd	M2,92	M2	D
Lu-Yb	M2,91	BAPD 4(3	D					Mo-Pm	M2	M2	D
Lu-Zn	M2	M2	D	Mg-W	M2	Binary Tungsten	T	Mo-Po	M2	[Molybdenum]	T
Lu-Zr	M2	M2	D	*Mg-Y	M2,92	Binary Magnesium	D	Mo-Pr	M2	M2	D
Md-Mo	M2	[Molybdenum]	D					*Mo-Pt	M2	BAPD 1(2)	D
*Mg-Mn	M2	Binary Magnesium	D					*Mo-Pu	M2	[Molybdenum]	D
								Mo-Ra	M2	[Molybdenum]	D
Mg-Mo	M2	Binary Magnesium	D					Mo-Rb	M2	M2	D
								Mo-Re	M2	M2	D

(continued)

System	Published	Data source	Data type	System	Published	Data source	Data type	System	Published	Data source	Data type
*Mo-Rh	M2	[Molybdenum]	D	Na-Np	No Data			*Nb-W	M2	Binary Tungsten	D
Mo-Rn	M2	[Molybdenum]	D	*Na-O	M2	BAPD 8(3)	D				
*Mo-Ru	M2	M2	D	Na-Os	M2	[81Loe]	T	Nb-Y	M2	JPE 12(2)	D
*Mo-S	M2	BAPD 1(2)	D	Na-P	No Data			Nb-Yb	M2	M2	D
Mo-Sb	M2	[Molybdenum]	D	*Na-Pb	M2	[Metals]	D	Nb-Zn	M2	JPE 13(4)	D
Mo-Sc	M2	[Molybdenum]	D	Na-Pd	M2	M2	D	*Nb-Zr	M2,92	BAPD 3(1)	D
Mo-Se	M2	[Molybdenum]	D	Na-Po	M2	M2	T	*Nd-Ni	M2,92	Binary Nickel	D
*Mo-Si	M2	JPE 12(4)	D	Na-Pr	No Data			Nd-Np	No Data		
Mo-Sm	M2	M2	D	Na-Pt	M2	M2	D	Nd-O	M2	M2	D
Mo-Sn	M2	BAPD 1(2)	D	Na-Pu	M2	M2	T	Nd-Os	M2	M2	T
Mo-Sr	M2	M2	D	*Na-Rb	M2	BAPD 3(3)	D	Nd-P	M2	M2	T
*Mo-Ta	M2	JAPD 2(3)	D	Na-Re	No Data			Nd-Pb	M2	M2	T
Mo-Tb	M2	M2	D	Na-Rh	M2	[81Loe]	T	Nd-Pd	M2,92	M2	D
Mo-Tc	M2	[Molybdenum]	D	Na-Ru	M2	[81Loe]	T	Nd-Pm	M2	M2	D
Mo-Te	M2	[Molybdenum]	D	*Na-S	M2	M2	D	Nd-Pr	M2	BAPD 3(2)	D
Mo-Th	M2	[Molybdenum]	D	*Na-Sb	M2	[06Mat]	D	*Nd-Pt	M2	M2	D
*Mo-Ti	M2	Binary Titanium	D	Na-Sc	No Data			Nd-Pu	M2	M2	D
Mo-Tl	M2	M2	D	*Na-Se	M2	M2	D	Nd-Rb	No Data		
Mo-Tm	M2	M2	D	Na-Si	M2	JPE 13(1)	T	Nd-Re	M2	M2	T
*Mo-U	M2	M2	D	Na-Sm	No Data			*Nd-Rh	M2	M2	D
*Mo-V	M2	JPE 13(1)	D	*Na-Sn	M2	M2	D	Nd-Ru	M2,91	M2	D
*Mo-W	M2	Binary Tungsten	D	*Na-Sr	M2	BAPD 6(1)	D	Nd-S	M2	M2	T
Mo-Xe	M2	[Molybdenum]	D	Na-Ta	M2,91	JAPD 6(1)	D	*Nd-Sb	M2	M2	D
Mo-Y	M2	[Molybdenum]	D	Na-Tb	No Data			Nd-Sc	M2	BAPD 3(3)	D
Mo-Yb	M2	M2	D	*Na-Te	M2	BAPD 11(5)	D	Nd-Se	M2	M2	T
Mo-Zn	M2	[Molybdenum]	D	Na-Th	M2	[42Gru]	D	*Nd-Si	M2	BAPD 10(3)	D
*Mo-Zr	M2	[Zirconium]	D	Na-Ti	M2	BAPD 10(2)	D	Nd-Sm	M2	BAPD 3(2)	D
N-Na	M2	M2	T	*Na-Tl	M2	[36Gru]	D	*Nd-Sn	M2	M2	D
*N-Nb	M2	[74Lev]	D	Na-Tm	No Data			Nd-Sr	M2	[78Esh]	D
N-Nd	M2	M2	T	Na-U	M2	M2	T	Nd-Ta	M2	[Moffatt]	D
*N-Ni	M2	Binary Nickel	D	Na-V	M2	Binary Vanadium	D	Nd-Tb	M2	M2	D
N-Np	M2	M2	T					*Nd-Te	M2	M2	D
N-Os	M2	M2	T	Na-W	M2	Binary Tungsten	T	Nd-Th	M2	[67Bad1]	D
N-Pa	M2	M2	T					*Nd-Ti	M2	Binary Titanium	D
N-Pb	M2	M2	T	Na-Y	No Data						
N-Pd	M2	[10Sie]	T	Na-Yb	No Data			*Nd-Tl	M2	Unpublished	D
N-Pr	M2	M2	T	Na-Zn	M2	BAPD 8(6)	D	Nd-Tm	M2	M2	D
N-Pt	No Data			Na-Zr	M2	BAPD 8(1)	D	Nd-U	M2	M2	D
N-Pu	M2	BAPD 10(5)	D	Nb-Nd	M2	M2	T	Nd-V	M2	Binary Vanadium	D
N-Rb	M2	M2	T	*Nb-Ni	M2	Binary Nickel	D				
N-Re	M2	M2	T	Nb-Np	No Data			Nd-W	M2	Binary Tungsten	D
N-Rh	No Data			Nb-O	M2	[59Ell,Shunk]	D				
N-Ru	No Data			*Nb-Os	M2	[77Wat]	D	Nd-Y	M2	BAPD 3(2)	D
N-Sb	No Data			Nb-P	M2	M2	T	Nd-Yb	M2	BAPD 3(2)	D
N-Sc	M2	M2	T	Nb-Pb	M2	M2	T	*Nd-Zn	M2	[72Mas]	D
N-Se	M2	M2	T	*Nb-Pd	M2	BAPD 9(4)	D	Nd-Zr	M2	[Shunk,Elliott]	T
N-Si	M2	BAPD 11(6)	D	Nb-Pr	No Data			Ne-W	M2	Binary Tungsten	T
N-Sm	M2	M2	T	*Nb-Pt	M2	M2	D				
N-Sn	M2	[08Fis,10Sie]	T	Nb-Pu	M2	M2	D	Ni-Np	M2	Binary Nickel	T
N-Sr	M2	M2	T	Nb-Rb	M2	BAPD 11(3)	D	*Ni-O	M2	Binary Nickel	D
*N-Ta	M2	[75Gat]	D	Nb-Re	M2	[60Gra]	D	*Ni-Os	M2	Binary Nickel	D
N-Tb	M2	M2	T	*Nb-Rh	M2	[64Rit]	D	*Ni-P	M2	Binary Nickel	D
N-Tc	M2	M2	T	*Nb-Ru	M2	M2	D	*Ni-Pb	M2	Binary Nickel	D
N-Te	M2	M2	T	Nb-S	M2	M2	D	*Ni-Pd	M2	Binary Nickel	D
*N-Th	M2	M2	D	Nb-Sb	M2	M2	D	Ni-Pm	M2	Binary Nickel	T
*N-Ti	M2	Binary Titanium	D	Nb-Sc	M2	M2	D	Ni-Po	M2	[Moffatt]	D
				Nb-Se	M2	M2	D	*Ni-Pr	M2	Binary Nickel	D
N-Tl	M2	M2	T	*Nb-Si	M2	Unpublished	D	*Ni-Pt	M2	Binary Nickel	D
N-Tm	M2	M2	T	Nb-Sm	M2,92	[Moffatt]	D	*Ni-Pu	M2	Binary Nickel	D
*N-U	M2	[Metals]	D	Nb-Sn	M2	[Shunk]	D	Ni-Rb	No Data		
N-V	M2	Binary Vanadium	D	Nb-Sr	No Data			*Ni-Re	M2,92	Binary Nickel	D
				*Nb-Ta	M2	JAPD 3(1)	D	*Ni-Rh	M2	Binary Nickel	D
N-W	M2	Binary Tungsten	D	Nb-Tb	No Data			*Ni-Ru	M2	Binary Nickel	D
				Nb-Tc	M2	M2	T	*Ni-S	M2	Binary Nickel	D
N-Y	M2	M2	D	Nb-Te	M2	M2	D	*Ni-Sb	M2	Binary Nickel	D
N-Yb	M2	M2	T	*Nb-Th	M2	[56Car]	D	*Ni-Sc	M2	Binary Nickel	D
N-Zn	M2	BAPD 9(3)	T	*Nb-Ti	M2	Binary Titanium	D	*Ni-Se	M2	Binary Nickel	D
*N-Zr	M2	[Zirconium]	D					*Ni-Si	M2	Binary Nickel	D
Na-Nb	M2	BAPD 9(4)	D	Nb-Tl	M2	M2	D	*Ni-Sm	M2	Binary Nickel	D
Na-Nd	No Data			Nb-Tm	No Data			*Ni-Sn	M2	Binary Nickel	D
Na-Ni	M2	Binary Nickel	T	*Nb-U	M2	M2	D	Ni-Sr	M2	Binary Nickel	D
				*Nb-V	M2	Binary Vanadium	D	*Ni-Ta	M2	Binary Nickel	D
								Ni-Tb	M2	Binary Nickel	T

(continued)

System	Published	Data source	Data type	System	Published	Data source	Data type	System	Published	Data source	Data type
Ni-Tc	M2	Binary Nickel	D	O-Te	M2	M2	D	P-Th	M2	M2	D
*Ni-Te	M2	Binary Nickel	D	O-Th	M2	[Smith]	D	*P-Ti	M2	Binary	D
Ni-Th	M2,91	Binary Nickel	D	*O-Ti	M2	Binary	D			Titanium	
*Ni-Ti	M2	Binary Nickel	D			Titanium		P-Tl	M2	Unpublished	D
Ni-Tl	M2	[08Vos]	D	O-Tl	M2	M2	T	P-Tm	M2	M2	T
Ni-Tm	M2	Binary Nickel	T	O-Tm	M2	M2	T	P-U	M2	M2	T
*Ni-U	M2	Binary Nickel	D	O-U	M2	[Elliott]	D	P-V	M2	JPE 12(4)	T
*Ni-V	M2	Binary Nickel	D	*O-V	M2	Binary	D	P-W	M2	Binary	T
*Ni-W	M2,91	Binary	D			Vanadium				Tungsten	
		Tungsten		*O-W	M2	Binary	D	P-Y	M2	M2	T
*Ni-Y	M2	Binary Nickel	D			Tungsten		P-Yb	M2	M2	T
*Ni-Yb	M2	Binary Nickel	D	*O-Y	M2	BAPD 11(1)	D	*P-Zn	M2	JPE 12(4)	D
*Ni-Zn	M2	Binary Nickel	D	O-Yb	M2	M2	T	P-Zr	M2	M2	T
*Ni-Zr	M2	Binary Nickel	D	O-Zn	M2	BAPD 8(2)	D	Pa-Pt	M2	BAPD 10(2)	T
Np-O	M2	M2	D	*O-Zr	M2	BAPD 7(2)	D	Pa-Rh	M2	M2	T
Np-Os	M2	M2	T	Os-P	M2	M2	D	Pa-Sb	M2	M2	T
Np-P	M2	M2	T	Os-Pb	No Data			Pa-Th	M2	M2	T
Np-Pb	No Data			Os-Pd	M2	[63Tyl]	D	Pa-W	M2	Binary	T
Np-Pd	M2	M2	T	Os-Pr	M2	M2	D			Tungsten	
Np-Pr	No Data			*Os-Pt	M2	M2	D	*Pb-Pd	M2	M2	D
Np-Pt	M2	BAPD 10(2)	T	*Os-Pu	M2	[55Kon]	D	Pb-Pm	M2	[63Wil]	T
*Np-Pu	M2	BAPD 6(3)	D	Os-Rb	M2	[81Loe]	T	Pb-Po	M2	M2	T
Np-Rb	No Data			*Os-Re	M2	[62Tyl1]	D	*Pb-Pr	M2	M2	D
Np-Re	M2	M2	T	*Os-Rh	M2	M2	D	*Pb-Pt	M2	[Hansen]	D
Np-Rh	M2	M2	T	*Os-Ru	M2	[62Tyl2]	D	*Pb-Pu	M2	BAPD 9(3)	D
Np-Ru	M2	M2	T	Os-S	M2	M2	D	*Pb-Rb	M2	[77Kuz,	D
Np-S	M2	M2	T	Os-Sb	M2	M2	T			64Hew]	
Np-Sb	M2	M2	T	Os-Sc	M2	M2	T	Pb-Re	No Data		
Np-Sc	No Data			Os-Se	M2	M2	D	*Pb-Rh	M2	M2	D
Np-Se	M2	M2	T	*Os-Si	M2	M2	D	Pb-Ru	M2	M2	T
Np-Si	M2	M2	T	Os-Sm	M2	[59Com,80Pal]	T	*Pb-S	M2	BAPD 7(4)	D
Np-Sm	No Data			Os-Sn	M2	M2	T	*Pb-Sb	M2	BAPD 2(1)	D
Np-Sn	M2	M2	T	Os-Sr	No Data			Pb-Sc	No Data		
Np-Sr	No Data			Os-Ta	M2	[60Kau]	D	*Pb-Se	M2	Unpublished	D
Np-Ta	No Data			Os-Tb	M2	[59Boz,80Pal]	T	Pb-Si	M2	BAPD 5(3)	D
Np-Tb	No Data			Os-Tc	M2	M2	T	Pb-Sm	M2	[Moffatt]	D
Np-Te	M2	M2	T	Os-Te	M2	M2	D	*Pb-Sn	M2	BAPD 9(2)	D
Np-Th	No Data			Os-Th	M2	M2	D	*Pb-Sr	M2	[81Bru]	D
Np-Ti	No Data			*Os-Ti	M2	Binary Nickel	D	Pb-Ta	No Data		
Np-Tl	M2	M2	T	Os-Tl	No Data			Pb-Tb	M2	M2	T
Np-Tm	No Data			Os-Tm	M2	M2	T	*Pb-Te	M2	BAPD 10(4)	D
*Np-U	M2	BAPD 6(3)	D	*Os-U	M2	[Shunk]	D	Pb-Th	M2	M2	D
Np-V	No Data			*Os-V	M2	Binary	D	Pb-Ti	M2	Binary	D
Np-W	M2	Binary	T			Vanadium				Titanium	
		Tungsten		*Os-W	M2,92	Binary	D	*Pb-Tl	M2	[Hultgren,B]	D
Np-Y	No Data					Tungsten		Pb-Tm	M2	M2	T
Np-Yb	No Data			Os-Y	M2	[73Sav]	D	Pb-U	M2	BAPD 8(6)	D
Np-Zn	No Data			Os-Yb	M2	M2	D	Pb-V	M2	Binary	T
Np-Zr	No Data			Os-Zn	M2	M2	T			Vanadium	
O-Os	M2	M2	T	*Os-Zr	M2	M2	D	Pb-W	M2	Binary	T
O-Pa	M2	M2	T	P-Pa	M2	M2	T			Tungsten	
*O-Pb	M2	BAPD 9(2)	D	P-Pb	M2	[1898Gra,	T	*Pb-Y	M2	[67Car]	D
O-Pd	M2	[Pearson3]	T			22Bru]		*Pb-Yb	M2	JPE 12(4)	D
O-Pm	M2	M2	T	*P-Pd	M2	Unpublished	D	*Pb-Zn	M2	[Hansen]	D
O-Po	M2	M2	T	*P-Pr	M2	[Moffatt]	D	Pb-Zr	M2	M2	D
*O-Pr	M2	M2	D	P-Pt	M2	BAPD 11(5)	D	Pd-Pr	M2	JAPD 6(2)	D
O-Pt	M2	[Pearson3]	T	P-Pu	M2	M2	T	*Pd-Pt	M2,91	M2	D
*O-Pu	M2	BAPD 11(2)	D	P-Rb	M2	M2	T	*Pd-Pu	M2	[67Kut1]	D
O-Rb	M2	M2	D	P-Re	M2	M2	T	Pd-Rb	M2	[81Loe]	T
O-Re	M2	[Pearson3]	T	P-Rh	M2	BAPD 11(4)	D	Pd-Re	M2	M2	D
O-Rh	M2	M2	T	*P-Ru	M2	M2	D	*Pd-Rh	M2	M2	D
O-Ru	M2	M2	T	P-S	91	[79Bla]	D	*Pd-Ru	M2	M2	D
O-Sb	M2	M2	D	P-Sb	M2,91	JPE 12(2)	D	*Pd-S	M2,92	[76Mat]	D
O-Sc	M2	M2	D	P-Sc	M2	M2	T	*Pd-Sb	M2,92	M2	D
O-Se	M2	M2	T	P-Se	M2	Unpublished	D	Pd-Sc	M2	M2	D
O-Si	M2	BAPD 11(1)	D	P-Si	M2	BAPD 6(2)	D	*Pd-Se	M2,91	JPE 13(1)	D
O-Sm	M2	M2	T	P-Sm	M2	M2	T	*Pd-Si	M2,91	JPE 12(3)	D
*O-Sn	M2	[Hansen]	D	*P-Sn	M2	[20Viv]	D	*Pd-Sm	M2	M2	D
O-Sr	M2	[56Swa,	T	P-Sr	M2	M2	T	*Pd-Sn	M2	M2	D
		63Sch]		P-Ta	M2	[Pearson3]	T	Pd-Sr	M2	M2	T
O-Ta	M2	[72Jeh]	D	P-Tb	M2	M2	T	Pd-Ta	M2	JAPD 6(2)	D
O-Tb	M2	M2	D	P-Tc	M2	M2	T	Pd-Tb	M2,91	M2	T
O-Tc	M2	M2	T	P-Te	M2	[42Mon]	T	Pd-Tc	M2	M2	D

(continued)

System	Published	Data source	Data type	System	Published	Data source	Data type	System	Published	Data source	Data type
*Pd-Te	M2	JPE 13(1)	D	Pr-W	M2	Binary Tungsten	D	Rb-Sc	No Data		
Pd-Th	M2	M2	D	Pr-Y	M2	M2	D	*Rb-Se	M2	M2	D
*Pd-Ti	M2	Binary Titanium	D	Pr-Yb	No Data			Rb-Si	M2	M2	T
*Pd-Tl	M2	M2	D	*Pr-Zn	M2	[70Mas]	D	Rb-Sm	No Data		
Pd-Tm	M2	M2	T	Pt-Pu	M2	BAPD 10(4a)	D	Rb-Sn	M2	M2	T
*Pd-U	M2,92	[56Cat, 63Pel]	D	Pt-Rb	M2	[81Loe]	T	Rb-Sr	M2	BAPD 6(1)	T
				Pt-Re	M2	M2	D	Rb-Ta	M2,91	JAPD 6(3)	D
*Pd-V	M2	Binary Vanadium	D	*Pt-Rh	M2,92	[Moffatt]	D	Rb-Tb	No Data		
				Pt-Ru	M2	[72Hut]	D	Rb-Te	M2	Unpublished	D
*Pd-W	M2,91,92	Binary Tungsten	D	Pt-S	M2	Unpublished	D	Rb-Th	No Data		
				Pt-Sb	M2,92	M2	D	Rb-Ti	M2	BAPD 10(2)	D
*Pd-Y	M2,91	M2	D	Pt-Sc	M2	M2	D	*Rb-Tl	M2	[70Thu]	D
*Pd-Yb	M2	[73Ian]	D	Pt-Se	M2	M2	T	Rb-Tm	No Data		
*Pd-Zn	M2	M2	D	*Pt-Si	M2	JPE 12(5)	D	Rb-U	No Data		
Pd-Zr	M2,92	JAPD 6(1)	D	Pt-Sm	M2	M2	D	Rb-V	M2	Binary Vanadium	D
Pm-Po	M2	M2	T	*Pt-Sn	M2	[Hansen]	D				
Pm-Pr	M2	M2	D	Pt-Sr	M2	M2	D	Rb-W	M2	Binary Tungsten	T
Pm-Pu	M2,92	M2	D	Pt-Ta	M2	[81Wat]	D				
Pm-Rh	M2	M2	D	Pt-Tb	M2	M2	D	Rb-Y	No Data		
Pm-Ru	M2	M2	D	Pt-Tc	M2	M2	D	Rb-Yb	No Data		
Pm-Sm	M2	M2	D	*Pt-Te	M2	M2	D	Rb-Zn	M2	BAPD 8(5)	D
Pm-Tb	M2	M2	D	Pt-Th	M2	BAPD 11(3)	D	Rb-Zr	M2	BAPD 8(1)	D
Pm-Th	M2	M2	D	*Pt-Ti	M2	Binary Titanium	D	Re-Rh	M2	[62Tyl3]	D
Pm-Tl	M2	[88Sac]	D					*Re-Ru	M2	[62Rud]	D
Pm-Tm	M2	M2	D	*Pt-Tl	M2	M2	D	Re-S	M2	M2	T
Pm-V	M2	Binary Vanadium	D	Pt-Tm	M2	M2	D	Re-Sb	M2	M2	D
				*Pt-U	M2	BAPD 11(3)	D	Re-Sc	M2	[66Sav]	D
Pm-W	M2	Binary Tungsten	T	*Pt-V	M2	Binary Vanadium	D	Re-Se	M2	M2	T
								*Re-Si	M2	Unpublished	D
Pm-Y	M2	M2	D	Pt-W	M2,91	Binary Tungsten	D	Re-Sm	M2	M2	T
Po-Pr	M2	[63Ker,Shunk]	T					Re-Sn	M2	M2	D
Po-Pt	M2	M2	T					Re-Sr	No Data		
Po-S	M2	[Hansen]	T	Pt-Y	M2	BAPD 11(5)	D	Re-Ta	M2	[60Bro]	D
Po-Sc	M2	M2	T	Pt-Yb	M2	M2	D	Re-Tb	M2	[68Sav]	D
Po-Sm	M2	M2	T	Pt-Zn	M2	JPE 12(4)	D	Re-Tc	M2	M2	D
Po-Sr	M2	M2	T	*Pt-Zr	M2	M2	D	*Re-Te	M2	[77Kur]	D
Po-Ta	M2	[60Wit]	T	Pu-Rb	No Data			Re-Th	M2	[77Gar]	D
Po-Tb	M2	M2	T	Pu-Re	M2	[67Bow]	D	Re-Ti	M2	Binary Titanium	D
Po-Ti	M2	M2	T	Pu-Rh	M2	[78Lan]	D				
Po-Tm	M2	M2	T	Pu-Ru	M2	[67Kut2]	D	Re-Tl	No Data		
Po-W	M2	Binary Tungsten	T	Pu-S	M2	M2	T	Re-Tm	M2	M2	T
				Pu-Sb	M2	M2	T	*Re-U	M2	M2	D
Po-Y	M2	M2	T	*Pu-Sc	M2	M2	D	*Re-V	M2	Binary Vanadium	D
Po-Yb	M2	M2	T	Pu-Se	M2	M2	T				
Po-Zn	M2	M2	T	Pu-Si	M2	[Shunk]	D	Re-W	M2,92	Binary Tungsten	D
Po-Zr	M2	M2	T	Pu-Sm	M2	M2	D				
Pr-Pt	M2	M2	D	Pu-Sn	M2	BAPD 9(2)	D	Re-Y	M2	[61Lun]	D
Pr-Pu	M2	M2	D	Pu-Sr	M2	M2	T	Re-Yb	M2	M2	T
Pr-Rb	No Data			Pu-Ta	M2	JPE 12(5)	D	Re-Zn	M2	M2	T
Pr-Re	M2	[64Ell]	D	Pu-Tb	M2	M2	D	Re-Zr	M2	M2	D
Pr-Rh	M2	M2	D	Pu-Te	M2	M2	T	Rh-Ru	M2	[84Pas]	D
Pr-Ru	M2	M2	D	Pu-Th	M2	BAPD 6(3)	D	Rh-S	M2,92	[Moffatt]	D
Pr-S	M2,91	M2	T	Pu-Ti	M2	Binary Titanium	D	Rh-Sb	M2	[Shunk]	D
*Pr-Sb	M2	M2	D					Rh-Sc	M2	[58Com, 61Dwi]	T
Pr-Sc	No Data			Pu-Tl	M2	[58Boc]	T				
*Pr-Se	M2	[70Yar]	D	Pu-Tm	M2	M2	D	*Rh-Se	M2	M2	D
*Pr-Si	M2	M2	D	*Pu-U	M2,92	BAPD 10(2)	D	Rh-Si	M2,92	JPE 13(1)	D
Pr-Sm	M2	M2	D	Pu-V	M2,91	JPE 12(5)	D	Rh-Sm	M2	M2	D
*Pr-Sn	M2	M2	D	Pu-W	M2	Binary Tungsten	D	Rh-Sn	M2	[Hansen]	D
Pr-Sr	No Data							Rh-Sr	M2	M2	T
Pr-Ta	M2	[Moffatt]	D	Pu-Y	M2	M2	D	*Rh-Ta	M2	[64Gie]	D
Pr-Tb	M2	M2	D	Pu-Yb	M2	M2	D	Rh-Tb	M2	M2	D
Pr-Tc	M2	[64Dar]	T	*Pu-Zn	M2	[Chiotti]	D	Rh-Tc	M2	M2	D
*Pr-Te	M2	[70Yar]	D	*Pu-Zr	M2	[Elliott]	D	Rh-Te	M2,91	M2	D
Pr-Th	M2	[67Bad1]	D	Ra-S	M2	M2	T	Rh-Th	M2	[63Tho]	D
Pr-Ti	M2	M2	D	Ra-Se	M2	M2	T	*Rh-Ti	M2	Binary Titanium	D
*Pr-Tl	M2	Unpublished	D	Ra-W	M2	Binary Tungsten	T				
Pr-Tm	M2	M2	D					Rh-Tl	No Data		
Pr-U	M2	M2	D	Rb-Re	No Data			Rh-Tm	M2	M2	T
Pr-V	M2	Binary Vanadium	D	Rb-Rh	M2	[81Loe]	T	*Rh-U	M2	[Ivanov]	D
				Rb-Ru	M2	[81Loe]	T	*Rh-V	M2	Binary Vanadium	D
				Rb-S	M2	M2	D				
				*Rb-Sb	M2	[61Dor2]	D				

(continued)

System	Published	Data source	Data type	System	Published	Data source	Data type	System	Published	Data source	Data type
Rh-W	M2	Binary Tungsten	D	Sb-V	M2	Binary Vanadium	T	*Sm-Sn	M2	[82Bor]	D
Rh-Y	M2	M2	D	Sb-W	M2,92	Binary Tungsten	T	Sm-Sr	No Data		
Rh-Yb	M2	[76Ian]	D					Sm-Ta	M2	[66Den2]	D
Rh-Zn	M2	M2	D	*Sb-Y	M2	[70Sch]	D	Sm-Tb	M2	M2	D
Rh-Zr	M2	Unpublished	D	Sb-Yb	M2	M2	D	Sm-Te	M2	M2	T
Rn-W		Binary Tungsten	T	*Sb-Zn	M2	[27Tak,66Vui]	D	Sm-Th	M2	M2	D
				Sb-Zr	M2	Unpublished	D	Sm-Ti	No Data		
Ru-S	M2,91	M2	D	Sc-Se	M2	M2	T	*Sm-Tl	M2	Unpublished	D
Ru-Sb	M2	M2	T	Sc-Si	M2	BAPD 7(4)	D	Sm-Tm	M2	M2	D
Ru-Sc	M2	[Moffatt]	D	Sc-Sm	No Data			Sm-U	M2	M2	D
Ru-Se	M2	M2	D	Sc-Sn	M2	M2	T	Sm-V	M2	Binary Vanadium	D
*Ru-Si	M2,92	M2	D	Sc-Sr	M2	M2	D				
Ru-Sm	M2,91	M2	D	Sc-Ta	M2	[66Den1]	D	Sm-W	M2	Binary Tungsten	D
Ru-Sn	M2	M2	D	Sc-Tb	M2	M2	D				
Ru-Sr	No Data			Sc-Tc	M2	M2	T	Sm-Y	M2	BAPD 4(2)	D
*Ru-Ta	M2,91	M2	D	Sc-Te	M2	M2	D	Sm-Yb	No Data		
Ru-Tb	M2	M2	D	Sc-Th	M2,91	[69Bad]	D	*Sm-Zn	M2	[Moffatt]	D
Ru-Tc	M2	M2	T	*Sc-Ti	M2	Binary Titanium	D	Sm-Zr	M2	[Elliott]	T
Ru-Te	M2	M2	D					Sn-Sr	M2	M2	D
*Ru-Th	M2	[63Tho]	D	Sc-Tl	No Data			Sn-Ta	M2	M2	T
*Ru-Ti	M2	Binary Titanium	D	Sc-Tm	No Data			Sn-Tb	No Data		
				Sc-U	M2	M2	D	Sn-Tc	M2	M2	T
Ru-Tl	No Data	D		Sc-V	M2	Binary Vanadium	D	*Sn-Te	M2	BAPD 7(1)	D
Ru-Tm	M2	M2	D					Sn-Th	M2	BAPD 10(4a)	D
*Ru-U	M2	BAPD 2(4)	D	Sc-W	M2	Binary Tungsten	D	*Sn-Ti	M2	Binary Titanium	D
*Ru-V	M2	Binary Vanadium	D	*Sc-Y	M2	BAPD 4(2)	D				
				Sc-Yb	M2	M2	D	*Sn-Tl	M2	M2	D
Ru-W	M2,92	Binary Tungsten	D	Sc-Zn	M2	[Pearson3]	T	Sn-Tm	M2	M2	T
				*Sc-Zr	M2	JPE 12(1)	D	*Sn-U	M2	BAPD 8(4)	D
Ru-Y	M2	M2	D	Se-Si	M2	M2	T	Sn-V	M2	Binary Vanadium	D
Ru-Yb	M2	[76Ian]	D	Se-Sm	M2	M2	T				
Ru-Zn	M2	M2	D	*Se-Sn	M2	BAPD 7(1)	D	Sn-W	M2	Binary Tungsten	T
Ru-Zr	M2	Unpublished	D	*Se-Sr	M2	[75Lys]	D				
S-Sb	M2	M2	D	Se-Ta	M2	[Pearson3]	T	*Sn-Y	M2	M2	D
S-Sc	M2	M2	T	Se-Tb	M2	M2	T	*Sn-Yb	M2	JPE 12(4)	D
*S-Se	M2	Unpublished	D	*Se-Te	M2	Unpublished	D	*Sn-Zn	M2	BAPD 6(4)	D
S-Si	M2	M2	D	Se-Th	M2	[Hansen]	D	*Sn-Zr	M2	BAPD 4(2)	D
S-Sm	M2	M2	D	Se-Ti	M2	Binary Titanium	T	Sr-Ta	No Data		
*S-Sn	M2	BAPD 7(3)	D					Sr-Tb	No Data		
S-Sr	M2	M2	T	*Se-Tl	M2	[81Mor]	D	*Sr-Te	M2	[75Lys]	D
S-Ta	No Data			*Se-Tm	M2	M2	D	Sr-Th	No Data		
S-Tb	M2	M2	T	*Se-U	M2	[75Ell]	D	Sr-Ti	M2	Binary Titanium	D
S-Tc	M2	M2	T	Se-V	M2	Binary Vanadium	D				
*S-Te	M2	BAPD 10(4)	D					*Sr-Tl	M2	M2	D
S-Th	M2	M2	T	Se-W	M2	Binary Tungsten	T	Sr-Tm	No Data		
*S-Ti	M2	Binary Titanium	D					Sr-U	M2	M2	T
				Se-Y	M2	M2	T	Sr-V	M2	Binary Vanadium	D
S-Tl	M2	M2	D	Se-Yb	M2	M2	D				
S-Tm	M2	M2	T	Se-Zn	M2	Unpublished	D	Sr-W	M2	Binary Tungsten	T
S-U	M2	M2	D	Se-Zr	M2	M2	T				
S-V	M2	Binary Vanadium	D	Si-Sm	M2	BAPD 9(5)	D	Sr-Y	M2	M2	D
				*Si-Sn	M2	BAPD 5(3)	D	Sr-Yb	No Data		
S-W	M2	Binary Tungsten	D	*Si-Sr	M2	BAPD 10(6)	D	*Sr-Zn	M2	M2	D
				*Si-Ta	M2	Unpublished	D	Sr-Zr	No Data		
S-Y	M2	M2	T	Si-Tb	M2	M2	T	T-Ta	92	[90Con]	D
S-Yb	M2	[78Eli]	D	Si-Tc	M2	M2	T	Ta-Tb	M2	[66Den1]	D
S-Zn	M2	Unpublished	D	*Si-Te	M2	[80Dav]	D	Ta-Tc	M2	M2	T
S-Zr	M2	M2	D	*Si-Th	M2	[Thorium]	D	Ta-Te	M2,92	JPE 13(3)	T
*Sb-Se	M2	M2	D	*Si-Ti	M2	Binary Titanium	D	*Ta-Th	M2	JAPD 5(1)	D
*Sb-Si	M2	BAPD 6(5)	D					*Ta-Ti	M2	Binary Titanium	D
*Sb-Sm	M2	M2	D	Si-Tl	M2	BAPD 6(6)	D				
*Sb-Sn	M2	[71Pre]	D	Si-Tm	M2	M2	D	Ta-Tl	M2	M2	D
*Sb-Sr	M2	[75Vak]	D	*Si-U	M2	M2	D	Ta-Tm	M2	[66Den1]	D
Sb-Ta	No Data			*Si-V	M2	Binary Vanadium	D	*Ta-U	M2	JAPD 4(3)	D
*Sb-Tb	M2	M2	D					*Ta-V	M2	Binary Vanadium	D
*Sb-Te	M2	M2	D	Si-W	M2	Binary Tungsten	D				
Sb-Th	M2	M2	T					*Ta-W	M2	Binary Tungsten	D
Sb-Ti	M2	Binary Titanium	D	Si-Y	M2,91	BAPD 7(5)	D				
				Si-Yb	M2	M2	D	Ta-Y	M2	M2	D
*Sb-Tl	M2	Unpublished	D	*Si-Zn	M2	BAPD 6(6)	D	Ta-Yb	M2	M2	D
Sb-Tm	M2	M2	T	*Si-Zr	M2	BAPD 11(5)	D	Ta-Zn	M2	[Pearson3]	T
*Sb-U	M2	BAPD 1(2)	D					*Ta-Zr	M2	JAPD 5(2)	D
								Tb-Tc	M2	M2	T

(continued)

System	Published	Data source	Data type	System	Published	Data source	Data type	System	Published	Data source	Data type
Tb-Te	M2	M2	T	*Th-Ti	M2	Binary Titanium	D	Tm-W	M2	Binary Tungsten	D
Tb-Th	M2	[67Bad2]	D								
Tb-Ti	M2	[83Kub]	D	*Th-Tl	M2	M2	D	Tm-Y	M2	M2	D
*Tb-Tl	M2	Unpublished	D	Th-Tm	M2	[Moffatt]	D	Tm-Yb	M2	M2	D
Tb-Tm	M2	M2	D	Th-U	M2	BAPD 6(5)	D	Tm-Zn	M2	M2	D
Tb-U	M2	[Elliott]	T	Th-V	M2	Binary Vanadium	D	Tm-Zr	M2	[Shunk]	T
Tb-V	M2	Binary Vanadium	D					U-V	M2	Binary Vanadium	D
Tb-W	M2	Binary Tungsten	D	Th-W	M2	Binary Tungsten	D				
				Th-Y	M2	[60Eas]	D	U-W	M2	Binary Tungsten	D
Tb-Y	M2	BAPD 4(2)	D	Th-Yb	M2	M2, D					
Tb-Yb	M2	M2	D	*Th-Zn	M2	[61Chi]	D	U-Y	M2	M2	D
Tb-Zn	M2	M2	D	*Th-Zr	M2	[58Gib,61Joh]	D	U-Yb	M2	M2	D
Tb-Zr	M2	[Moffatt]	D	Ti-Tm	M2	M2	D	U-Zn	M2	BAPD 1(2)	D
Tc-Te	M2	M2	D	*Ti-U	M2	Binary Titanium	D	*U-Zr	M2,92	BAPD 10(2)	D
Tc-Th	M2	[65Dar]	T					*V-W	M2	Binary Tungsten	D
Tc-Ti	M2	Binary Titanium	D	*Ti-V	M2	Binary Vanadium	D				
								V-Y	M2	Binary Vanadium	D
Tc-U	M2	[65Dar]	T	*Ti-W	M2	Binary Tungsten	D				
Tc-V	M2	Binary Vanadium	D					V-Yb	M2	Binary Vanadium	D
				*Ti-Y	M2	Binary Titanium	D				
Tc-W	M2	Binary Tungsten	D					V-Zn	M2	Binary Vanadium	D
				Ti-Yb	M2	M2	D	*V-Zr	M2	Binary Vanadium	D
Tc-Y	M2	M2	T	Ti-Zn	M2	Binary Titanium	D				
Tc-Zn	M2	[64Cha]	D					W-Xe		Binary Tungsten	T
Tc-Zr	M2	M2	T	*Ti-Zr	M2	Binary Titanium	D				
Te-Th	M2	M2	T					W-Y	M2	Binary Tungsten	D
Te-Ti	M2	Binary Titanium	T	Tl-Tm	M2	M2	D				
				Tl-U	M2	[52Ian,63Joh]	T	W-Yb	M2	Binary Tungsten	D
*Te-Tl	M2,91	M2	D	Tl-V	M2	Binary Vanadium	D				
Te-Tm	M2	M2	D					W-Zn	M2	Binary Tungsten	T
*Te-U	M2	[Moffatt]	D	Tl-W	M2	Binary Tungsten	T				
Te-V	M2	Binary Vanadium	D					*W-Zr	M2,92	Binary Tungsten	D
				Tl-Y	M2,91	M2	T				
Te-W	M2	Binary Tungsten	D	*Tl-Yb	M2	Unpublished	D	Y-Yb	M2	M2	D
				*Tl-Zn	M2	[07Veg,52Sei]	D	*Y-Zn	M2	M2	D
Te-Y	M2	M2	D	Tl-Zr	M2	M2	T	*Y-Zr	M2	JPE 12(4)	D
*Te-Yb	M2	M2	D	Tm-U	M2	M2	T	*Yb-Zn	M2	[68Mas]	D
*Te-Zn	M2	BAPD 8(1)	D	Tm-V	M2	Binary Vanadium	D	Yb-Zr	M2	M2	D
Te-Zr	M2	M2	D					Zn-Zr	M2	JPE 13(4)	D

References Cited in Binary Alloys Index

The references listed below represent the best available sources for the diagrams and data developed from them. They do not, however, represent *evaluations* conducted under the International Alloy Phase Diagram Programme.

1898Gra: A. Granger, *Ann. Chim. Phys., 14,* 5-90 (1898).

05Bol: W. v.Bolton, *Z. Elektrochem., 11,* 51 (1905).

05Emi: F. Emich, *Monatsh. Chem., 26,* 1013 (1905).

06Mat: C.H. Mathewson, *Z. Anorg. Allg. Chem., 50,* 192-195 (1906).

07Veg: A.V. Vegesack, *Z. Anorg. Allg. Chem., 52,* 30-34 (1907).

08Fis: F. Fisher and G. Iliovich, *Ber. Dtsch. Chem Ges., 41,* 3802, 4449 (1908); *42,* 527 (1909); quoted in [Elliott].

08Fri: K. Friedrich, *Metallurgie, 5,* 212-215 (1908).

08Vos: G. Voss, *Z. Anorg. Allg. Chem., 57,* 49-52 (1908).

10Sie: A. Sieverts and W. Krumbhaar, *Ber. Dtsch. Chem. Ges.,43,* 894 (1910) in German.

20Viv: A.C. Vivian, *J. Inst. Met. 23,* 325-366 (1920).

22Bru: A. Brukel, *Z. Anorg. Allg. Chem., 125,* 255-256 (1922).

27Tak: T. Takei, *Sci. Rep. Tohoku Univ., 16,* 1031-1056 (1927).

30Mul: L. Muller, *Ann. Phys., 7,* 9-47 (1930) in German.

34Gru: G. Grube and G. Schaufler, *Z. Elektrochem., 40,* 593-600 (1934).

36Gru: G. Grube and A. Schmidt, *Z. Elektrochem., 42,* 201-209 (1936).

38Gru: G. Grube and A. Dietrich, *Z. Elektrochem., 44,* 755-758 (1938).

40And: K.W. Andrews, H.E. Davies, W. Hume-Rothery, and C.R. Oswin, *Proc. Roy. Soc. (London), A177,* 149-167 (1940-1941).

41Rol: L. Rolla and A. Iandelli, *Ric. Sci., 20,* 1216-1226 (1941).

42Gru: G. Grube and L. Botzenhardt, *Z. Elektrochem., 48,* 418-425 (1942).

42Mon: E. Montignie, *Bull. Soc. Chim. Fr., 9,* 658-661 (1942).

50Ber: J. Berak and T. Heumann, *Z. Metallkd., 41,* 19-23 (1950).

52Ian: A. Iandelli and R. Ferro, *Ann. Chim. (Rome), 42,* 598-606 (1952).

52Now: H. Nowotny, E. Bauer, A. Stampfl, and H. Bittner, *Monatsh. Chem., 83,* 221-236 (1952).

52Kos: W. Koster and E. Horn, *Z. Metallkd., 43,* 444-449 (1952).

52Sei: W. Seith, H. Johnson, and J. Wagner, *Z. Metallkd., 46,* 773-779 (1952).

53Gea: G.A. Geach and R.A. Jettery, *J. Met., 5,* 1084 (1953).

54Vog: R. Vogel and H. Klose, *Z. Metallkd., 45,* 633-638 (1954).

55Kon: S.T. Konobeevsky, Conf. Acad. Sci. USSR. Peaceful Uses Atomic Energy, Div. Chem. Sci., 1 (1955).

55Rau: E. Raub and W. Mahler, *Z. Metallkd., 46,* 282-290 (1955).

56Car: O.N. Carlson, J.M. Dickenson, H.E. Lunt, and H.A. Wilhelm, *Trans. AIME, 206,* 132-136 (1956).

56Cat: J.A. Catterall, J.D. Grogan, and R.J. Pleasance, *J. Inst.Met., 85,* 63-67 (1956).

56Mul: R.N.P. Mulford and G.E. Sturdy, *J. Am. Chem. Soc., 78,* 3897-3901 (1956).

56Pel: E. Pelzel, *Metall, 10,* 717-718 (1956).

56Rau: E. Raub and W. Plate, *Z. Metallkd., 47,* 688-693 (1956) in German.

56Swa: H.E. Swanson, N.T. Gilfrich, and G.M. Ugrinic, NBS Circ. 539 (1956).

56Tru: F.A. Trumbore, C.D. Thurmond, and M. Kowalchik, *J. Chem. Phys., 24,* 1112 (1956).

58Boc: A.A. Bochvar *et al.,* Proc. U.N. Int Conf. Peaceful Uses At. Energy, 2nd, Geneva, Vol. 6, 184-193 (1958); quoted from [Shunk].

58Com: V.B. Compton, *Acta Crystallogr., 11,* 446 (1958).

58Dom: R.F. Domagala, R.P. Elliott, and W. Rostoker, *Trans. AIME, 212,* 393-395 (1958)

58Gib: E.D. Gibson, B.A. Loomis, and O.N. Carlson, *Trans. ASM, 50,* 348-369 (1958).

58Mul: R.N.R. Mulford, USAEC, AECU-3813 (1958).

59Boz: R.M. Bozworth, B.T. Matthias, H. Suhl, E. Corenzwit, and D.D. Davis, *Phys. Rev., 115,* 1595-1596 (1959).

59Com: V.B. Compton and B.T. Matthias, *Acta Crystallogr., 12,* 651-654 (1959).

59Ell: R.P. Elliott, *Trans. ASM, 52,* 990-1014 (1959).

59Hel: A. Hellawell, *J. Less-Common Met., 1,* 343-347 (1959).

59Sch: F.W. Schonfeld, E.M. Cramer, W.N. Miner, F.H. Elinger, and A.S. Coffinberry, *Progress in Nuclear Energy,* Ser. V, Vol. 2, Pergamon Press, New York, 579-599, (1959).

60Bec: R.L. Beck, USAEC, LAR-10, 93 p (1960).

60Bro: J.H. Brophy, P. Schwarzkopt, and J. Wulff, *Trans. AIME, 218,* 910-914 (1960).

60Cro: J. Croni, C.E. Armantrout, and H. Kato, U.S. Bur. Mines, Rep. Invest. 5688, 12 p (1960).

60Eas: D.T. Eash and O.N. Carlson, *Trans, ASM, 52,* 1097-1114 (1960).

60Gra: N.J. Grant and B.C. Giessen, WADD Tech. Rept., 60-132, 90-112 (1960); *J. Met., 13,* 87 (1961); as quoted in [Elliott].

60Kau: A.R. Kaufmann, E.J. Rapperport, and M.F. Smith, WADD Tech. Rep. 60-132, 33-39 (1960).

60Lya: V.S. Lyashenko and V. Bykov, *At.Energy (USSR), 8,* 146-148 (1960) in Russian; TR: *Sov. J. At. Energy, 8,* 132-134 (1960).

60Pet1: D.T. Peterson and M. Indig, *J. Am. Chem. Soc., 80,* 5645-5646 (1960).

60Pet2: D.T. Peterson and D.J. Beerntsen, *Trans. ASM, 52,* 763-777 (1960).

60Pre: B. Predel, *Z. Phys. Chem., 24,* 206-216 (1960).

60Sav: E.M. Savitskii and C.V. Kopetskii, *Zh. Neorg, Khim., 5,* 2638-2640 (1960) in Russian; TR: *Russ J. Inorg. Chem., 5,* 1274-1275 (1960).

60Wit: W.G. Witteman, A.L. Giorgi, and D.T. Vier, *J. Phys. Chem.,64,* 434-440 (1960).

61Chi: P. Chiotti and K.J. Gill, *Trans. AIME, 221,* 573-580 (1961).

61Dor1: F.W. Dorn, W. Klemm, and S. Lohmeyer, *Z. Anorg. Allg. Chem., 209,* 204-209 (1961).

61Dor2: F.W. Dorn and W. Klemm, *Z. Anorg. Allg. Chem., 309,* 189-203 (1961).

61Dwi: A.E. Dwight, J.W. Downey, and R.A. Conner, Jr., *Acta Crystallogr., 14,* 75-76 (1961).

61Joh: R.H. Johnson and R.W.K. Honeycombe, *J. Nucl. Mater, 4,* 66-69 (1961).

61Lun: C.E. Lundin, in *The Rare Earths,* F.H. Spedding and A.H. Daane, Ed., John Wiley & Sons, New York, 263-264 (1961).

61Lov: B. Love, WADD Tech. Rep., 61-123, 179p (1961); quoted in [Elliott].

61Poo: D.M. Poole, M.G. Bale, P.G. Mardon, J.A.C. Marples, and J.L. Nichols, *Plutonium 1960,* Cleaver-Humes Press, London, 267-280 (1961).

61Sav: E.M. Savitskii, M.A. Tylkina, R.V. Kirilenko, and C.V. Kopetskii, *Zh. Neorg. Khim., 6,* 1474-1476 (1961) in Russian; TR: *Russ. J. Inorg. Chem., 6,* 755-756 (1961).

62Lun: C.E. Lundin and D.T. Klodt, *Trans. AIME, 224,* 367-372 (1962).

62Rap: E.J. Rapperport and M.F. Smith, Tech. Rep., WADD-TR-60-132, Pt. II, 8-27 (1962).

62Rud: E. Rudy, B. Kietter, and H. Froelich, *Z. Metallkd., 53*, 90-92 (1962).

62Tyl1: M.A. Tylkina, V.P. Polyakova, and E.M. Savitskii, *Zh. Neorg. Khim., 7*, 1469-1470 (1962) in Russian; TR: *Russ. J. Inorg. Chem., 7*, 755-756 (1962).

62Tyl2: M.A. Tylkina, V.P. Polyakova, and E.M. Savitskii, *Zh. Neorg. Khim., 7*, 1467-1468 (1962) in Russian; TR: *Russ. J. Inorg. Chem., 7*, 755-756 (1962).

62Tyl3: M.A. Tylkina, V.P. Polyakova, and E.M. Savitskii, *Zh. Neorg. Khim., 7*, 1919-1927 (1962) in Russian; TR: *Russ. J. Inorg. Chem., 7*, 990-996 (1962).

63Joh: I. Johnson and M.G. Chasanov, *Trans. ASM, 56*, 272-277 (1963).

63Ker: C.J. Kershner and R.H. Steinmeyer, USAEC, MLM-1163, F1-F6 (1963).

63Liu: C.H. Liu, A.S. Pashinkin, and A.V. Novoselova, *Dokl. Akad. Nauk SSSR, 151*, 1335-1338 (1963) in Russian; TR: *Dokl. Chem., 151*, 662-664 (1963).

63Obr: W. Obrowski, *Metall, 17*, 108-112 (1963).

63Pel: G.P. Pells, *J. Inst. Met., 92*, 416-418 (1963-1964).

63Sch: S.J. Schneider, NBS Monograph 68, 31 pp (1963).

63Tay: A. Taylor, B.J. Kagle, and N.J. Doyle, *J. Less-Common Met., 5*, 26-40 (1963).

63Tho: J.R. Thompson, *J. Less-Common Met., 5*, 437-442 (1963).

63Tyl: M.A. Tylkina, V.P. Polyakova, and O.Kh. Khamidov, *Zh. Neorg. Khim., 8*, 776-778 (1963) in Russian; TR: *Russ. J. Inorg. Chem., 8*, 395-397 (1963).

63Wil: G.P. Williams and L. Slifkin, *Acta Metall., 11*, 319-322 (1963).

64Cha: M.G. Chasanov, I. Johnson, and R.V. Schablaske, *J. Less-Common Met., 7*, 127-132 (1964).

64Dar: J.B. Darby, Jr., L.J. Norton, and J.W. Downey, *J. Less-Common Met., 6*, 165-167 (1964).

64Ell: R.P. Elliott, in *Rare Earth Research III*, Proc. 4th Conf. Rare Earth Res., L. Eyring, Ed., Gordon and Breach, New York, 215-245 (1964).

64Gie: B.C. Giessen, H. Ibach, and N.J. Grant, *Trans. AIME, 230*, 113-122 (1964).

64Hew: I.F. Hewaidy, E. Busmann, and W. Klemm, *Z. Anorg. Allg. Chem., 328*, 283-293 (1964).

64Obi1: I. Obinata, Y. Takeuchi, K. Kurihara, and M. Watanabe, *Nippon Kinzoku Gakkaishi, 28*, 562-568 (1964).

64Obi2: I. Obinata, Y. Takeuchi, K. Kurihara, and M. Watanabe, *Nippon Kinzoku Gakkaishi, 28*, 568-576 (1964).

64Pec: W.H. Pechin, D.E. Williams, and W.L. Larsen, *Trans. ASM, 57*, 464-473 (1964).

64Pet: D.T. Peterson and R.P. Colburn, USAEC Comm. IS-613, 13 p (1964); quoted from [Shunk].

64Rhy: D.W. Rhys and E.G. Price, *Met. Ind., 105*, 243-247 (1964).

64Rit: D.L. Ritter, B.C. Giessen, and N.J. Grant, *Trans. AIME, 230*, 1250-1267 (1964).

64Wit: L.J. Wittenburg and G.R. Grove, USAEC, MLM-1208, 8-11 (1964); USAEC, MLM-1244, p 56 (1964); quoted in [Shunk].

65Dar: J.B. Darby, Jr., A.F. Berndt, and J.W. Downey, *J. Less-Common Met., 9*, 466-468 (1965).

65Ell: R.P. Elliott, in *Rare Earth Research III*, L. Eyring, Ed., Gordon and Breach, Science Publishers, New York, 215-245 (1965).

65Haa: D.J. Haase, H. Steinfink, and E.J. Weiss, in *Rare Earth Research 111*, Gordon and Breach, Science Publishers, New York, 535-544 (1965).

65Swi: J.H. Swisher, NASA Tech. Note, NASA-TN-D-2734, 18 p (1965); quoted in [Shunk].

66Bru: G. Bruzzone, *Ann. Chim. (Rome), 56*, 1306-1319 (1966).

66Den1: D.H. Dennison, M.J. Tschetter, and K.A. Gschneidner, Jr., *J. Less-Common Met., 10* (2), 108-115 (1966).

66Den2: D.H. Dennison, M.J. Tschetter, and K.A. Gschneidner, Jr., *J. Less-Common Met., 11*, 423-435 (1966).

66Ell: F.H. Ellinger, K.A. Johnson, and V.O. Struebing, *J. Nucl. Mat., 20*, 83-86 (1966).

66Sav: E.M. Savitskiy, M.A. Tylkina, and O.Kh. Khamidov, *Russ. Metall., 4*, 52-56 (1966).

66Vui: G. Vuillard and J.P. Piton, *Compt. Rend. C, 263*, 1018-1021 (1966).

67Bad1: T.A. Badayeva and R.I. Juznetsova, *Russ. Metall.,* (1), 89-92 (1967).

67Bad2: T.A. Badayeva and R.I. Juznetsova, *Russ. Metall., 6*, 99-100 (1967).

67Bow: D.F. Bowersox and J.A. Leary, *J. Nucl. Mater., 21*, 219-224 (1967).

67Car: O.N. Carlson, F.A. Schmidt, and D.E. Diesburg, *Trans. ASM, 60*(2), 119-124 (1967).

67Jan: G. Jangg, H.R. Kirchmayr, and W. Lugscheider, *Z. Metallkd., 58*, 724-726 (1967) in German.

67Kir1: H.R. Kirchmayr and W. Lugscheider, *Z. Metallkd., 58*(3), 185-188 (1967).

67Kir2: H.R. Kirchmayr and W. Lugscheider, *Z. Metallkd., 58*, 185-193 (1967) in German.

67Kut1: V.I. Kutaitsev, N.T. Chebortarev, I.G. Lebedev, M.A. Andrianov, V.N. Konev, and T.S. Menshikova, *Plutonium 1965*, Chapman & Hall, London, 420-449 (1967).

67Kut2: V.I. Kutaitsev, N.T. Chebortarev, I.G. Lebedev, M.A. Andrianov, V.N. Konev, and T.S. Menshikova, *Plutonium 1965*, Chapman & Hall, London, 420-447 (1967).

67Mcm: O.D. McMasters and K.A. Gschneidner, Jr., *J. Less-Common Met., 13*, 193-199 (1967).

67Par: J.K. Pargeter and W. Hume-Rothery, *J. Less-Common Met., 12*, 366-374 (1967).

67Rus: P.G. Rustamov, B.N. Mardakhaev, and M.G. Safarov, *Inorg. Mater., 3*(3), 429-433 (1967).

67Sto: E.K. Storms, *The Refractory Carbides*, Academic Press, New York (1967).

68Gul1: B.B. Gulyaev and G.F. Dvorshkaya, in *Phase Diagrams of Metallic Systems*, E.M. Savitskii, Ed., Akad. Nauk SSSR, 267-273 (1968) in Russian.

68Gul2: B.B. Gulyaev, in *Phase Diagrams of Metallic Systems*, E.M. Savitskii, Ed., Nauka, Moscow, 257-267 (1986) in Russian.

68Mas: J.T. Mason and P. Chiotti, *Trans. AIME, 242*, 1167-1171 (1968).

68Mcm: O.D. McMasters, T.J. O'Keefe, and K.A. Gschneidner, Jr., *Trans. Metall. Soc. AIME, 242*(5), 936-939 (1968).

68Sav: E.M. Savitskii and O.Kh. Khamidov, *Russ. Metall.,* (6), 108-111 (1968).

69Bad: T.A. Badayera and R.I. Kuznetsova, *Izv. Akad. Nauk SSSR, Met.,* (15), 156-193 (1969) in Russian; TR: *Russ. Metall.,* (5), 101-106 (1969).

69Ben1: R. Benz and P. L. Stone, *High Temp. Sci., 1*, 114-127 (1969).

69Ben2: R. Benz, C.G. Hoffman, and G.N. Rupert, *High. Temp. Sci., 1*, 342-359 (1969).

69Bor: V.A. Boryaleova, Ya Kh. Grinberg, E.G. Shukov, V.A. Koryazhkin, and Z.S. Medvedeva, *Inorg. Mater. J*, 397-399 (1969).

69Mcm: O.D. McMasters and K.A. Gschneidner, Jr., *J. Less-Common Met., 19*, 337-344 (1969).

70Kir: H.R. Kirchmayr and W. Lugscheider, *Z. Metallkd, 61*, 22-23 (1970).

70Mas: J.T. Mason and P. Chiotti, *Metall. Trans., 1*, 2119-2123 (1970).

70Sad: O.A. Sadovskaya and E.I. Yarembash, *Russ. Inorg. Mater., 6*(7), 1097-1101 (1970).

70Sch: F.A. Schmidt and O.D. McMasters, *J. Less-Common Met., 21*, 415-425 (1970).

70Thu: R. Thummel and W. Klemm, *Z. Anorg. Allg. Chem., 376*, 44-63 (1970) in German.

70Woo: D.H. Wood, E.M. Cramer, and P.L. Wallace, *Nucl. Metall., 17*, 707-719 (1970).

70Yar: E.I. Yarembach, *Colloq. Intern. CNRS (Paris), 1*, 472-481 (1970).

71Cun: P.T. Cunningham, S.A. Johnson, and E.J. Cairns, *J. Electrochem. Soc., 118*, 1941-1944 (1971).

71Gri: R.B. Griffin and K.A. Gschneidner, Jr., *Metall. Trans., 2*(9), 2517-2524 (1971).

71Pre: B. Predel and W. Schwermann, *J. Inst. Met., 99*, 169-173 (1971).

71Sve: V.N. Svechnikov, G.F. Kobzenko, and V.G. Ivanchenko, *Metallofizika,* (33), 93-95 (1971).

72Bor: J.D. Bornand and P. Feschotte, *J Less-Common Met., 29*, 81-91 (1972) in French.

72Hut: J.M. Hutchinson, Jr., *Platinum Met. Rev., 16*, 88-90 (1972).

72Jeh: H. Jehn and E. Olzi, *J. Less-Common Met., 27*, 297-309 (1972).

72Mas: J.T. Mason and P. Chiotti, *Metall. Trans., 3*, 2851-2855 (1972).

72Por: K.I. Portnoi and V.M. Romashov, *Sov. Powder Metall. Met. Ceram., 11*, 378-384 (1972).

72Shu: A.K. Shurin and V.V. Pet'kov, *Russ. Metall.,* (2), 122-144 (1972).

73Bus: K.H.J. Buschow, *J. Less-Common Met., 31*, 165-168 (1973).

73Gha: H. Ghassem and A. Raman, *Metall. Trans., 4*, 745-748 (1973).

73Ian: A. Iandelli and A. Palenzona, *Rev. Chim. Miner.*, 303-308 (1973).

73Loe: O. Loebich, Jr. and E. Raub, *J. Less-Common Met., 30*, 47-62 (1973).

73Sav: E. Savitskii, V. Polyakova, and E. Tsyganova, *Redkozemel. Met., Splavy Soedineniya,* Izd. Nauk, Moscow, 182-184 (1973).

73Sve: V.N. Svechnikov, G.F. Kobzenko, and V.G. Ivanchenko, *Dokl. Akad. Nauk SSSR, 213,* 1062-1064 (1973).

74Gsc1: K.A. Gschneidner, Jr. and M.E. Verkade, Document IS-RIC-7, Rare Earth Information Center, Iowa State Univ., Ames, IA, 40-41 (1974).

74Gsc2: K.A. Gschneidner, Jr. and M.E. Verkade, IS-RIC-7, Rare Earth Information Center, Iowa State Univ., Ames, IA, 30-31 (1974).

74Lev: Yu.V. Levinskiy, *Russ. Metall.,* (1), 34-37 (1974).

74Ray1: A.E. Ray, *Cobalt,* (1), 13-20 (1974).

74Ray2: A.E. Ray, *Cobalt,* (1), 3-20 (1974).

74Yar: E.I. Yarambash, E.S. Vigileva, A.A. Eliseev, A.V. Zachatskaya, T.G. Aminov, and M.A. Chernitsyna, *Inorg. Mater., 10*(8), 1212-1215 (1974).

75Ell: G.V. Ellert, V.G. Sevast'yanov, and V.K. Slovyanskikh, *Russ. J. Inorg. Chem., 20*(1), 120-124 (1975).

75Gat: J. Gatterer, D. Dufek, P. Ettmayer, and R. Kieffer, *Monatsh. Chem., 106,* 1137-1147 (1975).

75Lys: Yu.B. Lyskova and A.V. Vakhobov, *Inorg. Mater., 11,* 361-362 (1975).

75Sve: V.N. Svechnikov, G.F. Kobzenko, and V.G. Ivanchenko, *Metallofizika,* (59), 77-83 (1975).

75Vak: A.V. Vakhobov, Z.U. Niyazova, and B.N. Polev, *Inorg. Mater., 11,* 306-307 (1975).

76Ian: A. Iandelli and A. Palenzona, *Rev. Chim. Minerale, 13,* 55-61 (1976).

76Mat: P. Matkovic, M. El-Boragy, and K. Schubert, *J. Less-Common Met., 50,* 165-176 (1976).

76Vol: A.E. Vol and I.K Kagan, *Handbook of Binary Metallic Systems,* Nauka, Moscow (1976) in Russian; TR: NBS/NSF, 760-761 (1985).

77Ere: V.N. Eremenko, V.G. Batalin, Yu.I. Buyanov, and I.M. Obushenko, *Dop. Akad. Nauk Ukr. RSR, B,* (6) 516-521 (1977) in Russian.

77Gar: S.P. Garg and R.J. Ackermann, *J. Nucl. Mater., 64,* 265-274 (1977).

77Kur: T.Kh. Kurbanov, R.A. Dovlyatshina, I.A. Dzhavodova, and F.A. Akhmenov, *Russ J. Inorg. Chem., 22,* 622-624 (1977).

77Kuz: A.N. Kuznetsov, K.A. Chuntonov, and S.P. Yatsenko, *Russ. Metall.,* (5), 178-180 (1977).

77Wat: R.M. Waterstrat and R.C. Manuszewski, *J. Less Common Met.,* 51, 55-67 (1977).

77Yat: S.P. Yatsenko, *J. Chim. Phys., 74,* 836-843 (1977).

78Eli: A.A. Eliseev, G.M. Kuz'micheva, and V.I. Yushrov, *Zh. Neorg. Khim.,23,*(2), 492-296 (1978) in Russian; TR: *Russ J. Inorg. Chem., 23*(2), 273-276 (1978).

78Esh: K.K. Eshnov, M.A. Zukhuritdinov, A.V. Vakhobov, and T.D. Zhurayev, *Russ. Metall.,* (1), 171-173 (1978).

78Lan: C.C. Land, D.E. Peterson, and R.B. Root, *J. Nucl. Mat., 75,* 262-273 (1978).

78Yat: S.P. Yatsenko, B.G. Semenov, and K.A. Chuntonov, *Izv. Akad. Nauk SSSR, Met.,* (5) 222-224 (1978) in Russian; TR: *Russ. Metall.,* (51), 173-174 (1978).

79Bla: R. Blachnik and A. Hoppe, *Z. Anorg. Allg. Chem, 457,* 91-104 (1979) in German.

79Vol: A.E. Vol and I.K. Kagan, *Handbook of Binary Metallic Systems,* Vol. 4, Nauka, Moscow (1979) in Russian; translated by NBS and NSF, 588-605 (1986).

79Yat: S.P. Yatsenko, A.A. Semyannikov, B.G. Semenov, and K.A. Chuntonov, *J. Less-Common Met., 64,* 185-199 (1979).

80Ban: G. Banik, T. Schmitt, P. Ettmayer, and B. Lux, *Z. Metallkd.,71*(10), 644-645 (1980) in German.

80Dav: T.G. Davey and E.H. Baker, *J. Mater. Sci. Lett., 15,* 1601-1602 (1980).

80Ere: V.N. Eremenko, I.M. Obushenko, and Yu.I. Buyanov, *Dop. Akad. Nauk Ukr. RSRA,* (7), 87-91 (1980).

80Lu: X.S. Lu, J.K. Liang, and M.G. Zhou, *Acta Phys. Sin. (China),29,* 469-484 (1980).

80Pal: A. Palenzona, *J. Less-Common Met., 72*(1), P21-P24 (1980).

81Bru: G. Bruzzone, E. Franceschi, and F. Merlo, *J. Less-Common Met., 81,* 155-160 (1981).

81Bus: V.D. Busmanov and S.P. Yatsenko, *Russ. Metall.,* (5), 157-160 (1981).

81Ian: A. Iandelli and A. Palenzona, *J. Less-Common Met.,80,* P71-P82 (1981).

81Loe: O. Loebich, Jr. and C.J. Raub, *Platinum Met. Rev., 25*(3), 113-120 (1981).

81Mor: G. Morgaut, B. Legendre, S. Mareglier-Lacordaire, and C. Souleau, *Ann. Chim. Fr., 6,* 315-326 (1981).

81Wat: R.M. Waterstrat, *J. Less-Common Met., 80,* P31-P36 (1981).

82Bor: G. Borzone, A. Borsese, and R. Ferro, *J. Less-Common Met., 85,* 195-203 (1982).

82Pri: N.Yu. Pribyl'skii, I.G. Vasileva, and R.S. Gamidov, *Mater. Res. Bull., 17,* 1147-1153 (1982).

82Sub: P.R. Subramanian and J.F. Smith, *J. Less-Common Met., 87,* 205-213 (1982).

83Ere: V.N. Eremenko, K.A. Meleshevich, and Yu.I. Buyanov, *Dop. Akad. Nauk Ukr. RSRA,* (3), 83-88 (1983).

83Kub: O. Kubaschewski-Von Goldbeck, *Titanium: Physicochemical Properties of Its Compounds and Alloys,* Atomic Energy Review; Spec. Issue No. 9, O. Kubaschewski, Ed., IAEA, Vienna, 156 (1983).

84Pas: J.D.A. Paschoal, H. Kleykamp, and F. Thummler, *J. Less-Common Met., 98,* 279-284 (1984).

85Mur: J.L. Murray, *Int. Met. Rev.,30*(5), 211-233, 1985.

86Bar1: O.M. Barabash and Yu.N. Koval, *Crystal Structure of Metals and Alloys,* Naukova Dumka, Kiev, 211-212 (1986).

86Bar2: O.M. Barabash and Yu. N. Koval, *Crystal Structure of Metals and Alloys,* Naukova Dumka, Kiev, 247-248 (1986).

86Bar3: O.M. Barabash and Yu.N. Koval, *Crystal Structure of Metals and Alloys,* Naukova Dumka, Kiev, 296-297 (1986) in Russian.

87Gor: O.V. Gordiichuk, Author's Abstract of Candidate's Thesis, Chemical Sciences, Kiev (1987).

87Mel: L.Z. Melenkov, S.P. Yatsenko, K.A. Chuntonov, and Yu. N. Grin, *Izv. Akad. Nauk SSSR, Met.,* (2), 201-203 (1987) in Russian; TR: *Russ. Metall.,* (2), 2 08-211 (1987).

88Sac: A. Saccone, S. Delfino, and R. Ferro, *J. Less-Common Met., 143,* 1-23 (1988).

90Con: J.B. Condon, T. Schober, and R. Lasser, *J. Nucl. Mater., 170,* 24-30 (1990).

90Sac: A. Saccone, S. Delfino, and R. Ferro, *Calphad, 14*(2), 161 (1990).

Ag-Al

A.J. McAlister, 1987

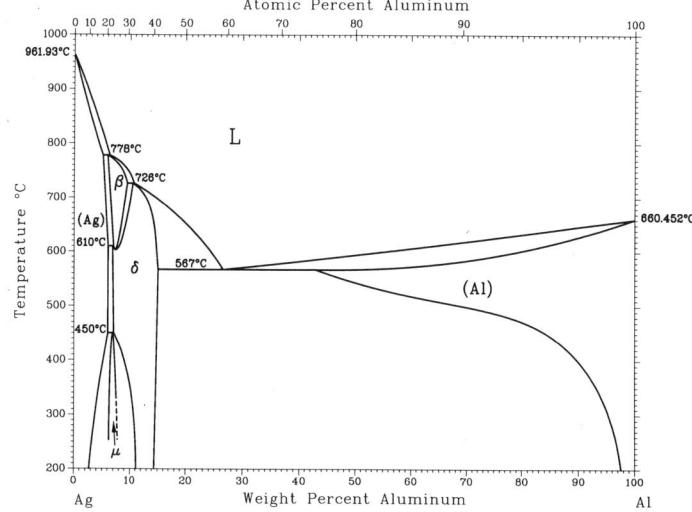

Phase	Composition, wt% Al	Pearson symbol	Space group
(Ag)	0.0	cF4	Fm$\bar{3}$m
β	6.1 to 7.4	cI2	Im$\bar{3}$m
δ	6.9 to 15.3	hP2	P6$_3$/mmc
μ	~6.2 to 7.3	cP20	P4$_1$32
			P2$_1$3(a)
(Al)	100	cF4	Fm$\bar{3}$m

(a) ~300 °C

Ag-As

M.R. Baren, 1990

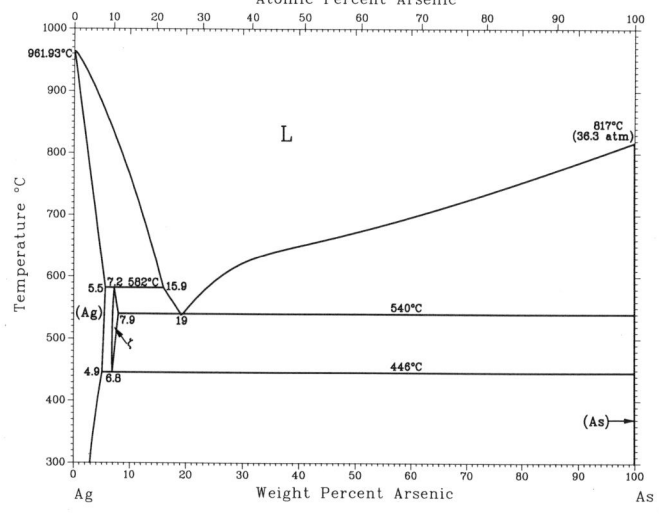

Phase	Composition, wt% As	Pearson symbol	Space group
(Ag)	0 to 5.5	cF4	Fm$\bar{3}$m
ζ	6.8 to 7.9	hP2	P6$_3$/mmc
(As)	100	hR2	R$\bar{3}$m

Ag-Au

H. Okamoto and T.B. Massalski, 1987

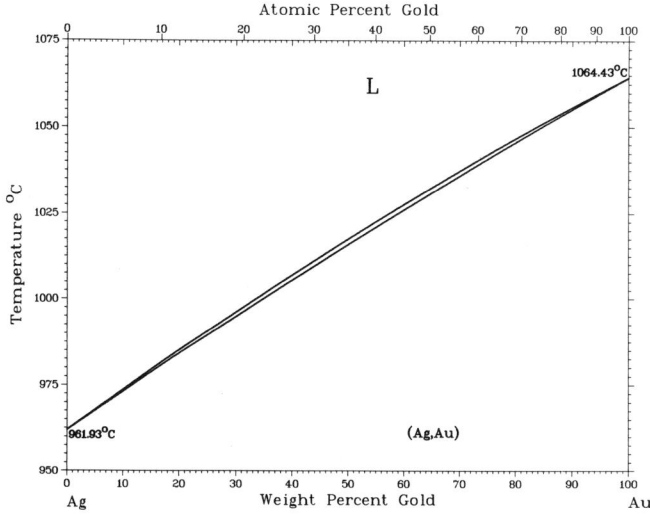

Phase	Composition, wt% Au	Pearson symbol	Space group
(Ag,Au)	0 to 100	cF4	Fm$\bar{3}$m

Ag-Be

H. Okamoto and L.E. Tanner, 1987

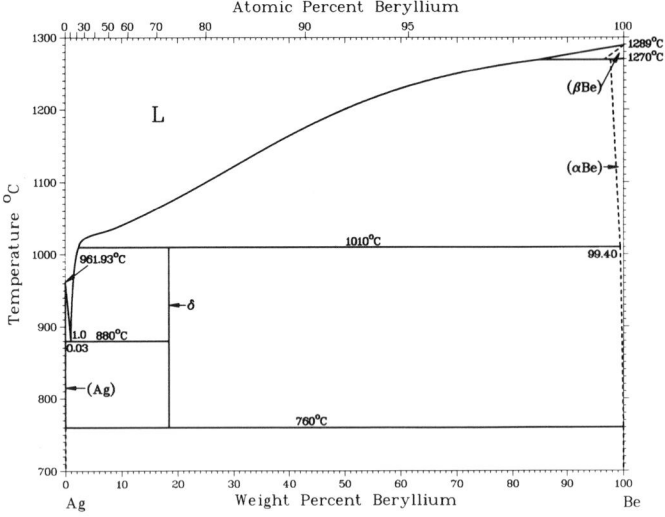

Phase	Composition, wt% Be	Pearson symbol	Space group
(Ag)	0 to 0.03	cF4	Fm$\bar{3}$m
δ or AgBe$_2$	~18?	cF24	Fd$\bar{3}$m
(αBe)	99.40 to 100	hP2	P6$_3$/mmc
(βBe)	100	cI2	Im$\bar{3}$m
Questionable phases (stable? metastable?)			
γ	~12	?	?
AgBe$_{12}$	50	tI26	I4/mmm

Ag-Bi

R.P. Elliott and F.A. Shunk, 1980

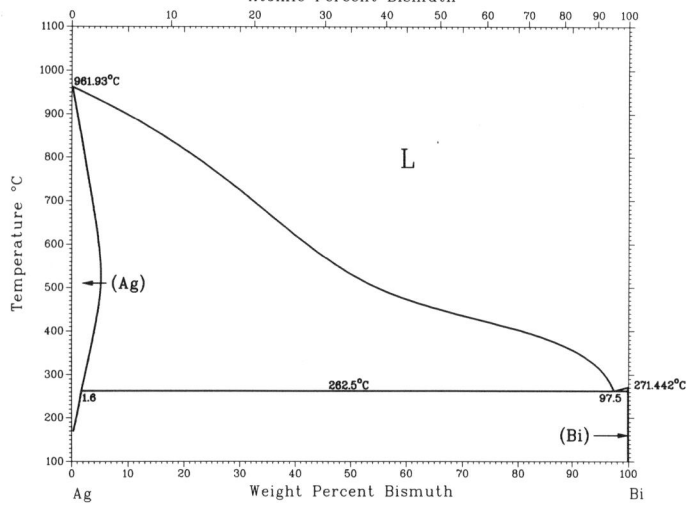

Phase	Composition, wt% Bi	Pearson symbol	Space group
(Ag)	0 to 4.945	cF4	Fm$\bar{3}$m
(Bi)	~100	hR2	R$\bar{3}$m

Ag-Ca

M.R. Baren, 1988

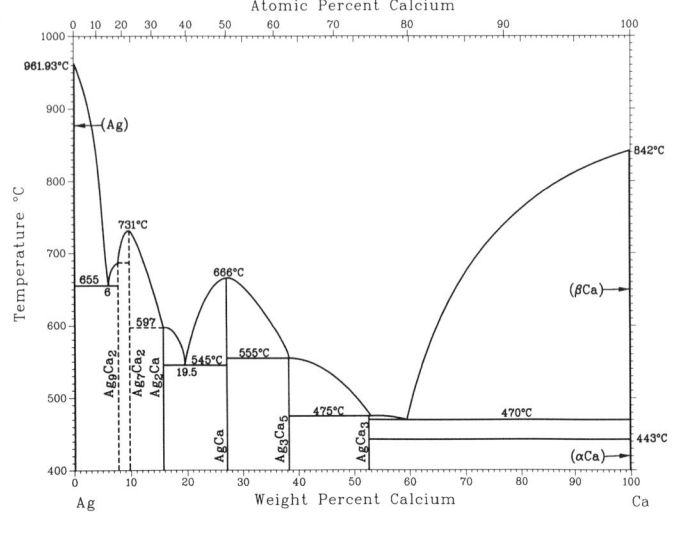

Phase	Composition, wt% Ca	Pearson symbol	Space group
(Ag)	0	cF4	Fm$\bar{3}$m
Ag$_9$Ca$_2$	7.7	…	…
Ag$_7$Ca$_2$	9.6	hP18	P6$_3$22
Ag$_2$Ca	15.6	oI12	Imma
AgCa	27.1	oC8	Cmcm
Ag$_3$Ca$_5$	38.2	tI32	I4/mcm
AgCa$_3$	52.7	…	…
(αCa)	100	cF4	Fm$\bar{3}$m
(βCa)(a)	100	cI2	Im$\bar{3}$m

(a) Above 443 °C

Ag-Cd

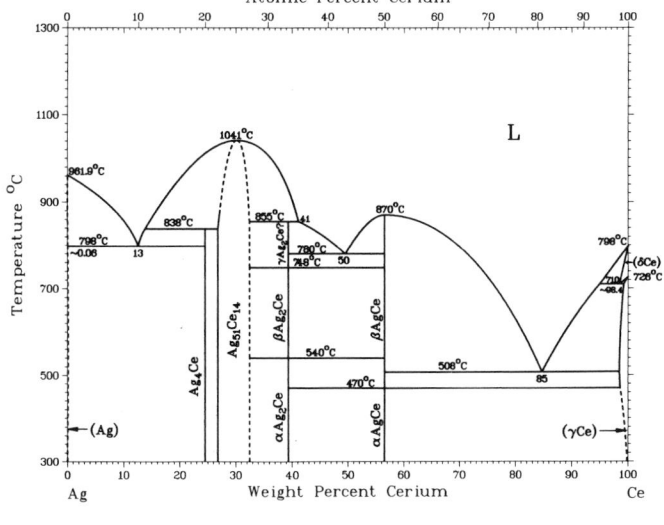

Phase	Composition, wt% Cd	Pearson symbol	Space group
(Ag)	0 to 43.2	cF4	Fm$\bar{3}$m
β	41 to 56	cI2	Im$\bar{3}$m
β′	49.5 to 51.0	(a)	…
ζ	50.5 to 57	(b)	…
γ	58 to 63.5	…	…
γ	58 to 63.5	cI52	I$\bar{4}$3m
ε	65.4 to 82	hP2	P6$_3$/mmc
(Cd)	93.3 to 100	hP2	P6$_3$/mmc

(a) Ordered bcc. (b) cph

Ag-Ce

K.A. Gschneidner, Jr. and F.W. Calderwood, 1985

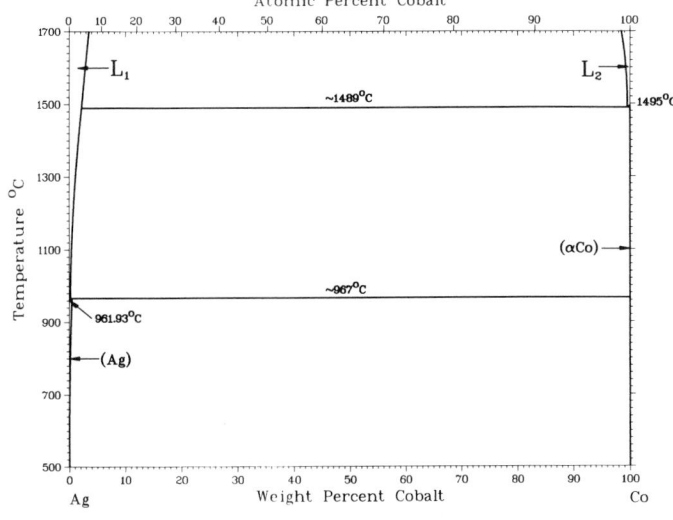

Phase	Composition, wt% Ce	Pearson symbol	Space group
(Ag)	0 to ~0.06	cF4	Fm$\bar{3}$m
Ag$_{51}$Ce$_{14}$	26.2 to ~30	hP65	P6/m
αAg$_2$Ce	39.3	oI12	Imma
AgCe	56.5	cP2	Pm$\bar{3}$m
(δCe)	~98 to 100	cI2	Im$\bar{3}$m
(γCe)	~98 to 100	cF4	Fm$\bar{3}$m
(βCe)	100	hP4	P6$_3$/mmc
(αCe)	100	cF4	Fm$\bar{3}$m

Ag-Co

I. Karakaya and W.T. Thompson, 1986

Phase	Composition, wt% Co	Pearson symbol	Space group
(Ag)	0 to 0.44	cF4	Fm$\bar{3}$m
(εCo)(a)	100	hP2	P6$_3$/mmc
(αCo)	~100	cF4	Fm$\bar{3}$m

(a) Below 422 °C

Ag-Cu

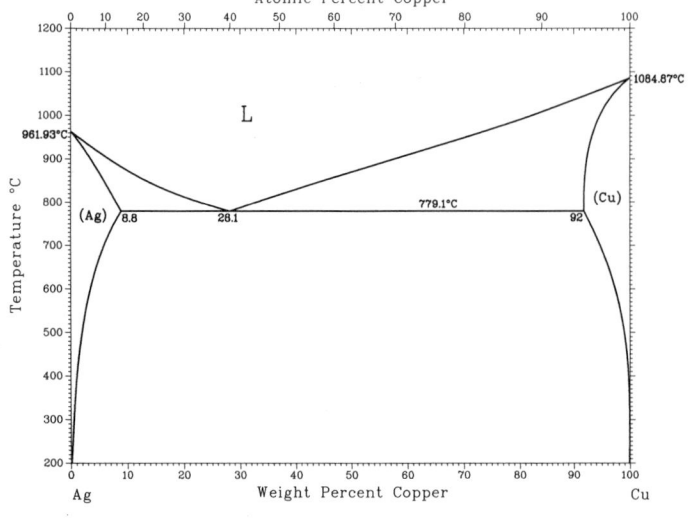

P.R. Subramanian and J.H. Perepezko, unpublished

Phase	Composition, wt% Cu	Pearson symbol	Space group
(Ag)	0 to 8.8	cF4	$Fm\overline{3}m$
(Cu)	92.0 to 100	cF4	$Fm\overline{3}m$

Ag-Dy

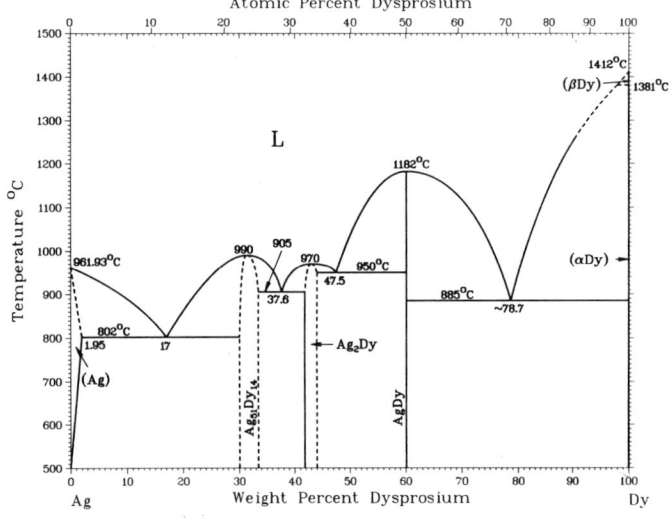

K.A. Gschneidner, Jr. and F.W. Calderwood, 1985

Phase	Composition, wt% Dy	Pearson symbol	Space group
(Ag)	0 to 1.95	cF4	$Fm\overline{3}m$
$Ag_{51}Dy_{14}$	29.2 to 34.0	hP65	...
Ag_2Dy	41.8 to 44.0	tI6	$I4/mmm$
AgDy	60.1	cP2	$Pm\overline{3}m$
(βDy)	100	cI2	$Im\overline{3}m$
(αDy)	100	hP2	$P6_3/mmc$
(α'Dy)(a)	100	...	...

(a) Below –187 °C

Ag-Er

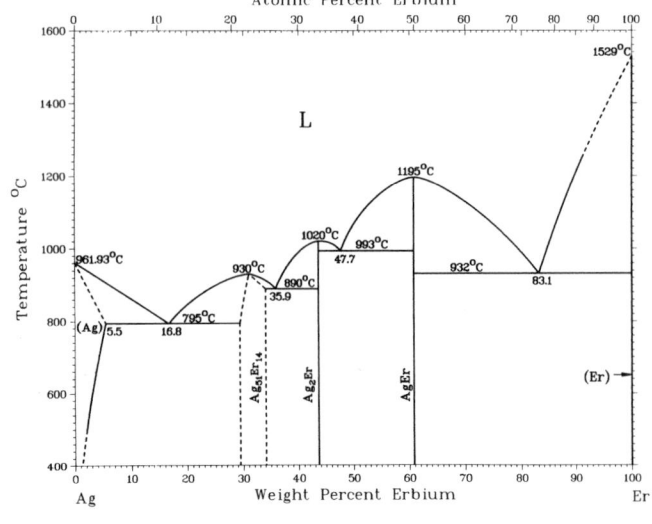

K.A. Gschneidner, Jr. and F.W. Calderwood, 1985

Phase	Composition, wt% Er	Pearson symbol	Space group
(Ag)	0 to 5.5	cF4	$Fm\overline{3}m$
$Ag_{51}Er_{14}$	29.8 to 34.7	hP65	...
Ag_2Er	43.6	tI6	$I4/mmm$
AgEr	60.8	cP2	$Pm\overline{3}m$
(Er)	100	hP2	$P6_3/mmc$

Ag-Eu

K.A. Gschneidner, Jr. and F.W. Calderwood, 1985

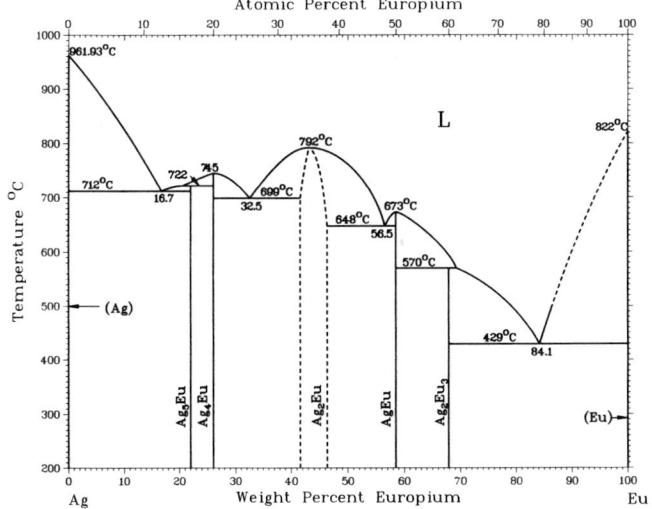

Phase	Composition, wt% Eu	Pearson symbol	Space group
(Ag)	0	cF4	Fm$\overline{3}$m
Ag₅Eu	22.0	hP6	P6/mmm
Ag₄Eu	26	tI10	I4/m
Ag₂Eu	41.3	oI12	Imma
AgEu	58.5	oP8	Pnma
Ag₂Eu₃	67.9	tP10	P4/mbm
(Eu)	100	cI2	Im$\overline{3}$m

Ag-Fe

L.J. Swartzendruber, 1984

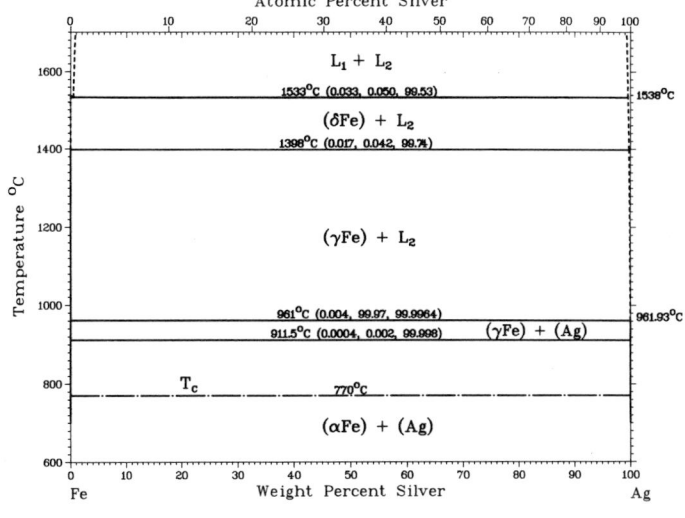

Phase	Composition, wt% Ag	Pearson symbol	Space group
δ or (δFe)	0 to 0.033	cI2	Im$\overline{3}$m
γ or (γFe)	0 to 0.042	cF4	Fm$\overline{3}$m
α or (αFe)	0 to 0.0004	cI2	Im$\overline{3}$m
(Ag)	99.99663 to 100	cF4	Fm$\overline{3}$m

Ag-Ga

H. Okamoto, 1992

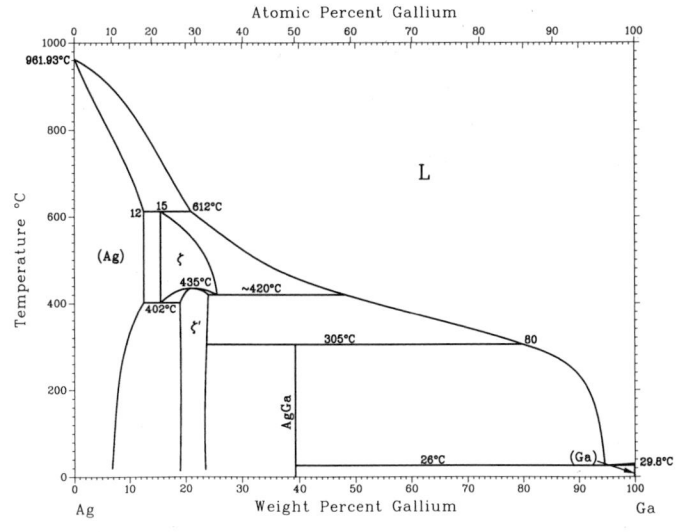

Phase	Composition, wt% Ga	Pearson symbol	Space group
(Ag)	0 to 12	cF4	Fm$\overline{3}$m
ζ	15 to 25	hP2	P6₃/mmc
ζ'	18 to 24	hP9	P$\overline{3}$
AgGa	39.2	cI2	Im$\overline{3}$m
(Ga)	100	oC8	Cmca

Ag-Gd

K.A. Gschneidner, Jr. and F.W. Calderwood, 1985

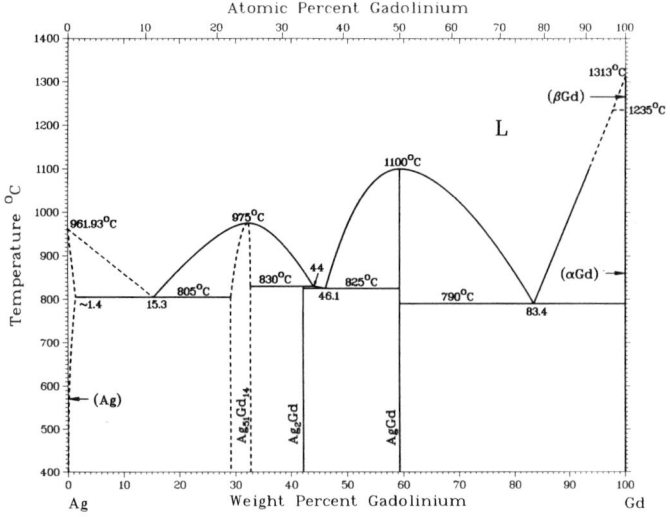

Phase	Composition, wt% Gd	Pearson symbol	Space group
(Ag)	0 to ~1.4	cF4	$Fm\overline{3}m$
Ag$_{51}$Gd$_{14}$	28.5	tP65	P6/m
Ag$_2$Gd	42.1	tI6	I4/mmm
AgGd	59.3	cP2	$Pm\overline{3}m$
(βGd)	100	cI2	$Im\overline{3}m$
(αGd)	100	hP2	P6$_3$/mmc

Ag-Ge

R.W. Olesinski and G.J. Abbaschian, 1988

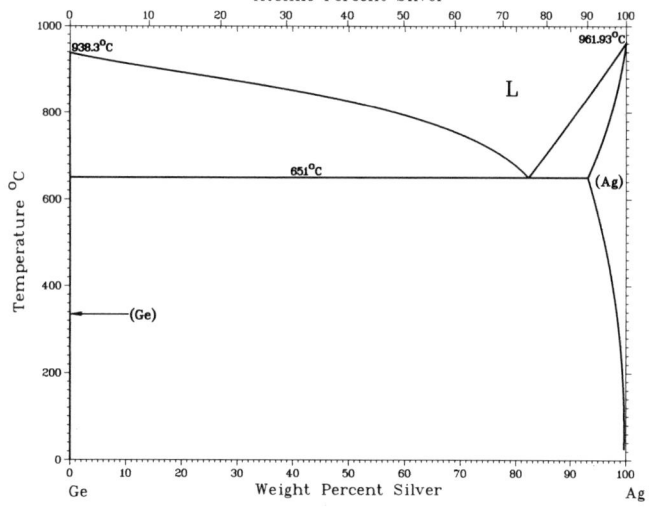

Phase	Composition, wt% Ag	Pearson symbol	Space group
(Ge)	~0	cF8	$Fd\overline{3}m$
GeII (HP)	0	tI4	I4$_1$/amd
(Ag)	93.3 to 100	cF4	$Fm\overline{3}m$
Metastable phases			
β (cph)	83 to 86	hP*	...
Tetragonal	85	t**	...

Ag-Hg

M.R. Baren, unpublished

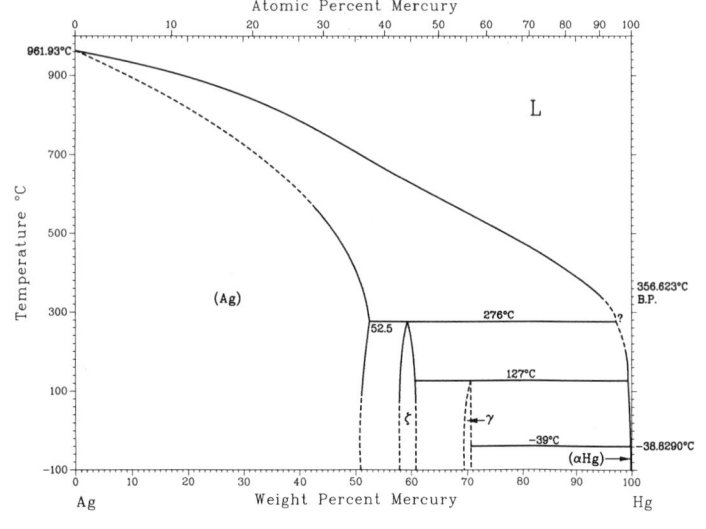

Phase	Composition, wt% Hg	Pearson symbol	Space group
(Ag)	0 to 52.5	cF4	$Fm\overline{3}m$
ζ	58.9 to 61.3	hP2	P6$_3$/mmc
γ	70.0 to 71.0	cI*	I23
(αHg)	100	hR1	$R\overline{3}m$

Ag-Ho

K.A. Gschneidner, Jr. and F.W. Calderwood, 1985

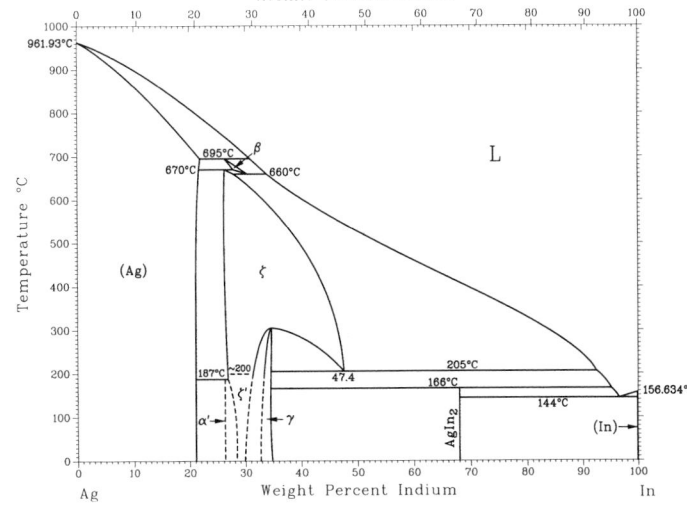

Phase	Composition, wt% Ho	Pearson symbol	Space group
(Ag)	0 to 2.4	cF4	$Fm\bar{3}m$
Ag$_{51}$Ho$_{14}$	29.5	hP65	P6/m
Ag$_2$Ho	43.3	tI6	I4/mmm
AgHo	60.5	cP2	$Pm\bar{3}m$
(Ho)	100	hP2	P6$_3$/mmc

Ag-In

M.R. Baren, 1992

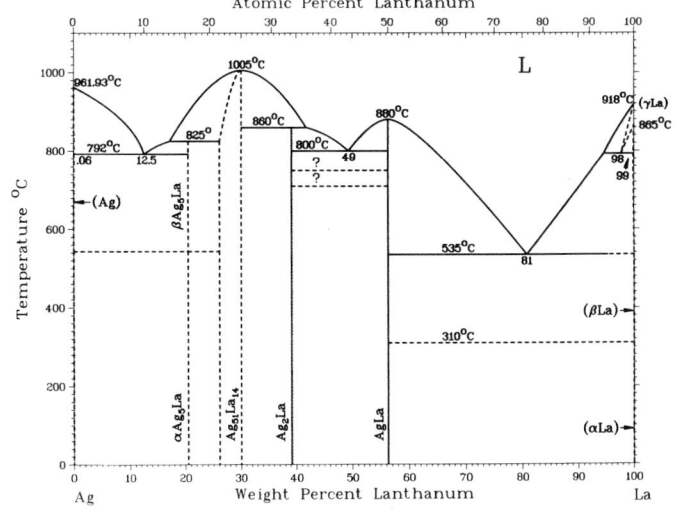

Phase	Composition, wt% In	Pearson symbol	Space group
α(Ag)	0 to 22.1	cF4	$Fm\bar{3}m$
β	26.2 to 31.3	cI2	$Im\bar{3}m$
α'(Ag$_3$In)	26	cP4?	$Pm\bar{3}m$?
ζ	26.2 to 47.6	hP*	…
ζ'	?	hP8	P6$_3$/mmc
γ(Ag$_2$In)	32.5 to 35.0	cP52	$P\bar{4}3m$
φ(AgIn$_2$)	68.1	tI12	I4/mcm
(In)	100	tI2	I4/mmm
Metastable phases			
…	19.4	hP2	P6$_3$/mmc
…	71 to 81	cF4	$Fm\bar{3}m$

Ag-La

K.A. Gschneidner, Jr. and F.W. Calderwood, 1983

Phase	Composition, wt% La	Pearson symbol	Space group
(Ag)	0	cF4	$Fm\bar{3}m$
Ag$_5$La	20.5	hP?	…
Ag$_{51}$La$_{14}$	26.1	hP65	…
Ag$_2$La	39.1	oI12	Imma
AgLa	56.3	cP2	$Pm\bar{3}m$
(γLa)	100	cI2	$Im\bar{3}m$
(βLa)	100	cF4	$Fm\bar{3}m$
(αLa)	100	hP4	P6$_3$/mmc

Ag-Li

A.D. Pelton, 1986

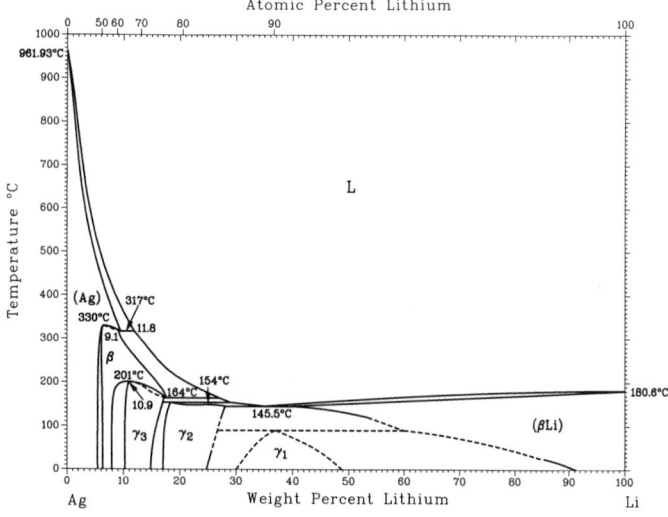

Phase	Composition, wt% Li	Pearson symbol	Space group
(Ag)	0 to 9.1	cF4	$Fm\bar{3}m$
β	6.1 to 18	cP2	$Pm\bar{3}m$
γ₃	10.9 to 17	Cubic (cP52?)	$P\bar{4}3m$?
γ₂	17 to 28	Cubic (cI52?)	$I\bar{4}3m$?
γ₁	32 to 43	Cubic	...
(βLi)	39 to 100	cI2	$Im\bar{3}m$
(αLi)	100	hP2	$P6_3/mmc$

Ag-Mg

A.A. Nayeb-Hashemi and J.B. Clark, 1988, with modifications

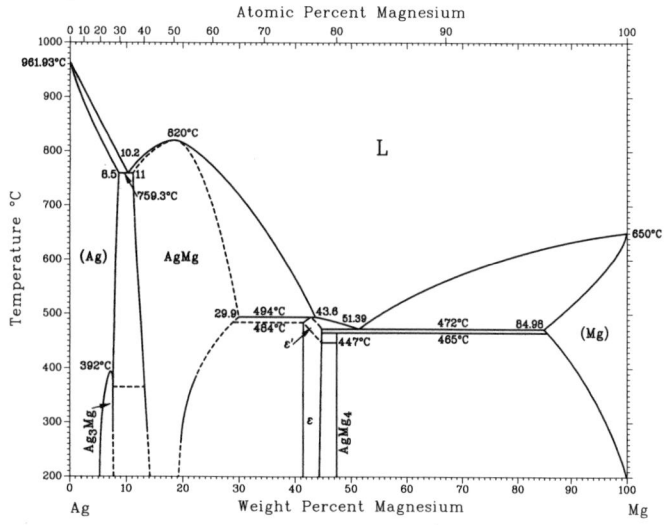

The two-phase region between (Ag) and Ag₃Mg (ordered) is not shown here.

Phase	Composition, wt% Mg	Pearson symbol	Space group
(Ag) or α	0 to 8.5	cF4	$Fm\bar{3}m$
Ag₃Mg ord or α′	7	cP4	$Pm\bar{3}m$
AgMg or β′	11 to 29.9	cP2	$Pm\bar{3}m$
ε′	41.4 to 44.7	tI*	...
ε	41.4 to 44.7	cF*	...
AgMg₄	47	hP*	...
(Mg) or δ	84.98 to 100	hP2	$P6_3/mmc$

Ag-Mo

M.R. Baren, 1990

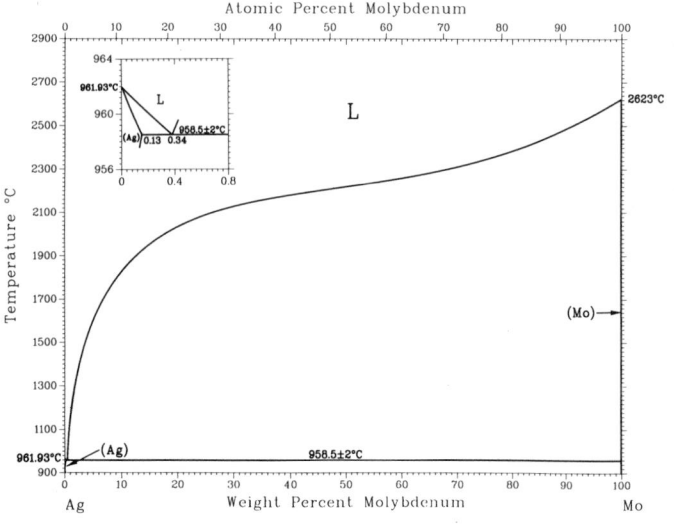

Phase	Composition, wt% Mo	Pearson symbol	Space group
(Ag)	0 to 0.13	cF4	$Fm\bar{3}m$
(Mo)	100	cI2	$Im\bar{3}m$

Ag-Na

A.D. Pelton, 1986

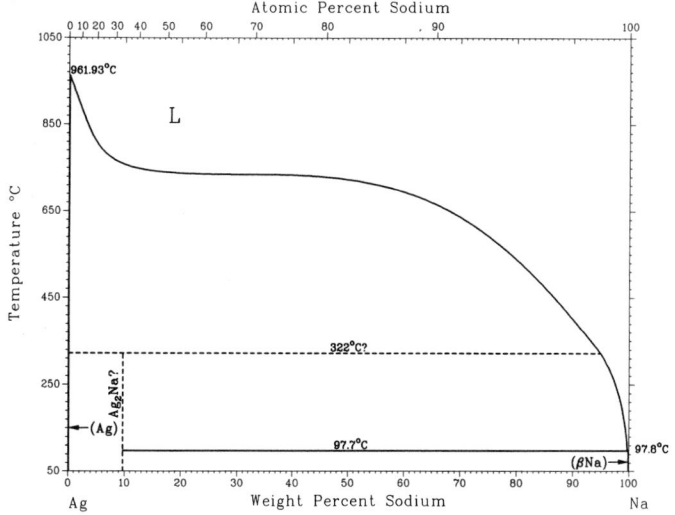

Phase	Composition, wt% Na	Pearson symbol	Space group
(Ag)	0	cF4	Fm$\overline{3}$m
Ag$_2$Na	9.6	cF24	Fd$\overline{3}$m
(βNa)	100	cI2	Im$\overline{3}$m

Ag-Nd

K.A. Gschneidner , Jr. and F.W. Calderwood, 1985

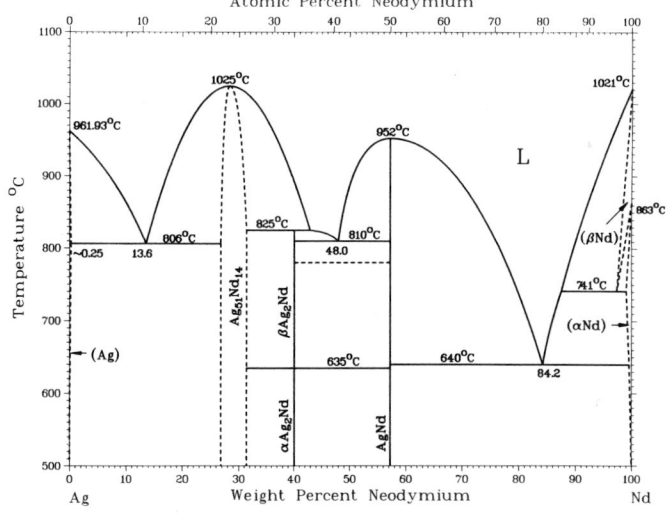

Phase	Composition, wt% Nd	Pearson symbol	Space group
(Ag)	0 to ~5	cF4	Fm$\overline{3}$m
Ag$_{51}$Nd$_{14}$	26.8 to 31.4	hP65	...
βAg$_2$Nd	40.0	hP?	...
αAg$_2$Nd	40.0	oI12	Imma
AgNd	57.2	cP2	Pm$\overline{3}$m
(βNd)	97.4 to 100	cI2	Im$\overline{3}$m
(αNd)	99.0 to 100	hP4	P6$_3$/mmc

Ag-Ni

M. Singleton and P. Nash, 1991

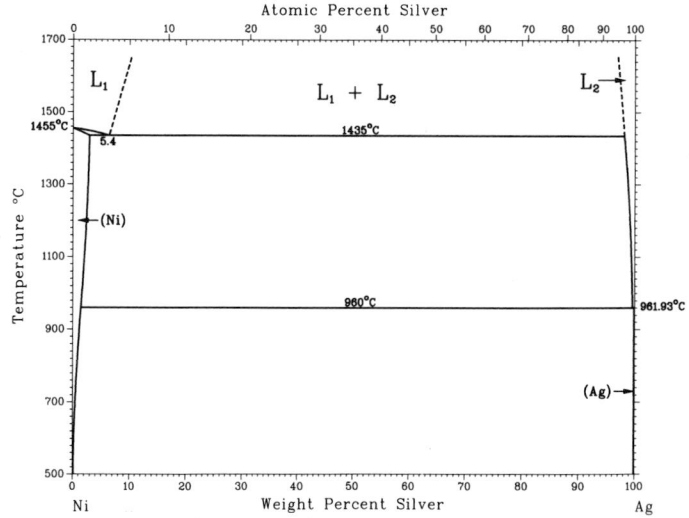

Phase	Composition, wt% Ag	Pearson symbol	Space group
(Ni)	0 to 1.8	cF4	Fm$\overline{3}$m
(Ag)	99.3 to 100	cF4	Fm$\overline{3}$m

Ag-P

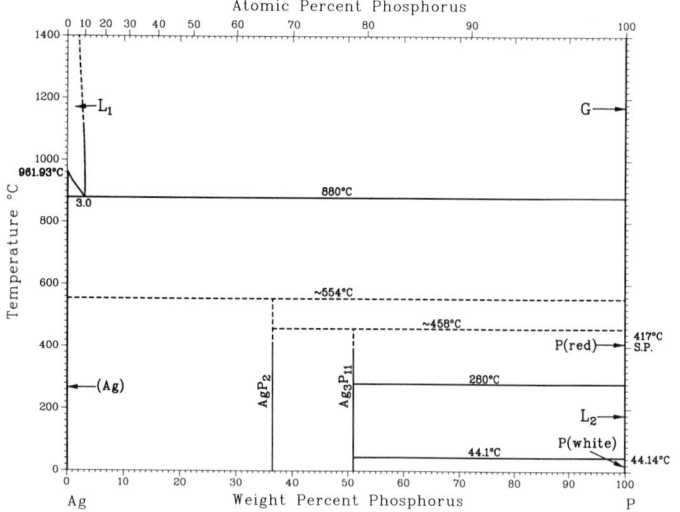

I. Karakaya and W.T. Thompson, 1988

Phase	Composition, wt% P	Pearson symbol	Space group
(Ag)	0	cF4	Fm$\overline{3}$m
AgP$_2$	36.5	(a)	...
Ag$_3$P$_{11}$	51.0	(b)	Cm
P(black)	100	oC8(c)	Cmca
P(white)	100	(d)	...
P(red)	100	(e)	...

(a) Monoclinic structure with β = 113.48°. (b) Monoclinic structure with β = 118.84°. (c) At high pressures black P transforms to a rhombohedral structure. (d) Cubic below –35 °C. (e) Cubic with 66 atoms per unit cell

Ag-Pb

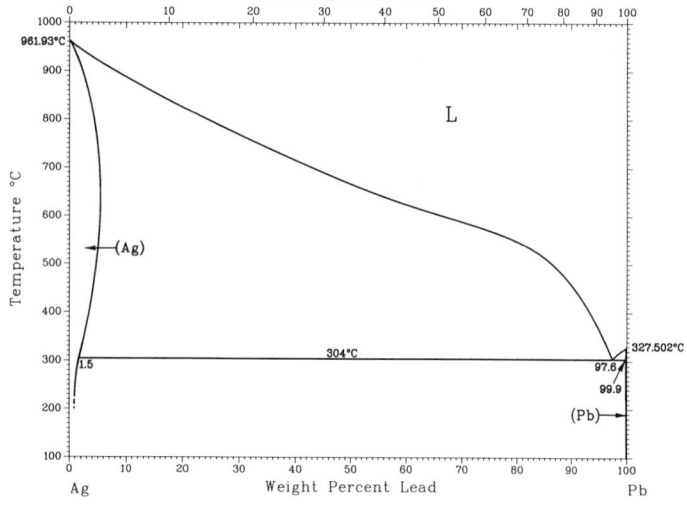

I. Karakaya and W.T. Thompson, 1987

Phase	Composition, wt% Pb	Pearson symbol	Space group
(Ag)	0 to 5.2	cF4	Fm$\overline{3}$m
(Pb)	99.9 to 100	cF4	Fm$\overline{3}$m

Ag-Pd

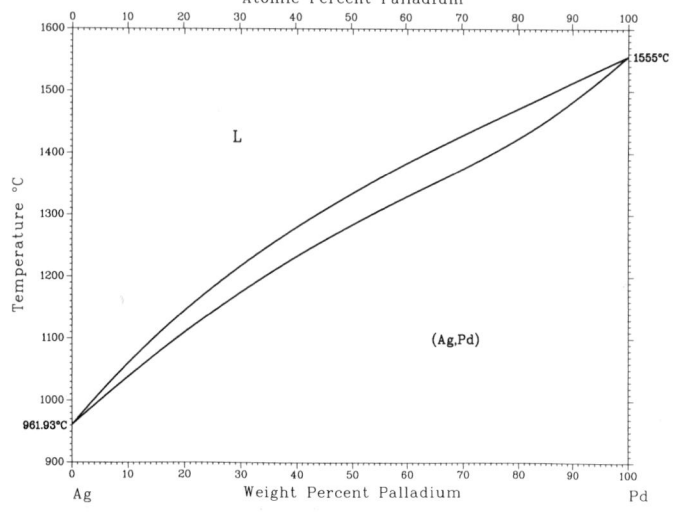

I. Karakaya and W.T. Thompson, 1988

Phase	Composition, wt% Pd	Pearson symbol	Space group
(Ag,Pd)	0 to 100	cF4	Fm$\overline{3}$m

Ag-Pr

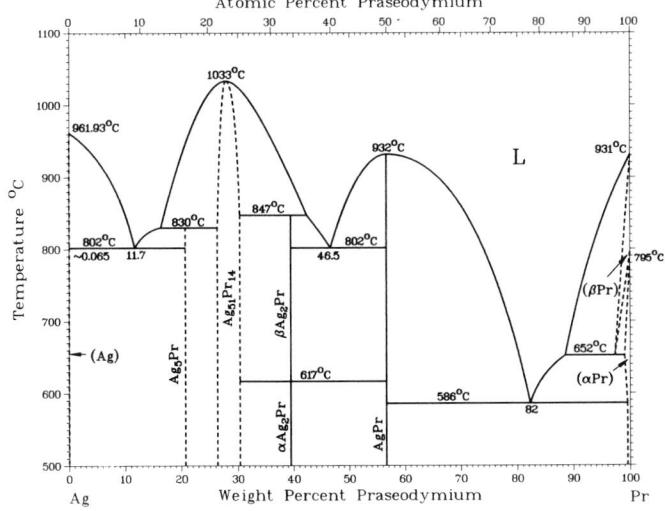

K.A. Gschneidner, Jr. and F.W. Calderwood, 1985

Phase	Composition, wt% Pr	Pearson symbol	Space group
(Ag)	0 to ~0.065	cF4	$Fm\bar{3}m$
Ag$_5$Pr	20.8	...	...
Ag$_{51}$Pr$_{14}$	26.4 to 30.3	hP65	...
βAg$_2$Pr	39.5	hP?	...
αAg$_2$Pr	39.5	oI12	Imma
AgPr	56.6	cP2	$Pm\bar{3}m$
(βPr)	97.3 to 100	cI2	$Im\bar{3}m$
(αPr)	99.0 to 100	hP4	$P6_3/mmc$

Ag-Pt

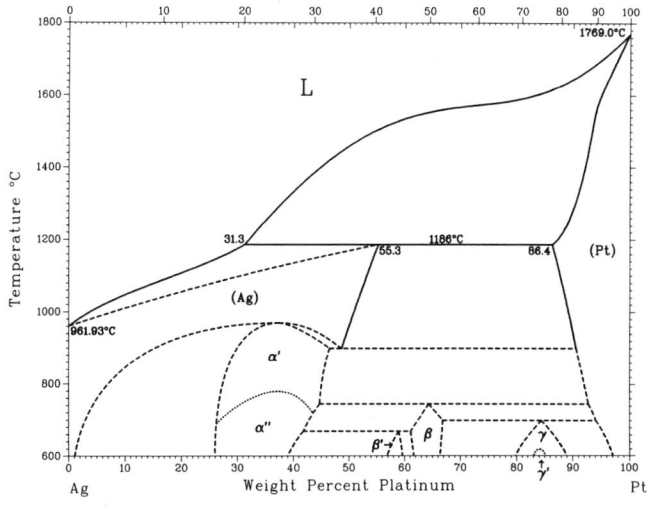

I. Karakaya and W.T. Thompson, 1987

Phase	Composition(a), wt% Pt	Pearson symbol	Space group
(Ag)	0 to 55.3	cF4	$Fm\bar{3}m$
(Pt)	86.4 to 100	cF4	$Fm\bar{3}m$
α′	26 to 47	cF4	$Fm\bar{3}m$
α″	26 to 43	cP4	$Pm\bar{3}m$
β(b)	61 to 67	...	...
β′(b)	57 to 60	...	...
γ	80 to 89	cP4	$Pm\bar{3}m$
γ′	83 to 85	cF*	...

Note: α′, α″, β, β′, γ, and γ′ phases are questionable. (a) Rough composition from phase diagram. (b) Rhombohedrally distorted cubic structure

Ag-S

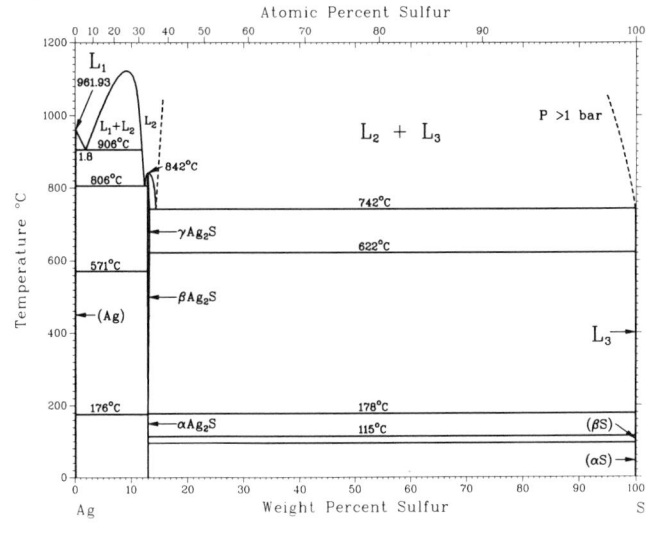

R.C. Sharma and Y.A. Chang, 1986

Phase	Composition, wt% S	Pearson symbol	Space group
(Ag)	0.04	cF4	$Fm\bar{3}m$
αAg$_2$S	12.9	mP24	$P2_1/c$
αAg$_2$S (acanthite)	12.9	mP12	$P2_1/n$
βAg$_{2+δ}$S	12.9	cI6	...
γAg$_{2+δ}$S	12.9	cF12	...
δAg$_2$S(a)	12.9	t**	...
(αS)	~100	oF128	Fddd
(βS)	~100	mP*	$P2_1/c$

(a) High-pressure phase

Ag-Sb

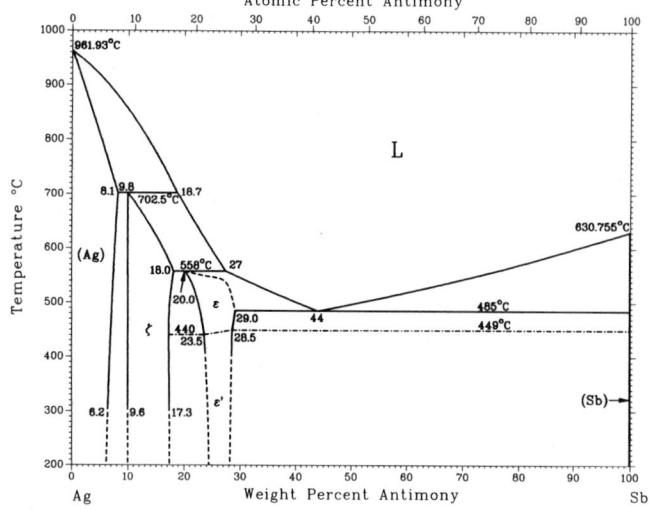

From [Hansen]

Phase	Composition, wt% Sb	Pearson symbol	Space group
(Ag)	0 to 8.1	cF4	$Fm\overline{3}m$
ζ	9.6 to 18.0	hP2	$P6_3/mmc$
ε	20.0 to 29.0	tP4	$P4/mmm$
ε′	23.5 to 28.5	(a)	…
(Sb)	100	hR2	$R\overline{3}m$

(a) Ordered orthorhombic, $L6_0$ related

Ag-Sc

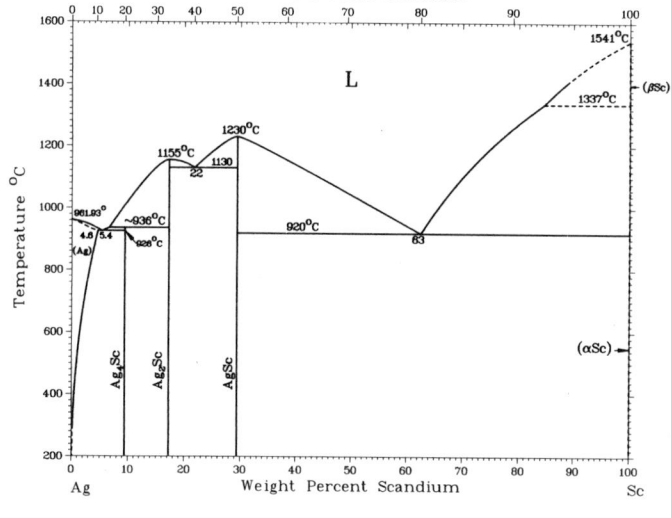

K.A. Gschneidner, Jr. and F.W. Calderwood, 1983

Phase	Composition, wt% Sc	Pearson symbol	Space group
(Ag)	0 to 4.6	cF4	$Fm\overline{3}m$
Ag₄Sc	9	tI10	$I4/m$
Ag₂Sc	17.2	tI6	$I4/mmm$
AgSc	29.4	cP2	$Pm\overline{3}m$
(βSc)	100	cI2	$Im\overline{3}m$
(αSc)	100	hP2	$P6_3/mmc$

Ag-Se

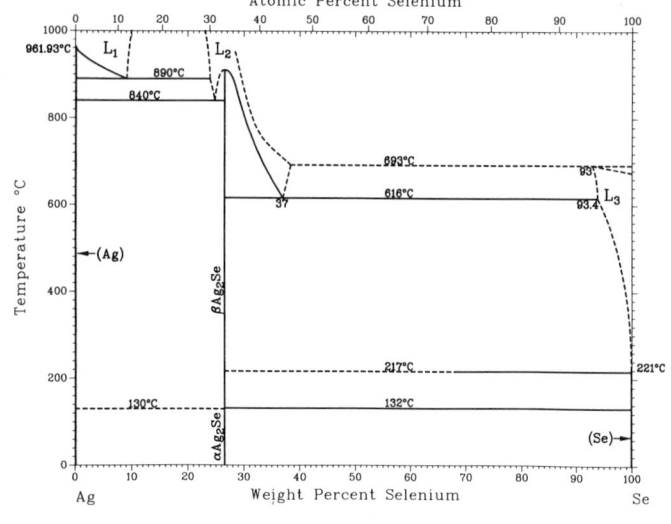

I. Karakaya and W.T. Thompson, 1990

Phase	Composition, wt% Se	Pearson symbol	Space group
(Ag)	0	cF4	$Fm\overline{3}m$
βAg₂Se	26.8	cI*	…
αAg₂Se	26.8	o**	…
(Se)	100	hP3	$P3_121$

Ag-Si

R.W. Olesinski and G.J. Abbaschian, 1989

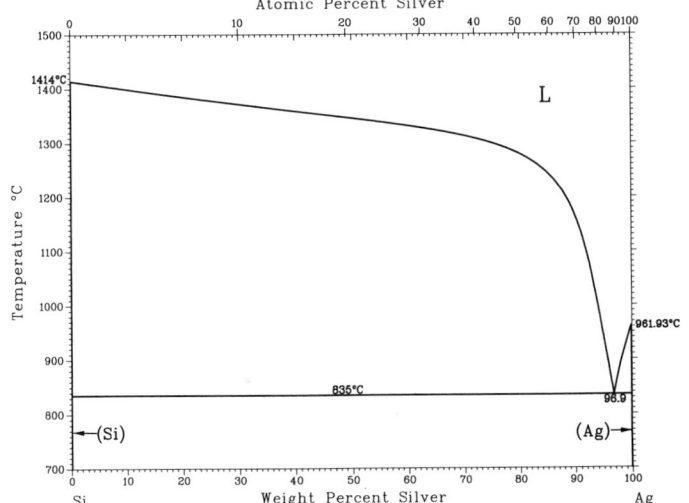

Phase	Composition, wt % Ag	Pearson symbol	Space group
(Si)	0	cF8	Fd3̄m
SiII(HP)	0	tI4	I4₁/amd
(Ag)	100	cF4	Fm3̄m
Metastable phases			
SiAg₂	~90	(a)	...
β	92 to 99	(b)	...

(a) Orthorhombic. (b) cph

Ag-Sm

K.A. Gschneidner, Jr. and F.W. Calderwood, 1985

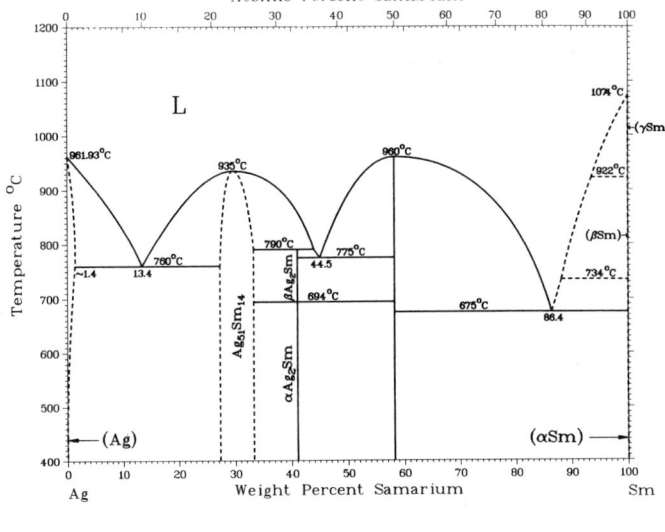

Phase	Composition, wt % Sm	Pearson symbol	Space group
(Ag)	0 to ~1.4	cF4	Fm3̄m
Ag₅₁Sm₁₄	~27.6 to 32.3	hP65	P6/m
βAg₂Sm	41.0	hP?	P6₃(?)
αAg₂Sm	41.0	...	...
AgSm	58.2	cP2	Pm3̄m
(γSm)	100	...	...
(βSm)	100	hP2	P6₃/mmc
(αSm)	100	hR3	R3̄m

Ag-Sn

I. Karakaya and W.T. Thompson, 1987

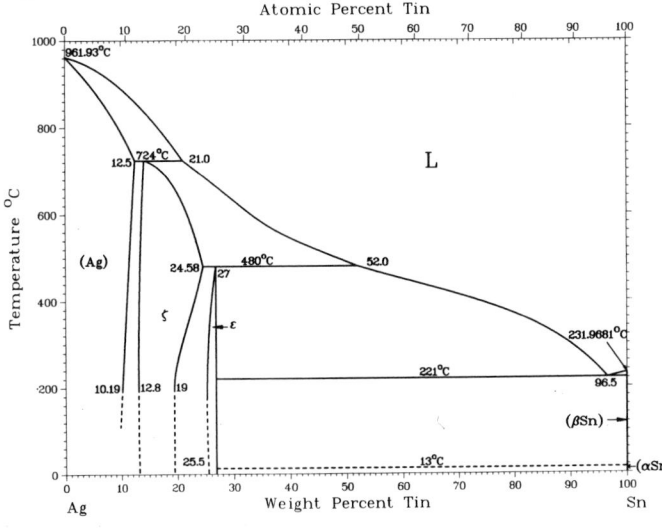

Phase	Composition, wt % Sn	Pearson symbol	Space group
(Ag)	0 to 12.5	cF4	Fm3̄m
ζ	12.8 to 24.58	hP2	P6₃/mmc
ε	25.5 to 27	oP8	Pmmn
(βSn)	99.92 to 100	tI4	I4₁/amd
(αSn)	100	cF8	Fd3̄m

Ag-Sr

M.R. Baren, 1990

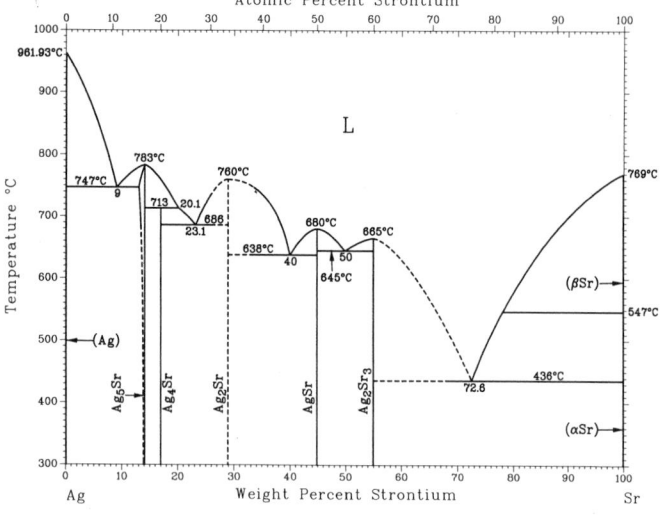

Phase	Composition, wt% Sr	Pearson symbol	Space group
(Ag)	0	cF4	Fm$\bar{3}$m
Ag$_5$Sr	12.9 to 14.1	hP6	P6/mmm
Ag$_4$Sr	17	…	…
Ag$_2$Sr	28.9	…	Imma
AgSr	44.8	oP8	Pnma
Ag$_2$Sr$_3$	55	hR45	R$\bar{3}$
Ag$_3$Sr$_7$(a)	65	hP20	P6$_3$mc
(αSr)	100	cF4	Fm$\bar{3}$m
(βSr)(b)	100	cI2	Im$\bar{3}$m

(a) Not shown on diagram; probably a peritectic reaction. (b) Above 547 °C

Ag-Te

I. Karakaya and W.T. Thompson, 1991

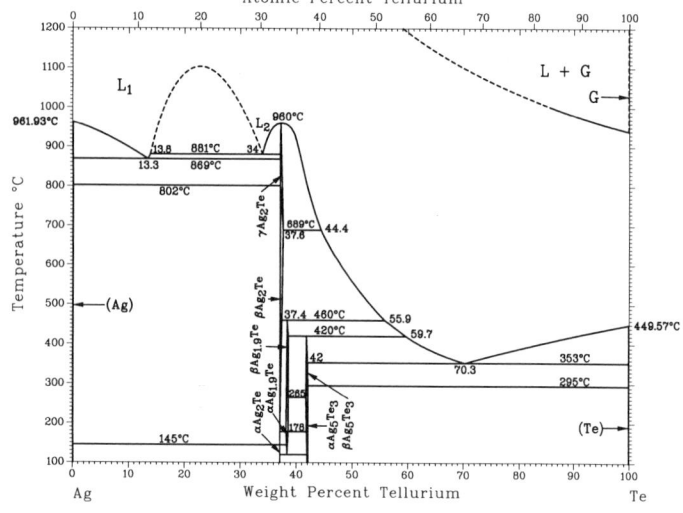

Phase	Composition(a), wt% Te	Pearson symbol	Space group
(Ag)	0	cF4	Fm$\bar{3}$m
αAg$_2$Te	37.1	mP12	P21/c
βAg$_2$Te(b)	37.1 to 37.6	cF12	…
γAg$_2$Te(c)	37.1 to 37.6	…	…
αAg$_{1.9}$Te	38.23 to 38.6	…	…
αAg$_5$Te$_3$(d)	41.67 to 42.06	hP55	P6/mmm
AgTe(e)	…	oP32	…
Ag$_2$TeII(f)	…	…	…
Ag$_2$TeIII(g)	…	…	…
AgTe$_4$-AgTe$_{2.33}$(h)	…	…	…
AgTe$_3$(j)	…	hR12	R$\bar{3}$m
(Te)	100	hP3	P3$_1$21

(a) Compositions are taken from the assessed diagram. (b) fcc structure. (c) bcc structure. (d) Referred to as Ag$_7$Te$_4$ by [Pearson2]. (e) Mineral empressite (regarded as metastable). (f) Tetragonal structure stable at pressures 2200 to 2500 kPa. Lattice parameters were measured at 2400 kPa pressure. (g) Tetragonal structure stable at pressures over 2500 kPa. Lattice parameters were measured at 4000 kPa pressure. (h) Simple cubic structure (metastable). (j) Stable at temperatures higher than 358 °C and pressures over 4.0 GPa

Ag-Ti

J.L. Murray and K.J. Bhansali, 1987

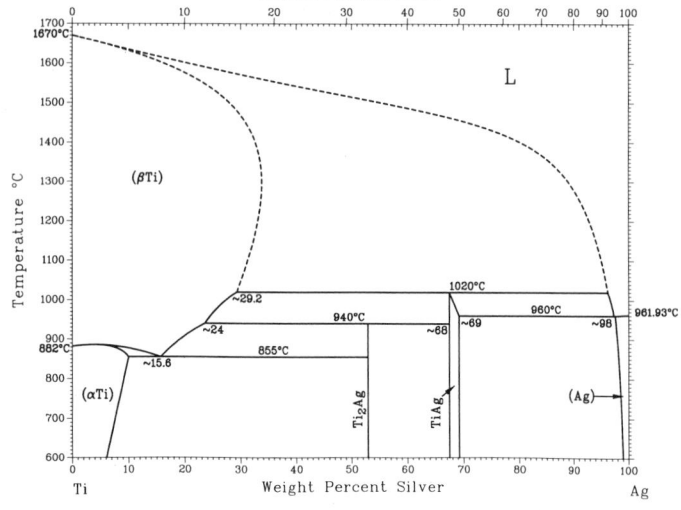

Phase	Composition, wt% Ag	Pearson symbol	Space group
(αTi)	0 to ~1.0	hP2	P6$_3$/mmc
(βTi)	0 to 29.2	cI2	Im$\bar{3}$m
Ti$_2$Ag	52.9	tI6	I4/mmm
TiAg	~68 to ~69	tP4	P4/nmn
(Ag)	~98 to 100	cF4	Fm$\bar{3}$m

Ag-Tl

M.R. Baren, 1989

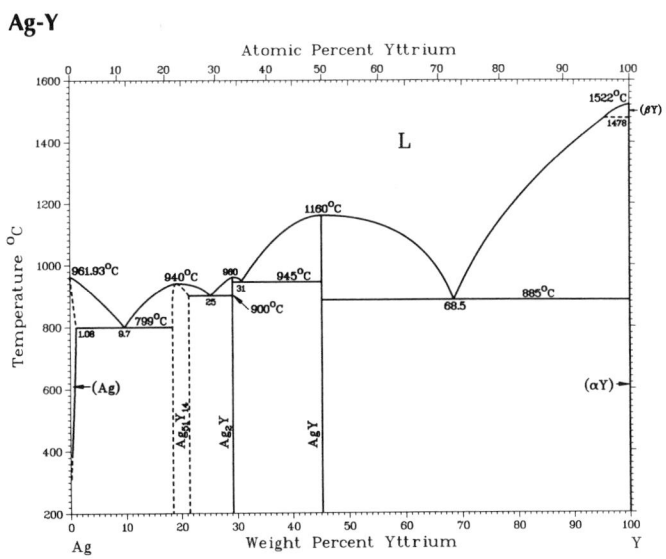

Phase	Composition, wt% Tl	Pearson symbol	Space group
(Ag)	0 to ~13.8	cF4	$Fm\overline{3}m$
(αTl)	100	hP2	$P6_3/mmc$
(βTl)(a)	? to 100	cI2	$Im\overline{3}m$

(a) Above 230 °C

Ag-Y

K.A. Gschneidner, Jr. and F.W. Calderwood, 1983

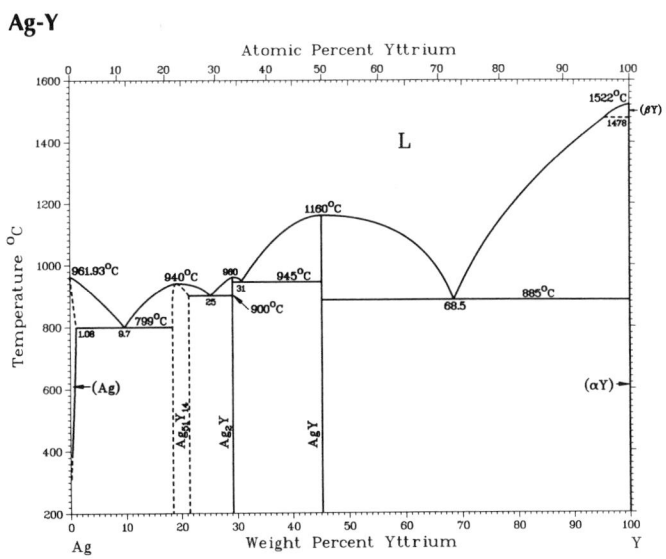

Phase	Composition, wt% Y	Pearson symbol	Space group
(Ag)	0 to 1.08	cF4	$Fm\overline{3}m$
Ag₅₁Y₁₄	18.4	hP65	...
Ag₂Y	29.2	tI6	$I4/mmm$
AgY	45.1	cP2	$Pm\overline{3}m$
(βY)	100	cI2	$Im\overline{3}m$
(αY)	100	hP2	$P6_3/mmc$

Ag-Yb

K.A. Gschneidner, Jr. and F.W. Calderwood, 1985

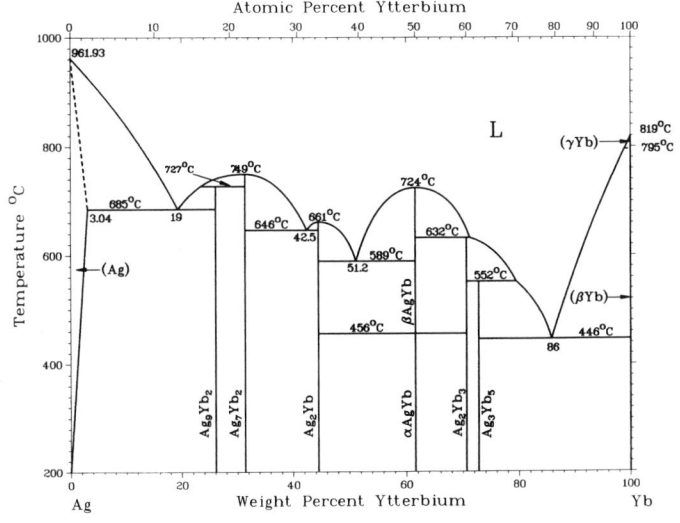

Phase	Composition, wt% Yb	Pearson symbol	Space group
(Ag)	0 to 3.04	cF4	$Fm\overline{3}m$
Ag₉Yb₂	...	...	...
Ag₇Yb₂	31.4	hP18	...
Ag₂Yb	44.5	oI2	$Imma$
βAgYb	61.6	cP2	$Pm\overline{3}m$
αAgYb	61.6	oP8	$Pnma$
Ag₂Yb₃	70.6	tP10	$P4/mbm$
Ag₃Yb₅	72.8	tI32	$I4/mcm$
(γYb)	100	cI2	$Im\overline{3}m$
(βYb)	100	cF4	$Fm\overline{3}m$
(αYb)	100	hP2	$P6_3/mmc$

Ag-Zn

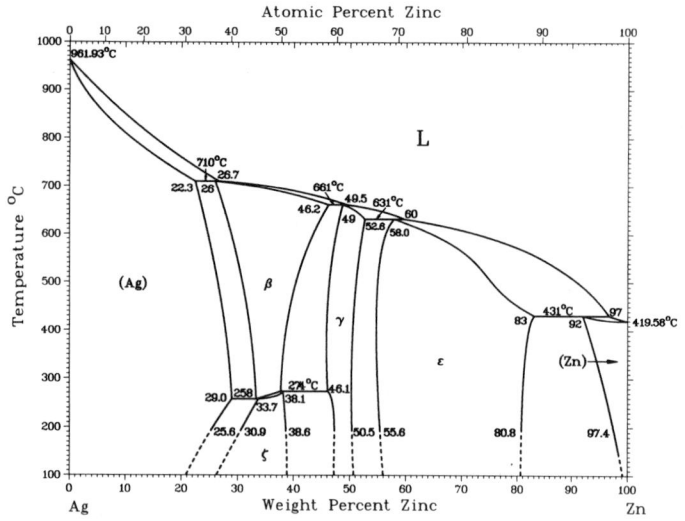

K.W. Andrews, H.E. Davies, W. Hume-Rothery, and
C.R. Oswin, 1940

Phase	Composition, wt% Zn	Pearson symbol	Space group
(Ag)	0 to 29.0	cF4	$Fm\overline{3}m$
ζ(AgZn)	26 to ~38.8	(a)	…
β(AgZn)	26 to 46.2	cI2	$Im\overline{3}m$
γ(Ag_5Zn_8)	46.1 to 52.6	cI52	$I\overline{3}m$
ε	~54.3 to 83	hP2	$P6_3/mmc$
(Zn)	92 to 100	hP2	$P6_3/mmc$

(a) Ordered hexagonal

Ag-Zr

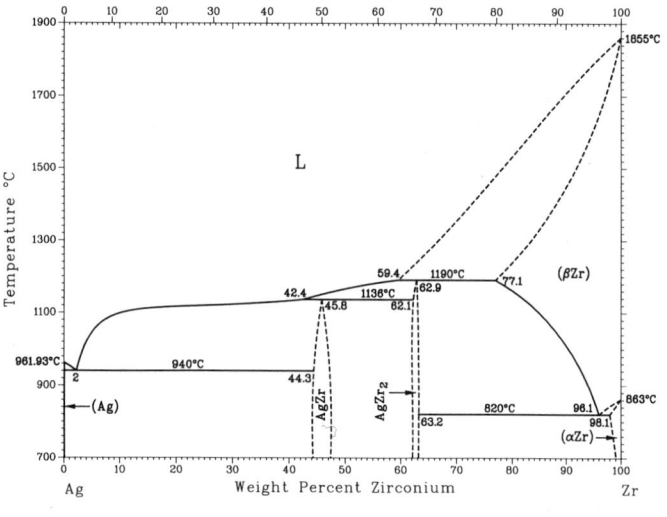

I. Karakaya and W.T. Thompson, 1992

Phase	Composition, wt% Zr	Pearson symbol	Space group
(Ag)	0 to 0.08	cF4	$Fm\overline{3}m$
AgZr	~45.8	tP4	$P4/nmm$
$AgZr_2$	~62.9	tI6	$I4/mmm$
(αZr)	98.1 to 100	hP2	$P6_3/mmc$
(βZr)	100	cI2	$Im\overline{3}m$

Al-As

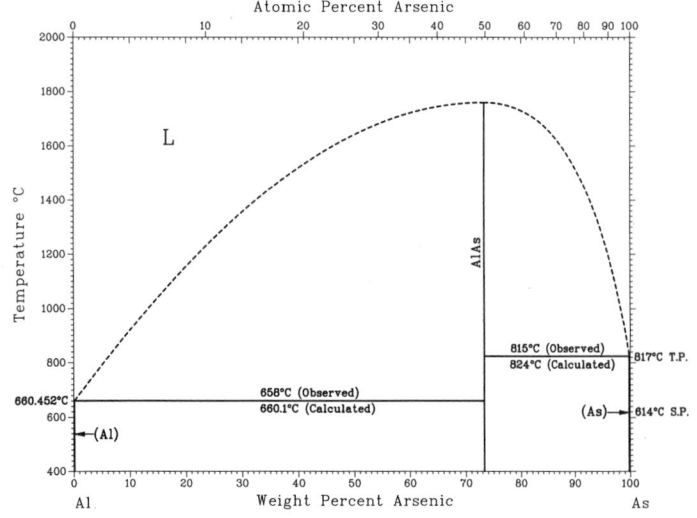

A.J. McAlister, 1984

Phase	Composition, wt% As	Pearson symbol	Space group
(Al)	0	cF4	$Fm\overline{3}m$
AlAs	73.5	cF8	$F\overline{4}3m$
(As)	100	hR2	$R\overline{3}m$

Al-Au

H. Okamoto, 1991

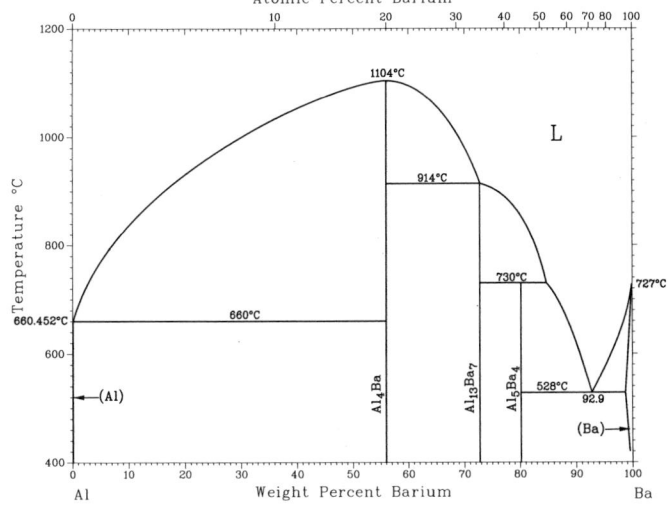

Phase	Composition, wt% Au	Pearson symbol	Space group
(Al)	0 to 0.44	cF4	$Fm\bar{3}m$
Al$_2$Au	78 to 79	cF12	$Fm\bar{3}m$
AlAu	88	mP28	$P2_1/m$
γAlAu$_2$	93 to 93.6	tI6	$I4/mmm$
βAlAu$_2$	93.2 to 93.4	oP32	Pnmn
αAlAu$_2$	93.5 to 93.6	oP12	Pnma
Al$_3$Au$_8$	95.1	hR132	$R\bar{3}c$
β	96.7 to 96.9	cI2	$Im\bar{3}m$
AlAu$_4$	96.7	cP20	$P2_13$
(Au)	98 to 100	cF4	$Fm\bar{3}m$

Al-Ba

H. Okamoto, 1992

Phase	Composition, wt% Ba	Pearson symbol	Space group
(Al)	0	cF4	$Fm\bar{3}m$
Al$_4$Ba	56	tI10	$I4/mmm$
Al$_{13}$Ba$_7$	73	hP20	$P\bar{3}m1$
Al$_5$Ba$_4$	80.3	hP18	$P6_3/mmc$
(Ba)	100	cI2	$Im\bar{3}m$
Other phases			
Al$_5$Ba$_3$	75.3	hP16	$P6_3/mmc$

Al-Be (calculated)

J.L. Murray and D.J. Kahan, 1988

Phase	Composition, wt% Be	Pearson symbol	Space group
(Al)	0 to 0.10	cF4	$Fm\bar{3}m$
(βBe)	99.979 to 100	cI2	$Im\bar{3}m$
(αBe)	99.979 to 100	hP2	$P6_3/mmc$

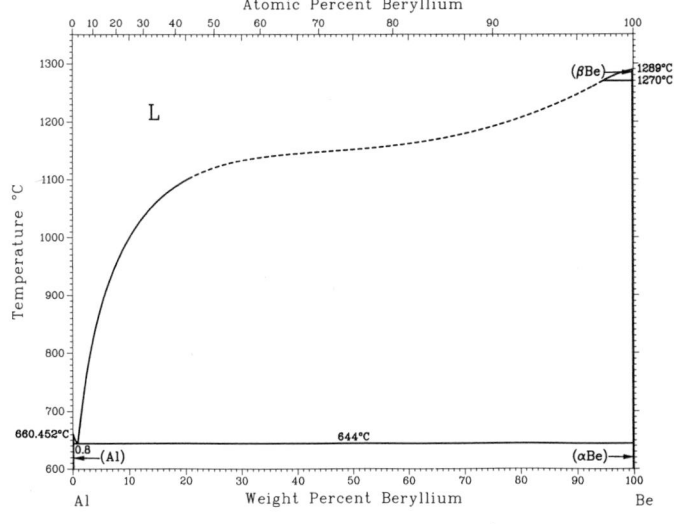

Al-Bi

A.J. McAllister, 1984

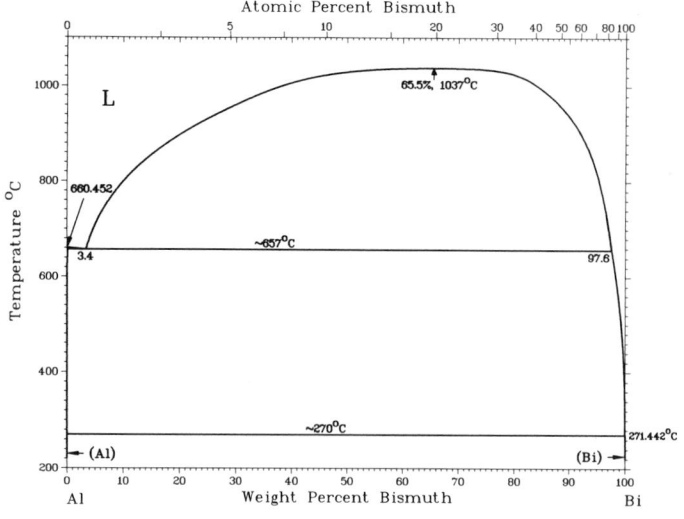

Phase	Composition, wt% Bi	Pearson symbol	Space group
(Al)	0 to ~0.23	cF4	Fm$\overline{3}$m
(Bi)	100	hR2	R$\overline{3}$m

Al-Ca

V.P. Itkin, C.B. Alcock, P.J. van Ekeren, and H.A.J. Oonk, 1988

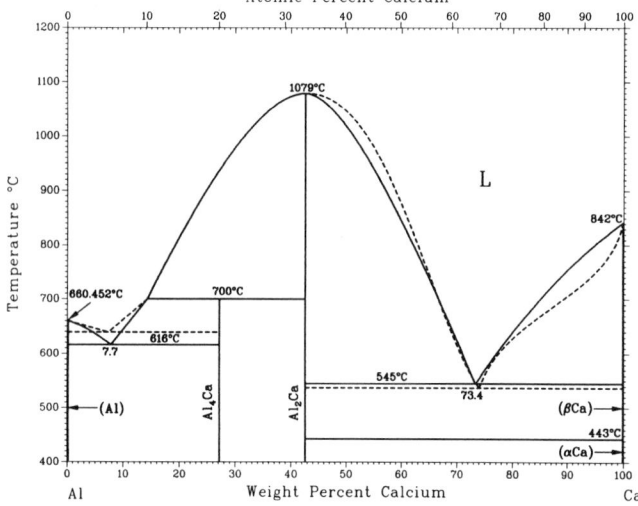

Dashed lines = calculated.

Phase	Composition, wt% Ca	Pearson symbol	Space group
(Al)	0	cF4	Fm$\overline{3}$m
Al$_4$Ca	27	tI10	I4/mmm
Al$_2$Ca	42.6	cF24	Fd$\overline{3}$m
(αCa)	100	cF4	Fm$\overline{3}$m
(βCa)	100	cI2	Im$\overline{3}$m

Al-Cd

A.J. McAlister, 1982

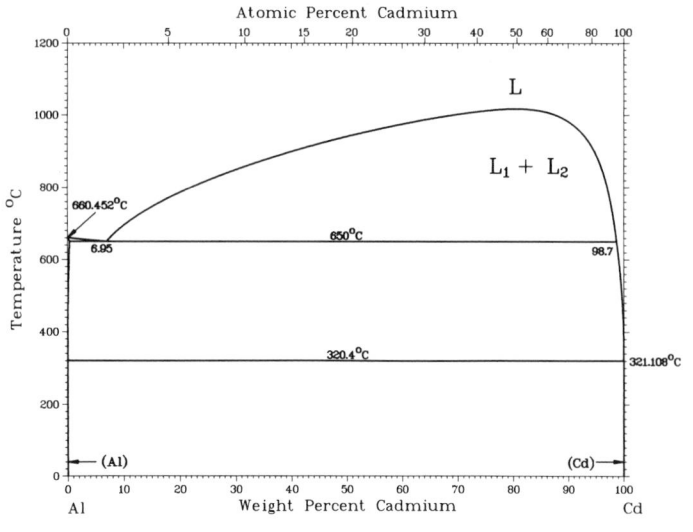

Phase	Composition, wt% Cd	Pearson symbol	Space group
(Al)	0	cF4	Fm$\overline{3}$m
(Cd)	100	hP2	P6$_3$/mmc

Al-Ce

K.A. Gschneidner, Jr. and F.W. Calderwood, 1988

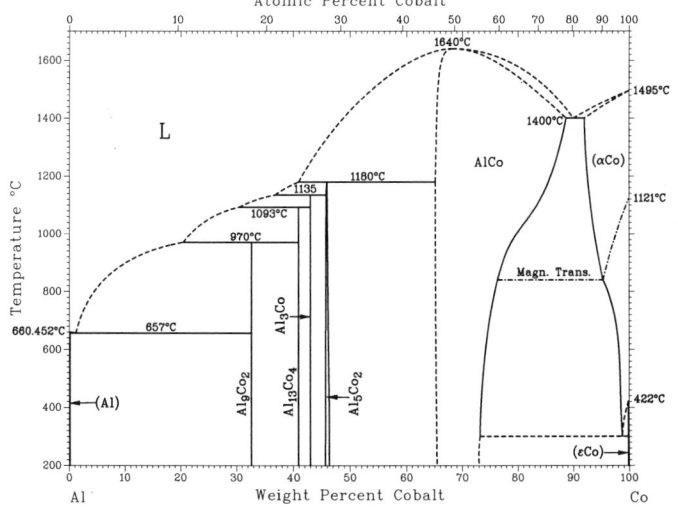

Phase	Composition, wt% Ce	Pearson symbol	Space group
(Al)	0	cF4	$Fm\overline{3}m$
$\alpha Al_{11}Ce_3$	58.6	oI28	Immm
$\beta Al_{11}Ce_3$	58.6	tI10	I_4/mmm
Al_3Ce	63	hP8	$P6_3/mmc$
Al_2Ce	72.2	cF24	$Fd\overline{3}m$
AlCe	83.9	oC16	Cmc2 or Cmcm
$\alpha AlCe_3$	94	hP8	$P6_3/mmc$
$\beta AlCe_3$	94	cP4	$Pm\overline{3}m$
(αCe)	100	cF4	$Fm\overline{3}m$
(βCe)	100	hP4	$P6_3/mmc$
(γCe)	100	cF4	$Fm\overline{3}m$
(δCe)	100	cI2	$Im\overline{3}m$

Al-Co

A.J. McAlister, 1989

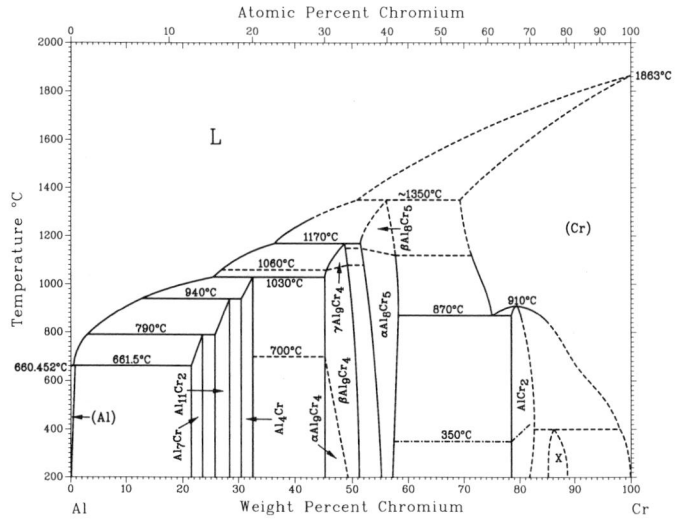

Phase	Composition, wt% Co	Pearson symbol	Space group
(Al)	~0	cF4	$Fm\overline{3}m$
Al_9Co_2	32.6	mP22	$P2_1/a$
$Al_{13}Co_4$	40.2	mC93	Cm
Al_3Co	42.9	...	(a)
Al_5Co_2	46.7	hP28	$P6_3/mmc$
AlCo	~67 to 88.9	cP2	$Pm\overline{3}m$
(ϵCo)	92 to 100	hP2	$P6_3/mmc$
(αCo)	~97 to 100	cF4	$Fm\overline{3}m$
Metastable phases			
α_{II}^{I}	95 to 98	...	(b)
α_{III}	93 to 94	...	(b)
α_{IV}	92 to 93	...	(b)
α	93 to 94	...	(b)

(a) Unknown. (b) Hexagonal

Al-Cr

J.L. Murray, unpublished

Phase	Composition, wt% Cr	Pearson symbol	Space group
(Al)	0 to 0.71	cF4	$Fm\overline{3}m$
Al_7Cr ($Al_{13}Cr_2$)	~21.4 to ~23.4	mC104	C2/m
$Al_{11}Cr_2$ (Al_5Cr)	~25.7 to ~28	mP48	P2
Al_4Cr	~30.4 to ~33	mP180	P2/m
αAl_9Cr_4	~45 to ~49.3	cI52	$I\overline{4}3m$
αAl_8Cr_5	~51.5 to ~58	hR26	$R\overline{3}m$
$AlCr_2$	~78.5 to ~82.8	tI6	I4/mmm
X(a)	~85	...	...
(Cr)	69 to 100	cI2	$Im\overline{3}m$

(a) It has been proposed that the structure is analogous to the ω phase seen in, for example, Zr at high pressure, but based on ordered bcc $AlCr_2$ rather than on the disordered bcc structure.

Al-Cu

J.L. Murray, 1985

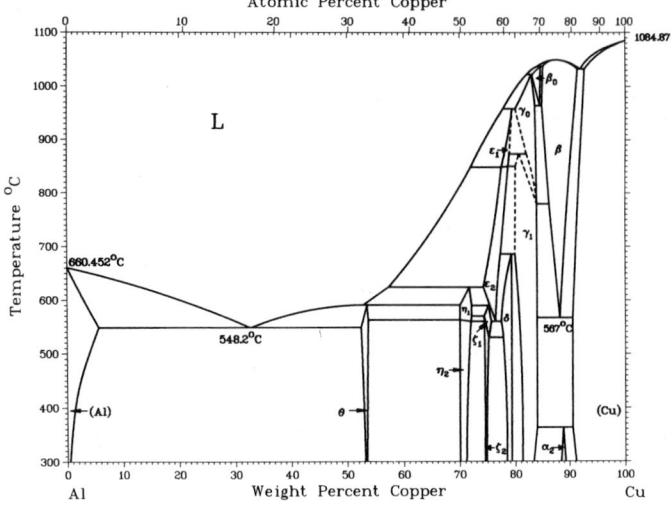

Phase	Composition, wt% Cu	Pearson symbol	Space group
(Al)	0 to 5.65	cF4	Fm$\bar{3}$m
θ	52.5 to 53.7	tI12	I4/mcm
η$_1$	70.0 to 72.2	oP16 or oC16	Pban or Cmmm
η$_2$	70.0 to 72.1	mC20	C2/m
ζ$_1$	74.4 to 77.8	hP42	P6/mmm
ζ$_2$	74.4 to 75.2	(a)	…
ε$_1$	77.5 to 79.4	(b)	…
ε$_2$	72.2 to 78.7	hP4	P6$_3$/mmc
δ	77.4 to 78.3	(c)	R$\bar{3}$m
γ$_0$	77.8 to 84	(d)	…
γ$_1$	79.7 to 84	cP52	P$\bar{4}$3m
β$_0$	83.1 to 84.7	(d)	…
β	85.0 to 91.5	cI2	Im$\bar{3}$m
α$_2$	88.5 to 89	(e)	…
(Cu)	90.6 to 100	cF4	Fm$\bar{3}$m
Metastable phases			
θ′	…	tP6	…
β′	…	cF16	Fm$\bar{3}$m
Al$_3$Cu$_2$	61 to 70	hP5	P$\bar{3}$m1

(a) Monoclinic? (b) Cubic? (c) Rhombohedral. (d) Unknown. (e) DO$_{22}$-type long-period superlattice

Al-Er

K.A. Gschneidner, Jr. and F.W. Calderwood, 1988

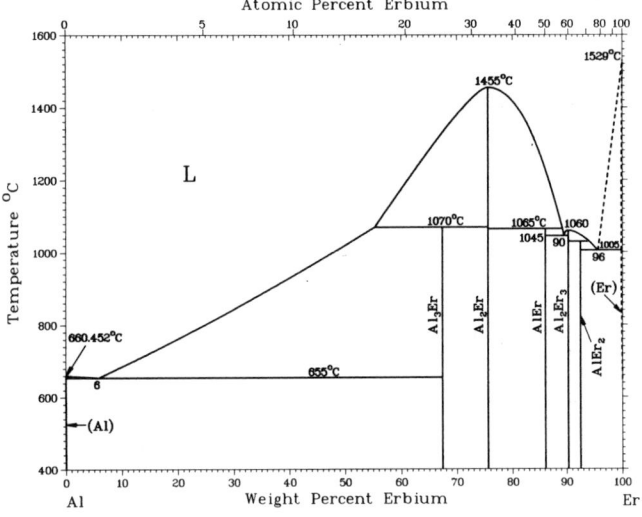

Phase	Composition, wt% Er	Pearson symbol	Space group
(Al)	0	cF4	Fm$\bar{3}$m
Al$_3$Er	67	cP4	Pm$\bar{3}$m
Al$_2$Er	75.6	cF24	Fd$\bar{3}$m
AlEr	86.1	oP16	Pmma
Al$_2$Er$_3$	90	tP20	P4$_2$/mnm
AlEr$_2$	92.6	oP12	Pnma
(Er)	100	hP2	P6$_3$/mmc

Al-Fe

U.R. Kattner and B.P. Burton, 1992

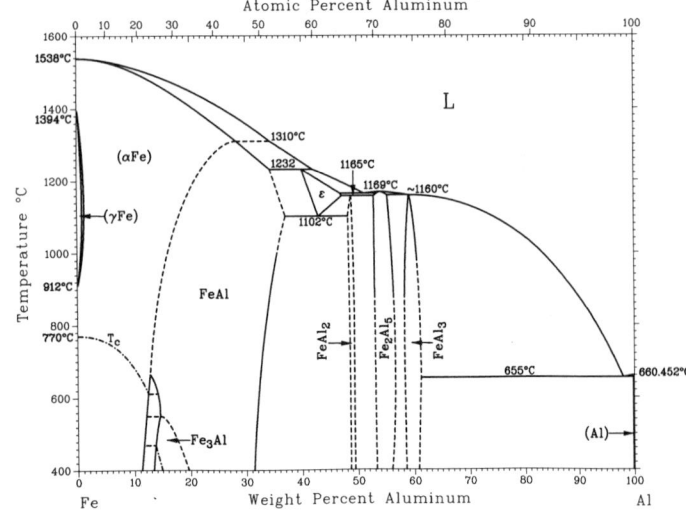

Phase	Composition, wt% Al	Pearson symbol	Space group
(αFe)	0 to ~28	cI2	Im$\bar{3}$m
(γFe)	0 to 0.6	cF4	Fm$\bar{3}$m
FeAl	12.8 to ~37	cP8	Pm$\bar{3}$m
Fe$_3$Al	~13 to ~20	cF16	Fm$\bar{3}$m
ε	~40 to ~47	cI16?	…
FeAl$_2$	48 to 49.4	aP18	P1
Fe$_2$Al$_5$	53 to 57	oC?	Cmcm
FeAl$_3$	58.5 to 61.3	mC102	C2/m
(Al)	100	cF4	Fm$\bar{3}$m
Metastable phases			
Fe$_2$Al$_9$	68.5	mP22	P2$_1$/c
FeAl$_6$	74.3	oC28	Cmc2$_1$

Al-Ga

J.L. Murray, 1983

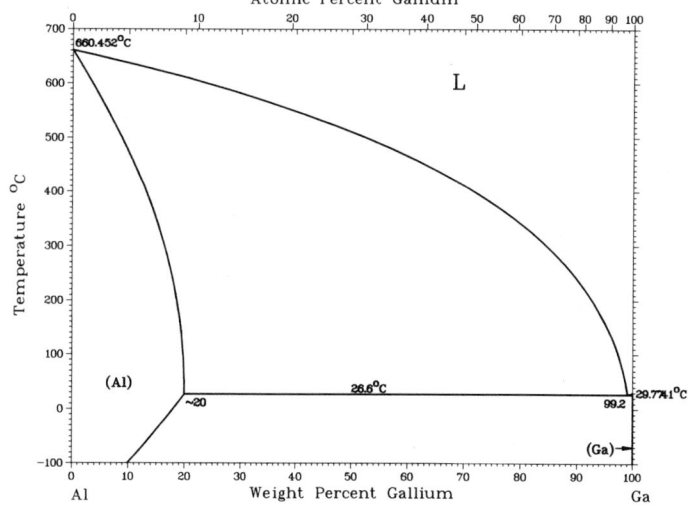

Phase	Composition, wt% Ga	Pearson symbol	Space group
(Al)	0 to ~20(a)	cF4	Fm$\bar{3}$m
(Ga)	100	oC8	Cmca
Metastable phases			
α′	83 to 92.4	tI2	I4/mmm
φ	94 to 95	(b)	(b)

(a) Can be extended to 83 wt% Ga by splat quenching. (b) Undetermined, low symmetry

Al-Gd

K.A. Gschneidner, Jr. and F.W. Calderwood, 1988

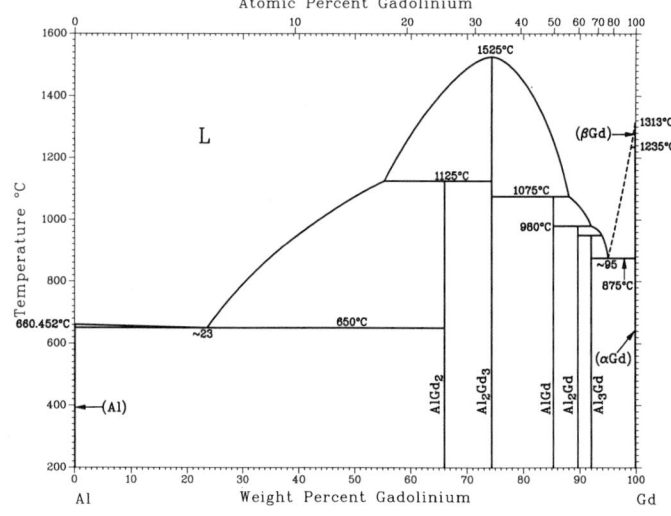

Phase	Composition, wt% Gd	Pearson symbol	Space group
(Al)	0	cF4	Fm$\bar{3}$m
Al$_3$Gd	66	hP8	P6$_3$/mmc
Al$_2$Gd	74.4	cF24	Fd$\bar{3}$m
AlGd	85.4	oP16	Pmma
Al$_2$Gd$_3$	90	tP20	P4$_2$/mnm
AlGd$_2$	92.1	oP12	Pnma
(βGd)	100	cI2	Im$\bar{3}$m
(αGd)	100	hP2	P6$_3$/mmc

Al-Ge

A.J. McAlister and J.L. Murray, 1984

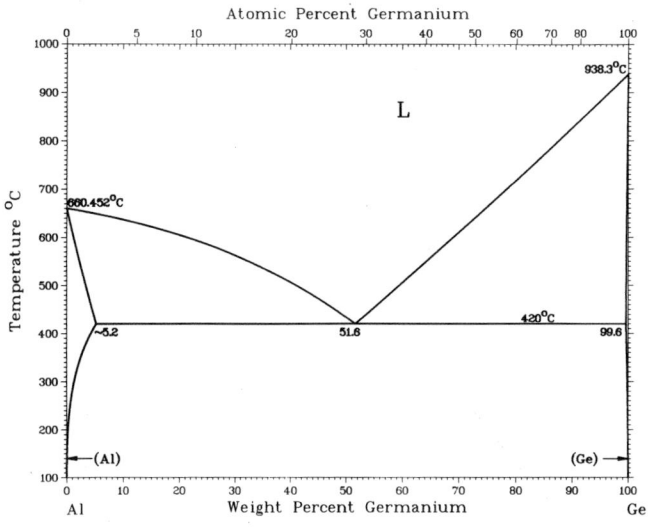

Phase	Composition, wt% Ge	Pearson symbol	Space group
(Al)	0 to ~5.2	cF4	Fm$\bar{3}$m
(Ge)	99.6 to 100	cF8	Fd$\bar{3}$m
Metastable phases			
γ$_1$	...	hR*	...
		t**	...
γ$_2$	...	cP*	...
		mC*	...
		t**	...
γ$_3$	...	cP*	...
		hP*	...

Al-H

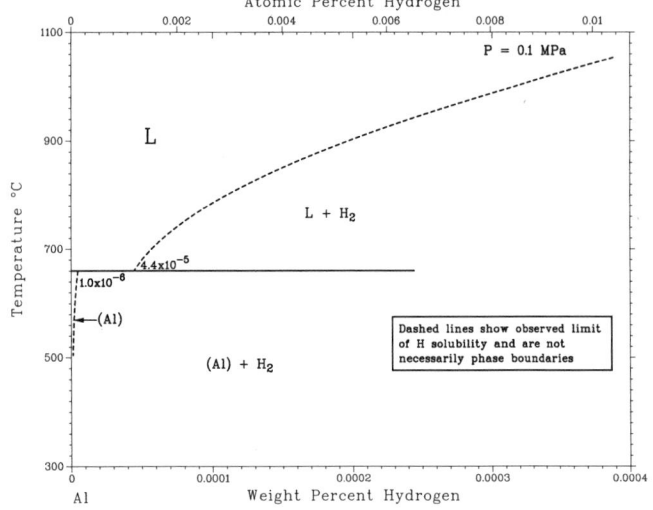

A. San-Martin and F.D. Manchester, 1992

Phase	Composition, wt% H	Pearson symbol	Space group
(Al)	0 to 4.48×10^{-6}(a)	cF4	$Fm\bar{3}m$
AlH₃(b)	10.1	...	$R\bar{3}c$

(a) At 660 °C and 0.1 MPa. (b) Produced by chemical reaction of organic solvents at atmospheric pressure

Al-Hg

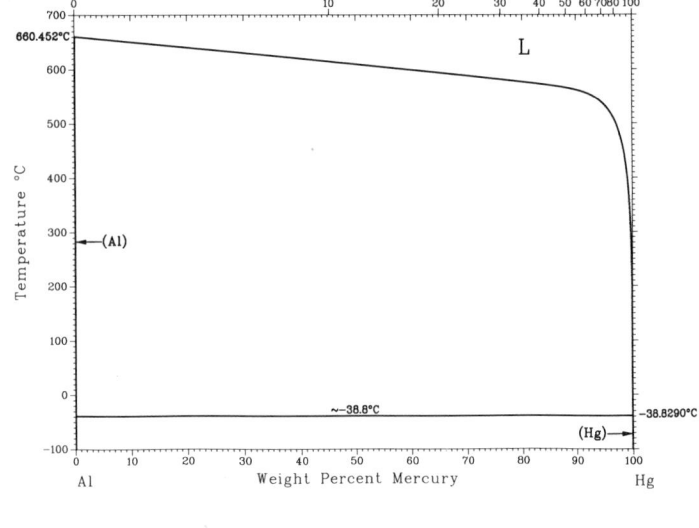

A.J. McAlister, 1985

Phase	Composition, wt% Hg	Pearson symbol	Space group
(Al)	0	cF4	$Fm\bar{3}m$
(Hg)	100	hR1	$R\bar{3}m$

Al-Ho

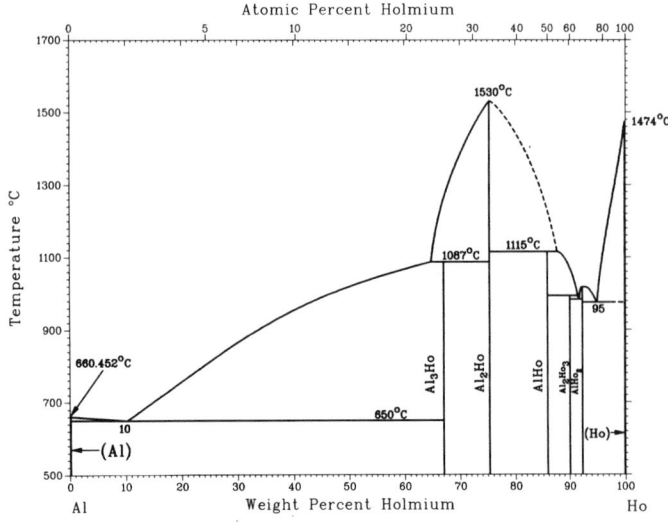

K.A. Gschneidner, Jr. and F.W. Calderwood, 1988

Phase	Composition, wt% Ho	Pearson symbol	Space group
(Al)	0	cF4	$Fm\bar{3}m$
Al₃Ho	67	hR20	$R\bar{3}m$
Al₂Ho	75.3	cF24	$Fd\bar{3}m$
AlHo	85.9	oP16	Pmma
Al₂Ho₃	90	tP20	$P4_2/mnm$
AlHo₂	92.5	oP12	Pnma
(Ho)	100	hP2	$P6_3/mmc$

Al-In

J.L. Murray, 1983

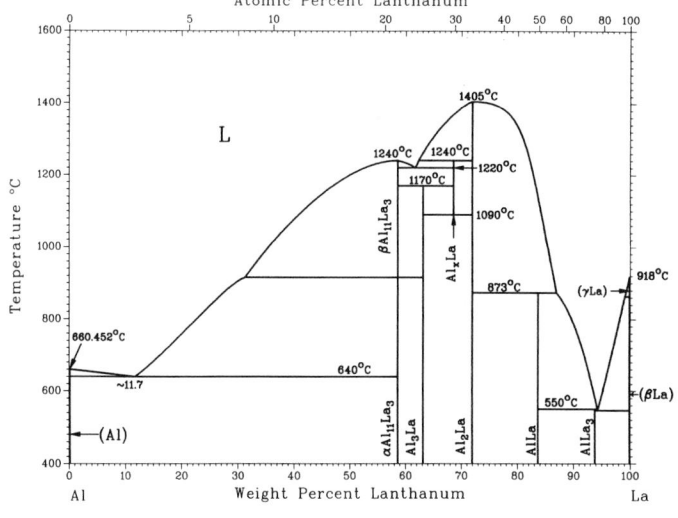

Phase	Composition, wt% In	Pearson symbol	Space group
(Al)	0 to 0.19	cF4	Fm$\bar{3}$m
(In)	~100	tI2	I4/mmm
Metastable phases			
(In′)	...	cF4	Fm$\bar{3}$m

Al-La

K.A. Gschneidner, Jr. and F.W. Calderwood, 1988

Phase	Composition, wt% La	Pearson symbol	Space group
(Al)	0 to ~0.05	cF4	Fm$\bar{3}$m
αAl$_{11}$La$_3$	58.4	oI28	Immm
βAl$_{11}$La$_3$	58.4	tI10	I4/mmm
Al$_3$La	63	hP8	P6$_3$/mmc
Al$_x$La	68.1	hP3	P6$_3$/mmm
Al$_2$La	72.0	cF24	Fd$\bar{3}$m
AlLa	83.7	oC16	Cmc2 or Cmcm
AlLa$_3$	94	hP8	P6$_3$/mmc
(αLa)	100	hP4	P6$_3$/mmc
(βLa)	100	cF4	Fm$\bar{3}$m
(γLa)	100	cI2	Im$\bar{3}$m

Al-Li

A.J. McAlister, 1991

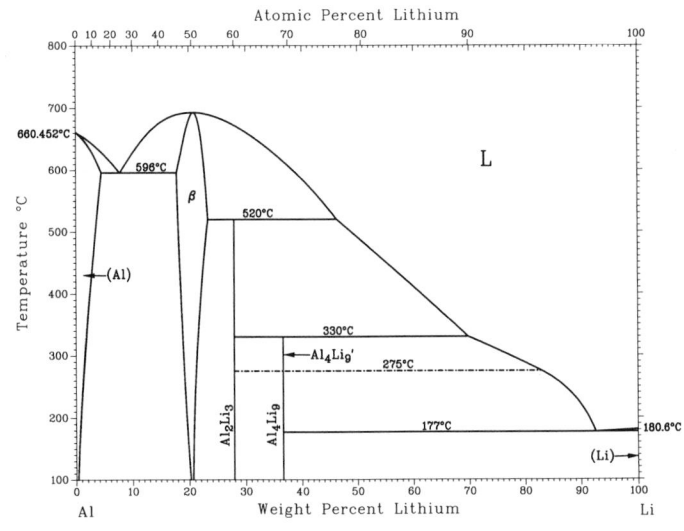

Phase	Composition, wt% Li	Pearson symbol	Space group
(Al)	0 to 4	cF4	Fm$\bar{3}$m
β	17 to 24	cF16	Fd$\bar{3}$m
Al$_2$Li$_3$	28 to 29	hR15	R$\bar{3}$m
Al$_4$Li$_9$	36.6	mC26	C2/m
Al$_4$Li$_9$′	36.6	...	...
(βLi)	100	cI2	Im$\bar{3}$m
(αLi)	100	hP2	P6$_3$/mmc
Metastable phases			
Al$_3$Li	...	cP4	Pm$\bar{3}$m

Al-Mg

J.L. Murray, 1988

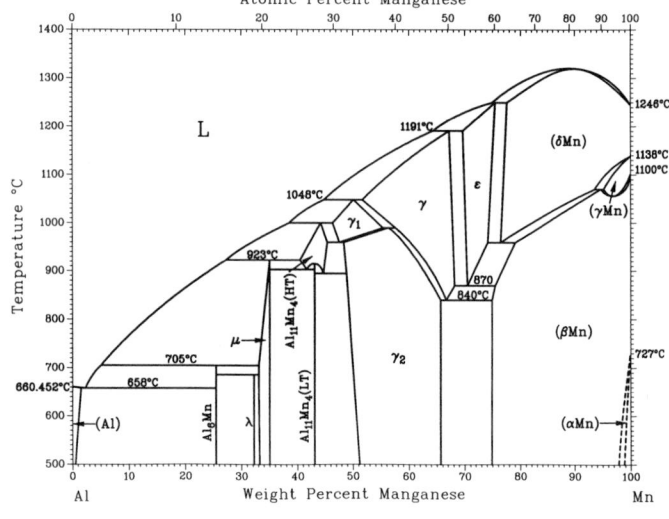

Phase	Composition, wt% Mg	Pearson symbol	Space group
(Al)	0 to 17.1	cF4	$Fm\bar{3}m$
β(Al₃Mg₂)	36.1 to 37.8	cF1168	$Fd\bar{3}m$
R	39	hR53	$R\bar{3}$
γ(Al₁₂Mg₁₇)	42 to 58.0	cI58	$I\bar{4}3m$
(Mg)	87.1 to 100	hP2	$P6_3/mmc$
Metastable phases			
Al₂Mg	31.0	tI24	$I4_1/amd$
γ′	38 to 56.2	(a)	…

(a) Tetragonal

Al-Mn

A.J. McAlister and J.L. Murray, 1987

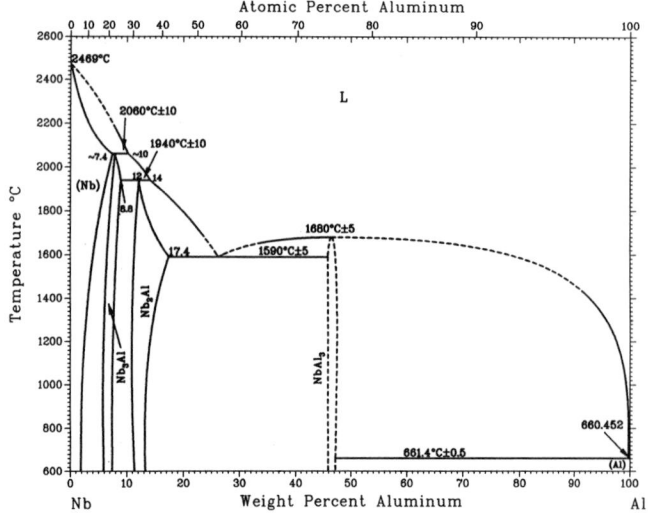

Phase	Composition, wt% Mn	Pearson symbol	Space group
(Al)	0 to 1.25	cF4	$Fm\bar{3}m$
G(a)	(b)	cI26	$Im\bar{3}m$
Al₆Mn	25.2	oC28	$Cmcm$
λ("Al₄Mn")(c)	~29.4 to ~32	(d)	…
μ	~32 to 34.8	(d)	…
Al₁₀Mn₃(φ)	(b)	hP28	$P6_3/mmc$
Al₁₁Mn₄(LT)(e)	43	aP30	$P\bar{1}$
Al₁₁Mn₄(HT)(e)	40 to 45.0	oP160	$Pnma$
γ₁	47 to 55.7	(f)	…
γ₂(g)	48.2 to 64	hR26	$R\bar{3}m$
γ	51.8 to 68.2	(f)	…
ε	69.8 to 75	hP2	$P6_3/mmc$
τ	(b)	tP2	$P4/mmm$
(δMn)	76.5 to 100	cI2	$Im\bar{3}m$
(γMn)	95.3 to 100	cF4	$Fm\bar{3}m$
(βMn)	75.0 to 100	cP20	$P4_132$
(αMn)	~99 to 100	cI58	$I\bar{4}3m$

(a) Several other structures have been ascribed to the G phase or variants of the G phase (G′, G″). (b) Metastable phase. (c) A simple orthorhombic structure was reported in an alloy described as "Al₄Mn." (d) Hexagonal. (e) Variants of this structure are described as complex stacking sequences along the b axis. (f) Unknown. (g) The structure has been described as distorted γ-brass type, cubic (bcc or fcc), and rhombohedral.

Al-Nb

U.R. Kattner, unpublished

Phase	Composition, wt% Al	Pearson symbol	Space group
(Nb)	0 to ~7.4	cI2	$Im\bar{3}m$
Nb₃Al	18.6 to 8.8	cP8	$Pm\bar{3}n$
Nb₂Al	11 to 17.4	tP30	$P4_2/mnm$
NbAl₃	47	tI8	$I4/mmm$
(Al)	100	cF4	$Fm\bar{3}m$

Al-Nd

H. Okamoto, 1991

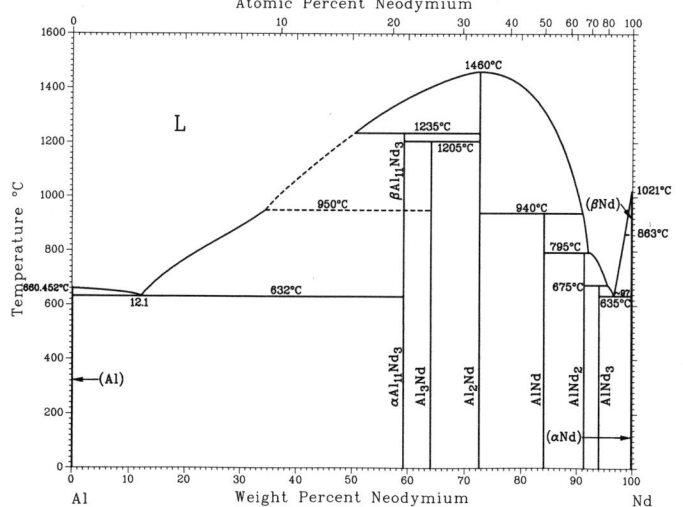

Phase	Composition, wt% Nd	Pearson symbol	Space group
(Al)	0 to 0.05	cF4	$Fm\bar{3}m$
$\alpha Al_{11}Nd_3$	59.3	oI28	Immm
$\beta Al_{11}Nd_3$	59.3	tI10	I4/mmm
Al_3Nd	64	hP8	$P6_3/mmc$
Al_2Nd	72.7	cF24	$Fd\bar{3}m$
AlNd	84.2	oP16	Pmma
$AlNd_2$	91.5	oP12	Pnma
$AlNd_3$	94	hP8	$P6_3/mmc$
(αNd)	100	hP4	$P6_3/mmc$
(βNd)	100	cI2	$Im\bar{3}m$

Al-Ni

P. Nash, M.F. Singleton, and J.L. Murray, 1991

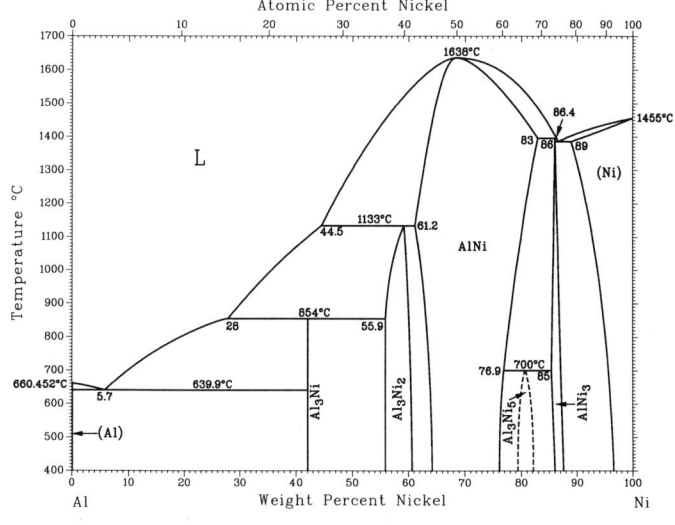

Phase	Composition, wt% Ni	Pearson symbol	Space group
(Al)	0 to 0.24	cF4	$Fm\bar{3}m$
Al_3Ni	42	oP16	Pnma
Al_3Ni_2	55.9 to 60.7	hP5	$P\bar{3}m1$
AlNi	61 to 83.0	cP2	$Pm\bar{3}m$
Al_3Ni_5	79 to ~82	...	Cmmm
$AlNi_3$	85 to 87	cP4	$Pm\bar{3}m$
(Ni)	89.0 to 100	cF4	$Fm\bar{3}m$

Al-Pb

A.J. McAlister, 1984

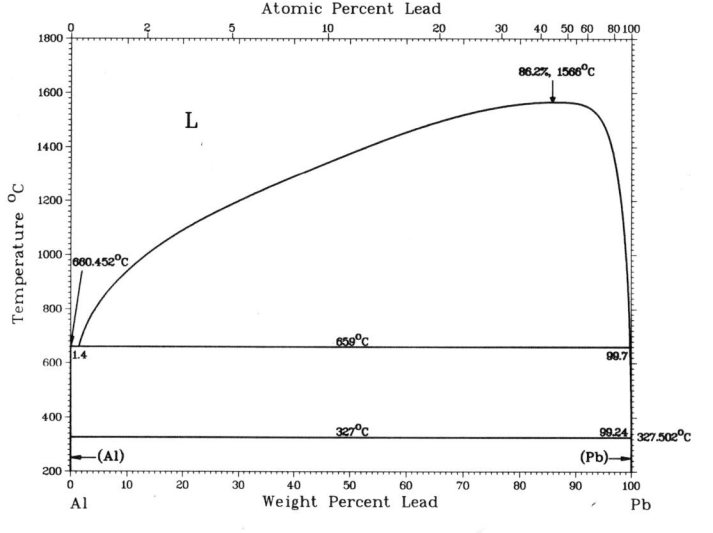

Phase	Composition, wt% Pb	Pearson symbol	Space group
(Al)	0	cF4	$Fm\bar{3}m$
(Pb)	99.7 to 100	cF4	$Fm\bar{3}m$

Al-Pd

A.J. McAlister, 1986

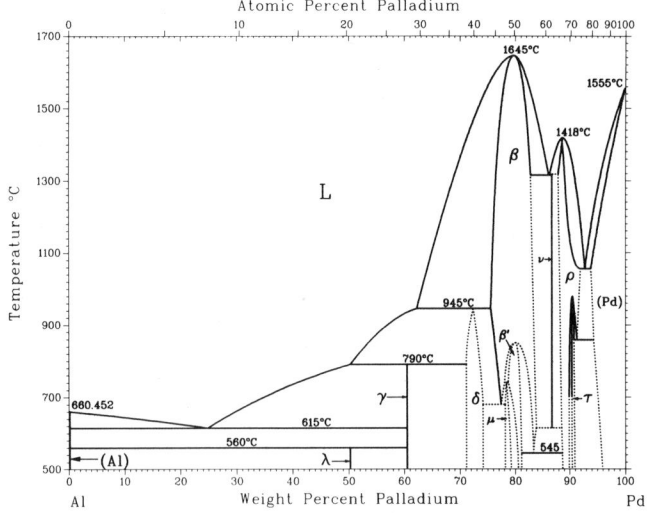

Phase	Composition, wt% Pd	Pearson symbol	Space group
(Al)	0	cF4	$Fm\bar{3}m$
λ	~50	(a)	...
γ	~60.1	(b)	...
δ	71 to 73.7	hP5	$P\bar{3}m1$
β	76 to 83	cP8	$Pm\bar{3}m$
β′	78.8 to 81.5	hR78	$R\bar{3}$
μ	78 to 79	cP8	$P2_13$
ν	86.8	oP16	Pbam
ρ	88 to 91	oP12	Pnma
τ	90.5 to 90.9	oP28	Pbmn
(Pd)	94 to 100	cF4	$Fm\bar{3}m$

(a) Hexagonal. (b) Orthorhombic

Al-Pr

K.A. Gschneidner, Jr. and F.W. Calderwood, 1989

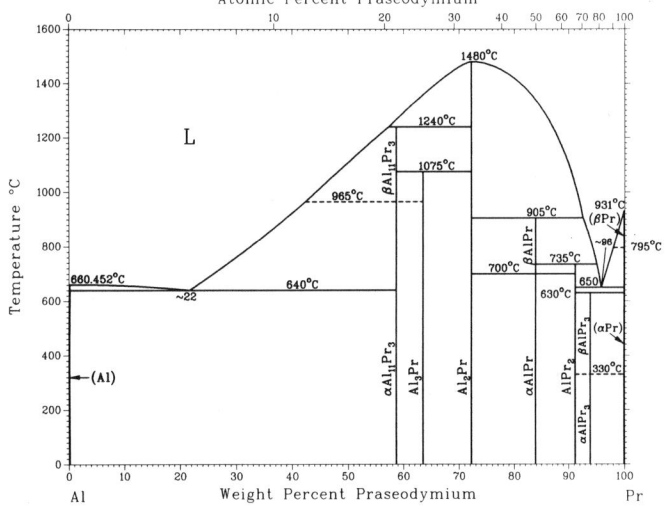

Phase	Composition, wt% Pr	Pearson symbol	Space group
(Al)	0 to ~0.5	cF4	$Fm\bar{3}m$
$\alpha Al_{11}Pr_3$	58.7	oI28	Immm
$\beta Al_{11}Pr_3$	58.7	tI10	I4/mmm
Al_3Pr	64	hP8	$P6_3/mmc$
Al_2Pr	72.3	cF24	$Fd\bar{3}m$
$\alpha AlPr$	83.9	oP16	Pmma
$\beta AlPr$	83.9	oC16	Cmc2 or Cmcm
$AlPr_2$	91.3	oP12	Pnma
$\alpha AlPr_3$	94	hP8	$P6_3/mmc$
$\beta AlPr_3$	94	cP4	$Pm\bar{3}m$
(αPr)	100	hP4	$P6_3/mmc$
(βPr)	100	cI2	$Im\bar{3}m$

Al-Pt

A.J. McAlister and D.J. Kahan, 1986

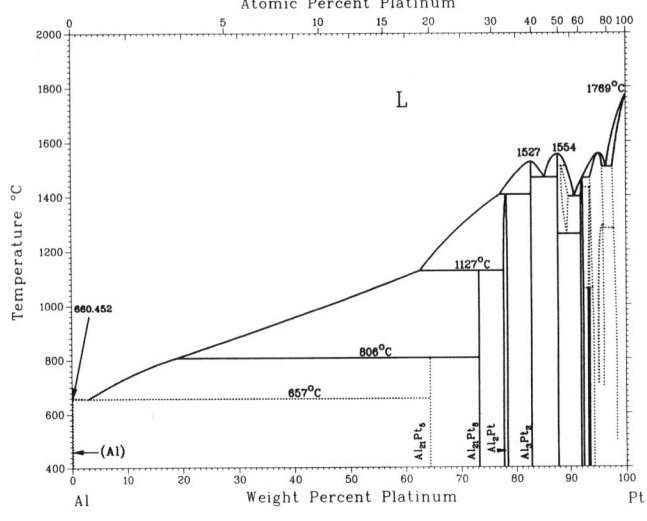

Phase	Composition, wt% Pt	Pearson symbol	Space group
(Al)	0	cF4	$Fm\bar{3}m$
$Al_{21}Pt_5$	63.2	c**	...
$Al_{21}Pt_8$	72.8	tI116	$I4_1a$
Al_2Pt	76.9 to 78.5	cF12	$Fm\bar{3}m$
Al_3Pt_2	82.8	hP5	P3m1
AlPt	87.9	cP8	$P2_13$
β	~89 to ~90	cP2	$Pm\bar{3}m$
Al_3Pt_5	~92.0 to ~92.5	oP16	Pbam
$AlPt_2$	~93 to ~94	oP12	Pnma
$AlPt_2(LT)$	~93 to ~94	oP24	Pmma
$AlPt_3$	~93.7 to ~96.18	cP4	$Pm\bar{3}m$
$AlPt_3(LT)$	~95.3 to ~96.25	tP16	P4/mbm
(Pt)	~97.4 to 100	cF4	$Fm\bar{3}m$
Metastable phases			
α′	...	cF4	$Fm\bar{3}m$
Al_4Pt	~64	hP*	...
Al_6Pt	~54	o**	...
ε′	...	c**	...
λ′	45 to 71	...	...

Al-S

R.C. Sharma and Y.A. Chang, 1991

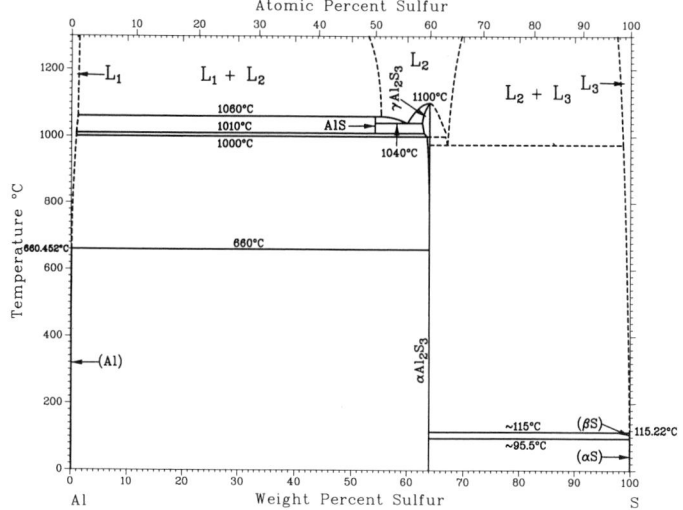

Phase	Composition, wt% S	Pearson symbol	Space group
(Al)	0	cF4	$Fm\bar{3}m$
αAl_2S_3	64	hP30	...
βAl_2S_3(a)	64	(b)	$P6_3mc$
γAl_2S_3	63 to 64	hR10	R3c
Al_2S_3(c)	64	(d)	$I4_1/amd$
Al_2S_3(e)	64	(f)	$Fd\bar{3}m$
(αS)	100	oF128	Fddd
(βS)	100	mP*	$P2_1/c$

(a) Stable in the presence of Al_4C_3 between 1000 and 1100 °C. (b) Hexagonal. (c) High pressure, formed at 2 to 65 kbar and 1000 to 1200 °C. (d) Tetragonal. (e) High pressure, formed at 40 kbar and 400 °C. (f) Cubic

Al-Sb

A.J. McAlister, 1984

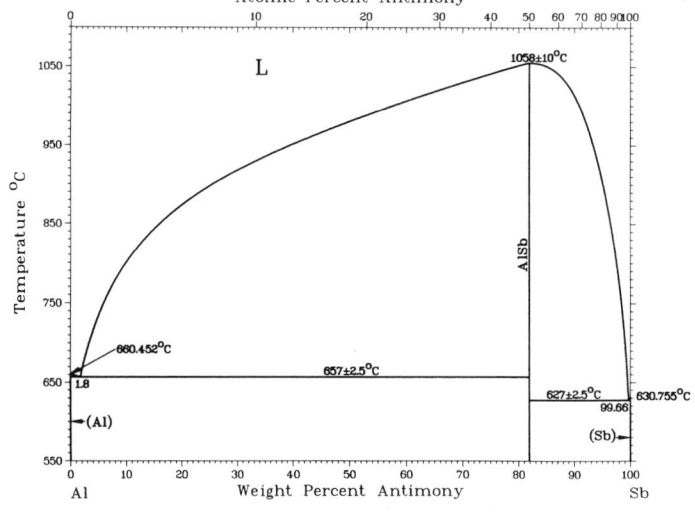

Phase	Composition, wt% Sb	Pearson symbol	Space group
(Al)	0	cF4	$Fm\bar{3}m$
AlSb	81.9	cF8	$F\bar{4}3m$
(Sb)	100	hR2	$R\bar{3}m$
High-pressure phase			
AlSb(a)	81.9	tI4	$I4_1/amd$

(a) At 120 kbar

Al-Se

J.M. Howe, 1989

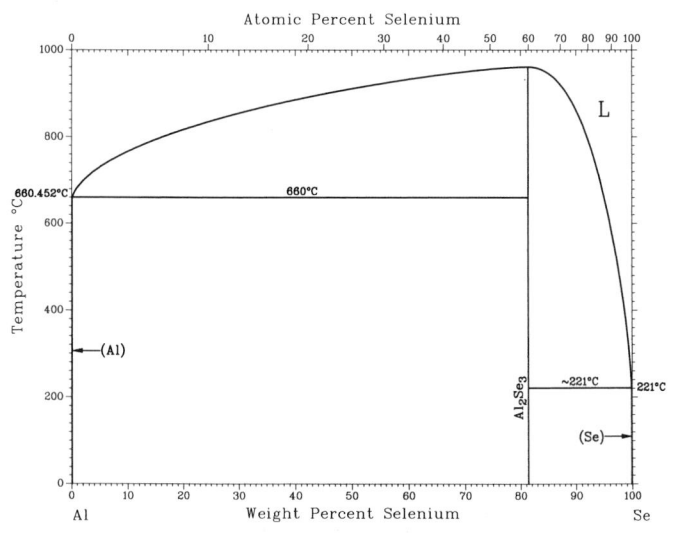

Phase	Composition, wt% Se	Pearson symbol	Space group
(Al)	<0.009	cF4	$Fm\bar{3}m$
Al_2Se_3	81	mC20	C_c
(Se)	100	hP3	$P3_121$

Al-Si

J.L. Murray and A.J. McAlister, 1984

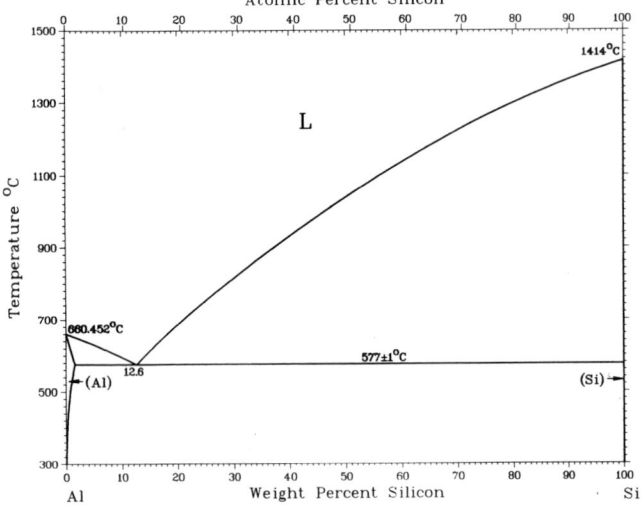

Phase	Composition, wt% Si	Pearson symbol	Space group
(Al)	0 to 1.6	cF4	Fm$\overline{3}$m
(Si)	99.985 to 100	cF8	Fd$\overline{3}$m

Al-Sn

A.J. McAlister and D.J. Kahan, 1983

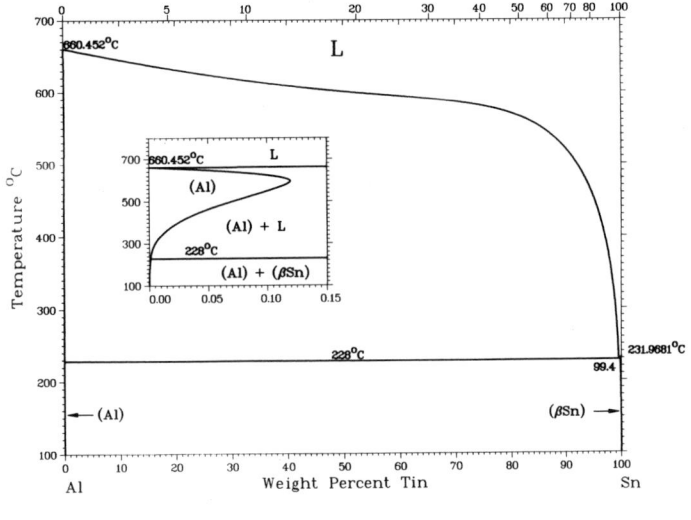

Phase	Composition, wt% Sn	Pearson symbol	Space group
(Al)	0	cF4	Fm$\overline{3}$m
(βSn)	100	tI4	I4$_1$/amd
(αSn)	100	cF8	Fd$\overline{3}$m
Metastable phase			
Γ	>81.5	hP1	P6/mmm

Al-Sr

C.B. Alcock and V.P. Itkin, 1989

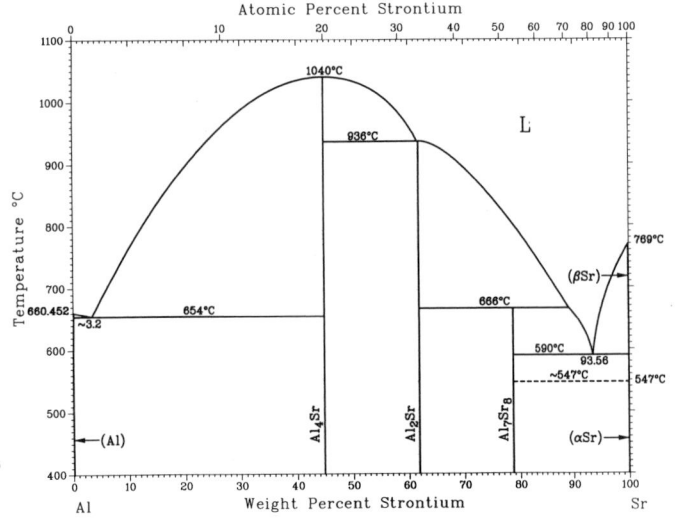

Phase	Composition, wt% Sr	Pearson symbol	Space group
(Al)	0	cF4	Fm$\overline{3}$m
Al$_4$Sr	45	tI10	I4/mmm
Al$_2$Sr	61.9	oI12	Imma
Al$_7$Sr$_8$	78.8	cP60	P2$_1$3
(βSr)	100	cI2	Im$\overline{3}$m
(αSr)	100	cF4	Fm$\overline{3}$m

Al-Ta

U.R. Kattner, unpublished

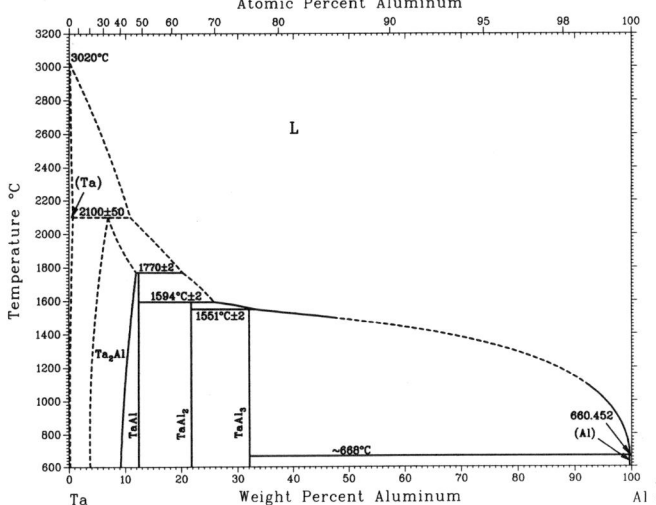

Phase	Composition, wt% Al	Pearson symbol	Space group
(Ta)	0 to 0.6	cI2	Im$\bar{3}$m
Ta₂Al	4 to 9	tP30	P4₂/mnm
TaAl	12.3	...	...
TaAl₂	22	c, h, or o	...
TaAl₃	32	tI8	I4/mmm
(Al)	100	cF4	Fm$\bar{3}$m

Note: Different unit cells are proposed for TaAl₂.

Al-Te

N. Prabhu and J.M. Howe, 1990

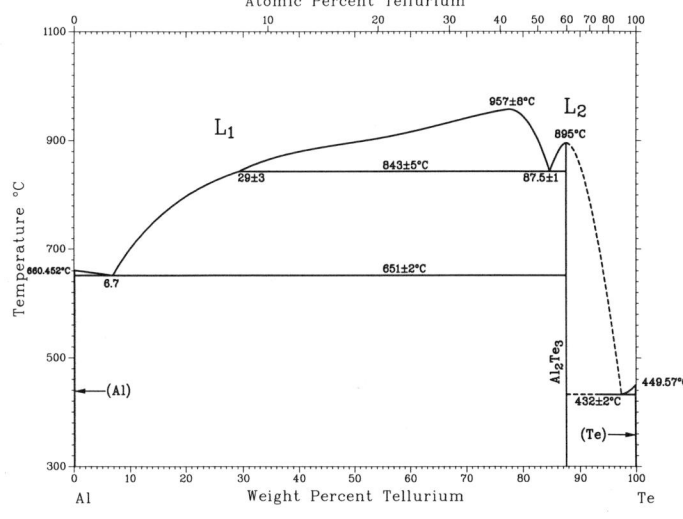

Phase	Composition, wt% Te	Pearson symbol	Space group
(Al)	0	cF4	Fm$\bar{3}$m
Al₂Te₃	88	hP4	P6₃/mc
(Te)	100	hP3	P3₁21

Al-Th

M.E. Kassner and D.E. Peterson, 1989

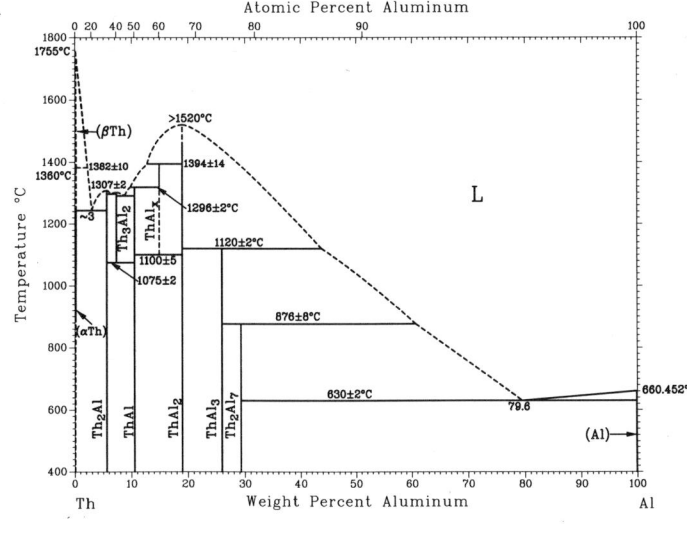

Phase	Composition, wt% Al	Pearson symbol	Space group
(αTh)	0 to 0.10	cF4	Fm$\bar{3}$m
(βTh)	0	cI2	Im$\bar{3}$m
Th₂Al	5.5	tI12	I4/mcm
Th₃Al₂	7	tP10	P4/mbm
ThAl	10.4	oC8	Cmcm
ThAlₓ	15.6 to 16.2	(a)	...
Th₂Al₃(b)	15	(a)	...
Th₄Al₇(b)	16.9	(a)	...
ThAl₂	18.9	hP3	P6/mmm
ThAl₃	26	hP8	P6₃/mmc
Th₂Al₇	29.0	oP18	Pbam
(Al)	100	cF4	Fm$\bar{3}$m

(a) Tetragonal. (b) Considered same as ThAlₓ

Al-Ti

J.L. Murray, 1987

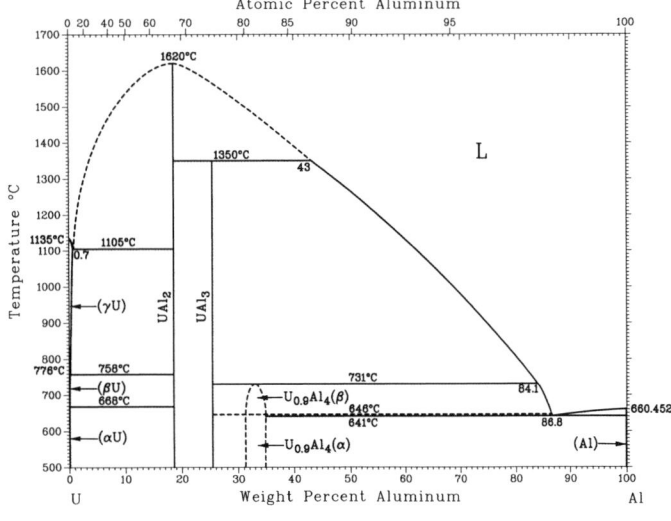

Phase	Composition, wt% Al	Pearson symbol	Space group
(βTi)	0 to 33.8	cI2	Im$\bar{3}$m
(αTi)	0 to 32	hP2	P6$_3$/mmc
Ti$_3$Al	14 to 26	hP8	P6$_3$/mmc
TiAl	34 to 56.2	tP4	P4/mmm
Ti$_3$Al$_5$(a)	44 to 49	tP32	I4/mbm
TiAl$_2$	51 to 54	tI24	I4$_1$/amd
αTiAl$_2$(b)	...	oC12	Cmmm
δ	57 to 59.8	(c)	...
TiAl$_3$	~63	tI8	I4/mmm
αTiAl$_2$	63	(d)	...
(Al)	98.8 to 100	cF4	Fm$\bar{3}$m

(a) Not an equilibrium phase. (b) Not shown on the assessed diagram. (c) Long-period superlattice structures. (d) Tetragonal; a superstructure of the D0$_{22}$ lattice

Al-U

M.E. Kassner, M.G. Adamson, P.H. Adler, and D.E. Peterson, 1990

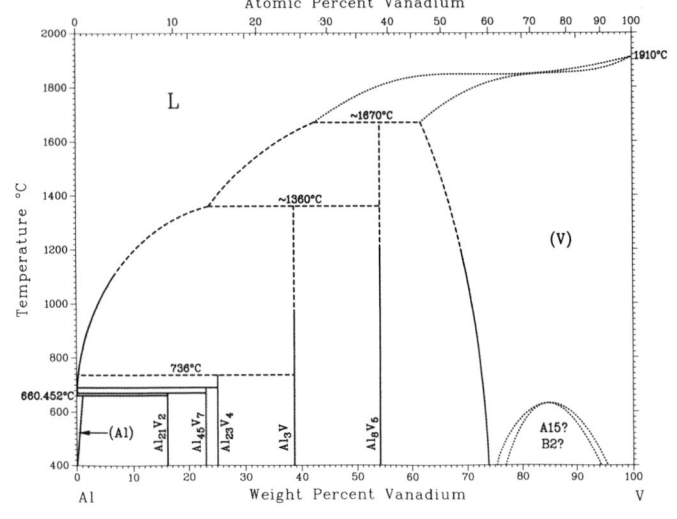

Phase	Composition, wt% Al	Pearson symbol	Space group
(γU)	0 to 0.6	cI2	Im$\bar{3}$m
(βU)	0 to 0.06	tP30	P$\bar{4}$n2
(αU)	0	oC4	Cmcm
UAl$_2$	18.5	(a)	...
	18.5	cF24	Fd$\bar{3}$m
UAl$_3$	25	cP4	Pm$\bar{3}$m
	25	(a)	...
UAl$_4$(b)	31	oI20	I2ma or Imma
U$_{0.9}$Al$_4$(α)	33.5	oI20	Imma
U$_{0.9}$Al$_4$(β)	33.5	oI20	Imma
UAl$_5$	...	(c)	...
(Al)	100	cF4	Fm$\bar{3}$m

(a) Cubic. (b) Considered same as U$_{0.9}$Al$_4$(α). (c) Unknown

Al-V

J.L. Murray, 1989

Phase	Composition, wt% V	Pearson symbol	Space group
(Al)	0 to 0.6	cF4	Fm$\bar{3}$m
Al$_{21}$V$_2$	~15.3 to 15.9	cF176	Fd$\bar{3}$m
Al$_{45}$V$_7$	~23.1	mC104	C2/m
Al$_{23}$V$_4$	~24.7	hP54	P6$_3$/mmc
Al$_3$V	~39	tI8	I4/mmm
Al$_8$V$_5$	54.2	cI52	I4$\bar{3}$m
(V)	~65 to 100	cI2	Im$\bar{3}$m
AlV$_3$	(a)	cP8	Pm$\bar{3}$m
βAlV$_3$	(a)	h**	...
αAlV$_3$	(a)	t**	...

Note: The structure of Al$_{23}$V$_4$ is related to that of Co$_2$Al$_5$ (φ). It contains nearly regular icosahedra as structural elements. (a) Unknown

Al-W

From [Metals]

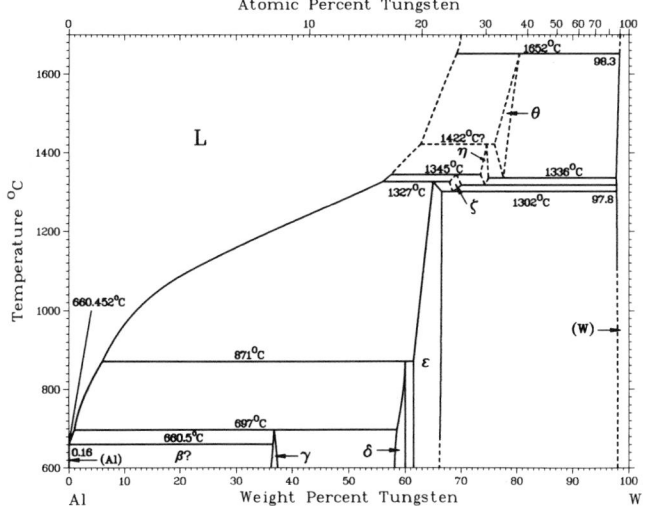

Phase	Composition, wt% W	Pearson symbol	Space group
(Al)	0	cF4	Fm$\overline{3}$m
γ	~37	cI26	Im3
δ	~58 to 60	hP12	P6₃
ε	~62 to 66	mC30	Cm
(W)	100	cI2	Im$\overline{3}$m

Al-Y

K.A. Gschneidner, Jr. and F.W. Calderwood, 1989

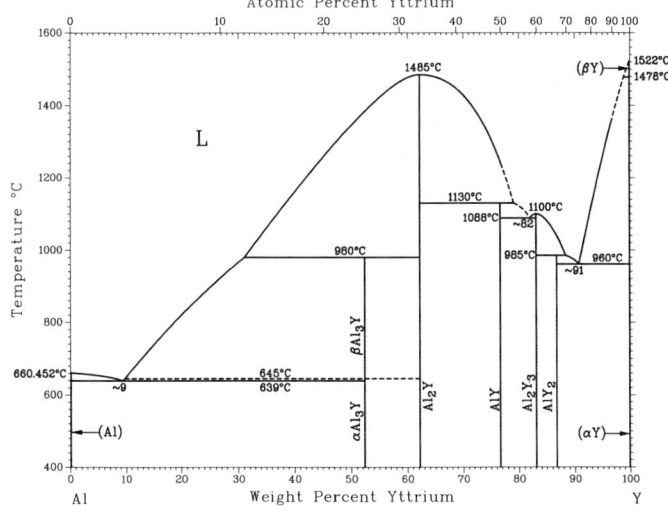

Phase	Composition, wt% Y	Pearson symbol	Space group
(Al)	0 to ~0.17	cF4	Fm$\overline{3}$m
αAl₃Y	52	hP8	P6₃/mmc
βAl₃Y	52	hR12	R$\overline{3}$m
Al₂Y	62.2	cF24	Fd$\overline{3}$m
AlY	76.7	oC8	Cmcm
Al₂Y₃	83	tP20	P4₂/mnm
AlY₂	86.8	oP12	Pnma
AlY₃(a)	91	cP4	Pm$\overline{3}$m
(αY)	100	hP2	P6₃/mmc
(βY)	100	cI2	Im$\overline{3}$m

(a) Metastable

Al-Yb

K.A. Gschneidner, Jr. and F.W. Calderwood, 1989

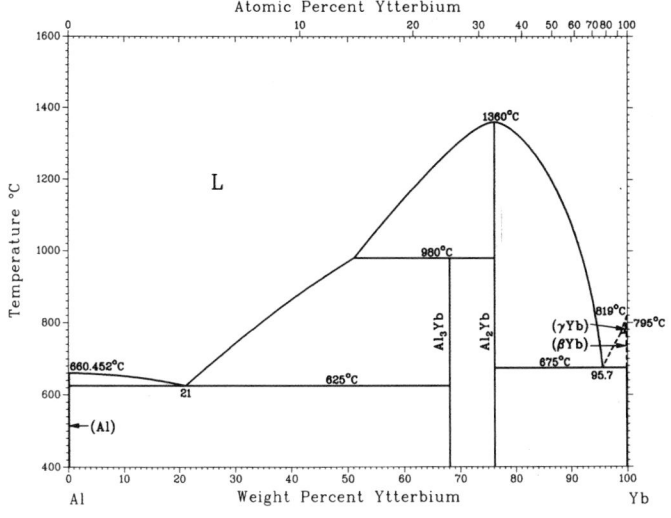

Phase	Composition, wt% Yb	Pearson symbol	Space group
(Al)	0	cF4	Fm$\overline{3}$m
Al₃Yb	68	cP4	Pm$\overline{3}$m
Al₂Yb	76.2	cF24	Fd$\overline{3}$m
(γYb)	99.6 to 100	cI2	Im$\overline{3}$m
(βYb)	99.9 to 100	cF4	Fm$\overline{3}$m

Al-Zn

J.L. Murray, 1983

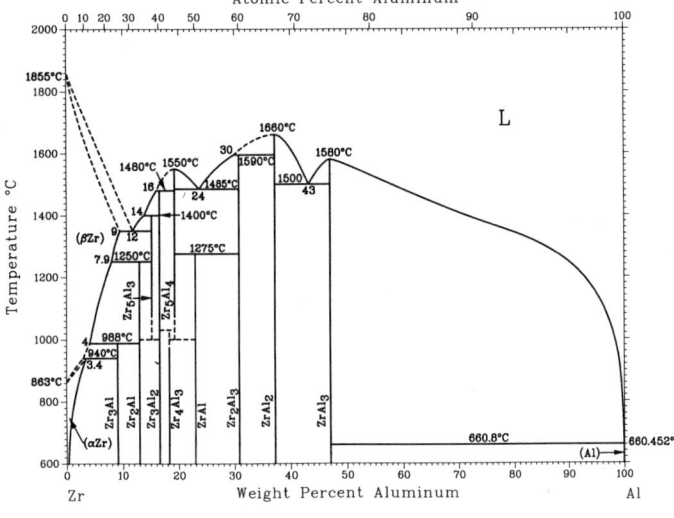

Phase	Composition, wt% Zn	Pearson symbol	Space group
(Al)	0 to 83.1	cF4	$Fm\bar{3}m$
(Zn)	98.8 to 100	hP2	$P6_3/mmc$
Metastable phases			
$(\alpha'\text{Al})_R$	78 to ~85	...	$R\bar{3}m$
"R"	(a)	...	...
Y	...	...	...

(a) Coherent precipitate

Al-Zr

J. Murray, A. Peruzzi, and J.P. Abriata, 1992

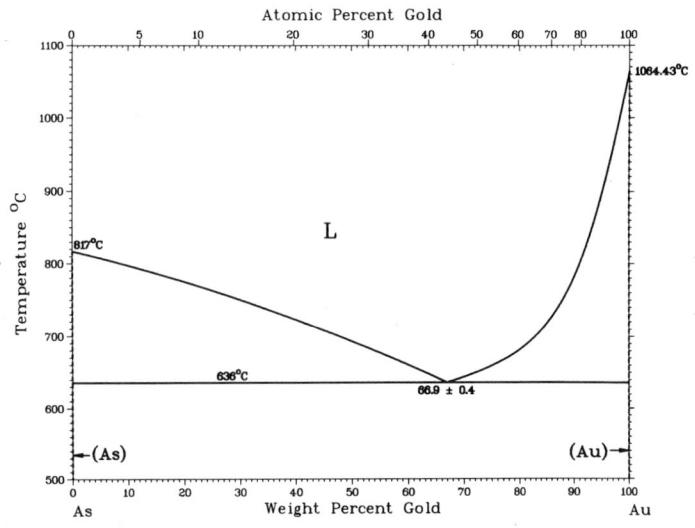

Phase	Composition, wt% Al	Pearson symbol	Space group
(αZr)	0 to 3.4	hP2	$P6_3/mmc$
(βZr)	0 to 9.4	cI2	$Im\bar{3}m$
Zr_3Al	9.0	cP4	$Pm\bar{3}m$
Zr_2Al	12.9	hP6	$P6_3/mmc$
Zr_5Al_3	15.1	tI32	$I4/mcm$
Zr_3Al_2	16	tP20	$P4_2/mmm$
Zr_4Al_3	18.2	hP7	$P\bar{6}$
Zr_5Al_4	19.1	hP18	$P6_3/mcm$
ZrAl	22.8	oC8	$Cmcm$
Zr_2Al_3	31	oF40	$Fdd2$
$ZrAl_2$	37.2	hP12	$P6_3/mmc$
$ZrAl_3$	47	tI16	$I4/mmm$
(Al)	99.86 to 100	cF4	$Fm\bar{3}m$

As-Au

H. Okamoto and T.B. Massalski, 1987

Phase	Composition, wt% Au	Pearson symbol	Space group
(As)	0	hR2	$R\bar{3}m$
(Au)	100	cF4	$Fm\bar{3}m$

As-Bi

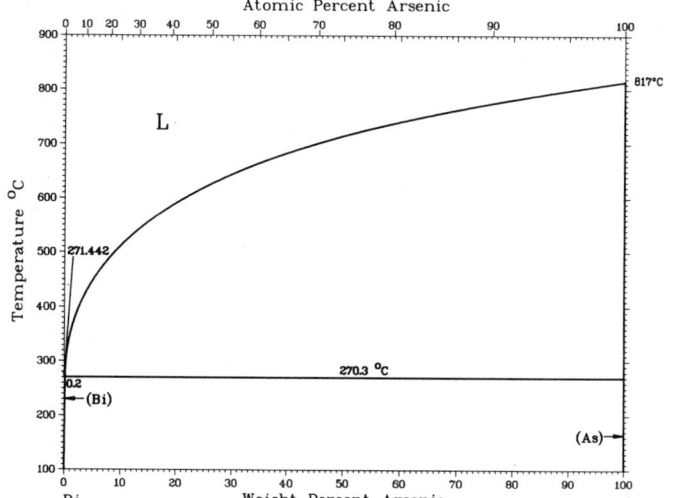

Phase	Composition, wt% As	Pearson symbol	Space group
(Bi)	0 to ~0.2	$hR2$	$R\overline{3}m$
(As)	~100	$hR2$	$R\overline{3}m$

As-Cd

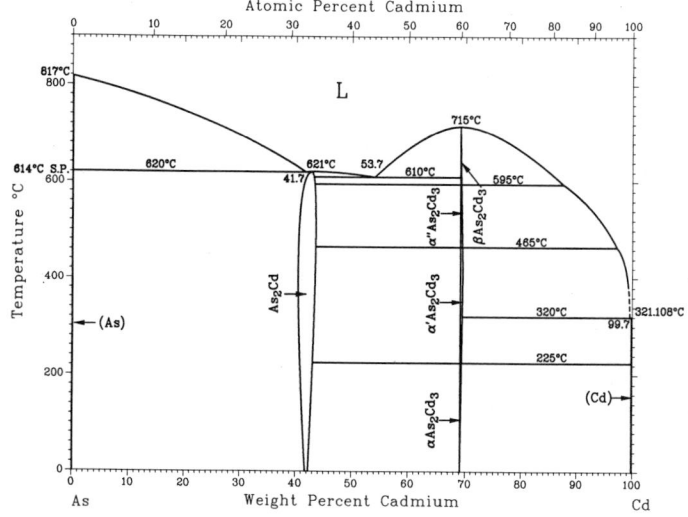

Phase	Composition, wt% Cd	Pearson symbol	Space group
(As)	0	$hR2$	$R\overline{3}m$
As$_2$Cd	42.8	$tI12$	$I4_122$
βAs$_2$Cd$_3$	69	$cF12$	$Fm\overline{3}m$
α″As$_2$Cd$_3$	69	$tP40$	$P4_2/nmc$
α′As$_2$Cd$_3$	69	$tP160$	$P4_2/nbc$
αAs$_2$Cd$_3$	69	$tI160$	$I4_1cd$
(Cd)	100	$hP2$	$P6_3/mmc$
High-pressure phases			
As$_2$CdII	42.8	...	...
AS$_2$CdIII(a)	42.8	...	...
AsCd	60	$oP16$	$Pbca$
As$_2$Cd$_3$(b)	69	$hP30$	...
As$_2$Cd$_3$II(c)	69	$hP5$	$P\overline{3}m1$
		$oP*$	$Pmmn$
As$_2$Cd$_3$II′	69	...	...
As$_2$Cd$_3$III	69	...	...
As$_2$Cd$_3$III′	69	...	...
Metastable phase			
As$_4$Cd	27	$t*20$	...
Other phases			
As$_2$Cd$_3$(d)	69	$tI160$	$I4_1/acd$
As$_2$Cd$_3$(e)	69	$tI160$	$I4_1\alpha$
As$_2$Cd$_3$(f)	69	$tI160$	$Iacd$

(a) >46 kbar. (b) 55 kbar. (c) 30 kbar. (d) Also might be βAs$_2$Cd$_3$. (e) Vapor deposition. (f) Synthesis at 675 °C

As-Co

K. Ishida and T. Nishizawa, 1990

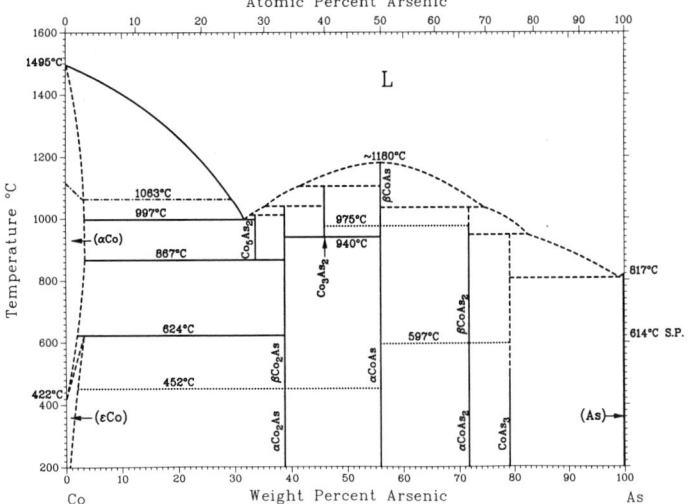

Phase	Composition, wt% As	Pearson symbol	Space group
(αCo)	0 to ~3.2	cF4	Fm$\overline{3}$m
(εCo)	0 to ~3	hP2	P6₃/mmc
Co₅As₂	33.7	hP42	P6₃cm
βCo₂As(a)	38.8 to 39.2	hP9	P62m
αCo₂As(a)	38.8	...	...
Co₃As₂	46	?	?
βCoAs	55.9	hP4	P6₃/mmc
αCoAs	55.9	oP8	Pna2₁
βCoAs₂	71.8	oP6	Pnnm
αCoAs₂	71.8	mP12	P2₁/c
CoAs₃	79 to 79.2	cI32	Im$\overline{3}$
(As)	~100	hR2	R$\overline{3}$m

(a) αCo₂As (low-temperature form) transforms into βCo₂As (high-temperature form) at 452 °C

As-Cu

P.R. Subramanian and D.E. Laughlin, 1988

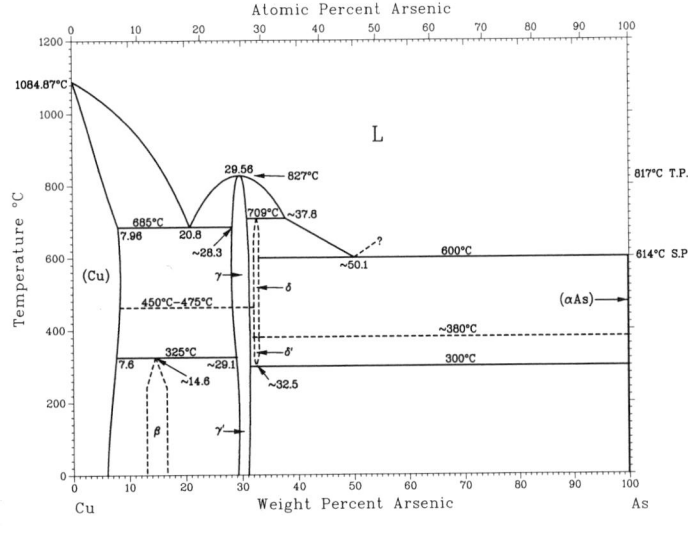

Phase	Composition, wt% As	Pearson symbol	Space group
(Cu)	0 to ~7.96	cF4	Fm$\overline{3}$m
β	12.8 to 16.4	hP2	P6₃/mmc
γ(HT)	28.2 to 31.2	hP8	P6₃/mmc
γ'(LT)	28.8 to 31.2	hP24	P$\overline{3}$c1
δ(HT)	32.1 to 33.1	cF16	Fm$\overline{3}$m
δ'(LT)	32.1 to 33.1	oI28	Ibam
(As)	100	hR2	R$\overline{3}$m
Metastable phases			
Cu₂As	~37.1	tP6	P4/nmm
Cu₃As₄	~61.12	oI28	Immm

As-Fe

H. Okamoto, 1992

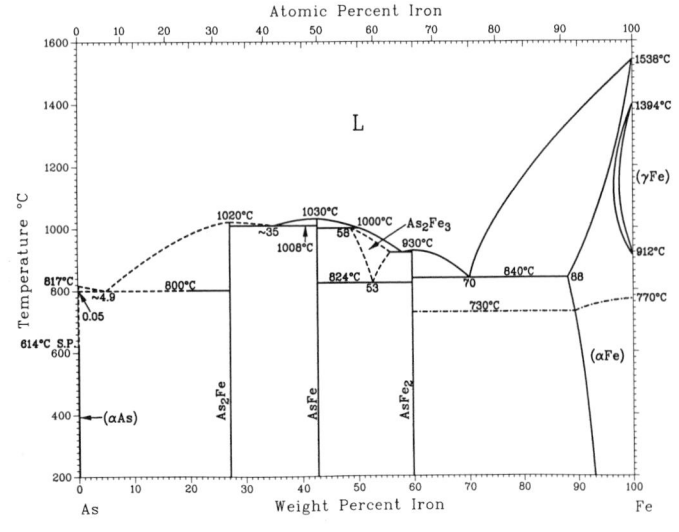

Phase	Composition, wt% Fe	Pearson symbol	Space group
(αAs)	0 to 0.05	hR2	R$\overline{3}$m
As₂Fe	27.1	oP6	Pnnm
AsFe	42.7	oP8	Pnma
As₂Fe₃	50 to 55	...	...
AsFe₂	59.9	tP6	P4/nmm
(αFe)	88 to 100	cI2	Im$\overline{3}$m
(γFe)	98.7 to 100	cF4	Fm$\overline{3}$m
High-pressure phase			
As₅Fe₁₂	64.2	hR17	R3

As-Ga

H. Okamoto, 1990

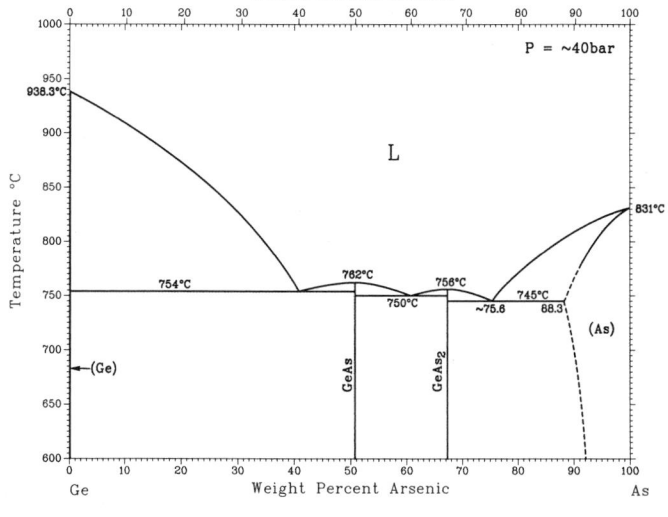

Phase	Composition, wt% As	Pearson symbol	Space group
(Ga)	0	oC8	Cmca
GaAs	51.8	cF8	F$\bar{4}$3m
(As)	100	hR2	R$\bar{3}$m

As-Ge

H. Okamoto, 1991

Phase	Composition, wt% As	Pearson symbol	Space group
(Ge)	0 to 0.19	cF8	Fm$\bar{3}$m
GeAs	50.8	mC24	C2/m
GeAs(a)	50.8	tI4	I4mm
GeAs$_2$	67.4	oP24	Pbam
(As)	88 to 100	hR2	R$\bar{3}$m

(a) High-pressure phase

As-In

H. Okamoto, 1992

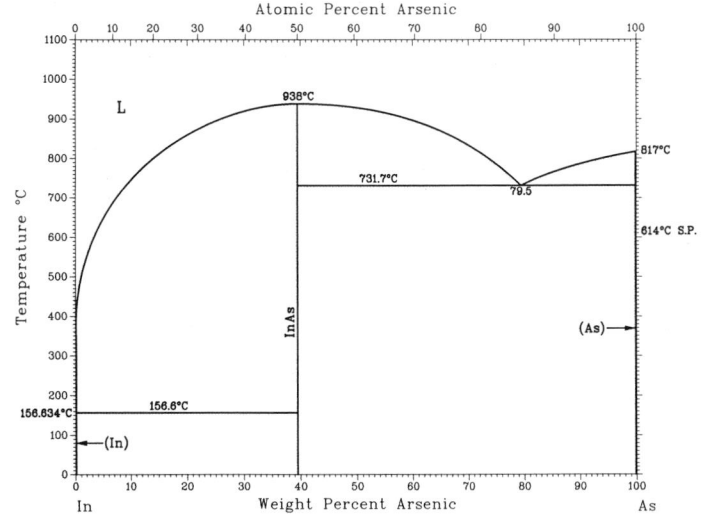

Phase	Composition, wt% As	Pearson symbol	Space group
Stable phases			
(In)	0	tI2	I4/mmm
InAs	39.5	cF8	F$\bar{4}$3m
(As)	100	hR2	R$\bar{3}$m
High-pressure phases			
InAs II(a)	39.5	cF8	Fm$\bar{3}$m
InAs III(b)	39.5	tI4	I4/amd

(a) Between 7 and 15 GPa. (b) Above 17 GPa (hysteresis between 15 and 17 GPa)

As-K

F.W. Dorn, W. Klemm, and S. Lohmeyer, 1961

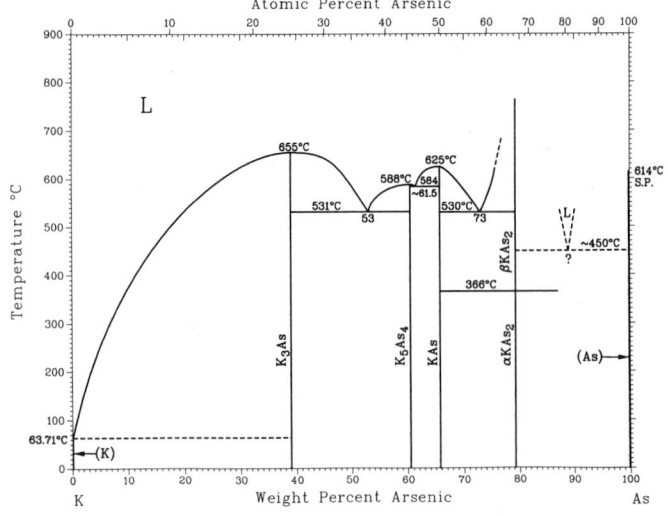

Phase	Composition, wt% As	Pearson symbol	Space group
(K)	~0	cI2	$Im\bar{3}m$
K_3As	39	hP8	$P6_3/mmc$
K_5As_4	60.5	...	...
KAs	65.7	oP16	$P2_12_12_1$
βKAs_2	79.3	...	...
αKAs_2	79.3	...	...
(As)	~100	hR8	$R\bar{3}m$

As-Mn

H. Okamoto, 1989

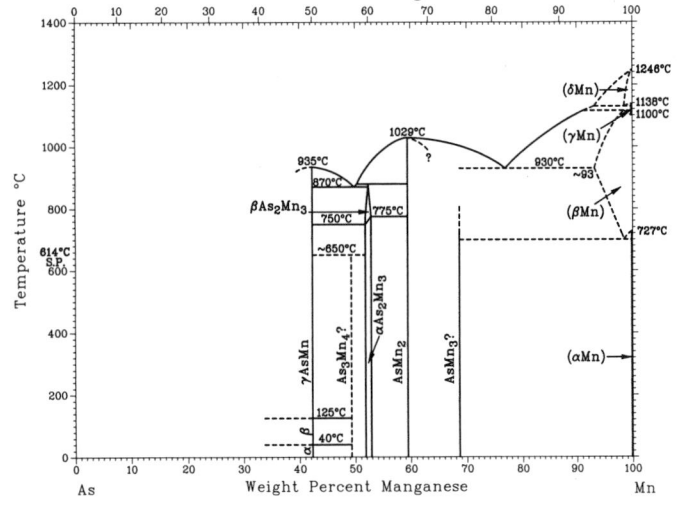

Phase	Composition, wt% Mn	Pearson symbol	Space group
(As)	0	hR2	$R\bar{3}m$
$\gamma AsMn$	42.3	hP4	$P6_3/mmc$
$\beta AsMn$	42.3	oP8	Pnma
$\alpha AsMn$	42.3	hP4	$P6_3/mmc$
As_3Mn_4	49.4	tI*	...
βAs_2Mn_3	52	...	...
αAs_2Mn_3	52	(a)	...
$AsMn_2$	59.5	tP6	P4/nmm
$AsMn_3$	69	oP16	Pmmn
(δMn)	100	cI2	$Im\bar{3}m$
(γMn)	100	cF4	$Fm\bar{3}m$
(βMn)	~ 93 to 100	cP20	$P4_132$
(αMn)	100	cI58	$I\bar{4}3m$
High-pressure phase			
$AsMn_2$	59.5	hP9	$P\bar{6}2m$

(a) Distorted cubic

As-Nd

K.A. Gschneidner, Jr. and F.W. Calderwood, 1986

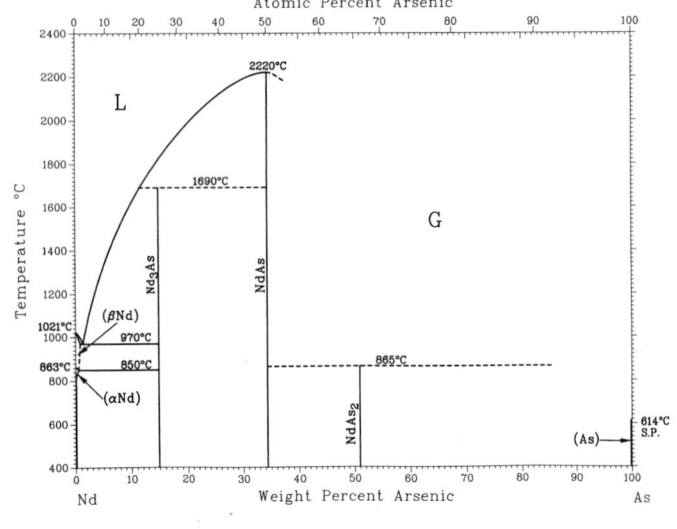

Phase	Composition, wt% As	Pearson symbol	Space group
(αNd)	0	hP4	$P6_3/mmc$
(βNd)	0	cI2	$Im\bar{3}m$
Nd_3As	15	(a)	...
NdAs	34.2	cF8	$Fm\bar{3}m$
$NdAs_2$	51.0	mP12	$P2_1/c$
(As)	100	hR2	$R\bar{3}m$

(a) Structure not known

As-Ni

M. Singleton and P. Nash, 1991

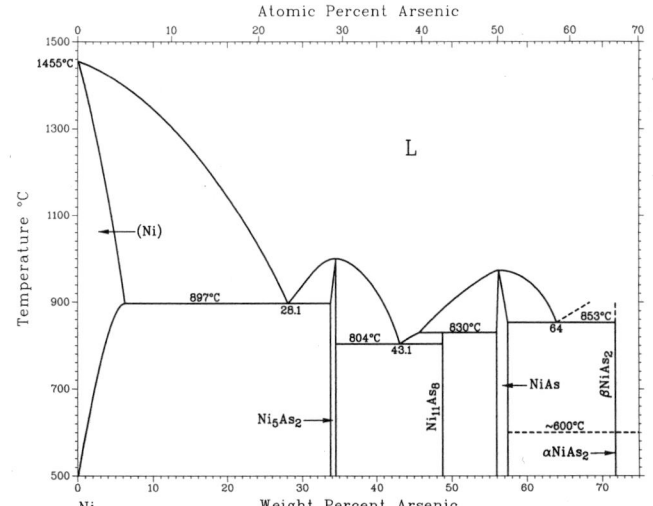

Phase	Composition, wt% As	Pearson symbol	Space group
(Ni)	0 to 6.30	cF4	Fm$\bar{3}$m
Ni$_5$As$_2$	33.27 to 33.99	hP42	Pb$_3$cm
Ni$_{11}$As$_8$	48.1	tP76	P4$_1$2$_1$2
NiAs	56.1 to 57.4	hP4	P6$_3$/mmc
αNiAs$_2$	71.86(a)	oP24	Pbca
βNiAs$_2$	71.86	oP6	Pnnm

(a) Up to 600 °C

As-P

I. Karakaya and W.T. Thompson, 1991

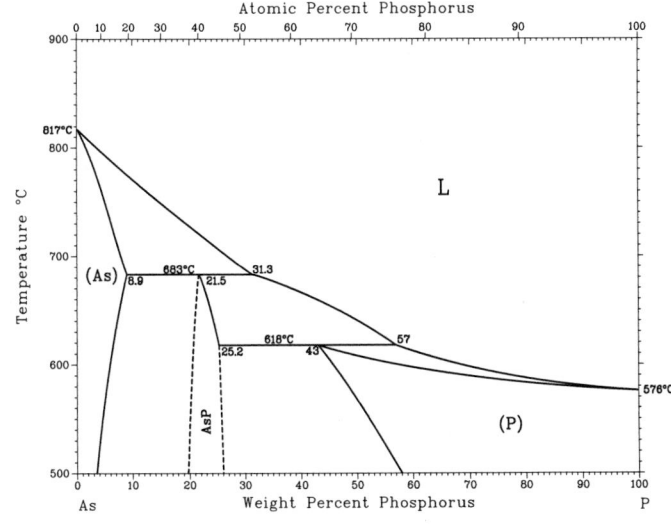

Phase	Composition, wt% P	Pearson symbol	Space group
(As)	0 to 8.9	hR2	R$\bar{3}$m
AsP	~21.5	...	...
P (black)	100	oC8(a)	Cmca
P (white)	43 to 100	(b)	...
P (red)	100	(c)	...

(a) At high pressures, transforms to a rhombohedral structure. (b) Cubic at 35 °C. (c) Cubic with 66 atoms per unit cell

As-Pb

N.A. Gokcen, 1990

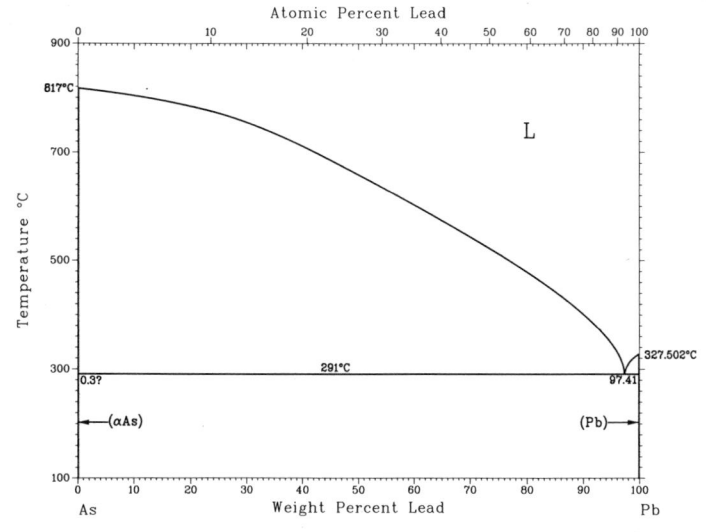

Phase	Composition, wt% Pb	Pearson symbol	Space group
(As)	0	hR2	R$\bar{3}$m
(Pb)	100	cF4	Fm$\bar{3}$m

As-Pd

H. Okamoto, 1992

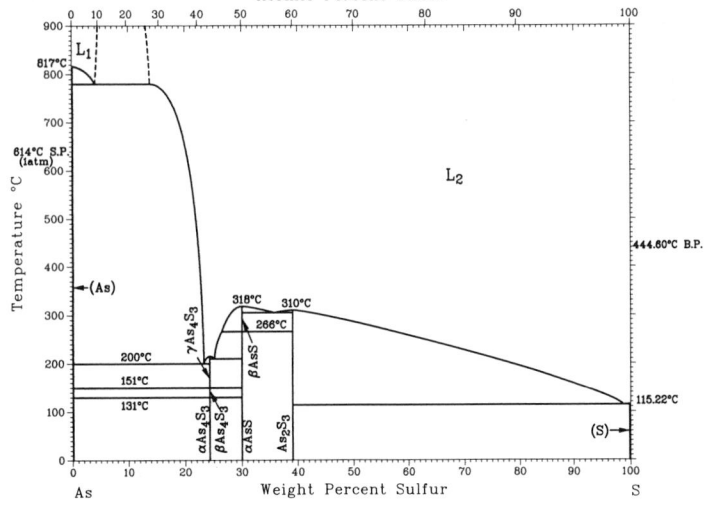

Phase	Composition, wt% Pb	Pearson symbol	Space group
(As)	0	hR2	$R\bar{3}m$
As₂Pd	41.5	cP12	$Pa\bar{3}$
βAsPd₂	74.0	hP9	$P\bar{6}2m$
αAsPd₂	74.0	mP54	$P2/m$
As₂Pd₅	78.0	hP*	...
As₂Pd₅	78.0	hP84	$P\bar{3}m1$
As₂Pd₅	78.0	hP*	$P6_322$
As₂Pd₅	78.0	hP*	$P\bar{3}m1$
As₃Pd₈	79.1	hP33	$P3$
AsPd₃	81	tI32	$I\bar{4}$
AsPd₅	87.6	mC24	$C2$
(Pd)	100	cF4	$Fm\bar{3}m$
Metastable phase			
AsPd₅	87.6	cI2	$Im\bar{3}m$
Questionable phases			
αAsPd₂	74.0	oC24	$Cmc2_1$
αAsPd₂	74.0	hP*	...
As₂Pd₅	78.0	o**	...
AsPd₇	90.9	...	...

As-S

H. Okamoto, 1990

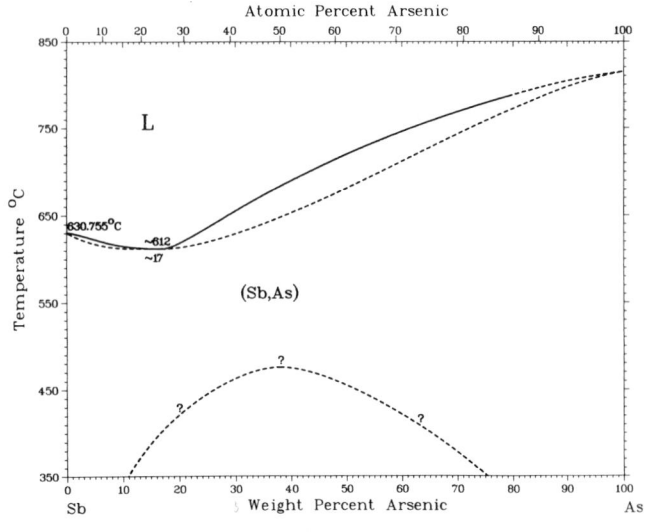

Phase	Composition, wt% S	Pearson symbol	Space group
(αAs)	0	hR2	$R\bar{3}m$
γAs₄S₃	24.3	...	...
βAs₄S₃	24.3	t**	...
αAs₄S₃	24.3	oP28	$Pnma$
βAsS	30.0	mP32	$P2_1/n$
αAsS	30.0	mP32	$P2_1/c$
As₂S₃	39	mP20	$P2_1/c$
(S)	100	oF128	$Fddd$

As-Sb

H. Okamoto, 1990

Phase	Composition, wt% As	Pearson symbol	Space group
(Sb,As)	0 to 100	hR2	$R\bar{3}m$

As-Se

H. Okamoto, 1990

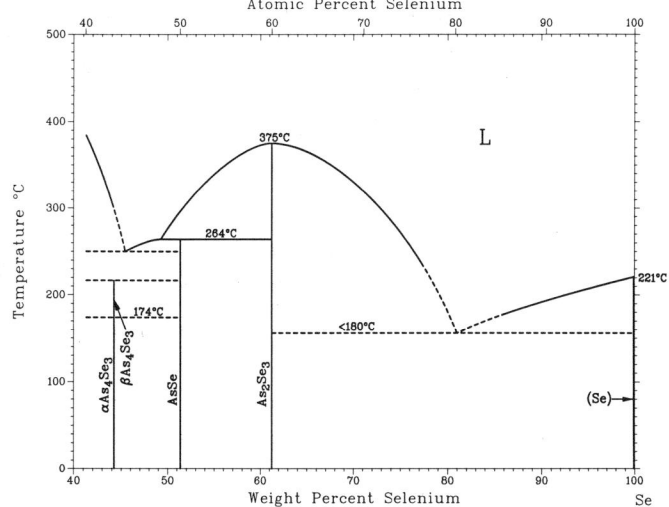

Phase	Composition, wt% Se	Pearson symbol	Space group
(As)	0	hR2	$R\bar{3}m$
βAs_4Se_3	44.2	mC112	C2/c
αAs_4Se_3	44.2	oP28	Pnma
AsSe	51.3	mP32	$P2_1/c$
As_2Se_3	61	mP20	$P2_1/c$
(γSe)	100	hP3	$P3_121$

As-Si

R. W. Olesinski and G.J. Abbaschian, 1985

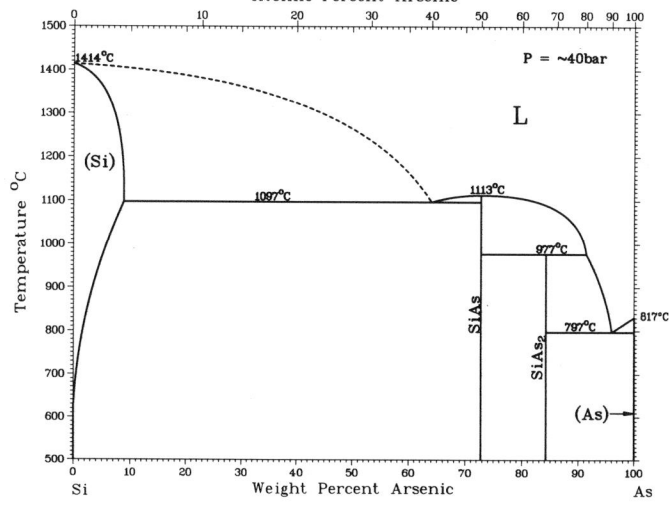

Phase	Composition, wt% As	Pearson symbol	Space group
(Si)	0 to 8.8	cF8	$Fd\bar{3}m$
SiAs	72.7	o**	...
$SiAs_2$	84.2	oP*	Pbam
$SiAs_2$(a)	84.2	cP12	$Pa\bar{3}$
(As)	~100	hR2	$R\bar{3}m$

(a) High-pressure phase

As-Sn

N.A. Gokcen, 1990

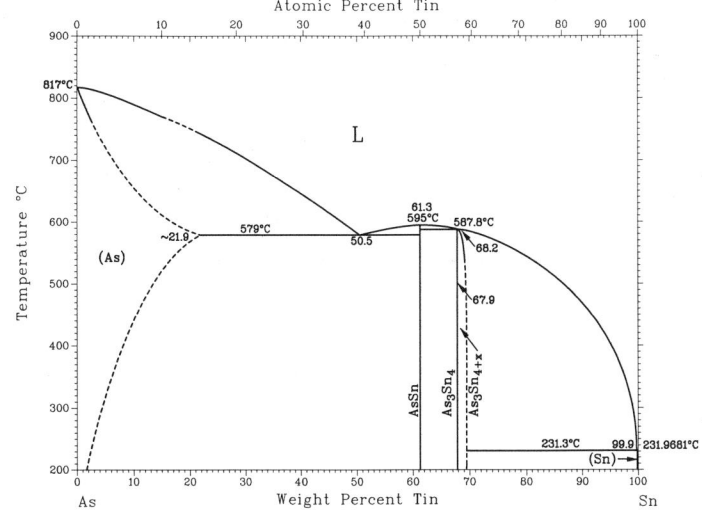

Phase	Composition, wt% Sn	Pearson symbol	Space group
(As)	0 to ~21.9	hR2	$R\bar{3}m$
AsSn	61.3	cF8	$Fm\bar{3}m$
As_3Sn_4	67.87 to 70?	hR7	$R\bar{3}m$
(βSn)(a)	99.9 to 100	tI4	$I4_1/amd$
(αSn)(b)	100	cF8	$Fm\bar{3}m$

(a) White tin, stable above 13 °C. (b) Grey tin, stable below 13 °C

As-Te

H. Okamoto, 1990

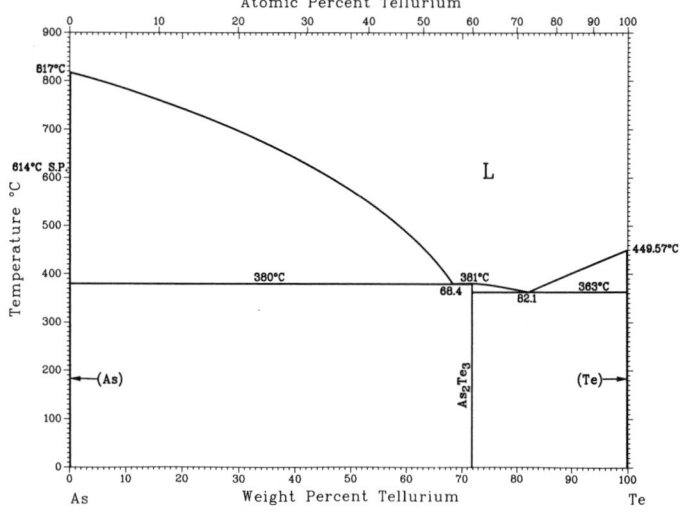

Phase	Composition, wt% Te	Pearson symbol	Space group
(As)	0	hR2	R$\bar{3}$m
As$_2$Te$_3$	72	mC20	Cm/2
(Te)	100	hP3	P3$_1$21

As-Tl

R.C. Sharma and Y.A. Chang, unpublished

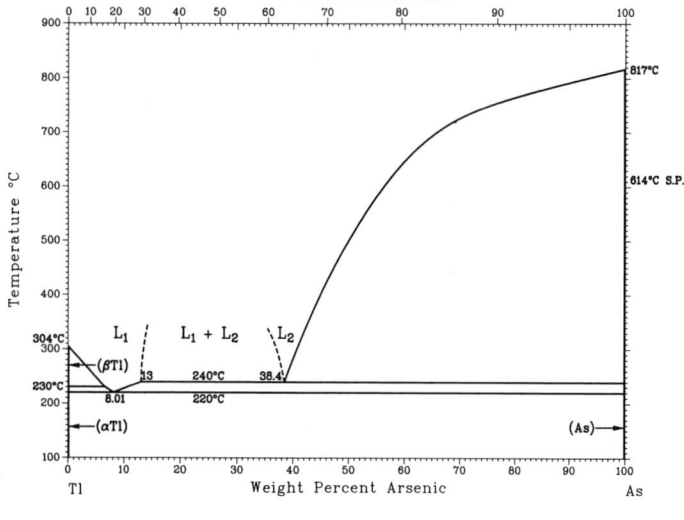

Phase	Composition, wt% As	Pearson symbol	Space group
(αTl)	0	hP2	P6$_3$/mmc
(βTl)	0	cI2	Im$\bar{3}$m
(As)	100	hR2	R$\bar{3}$m

As-Yb

H. Okamoto, 1990

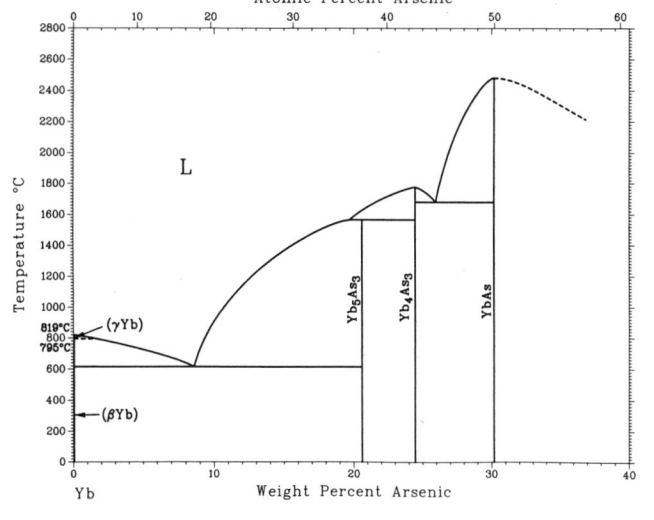

Phase	Composition, wt% As	Pearson symbol	Space group
(αYb)	0	hP2	P6$_3$/mmc
(βYb)	0	cF4	Fm$\bar{3}$m
(γYb)	0	cI2	Im$\bar{3}$m
Yb$_5$As$_3$	20.6	hP16	P6$_3$/mcm
αYb$_4$As$_3$	24.5	hR28	R3
βYb$_4$As$_3$	24.5	cI28	I$\bar{4}$3d
YbAs	30.2	cF8	Fm$\bar{3}$m
(As)	100	hR2	R$\bar{3}$m

As-Zn

H. Okamoto, 1992

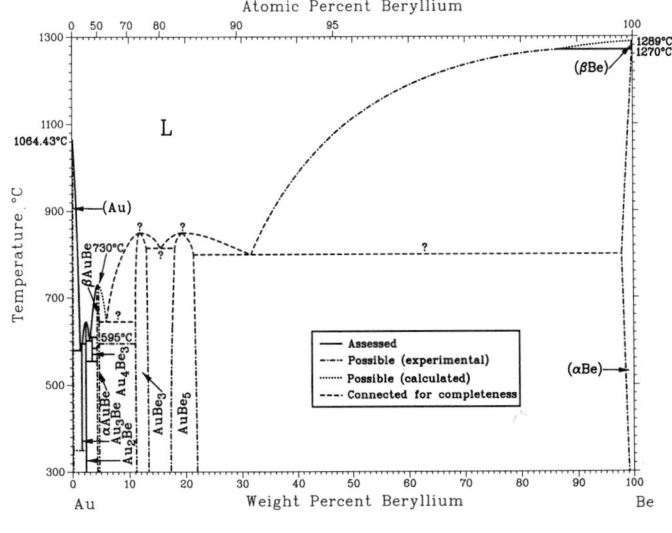

Phase	Composition, wt% Zn	Pearson symbol	Space group
(αAs)	0	hR2	$R\bar{3}m$
As₂Zn	30.3	mP24	$P2_1/c$
βAs₂Zn₃	56.7	cF12	$Fm\bar{3}m$
α'As₂Zn₃	56.7	tP160	$P4_2/nbc$
αAs₂Zn₃	56.7	tI160	$I4_1cd$
(Zn)	100	hP2	$P6_3/mmc$
High-pressure phases			
AsZn	46.6	oP16	Pbca
As₂Zn₃II(a)	56.7	cF*	...
As₂Zn₃II'	56.7	oP*	Pmmn
As₂Zn₃III	56.7	...	...
As₂Zn₃(b)	56.7	hP30	...
Other phases			
As₂Zn	30.39	o*32	...
As₂Zn₃	56.7	cI80	Ia3
	56.7	tP40	$P4_2/mmc$

(a) At 55 kbar. (b) At 70 kbar

Au-Be

H. Okamoto and T.B. Massalski, 1987

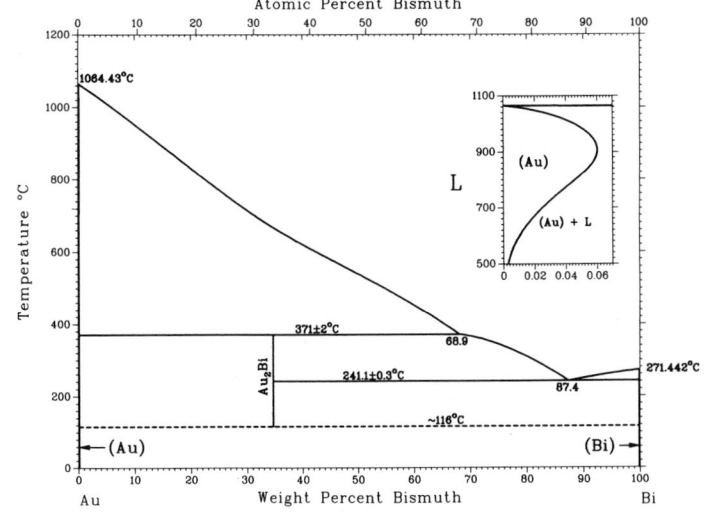

Phase	Composition, wt% Be	Pearson symbol	Space group
(Au)	0 to 0.009	cF4	$Fm\bar{3}m$
Au₃Be	2	o**	...
Au₂Be	2.2	tI6	I4/mmm
Au₄Be₃	3.3	...	...
βAuBe	4.2 to 4.6	...	...
αAuBe	4.2 to 4.6	cP8	$P2_13$
AuBe₃	11 to 13	cF16	$Fd\bar{3}m$
AuBe₅	17 to 22	cF24	$F\bar{4}3m$
(βBe)	? to 100	cI2	$Im\bar{3}m$
(αBe)	94.81 to 100	hP2	$P6_3/mmc$

Au-Bi

H. Okamoto, 1990

Phase	Composition, wt% Bi	Pearson symbol	Space group
(Au)	0	cF4	$Fm\bar{3}m$
Au₂Bi	34.6	cF24	$Fd\bar{3}m$
(Bi)	100	hR2	$R\bar{3}m$
Metastable phases			
π	76 to 81	cP1	$Pm\bar{3}m$
π'	61	hR1	$R\bar{3}m$
Microcrystalline	46 to 71	~200 π'-like unit cells	...
(AuBi)?	56	Complex	...

Au-Ca

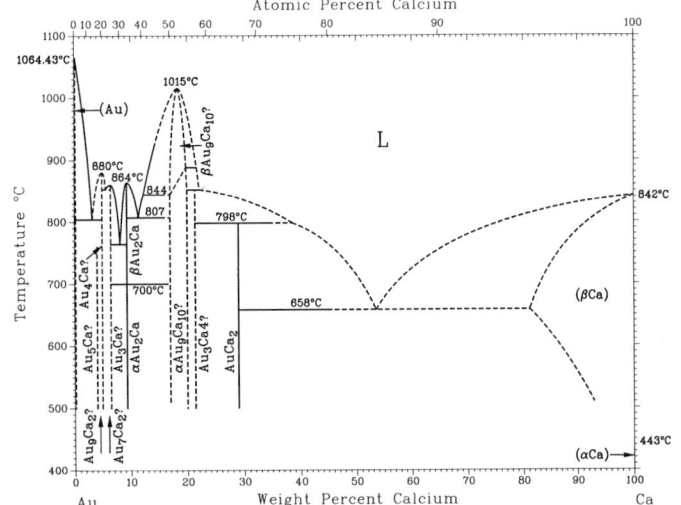

H. Okamoto, T.B. Massalski, C.B. Alcock, and V.P. Itkin, 1987

Phase	Composition, wt% Ca	Pearson symbol	Space group
(Au)	0 to <4.3	cF4	$Fm\bar{3}m$
Au$_5$Ca	3.9	cF24	$F\bar{4}3m$
Au$_9$Ca$_2$	4.3	?	…
Au$_4$Ca	5	?	…
Au$_7$Ca$_2$	5.5	(a)	…
Au$_3$Ca	6	(b)	…
βAu$_2$Ca	9.2	?	…
αAu$_2$Ca	9.2	?	…
AuCa	17	oC8	Cmcm
βAu$_9$Ca$_{10}$	17 to 20	(c)	…
αAu$_9$Ca$_{10}$	16 to 20.2	?	…
Au$_3$Ca$_4$	21.3	?	…
AuCa$_2$	29.0	?	…
(βCa)	81.2 to 100	cI2	$Im\bar{3}m$
(αCa)	? to 100	cF4	$Fm\bar{3}m$

(a) Same as Au$_3$Ca? (b) Not cubic. (c) Same as AuCa?

Au-Cd

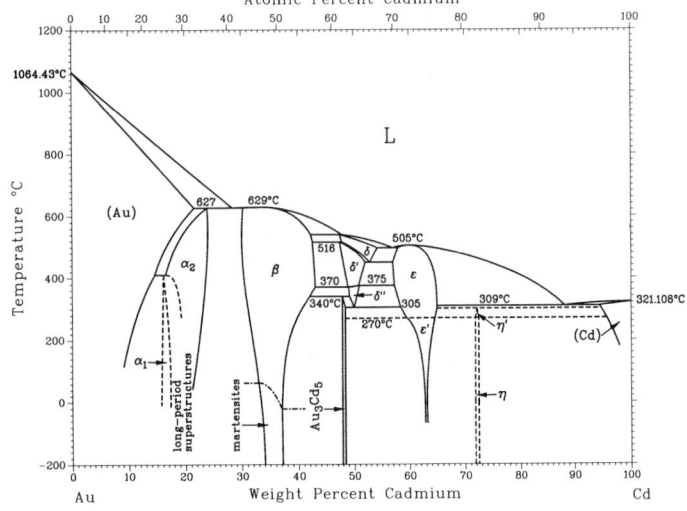

H. Okamoto and T.B. Massalski, 1987

Phase	Composition, wt% Cd	Pearson symbol	Space group
(Au)	0 to 21.6	cF4	$Fm\bar{3}m$
α$_1$	~16	…	$Pm\bar{3}m$
Long-period superstructures			
Au$_3$Cd	15	tI16	I4/mmm
4H (1d)	~16	hP4	P6$_3$/mmc
9a$_0$-4H (2d)	15 to 17	(a)	…
6H	16 to 17	hR3	$R\bar{3}m$
9R	16.7 to 19.7	(b)	…
7a$_0$ - 2H	19 to 23	(a)	P6$_3$/mmc
9a$_0$ - 2H	19 to 23	(a)	P6$_3$/mcm
α$_2$	16.3 to 24.0	hP*	…
β	30 to 43	cP2	$Pm\bar{3}m$
β' (c)	34.1	oP4	Pmma
β" (c)	36	(a)	…
δ	47.3 to 54.3	cI52	$I\bar{4}3m$
δ'	47.3 to 52.0	cI52	$I\bar{4}3m$
δ"	48.9 to 50.9	…	…
Au$_3$Cd$_5$	47.8 to 48.3	tI32	I4/mcm
ε	57.0 to 65.0	(d)	…
ε'	59 to 64	(d)	…
η	72	…	…
η'	72	…	…
(Cd)	94.0 to 100	hP2	P6$_3$/mmc

Note: d = dimensional. (a) Hexagonal. (b) Rhombohedral. (c) Not shown in the assessed diagram. (d) bct

Au-Ce

K.A. Gschneidner, Jr., F.W. Calderwood, H. Okamoto, and T.B. Massalski, 1987

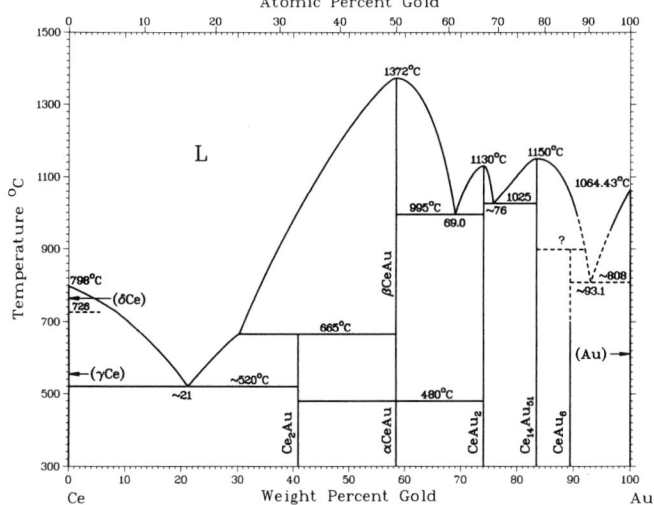

Phase	Composition, wt% Au	Pearson symbol	Space group
(αCe)	0	cF4	Fm$\bar{3}$m
(βCe)	0	hP4	P6$_3$/mmc
(γCe)	0	cF4	Fm$\bar{3}$m
(δCe)	0	cI2	Im$\bar{3}$m
Ce$_2$Au	41.2	oP12	Pnma
αCeAu	58	oP8	Pnma
βCeAu	58	oC8	Cmcm
CeAu$_2$	73.8	oI12	Imma
Ce$_{14}$Au$_{51}$	81 to 83.7	hP65	P6/m
CeAu$_6$	89.4	mC28	C/2c
(Au)	100	cF4	Fm$\bar{3}$m

Au-Co

H. Okamoto, T.B. Massalski, M. Hasebe, and T. Nishizawa, 1987

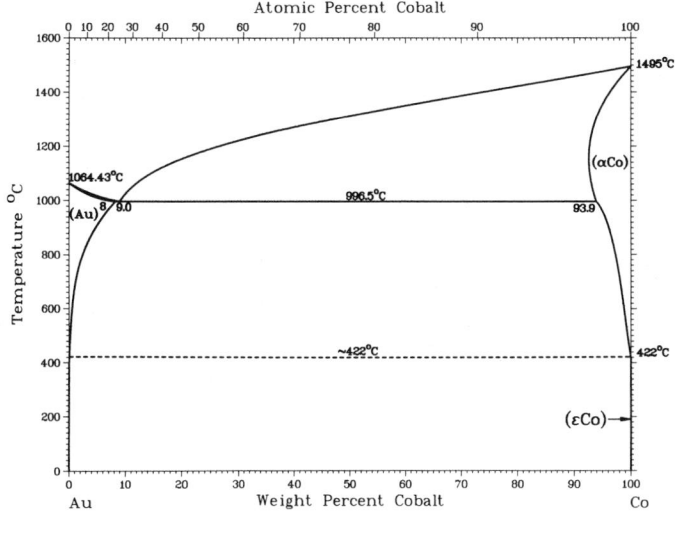

Phase	Composition, wt% Co	Pearson symbol	Space group
(Au)	0 to 8	cF4	Fm$\bar{3}$m
(αCo)	92.1 to 100	cF4	Fm$\bar{3}$m
(εCo)	? to 100	hP2	P6$_3$/mmc

Au-Cr

H. Okamoto and T.B. Massalski, 1987

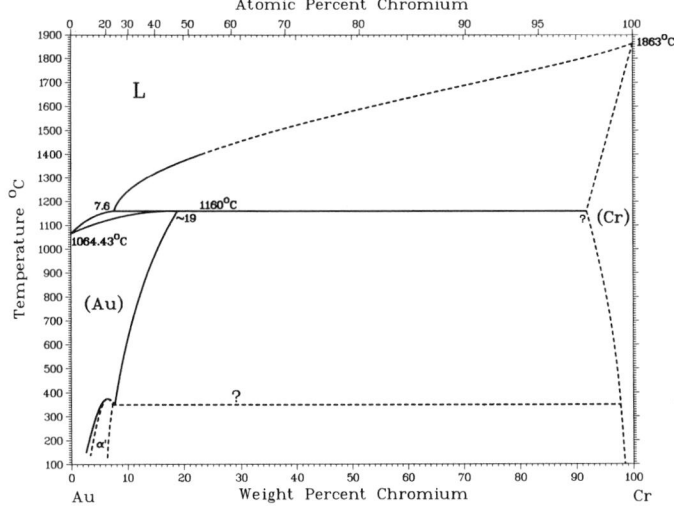

Phase	Composition, wt% Cr	Pearson symbol	Space group
(Au)	0 to ~19	cF4	Fm$\bar{3}$m
α'	2 to 8	tI10	I4/m
(Cr)	~90 to 100	cI2	Im$\bar{3}$m

Au-Cu

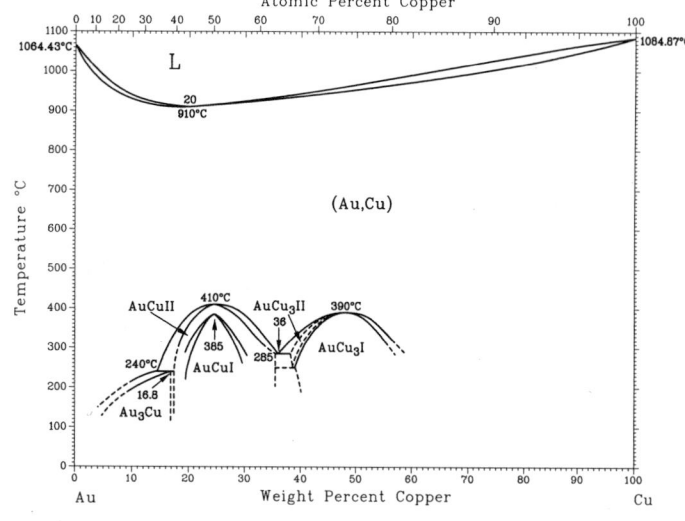

H. Okamoto, D.J. Chakrabarti, D.E. Laughlin, and
T.B. Massalski, 1987

Phase	Composition, wt% Cu	Pearson symbol	Space group
(Au,Cu)	0 to 100	cF4	$Fm\overline{3}m$
Au₃Cu	3 to 16.8	cP4	$Pm\overline{3}m$
AuCu(I)	19 to 30	tP4	$P4/mmm$
AuCu(II)	16.8 to 35	oI40	$Imma$
AuCu₃(I)	40 to 58	cP4	$Pm\overline{3}m$
AuCu₃(II)	39 to ?	tP28	$P\overline{4}mm$

Au-Dy

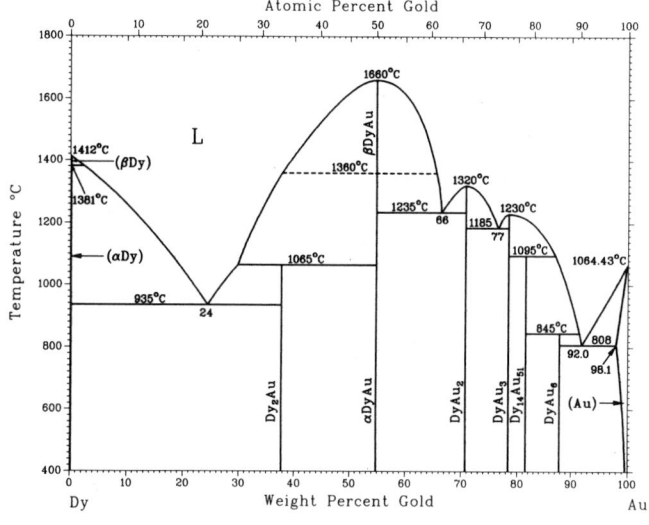

K.A. Gschneidner, Jr., F.W. Calderwood, H. Okamoto, and
T.B. Massalski, 1987

Phase	Composition, wt% Au	Pearson symbol	Space group
(αDy)	0	hP2	$P6_3/mmc$
(α'Dy)	0	(a)	...
(βDy)	0	cI2	$Im\overline{3}m$
Dy₂Au	37.7	oP12	$Pnma$
αDyAu	55	oC8	$Cmcm$
βDyAu	56	cP2	$Pm\overline{3}m$
DyAu₂	70.8	tI6	$I4/mmm$
DyAu₃	78	oP8	$Pmmn$
Dy₁₄Au₅₁	~81.6	hP65	$P6/m$
DyAu₆	87.9	tP56	$P4_2/ncm$
(Au)	98.1 to 100	cF4	$Fm\overline{3}m$

(a) Orthorhombic distortion, $T \leq 86$ K

Au-Eu

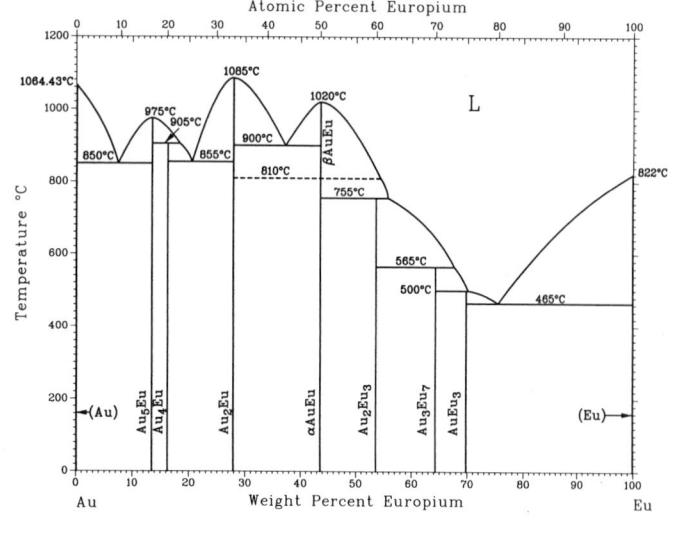

H. Okamoto, 1990

Phase	Composition, wt% Eu	Pearson symbol	Space group
(Au)	0	cF4	$Fm\overline{3}m$
Au₅Eu	13.4	hP6	$P6/mmm$
Au₄Eu	16	...	...
Au₂Eu	27.8	oI12	$Imma$
βAuEu	43.6	...	...
αAuEu	43.6	oP8	$Pnma$
Au₂Eu₃	54	hR45	$R\overline{3}$
Au₃Eu₇	64	hP20	$P6_3/mc$
AuEu₃	70	oP16	$Pnma$
(Eu)	100	cI2	$Im\overline{3}m$

Au-Fe

H. Okamoto, T.B. Massalski, L.J. Swartzendruber, and P.A. Beck, 1987

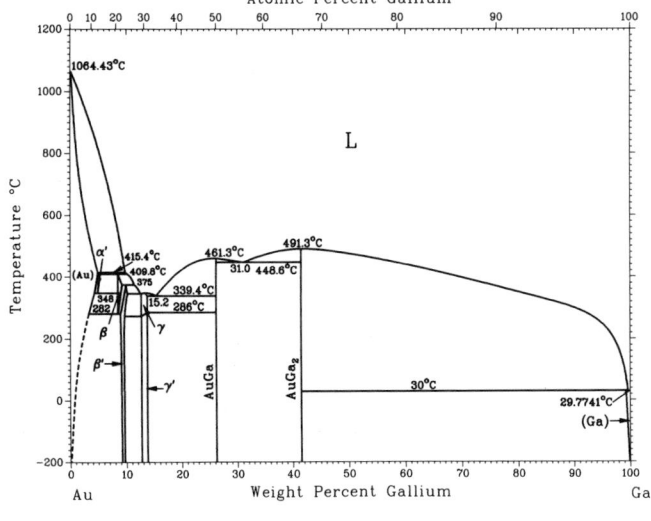

Phase	Composition, wt% Fe	Pearson symbol	Space group
(Au)	0 to 45	cF4	Fm$\bar{3}$m
(γFe)	77 to 100	cF4	Fm$\bar{3}$m
(αFe)	96 to 100	cI2	Im$\bar{3}$m
(δFe)	93 to 100	cI2	Im$\bar{3}$m
Metastable phases			
Amorphous(a)	19 to 72	...	...
fcc(b)	30 to 32	...	...
bcc(b)	32 to 53	...	...

(a) Found in thin films deposited at liquid nitrogen temperature or below. (b) Formed by crystallization on heating amorphous phase

Au-Ga

T.B. Massalski and H. Okamoto, 1987

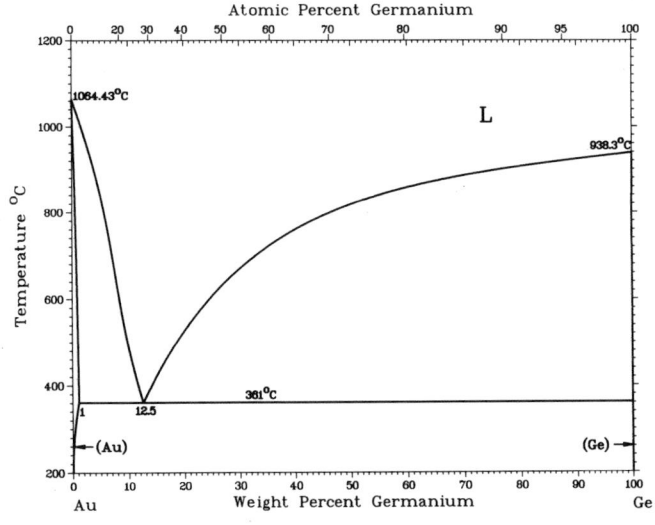

Phase	Composition, wt% Ga	Pearson symbol	Space group
(Au)	0 to 4.8	cF4	Fm$\bar{3}$m
α'	4.9 to 5.5	hP16	P6$_3$/mmc
β	8.3 to 9.1	(a)	...
β'	8.7 to 10.5	(b)	...
γ	13.1 to 14	(b)	...
γ'	13.1 to 14	(b)	...
AuGa	26.1	oP8	Pnma
AuGa$_2$	41.5	cF12	Fm$\bar{3}$m
(Ga)	100	oC8	Cmca

(a) Hexagonal. (b) Orthorhombic

Au-Ge

H. Okamoto and T.B. Massalski, 1987

Phase	Composition, wt% Ge	Pearson symbol	Space group
(Au)	0 to 1(a)	cF4	Fm$\bar{3}$m
(Ge)	100(a)	cF8	Fd$\bar{3}$m
Metastable phases			
β	7 to 11(a)	hP2	P6$_3$/mmc
γ	11 to 29(a)	tI*	...

(a) Approximate composition from the phase diagram

Au-Hg

H. Okamoto and T.B. Massalski, 1989

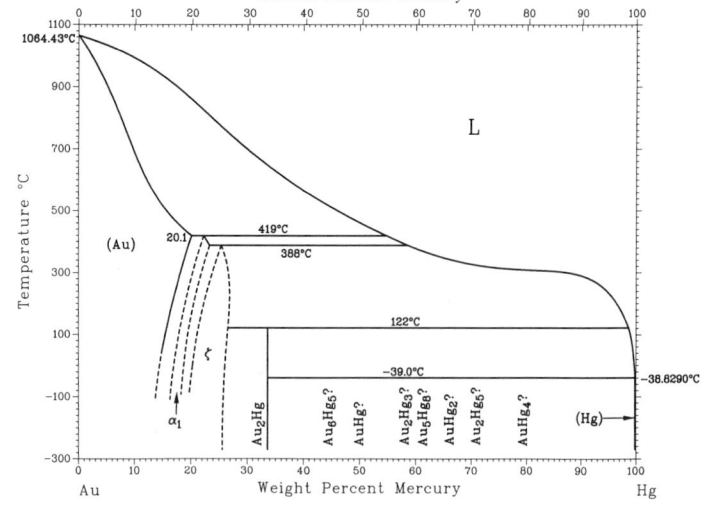

Phase	Composition, wt % Hg	Pearson symbol	Space group
(Au)	0 to 20.1	$cF4$	$Fm\bar{3}m$
α_1	16.2 to 23	$hP36$	$P6_3/mmc$
ζ	21 to 26	$hP2$	$P6_3/mmc$
Au_2Hg	33.7	$hP{\sim}150$	...
		$hP22$	$P6_3/mcm$
Au_6Hg_5	46.0	$hP22$	$P6_3/mcm$
Au_5Hg_8	62.0	$cI52$	$I\bar{4}3m$
(Hg)	100	$hR1$	$R\bar{3}m$

Au-In

H. Okamoto and T.B. Massalski, 1992

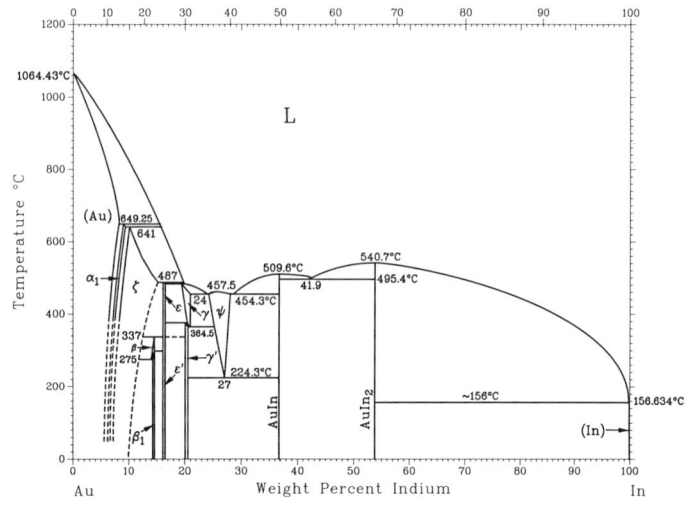

Phase	Composition, wt % In	Pearson symbol	Space group
(Au)	0 to 7.8	$cF4$	$Fm\bar{3}m$
α_1	7.4 to 8.9	$hP16$	$P6_3/mmc$
		$hP4$	$P6_3/mmc$
ζ	8 to 14.8	$hP2$	$P6_3/mmc$
β	13.8 to 14.3	(a)	...
		(b)	...
β_1	13.9 to 14.5	$hP26$	$P3$
		(a)	...
ε	15.9 to 16.3	(b)	...
ε'	15.9 to 16.3	$oP8$	$Pmmn$
γ	19.1 to 21.1	$cP52$	$P\bar{4}3m$
		$cP76$	$P\bar{4}3m$
γ'	19.8 to 20.5	$hP60$	$P3$
ψ	24.1 to 27.6	$hP5$	$P\bar{3}m1$
AuIn	37 to 36.9	(c)	...
$AuIn_2$	53.9	$cF12$	$Fm\bar{3}m$
(In)	100	$tI2$	$I4/mmm$

(a) Hexagonal. (b) Orthorhombic. (c) Triclinic

Au-K

A.D. Pelton, 1987

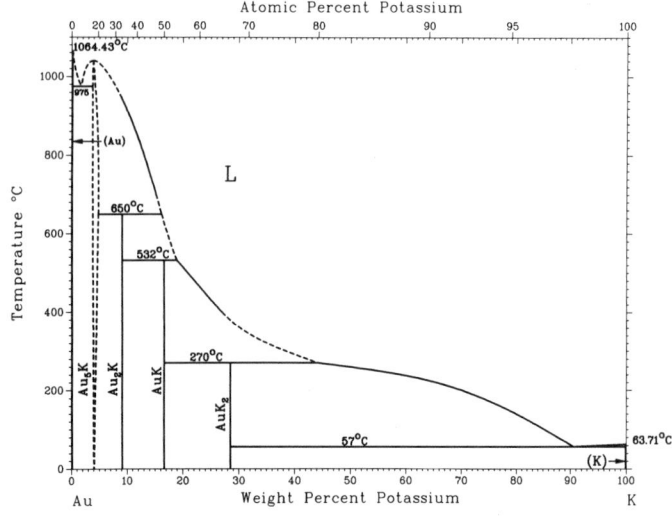

Phase	Composition, wt % K	Pearson symbol	Space group
(Au)	0	$cF4$	$Fm\bar{3}m$
Au_5K	3.8	$hP6$	$P6/mmm$
Au_2K	9.0	...	...
AuK	16.6	...	...
AuK_2	28.5	...	...
(K)	100	$cI2$	$Im\bar{3}m$

Note: At 25 °C

Au-La

K.A. Gschneidner, Jr., F.W. Calderwood, H. Okamoto, and
T.B. Massalski, 1987

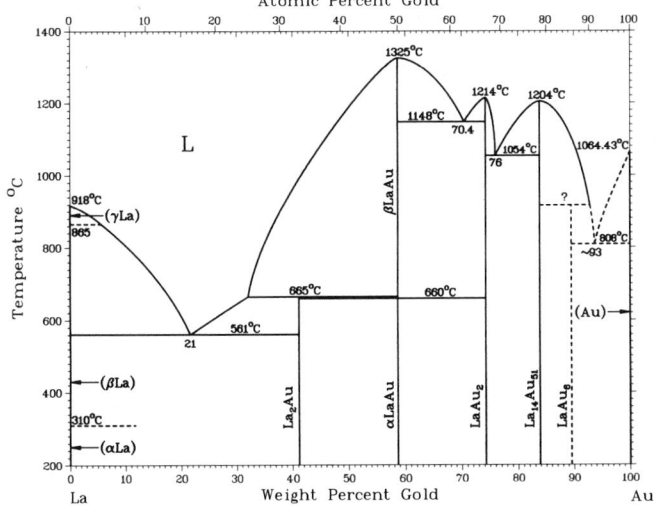

Phase	Composition, wt% Au	Pearson symbol	Space group
(αLa)	0	hP4	P6₃/mmc
(βLa)	0	cF4	Fm3̄m
(γLa)	0	cI2	Im3̄m
La₂Au	41.5	oP12	Pnma
αLaAu	59	oP8	Pnma
βLaAu	59	oC8	Cmcm
LaAu₂	74.0	oI12	Imma
La₁₄Au₅₁	~81 to ~83.8	hP65	P6/m
LaAu₆	89.5	mC28	C2/c
(Au)	100	cF4	Fm3̄m

Au-Li

A.D. Pelton, 1987

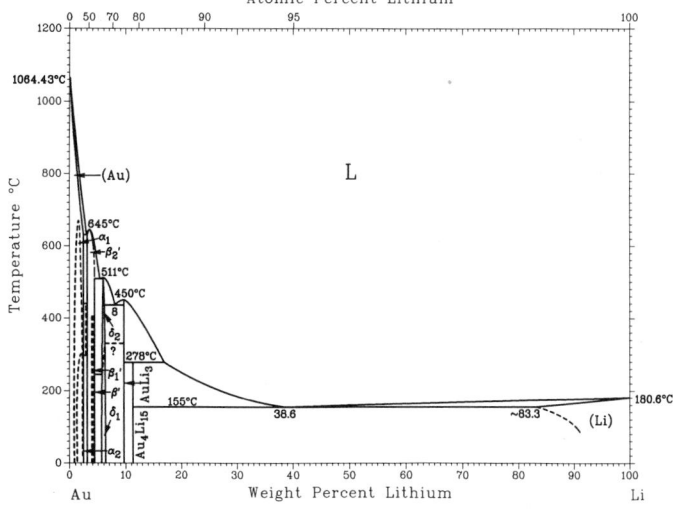

Phase	Composition, wt% Li	Pearson symbol	Space group
(αAu)	0 to 0.7	cF4	Fm3̄m
(α₁Au)	0.7 to 1	cP4	Pm3̄m
(α₂Au)	2 to 2.3	(b)	...
Au₅Li₄	2.7	(c)	...
β′₂	3 to 4	oP2	...
β′₁	4 to 4.1	tP2?(b)	...
β′	4.1 to 4.3	cP2	Pm3̄m
δ₂(HT)	5.6 to 6.3	hP9	...
δ₁(LT)	5.6 to 6.3	(d)	...
AuLi₃	10	cF16	Fm3̄m
Au₄Li₁₅	12	cI76	I4̄3d
(βLi)(a)	100	cI2	Im3̄m
(αLi)(e)	100	hP2	P6₃/mmc

(a) At 25 °C. (b) Complex. (c) Hexagonal. (d) Similar to δ₂. (e) T less than −201 °C

Au-Mg

A.A. Nayeb-Hashemi and J.B. Clark, 1988

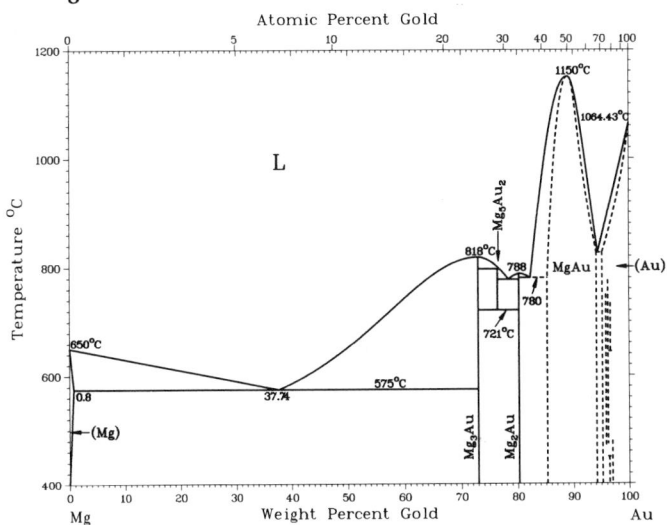

Phase	Composition, wt% Au	Pearson symbol	Space group
(Mg)	0 to 0.8	hP2	P6₃/mmc
Mg₃Au	73	hP8	P6₃/mmc or P3̄c1
Mg₅Au₂	76.42	...	...
Mg₂Au	80.20	...	Pnam or Pna2₁
(MgAu)	89.5	cP2	Pm3̄m
Mg₂₆Au₇₄	96	oC160	Cm2m
Mg₂₄Au₇₆	96.3	oC64	Cmcm
Mg₂₃Au₇₇	96.6	hP108	P6₃/mcm
Mg₂₂Au₇₈	96.64	tI16	I4/mmm
Mg₄Au₁₅	96.81	mP38	B2/m
MgAu₄	97	(a)	...
(Au)	100	cF4	Fm3̄m

(a) Structure reportedly is related to that of the "X-phase," Mg₄Au₁₅.

Au-Mn

T.B. Massalski and H. Okamoto, 1987

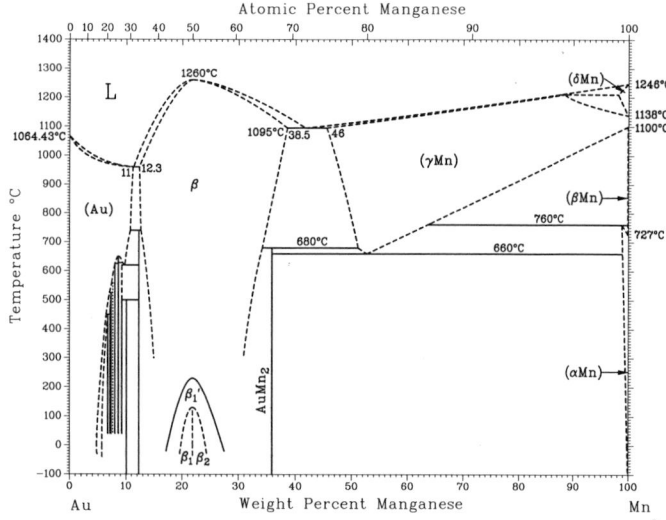

Au-rich region of the Au-Mn phase diagram

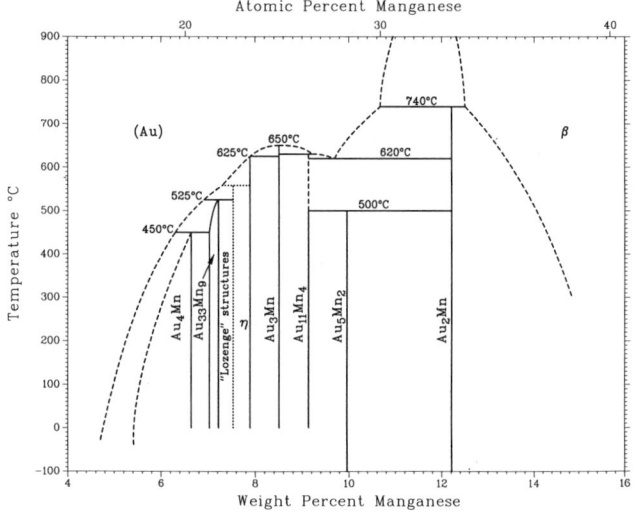

Phase	Composition, wt% Mn	Pearson symbol	Space group
(Au)	0 to 11	cF4	Fm$\bar{3}$m
Au$_4$Mn	5 to 6	tI10	I4/m
Au$_{33}$Mn$_9$	7.07	(a)	P2$_1$/b
Au$_{22}$Mn$_6$	7.07	(a)	...
α″	7.0 to 7.2	(b)	Pnnm
Au$_{31}$Mn$_9$	7.49	(c)	...
2d-APS(I)	...	(d)	...
Au$_{31}$Mn$_9$	7.49	(b)	...
Au$_{72}$Mn$_{21}$	5.52	(a)	...
Au$_{41}$Mn$_{12}$	7.55	(a)	...
Au$_{167}$Mn$_{49}$	7.57	(b)	...
Au$_{95}$Mn$_{28}$	7.59	(a)	...
Au$_{27}$Mn$_8$	7.63	(b)	...
Au$_{13}$Mn$_4$ (η)	7.50	(b)	...
2d-Au$_3$Mn(e)	~8	oP32	Pnnm
Au$_3$Mn	7.2 to 10	...	...
5H	...	(b)	...
3R	...	(b)	...
M = 1	...	tI8	I4/mmm
6H$_1$	...	(b)	...
6H$_2$	...	(b)	...
Au$_3$Mn(I)(f)	9	(b)	...
X (Au$_{11}$Mn$_4$)	9.21	(a)	...
(f)	~9.2	(b)	...
Au$_{11}$Mn$_4$I(f)	~9.2	(a)	...
Au$_{11}$Mn$_4$II(f)	~9.2	(a)	...
Au$_{11}$Mn$_4$III(e,f)	~9.2	(a)	...
AABB(f)	~9.2	(g)	...
Au$_5$Mn$_2$	10.04	(a)	C2/m
Au$_2$Mn	12.24	tI6	I4/mmm
β	12.3 to 38.5	cP2	Pm$\bar{3}$m
β′$_1$	16 to 29	(g)	...
β$_1$	19 to 22	(g)	...
β$_2$	22 to 25	(g)	...
	23	(b)	...
AuMn$_2$	36	tI6	I4/mmm
(δMn)	100	cI12	Im$\bar{3}$m
(γMn)	46 to 100	cF4	Fm$\bar{3}$m
(βMn)	100	cP20	P4$_1$32
(αMn)	100	cI58	I$\bar{4}$3m
(γMn1)(f)	67 to 100	(g)	...
(γMn2)(f)	60.5 to 75.3	(g)	...
(f)	73.4	(b)	...

Note: 2d = two dimensional. APS = antiphase structure. (a) Monoclinic. (b) Orthorhombic. (c) Square island. (d) Lozenge island. (e) Thin film. (f) Metastable. (g) Tetragonal

Au-Na

A.D. Pelton, 1987

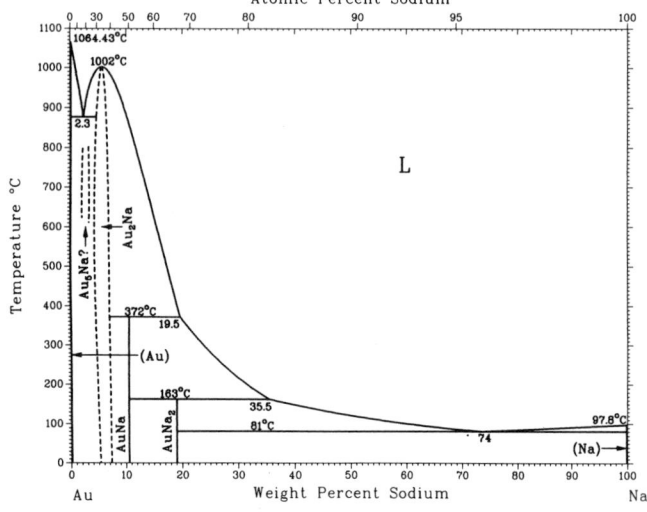

Phase	Composition, wt% Na	Pearson symbol	Space group
(Au)	0	cF4	Fm$\bar{3}$m
Au$_5$Na(a)	2 to 3	...	...
Au$_2$Na	5 to 7	cF24	Fd$\bar{3}$m
AuNa	10	(b)	...
AuNa$_2$	18.9 to 21	tI12	I4/mcm
(Na)	100	cI2	Im$\bar{3}$m
(Na)(c)	100	hP2	P6$_3$/mmc

(a) Existence requires verification; T = 775 °C. (b) Complex structure. (c) T is less than −237 °C.

Au-Nb

H. Okamoto and T.B. Massalski, 1987

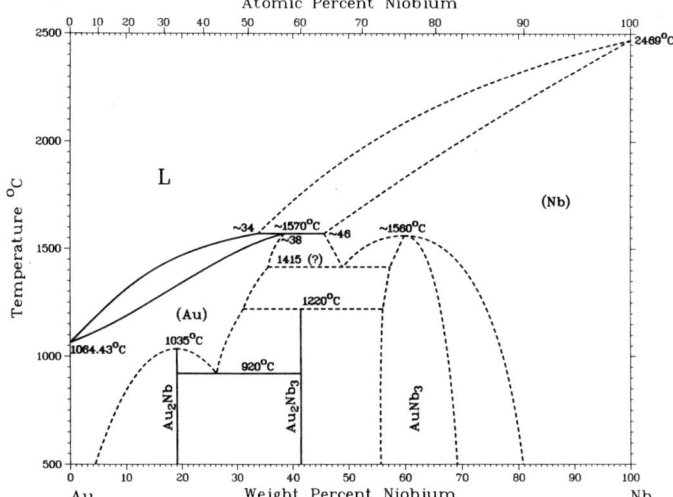

Phase	Composition, wt% Nb	Pearson symbol	Space group
(Au)	0 to ~38	$cF4$	$Fm\bar{3}m$
Au₂Nb	19.1	$hP3$	$P6/mmm$
Au₂Nb₃	41	$tI10$	$I4/mmm$
AuNb₃	56 to 70	$cP8$	$Pm\bar{3}n$
(Nb)	~46 to 100	$cI2$	$Im\bar{3}m$

Au-Ni

H. Okamoto and T.B. Massalski, 1991

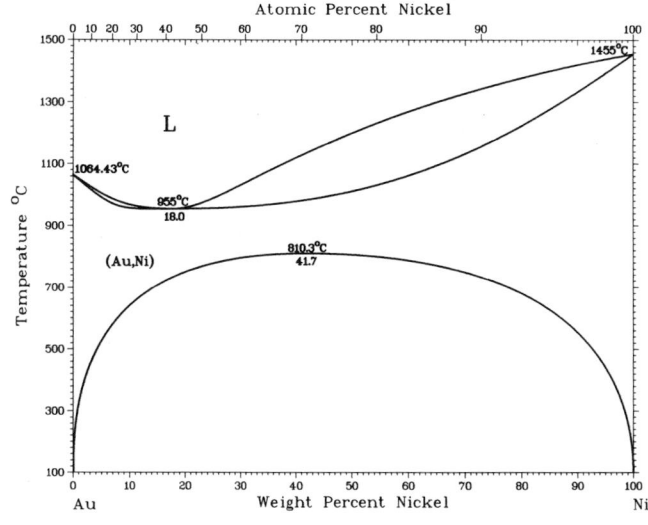

Phase	Composition, wt% Ni	Pearson symbol	Space group
(Au,Ni)	0 to 100	$cF4$	$Fm\bar{3}m$

Au-Pb

H. Okamoto and T.B. Massalski, 1987

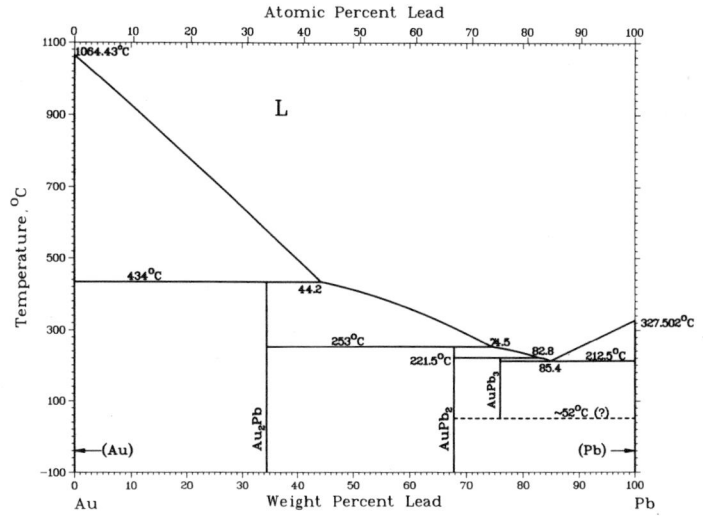

Phase	Composition, wt% Pb	Pearson symbol	Space group
(Au)	0 to 0.12	$cF4$	$Fm\bar{3}m$
Au₂Pb	34.4	$cF24$	$Fd\bar{3}m$
AuPb₂	67.8	$tI12$	$I4/mcm$
AuPb₃	75.9	$tI32$	$I\bar{4}2m$
(Pb)	99.81 to 100	$cF4$	$Fm\bar{3}m$

Au-Pd

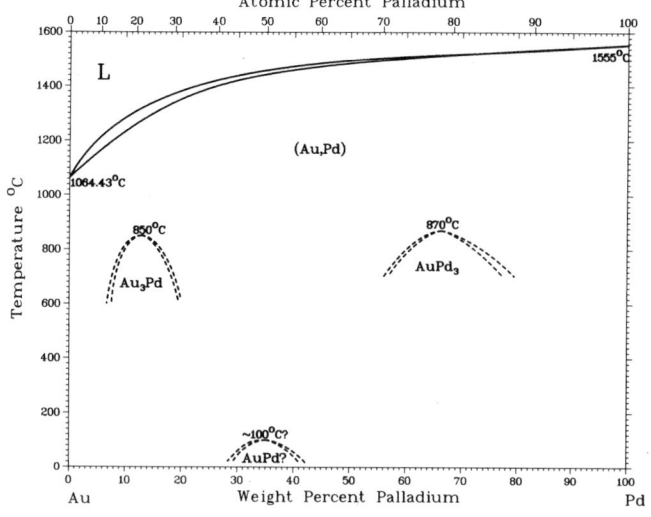

H. Okamoto and T.B. Massalski, 1987

Phase	Composition, wt% Pd	Pearson symbol	Space group
(Au,Pd)	0 to 100	cF4	Fm3̄m
Au₃Pd	7 to 20	cP4	Pm3̄m
AuPd	?	(a)	…
AuPd₃	53 to 83	cP4(?)	Pm3̄m

(a) Long period?

Au-Pr

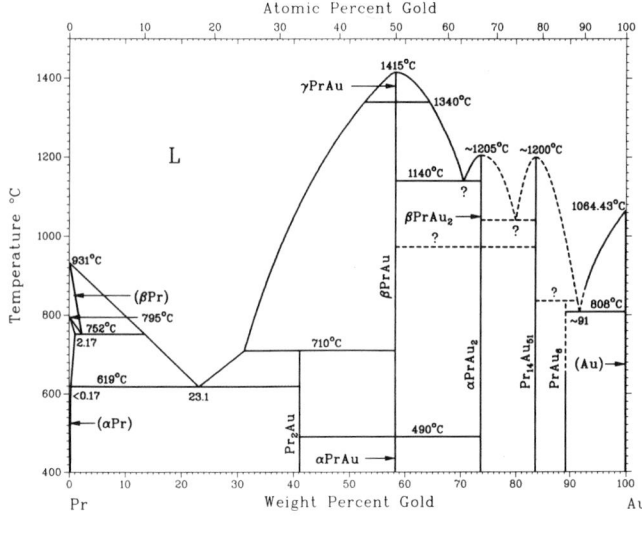

K.A. Gschneidner, Jr., F.W. Calderwood, H. Okamoto, and T.B. Massalski, 1987

Phase	Composition, wt% Au	Pearson symbol	Space group
(αPr)	0 to <0.17	hP4	P6₃/mmc
(βPr)	0 to 2.17	cI2	Im3̄m
Pr₂Au	41.1	oP12	Pnma
αPrAu	58	oP8	Pnma
βPrAu	58	oC8	Cmcm
γPrAu	58	cP2	Pm3̄m
αPrAu₂	73.7	oI12	Imma
βPrAu₂	73.7	tP108	P4/nmm
Pr₁₄Au₅₁	~81 to ~83.6	hP65	P6/m
PrAu₆	89.3	mC28	C2/c
(Au)	~99.93 to 100	cF4	Fm3̄m

Au-Pt

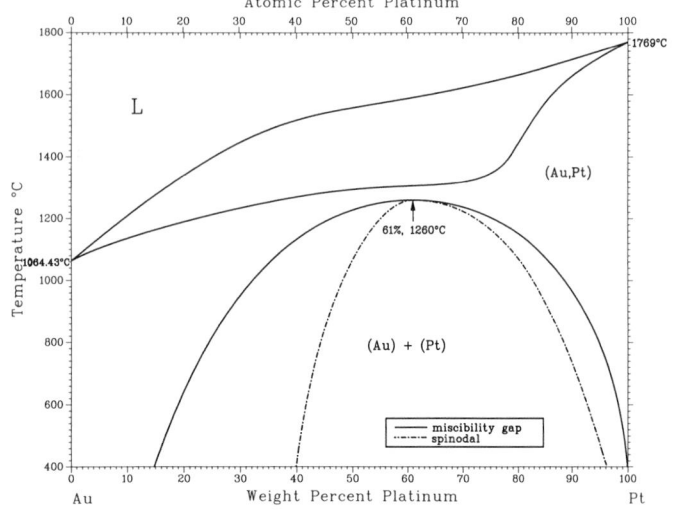

H. Okamoto and T.B. Massalski, 1987

Phase	Composition, wt% Pt	Pearson symbol	Space group
(Au,Pt)	0 to 100	cF4	Fm3̄m
Metastable phases			
Au₃Pt	4.9 to 39.8	…	…
AuPt	49.8	(a)	…
AuPt₃	74.8	…	…

(a) Tetragonal

Au-Pu

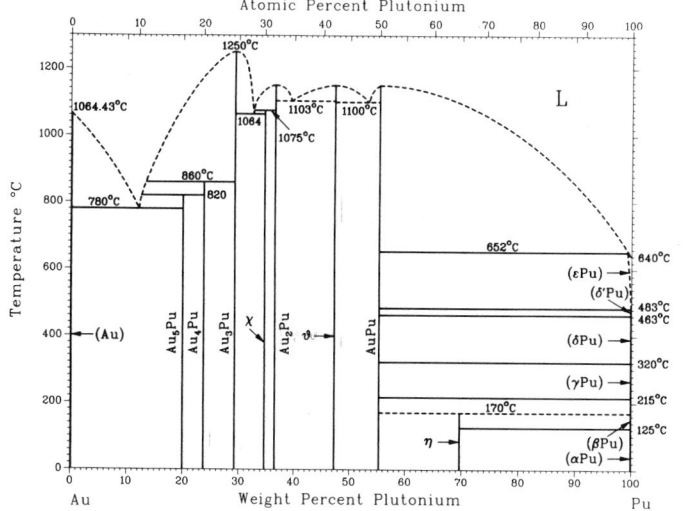

H. Okamoto, T.B. Massalski, and D.E. Peterson, 1987

Phase	Composition, wt% Pu	Pearson symbol	Space group
(Au)	0	cF4	Fm$\overline{3}$m
Au₅Pu (ν)	19.9	Unknown	...
Au₄Pu (μ)	23.7	Unknown	...
Au₃Pu (λ)	29.2	(a)	...
χ	35	Unknown	...
Au₂Pu(ι)	38.2	Unknown	...
θ	47	Unknown	...?
AuPu(ζ)	55.3	Unknown	...
η	70	Unknown	...
(εPu)	99.2 to 100	cI2	Im$\overline{3}$m
(δ'Pu)	100	tI2	I4/mmm
(δPu)	100	cF4	Fm$\overline{3}$m
(γPu)	100	oF8	Fddd
(βPu)	100	mC34	C2/m
(αPu)	100	mP16	P2₁/m

(a) Hexagonal

Au-Rb

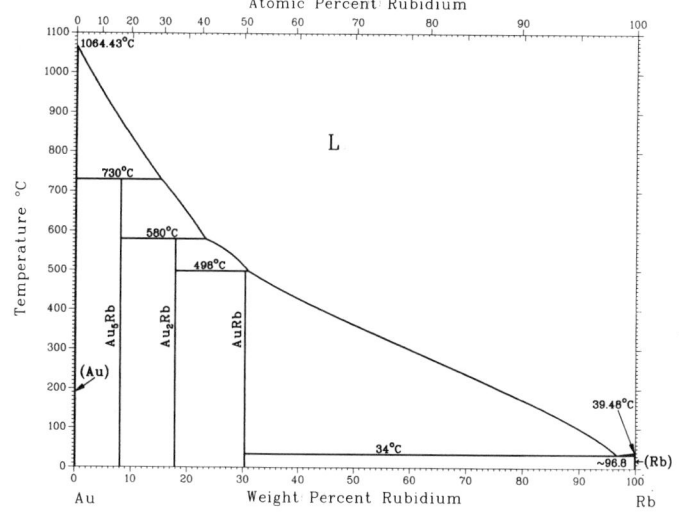

A.D. Pelton, 1987

Phase	Composition, wt% Rb	Pearson symbol	Space group
(Au)	0	cF4	Fm$\overline{3}$m
Au₅Rb	8.0	hP6	P6/mmm
Au₂Rb	17.8	...	...
AuRb	30.3	cP2	Pm$\overline{3}$m
(Rb)	100	cI2	Im$\overline{3}$m

Au-Sb

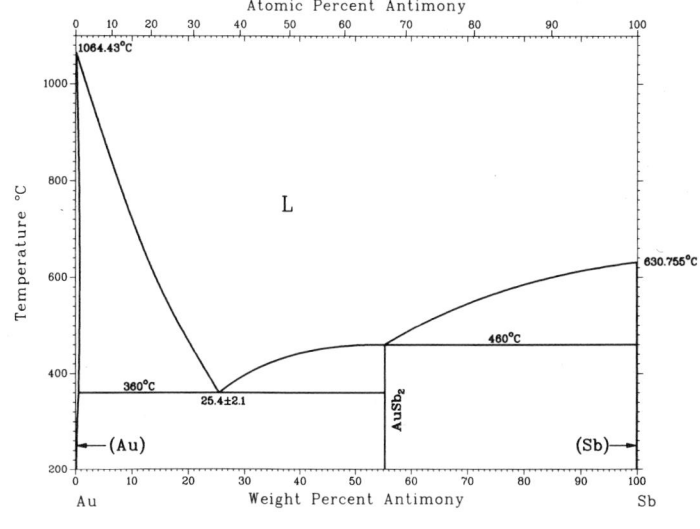

H. Okamoto and T.B. Massalski, 1987

Phase	Composition, wt% Sb	Pearson symbol	Space group
(Au)	0 to 0.75	cF4	Fm$\overline{3}$m
AuSb₂	55.3	cP12	Pa3
(Sb)	100	hR2	R$\overline{3}$m
Metastable phases			
...	8 to 10	hP2	P6₃/mmc
...	61 to 76	cP1	Pm$\overline{3}$m

Au-Se

H. Okamoto and T.B. Massalski, 1987

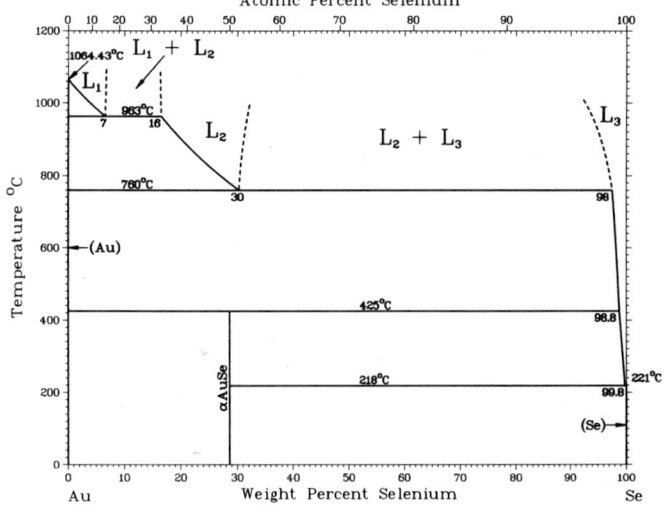

Phase	Composition, wt% Se	Pearson symbol	Space group
(Au)	0	cF4	Fm$\overline{3}$m
αAuSe	29	mC24	C2/m
βAuSe(a)	29	mC12	C2/m
(γSe)	100	hP3	P3$_1$21

(a) Metastable

Au-Si

H. Okamoto and T.B. Massalski, 1987

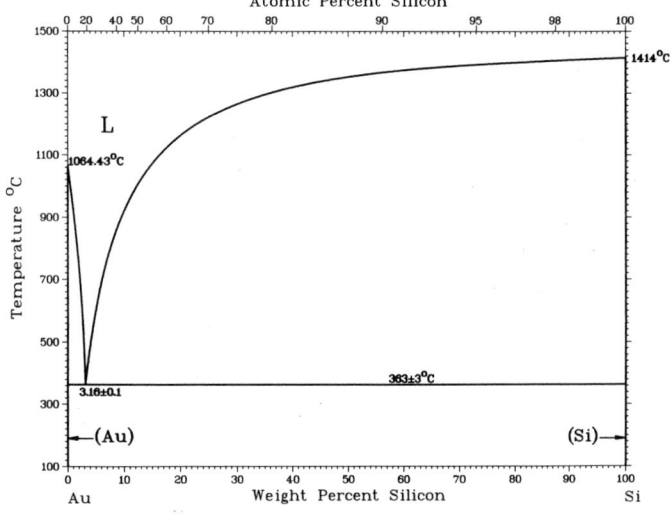

Phase	Composition, wt% Si	Pearson symbol	Space group
(Au)	0	cF4	Fm$\overline{3}$m
(Si)	100	cF8	Fd$\overline{3}$m

Au-Sn

H. Okamoto and T.B. Massalski, 1987

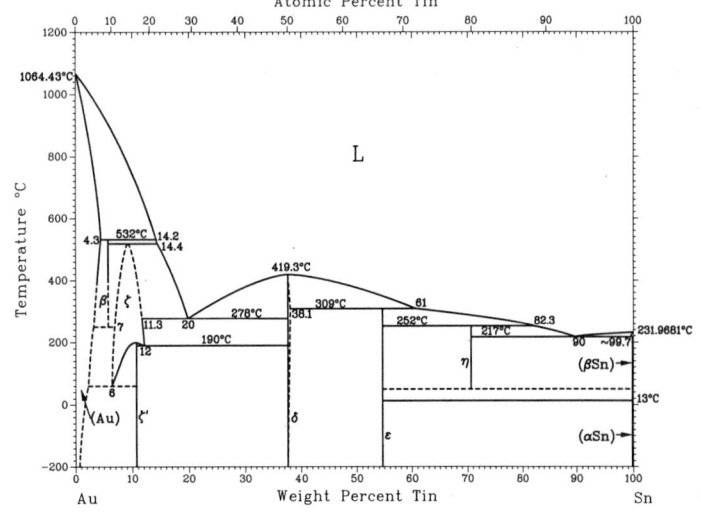

Phase	Composition, wt% Sn	Pearson symbol	Space group
(Au)	0 to 4.3	cF4	Fm$\overline{3}$m
β or Au$_{10}$Sn	5.7	hP16	P6$_3$/mmc
ζ	7 to 12	hP2	P6$_3$/mmc
ζ′ or Au$_5$Sn	10.8	(a)	R$\overline{3}$
δ or AuSn	38 to 38.08	hP4	P6$_3$/mmc
ε or AuSn$_2$	54.7	(b)	Pbca
η or AuSn$_4$	71	oC20	Aba2
(βSn)	99.7 to 100	tI4	I4$_1$/amd
(αSn)	99.990 to 100	cF8	Fm$\overline{3}$m

(a) Hexagonal. (b) Orthorhombic

Au-Sr

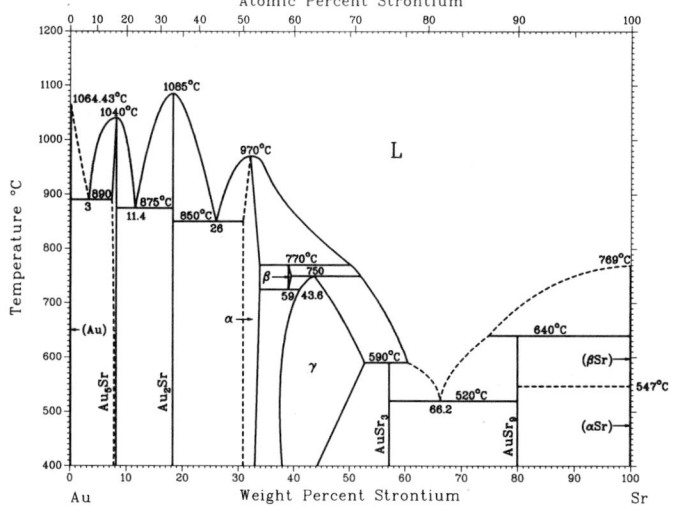

C.B. Alcock, V.P. Itkin, H. Okamoto, and T.B. Massalski, 1987

Phase	Composition, wt% Sr	Pearson symbol	Space group
(Au)	0	cF4	Fm3̄m
Au₅Sr	7.7 to 8.2	hP6	P6/mmm
Au₂Sr	18.2	oI12	Imma
α	32.1 to 34	?	…
β	39 to 40	?	…
γ	38.5 to 52.7	?	…
AuSr₃	57	?(a)	…
AuSr₉	80	?(b)	…
(βSr)	100	cI2	Im3̄m
(αSr)	100	cF4	Fm3̄m

(a) Complex. (b) Hexagonal

Au-Te

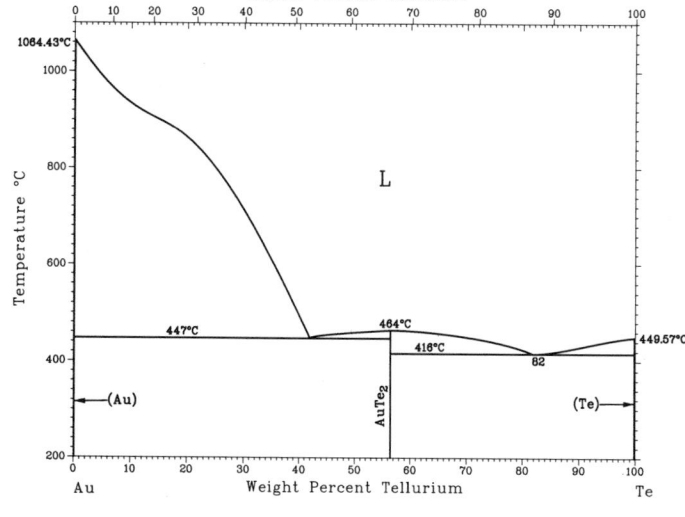

H. Okamoto and T.B. Massalski, 1987

Phase	Composition, wt% Te	Pearson symbol	Space group
(Au)	0 to 0.10	cF4	Fm3̄m
AuTe₂ (calaverite)	56.5	mC6	C2/m
(Te)	100	hP3	P3₁21
Metastable phases and other phases			
Petzite (a)	24.4	…	…
Montbrayite(a)	49	aP60	P1
Krennerite(a)	56.5	oP24	Pma2
(b)	48.9 to 79	cP1	Pm3̄m
(c)	91.8	(d)	…
(e)	60 to 100	(f)	…

(a) Natural ore. May be stable only with additional impurities. (b) Splat cooled at room temperature. Complete decomposition in 10 min at 165 °C (>69.6 at.% Te), 8 min at 260 °C or 10 h at 175 °C (62.5 at.% Te). (c) Splat cooled at room temperature. (d) Unidentified structure. (e) Vapor deposition of Te on Au at room temperature. (f) Amorphous

Au-Th

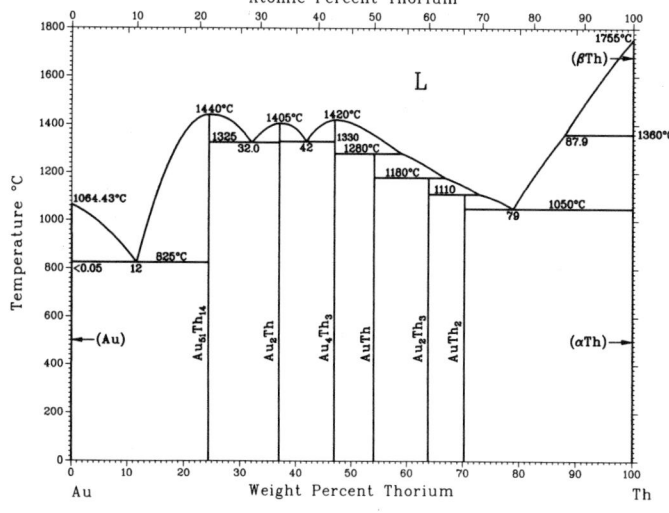

H. Okamoto, T.B. Massalski, and D.E. Peterson, 1991

Phase	Composition, wt% Th	Pearson symbol	Space group
(Au)	~0	cF4	Fm3̄m
Au₅₁Th₁₄	24.44	hP65	P6/m
Au₂Th	37.08	hP3	P6/mmm
Au₄Th₃	46.91	hR42	R3
AuTh	54	oC8	Cmcm
Au₂Th₃	64	(a)	…
AuTh₂	70.21	tI12	I4/mcm
(βTh)	100	cI2	Im3̄m
(αTh)	~100	cF4	Fm3̄m

(a) Cubic?

Au-Ti

J.L. Murray, 1987

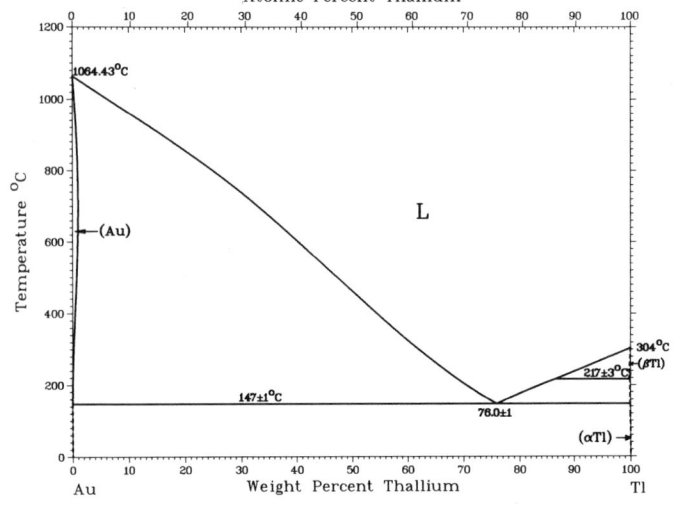

Phase	Composition, wt% Au	Pearson symbol	Space group
(αTi)	0 to 6.6	hP2	$P6_3/mmc$
(βTi)	0 to 42	cI2	$Im\bar{3}m$
Ti$_3$Au	58	cP8	$Pm\bar{3}n$
γTiAu	72 to 82	cP2	$Pm\bar{3}m$
βTiAu	80 to 80.4	oP4	Pmma
αTiAu	80.4	tP4	P4/nmm
TiAu$_2$	89.2	tI6	I4/mmm
TiAu$_4$	94 to 95	tI10	I4/m
(Au)	97 to 100	cF4	$Fm\bar{3}m$

Au-Tl

H. Okamoto and T.B. Massalski, 1987

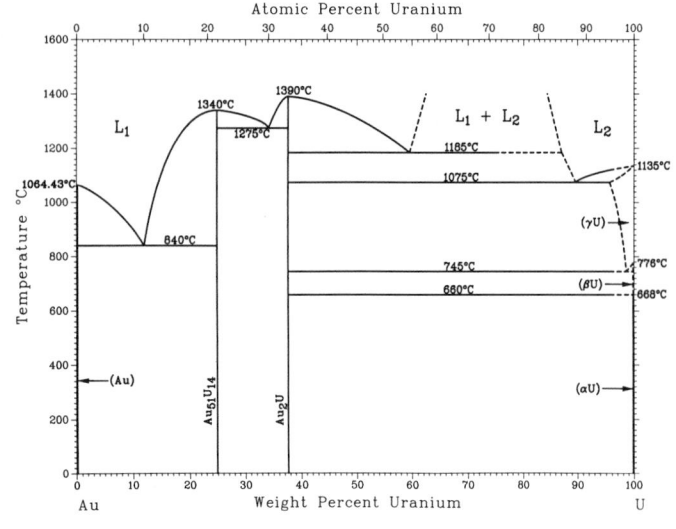

Phase	Composition, wt% Tl	Pearson symbol	Space group
(Au)	0 to 1.04	cF4	$Fm\bar{3}m$
(αTl)	100	hP2	$P6_3/mmc$
(βTl)	100	cI2	$Im\bar{3}m$

Au-U

H. Okamoto, 1990

Phase	Composition, wt% U	Pearson symbol	Space group
(Au)	0	cF4	$Fm\bar{3}m$
Au$_{51}$U$_{14}$	24.9	hP65	P6/m
Au$_2$U	37.6	hP3	P6/mmm
(γU)	100	cI2	$Im\bar{3}m$
(βU)	100	tP30	$P4_2/mnm$
(αU)	100	oC4	Cmcm

Au-V

J.F. Smith, 1989

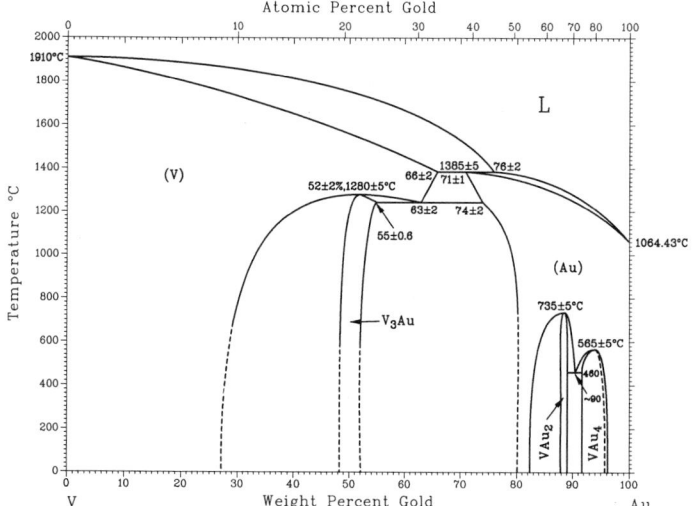

Phase	Composition, wt% Au	Pearson symbol	Space group
(V)	0 to ~66	cI2	Im$\overline{3}$m
V$_3$Au(a)	48 to 55	cP8	Pm$\overline{3}$n
VAu$_2$	88 to 89	oC12	(b)
VAu$_4$	92 to 96	tI10	I4/m
(Au)	~71 to 100	cF4	Fm$\overline{3}$m

(a) In the presence of small amounts of O or N, a second phase with the Cu$_3$Au-type structure may co-exist with the Cr$_3$Si-type structure. (b) Crystal structure related to the MoSi$_2$-type structure, but with a unit cell of twice the size.

Au-Yb

K.A. Gschneidner, Jr., F.W. Calderwood, H. Okamoto, and T.B. Massalski, 1987

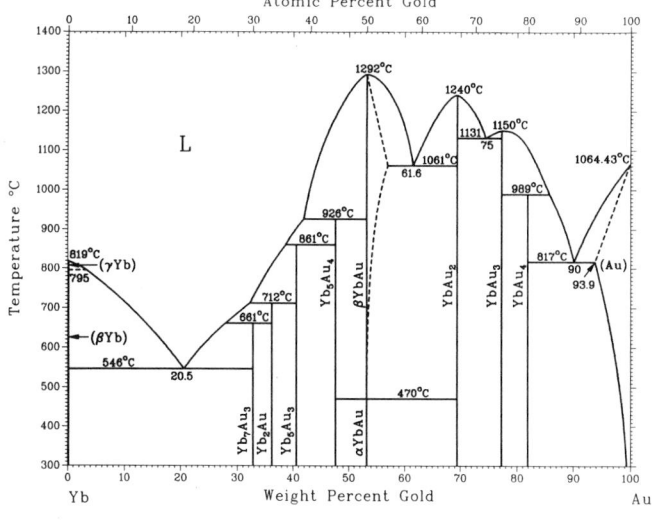

Phase	Composition, wt% Au	Pearson symbol	Space group
(βYb)	0	cF4	Fm$\overline{3}$m
(γYb)	0	cI2	Im$\overline{3}$m
Yb$_7$Au$_3$	33	hP20	P6$_3$/mc
Yb$_2$Au	36.2	oP12	Pnma
Yb$_5$Au$_3$	40.6	tI32	I4/mcm
Tb$_5$Au$_4$	47.6	oP36	Pnma
αYbAu	53	oP8	Pnma
βYbAu	53	cP2	Pm$\overline{3}$m
YbAu$_2$	69.5	tI6	I4/mmm
YbAu$_2$	77	oP8	Pmmn
YbAu$_4$	82	tI10	I4/m
(Au)	93.9 to 100	cF4	Fm$\overline{3}$m

Au-Zn

H. Okamoto and T.B. Massalski, 1990

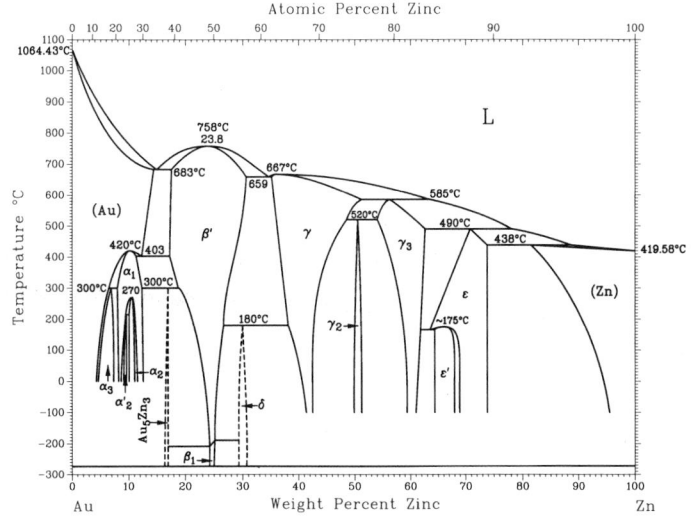

Phase	Composition, wt% Zn	Pearson symbol	Space group
(Au)	0 to 14	cF4	Fm$\overline{3}$m
α$_3$	~4 to 7.4	(a)	Pn$\overline{2}$n or Pnmn
α$_1$	~7.9 to 11.7	(b)	
α'$_2$	9.0 to 9.5	(b)	I4$_1$/acd
α$_2$	~9.7 to 10.2	(a)	Abam (Cmca)
Au$_5$Zn$_3$	16.6	(a)	...
β'	17 to 31	cP2	Pm$\overline{3}$m
β$_1$	24 to 26	?	...
δ	30	?	...
γ	38 to 51	cI52	...
γ$_2$	50 to 51	cP32	Pm$\overline{3}$m
γ$_3$	54 to 62.7	hP*	P6/mmm
ε	64 to 73	hP2	P6$_3$/mmc
ε'	64 to 67	(c)	...
(Zn)	80.4 to 100	hP2	P6$_3$/mmc

(a) Orthorhombic, antiphase domain. (b) Tetragonal, antiphase domain. (c) Orthorhombic, pseudocell

Au-Zr

T.B. Massalski, H. Okamoto, and J.P. Abriata, 1987

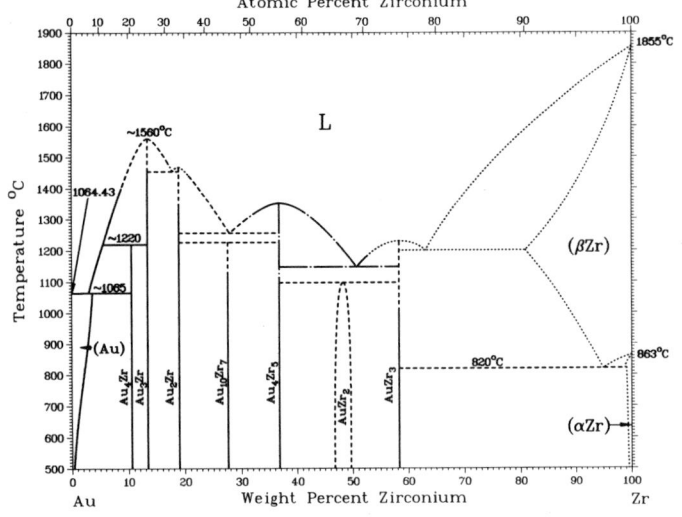

Phase	Composition, wt% Zr	Pearson symbol	Space group
(Au)	0	$cF4$	$Fm\overline{3}m$
Au_4Zr	10	$oP20$	$Pnma$
Au_3Zr	13	$oP8$	$Pmmn$
Au_2Zr	18.8	$tI6$	$I4/mmm$
$Au_{10}Zr_7$	27	$tI34$	?
Au_4Zr_5	36.7	...	...
$AuZr_2$	48.1	$tI6$	$I4/mmm$
$AuZr_3$	58	$cP8$	$Pm\overline{3}n$
(βZr)	100	$cI2$	$Im\overline{3}m$
(αZr)	100	$hP2$	$P6_3/mmc$

B-C

H. Okamoto, 1992

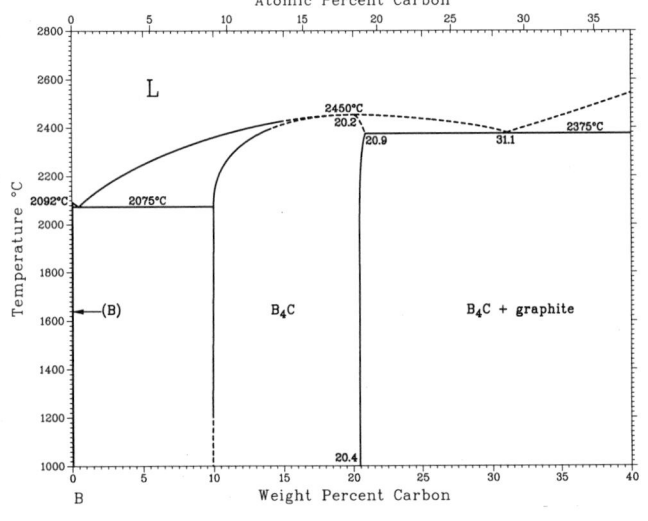

Phase	Composition, wt% C	Pearson symbol	Space group
(βB)	0	$hR108$	$R\overline{3}m$
"B_4C"	10 to 20.9	$hR15$	$R\overline{3}m$
(C)	100	$hP4$	$P6_3/mmc$

B-Co

P.K. Liao and K.E. Spear, 1988

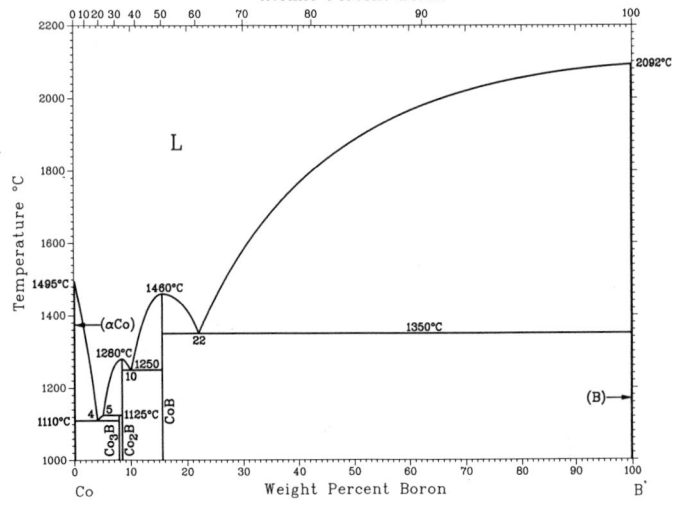

Phase	Composition, wt% B	Pearson symbol	Space group
(αCo)	~0	$cF4$	$Fm\overline{3}m$
(εCo)	~0	$hP2$	$P6_3/mmc$
Co_3B	7.8	...	$Pbnm$
Co_2B	8.4	$tI12$	$I4/mcm$
CoB	15.5	$oP8$	$Pnma$
(βB)	100	$hR108$	$R\overline{3}m$

B-Cr

P.K. Liao and K.E. Spear, 1986

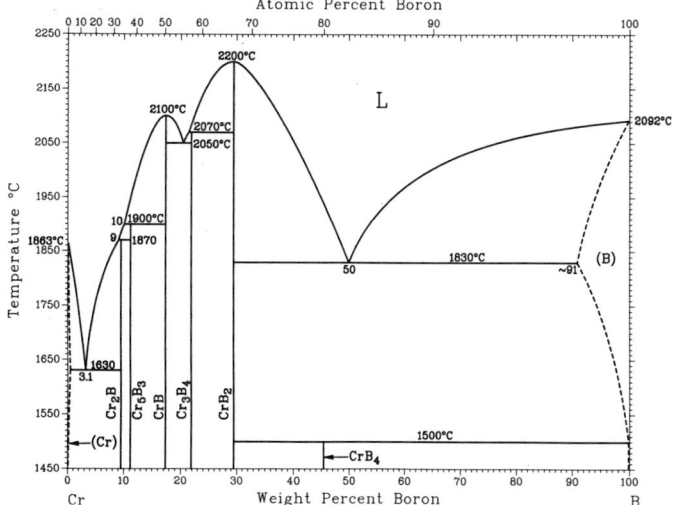

Phase	Composition, wt% B	Pearson symbol	Space group
(αCr)	0 to ~ 0.2	cI2	Im$\overline{3}$m
Cr$_4$B(a)	5	oF40	Fddd
Cr$_2$B	9.4	oF40	Fddd
Cr$_2$B(a)	9.4	(b)	Abmm
Cr$_2$B(a)	9.4	tI12	I4/mcm
Cr$_5$B$_3$	11.1	tI32	I4/mcm
CrB	17.2	oC8	Cmcm
Cr$_3$B$_4$	21.7	oI14	Immm
CrB$_2$	29.4	hP3	P6/mmm
CrB$_4$	45	(b)	...
CrB$_6$(a)	55.5	(c)	...
(βB)	~91 to 100	hR108	R$\overline{3}$m

(a) Unstable or stability is uncertain. (b) Orthorhombic. (c) Tetragonal

B-Cu

D.J. Chakrabarti and D.E. Laughlin, 1982

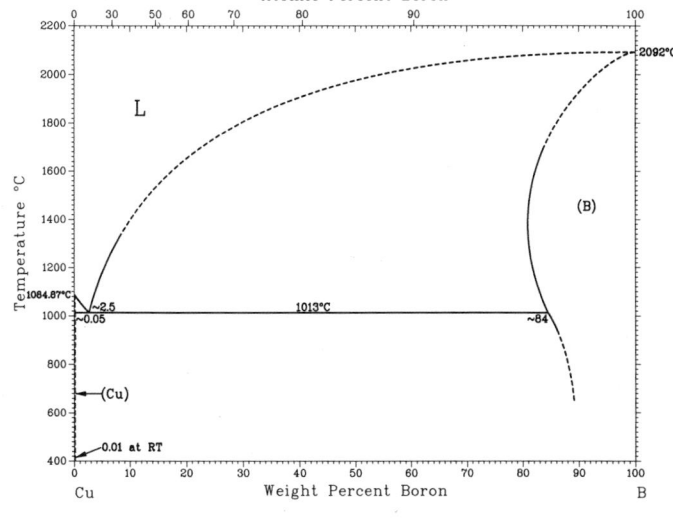

Phase	Composition, wt% B	Pearson symbol	Space group
(Cu)	0 to ~0.05	cF4	Fm$\overline{3}$m
(B)	>80	tP192	P4$_1$2$_1$2 or P4$_3$2$_1$2(?)
(B)	>80	hR105	R$\overline{3}$m
(βB)	100	hR108	R$\overline{3}$m
(αB)	100	hR12	R$\overline{3}$m
		tP192	P4$_1$2$_1$2 or P4$_3$2$_1$2(?)

B-Fe

P.K. Liao and K.E. Spear, unpublished

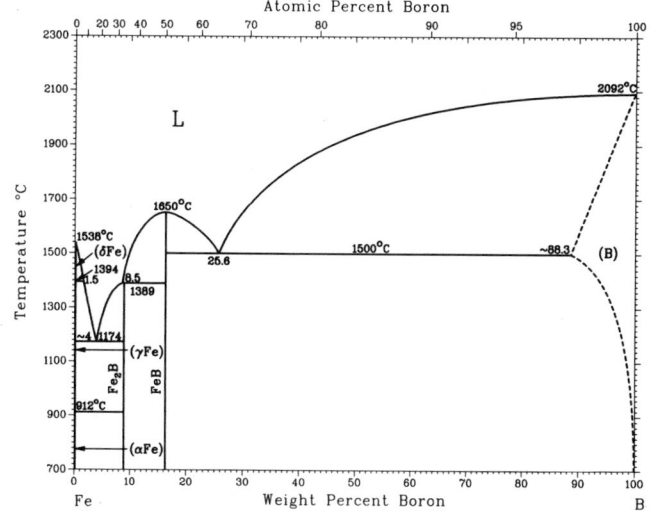

Phase	Composition, wt% Fe	Pearson symbol	Space group
(αFe)	0	cI2	Im$\overline{3}$m
Fe$_2$B	8.8	tI12	I4/mcm
FeB	16.0 to 16.2	oP8	Pbmn
(βB)	100	hR108	R$\overline{3}$m
Metastable phases			
Fe$_3$B	~6	oP16	Pnma
Fe$_3$B(HT)	~6	(a)	...
Fe$_3$B(LT)	~6	(b)	...

(a) bct. (b) Tetragonal

B-Mn

P.K. Liao and K.E. Spear, 1986

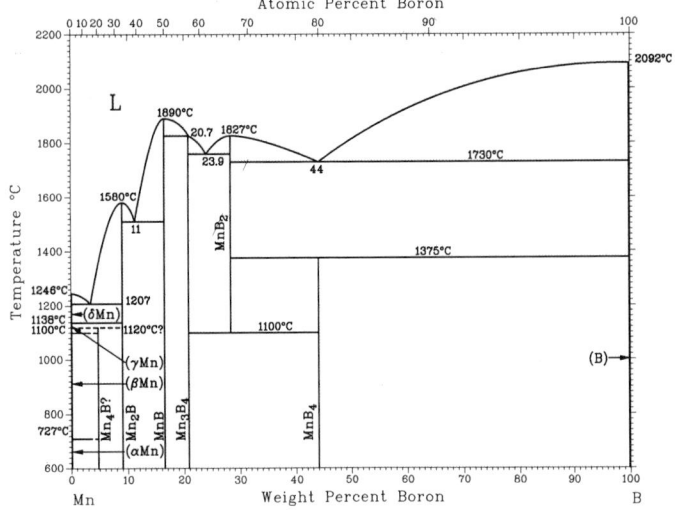

Phase	Composition, wt% B	Pearson symbol	Space group
(δMn)	0	cI2	Im$\bar{3}$m
Mn$_4$B(a)	5	oF40	Fddd
Mn$_2$B(a)	9.0	(b)	Fddd
	9.0	tI12	I4/mcm
MnB	16	oP	Pnma
Mn$_3$B$_4$	20.8	oI14	Immm
MnB$_2$	28.3	hP3	P6/mmm
MnB$_4$	44	(c)	C2/m
MnB$_{\sim23}$(d)	...	hR108	R$\bar{3}$m
(βB)	100	hR108	R$\bar{3}$m

(a) Probably not thermodynamically stable. Also, orthorhombic Mn$_4$B and Mn$_2$B may refer to the same phase. (b) Orthorhombic. (c) Monoclinic. (d) Probably the Mn-rich boundary or rhombohedral B

B-Mo

K.E. Spear and P.K. Liao, 1988

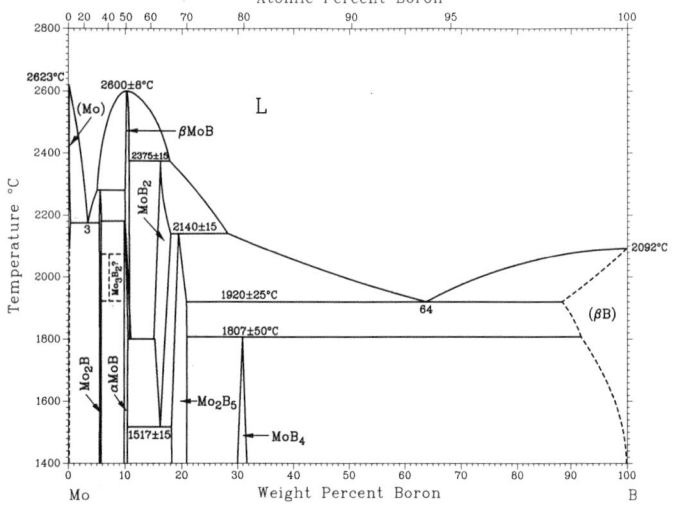

Phase	Composition, wt% B	Pearson symbol	Space group
(Mo)	0 to <0.1	cI2	Im$\bar{3}$m
Mo$_2$B	~5	tI12	I4/mcm
αMoB	9 to 10	tI16	I4$_1$/amd
βMoB	9 to 10.4	oC8	Cmcm
MoB$_2$	16 to 18	hP3	P6/mmm
Mo$_2$B$_5$	18.6 to 20	hR21	R$\bar{3}$m
MoB$_4$	~30	hP20	P6$_3$/mmc
(βB)	>92 to 100	hR108	R$\bar{3}$m

B-Nb

H. Okamoto, 1990

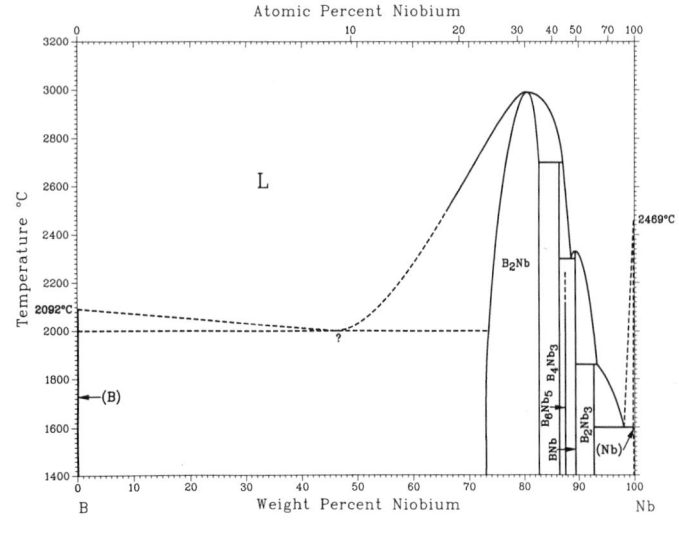

Phase	Composition, wt% Nb	Pearson symbol	Space group
(βB)	0	hR108	R$\bar{3}$m
B$_2$Nb	73 to 83	hP3	P6/mmm
B$_4$Nb$_3$	86.6	oI14	Immm
B$_6$Nb$_5$	87.8	oC*	Cmmm
BNb	90	oC8	Cmcm
B$_2$Nb$_3$	93	tP10	P4/mbm
(Nb)	100	cI2	Im$\bar{3}$m

B-Ni

P.K. Liao and K.E. Spear, 1991

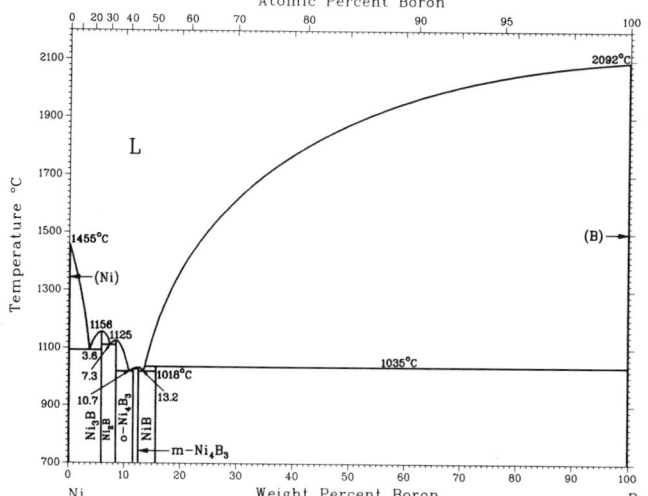

Phase	Composition, wt% B	Pearson symbol	Space group
(Ni)	0	cF4	$Fm\bar{3}m$
Ni₃B	6	oP6	Pnma
Ni₂B	8.4	tI12	I4/mcm
Ni₄B₃	11.5	(a)	Pnma
Ni₄B₃	12.5	(b)	C2/c
NiB	16	oC8	Cmcm
NiB₂(c)	26.9	(d)	...
NiB₁₂(c)	68.8	(d)	...
(βB)	100	hR108	$R\bar{3}m$

(a) Orthorhombic. (b) Monoclinic. (c) Existence of these compounds has been reported, but is highly unlikely. (d) Cubic

B-Pd

P.K. Liao and K.E. Spear, unpublished

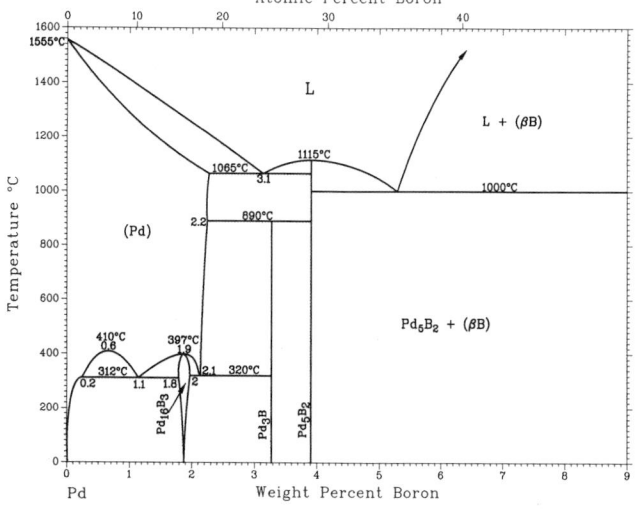

Phase	Composition, wt% B	Pearson symbol	Space group
(Pd)	0.00 to 2.2	cF4	$Fm\bar{3}m$
Pd₁₆B₃	1.9	...	...
Pd₃B	3.4	oP16	Pnma
Pd₅B₂	3.9	mC28	C2/c
(βB)	100	hR105	$FR\bar{3}m$

B-Pt

H. Okamoto, 1990

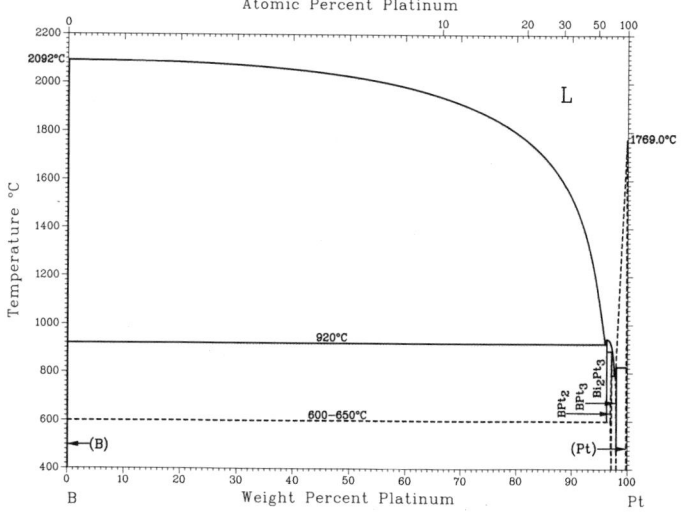

Phase	Composition, wt% Pt	Pearson symbol	Space group
(βB)	0	hR108	$R\bar{3}m$
B₂Pt₃	96	...	...
BPt₂	97.3	hP6	P6₃/mmc
BPt₃	98	t**	...
(Pt)	100	cF4	$Fm\bar{3}m$

B-Re

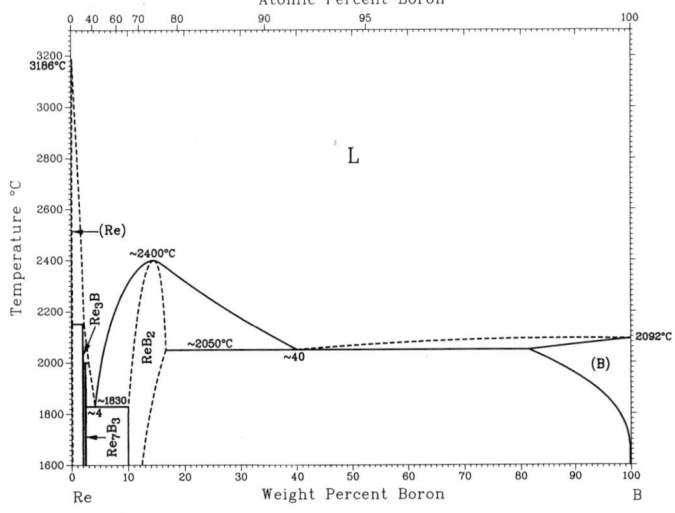

K.I. Portnoi and V.M. Romashov, 1972

Phase	Composition, wt% B	Pearson symbol	Space group
(Re)	0 to ~0.06	hP2	$P6_3/mmc$
Re₃B	~2	oC16	Cmcm
Re₇B₃	~2.4	hP20	$P6_3/mc$
ReB₂	~10 to ~17	hP6	$P6_3/mmc$
(B)	~85 to 100	hR105	$R\overline{3}m$

B-Ru

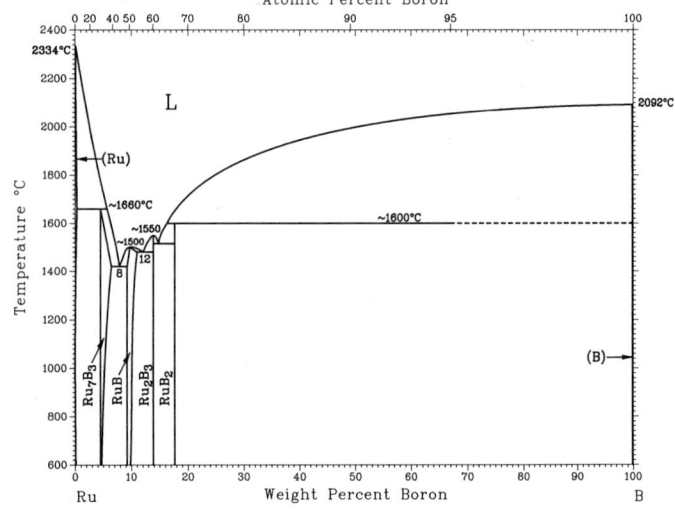

W. Obrowski, 1963

Phase	Composition, wt% B	Pearson symbol	Space group
(Ru)	0 to ~0.2	hP2	$P6_3/mmc$
Ru₇B₃	~4 to 6	hP20	$P6_3/mc$
RuB	~9 to 11	hP2	$P6m2$
Ru₂B₃	14	hP12	$P6_3/mmc$
RuB₂	17.6	oP6	Pmmn
(B)	~100	hR105	$R\overline{3}m$

B-Sc

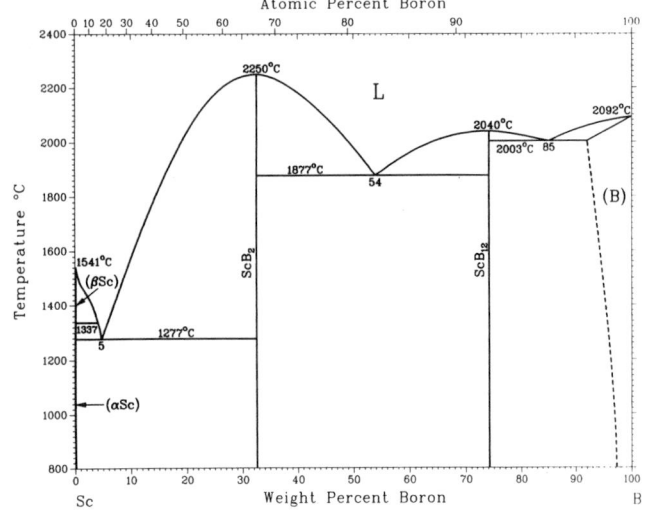

K.E. Spear and P.K. Liao, 1990

Phase	Composition, wt% B	Pearson symbol	Space group
(αSc)	0	hP2	$P6_3/mmc$
(βSc)	0	cI2	$Im\overline{3}m$
ScB₂	33	hP3	P6/mmm
ScB₁₂	73	tI26	I4/mmm
ScB₂₀	(a)	...	...
(βB)	100	hR108	$R\overline{3}m$

(a) Metastable, rhombohedral (βB)

B-Si

R.W. Olesinski and G.J. Abbaschian, 1984

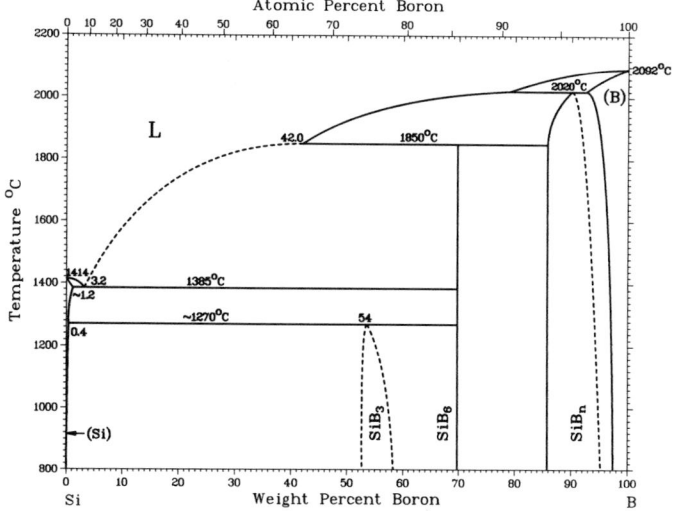

Phase	Composition, wt % B	Pearson symbol	Space group
(αSi)	0 to ~1.2	cF8	Fd$\bar{3}$m
(βSi) (HP)	0	tI4	I4$_1$/amd
SiB$_3$	52.7 to 58.4	hR15	R$\bar{3}$m
SiB$_6$	69.8	oP280	Pnnm
SiB$_n$	84.3 to ~93	hR12	R$\bar{3}$m
(B)	~93 to ~100	hR12	R$\bar{3}$m
(βB)(a)	100	hR105	R$\bar{3}$m

(a) Assumed to be the only stable phase of pure B

B-Ta

H. Okamoto, 1990

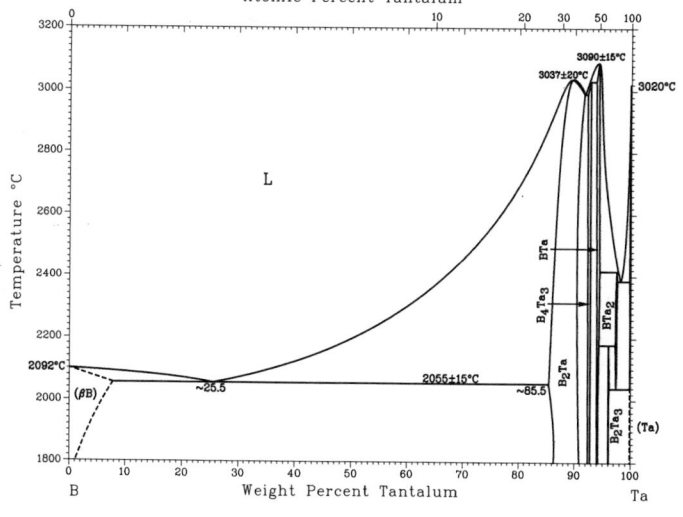

Phase	Composition, wt% Ta	Pearson symbol	Space group
(βB)	0 to ~2	hR108	R$\bar{3}$m
B$_2$Ta	~85.5 to 91	hP3	P6/mmm
B$_4$Ta$_3$	92.4 to 92.9	oI14	Immm
BTa	94 to 95	oC8	Cmcm
B$_2$Ta$_3$	96.0 to 96.3	tP10	P4/mbm
BTa$_2$	97.4 to 97.7	tI12	I4/mcm
(Ta)	100	cI2	Im$\bar{3}$m

B-Ti

J.L. Murray, P.K Liao, and K.E. Spear, 1987

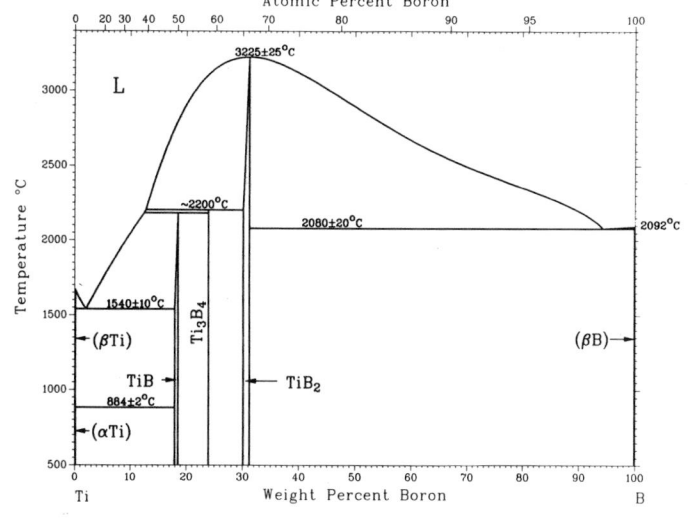

Phase	Composition, wt% B	Pearson symbol	Space group
(αTi)	0 to <0.05	hP2	P6$_3$/mmc
(βTi)	0 to <0.05	cI2	Im$\bar{3}$m
TiB	18 to 18.4	oP8	Pnma
Ti$_3$B$_4$	22.4	oI14	Immm
TiB$_2$	30.1 to 31.1	hP3	P6/mmm
(βB)	~100	hR108	R$\bar{3}$m

B-V

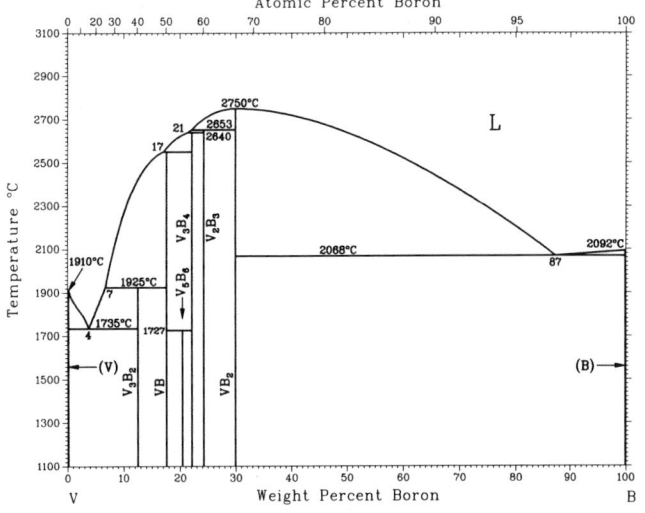

K.E. Spear, P.K. Liao, and J.F. Smith, 1991

Phase	Composition, wt% B	Pearson symbol	Space group
(V)	0	cI2	Im$\bar{3}$m
V$_3$B$_2$	12	tP10	P4/mbm
VB	18	oC8	Cmcm
V$_5$B$_6$	20.3	(a)	Ammm
V$_3$B$_4$	22	oI14	Immm
V$_2$B$_3$	24	(a)	Cmcm
VB$_2$	30	hP3	P6/mmm
(βB)	100	hR108	R$\bar{3}$m

(a) Orthorhombic

B-W

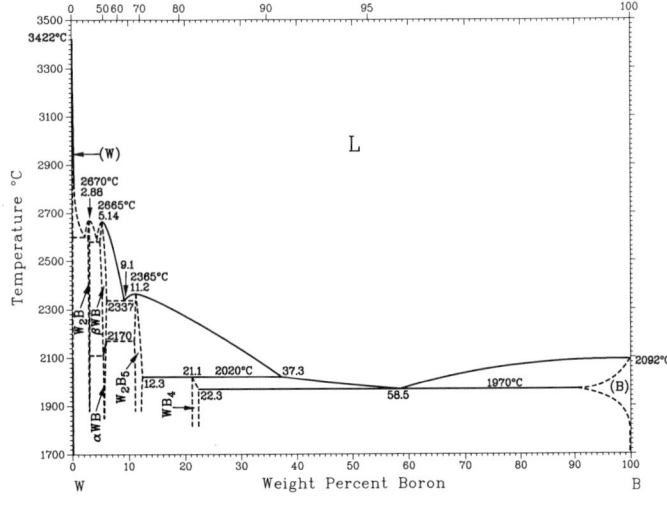

S.V. Nagender Naidu and P. Rama Rao, 1991

Phase	Composition, wt% B	Pearson symbol	Space group
(W)	0	cI2	Im$\bar{3}$m
W$_2$B	2.9	tI12	I4/mcm
βWB	5.2	oC8	Cmcm
αWB	5.4	tI16	I4$_1$/amd
W$_2$B$_5$	11.1	hP14	P6$_3$/mmc
WB$_4$	21.1	hP20	P6$_3$/mmc
(B)	100	hR12	R$\bar{3}$m
		tP50	P4$_2$/nnm

B-Y

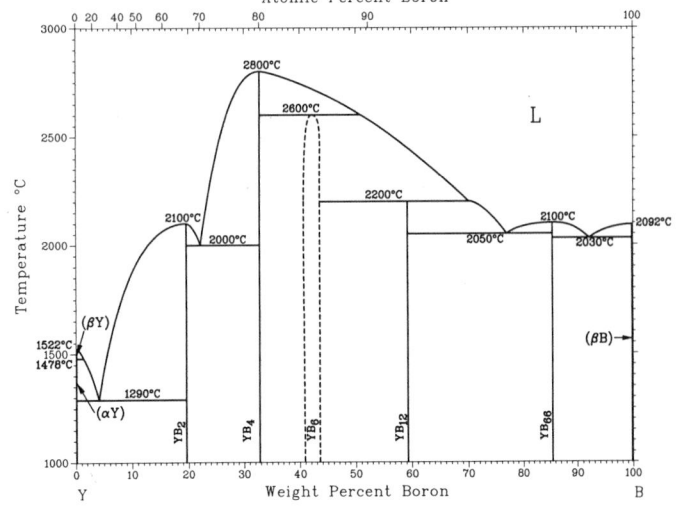

P.K. Liao and K.E. Spear, unpublished

Phase	Composition, wt% B	Pearson symbol	Space group
(βY)	0	cI2	Im$\bar{3}$m
(α)Y	0	hP2	P6$_3$/mmc
YB$_2$	19.6	hP3	P6/mmm
YB$_4$	32.7	tP20	P4/mbm
YB$_6$	42.2	cP7	Pm$\bar{3}$m
YB$_{12}$	59.3	cF52	Fm$\bar{3}$m
YB$_{66}$	85.6	...	Fm$\bar{3}$c
(βB)	100	hR108	R$\bar{3}$m

B-Zr

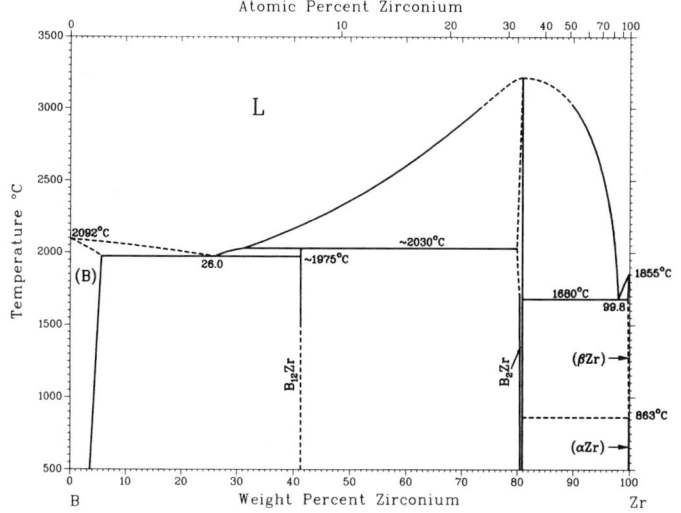

Phase	Composition, wt% Zr	Pearson symbol	Space group
(B)	~0	hR105	$R\overline{3}m$
B$_{12}$Zr	40.9	cF52	$Fm\overline{3}m$
B$_2$Zr	80 to 83.8	hP3	$P6/mmm$
(βZr)	99.8 to 100	cI2	$Im\overline{3}m$
(αZr)	~100	hP2	$P6_3/mmc$

Ba-Ca

C.B. Alcock and V.P. Itkin, 1986

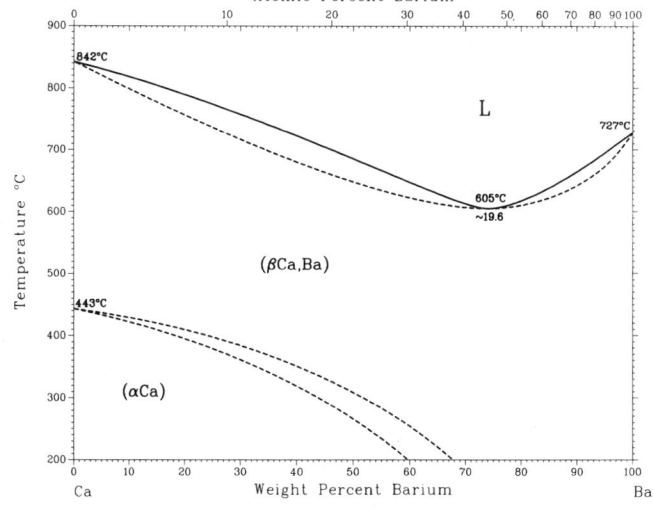

Phase	Composition, wt% Ba	Pearson symbol	Space group
(αCa)	0 to 60	cF4	$Fm\overline{3}m$
(βCa,Ba)	0 to 100	cI2	$Im\overline{3}m$

Ba-Cd

H. Okamoto, 1990

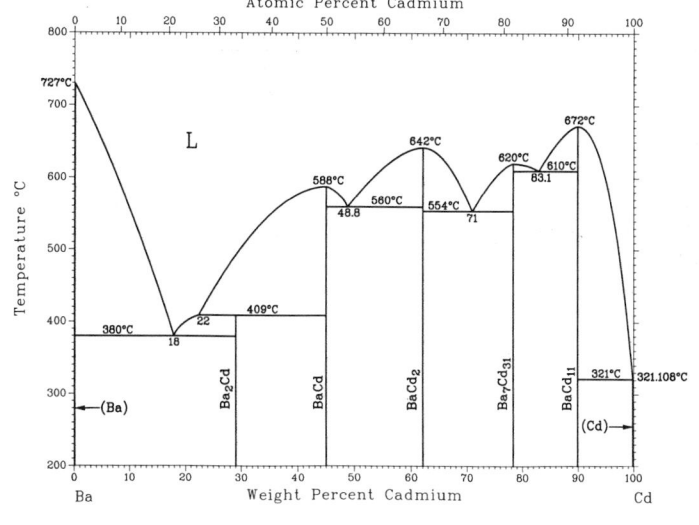

Phase	Composition, wt% Cd	Pearson symbol	Space group
(Ba)	0	cI2	$Im\overline{3}m$
Ba$_2$Cd	29.0	tI6	$I4/mmm$
BaCd	45	cP2	$Pm\overline{3}m$
BaCd$_2$	62.1	oI12	$Imma$
Ba$_7$Cd$_{31}$	78.4	hP41	$P6/mmm$
BaCd$_{11}$	90.0	tI48	$I4_1/amd$
(Cd)	100	hP2	$P6_3/mmc$

Ba-Cu

D.J. Chakrabarti and D.E. Laughlin, 1984

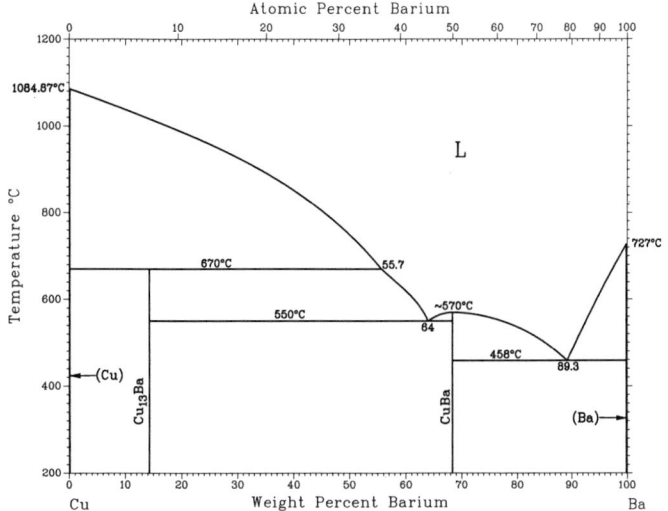

Phase	Composition, wt% Ba	Pearson symbol	Space group
(Cu)	0	cF4	$Fm\overline{3}m$
Cu$_{13}$Ba	14.25	cF112	$Fm\overline{3}c$
CuBa	68.3	hP8	$P6_3/mmc$
(Ba)	100	cI2	$Im\overline{3}m$
Pressure-stabilized phase			
Ba	100	hP2	$P6_3/mmc$

Ba-Ga

V.P. Itkin and C.B. Alcock, 1991

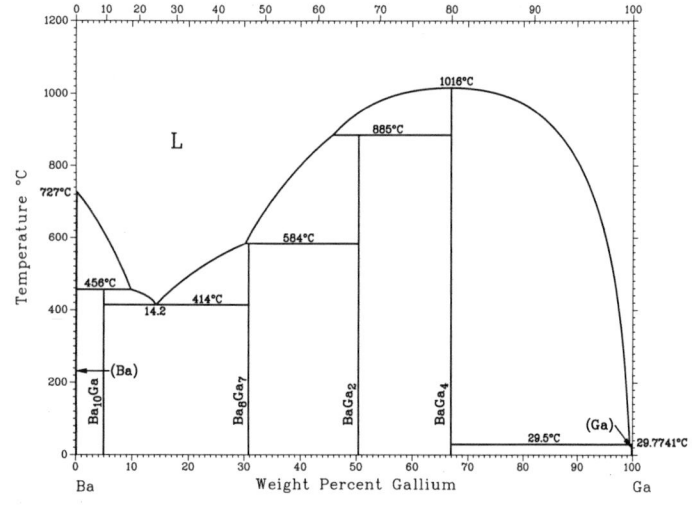

Phase	Composition, wt% Ga	Pearson symbol	Space group
(Ba)	0	cI2	$Im\overline{3}m$
Ba$_{10}$Ga	4.8	cF176	$Fd\overline{3}m$
Ba$_8$Ga$_7$	30.8	cP60	$P2_13$
BaGa$_2$	50.4	hP3	$P6/mmm$
BaGa$_4$	67	tI10	$I4/mmm$
(Ga)	100	hP2	$P6_3/mmc$

Ba-Ge

P.R. Subramanian, 1990

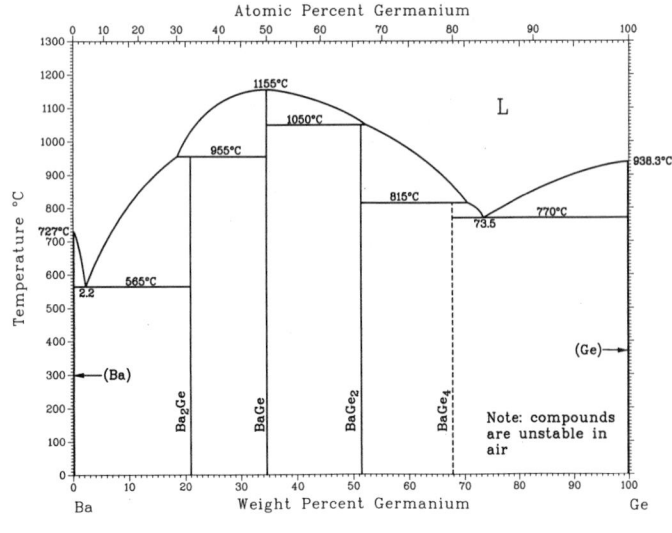

Phase	Composition, wt% Ge	Pearson symbol	Space group
(Ba)	0	cI2	$Im\overline{3}m$
Ba$_2$Ge	20.9	oP12	Pnma
BaGe	35	oC8	Cmcm
BaGe$_2$	51.4	cP84	$P4_132$
		oP24	Pmna
BaGe$_4$	68	...	...
(Ge)	100	cF8	$Fd\overline{3}m$
High-pressure phase			
BaGe$_2$(a)	51.4	tI12	$I4_1/amd$

(a) Prepared at 1000 °C, 40 kbar pressure

Ba-H

D.T. Peterson and M. Indig, 1960

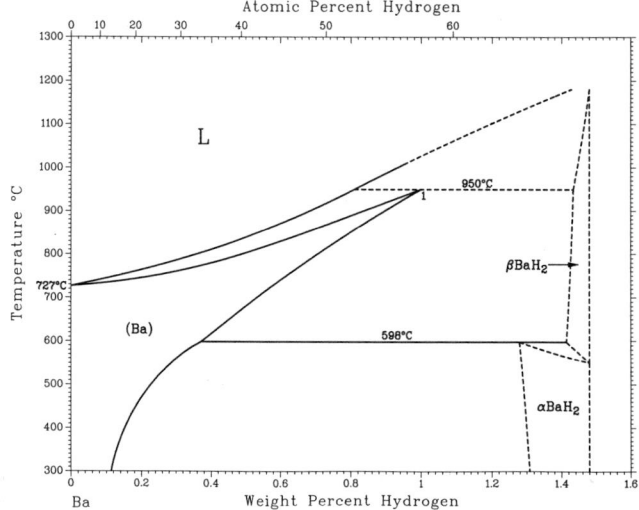

Phase	Composition, wt% H	Pearson symbol	Space group
(Ba)	0 to 1	cI2	Im$\bar{3}$m
αBaH$_2$	~1.3 to 1.5	oP12	Pnma
βBaH$_2$	~1.4 to 1.5	cI*	...

Ba-Hg

P.R. Subramanian, 1990

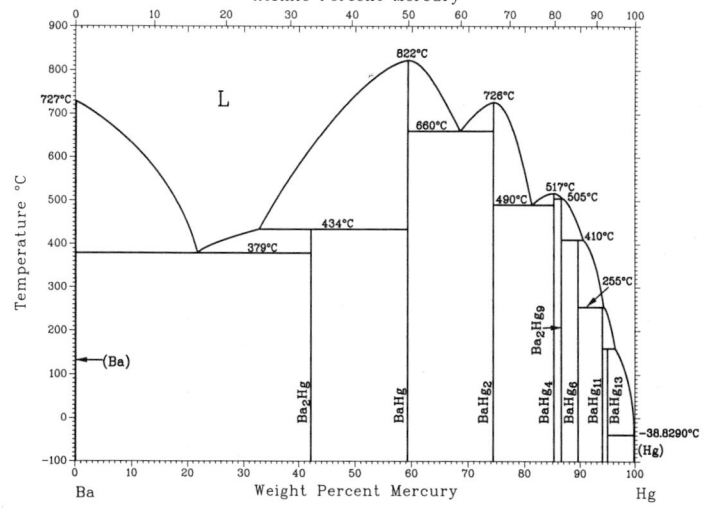

Phase	Composition, wt% Hg	Pearson symbol	Space group
(Ba)	0	cI2	Im$\bar{3}$m
Ba$_2$Hg	42.2	tI6	I4/mmm
BaHg	59	cP2	Pm$\bar{3}$m
BaHg$_2$	74.5	oI12	Imma
BaHg$_4$	85	...	...
Ba$_2$Hg$_9$	~86.7	hP38	P6/mmm
BaHg$_6$	~89.8	...	...
BaHg$_{11}$	~94.1	cP36	Pm$\bar{3}$m
BaHg$_{13}$	~95	...	...
(Hg)	~100	hR1	R$\bar{3}$m

Ba-In

H. Okamoto, 1992

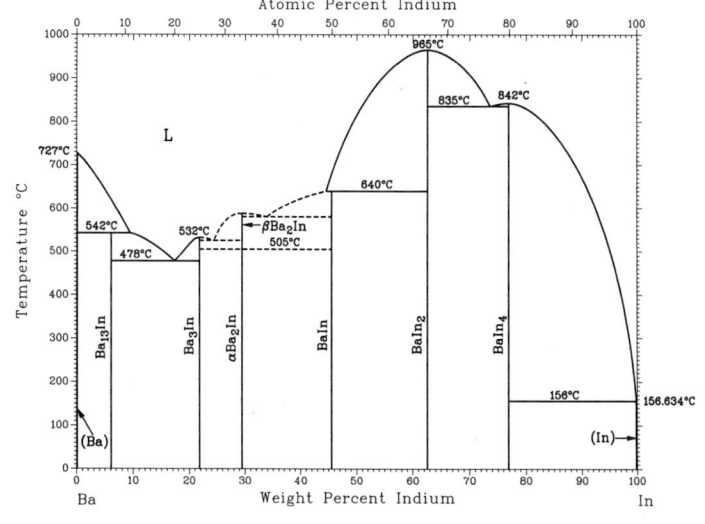

Phase	Composition, wt% In	Pearson symbol	Space group
(Ba)	0	cI2	Im$\bar{3}$m
Ba$_{13}$In	6.0	...	...
Ba$_3$In	22	...	...
βBa$_2$In	29.5	...	...
αBa$_2$In	29.5	...	...
BaIn	46	(a)	...
BaIn$_2$	62.6	oI12	Imma
BaIn$_4$	77	tI10	I4/mmm
(In)	100	tI2	I4/mmm

Ba-Li

A.D. Pelton, 1984

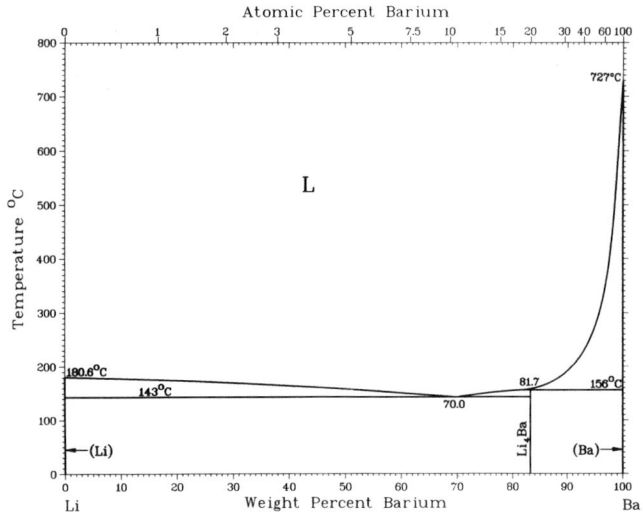

Phase	Composition, wt% Ba	Pearson symbol	Space group
(βLi)	0	cI2	Im$\bar{3}$m
(αLi)(a)	0	hP2	P6$_3$/mmc
Li$_4$Ba	83	hP30	P6$_3$/mmc
(Ba)	100	cI2	Im$\bar{3}$m

(a) Exists below −201 °C

Ba-Mg

A.A. Nayeb-Hashemi and J.B. Clark, 1988

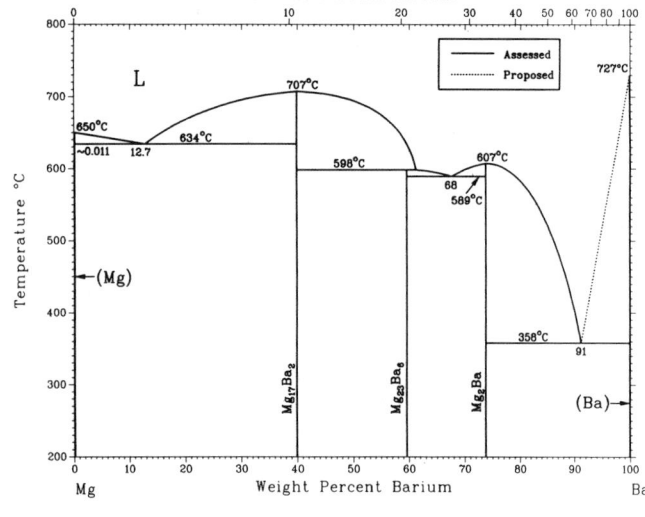

Phase	Composition, wt% Ba	Pearson symbol	Space group
(Mg)	0.0 to ~0.011	hP2	P6$_3$/mmc
Mg$_{17}$Ba$_2$	39.94	hR19	R$\bar{3}$m
Mg$_{23}$Ba$_6$	59.58	cF116	Fm$\bar{3}$m
Mg$_2$Ba	73.85	hP12	P6$_3$/mmc
(αBa)	100	cI2	Im$\bar{3}$m

Ba-Na

A.D. Pelton, 1985

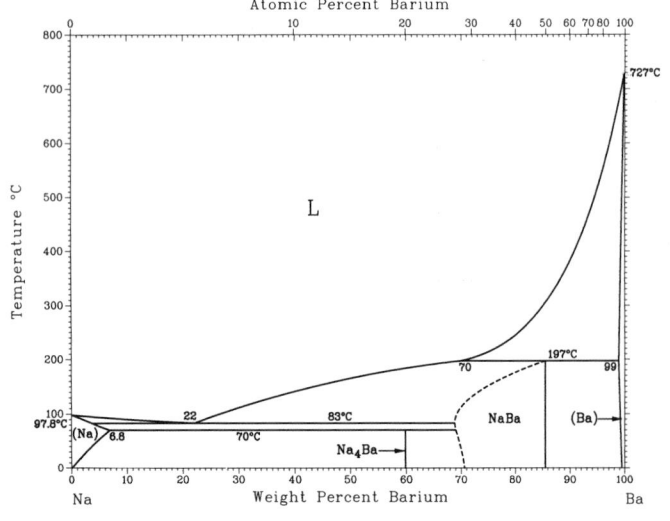

Phase	Composition, wt% Ba	Pearson symbol	Space group
(αNa)	0	hP2	P6$_3$/mmc
(βNa)	0 to 6.8	cI2	Im$\bar{3}$m
Na$_4$Ba	60	(a)	…
NaBa	69 to 86	(b)	…
(Ba)	99 to 100	cI2	Im$\bar{3}$m

(a) Tetragonal. (b) Orthorhombic

Ba-P

P.R. Subramanian, 1990

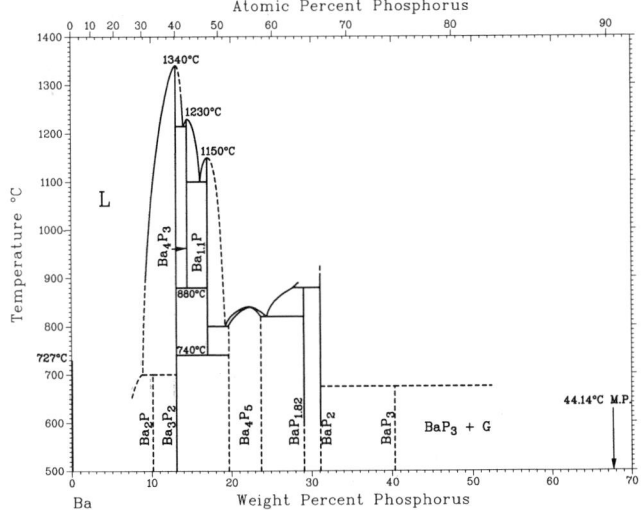

Phase	Composition, wt% P	Pearson symbol	Space group
(Ba)	0	cI2	Im$\bar{3}$m
Ba$_2$P	10.1	...	...
Ba$_3$P$_2$	13	cI28	I$\bar{4}$3d
Ba$_4$P$_3$	~14.5	...	...
Ba$_{1.1}$P	~17.0	...	...
Ba$_4$P$_5$	~22.0	...	...
BaP$_{1.82}$	~29.0	...	...
BaP$_2$	31.1	...	...
BaP$_3$	40	mC16	C2/m
Ba$_3$P$_{14}$	~51.3	mP34	P21/a
BaP$_{10}$	~69.3	oC44	Cmc2$_1$

Ba-Pb

From [Hansen]

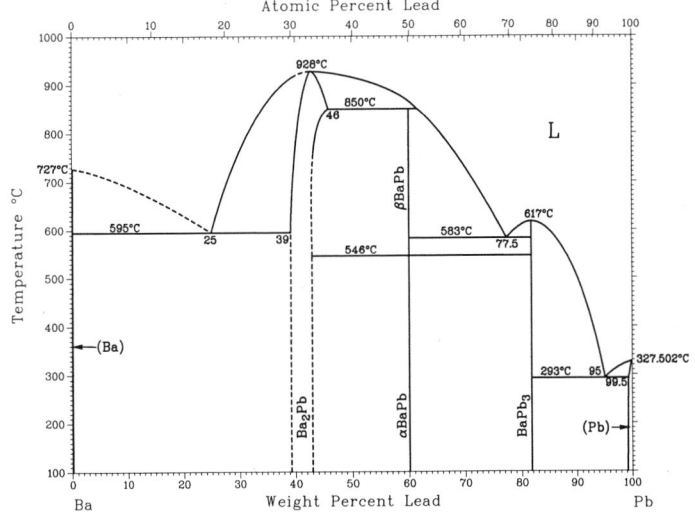

Phase	Composition, wt% Pb	Pearson symbol	Space group
(Ba)	0	cI2	Im$\bar{3}$m
Ba$_2$Pb	~39 to 43.0	oP12	Pnma
βBaPb	60	...	...
αBaPb	60	oC8	Cmcm
BaPb$_3$	82	hR12	R$\bar{3}$m
(Pb)	99.5 to 100	cF4	Fm$\bar{3}$m

Ba-Se

H. Okamoto, 1991

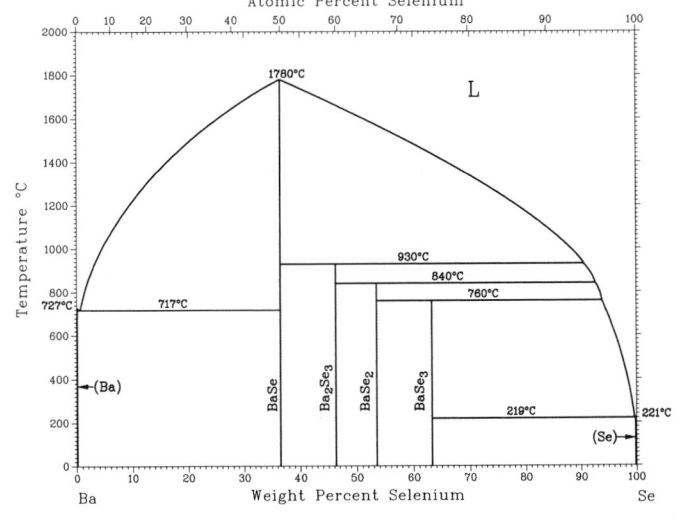

Phase	Composition, wt% Se	Pearson symbol	Space group
(Ba)	0	cI2	Im$\bar{3}$m
BaSe	37	cF8	Fm$\bar{3}$m
Ba$_2$Se$_3$	46	...	...
BaSe$_2$	53.5	mC12	C2/c
BaSe$_3$	63	tP8	P$\bar{4}$2$_1$m
(γSe)	100	hP3	P3$_1$21

Ba-Si

I. Obinata, Y. Takeuchi, K. Kurihara, and M. Watanabe, 1964

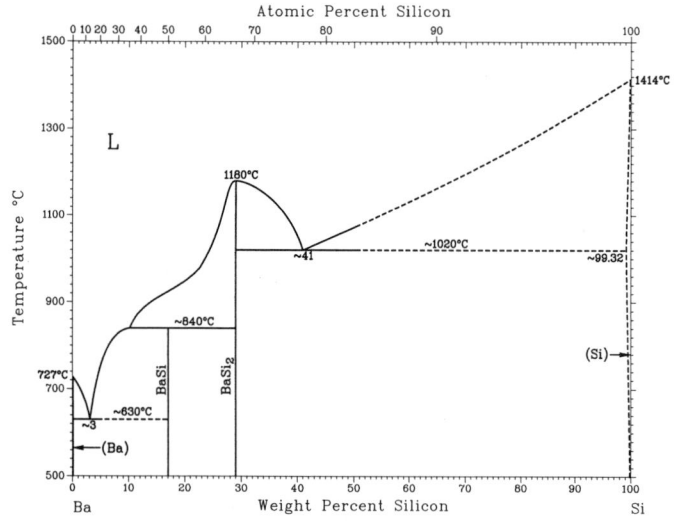

Phase	Composition, wt% Si	Pearson symbol	Space group
(Ba)	~0	cI2	$Im\bar{3}m$
Ba₂Si(a)	9.3	oP12	Pnma
Ba₅Si₃(a)	10.9	tP32	P4/ncc
BaSi	17	oC8	Cmcm
Ba₃Si₄	21.4	tP28	P4₂/mnm
BaSi₂	29.1	oP24	Pnma
		hP3	P6/mmm
(Si)	~100	cF8	$Fd\bar{3}m$

(a) Found after the diagram was constructed

Ba-Te

H. Okamoto, unpublished

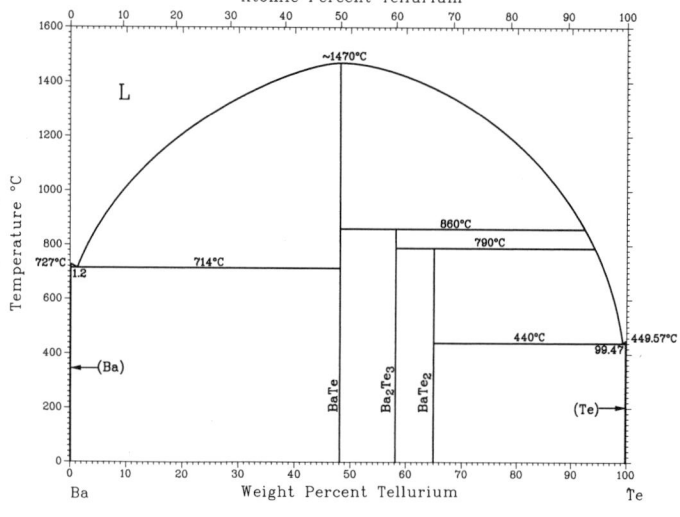

Phase	Composition, wt% Te	Pearson symbol	Space group
(Ba)	0	cI2	$Im\bar{3}m$
BaTe	48	cF8	$Fm\bar{3}m$
Ba₂Te₃	58	...	...
BaTe₂	65.1	...	...
(Te)	100	hP3	P3₁21

Ba-Tl

G. Bruzzone, 1966

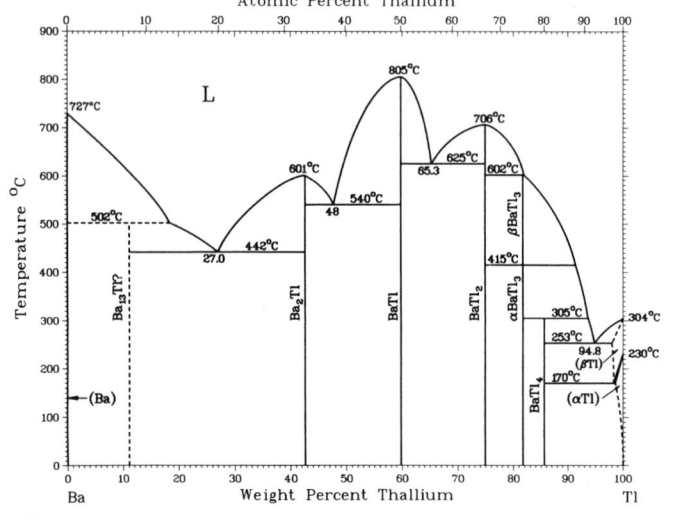

Phase	Composition, wt% Tl	Pearson symbol	Space group
(Ba)	~0	cI2	$Im\bar{3}m$
Ba₁₃Tl	~10	...	...
Ba₂Tl	42.6	...	...
BaTl	60	...	...
BaTl₂	74.9	hP6	P6₃/mmc
αBaTl₃	79	...	...
βBaTl₃	79	...	...
BaTl₄	86	...	...
(βTl)	~98 to 100	hP2	P6₃/mmc
(αTl)	~98.7 to 100	cI2	$Im\bar{3}m$

Ba-Zn

H. Okamoto, 1991

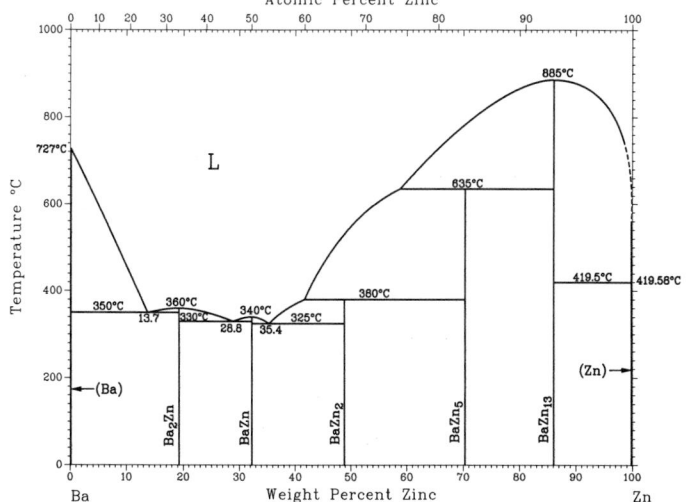

Phase	Composition, wt% Zn	Pearson symbol	Space group
(Ba)	0	cI2	Im$\bar{3}$m
Ba$_2$Zn	19.2	tI6	I4/mmm
BaZn	32	cP2	Pm$\bar{3}$m
BaZn$_2$	48.8	oI12	Imma
BaZn$_5$	70.4	oC25	Cmcm
BaZn$_{13}$	86.2	cF112	Fm$\bar{3}$c
(Zn)	100	hP2	P6$_3$/mmc

Be-Co

H. Okamoto, L.E. Tanner, and T. Nishizawa, 1988

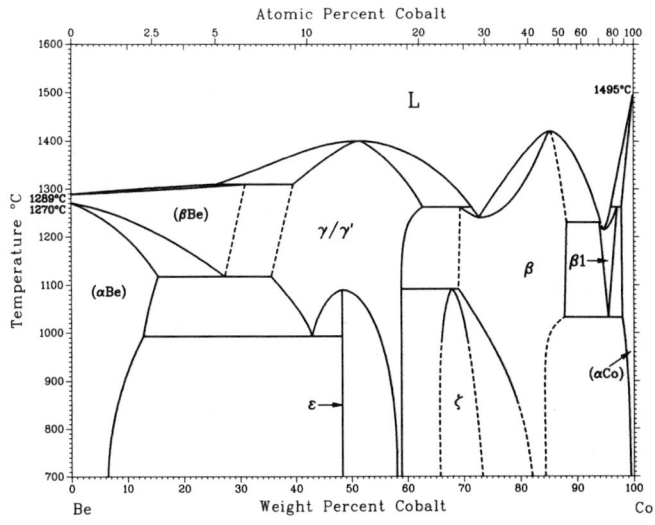

Phase	Composition, wt% Co	Pearson symbol	Space group
(βBe)	0 to 29	cI2	Im$\bar{3}$m
(αBe)	0 to 15.61	hP2	P6$_3$/mmc
Be$_{12}$Co	(a)	tI26	I4/mmm
γ	34.7 to ?	cI52	Im$\bar{3}$m
γ'	? to 62	cF416	Fm$\bar{3}$m
ε	~47	hP19	P6m2
		hP48	P6$_3$/mcm
δ	(a)	cF24	F4$\bar{3}$m
β'	(a)	cI2	Im$\bar{3}$m
ζ'	(a)	(b)	?
ζ	66 to 70	hP96	P6$_3$/mcm
β	70 to 88	cP2	Pm$\bar{3}$m
β1	94 to 97	cI2?	Im$\bar{3}$m
(αCo)	98 to 100	cF4	Fm$\bar{3}$m
(εCo)	99.9 to 100(a)	hP2	P6$_3$/mmc
Metastable phases			
...	~86.7	(c)	?
...	91 to 97	(d)	?

(a) Not shown in the assessed diagram. (b) Orthorhombic. (c) bct. (d) Tetragonal

Be-Cr

M. Venkatraman and J.P. Neumann, 1987

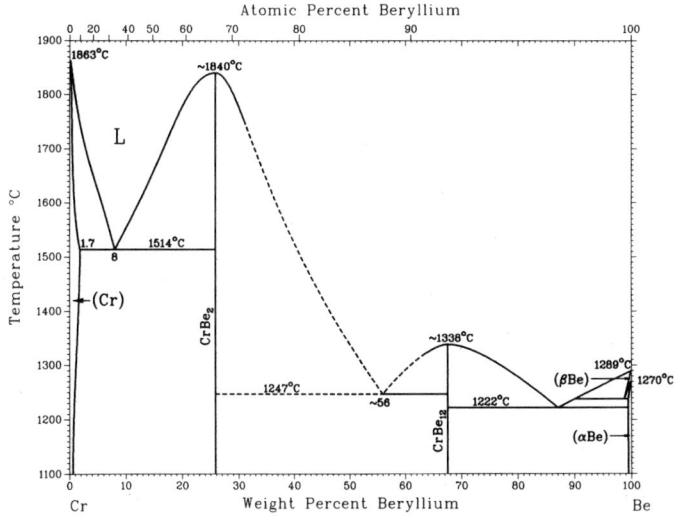

Phase	Composition, wt% Be	Pearson symbol	Space group
(Cr)	0 to 1.7	cI2	Im$\bar{3}$m
CrBe$_2$	25.8 to ~26	hP12	P6$_3$/mmc
CrBe$_{12}$	67.5	tI26	I4/mmm
(βBe)	~98.9 to 100	cI2	Im$\bar{3}$m
(αBe)	~99.54 to 100	hP2	P6$_3$/mmc

Be-Cu

H. Okamoto, 1992

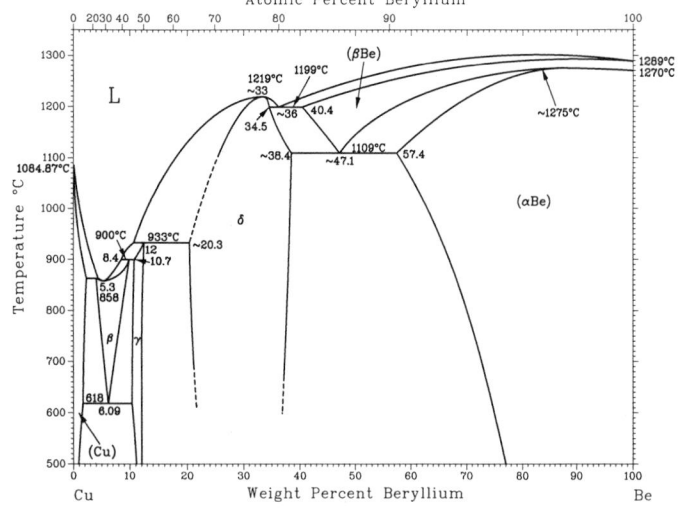

Phase	Composition, wt% Be	Pearson symbol	Space group
(Cu)	0 to 2.2	cF4	$Fm\bar{3}m$
β	4.3 to 9.8	cI2	$Im\bar{3}m$
γ	10.3 to 12.4	cP2	$Pm\bar{3}m$
δ	~20.4 to ~38.5	cF24	$Fd\bar{3}m$
(βBe)	40.4 to 100	cI2	$Im\bar{3}m$
(αBe)	57.5 to 100	hP2	$P6_3/mmc$

Be•Fe

H. Okamoto and L.E. Tanner, 1992

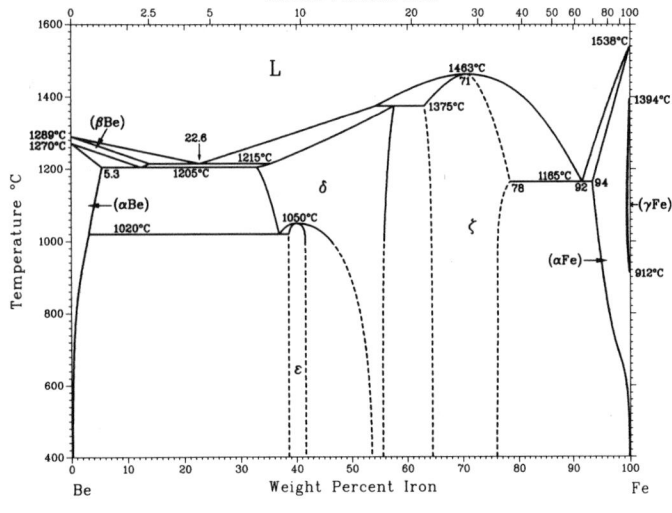

Phase	Composition, wt% Fe	Pearson symbol	Space group
(βBe)	0 to 11	cI2	$Im\bar{3}m$
(αBe)	0 to 5.3	hP2	$P6_3/mmc$
ε	~35 to 41	hP19	$P6m2$
		hP48	$P6_3/mcm$
δ	32 to 58	cF24	$Fd\bar{3}m$
ζ	62 to 78	hP12	$P6_3/mmc$
(γFe)	99.7 to 100	cF4	$Fm\bar{3}m$
(αFe)	94 to 100	cI2	$Im\bar{3}m$
Metastable phases			
...	~86	cF16	$Fd\bar{3}m$
β	?	cP2	$Pm\bar{3}m$
BeF₃	~95	cF16	$Fm\bar{3}m$

Be-rich portion of the Be-Fe phase diagram

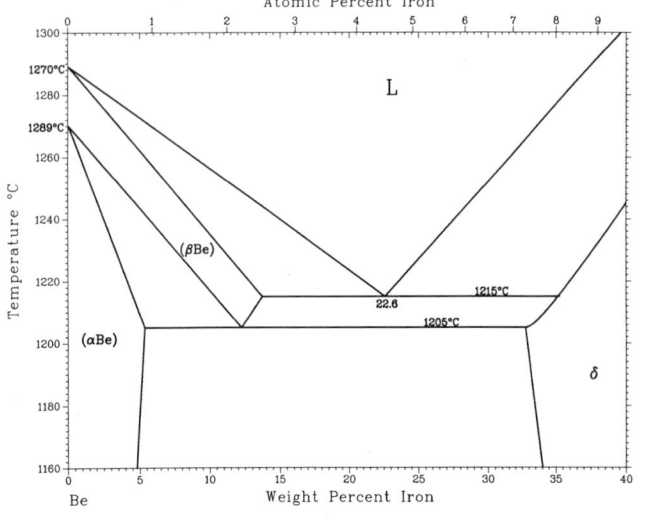

Be-Hf

H. Okamoto and L.E. Tanner, 1987

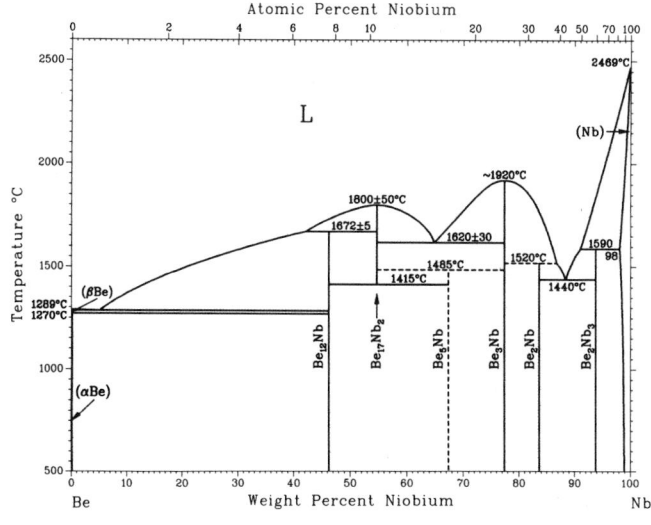

Phase	Composition, wt% Hf	Pearson symbol	Space group
(βBe)	0	cI2	Im$\bar{3}$m
(αBe)	0	hP2	P6$_3$/mmc
Be$_{13}$Hf	60.2	cF112	Fm$\bar{3}$c
Be$_{17}$Hf	69.9	hP*	P$\bar{6}$m2
(αBe$_{17}$Hf$_2$)	(a)	hR19	R$\bar{3}$m
(βBe$_{17}$Hf$_2$)	(b)	hP38	P6$_3$/mmc
Be$_5$Hf	79.9	hP6	P6/mmm
Be$_2$Hf	90.8	hP3	P6/mmm
(βHf)	100	cI2	Im$\bar{3}$m
(αHf)	100	hP2	P6$_3$/mmc
Metastable phases			
BeHf	95	oC8	Cmcm
α'	99.7 to 100(c)	...	...

(a) Be-poor side. (b) Be-rich side. (c) Acicular martensite

Be-Nb

H. Okamoto and L.E. Tanner, 1987

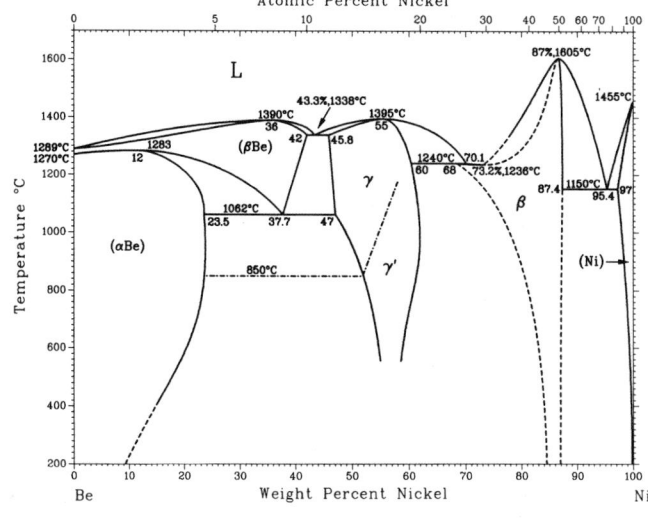

Phase	Composition, wt% Nb	Pearson symbol	Space group
(βB)	0	cI2	Im$\bar{3}$m
(αBe)	0	hP2	P6$_3$/mmc
Be$_{12}$Nb	46.2	tI26	I4/mmm
Be$_{17}$Nb$_2$	54.7	hR19	R$\bar{3}$m
(a)	56.3	hP*	...
Be$_5$Nb	67.4	hP6	P6/mmm
Be$_3$Nb	77	hR12	R$\bar{3}$m
(b)	83.7	hP*	...
Be$_2$Nb	83.7	cF24	Fd$\bar{3}$m
	83.7	...	...
	83.73	...	...
Be$_2$Nb$_3$	94	tP10	P4/mbm
(Nb)	98.6 to 100	cI2	Im$\bar{3}$m

(a) Proposed as Be$_8$Nb. (b) Reported as Be$_2$Nb

Be-Ni

H. Okamoto and L.E. Tanner, 1991

Phase	Composition, wt% Ni	Pearson symbol	Space group
(βBe)	0 to 23.5	cI2	Im$\bar{3}$m
(αBe)	0 to 42	hP2	P6$_3$/mmc
γ	45.8 to >51	cI52	I$\bar{4}$3m
γ'	51 to 62	cF416	F23
β	68 to 87.4	cP2	Pm$\bar{3}$m
(Ni)	95.4 to 100	cF4	Fm$\bar{3}$m
Metastable phases			
?	92.2 to 93.4	o**	?
β'	>87 to <95	tI*	?
γ'BeNi$_3$	95	?	?

Be-Pd

H. Okamoto and L.E. Tanner, 1987

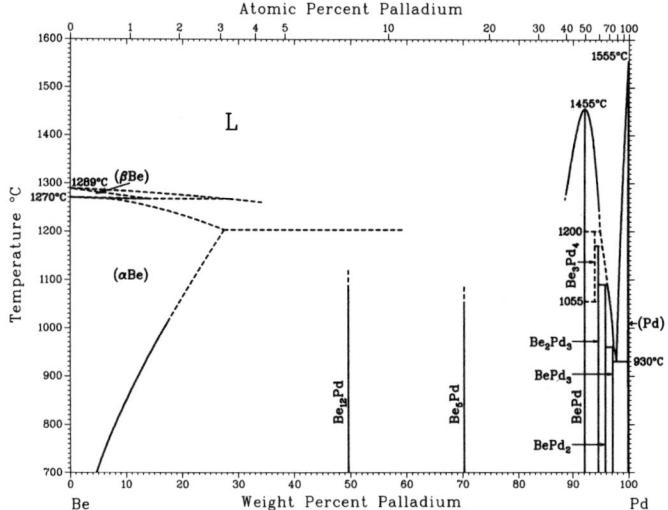

Phase	Composition, wt% Pd	Pearson symbol	Space group
(βBe)	0	cI2	Im$\bar{3}$m
(αBe)	0 to 38	hP2	P6$_3$/mmc
Be$_{12}$Pd	49.6	tI26	I4/mmm
Be$_5$Pd	70.3	cF24	F$\bar{4}$3m
BePd	92	cP2	Pm$\bar{3}$m
Be$_3$Pd$_4$	94.0	?	?
Be$_2$Pd$_3$	95	?	?
BePd$_2$	95.9	?	?
BePd$_3$	97	(a)	?
(Pd)	99.9 to 100	cF4	Fm$\bar{3}$m

(a) Orthorhombic

Be-Si

H. Okamoto and L.E. Tanner, 1987

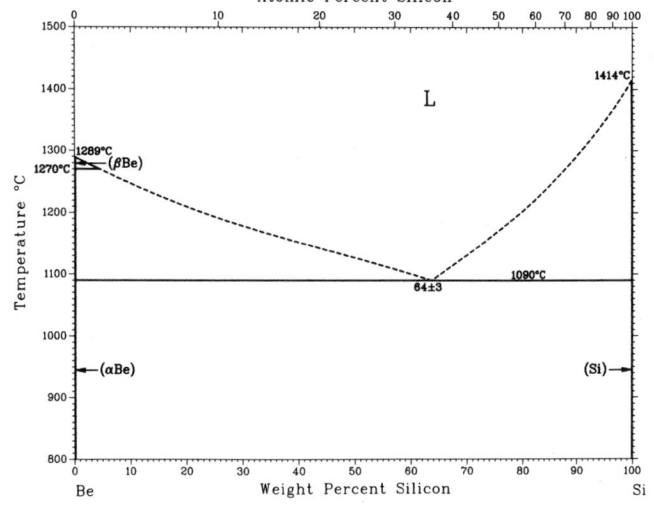

Phase	Composition, wt% Si	Pearson symbol	Space group
(βBe)	0	cI2	Im$\bar{3}$m
(αBe)	0	hP2	P6$_3$/mmc
(Si)	100	cF8	Fd$\bar{3}$m

Be-Th

H. Okamoto, L.E. Tanner, and D.E. Peterson, 1987

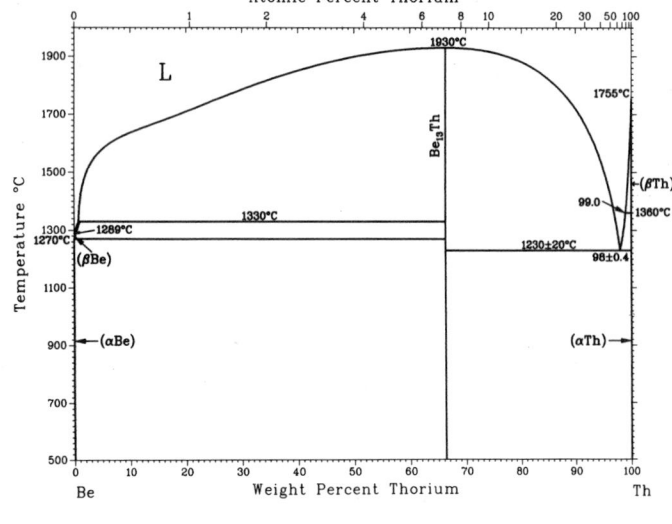

Phase	Composition, wt% Th	Pearson symbol	Space group
(βBe)	0	cI2	Im$\bar{3}$m
(αBe)	0	hP2	P6$_3$/mmc
Be$_{13}$Th	66.44	cF112	Fm$\bar{3}$c
(βTh)	100	cI2	Im$\bar{3}$m
(αTh)	100	cF4	Fm$\bar{3}$m

Be-Ti

J.L. Murray, 1987

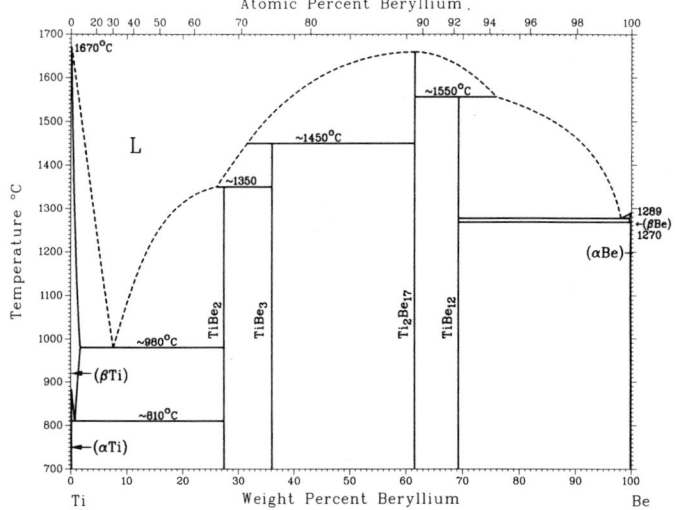

Phase	Composition, wt% Be	Pearson symbol	Space group
(βTi)	0 to ~1.5	cI2	Im$\bar{3}$m
(αTi)	~0	hP2	P6$_3$/mmc
TiBe$_2$	27.4	cF24	Fd$\bar{3}$m
TiBe$_3$	36	hR12	R$\bar{3}$m
αTi$_2$Be$_{17}$	61.6	hR19	R$\bar{3}$m
βTi$_2$Be$_{17}$	61.6	hP38	P6$_3$/mmc
TiBe$_{12}$	69.3	tI26	I4/mmm
TiBe(a)	~16	cP2	Pm$\bar{3}$m
(βBe)	~100	cI2	Im$\bar{3}$m
(αBe)	~100	hP2	P6$_3$/mmc

(a) Metastable

Be-W

H. Okamoto and L.E. Tanner, 1987

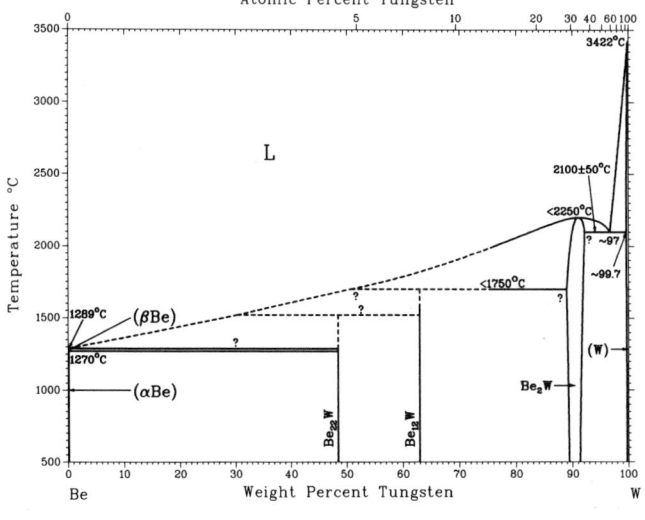

Phase	Composition, wt% W	Pearson symbol	Space group
(βBe)	0	cI2	Im$\bar{3}$m
(αBe)	0	hP2	P6$_3$/mmc
Be$_{24}$W?(a)	46	(b)	...
Be$_{22}$W	47.8	cF184	Fd$\bar{3}$m
Be$_{12}$W	63.0	tI26	I4/mmm
Be$_2$W	~89 to ~92	hP12	P6$_3$/mmc
(W)	~99.7 to 100	cI2	Im$\bar{3}$m

(a) Not accepted in the assessed phase diagram. (b) Tetragonal

Be-Zr

H. Okamoto, L.E. Tanner, and J.P. Abriata, 1987

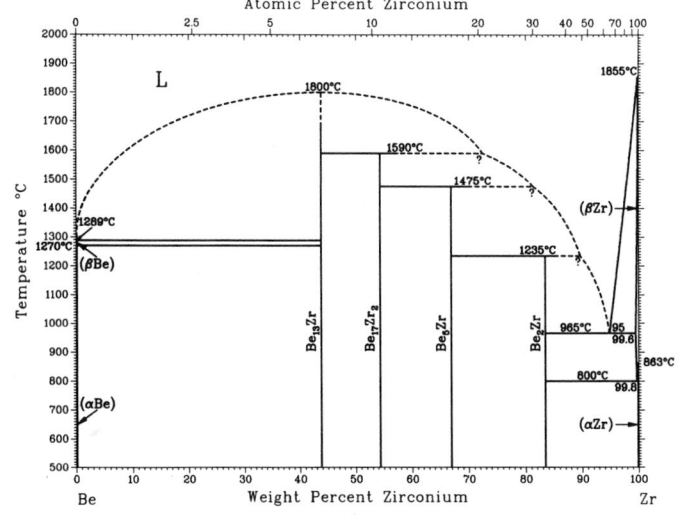

Phase	Composition, wt% Zr	Pearson symbol	Space group
(βBe)	0	cI2	Im$\bar{3}$m
(αBe)	0	hP2	P6$_3$/mmc
Be$_{13}$Zr	43.6	cF112	Fm$\bar{3}$c
Be$_{12}$Zr(a)	43.6	tI*	...
Be$_{17}$Zr$_2$	54.3	hR19	R$\bar{3}$m
Be$_5$Zr	67.0	hP6	P6/mmm
Be$_2$Zr	83.5	hP3	P6/mmm
(βZr)	100	cI2	Im$\bar{3}$m
(αZr)	100	hP2	P6$_3$/mmc
Metastable phases			
BeZr	91	oC8	Cmcm
α'	99 to 100	(b)	...

(a) Not accepted in the assessed diagram. (b) Acicular martensite

Bi-Ca

H. Okamoto, 1991

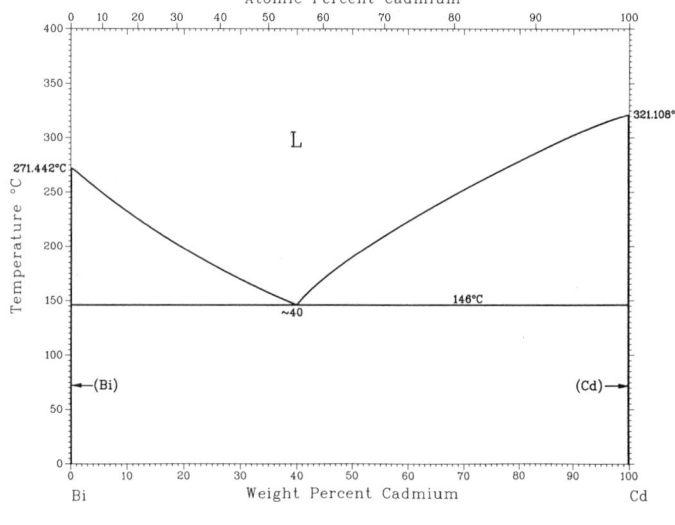

Phase	Composition, wt% Ca	Pearson symbol	Space group
(Bi)	0	$hR2$	$R\bar{3}m$
Bi_3Ca	6	...	...
$Bi_{10}Ca_{11}$	17.4	$tI84$	$I4/mmm$
Bi_3Ca_5	24.2	$oP32$	$Pnma$
$BiCa_2$	27.8	$tI12$	$I4/mmm$
(αCa)	100	$cF4$	$Fm\bar{3}m$
(βCa)	100	$cI2$	$Im\bar{3}m$

Bi-Cd

Z. Moser, J. Dutkiewicz, L. Zabdyr, and J. Salawa, 1988

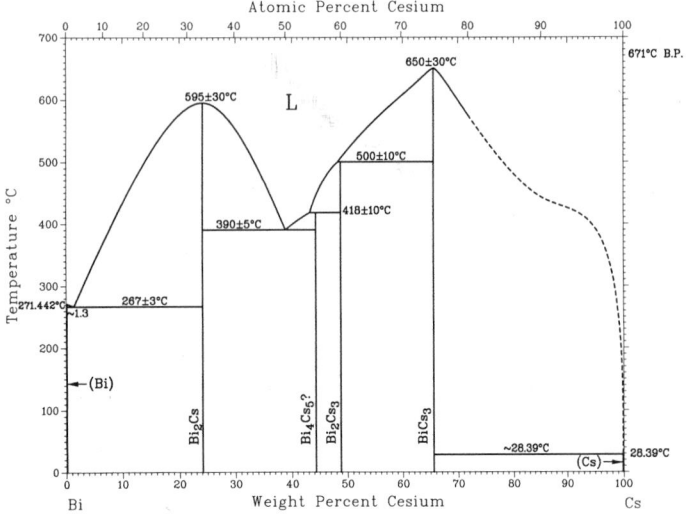

Phase	Composition, wt% Cd	Pearson symbol	Space group
(Bi)	0	$hR2$	$R\bar{3}m$
(Cd)	100	$hP2$	$P6_3/mmc$

Bi-Cs

J. Sangster and A.D. Pelton, 1991

Phase	Composition, wt% Cs	Pearson symbol	Space group
(αBi)	0	$hR2$	$R\bar{3}m$
Bi_2Cs	24.1	$cF24$	$Fd\bar{3}m$
$Bi_4Cs_5(?)$	44.3	...	...
Bi_2Cs_3	49	...	...
$BiCs_3$	66	$cF16$	$Fd\bar{3}m$
	66	$cF16$	$Fm\bar{3}m$
(Cs)	100	$cI2$	$Im\bar{3}m$

Bi-Cu

D.J. Chakrabarti and D.E. Laughlin, 1984

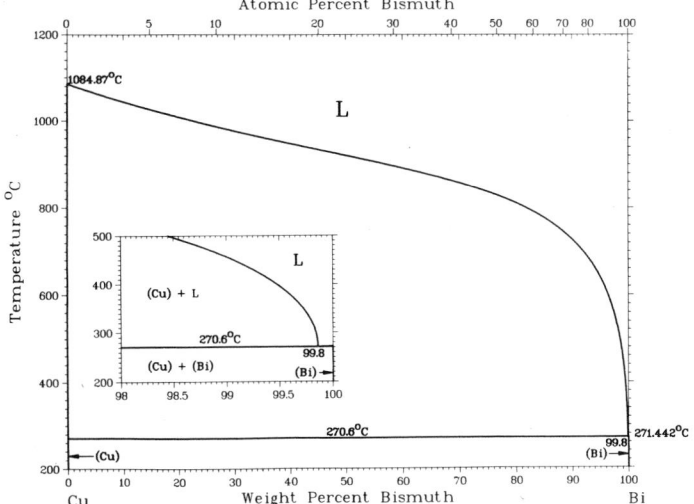

Phase	Composition, wt% Bi	Pearson symbol	Space group
(Cu)	0 to 0.010	cF4	$Fm\bar{3}m$
(Bi)	100	hR2	$R\bar{3}m$
Metastable phase			
Cu₅Bi₂	57	...	...

Bi-Ga

H. Okamoto, 1990

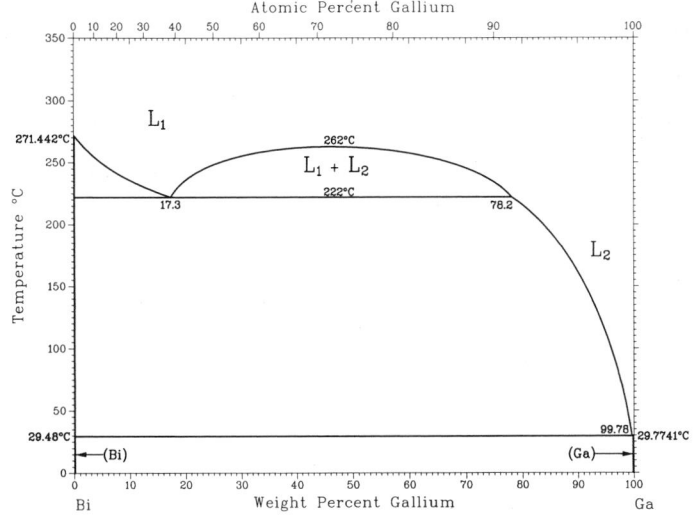

Phase	Composition, wt% Ga	Pearson symbol	Space group
(Bi)	~0	hR2	$R\bar{3}m$
(Ga)	~100	oC8	Cmca

Bi-Ge

R.W. Olesinski and G.J. Abbaschian, 1986

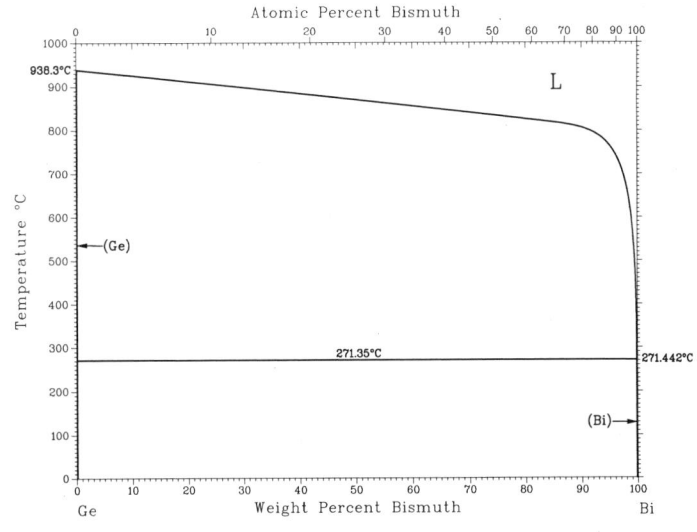

Phase	Composition, wt% Bi	Pearson symbol	Space group
(Ge)	0	cF8	$Fd\bar{3}m$
(GeII)(HP)	0	tI4	$I4_1/amd$
(Bi)	100	hR2	$R\bar{3}m$

Bi-Hg

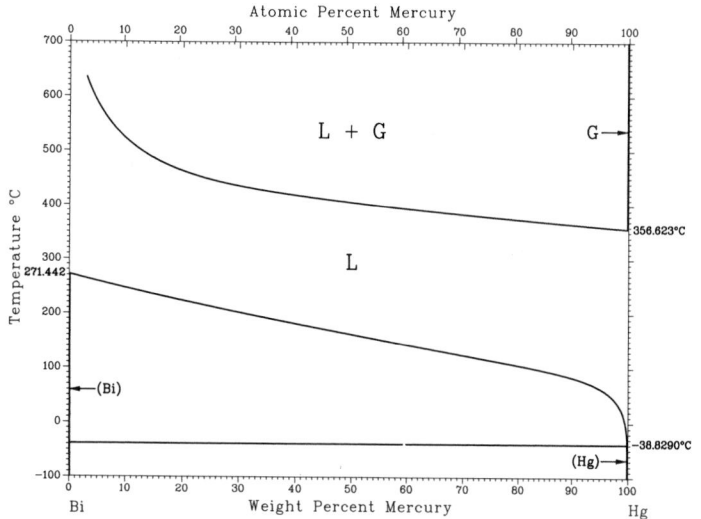

L. Zabdyr and C. Guminski, unpublished

Phase	Composition, wt% Hg	Pearson symbol	Space group
(Bi)	0	hR2	R$\bar{3}$m
(Hg)	100	hR1	R$\bar{3}$m

Bi-In

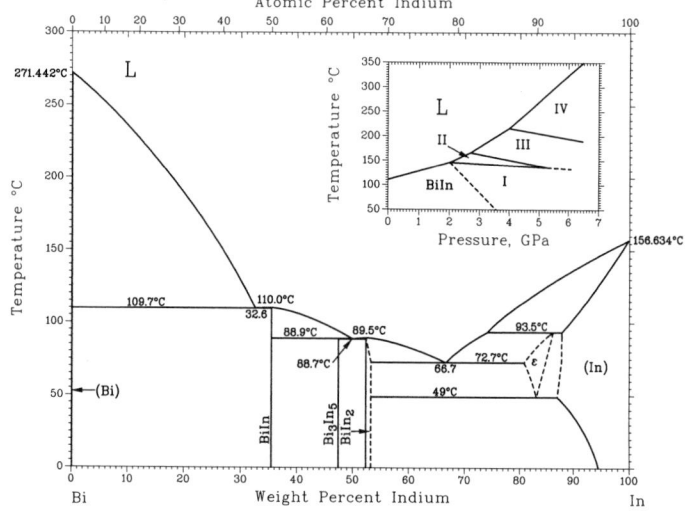

H. Okamoto, 1992

Phase	Composition, wt% In	Pearson symbol	Space group
Stable phases			
(αBi)	0 to 0.005	hR2	R$\bar{3}$m
BiIn	35.4	tP4	P4/nmm
Bi$_3$In$_5$	47.5 to 47.97	tI32	I4/mcm
BiIn$_2$	52.5 to 53.5	hP6	P6$_3$/mmc
ε	80 to 86	tI2	...
(In)	~86 to 100	tI2	I4/mmm
High-pressure/metastable phases			
(γBi)	0	mP4	P2$_1$/m
(βBi)	0	mC4	C2/m
Bi$_4$In	12	tI4	I4$_1$/amd
Bi$_3$In	~15	oI*	Immb
γ	12 to 42	hP1	P6/mmm
γ$_1$	21 to 35.4	oI*	...
γ$_2$	21 to 35.4	oI*	...
X	21 to 35.4	oI*	...
BiIn′	35.4	tP4	P4/nmm
α$_2$	45 to 51	cI2	Im$\bar{3}$m
Bi$_2$In$_3$(a)	45	hP*	...
BiIn$_3$	62	t**	...
α$_1$	~59 to 76	tI2	I4/mmm

(a) Thin film. Probably Bi$_3$In$_5$

Bi-K

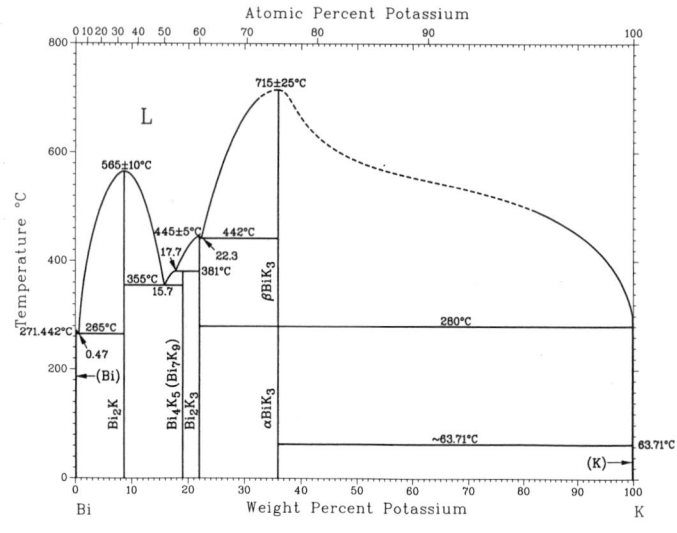

A. Petric and A.D. Pelton, 1991

Phase	Composition, wt% K	Pearson symbol	Space group
(Bi)	0	hR2	R$\bar{3}$m
Bi$_2$K	8.5	cF24	Fd$\bar{3}$m
BiK$_5$(a)	19.0	...	...
Bi$_2$K$_3$	22	...	...
αBiK$_3$(b)	36	hP8	P6$_3$/mmc
βBiK$_3$(c)	36	cF16	Fm$\bar{3}$m
(K)	100	cI2	Im$\bar{3}$m

(a) Might be Bi$_7$K$_9$. (b) Stable below 280 °C. (c) Stable above 280 °C

Bi-La

K.A. Gschneidner, Jr. and F.W. Calderwood, 1989

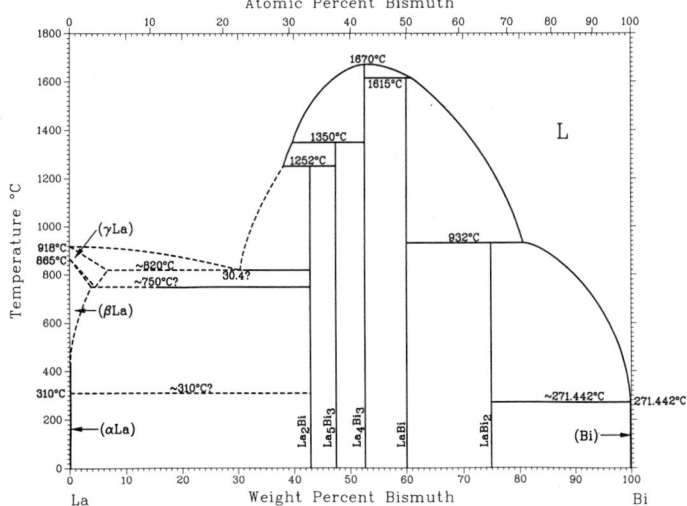

Phase	Composition, wt% Bi	Pearson symbol	Space group
(γLa)	0 to ?	cI2	Im$\bar{3}$m
(βLa)	0 to ?	cF4	Fm$\bar{3}$m
(αLa)	0	hP4	P6$_3$/mmc
La$_2$Bi	42.9	tI12	I4/mmm
La$_5$Bi$_3$	47.4	hP16	P6$_3$/mcm
La$_4$Bi$_3$	53.1	cI28	I$\bar{4}$3d
LaBi	60.1	cF8	Fm$\bar{3}$m
LaBi$_2$(a)	75.1	o?12	...
LaBi$_2$(a)	75.1	aP27(?)	P1 or P$\bar{1}$
(αBi)	100	hR2	R$\bar{3}$m

(a) Conflicting reports regarding LaBi$_2$ structure

Bi-Li

J. Sangster and A.D. Pelton, 1991

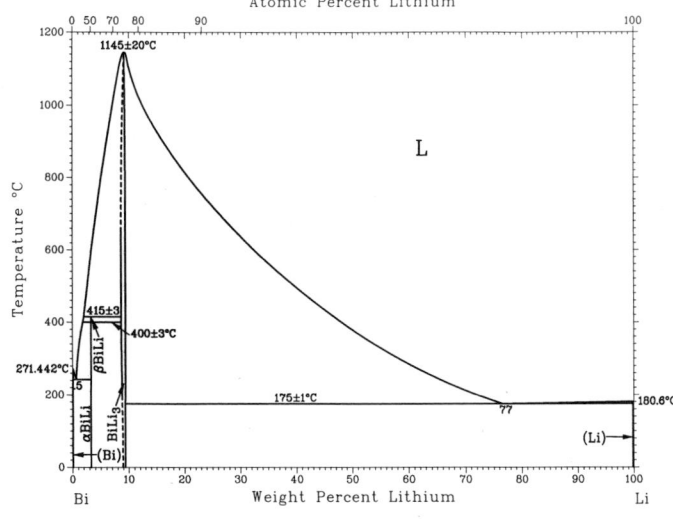

Phase	Composition, wt% Li	Pearson symbol	Space group
(αLi)	0	hR2	R$\bar{3}$m
αBiLi(a)	3.2	tP4	P4/mmm
βBiLi	3.2	...	...
BiLi$_3$	8.6 to 9.2(b)	cF16	Fm$\bar{3}$m
(αLi)(c)	100	hP2	P6$_3$/mmc
(βLi)	100	cI2	Im$\bar{3}$m

(a) Below 415 °C. (b) At 380 °C. (c) Below −201 °C

Bi-Mg

A.A. Nayeb-Hashemi and J.B. Clark, 1988

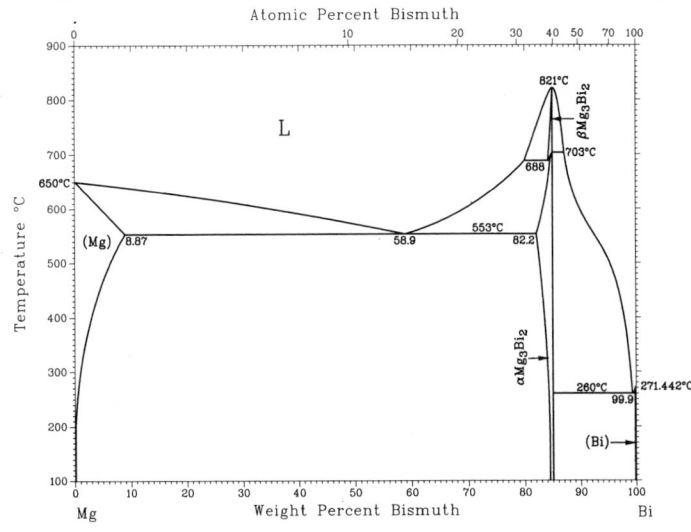

Phase	Composition, wt% Bi	Pearson symbol	Space group
(Mg)	0 to 8.87	hP2	P6$_3$/mmc
Mg$_3$Bi$_2$(LT) or αMg$_3$Bi$_2$	82.2 to 85	hP5	P$\bar{3}$m1
Mg$_3$Bi$_2$(HT) or βMg$_3$Bi$_2$	85	(a)	...
(Bi)	100	hR2	R$\bar{3}$m

(a) The structure of the high-temperature Mg$_3$Bi$_2$ is unknown.

Bi-Mn

H. Okamoto, 1990

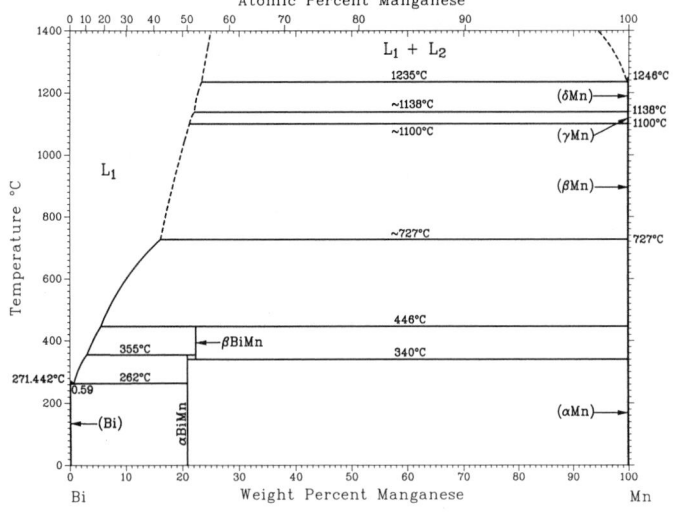

Phase	Composition, wt% Mn	Pearson symbol	Space group
(Bi)	0	$hR2$	$R\bar{3}m$
βBiMn	22.1	$o*32$	...
αBiMn	20.8	$hP4$	$P6_3/mmc$
(δMn)	100	$cI2$	$Im\bar{3}m$
(γMn)	100	$cF4$	$Fm\bar{3}m$
(βMn)	100	$cP20$	$P4_132$
(αMn)	100	$cI58$	$I43m$

Bi-Na

J. Sangster and A.D. Pelton, 1991

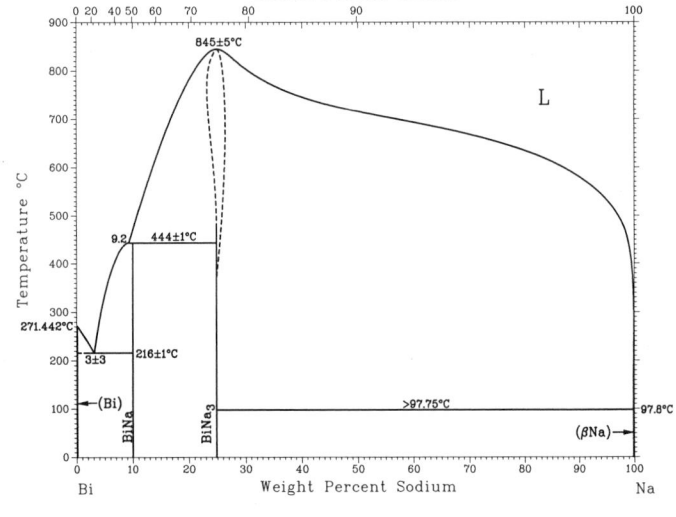

Phase	Composition, wt% Na	Pearson symbol	Space group
(αBi)	0	$hR2$	$R\bar{3}m$
BiNa	10.1	$tP4$	$P4/mmm$
BiNa$_3$(a)	23.4 to 27.5(b)	$hP8$	$P6_3/mmc$
(αNa)(c)	100	$hP2$	$P6_3/mmc$
(βNa)	100	$cI2$	$Im\bar{3}m$

(a) Might be $hP24$, Cu$_3$As prototype. (b) At 800 °C. (c) Below −237 °C

Bi-Nd

K.A. Gschneidner, Jr. and F.W. Calderwood, 1989

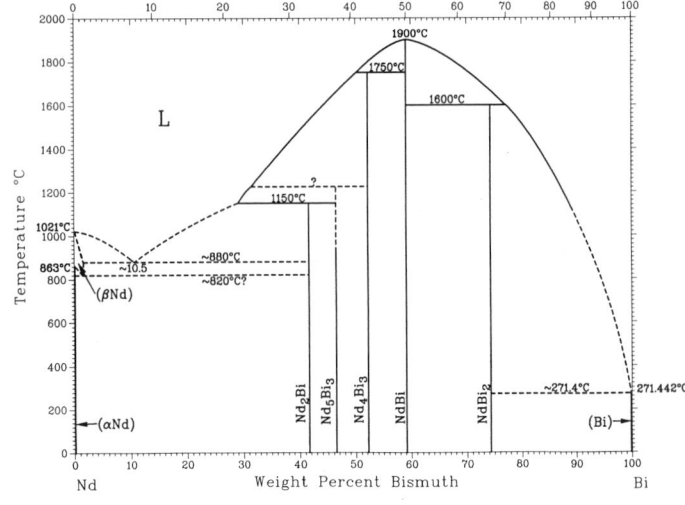

Phase	Composition, wt% Bi	Pearson symbol	Space group
(αNd)	0	$hP4$	$P6_3/mmc$
Nd$_2$Bi	42.0	$tI12$	$I4/mmm$
Nd$_5$Bi$_3$	46.5	$hP16$	$P6_3/mcm$
Nd$_4$Bi$_3$	52.1	$cI28$	$I\bar{4}3d$
NdBi	59.1	$cF8$	$Fm\bar{3}m$
NdBi$_2$	74.4	$aP27(?)$	$P1$ or $P\bar{1}$
(αBi)	100	$hR2$	$R\bar{3}m$

Bi-Ni

P. Nash, 1991

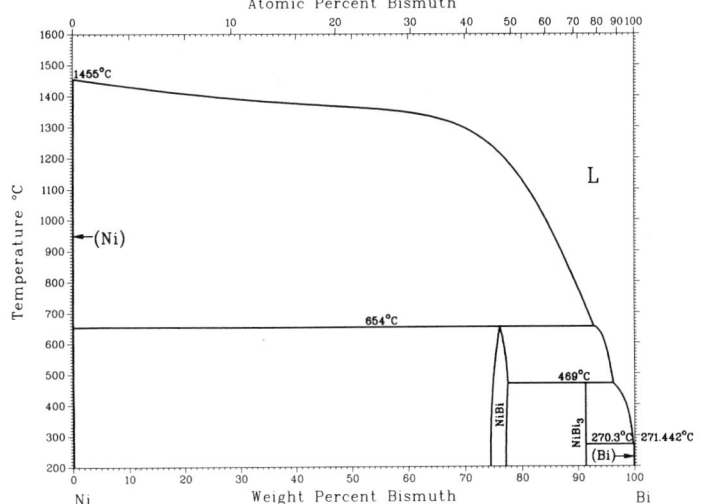

Phase	Composition, wt% Bi	Pearson symbol	Space group
(Ni)	0	cF4	$Fm\overline{3}m$
NiBi	74 to 77	hP4	$P6_3/mmc$
NiBi$_3$	91	…	…
(Bi)	100	hR2	$R\overline{3}m$

Bi-Pb

N.A. Gokcen, 1992

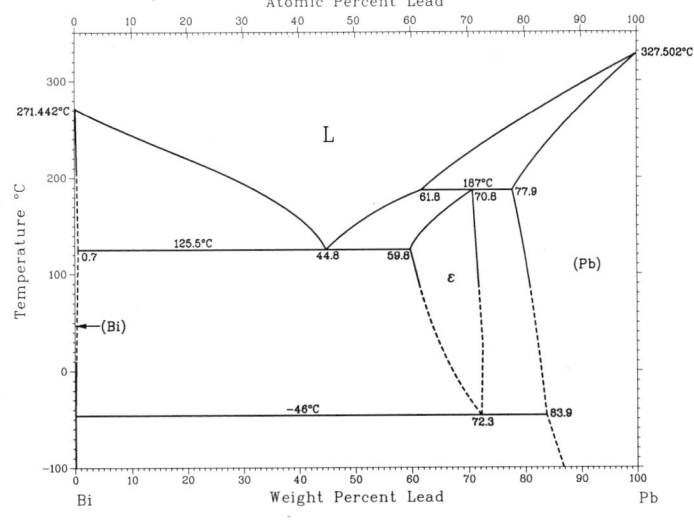

Phase	Composition, wt% Pb	Pearson symbol	Space group
(Bi)	0 to 0.7	hR2	$R\overline{3}m$
ε	59.8 to 73	hP2	$P6_3/mmc$
(Pb)	77.9 to 100	cF4	$Fm\overline{3}m$

Bi-Pd

H. Okamoto, unpublished

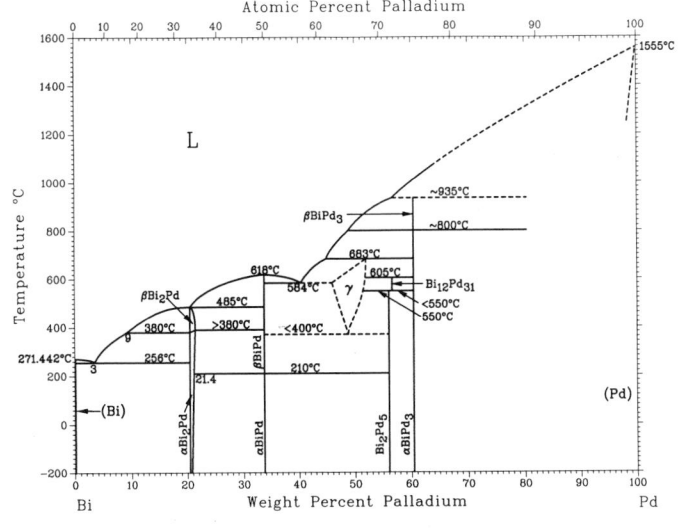

Phase	Composition, wt% Pd	Pearson symbol	Space group
(αBi)	0	hR2	$R\overline{3}m$
βBi$_2$Pd	20.3	tI6	$I4/mmm$
αBi$_2$Pd	20.3	mC12	$C2/m$
βBiPd	33.7	oC32	$Cmc2_1$
αBiPd	33.7	mP32	$P2_1$
γ(a)	45.9	hP16	…
Bi$_2$Pd$_5$	56.0	mC28	$C2/m$
Bi$_{12}$Pd$_{31}$	56.8	hR44	$R3$
βBiPd$_3$	60	…	…
αBiPd$_3$	60	oP16	$Pmma$
(Pd)	100	cF4	$Fm\overline{3}m$

(a) Superlattice of NiAs type

Bi-Pt

H. Okamoto, 1991

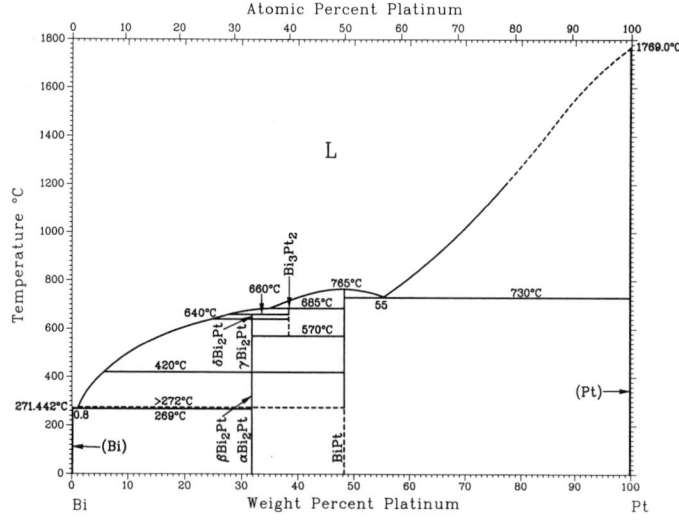

Phase	Composition, wt% Pt	Pearson symbol	Space group
(αBi)	0	hR2	$R\bar{3}m$
δBi₂Pt	31.8	oP6	Pnnm
γBi₂Pt	31.8	hP9	$P\bar{3}$
βBi₂Pt	31.8	cP12	Pa3
αBi₂Pt	31.8	oP24	Pbca
Bi₃Pt₂	38	o**	...
		hP4	P6₃/mmc
BiPt	48.2	hP4	P6₃/mmc
(Pt)	100	cF4	$Fm\bar{3}m$

Bi-Rb

A.D. Pelton and A. Petric, unpublished

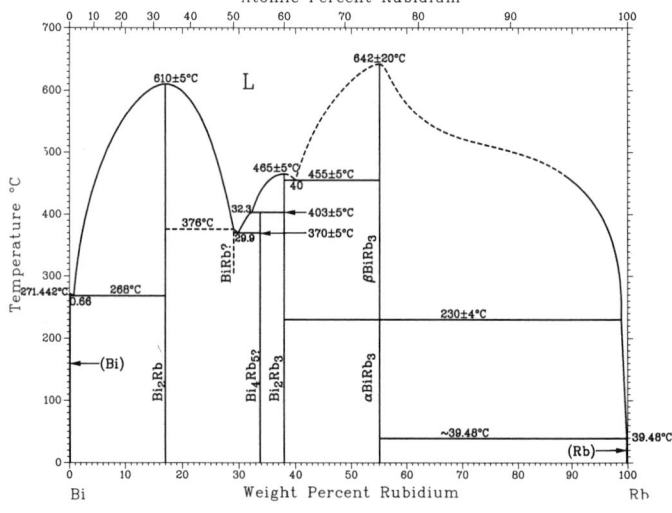

Phase	Composition, wt% Rb	Pearson symbol	Space group
(Bi)	0	hR2	$R\bar{3}m$
Bi₂Rb	17.0	cF24	$Fd\bar{3}m$
BiRb(?)	29.0	...	...
Bi₄Rb₅	33.9	...	...
Bi₂Rb₃	38	...	...
αBiRb₃(a)	55	hP8	P6₃/mmc
βBiRb₃(b)	55	cF16	$Fm\bar{3}m$
(Rb)	100	cI2	$Im\bar{3}m$

(a) Stable below 230 °C. (b) Stable above 230 °C

Bi-S

J.-C. Lin, R.C. Sharma, and Y.A. Chang, unpublished

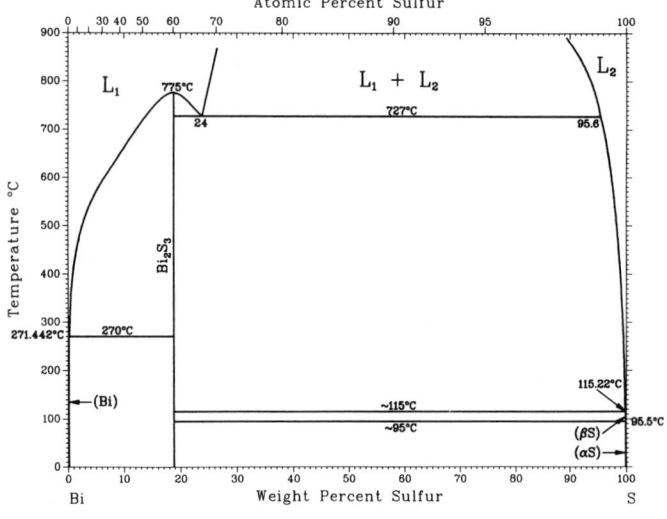

Phase	Composition, wt% S	Pearson symbol	Space group
(αBi)	0	hR2	$R\bar{3}m$
Bi₂S₃	19	oP20	Pnma
(αS)	100	oF128	Fddd
(βS)	100	mP*	P2₁/c

Bi-Sb

H. Okamoto, unpublished

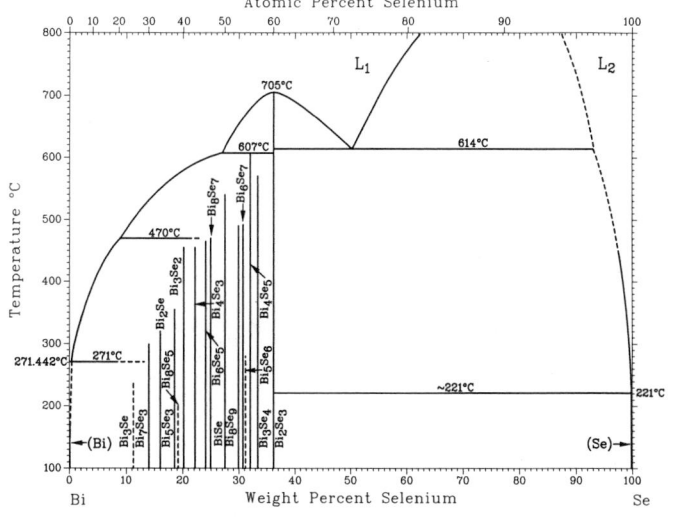

Phase	Composition, wt% Sb	Pearson symbol	Space group
(Bi,Sb)	0 to 100	$hR2$	$R\bar{3}m$
High-pressure phases			
(BiII)	0 to 2.1	$mC4$	$C2/m$
(Bi,SbIII)	0 to 100	$mP4$	$P2_1/m$
(BiIII′)	0 to ?	...	...
(BiIV)	0 to ?	$m*8$	...
(BiV)	0 to ?	$cI2$	$Im\bar{3}m$
(SbII)	70 to 100	$cP1$	$Pm\bar{3}m$
(SbIII)	? to 100	$hP2$	$P6_3/mmc$

(a) At room temperature. (b) High-temperature, high-pressure phase

Bi-Se

H. Okamoto, unpublished

Phase	Composition, wt% Se	Pearson symbol	Space group
(αBi)	0	$hR2$	$R\bar{3}m$
Bi_7Se_3	14	$hR20$	$R\bar{3}m$
Bi_2Se	15.9	$hP9$	$P\bar{3}m1$
Bi_5Se_3(a)	18.5	$hP48$	$P\bar{3}m1$
Bi_3Se_2	20	$hP30$	$P\bar{3}m1$
Bi_4Se_3	22.1	$hR7$	$R\bar{3}m$
Bi_6Se_5	24.0	$hP33$	$P\bar{3}m1$
Bi_8Se_7	24.8	$hP45$	$P\bar{3}m1$
BiSe	27.4	$hP12$	$P\bar{3}m1$
Bi_8Se_9	29.8	$hP17$	$R\bar{3}m$
Bi_6Se_7	30.6	$hP39$	$P\bar{3}m1$
Bi_4Se_5	32.1	$hP27$	$P\bar{3}m1$
Bi_3Se_4	33.5	$hP42$	$P\bar{3}m1$
Bi_2Se_3	36	$hR5$	$R\bar{3}m$
(Se)	100	$hP3$	$P3_121$
Metastable phases			
BiSe(b)	27.4	$cF8$	$Fm\bar{3}m$
Bi_2Se_3IIIa	36	$c**$	...
High-pressure phases			
Bi_2Se_3II(c)	36	$oP20$	$Pnma$
Bi_2Se_3III	36	$tP40$	$P4_2/nmc$
$BiSe_2$	43.1	...	...

(a) Laitakarite. (b) Thin film. (c) Bismuthite

Bi-Sm

H. Okamoto, 1990

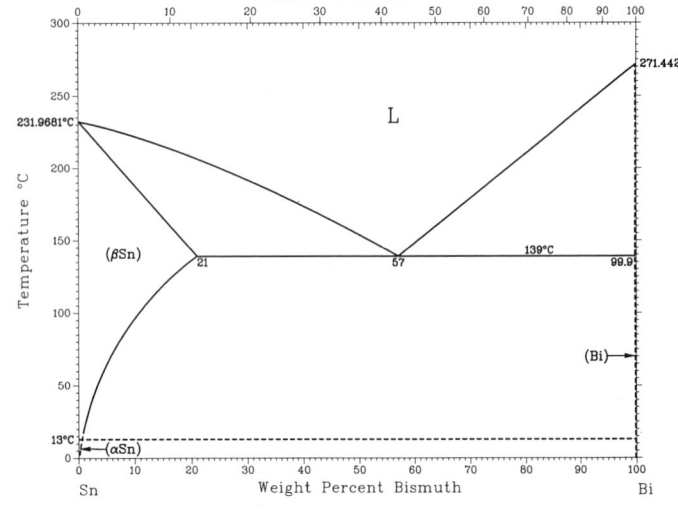

Phase	Composition, wt% Sm	Pearson symbol	Space group
(αBi)	0	hR2	$R\bar{3}m$
Bi₂Sm	26.5	oP12	Pmmm
BiSm	42.8	cF8	$Fm\bar{3}m$
Bi₃Sm₄	48.9	cI28	$I\bar{4}3d$
Bi₃Sm₅	54.5	hP16	P6₃/mcm
BiSm₂	59.0	tI6	I4/mmm
(γSm)	100	cI2	$Im\bar{3}m$
(βSm)	100	hP2	P6₃/mmc
(αSm)	100	hR3	$Rm\bar{3}m$

Bi-Sn

H. Okamoto, 1990

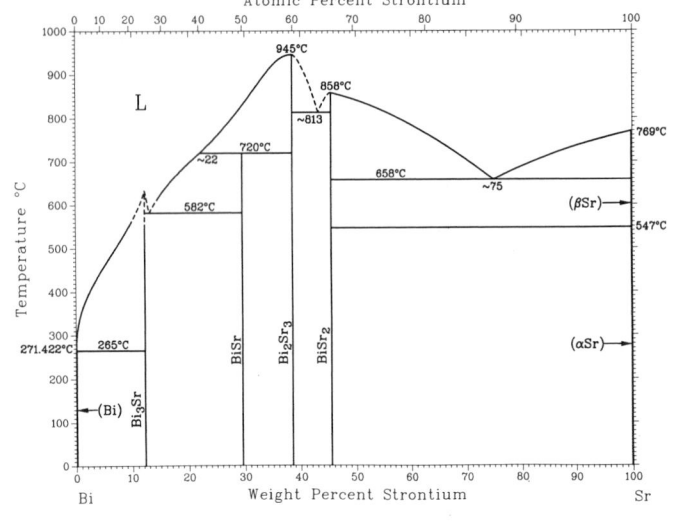

Phase	Composition, wt% Bi	Pearson symbol	Space group
(βSn)	0 to 21	tI4	I4₁/amd
(αSn)	0 to ?	cF8	$Fd\bar{3}m$
(Bi)	99.9 to 100	hR2	$R\bar{3}m$

Bi-Sr

From [Elliott]

Phase	Composition, wt% Sr	Pearson symbol	Space group
(Bi)	0	hR2	$R\bar{3}m$
Bi₃Sr	12	cP4	$Pm\bar{3}m$
BiSr	29.5	...	...
Bi₂Sr₃	39	...	...
BiSr₂	45.7	tI12	I4/mmm
(βSr)	100	cI2	$Im\bar{3}m$
(αSr)	100	cF	$Fm\bar{3}m$

Bi-Te

H. Okamoto and L.E. Tanner, unpublished

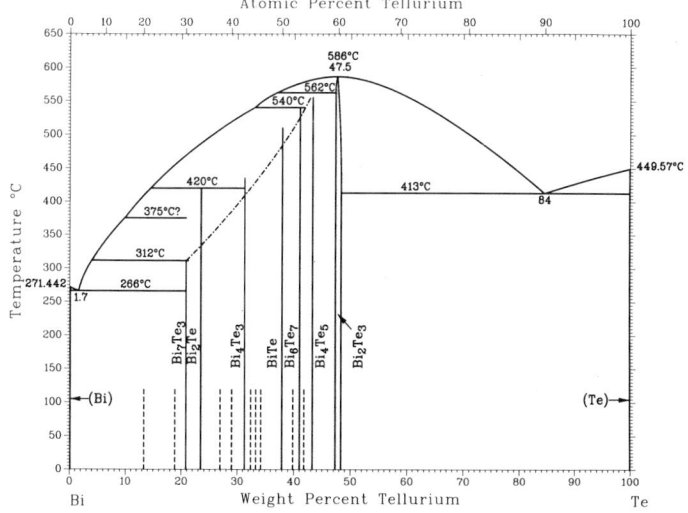

Phase	Composition, wt% Te	Pearson symbol	Space group
(αBi)	0 to ?	hR2	$R\bar{3}m$
Bi$_2$Te$_3$	47.5 to 48.0	hR5	$R\bar{3}m$
(αTe)	99.992 to 100	hP3	$P3_121$
Stacking variants			
Bi$_7$Te$_3$	21	hR20	$R\bar{3}m$
Bi$_2$Te	23.4	...	$P\bar{3}m1$
Bi$_4$Te$_3$	31.5	hR7	$R\bar{3}m$
BiTe	37.9	hP12	$P\bar{3}m1$
Bi$_6$Te$_7$	41.6	hP39	$P\bar{3}m1$
Bi$_4$Te$_5$	43.3	hP27	$P\bar{3}m1$
Metastable phases			
BiTe(a)	37.9	cF8	$Fm\bar{3}m$
Bi$_2$Te$_5$	60.4	...	...
High-pressure phase			
Bi$_2$Te$_3$II	48	hR5	$R\bar{3}m$

(a) Thin film

Bi-Tl

H. Okamoto, unpublished

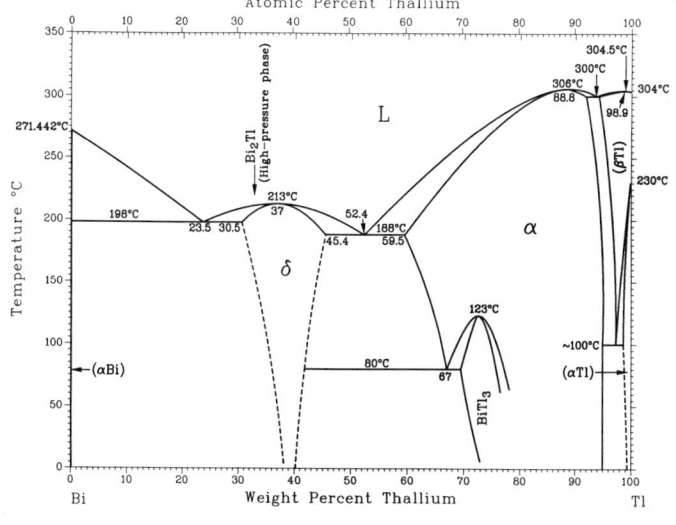

Phase	Composition, wt% Tl	Pearson symbol	Space group
(γBi)(a)	0	mP4	$P2_1/m$
(βBi)(a)	0	mC4	$C2/m$
(αBi)	0	hR2	$R\bar{3}m$
Bi$_2$Tl(a)	33.8	...	...
δ	30.5 to 45.4	hP3	$P6/mmm$
BiTl(b)	...	cP2	$Pm\bar{3}m$
α	59.5 to 94.9	cF4	$Fm\bar{3}m$
BiTl$_3$	69.5 to 76.6	hP2	$P6_3/mmc$
		hP8	$P6_3/mmc$
BiTl$_3$I(c)	~74.6	cP4	$Pm\bar{3}m$
BiTl$_7$(a)	87.3	...	...
(γTl)(a)	100	cF4	$Fm\bar{3}m$
(βTl)	94.9 to 100	cI2	$Im\bar{3}m$
(αTl)	98.98 to 100	hP2	$P6_3/mmc$

Note: Not all high-pressure phases of Bi are listed. (a) High-pressure phase. (b) Not accepted in the assessed diagram. (c) Metastable?

Bi-U

From [Chiotti]

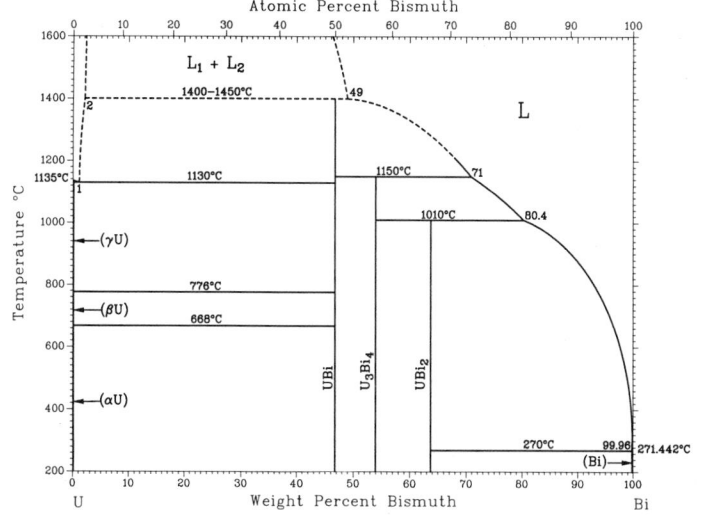

Phase	Composition, wt% Bi	Pearson symbol	Space group
(γU)	0	cI2	$Im\bar{3}m$
(βU)	0	tP30	$P4_2/mnm$
(αU)	0	oC4	$Cmcm$
UBi	46.7	cF8	$Fm\bar{3}m$
U$_3$Bi$_4$	53.9	cI28	$I\bar{4}3d$
UBi$_2$	63.8	tP6	$P4/nmm$
(Bi)	100	hR2	$R\bar{3}m$

Bi-Y

K.A. Gschneidner, Jr. and F.W. Calderwood, 1989

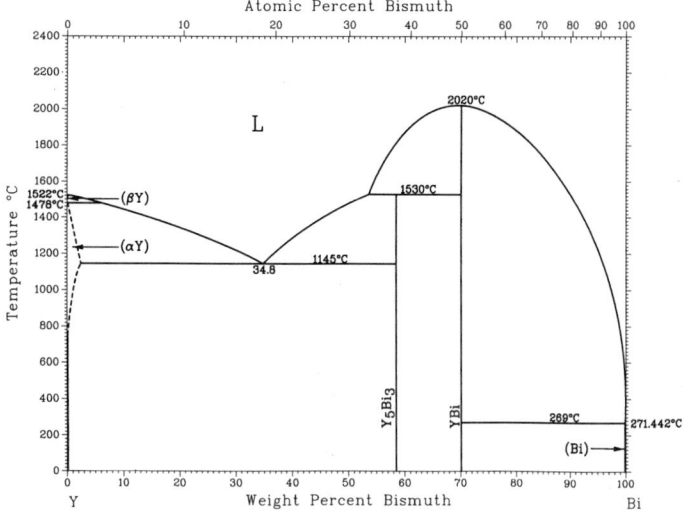

Phase	Composition, wt% Bi	Pearson symbol	Space group
(αY)	0	hP2	P6$_3$/mmc
Y$_5$Bi$_3$	58.5	oP32	Pnma
YBi	70.1	cF8	Fm$\overline{3}$m
(αBi)	100	hR2	R$\overline{3}$m

Bi-Yb

H. Okamoto, 1990

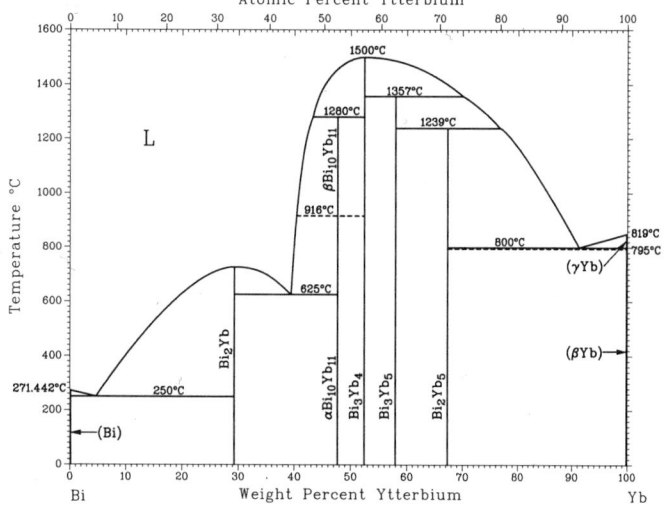

Phase	Composition, wt% Yb	Pearson symbol	Space group
(αBi)	0	hR2	R$\overline{3}$m
Bi$_2$Yb	29.3	oC12	Cmcm
βBi$_{10}$Yb$_{11}$	47.7	tI84	I4/mmm
αBi$_{10}$Yb$_{11}$	47.7	...	...
Bi$_3$Yb$_4$	52.4	cI28	I$\overline{4}$3d
Bi$_3$Yb$_5$	58	oP32	Pnma
Bi$_2$Yb$_5$	67.4	oP*	Pn2$_1$a
(γYb)	100	cI2	Im$\overline{3}$m
(βYb)	100	cF4	Fm$\overline{3}$m
(αYb)	100	hP2	P6$_3$/mmc

Bi-Zn

H. Okamoto, 1990

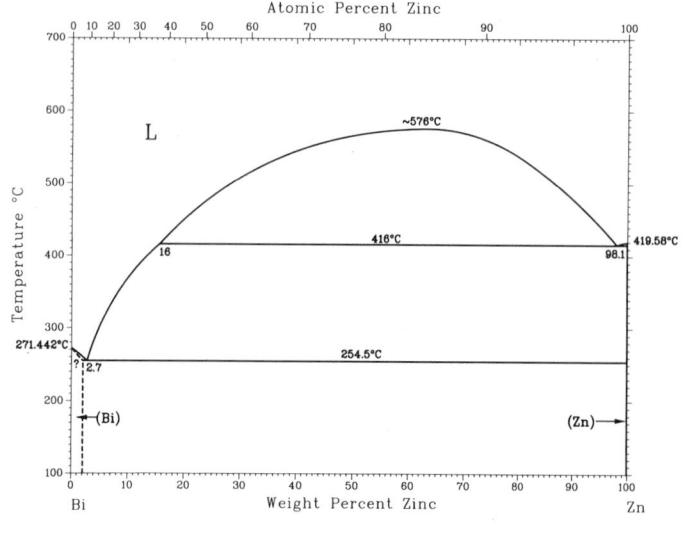

Phase	Composition, wt% Zn	Pearson symbol	Space group
(Bi)	0 to ?	hR2	R$\overline{3}$m
(Zn)	~100	hP2	P6$_3$/mmc

Bi-Zr

H. Okamoto, 1990

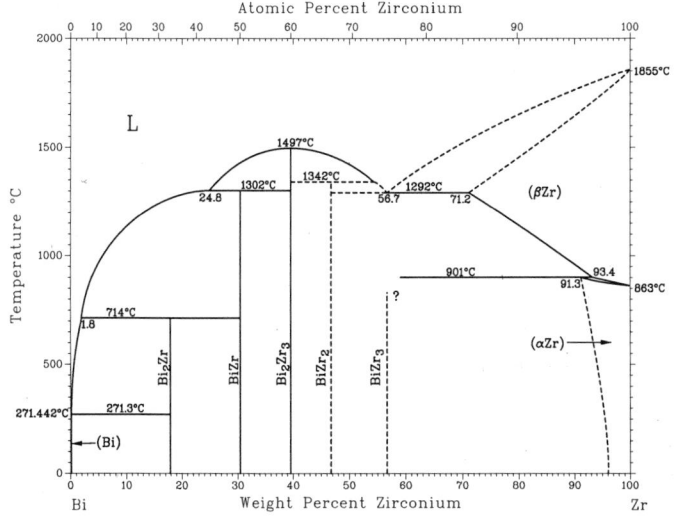

Phase	Composition, wt% Zr	Pearson symbol	Space group
(αBi)	0	hR2	R$\overline{3}$m
Bi₂Zr	17.9	oP24	Pnnm
BiZr	30.4	...	...
Bi₂Zr₃	39.6	...	...
BiZr₂	46.7	...	...
BiZr₃	56.7	tI32	I$\overline{4}$
(βZr)	71.2 to 100	cI2	Im$\overline{3}$m
(αZr)	91.3 to 100	hP2	P6₃/mmc

C-Co

K. Ishida and T. Nishizawa, 1991

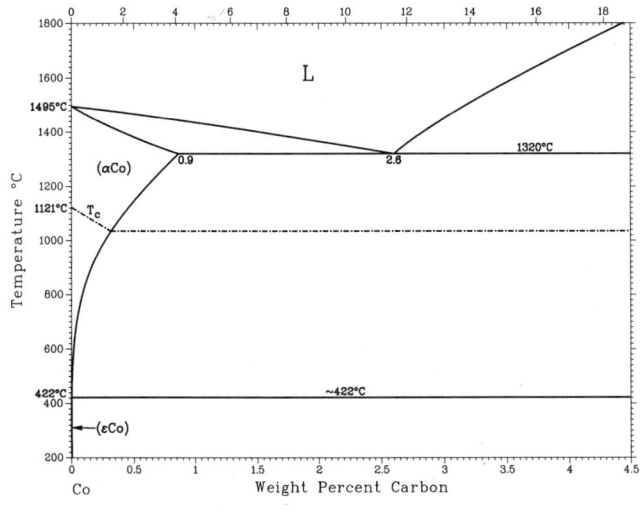

Phase	Composition, wt% C	Pearson symbol	Space group
(αCo)	0 to 0.9	cF4	Fm$\overline{3}$m
(εCo)	~0	hP2	P6₃/mmc
C	~100	hP4	P6₃/mmc
Metastable phases			
(ε'Co)	~0.3 to ~0.4	(a)	...
Co₃C	6	oP6	Pnma
Co₂C	9	oP6	Pnnm

(a) Hexagonal

C-Cr

M. Venkatraman and J.P. Neumann, 1990

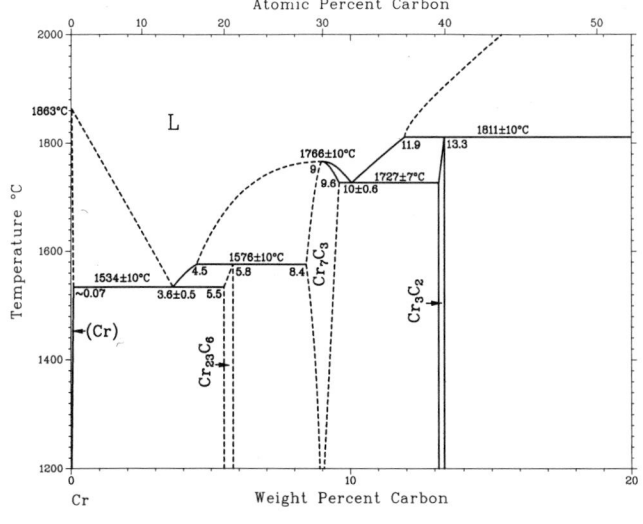

Phase	Composition, wt% C	Pearson symbol	Space group
(Cr)	0 to ~0.07	cI2	Im$\overline{3}$m
Cr₂₃C₆	5.5 to 5.8	cF116	Fm$\overline{3}$m
Cr₃C(a)	~7	oP16	Pnma
Cr₇C₃	~9	oP40	Pnma
Cr₃C₂	~13	oP20	Pnma
CrC(?)	~19	...	...
(C)	~100	hP4	P6₃/mmc

(a) Metastable

C-Cu

P.R. Subramanian and D.E. Laughlin, unpublished

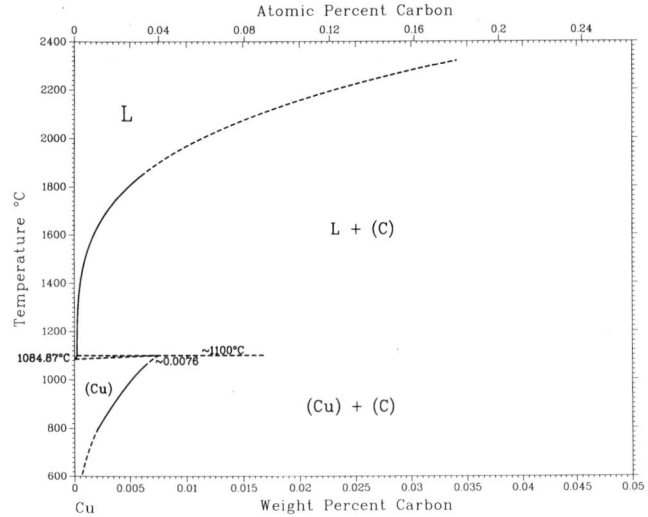

Phase	Composition, wt% C	Pearson symbol	Space group
(Cu)	0 to 0.01	cF4	$Fm\bar{3}m$
(C)	100	hP4	$P6_3/mmc$

H. Okamoto, 1992

Phase	Composition, wt% C	Pearson symbol	Space group
(δFe)	0 to 0.09	cI2	$Im\bar{3}m$
(γFe)	0 to 2.1	cF4	$Fm\bar{3}m$
(αFe)	0 to 0.021	cI2	$Im\bar{3}m$
(C)	100	hP4	$P6_3/mmc$
Metastable/high-pressure phases			
(εFe)	0	hP2	$P6_3/mmc$
Martensite	<2.1	tI4	$I4/mmm$
Fe_4C	5.1	cP5	$P\bar{4}3m$
Fe_3C (θ)	6.7	oP16	$Pnma$
Fe_5C_2 (χ)	7.9	mC28	$C2/c$
Fe_7C_3	8.4	hP20	$P6_3mc$
Fe_7C_3	8.4	oP40	$Pnma$
Fe_2C (η)	9.7	oP6	$Pnnm$
Fe_2C (ε)	9.7	hP*	$P6_322$
Fe_2C	9.7	hP*	$P\bar{3}m1$
(C)	100	cF8	$Fd\bar{3}m$

C-Fe

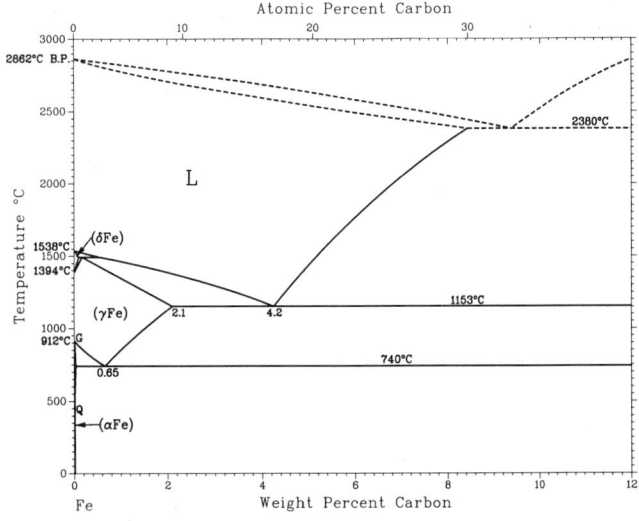

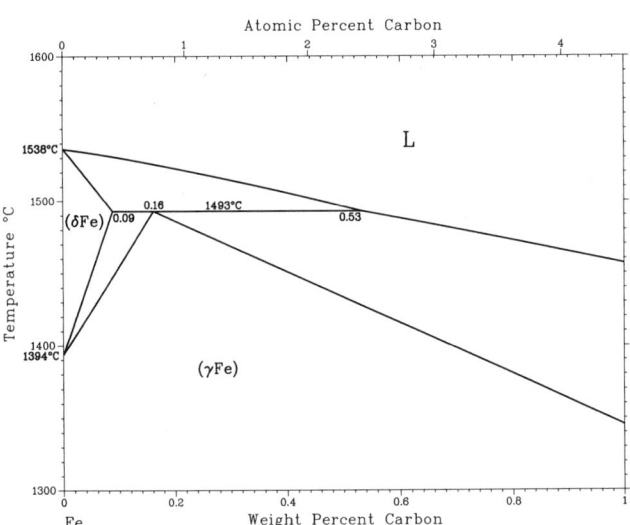

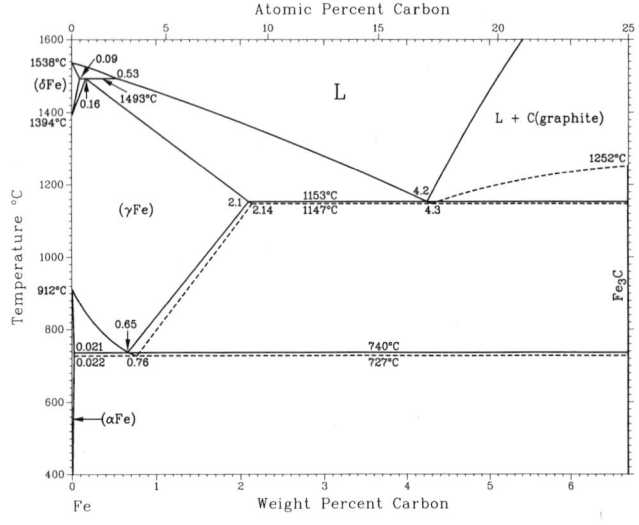

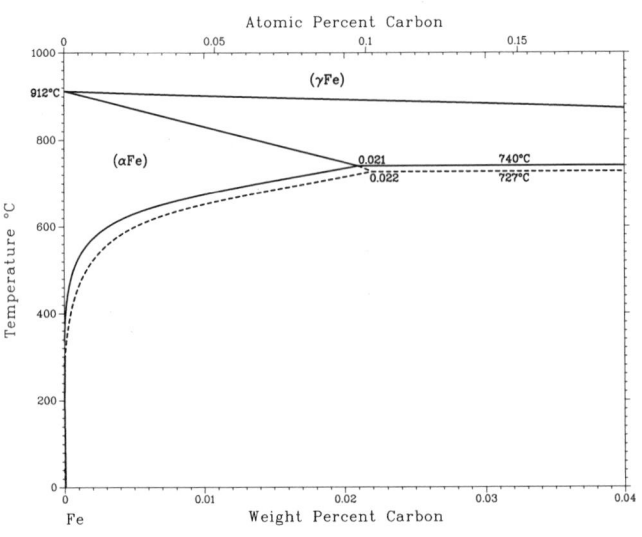

C-Hf

H. Okamoto, 1990

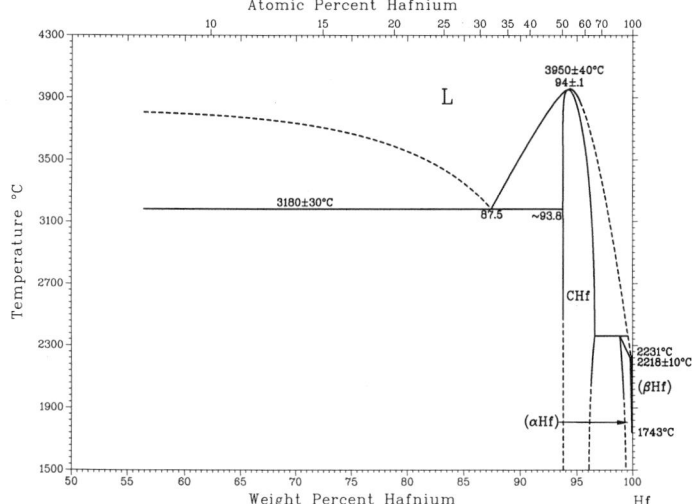

Phase	Composition, wt% Hf	Pearson symbol	Space group
(C)	0	hP4	$P6_3/mmc$
CHf	~93.8 to 96.6	cF8	$Fm\overline{3}m$
(βHf)	99.9 to 100	cI2	$Im\overline{3}m$
(αHf)	98.9 to 100	hP2	$P6_3/mmc$

C-La

K.A. Gschneidner, Jr. and F.W. Calderwood, 1986

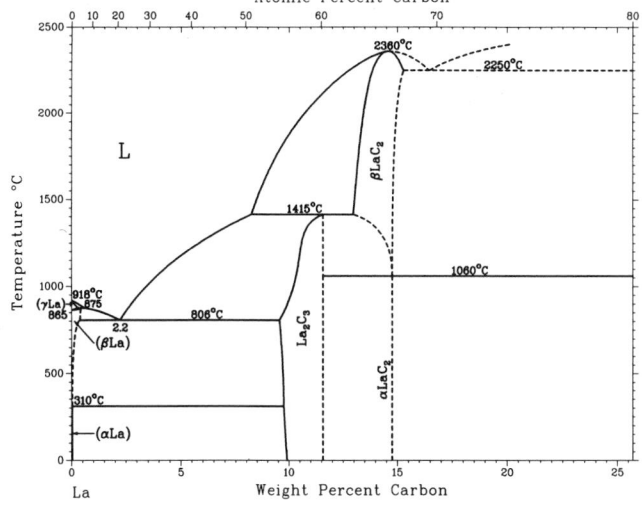

Phase	Composition, wt% C	Pearson symbol	Space group
(αLa)	0	hP4	$P6_3/mmc$
(βLa)	0 to ~0.3	cF4	$Fm\overline{3}m$
(γLa)	0 to ~0.2	cI2	$Im\overline{3}m$
La₂C₃	~9 to ~11	cI40	$I\overline{4}3d$
αLaC₂	~15	tI6	$I4/mmm$
βLaC₂	~13 to ~15	cF12	$Fm\overline{3}m$
(C)	100	hP4	$P6_3/mmc$

C-Mn

H. Okamoto, 1990

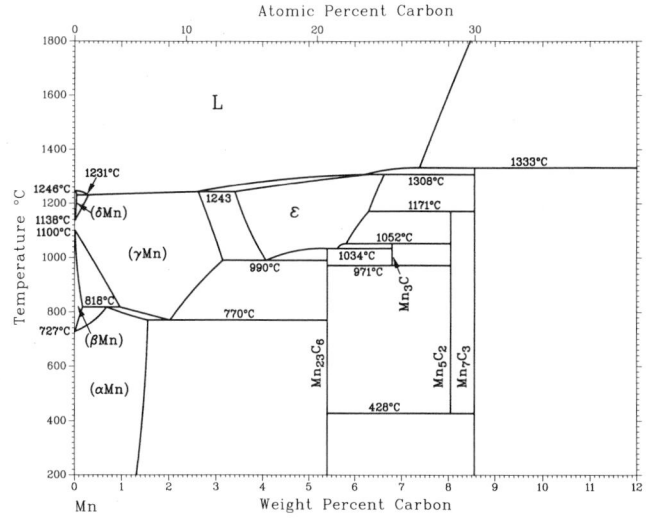

Phase	Composition, wt% C	Pearson symbol	Space group
(δMn)	0 to 0.02	cI2	$Im\overline{3}m$
(γMn)	0 to 3	cF4	$Fm\overline{3}m$
(βMn)	0 to 0.1	cP20	$P4_132$
(αMn)	0 to 1.5	cI58	$I\overline{4}3m$
ε	3.3 to 6.6	...	...
Mn₂₃C₆	5.4	cF116	$Fm\overline{3}m$
Mn₃C	6.8	oP16	Pnma
Mn₅C₂	8.1	mC28	C2/c
Mn₇C₃	8.6	oP40	Pnma
(C)	100	hP4	$P6_3/mmc$

C-Mo

H. Okamoto, 1990

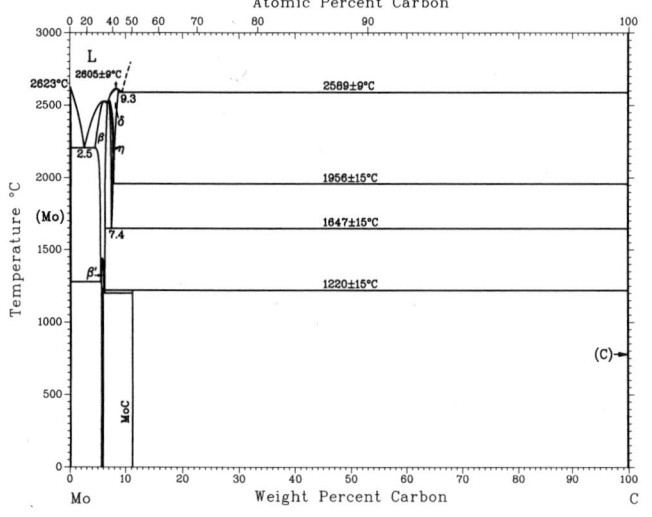

Phase	Composition, wt% C	Pearson symbol	Space group
(Mo)	0 to 0.14	cI2	Im$\bar{3}$m
β	4.4 to 6.6	hP3	P6$_3$/mmc
β′	~5.7	oP12	Pbcn
β″	~5.9	...	...
η	6.8 to 7.7	hP8	P6$_3$/mmc
δ	6.8 to 8.6	oF8	Fm$\bar{3}$m
MoC	11	hP2	P$\bar{6}$m2
(C)	100	hP4	P6$_3$/mmc

C-Ni

M.F. Singleton and P. Nash, 1991

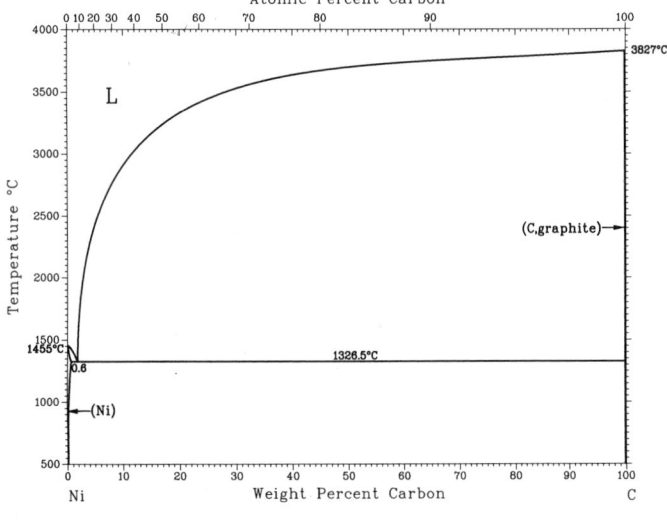

Phase	Composition, wt% C	Pearson symbol	Space group
(Ni)	0 to 0.6(a)	cF4	Fm$\bar{3}$m
(C, graphite)	~100	hP4	P6$_3$/mmc
Metastable phase			
Ni$_3$C	...	oP16	Pnma

(a) Can be extended to 1.6 wt% C at 1314 °C

C-Pr

H. Okamoto, 1990

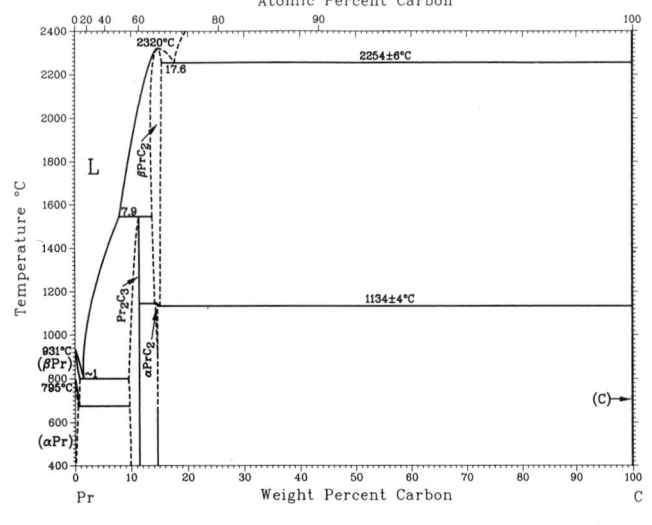

Phase	Composition, wt% C	Pearson symbol	Space group
(αPr)	0 to ?	hP4	P6$_3$/mmc
(βPr)	0 to ?	cI2	Im$\bar{3}$m
Pr$_2$C$_3$	~9 to ~11	cI40	I$\bar{4}$3d
αPrC$_2$	~14.6	tI6	I4/mmm
βPrC$_2$	...	c**	...
(C)	100	hP4	P6$_3$/mmc

C-Sc

O.V. Gordiichuk, 1987

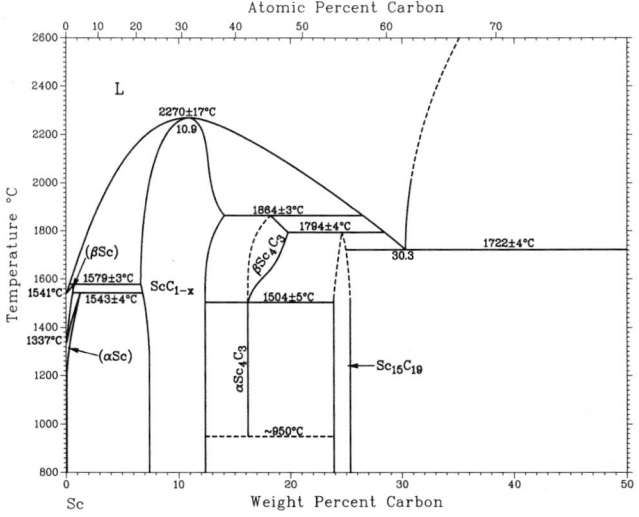

Phase	Composition, wt% C	Pearson symbol	Space group
(αSc)	0	hP2	$P6_3/mmc$
(βSc)	0	cI2	$Im\bar{3}m$
Sc₂C	~12	hR3	$R\bar{3}m$
Sc₄C₃	16.7	cI28	$I\bar{4}3d$
Sc₁₃C₁₀	17.1	c**	...
Sc₁₅C₁₉	25.3	tP68	$P\bar{4}2_1c$
(C)	100	hP4	$P6_3/mmc$

C-Si

R.W. Olesinski and G.J. Abbaschian, 1984

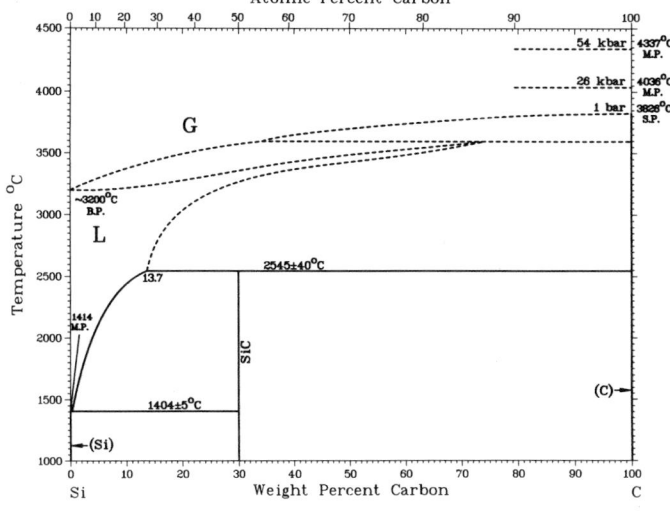

Phase	Composition, wt% C	Pearson symbol	Space group
(Si)	0	cF8	$Fd\bar{3}m$
SiC or βSiC	30	cF8	$F\bar{4}3m$
(C)	100	hP4	$P6_3/mmc$
Metastable			
αSiC(a)	30	(b)	...
Amorphous	22 to 40	...	...
High pressure			
SiC II	...	tI4	$I4_1/amd$

(a) Other SiC polytypes have been reported. (b) Hexagonal

C-Ta

O.M. Barabash and Yu. N. Koval, 1986

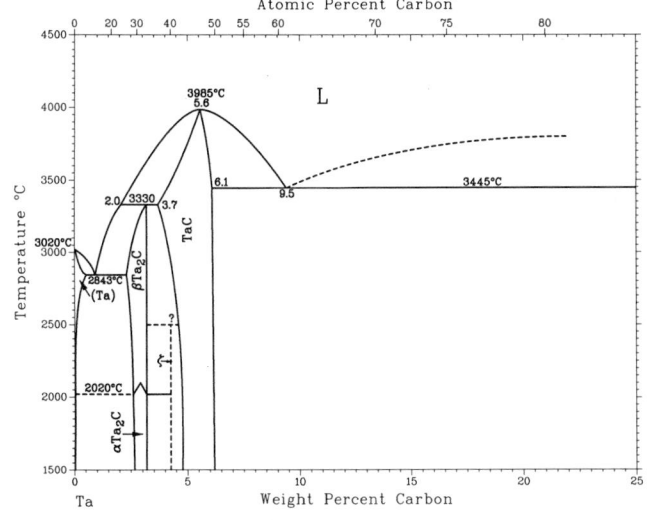

Phase	Composition, wt% C	Pearson symbol	Space group
(Ta)	0 to 0.5	cI2	$Im\bar{3}m$
αTa₂C	2.6 to 3.2	hP3	$P\bar{3}m1$
βTa₂C	2.3 to 3.2	hP3	$P6_3/mmc$
ζ	~4.2	hR20	$R\bar{3}m$
TaC	3.7 to 6.1	cF8	$Fm\bar{3}m$
(C)	100	hP4	$P6_3/mmc$

C-Th

R. Benz and P.L. Stone, 1969

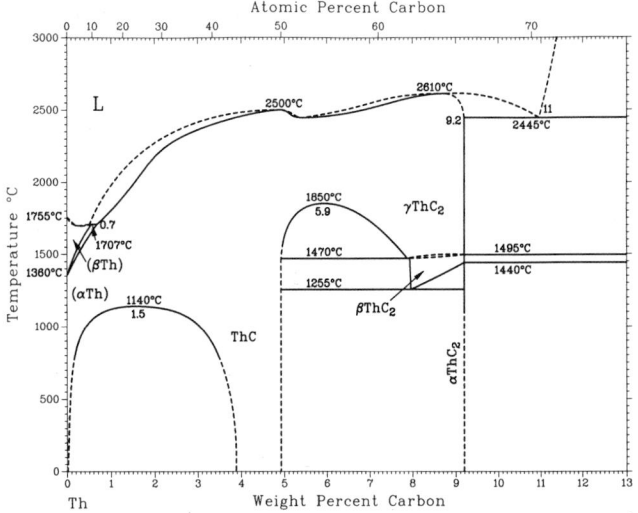

Phase	Composition, wt% C	Pearson symbol	Space group
(βTh)	0 to 0.3	cI2	Im$\bar{3}$m
(αTh)	0 to ?	cF4	Fm$\bar{3}$m
ThC	?	cF8	Fm$\bar{3}$m
γThC$_2$	? to 9.2	cP12	Pa$\bar{3}$
βThC$_2$	~8.1 to 9.1	tP6	P4$_2$/mmc
αThC$_2$	~9.1	mC12	C2/c

C-Ti

J.L. Murray, 1987

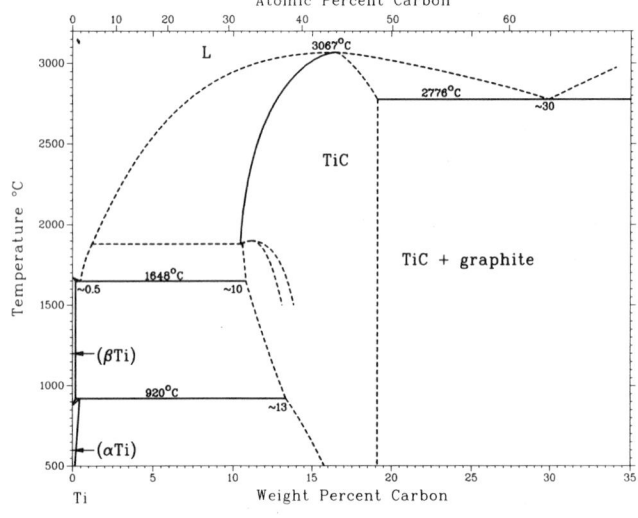

Phase	Composition, wt% C	Pearson symbol	Space group
(βTi)	0 to 0.2	cI2	Im$\bar{3}$m
(αTi)	0 to 0.4	hP2	P6$_3$/mmc
TiC	~10 to 19.3	cF8	Fm$\bar{3}$m
Ti$_2$C	~10 to 12.4	cF48	Fd$\bar{3}$m
(C)	100	hP4	P6$_3$/mmc

C-U

E.K. Storms, 1967; and R. Benz, 1969

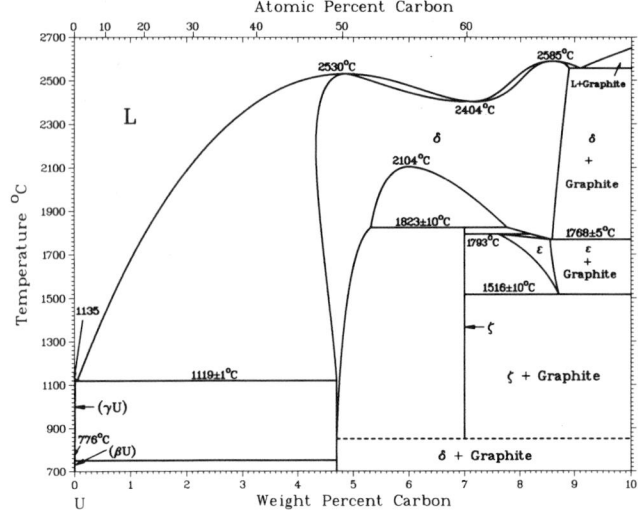

Phase	Composition, wt% C	Pearson symbol	Space group
(γU)	0	cI2	Im$\bar{3}$m
(βU)	0	tP30	P4$_2$/mnm
δ	~4.3 to 8.9	cF8	Fm$\bar{3}$m
ε	~7.6 to 8.7	tI6	I4/mmm
ζ	7.0	cI40	I$\bar{4}$3d

C-V

H. Okamoto, 1991

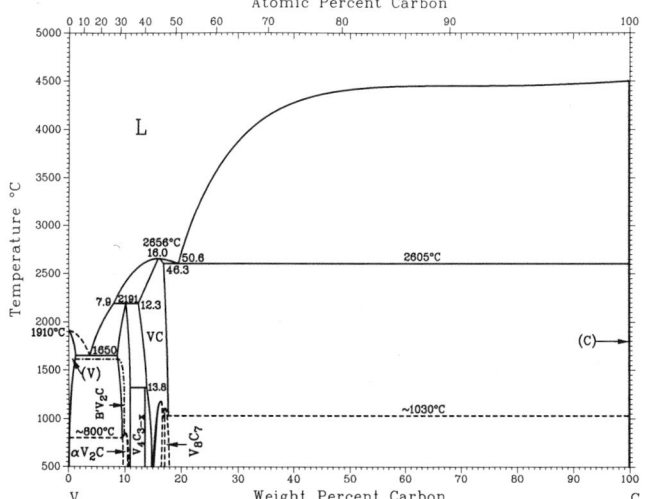

Phase	Composition, wt% C	Pearson symbol	Space group
(V)	0 to 1.0	cI2	Im$\bar{3}$m
αV₂C	9.6 to 10.4	oP12	Pbcn
βV₂C	8.5 to 10.8	hP3	P6₃/mmc(b)
β'V₂C(a)	~8.6 to 9.9	hP9	P$\bar{3}$1m(b)
V₄C₃₋ₓ	~13.6	hR20	R$\bar{3}$m
VC(b)	12.3 to 17.9	cF8	Fm$\bar{3}$m
V₆C₅(c)	15.1 to 16.7	C44	B2
V₈C₇	16.7 to 17.9	cP60	P4₁32
			P4₃32
(C)	100	hP4	P6₃/mmc

(a) High-temperature form. (b) Either one or the other of these two space groups is in error, or the postulated transition in the diagram is in error, with the transition being first order, requiring a two-phase region between the ordered and disordered structures. (c) Enantiomorphic and twinned forms have been described with other lattice parameters and/or space groups.

C-W

R.V. Sara, 1965; and E. Rudy and J.R. Hoffman, 1967

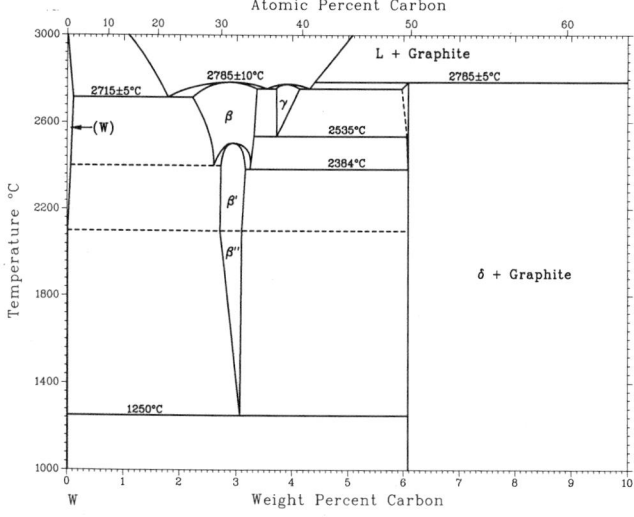

Phase	Composition, wt% C	Pearson symbol	Space group
(W)	0	cI2	Im$\bar{3}$m
β	~2.2 to 3.3	…	…
β'	~2.7 to 3.1	hP3	P6₃/mmc
β"	~2.7 to 3.05	hP3	P$\bar{3}$m1
γ	~3.7 to 4.1	cF8	Fm$\bar{3}$m
δ	6.1	hP2	P$\bar{6}$m2

C-Y

K.A. Gschneidner, Jr. and F.W. Calderwood, 1986

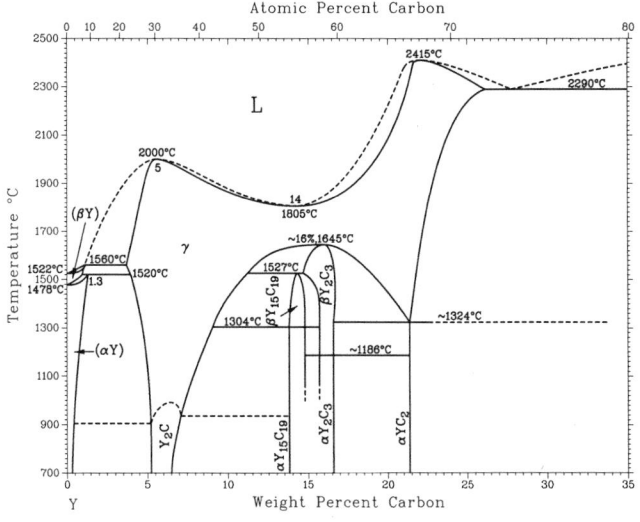

Phase	Composition, wt% C	Pearson symbol	Space group
(αY)	0 to 1.3	hP2	P6₃/mmc
(βY)	0 to 1.0	cI2	Im$\bar{3}$m
Y₂C	~6.2	hR3	R$\bar{3}$m
γ	~3.7 to 25.8	cF5	Fm$\bar{3}$m
αY₁₅C₁₉	14.6	tP68	P$\bar{4}$2₁c
Y₂C₃(a)	~17	cI40	I$\bar{4}$3d
αYC₂	~21.3	tI6	I4/mmm
"βYC₂"	~21.3	cF12	Fm$\bar{3}$m
(C)	100	hP4	P6₃/mmc

(a) Metastable form produced under pressure at high temperature

C-Zr

H. Okamoto, 1990

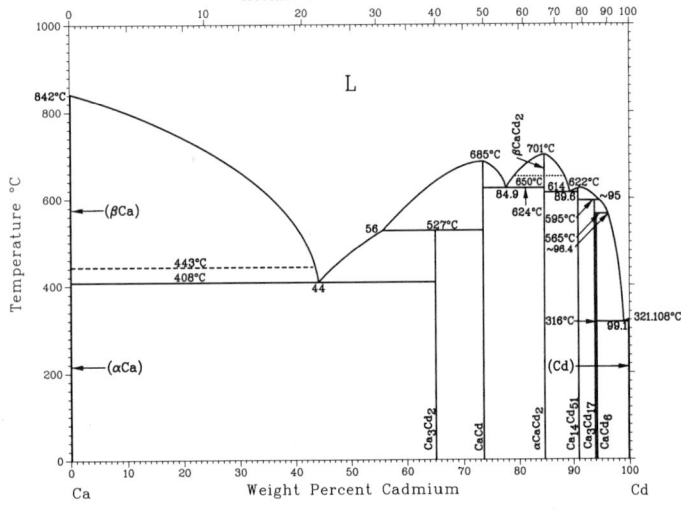

Phase	Composition, wt% C	Pearson symbol	Space group
(βZr)	0	cI2	$Im\bar{3}m$
(αZr)	0	hP2	$P6_3/mmc$
ZrC	~6 to 12	cF8	$Fm\bar{3}m$
(C)	0	hP4	$P6_3/mmc$

Ca-Cd

P.R. Subramanian, 1990

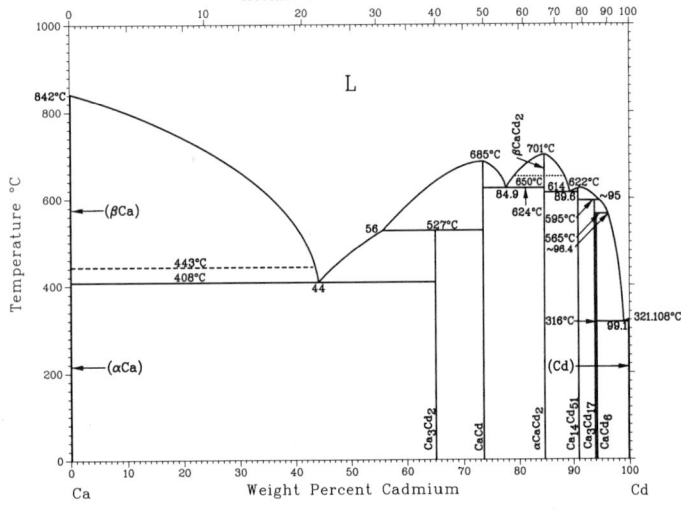

Phase	Composition, wt% Cd	Pearson symbol	Space group
(αCa)(a)	0	cF4	$Fm\bar{3}m$
(βCa)(b)	0	cI2	$Im\bar{3}m$
Ca_3Cd_2	65	tP20	$P4_2nm$
CaCd	73.7	cP2	$Pm\bar{3}m$
$\alpha CaCd_2(c)$	84.9	hP12	$P6_3/mmc$
$\beta CaCd_2$	84.9	oI12	Imma
$Ca_{14}Cd_{51}$	91.1	hP68	P6/m
Ca_3Cd_{17}	94	...	...
$CaCd_6$	94.4	cI184	$Im\bar{3}$
(Cd)	100	hP2	$P6_3/mmc$

(a) Below 443 °C. (b) From 443 to 842 °C. (c) From 0 to 650 °C. (d) From 650 to 701 °C

Ca-Cu

D.J. Chakrabarti and D.E. Laughlin, 1984

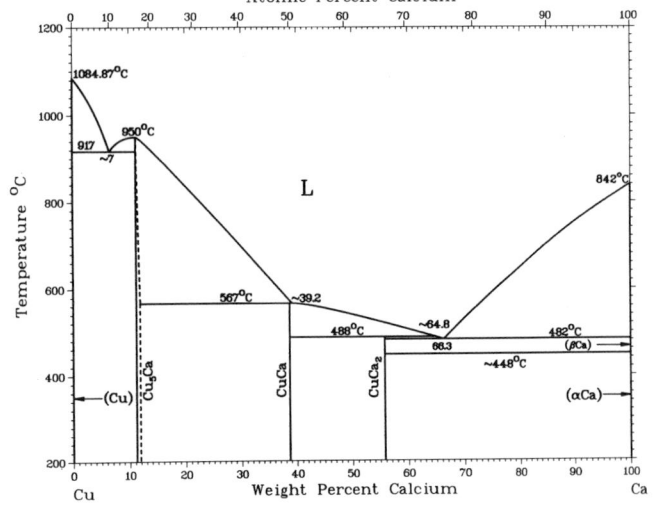

Phase	Composition, wt% Ca	Pearson symbol	Space group
(Cu)	0	cF4	$Fm\bar{3}m$
Cu_5Ca	10.7 to 11.4	hP6	P6/mmm
αCuCa(b)	38.7	mP20	$P2_1/c$
βCuCa(c)	38.7	oP40	Pnma
$CuCa_2$	55.8	oP12	Pnma
(αCa)	100	cF4	$Fm\bar{3}m$
(βCa)	100	cI2	$Im\bar{3}m$

(a) A much wider homogeneity range (approximately 14.1 to 20 at.% Ca) indicated. (b) High temperature; 94.3° interaxial angle. (c) Low temperature

Ca-Ga

V.P. Itkin and C.B. Alcock, 1992

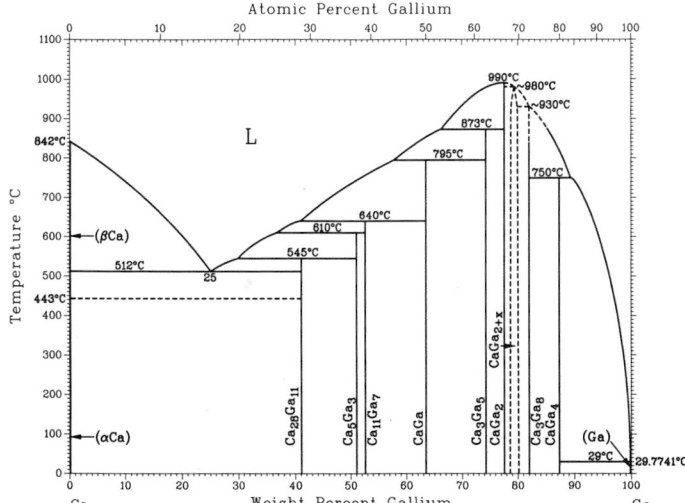

Phase	Composition, wt% Ga	Pearson symbol	Space group
(αCa)(a)	0	cF4	Fm$\bar{3}$m
(βCa)(b)	0	cI2	Im$\bar{3}$m
Ca$_{28}$Ga$_{11}$	40.6	oI78	Imm2
Ca$_5$Ga$_3$	51.1	tI32	I4/mcm
Ca$_{11}$Ga$_7$	52.6	cF144	Fm$\bar{3}$m
CaGa	63.5	oC8	Cmcm
Ca$_3$Ga$_5$	72.4	oC32	Cmcm
CaGa$_2$	77.7	hP6	P6$_3$/mmc
CaGa$_{2+x}$	78.5 to 80.7	hP3	P6/mmm
Ca$_3$Ga$_8$	82.2	oI22	Immm
CaGa$_4$	87	mC10	C2/m
(Ga)	100	oC8	Cmca

(a) <443 °C. (b) From 443 to 842 °C

Ca-Ge

H. Okamoto, 1990

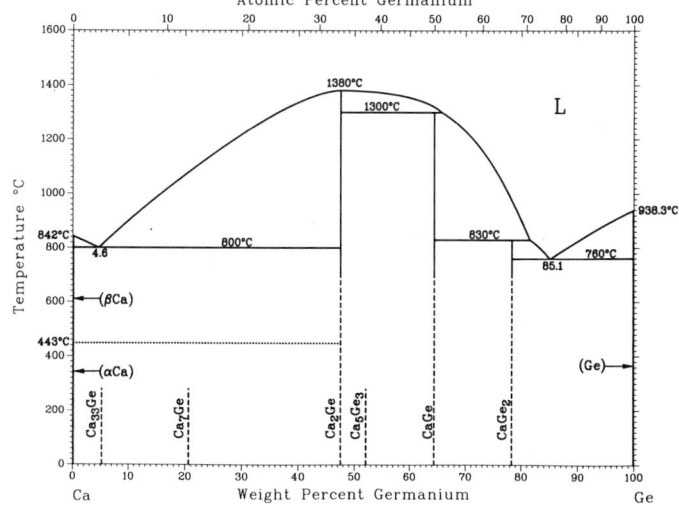

Phase	Composition, wt% Ge	Pearson symbol	Space group
(βCa)	0	cI2	Im$\bar{3}$m
(αCa)	0	cF4	Fm$\bar{3}$m
Ca$_{33}$Ge	5.1	cF48	Fd$\bar{3}$m
Ca$_7$Ge	20.6	cF32	Fm$\bar{3}$m
Ca$_2$Ge	47.5	oP12	Pnma
Ca$_5$Ge$_3$	52.1	tI32	I4/mcm
CaGe	64.4	oC8	Cmcm
CaGe$_2$	78.4	hR6	R$\bar{3}$m
(Ge)	100	cF8	Fm$\bar{3}$m

Ca-Hg

P.R. Subramanian, 1990

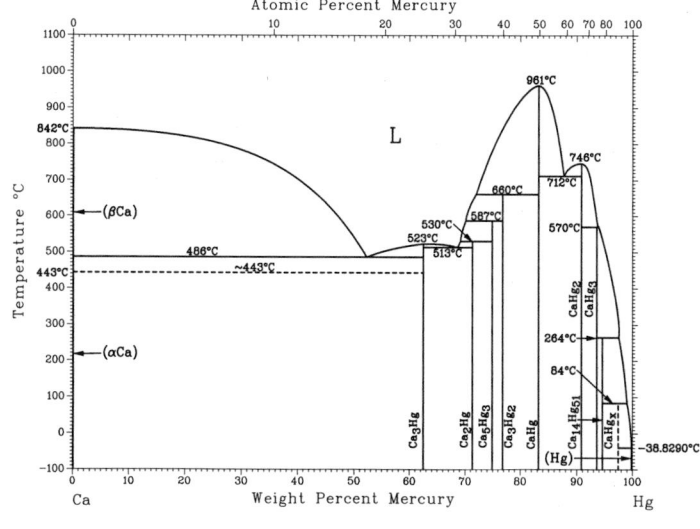

Phase	Composition, wt% Hg	Pearson symbol	Space group
(αCa)(a)	0	cF4	Fm$\bar{3}$m
(βCa)(b)	0	cI2	Im$\bar{3}$m
Ca$_3$Hg	63	oP16	Pnma
		cI32	I$\bar{4}$3m
Ca$_2$Hg	71.4	oP12	Pnma
Ca$_5$Hg$_3$	75.0	tI32	I4/mcm
Ca$_3$Hg$_2$	77	tP10	P4/mbm
CaHg	83.3	cP2	Pm$\bar{3}$m
CaHg$_2$	90.9	hP3	P$\bar{3}$m1
		hP3	P6/mmm
CaHg$_3$	94	hP8	P6$_3$/mmc
Ca$_{14}$Hg$_{51}$	94.8	hP68	P6/m
CaHg$_x$	~98	...	...
(Hg)	100	hR1	R$\bar{3}$m

(a) Below 443 °C. (b) From 443 to 842 °C

Ca-In

H. Okamoto, V.P. Itkin, and C.B. Alcock, 1992

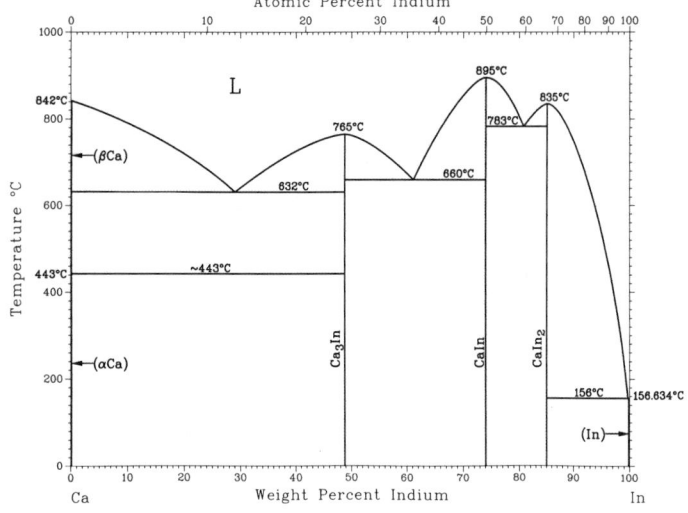

Phase	Composition, wt% In	Pearson symbol	Space group
(βCa)	0	cI2	Im$\overline{3}$m
(αCa)	0	cF4	Fm$\overline{3}$m
Ca₃In	49	cF16	Fm$\overline{3}$m
CaIn	74.1	cP2	Pm$\overline{3}$m
CaIn₂	85.2	hP6	P6₃/mmc
(In)	100	tI2	I4/mmm

Ca-Li

C.W. Bale and A.D. Pelton, 1987

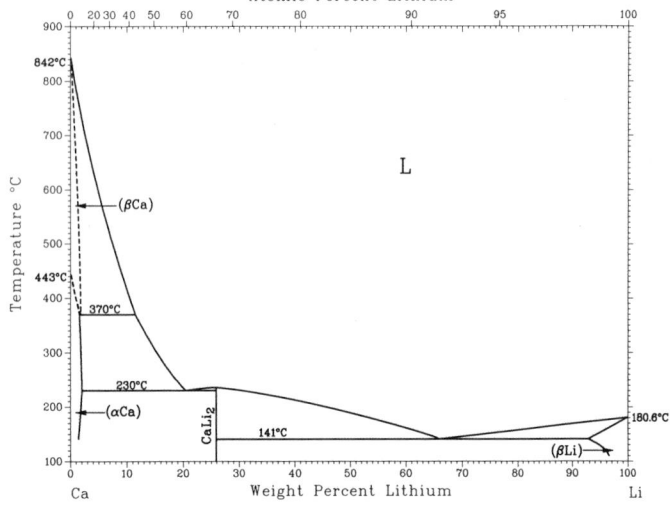

Phase	Composition, wt% Li	Pearson symbol	Space group
(αCa)	0	cF4	Fm$\overline{3}$m
(βCa)	0	cI2	Im$\overline{3}$m
CaLi₂	87.4	hP12	P6₃/mmc
(αLi)	100	hP2	P6₃/mmc
(βLi)	100	cI2	Im$\overline{3}$m

Ca-Mg

A.A. Nayeb-Hashemi and J.B. Clark, 1988

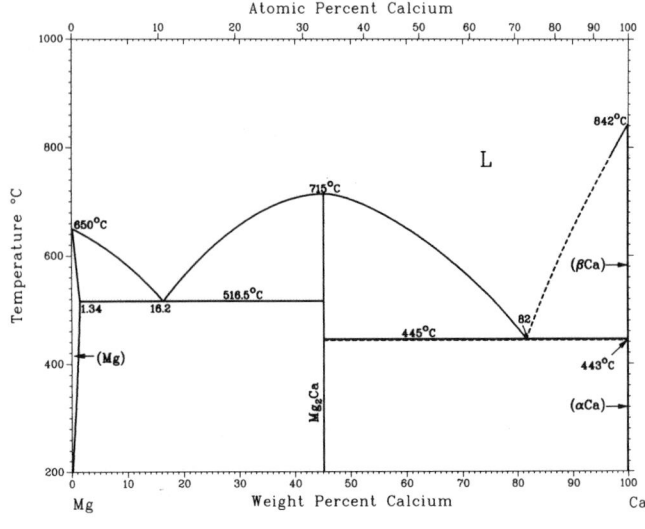

Phase	Composition, wt% Ca	Pearson symbol	Space group
(Mg)	0	hP2	P6₃/mmc
Mg₂Ca	45.2	hP12	P6₃/mmc
(βCa)	100	cI2	Im$\overline{3}$m
(αCa)	100	cF4	Fm$\overline{3}$m

Ca-Na

A.D. Pelton, 1985

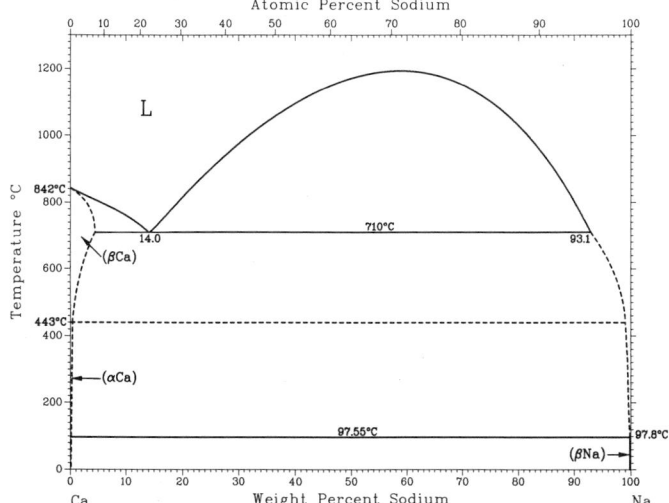

Phase	Composition, wt% Na	Pearson symbol	Space group
(αCa)	0	cF4	Fm$\bar{3}$m
(βCa)	0 to ~4.4	cI2	Im$\bar{3}$m
(βNa)	100	cI2	Im$\bar{3}$m
(αNa)	100	hP2	P6₃/mmc

Ca-Nd

K.A. Gschneidner, Jr. and F.W. Calderwood, 1987

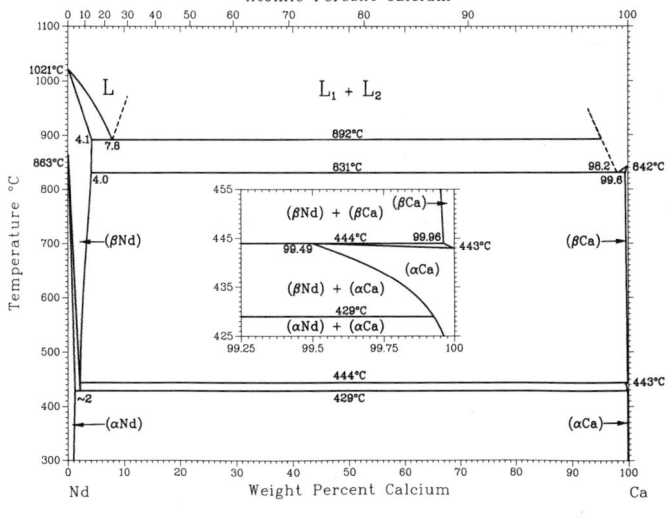

Phase	Composition, wt% Ca	Pearson symbol	Space group
(αNd)	0	hP4	P6₃/mmc
(βNd)	0	cI2	Im$\bar{3}$m
(αCa)	99.5 to 100	cF4	Fm$\bar{3}$m
(βCa)	99.6 to 100	cI2	Im$\bar{3}$m

Ca-Ni

H. Okamoto, 1991

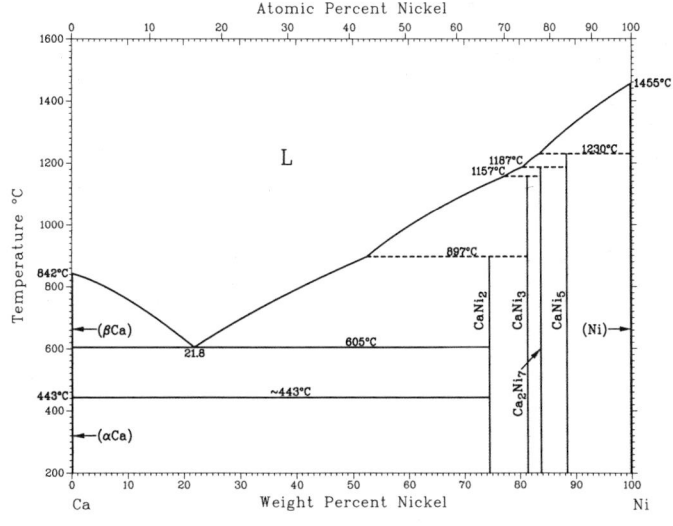

Phase	Composition, wt% Ni	Pearson symbol	Space group
(βCa)	0	cI2	Im$\bar{3}$m
(αCa)	0	cF4	Fm$\bar{3}$m
CaNi₂	74.6	cF24	Fd$\bar{3}$m
Ca₂Ni₅(a)	78.5	hP*	?
CaNi₃	82	hR12	R$\bar{3}$m
Ca₂Ni₇	83.7	hR18	R$\bar{3}$m
CaNi₅	88.0	hP6	P6/mmm
(Ni)	100	cF4	Fm$\bar{3}$m

(a) Not shown on diagram

Ca-O (condensed system)

H.A. Wriedt, 1985

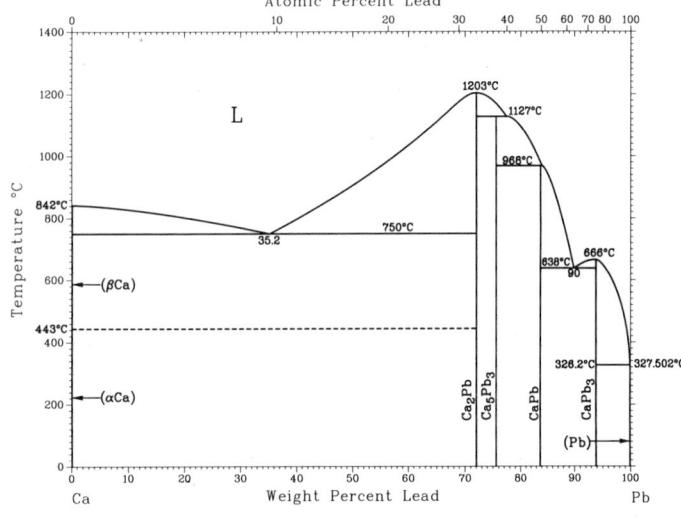

Phase	Composition, wt% O	Pearson symbol	Space group
(αCa)	0	cF4	Fm$\bar{3}$m
(βCa)	0	cI2	Im$\bar{3}$m
CaO	28.5	cF8	Fm$\bar{3}$m
CaO$_2$	44.4	tI6	I4/mmm
CaO$_4$	62	...	...
CaO$_6$	66.6	...	...

Ca-Pb

V.P. Itkin and C.B. Alcock, 1992

Phase	Composition, wt% Pb	Pearson symbol	Space group
(αCa)	0	cF4	Fm$\bar{3}$m
(βCa)	0	cI2	Im$\bar{3}$m
Ca$_2$Pb	72.1	oP12	Pnma
Ca$_5$Pb$_3$	75.6	hP48	P6$_3$/mc
CaPb	83.8	tP4	P4/mmm
CaPb$_3$	94	cP4	Pm$\bar{3}$m
(Pb)	99.9 to 100	cF4	Fm$\bar{3}$m

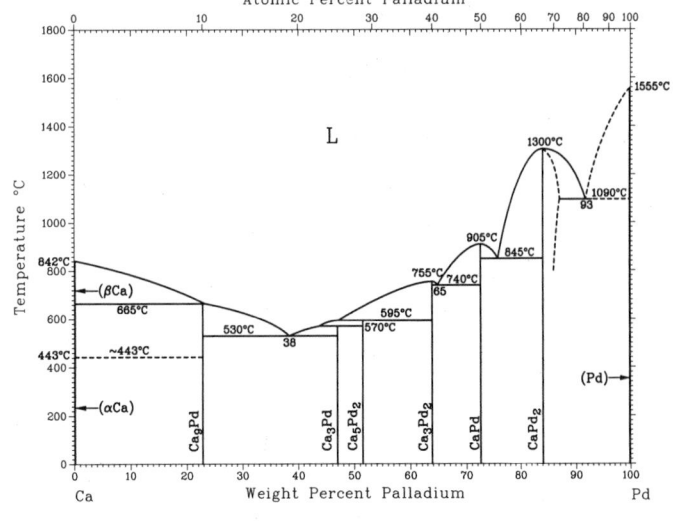

Ca-Pd

H. Okamoto, 1992

Phase	Composition, wt% Pd	Pearson symbol	Space group
(βCa)	0	cI2	Im$\bar{3}$m
(αCa)	0	cF4	Fm$\bar{3}$m
Ca$_9$Pd	~23	...	...
Ca$_3$Pd	47	oP16	Pnma
Ca$_5$Pd$_2$	51.5	mC28	C2/c
Ca$_3$Pd$_2$	64	hR45	R$\bar{3}$
CaPd	72.6	cP2	Pm$\bar{3}$m
CaPd$_2$	84.2	cF24	Fd$\bar{3}$m
CaPd$_5$	93.2	hP6	P6/mmm
(Pd)	100	cF4	Fm$\bar{3}$m

Ca-Pt

P.R. Subramanian, 1990

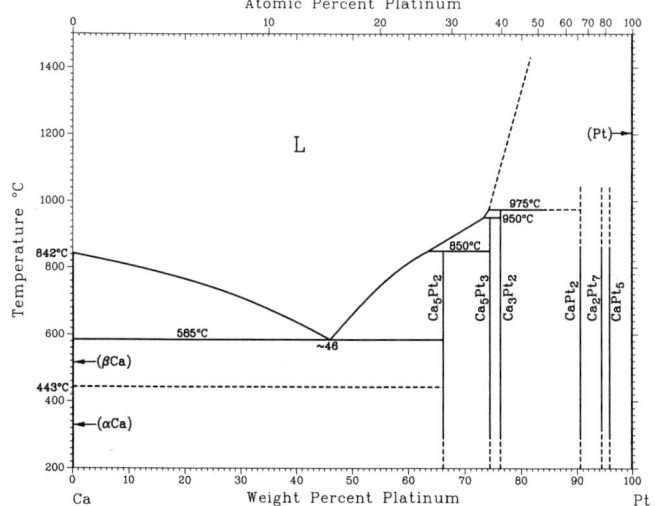

Phase	Composition, wt% Pt	Pearson symbol	Space group
(αCa)	0	cF4	Fm$\overline{3}$m
(βCa)	0	cI2	Im$\overline{3}$m
Ca$_5$Pt$_2$	66.1	mC28	C2/c
Ca$_5$Pt$_3$	74.5	tI32	I4/mcm
Ca$_3$Pt$_2$	76	hR45	R$\overline{3}$
CaPt$_2$	~90.7	cF24	Fd$\overline{3}$m
Ca$_2$Pt$_7$	~94.5	hP36	P6$_3$/mmc
CaPt$_5$	96.0	hP6	P6/mmm
(Pt)	100	cF4	Fm$\overline{3}$m

Ca-Sb

P.R. Subramanian, 1990

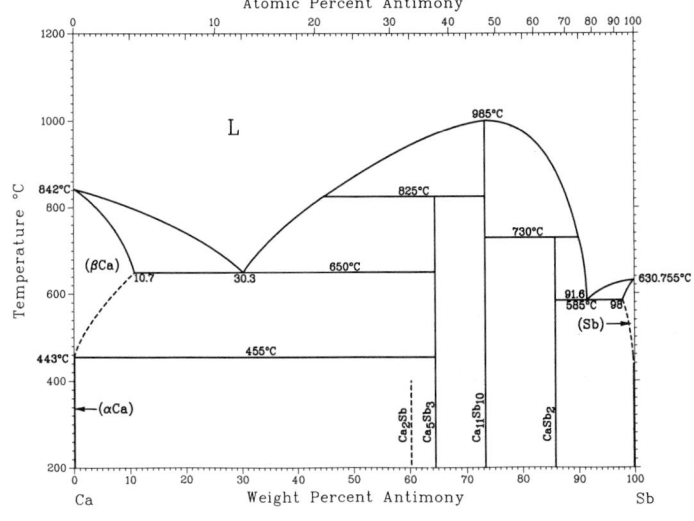

Phase	Composition, wt% Sb	Pearson symbol	Space group
(αCa)	0	cF4	Fm$\overline{3}$m
(βCa)	0 to 10.7	cI2	Im$\overline{3}$m
Ca$_2$Sb	~60.3	tI12	I4/mmm
		tI14(b)	I4/mmm
αCa$_5$Sb$_3$(a)	~64.6	oP32	Pnma
βCa$_5$Sb$_3$(b)	~64.6	hP16	P6$_3$/mcm
Ca$_{11}$Sb$_{10}$	~73.4	tI84	I4/mmm
CaSb(c)	75	cF8	F$\overline{4}$3m
CaSb$_2$	~85.9	mP6	P2$_1$/m
(Sb)	98.0 to 100	hR2	R$\overline{3}$m

(a) Room temperature modification. (b) High-temperature modification; allotropic transformation temperature unknown. (c) Not shown on diagram

Ca-Si

P.R. Subramanian, 1990

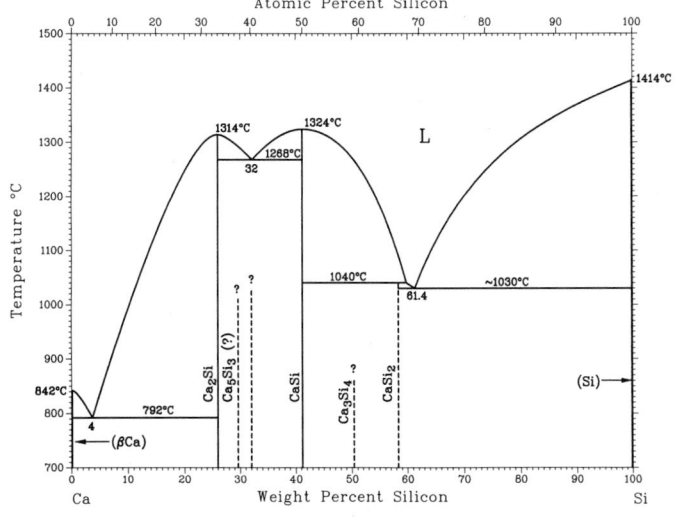

Phase	Composition, wt% Si	Pearson symbol	Space group
(αCa)	0	cF4	Fm$\overline{3}$m
(βCa)	0	cI2	Im$\overline{3}$m
Ca$_2$Si	25.9	oP12	Pnma
Ca$_5$Si$_3$	29.6	tI32	I4/mcm
CaSi	41.2	oC8	Cmcm
Ca$_3$Si$_4$	~48.3	...	...
CaSi$_2$	58.4	hR6	R$\overline{3}$m
(Si)	100	cF8	Fd$\overline{3}$m
High-pressure phase			
CaSi$_2$(a)	58.4	tI12	I4$_1$/amd

(a) Prepared by high-temperature/high-pressure treatment of rhombohedral CaSi$_2$ at 1000 to 1500 °C and 40 kbar, followed by quenching to ambient conditions

Ca-Sr

C.B. Alcock and V.P. Itkin, 1986

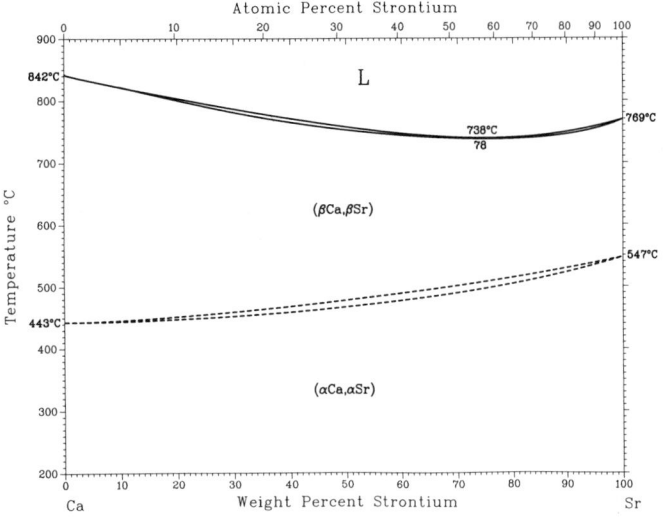

Phase	Composition, wt% Sr	Pearson symbol	Space group
(αCa, αSr)	0 to 100	cF4	Fm$\bar{3}$m
(βCa, βSr)	0 to 100	cI2	Im$\bar{3}$m

Ca-Tl

P.R. Subramanian, 1990

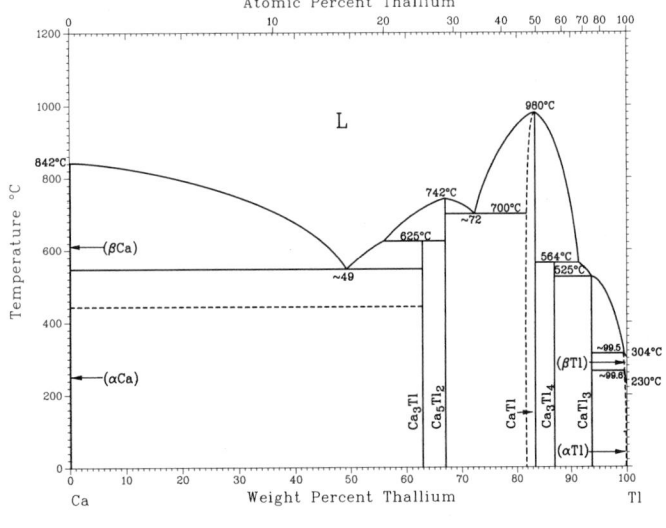

Phase	Composition, wt% Tl	Pearson symbol	Space group
(αCa)	0	cF4	Fm$\bar{3}$m
(βCa)	0	cI2	Im$\bar{3}$m
Ca$_3$Tl	63	cF16	Fm$\bar{3}$m
Ca$_5$Tl$_2$	~67.1	...	...
CaTl	83.6	cP2	Pm$\bar{3}$m
Ca$_3$Tl$_4$	~87.2	...	...
CaTl$_3$	94	cP4	Pm$\bar{3}$m
(αTl)	~99.6	hP2	P6$_3$/mmc
(βTl)	~99.5	cI2	Im$\bar{3}$m

Ca-Yb

K.A. Gschneidner, Jr. and F.W. Calderwood, 1987

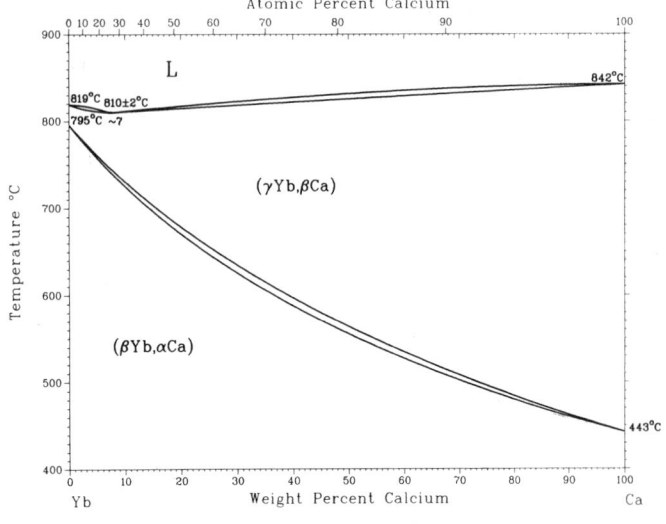

Phase	Composition, wt% Ca	Pearson symbol	Space group
(αYb)	0	hP2	P6$_3$/mmc
(βYb)	0	cF4	Fm$\bar{3}$m
(γYb)	0	cI2	Im$\bar{3}$m
(αCa)	100	cF4	Fm$\bar{3}$m
(βCa)	100	cI2	Im$\bar{3}$m

Ca-Zn

V.P. Itkin and C.B. Alcock, 1990

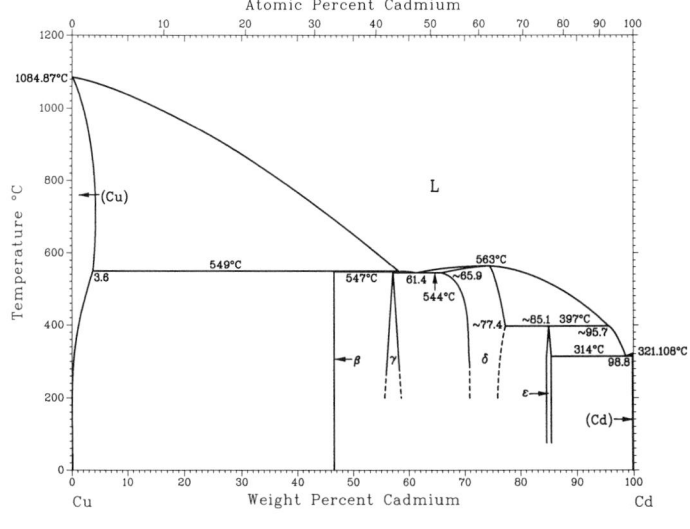

Phase	Composition, wt% Zn	Pearson symbol	Space group
(βCa)	0	cI2	$Im\bar{3}m$
(αCa)	0	cF4	$Fm\bar{3}m$
Ca₃Zn	35	oC16	Cmcm
Ca₅Zn₃	46.5	tI32	I4/mcm
CaZn	62.0	oC8	Cmcm
CaZn₂	76.6	oI12	Imma
CaZn₃	82 to 83	hP32	P6₃/mmc
CaZn₅	81.7 to 89.5	hP6	P6/mmm
CaZn₁₁	94.7	tI48	I4₁/amd
CaZn₁₃	95.5	cF112	$Fm\bar{3}c$
(Zn)	100	hP2	P6₃/mmc

Cd-Cu

P.R. Subramanian and D.E. Laughlin, 1990

Phase	Composition, wt% Cd	Pearson symbol	Space group
(Cu)	0 to 3.6	cF4	$Fm\bar{3}m$
β	46.9	hP24	P6₃/mmc
γ	56.0 to 58.3(a)	cF1124	$F\bar{4}3m$
δ	65.9 to 77	cI52	...
ε	84.6 to 85.9	hP28	P6₃/mmc
(Cd)	~99.9 to 100	hP2	P6₃/mmc

(a) At 300 °C

Cd-Eu

K.A. Gschneidner, Jr. and F.W. Calderwood, 1988

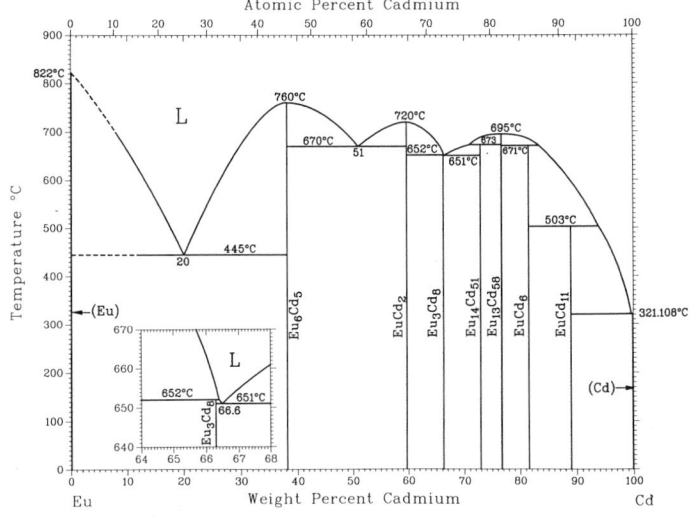

Phase	Composition, wt% Cd	Pearson symbol	Space group
(Eu)	0	cI2	$Im\bar{3}m$
EuCd(a)	38.2	cP2	$Pm\bar{3}m$
EuCd₂	59.7	oI12	Imma
Eu₃Cd₈	66.3	(b)	...
Eu₁₄Cd₅₁	73.0	hP65	P6/m
Eu₁₃Cd₅₈	76.8	hP142	P6₃/mmc
EuCd₆	81.6	cI168	$Im\bar{3}$
EuCd₁₁	88.8 to 89.1	tI48	I4/amd
(Cd)	100	hP2	P6₃/mmc

(a) Defect structure reported as Eu₆Cd₅. (b) Structure not known

Cd-Ga

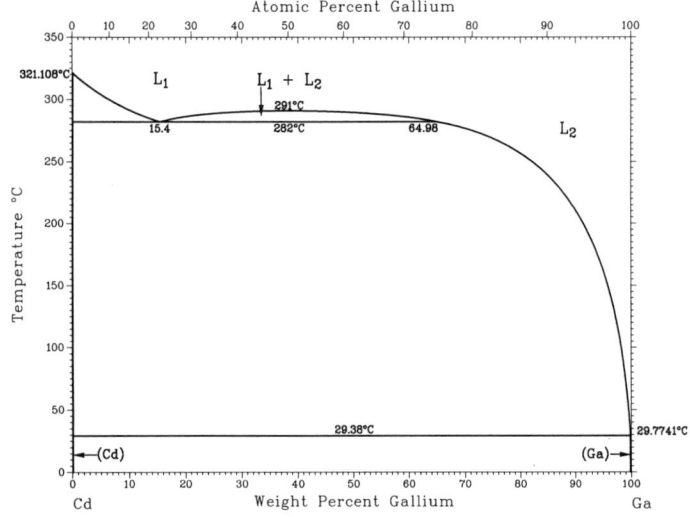

Z. Moser, J. Dutkiewicz, W. Gasior, and J. Salawa, unpublished

Phase	Composition, wt% Ga	Pearson symbol	Space group
(Cd)	0	hP2	P6₃/mmc
(Ga)	100	oC8	Cmca

Cd-Gd

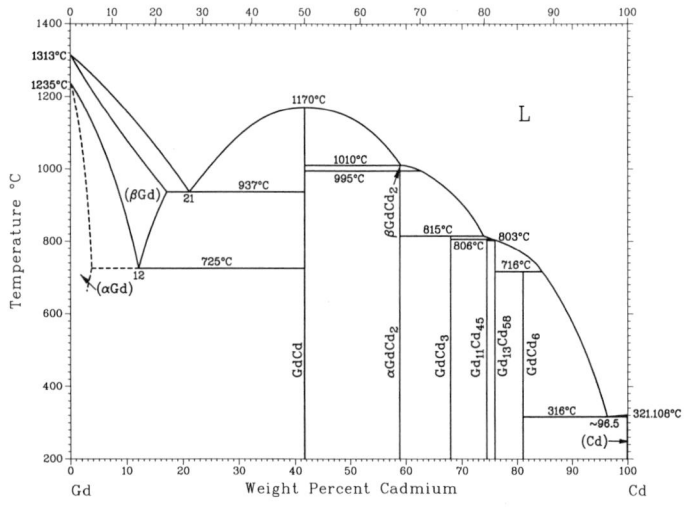

K.A. Gschneidner, Jr. and F.W. Calderwood, 1988

Phase	Composition, wt% Cd	Pearson symbol	Space group
(αGd)	0 to ~3.6	hP2	P6₃/mmc
(βGd)	0 to ~17	cI2	Im3̄m
GdCd	41.7	cP2	Pm3̄m
GdCd₂	58.9	hP3	P3m1
GdCd₃	68	hP8	P6₃/mmc
Gd₁₁Cd₄₅	74.6	cF448	F4̄3m
Gd₁₃Cd₅₈	76.1	hP142	P6₃/mmc
GdCd₆	81.1	cI168	Im3̄
(Cd)	100	hP2	P6₃/mmc

Cd-Ge

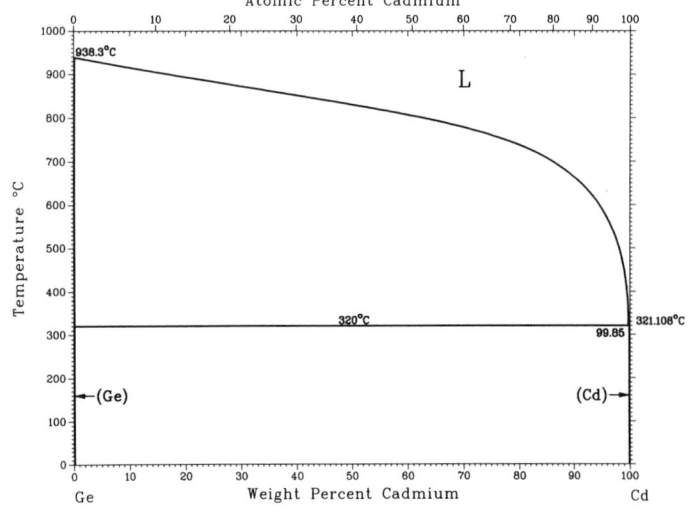

R.W. Olesinski and G.J. Abbaschian, 1986

Phase	Composition, wt% Cd	Pearson symbol	Space group
(Ge)	0.0	cF8	Fd3̄m
GeII(HP)	0.0	tI4	I4₁/amd
(Cd)	100	hP2	P6₃/mmc

Cd-Hg

C. Guminski and L.A. Zabdyr, 1992

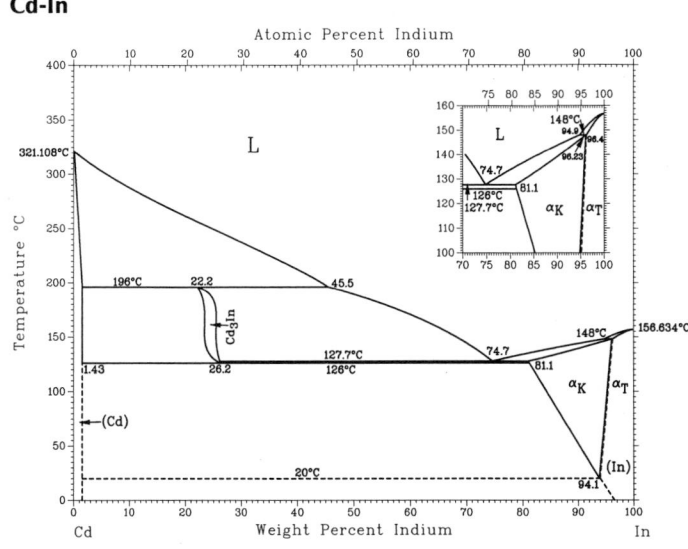

Phase	Composition, wt% Hg	Pearson symbol	Space group
(Cd)	0 to 37	hP2	$P6_3/mmc$
ω	42 to 94	tI2	I4/mmm
ω'	47 to 56	tI6	I4/mmm
ω''	71 to 81	tI6	I4/mmm
(αHg)(a)	98 to 100	hR1	$R\bar{3}m$
(βHg)(b)	~100	tI2	I4/mmm

(a) From –38.8290 to –193 °C at 100 wt% Hg. (b) Below –193 °C

Cd-In

J. Dutkiewicz, L.A. Zabdyr, W. Zakulski, Z. Moser, J. Salawa, P.J. Horrocks, F.H. Hayes, and M.H. Rand, 1992

Phase	Composition, wt% In	Pearson symbol	Space group
(Cd)	0 to 1.4	hP2	$P6_3/mmc$
Cd₃In	22.2 to 26.2	cP4	$Pm\bar{3}m$
α_K	81.1 to 94	cF4	$Fm\bar{3}m$
(In)(α_T)	94 to 100	tI2	I4/mmm

Cd-La

K.A. Gschneidner, Jr. and F.W. Calderwood, 1988

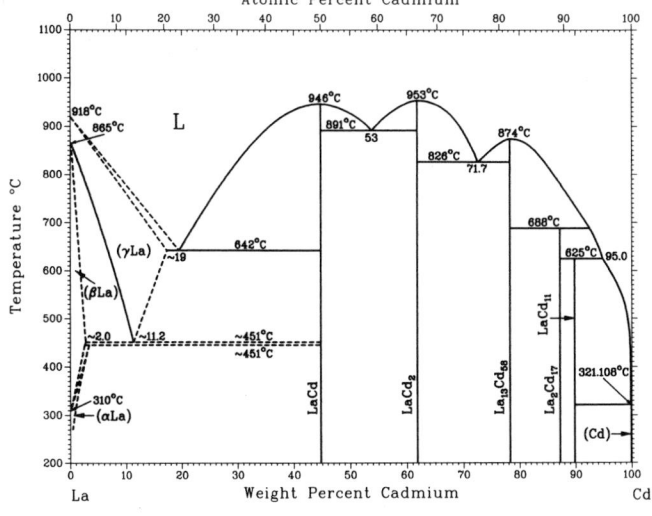

Phase	Composition, wt% Cd	Pearson symbol	Space group
(αLa)	0	hP4	$P6_3/mmc$
(βLa)	0 to ~2.0	cF4	$Fm\bar{3}m$
(γLa)	0 to ~18	cI2	$Im\bar{3}m$
LaCd	44.7	cP2	$Pm\bar{3}m$
LaCd₂	61.8	hP3	P3m1
La₁₃Cd₅₈	78.3	hP142	$P6_3/mmc$
La₂Cd₁₇	85.8	hP38	$P6_3/mmc$
LaCd₁₁	89.9	cP36	$Pm\bar{3}m$
(Cd)	100	hP2	$P6_3/mmc$

Cd-Li

A.D. Pelton, 1988

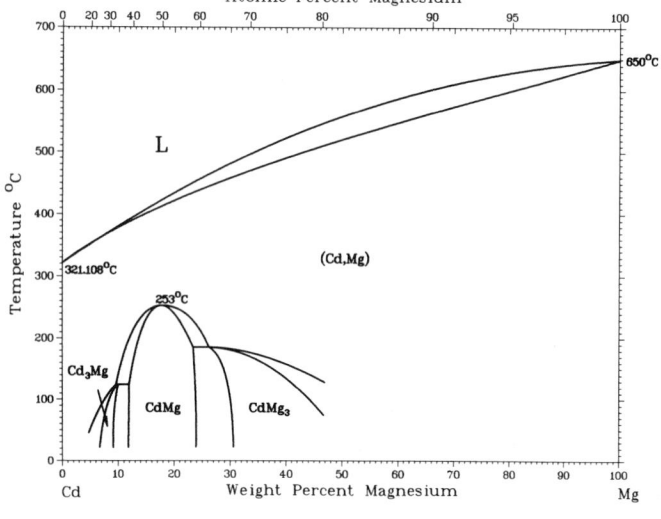

Phase	Composition, wt% Li	Pearson symbol	Space group
(Cd)	0 to 2.6	hP2	P6₃/mmc
Cd₃Li	2? to 2.5	hP2	P6₃/mmc
CdLi	3.6 to 18	cF16	Fd3̄m
CdLi₃	10? to 18	cF4	Fm3̄m
(βLi)	22 to 100	cI2	Im3̄m
(αLi)(a)	100	hP2	P6₃/mmc

(a) Below −193 °C

Cd-Mg

Z. Moser, W. Gasior, J. Wypartowicz, and L. Zabdyr, 1984

Phase	Composition, wt% Mg	Pearson symbol	Space group
(Cd, Mg)	0 to 100	hP2	P6₃/mmc
α' or Cd₃Mg	7 to 9	hP8	P6₃/mmc
α'' or CdMg	12 to 25	oP4	Pmma
α''' or CdMg₃	29 to 50	hP8	P6₃/mmc

Cd-Na

A.D. Pelton, 1988

Phase	Composition, wt% Na	Pearson symbol	Space group
(Cd)	0	hP2	P6₃/mmc
Cd₁₁Na₂	3.9	cP39	Pm3̄
Cd₂Na(a)	9.3	cF1192	...
(βNa)	100	cI2	Im3̄m
(αNa)(b)	100	hP2	P6₃/mmc

(a) Complex cubic structure that corresponds to the formula Cd₁.₉₂Na at 0.070 wt% Na. (b) Below −237 °C

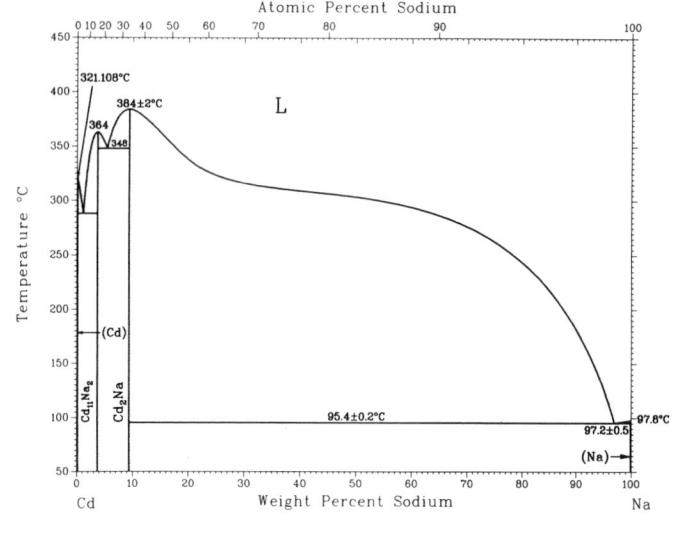

Cd-Ni

F.A. Shunk and P. Nash, 1991

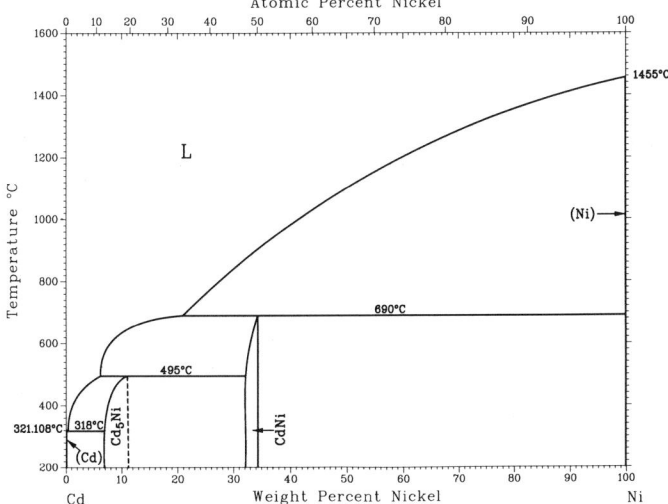

Phase	Composition, wt% Ni	Pearson symbol	Space group
(Cd)	0	$hP2$	$P6_3/mmc$
Cd_5Ni	9 to 10.6	$cP52$	$P\bar{4}3m$
CdNi	31.9 to 34.3	$cF112$	$Fd\bar{3}m$
(Ni)	100	$cF4$	$Fm\bar{3}m$

Cd-P

H. Okamoto, 1990

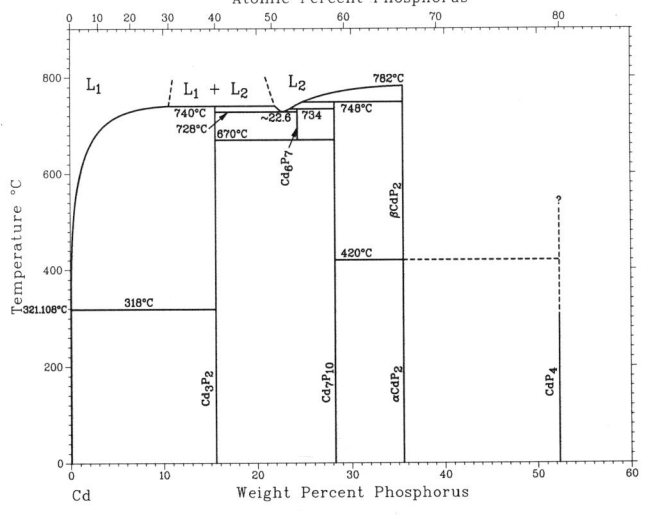

Phase	Composition, wt% P	Pearson symbol	Space group
(Cd)	0	$hP2$	$P6_3/mmc$
Cd_3P_2	16	$tI40$	$P4_2/nmc$
Cd_6P_7	24.3	$c*52$	...
Cd_7P_{10}	24.3	$oF136$	$Fdd2$
βCdP_2	55.6	$tP24$	$P4_32_12$
αCdP_2	35.6	$oP12$	$Pna2_1$
CdP_4	52.4	$mP10$	$P2_1/c$

Cd-Pb

J. Dutkiewicz, Z. Moser, and W. Zakulski, 1988

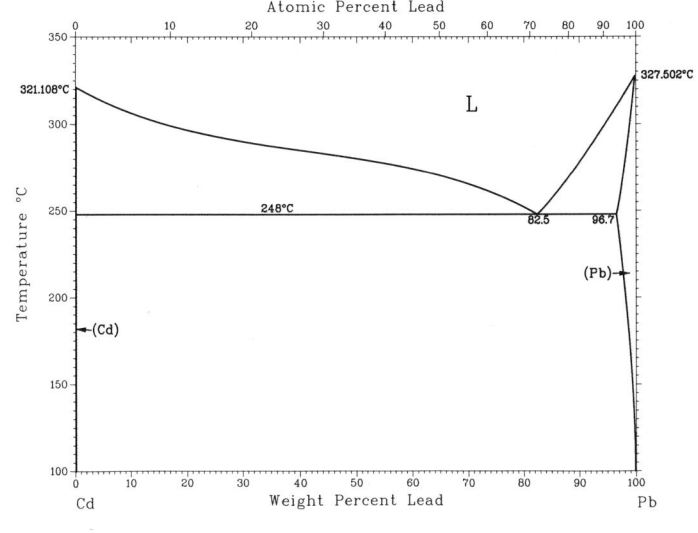

Phase	Composition, wt% Pb	Pearson symbol	Space group
(Cd)	0	$hP2$	$P6_3/mmc$
(Pb)	96.7 to 100	$cF4$	$Fm\bar{3}m$

Cd-Sb

H. Okamoto, 1990

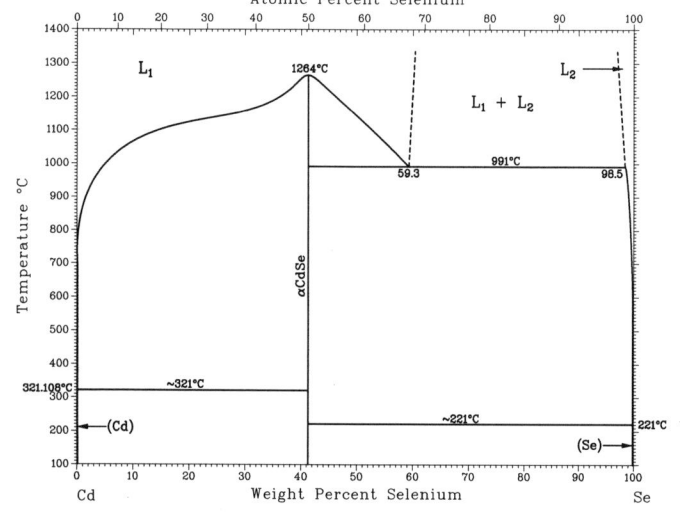

Phase	Composition, wt% Sb	Pearson symbol	Space group
(Cd)	0	hP2	$P6_3/mmc$
CdSb	52.0 to 53	oP16	Pbca
(Sb)	100	hR2	$R\bar{3}m$
Metastable phases			
Cd_3Sb_2	42	m*20	...
Cd_4Sb_3	44.9	hR*	...

Cd-Se

R.C. Sharma and Y.A. Chang, unpublished

Phase	Composition, wt% Se	Pearson symbol	Space group
(Cd)	0	hP2	$P6_3/mmc$
αCdSe	41.3	hP4	$P6_3mc$
βCdSe(a)	41.3	cF8	$Fm\bar{3}m$
(Se)	100	hP3	$P3_121$

(a) High-pressure phase

Cd-Sm

K.A. Gschneidner, Jr. and F.W. Calderwood, 1988

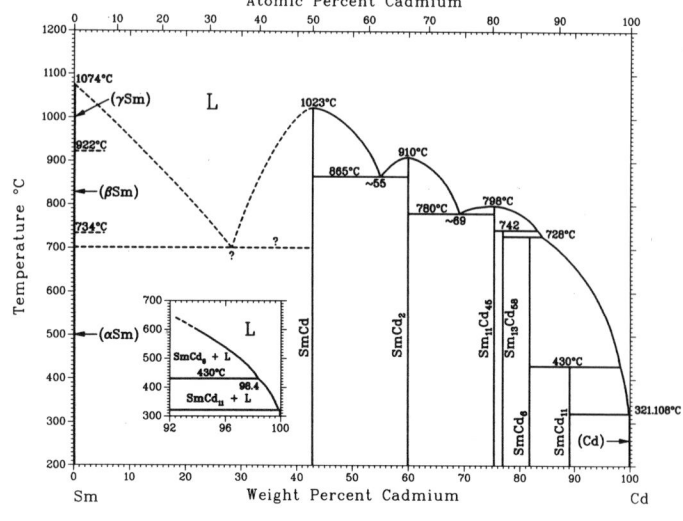

Phase	Composition, wt% Cd	Pearson symbol	Space group
(αSm)	0	hR3	$R\bar{3}m$
(βSm)	0	hP2	$P6_3/mmc$
(γSm)	0	cI2	$Im\bar{3}m$
SmCd	42.8	cP2	$Pm\bar{3}m$
$SmCd_2$	60.0	hP3	$P3m1$
$Sm_{11}Cd_{45}$	75.4	cF448	$F\bar{4}3m$
$Sm_{13}Cd_{58}$	76.9	hP142	$P6_3/mmc$
$SmCd_6$	81.8	cI168	$Im\bar{3}$
$SmCd_{11}$	89.2	cP36	$Pm\bar{3}m$
(Cd)	100	hP2	$P6_3/mmc$

Cd-Sn

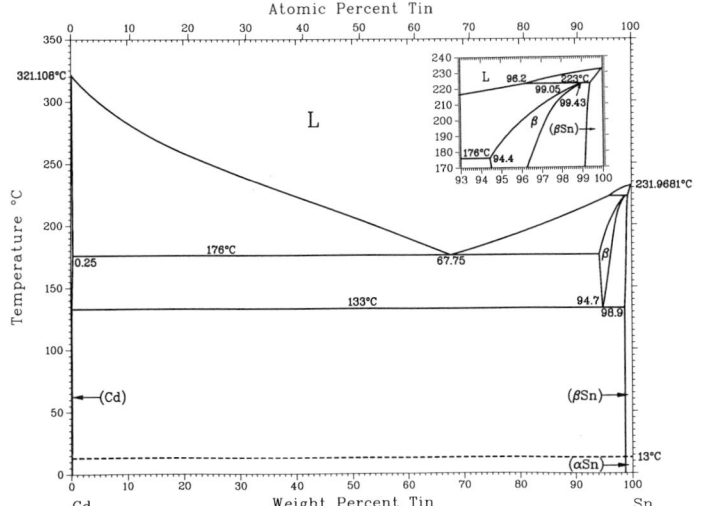

J. Dutkiewicz, L.A. Zabdyr, Z. Moser, and J. Salawa, 1989

Phase	Composition, wt% Sn	Pearson symbol	Space group
(Cd)	0 to 0.25	hP2	P6₃/mmc
β	94.3 to 99.1	hP2	P6₃/mmc
(Sn)	98.9 to 100	tI4	I4₁/amd

Cd-Sr

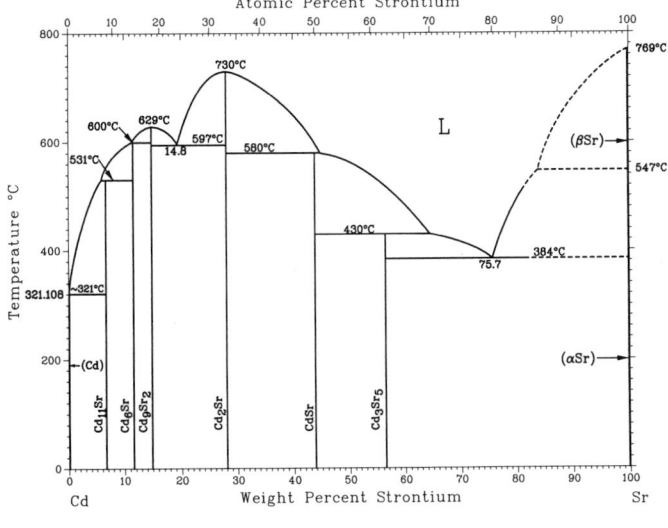

H. Okamoto, 1990

Phase	Composition, wt% Sr	Pearson symbol	Space group
(Cd)	0	hP2	P6₃/mmc
Cd₁₁Sr	6.6	tI48	I4₁/amd
Cd₆Sr	11.5	…	…
Cd₉Sr₂	14.8	…	…
Cd₂Sr	28.0	oI12	Imma
CdSr	43.8	cP2	Pm3m
Cd₃Sr₅	56.5	tI32	I4/mcm
(βSr)	100	cI2	Im3m
(αSr)	100	cF4	Fm3m

Cd-Te

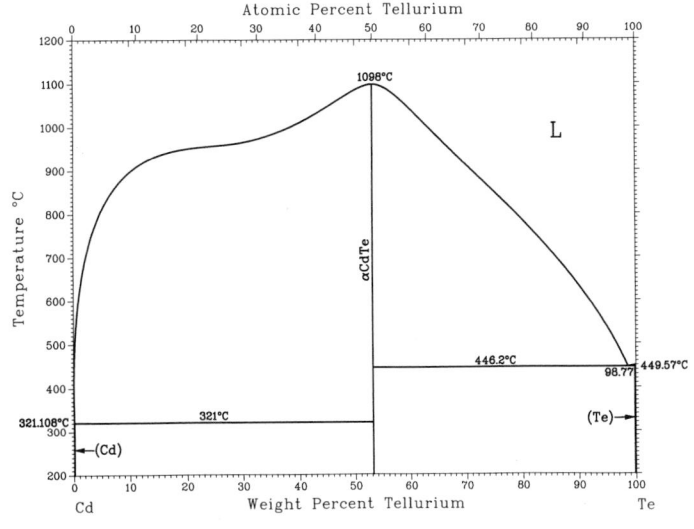

R.C. Sharma and Y.A. Chang, 1989

Phase	Composition, wt% Te	Pearson symbol	Space group
(Cd)	0	hP2	P6₃/mmc
αCdTe	53.2	cF8	F43m
βCdTe(a)	53.2	cF8	Fm3m
γCdTe(a)	53.2	tI4	I4₁/amd
(Te)	100	hP3	P3₁21

(a) High-pressure phase

Cd-Th

J. Dutkiewicz, unpublished

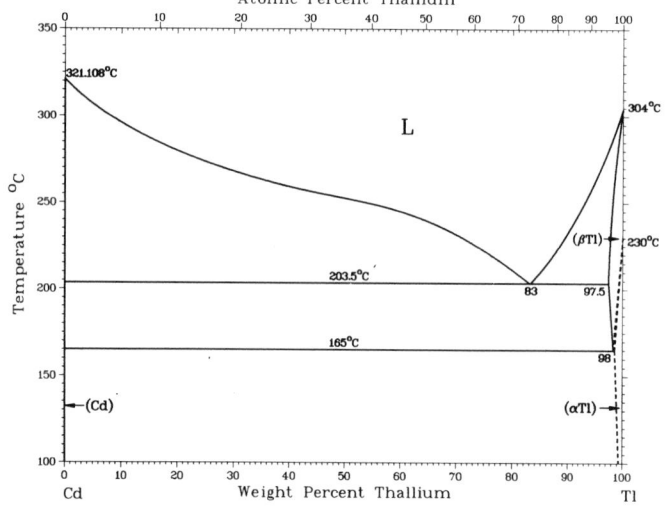

Phase	Composition, wt% Th	Pearson symbol	Space group
(Cd)	0	hP2	$P6_3/mmc$
$Cd_{11}Th$	15.79	cP36	$Pm\overline{3}m$
Cd_5Th	29.21	hP36	$P6_3/mmc$
$Cd_{23}Th_6$	35.00	cF116	$Fm\overline{3}m$
Cd_3Th	41	hP8	$P6_3/mmc$
Cd_2Th	50.79	hP3	$P6/mmm$
CdTh	67.4	oP24	...
(αTh)	100	cF4	$Fm\overline{3}m$
(βTh)	100	cI2	$Im\overline{3}m$

Cd-Tl

H. Okamoto, 1990

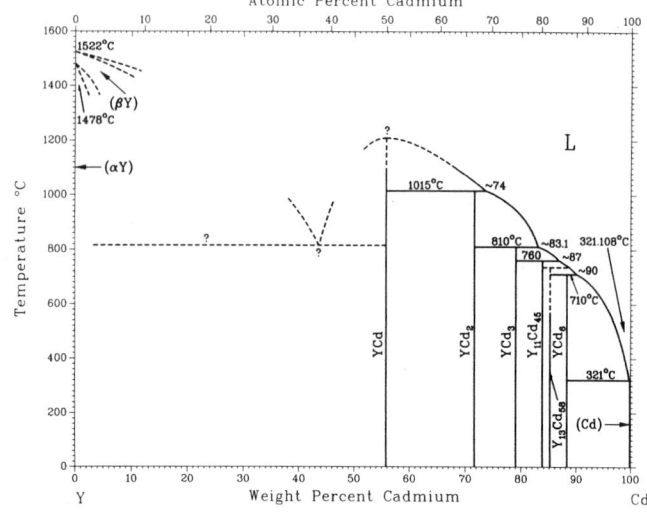

Phase	Composition, wt% Tl	Pearson symbol	Space group
(Cd)	0	hP2	$P6_3/mmc$
(βTl)	97.5 to 100	cI2	$Im\overline{3}m$
(αTl)	~98 to 100	hP2	$P6_3/mmc$

Cd-Y

K.A. Gschneidner, Jr. and F.W. Calderwood, 1988

Phase	Composition, wt% Cd	Pearson symbol	Space group
(αY)	0	hP2	$P6_3/mmc$
(βY)	0	cI2	$Im\overline{3}m$
YCd	55.8	cP2	$Pm\overline{3}m$
YCd_2	71.7	hP3	$P3m1$
YCd_3	79	oC16	Cmcm
$Y_{11}Cd_{45}$	83.8	cF448	$F\overline{4}3m$
$Y_{13}Cd_{58}$	85.0	hP142	$P6_3/mmc$
YCd_6	88.3	cI168	$Im\overline{3}$
(Cd)	100	hP2	$P6_3/mmc$

Cd-Yb

K.A. Gschneidner, Jr. and F.W. Calderwood, 1988

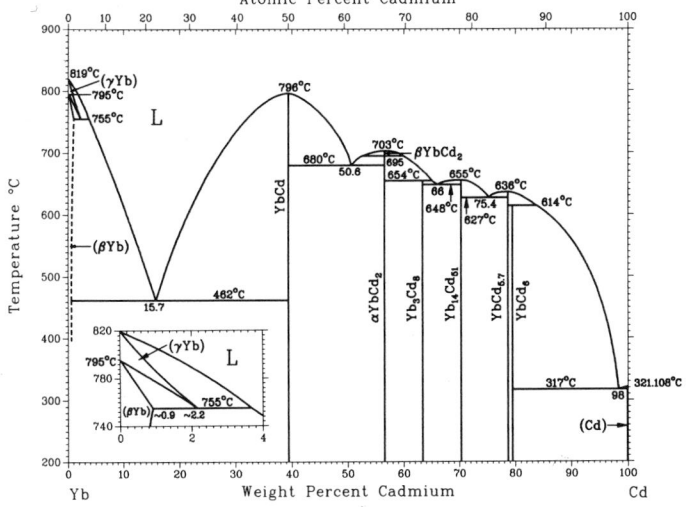

Phase	Composition, wt% Cd	Pearson symbol	Space group
(βYb)	0 to ~0.91	cF4	$Fm\overline{3}m$
(γYb)	0 to ~2.2	cI2	$Im\overline{3}m$
YbCd	39.4	cP2	$Pm\overline{3}m$
YbCd$_2$	56.5	hP12	$P6_3/mmc$
Yb$_3$Cd$_8$	63.4	...	...
Yb$_{14}$Cd$_{51}$	70.3	hP65	$P6/m$
YbCd$_{5.7}$	78.8	...	...
YbCd$_6$	79.6	cI168	$Im\overline{3}$
(Cd)	100	hP2	$P6_3/mmc$

Cd-Zn

J. Dutkiewicz and W. Zakulski, 1984

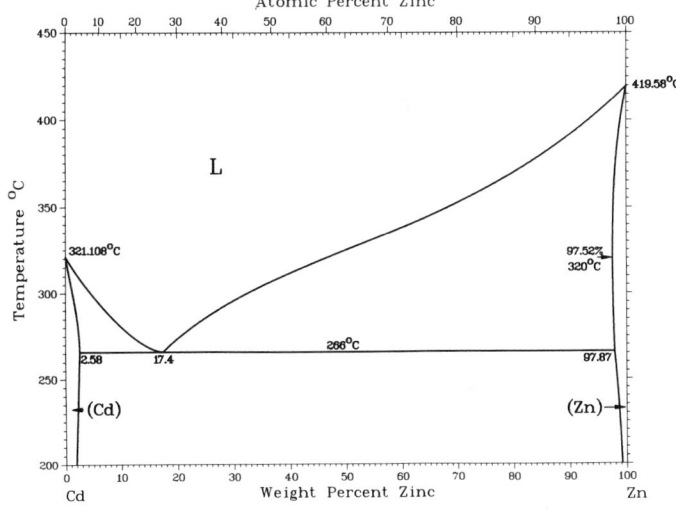

Phase	Composition, wt% Zn	Pearson symbol	Space group
(Cd)	0 to 2.58	hP2	$P6_3/mmc$
(Zn)	97.52 to 100	hP2	$P6_3/mmc$

Ce-Co

K.A. Gschneidner, Jr. and M.E. Verkade, 1974

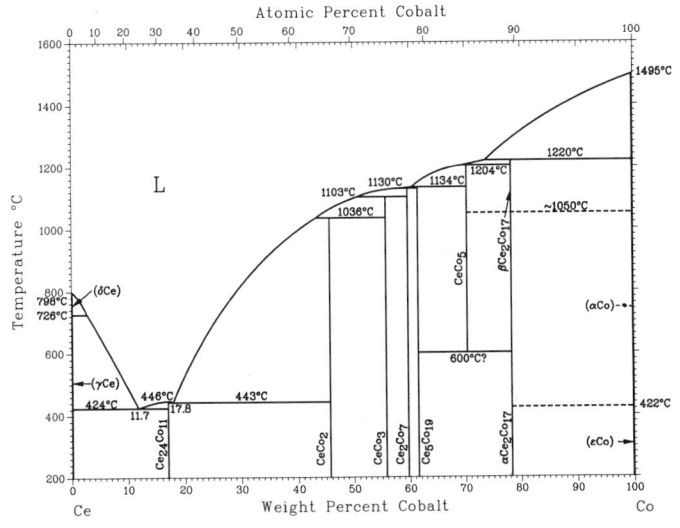

Phase	Composition, wt% Co	Pearson symbol	Space group
(δCe)	0	cI2	$Im\overline{3}m$
(γCe)	0	cF4	$Fm\overline{3}m$
Ce$_{24}$Co$_{11}$	16.1	hP70	$P6_3/mc$
CeCo$_2$	45.7	cF24	$Fd\overline{3}m$
CeCo$_3$	56	hR12	$R\overline{3}m$
Ce$_2$Co$_7$	59.6	hP36	$P6_3/mmc$
Ce$_5$Co$_{19}$	61.1	hR24	$R\overline{3}m$
CeCo$_5$	67.7	hP6	$P6/mmm$
βCe$_2$Co$_{17}$	78.2	hP38	$P6_3/mmc$
αCe$_2$Co$_{17}$	78.2	hR19	$R\overline{3}m$
(αCo)	100	cF4	$Fm\overline{3}m$
(εCo)	100	hP2	$P6_3/mmc$

Ce-Cu

P.R. Subramanian and D.E. Laughlin, 1988

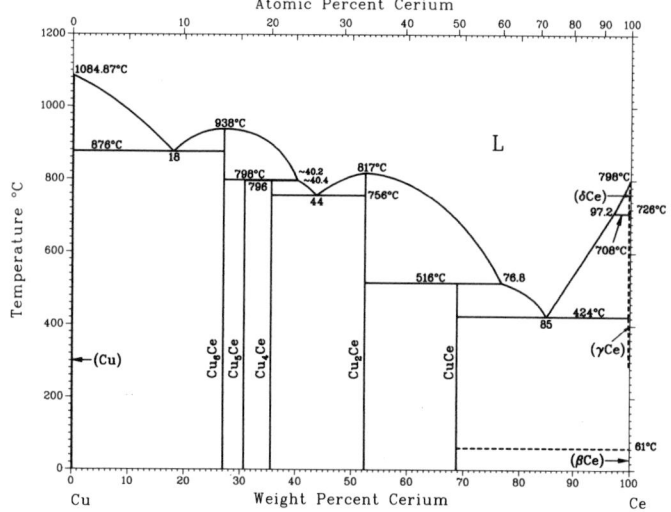

Phase	Composition, wt% Ce	Pearson symbol	Space group
(Cu)	0	cF4	Fm$\overline{3}$m
Cu$_6$Ce	~26.88	oP28	Pnma
Cu$_5$Ce	~30.61	hP6	P6/mmm
Cu$_4$Ce	~35.5	oP20	Pnnm
Cu$_2$Ce	~52.4	oI12	Imma
CuCe	~68.8	oP8	Pnma
(δCe)	100	cI2	Im$\overline{3}$m
(γCe)	100	cF4	Fm$\overline{3}$m
(βCe)	100	hP2	P6$_3$/mmc
(αCe)	100	cF4	Fm$\overline{3}$m

Ce-Fe

W. Zhang, G. Liu, and K. Han, 1992

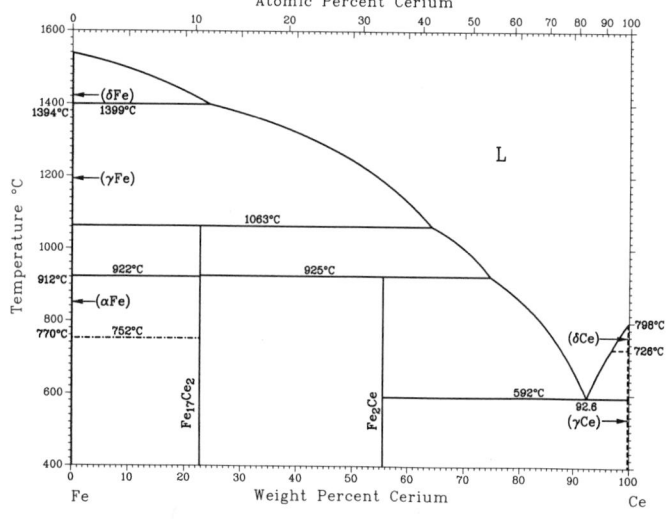

Phase	Composition, wt% Ce	Pearson symbol	Space group
(δFe)	0	cI2	Im$\overline{3}$m
(γFe)	0	cF4	Fm$\overline{3}$m
(αFe)	0	cI2	Im$\overline{3}$m
αFe$_{17}$Ce$_2$	22.7	hP38	P6/mmm
βFe$_{17}$Ce$_2$	22.7	hR19	R$\overline{3}$m
Fe$_2$Ce	55.6	cF24	Fd$\overline{3}$m
(δCe)	100	cF4	Fm$\overline{3}$m
(βCe)	100	hP2	P6$_3$/mmc
(αCe)	100	cF4	Fm$\overline{3}$m

Enlargement of the Ce-rich portion of the Fe-Ce phase diagram

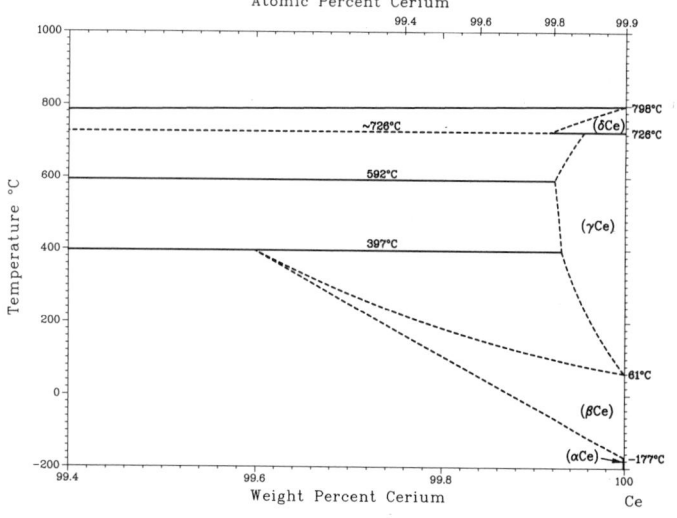

Ce-Ga

H. Okamoto, 1990

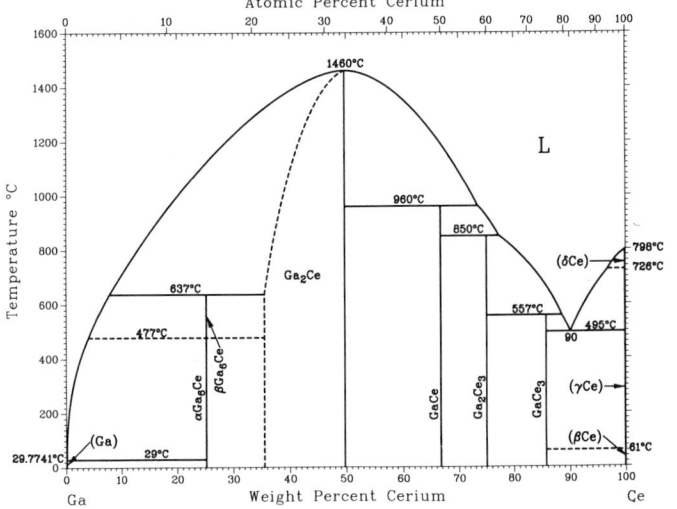

Phase	Composition, wt% Ce	Pearson symbol	Space group
(Ga)	0	oC8	Cmca
βGa₆Ce	21.1	...	...
αGa₆Ce	21.1	tI14	P4/nbm
Ga₂Ce	? to 44.6	hP3	P6/mmm
GaCe	61.7	oC8	Cmcm
Ga₂Ce₃	71	tP20	P4₂/mnm
Ga₃Ce₅(a)	73	tI32	I4/mcm
GaCe₃	83	cP4	Pm̄3m
(δCe)	100	cI2	Im̄3m
(γCe)	100	cF4	Fm̄3m
(βCe)	100	hP4	P6₃/mmc

(a) Not shown in the diagram

Ce-Ge

A.B. Gokhale and G.J. Abbaschian, 1989

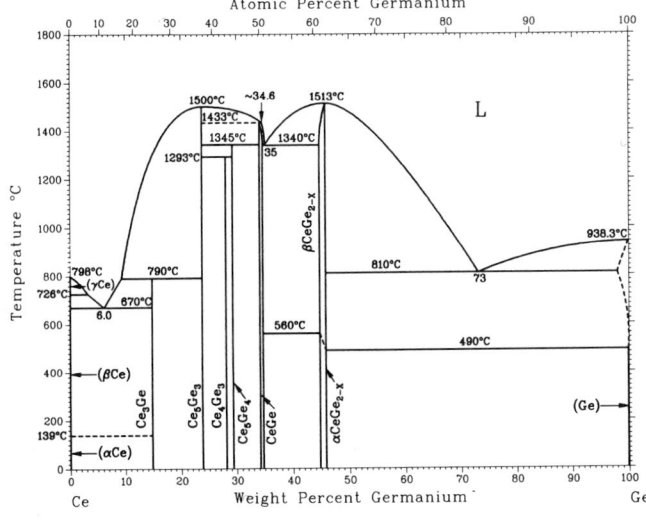

Phase	Composition, wt% Ge	Pearson symbol	Space group
(δCe)(a)	0	cI2	Im̄3m
(γCe)(b)	0	cF4	Fm̄3m
(βCe)(c)	0	hP4	P6₃/mmc
(αCe)(d)	0	cF4	Fm̄3m
Ce₃Ge	15	...	...
Ce₅Ge₃	23.7	hP16	P6₃/mcm
Ce₄Ge₃	28.0	cI28	I43d
Ce₅Ge₄	29.4	...	Pnma
CeGe	34.1	oP8	Pnma
αCeGe₂₋ₓ	44.9 to 45.94	(e)	Imma
βCeGe₂₋ₓ	44.9 to 45.94	tI12	I4₁/amd
(Ge)	100	cF8	Fd̄3m

(a) From 798 to 726 °C. (b) From 726 to 61 °C (139 °C on heating, 16 °C on cooling). (c) From 61 to −177 °C. (d) Below −177 °C. (e) Orthorhombic

Ce-In

H. Okamoto, 1992

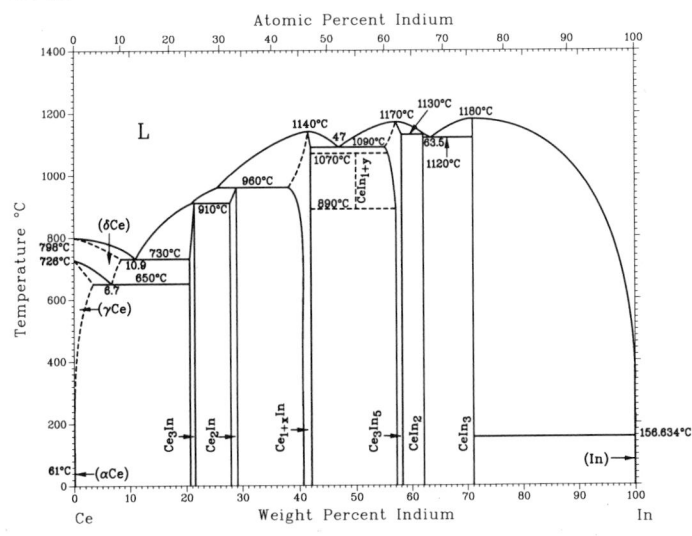

Phase	Composition, wt% In	Pearson symbol	Space group
(δCe)	0 to 8	cI2	Im̄3m
(γCe)	0 to 3	cF4	Fm̄3m
(βCe)	0	hP4	P6₃/mmc
(αCe)	0	cF4	Fm̄3m
βCe₃In	22	cF4	Fm̄3m
αCe₃In	21 to 22	cP4	Pm̄3m
Ce₂In	28 to 29.0	hP6	P6₃/mmc
Ce₁₊ₓIn	38 to 42	...	...
CeIn₁₊ᵧ	...	...	...
Ce₃In₅	55 to 58	oC32	Cmcm
CeIn₂	62.1	oI12	Imma
CeIn₃	71	cP4	Pm̄3m
(In)	100	tI2	I4/mmm

Ce-Ir

H. Okamoto, 1991

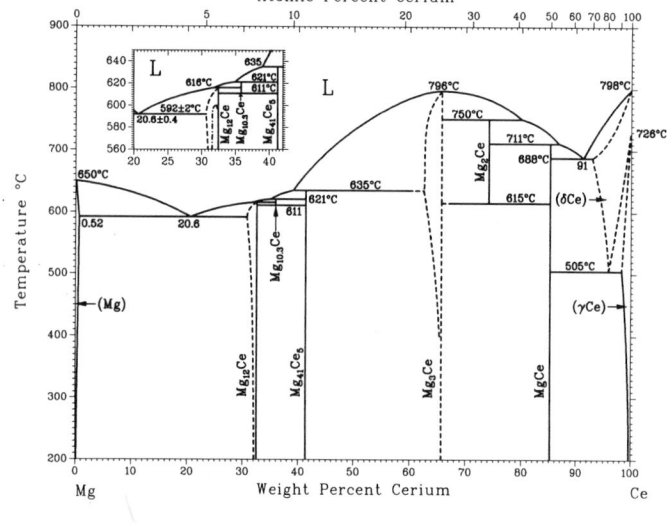

Phase	Composition, wt% Ir	Pearson symbol	Space group
(δCe)	0	cI2	$Im\bar{3}m$
(γCe)	0	cF4	$Fm\bar{3}m$
(βCe)	0	hP4	$P6_3/mmc$
(αCe)	0	cF4	$Fm\bar{3}m$
Ce₄Ir	26	...	...
Ce₃Ir	31	...	...
Ce₇Ir₃	37	hP20	$P6_3mc$
Ce₅Ir₃	45.1	tP32	$P4/ncc$
Ce₅Ir₄	52.3	oP36	$Pnma$
CeIr₂	70 to 76	cF24	$Fd\bar{3}m$
CeIr₃	81	hR12	$R\bar{3}m$
Ce₂Ir₇	82.8	hR18	$R\bar{3}m$
CeIr₅	87.2	cF24	$F\bar{4}3m$
(Ir)	100	cF4	$Fm\bar{3}m$

Ce-Mg

A.A. Nayeb-Hashemi and J.B. Clark, 1988

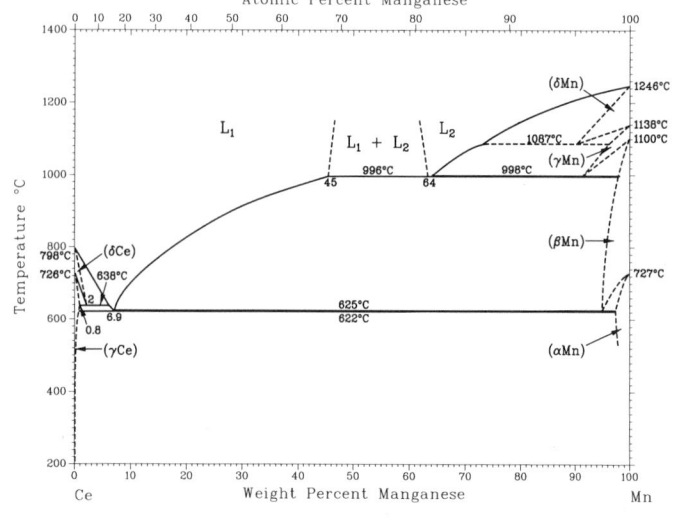

Phase	Composition, wt% Ce	Pearson symbol	Space group
(Mg)	0 to 0.52	hP2	$P6_3/mmc$
Mg₁₂Ce(I)	32.44(a,b)	tI26	$I4/mmm$
Mg₁₂Ce(II)	32.44(b)	oI338	$(Immm)$
Mg₁₀.₃Ce	35.89(a)	hP38	$P6_3/mmc$
Mg₄₁Ce₅	41.28(a)	tI92	$I4/m$
Mg₃Ce	? to 66	cF16	$Fm\bar{3}m$
Mg₂Ce	74.24(a)	cF24	$Fd\bar{3}m$
MgCe	85.22	cP2	$Pm\bar{3}m$
(δCe)	? to 100	cI2	$Im\bar{3}m$
(γCe)	98.5 to 100	cF4	$Fm\bar{3}m$

(a) Appears to be a line compound. The composition range, if any, is unknown. (b) Composition has not been established with certainty. (c) The Ni₁₇Th₂ structure type is taken from the homologous Mg-Nd system. In the Mg-Ce system, the Ni₁₇Th₂ structure has not yet been found.

Ce-Mn

A. Palenzona and S. Cirafici, unpublished

Phase	Composition, wt% Mn	Pearson symbol	Space group
(δCe)	0 to 2	cI2	$Im\bar{3}m$
(γCe)	0 to 0.8	cF4	$Fm\bar{3}m$
(βCe)	0	hP4	$P6_3/mmc$
(αCe)	0	cF4	$Fm\bar{3}m$
(δMn)	~100	cI2	$Im\bar{3}m$
(γMn)	~100	cF4	$Fm\bar{3}m$
(βMn)	~100	cP20	$P4_132$
(αMn)	~100	cI58	$I\bar{4}3m$

Ce-Ni

P. Nash and C.H. Tung, 1991

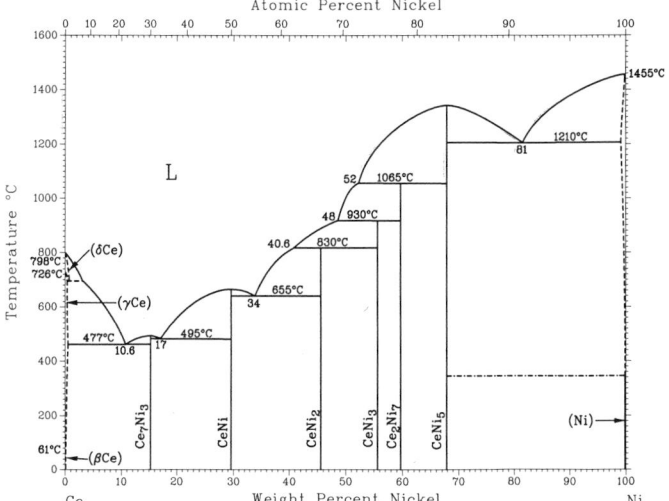

Phase	Composition, wt% Ni	Pearson symbol	Space group
(γCe)	~0	cF4	Fm$\bar{3}$m
(δCe)	~0	cI2	Im$\bar{3}$m
Ce$_7$Ni$_3$	15	hP20	P6$_3$mc
CeNi	29.5	oC8	Cmcm
CeNi$_2$	45.6	cF24	Fd$\bar{3}$m
CeNi$_3$	55.7	(a)	P6$_3$/mmc
Ce$_2$Ni$_7$	59.5	(c)	P6$_3$/mmc
CeNi$_5$	67.6	hP6	P6/mmm
(Ni)(b)	99.90 to 100	cF4	Fm$\bar{3}$m

(a) Hexagonal. (b) Solubility of Ce in Ni is 0.05 at.% Ce at 1200 °C and 0.04 at.% Ce at room temperature; data were obtained from pure Ni.

Ce-O

P.R. Subramanian, 1990

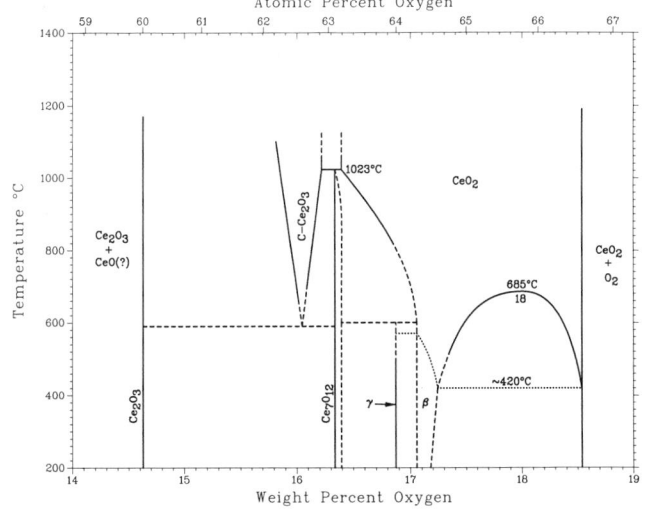

Phase	Composition, wt% O	Pearson symbol	Space group
(αCe)(a)	~0	cF4	Fm$\bar{3}$m
(βCe)(b)	~0	hP4	P6$_3$/mmc
(γCe)(c)	~0	cF4	Fm$\bar{3}$m
(δCe)(d)	~0	cI2	Im$\bar{3}$m
CeO	~10.2	cF8	Fm$\bar{3}$m
Ce$_2$O$_3$	~15	hP5	P$\bar{3}$m1
"C-C$_2$O$_3$" (e, g)	15.86 to 16.16	cI80	Ia$\bar{3}$
Ce$_7$O$_{12}$	16.3 to 16.43	hR22	R$\bar{3}$
γ(f)	~16.90	hR?	...
β(g)	~17.1 to 17.2	hR?	...
Ce$_6$O$_{11}$(h)	~17.3	mP?	P2$_1$/n
CeO$_2$	~18.6	cF12	Fm$\bar{3}$m
CeO$_2$(j)	~18.6	hP48	...
High-pressure phase			
CeO(k)	~10.2	cF?	...

(a) Below room temperature. (b) Up to 61 °C. (c) From 61 to 726 °C. (d) From 726 to 798 °C. (e) High-temperature phase; stable above ~590 °C. (f) Reported to be γ form of Ce$_2$O$_3$, perhaps a compound with stoichiometry Ce$_9$O$_{16}$, with monoclinic or lower symmetry. (g) Reported to be β form of Ce$_2$O$_3$, perhaps a compound with stoichiometry Ce$_{10}$O$_{18}$, with monoclinic or lower symmetry. (h) High-temperature phase; reported to be stable between 790 and 850 °C. (j) Reported to be high-temperature phase, observed at 1340 °C. (k) High-pressure phase, formed by reaction of Ce and CeO$_2$ at 700 °C and 15 kbar pressure

Ce-Pd

H. Okamoto, 1991

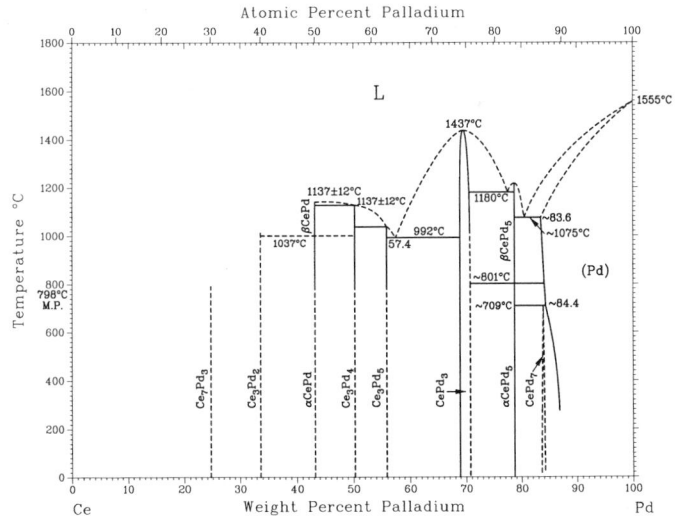

Phase	Composition, wt% Pd	Pearson symbol	Space group
(δCe)	0	cI2	Im$\bar{3}$m
(γCe)	0	cF4	Fm$\bar{3}$m
(βCe)	0	hP4	P6$_3$/mmc
(αCe)	0	cF4	Fm$\bar{3}$m
Ce$_7$Pd$_3$	25	hP20	P6$_3$mc
Ce$_3$Pd$_2$	34	...	...
βCePd	43.2	oP8	Pnma
αCePd	43.2	oC8	Cmcm
Ce$_3$Pd$_4$	50.3	hR14	R$\bar{3}$m
Ce$_3$Pd$_5$	55.9	hP8	P$\bar{6}$2m
CePd$_3$	69.3 to 70.9	cP4	Pm$\bar{3}$m
βCePd$_5$	79.1	hR*	...
αCePd$_5$	79.1	cF*	...
CePd$_7$	84.2	cF*	Fm$\bar{3}$m
(Pd)	84 to 100	cF4	Fm$\bar{3}$m

Ce-Pu

J.E. Selle and D.E. Etter, 1964

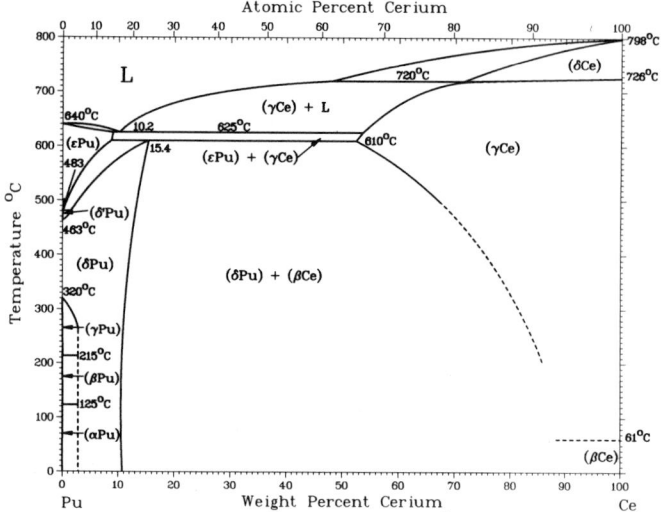

Phase	Composition, wt% Ce	Pearson symbol	Space group
(εPu)	0 to 9	cI2	$Im\bar{3}m$
(δ'Pu)	0	tI2	I4/mmm
(δPu)	0 to 15.4	cF4	$Fm\bar{3}m$
(γPu)	0	oF8	Fddd
(βPu)	0	mC34	I2/m
(αPu)	0	mP16	$P2_1/m$
(δCe)	72 to 100	cI2	$Im\bar{3}m$
(γCe)	53 to 100	cF4	$Fm\bar{3}m$
(βCe)	100	hP4	$P6_3/mmc$
(αCe)	100	cF4	$Fm\bar{3}m$

Ce-S

K.A. Gschneidner, Jr. and M.E. Verkade, 1974

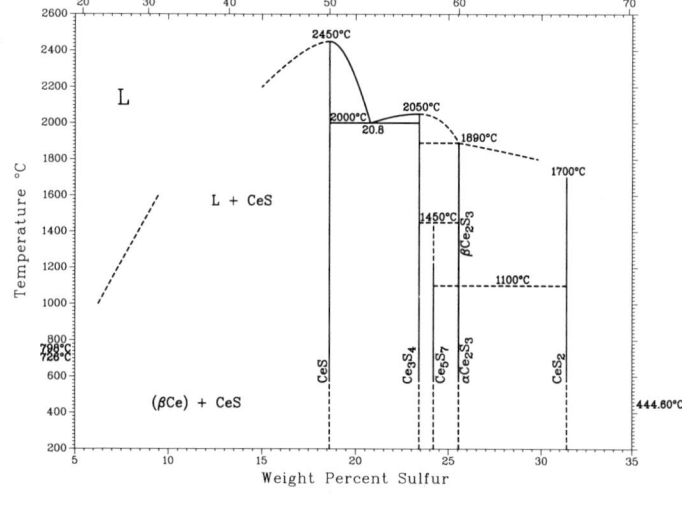

Phase	Composition, wt% S	Pearson symbol	Space group
(γCe)	0	cF4	$Fm\bar{3}m$
CeS	18.6	cF8	$Fm\bar{3}m$
Ce_3S_4	23.3	...	...
Ce_5S_7	24.2	tI92	$I4_1/acd$
βCe_2S_3	26	cI28	$I\bar{4}3d$
αCe_2S_3	26	oP20	Pnma
CeS_2	31.4	tP24	P4/nmm

Ce-Si

A. Munitz, A.B. Gokhale, and G.J. Abbaschian, 1989

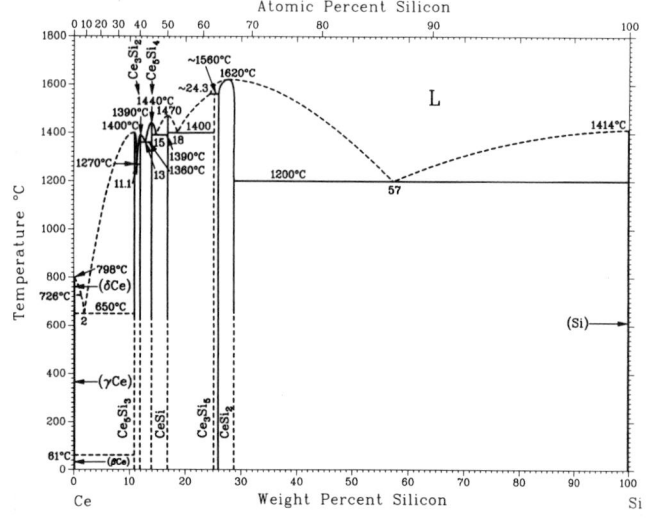

Phase	Composition, wt% Si	Pearson symbol	Space group
δCe(a)	0	cI2	$Im\bar{3}m$
γCe(b)	0	cF4	$Fm\bar{3}m$
βCe(c)	0	hP4	$P6_3/mmc$
αCe(d)	0	cF4	$Fm\bar{3}m$
Ce_5Si_3	10.7	tI32	I4/mcm
Ce_3Si_2	12	tP10	P4/mbm
Ce_5Si_4	13.8	(e)	...
CeSi	16.7	oP8	Pnma
Ce_3Si_5	25.0	(f)	Imma
$CeSi_2$	26 to 28.62	tI12	$I4_1/amd$
Si	100	cF8	$Fd\bar{3}m$
SiII(H.P.)	100	tI4	$I4_1/amd$

(a) From 798 to >726 °C. (b) From 726 to >61 °C (139 °C on heating, 16 °C on cooling). (c) From 61 °C to ? (d) <177 °C. (e) Tetragonal. (f) Orthorhombic

Ce-Sn

H. Okamoto, 1990

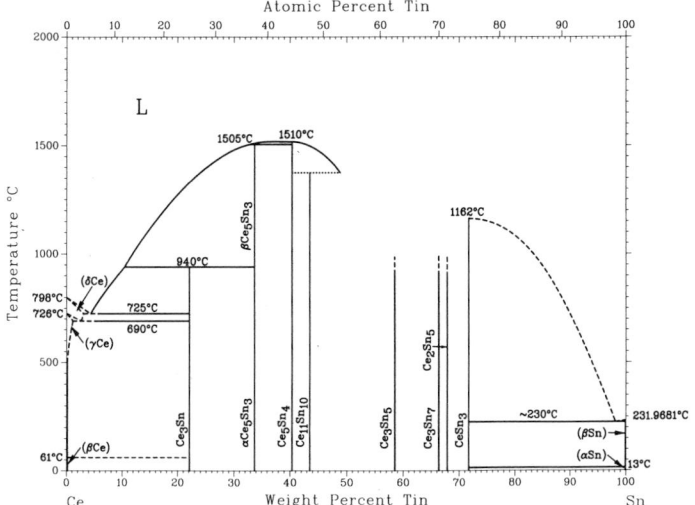

Phase	Composition, wt% Sn	Pearson symbol	Space group
(δCe)	0 to ?	cI2	Im$\bar{3}$m
(γCe)	0 to ?	cF4	Fm$\bar{3}$m
(βCe)	0	hP4	P6$_3$/mmc
(αCe)	0	cF4	Fm$\bar{3}$m
Ce$_3$Sn	22	cP4	Pm$\bar{3}$m
βCe$_5$Sn$_3$	33.7	hP16	P6$_3$/mcm
αCe$_5$Sn$_3$	33.7	tI32	I4/mcm
Ce$_5$Sn$_4$	40.4	oP36	Pnma
Ce$_{11}$Sn$_{10}$	43.5	tI84	I4/mmm
Ce$_3$Sn$_5$	58.5	oC32	Cmcm
Ce$_3$Sn$_7$	66	o**	...
Ce$_2$Sn$_5$	67.9	o**	...
CeSn$_3$	72	cP4	Pm$\bar{3}$m
(βSn)	100	tI4	I4$_1$/amd
(αSn)	100	cF8	Fm$\bar{3}$m

Ce-Te

H. Okamoto, 1990

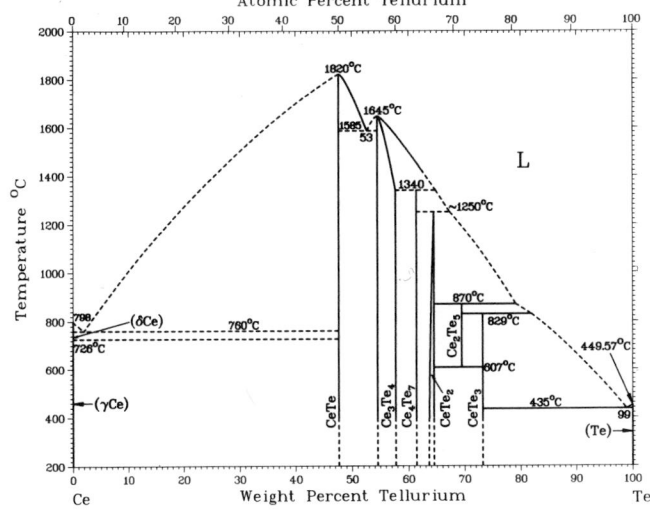

Phase	Composition, wt% Te	Pearson symbol	Space group
(δCe)	0	cI2	Im$\bar{3}$m
(γCe)	0	cF4	Fm$\bar{3}$m
(βCe)	0	hP4	P6$_3$/mmc
CeTe	47.7	cF8	Fm$\bar{3}$m
Ce$_3$Te$_4$	54.8 to 58	cI28	I$\bar{4}$3d
Ce$_4$Te$_7$	61.1	tP*	...
CeTe$_2$	64.6	tP6	P4/nmm
Ce$_2$Te$_5$	69.5	oC28	Cmcm
CeTe$_3$	73	oC16	Cmcm
(Te)	100	hP3	P3$_1$21

Ce-Ti

J.L. Murray, 1987

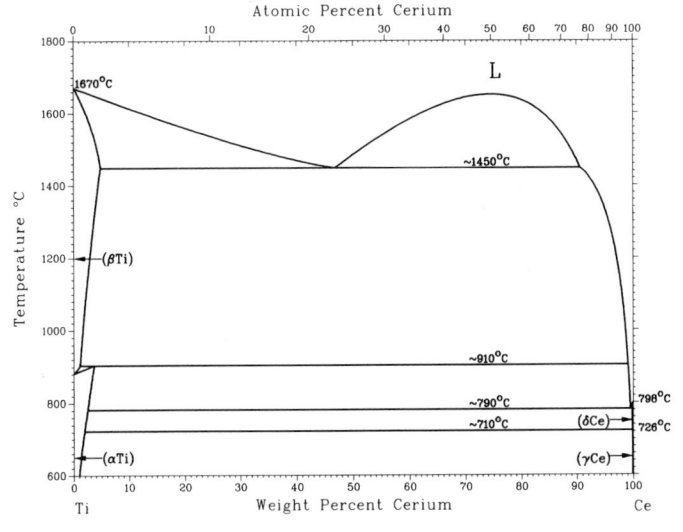

Phase	Composition, wt% Ce	Pearson symbol	Space group
(αTi)	0 to 3.4	hP2	P6$_3$/mmc
(βTi)	0 to 4.8	cI2	Im$\bar{3}$m
(δCe)	99.9 to 100	cI2	Im$\bar{3}$m
(γCe)	100	cF4	Fm$\bar{3}$m
(βCe)	100	hP4	P6$_3$/mmc

Ce-Tl

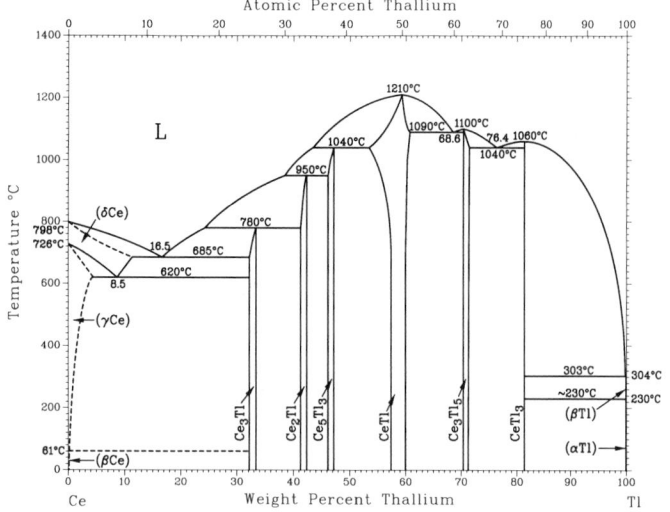

S. Delfino, A. Saccone, A. Palenzona, and R. Ferro, unpublished

Phase	Composition, wt% Tl	Pearson symbol	Space group
(αCe)	0	cF4	$Fm\bar{3}m$
(βCe)	0	hP4	$P6_3/mmc$
(γCe)	0 to 4	cF4	$Fm\bar{3}m$
(δCe)	0 to 13	cI2	$Im\bar{3}m$
Ce₃Tl(a)	~32.1 to ~33.3	cP4	$Pm\bar{3}m$
	~33	cF4	$Fm\bar{3}m$
Ce₂Tl	~42	...	...
Ce₅Tl₃	~46 to ~47	tI32	$I4/mcm$
CeTl(b)	~53 to ~60	cP2	$Pm\bar{3}m$
		(or cI2)	$Im\bar{3}m$
CeTl(c)	~53 to ~60	tP2	$P4/mmm$
Ce₃Tl₅	~70 to ~71	oC32	$Cmcm$
CeTl₃	81	cP4	$Pm\bar{3}m$
(βTl)	100	cI2	$Im\bar{3}m$
(αTl)	100	hP2	$P6_3/mmc$

(a) A cP4-cF4 order-disorder transformation in this phase has been suggested. (b) Cubic structure presumed to be room- and high-temperature phases. (c) Tetragonal structure presumed to be low-temperature phase

Ce-Zn

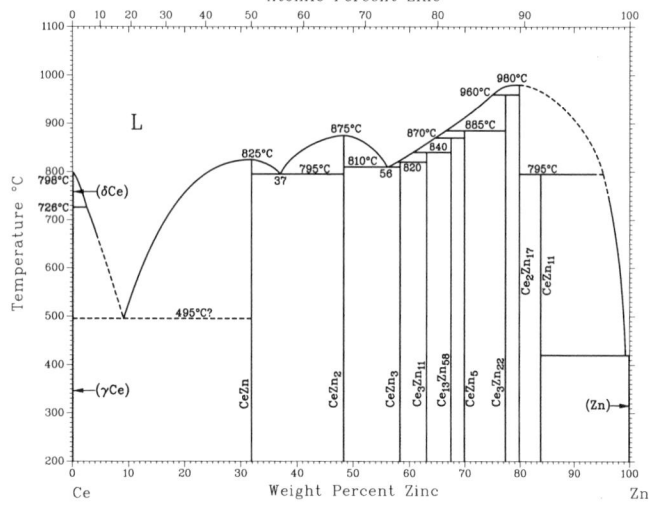

H. Okamoto, 1990

Phase	Composition, wt% Zn	Pearson symbol	Space group
(δCe)	0	cI2	$Im\bar{3}m$
(γCe)	0	cF4	$Fm\bar{3}m$
(βCe)	0	hP4	$P6_3/mmc$
(αCe)	0	cF4	$Fm\bar{3}m$
CeZn	31.8	cP2	$Pm\bar{3}m$
CeZn₂	48.3	oI12	$Imma$
CeZn₃	58	oC16	$Cmcm$
Ce₃Zn₁₁	63.2	oI28	$Immm$
Ce₁₃Zn₅₈	67.6	hP142	$P6_3mc$
CeZn₅	70.0	hP6	$P6/mmm$
Ce₃Zn₂₂	77	tI100	$I4_1/amd$
Ce₂Zn₁₇	79.9	hR19	$R\bar{3}m$
CeZn₁₁	83.8	tI48	$I4_1/amd$
(Zn)	100	hP2	$P6_3/mmc$

Cl-Cs

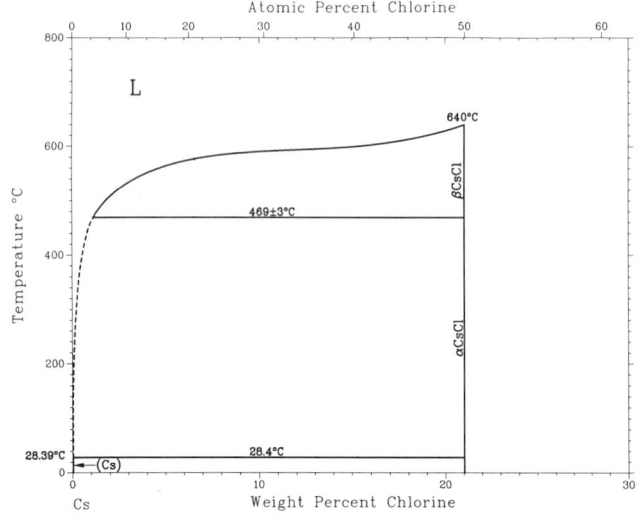

H. Okamoto, 1990

Phase	Composition, wt% Cl	Pearson symbol	Space group
(Cs)	0	cI2	$Im\bar{3}m$
βCsCl	21.1	cF8	$Fm\bar{3}m$
αCsCl	21.1	cP2	$Pm\bar{3}m$
(Cl)	100	oC8	$Cmca$

Cl-Ga

H. Okamoto, 1990

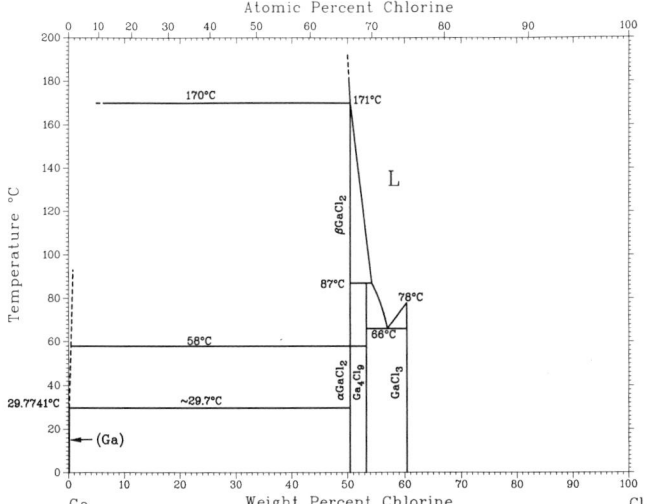

Phase	Composition, wt% Cl	Pearson symbol	Space group
(Ga)	0	oC8	Cmca
βGaCl₂	50.5	tP24	Pnma
αGaCl₂	50.5	...	...
Ga₄Cl₉	53.3	...	...
GaCl₃	60	aP8	$P\bar{1}$
(Cl)	100	oC8	Cmca

Cl-Hg

H. Okamoto, 1990

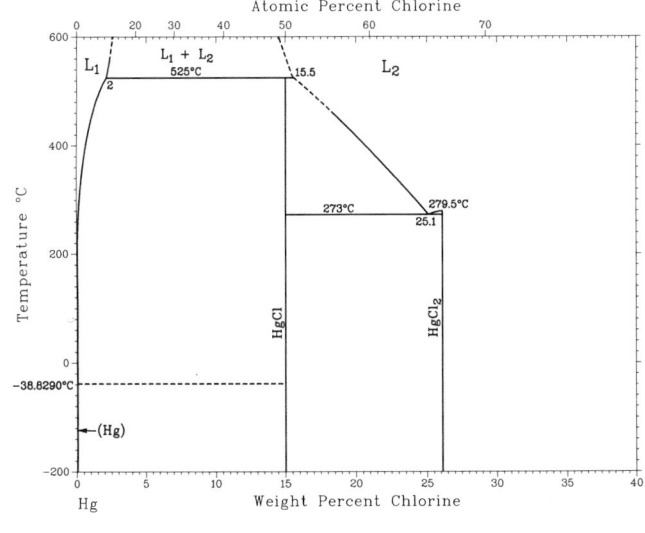

Phase	Composition, wt% Cl	Pearson symbol	Space group
(αHg)	0	hR1	$R\bar{3}m$
HgCl	15.0	tI8	I4/mmm
HgCl₂	26.1	oP12	Pmnb
(Cl)	100	oC8	Cmca

Cl-In

H. Okamoto, 1992

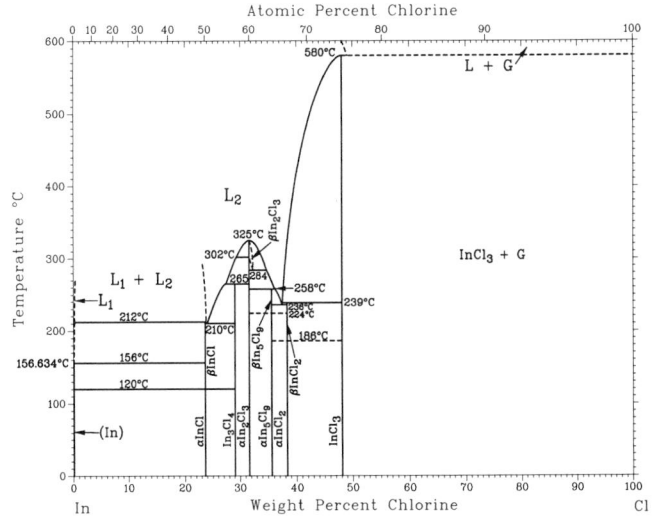

Phase	Composition, wt% Cl	Pearson symbol	Space group
(In)	0	tI2	I4/mmm
βInCl	23.6	oC8	Cmcm
αInCl	23.6	cP64	P2₁3
In₃Cl₄	29.1	...	...
In₂Cl₃(I)	32	o*30	...
In₂Cl₃(II)	32	t*45	...
In₂Cl₃(III)	32	hP*	...
βIn₅Cl₉(a)	35.7	...	...
αIn₅Cl₉(a)	35.7	...	...
βInCl₂	38.2	oP24	Pnna
αInCl₂	38.2	m**	...
InCl₃	48	mC16	C2/m
(Cl)	100	oC8	Cmca

(a) Or In₄Cl₇

Cl-Na

H. Okamoto, 1990

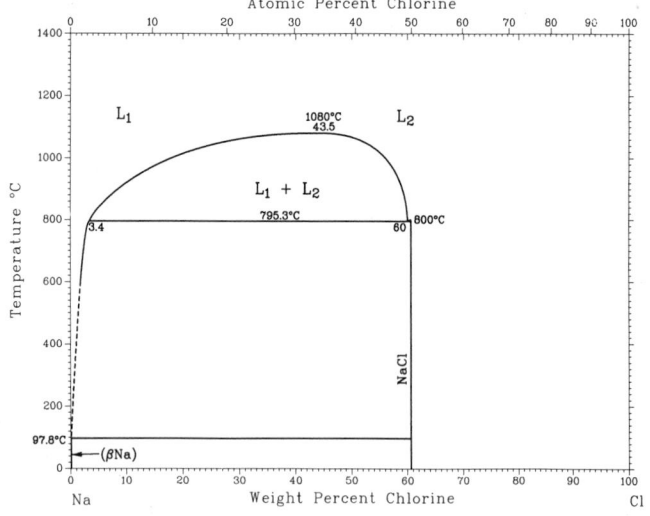

Phase	Composition, wt% Cl	Pearson symbol	Space group
(Na)	0	cI2	Im$\bar{3}$m
NaCl	60.7	cF8	Fm$\bar{3}$m
(Cl)	100	oC8	Cmca

Co-Cr

K. Ishida and T. Nishizawa, 1990

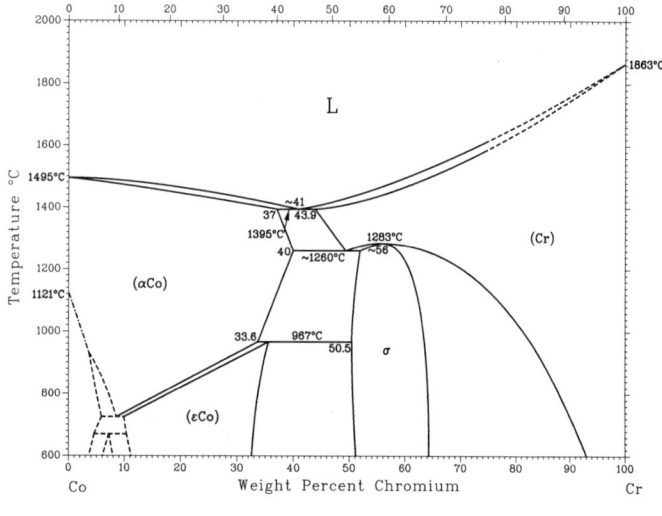

Phase	Composition, wt% Cr	Pearson symbol	Space group
(αCo)	0 to 40	cF4	Fm$\bar{3}$m
(εCo)	0 to 36	hP2	P6$_3$/mmc
(αCr)	43.9 to 100	cI2	Im$\bar{3}$m
σ	50.5 to 63	tP30	P4$_2$/mnm
Metastable phases			
(αCr)	~16	cI2	Im$\bar{3}$m
(αCo)	40 to 62.9	cF4	Fm$\bar{3}$m
(δCr)	54 to 100	cP8	Pm$\bar{3}$n
Co$_3$Cr?	23	hP8	P6$_3$/mmc

Co-Cu

T. Nishizawa and K. Ishida, 1984

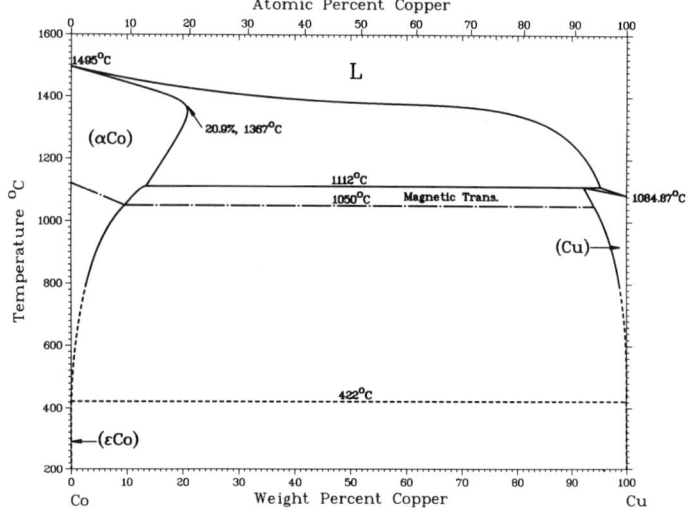

Phase	Composition, wt% Cu	Pearson symbol	Space group
(αCo)	0 to 20.9	cF4	Fm$\bar{3}$m
(εCo)	0 to 9(a)	hP2	P6$_3$/mmc
(Cu)	93 to 100	cF4	Fm$\bar{3}$m
Metastable phase			
ε′	9 to 10	hR1	R$\bar{3}$m

(a) The composition of (εCo) is between 0 and 0.3 wt% Cu in equilibrium, but is 0 to 9 wt% Cu in the metastable state, which is obtained by quenching from high temperatures.

Co-Dy

H. Okamoto, 1990

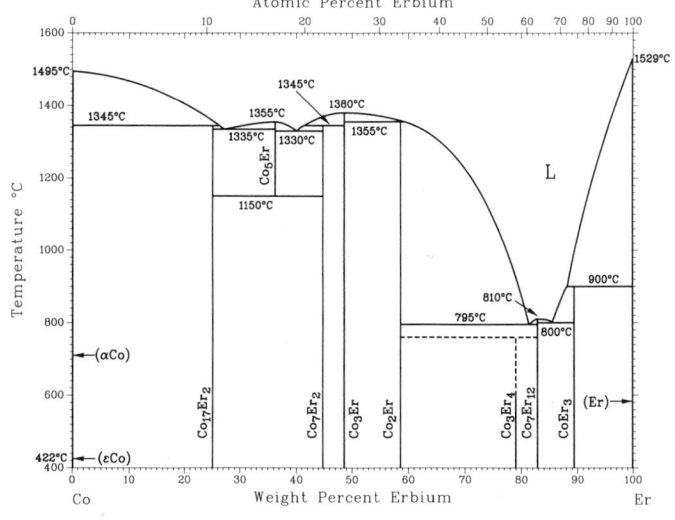

Phase	Composition, wt% Dy	Pearson symbol	Space group
(αCo)	~0	cF4	$Fm\bar{3}m$
(εCo)	~0	hP2	$P6_3/mmc$
$\beta Co_{17}Dy_2$	24.4	hP38	$P6_3/mmc$
$\alpha Co_{17}Dy_2$	24.4	hR19	$R\bar{3}m$
Co_5Dy	35.6	hP6	$P6/mmm$
Co_7Dy_2	44.0	hR18	$R\bar{3}m$
Co_3Dy	48	hR12	$R\bar{3}m$
Co_2Dy	57.9	cF24	$Fd\bar{3}m$
Co_3Dy_4	78.6	hP22	$P6_3/m$
Co_7Dy_{12}	82.6	mP38	$P2_1/c$
$CoDy_3$	89	oP16	$Pnma$
(βDy)	~100	cI2	$Im\bar{3}m$
(αDy)	~100	hP2	$P6_3/mmc$

Co-Er

H. Okamoto, 1990

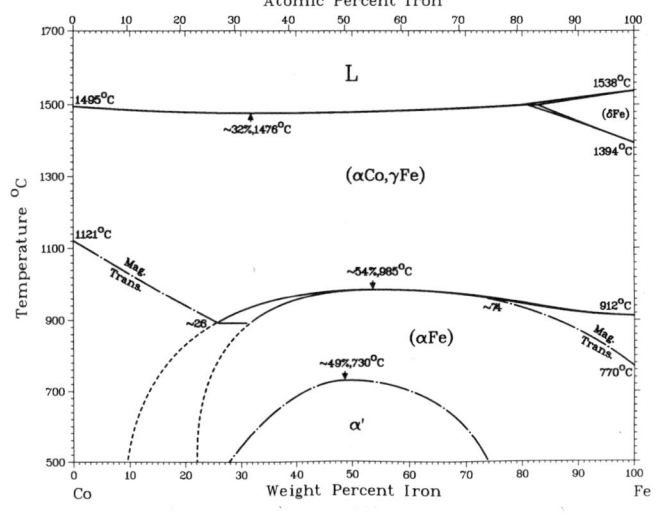

Phase	Composition, wt% Er	Pearson symbol	Space group
(αCo)	~0	cF4	$Fm\bar{3}m$
$Co_{17}Er_2$	25.0	hP38	$P6_3/mmc$
Co_5Er	36.3	hP6	$P6/mmm$
Co_7Er_2	44.7	hR18	$R\bar{3}m$
Co_3Er	49	hR12	$R\bar{3}m$
Co_2Er	58.6	cF24	$Fd\bar{3}m$
Co_3Er_4	79.1	hP22	$P6_3/m$
Co_7Er_{12}	83.0	mP38	$P2_1/c$
$CoEr_3$	99.5	oP16	$Pnma$
(Er)	~100	hP2	$P6_3/mmc$

Co-Fe

T. Nishizawa and K. Ishida, 1984

Phase	Composition, wt% Fe	Pearson symbol	Space group
(αCo, γFe)	0 to 100	cF4	$Fm\bar{3}m$
α'	~28 to ~74	cP2	$Pm\bar{3}m$
(αFe)	~22 to 100	cI2	$Im\bar{3}m$
(δFe)	82 to 100	cI2	$Im\bar{3}m$
Metastable phase			
η	0.5 to 5.7	hP4	$P6_3/mmc$

Co-Ga

H. Okamoto, 1990

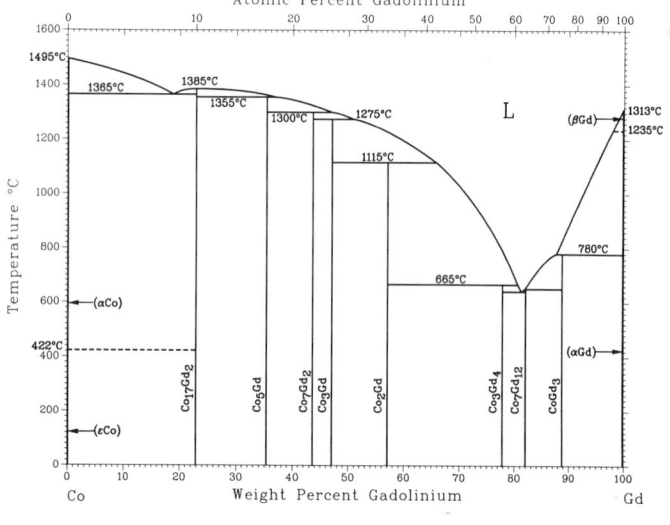

Phase	Composition, wt% Ga	Pearson symbol	Space group
(αCo)	0 to 22	cF4	Fm$\bar{3}$m
(εCo)	0 to 17	hP2	P6₃/mmc
β	33 to 67.7	cP2	Pm$\bar{3}$m
CoGa₃	78	tP16	P$\bar{4}$n2
(Ga)	100	oC8	Cmca

Co-Gd

H. Okamoto, 1990

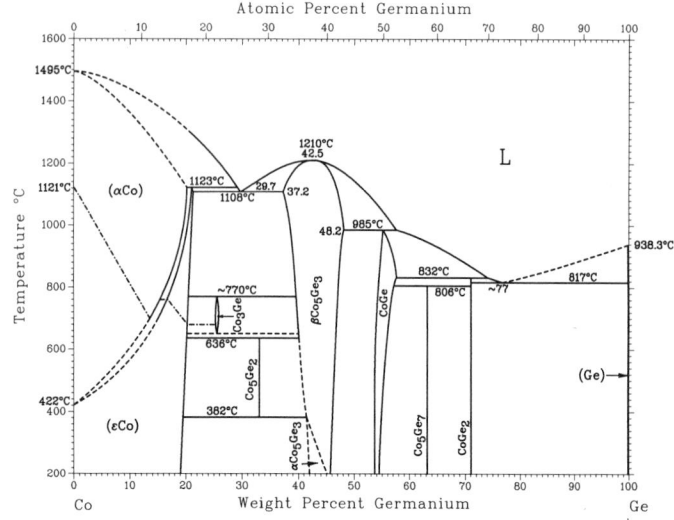

Phase	Composition, wt% Gd	Pearson symbol	Space group
(αCo)	~0	cF4	Fm$\bar{3}$m
(εCo)	~0	hP2	P6₃/mmc
Co₁₇Gd₂	~23.8	hP38	P6₃/mmc
		hR19	R$\bar{3}$m
Co₅Gd	~34.9	hP6	P6/mmm
Co₇Gd₂	~43.2	hR18	R$\bar{3}$m
		hP36	P6₃/mmc
Co₃Gd	47	hR12	R$\bar{3}$m
Co₂Gd	57.1	cF24	Fd$\bar{3}$m
Co₃Gd₄	78.0	hP22	P6₃/m
Co₇Gd₁₂	~82.1	mP38	P2₁/c
CoGd₃	89	oP16	Pnma
(βGd)	~100	cI2	Im$\bar{3}$m
(αGd)	100	hP2	P6₃/mmc
Other reported phases			
Co₈Gd	~25.0	hP8	P6/mmm
CoGd	72.7	oP8	Pnma
Co₃Gd₇	86	o*	...
CoGd₉	96	o*	...

Co-Ge

K. Ishida and T. Nishizawa, 1991

Phase	Composition, wt% Ge	Pearson symbol	Space group
(αCo)	0 to 20.7	cF4	Fm$\bar{3}$m
(εCo)	0 to 21	hP2	P6₃/mmc
Co₃Ge	25.2 to 26	cP8	Pm$\bar{3}$n?
Co₅Ge₂	33.0	(a)	...
αCo₅Ge₃	~41.5 to ~45	(b)?	Pbnm?
βCo₅Ge₃	37.2 to 48.2	hP6	P6₃/mmc
CoGe	53.7 to 57.7	mC16	C2/m
		cP8	P2₁3
Co₅Ge₇	63.3	tI24	I4mm
CoGe₂	71.2	oC24	Aba2
(Ge)	~100	cF8	Fd$\bar{3}$m

Co-Hf

K. Ishida and T. Nishizawa, 1991

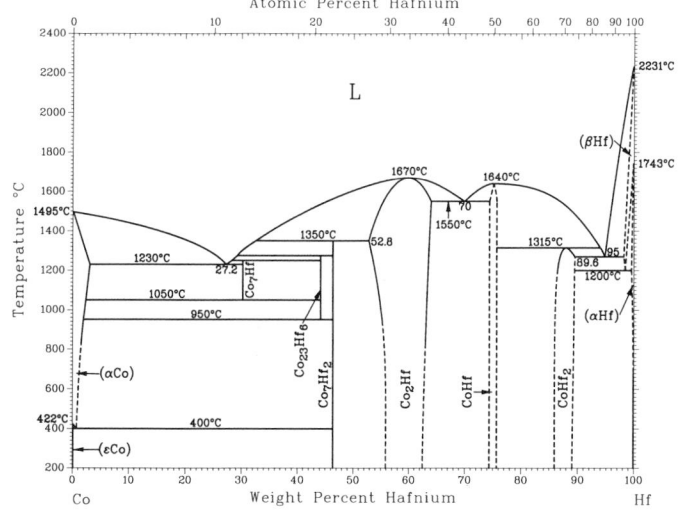

Phase	Composition, wt% Hf	Pearson symbol	Space group
(αCo)	0 to ~6	cF4	Fm3̄m
(εCo)	0 to ~1.5	hP2	P6₃/mmc
Co₇Hf	30.2	tP32	...
Co₂₃Hf₆	44.2	cF116	Fm3̄m
Co₇Hf₂	46.4	o**	...
Co₂Hf	52.8 to ~63	cF24	Fd3̄m
CoHf	~74 to ~76	cP2	Pm3̄m
CoHf₂	~86 to 89.6	cF112	Fd3̄m
(βHf)	~98 to 100	cI2	Im3̄m
(αHf)	~99 to 100	hP2	P6₃/mmc

Co-Ho

H. Okamoto, 1990

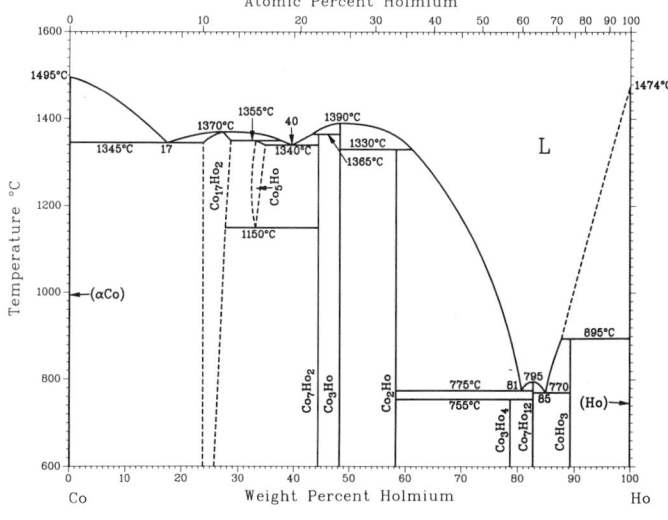

Phase	Composition, wt% Ho	Pearson symbol	Space group
(αCo)	~0	cF4	Fm3̄m
(εCo)	~0	hP2	P6₃/mmc
Co₁₇Ho₂	24.7	hR19	R3̄m
		hP38	P6₃/mmc
		hP52	P6₃/mmc
Co₅Ho	35.9	hP6	P6/mmm
Co₇Ho₂	44.4	hR18	R3̄m
Co₃Ho	48	hR12	R3̄m
Co₂Ho	58.3	cF24	Fd3̄m
Co₃Ho₄	78.8	hP22	P6₃/m
Co₇Ho₁₂	82.8	mP38	P2₁/c
CoHo₃	89	oP16	Pnma
(Ho)	~100	hP2	P6₃/mmc

Co-Mn

K. Ishida and T. Nishizawa, 1990

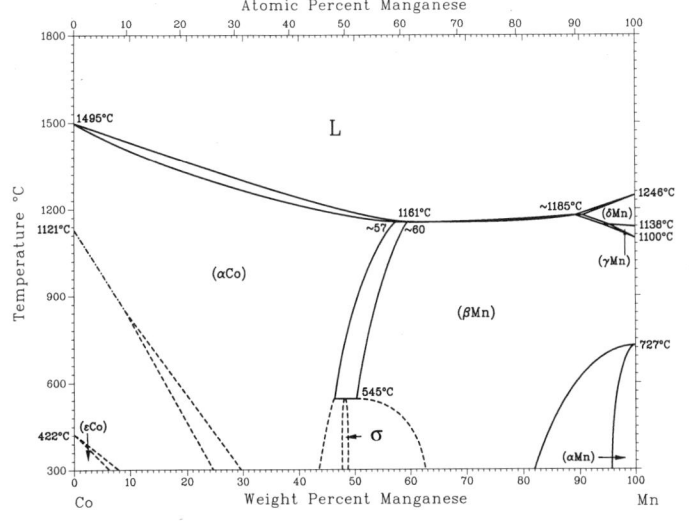

Phase	Composition, wt% Mn	Pearson symbol	Space group
(εCo)	0 to ~19	hP2	P6₃/mmc
(αCo)	0 to ~57	cF4	Fm3̄m
σ	~48	tP30	P4₂/mnm
(αMn)	97 to 100	cI58	I4̄3m
(βMn)	49 to 100	cP20	P4₁32
(γMn)	95 to 100	cF4	Fm3̄m
(δMn)	90 to 100	cI2	Im3̄m
(γ'Mn)(a)	90 to 100	tI2	I4/mmm

(a) Splat quenched from the liquid state or rapid quenched from the high-temperature solid field

Co-Mo

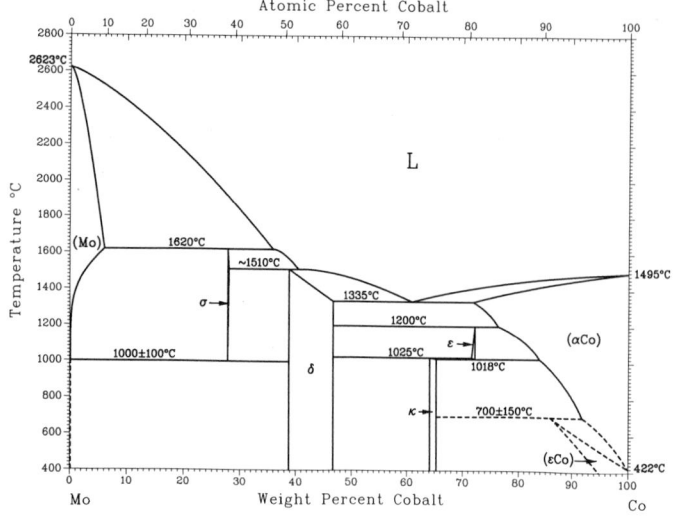

Phase	Composition, wt% Co	Pearson symbol	Space group
(Mo)	0 to ~6	cI2	$Im\bar{3}m$
σ	~27.8 to 28	tP30	$P4_2/mnm$
ε	~38.8 to ~46.7	hR13	$R\bar{3}m$
κ	~64.2 to ~65.4	hP8	$P6_3/mmc$
cph	~72	hP2	$P6_3/mmc$
(αCo)	~72 to 100	cF4	$Fm\bar{3}m$
(εCo)	~86 to 100	hP2	$P6_3/mmc$

Co-Nb

J.K. Pargeter and W. Hume-Rothery, 1967

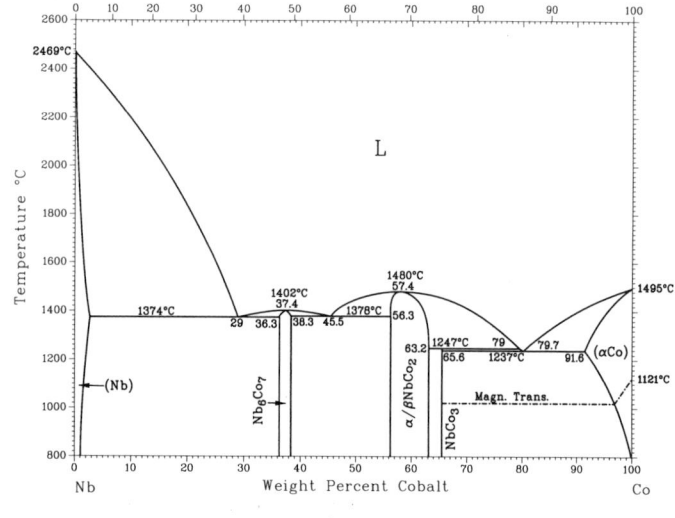

Phase	Composition, wt% Co	Pearson symbol	Space group
(Nb)	0 to ~3	cI2	$Im\bar{3}m$
Nb_6Co_7	36.3 to 38.3	hR13	$R\bar{3}m$
$\beta NbCo_2$(a)	56.3 to ?	hP12	$P6_3/mmc$
$\alpha NbCo_2$	56.3 to 63.2	cF24	$Fd\bar{3}m$
$NbCo_3$	65.3	hP24	$P6_3/mmc$
(αCo)	91.6 to 100	cF4	$Fm\bar{3}m$

(a) $\beta NbCo_2$ is stable above ~1200 °C.

Co-Nd

A.E. Ray, 1974

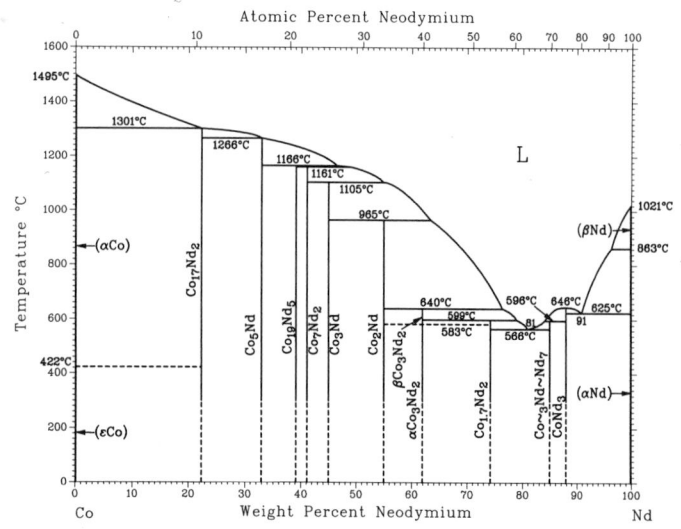

Phase	Composition, wt% Nd	Pearson symbol	Space group
(αCo)	~0	cF4	$Fm\bar{3}m$
(εCo)	~0	hP2	$P6_3/mmc$
$Co_{17}Nd_2$	~22.3	hR19	$R\bar{3}m$
Co_5Nd	~32.9	hP6	$P6/mmm$
$Co_{19}Nd_5$	~39.1	hR24	$R\bar{3}m$
βCo_7Nd_2	~41.1	hR18	$R\bar{3}m$
αCo_7Nd_2	~41.1	hP36	$P6_3/mmc$
Co_3Nd	45	hR12	$R\bar{3}m$
Co_2Nd	55.0	cF24	$Fd\bar{3}m$
Co_3Nd_2	62	o**	...
$Co_{1.7}Nd_2$	~74.3	h**	...
$Co_{~3}Nd_{~7}$	~85	hP20	$P6_3mc$
$CoNd_3$	88	oP16	Pnma
(βNd)	~100	cI2	$Im\bar{3}m$
(αNd)	~100	hP4	$P6_3/mmc$
Other reported phases			
Co_3Nd_4	~76.5	hP7	$P\bar{6}$
$Co_{11}Nd_{24}$	~84.2	hP70	$P6_3mc$
Co_2Nd_5	~85.9	mC28	C2/c

Co-Ni

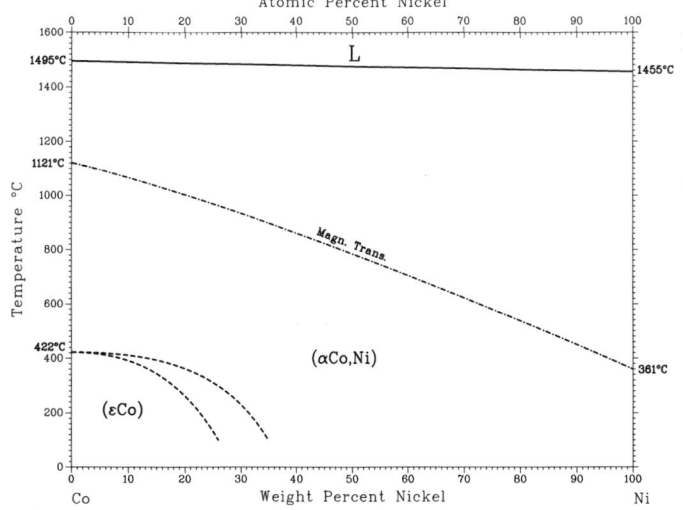

Phase	Composition, wt% Ni	Pearson symbol	Space group
(αCo,Ni)	0 to 100	cF4	Fm$\overline{3}$m
(εCo)	0 to 35	hP2	P6$_3$/mmc

Co-P

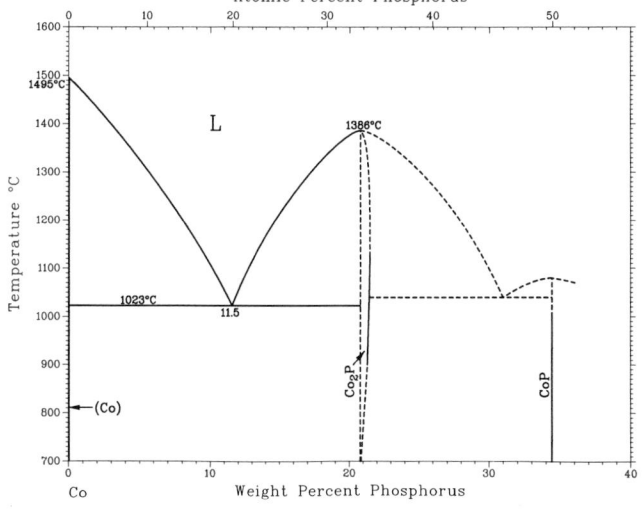

Phase	Composition, wt% P	Pearson symbol	Space group
(αCo)	~0	cF4	Fm$\overline{3}$m
(εCo)	~0	hP2	P6$_3$/mmc
Co$_2$P	~20.6 to 21.3	oP12	Pnma
CoP	34.5	oP8	Pnma
CoP$_2$	51.3	(a)	...
CoP$_3$	61	cI32	Im$\overline{3}$m
Red (P)	100	(b)	...
White (P)	100	(b)	...
Black (P)	100	oC8	Cmca

(a) Monoclinic. (b) Cubic

Co-Pd

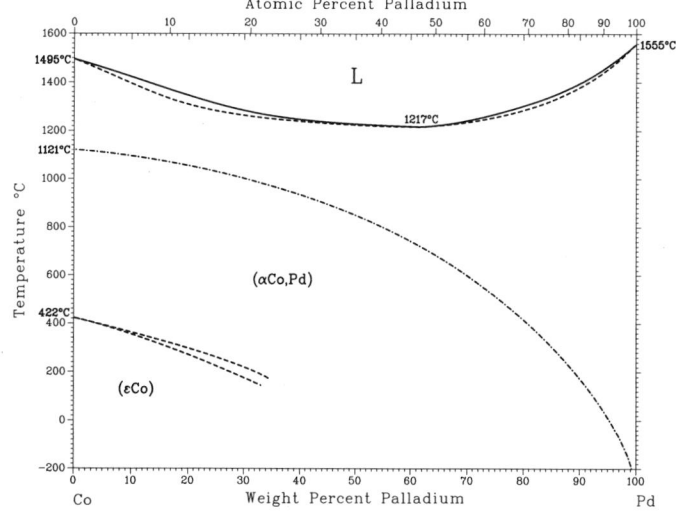

Phase	Composition, wt% Pd	Pearson symbol	Space group
(αCo,Pd)	0 to 100	cF4	Fm$\overline{3}$m
(εCo)	0 to ~31	hR2	P6$_3$/mmc
Metastable phases			
α″	~63 to ~66	tP4	P4/mmm
α′	73 to 94	cP4	Pm$\overline{3}$m

Co-Pr

A.E. Ray, 1974

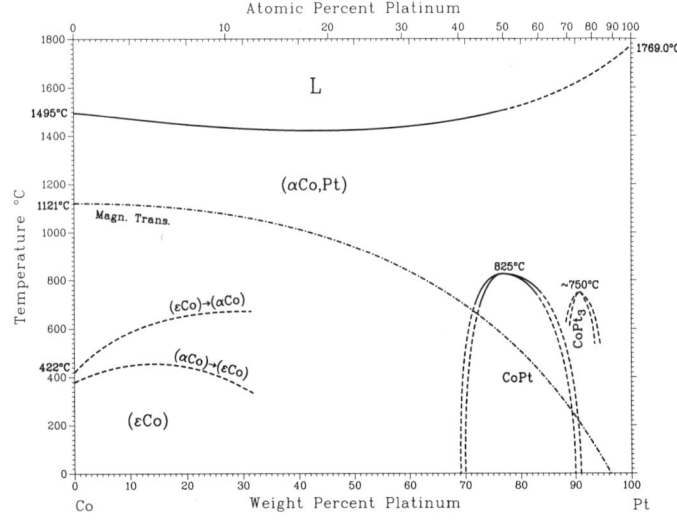

Phase	Composition, wt% Pr	Pearson symbol	Space group
(αCo)	~0	cF4	Fm$\bar{3}$m
(εCo)	~0	hP2	P6$_2$/mmc
Co$_{17}$Pr$_2$	21.9	hR19	R$\bar{3}$m
Co$_5$Pr	32.4	hP6	P6/mmm
Co$_{19}$Pr$_5$	38.6	hR24	R$\bar{3}$m
βCo$_7$Pr$_2$	40.6	hR18	R3m
αCo$_7$Pr$_2$	40.6	hP36	P6$_3$/mmc
Co$_3$Pr	44	hR12	R$\bar{3}$m
Co$_2$Pr	54.4	cF24	Fd$\bar{3}$m
Co$_{1.7}$Pr$_2$	~73.8	hP*	...
Co$_2$Pr$_5$	~85.7	mC28	C2/c
CoPr$_3$	88	oP16	Pnma
(βPr)	~100	cI2	Im$\bar{3}$m
(αPr)	~100	hP4	P6$_3$/mmc

Co-Pt

H. Okamoto, 1990

Phase	Composition, wt% Pt	Pearson symbol	Space group
(αCo, Pt)	0 to 100	cF4	Fm$\bar{3}$m
(εCo)	0 to ?	hP2	P6$_3$/mmc
CoPt	~76.8	tP4	P4/mmm
CoPt$_3$	~91	cP4	Pm$\bar{3}$m

Co-Pu

D.M. Poole, M.G. Bale, P.G. Mardon, J.A.C. Marples, and J.L. Nichols, 1961

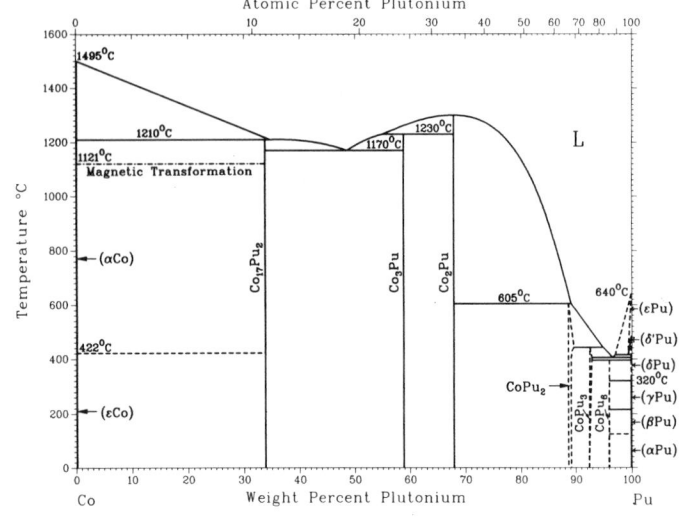

Phase	Composition, wt% Pu	Pearson symbol	Space group
(αCo)	~0	cF4	Fm$\bar{3}$m
(εCo)	~0	hP2	P6$_3$/mmc
Co$_{17}$Pu$_2$	34	hP38	P6$_3$/mmc
Co$_3$Pu	~58.9	hR12	R$\bar{3}$m
Co$_2$Pu	~67.4	cF24	Fd$\bar{3}$m
CuPu$_2$	~88.7 to 90	hP9	P6$_3$/mmc
CoPu$_3$	~92.6 to 93	oC16	Cmcm
CoPu$_6$	96.1	tI28	I4/mcm
(εPu)	~99.5 to 100	cI2	Im$\bar{3}$m
(δ'Pu)	~100	tI2	I4/mmm
(δPu)	~100	cF4	Fm$\bar{3}$m
(γPu)	~100	oF8	Fddd
(βPu)	~100	mC34	C2/m
(αPu)	~100	mP16	P2$_1$/m

Co-Re

H. Okamoto, 1990

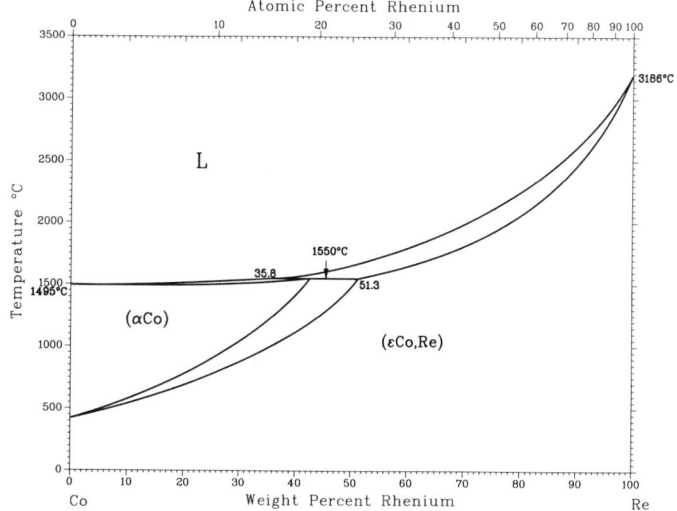

Phase	Composition, wt% Re	Pearson symbol	Space group
(αCo)	0 to 43	cF4	Fm$\overline{3}$m
(εCo,Re)	0 to 100	hP2	P6$_3$/mmc

Co-S

K. Friedrich, 1908

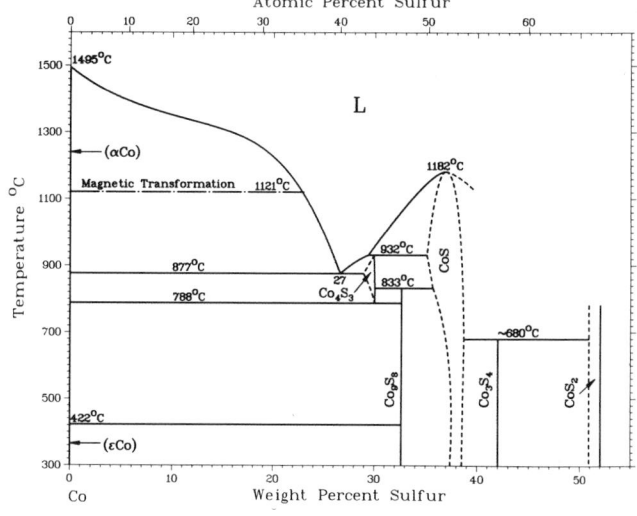

Phase	Composition, wt% S	Pearson symbol	Space group
(αCo)	0	cF4	Fm$\overline{3}$m
(εCo)	0	hP2	P6$_3$/mmc
Co$_4$S$_3$	~29.0	...	...
Co$_9$S$_8$	32.6	cF68	Fm$\overline{3}$m
CoS	35.2 to 40	hP4	P6$_3$/mmc
Co$_3$S$_4$	42.0	cF56	Fd$\overline{3}$m
CoS$_2$	52.1	cP12	Pa$\overline{3}$
(S)	100	oF128	Fddd

Co-Sb

H. Okamoto, 1991

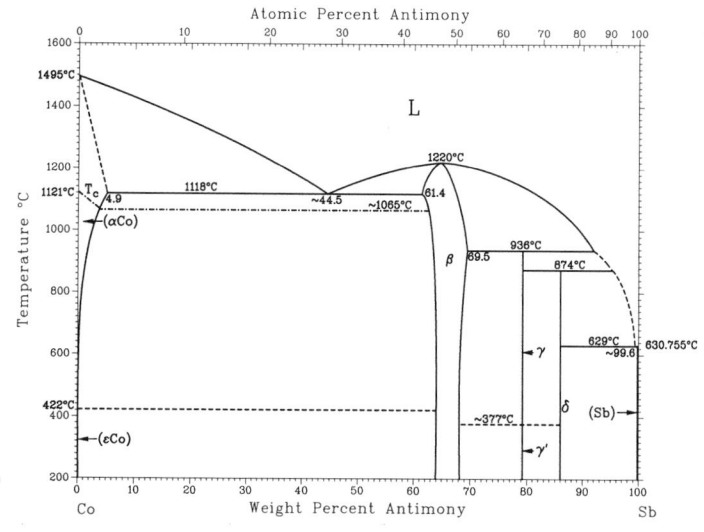

Phase	Composition, wt% Sb	Pearson symbol	Space group
(αCo)	0 to ~5.0	cF4	Fm$\overline{3}$m
(εCo)	0	hP2	P6$_3$/mmc
β	61.4 to ~69	hP4	P6$_3$/mmc
γ	79	oP6	Pnnm
γ	79	mP12	P2$_1$/c
δ	~86	cI32	Im$\overline{3}$
(Sb)	~100	hR2	R$\overline{3}$m

Co-Se

H. Okamoto, 1990

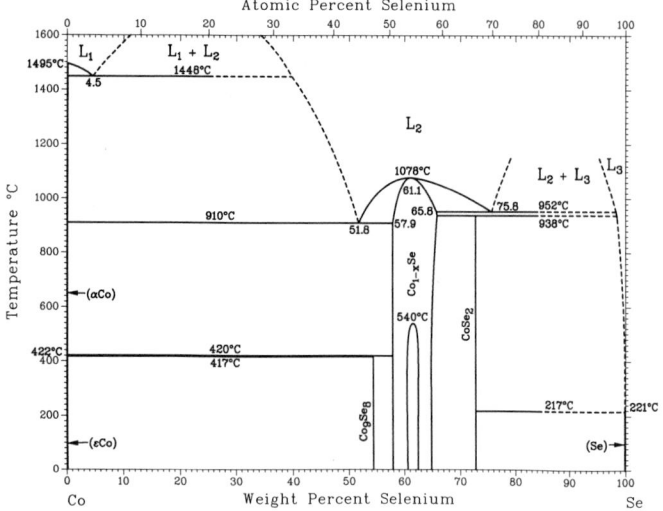

Phase	Composition, wt% Se	Pearson symbol	Space group
(αCo)	0	cF4	$Fm\overline{3}m$
(εCo)	0	hP2	$P6_3/mmc$
Co₉Se₈	54.4	cF68	$Fm\overline{3}m$
Co₁₋ₓSe	57.9 to 65.8	m**	...
CoSe₂	72.9	cP12	$Pa\overline{3}$
(Se)	100	oC8	Cmca

Co-Si

K. Ishida and T. Nishizawa, 1991

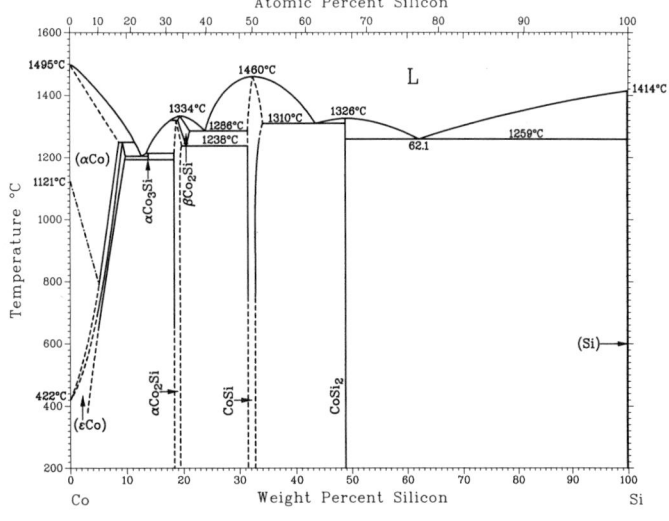

Phase	Composition, wt% Si	Pearson symbol	Space group
(αCo)	0 to 8.5	cF4	$Fm\overline{3}m$
(εCo)	0 to 9.7	hP2	$P6_3/mmc$
Co₃Si	14	t**	...
αCo₂Si	~18 to ~20	oP12	Pnma
βCo₂Si	~18 to 21.0	...	...
CoSi	31 to ~34	cP8	$P2_13$
CoSi₂	48.8	cF12	$Fm\overline{3}m$
(Si)	~100	cF8	$Fd\overline{3}m$
Metastable phases			
Co₃Si	~4 to 14	hP8	$P6_3/mmc$
Co₄Si	~11	...	...
γCo₂Si(a)	~14	o**	...
Co₂Si₃	42	tP20	$P\overline{4}c2$

(a) Formed by massive transformation

Co-Sm

From [Moffatt]

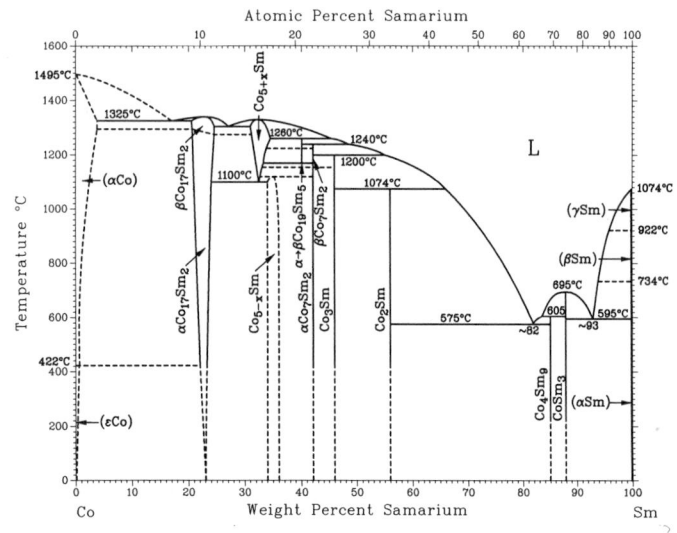

Phase	Composition, wt% Sm	Pearson symbol	Space group
(αCo)	0 to ~3.7	cF4	$Fm\overline{3}m$
(εCo)	~0	hP2	$P6_3/mmc$
βCo₁₇Sm₂	~23.0	hP38	$P6_3/mmc$
αCo₁₇Sm₂	~23.0	hR19	$R\overline{3}m$
		hP8	$P6/mmm$
Co₅₊ₓSm	~33 to 34	...	...
Co₅₋ₓSm	~34 to 35	...	...
Co₁₉Sm₅	~40.1	hR24	$R\overline{3}m$
		hP48	$P6_3/mmc$
αCo₇Sm₂	~42.1	hR18	$R\overline{3}m$
βCo₇Sm₂	~42.1	hP36	$P6_3/mmc$
Co₃Sm	46	hR12	$R\overline{3}m$
Co₂Sm	56.0	hR4	$R\overline{3}m$
		cF24	$Fd\overline{3}m$
Co₄Sm₉	~85.1	o**	...
CoSm₃	88	oP16	Pnma
(γSm)	~100	cI2	$Im\overline{3}m$
(βSm)	~100	hP2	$P6_3/mmc$
(αSm)	~100	hR3	$R\overline{3}m$
Other reported phases			
Co₅Sm	~33.8	hP6	$P6/mmm$
		hP*	...
Co₂Sm₅	~86.4	mC28	C2/c

Co-Sn

K. Ishida and T. Nishizawa, 1991

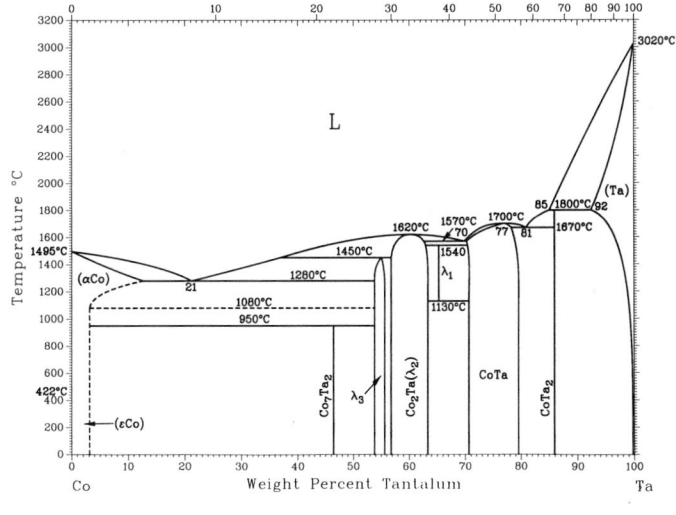

Phase	Composition, wt% Sn	Pearson symbol	Space group
(αCo)	0 to ~4	cF4	$Fm\bar{3}m$
(εCo)	0 to ~0.4	hP2	$P6_3/mmc$
βCo₃Sn₂	~52 to ~59	hP4	$P6_3/mmc$
αCo₃Sn₂	~58 to ~59	oP20	$Pnma$
CoSn	66.8	hP6	$P6/mmm$
CoSn₂	80.1	tI12	$I4/m$
(βSn)	~100	tI4	$I4_1/amd$
Metastable phases			
(ε'Co)	3.0 to 15.1	...	$R\bar{3}m$
Co₃Sn	40.2	cI2	$Im\bar{3}m$
		cP2	$Pm\bar{3}m$

Co-Ta

H. Okamoto, 1991

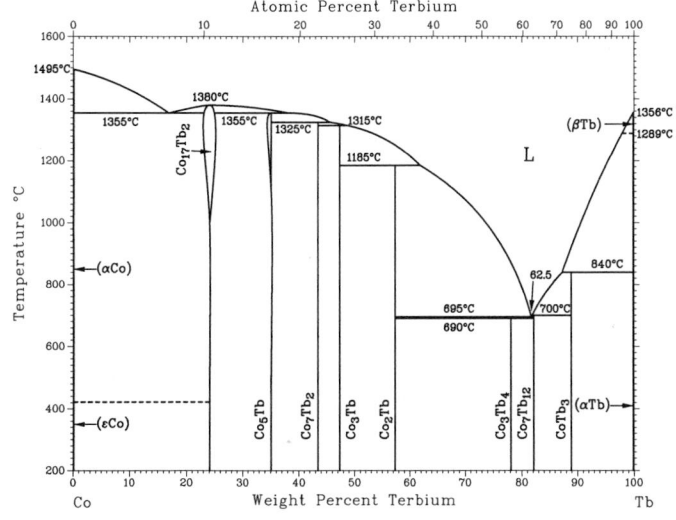

Phase	Composition, wt% Ta	Pearson symbol	Space group
(αCo)	0 to 11	cF4	$Fm\bar{3}m$
Co₇Ta₂	46.7	...	...
λ₃	53.8 to 56	hP24	$P6_3/mmc$
λ₂	56.2 to 63	cF24	$Fd\bar{3}m$
λ₁	~64	hP12	$P6_3/mmc$
Co₆Ta₇	71 to 80	hR13	$R\bar{3}m$
CoTa₂	86.0	tI12	$I4/mcm$
(Ta)	92 to 100	cI2	$Im\bar{3}m$

Co-Tb

H. Okamoto, 1990

Phase	Composition, wt% Tb	Pearson symbol	Space group
(αCo)	0	cF4	$Fm\bar{3}m$
(εCo)	0	hP2	$P6_3/mmc$
βCo₁₇Tb₂	24.0	hP38	$P6_3/mmc$
αCo₁₇Tb₂	24.0	hR19	$R\bar{3}m$
Co₅Tb	35.1	hP6	$P6/mmm$
Co₇Tb₂	43.5	hR18	$R\bar{3}m$
Co₃Tb	47	hR12	$R\bar{3}m$
Co₂Tb	57.4	cF24	$Fd3m$
Co₃Tb₄	78.2	hP22	$P6_3/m$
Co₇Tb₁₂	82.2	mP38	$P2_1/c$
CoTb₃	89	oP16	$Pnma$
(Tb)	100	hP2	$P6_3/mmc$

Co-Te

K. Ishida and T. Nishizawa, unpublished

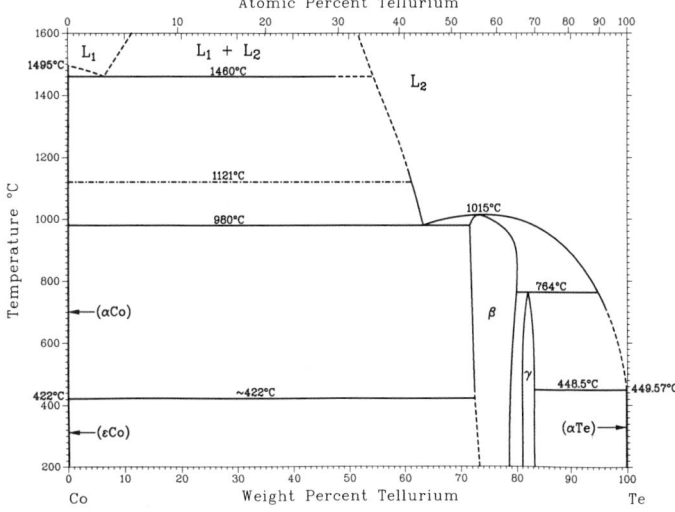

Phase	Composition, wt% Te	Pearson symbol	Space group
(αCo)	~0	cF4	Fm$\overline{3}$m
(εCo)	~0	hP2	P6₃/mmc
β(Co₂Te₃)	73 to 80	hP4	P6₃/mmc
γ(CoTe₂)	81.1 to 83.3	oP6	Pnn2
CoTe₂(a)	81.3	hP3	P$\overline{3}$m1
CoTe₂(b)	81.3	cP12	Pa$\overline{3}$
(αTe)	~100	hP3	P3₁21

(a) Metastable? (b) Under high pressure

Co-Th

K. Ishida, T. Nishizawa, and H. Okamoto, unpublished

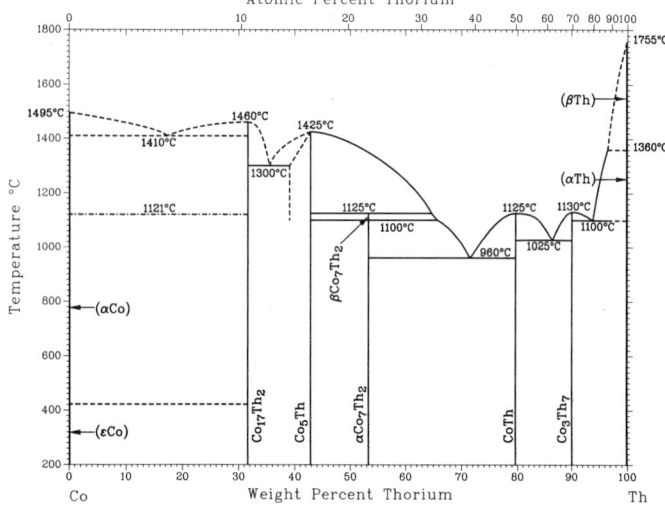

Phase	Composition, wt% Th	Pearson symbol	Space group
(αCo)	~0	cF4	Fm$\overline{3}$m
(εCo)	~0	hP2	P6₃/mmc
Co₁₇Th₂	31.6	hR19	R$\overline{3}$m
Co₅Th	44.1	hP6	P6/mmm
αCo₇Th₂	52.9	hP36	P6₃/mmc
βCo₇Th₂	52.9	hR18	R$\overline{3}$m
CoTh	79.7	oC8	Cmcm
Co₃Th₇	90	hP20	P6₃mc
(βTh)	~100	cI2	Im$\overline{3}$m
(αTh)	~100	cF4	Fm$\overline{3}$m

Co-Ti

J.L. Murray, 1987

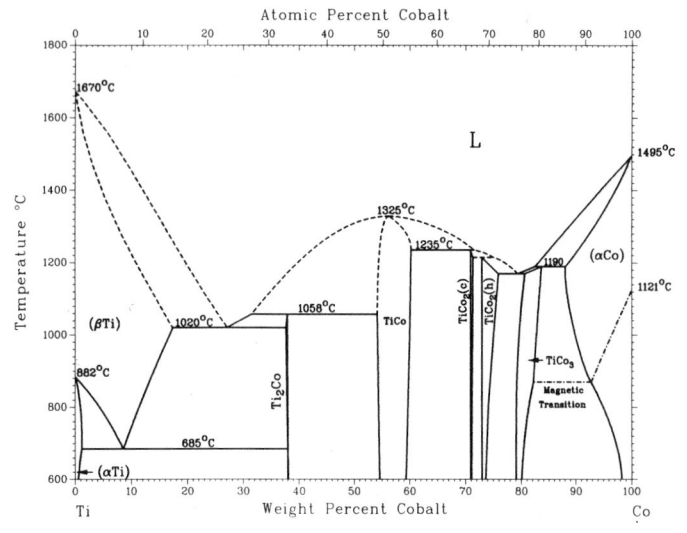

Phase	Composition, wt% Co	Pearson symbol	Space group
(αTi)	0 to 1.0	hP2	P6₃/mmc
(βTi)	0 to 17.3	cI2	Im$\overline{3}$m
Ti₂Co	37.6 to 38.1	cF96	Fd$\overline{3}$m
TiCo	54 to 60	cP2	Pm$\overline{3}$m
TiCo₂ (cubic)	71.0 to 71	cF24	Fd$\overline{3}$m
TiCo₂ (hexagonal)	73.0 to 76	hP24	P6₃/mmc
TiCo₃	79.1 to 83.7	cP4	Pm$\overline{3}$m
(εCo)	~99.2 to 100	hP2	P6₃/mmc
(αCo)	88.0 to 100	cF4	Fm$\overline{3}$m
Metastable phases			
ω	...	(a)	P6/mmm
(α″Co)	...	(b)	...

(a) The "ideal" ω structure is hexagonal, but a distorted trigonal form has also been observed in some Ti systems. The structure of ω in Ti-Co has not been definitively established. (b) Rhombohedral

Co-V

J.F. Smith, 1989

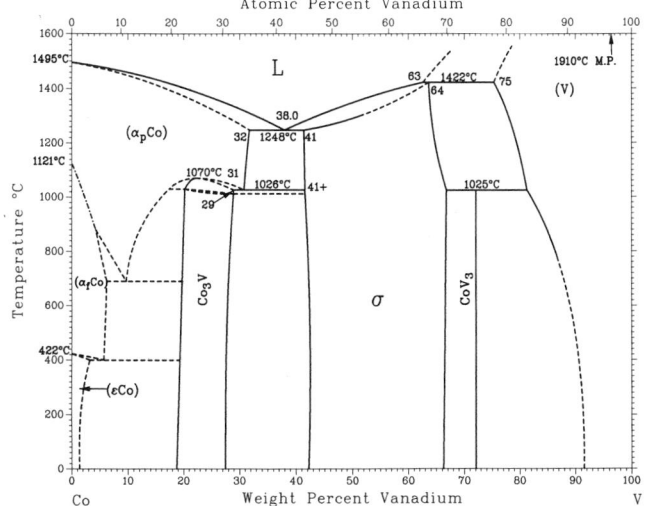

Phase	Composition, wt% V	Pearson symbol	Space group
(αCo)	0 to 32	cF4	$Fm\bar{3}m$
(εCo)	0 to ?	hP2	$P6_3/mmc$
Co₃V(hex)	~21 to 29	hP24	$P\bar{6}m2$
Co₃V(fcc)	~19 to 28	cP4	$Pm\bar{3}m$
σ	41 to ~67	tP30	$P4_2/mnm$
CoV₃	~72	cP8	$Pm\bar{3}n$
(V)	75 to 100	cI2	$Im\bar{3}m$

Co-W

S.V. Nagender Naidu, A.M. Sriramamurthy, and P. Rama Rao, 1986

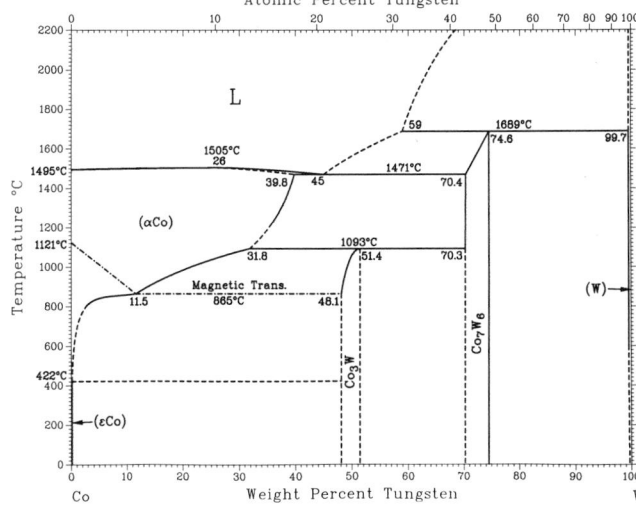

Phase	Composition, wt% W	Pearson symbol	Space group
(αCo)	0 to 39.8	cF4	$Fm\bar{3}m$
(εCo)	0	hP2	$P6_3/mmc$
Co₃W	48.1 to 51.4	hP8	$P6_3/mmc$
Co₇W₆	70.3 to 74.6	hR13	$R\bar{3}m$
(W)	99.7 to 100	cI2	$Im\bar{3}m$

Co-Y

H. Okamoto, 1992

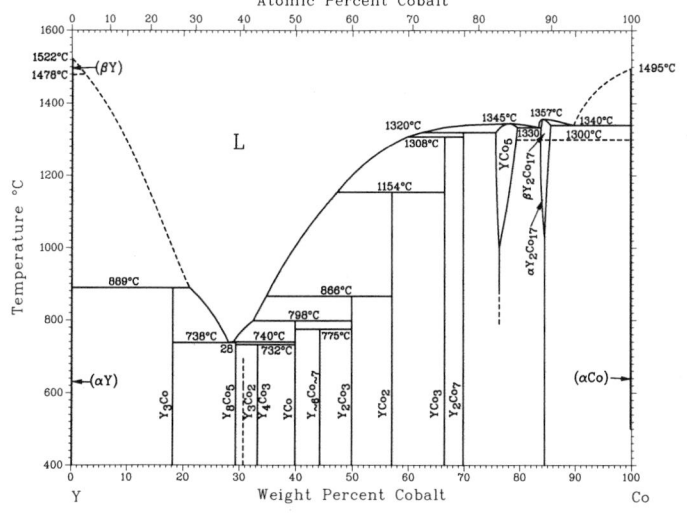

Phase	Composition, wt% Co	Pearson symbol	Space group
(βY)	0	cI2	$Im\bar{3}m$
(αY)	0	hP2	$P6_3/mmc$
Y₃Co	18	oP16	$Pnma$
Y₈Co₅	29.3	mP52	$P2_1/c$
Y₄Co₃	33.2	hP22	$P6_3/m$
YCo	39.9	oC8	$Cmcm$
Y₆Co₇	44.4	...	...
Y₂Co₃	49.9	cP*	...
YCo₂	57.0	cF24	$Fd\bar{3}m$
YCo₃	67	hR12	$R\bar{3}m$
		hP24	$P6_3/mmc$
Y₂Co₇	69.9	hR18	$R\bar{3}m$
YCo₅	75.8 to 80	hP6	$P6/mmm$
βY₂Co₁₇	84 to 86	hP38	$P6_3/mmc$
αY₂Co₁₇	~84	hP19	$R\bar{3}m$
(αCo)	100	cF4	$Fm\bar{3}m$
(εCo)	100	hP2	$P6_3/mmc$
Metastable phase			
Y₃Co₂	31	oP20	$Pnnm$

Co-Zn

H. Okamoto, 1990

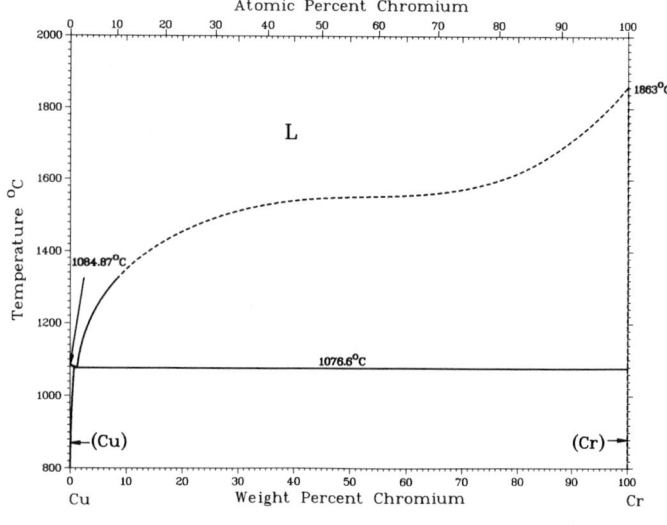

Phase	Composition, wt% Zn	Pearson symbol	Space group
(αCo)	0 to 40	cF4	$Fm\bar{3}m$
(εCo)	0 to ?	hP2	$P6_3/mmc$
β	~52 to 54	cI2?	$Im\bar{3}m$
$β_1$	50.5 to 59.0	cP20	$P4_132$
γ	71 to 86.6	cP52	$P\bar{4}3m$
$γ_1$	88.5 to 89.6	...	...
δ	~89 to <91	...	...
$γ_2$	92 to 93.5	mC28	C2/m
(Zn)	~100	hP2	$P6_3/mmc$

Cr-Cu

D.J. Chakrabarti and D.E Laughlin, 1984

Phase	Composition, wt% Cr	Pearson symbol	Space group
(Cu)	0 to 0.73	cF4	$Fm\bar{3}m$
(Cr)	99.8 to 100	cI2	$Im\bar{3}m$

Cr-Fe

H. Okamoto, 1990

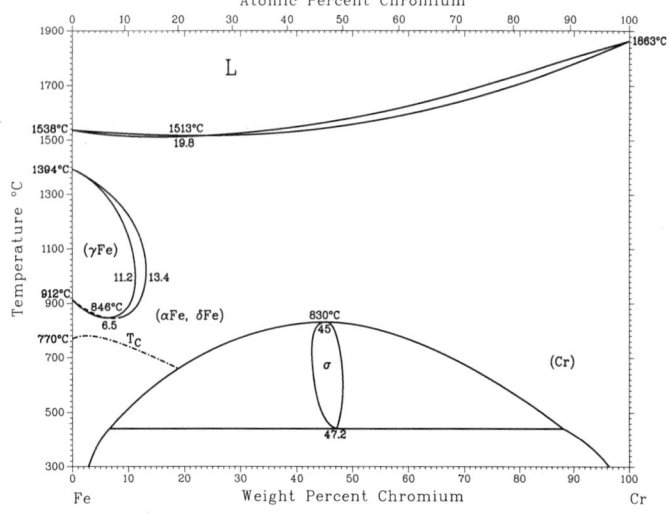

Phase	Composition, wt% Cr	Pearson symbol	Space group
(αFe,Cr)	0 to 100	cI2	$Im\bar{3}m$
(γFe)	0 to 11.2	cF4	$Fm\bar{3}m$
σ	42.7 to 48.2	tP30	$P4_2/mnm$

Cr-Ga

J.-D. Bornand and P. Feschotte, 1972

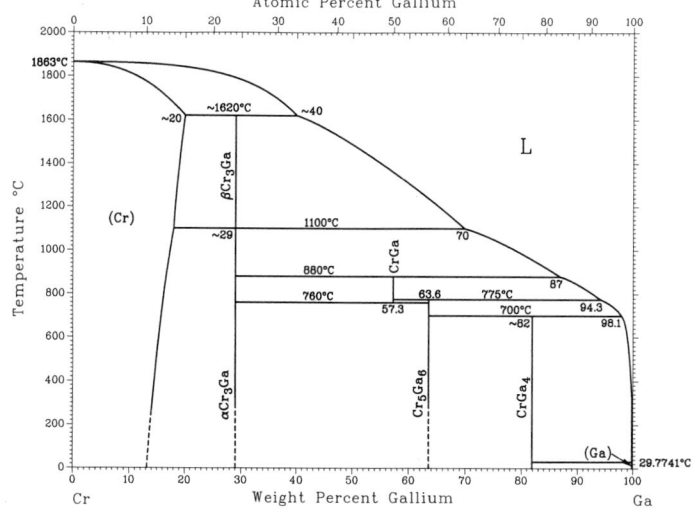

Phase	Composition, wt% Ga	Pearson symbol	Space group
(Cr)	0 to ~20	cI2	$Im\bar{3}m$
βCr$_3$Ga	~29	...	...
αCr$_3$Ga	~29	cP8	$Pm\bar{3}n$
CrGa	57.3	hR26	$R\bar{3}m$
Cr$_5$Ga$_6$	63.6	...	...
CrGa$_4$	~82	cI10	$I432$
(Ga)	~100	oC8	$Cmca$

Cr-Ge

A.B. Gokhale and G.J. Abbaschian, 1986

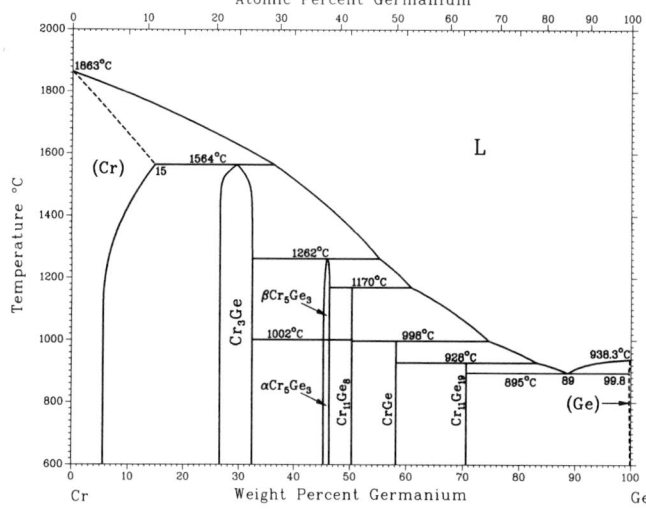

Phase	Composition, wt% Ge	Pearson symbol	Space group
(Cr)	0 to 15	cI2	$Im\bar{3}m$
Cr$_3$Ge	26.5 to 31.9	cP8	$Pm\bar{3}n$
Cr$_5$Ge$_3$	45.5 to 46.3	hP16	$I4/mcm$
Cr$_{11}$Ge$_8$	50.4	oP76	$Pnam$
CrGe	58.3	cP8	$P2_13$
Cr$_{11}$Ge$_{19}$	70.7	(a)	$P\bar{4}n2$
(Ge)	100	cF8	$Fd\bar{3}m$

(a) Tetragonal

Cr-Hf

M. Venkatraman and J.P. Neumann, 1986

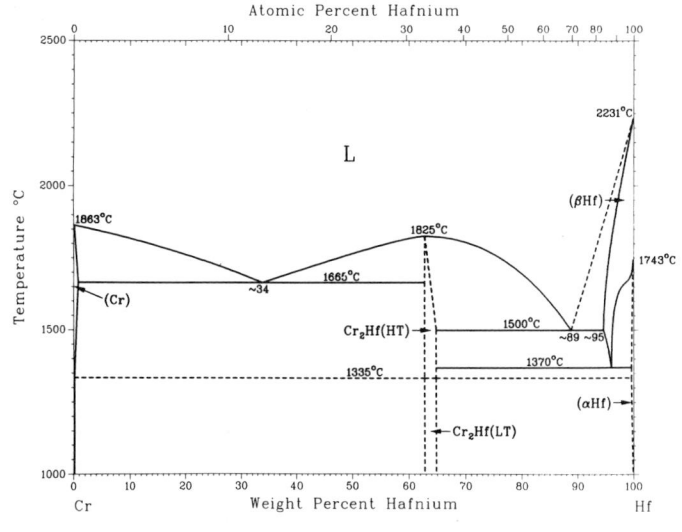

Phase	Composition, wt% Hf	Pearson symbol	Space group
(Cr)(a)	~0	cI2	$Im\bar{3}m$
Cr$_2$Hf(HT)(b)	63 to 65	hP12	$P6_3/mmc$
Cr$_2$Hf(LT)(c)	63 to 65	cF24	$Fd\bar{3}m$
(βHf)(d)	~95 to 100	cI2	$Im\bar{3}m$
(αHf)(e)	98 to 100	hP2	$P6_3/mmc$

(a) Stable at <1863 °C. (b) Stable at 1335 to 1825 °C. (c) Stable at <1335 °C. (d) Stable at 1740 to 2224 °C. (e) Stable at <1740 °C

Cr-Ir

M. Venkatraman and J.P. Neumann, 1990

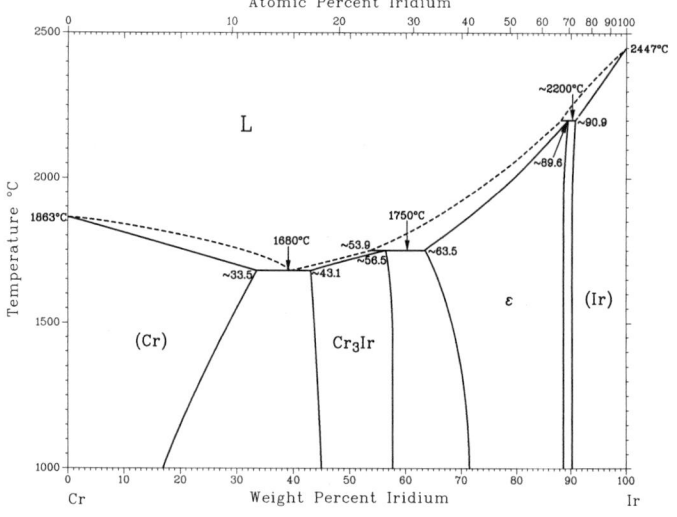

Phase	Composition, wt% Ir	Pearson symbol	Space group
(Cr)	0 to ~33.5	cI2	Im$\bar{3}$m
Cr$_3$Ir	~43.1 to 58	cP8	Pm$\bar{3}$n
ε	~63.5 to ~89.6	hP2	P6$_3$/mmc
CrIr$_3$(a)	~90 to ~95	cP4	Pm$\bar{3}$m
(Ir)	91 to 100	cF4	Fm$\bar{3}$m

(a) Order-disorder temperature has not been determined, but because it is presumably below 1000 °C, the phase is not shown in the diagram.

Cr-Lu

H. Okamoto, 1992

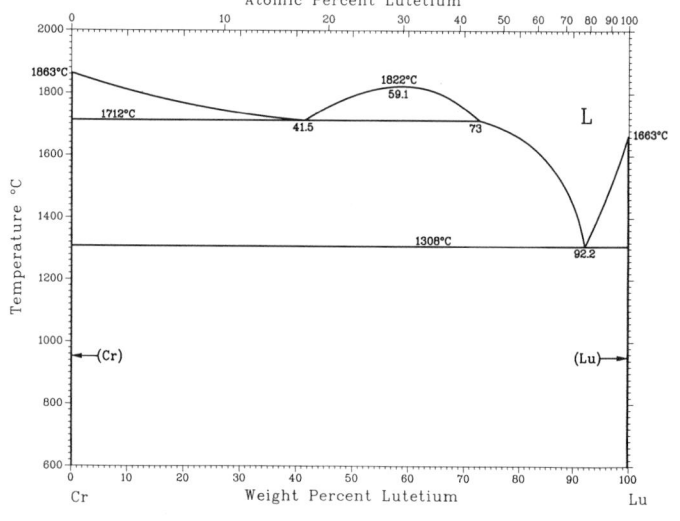

Phase	Composition, wt% Lu	Pearson symbol	Space group
(Cr)	0	cI2	Im$\bar{3}$m
(Lu)	100	hP2	P6$_3$/mmc

Cr-Mn

M. Venkatraman and J.P. Neumann, 1986

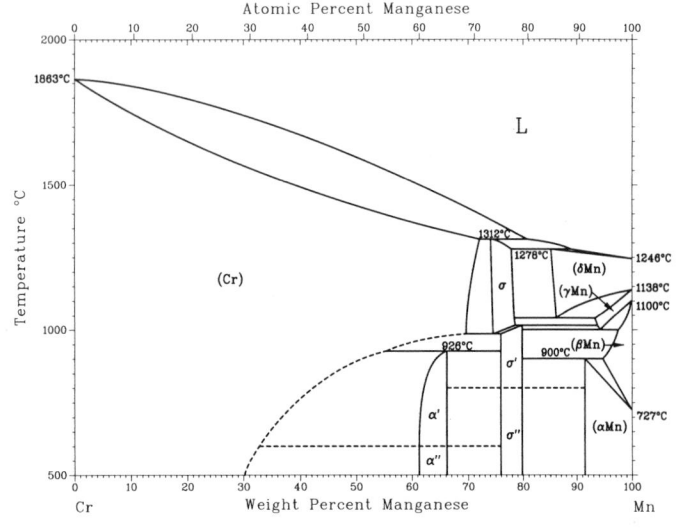

Phase	Composition, wt% Mn	Pearson symbol	Space group
(Cr)(a)	0 to 72.5	cI2	Im$\bar{3}$m
α' (HT)(b)	61.5 to 66.5	...	...
α" (LT)(c)	61.5 to 66.5	...	...
σ (HT)(d)	74 to 79	tP30	P4$_2$/mnm
σ' (MT)(e)	76 to 80	tP30	P4$_2$/mnm
σ" (LT)(f)	76 to 80	tP30	P4$_2$/mnm
(δMn)(g)	86 to 100	cI2	Im$\bar{3}$m
(αMn)(h)	91 to 100	cI58	I$\bar{4}$3m
(γMn)(j)	93 to 100	cF4	Fm$\bar{3}$m
(βMn)(k)	94 to 100	cP20	P4$_1$32
Metastable phases			
"(δMn)"	73 to 84	cI2	Im$\bar{3}$m
"(γMn)"	85 to 100	tI2	I4/mmm

(a) Below 1863 °C. (b) From 600 to 926 °C. (c) Below 600 °C. (d) From 999 to 1312 °C. (e) From ~800 to 1006 °C. (f) Below ~800 °C. (g) From 1140 to 1246 °C. (h) Below 707 °C. (j) From 1088 to 1140 °C. (k) From 707 to 1088 °C

Cr-Mo

M. Venkatraman and J.P. Neumann, 1987

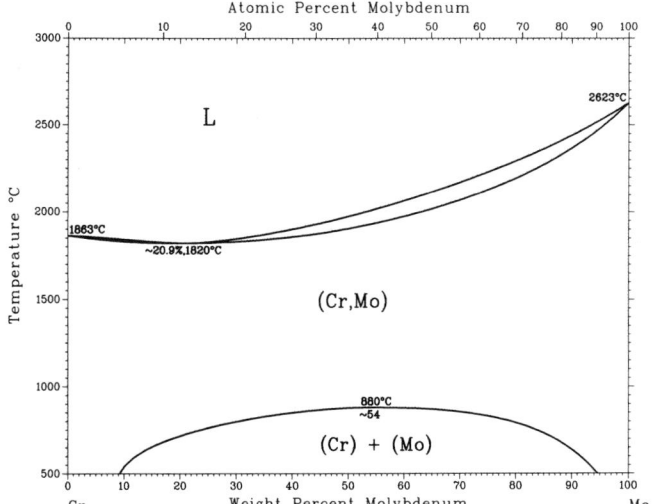

Phase	Composition, wt% Mo	Pearson symbol	Space group
(Cr,Mo)	0 to 100	cI2	Im$\bar{3}$m

Cr-Nb

M. Venkatraman and J.P. Neumann, 1986

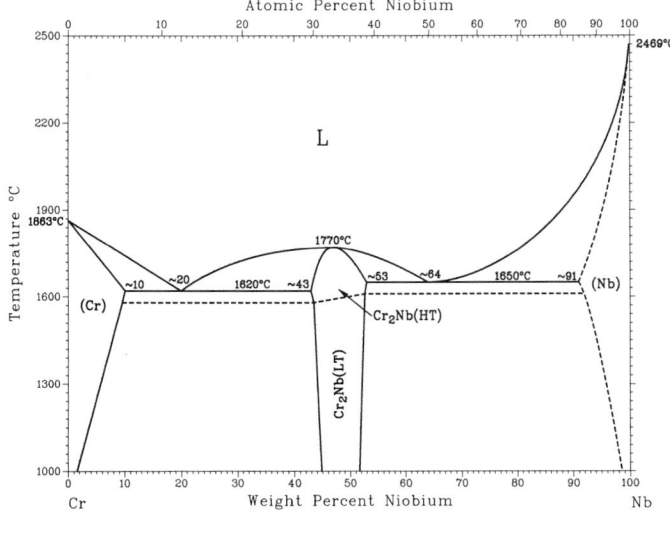

Phase	Composition, wt% Nb	Pearson symbol	Space group
(Cr)	0 to ~10	cI2	Im$\bar{3}$m
Cr$_2$Nb (HT)	~43 to ~53	hP12	P6$_3$/mmc
Cr$_2$Nb (LT)	43 to 53	cF24	Fd$\bar{3}$m
(Nb)	~91 to 100	cI2	Im$\bar{3}$m

Cr-Ni

P. Nash, 1991

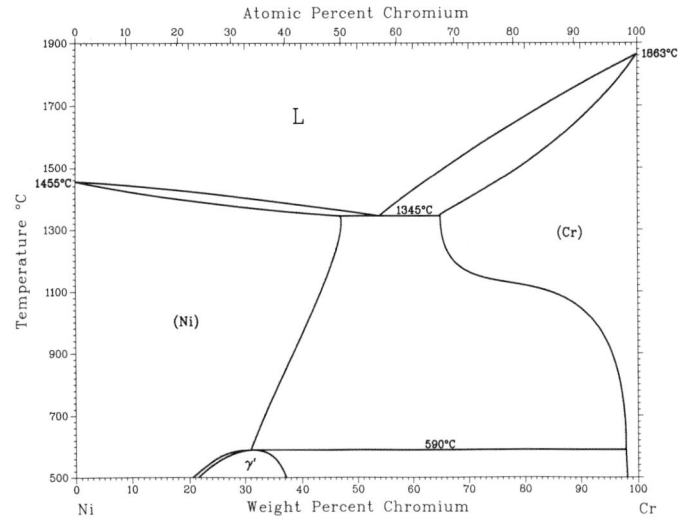

Phase	Composition, wt% Cr	Pearson symbol	Space group
(Ni)	0 to 47.0	cF4	Fm$\bar{3}$m
Ni$_2$Cr or γ	21 to 37	oI6	Immm
(Cr)	65 to 100	cI2	Im$\bar{3}$m
Metastable phases			
σ	~28	tP30	P4$_2$/mnm
δ	100	cP8	Pm$\bar{3}$m

Cr-O

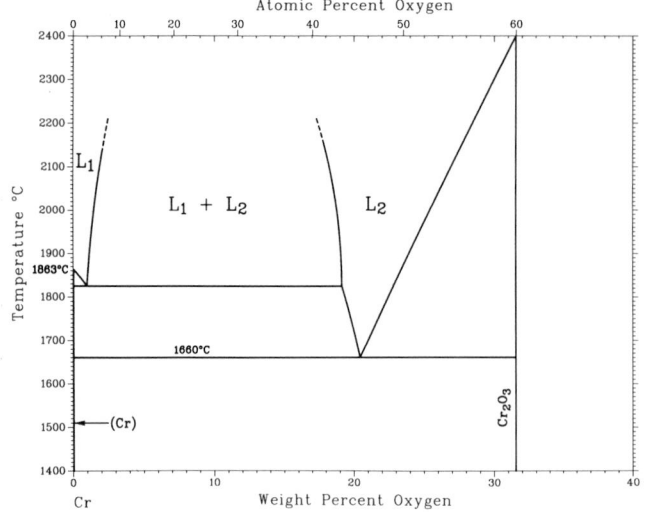

C. Banik, T. Schmitt, P. Ettmayer, and B. Lux, 1980

Phase	Composition, wt% O	Pearson symbol	Space group
(Cr)	0	cI2	$Im\bar{3}m$
Cr$_3$O$_4$(a)	29.1	tI28	$I4_1/amd$
Cr$_2$O$_3$	32	hR10	$R\bar{3}c$
CrO$_2$	38.1	tP6	$P4_2/mnm$
Cr$_5$O$_{12}$	42.5	oP68	Pbcn
Cr$_6$O$_{15}$	43.4	oC84	Cmcm
CrO$_3$	48	oC16	Ama2

(a) Metastable or high-pressure phase

Cr-Os

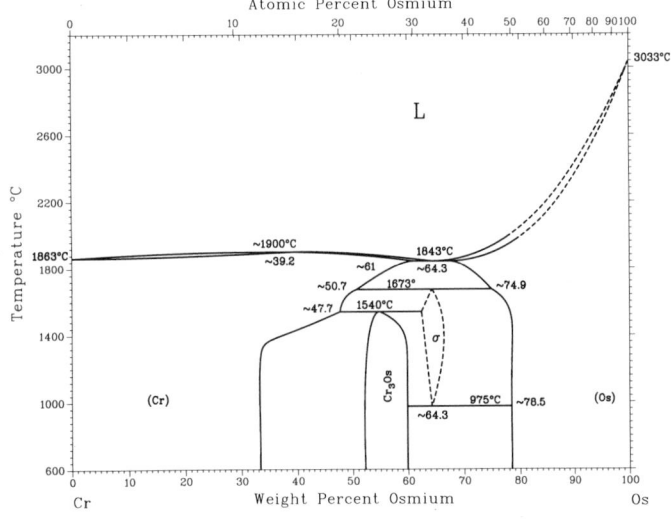

M. Venkatraman and J.P. Neumann, 1990

Phase	Composition, wt% Os	Pearson symbol	Space group
(Cr)(a)	0 to ~61	cI2	$Im\bar{3}m$
Cr$_3$Os(b)	~52 to ~60	cP8	$Pm\bar{3}n$
σ(c)	~61 to ~81	tP30	$P4_2/mnm$
(Os)(d)	~66 to 100	hP2	$P6_3/mmc$

(a) Below 1900 °C. (b) Below 1540 °C. (c) 975 to 1673 °C. (d) Below 3033 °C

Cr-Pd

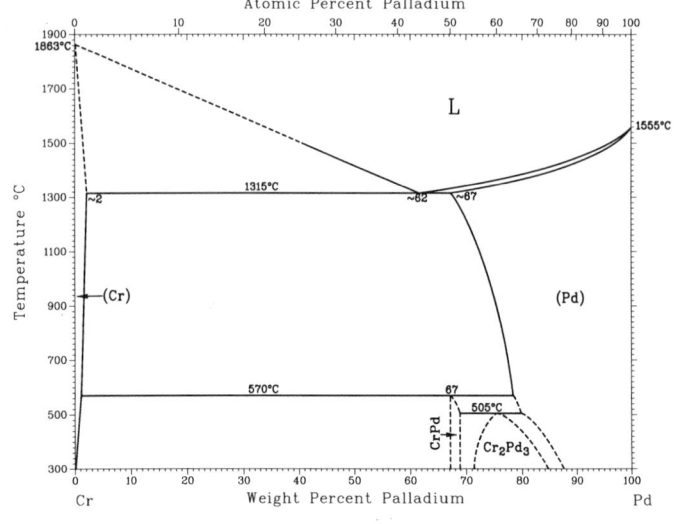

M. Venkatraman and J.P. Neumann, 1990

Phase	Composition, wt% Pd	Pearson symbol	Space group
(Cr)	0 to ~2	cI2	$Im\bar{3}m$
CrPd	67 to ~69	tP2	P4/mmm
Cr$_2$Pd$_3$	~71 to ~86	cP4	$Pm\bar{3}m$
(Pd)	~67 to 100	cF4	$Fm\bar{3}m$

Cr-Pt

M. Venkatraman and J.P. Neumann, 1990

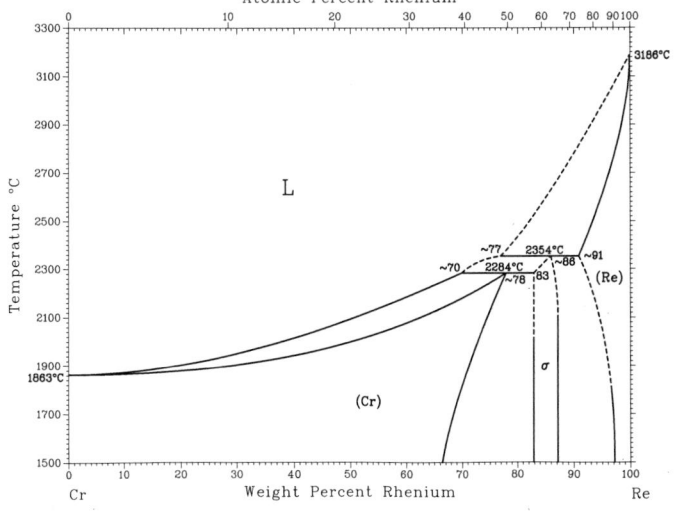

Phase	Composition, wt% Pt	Pearson symbol	Space group
(Cr)	0 to ~29	cI2	$Im\bar{3}m$
Cr$_3$Pt	44 to ~53	cP8	$Pm\bar{3}n$
CrPt	~78 to ~80	tP2	$P4/mmm$
CrPt$_3$	~66 to 96	cP4	$Pm\bar{3}m$
(Pt)	~61 to 100	cF4	$Fm\bar{3}m$

Cr-Re

M. Venkatraman and J.P. Neumann, 1987

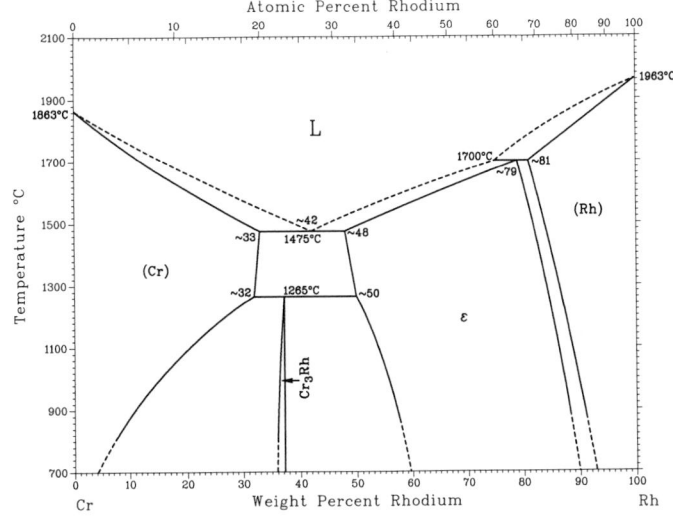

Phase	Composition, wt% Re	Pearson symbol	Space group
(Cr)	0 to ~78	cI2	$Im\bar{3}m$
σ (Cr$_2$Re$_3$)	83 to 87	tP30	$P4_2/mnm$
(Re)	~91 to 100	hP2	$P6_3/mmc$

Cr-Rh

M. Venkatraman and J.P. Neumann, 1987

Phase	Composition, wt% Rh	Pearson symbol	Space group
(Cr)	0 to ~33	cI2	$Im\bar{3}m$
Cr$_3$Rh	36 to 37	cP8	$Pm\bar{3}n$
ε	~48 to 81	hP2	$P6_3/mmc$
(Rh)	~81 to 100	cF4	$Fm\bar{3}m$

Cr-Ru

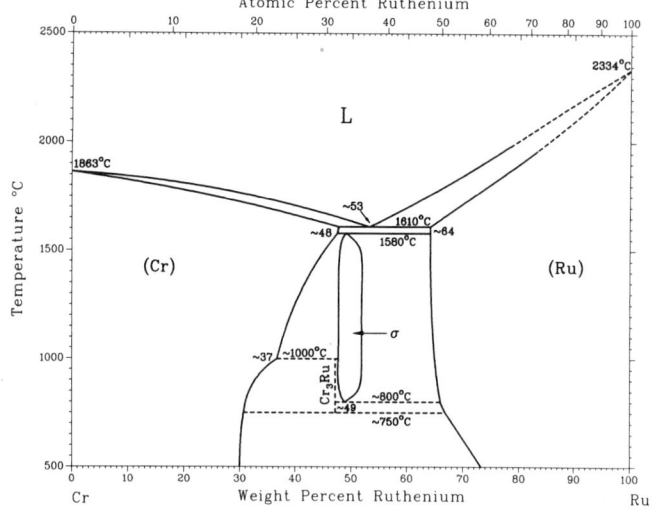

M. Venkatraman and J.P. Neumann, 1987

Phase	Composition, wt% Ru	Pearson symbol	Space group
(Cr)(a)	0 to ~48	cI2	$Im\bar{3}m$
Cr₃Ru(b)	47.2	cP8	$Pm\bar{3}n$
σCr₂Ru(c)	48 to 52	tP30	$P4_2/mnm$
(Ru)(d)	~64 to 100	hP2	$P6_3/mmc$

(a) Stable below 1863 °C. (b) Stable from 750 to 1000 °C; might be located at ~39.3 wt%, instead. (c) Stable from 800 to 1580 °C. (d) Stable below 2334 °C

Cr-S

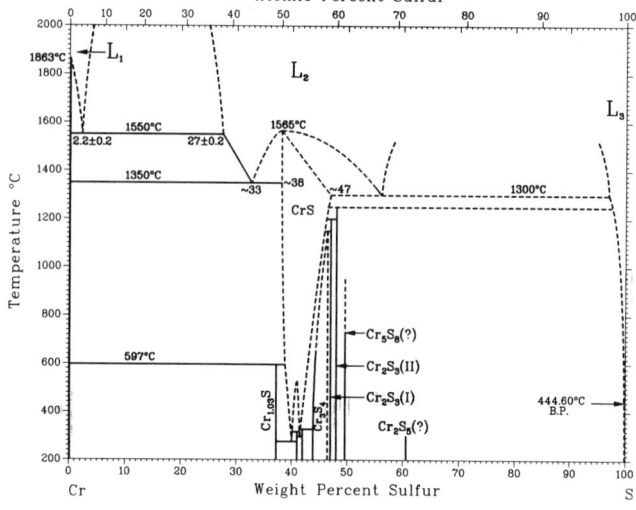

M. Venkatraman and J.P. Neumann, unpublished

Phase	Composition, wt% S	Pearson symbol	Space group
(Cr)	0 to ~0.001	cI2	$Im\bar{3}m$
Cr₁.₀₃S	~37.5	mC8	C2/c
CrS	~38 to ~47	hP4	$P6_3/mmc$
Cr₇S₈	41.2 to 41.5	hP4	$P\bar{3}m1$
Cr₅S₆	~42	hP22	$P\bar{3}1c$
Cr₃S₄	44 to ~46.2	mC14	C2/m
Cr₂S₃(I)	46.5 to 47.5	hP20	$P\bar{3}1c$
Cr₂S₃(II)	47.8 to 48.7	hR10	$R\bar{3}$
Cr₅S₈(a)	49.6	mC*	C2/m
Cr₂S₅(?)	60.6	(b)	...

(a) High-pressure phase. (b) Unknown

Cr-Sb

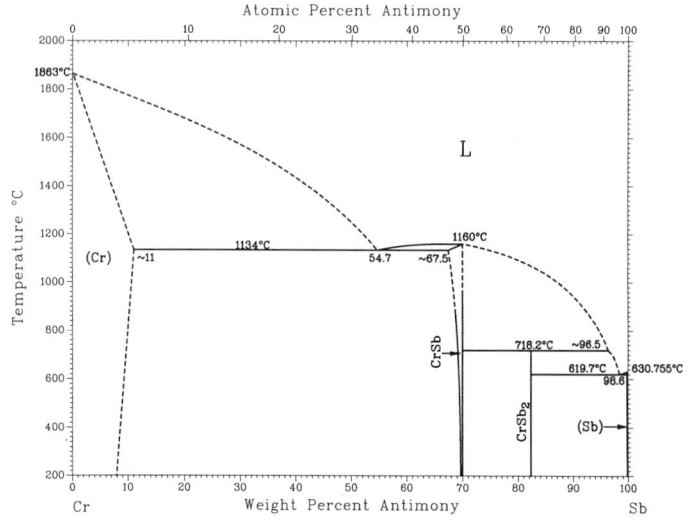

H. Okamoto, 1992

Phase	Composition, wt% Sb	Pearson symbol	Space group
(Cr)	0 to ~11	cI2	$Im\bar{3}m$
CrSb	~67.5 to 70.1	hP4	$P6_3/mmc$
CrSb₂	82.4	oP6	Pnnm
(Sb)	100	hR2	$R\bar{3}m$

Cr-Sc

M. Venkatraman and J.P. Neumann, 1985

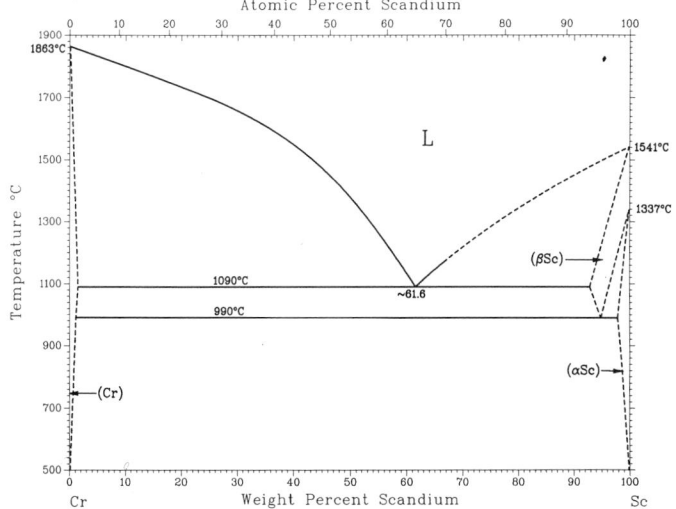

Phase	Composition, wt% Sc	Pearson symbol	Space group
(Cr)	0 to <0.09	cI2	$Im\bar{3}m$
(βSc)	>89 to 100	cI2	$Im\bar{3}m$
(αSc)	~100	hP2	$P6_3/mmc$
Metastable phase			
$Cr_{0.85}Sc_{2.15}B_x$	~69.0	cF112	$Fd\bar{3}m$

Cr-Se

M. Venkatraman and J.P. Neumann, unpublished

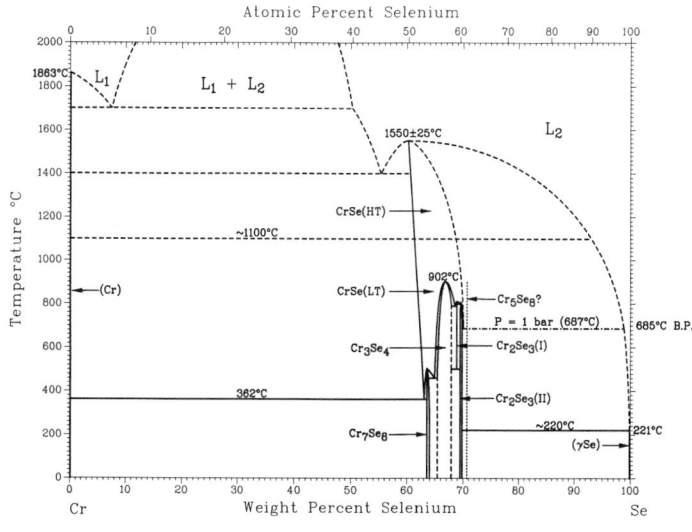

Phase	Composition, wt% Se	Pearson symbol	Space group
(Cr)	~0	cI2	$Im\bar{3}m$
CrSe(HT)	60.3 to ~69.5	hP4	$P6_3/mmc$
CrSe(LT)	~61 to ~69.9	hP4	$P\bar{3}m1$
Cr_7Se_8	63.6 to 64.1	mF60	$F2/m$
Cr_3Se_4	65.4 to 68.0	mI14	$I2/m$
$Cr_2Se_3(I)$	~69.0	hP20	$P\bar{3}1c$
$Cr_2Se_3(II)$	69.3 to 69.7	hR10	$R\bar{3}$
$Cr_2Se_3(III)(a)$	69.9 to 70.4	mI15	$I2/m$
$Cr_5Se_8(b)$	70.8	mF52	$F2/m$
$CrSe_2(a)$	75.3	hP3	$P\bar{3}m1$
(γSe)	~100	hP3	$P3_121$

(a) Metastable. (b) Stable at high pressure

Detailed view of the Cr-Se phase diagram in the region 59.9 to 70.5 wt% Se

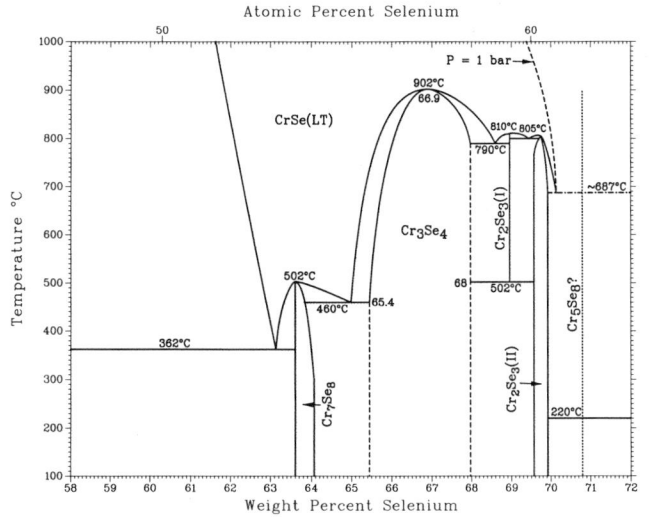

Cr-Si

A.B. Gokhale and G.J. Abbaschian, 1987

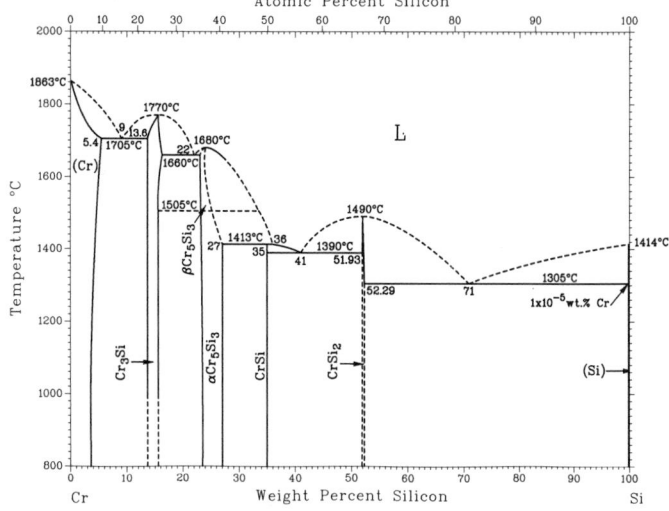

Phase	Composition, wt% Si	Pearson symbol	Space group
(Cr)	0 to 5.4	cI2	Im$\bar{3}$m
Cr$_3$Si	13.6 to 16.2	cP8	Pm$\bar{3}$n
αCr$_5$Si$_3$	23 to 27	tI38	I4/mcm
CrSi	35	cP8	P2$_1$3
CrSi$_2$	51.9 to 52.29	hP9	P6$_2$22
(Si)	~100	cF8	Fd$\bar{3}$m

Cr-Sn

M. Venkatraman and J.P. Neumann, 1988

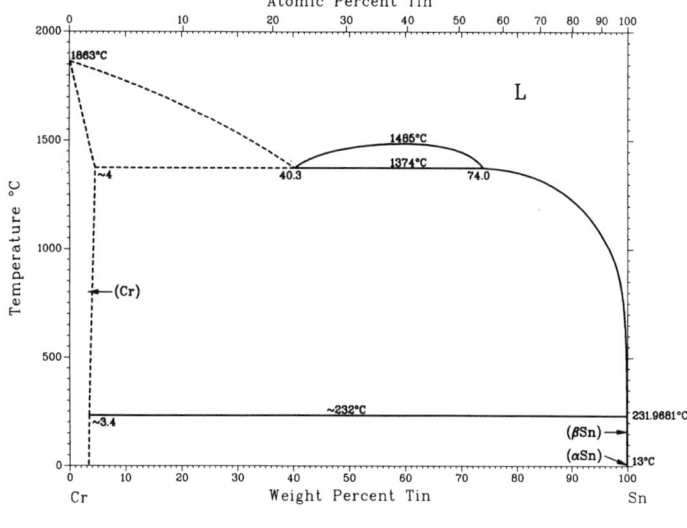

Phase	Composition, wt% Sn	Pearson symbol	Space group
(Cr)	0 to ~4	cI2	Im$\bar{3}$m
(βSn)	~100	tI4	I4$_1$/amd
(αSn)	~100	cF8	Fd$\bar{3}$m
Metastable phase			
Cr$_2$Sn$_3$	77 to 78	oF48	Fddd

Cr-Ta

M. Venkatraman and J.P. Neumann, 1987

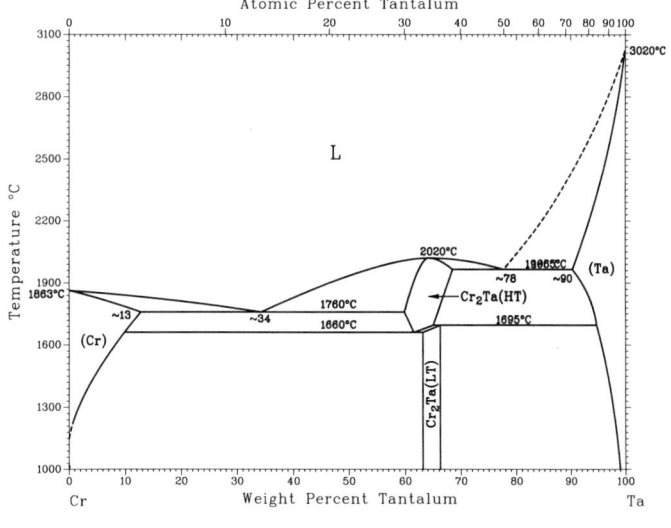

Phase	Composition, wt% Ta	Pearson symbol	Space group
(Cr)	0 to ~13	cI2	Im$\bar{3}$m
Cr$_2$Ta(HT)	60 to 68	hP12	P6$_3$/mmc
Cr$_2$Ta(LT)	63 to 66	cF24	Fd$\bar{3}$m
(Ta)	~90 to 100	cI2	Im$\bar{3}$m

Cr-Te

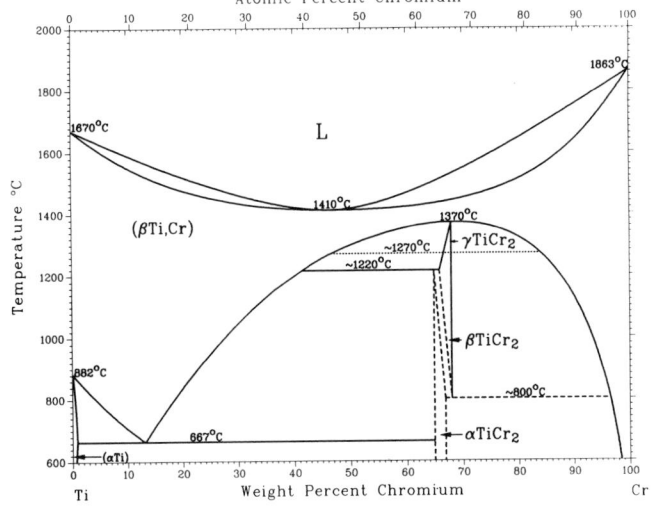

M. Venkatraman and J.P. Neumann, unpublished

Phase	Composition, wt% Te	Pearson symbol	Space group
(Cr)	~0	cI2	Im$\bar{3}$m
Cr$_{1-x}$Te	73.1 to 73.8	hP4	P6$_3$/mmc
Cr$_3$Te$_4$(HT)	~73.9 to ~80.0	mC14	C2/m
Cr$_3$Te$_4$(LT)	~76 to 77.5	...	...
Cr$_5$Te$_8$-I(a)	78.4 to ~78.9	mC26	C2/m
Cr$_5$Te$_8$-II(a)	~79.7 to ~80.0	...	P$\bar{3}$c1 (?)
Cr$_2$Te$_3$	78.3 to 78.6	hP20	P$\bar{3}$1c
CrTe$_3$	~88	mP32	P2$_1$/c
(Te)	~100	hP3	P3$_1$21

(a) Not shown in diagram

Cr-Ti

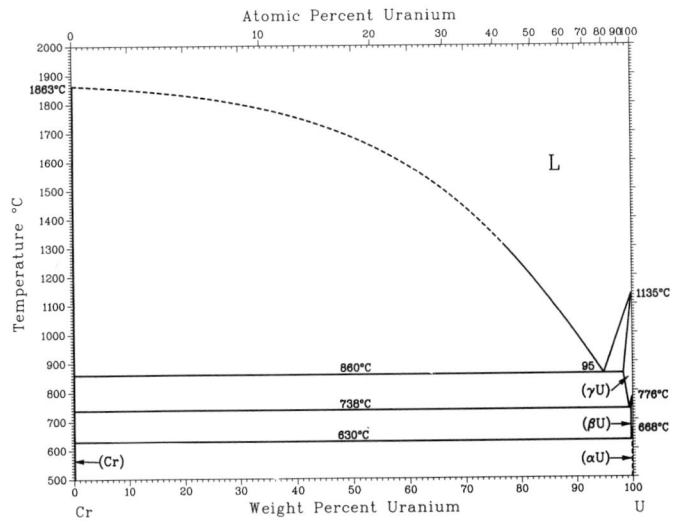

J.L. Murray, 1987

Phase	Composition, wt% Cr	Pearson symbol	Space group
(βTi,Cr)	0 to 100	cI2	Im$\bar{3}$m
(αTi)	0 to 0.2	hP2	P6$_3$/mmc
αTiCr$_2$	65 to 67	cF24	Fd$\bar{3}$m
βTiCr$_2$	66 to 68	hP12	P6$_3$/mmc
γTiCr$_2$	66 to 68	hP24	P6$_3$/mmc
Metastable phase			
ω	...	hP3	P$\bar{3}$m1

Cr-U

M. Venkatraman, J.P. Neumann, and D.E. Peterson, 1985

Phase	Composition, wt% U	Pearson symbol	Space group
(Cr)(a)	~0	cI2	Im$\bar{3}$m
(γU)(b)	99 to 100	cI2	Im$\bar{3}$m
(βU)(c)	99.8 to 100	tP30	P4$_2$/mnm
(αU)(d)	~100	oC4	Cmcm

(a) Stable below 1863 °C. (b) Stable from 775 to 1135 °C. (c) Stable from 668 to 775 °C. (d) Stable below 668 °C.

Cr-V

J.F. Smith, 1989

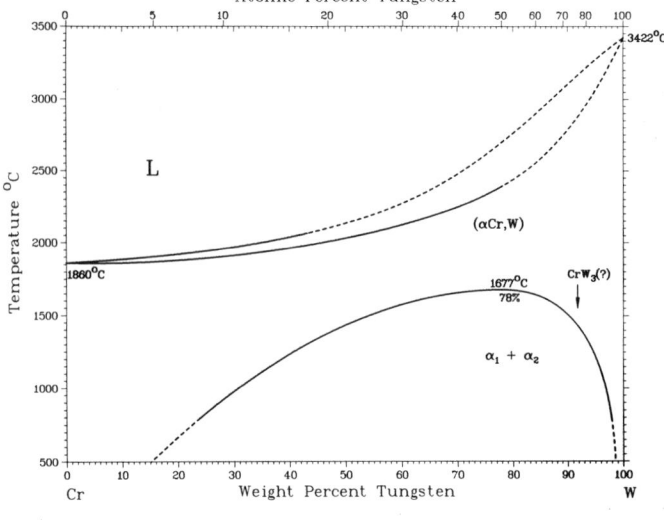

Phase	Composition, wt% Cr	Pearson symbol	Space group
(V,Cr)	0 to 100	cI2	$Im\bar{3}m$

Cr-W

S.V. Nagender Naidu, A.M. Sriramamurthy, and P. Rama Rao, 1984

Phase	Composition, wt% W	Pearson symbol	Space group
(βCr)(a)	0	cF4	$Fm\bar{3}m$
(γCr)(b)	0	cI58	$I\bar{4}3m$
(δCr)	0	cP8	$Pm\bar{3}n$
(εCr)	0	hP2	$P6_3/mmc$
(αCr,W)	0 to 100	cI2	$Im\bar{3}m$
CrW₃(?)	91	tI*	...

(a) Above 1840 °C. (b) Electrolytic

Cr-Zr

D. Arias and J.P. Abriata, 1986

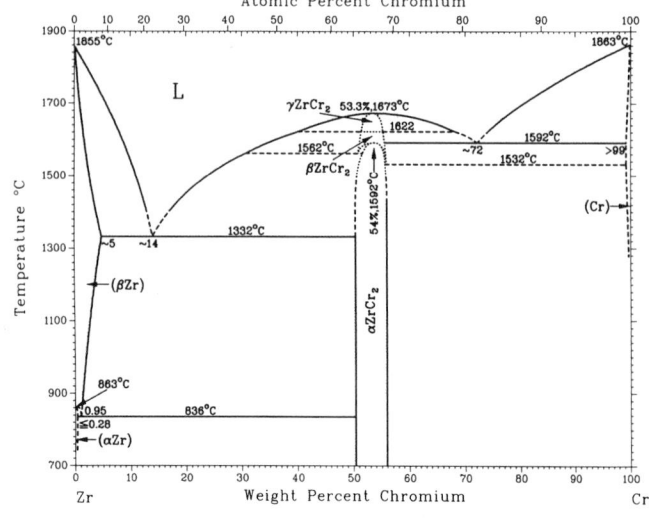

Phase	Composition, wt% Cr	Pearson symbol	Space group
(αZr)	0 to 0.28	hP2	$P6_3/mmc$
(βZr)	0 to ~5	cI2	$Im\bar{3}m$
γZrCr₂	50 to 56	hP12	$P6_3/mmc$
βZrCr₂	50 to 56	hP24	$P6_3/mmc$
αZrCr₂	50 to 56	cF24	$Fd\bar{3}m$
(Cr)	>99 to 100	cI2	$Im\bar{3}m$
Metastable phases			
ω	...	hP3	$P\bar{3}m1$ ($P6/mmm$?)

Cs-Ge

H. Okamoto, 1990

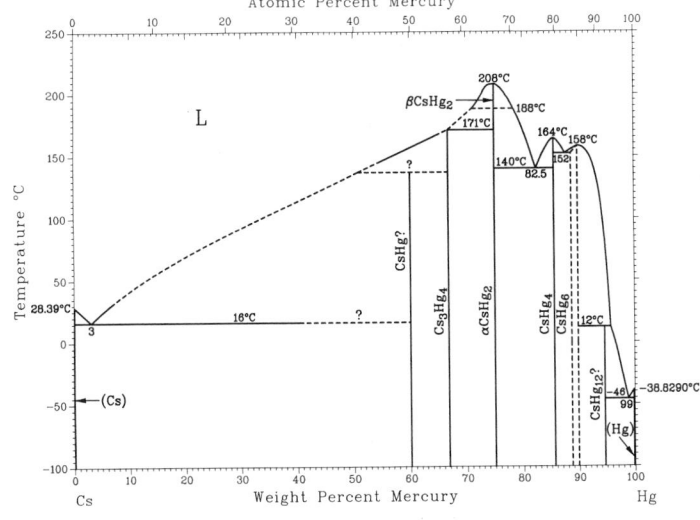

Phase	Composition, wt% Ge	Pearson symbol	Space group
(Cs)	0	cI2	Im$\bar{3}$m
Cs$_3$Ge	15	...	...
Cs$_3$Ge$_2$	27	...	...
CsGe	35.3	cP64	P$\bar{4}$3n
CsGe$_4$	69	cP*	Pm$\bar{3}$n
(Ge)	100	cF8	Fm$\bar{3}$m

Cs-Hg

From [Hansen]

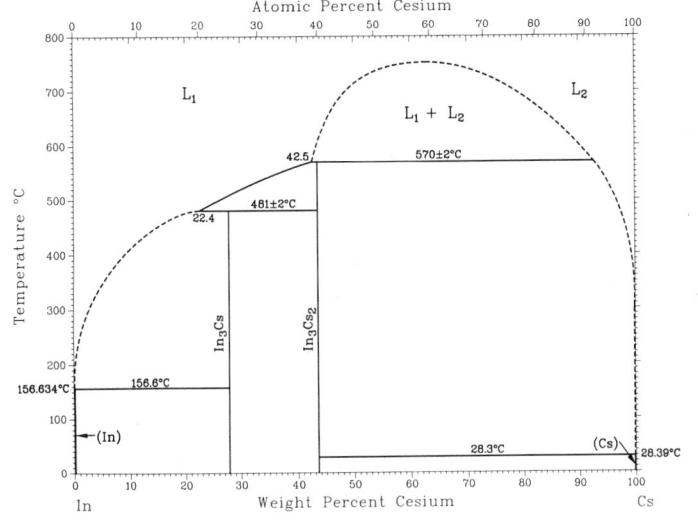

Phase	Composition, wt% Hg	Pearson symbol	Space group
(Cs)	0	cI2	Im$\bar{3}$m
CsHg?	60.1	...	...
Cs$_3$Hg$_4$	66.8	...	...
βCsHg$_2$	75.1	...	...
αCsHg$_2$	75.1	...	...
CsHg$_4$	86	...	...
CsHg$_6$	90.0	...	...
CsHg$_{12}$?	~95	c**	...
(Hg)	100	hR1	R$\bar{3}$m

Cs-In

A.D. Pelton and S. LaRose, 1990

Phase	Composition, wt% Cs	Pearson symbol	Space group
(In)	0	tI2	I4/mmm
In$_3$Cs	28	tI24	I4m2
In$_3$Cs$_2$	44	...	I4m2
(Cs)	100	cI2	Im$\bar{3}$m

Cs-K

C.W. Bale and A.D. Pelton, 1983

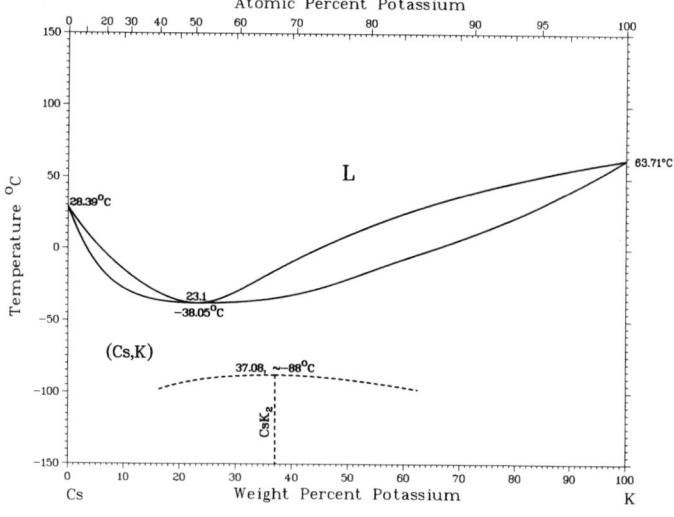

Phase	Composition, wt% K	Pearson symbol	Space group
(Cs,K)	0 to 100	cI2	$Im\bar{3}m$
CsK$_2$	37.0	hP2?	...
Other reported phase			
Cs$_6$K$_7$	?	...	...

Cs-Na

C.W. Bale, 1982

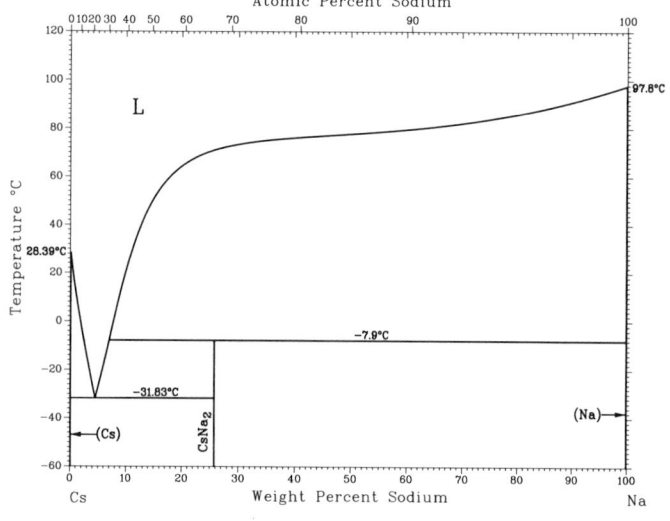

Phase	Composition, wt% Na	Pearson symbol	Space group
(Cs)	0	cI2	$Im\bar{3}m$
CsNa$_2$	25.7	...	...
(Na)	100	cI2	$Im\bar{3}m$

Cs-O

P.R. Subramanian, 1990

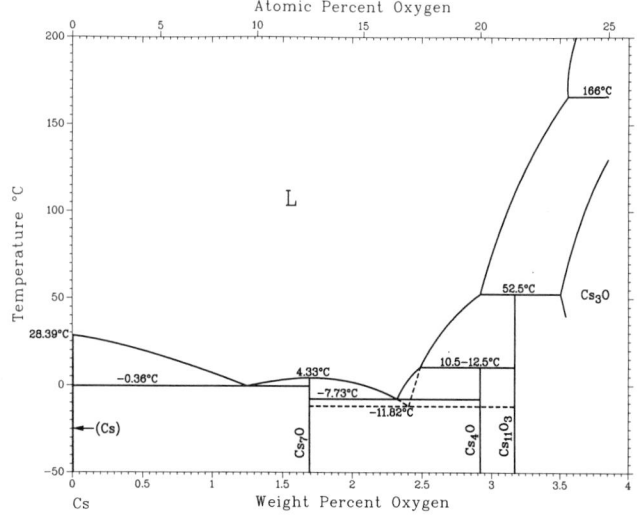

Phase	Composition, wt% O	Pearson symbol	Space group
(Cs)	~0	cI2	$Im\bar{3}m$
Cs$_7$O	~1.7	hP24	$P\bar{6}m2$
Cs$_4$O	3	...	...
Cs$_{11}$O$_3$(a)	~3.2	mP56	$P2_1/c$
Cs$_3$O	~4	...	...
Cs$_2$O	~5.7	hR3	$R\bar{3}m$
CsO	~10.7	oI8	Immm
Cs$_2$O$_3$	~15	cI28	$I\bar{4}3d$
CsO$_2$(LT)	~19.4	tI6	I4/mmm
CsO$_2$(HT)(b)	~19.4	cF8	$Fm\bar{3}m$

(a) Also reported as Cs$_7$O$_2$. (b) Above ~200 °C

Cs-Rb

C.W. Bale and A.D. Pelton, 1983

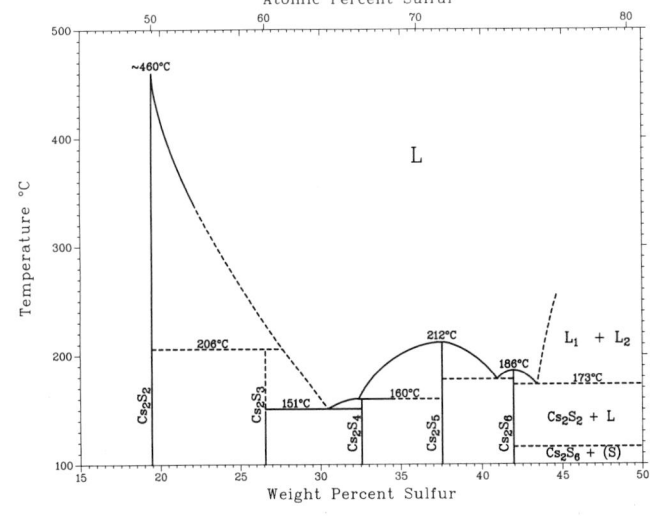

Phase	Composition, wt% Rb	Pearson symbol	Space group
(Cs,Rb)	0 to 100	$cI2$	$Im\bar{3}m$

Cs-S

From [Smithells]

Phase	Composition, wt% S	Pearson symbol	Space group
Cs_2S_2	19.4	$oI8$	...
Cs_2S_3	27	$oC20$	$Cmc2_1$
Cs_2S_4	~34.7	...	...
Cs_2S_5	~40.0	...	...
Cs_2S_6	~42.5	...	...

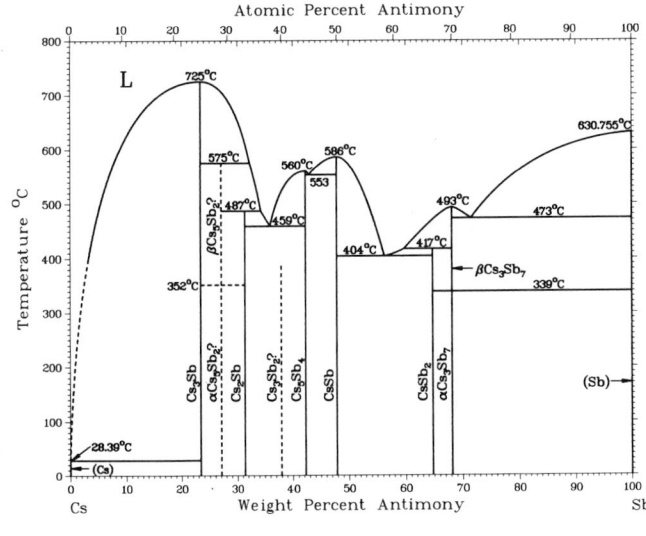

Cs-Sb

F.W. Dorn and W. Klemm, 1961

Phase	Composition, wt% Sb	Pearson symbol	Space group
(Cs)	0	$cI2$	$Im\bar{3}m$
Cs_3Sb	23	$cF16$	$Fd\bar{3}m$
αCs_5Sb_2	26.8	...	...
βCs_5Sb_2	26.8	...	...
Cs_2Sb	31	...	...
Cs_3Sb_2	38	...	...
Cs_5Sb_4	42.2	...	...
$CsSb$	47.8	$oP16$	$P2_12_12_1$
$CsSb_2$	64.7	...	...
αCs_3Sb_7	68	...	...
βCs_3Sb_7	68	...	...
(Sb)	100	$hR2$	$R\bar{3}m$

Cs-Se

H. Okamoto, 1990

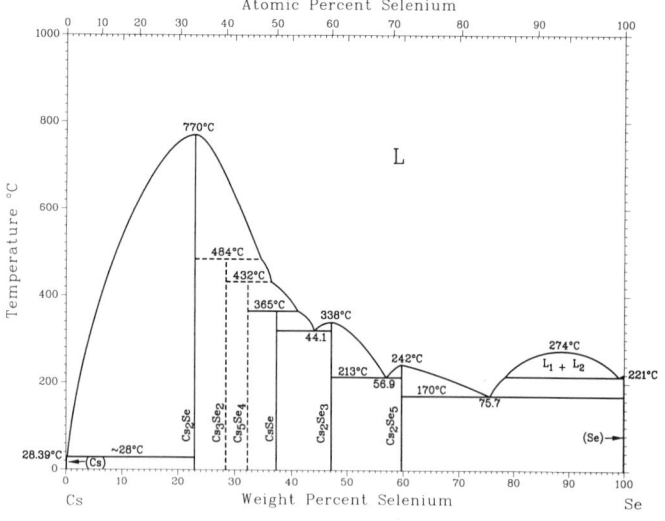

Phase	Composition, wt% Se	Pearson symbol	Space group
(Cs)	0	cI2	Im$\bar{3}$m
Cs$_2$Se	22.9	oP12	Pnma
Cs$_3$Se$_2$	28	...	...
Cs$_5$Se$_4$	32.2	...	...
CsSe	37.3	...	...
Cs$_2$Se$_3$	47	oC20	Cmc2$_1$
Cs$_2$Se$_5$	59.7	oP28	P2$_1$2$_1$2$_1$
(Se)	100	hP3	P3$_1$2$_1$
High-pressure phase			
Cs$_2$Se	22.9	oF24	Fdd2

Cs-Sn

L.Z. Melenkov, S.P. Yatsenko, K.A. Chantonov, and Yu.N. Grin, 1987

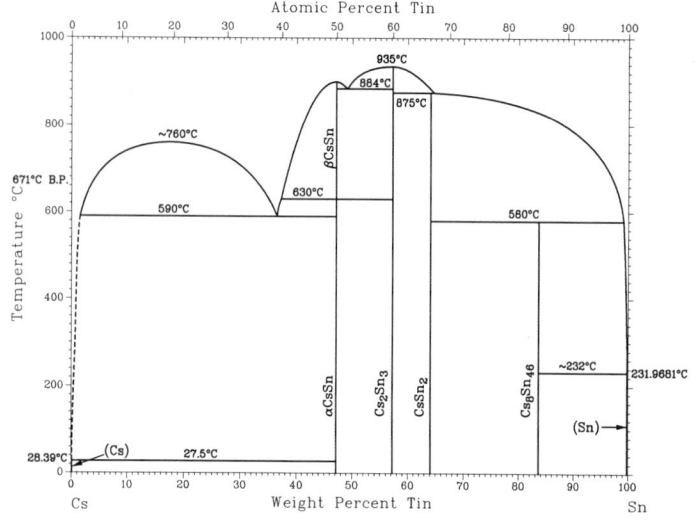

Phase	Composition, wt% Sn	Pearson symbol	Space group
(Cs)	~0	cI2	Im$\bar{3}$m
βCsSn	47.2	...	...
αCsSn	47.2	tI64	I4$_1$/acd
Cs$_2$Sn$_3$	57	...	...
CsSn$_2$	64.1	...	...
Cs$_8$Sn$_{46}$	84	...	Pm$\bar{3}$n
(βSn)(a)	~100	tI4	I4$_1$/amd
(αSn)(b)	~100	cF8	Fd$\bar{3}$m

(a) Between 13 and 231.9681 °C. (b) Below 13 °C

Cs-Te

J. Sangster and A.D. Pelton, unpublished

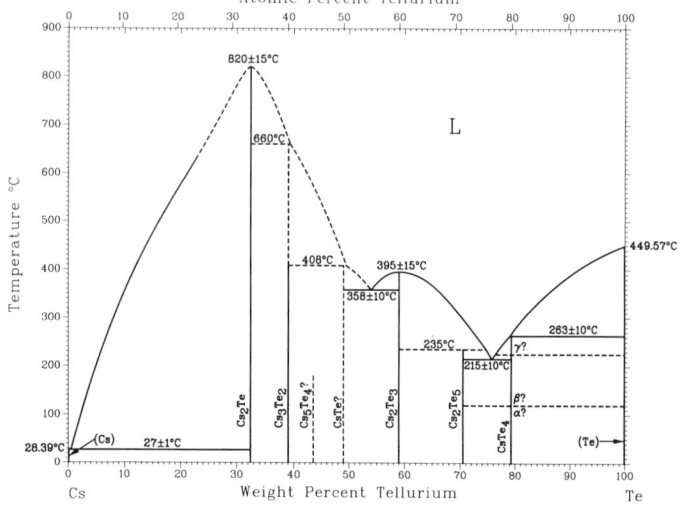

Phase	Composition, wt% Te	Pearson symbol	Space group
(Cs)	0	cI2	Im$\bar{3}$m
Cs$_2$Te	32.4	oP12	P2$_1$2$_1$2$_1$
Cs$_3$Te$_2$	39.0	...	...
Cs$_5$Te$_4$(a)	43.4	...	...
CsTe(a)	49.0	...	...
Cs$_2$Te$_3$	59	oC20	Cmc2$_1$
Cs$_2$Te$_5$	70.6	oC28	Cmcm
CsTe$_4$(b)	79	mP20	P2$_1$/c
(Te)	100	hP3	P3$_1$2$_1$

(a) Might not exist. (b) Three allotropic forms have been reported to exist. If so, this is the structure of a metastable high-temperature allotrope.

Cs-Tl

V.D. Busmanov and S.P. Yatsenko, 1981

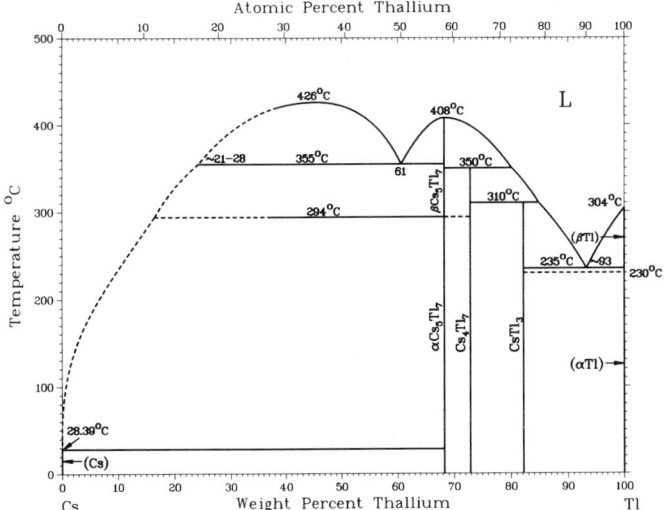

Phase	Composition, wt% Tl	Pearson symbol	Space group
(Cs)	0	cI2	$Im\bar{3}m$
αCs_5Tl_7	68.3	...	...
βCs_5Tl_7	68.3	...	...
Cs_4Tl_7	62.9	...	...
$CsTl_3$	82	...	...
(αTl)	100	hP2	$P6_3/mmc$
(βTl)	100	cI2	$Im\bar{3}m$

Cu-Dy

P.R. Subramanian and D.E. Laughlin, 1988

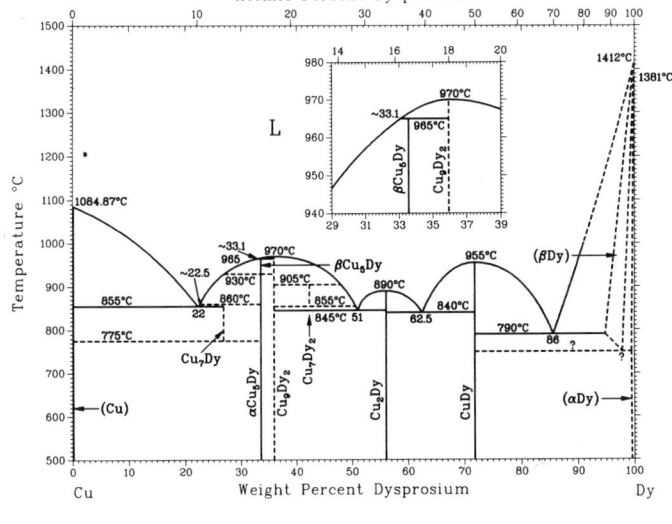

Phase	Composition, wt% Dy	Pearson symbol	Space group
(Cu)	0	cF4	$Fm\bar{3}m$
βCu_5Dy	~33.84	hP6	$P6/mmm$
αCu_5Dy	~33.84	cF24	$F\bar{4}3m$
Cu_2Dy	~56.1	oI12	$Imma$
$CuDy$	~72	cP2	$Pm\bar{3}m$
$(\alpha'Dy)$	100	oC4	$Cmcm$
(αDy)	100	hP2	$P6_3/mmc$
(βDy)	100	cI2	$Im\bar{3}m$

Cu-Er

P.R. Subramanian and D.E. Laughlin, 1988

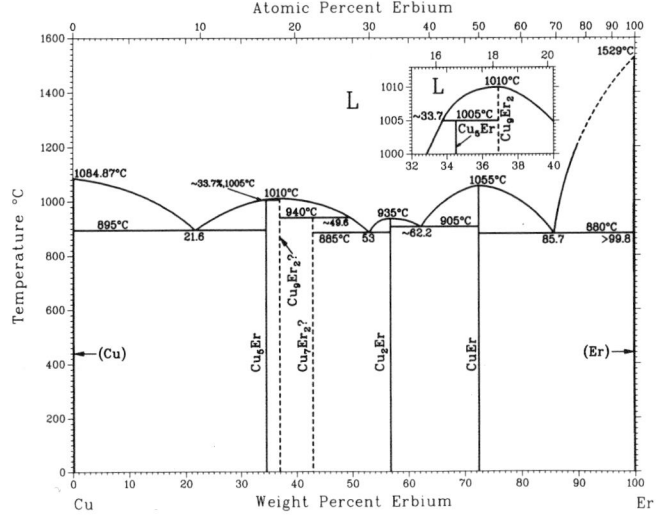

Phase	Composition, wt% Er	Pearson symbol	Space group
(Cu)	0	cF4	$Fm\bar{3}m$
Cu_5Er	~34.49	cF24	$F\bar{4}3m$
Cu_2Er	~56.8	oI12	$Imma$
$CuEr$	~73	cP2	$Pm\bar{3}m$
(Er)	100	hP2	$P6_3/mmc$

Cu-Eu

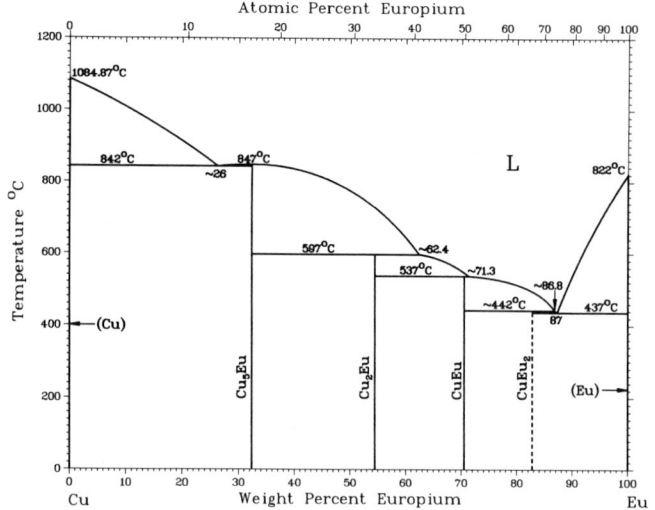

P.R. Subramanian and D.E. Laughlin, 1988

Phase	Composition, wt% Eu	Pearson symbol	Space group
(Cu)	0	cF4	Fm$\bar{3}$m
Cu$_5$Eu	~35.24	hP6	P6/mmm
Cu$_2$Eu	~57.6	oI12	Imma
CuEu	~73	oP8	Pnma
CuEu$_2$	~84.48	oP12	Pnma
(Eu)	100	cI2	Im$\bar{3}$m

Cu-Fe

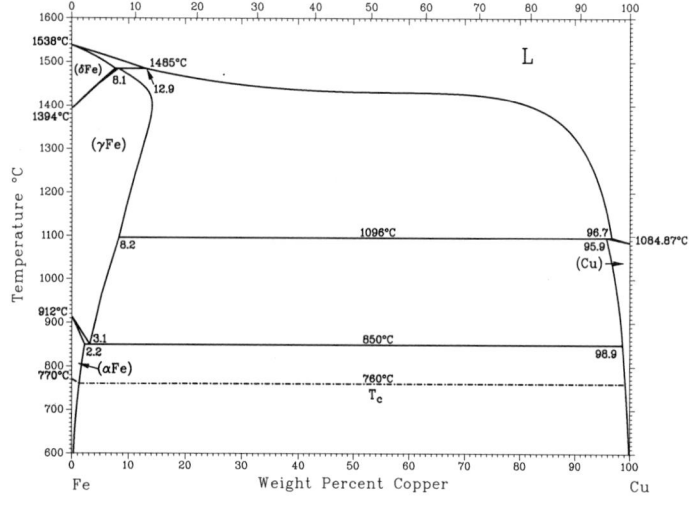

L.J. Swartzendruber, 1992

Phase	Composition, wt% Cu	Pearson symbol	Space group
(δFe)	0 to 7.6	cI2	Im$\bar{3}$m
(γFe)	0 to 13	cF4	Fm$\bar{3}$m
(αFe)	0 to 2.2	cI2	Im$\bar{3}$m
(Cu)	95.9 to 100	cF4	Fm$\bar{3}$m

Cu-Ga

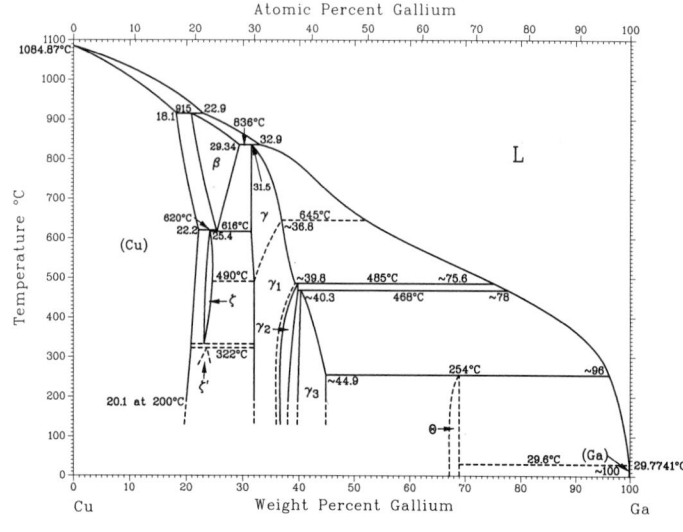

P.R. Subramanian and D.E. Laughlin, unpublished

Phase	Composition, wt% Ga	Pearson symbol	Space group
(Cu)	0 to 22.2	cF4	Fm$\bar{3}$m
β	20.8 to 29.34	cI2	Im$\bar{3}$m
γ	31.5 to ~36.8	cP52	P$\bar{4}$3m
γ$_1$	31.8 to 39.8	cP52	P$\bar{4}$3m
γ$_2$	36.0 to 39.9	cP?(a)	P$\bar{4}$3m
γ$_3$	39.7 to ~44.9	cP?(a)	P$\bar{4}$3m
ζ	22.1 to 24.2	hP2	P6$_3$/mmc
ζ'	22.6 to 24.1	...	...
θ	66.7 to 68.70	tP3	P4/mmm
(Ga)	~100	oC8	Cmca

(a) The number of atoms/cell decreases from 52 to ~47, as the Ga contents decrease from 32.0 to 44.6 wt%.

Cu-Gd

P.R. Subramanian and D.E. Laughlin, 1988

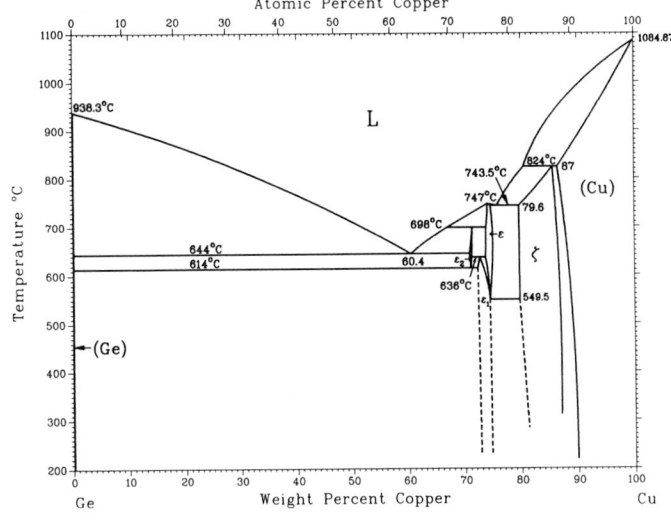

Phase	Composition, wt% Gd	Pearson symbol	Space group
(Cu)	0	cF4	$Fm\overline{3}m$
Cu₆Gd	~29.21	oP28	Pnma
βCu₅Gd	~32 to ~34.1	hP6	P6/mmm
αCu₅Gd	~32 to ~34.1	cF4	$F\overline{4}3m$
Cu₂Gd	~55.3	oI12	Imma
CuGd	~71.2	cP2	$Pm\overline{3}m$
(αGd)	~99.3 to 100	hP2	P6₃/mmc
(βGd)	~93 to 100	cI2	$Im\overline{3}m$

Cu-Ge

R.W. Olesinski and G.J. Abbaschian, 1986

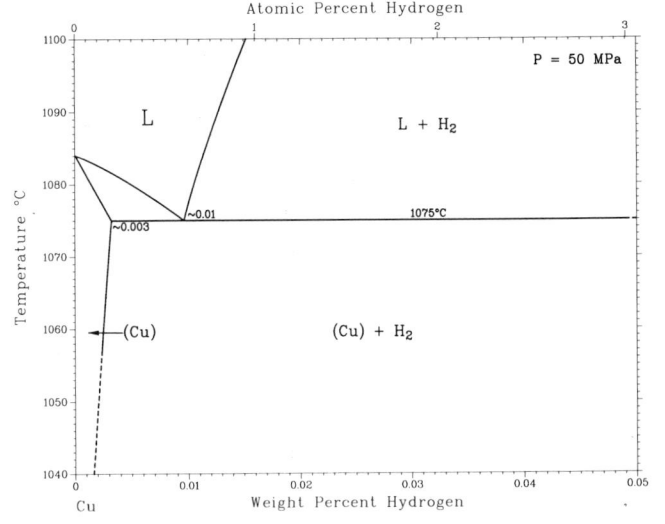

Phase	Composition, wt% Cu	Pearson symbol	Space group
(Ge)	0	cF8	$Fd\overline{3}m$
GeII (HP)	...	tI4	I4₁/amd
ε₂(a)	70.8 to 71.3	cI2	$Im\overline{3}m$
ε₁(a)	72.3 to 74.4	oP8	Pmnm
ε(a)	73.7 to 74.4	(b)	...
ζ(c)	79.6 to 87.1	hP2	P6₃/mmc
(Cu)	87 to 100	cF4	$Fm\overline{3}m$
Other reported phase			
γ'''	75.6	(d)	...

(a) Also denoted as Cu₃Ge. (b) Rhombohedral. (c) Also denoted as Cu₅Ge. (d) Cubic

Cu-H

O.M. Barabash and Yu.N. Koval, 1986

Phase	Composition, wt% H	Pearson symbol	Space group
(Cu)	0 to ~0.003	cF4	$Fm\overline{3}m$

Cu-Hf

P.R. Subramanian and D.E. Laughlin, 1988

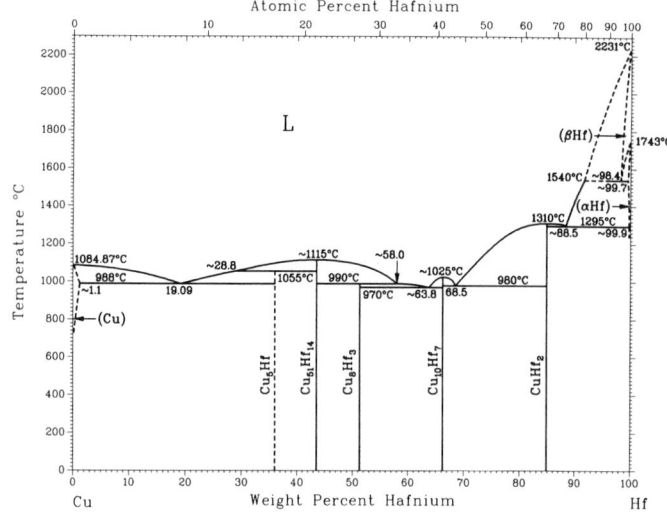

Phase	Composition, wt% Hf	Pearson symbol	Space group
(Cu)	0 to ~1.1	cF4	$Fm\bar{3}m$
Cu$_{51}$Hf$_{14}$	43.54	hP68	$P6/m$
Cu$_8$Hf$_3$	51.29	oP44	$Pnma$
Cu$_{10}$Hf$_7$	66.29	oC68	...
CuHf$_2$	84.89	tI6	$I4/mmm$
(αHf)	~99.7 to 100	hP2	$P6_3/mmc$
(βHf)	~98.4 to 100	cI2	$Im\bar{3}m$

Cu-Hg

D.J. Chakrabarti and D.E. Laughlin, 1985

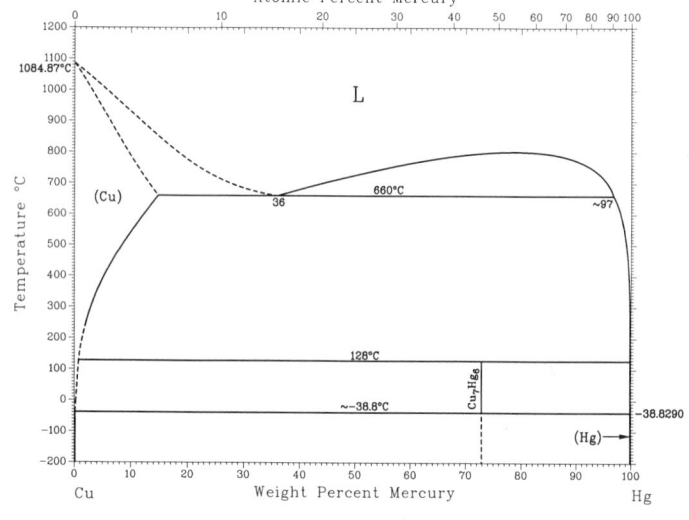

Phase	Composition, wt% Hg	Pearson symbol	Space group
(Cu)	0 to ?	cF4	$Fm\bar{3}m$
γ(a)	73	hR52	$R\bar{3}m$
(αHg)	100	hR1	$R\bar{3}m$
(βHg)	100	tI2	$I4/mmm$
(γHg)(b)	100	...	...

(a) Composition of the γ phase corresponds to stoichiometry Cu$_7$Hg$_6$. (b) Formed from αHg by strain-induced (martensitic) transformation at 4.2 K, reverting to αHg at 50 K

Cu-In

H. Okamoto, 1991

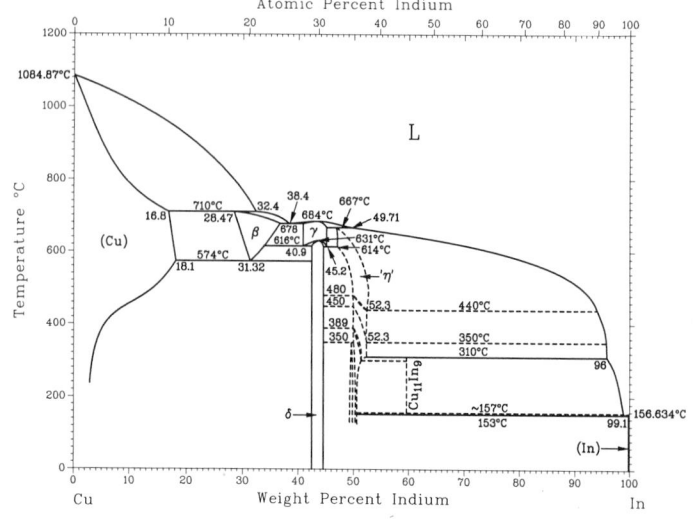

Phase	Composition, wt% In	Pearson symbol	Space group
(Cu)	0 to 18.1	cF4	$Fm\bar{3}m$
β	28.47 to 37.0	cI2	$Im\bar{3}m$
γ	40.9 to 45.2	cP52	$P\bar{4}3m$
δ	42.52 to 44.3	aP40	$P\bar{1}$
"η"	47.00 to 52.3	hP4	$P6_3/mmc$
		hP6	$P6_3/mmc$
	49.5 to 52.3	o**	...
Cu$_{11}$In$_9$	~59	mC20	$C2/m$
(In)	~100	tI2	$I4/mmm$

Cu-Ir

D.J. Chakrabarti and D.E. Laughlin, 1987

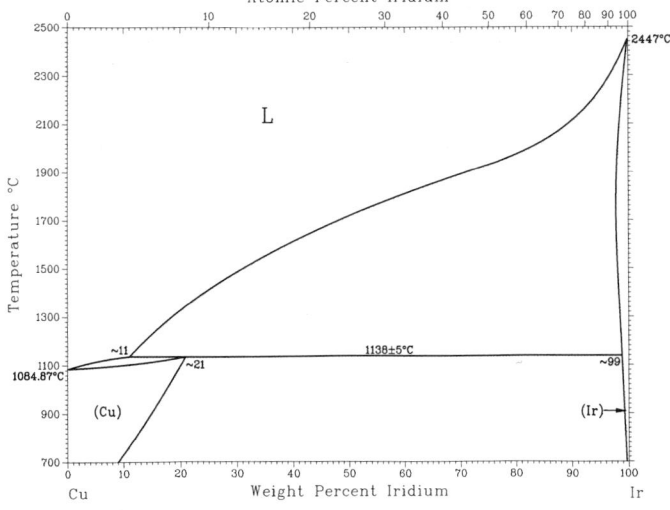

Phase	Composition, wt% Ir	Pearson symbol	Space group
(Cu)	0 to ~21	cF4	$Fm\overline{3}m$
(Ir)	~97.8 to 100	cF4	$Fm\overline{3}m$

Cu-La

H. Okamoto, 1991

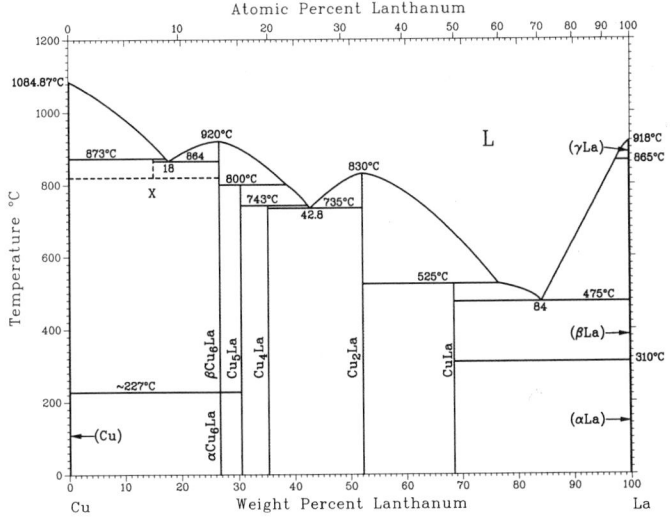

Phase	Composition, wt% La	Pearson symbol	Space group
(Cu)	~0	cF4	$Fm\overline{3}m$
X	15.1	...	...
βCu_6La	26.7	oP28	Pnma
αCu_6La(a)	26.7	mP*	...
Cu_5La	30.3	hP6	P6/mmm
Cu_4La	35	tI90	$I\overline{4}m2$
Cu_2La	52.2	hP3	P6/mmm
CuLa	69	oP8	Pnma
(γLa)	~100	cI2	$Im\overline{3}m$
(βLa)	~100	cF4	$Fm\overline{3}m$
(αLa)	~100	hP4	$P6_3/mmc$

(a) Below ~227 °C

Cu-Li

A.D. Pelton, 1986

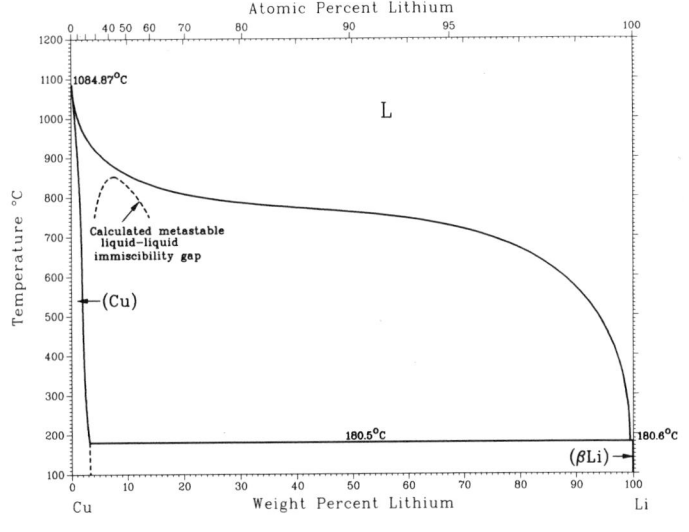

Phase	Composition, wt% Li	Pearson symbol	Space group
(Cu)	0 to 3	cF4	$Fm\overline{3}m$
(βLi)	100	cI2	$Im\overline{3}m$
(αLi)(a)	100	hP2	$P6_3/mmc$

(a) Below −193 °C

Cu-Mg

H. Okamoto, 1992

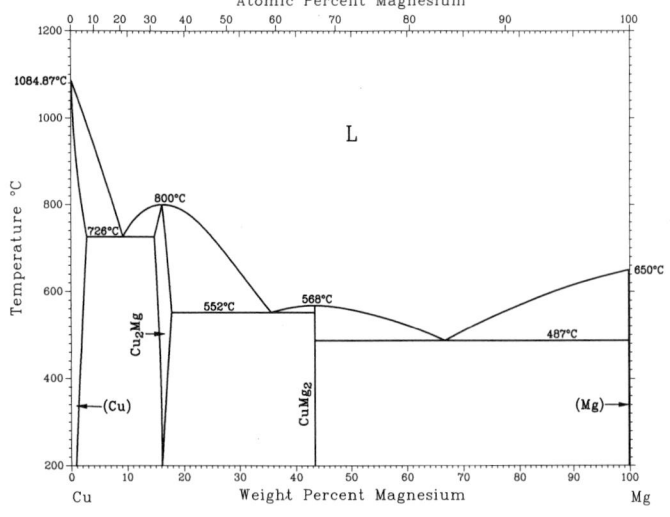

Phase	Composition, wt% Mg	Pearson symbol	Space group
(Cu)	0 to 2.77	cF4	$Fm\overline{3}m$
Cu$_2$Mg	15 to 18	cF24	$Fd\overline{3}m$
CuMg$_2$	43.4	oF48	Fddd
(Mg)	100	hP2	$P6_3/mmc$

Cu-Mn

N.A. Gokcen, unpublished

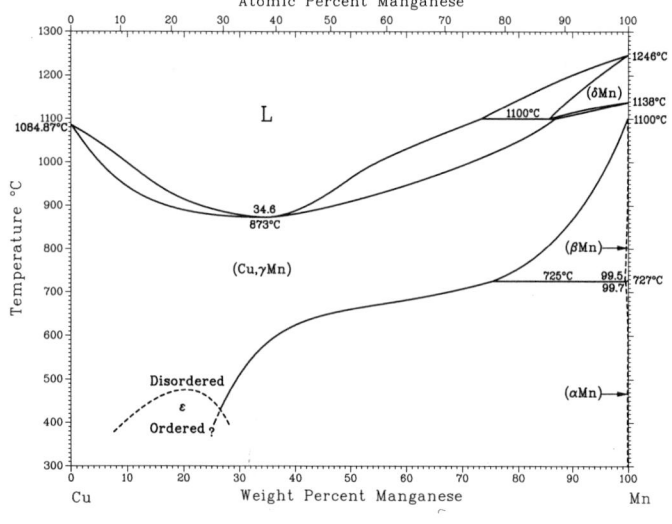

Phase	Composition, wt% Mn	Pearson symbol	Space group
(Cu,γMn)	0 to 100	cF4	$Fm\overline{3}m$
(δMn)	85.8 to 100	cI2	$Im\overline{3}m$
(βMn)	99.5 to 100	cP20	$P4_132$
(αMn)	99.7 to 100	cI58	$I\overline{4}3m$

Cu-Nb

H. Okamoto, 1991

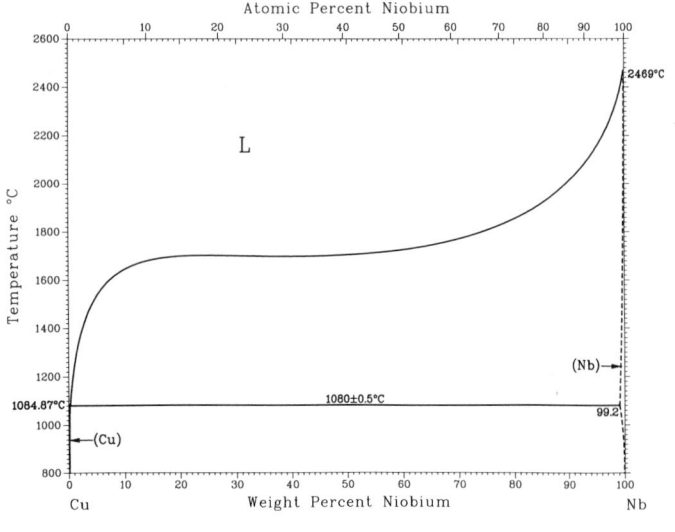

Phase	Composition, wt% Nb	Pearson symbol	Space group
(Cu) or α	0 to 0.15	cF4	$Fm\overline{3}m$
(Nb) or β	99.2 to 100	cI2	$Im\overline{3}m$

Enlargement of the Cu-rich portion of the Cu-Nb system

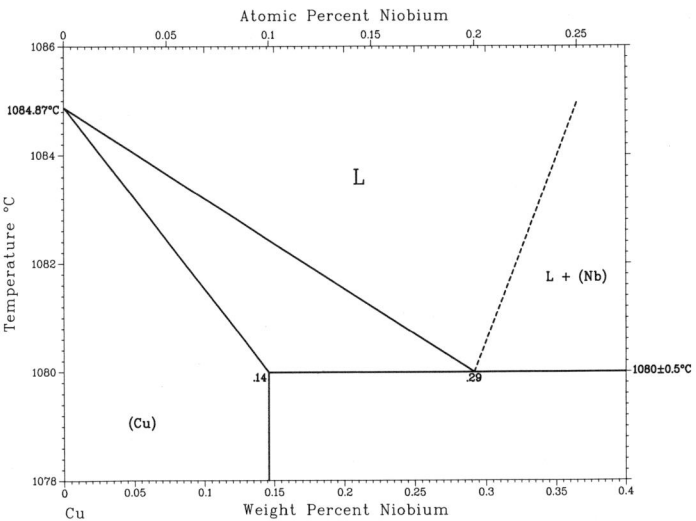

Cu-Nd

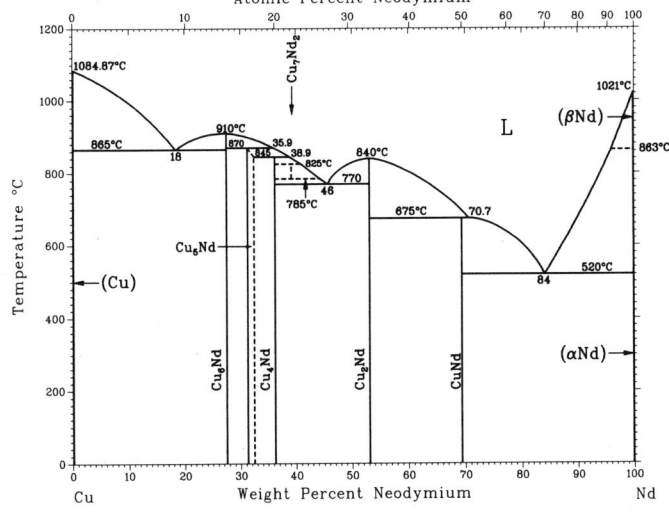

P.R. Subramanian and D.E. Laughlin, 1988

Phase	Composition, wt% Nd	Pearson symbol	Space group
(Cu)	0	cF4	Fm$\overline{3}$m
Cu$_6$Nd	~27.45	oP28	Pnma
Cu$_5$Nd	~31.23	hP6	P6/mmm
Cu$_4$Nd	~36.2	...	Pnnm
Cu$_2$Nd	~53.1	oI12	Imma
CuNd	~69	oP8	Pnma
(βNd)	100	cI2	Im$\overline{3}$m
(αNd)	100	hP4	P6$_3$/mmc

Cu-Ni

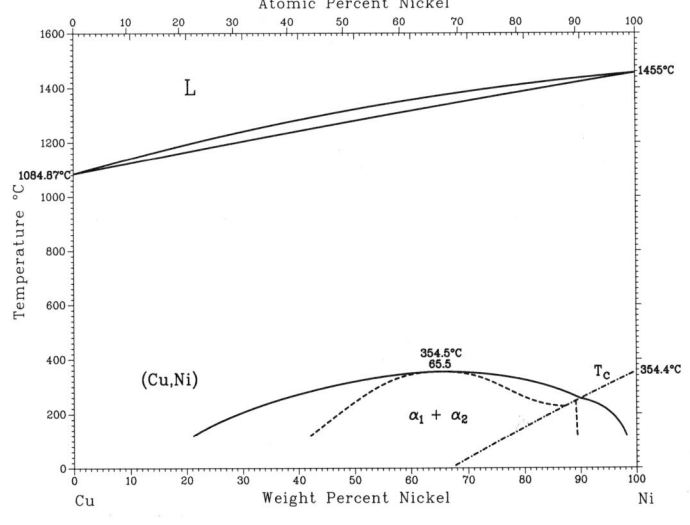

D.J. Chakrabarti, D.E. Laughlin, S.W. Chen, and Y.A. Chang, 1991

Phase	Composition, wt% Ni	Pearson symbol	Space group
(Cu,Ni)	0 to 100(a)	cF4	Fm$\overline{3}$m

(a) Above 354.5 °C

Cu-O

J.P. Neumann, T. Zhong, and Y.A. Chang, 1984

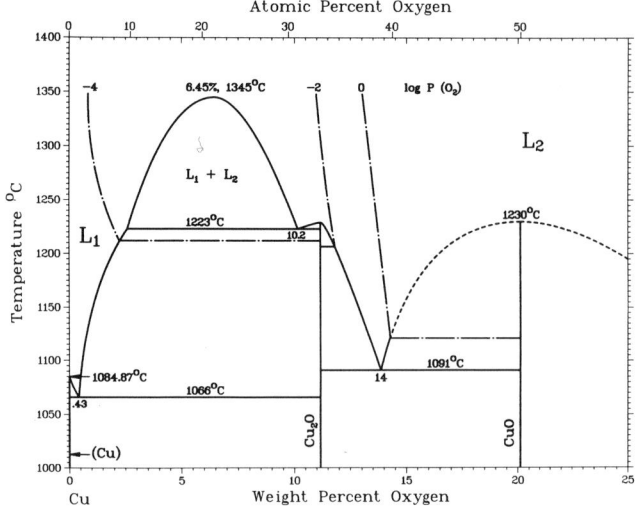

Phase	Composition, wt% O	Pearson symbol	Space group
(Cu)	0 to 0.008	cF4	$Fm\bar{3}m$
Cu₂O(a)	11.2	cP6	$Pn\bar{3}m$
CuO(b)	20	mC8	...
Cu₄O₃(c)	15.9	tI28	$I4/mcm$

(a) κ or cuprite. (b) τ or tenorite. (c) Additional possible phase, π or paramelaconite

Cu-O stability diagram

Solubility of O in (Cu)

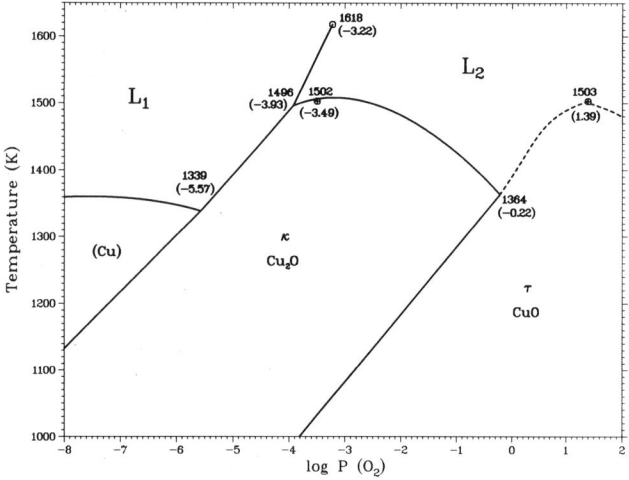

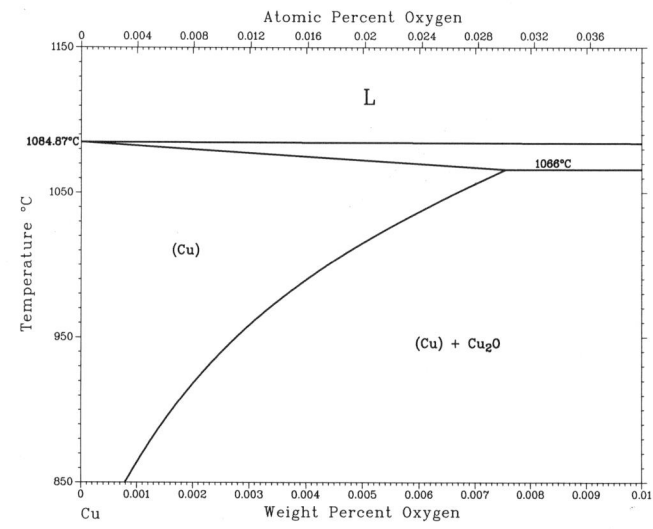

Cu-P

H. Okamoto, 1990

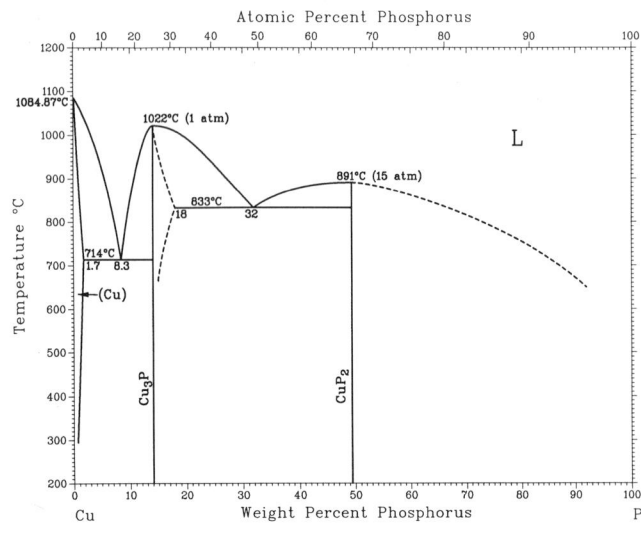

Phase	Composition, wt% P	Pearson symbol	Space group
(Cu)	0 to 1.7	cF4	$Fm\bar{3}m$
Cu₃P	14 to 18	hP24	$P6_3cm$
CuP₂	49.4	mP12	$P2_1/c$
Cu₂P₇(a)	63.1	mC72	$C2/m$

(a) Not shown in the diagram

Cu-Pb

D.J. Chakrabarti and D.E. Laughlin, 1984

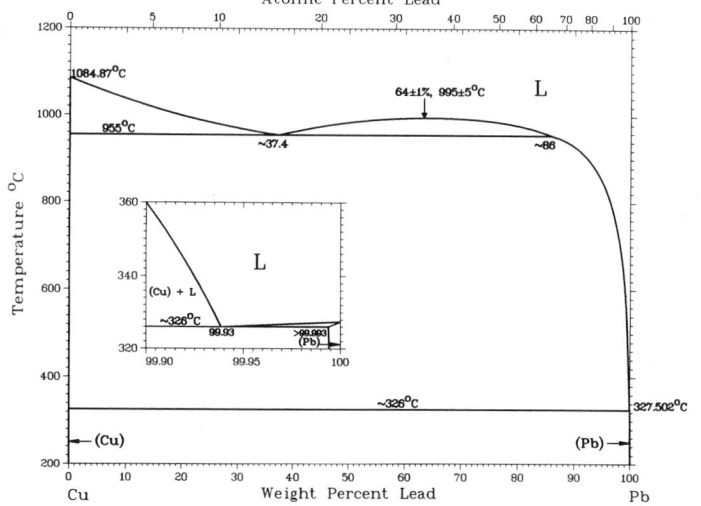

Phase	Composition, wt% Pb	Pearson symbol	Space group
(Cu)	0(a)	cF4	$Fm\bar{3}m$
(αPb)	100	cF4	$Fm\bar{3}m$
(βPb)(b)	100	hP2	$P6_3/mmc$

(a) Metastable solid solubility may extend up to 10.0 to 12.0 wt% Pb. (b) Above 10.3 GPa

Cu-Pd

P.R. Subramanian and D.E. Laughlin, 1991

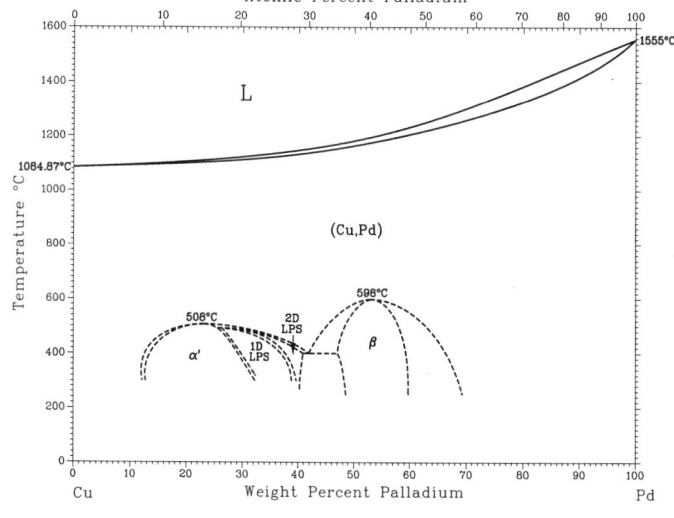

Phase	Composition, wt% Pd	Pearson symbol	Space group
(Cu,Pd)	0 to 100	cF4	$Fm\bar{3}m$
Cu₃Pd (α′)	~12.1 to ~32	cP4	$Pm\bar{3}m$
Cu₃Pd (α″)			
1D-LPS	~26 to ~39	tP28	P4mm
2D-LPS	~28 to ~43	...	...
CuPd (β)	~49 to ~60	cP2	$Pm\bar{3}m$

Cu-Pt

P.R. Subramanian and D.E. Laughlin, unpublished

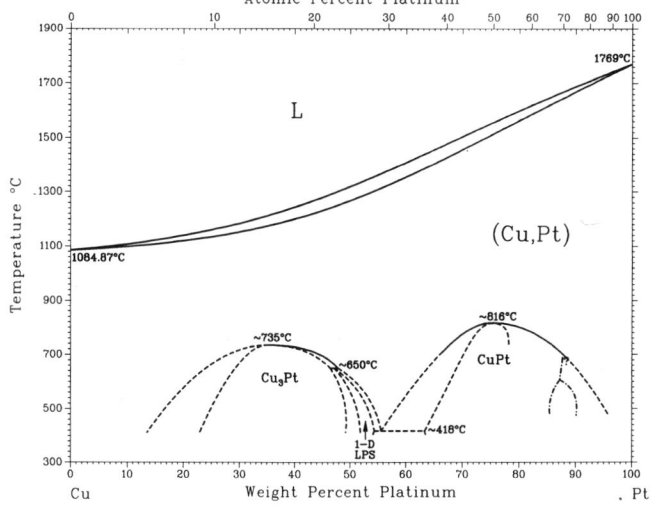

Phase	Composition, wt% Pt	Pearson symbol	Space group
(Cu,Pt)	0 to 100	cF4	$Fm\bar{3}m$
Cu₃Pt	~16 to ~52	cP4	$Pm\bar{3}m$
1D-LPS	~43 to ~56	tP28	P4mm
CuPt	~63 to ~81	hR32	$R\bar{3}m$
Cu₃Pt₅	~85 to ~88	hR*	?
CuPt₃	~87 to ~90	o**	?
CuPt₃	~88 to ?	c**	?

Cu-Pu

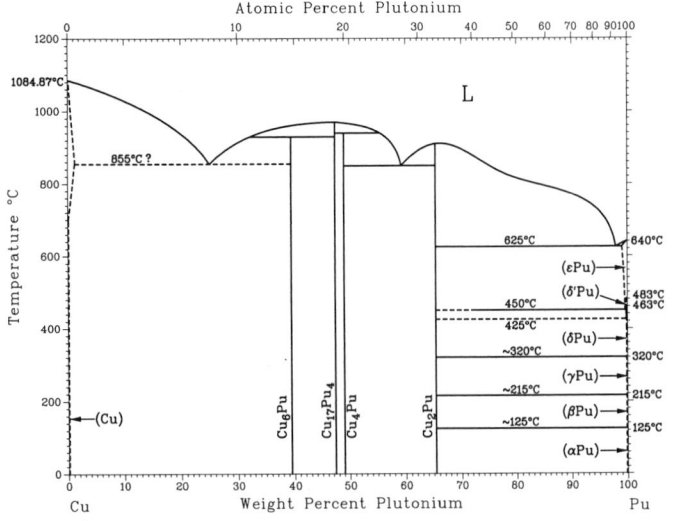

V.I. Kutaitsev, N.T. Chebotarev, I.G. Lebedev, M.A. Andrianov, V.N. Konev, and T.S. Menshikova, 1967

Phase	Composition, wt% Pu	Pearson symbol	Space group
(Cu)	0	cF4	$Fm\bar{3}m$
Cu₆Pu	39.1	...	...
Cu₁₇Pu₄	47.4	...	...
Cu₄Pu	49	o*20	...
Cu₂Pu	65.7	oI12	Imma
(εPu)	? to 100	cI2	$Im\bar{3}m$
(δ'Pu)	100	tI2	I4/mmm
(δPu)	100	cF4	$Fm\bar{3}m$
(γPu)	100	oF8	Fddd
(βPu)	100	mC34	C2/m
(αPu)	100	mP16	P2₁/m

Cu-Rh

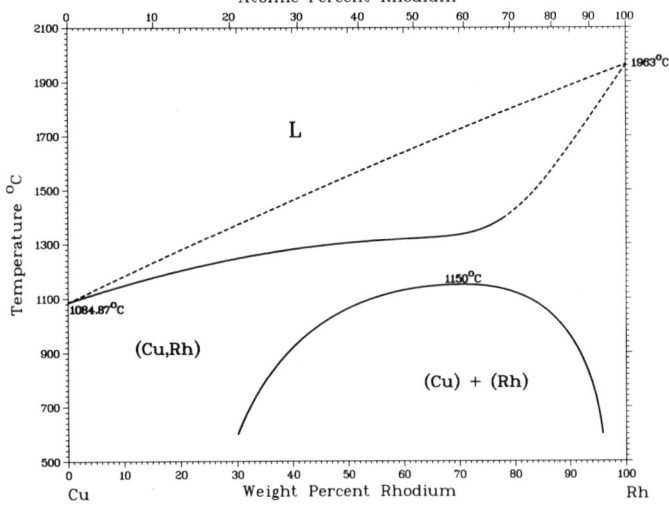

D.J. Chakrabarti and D.E. Laughlin, 1982

Phase	Composition, wt% Rh	Pearson symbol	Space group
(Cu,Rh)	0 to 100	cF4	$Fm\bar{3}m$

Cu-S

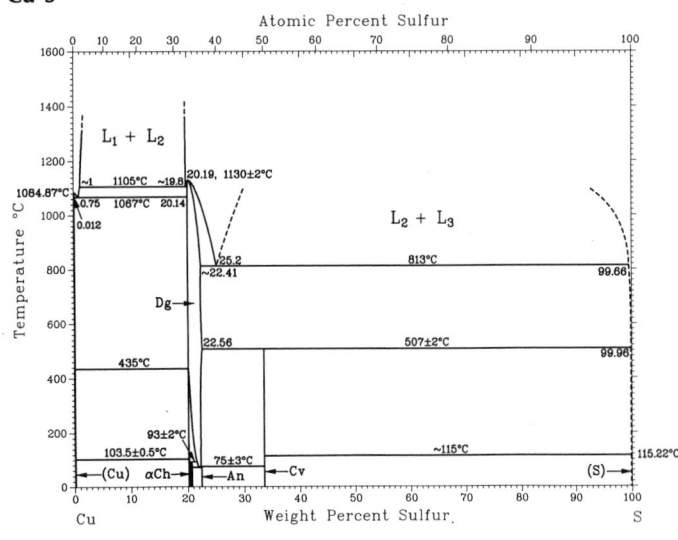

Enlargement of the Cu-S diagram from 0 to 160 °C

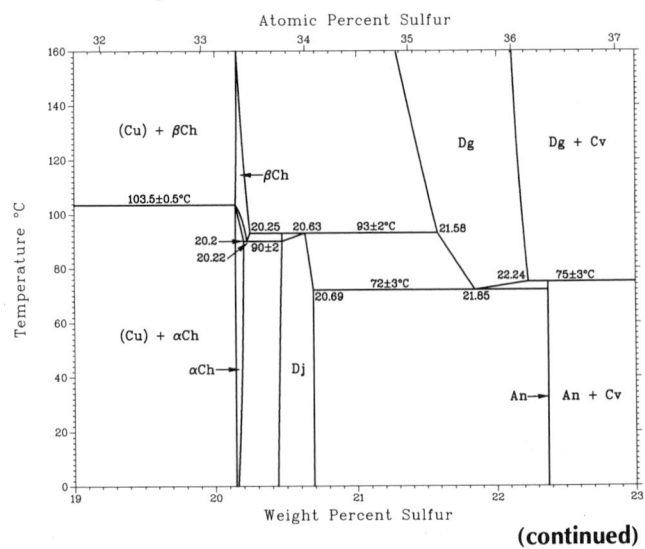

(continued)

Cu-saturated boundary of digenite

D.J. Chakrabarti and D.E. Laughlin, 1983

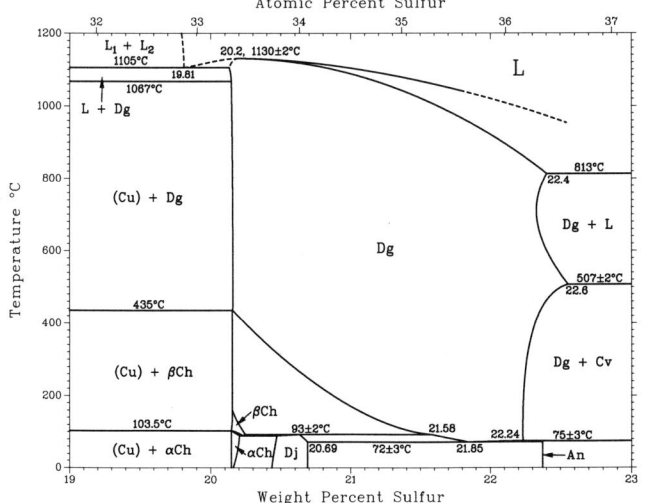

Phase	Composition, wt% S(Cu/S)	Pearson symbol	Space group
(Cu)	0 to 0.012	cF4	Fm$\bar{3}$m
α chalcocite (αCu$_2$S)	20.14 to 20.01	mP144(?)	P2$_1$/c
β chalcocite (βCu$_2$S)	20.14 to 20.22	hP6	P6$_3$/mmc
Djurleite (Cu$_{\sim 1.96}$S)	20.4 to 20.69	oP380(?)	Pmnm
			P2$_1$nm(?)
			Pmn2$_1$
Digenite (Cu$_{2-\delta}$S)	20.14 to 22.24	cF12	Fm$\bar{3}$m
Anilite(Cu$_{1.75}$S)	22.38 ± 0.03	oP44(?)	Pnma
Covellite (CuS)	33.5	hP12	P6$_3$/mmc
(S)	~100	oF128	Fddd
		mP48	P2$_1$/a
		hR6	R$\bar{3}$
Metastable phases			
Protodjurleite	20.4 (1.00)(a)	...	...
	20.5 (0.999)(b)		
Tetragonal	20.5 (0.999)	tP12	P4$_3$2$_1$2
Hexagonal-tetragonal Cu$_x$S	20.7 to 22.4 (0.98 to 0.89)	...	...
Low digenite (αDg)	21.99 to 22.22 (0.911 to 0.899)(c)	...	R$\bar{3}$m
Blaubleibender covellite I	26.5 ± 1.4 (0.71 ± 0.5)	...	...
Blaubleibender covellite II	31.6 ± 1.95 (0.6 ± 0.1)	...	...
CuS$_2$	50.23 (0.3)	...	Pa3(?)

(a) At 75 °C. (b) At 93 °C. (c) At 25 °C.

Cu-Sb

P.R. Subramanian, 1990

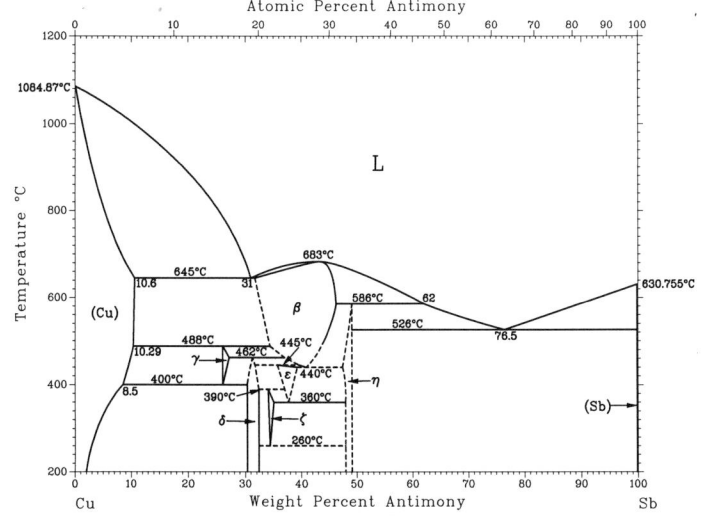

Phase	Composition, wt% Sb	Pearson symbol	Space group
(Cu)	0 to 10.6	cF4	Fm$\bar{3}$m
β	31.6 to 46.0	oF16	Fm$\bar{3}$m
γ	~26.0 to 26.7	hP2	P6$_3$/mmc
δ	30.3 to 32	hP?	P6$_3$/mmc
ε	~36.1 to 39.4	oP8	Pmmn
ζ	~34.1 to 34.5	hP26	P$\bar{3}$
η	~47.4 to 48.9	tP6	P4/nmm
(Sb)	~100	hR2	R$\bar{3}$m

Cu-Se

D.J. Chakrabarti and D.E. Laughlin, 1981

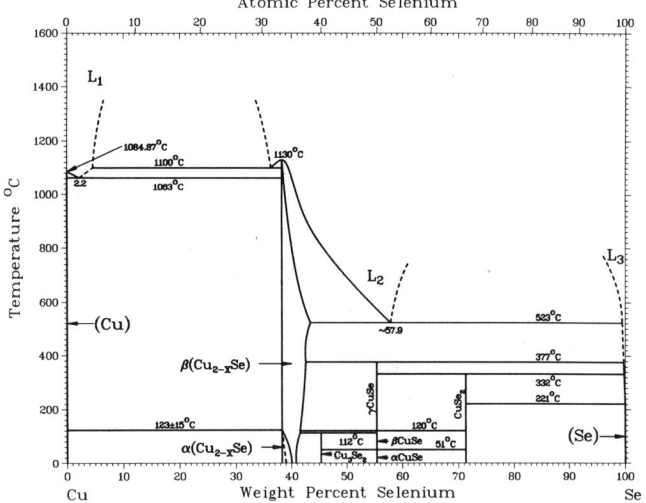

Phase	Composition, wt% Se	Pearson symbol	Space group
(Cu)	~0	cF4	Fm$\bar{3}$m
αCu$_{2-x}$Se	~38.3 to 38.8	(a)	...
βCu$_{2-x}$Se	~38.3 to 41.6(b)	cF12	Fm$\bar{3}$m
Cu$_3$Se$_2$	45	...	P$\bar{4}$2$_1$m
αCuSe	55.4	...	P6$_3$/mmc
βCuSe	55.4	...	...
γCuSe	55.4	...	P6$_3$/mmc
CuSe$_2$	71.3	oP6	Pnnm
(Se)	~100	hP3	P3$_1$21

(a) Monoclinic. (b) Homogeneity range at room temperature, $0.18 \leq x \leq 0.22$, and at 500 °C, $x = 0$ to ~0.26

Cu-Si

R.W. Olesinski and G.J. Abbaschian, 1986

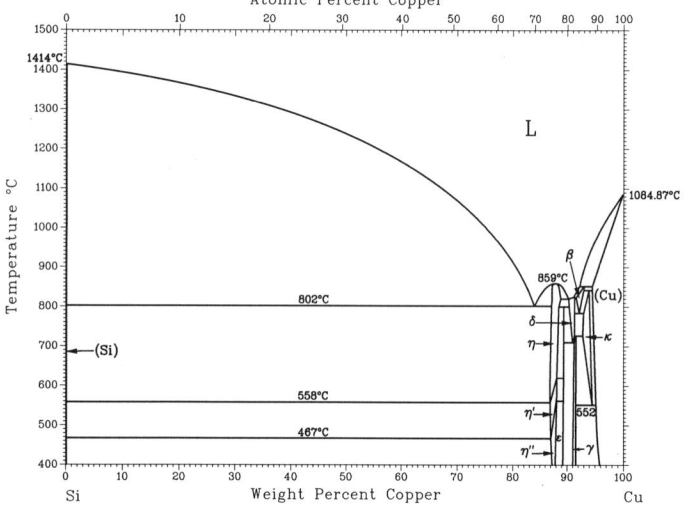

Phase	Composition, wt% Cu	Pearson symbol	Space group
(Si)	0	cF8	Fd$\bar{3}$m
SiII (HP)	0	tI4	I4$_1$/amd
η″(a)	87.2 to 88.16	(b)	...
η′(a)	87.0 to 88.22	(c)	R$\bar{3}$
η(a)	87.2 to 88.8	(c)	R$\bar{3}$m
ε(d)	89.3 to 89.4	(e)	...
δ	90.3 to 91.4	(f)	...
γ(g)	91.4 to 91.62	cP20	P4$_1$32
β	91.6 to 93.2	cI2	Im$\bar{3}$m
κ(h)	93.0 to 94.80	hP2	P6$_3$/mmc
(Cu)	94.6 to ~100	cF4	Fm$\bar{3}$m
Other reported phases			
η‴(j)	...	(f)	...
Metastable	...	(f)	...

(a) Also denoted Cu$_3$Si. (b) Orthorhombic. (c) Rhombohedral. (d) Also denoted Cu$_{15}$Si$_4$. (e) Cubic. (f) Tetragonal. (g) Also denoted Cu$_5$Si. (h) Also denoted Cu$_7$Si. (j) Originally denoted η′

Cu-Sn

N. Saunders and A.P. Miodownik, 1990

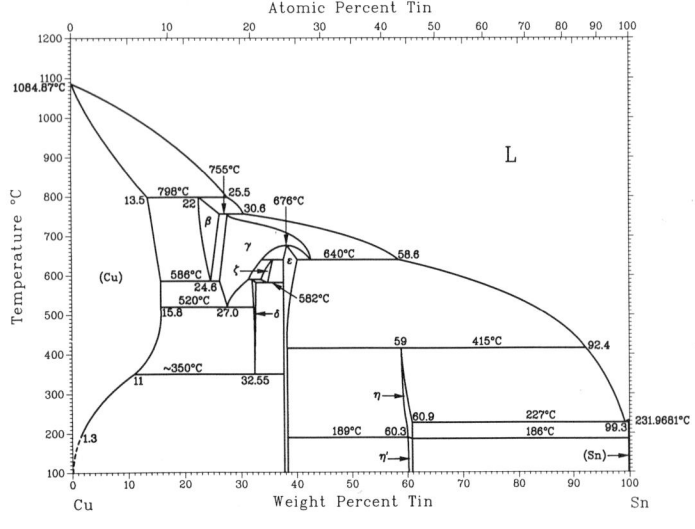

Phase	Composition, wt% Sn	Pearson symbol	Space group
α	0 to 15.8	cF4	Fm$\bar{3}$m
β	22.0 to 27.0	cI2	Im$\bar{3}$m
γ	25.5 to 41.5	cF16	Fm$\bar{3}$m
δ	32 to 33	cF416	F$\bar{4}$3m
ζ	32.2 to 35.2	hP26	P6$_3$
ε	27.7 to 39.5	oC80	Cmcm
η	59.0 to 60.9	hP4	P6$_3$/mmc
η′	44.8 to 60.9	(a)	...
(βSn)	~100	tI4	I4$_1$/amd
(αSn)	100	cF8	Fd$\bar{3}$m

Note: Lattice parameter data can be found in [Pearson3]. (a) Hexagonal; superlattice based on NiAs-type structure

Cu-Sr

D.J. Chakrabarti and D.E. Laughlin, 1984

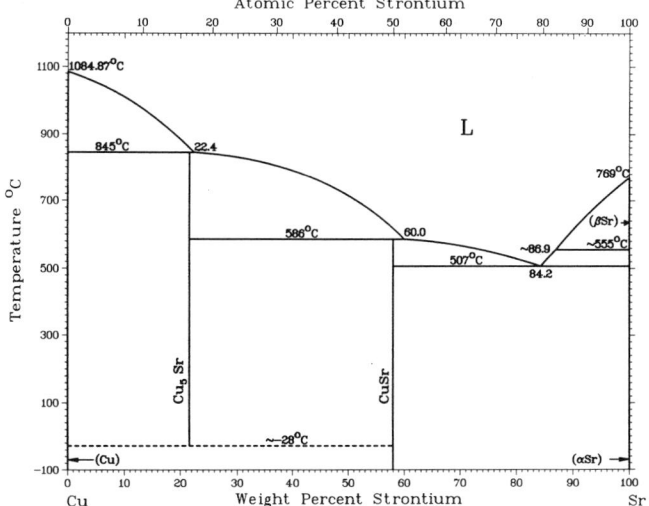

Phase	Composition, wt% Sr	Pearson symbol	Space group
(Cu)	0	cF4	Fm$\bar{3}$m
Cu$_5$Sr	21.62	hP6	P6/mmm
CuSr	58.0	hP8(?)	P6$_3$/mmc
(βSr)	100	cI2	Im$\bar{3}$m
(αSr)	100	cF4	Fm$\bar{3}$m
Pressure-stabilized form			
βSr or Sr-II	100	cI2	Im$\bar{3}$m

Cu-Te

H. Okamoto, unpublished

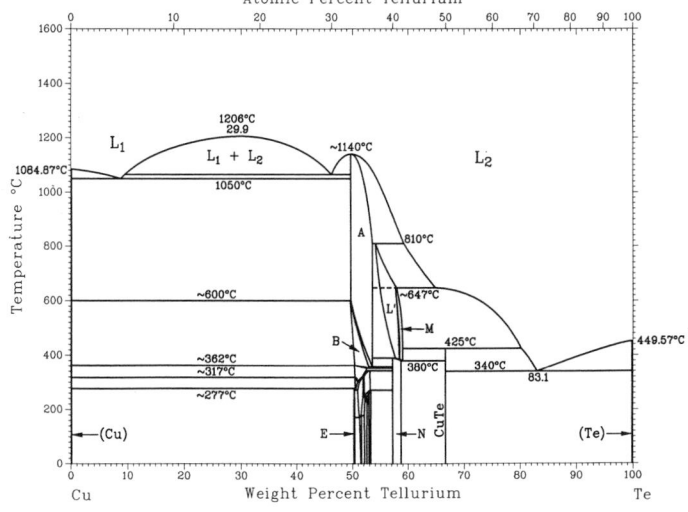

Phase	Composition, wt% Te	Pearson symbol	Space group
(Cu)	0	cF2	Fm$\bar{3}$m
Cu$_2$Te group			
A	50 to 53.6	cF12	Fd$\bar{3}$m
B	50 to 52.99	hP6	P6/mmm
C	50.4 to 52.5	hP*	…
D	50.46 to 51.1	o**	…
E	50.3 to 50.46	o**	…
F	51.0 to 52	o**	…
G	51.3 to 51.6	o**	…
H	52.12 to 53.1	hP72	P3m1
I	52.23 to 52.88	…	…
J	52.23 to 52.6	hP*	…
K	52.9 to 53.3	hP22	P3m1
L	54 to 58	tP6	P4/nmm
L'	55 to 58	…	…
M	58 to 59	…	…
N	57 to 58.8	…	…
CuTe	67	oP4	Pmmn
(Te)	100	hP3	P3$_1$21
High-pressure phase			
CuTe$_2$	50.1	cP12	Pa3

Details of the Cu-Te phase diagram from 49.7 to 53.6 wt% Te

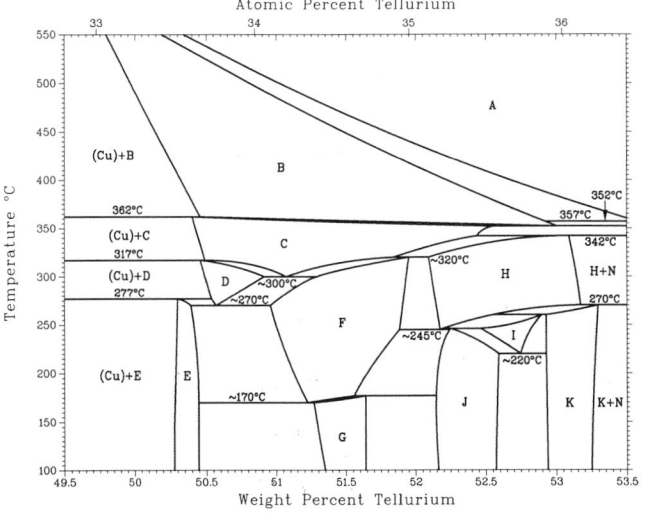

Cu-Th

D.J. Chakrabarti, D.E. Laughlin, and D.E. Peterson, 1986

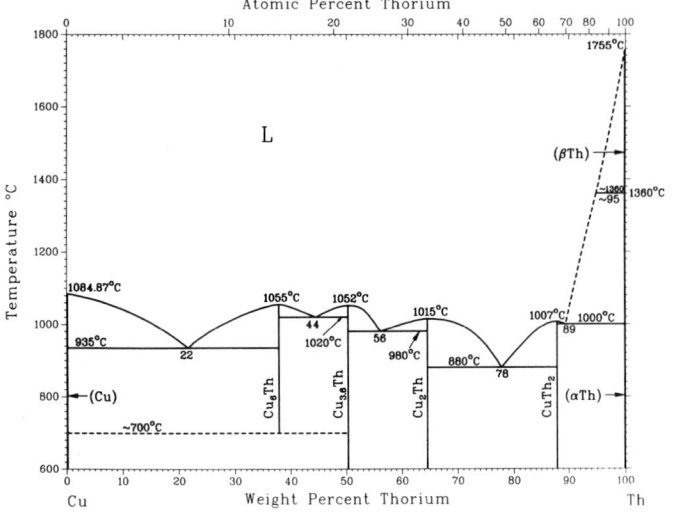

Phase	Composition, wt% Th	Pearson symbol	Space group
(Cu)	0	cF4	Fm$\bar{3}$m
Cu$_6$Th	37.84	oP28?	Pnma
Cu$_{3.6}$Th	50.36	(a)	P6/m
Cu$_2$Th	64.61	hP3	P6/mmm
CuTh(b)	79	oC8	Cmcm
CuTh$_2$	87.96	tI12	I4/mcm
(βTh)	100	cI2	Im$\bar{3}$m
(αTh)	100	cF4	Fm$\bar{3}$m

(a) Hexagonal. (b) Metastable

Cu-Ti

J.L. Murray, 1987

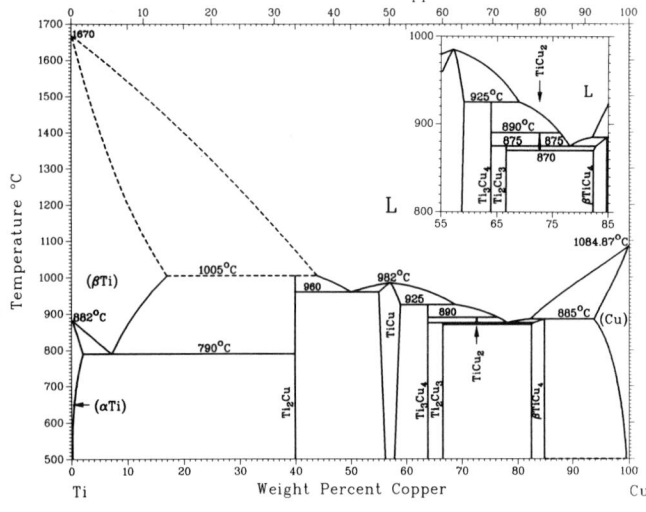

Phase	Composition, wt% Cu	Pearson symbol	Space group
(αTi)	0 to 2.1	hP2	P6$_3$/mmc
(βTi)	0 to 17.2	cI2	Im$\bar{3}$m
Ti$_2$Cu	39.9	tI6	I4/mmm
TiCu	55 to 59	tP4	P4/nmm
Ti$_3$Cu$_4$	63.9	tI14	I4/mmm
Ti$_2$Cu$_3$	67	tP10	P4/nmm
TiCu$_2$	72.7	oC12	Amm2
TiCu$_4$	83 to 84.9	oP20	Pnma
αTiCu$_4$	~83 to 84.9	tI10	I4/m
(Cu)	94 to 100	cF4	Fm$\bar{3}$m
Metastable phases			
TiCu$_3$	...	oP8	Pmnm
β''	...	tP2	P4/mmm

Transformation of βTiCu$_4$ ↔ αTiCu$_4$

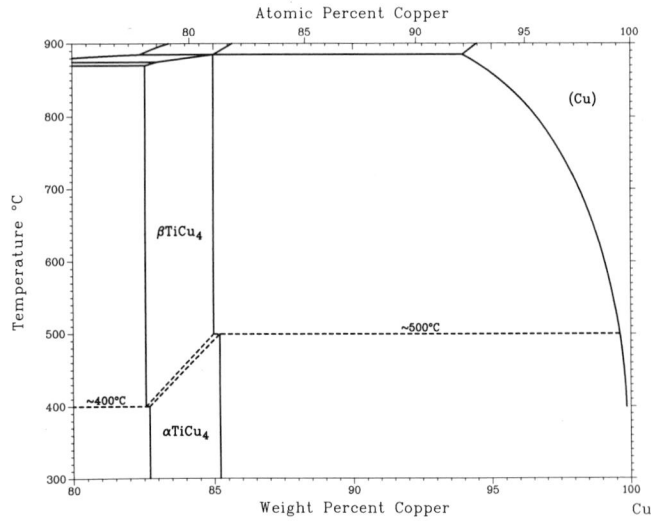

Cu-Tl

D.J. Chakrabarti and D.E. Laughlin, 1984

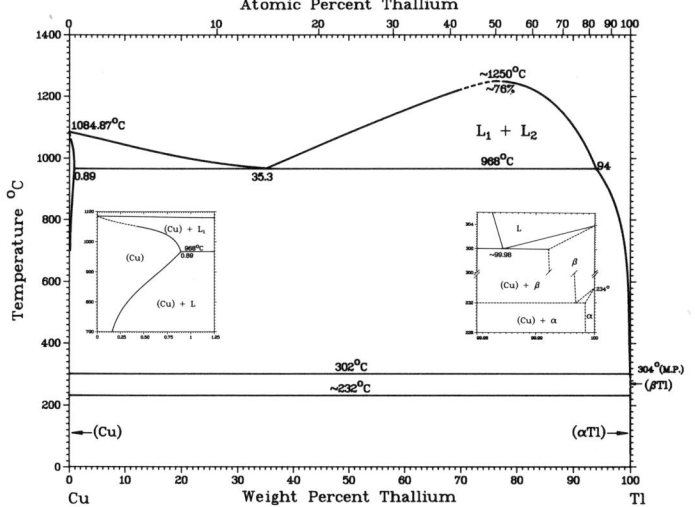

Phase	Composition, wt% Tl	Pearson symbol	Space group
(Cu)	0 to 0.89	cF4	Fm$\overline{3}$m
(αTl)	100	hP2	P6$_3$/mmc
(βTl)	100	cI2	Im$\overline{3}$m
Pressure-stabilized phase			
γTl	100	cF?	…

Cu-V

J.F. Smith and O.N. Carlson, 1989

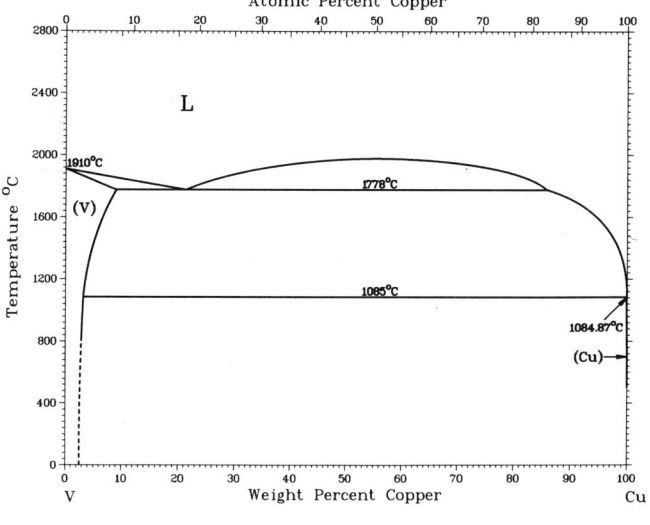

Phase	Composition, wt% Cu	Pearson symbol	Space group
(V)	0 to 9.2	cI2	Im$\overline{3}$m
(Cu)	99.9 to 100	cF4	Fm$\overline{3}$m

Cu-Yb

P.R. Subramanian and D.E. Laughlin, 1988

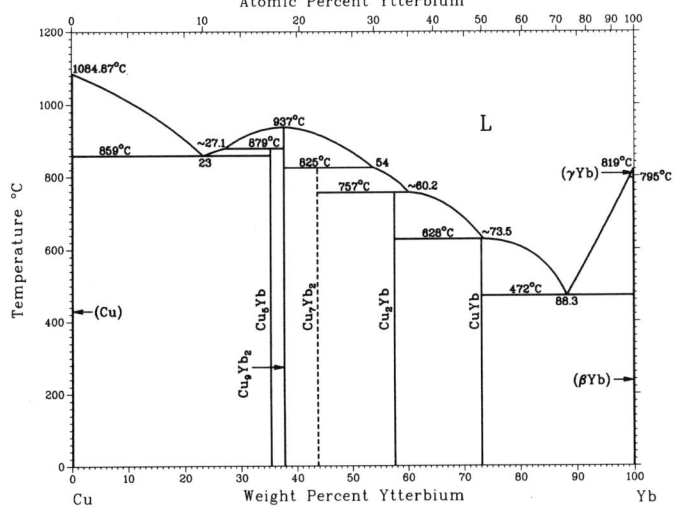

Phase	Composition, wt% Yb	Pearson symbol	Space group
(Cu)	0	cF4	Fm$\overline{3}$m
Cu$_5$Yb	~35.26	hP6	P6/mmm
Cu$_2$Yb	~57.6	oI12	Imma
CuYb	~73.1	oP8	Pnma
(γYb)	100	cI2	Im$\overline{3}$m
(βYb)	~99.99 to 100	cF4	Fm$\overline{3}$m
(αYb)	100	hP2	P6$_3$/mmc

Cu-Zn

A.P. Miodownik, unpublished

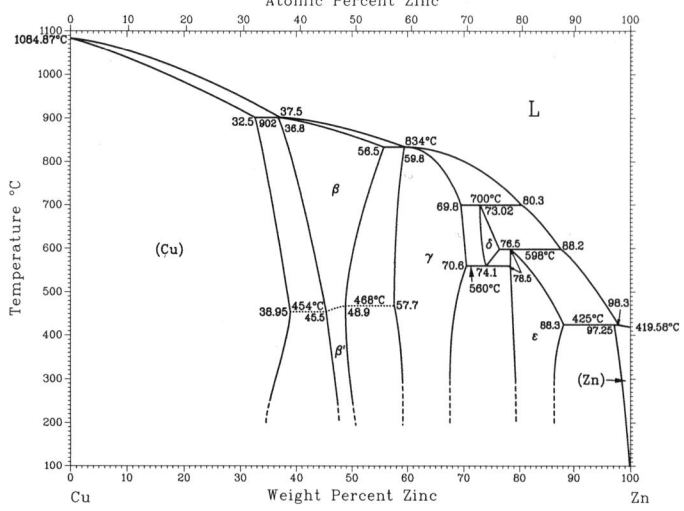

Phase	Composition, wt% Zn	Pearson symbol	Space group
α or (Cu)	0 to 38.95	cF4	$Fm\bar{3}m$
β	36.8 to 56.5	cI2	$Im\bar{3}m$
β′	45.5 to 50.7	cP2	$Pm\bar{3}m$
γ	57.7 to 70.6	cI52	$I\bar{4}3m$
δ	73.02 to 76.5	hP3	$P\bar{6}$
ε	78.5 to 88.3	hP2	$P6_3/mmc$
η or (Zn)	97.25 to 100	hP2	$P6_3/mmc$

Cu-Zr

D. Arias and J.P. Abriata, 1990

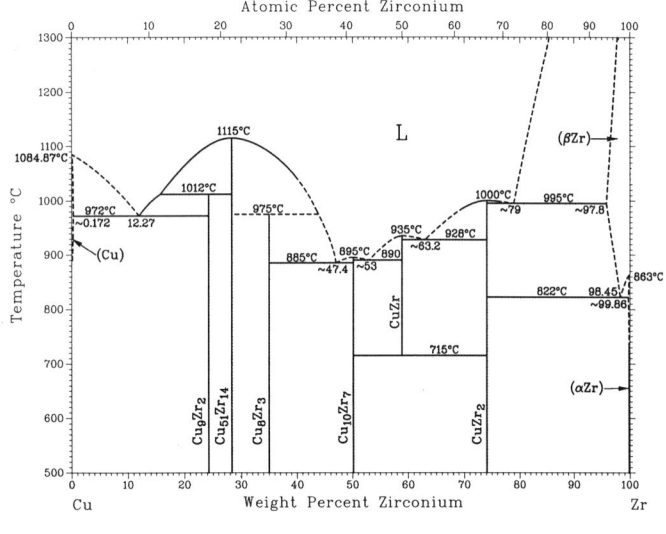

Phase	Composition, wt% Zr	Pearson symbol	Space group
(Cu)	0 to ~0.172	cF4	$Fm\bar{3}m$
Cu₉Zr₂(a)	24.18	tP24	$P4/m$
Cu₅₁Zr₁₄	28.27	hP65	$P6/m$
Cu₈Zr₃	34.99	oP44	$Pnma$
Cu₁₀Zr₇	50.13	oC68	...
CuZr	58.9	cP2	$Pm\bar{3}m$
CuZr₂	74.17	tI6	$I4/mmm$
(βZr)	~97.8 to 100	cI2	$Im\bar{3}m$
(αZr)	~99.86 to 100	hP2	$P6_3/mmc$

(a) Tetragonal long-period superlattice derived from the AuBe₅-type structure

Dy-Fe

H. Okamoto, 1992

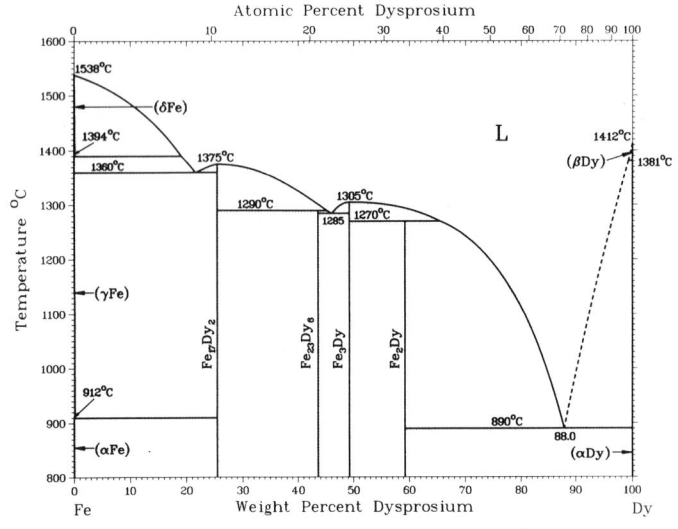

Phase	Composition, wt% Dy	Pearson symbol	Space group
(δFe)	0	cI2	$Im\bar{3}m$
(γFe)	0	cF4	$Fm\bar{3}m$
(αFe)	0	cI2	$Im\bar{3}m$
Fe₁₇Dy₂	25.4	hP38	$P6_3/mmc$
Fe₂₃Dy₆	43.2	cF116	$Fm\bar{3}m$
Fe₃Dy	49	hR12	$R\bar{3}m$
Fe₂Dy	59.2	cF24	$Fd\bar{3}m$
Fe₂Dy′(a)	59.2	t**	...
(βDy)	100	cI2	$Im\bar{3}m$
(αDy)	100	hP2	$P6_3/mmc$

(a) Below −23 °C

Dy-Ga

From [Moffatt]

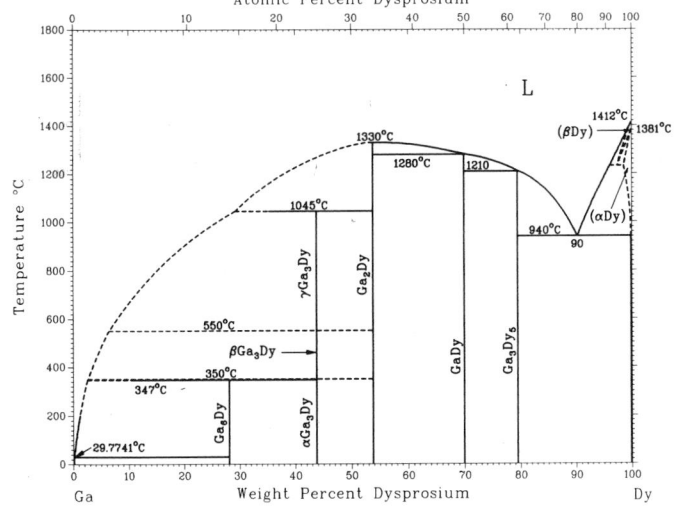

Phase	Composition, wt% Dy	Pearson symbol	Space group
(Ga)	0	oC8	Cmca
Ga₆Dy	28.0	tP14	P4/nbm
γGa₃Dy	44	cP4	Pm$\bar{3}$m
βGa₃Dy	44	hP40	P6₃/mmc
αGa₃Dy	44	hP16	R$\bar{3}$m
Ga₂Dy	53.8	hP3	P6/mmm
GaDy	70.0	oC8	Cmcm
Ga₃Dy₅	79.5	tI32	I4/mcm
(βDy)	100	cI2	Im$\bar{3}$m
(αDy)	100	hP2	P6₃/mmc

Dy-Ge

V.N. Eremenko, V.G. Batalin, Yu.I. Buyanov, and I.M. Obushenko, 1977

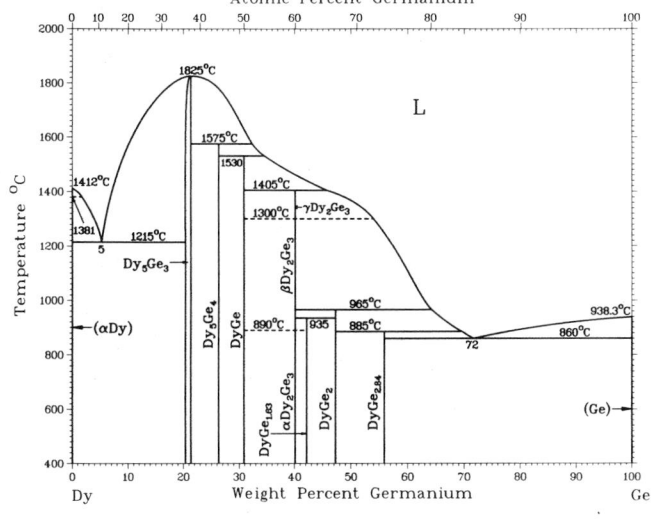

Phase	Composition, wt% Ge	Pearson symbol	Space group
(βDy)	0	cI2	Im$\bar{3}$m
(αDy)	0	hP2	P6₃/mmc
Dy₅Ge₃	~21.4	hP16	P6₃/mcm
Dy₅Ge₄	26.3	oP36	Pnma
DyGe	30.9	oC8	Cmcm
γDy₂Ge₃	40	...	...
βDy₂Ge₃	40	...	...
αDy₂Ge₃	40	hP3	P6/mmm
DyGe₁.₈₃	45.0	...	...
DyGe₂	47.2	tI12	I4₁/amd
		o**(a)	...
DyGe₂.₈₄	56	...	...
(Ge)	100	hR2	R$\bar{3}$m

(a) High-temperature (>750 °C) phase?

Dy-In

H. Okamoto, 1992

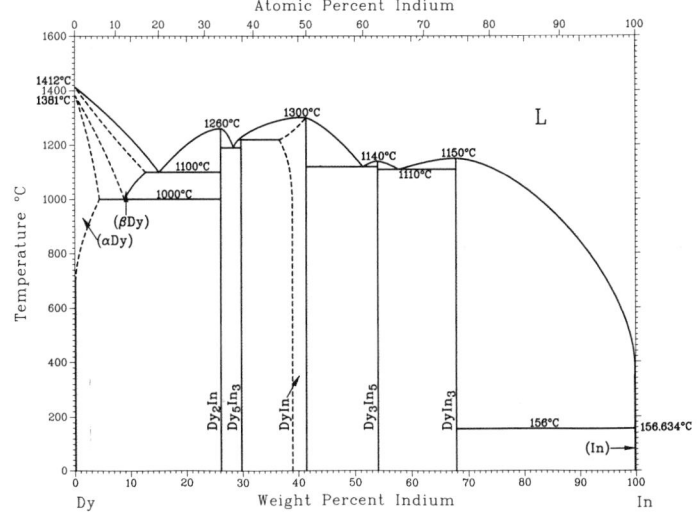

Phase	Composition, wt% In	Pearson symbol	Space group
(βDy)	0 to 13	cI2	Im$\bar{3}$m
(αDy)	0 to 6	hP2	P6₃/mmc
Dy₃In(a)	19	tP4	P4/mmm
Dy₂In	26.1	hP6	P6₃/mmc
Dy₅In₃	29.8	tI32	I4/mcm
DyIn	37 to 41	cP2	Pm$\bar{3}$m
Dy₃In₅	54.1	oC32	Cmcm
DyIn₃	68	cP4	Pm$\bar{3}$m
(In)	100	tI2	I4/mmm

(a) Not accepted in the assessed diagram

Dy-Mn

H.R. Kirchmayr and W. Lugscheider, 1967

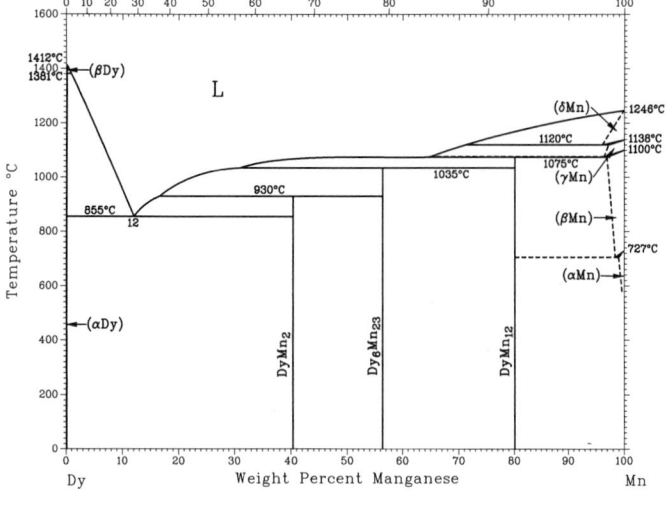

Phase	Composition, wt% Mn	Pearson symbol	Space group
(βDy)	0	cI2	Im$\overline{3}$m
(αDy)	0	hP2	P6$_3$/mmc
DyMn$_2$	40.4	cF24	Fd$\overline{3}$m
Dy$_6$Mn$_{23}$	56.4	cF116	Fm$\overline{3}$m
DyMn$_{12}$	80.2	tI26	I4/mmm
(δMn)	100	cI2	Im$\overline{3}$m
(γMn)	100	cF4	Fm$\overline{3}$m
(βMn)	100	cP20	P4$_1$32
(αMn)	100	cI58	I$\overline{4}$3m

Dy-Ni

Y.Y. Pan and P. Nash, 1991

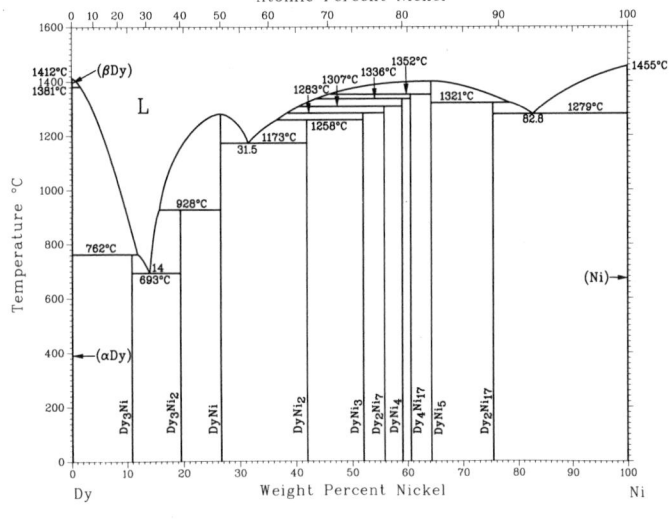

Phase	Composition, wt% Ni	Pearson symbol	Space group
(αDy)	0	cI2	Im$\overline{3}$m
(βDy)	0	hP2	P6$_3$/mmc
(α'Dy)	0	oC4	Cmcm
Dy$_3$Ni	10.7	oP16	Pnma
Dy$_3$Ni$_2$	19.4	mC20	C2/m
DyNi	26.5	oP8	Pbnm
DyNi$_2$	42.0	cF24	Fd$\overline{3}$m
DyNi$_3$	52.0	hR24	R$\overline{3}$m
Dy$_2$Ni$_7$	55.9	hR54	R$\overline{3}$m
		hP36	P6$_3$/mmc
DyNi$_4$	59.1	...	...
Dy$_4$Ni$_{17}$	61	...	...
DyNi$_5$	64.3	hP6	P6$_3$/mmm
Dy$_2$Ni$_{17}$	75.5	hP38	P6$_3$/mmc
(Ni)	100	cF4	Fm$\overline{3}$m

(a) Low-temperature form. (b) High-temperature form

Dy-Pb

O.D. McMasters, T.J. O'Keefe, and K.A. Gschneidner, Jr., 1968

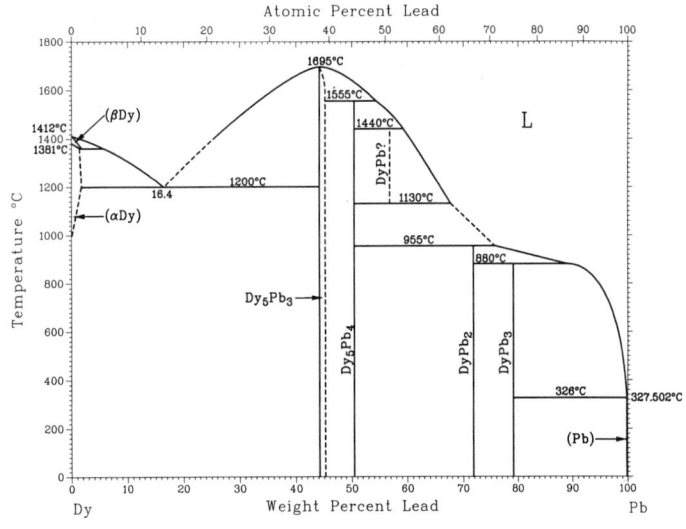

Phase	Composition, wt% Pb	Pearson symbol	Space group
(βDy)	0	cI2	Im$\overline{3}$m
(αDy)	0	hP2	P6$_3$/mmc
Dy$_5$Pb$_3$	43.3	hP16	P6$_3$/mcm
Dy$_5$Pb$_4$	50.5	oP36	Pnma
DyPb	56.0	...	...
DyPb$_2$	71.9	...	...
DyPb$_3$	79	cP4	Pm$\overline{3}$m
(Pb)	100	cF4	Fm$\overline{3}$m

Dy-Pd

H. Okamoto, 1990

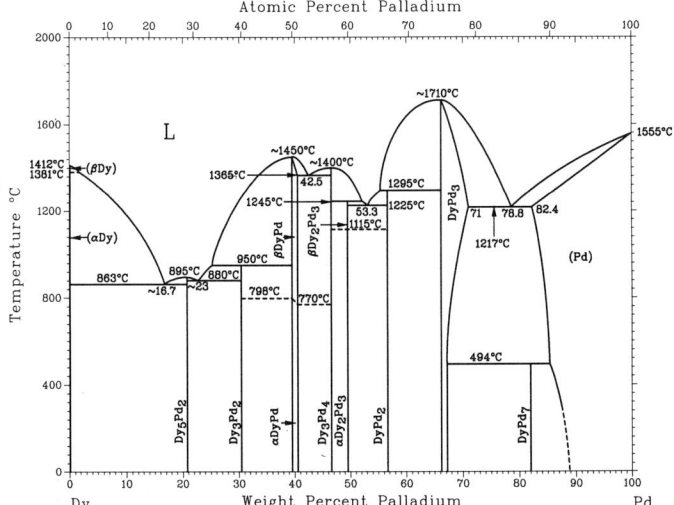

Phase	Composition, wt% Pd	Pearson symbol	Space group
(βDy)	0	cI2	$Im\bar{3}m$
(αDy)	0	hP2	$P6_3/mmc$
Dy$_5$Pd$_2$	20.8	tI49	$I4_1/a$
Dy$_3$Pd$_2$	30	tP10	$P4/mbm$
βDyPd	39.6	cP2	$Pm\bar{3}m$
αDyPd	39.6	oP8	$Pnma$
Dy$_3$Pd$_4$	46.6	hR14	$R\bar{3}$
βDy$_2$Pd$_3$	50	...	...
αDy$_2$Pd$_3$	50	...	...
DyPd$_2$	56.7	...	...
DyPd$_3$	66 to 71	cP4	$Pm\bar{3}m$
DyPd$_7$	82	c**	...
(Pd)	82.4 to 100	cF4	$Fm\bar{3}m$

Dy-S

H. Okamoto, 1990

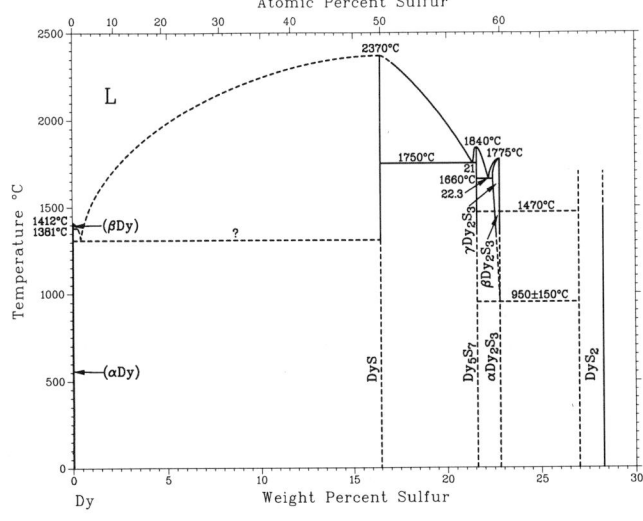

Phase	Composition, wt% S	Pearson symbol	Space group
(βDy)	0	cI2	$Im\bar{3}m$
(αDy)	0	hP2	$P6_3/mmc$
DyS	17	cF8	$Fm\bar{3}m$
Dy$_5$S$_7$	21.6	mC24	$C2/m$
γDy$_2$S$_3$	23	cI28	$I\bar{4}3d$
βDy$_2$S$_3$	23	oP20	$Pnma$
αDy$_2$S$_3$	23	m**	...
DyS$_2$	27.2 to 28.3	cF24	$Fd\bar{3}m$
(S)	100	oF128	$Fddd$

Dy-Sb

H. Okamoto, 1990

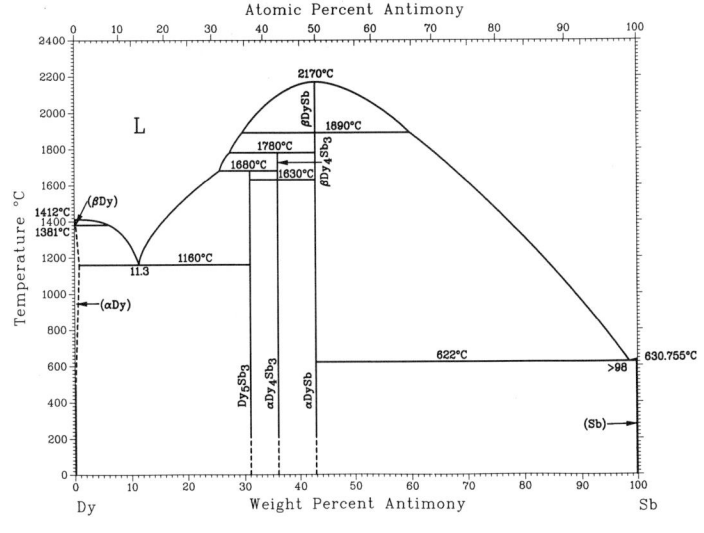

Phase	Composition, wt% Sb	Pearson symbol	Space group
(βDy)	0	cI2	$Im\bar{3}m$
(αDy)	0	hP2	$P6_3/mmc$
Dy$_5$Sb$_3$	31.0	hP16	$P6_3/mcm$
βDy$_4$Sb$_3$	36.0	...	...
αDy$_4$Sb$_3$	36.0	cI28	$I\bar{4}3d$
βDySb	42.8	...	...
αDySb	42.8	cF8	$Fm\bar{3}m$
α'DySb(a)	42.8	tI4	$I4/mmm$
(Sb)	100	hR2	$R\bar{3}m$
High-pressure phase			
DySb$_2$	60.1	o*6	...

(a) Below 11 K

Dy-Sn

H. Okamoto, 1990

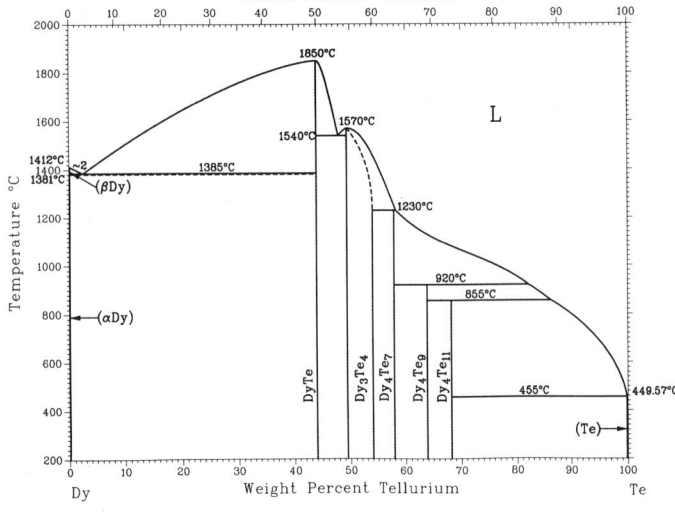

Phase	Composition, wt% Sn	Pearson symbol	Space group
(βDy)	0	cI2	Im$\overline{3}$m
(αDy)	0	hP2	P6₃/mmc
Dy₂Sn	26.7	...	...
Dy₅Sn₃	30.5	hP16	P6₃/mcm
Dy₅Sn₄	36.8	oP36	Pnma
Dy₁₁Sn₁₀	39.9	tI84	I4/mmm
DySn	42.2	...	...
Dy₈Sn₉	45.1	...	...
Dy₄Sn₇	56.1	...	...
DySn₂	59.4	oC12	Cmcm
DySn₄	75	...	...
(βSn)	100	tI4	I4₁/amd
(αSn)	100	cF8	Fd$\overline{3}$m
High-pressure phase			
DySn₃	69	cP4	Pm$\overline{3}$m

Dy-Te

H. Okamoto, 1990

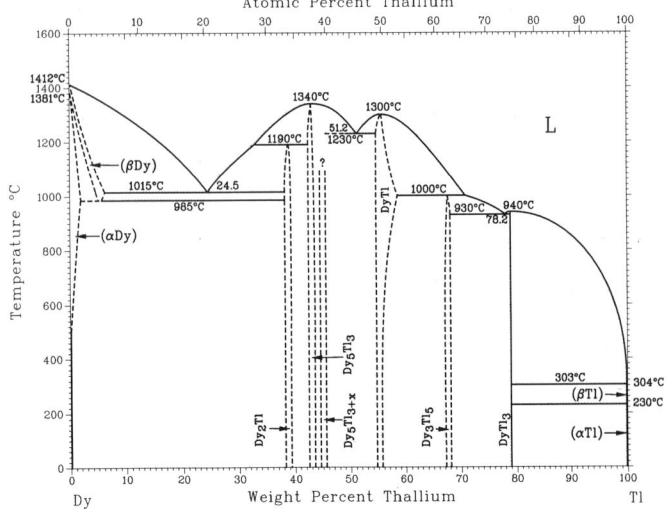

Phase	Composition, wt% Te	Pearson symbol	Space group
(βDy)	0	cI2	Im$\overline{3}$m
(αDy)	0	hP2	P6₃/mmc
DyTe	44.0	cF8	Fm$\overline{3}$m
Dy₃Te₄	51.1 to 54	oF80	Fddd
Dy₄Te₇	57.8	tP6	P4/nmm
Dy₄Te₉	63.8	...	...
Dy₄Te₁₁	68.3	...	...
(Te)	100	hP3	P3₁21
Other phases			
Dy₂Te₅	66.2	oC28	Cmcm
DyTe₃	70	oC16	Cmcm
		tP16	P4₂/n

Dy-Tl

S. Delfino, A. Saccone, A. Palenzona, and R. Ferro, unpublished

Phase	Composition, wt% Tl	Pearson symbol	Space group
(βDy)	0 to ~6	cI2	Im$\overline{3}$m
(αDy)	0 to ?	hP2	P6₃/mmc
Dy₂Tl	~38 to ~39	hP6	P6₃/mmc
Dy₅Tl₃	~43 to ~44	hP16	P6₃/mcm
Dy₅Tl₃₊ₓ	?	tI32	I4/mcm
DyTl(a)	55 to ~59	cP2	Pm$\overline{3}$m
		(or cI2)	Im$\overline{3}$m
DyTl(b)	~55 to ~59	tP2	P4/mmm
Dy₃Tl₅	~67 to ~68	oC32	Cmcm
DyTl₃	79	cP4	Pm$\overline{3}$m
(βTl)	100	cI2	Im$\overline{3}$m
(αTl)	100	hP2	P6₃/mmc

(a) Cubic structure presumed to be room- and higher temperature phases. (b) Tetragonal structure presumed to be lower temperature phase

Dy-Zr

J. Croni, C.E. Armantrout, and H. Kato, 1960

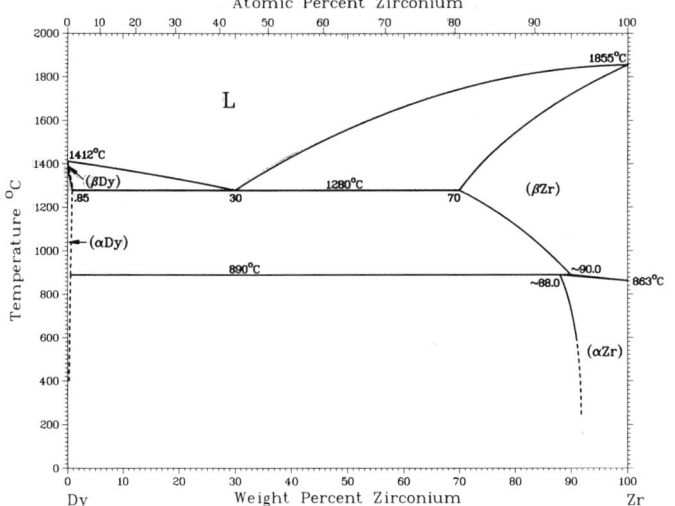

Phase	Composition, wt% Zr	Pearson symbol	Space group
(βDy)	0 to ?	cI2	Im$\bar{3}$m
(αDy)	0 to 0.85	hP2	P6₃/mmc
(βZr)	70 to 100	cI2	Im$\bar{3}$m
(αZr)	~88 to 100	hP2	P6₃/mmc

Er-Fe

H. Okamoto, 1992

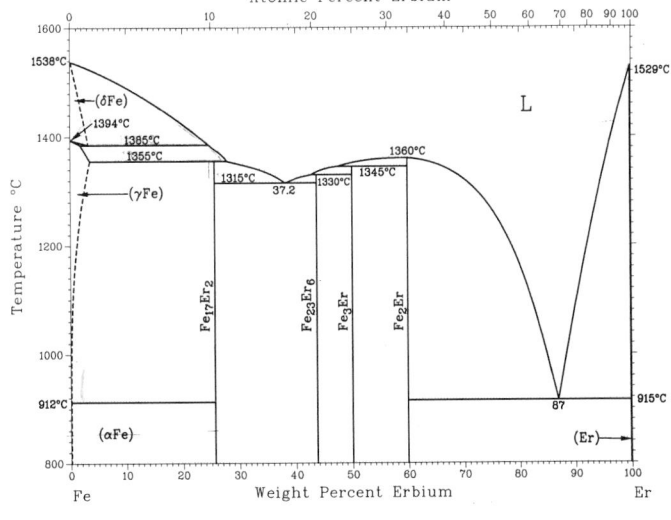

Phase	Composition, wt% Er	Pearson symbol	Space group
(δFe)	0	cI2	Im$\bar{3}$m
(γFe)	0	cF4	Fm$\bar{3}$m
(αFe)	0	cI2	Im$\bar{3}$m
Fe₁₇Er₂	26.0	hP38	P6₃/mmc
Fe₂₃Er₆	43.9	cF116	Fm$\bar{3}$m
Fe₃Er	50	hR12	R$\bar{3}$m
Fe₂Er	59.9	cF24	Fd$\bar{3}$m
(Er)	100	hP2	P6₃/mmc
Metastable phase			
	~75.0	hP12	P6₃/mmc

Er-Ga

H. Okamoto, 1990

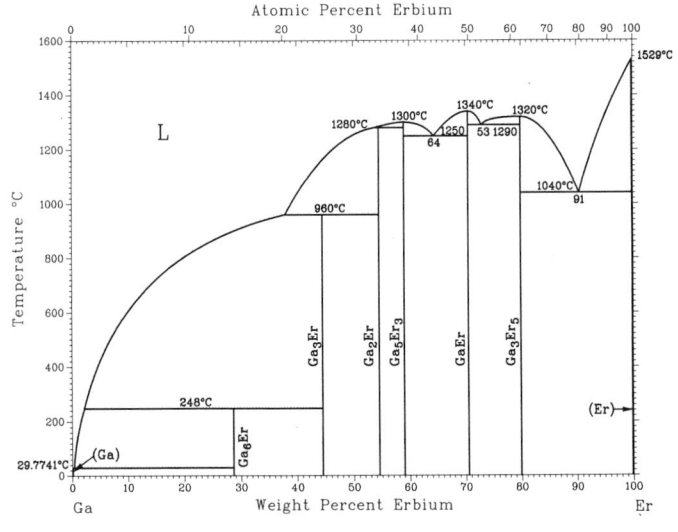

Phase	Composition, wt% Er	Pearson symbol	Space group
(Ga)	0	oC8	Cmca
Ga₆Er	28.6	tP14	P4/nbm
Ga₃Er	44	cP4	Pm$\bar{3}$m
Ga₂Er	54.5	hP3	P6/mmm
Ga₅Er₃	59.0	oP32	Pnma
GaEr	70.6	oC8	Cmcm
Ga₃Er₅	80.0	hP16	P6₃/mcm
(Er)	100	hP2	P6₃/mmc

Er-Ge

H. Okamoto, 1990

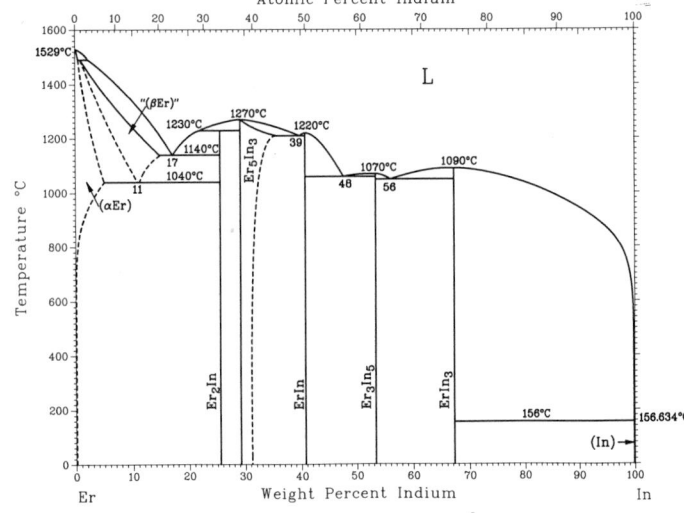

Phase	Composition, wt% Ge	Pearson symbol	Space group
(Er)	0	hP2	P6₃/mmc
Er₅Ge₃	~20.7	hP16	P6₃/mcm
Er₅Ge₄	25.7	oP36	Pnma
Er₁₁Ge₁₀	28.3	tI84	I4/mmm
ErGe	30.3	oC8	Cmcm
Er₄Ge₅	35.2	...	...
βEr₂Ge₃	39	hP3	P6/mmm
αEr₂Ge₃	39	...	...
γErGe₂	46.5	...	...
βErGe₂	46.5	...	...
αErGe₂	46.5	...	...
ErGe₃₋ₓ	55	oC16	C222₁
(Ge)	100	cF8	Fd̄3m

Er-In

H. Okamoto, 1992

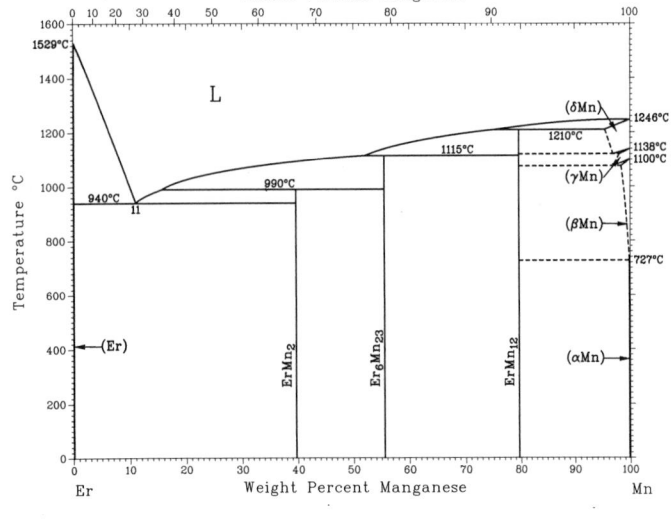

Phase	Composition, wt% In	Pearson symbol	Space group
(αEr)	0 to 5	hP2	P6₃/mmc
"(βEr)"	? to 15	cI2	Im̄3m
Er₂In	25.5	hP6	P6₃/mmc
Er₅In₃	29.2 to 36	hP16	P6₃/mcm
ErIn	40.7	cP2	Pm̄3m
Er₃In₅	53.4	oC32	Cmcm
ErIn₃	67	cP4	Pm̄3m
(In)	100	tI2	I4/mmm

Er-Mn

H.R. Kirchmayr and W. Lugscheider, 1967

Phase	Composition, wt% Mn	Pearson symbol	Space group
(Er)	0	hP2	P6₃/mmc
ErMn₂	39.7	...	...
Er₆Mn₂₃	55.7	cF116	Fm̄3m
ErMn₁₂	79.7	tI26	I4/mmm
(δMn)	100	cI2	Im̄3m
(γMn)	100	cF4	Fm̄3m
(βMn)	100	cP20	P4₁32
(αMn)	100	cI58	Ī43m

Er-Ni

Y.Y. Pan and P. Nash, 1991

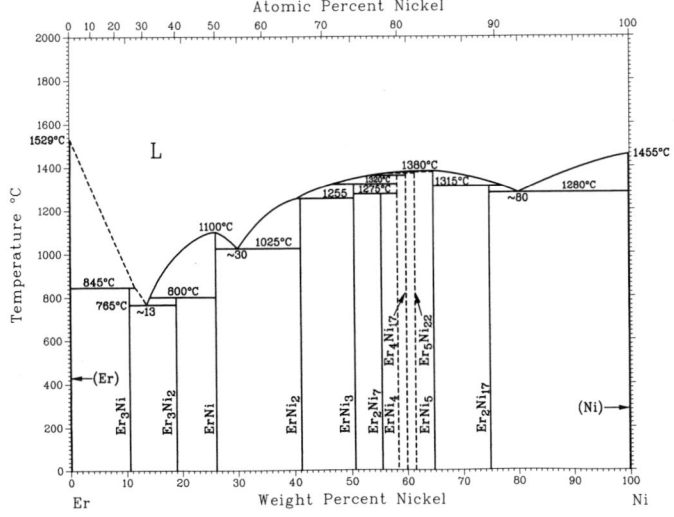

Phase	Composition, wt% Ni	Pearson symbol	Space group
(Er)	0	hP2	$P6_3/mmc$
Er₃Ni	10.5	oP16	Pnma
Er₃Ni₂	19.0	hR5	$R\overline{3}$
ErNi	26.0	oP8	Pnma
ErNi₂	41.3	cF24	$Fd\overline{3}m$
ErNi₃	51.3	hR24	$R\overline{3}m$
Er₂Ni₇	55.2	hR54	$R\overline{3}m$
ErNi₄	58.4	hP36	$P6_3/mmc$
Er₄Ni₁₇	60	...	...
Er₅Ni₂₂	60.7	...	...
ErNi₅	63.6	hP6	P6/mmm
Er₂Ni₁₇	74.9	hP38	$P6_3/mmc$
(Ni)	100	cF4	$Fm\overline{3}m$

Er-Pd

H. Okamoto, 1991

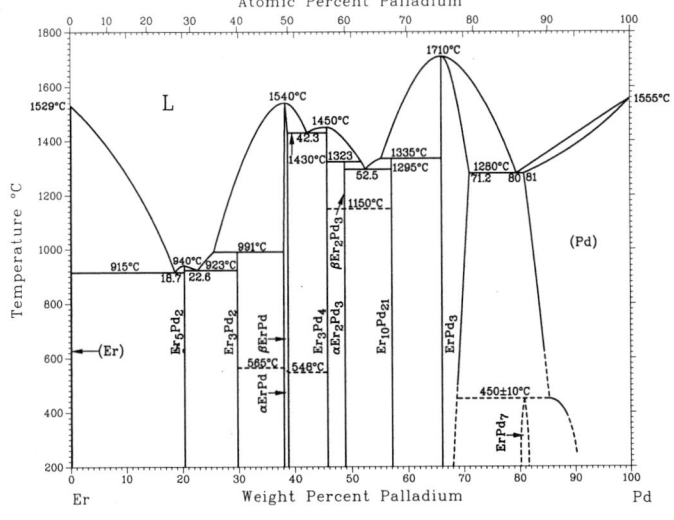

Phase	Composition, wt% Pd	Pearson symbol	Space group
(Er)	0	hP2	$P6_3/mmc$
Er₅Pd₂	20.3	cF96	$Fd\overline{3}m$
		tI49	$I4_1/a$
Er₃Pd₂	30	tP10	P4/mbm
βErPd	~38.9	cP2	$Pm\overline{3}m$
αErPd	~38.9	...	...
Er₃Pd₄	45.9	hR14	$R\overline{3}$
βEr₂Pd₃	49	...	...
αEr₂Pd₃	49	...	...
Er₁₀Pd₂₁(a)	57.1	mC124	C2/m
ErPd₃	66 to 71.2	cP4	$Pm\overline{3}m$
ErPd₇	81.7	c**	...
(Pd)	100	cF4	$Fm\overline{3}m$

(a) Similarity to Sm₁₀Pd₂₁ is assumed.

Er-Pt

H. Okamoto, 1990

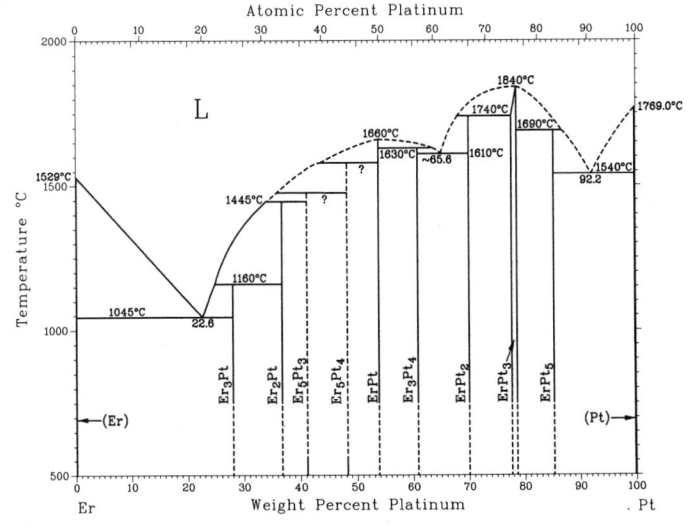

Phase	Composition, wt% Pt	Pearson symbol	Space group
(Er)	0	hP2	$P6_3/mmc$
Er₃Pt	28	oP16	Pnma
Er₂Pt	36.8	oP12	Pnma
Er₅Pt₃	41.2	hP16	$P6_3/mcm$
Er₅Pt₄	48.2	oP36	Pnma
ErPt	53.8	oP8	Pnma
Er₃Pt₄	60.8	hR14	$R\overline{3}m$
ErPt₂	70.0	cF24	$Fd\overline{3}m$
ErPt₃	~78	cP4	$Pm\overline{3}m$
ErPt₅	85.3	o*72	...
(Pt)	100	cF4	$Fm\overline{3}m$

Er-Ru

H. Okamoto, 1990

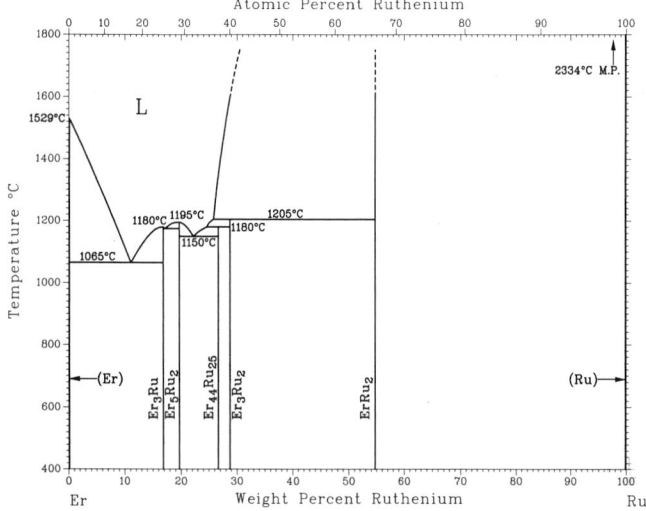

Phase	Composition, wt% Ru	Pearson symbol	Space group
(Er)	0	hP2	P6₃/mmc
Er₃Ru	17	oP16	Pnma
Er₅Ru₂	19.5	mC28	C2/c
Er₄₄Ru₂₅	25.4	oP276	Pnma
Er₃Ru₂	29	hP10	P6₃/m
ErRu₂	54.8	hP12	P6₃/mmc
(Ru)	100	hP2	P6₃/mmc

Er-Se

H. Okamoto, 1990

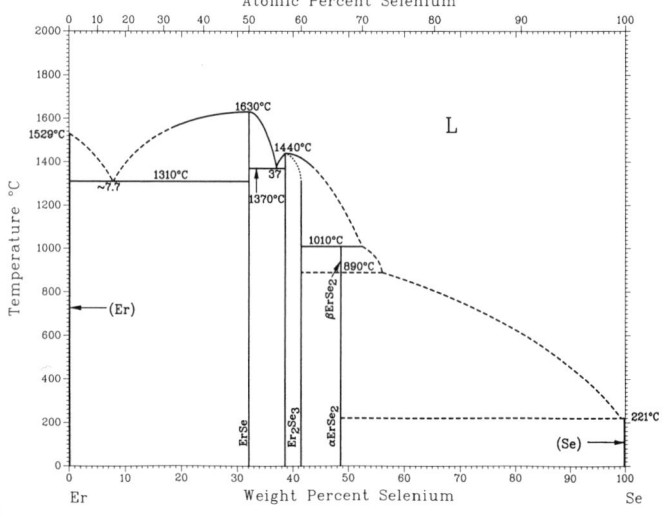

Phase	Composition, wt% Se	Pearson symbol	Space group
(Er)	0	hP2	P6₃/mmc
ErSe	32.1	cF8	Fm$\bar{3}$m
Er₂Se₃	38.6 to 42	oF80	Fddd
βErSe₂	48.6	oC132	Cmma
αErSe₂	48.6	oI12	Immm
(Se)	100	hP3	P3₁21

Er-Te

H. Okamoto, 1990

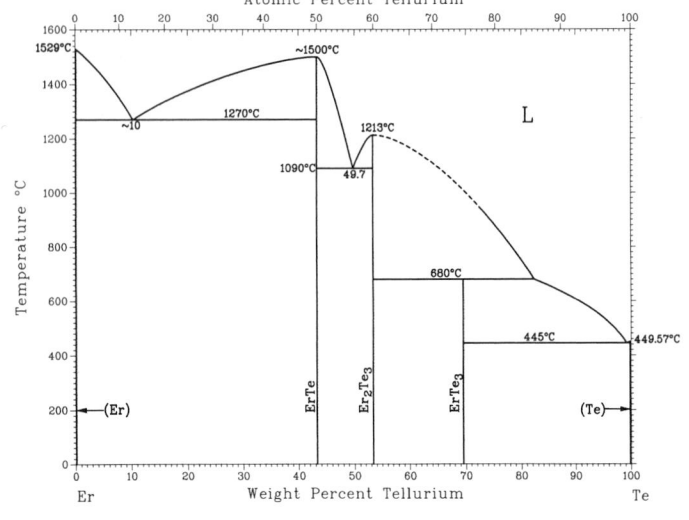

Phase	Composition, wt% Te	Pearson symbol	Space group
(Er)	0	hP2	P6₃/mmc
ErTe	43.3	cF8	Fm$\bar{3}$m
Er₂Te₃	53	oF80	Fddd
ErTe₃	70	oC16	Cmcm
(Te)	100	hP3	P3₁21
High-temperature, high-pressure phase			
ErTe₂	60.4	tP6	P4/nmm

Er-Ti

J.L. Murray, 1987

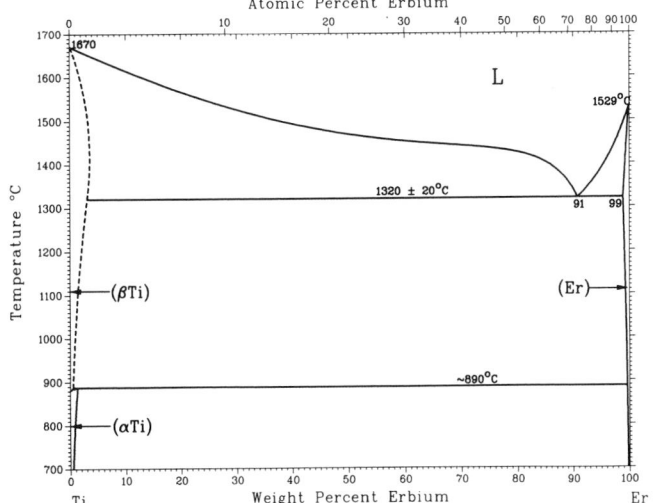

Phase	Composition, wt% Er	Pearson symbol	Space group
(βTi)	0 to ~3.1	cI2	$Im\overline{3}m$
(αTi)	0 to ~1.0	hP2	$P6_3/mmc$
(Er)	99.7 to 100	hP2	$P6_3/mmc$

Er-Tl

S. Delfino, A. Saccone, A. Palenzona, and R. Ferro, unpublished

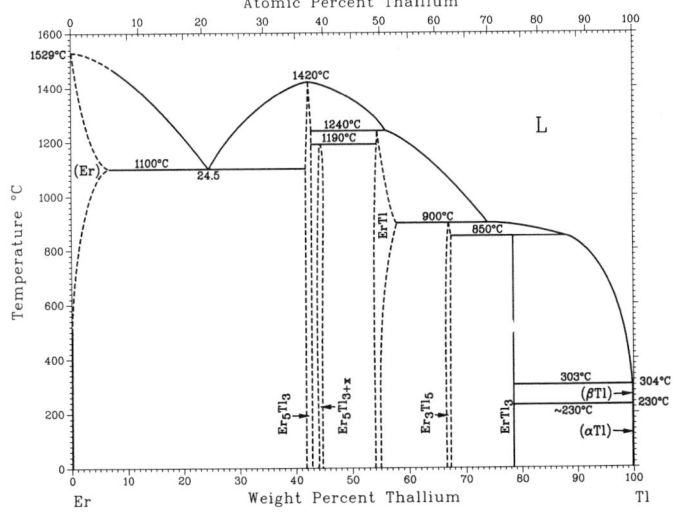

Phase	Composition, wt% Tl	Pearson symbol	Space group
(Er)	0 to ~7	hP2	$P6_3/mmc$
Er$_5$Tl$_3$	~42 to 43	hP16	$P6_3/mcm$
Er$_5$Tl$_{3+x}$	?	tI32	I4/mcm
ErTl(a)	~54 to ~58	cP2	$Pm\overline{3}m$
		(or cI2)	$Im\overline{3}m$
ErTl(b)	~54 to ~58	tP2	P4/mmm
Er$_3$Tl$_5$	~67 to ~68	oC32	Cmcm
ErTl$_3$	79	cP4	$Pm\overline{3}m$
(βTl)	100	cI2	$Im\overline{3}m$
(αTl)	100	hP2	$P6_3/mmc$

(a) Cubic structure presumed to be room-temperature and higher temperature phases. (b) Tetragonal structure presumed to be lower temperature phase

Eu-Ga

S.P. Yatsenko, B.G. Semenov, and K.A. Chuntonov, 1978

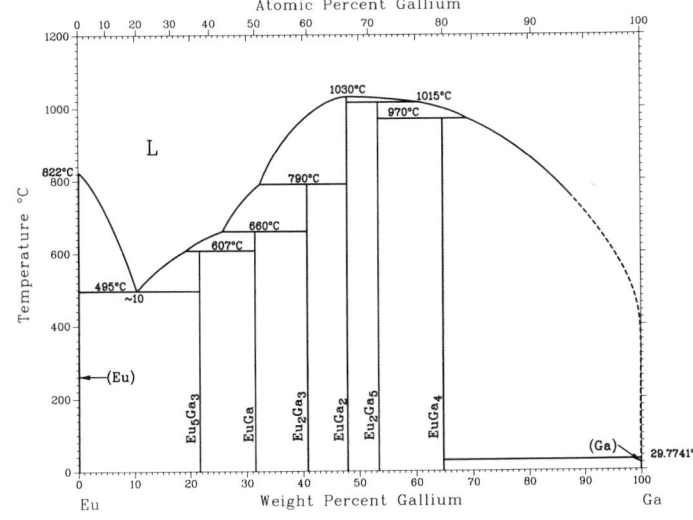

Phase	Composition, wt% Ga	Pearson symbol	Space group
(Eu)	0	cI2	$Im\overline{3}m$
Eu$_5$Ga$_3$	21.6	...	...
EuGa	31.5	...	...
Eu$_2$Ga$_3$	41	...	...
βEuGa$_2$(a)	47.9	hP3	P6/mmm
αEuGa$_2$(b)	47.9	oI12	Imma
Eu$_2$Ga$_5$	53.4	...	...
EuGa$_4$	65	tI10	I4/mmm
(Ga)	100	oC8	Cmca

(a) Hexagonal structure presumed to be lower temperature phase. (b) Cubic structure presumed to be higher temperature phase

Eu-Ge

A.B. Gokhale and G.J. Abbaschian, 1991

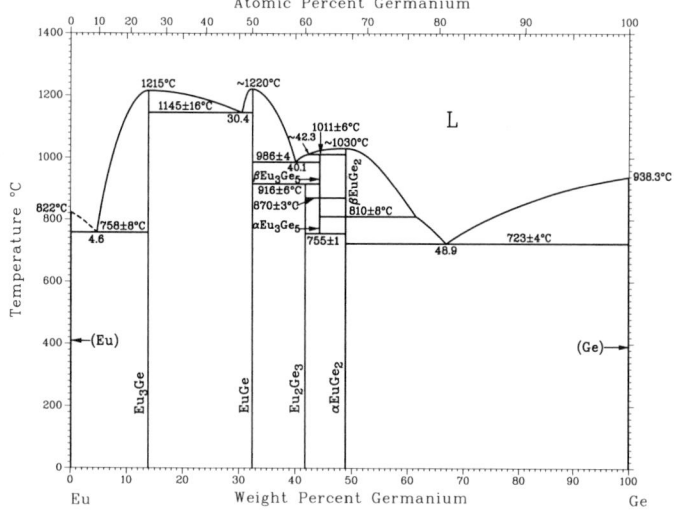

Phase	Composition, wt% Ge	Pearson symbol	Space group
(Eu)	0	$cI2$	$Im\bar{3}m$
EuGe	32.3	$oC8$	$Cmcm$
Eu_2Ge_3	41.7	…	…
βEu_3Ge_5	44.3	…	…
αEu_3Ge_5	44.3	(a)	…
$\beta EuGe_2$	48.82	…	…
$\alpha EuGe_2$	48.9	$hP3$	$P\bar{3}m1$
(Ge)	100	$cF8$	$Fd\bar{3}m$

(a) Hexagonal structure

Eu-In

H. Okamoto, 1990

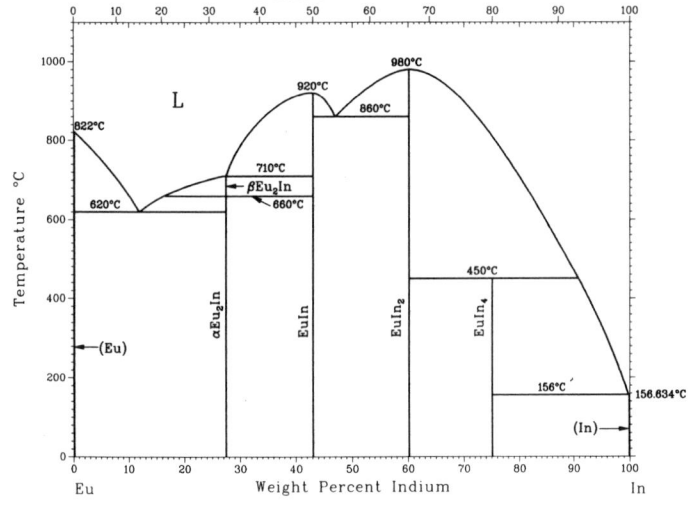

Phase	Composition, wt% In	Pearson symbol	Space group
(Eu)	0	$cI2$	$Im\bar{3}m$
βEu_2In	27.4	…	…
αEu_2In	27.4	…	…
EuIn	43.0	…	…
$EuIn_2$	60.1	$hP6$	$P6_3/mmc$
$EuIn_4$	75.1	…	…
(In)	100	$tI2$	$I4/mmm$

Eu-Mg

H. Okamoto, 1992

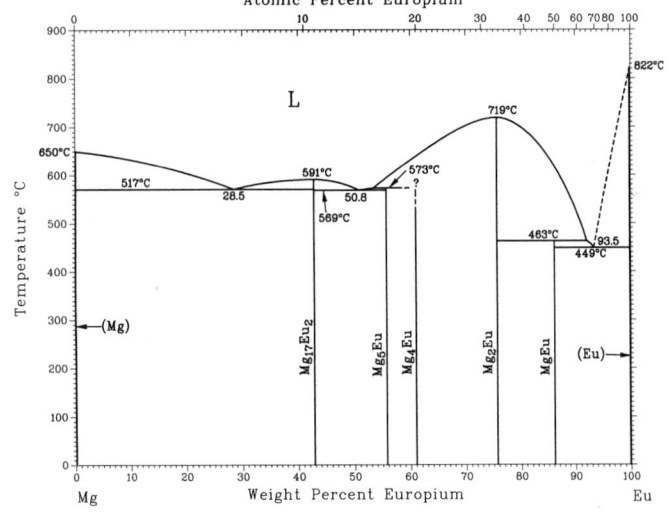

Phase	Composition, wt% Eu	Pearson symbol	Space group
(Mg)	0	$hP2$	$P6_3/mmc$
$Mg_{17}Eu_2$	42.3	$hP38$	$P6_3/mmc$
Mg_5Eu	55.6	$hP36$	$P6_3/mmc$
Mg_4Eu	61	$hP90$	$P6_3/mmc$
Mg_2Eu	75.7	$hP12$	$P6_3/mmc$
MgEu	86.2	$cP2$	$Pm\bar{3}m$
(Eu)	100	$cI2$	$Im\bar{3}m$

Eu-Pb

O.D. McMasters and K.A. Gschneidner, Jr., 1967

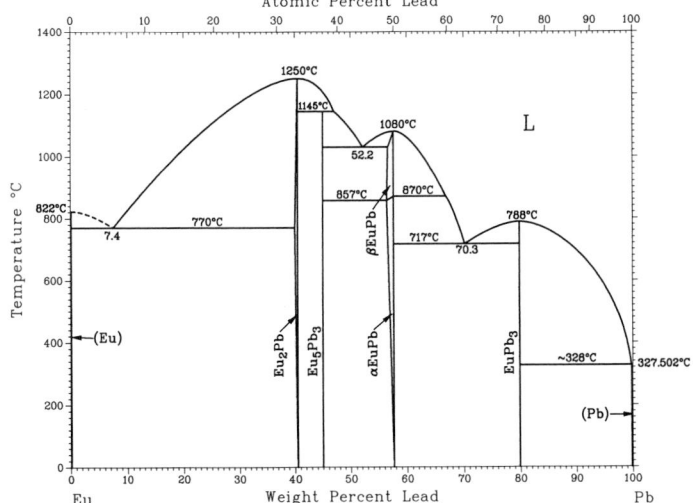

Phase	Composition, wt% Pb	Pearson symbol	Space group
(Eu)	0	cI2	Im3̄m
Eu₂Pb	~40 to 40.5	oP12	Pnma
Eu₅Pb₃	45.0	tI32	I4/mcm
βEuPb	~57.7	...	...
αEuPb(a)	~57.7	tP2	P4/mmm
EuPb₃	80	cP4	Pm3̄m
(Pb)	100	cF4	Fm3̄m

(a) Crystal structure data might be for βEuPb.

Eu-Pd

H. Okamoto, 1990

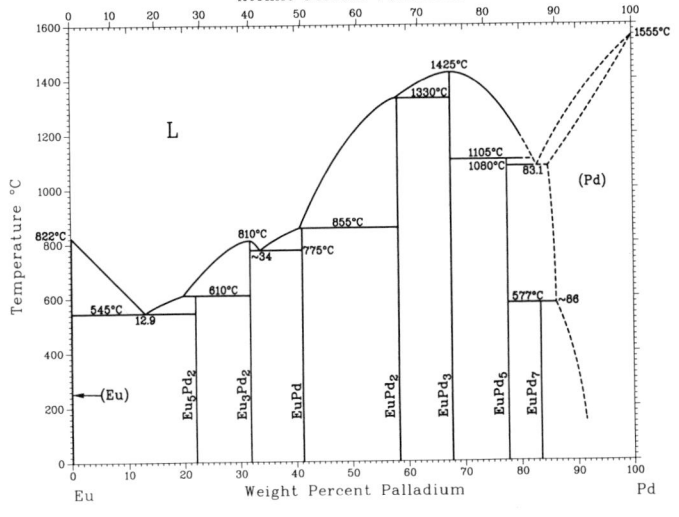

Phase	Composition, wt% Pd	Pearson symbol	Space group
(Eu)	0	cI2	Im3̄m
Eu₅Pd₂	21.8	mC28	C2/c
Eu₃Pd₂	32	hR15	R3̄
EuPd	41.2	oC8	Cmcm
EuPd₂	58.4	...	...
EuPd₃	68	cP4	Pm3̄m
EuPd₅	77.7	o*72	...
EuPd₇	~83.1	c**	...
(Pd)	~86 to 100	cF4	Fm3̄m

Eu-Pt

A. Iandelli and A. Palenzona, 1981

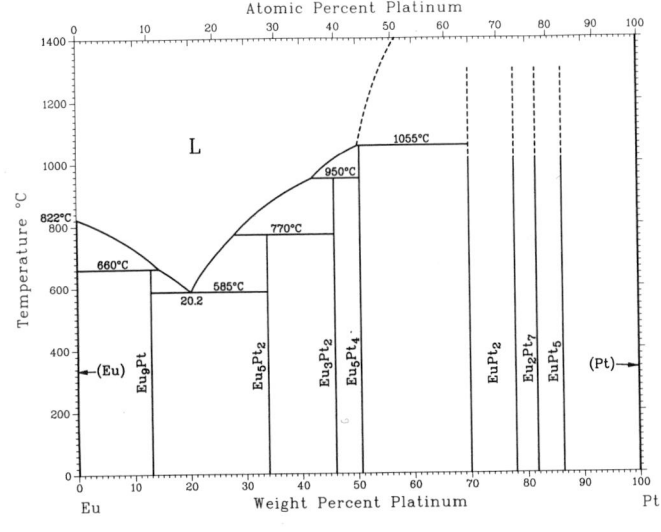

Phase	Composition, wt% Pt	Pearson symbol	Space group
(Eu)	0	cI2	Im3̄m
Eu₉Pt	13	cF*	...
Eu₅Pt₂	34.0	mC28	C2/c
Eu₃Pt₂	46	hR15	R3̄
Eu₅Pt₄	50.6	oP36	Pnma
EuPt₂	70 to 78	cF24	Fd3̄m
Eu₂Pt₇	81.8	hP36	P6₃/mmc
EuPt₅	86.5	o**	...
(Pt)	100	cF4	Fm3̄m

Eu-Te

O.A. Sadovskaya and E.I. Yarembash, 1970

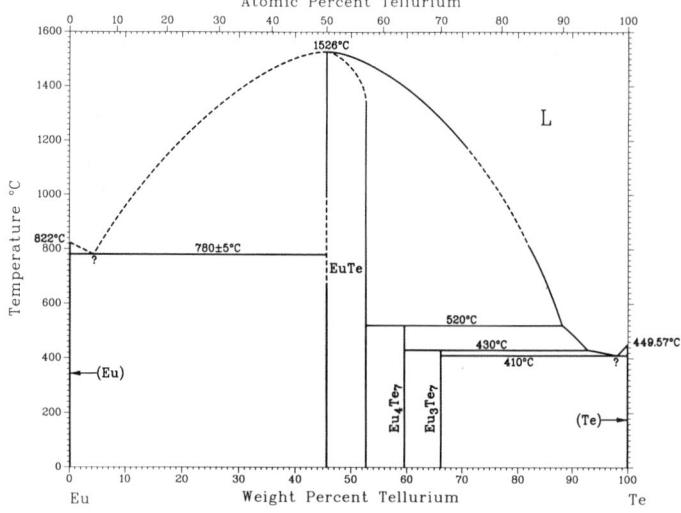

Phase	Composition, wt% Te	Pearson symbol	Space group
(Eu)	0	cI2	$Im\bar{3}m$
EuTe	46 to 52.8	cF8	$Fm\bar{3}m$
Eu₄Te₇	59.5	...	...
Eu₃Te₇	66	...	...
(Te)	100	hP3	$P3_121$

Fe-Ga

H. Okamoto, 1992

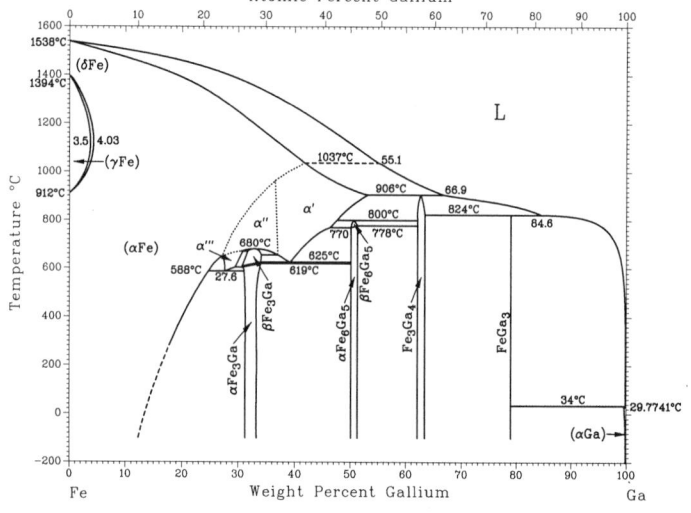

Phase	Composition, wt% Ga	Pearson symbol	Space group
(γFe)	0 to 3.5	cF4	$Fm\bar{3}m$
(αFe)	0 to 41	cI2	$Im\bar{3}m$
α'	36.5 to 53.0	cP2	$Pm\bar{3}m$
α''	26.9 to 37.1	cF16	$Fm\bar{3}m$
α'''	26.9 to 30.4	cF16	$Fm\bar{3}m$
βFe₃Ga	30.5 to 33.8	hP8	$P6_3/mmc$
αFe₃Ga	30.7 to 34.0	cP4	$Pm\bar{3}m$
βFe₆Ga₅	50.0 to 51.0	hR26	$R\bar{3}m$
αFe₆Ga₅	50.0 to 51.0	mC44	$C2/m$
Fe₃Ga₄	61.9 to 63.3	mC42	$C2/m$
		t*63	...
FeGa₃	79	tP16	$P\bar{4}n2$
		tP16	$P4_2/mnm$
(αGa)	100	oC8	$Cmca$
Metastable phase			
Fe₁₃Ga₉	46.4	...	...

Fe-Gd

H. Okamoto, 1992

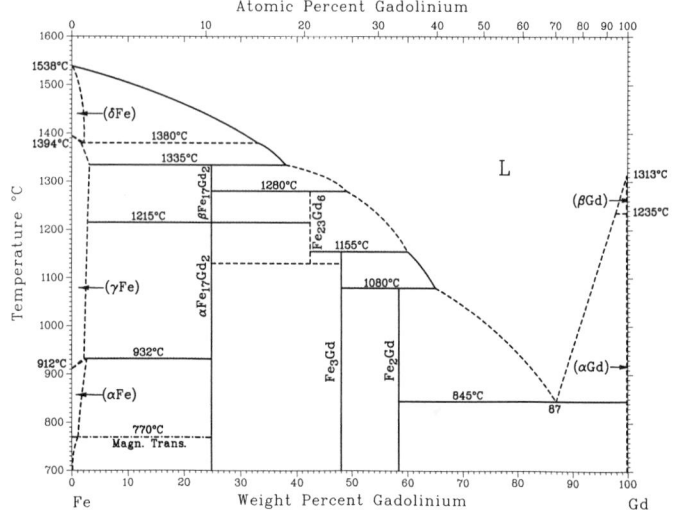

Phase	Composition, wt% Gd	Pearson symbol	Space group
(δFe)	0	cI2	$Im\bar{3}m$
(γFe)	0	cF4	$Fm\bar{3}m$
(αFe)	0	cI2	$Im\bar{3}m$
βFe₁₇Gd₂	24.8	hP38	$P6_3/mmc$
αFe₁₇Gd₂	24.8	hR19	$R\bar{3}m$
Fe₂₃Gd₆	42.4	cF116	$Fm\bar{3}m$
Fe₃Gd	48	hR12	$R\bar{3}m$
Fe₂Gd	58.4	cF24	$Fd\bar{3}m$
(βGd)	100	cI2	$Im\bar{3}m$
(αGd)	100	hP2	$P6_3/mmc$
Questionable phases			
Fe₅Gd	24	hP*	...
Fe₁₇Gd₂	24.8	hP8	$P6/mmm$
Fe₅Gd	36.1	hP6	$P6/mmm$
Fe₄Gd	41	hP10	...
Fe₇Gd₂	44.6	o*18	...
Fe₃Gd₂	65	c*30	...

Fe-Ge

E. Kato and S. Nunoue, 1992

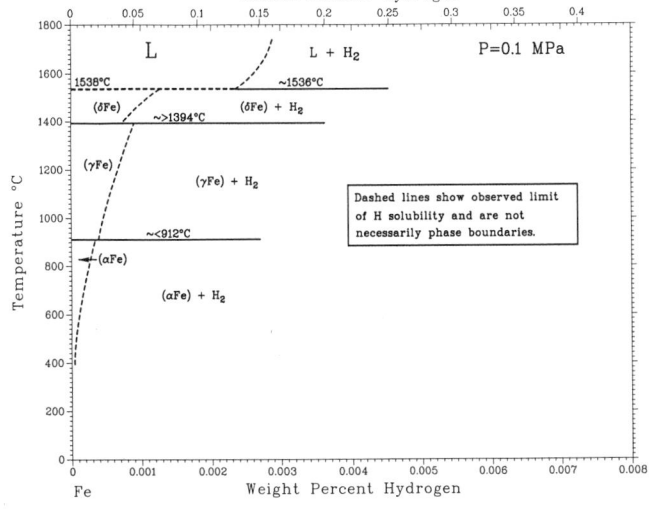

Phase	Composition, wt% Ge	Pearson symbol	Space group
(γFe)	0 to 4.4	cF4	$Fm\bar{3}m$
(αFe)	0 to 21.6	cI2	$Im\bar{3}m$
α_2	12.6 to 26.8	cP2	$Pm\bar{3}m$
α_1	18.9 to 25.7	cF16	$Fm\bar{3}m$
ε(Fe₃Ge)	28.8 to 31.0	hP8	$P6_3/mmc$
ε'(Fe₃Ge)	28.8 to 31.0	cP4	$Pm\bar{3}m$
β	39.6 to 47.5	hP4	$P6_3/mmc$
η	47.3 to 50.0	hP6	$P6_3/mmc$
Fe₆Ge₅	52.0	…	$C2/m$
FeGe	56.5	…	$C2/m$
		hP6	$P6/mmm$
		cP8	$P2_13$
FeGe₂	72.3	tI2	$I4/mcm$
(Ge)	100	cF8	$Fd\bar{3}m$

Fe-H

A. San-Martin and F.D. Manchester, 1992

Phase	Composition, wt% H	Pearson symbol	Space group
(δFe) or δ	0 to 0.0013	cI2	$Im\bar{3}m$
(γFe) or γ	0 to 0.0008	cF4	$Fm\bar{3}m$
(αFe) or α	0 to 0.0003	cI2	$Im\bar{3}m$
Metastable phases			
ε	1.2 to 1.4(a)	hP2	$P6_3/mmc$
		hP4	$P6_3/mmc$
		hP4	$P6_3mc$

(a) Produced under a pressure of 6.7 GPa at 250 °C

Fe-Hf

H. Okamoto, 1992

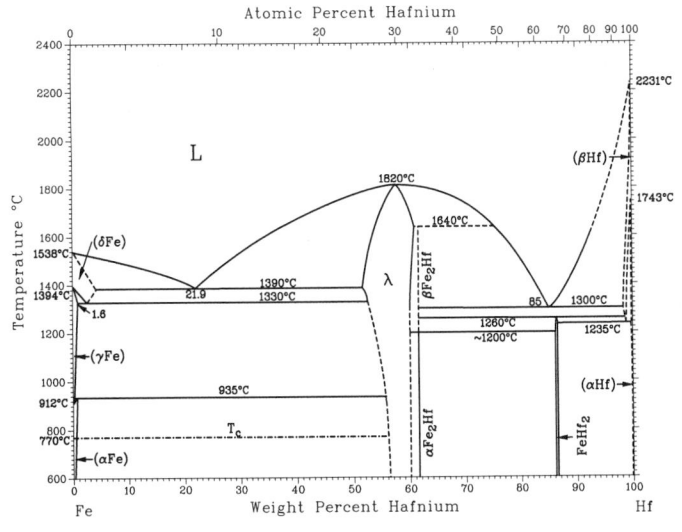

Phase	Composition, wt% Hf	Pearson symbol	Space group
(δFe)	0 to 6	cI2	$Im\bar{3}m$
(γFe)	0 to 1.6	cF4	$Fm\bar{3}m$
(αFe)	0 to 0.70	cI2	$Im\bar{3}m$
λ	52 to 61.2	hP12	$P6_3/mmc$
βFe₂Hf	61.5	hP24	$P6_3/mmc$
αFe₂Hf	61.5	cF24	$Fd\bar{3}m$
FeHf₂	85.6 to 86.6	cF96	$Fd\bar{3}m$
(βHf)	? to 100	cI2	$Im\bar{3}m$
(αHf)	? to 100	hP2	$P6_3/mmc$

Fe-Ho

H. Okamoto, 1992

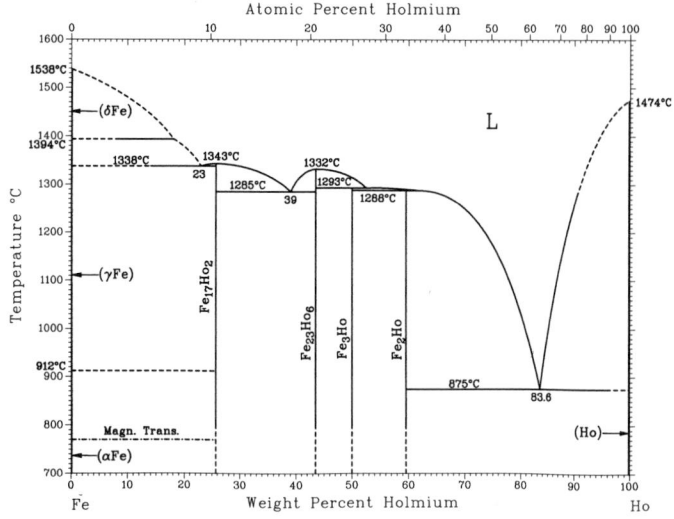

Phase	Composition, wt% Ho	Pearson symbol	Space group
(δFe)	0	cI2	Im$\overline{3}$m
(γFe)	0	cF4	Fm$\overline{3}$m
(αFe)	0	cI2	Im$\overline{3}$m
Fe$_{17}$Ho$_2$	25.7	hP38	P6$_3$/mmc
Fe$_{23}$Ho$_6$	43.5	cF116	Fm$\overline{3}$m
Fe$_3$Ho	50	hR12	R$\overline{3}$m
Fe$_2$Ho	59.6	cF24	Fd$\overline{3}$m
(Ho)	100	hP2	P6$_3$/mmc
Metastable phase			
...	~75	hP12	P6$_3$/mmc

Fe-Ir

L.J. Swartzendruber, 1992

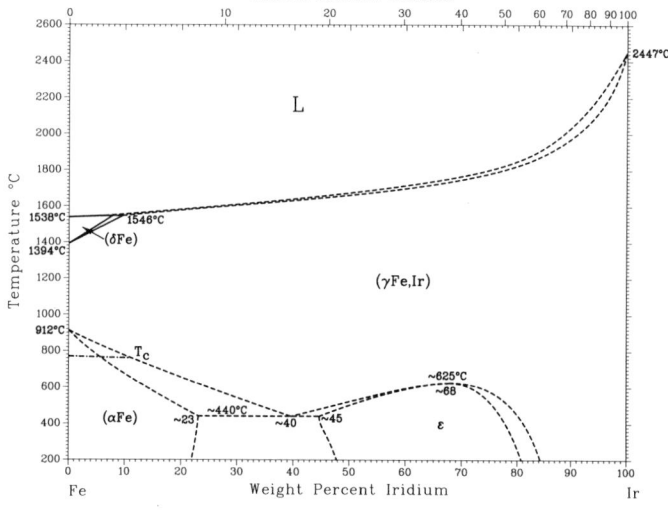

Phase	Composition, wt% Ir	Pearson symbol	Space group
(αFe)	0 to ~23	cI2	Im$\overline{3}$m
(γFe,Ir)	0 to 100	cF4	Fm$\overline{3}$m
(δFe)	0 to 7	cI2	Im$\overline{3}$m
ε	~45 to 80	hP2	P6$_3$/mmc

Fe-La

H. Okamoto, 1992

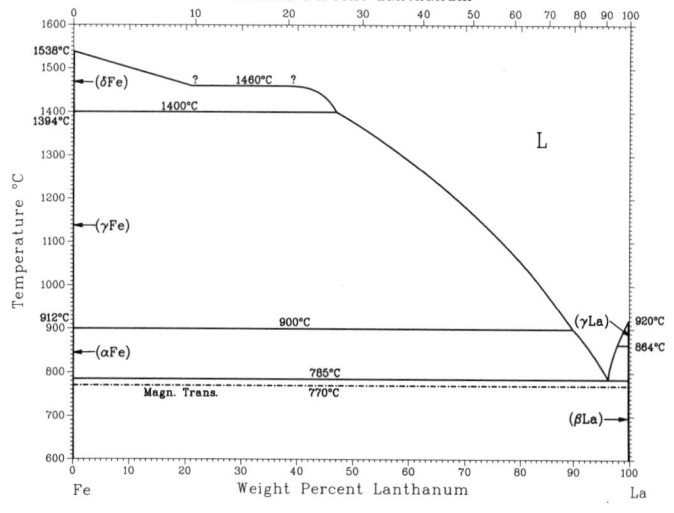

Phase	Composition, wt% La	Pearson symbol	Space group
(δFe)	0	cI2	Im$\overline{3}$m
(γFe)	0	cF4	Fm$\overline{3}$m
(αFe)	0	cI2	Im$\overline{3}$m
(γLa)	100	cI2	Im$\overline{3}$m
(βLa)	100	cF4	Fm$\overline{3}$m
(αLa)	100	hP4	P6$_3$/mmc

Fe-Lu

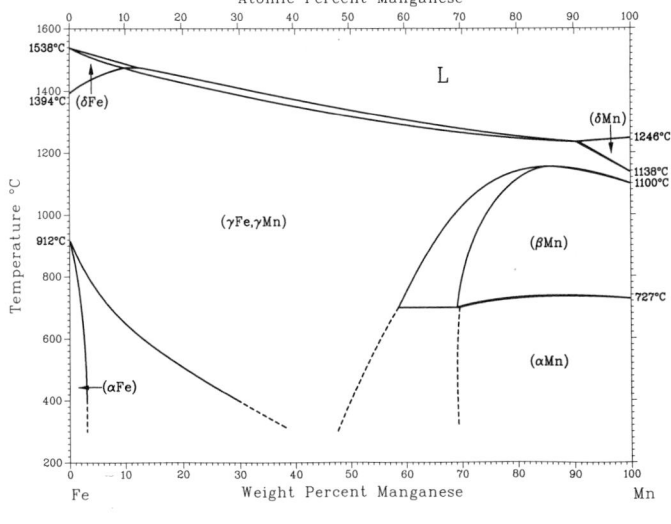

Phase	Composition, wt% Lu	Pearson symbol	Space group
(δFe)	0	cI2	$Im\bar{3}m$
(γFe)	0	cF4	$Fm\bar{3}m$
(αFe)	0	cI2	$Im\bar{3}m$
Fe₁₇Lu₂	24.7 to 26.9	hP38	$P6_3/mmc$
Fe₂₃Lu₆	45.0	cF116	$Fm\bar{3}m$
Fe₃Lu	51	hR12	$R\bar{3}m$
Fe₂Lu	61.0	cF24	$Fd\bar{3}m$
(Lu)	100	hP2	$P6_3/mmc$
Metastable phase			
...	~76	hP12	$P6_3/mmc$

Fe-Mn

Phase	Composition, wt% Mn	Pearson symbol	Space group
(δFe)	0 to 10	cI2	$Im\bar{3}m$
(γFe,γMn)	0 to 100	cF4	$Fm\bar{3}m$
(αFe)	0 to 3	cI2	$Im\bar{3}m$
(δMn)	91 to 100	cI2	$Im\bar{3}m$
(βMn)	69.2 to 100	cP20	$P4_132$
(αMn)	~70 to 100	cI58	$I\bar{4}3m$
Metastable phases			
α′	3 to 18	tI2	$I4/mmm$
ε	12 to 30	hP2	$P6_3/mmc$
γ′	?	t**	...

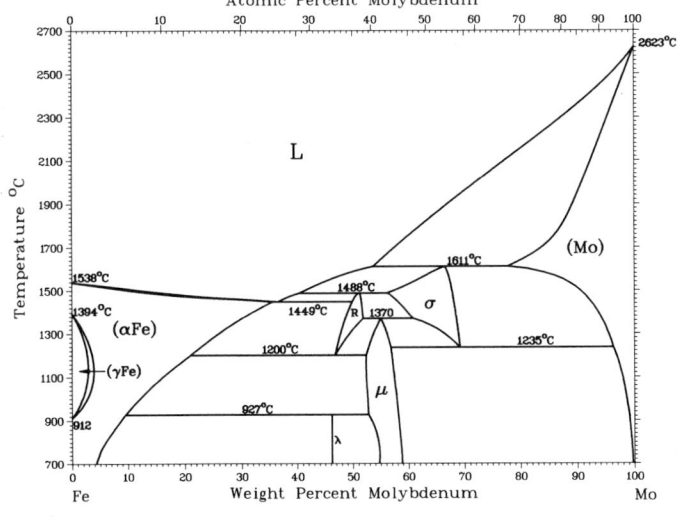

Fe-Mo

Phase	Composition, wt% Mo	Pearson symbol	Space group
(αFe)	0 to 35.7	cI2	$Im\bar{3}m$
(γFe)	0 to 2.9	cF4	$Fm\bar{3}m$
λ	46.2	hP12	$P6_3/mmc$
R	46.8 to 51.8	hR53	...
μ	52.3 to 57.4	hR13	$R\bar{3}m$
σ	56.3 to 69.2	tP30	$P4_2/mnm$
(Mo)	79.0 to 100	cI2	$Im\bar{3}m$

Fe-N

H.A. Wriedt, N.A. Gokcen, and R.H. Nafziger, 1992

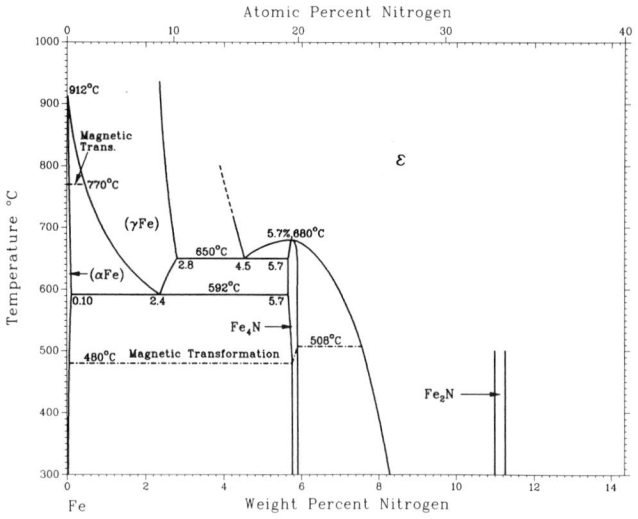

Phase	Composition, wt% N	Pearson symbol	Space group
Stable at 0.1 MPa			
(δFe)	0 to ~0.9	cI2	Im$\bar{3}$m
(γFe)	0 to 2.8	cF4	Fm$\bar{3}$m
(αFe)	0 to 0.10	cI2	Im$\bar{3}$m
Fe$_4$N	5.7 to 5.9	cP5	Pm$\bar{3}$m or P$\bar{4}$3m
ε	~4 to ~11	hP3	P6$_3$/mmc
Fe$_2$N	~11.1	o**	...
FeN$_6$	~61	...	...
FeN$_9$	~69	...	...
Other phases			
(εFe)(a)	0 to ?	hP2	P6$_3$/mmc
Martensite	0 to 0.6	cI2	Im$\bar{3}$m
	0.7 to 2.6	(b)	...
Fe$_{16}$N$_2$	~3.0	(b)	I4/mmm

(a) Stable at pressures >13 GPa. (b) bct

Fe-Nb

E. Paul and L.J. Swartzendruber, 1992

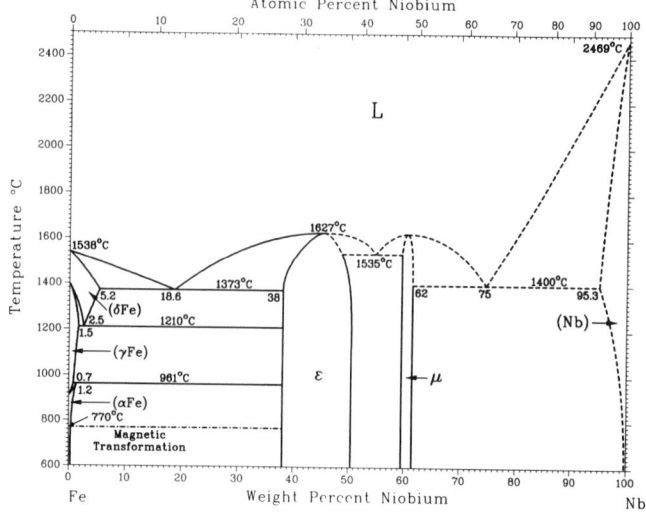

Phase	Composition, wt% Nb	Pearson symbol	Space group
δ or (δFe)	0 to 5.2	cI2	Im$\bar{3}$m
γ or (γFe)	0 to 1.5	cF4	Fm$\bar{3}$m
α or (αFe)	0 to 1.2	cI2	Im$\bar{3}$m
ε or Fe$_2$Nb	38 to 51	hP12	P6$_3$/mmc
μ or FeNb	60 to 62	hR13	R$\bar{3}$m
(Nb)	95.3 to 100	cI2	Im$\bar{3}$m

Fe-Nd

W. Zhang, G. Liu, and K. Han, 1992

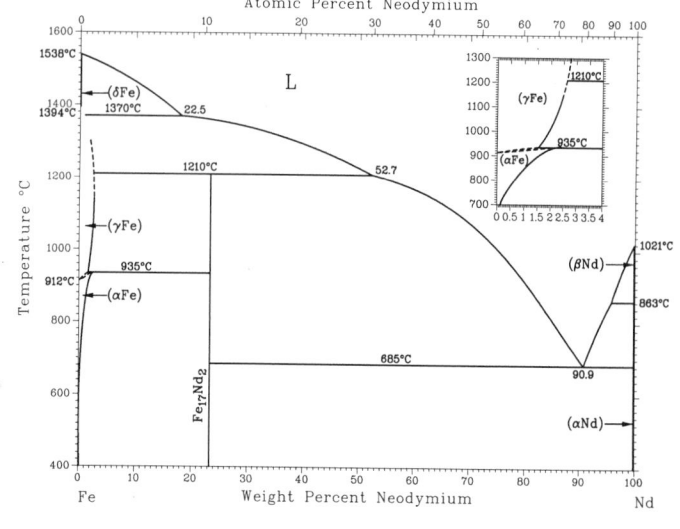

Phase	Composition, wt% Nd	Pearson symbol	Space group
(δFe)(a)	0	cI2	Im$\bar{3}$m
(γFe)(b)	0 to ~1	cF4	Fm$\bar{3}$m
(αFe)(c)	0 to ~1.1	cI2	Im$\bar{3}$m
Fe$_{17}$Nd$_2$	23.3	(d)	R$\bar{3}$m
(βNd)(e)	100	cI2	Im$\bar{3}$m
(αNd)(f)	100	hP4	P6$_3$/mmc
Metastable phase			
Fe$_{5+x}$Nd	...	hP6	P6/mmm

(a) From 1538 to 1394 °C. (b) From <1394 to 912 °C. (c) Below 912 °C. (d) Rhombohedral. (e) From 1021 to 863 °C. (f) Below 863 °C

Fe-Ni

L.J. Swartzendruber, V.P. Itkin, and C.B. Alcock, 1992

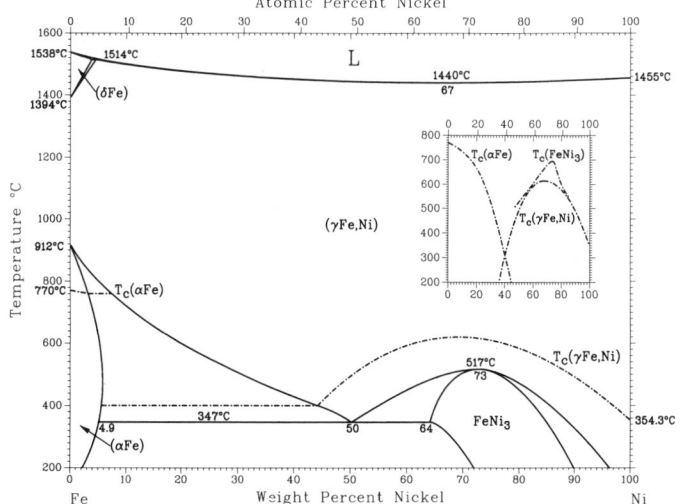

Phase	Composition, wt% Ni	Pearson symbol	Space group
(δFe)	0 to 3.7	cI2	Im$\bar{3}$m
(γFe, Ni)	0 to 100	cF4	Fm$\bar{3}$m
(αFe)	0 to 5.8	cI2	Im$\bar{3}$m
Fe$_3$Ni(a)	26	cP4	Pm$\bar{3}$m
FeNi(a)	51	tP2	P4/mmm
FeNi$_3$	64 to ~90	cP4	Pm$\bar{3}$m

(a) Metastable

Fe-O

H.A. Wriedt, 1992

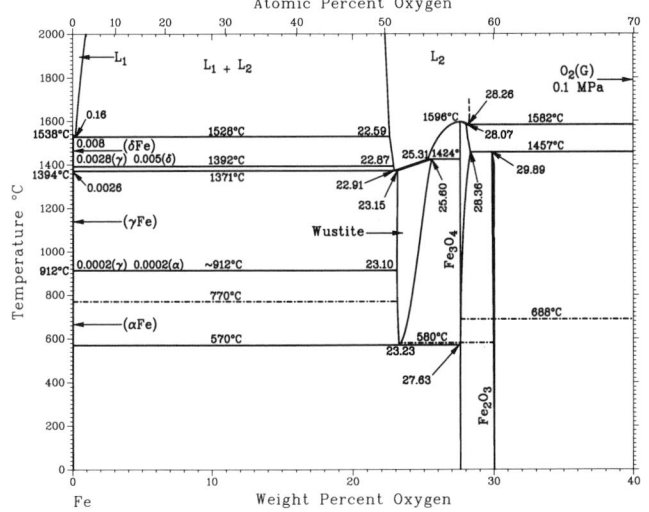

Phase	Composition, wt% O	Pearson symbol	Space group
Stable phases			
(δFe)	~0	cI2	Im$\bar{3}$m
(γFe)	~0	cF4	Fm$\bar{3}$m
(αFe)	~0	cI2	Im$\bar{3}$m
Wustite	23.15 to 25.60	cF8	Fm$\bar{3}$m
Fe$_3$O$_4$(LT)	~27.6	mC224	Cc
Fe$_3$O$_4$	27.56 to 28.36	cF56	Fd$\bar{3}$m
αFe$_2$O$_3$	~30.1	hR10	R$\bar{3}$c
Other phases			
(εFe)(a)	0 to ?	hP2	P6$_3$/mmc
P′(wustite)	~23.2 to ~24.8	c**(?)(b)	...
P″(wustite)	~24 to ~25	mP500(?)	P2$_1$/m
P‴(wustite)	...	...	...
Wustite(LT)	23.2 to 24.6	hR2(c)	R$\bar{3}$
Fe$_3$O$_4$(P)(d)	~27.6	m*14	...
βFe$_2$O$_3$	~30.1	cI80	Ia$\bar{3}$
γFe$_2$O$_3$	~30.1	tP60	P4$_3$2$_1$2
εFe$_2$O$_3$	~30.1	m*100	...

(a) Stable at pressures >13 GPa. (b) Incommensurate or orthorhombic. (c) Magnetic reflections might indicate linear cell dimensions are doubled, corresponding to hR16. (d) Stable at pressures >25 GPa

Fe-O phase diagram from 22 to 31 wt% O

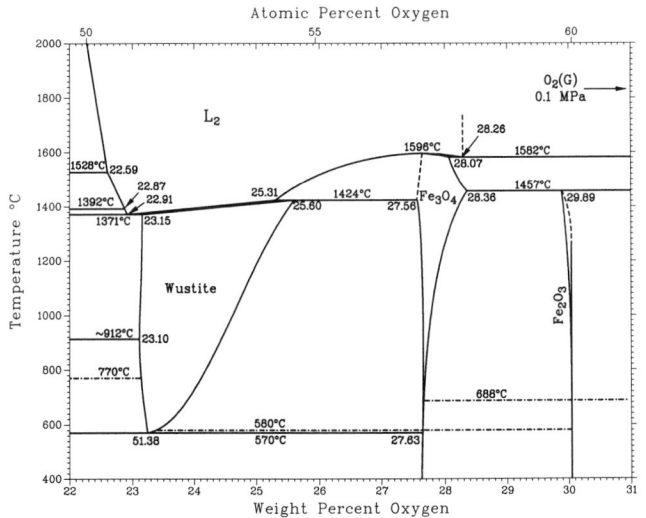

Fe-P

H. Okamoto, 1992

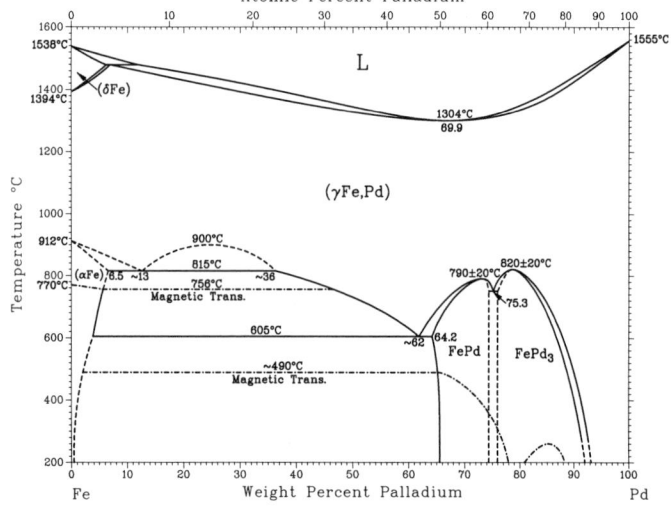

Phase	Composition, wt% P	Pearson symbol	Space group
(γFe)	0 to 0.31	cF4	Fm$\bar{3}$m
(αFe)	0 to 2.8	cI2	Im$\bar{3}$m
Fe₃P	16	tI32	I4
Fe₂P	21.7 to 22.2	hP9	P$\bar{6}$2m
FeP	36	oP8	Pna2₁
FeP₂	52 to 53	oP6	Pnnm
FeP₄	69	mP30	P2₁/c
(P) (white)	100	c**	...
Metastable phases			
Fe₄₊P	<12	o**	...
(P) black	100	oC8	Cmca
High-pressure phases			
Fe₂P	21.7	oP12	Pnma
FeP₄	69	oC20	C222₁

Fe-Pd

H. Okamoto, 1992

Phase	Composition, wt% Pd	Pearson symbol	Space group
(δFe)	0 to 6.1	cI2	Im$\bar{3}$m
(γFe, Pd)	0 to 100	cF4	Fm$\bar{3}$m
(αFe)	0 to 6.5	cI2	Im$\bar{3}$m
FePd	64.2 to 74	tP2	P4/mmm
FePd₃	76 to ?	cP4	Pm$\bar{3}$m

Fe-Pu

H. Okamoto, 1992

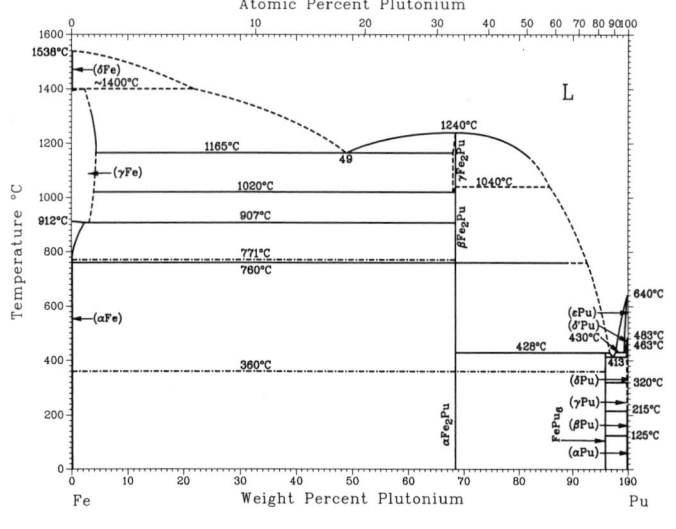

Phase	Composition, wt% Pu	Pearson symbol	Space group
(δFe)	0	cI2	Im$\bar{3}$m
(γFe)	0 to ~4	cF4	Fm$\bar{3}$m
(αFe)	0	cI2	Im$\bar{3}$m
γFe₂Pu	68.6	c**	...
βFe₂Pu	68.6	hP24	P6₃/mmc
αFe₂Pu	68.6	cF24	Fd$\bar{3}$m
FePu₆	96.3	tI28	I4/mcm
(εPu)	99.5 to 100	cI2	Im$\bar{3}$m
(δ'Pu)	~100	tI2	I4/mmm
(δPu)	99.9 to 100	cF4	Fm$\bar{3}$m
(γPu)	100	oF8	Fddd
(βPu)	100	mC34	C2/m
(αPu)	100	mP16	P2₁/m

Fe-Rh

L.J. Swartzendruber, 1992

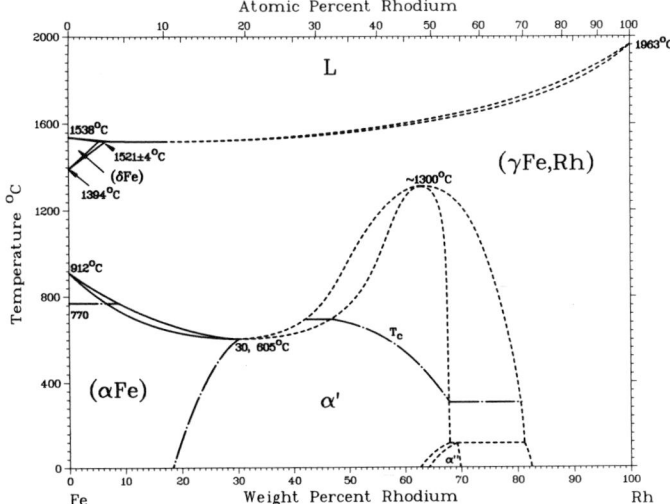

Phase	Composition, wt% Rh	Pearson symbol	Space group
(δFe)	0 to 5	cI2	Im$\bar{3}$m
(γFe,Rh)	0 to 100	cF4	Fm$\bar{3}$m
(αFe)	0 to 30	cI2	Im$\bar{3}$m
α'	19 to 69	cP2	Pm$\bar{3}$m
α" (chemical cell)	63 to 69	cP2	Pm$\bar{3}$m
α" (magnetic cell)	63 to 69	cF16	Fm$\bar{3}$m

Fe-S

From [Kubaschewski]

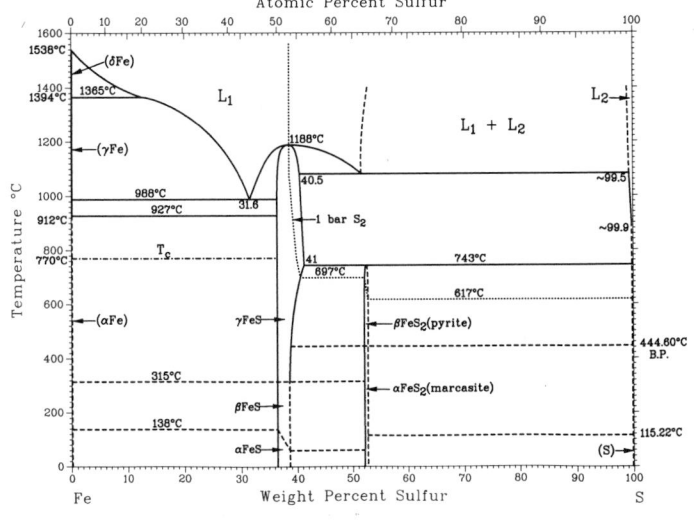

Phase	Composition, wt% S	Pearson symbol	Space group
(δFe)	0 to ~0.14	cI2	Im$\bar{3}$m
(γFe)	0 to ~0.05	cF4	Fm$\bar{3}$m
(αFe)	0 to 0.019	cI2	Im$\bar{3}$m
γFeS	36.5 to 41	hP4	P6$_3$/mmc
βFeS	36.5 to ~38	hP24	P$\bar{6}$2c
αFeS	36.5 to ~38	...	...
βFeS$_2$	~53.5	cP12	Pa3
αFeS$_2$	~53.5	oP6	Pnnm

Fe-rich region of the Fe-S system

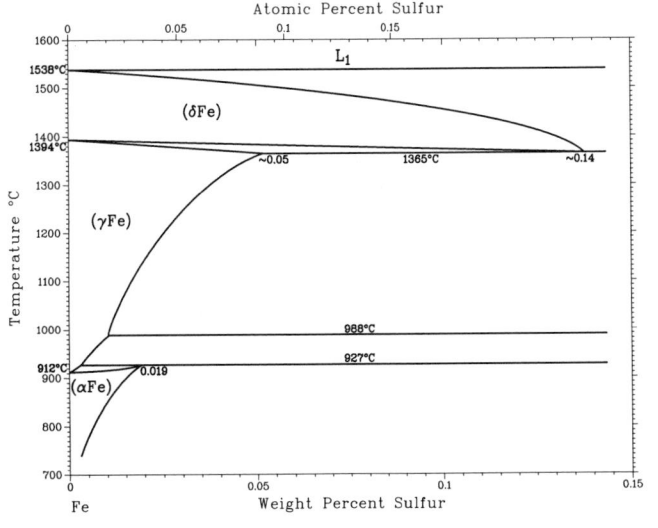

Fe-Sb

H. Okamoto, 1992

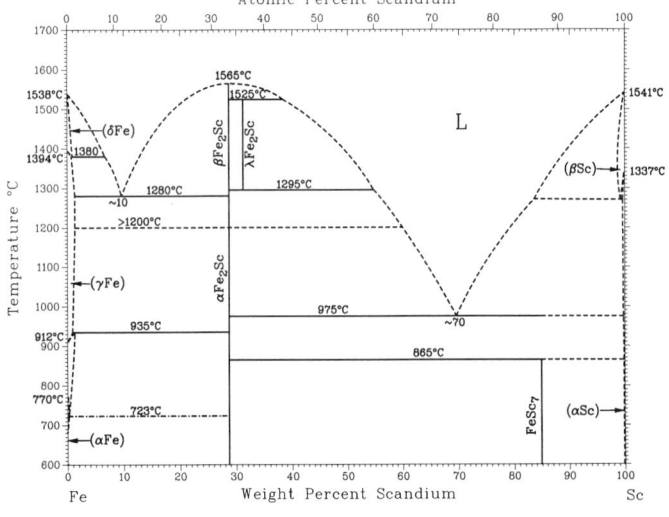

Phase	Composition, wt% Sb	Pearson symbol	Space group
(αFe)	0 to 10.3	cI2	Im$\bar{3}$m
(γFe)	0 to 2.4	cF4	Fm$\bar{3}$m
ε	59.2 to 65.9	hP4	P6₃/mmc
FeSb₂	81.4	oP6	Pnn2
(αSb)	100	hR2	R$\bar{3}$m
Metastable phase			
FeSb₄	90	cP1	Pm$\bar{3}$m

Fe-Sc

H. Okamoto, 1992

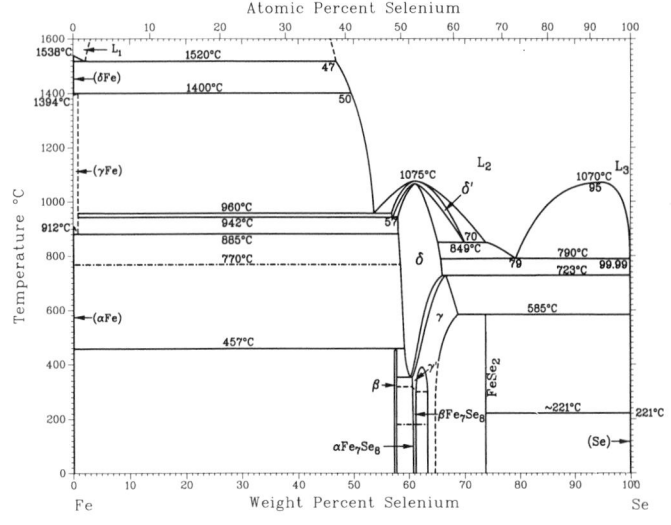

Phase	Composition, wt% Sc	Pearson symbol	Space group
(δFe)	~0	cI2	Im$\bar{3}$m
(γFe)	~0	cF4	Fm$\bar{3}$m
(αFe)	~0	cI2	Im$\bar{3}$m
βFe₂Sc	28.7	hP24	P6₃/mmc
αFe₂Sc	28.7	hP12	P6₃/mmc
λFe₂Sc	~31	cF24	Fd$\bar{3}$m
FeSc₇	84.9	...	...
(βSc)	~100	cI2	Im$\bar{3}$m
(αSc)	~100	hP2	P6₃/mmc

Fe-Se

H. Okamoto, 1992

Phase	Composition, wt% Se	Pearson symbol	Space group
(δFe)	~0	cI12	Im$\bar{3}$m
(γFe)	~0	cF4	Fm$\bar{3}$m
(αFe)	~0	cI2	Im$\bar{3}$m
β	57.6 to 58.0	tP4	P4/nmm
δ′	57 to 70	...	...
δ	58.1 to 66	hP4	P6₃/mmc
γ′	?	mC7	C2/m
γ	? to 69	mC14	C2/m
βFe₇Se₈	61.7	hP45	P3₁21
αFe₇Se₈	61.7	aP120	
FeSe₂	73.9	oP6	Pnnm
(γSe)	100	hP3	P3₁21
Metastable phases			
FeSe	58.6	c**	...
FeSe	58.6	hP4	P6₃/mmc
FeSe	58.6	tP2	P4/mmm
High-pressure phase			
FeSe₂	73.9	cP12	Pa3

Fe-Si

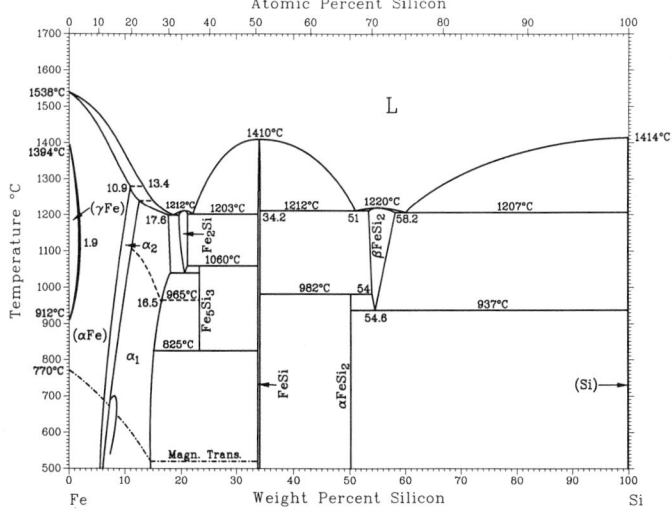

Phase	Composition, wt% Si	Pearson symbol	Space group
(γFe)	0 to 10.9	cF4	$Fm\bar{3}m$
(αFe)	0 to 1.63	cI2	$Im\bar{3}m$
α_2	~5 to 12	cP2	$Pm\bar{3}m$
α_1	~5 to 18	cF16	$Fm\bar{3}m$
Fe_2Si	~20.1	hP6	$P\bar{3}m1$
Fe_5Si_3	23.2	hP16	$P6_3/mcm$
FeSi	~34	cP8	$P2_13$
$\beta FeSi_2$	53.4 to 58.2	tP3	$P4/mmm$
$\alpha FeSi_2$	50.2	oC48	$Cmca$
(Si)	100	cF8	$Fd\bar{3}m$

Fe-Sm

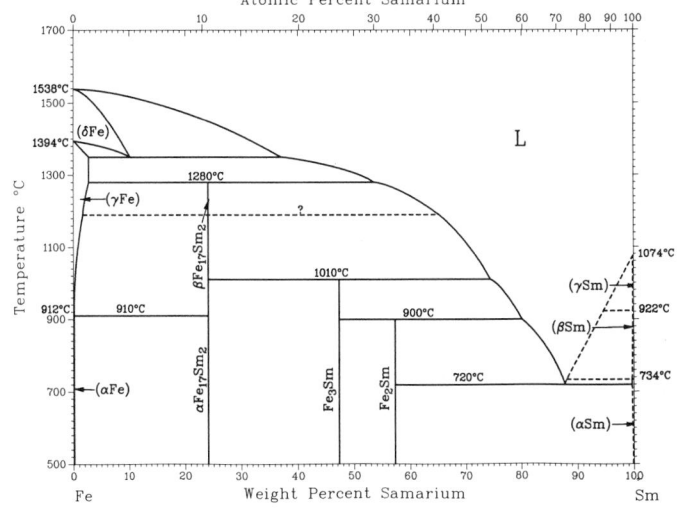

Phase	Composition, wt% Sm	Pearson symbol	Space group
(δFe)	~0	cI2	$Im\bar{3}m$
(γFe)	~0	cF4	$Fm\bar{3}m$
(αFe)	0	cI2	$Im\bar{3}m$
$\beta Fe_{17}Sm_2$	24.0	hP38	$P6_3/mmc$
$\alpha Fe_{17}Sm_2$	24.0	hR19	$R\bar{3}m$
Fe_3Sm	47	hR12	$R\bar{3}m$
Fe_2Sm	57.3	cF24	$Fd\bar{3}m$
(γSm)	100	cI2	$Im\bar{3}m$
(βSm)	~100	hP2	$P6_3/mmc$
(αSm)	>99.8 to 100	hR3	$R\bar{3}m$
Questionable phase			
Fe_5Sm	35.1	hP6	$P6/mmm$

Fe-Sn

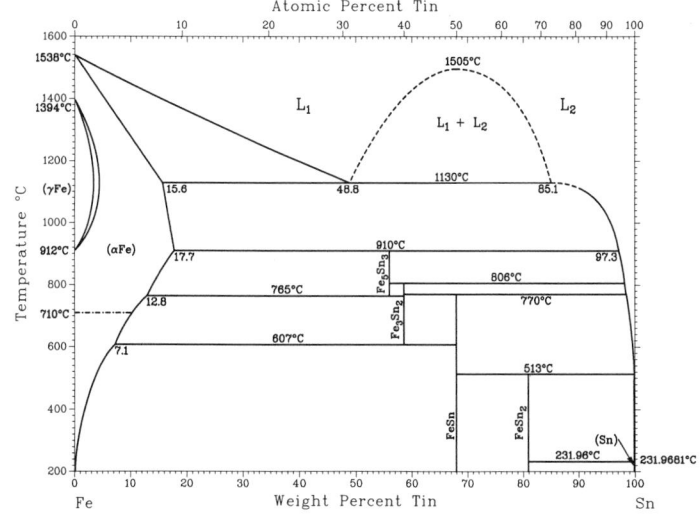

Phase	Composition, wt% Sn	Pearson symbol	Space group
(γFe)	0 to 1.7	cF4	$Fm\bar{3}m$
(αFe)	0 to 17.7	cI2	$Im\bar{3}m$
Fe_5Sn_3	56.1	hP6	$P6_3/mmc$
Fe_3Sn_2	59	hR10	$R\bar{3}m$
FeSn	68.0	hP6	$P6/mmm$
$FeSn_2$	81.0	tI2	$I4/mcm$
(βSn)	100	tI4	$I4_1/amd$
(αSn)	100	cF8	$Fm\bar{3}m$
Oxygen stabilized phase			
"Fe_3Sn"	42	hP8	$P6_3/mmc$

Fe-Tb

H. Okamoto, 1992

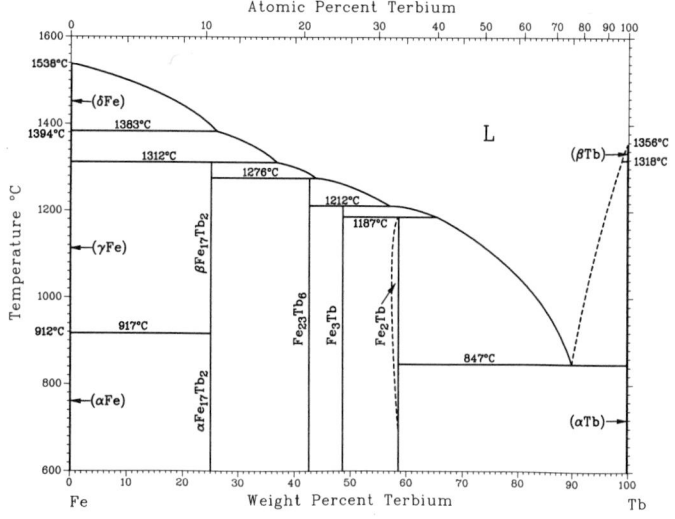

Phase	Composition, wt % Tb	Pearson symbol	Space group
(δFe)	0	cI2	Im$\bar{3}$m
(γFe)	~0	cF4	Fm$\bar{3}$m
(αFe)	0	cI2	Im$\bar{3}$m
βFe$_{17}$Tb$_2$	25.0	hP38	P6$_3$/mmc
αFe$_{17}$Tb$_2$	25.0	hR19	R$\bar{3}$m
Fe$_{23}$Tb$_6$	42.6	cF116	Fm$\bar{3}$m
Fe$_3$Tb	49	hR12	R$\bar{3}$m
Fe$_2$Tb	58.7	cF24	Fd$\bar{3}$m
Fe$_2$Tb(a)	58.7	hR6	R$\bar{3}$m
(βTb)	100	cI2	Im$\bar{3}$m
(αTb)	100	hP2	P6$_3$/mmc

(a) Distorted Cu$_2$Mg type due to magnetostriction at low temperatures

Fe-Te

H. Okamoto and L.E. Tanner, 1992

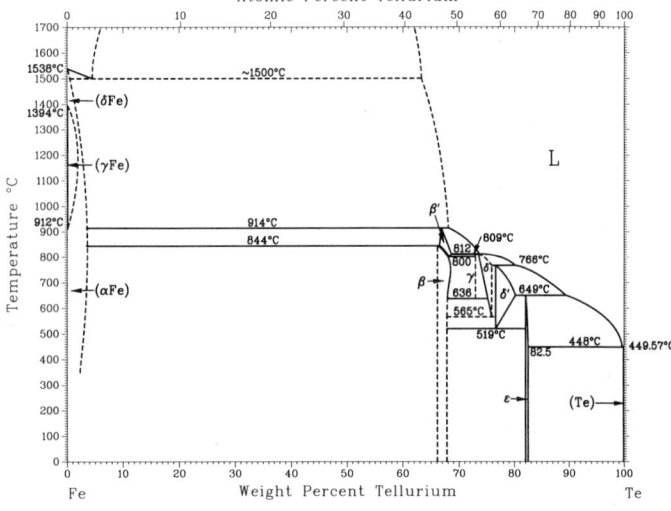

Phase	Composition, wt % Te	Pearson symbol	Space group
(δFe, αFe)	0 to 3.4	cI2	Im$\bar{3}$m
(γFe)	~0	cF4	Fm$\bar{3}$m
β'	66.5 to 68.3	hR*	...
β	66 to 68.3	tP4	P4/nmm
		tP6	P4/nmm
β$_1$(a)	66 to 68	mP*	P2$_1$/m
γ	73.0	...	...
δ	74 to 77	mC14	C2/m
δ'	76.9 to 80.6	hP4	P6$_3$/mmc
ε	82.0 to 82.5	oP6	Pnn2
FeTe$_2$I(b)	82.1	cP12	Pa$\bar{3}$
(Te)	100	hP3	P3$_1$21

(a) Low-temperature phase. (b) High-pressure phase

Fe-Th

H. Okamoto, 1992

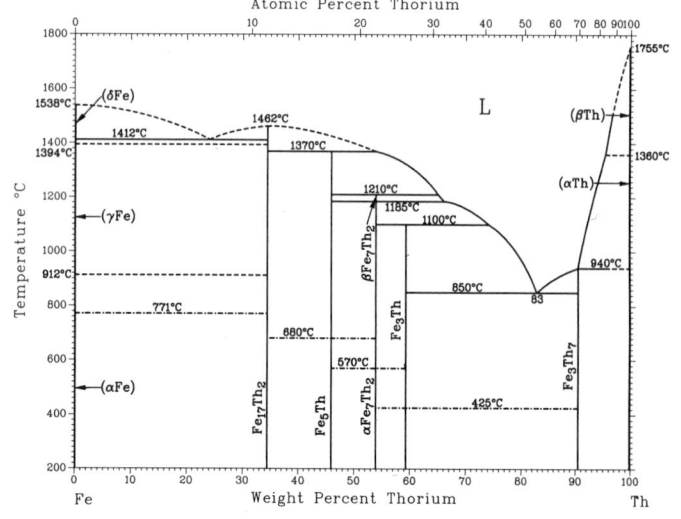

Phase	Composition, wt % Th	Pearson symbol	Space group
(δFe)	0	cI2	Im$\bar{3}$m
(γFe)	0	cF4	Fm$\bar{3}$m
(αFe)	0	cI2	Im$\bar{3}$m
Fe$_{17}$Th$_2$	32.8	hR19	R$\bar{3}$m
Fe$_5$Th	45.4	hP6	P6/mmm
βFe$_7$Th$_2$	54.2	hR18	R$\bar{3}$m
αFe$_7$Th$_2$	54.2	hP36	P6$_3$/mmc
Fe$_3$Th	58	hR12	R$\bar{3}$m
Fe$_3$Th$_7$	91	hP20	P6$_3$mc
(βTh)	100	cI2	Im$\bar{3}$m
(αTh)	100	cF4	Fm$\bar{3}$m

Fe-Ti

J.L. Murray, 1992

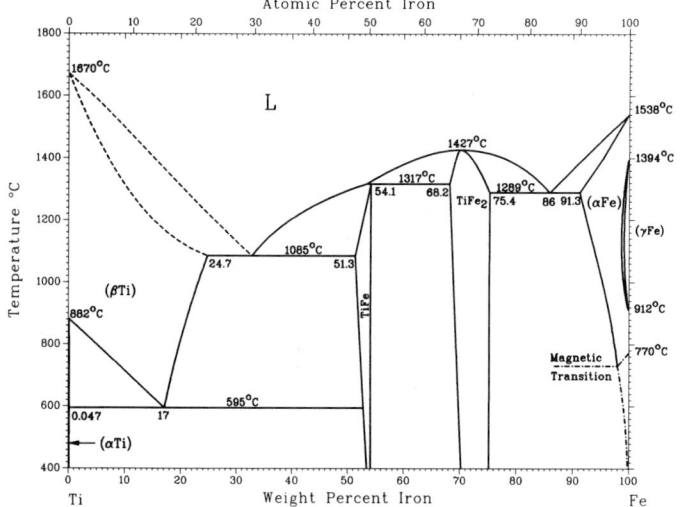

Phase	Composition, wt% Fe	Pearson symbol	Space group
(αTi)	0 to 0.047	hP2	$P6_3/mmc$
(βTi)	0 to 24.7	cI2	$Im\overline{3}m$
TiFe	51.3 to 54.1	cP2	$Pm\overline{3}m$
TiFe$_2$	68.2 to 75.4	hP12	$P6_3/mmc$
(αFe)	91.3 to 100	cI2	$Im\overline{3}m$
(γFe)	99.5 to 100	cF4	$Fm\overline{3}m$
ω	(a)	hP3	$P6/mmm$

(a) Metastable phase

Fe-Tm

H. Okamoto, 1992

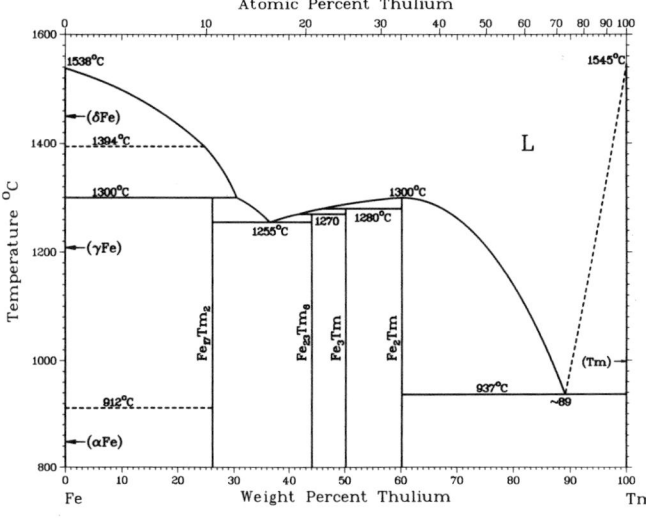

Phase	Composition, wt% Tm	Pearson symbol	Space group
(δFe)	0	cI2	$Im\overline{3}m$
(γFe)	0	cF4	$Fm\overline{3}m$
(αFe)	0	cI2	$Im\overline{3}m$
Fe$_{17}$Tm$_2$	26.2	hP38	$P6_3/mmc$
Fe$_{23}$Tm$_6$	44.1	cF114	$Fm\overline{3}m$
Fe$_3$Tm	50.2	hR12	$R\overline{3}m$
Fe$_2$Tm	60.2	cF24	$Fd\overline{3}m$
(Tm)	100	hP2	$P6_3/mmc$
Metastable phase			
...	~75	hP12	$P6_3/mmc$

Fe-U

H. Okamoto, 1992

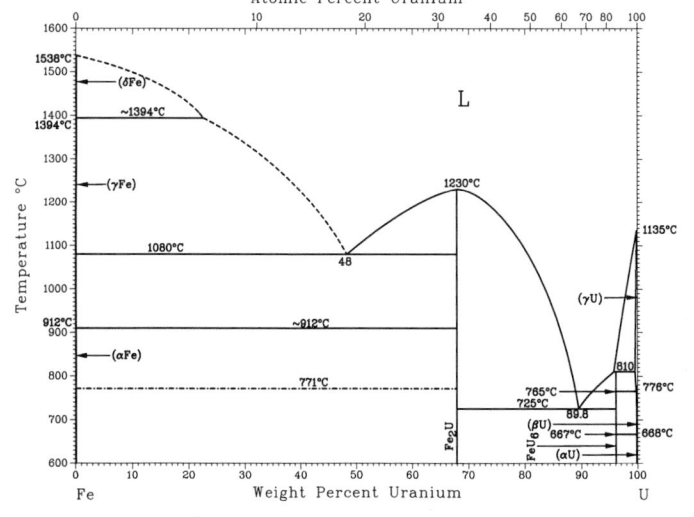

Phase	Composition, wt% U	Pearson symbol	Space group
(δFe)	0	cI2	$Im\overline{3}m$
(γFe)	0	cF4	$Fm\overline{3}m$
(αFe)	0	cI2	$Im\overline{3}m$
Fe$_2$U	68.0	cF24	$Fd\overline{3}m$
FeU$_6$	96.2	tI28	$I4/mcm$
(γU)	99.7 to 100	cI2	$Im\overline{3}m$
(βU)	99.9 to 100	tP30	$P4_2/mnm$
(αU)	99.99 to 100	oC4	$Cmcm$

Fe-V

J.F. Smith, 1992

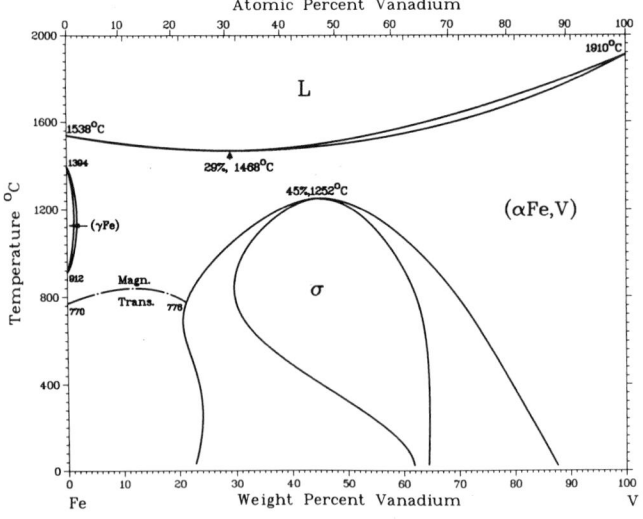

Phase	Composition, wt% V	Pearson symbol	Space group
(αFe,V)	0 to 100	cI2	Im$\bar{3}$m
(γFe)	0 to 1.2	cF4	Fm$\bar{3}$m
σ	30 to 65	tP30	P4$_2$/mnm
Metastable phase			
α′	47.7	cP2	Pm$\bar{3}$m

Fe-W

S.V. Nagender Naidu, A.M. Sriramamurthy, and P. Rama Rao, 1992

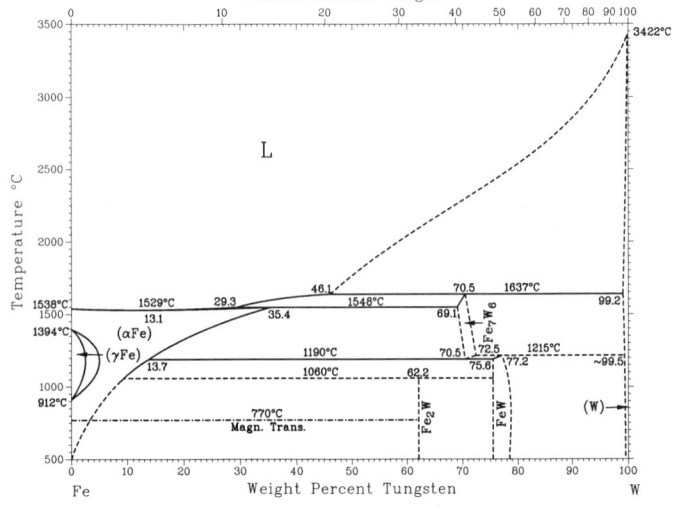

Phase	Composition, wt% W	Pearson symbol	Space group
(γFe)	0	cF4	Fm$\bar{3}$m
(αFe)	0	cI2	Im$\bar{3}$m
Fe$_7$W$_6$ (μ)	~70.5	hR13	R$\bar{3}$m
FeW (δ)	~77.2	(a)	P2$_1$2$_1$2$_1$
(W)	100	cI2	Im$\bar{3}$m
Metastable phase			
Fe$_2$W (λ)	62.2	hP12	P6$_3$/mmc

(a) Orthorhombic

Fe-Zn

B.P. Burton and P. Perrot, 1992

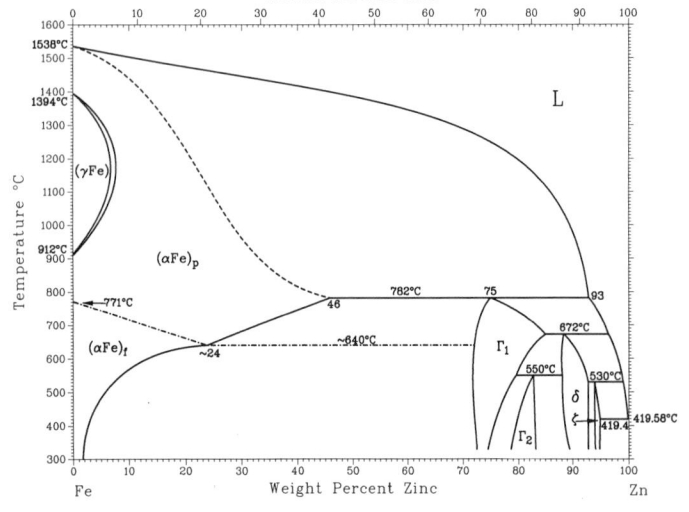

Phase	Composition, wt% Zn	Pearson symbol	Space group
(γFe)	0 to 6.59	cF4	Fm$\bar{3}$m
(αFe,δFe)	0 to 46	cI2	Im$\bar{3}$m
Γ$_1$	~72 to ~85	cI52	I$\bar{4}$3m
Γ$_2$	0.91 to 83	cF408	F$\bar{4}$3m
δ-FeZn$_{10}$	88.5 to 93.0	hP555	P6$_3$mc
ζ-FeZn$_{13}$	~94 to 94.8?	mC28	C2/m
(Zn)	~100	hP2	P6$_3$/mmc

Fe-Zr

D. Arias and J.P. Abriata, 1992

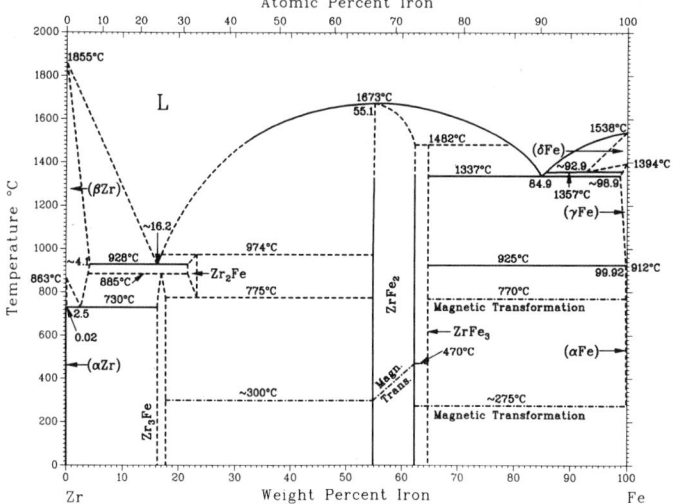

Phase	Composition, wt% Fe	Pearson symbol	Space group
(βZr)	0 to ~4.1	cI2	Im$\bar{3}$m
(αZr)	0 to 0.02	hP2	P6$_3$/mmc
Zr$_3$Fe	16.2 to 18.3	oC16	Cmcm
Zr$_2$Fe	21.6 to 23.4	tI12	I4/mcm
ZrFe$_2$	54.3 to 62.2	cF24	Fd$\bar{3}$m
ZrFe$_3$	64.7	cF116	Fm$\bar{3}$m
(δFe)	~92.9 to 100	cI2	Im$\bar{3}$m
(γFe)	~98.9 to 100	cF4	Fm$\bar{3}$m
(αFe)	99.91 to 100	cI2	Im$\bar{3}$m

Ga-Gd

A. Palenzona and S. Cirafici, 1990

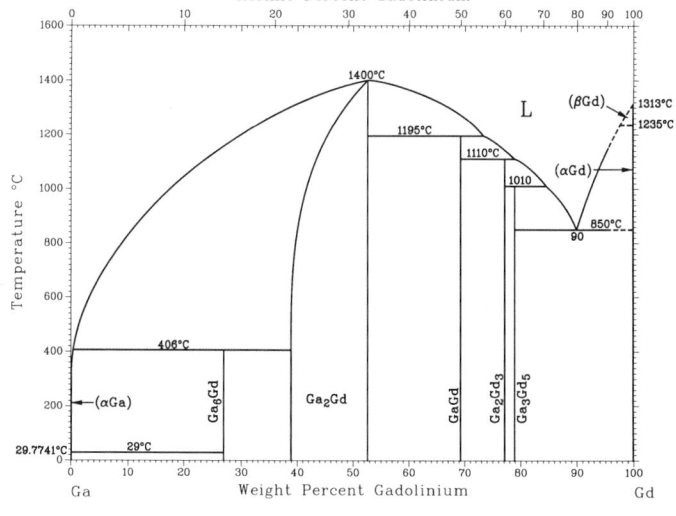

Phase	Composition, wt% Gd	Pearson symbol	Space group
(Ga)	~0	oC8	Cmca
Ga$_6$Gd	27.32	tP14	P4/nbm
Ga$_2$Gd	39 to 53.0	hP3	P6/mmm
GaGd	69.3	oC8	Cmcm
Ga$_2$Gd$_3$	77	tI80	I4/mcm
Ga$_3$Gd$_5$	79.0	tI32	I4/mcm
(βGd)	~100	cI2	Im$\bar{3}$m
(αGd)	~100	hP2	P6$_3$/mmc

Ga-Ho

H. Okamoto, 1990

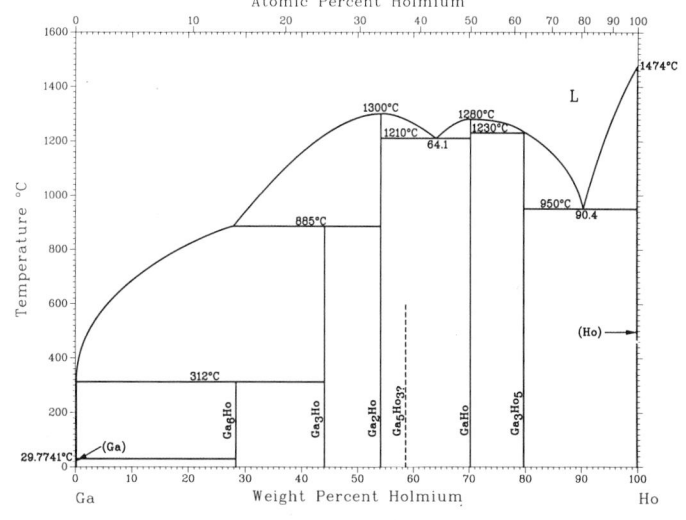

Phase	Composition, wt% Ho	Pearson symbol	Space group
(Ga)	0	oC8	Cmca
Ga$_6$Ho	28.3	tP14	P4/nbm
Ga$_3$Ho	44	cP4	Pm$\bar{3}$m
Ga$_2$Ho	54.1	hP3	P6/mmm
Ga$_5$Ho$_3$	58.7	oP32	P4/nbm
GaHo	70.3	oC8	Cmcm
Ga$_3$Ho$_5$	79.8	hP16	P6$_3$/mcm
(Ho)	100	hP2	P6$_3$/mmc

Ga-In

T.J. Anderson and I. Ansara, 1992

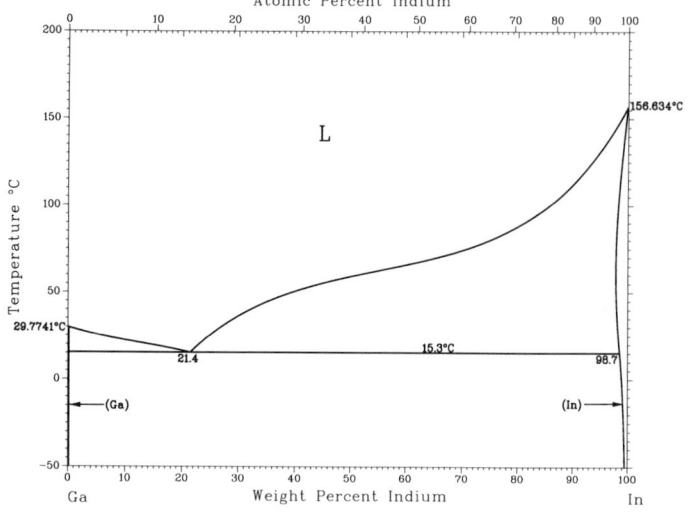

Phase	Composition, wt% In	Pearson symbol	Space group
(αGa)	0	oC8	Cmca
(βGa)	0	...	C2/c
(In)	98.6 to 100	tI2	I4/mmm

Ga-La

A. Palenzona and S. Cirafici, 1990

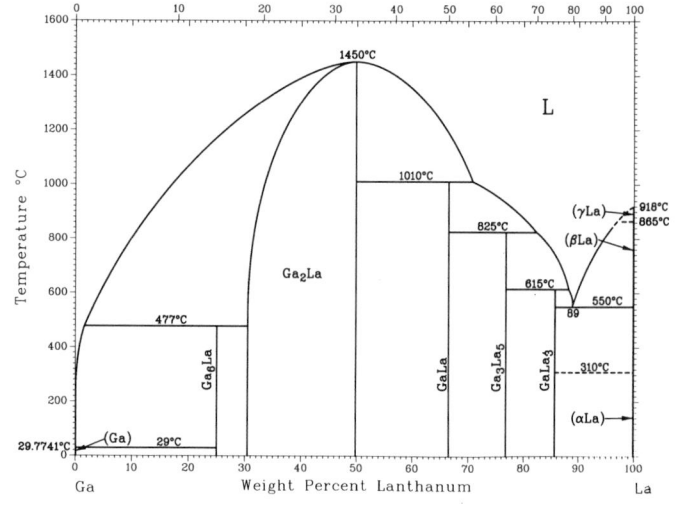

Phase	Composition, wt% La	Pearson symbol	Space group
(αGa)	~0	oC8	Cmca
Ga6La	24.9	tP14	P4/nbm
Ga4La(a)	33	o**	...
Ga2La	30 to 49.9	hP3	P6/mmm
GaLa	66.6	oC8	Cmcm
Ga3La5	76.9	tI32	I4/mcm
GaLa3	86	cP4	Pm$\overline{3}$m
(γLa)	~100	cI2	Im$\overline{3}$m
(βLa)	~100	cF4	Fm$\overline{3}$m
(αLa)	~100	hP4	P6₃/mmc

(a) Not shown on diagram; needs further confirmation

Ga-Li

J. Sangster and A.D. Pelton, 1991

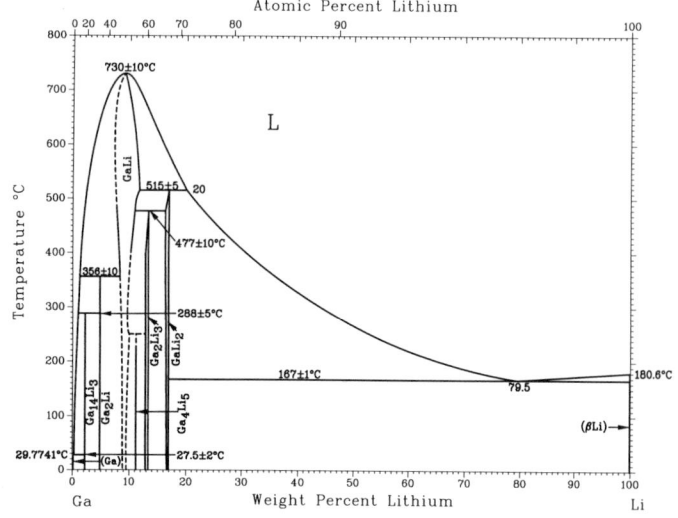

Phase	Composition, wt% Li	Pearson symbol	Space group
(Ga)	0	oC8	Cmca
Ga14Li3	2.1	hR51	R$\overline{3}$m
Ga2Li	4.7(a)	...	...
GaLi	8 to 11(b)	cF16	Fd$\overline{3}$m
Ga4Li5	11.1	hR9	P$\overline{3}$m1
Ga2Li3	12.8 to 13.2	hR15	R$\overline{3}$m
GaLi2	16 to 17	oC12	Cmcm
(βLi)	100	cI2	Im$\overline{3}$m
(αLi)(c)	100	hP2	P6₃/mmc

(a) Stoichiometry uncertain. (b) Near 400 °C. (c) Below −193 °C

Ga-Lu

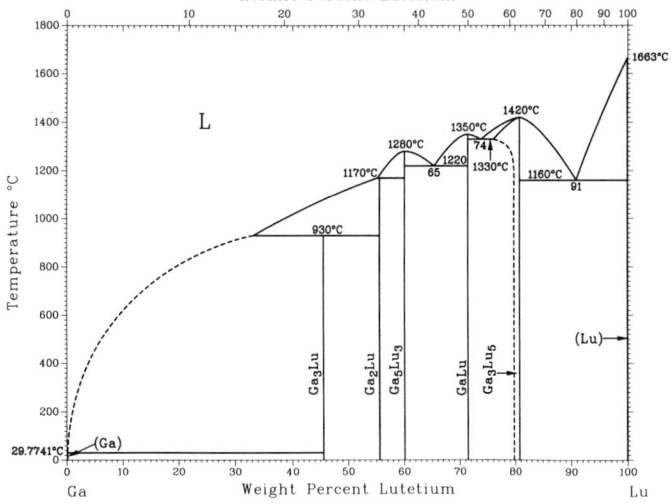

S.P. Yatsenko, A.A. Semyannikov, B.G. Semenov, and K.A. Chuntonov, 1979

Phase	Composition, wt% Lu	Pearson symbol	Space group
(Ga)	0	oC8	Cmca
Ga₃Lu	46	cP4	Pm$\bar{3}$m
Ga₂Lu	55.6	oI12	Imma
Ga₅Lu₃	60.1	oP32	Pnma
GaLu	71.5	oC8	Cmcm
Ga₃Lu₅	? to 80.7	hP16	P6₃/mcm
(Lu)	100	hP2	P6₃/mmc

Ga-Mg

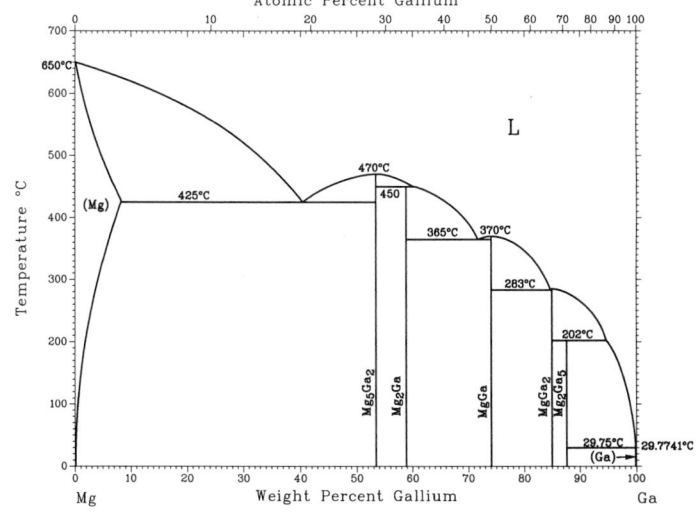

H. Okamoto, 1991

Phase	Composition, wt% Ga	Pearson symbol	Space group
(Mg)	0 to 9.4	hP2	P6₃/mmc
Mg₅Ga₂	53.43	oI28	Ibam
Mg₂Ga(a)	58.9	hP18	P$\bar{6}$2c
MgGa	74.2	tI32	I4₁/a
MgGa₂	85.15	oP24	Pbam
Mg₂Ga₅	87.76	tI28	I4/mmm
(Ga)	100	oC8	Cmca

(a) The structure is closely related to the Fe₂P (hP9) type with a small deviation.

Ga-Mn

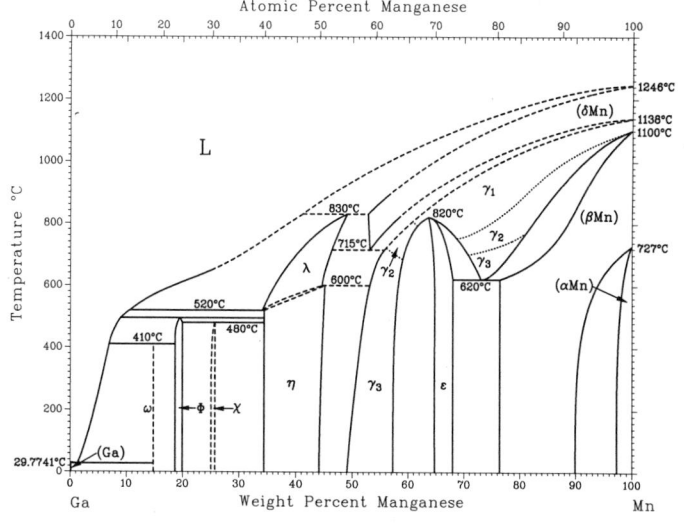

X.S. Lu, J.K. Liang, and M.G. Zhou, 1980

Phase	Composition, wt% Mn	Pearson symbol	Space group
(Ga)	0	oC8	Cmca
ω	15	oC28	Cmcm
φ	~18 to 20	...	...
χ	~25	tP14	P4/mbm
λ	~34.0 to 49	hR26	R$\bar{3}$m
η	34.4 to 44	...	...
(δMn)	~53 to 100	cI2	Im$\bar{3}$m
γ₁(γMn)	~62 to 100	cF4	Fm$\bar{3}$m
γ₂	~56 to 100	tI8	I4/mmm
γ₃	~49 to ~59	tP4	P4/mmm
ε	64 to 68	...	...
(βMn)	76.3 to ~100	cP20	P4₁32
(αMn)	97.3 to ~100	cI58	I$\bar{4}$3m

Ga-Mo

From [Molybdenum]

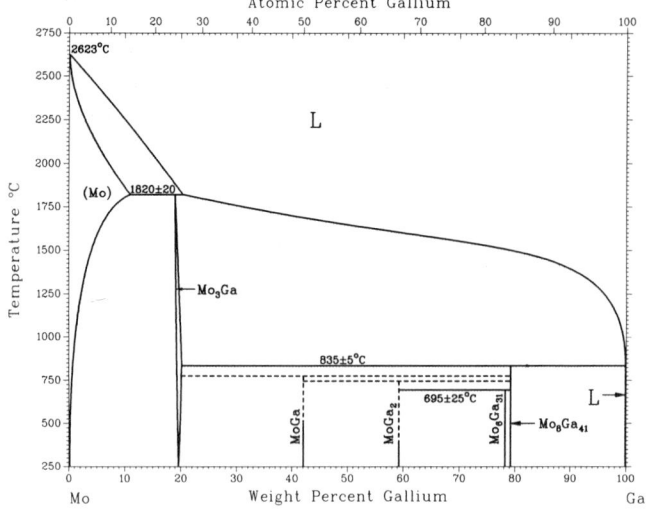

Phase	Composition, wt% Ga	Pearson symbol	Space group
(Mo)	0 to 11	cI2	$Im\bar{3}m$
Mo₃Ga	~20	cP8	$Pm\bar{3}n$
MoGa	42.1	...	...
MoGa₂	59.3	...	...
Mo₆Ga₃₁	~78	mP148	$P2_1/c$
Mo₈Ga₄₁	~79	hR49	$R\bar{3}$
(Ga)	100	oC8	Cmca

Ga-Na

A.D. Pelton and S. Larose, 1990

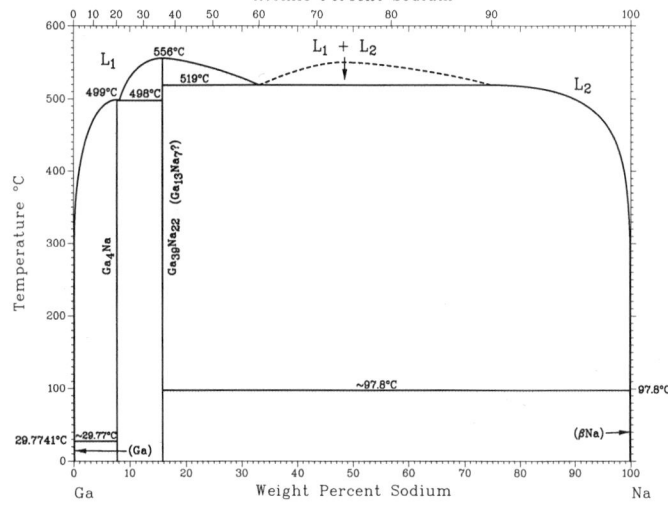

Phase	Composition, wt% Na	Pearson symbol	Space group
(Ga)	0	oC8	Cmca
Ga₄Na	8	tI10	I4/mmm
Ga₁₃Na₇(a)	15	hR360	$R\bar{3}m$
Ga₁₃Na₇(b, c)	15	oP240	Pnma
Ga₃₉Na₂₂(c)	15.7	oP244	Pnma
(βNa)	100	cI2	$Im\bar{3}m$
(αNa)	100	hP2	P6₃/mmc

(a) Structure observed when compound prepared with excess Ga. (b) Structure observed when compound prepared with excess Na. (c) Same compound with same diffractogram, although different stoichiometries have been reported

Ga-Nb

H. Okamoto, 1990

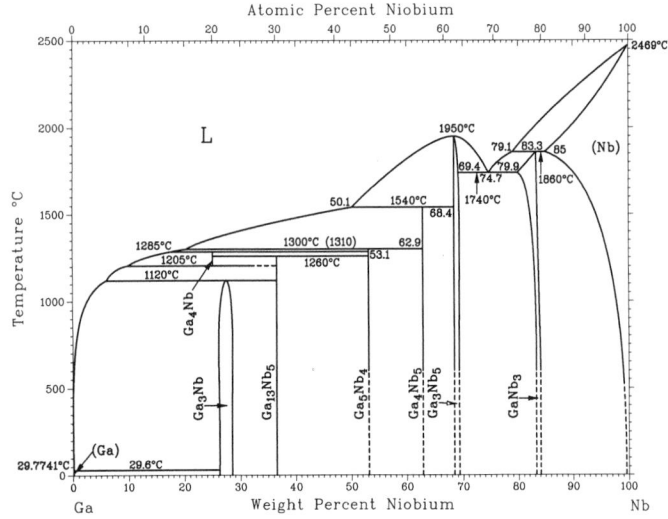

Phase	Composition, wt% Nb	Pearson symbol	Space group
(Ga)	0	oC8	Cmca
Ga₄Nb	25	...	...
Ga₃Nb	26 to 29	tI8	I4/mmm
Ga₁₃Nb₅	36.7	oC36	Cmmm
Ga₅Nb₄	53.3	t**	...
Ga₄Nb₅	62.5	hP18	P6₃/mcm
Ga₂Nb₃(a)	67	tP10	P4/mbm
Ga₃Nb₅	68.4 to 69.4	tI32	I4/mcm
GaNb₃	79.9 to 84	cP8	$Pm\bar{3}n$
(Nb)	100	cI2	$Im\bar{3}m$

(a) Not in phase diagram

Ga-Nd

From [Moffatt]

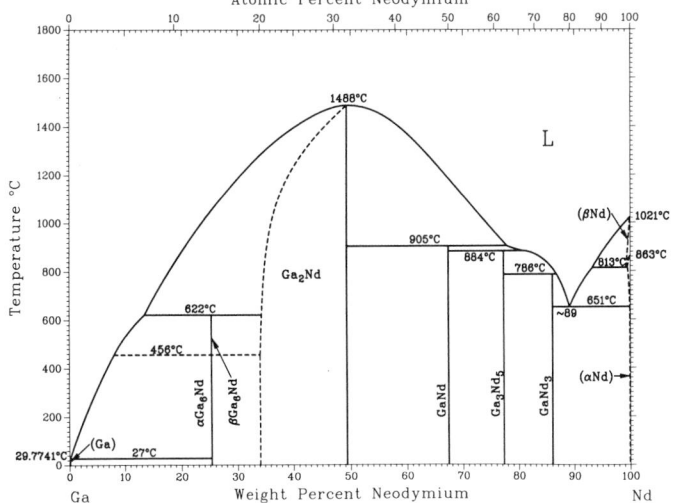

Phase	Composition, wt% Nd	Pearson symbol	Space group
(Ga)	0	oC8	Cmca
βGa₆Nd	25.7	…	…
αGa₆Nd	25.7	tP14	P4/nbm
Ga₂Nd	~34 to 50.8	hP3	P6/mmm
GaNd	67.4	oC8	Cmcm
Ga₃Nd₅	77.5	tI32	I4/mcm
GaNd₃	86	cP4	Pm$\bar{3}$m
(βNd)	? to 100	cI2	Im$\bar{3}$m
(αNd)	? to 100	hP4	P6₃/mmc

Ga-Ni

S.Y. Lee and P. Nash, 1991

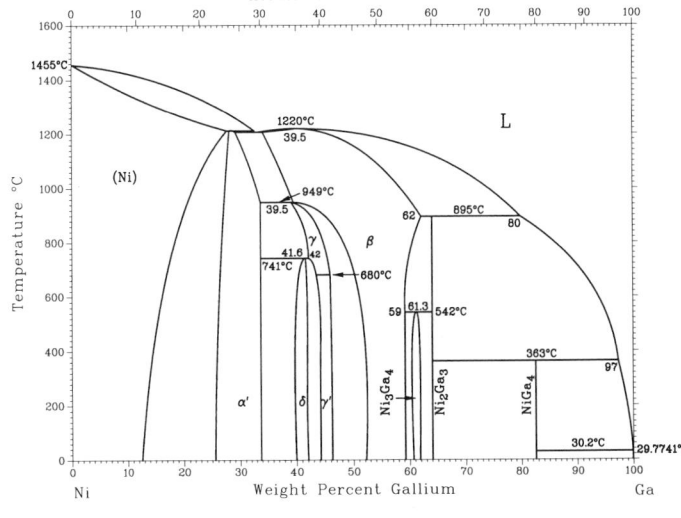

Phase	Composition, wt% Ga	Pearson symbol	Space group
(Ni)	0 to 27.6	cF4	Fm$\bar{3}$m
α′(Ni₃Ga)	25.8 to 34	cP4	Pm$\bar{3}$m
β(NiGa)	34.2 to 62	cP2	Pm$\bar{3}$m
γ(Ni₃Ga₂)	39.5 to 46	hP4	P6₃/mmc
δ(Ni₅Ga₃)	40.3 to 42	oC16	Cmmm
γ′(Ni₃Ga₂)	~43.4 to ~46.4	…	…
Ni₃Ga₄	~60.8 to 61.7	cI112	Ia3d
β′(Ni₂Ga₃)	64	hP5	P$\bar{3}$m1
ε(NiGa₄)	83	cI52	I$\bar{4}$3m
(Ga)	100	oC8	Cmca

Ga-Pb

I. Ansara and F. Ajersch, 1991

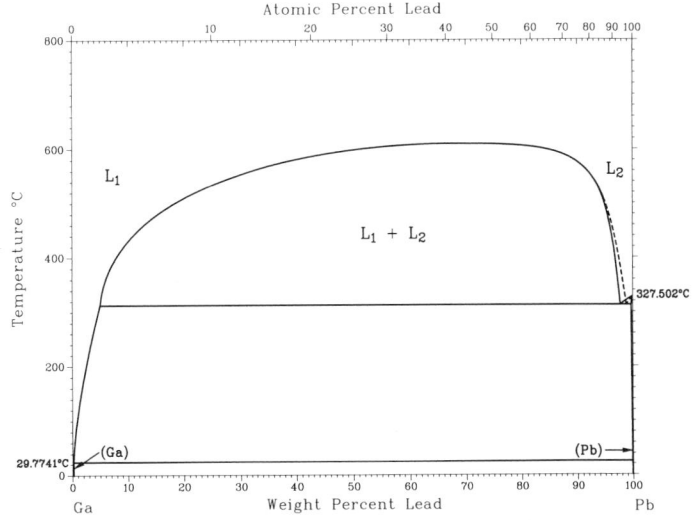

Phase	Composition, wt% Pb	Pearson symbol	Space group
(Ga)	0	oC8	Cmca
(Pb)	100	cF4	Fm$\bar{3}$m

Ga-Pd

H. Okamoto, 1990

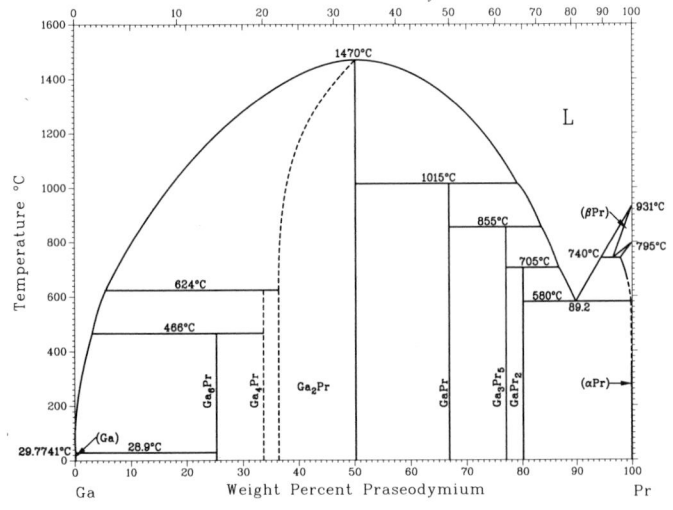

Phase	Composition, wt% Pd	Pearson symbol	Space group
(Ga)	0	oC8	Cmca
Ga₅Pd	~23.4	tI24	I4/mcm
Ga₇Pd₃	~40	cI40	Im$\bar{3}$m
GaPd	~60.4	cP8	P2₁3
Ga₄Pd₅	~65.7	cP2	Pm$\bar{3}$m
Ga₃Pd₅	~71.8	oP16	Pbam
GaPd₂	73 to 82	oP12	Pnma
Ga₃Pd₇	~78	...	...
Ga₂Pd₅	79 to 80.5	oP28	Pnma
Ga₅Pd₁₃	80 to 81.3	o**	...
(Pd)	? to 100	cF4	Fm$\bar{3}$m

Ga-Pr

H. Okamoto, 1990

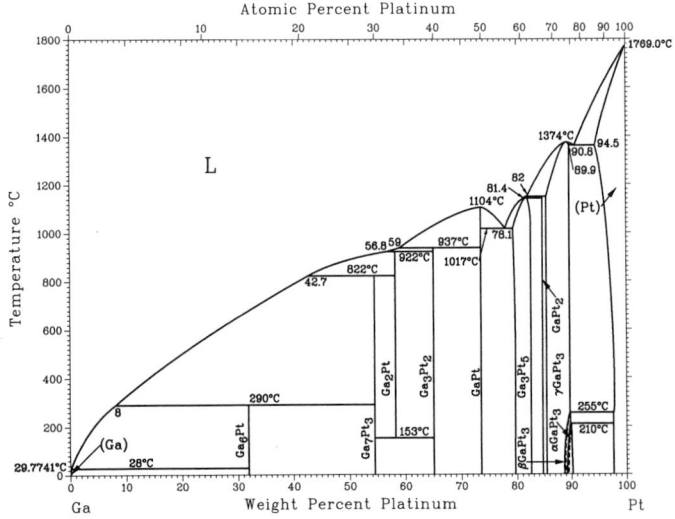

Phase	Composition, wt% Pr	Pearson symbol	Space group
(Ga)	0	oC8	Cmca
Ga₆Pr	25.2	tI14	P4/nbm
Ga₄Pr	34	...	...
Ga₂Pr	36 to 50.2	hP3	P6/mmm
GaPr	66.9	oC8	Cmcm
Ga₃Pr₅	77.1	tP32	P4/ncc
GaPr₂	80.2	oP12	Pnma
(βPr)	96 to 100	cI2	Im$\bar{3}$m
(αPr)	98 to 100	hP4	P6₃/mmc

Ga-Pt

H. Okamoto, 1990

Phase	Composition, wt% Pt	Pearson symbol	Space group
(Ga)	0	oC8	Cmca
Ga₆Pt	31.8	o**	...
Ga₇Pt₃	55	cI40	Im$\bar{3}$m
Ga₂Pt	58.3	cF12	Fm$\bar{3}$m
Ga₃Pt₂	65	hP5	P$\bar{3}$m1
GaPt	73.7	cP8	P2₁3
Ga₃Pt₅	79 to 83	oC16	Cmmm
γGaPt₂	84.9	oP16	Pbam
βGaPt₂	84.9	o**	...
αGaPt₂	84.9	tP2	P4/mmm
γGaPt₃	85 to 90	cP4	Pm$\bar{3}$m
βGaPt₃	~89	tI16	I4/mcm
αGaPt₃	~89	tP16	P4/mbm
(Pt)	94.5 to 100	cF4	Fm$\bar{3}$m

Ga-Pu

D.E. Peterson and M.E. Kassner, 1988

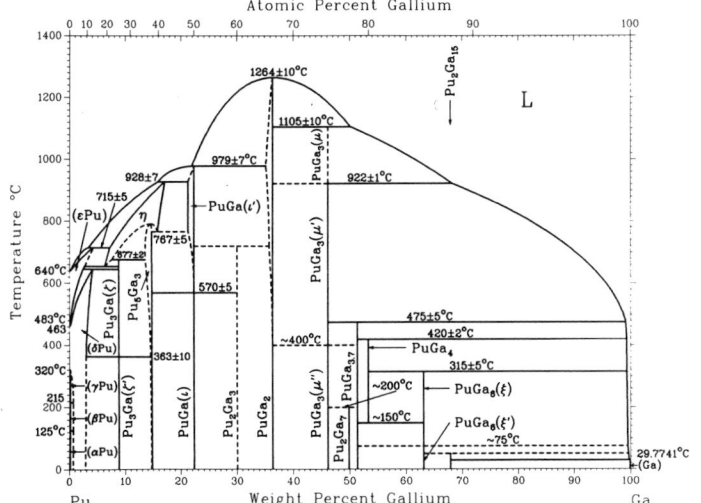

Phase	Composition, wt% Ga	Pearson symbol	Space group
(εPu)	0 to 4	cI2	Im$\bar{3}$m
(δ'Pu)	0 to 0.07	tI2	I4/mmm
(δPu)	0 to 3.9	cF4	Fm$\bar{3}$m
(γPu)	0	oF8	Fddd
(βPu)	0	mC34	C2/m
(αPu)	0	mP16	P2₁/m
η	6.1 to 17	...	I2₁3(a)
Pu₃Ga(ζ)	9	cP4	Pm$\bar{3}$m
Pu₃Ga(ζ')	9	tP4	P4/mmm
Pu₅Ga₃	14 to 14.6	tI38	I4/mcm
		(b)	...
PuGa(ι')	~22.2	cI2	Im$\bar{3}$m
		(c)	I4/mmm
PuGa(ι)	22.2	(d)	I4mm
Pu₂Ga₃	30	(e)	...
PuGa₂	36.4	hP3	P6/mmm
PuGa₃(μ)	46	...	...
PuGa₃(μ')	46	hP8	P6₃/mmc
PuGa₃(μ")	46	...	R$\bar{3}$m
Pu₂Ga₇	50.0	(c)	...
PuGa₃.₇	51.4	...	...
PuGa₄	53	oI20	Imma
PuGa₆(ξ)	63.1	...	P4/nbm
			P4/mbm
PuGa₆(ξ')	63.1	...	...
Pu₂Ga₁₅	68.1	(c)	...
(Ga)	100	oC8	Cmca

(a) Partially ordered. (b) Face-centered cubic. (c) Tetragonal. (d) Body-centered tetragonal. (e) Hexagonal

Ga-S

P.G. Rustamov, B.N. Mardakhaev, and M.G. Safarov, 1967

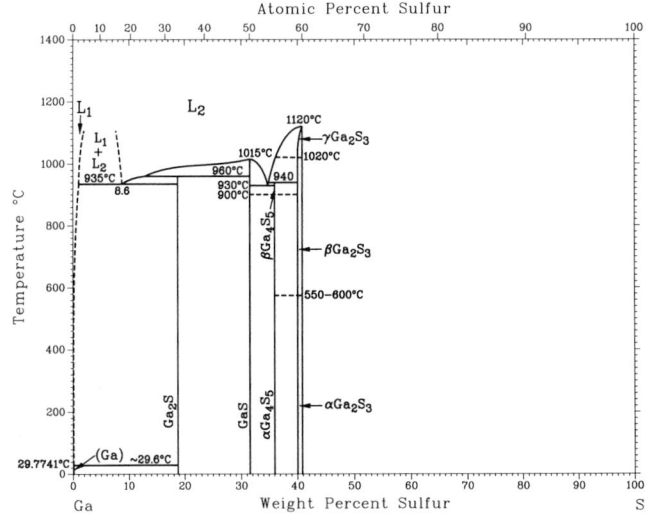

Phase	Composition, wt% S	Pearson symbol	Space group
(Ga)	0	oC8	Cmca
Ga₂S	18.7	...	...
GaS	31.5	hP8	P6₃/mmc
βGa₄S₅	36.5	...	...
αGa₄S₅	36.5	...	...
γGa₂S₃	41	hP4	P6₃mc
βGa₂S₃	41	mC20	Cc
αGa₂S₃	41	cF8	F$\bar{4}$3m

Ga-Sb

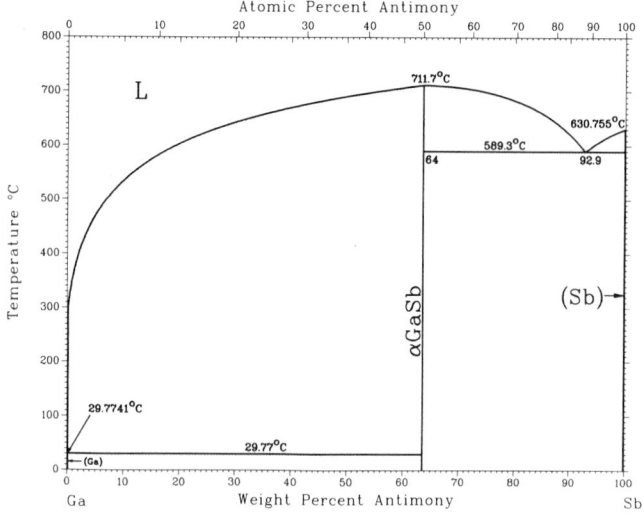

T.I. Ngai, R.C. Sharma, and Y.A. Chang, 1988

Phase	Composition, wt% Sb	Pearson symbol	Space group
(Ga)	0	oC8	Cmca
αGaSb	63.6	cF8	F$\overline{4}$3m
βGaSb(a)	63.6	tI4	I4₁/amd
(Sb)	100	hR2	R$\overline{3}$m

(a) At high pressure

Ga-Sc

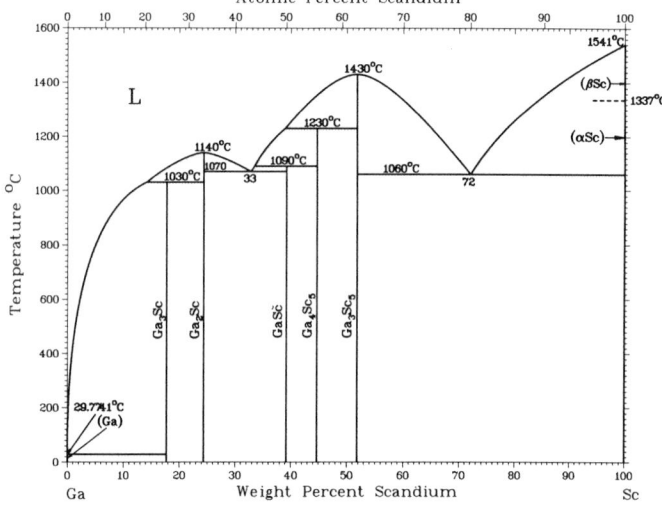

S.P. Yatsenko, A.A. Semyannikov, G.B. Semenov, and K.A. Chuntonov, 1979

Phase	Composition, wt% Sc	Pearson symbol	Space group
(Ga)	0	oC8	Cmca
Ga₃Sc	18	cP4	Pm$\overline{3}$m
Ga₂Sc	24.4	oI12	Imma
GaSc	39.2	oC8	Cmcm
Ga₄Sc₅	44.7	tI84	I4/mmm
Ga₃Sc₅	51.8	hP16	P6₃/mcm
(βSc)	100	cI2	Im$\overline{3}$m
(αSc)	100	hP2	P6₃/mmc
Other reported phase			
Ga₃Sc₂	30	oP32	Pnma

Ga-Se

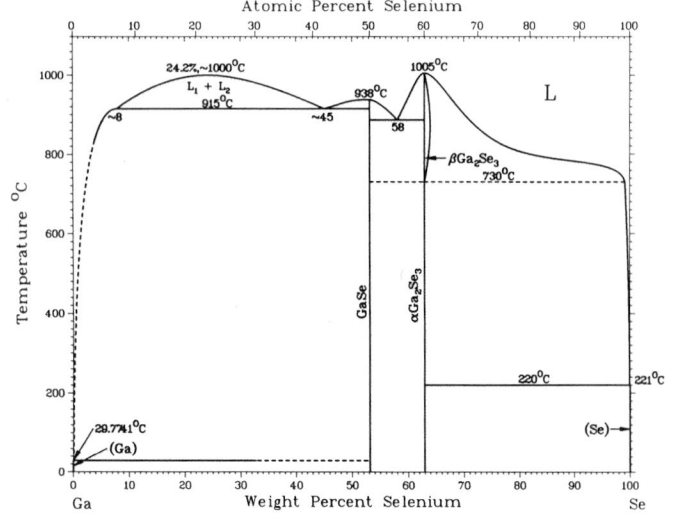

From [Moffatt]

Phase	Composition, wt% Se	Pearson symbol	Space group
(Ga)	0	oC8	Cmca
GaSe	53.1	hR4	R$\overline{3}$m
		hP8	P$\overline{6}$
		hP16	P6₃mc
βGa₂Se₃	~63	c**	...
αGa₂Se₃	63	mC20	Cc
(Se)	100	hP3	P3₁21

Ga-Sm

From [Moffatt]

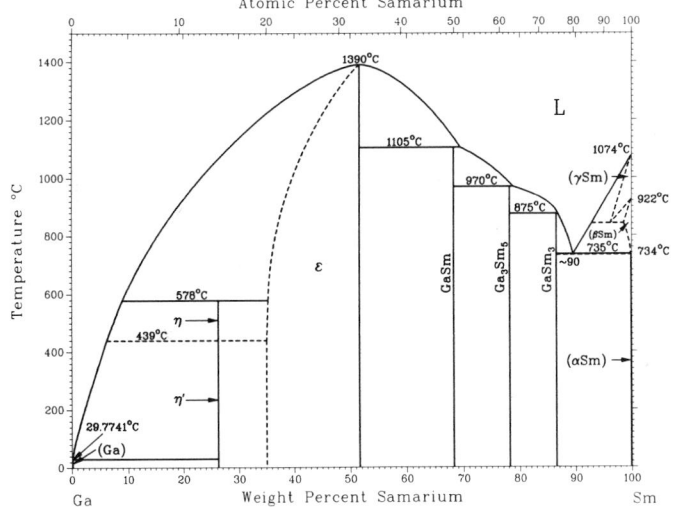

Phase	Composition, wt% Sm	Pearson symbol	Space group
(Ga)	0	oC8	Cmca
βGa₆Sm(η)	26.5	...	...
αGa₆Sm(η′)	26.5	tP14	P4/nbm
Ga₂Sm(ε)	~35 to 51.8	hP3	P6/mmm
GaSm	68.3	oC8	Cmcm
Ga₃Sm₅	78.2	tI32	I4/mcm
GaSm₃	87	cP4	Pm$\overline{3}$m
(γSm)	? to 100	cI2	Im$\overline{3}$m
(βSm)	? to 100	hP2	P6₃/mmc
(αSm)	100	hR3	R$\overline{3}$m

Ga-Sn

T.J. Anderson and I. Ansara, 1992

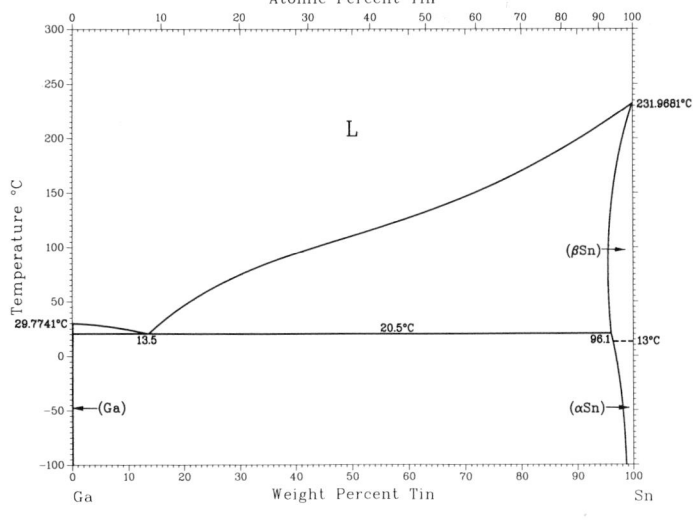

Phase	Composition, wt% Sn	Pearson symbol	Space group
(αGa)	0	oC8	Cmca
(βGa)(a)	0 to 0.03	tI2	I4/mmm
(βSn)	96.1 to 100	tI4	I4₁/amd

(a) Above 1.2 GPa

Ga-Sr

V.P. Itkin and C.B. Alcock, 1992

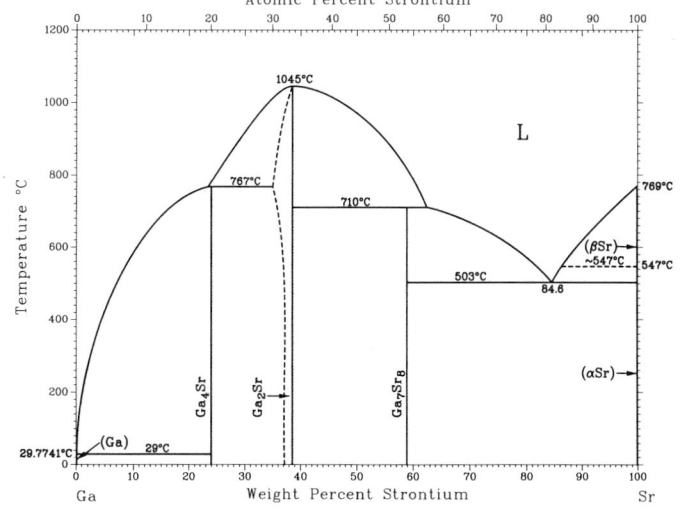

Phase	Composition, wt% Sr	Pearson symbol	Space group
(Ga)	0	hP2	P6₃/mmc
Ga₄Sr	24	tI10	I4/mmm
Ga₂Sr	35 to 38.6 (a)	hP3	P6/mmm
Ga₇Sr₈	58.9	cP60	P2₁3
(αSr)	100	cF4	Fm$\overline{3}$m
(βSr)	100	cI2	Im$\overline{3}$m

(a) After annealing at 900 °C

Ga-Tb

From [Moffatt]

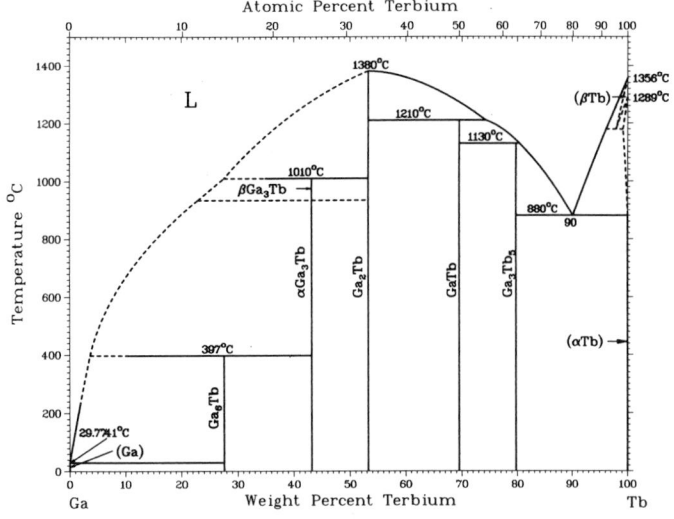

Phase	Composition, wt% Tb	Pearson symbol	Space group
(Ga)	0	oC8	Cmca
Ga₆Tb	27.6	tP14	P4/nbm
βGa₃Tb	43	cP4	Pm$\overline{3}$m
αGa₃Tb	43	hP8	P6₃/mmc
Ga₂Tb	53.2	hP3	P6/mmm
GaTb	69.5	oC8	Cmcm
Ga₃Tb₅	79.2	tI32	I4/mcm
(βTb)	? to 100	cI2	Im$\overline{3}$m
(αTb)	? to 100	hP2	P6₃/mmc

Ga-Te

U.R. Kattner, unpublished

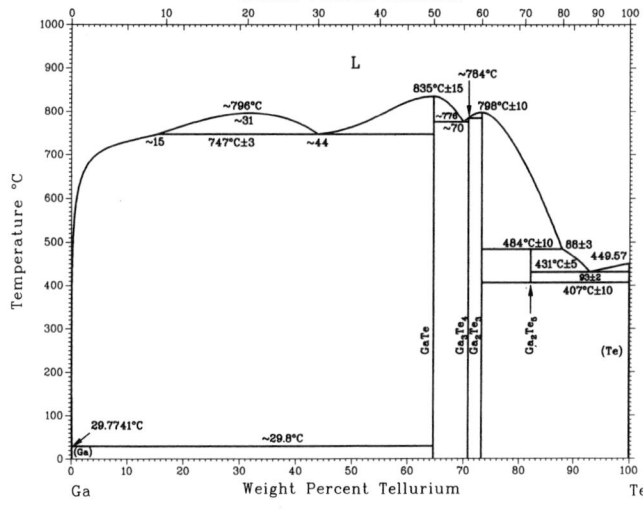

Phase	Composition, wt% Te	Pearson symbol	Space group
(Ga)	0	oC8	Cmca
GaTe	64.7	mC24	C2/m
Ga₃Te₄	70.9	hP*	...
Ga₂Te₃	73	cF8	F$\overline{4}$3m
Ga₂Te₅	82.1	tI14	I4/m
(Te)	100	hP3	P3₁21
Metastable (thin film)			
GaTe	64.7	hP8	P6₃/mmc

Ga-Tl

J. Klingbeil and R. Schmid-Fetzer, 1991

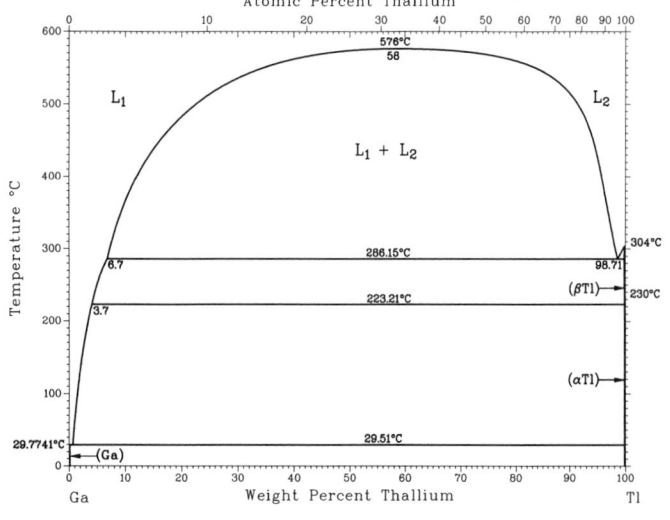

Phase	Composition, wt% Tl	Pearson symbol	Space group
(Ga)	0	oC	Cmca
(αTl)	100	hP2	P6₃/mmc
(βTl)	100	cI2	Im$\overline{3}$m

Ga-Tm

From [Moffatt]

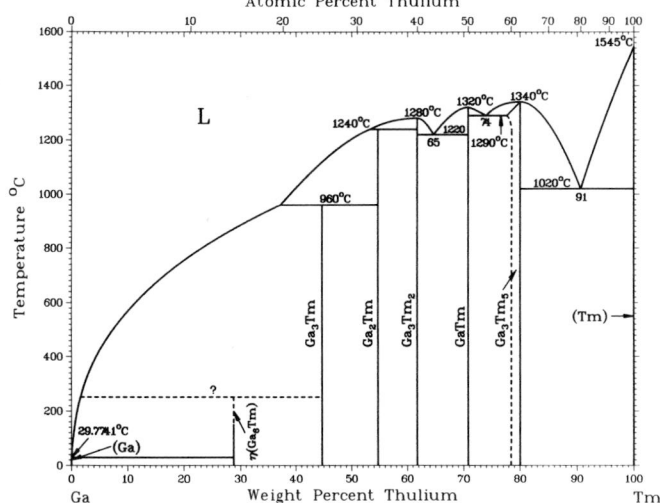

Phase	Composition, wt% Tm	Pearson symbol	Space group
(Ga)	0	oC8	Cmca
Ga₆Tm	28.8	tP14	P4/nbm
Ga₃Tm	45	cP4	Pm3̄m
Ga₂Tm	54.7	oI12	Imma
Ga₃Tm₂	62	…	…
GaTm	70.8	oC8	Cmcm
Ga₃Tm₅	? to 80.2	oP32	Pnma
(Tm)	100	hP2	P6₃/mmc

Ga-U

K.H.J. Buschow, 1973

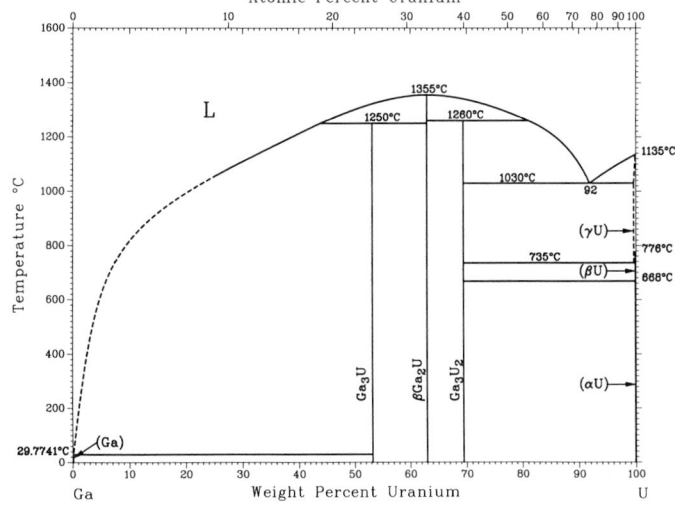

Phase	Composition, wt% U	Pearson symbol	Space group
(Ga)	0	oC8	Cmca
Ga₃U	53	cP4	Pm3̄m
βGa₂U	63.0	hP3	P6/mmm
αGa₂U(a)	63.0	oC*	Cmmm
Ga₃U₂	70	oC32	Cmcm
(γU)	? to 100	cI2	Im3̄m
(βU)	100	tP30	P4₂/mnm
(αU)	100	oC4	Cmcm

(a) Below the Curie temperature (−148 °C)

Ga-V

J.F. Smith, 1989

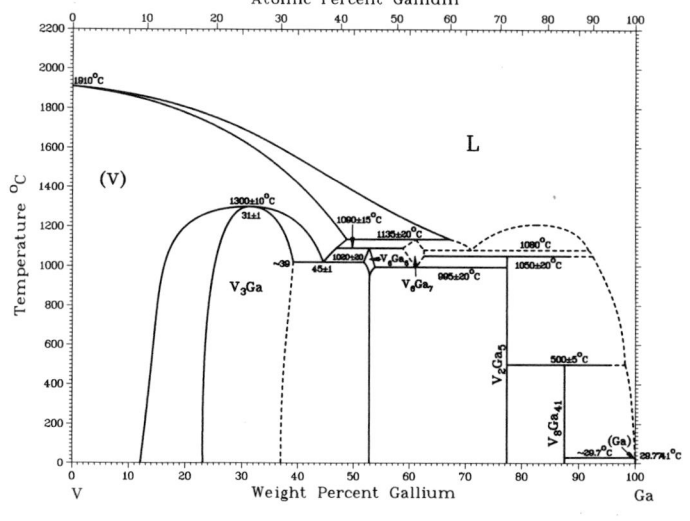

Phase	Composition, wt% Ga	Pearson symbol	Space group
(V)	0 to 49	cI2	Im3̄m
V₃Ga	~23 to ~39	cP8	Pm3̄n
V₆Ga₅	~53.3	hP22	…
V₆Ga₇	~59 to ~63	cI52	I4̄3m
V₂Ga₅	77.4	tP14	…
V₈Ga₄₁	87.5	hR49	…
(Ga)	100	oC8	Cmca

Ga-Y

S.P. Yatsenko, 1977

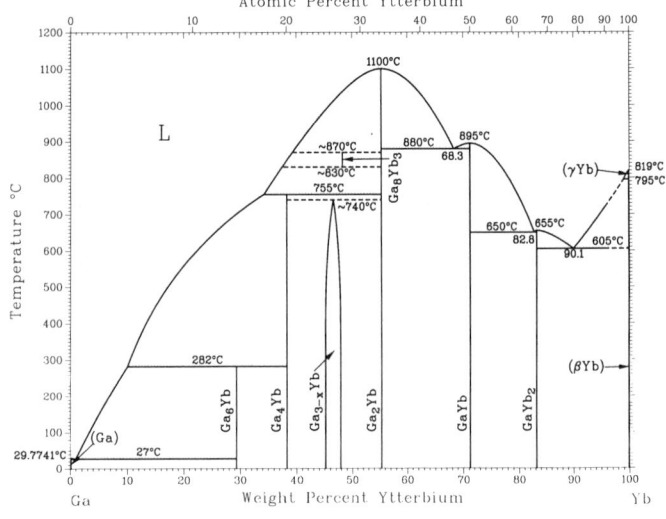

Phase	Composition, wt % Y	Pearson symbol	Space group
(Ga)	0	$oC8$	$Cmca$
Ga_2Y	38.9	...	...
Ga_5Y_3	43.3	$oP32$	$Pnma$
GaY	56.0	$oC8$	$Cmcm$
Ga_3Y_5	68.0	$hP16$	$P6_3/mcm$
(βY)	100	$cI2$	$Im\overline{3}m$
(αY)	100	$hP2$	$P6_3/mmc$

Ga-Yb

A. Palenzona and S. Cirafici, 1992

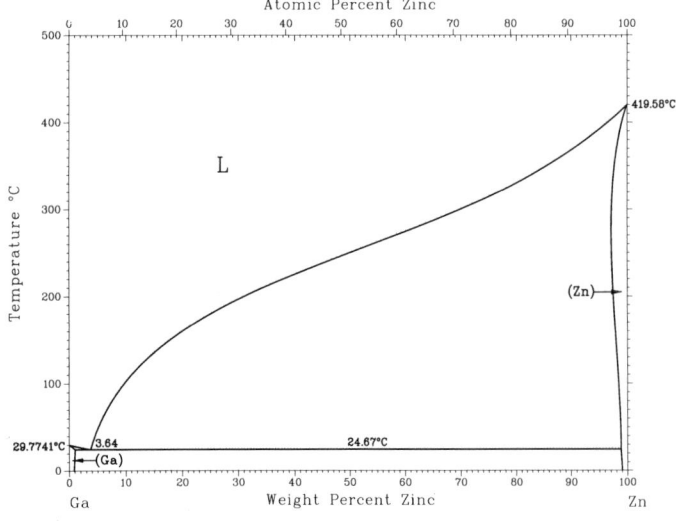

Phase	Composition, wt % Yb	Pearson symbol	Space group
(αGa)	~0	$oC8$	$Cmca$
Ga_6Yb	29.3	$tP14$	$P4/nbm$
Ga_4Yb	38	$mC10$	$C2/m$
$Ga_{3-x}Yb$(a)	45 to 48.5	$hP54.3$	$P6/mmm$
Ga_8Yb_3	48.2	$oI22$	$Immm$
Ga_2Yb	55.3	$hP6$	$P6_3/mmc$
GaYb	71.3	$tP4$	$P4/mmm$
$GaYb_2$	83.3	$oP12$	$Pnma$
(γYb)	~100	$cI2$	$Im\overline{3}m$
(βYb)	~100	$cF4$	$Fm\overline{3}m$
(αYb)	~100	$hP2$	$P6_3/mmc$

(a) $0 \leq x \leq 0.36$

Ga-Zn

J. Dutkiewicz, Z. Moser, L. Zabdyr, D.D. Gohil, T.G. Chart, I. Ansara, and C. Girard, 1990

Phase	Composition, wt % Zn	Pearson symbol	Space group
(Ga)	0 to 0.75	$oC8$	$Cmca$
(Zn)	97.49 to 100	$hP2$	$P6_3/mmc$

Ga-Zr

From [Shunk]

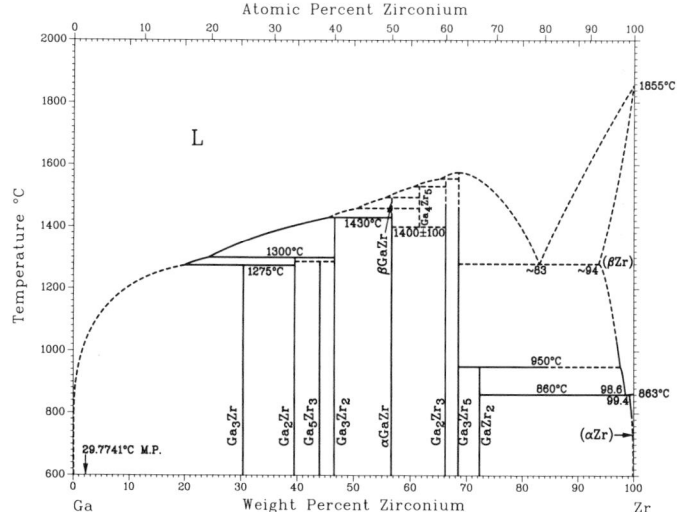

Phase	Composition, wt% Zr	Pearson symbol	Space group
(Ga)	0	oC8	Cmca
Ga₃Zr	30	tI16	I4/mmm
Ga₂Zr	39.5	oC12	Cmmm
Ga₅Zr₃	44.0	oC32	Cmcm
Ga₃Zr₂	47	oF40	Fdd2
βGaZr	56.7	…	…
αGaZr	56.7	tI16	I4₁/amd
Ga₄Zr₅	62.1	hP18	P6₃/mcm
Ga₂Zr₃	66	tP10	P4/mbm
Ga₃Zr₅	68.6	hP16	P6₃/mcm
GaZr₂	72.4	tI12	I4/mcm
(βZr)	~94 to 100	cI2	Im3̄m
(αZr)	99.4 to 100	hP2	P6₃/mmc

Gd-Ge

A.B. Gokhale and G.J. Abbaschian, 1989

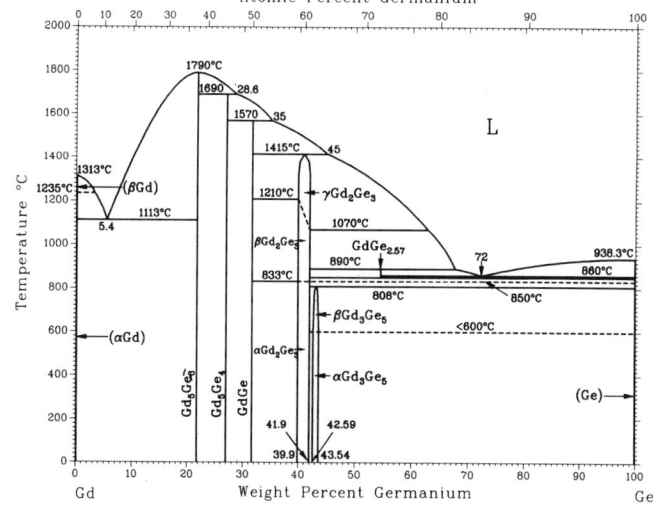

Phase	Composition, wt% Ge	Pearson symbol	Space group
βGd(a)	0	cI2	Im3̄m
αGd(b)	0	hP2	P6₃/mmc
Gd₅Ge₃	21.7	hP16	P6₃/mcm
Gd₅Ge₄	27.0	(c)	Pnma
GdGe	31.6	oC8	Cmcm
γGd₂Ge₃	40 to 42	…	…
βGd₂Ge₃	40 to 42	…	…
αGd₂Ge₃	40 to 42	hP3	P6/mmm
βGd₃Ge₅(d)	42.59 to 43.54	(c)	Imma
αGd₃Ge₅	42.59 to 43.54	tI12	I4̄₁/amd
GdGe₂.₅₇	54	(c)	C222₁
Ge	100	cF8	Fd3̄m

(a) From 1313 to >1235 °C. (b) From 1235 °C. (c) Orthorhombic. (d) Also designated "GdGe₂₋ₙ"

Gd-In

A. Palenzona and S. Cirafici, 1992

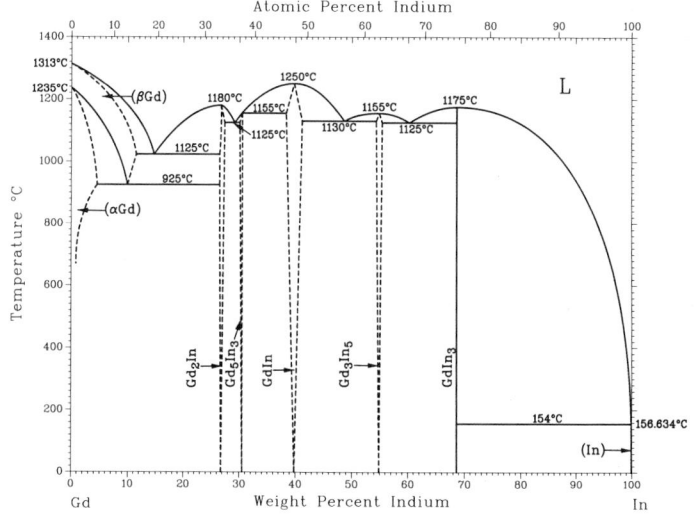

Phase	Composition, wt% In	Pearson symbol	Space group
(βGd)	0 to ~11	cI2	Im3̄m
(αGd)	0 to ~5	hP2	P6₃/mmc
Gd₂In	26.7 ± ~1	hP6	P6₃/mmc
Gd₅In₃	30.5 ± ~1	tI32	I4/mcm
GdIn(a)	39 ± ~2	cP2 or cI2	Pm3̄m or Im3̄m
Gd₃In₅	54.9 ± ~1	oC32	Cmcm
GdIn₃	69	cP4	Pm3̄m
(In)	~100	tI2	I4/mmm

(a) Possibly metastable

Gd-Mg

A.A. Nayeb-Hashemi and J.B. Clark, 1988

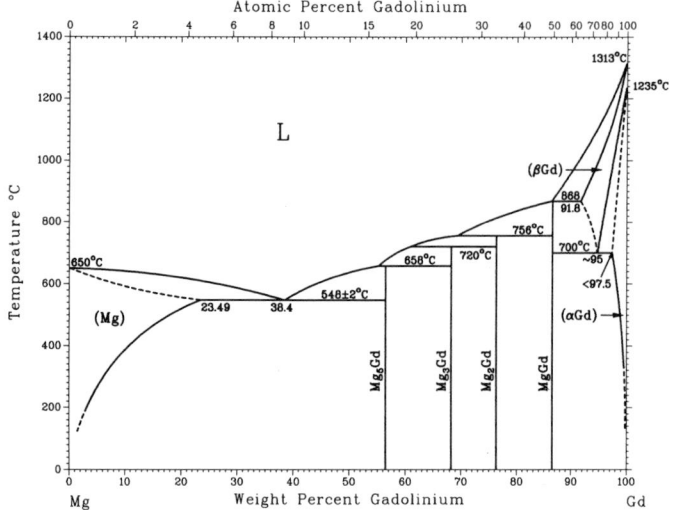

Phase	Composition, wt% Gd	Pearson symbol	Space group
(Mg)	0 to 23.49	hP2	P6₃/mmc
Mg₅Gd	56.41(a)	(b)	F̄43m
Mg₃Gd	68	cF16	Fm̄3m
Mg₂Gd	76.38	cF24	Fd̄3m
MgGd	86.6	cP2	Pm̄3m
(βGd)	? to 100	cI2	Im̄3m
(αGd)	<97.5 to 100	hP2	P6₃/mmc

(a) There may be a small homogeneity range. The exact stoichiometry was reported as Mg₅.₀₅Gd, closely related to that of Sm₁₁Cd₄₅. (b) Cubic

Gd-Mn

H. Okamoto, 1990

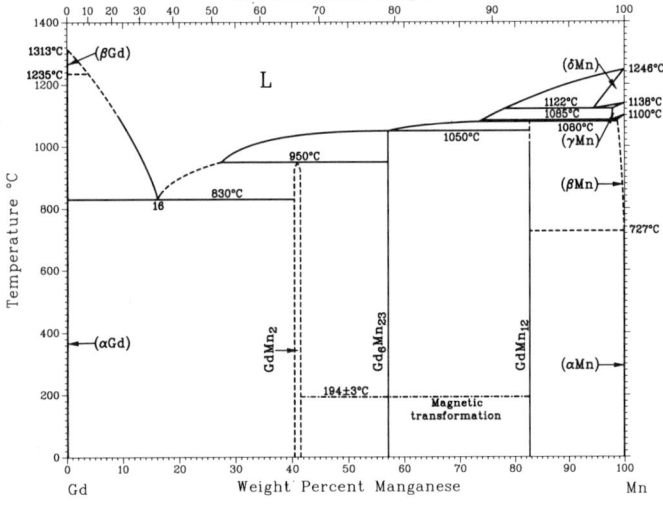

Phase	Composition, wt% Mn	Pearson symbol	Space group
(βGd)	0	cI2	Im̄3m
(αGd)	0	hP2	P6₃/mmc
GdMn₂	~41.2	cF24	Fd̄3m
Gd₆Mn₂₃	57.2	cF116	Fm̄3m
GdMn₁₂	80.7	tI26	I4/mmm
(δMn)	~95 to 100	cI2	Im̄3m
(γMn)	~97 to 100	cF4	Fm̄3m
(βMn)	~100	cP20	P4₁32
(αMn)	100	cI58	Ī43m

Gd-Ni

Y.Y. Pan and P. Nash, 1991

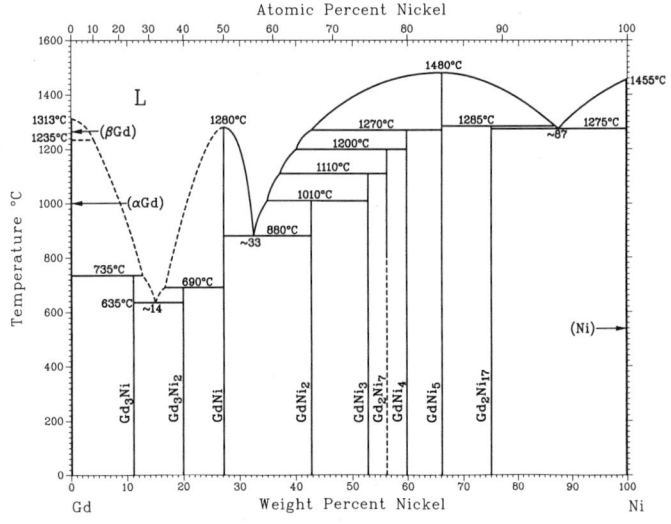

Phase	Composition, wt% Ni	Pearson symbol	Space group
(βGd)	0	hP2	P6₃/mmc
(αGd)	0	cI2	Im̄3m
Gd₃Ni	11.1	oP16	P6/mmm
Gd₃Ni₂	19.9	t**	...
GdNi	27.2	oC8	Cmcm
GdNi₂	42.8	cF24	Fd̄3m
GdNi₃	52.8	hP24	R̄3m
Gd₂Ni₇	56.7	hP36(a)	P6₃/mmc
		hR54(b)	R̄3m
GdNi₄	59.9	hP6	...
GdNi₅	65.1	hP6	P6/mmm
Gd₂Ni₁₇	76.1	hP38	P6₃/mmc
(Ni)	100	cF4	Fm̄3m

(a) High-temperature form. (b) Low-temperature form

Gd-Pb

A. Palenzona and S. Cirafici, 1991

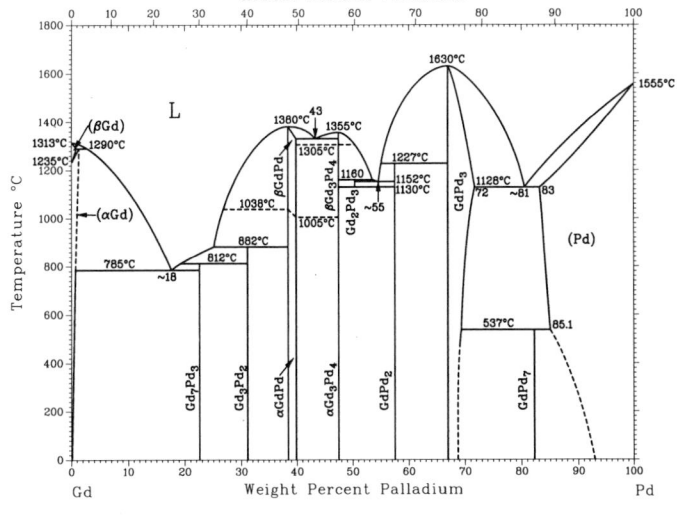

Phase	Composition, wt% Pb	Pearson symbol	Space group
(βGd)	0 to 3	cI2	Im$\overline{3}$m
(αGd)	0 to 1	hP2	P6₃/mmc
Gd₅Pb₃	44.2 to 46	hP16	P6₃/mcm
Gd₅Pb₄	51.3	oP36	Pnma
Gd₁₁Pb₁₀	54.5	...	...
GdPb₃	80	cP4	Pm$\overline{3}$m
(Pb)	>99.6 to 100	cF4	Fm$\overline{3}$m

Gd-Pd

H. Okamoto, 1990

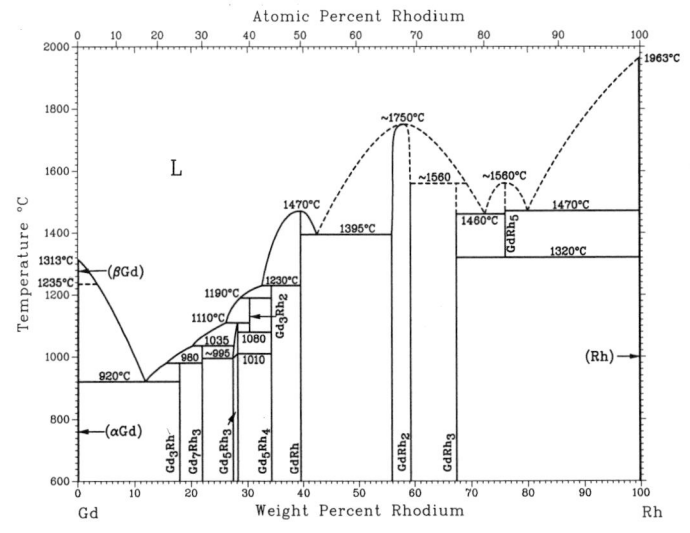

Phase	Composition, wt% Pd	Pearson symbol	Space group
(βGd)	0 to 0.68	cI2	Im$\overline{3}$m
(αGd)	0 to 1.4	hP2	P6₃/mmc
Gd₇Pd₃	23	hP20	P6₃mc
Gd₃Pd₂	31	tP10	P4/mbm
βGdPd	~40.4	...	...
αGdPd	~40.4	oC8	Cmcm
Gd₃Pd₄	47.4	hR14	R$\overline{3}$
Gd₂Pd₃	50	...	...
GdPd₃	67 to 72	cP4	Pm$\overline{3}$m
GdPd₇	82.6	...	...
(Pd)	83 to 100	cF4	Fm$\overline{3}$m

Gd-Rh

H. Okamoto, 1990

Phase	Composition, wt% Rh	Pearson symbol	Space group
(βGd)	0	cI2	Im$\overline{3}$m
(αGd)	0	hP2	P6₃/mmc
Gd₃Rh	18	oP16	Pnma
Gd₇Rh₃	22	hP20	P6₃mc
βGd₅Rh₃	~28.2	hP16	P6₃/mcm
αGd₅Rh₃	~28.2	...	...
Gd₃Rh₂	30	tI140	I4/mcm
Gd₅Rh₄	34.3	oP36	Pnma
GdRh	39.6	cP2	Pd$\overline{3}$m
GdRh₂	56 to 59	cF24	Fd$\overline{3}$m
GdRh₃	66	hP24	P6₃/mmc
GdRh₅	76.5	hP6	P6/mmm
(Rh)	100	cF4	Fm$\overline{3}$m

Gd-Sb

H. Okamoto, 1990

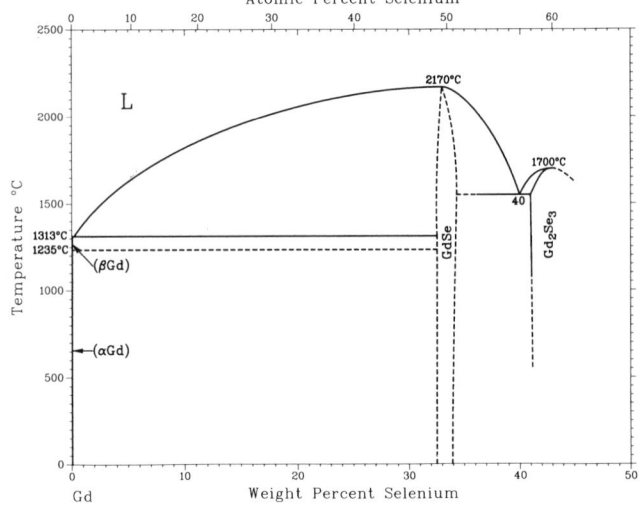

Phase	Composition, wt% Sb	Pearson symbol	Space group
(βGd)	0	cI2	Im$\overline{3}$m
(αGd)	0	hP2	P6₃/mmc
Gd₅Sb₃	31.7	hP16	P6₃/mcm
Gd₄Sb₃	36.8	cI28	I$\overline{4}$3d
βGdSb	43.6	...	...
αGdSb	43.6	cF8	Fm$\overline{3}$m
GdSb₂	60.8	hP12	P6₃/mmc
(Sb)	100	hR2	R$\overline{3}$m

Gd-Se

N.Yu. Pribyl'skiĭ, I.G. Vasileva, and R.S. Gamidov, 1982

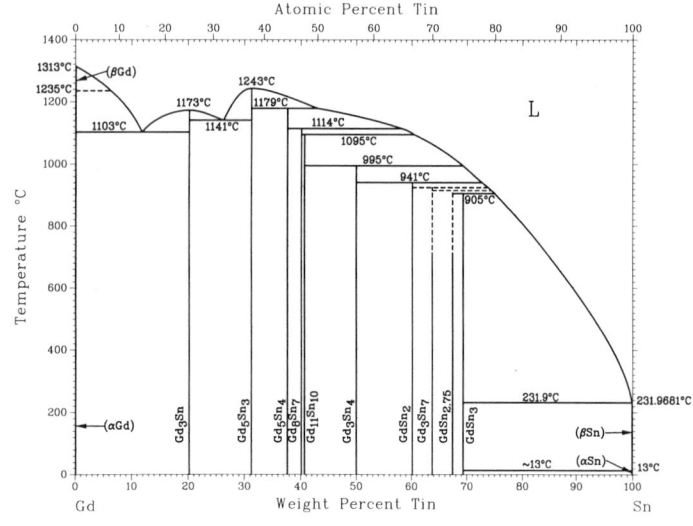

Phase	Composition, wt% Se	Pearson symbol	Space group
(βGd)	0	cI2	Im$\overline{3}$m
(αGd)	0	hP2	P6₃/mmc
GdSe	33 to 34	cF8	Fm$\overline{3}$m
Gd₂Se₃	~41 to ?	oP20	Pnma
		or cI28	I$\overline{4}$3d
Other reported phases			
GdSe₂	50.1	oP12	Pnma
		o*144	...
		t*24	...

Gd-Sn

A. Palenzona and S. Cirafici, 1991

Phase	Composition, wt% Sn	Pearson symbol	Space group
(βGd)	0	hP2	P6₃/mmc
(αGd)	0	cI2	Im$\overline{3}$m
Gd₃Sn	20	...	...
Gd₅Sn₃	31.2	hP16	P6₃/mcm
Gd₅Sn₄	37.6	oP36	Pnma
Gd₈Sn₇	39.8	...	...
Gd₁₁Sn₁₀	40.7	tI84	I4/mmm
Gd₃Sn₄	50.1	...	...
GdSn₂	60.2	oC12	Cmcm
Gd₃Sn₇	64	oC20	Cmmm
GdSn₂.₇₅	67.5	oC15	Amm2
GdSn₃	69	cP4	Pm$\overline{3}$m
(βSn)	100	tI4	I4₁/amd
(αSn)	100	cF8	Fd$\overline{3}$m

Gd-Te

H. Okamoto, 1990

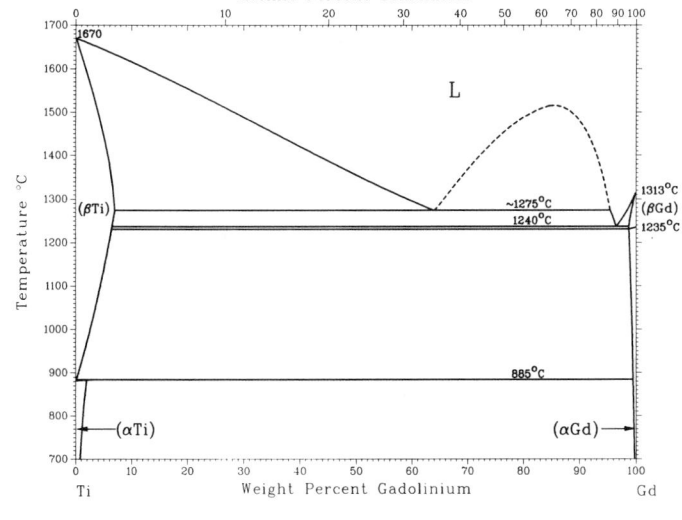

Phase	Composition, wt% Te	Pearson symbol	Space group
(βGd)	0	cI2	Im$\overline{3}$m
(αGd)	0	hP2	P6₃/mmc
GdTe	44.8	cF8	Fm$\overline{3}$m
Gd₂Te₃	52 to 55	oP20	Pnma
Gd₄Te₇	58.6	tP6	P4/nmm
GdTe₂	61.9	...	...
Gd₂Te₅	67.0	oC28	Cmcm
GdTe₃	71	oC16	Cmcm
(Te)	100	hP3	P3₁21

Gd-Ti

J.L. Murray, 1987

Phase	Composition, wt% Gd	Pearson symbol	Space group
(βTi)	0 to ~6	cI2	Im$\overline{3}$m
(αTi)	0 to ~1.9	hP2	P6₃/mmc
(βGd)	~99 to 100	cI2	Im$\overline{3}$m
(αGd)	~99 to 100	hP2	P6₃/mmc

Gd-Tl

S. Delfino, A. Saccone, A. Palenzona, and R. Ferro, unpublished

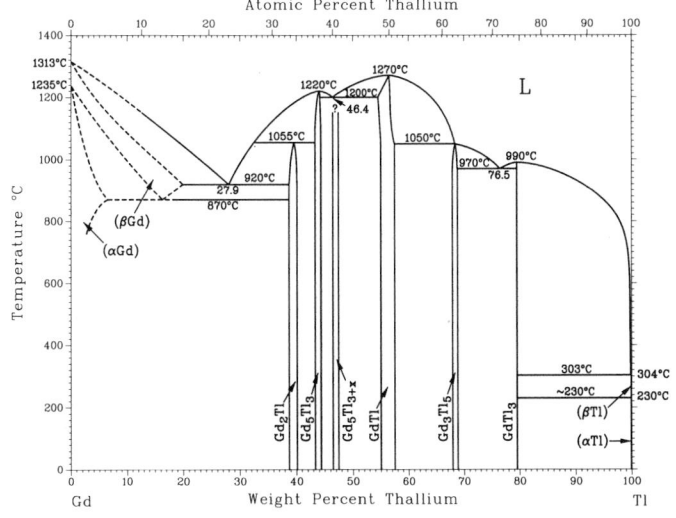

Phase	Composition, wt% Tl	Pearson symbol	Space group
(αGd)	0 to ?	hP2	P6₃/mmc
(βGd)	0 to ~20	cI2	Im$\overline{3}$m
Gd₂Tl	~39 to ~40	hP6	P6₃/mmc
Gd₅Tl₃	~43 to ~44	hP16	P6₃/mcm
Gd₅Tl₃₊ₓ	?	tI32	I4/mcm
GdTl(a)	~55 to ~58	cP2	Pm$\overline{3}$m
		(or cI2)	Im$\overline{3}$m
GdTl(b)	~55 to ~58	tP2	P4/mmm
Gd₃Tl₅	~68 to ~69	oC32	Cmcm
GdTl₃	80	cP4	Pm$\overline{3}$m
(βTl)	100	cI2	Im$\overline{3}$m
(αTl)	100	hP2	P6₃/mmc

(a) Cubic structure presumed to be room- and higher-temperature phases. (b) Tetragonal structure presumed to be lower-temperature phase.

Ge-Ho

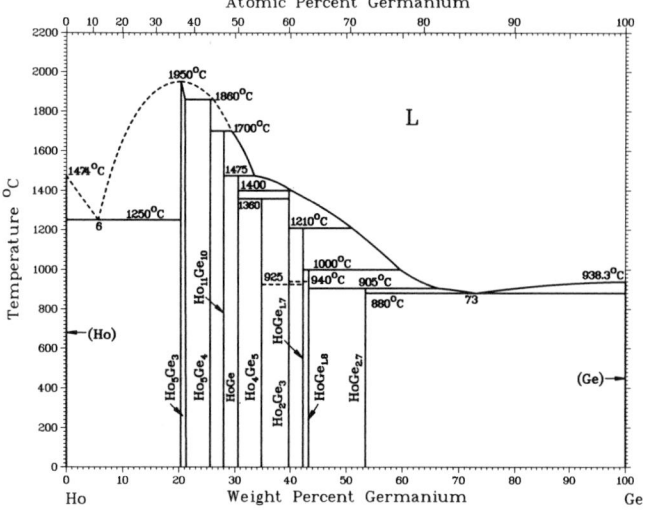

V.N. Eremenko, I.M. Obushenko, and Yu.I. Buyanov, 1980

Phase	Composition, wt% Ge	Pearson symbol	Space group
(Ho)	0	hP2	P6₃/mmc
Ho₅Ge₃	~20.9	hP16	P6₃/mcm
Ho₅Ge₄	26.0	oP36	Pnma
Ho₁₁Ge₁₀	28.6	tI84	I4/mmm
HoGe	30.6	oC8	Cmcm
Ho₄Ge₅	35.5	...	...
βHo₂Ge₃	40	oC12	Cmmm
αHo₂Ge₃	40	hP3	P6/mmm
βHoGe₁.₇	43	...	...
αHoGe₁.₇	43	tI12	I4₁/amd
HoGe₁.₈	44.2	...	...
HoGe₂.₇	54	o**	...
(Ge)	100	cF8	Fd3̄m

Ge-In

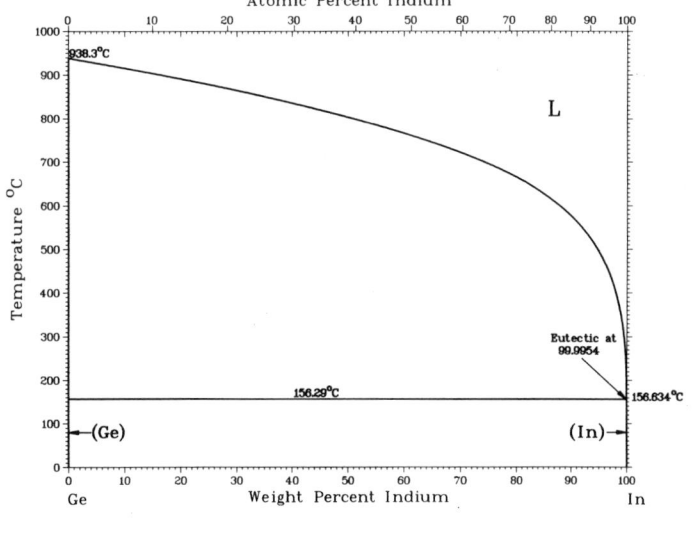

R.W. Olesinski, N. Kanani, and G.J. Abbaschian, 1992

Phase	Composition, wt% In	Pearson symbol	Space group
(Ge)	0	cF8	Fd3̄m
(In)	100	tI2	I4/mmm

Ge-K

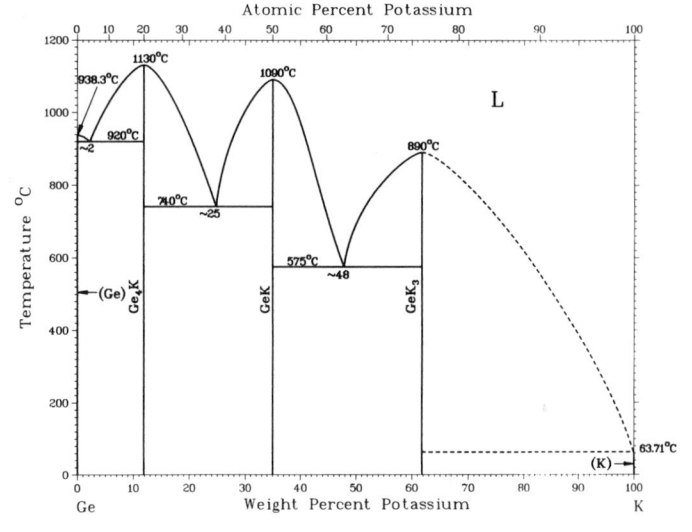

H. Okamoto, 1990

Phase	Composition, wt% K	Pearson symbol	Space group
(Ge)	0	cF8	Fd3̄m
Ge₂₃K₄(a)	8.6	cP54	Pm3̄n
Ge₄K	12	...	...
GeK	35.0	cP64	P4̄3m
GeK₃	62	...	...
(K)	100	cI2	Im3̄m

(a) Not shown in the phase diagram

Ge-La

A.B. Gokhale and G.J. Abbaschian, 1989

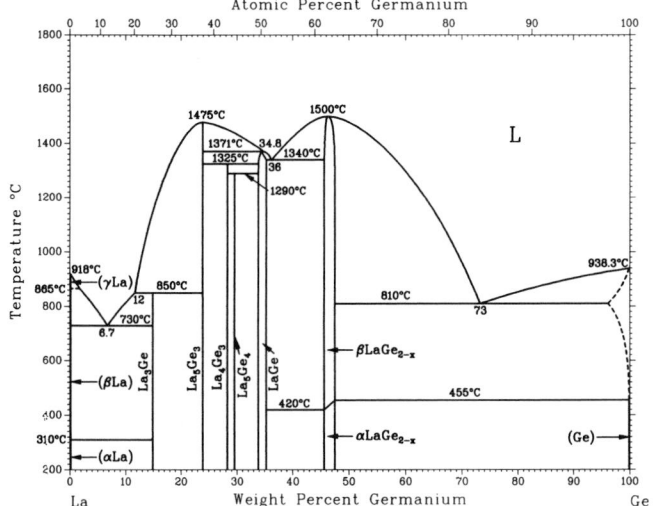

Phase	Composition, wt% Ge	Pearson symbol	Space group
(γLa)	0	cI2	Im$\bar{3}$m
(βLa)	0	cF4	Fm$\bar{3}$m
(αLa)	0	hP4	P6$_3$/mmc
La$_3$Ge	15	t**	...
La$_5$Ge$_3$	23.9	hP16	P6$_3$/mcm
La$_4$Ge$_3$	28.2	cI28	I$\bar{4}$3d
La$_5$Ge$_4$	29.5	oP*	Pnma
LaGe	33 to 35	oP8	Pnma
αLaGe$_{2-x}$	45.5 to 46.4	oI*	Imma
βLaGe$_{2-x}$	45.5 to 46.4	tI12	I4$_1$/amd
(Ge)	? to 100	cF8	Fd$\bar{3}$m

Ge-Li

H. Okamoto, 1990

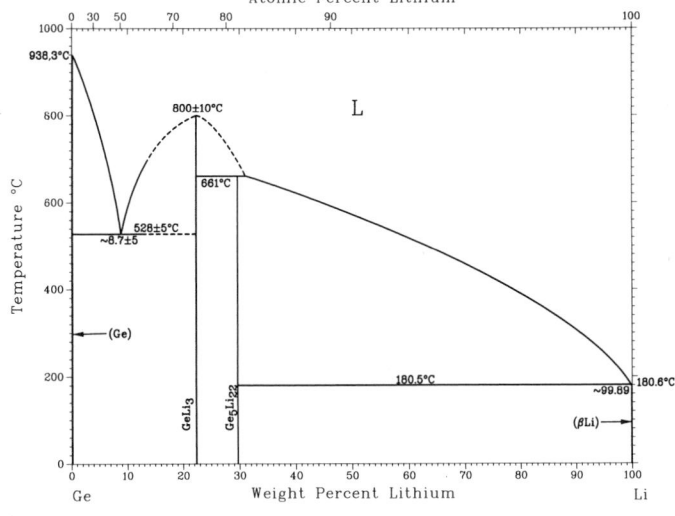

Phase	Composition, wt% Li	Pearson symbol	Space group
(Ge)	0	cF8	Fd$\bar{3}$m
GeLi(a)	8.7	tI32	I4$_1$/a
Ge$_6$Li$_{11}$(a)	14.9	oC68	Cmcm
GeLi$_2$(a)	16.1	...	...
GeLi$_3$	22	...	...
Ge$_2$Li$_7$(a)	25.1	oC36	Cmmm
Ge$_4$Li$_{15}$(a)	26.3	cI76	I$\bar{4}$3d
Ge$_5$Li$_{22}$	29.6	cF432	F23
(βLi)	100	cI2	Im$\bar{3}$m

(a) Not shown in the diagram

Ge-Lu

H. Okamoto, 1990

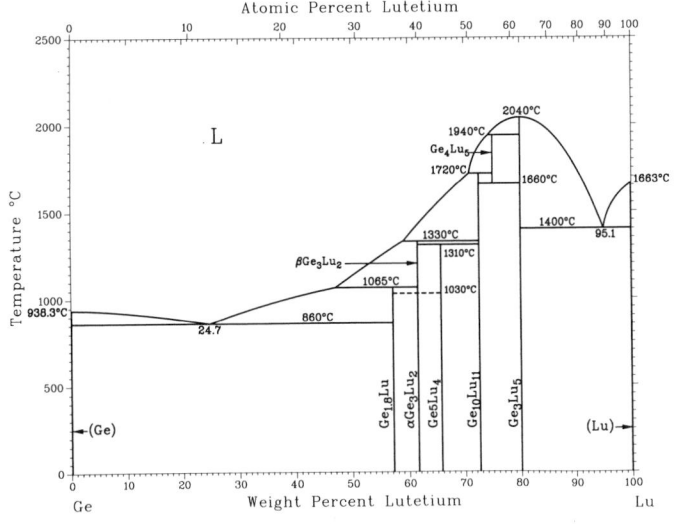

Phase	Composition, wt% Lu	Pearson symbol	Space group
(Ge)	0	cF8	Fd$\bar{3}$m
Ge$_{1.8}$Lu	57.2	oC12	Cmcm
Ge$_3$Lu$_2$	62	hP3	P6/mmm
Ge$_5$Lu$_4$	65.8	...	...
Ge$_{10}$Lu$_{11}$	72.6	tI84	I4/mmm
Ge$_4$Lu$_5$	75.1	oP36	Pnma
Ge$_3$Lu$_5$	80.1	hP16	P6$_3$/mcm
(Lu)	100	hP2	P6$_3$/mmc

Ge-Mg

A.A. Nayeb-Hashemi, R.W. Olesinski, G.J. Abbaschian, and
J.B. Clark, 1988

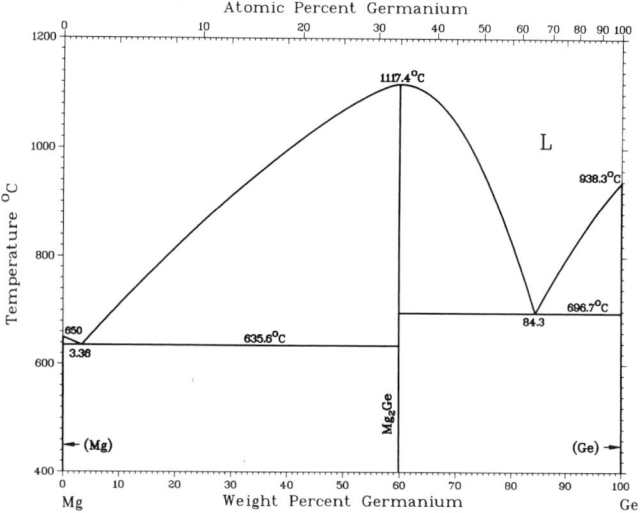

Phase	Composition, wt% Ge	Pearson symbol	Space group
(Mg)	~0	hP2	$P6_3/mmc$
Mg$_2$Ge	59.90	cF12	$Fm\overline{3}m$
(Ge)	~100	cF8	$Fd\overline{3}m$

Ge-Mn

A.B. Gokhale and G.J. Abbaschian, 1990

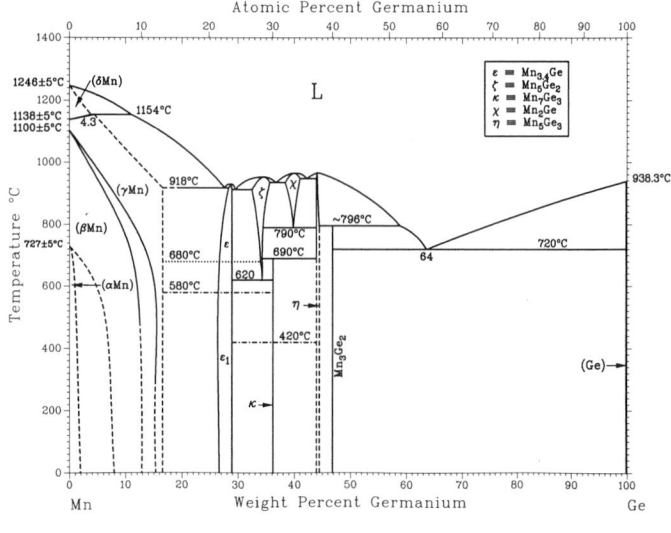

Phase	Composition, wt% Ge	Pearson symbol	Space group
(δMn)	0 to 4.3	cI2	$Im\overline{3}m$
(γMn)	0 to ~16	cF4	$Fm\overline{3}m$
(βMn)	0 to ~13	cP20	$P4_132$
(αMn)	0 to ~2.0	cI58	$I\overline{4}3m$
ε	~28.0	hP8	$P6_3/mmc$
ε$_1$	~28.0	tI8	$I4/mmm$
ξ	~34.6	hP128	$P3c1$
κ	36	o**	...
χ	~39.9	hP6	$P6_3/mmc$
η	~44.2	hP16	$P6_3/mcm$
Mn$_3$Ge$_2$	47	...	...
(Ge)	100	cF8	$Fd\overline{3}m$

Enlarged region of the Mn-Ge system

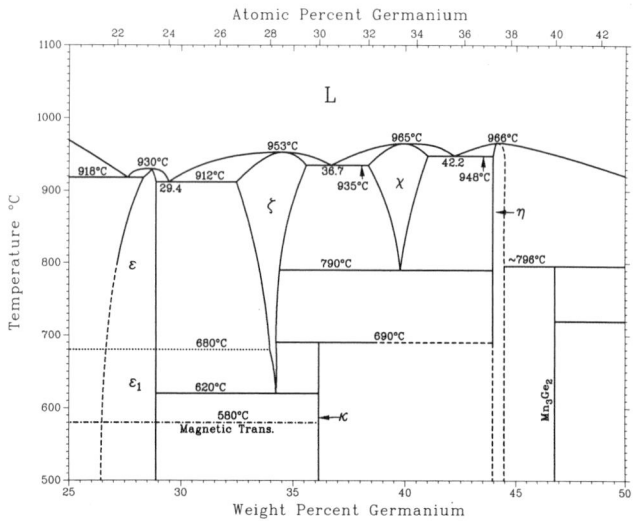

Ge-Mo

R.W. Olesinski and G.J. Abbaschian, 1987

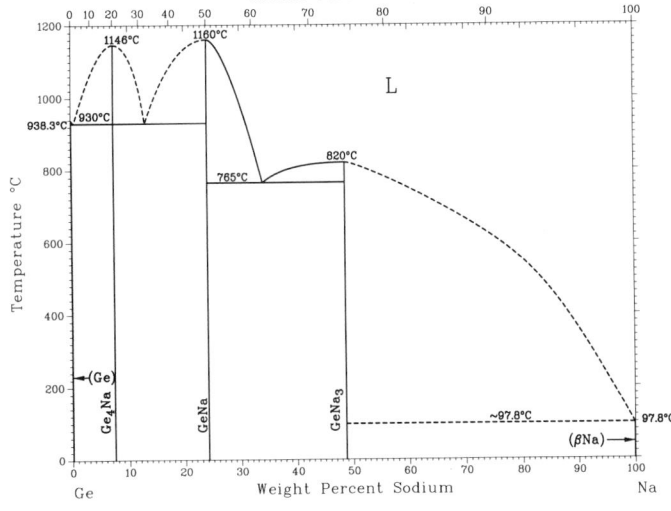

Phase	Composition, wt% Mo	Pearson symbol	Space group
(Ge)	0	cF8	$Fd\overline{3}m$
αGe$_2$Mo	39.7	oP12	Pnma
βGe$_2$Mo (HP)	...	tI8	I4/mmm
Ge$_{23}$Mo$_{13}$(a)	43 to 44	tP144	$P\overline{4}n2$
Ge$_3$Mo$_5$	68.8	hP16	$P6_3/mcm$
GeMo$_3$	80	cP8	$Pm\overline{3}n$
(Mo)	? to 100	cI2	$Im\overline{3}m$

(a) Also reported as Ge$_{41}$Mo$_{23}$ and Ge$_{16}$Mo$_9$ or Ge$_{1.7}$Mo

Ge-Na

H. Okamoto, 1990

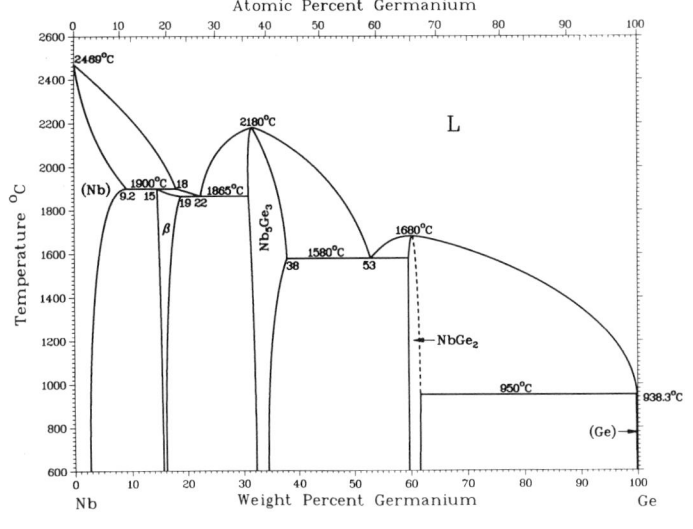

Phase	Composition, wt% Na	Pearson symbol	Space group
(Ge)	0	cF8	$Fd\overline{3}m$
Ge$_4$Na	7	cP*	$Pm\overline{3}n$
GeNa	24.0	mP32	$P2_1/c$
GeNa$_3$	49	...	...
(Na)	100	cI2	$Im\overline{3}m$

Ge-Nb

From [Moffatt]

Phase	Composition, wt% Ge	Pearson symbol	Space group
(Nb)	0 to 9.2	cI2	$Im\overline{3}m$
β	15 to 19	cP8	$Pm\overline{3}n$
Nb$_5$Ge$_3$	32 to 38	tI32	I4/mcm
NbGe$_2$	~61.0	hP9	$P6_222$
(Ge)	100	cF8	$Fd\overline{3}m$

Ge-Nd

A.B. Gokhale and G.J. Abbaschian, 1989

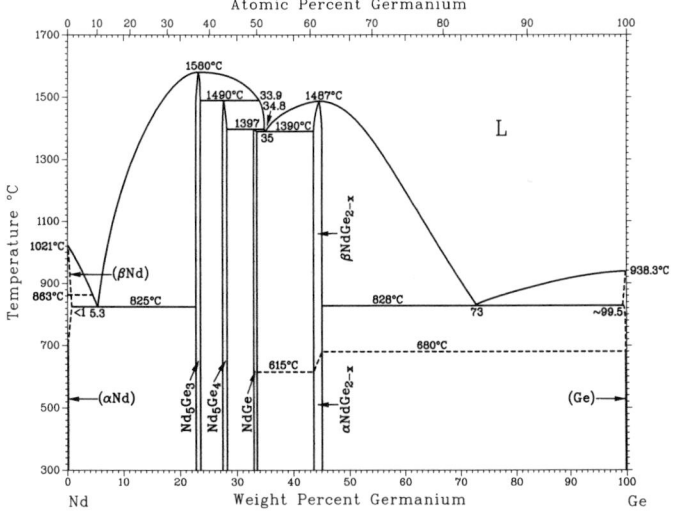

Phase	Composition, wt% Ge	Pearson symbol	Space group
(αNd)	0 to <1	hP4	P6₃/mmc
(βNd)	0 to <1	cI2	Im3̄m
Nd₅Ge₃	22.3 to 23.2	hP16	P6₃/mcm
Nd₅Ge₄	26.8 to 27.8	oP*	Pnma
NdGe	33 to 33.5	oC8	Cmcm
αNdGe₂₋ₓ	43 to 44.7	oI*	Imma
βNdGe₂₋ₓ	43 to 44.7	tI12	I4₁/amd
(Ge)	100	cF8	Fd3̄m

Ge-Ni

A. Nash and P. Nash, 1991

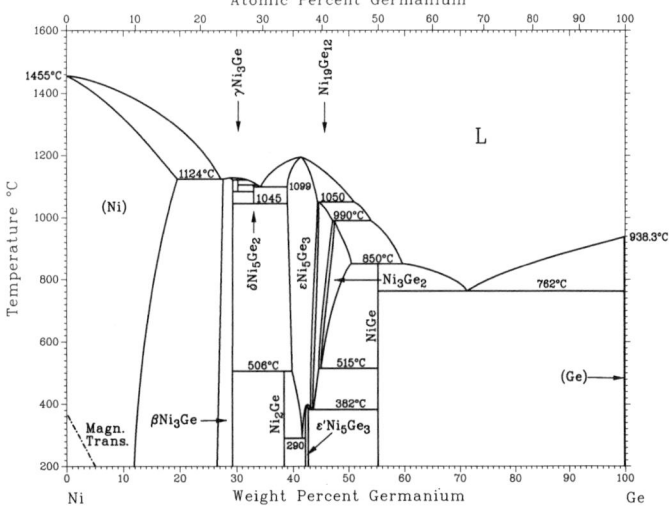

Phase	Composition, wt% Ge	Pearson symbol	Space group
(Ni)	0 to 19	cF4	Fm3̄m
βNi₃Ge	26.4 to 29	cP4	Pm3̄m
γNi₃Ge	29.9	...	...
δNi₅Ge₂	33	hP84	P6₃/mmc
Ni₂Ge	38.4	oP12	Pnma
ε′Ni₅Ge₃	~42	mC32	C2
εNi₅Ge₃	40 to 49	hP4	P6₃/mmc
Ni₁₉Ge₁₂	43 to 46	mC62	C2
Ni₃Ge₂(a)	46 to 48	hP4	P6₃/mmc
NiGe	55.3	oP8	Pnma
(Ge)	100	cF8	Fd3̄m

Ge-P

R.W. Olesinski, N. Kanani, and G.J. Abbaschian, 1985

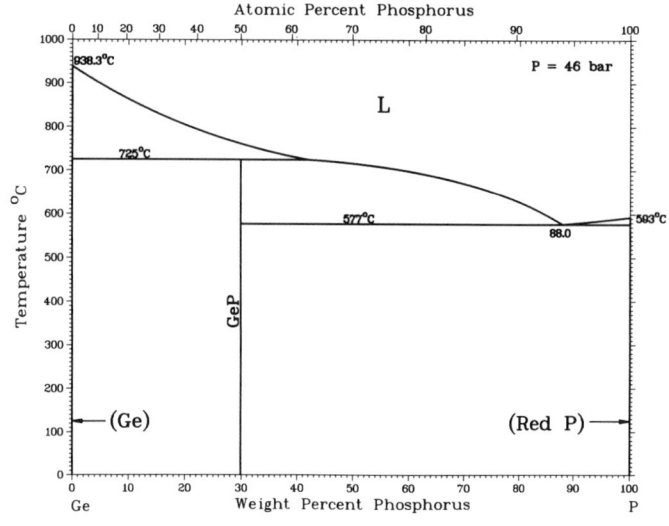

Phase	Composition, wt% P	Pearson symbol	Space group
(Ge)	0 to 0.07	cF8	Fd3̄m
GeP	29.9	(a)	C2/m
GeP(b)	29.9	(c)	I4/mm
GeP₃(b)	56	hR2	R3̄m
GeP₅(b)	68.0	hR2	R3̄m
Red P	100	...	...
White P	100	...	...
Black P	100	oC8	Cmca

(a) Orthorhombic. (b) High-temperature phase. (c) Tetragonal

Ge-Pb

R.W. Olesinski and G.J. Abbaschian, 1984

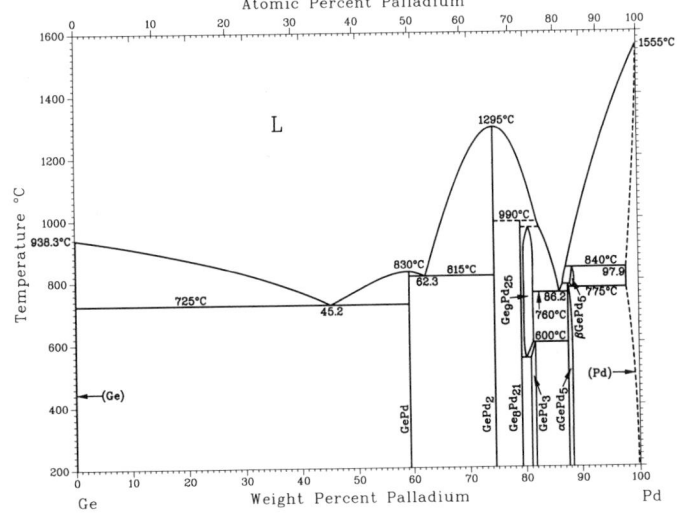

Phase	Composition, wt% Pb	Pearson symbol	Space group
(Ge)	0	cF8	$Fd\bar{3}m$
(Pb)	100	cF4	$Fm\bar{3}m$

Ge-Pd

H. Okamoto, 1992

Phase	Composition, wt% Pd	Pearson symbol	Space group
(Ge)	0	cF8	$Fd\bar{3}m$
GePd	59.4	oP8	Pnma
GePd$_2$	74.6	hP9	$P\bar{6}2m$
Ge$_8$Pd$_{21}$	79.4	tI116	$I4_1/a$
Ge$_9$Pd$_{25}$	80 to 81.5	hP34	$P\bar{3}$
GePd$_3$	81.1 to 81.9	...	...
βGePd$_5$	88.1 to 88.9	cI2	$Im\bar{3}m$
αGePd$_5$	87.7 to 88.5	mC24	C2
(Pd)	97.9 to 100	cF4	$Fm\bar{3}m$

Ge-Pr

A.B. Gokhale, A. Munitz, and G.J. Abbaschian, 1989

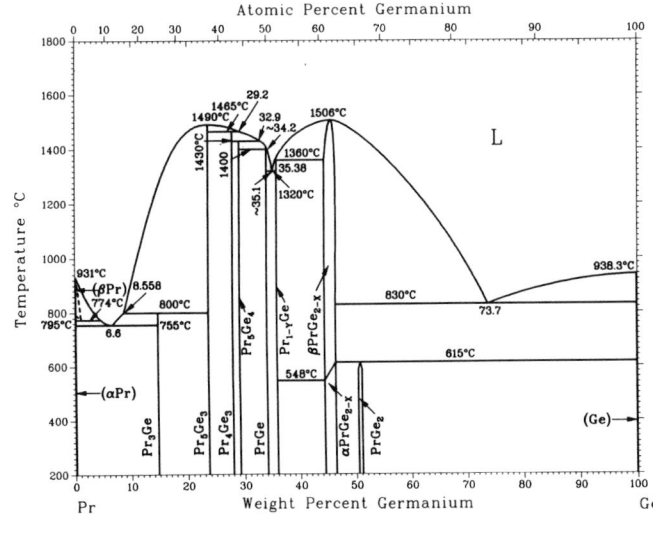

Phase	Composition, wt% Ge	Pearson symbol	Space group
(βPr)	0 to ?	cI2	$Im\bar{3}m$
(αPr)	0	hP4	$P6_3/mmc$
Pr$_3$Ge	15	t**	...
Pr$_5$Ge$_3$	23.8	hP16	$P6_3/mcm$
Pr$_4$Ge$_3$	27.9	cI28	$I\bar{4}d$
Pr$_5$Ge$_4$	29.2	oP*	Pnma
PrGe	34.0	oC8	Cmcm
Pr$_{1-y}$Ge	35.5	oP8	Pnma
αPrGe$_{2-x}$	45 to 46.3	oI*	Imma
βPrGe$_{2-x}$	45 to 46.3	tI12	$I4_1/amd$
PrGe$_2$	~50.8	tI12	$I4_1/amd$
(Ge)	100	cF8	$Fd\bar{3}m$

Ge-Pt

H. Okamoto, 1992

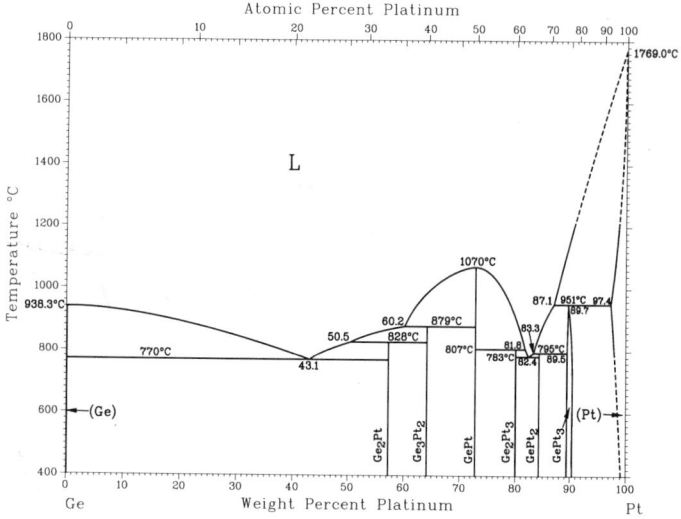

Phase	Composition, wt% Pt	Pearson symbol	Space group
(Ge)	0	cF8	$Fd\bar{3}m$
Ge$_2$Pt	57.3	oP6	Pnnm
Ge$_3$Pt$_2$	64	oP20	Pnma
GePt	72.9	oP8	Pnma
Ge$_2$Pt$_3$	80	oP40	Pnma
GePt$_2$	84.3	hP9	$P\bar{6}2m$
GePt$_3$	90 to 91	mC16	C2/m
(Pt)	97.4 to 100	cF4	$Fm\bar{3}m$
Metastable phase			
GePt$_3$	90 to 91	tI16	I4/mcm

Ge-S

C.H. Lin, A.S. Pashinkin, and A.V. Novoselova, 1963

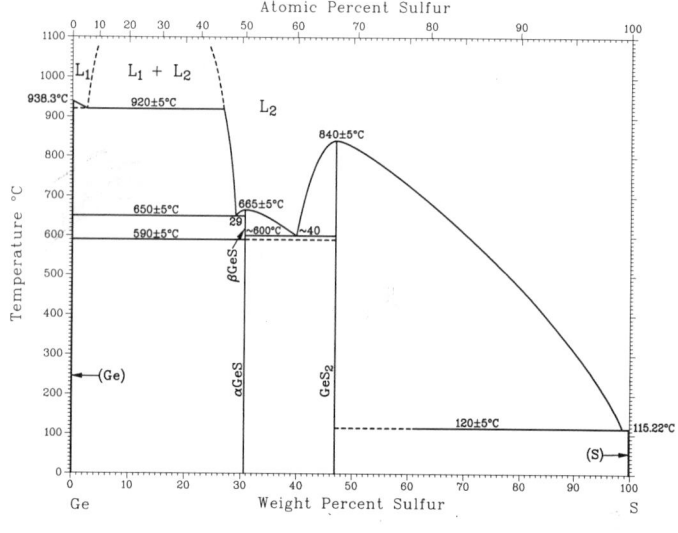

Phase	Composition, wt% S	Pearson symbol	Space group
(Ge)	0	cF8	$Fd\bar{3}m$
βGeS	30.6	h**	...
αGeS	30.6	oP8	Pnma
GeS$_2$	46.9	oF72	Fdd2

Ge-Sb

R.W. Olesinski and G.J. Abbaschian, 1986

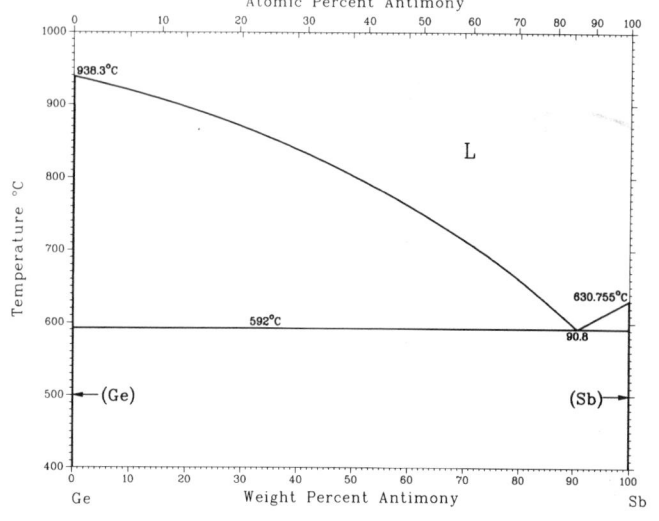

Phase	Composition, wt% Sb	Pearson symbol	Space group
(Ge)	0	cF8	$Fd\bar{3}m$
(Sb)	100	hR2	$R\bar{3}m$

Ge-Sc

A.B. Gokhale and G.J. Abbaschian, 1986

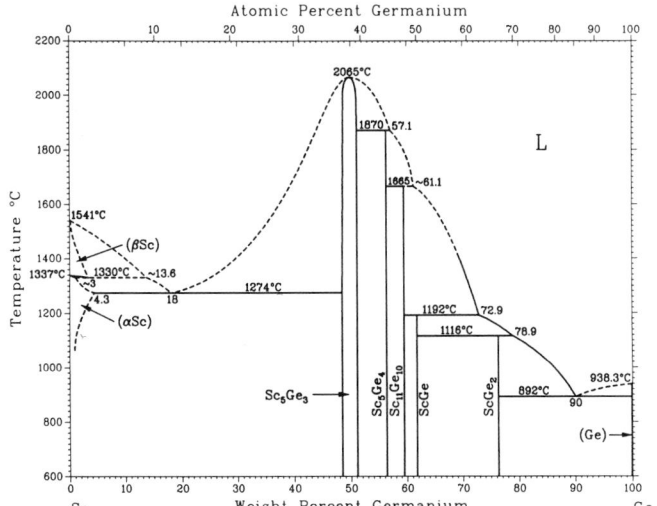

Phase	Composition, wt% Ge	Pearson symbol	Space group
(βSc)	0 to ~3	cI2	$Im\bar{3}m$
(αSc)	0 to 4.3	hP2	$P6_3/mmc$
Sc_5Ge_3	48.1 to 50.3	hP16	$P6_3/mcm$
Sc_5Ge_4	56.4	...	...
$Sc_{11}Ge_{10}$	59.5	...	I4/mmm
ScGe	61.8	oC8	Cmcm
$ScGe_2$	76.4	oC12	Cmcm
(Ge)	~100	cF8	$Fd\bar{3}m$

Ge-Se

A.B. Gokhale and G.J. Abbaschian, 1990

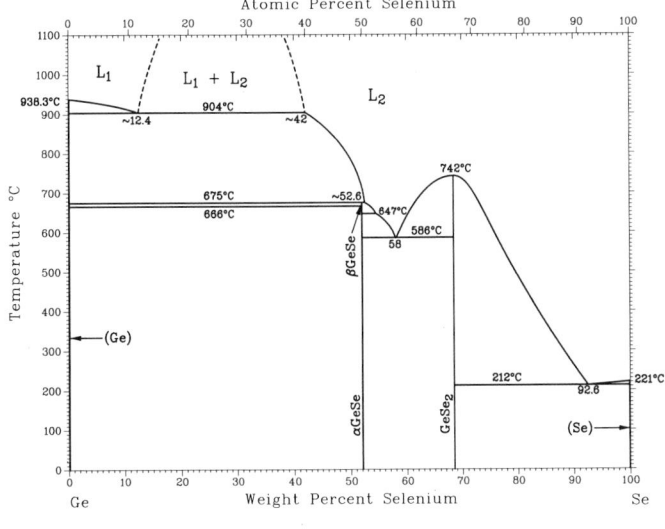

Phase	Composition, wt% Se	Pearson symbol	Space group
(Ge)	0	cF8	$Fd\bar{3}m$
αGeSe	52.1	oC8	Cmca
βGeSe	52.1	cF8	$Fm\bar{3}m$
$GeSe_2$	68.51	...	...
(γSe)	100	hP3	$P3_121$

Note: Crystal structures of the low-temperature α and β forms of Se are not known.

Ge-Si

R.W. Olesinski and G.J. Abbaschian, 1984

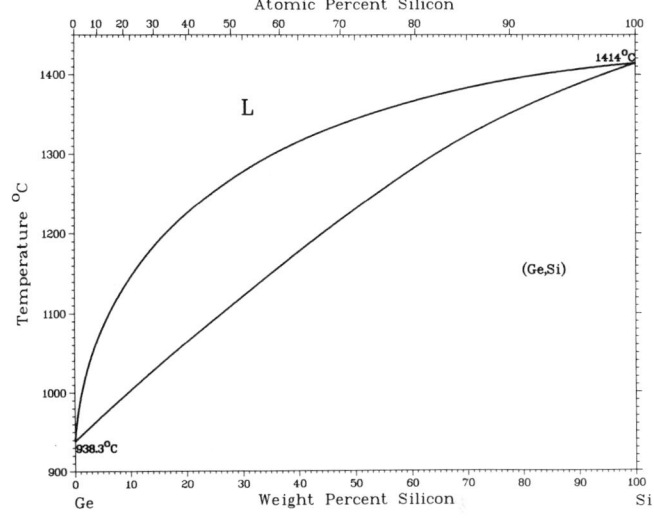

Phase	Composition, wt% Si	Pearson symbol	Space group
(Ge,Si)	0 to 100	cF8	$Fd\bar{3}m$
High-pressure phases			
GeII	...	tI4	$I4_1/amd$
SiII	...	tI4	$I4_1/amd$

Ge-Sm

A.B. Gokhale and G.J. Abbaschian, 1988

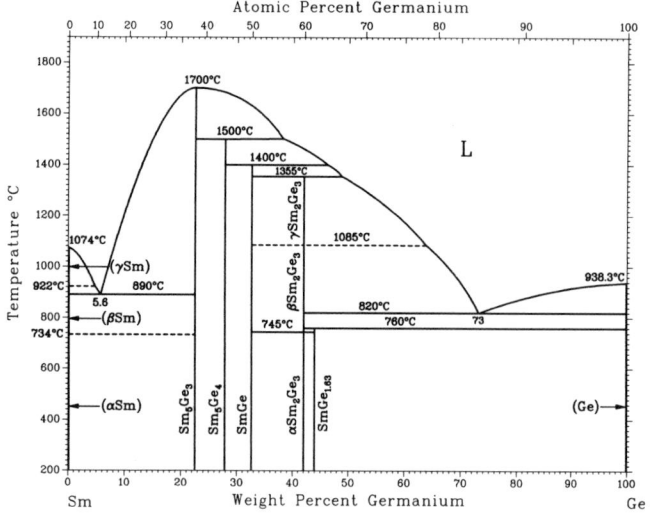

Phase	Composition, wt% Ge	Pearson symbol	Space group
(γSm)	0	cI2	Im$\bar{3}$m
(βSm)	0	hP2	P6₃/mmc
(αSm)	0	hR3	R$\bar{3}$m
Sm₅Ge₃	22.5	hP16	P6₃/mcm
Sm₅Ge₄	27.9	oP*	Pnma
SmGe	32.6	oC8	Cmcm
γSm₂Ge₃	42	...	...
βSm₂Ge₃	42	...	...
αSm₂Ge₃	42	hP3	P6/mmm
SmGe₁.₆₃	44	tI12	I4₁/amd
(Ge)	100	cF8	Fd$\bar{3}$m

Ge-Sn

R.W. Olesinski and G.J. Abbaschian, 1984

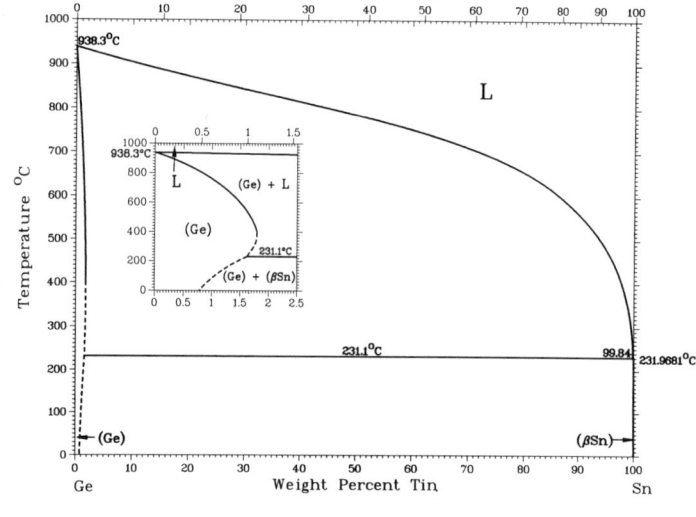

Phase	Composition, wt% Sn	Pearson symbol	Space group
(Ge)	0 to 1.8	cF8	Fd$\bar{3}$m
(βSn)	100	tI4	I4₁/amd
(αSn)	100	cF8	Fd$\bar{3}$m
Pressure stabilized phase			
GeII	0 to 15	tI4	I4₁/amd
Crystallized from amorphous phase			
Ge$_y$Sn$_{1-y}$	42 to 62.0	cF8	F$\bar{4}$3m

Ge-Sr

P.R. Subramanian, 1990

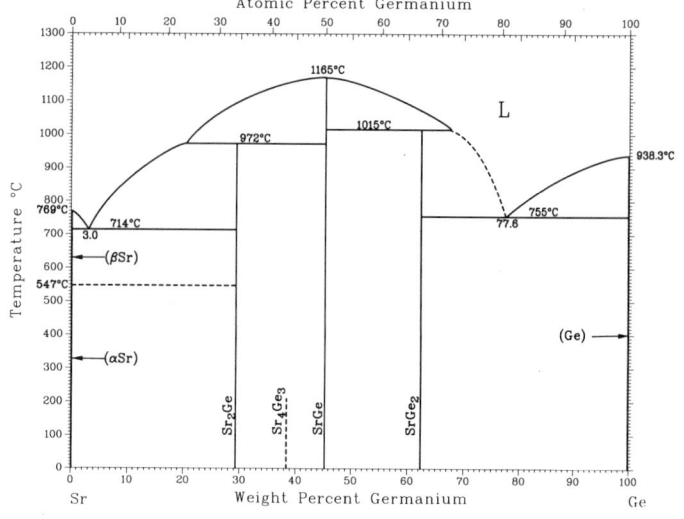

Phase	Composition, wt% Ge	Pearson symbol	Space group
(αSr)	0	cF4	Fm$\bar{3}$m
(βSr)	0	cI2	Im$\bar{3}$m
Sr₂Ge	29.3	oP12	Pnma
Sr₄Ge₃	~38.4	oI40	Immm
SrGe	45.3	oC8	Cmcm
SrGe₂	62.4	oP24	Pnma
(Ge)	100	cF8	Fd$\bar{3}$m

Ge-Tb

H. Okamoto, 1990

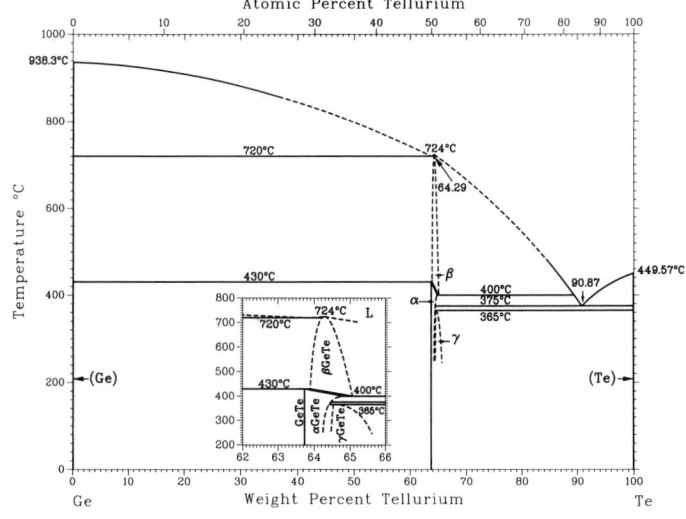

Phase	Composition, wt% Tb	Pearson symbol	Space group
(Ge)	0	cF8	$Fd\overline{3}m$
Ge$_{3-\alpha}$Tb	45	oC18	$C222_1$
Ge$_2$Tb	52.2	...	...
βGe$_{2-x}$Tb	56	tI12	$I4_1/amd$
αGe$_{2-x}$Tb	56	...	...
βGe$_{2-y}$Tb	59	hP3	$P6/mmm$
αGe$_{2-y}$Tb	59	...	...
GeTb	68.6	oC8	$Cmcm$
Ge$_{10}$Tb$_{11}$	70.7	tI84	$I4/mmm$
Ge$_4$Tb$_5$	73.3	oP36	$Pnma$
GeTb$_5$	78.5	hP16	$P6_3/mcm$
(Tb)	100	hP2	$P6_3/mmc$

Ge-Te

H. Okamoto, 1990

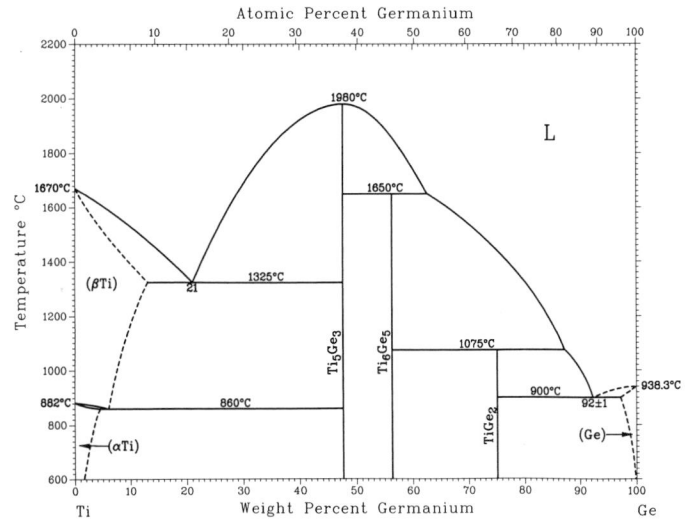

Phase	Composition, wt% Te	Pearson symbol	Space group
(Ge)	0	cF8	$Fd\overline{3}m$
βGeTe	~63.7 to 65	cF8	$Fm\overline{3}m$
αGeTe	~63.7 to 65	hR*	$R\overline{3}m$
γGeTe	~65 to 66	o**	...
(Te)	100	hP3	$P3_121$

Ge-Ti

H. Okamoto, 1990

Phase	Composition, wt% Ge	Pearson symbol	Space group
(βTi)	0 to ?	cI2	$Im\overline{3}m$
(αTi)	0 to ?	hP2	$P6_3/mmc$
Ti$_5$Ge$_3$	47.6	hP16	$P6_3/mcm$
Ti$_6$Ge$_5$	55.9	oI44	$Immm$
TiGe$_2$	75.2	oF24	$Fddd$
(Ge)	100	cF8	$Fm\overline{3}m$

Ge-Tl

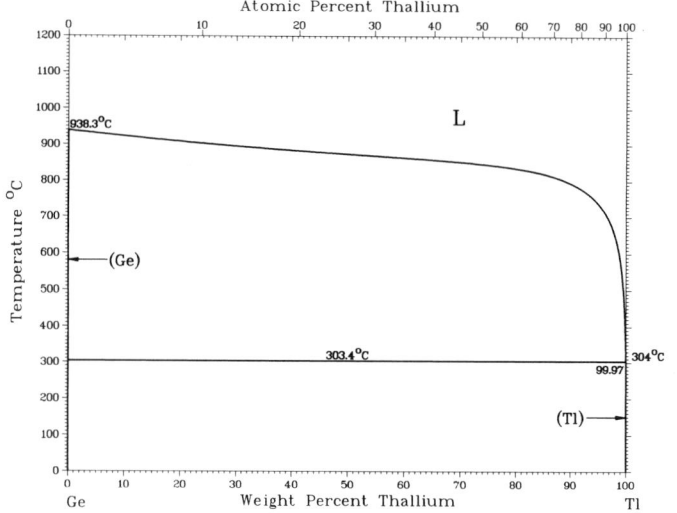

R.W. Olesinski and G.J. Abbaschian, 1985

Phase	Composition, wt% Tl	Pearson symbol	Space group
(Ge)	0.06	cF8	Fd3̄m
(βTl)	99.97 to 100	cI2	Im3̄m
(αTl)	100	hP2	P6₃/mmc

Ge-Tm

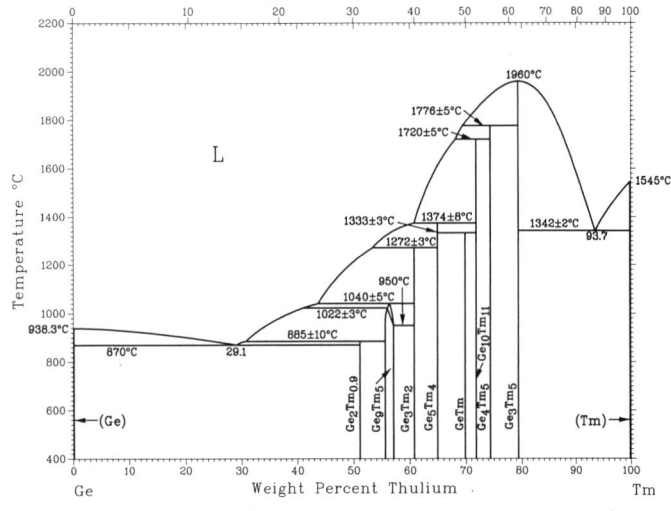

H. Okamoto, 1990

Phase	Composition, wt% Tm	Pearson symbol	Space group
(Ge)	0	cF8	Fd3̄m
Ge₂Tm₀.₉	51	oC12	Cmcm
βGe₉Tm₅	56.4	...	...
αGe₉Tm₅	56.4	...	...
Ge₃Tm₂	61	hP3	P6/mmm
Ge₅Tm₄	65.0	...	...
GeTm	69.9	oC8	Cmcm
Ge₁₀Tm₁₁	71.9	tI84	I4/mmm
Ge₄Tm₅	74.4	oP36	Pnma
Ge₃Tm₅	79.5	hP16	P6₃/mcm
(Tm)	100	hP2	P6₃/mmc

Ge-U

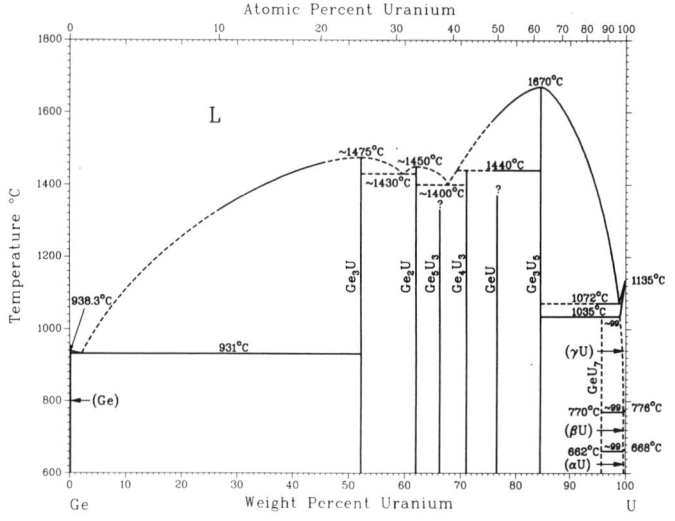

V.S. Lyashenko and V. Bykov, 1960

Phase	Composition, wt% U	Pearson symbol	Space group
(Ge)	0	cF8	Fd3̄m
Ge₃U	52	cP4	Pm3̄m
Ge₂U	62.1	hP3	P6/mmm
		or oC12	Cmcm
Ge₅U₃	66.3	...	...
Ge₄U₃	71.1	o**	...
GeU	76.6	...	...
Ge₃U₅	84.5	hP16	P6₃/mcm
GeU₇	95.8	...	...
(γU)	~99 to 100	cI2	Im3̄m
(βU)	~99 to 100	tP30	P4₂/mnm
(αU)	~99 to 100	oC4	Cmcm

Ge-Y

A.B. Gokhale and G.J. Abbaschian, 1988

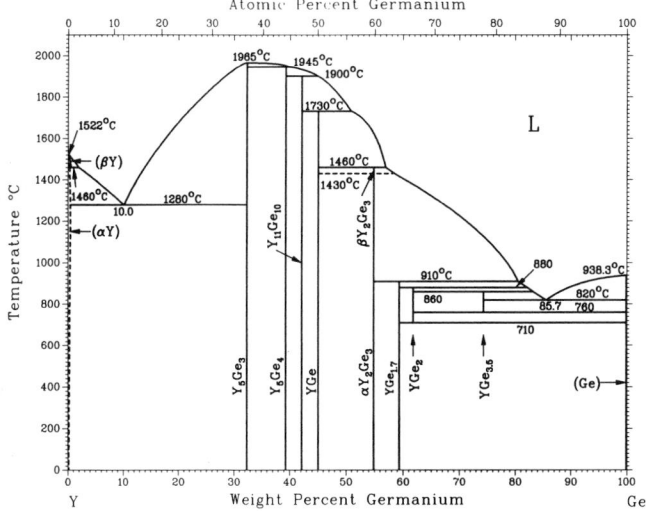

Phase	Composition, wt% Ge	Pearson symbol	Space group
(αY)	0 to ~0.81	hP2	P6₃/mmc
(βY)	0 to ~0.81	cI2	Im3̄m
Y₅Ge₃	32.9	hP16	P6₃/mcm
Y₅Ge₄	39.52	oP36	Pnma
Y₁₁Ge₁₀	42.6	tI84	I4/mmm
YGe	45.0	oC8	Cmcm
βY₂Ge₃	55	…	Pccm(a)
αY₂Ge₃	55	hP3	P6/mmm
βY₃Ge₅	57.6	oF72	Fdd2
αY₃Ge₅	57.6	tI12	I4₁/amd
YGe₂	62.03	oC12	Cmcm
Y₂Ge₇	74.09	…	C222₁(a)
(Ge)	0 to ~0.4	cF8	Fd3̄m

(a) Tentative

Ge-Yb

V.N. Eremenko, K.A. Meleshevich, and Yu.I. Buyanov, 1983

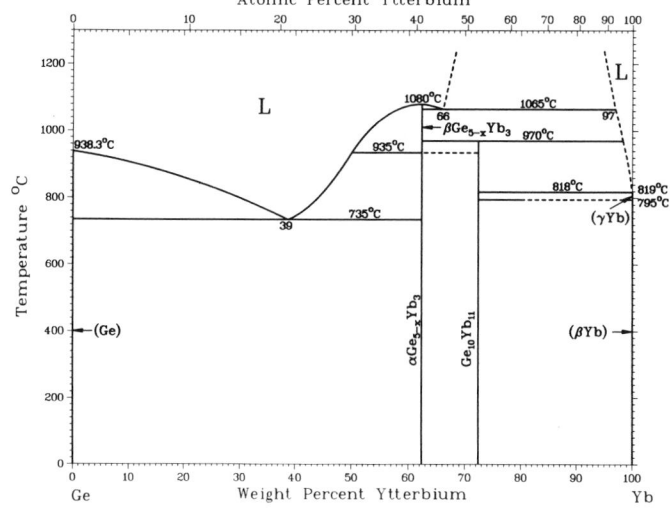

Phase	Composition, wt% Yb	Pearson symbol	Space group
(Ge)	0	cF8	Fd3̄m
βGe₅₋ₓYb	~61	hP3	P6/mmm
αGe₅₋ₓYb	~61	hP8	P6̄2m
Ge₁₀Yb₁₁	72.3	tI84	I4/mmm
(γYb)	100	cI2	Im3̄m
(βYb)	100	cF4	Fm3̄m
(αYb)	100	hP2	P6₃/mmc

Ge-Zn

R.W. Olesinski and G.J. Abbaschian, 1985

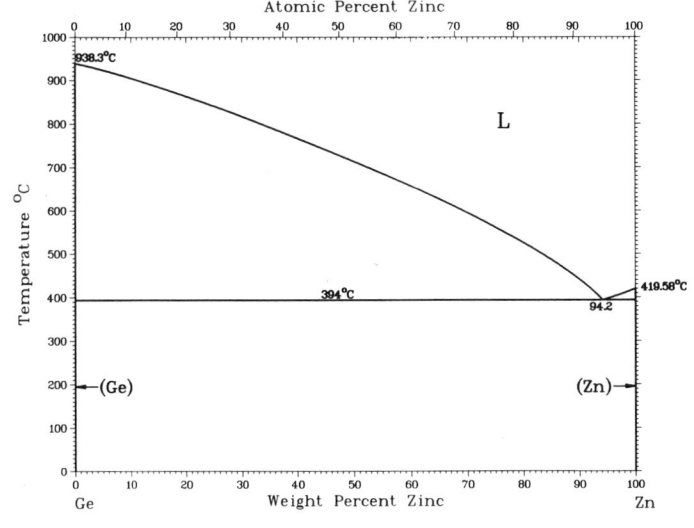

Phase	Composition, wt% Zn	Pearson symbol	Space group
(Ge)	0	cF8	Fd3̄m
(Zn)	100	hP2	P6₃/mmc

H-La

D. Khatamian and F.D. Manchester, 1990

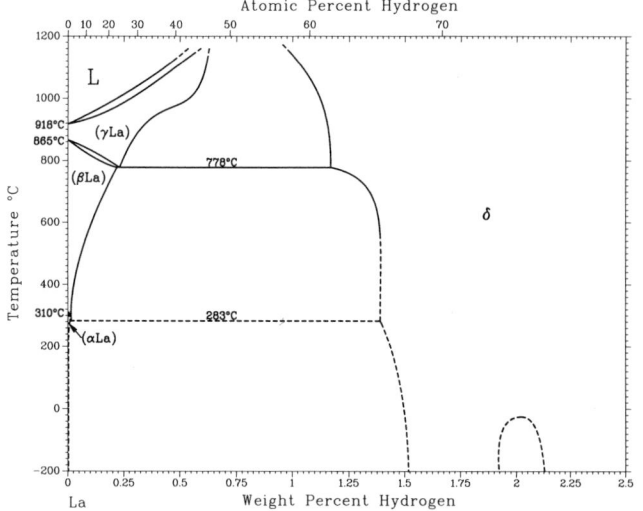

Phase	Composition, wt% H	Pearson symbol	Space group
(γLa)(a)	0 to 0.6	cI2	Im$\bar{3}$m
(βLa)(b)	0 to 0.2	cF4	Fm$\bar{3}$m
(αLa)(c)	0 to 0.01	hP4	P6$_3$/mmc
δ	1 to 2	cF16	Fm$\bar{3}$m

(a) From 865 to 918 °C at 0 at.% H. (b) From 310 to <865 °C at 0 at.% H. (c) Up to <310 °C at 0 at.% H

H-Nb

J.F. Smith, 1983

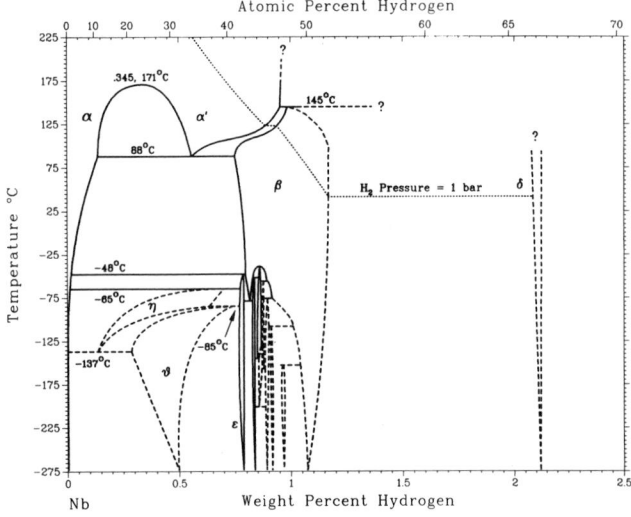

Phase	Composition, wt% H	Pearson symbol	Space group
α, α'	0 to 470.95	cI2	Im$\bar{3}$m
η	~0.13 to ~0.69	...	...
θ	~0.29 to ~0.75	...	...
β	0.75 to ~1.2	oP8	...
ε	~0.78	oP28	...
			'
ι, λ, o, μ, υ, ξ	~0.83 to 0.92	(a)	...
γ	~0.96	(b)	...
δ	~2.13	cF12	Fm$\bar{3}$m

(a) H-deficient β structure having ordering of H atoms. (b) Possibly a face-centered tetragonal structure

Peritectoid cascade region of the Nb-H phase diagram

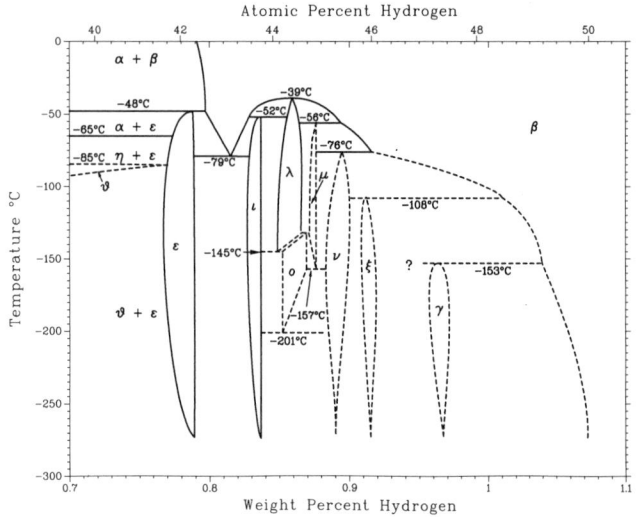

H-Nd

P.R. Subramanian, 1990

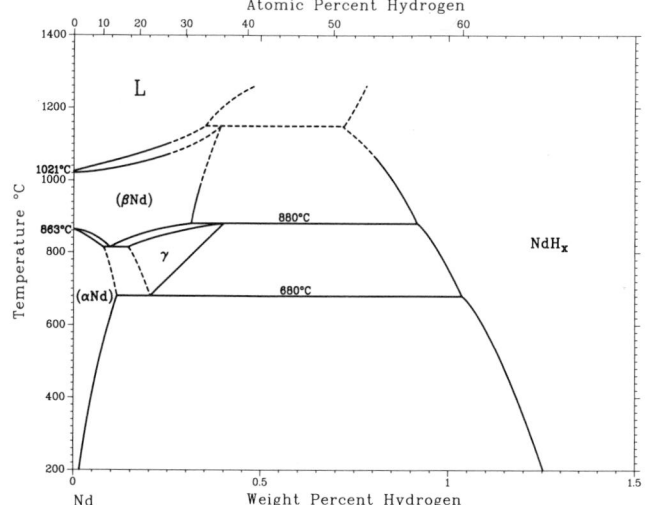

Phase	Composition, wt% H	Pearson symbol	Space group
(βNd)	0	cI2	Im3̄m
(αNd)	0	hP4	P6₃/mmc
γ(a)	~0.15 to 0.43	…	…
NdH₂(b)	~1.38	cF12	Fm3̄m
Nd₂H₅(b, c)	~1.6	tI28	I4₁md
		tI40	I4₁md

(a) High-temperature phase; exists between 680 and 880 °C. (b) Not shown in the phase diagram. (c) Ideal stoichiometry; structure based on neutron-diffraction studies on samples with the composition NdD₂.₃₆

H-Ni

M.L. Wayman and G.C. Weatherly, 1991

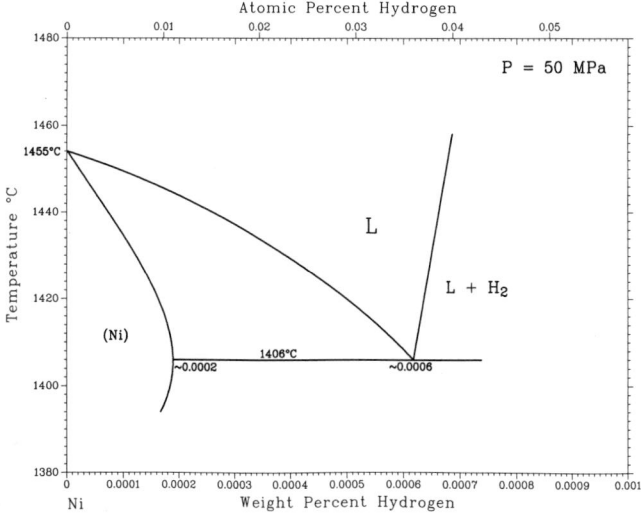

Phase	Composition, wt% H	Pearson symbol	Space group
(Ni)	0 to ~0.0002	cF4	Fm3̄m

H-Pd

A. San-Martin and F.D. Manchester, unpublished

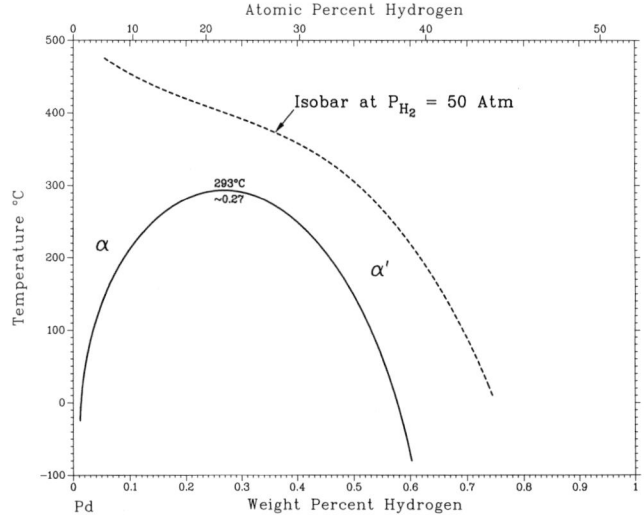

Phase	Composition, wt% H	Pearson symbol	Space group
(Pd)	0	cF4	Fm3̄m
α or (Pd)	0 to 0.019(a)	cF8	Fm3̄m
α' or (Pd)	~0.567(a)	(b)	…
Low-temperature phases(c)			
A₂B₂	0.601	…	I4₁/amd
A₄B	0.715	tI10	I4/m

(a) At 25 °C. (b) fcc. (c) Below 100 K

H-Sr

D.T. Peterson and R.P. Colburn, 1964

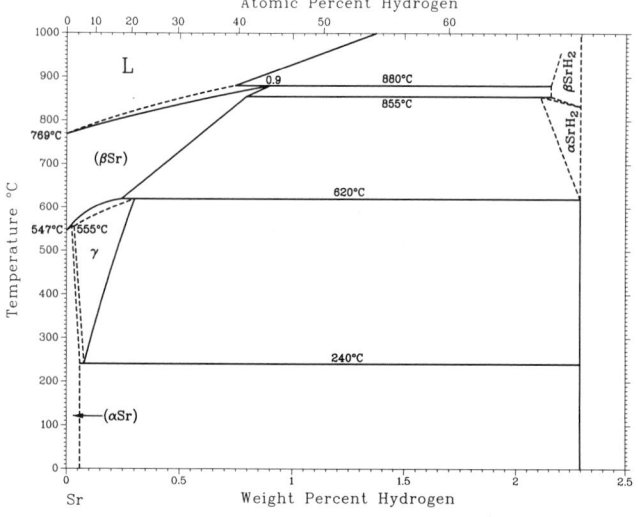

Phase	Composition, wt% H	Pearson symbol	Space group
(βSr)	0 to 0.9	cI2	Im$\bar{3}$m
(αSr)	0 to ?	cF4	Fm$\bar{3}$m
γ	? to 0.3	hP*	…
βSrH$_2$	2.3	…	…
αSrH$_2$	2.3	oP12	Pnma

H-Ta

A.San-Martin and F.D. Manchester, 1991

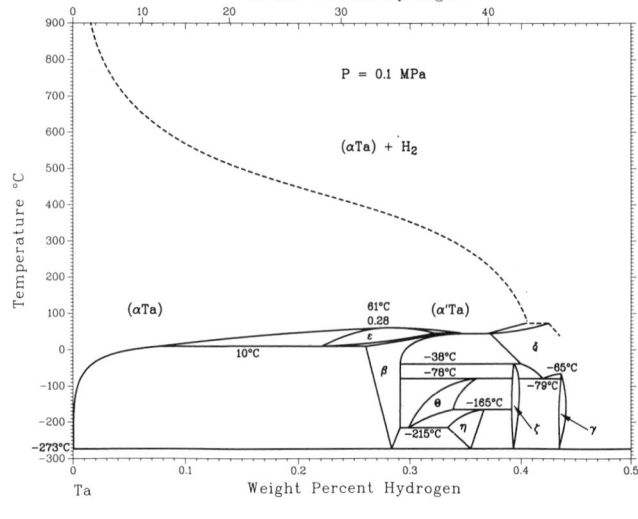

Phase	Composition, wt% H	Pearson symbol	Space group
(αTa)	0 to 0.28	cI2	Im$\bar{3}$m
(α'Ta)	0.28 to 0.42	cI2	Im$\bar{3}$m
ε	0.22 to 0.32	mC*	C222
β	0.26 to 0.35	mC*	C222
θ	0.30 to 0.36	…	…
η	0.34 to ~0.37	…	…
δ	0.37 to 0.438	oP*	Pnnm
ζ	0.395 to ~0.398	…	…
γ	0.436 to 0.439	…	…

H-Ti

H. Okamoto, 1992

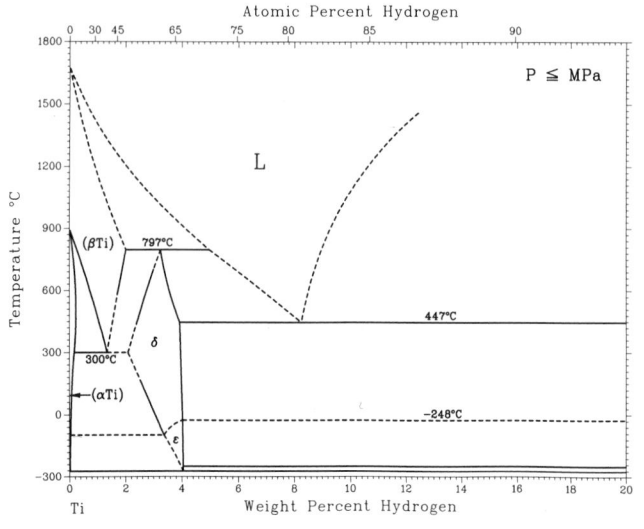

Phase	Composition, wt% H	Pearson symbol	Space group
(βTi)	0 to 2.06	cI2	Im$\bar{3}$m
(αTi)	0 to 0.2	hP2	P6$_3$/mmc
δ	2.06 to 4.05	cF12	Fm$\bar{3}$m
ε	3.06 to 4.05	tI6	I4/mmm

H-U

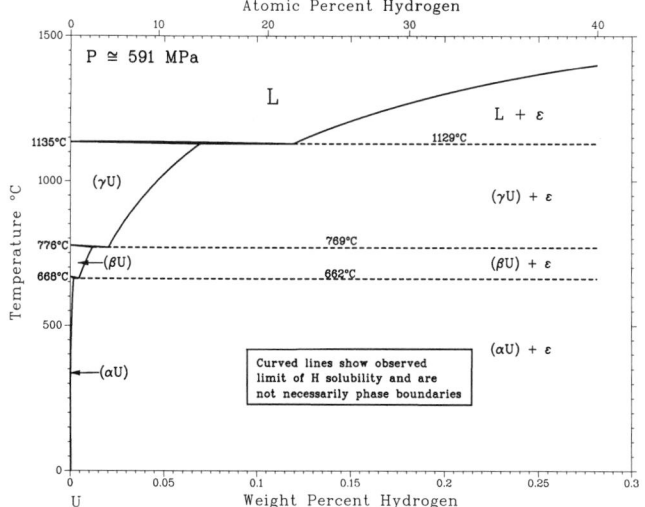

A. San-Martin and F.D. Manchester, unpublished

Phase	Composition, wt% H	Pearson symbol	Space group
(γU)	0 to 0.069	cI2	Im$\overline{3}$m
(βU)	0 to 0.011	tP30	P4$_2$/mnm
(αU)	0 to 0.0014	oC4	Cmcm
ε	1.25	cP32	Pm3n
δ(a)	1.25	cP8	Pm3n

(a) Metastable phase

H-V

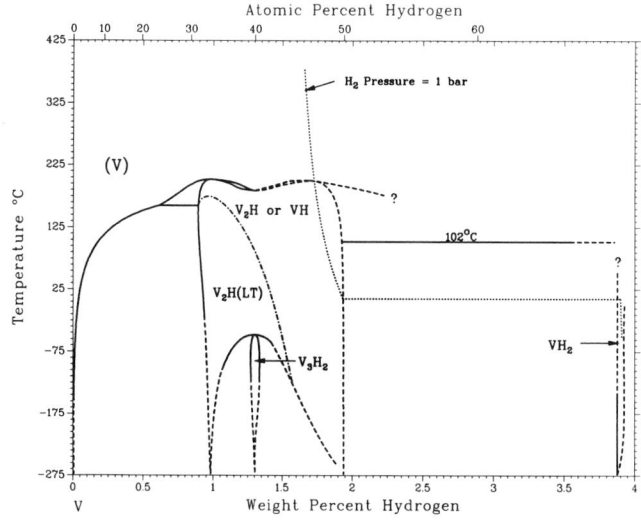

J.F. Smith and D.T. Peterson, 1989

Phase	Composition, wt% H	Pearson symbol	Space group
α or (V)	0 to ?	cI2	Im$\overline{3}$m
β$_1$ or V$_2$H(LT)	~0.97	mC6	C2/m
β$_2$ or V$_2$H or VH	~0.97 to 1.94	tF6, tF8?	…
δ or V$_3$H$_2$	~1.30	mC10	…
γ or VH$_2$	3.81	cF12	Fm$\overline{3}$m

H-Zr

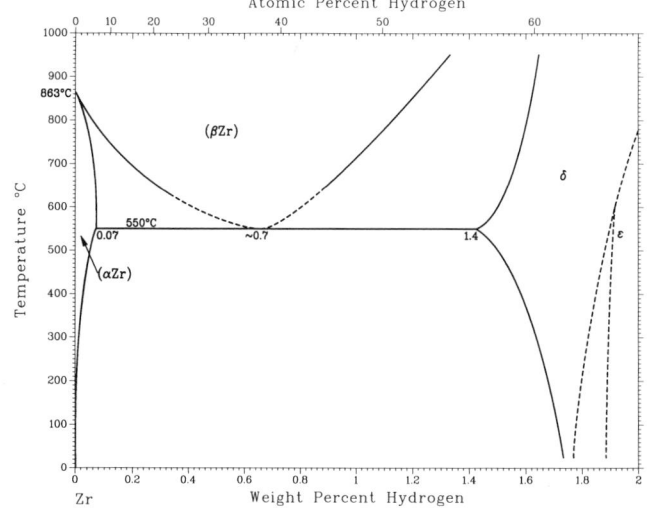

E. Zuzek, J.P. Abriata, A.San-Martin, and F.D. Manchester, 1990

Phase	Composition, wt% H	Pearson symbol	Space group
α or (αZr)	0 to 0.07	hP2	P6$_3$/mmc
β or (βZr)	0 to ~1.28?	cI2	Im$\overline{3}$m
δ	1.4 to ~2.1?	cF12	Fm$\overline{3}$m
ε	1.89	tI6	I4/mmm
Metastable phase			
γ	~0.011	tP6	P4$_2$/n

Hf-Ir

H. Okamoto, 1990

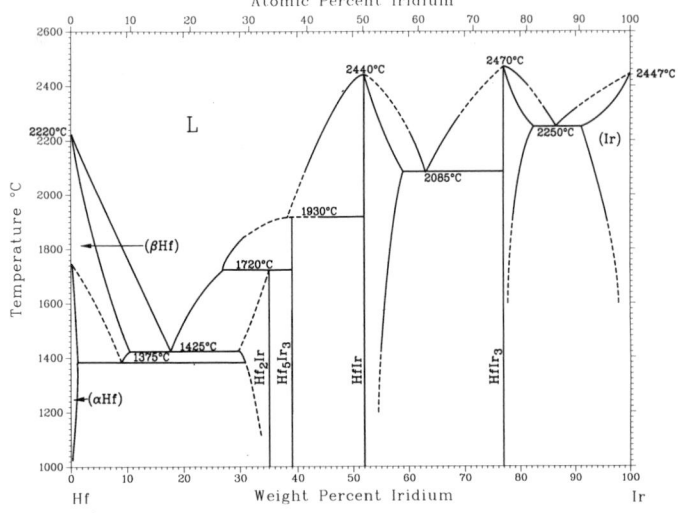

Phase	Composition, wt% Ir	Pearson symbol	Space group
(βHf)	0 to ~10.5	cI2	Im$\overline{3}$m
(αHf)	0 to ~ 1.5	hP2	P6$_3$/mmc
Hf$_2$Ir	~28 to 35.0	cF96	Fd$\overline{3}$m
Hf$_5$Ir$_3$	39.3	hP16	P6$_3$/mcm
HfIr	51.9 to 59	o**	...
HfIr$_3$	76 to 82	cP4	Pm$\overline{3}$m
(Ir)	~91 to 100	cF4	Fm$\overline{3}$m

Hf-Mn

H. Okamoto, unpublished

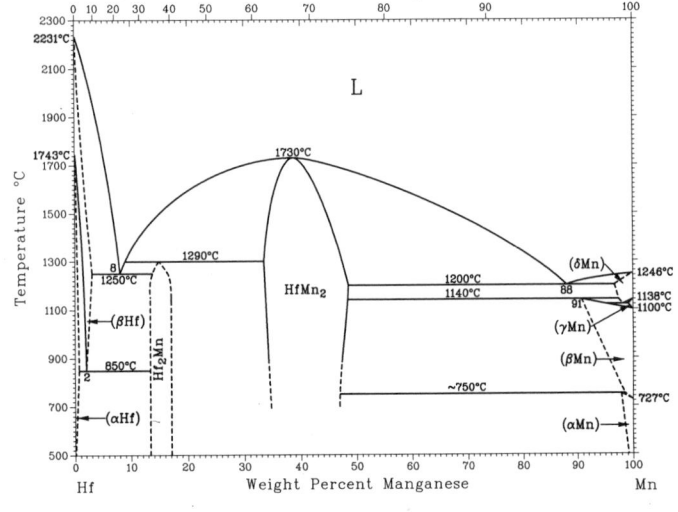

Phase	Composition, wt% Mn	Pearson symbol	Space group
(βHf)	0 to 3	cI2	Im$\overline{3}$m
(αHf)	0 to 0.62	hP2	P6$_3$/mmc
Hf$_2$Mn	13.3 to ?	cF96	Fd$\overline{3}$m
βHfMn$_2$	?	hP24	P6$_3$/mmc
αHfMn$_2$	33 to 48.7	hP12	P6$_3$/mmc
(δMn)	97.4 to 100	cI2	Im$\overline{3}$m
(γMn)	99.0 to 100	cF4	Fm$\overline{3}$m
(βMn)	91 to 100	cP20	P4$_1$32
(αMn)	99.4 to 100	cI58	I$\overline{4}$3m

Hf-Mo

From [Molybdenum]

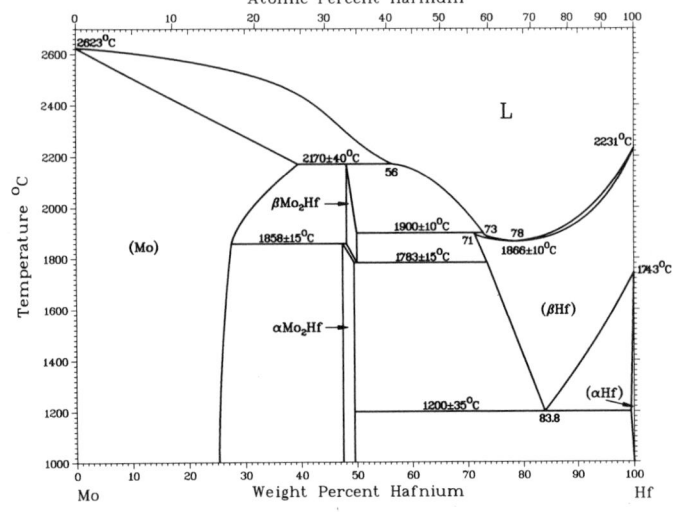

Phase	Composition, wt% Hf	Pearson symbol	Space group
(Mo)	0 to 38	cI2	Im$\overline{3}$m
βMo$_2$Hf	~48.2	hP24	P6$_3$/mmc
αMo$_2$Hf	~48.2	cF25	Fd$\overline{3}$m
(βHf)	71 to 100	cI2	Im$\overline{3}$m
(αHf)	~100	hP2	P6$_3$/mmc

Hf-N

H. Okamoto, 1990

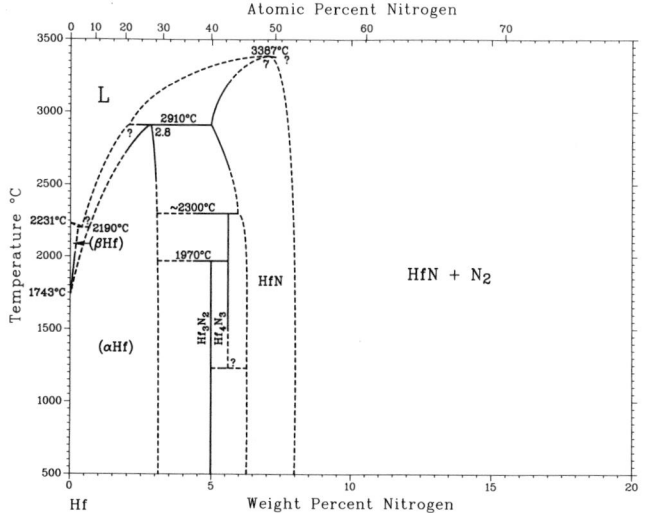

Phase	Composition, wt% N	Pearson symbol	Space group
(βHf)	0 to ?	cI2	$Im\bar{3}m$
(αHf)	0 to 3.1	hP2	$P6_3/mmc$
Hf$_3$N$_2$	4.97	hR6	$R\bar{3}m$
Hf$_4$N$_3$	5.57	hR8	$R\bar{3}m$
HfN	4.59 to 7.98	cF8	$Fm\bar{3}m$
(αN)	100	cP8	$Pa3$

Hf-Nb

H. Okamoto, 1991

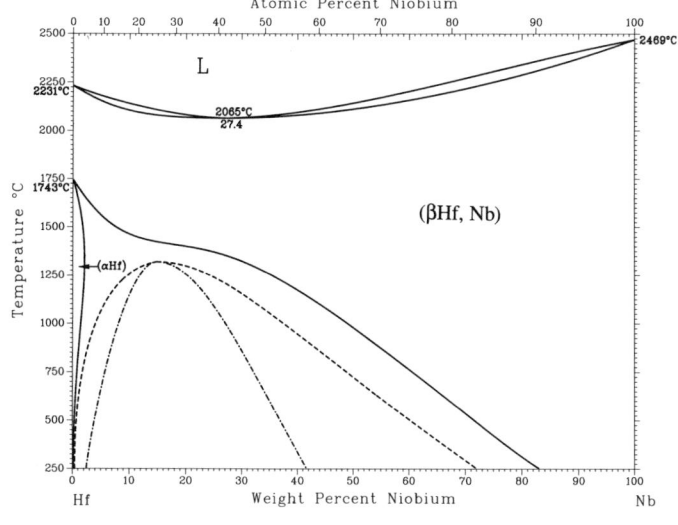

Phase	Composition, wt% Nb	Pearson symbol	Space group
(βHf,Nb)	0 to 100	cI2	$Im\bar{3}m$
(αHf)	0 to 2.4	hP2	$P6_3/mmc$

Hf-Ni

P. Nash and A. Nash, 1991

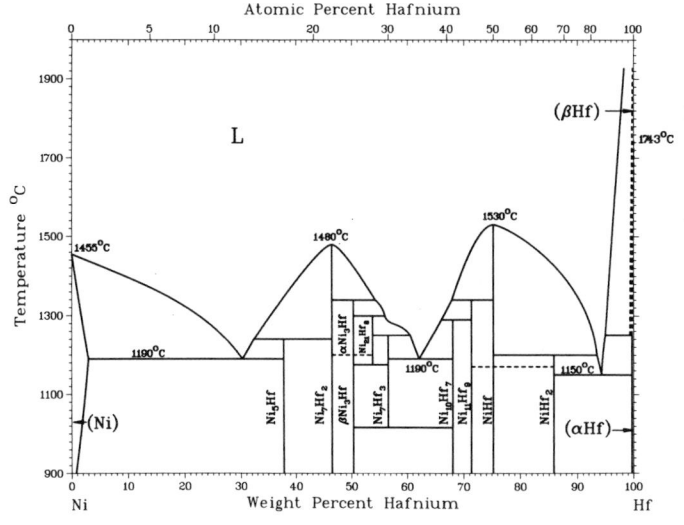

Phase	Composition, wt% Hf	Pearson symbol	Space group
(Ni)	0 to 3	cF4	$Fm\bar{3}m$
Ni$_5$Hf	37.9	cF24	$F\bar{4}3m$
Ni$_7$Hf$_2$	46.5	m**	...
βNi$_3$Hf	50	hP40	$P6_3/mmc$
αNi$_3$Hf	50	hR12	$R\bar{3}m$
Ni$_{21}$Hf$_6$	53.7	aP29	$P\bar{1}$
Ni$_7$Hf$_3$	57	aP20	$P\bar{1}$
Ni$_{10}$Hf$_7$	68.1	oC68	$C2ca$
Ni$_{11}$Hf$_9$	71	tI*	$I4/m$
NiHf	75.3	oC8	$Cmcm$
NiHf$_2$	85.9	tI12	$I4/mcm$
(βHf)	99.3 to 100	cI2	$Im\bar{3}m$
(αHf)	99.7 to 100	hP2	$P6_3/mmc$

Hf-O

From [Hafnium]

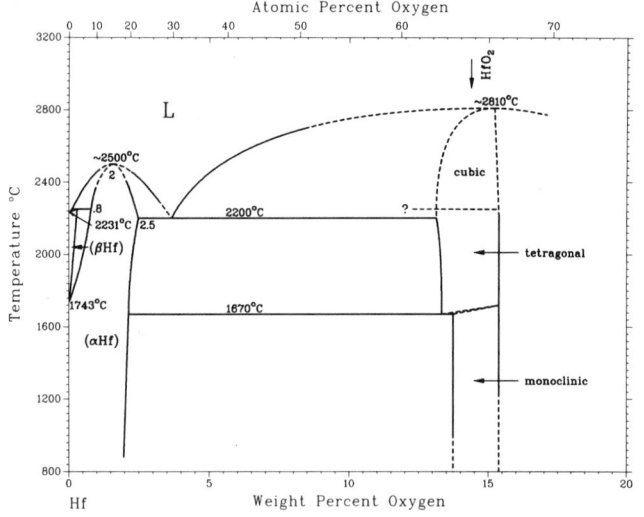

Phase	Composition, wt% O	Pearson symbol	Space group
(βHf)	0 to 0.8	cI2	Im3̄m
(αHf)	0 to 2.5	hP2	P6₃/mmc
HfO₂	~13.2 to 15.4	cF12	Fm3̄m
	~13.2 to 15.4	t**	...
	~13.7 to 15.4	mP12	P2₁/c

Hf-Os

H. Okamoto, 1990

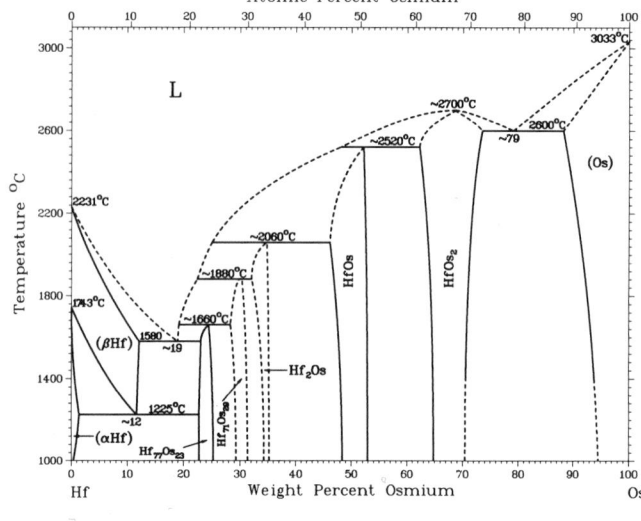

Phase	Composition, wt% Os	Pearson symbol	Space group
(βHf)	0 to 13	cI2	Im3̄m
(αHf)	0 to 2	hP2	P6₃/mmc
θ	~24	...	...
ζ	~30	...	...
Hf₂Os	~35	...	...
HfOs	~47 to 54	cP2	Pm3̄m
HfOs₂	~64 to 73	cF96	Fd3̄m
		hP12	P6₃/mmc
(Os)	100	hP2	P6₃/mmc

Hf-Rh

H. Okamoto, 1990

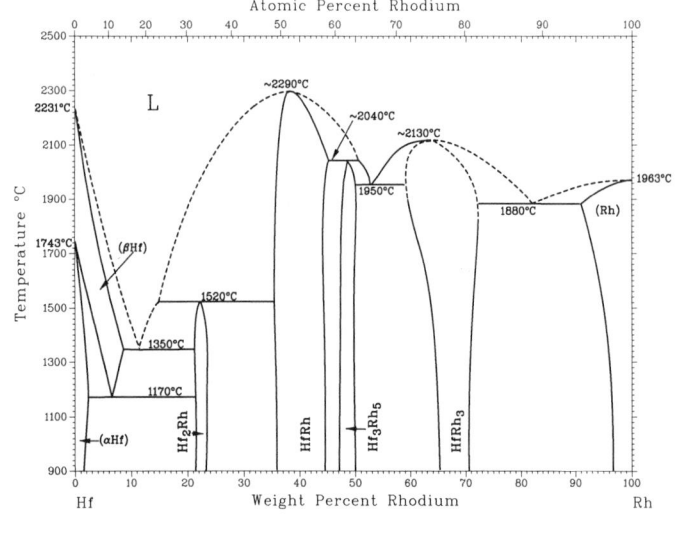

Phase	Composition, wt% Rh	Pearson symbol	Space group
(βHf)	0	cI2	Im3̄m
(αHf)	0	hP2	P6₃/mmc
Hf₂Rh	22 to 23	cF96	Fd3̄m
HfRh	36 to 44	cP2	Pm3̄m
Hf₃Rh₅	47 to 51	oP16	Pbam
HfRh₃	59 to 72	cP4	Pm3̄m
(Rh)	100	cF4	Fm3̄m

Hf-Si

A.B. Gokhale and G.J. Abbaschian, 1989

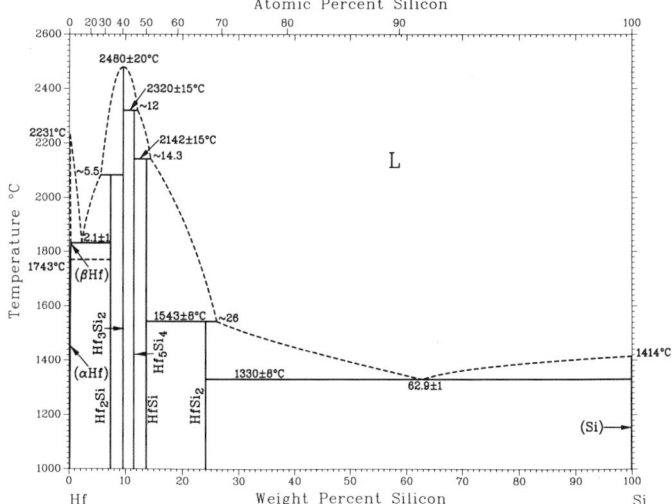

Phase	Composition, wt% Si	Pearson symbol	Space group
(αHf)	~0	hP2	P6₃/mmc
(βHf)	~0	cI2	Im$\bar{3}$m
Hf₂Si	7.3	tI12	I4/mcm
Hf₃Si₂	9	tP10	P4/mbm
Hf₅Si₄	11.2	...	P4₁2₁2
HfSi	13.6	oP8	Pnma
HfSi₂	24.0	oC12	Cmcm
(Si)	100	cF8	Fd$\bar{3}$m

Note: The presence of Mn₅Si₃-type (D8₈) Hf₅Si₃ has been reported. However, the phase occurs only in the presence of interstitial impurities.

Hf-Ta

R. Krishnan, S.P. Garg, and N. Krishnamurthy, 1989

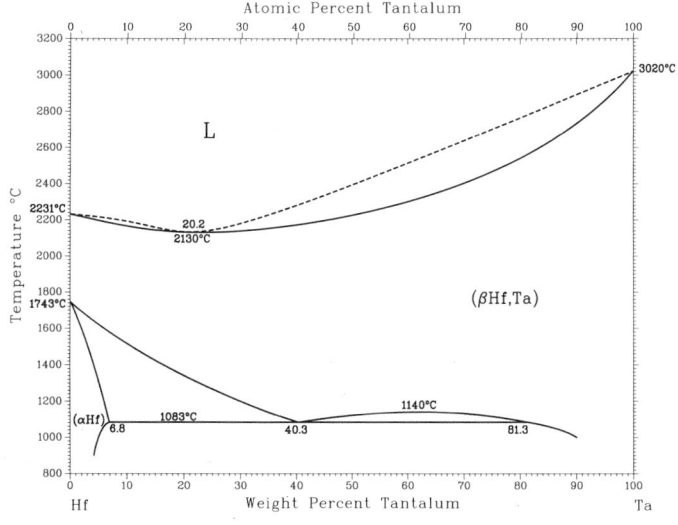

Phase	Composition, wt% Ta	Pearson symbol	Space group
(βHf,Ta)	0 to 100	cI2	Im$\bar{3}$m
(αHf)	0 to 6.8	hP2	P6₃/mmc

Hf-U

D.T. Peterson and D.J. Beerntsen, 1960

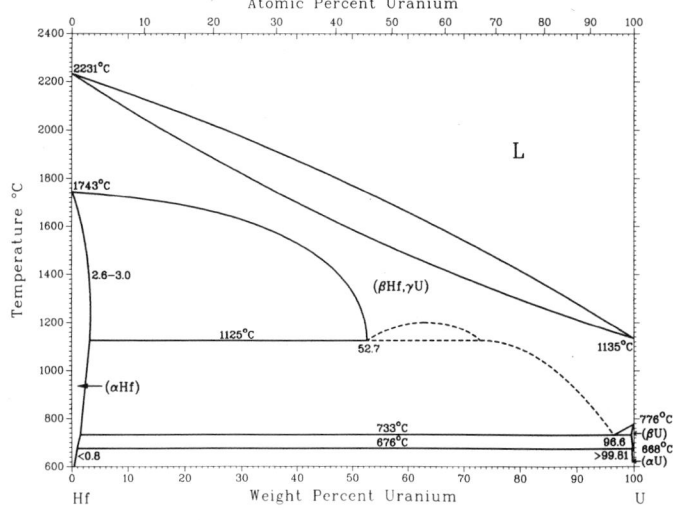

Phase	Composition, wt% U	Pearson symbol	Space group
(βHf,γU)	0 to 100	cI2	Im$\bar{3}$m
(αHf)	0 to ~3	hP2	P6₃/mmc
(βU)	100	tP30	P4₂/mnm
(αU)	100	oC4	Cmcm

Hf-V

J.F. Smith, 1989

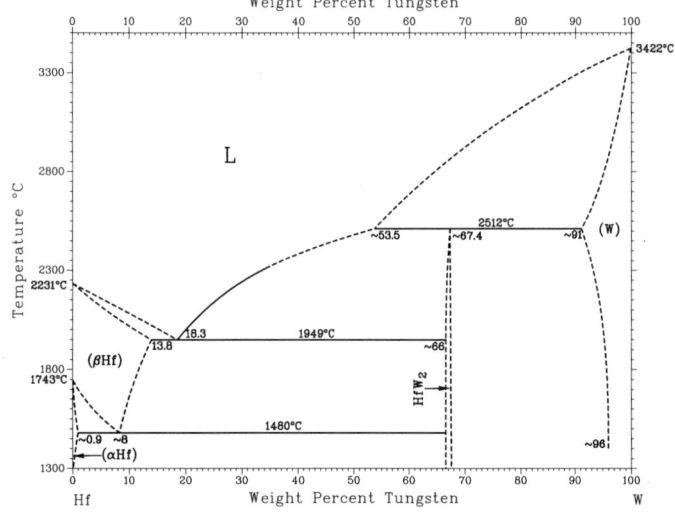

Phase	Composition, wt% V	Pearson symbol	Space group
(βHf)	0 to 8.1	cI2	$Im\bar{3}m$
(αHf)	0 to <1	hP2	$P6_3/mmc$
HfV$_2$	~36.4	cF24	$Fd\bar{3}m$
(V)	~87 to 100	cI2	$Im\bar{3}m$

Hf-W

S.V. Nagender Naidu and P. Rama Rao, 1991

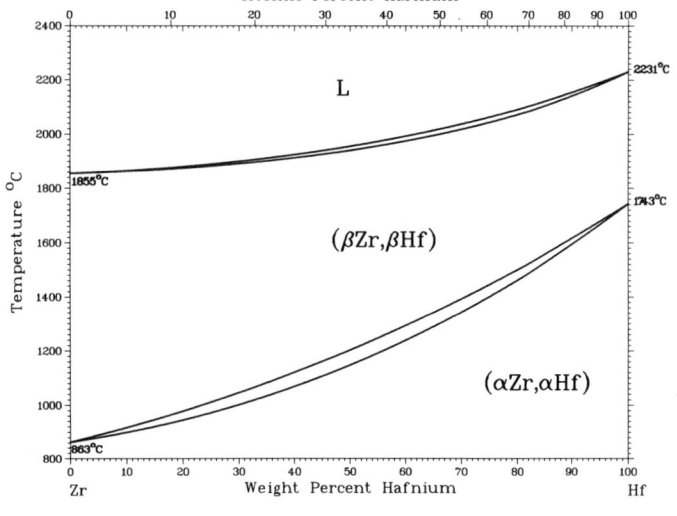

Phase	Composition, wt% W	Pearson symbol	Space group
(βHf)	0 to 13.8	cI2	$Im\bar{3}m$
(αHf)	0 to ~0.9	hP2	$P6_3/mmc$
HfW$_2$	~67.4	cF24	$Fd3n$
(W)	~91 to 100	cI2	$Im\bar{3}m$

Hf-Zr

J.P. Abriata, J.C. Bolcich, and H.A. Peretti, 1982

Phase	Composition, wt% Hf	Pearson symbol	Space group
(αZr,αHf)	0 to 100	hP2	$P6_3/mmc$
(βZr,βHf)	0 to 100	cI2	$Im\bar{3}m$
ω(a)	0 to 100	hP3	$P\bar{3}m1$ ($P6/mmm$?)

(a) Metastable at room temperature and zero pressure

Hg-In

H. Okamoto, 1992

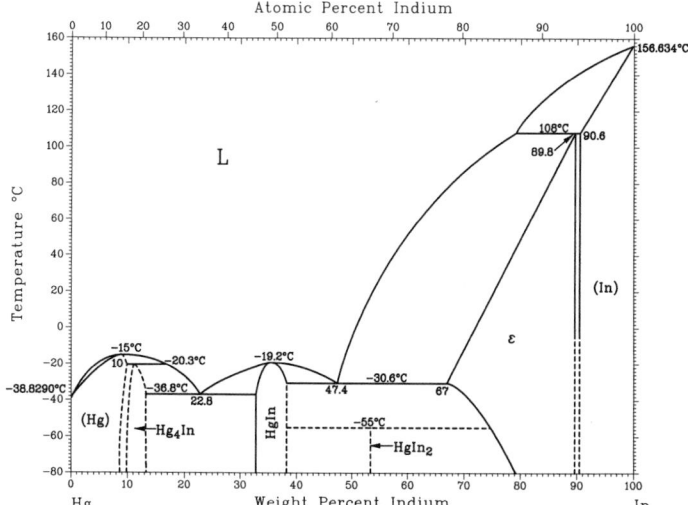

Phase	Composition, wt% In	Pearson symbol	Space group
(Hg)	0 to ~10	hR1	R$\overline{3}$m
Hg$_4$In	10 to 14	oF8	Fddd
HgIn	33 to 38	hR2	R$\overline{3}$m
HgIn$_2$	53.4	...	...
ε	67 to 89.8	cF4	Fm$\overline{3}$m
(In)	90.6 to 100	tI2	I4/mmm

Hg-K

A.E. Vol and I.K. Kagan, 1979

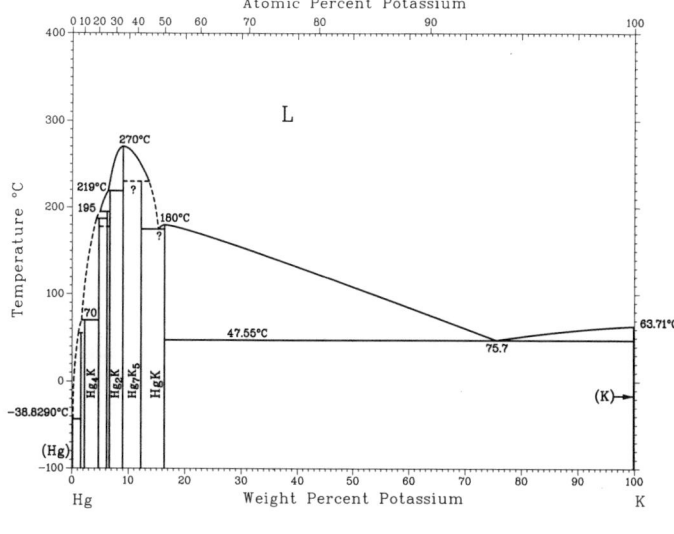

Phase	Composition, wt% K	Pearson symbol	Space group
(Hg)	0	hR1	R$\overline{3}$m
Hg$_{11}$K	1.7	cP36	Pm$\overline{3}$m
Hg$_9$K	2.1	...	...
Hg$_4$K	4.6	...	...
Hg$_3$K	6.1	...	...
Hg$_{2.7}$K	6.7	...	...
Hg$_2$K	8.9	oI12	Imma
Hg$_7$K$_5$	12.2	oP48	Pbcm
HgK	16.3	aP8	P1
(K)	100	cI2	Im$\overline{3}$m

Hg-La

C. Guminski, unpublished

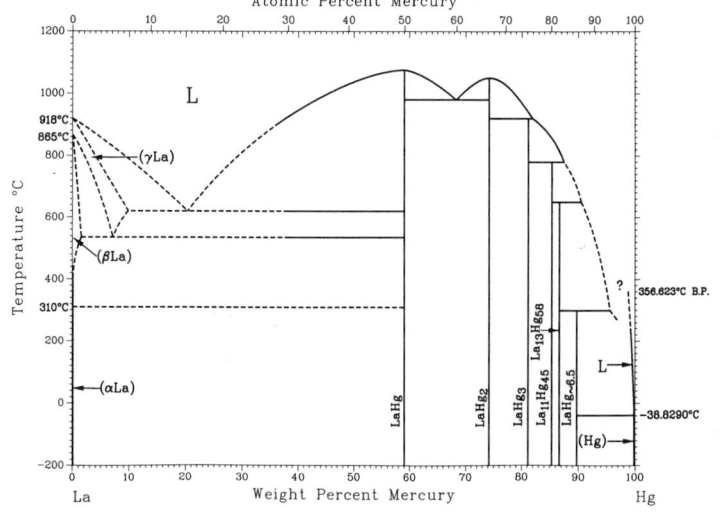

Phase	Composition, wt% Hg	Pearson symbol	Space group
(γLa)	0	cI2	Im$\overline{3}$m
(βLa)	0	cF4	Fm$\overline{3}$m
(αLa)	0	hP4	P6$_3$/mmc
LaHg	59.1	cP2	Pm$\overline{3}$m
LaHg$_2$	74.3	hP3	P6/mmm
LaHg$_3$	81	hP8	P6$_3$/mmc
La$_{11}$Hg$_{45}$	85.5	cF448	F$\overline{4}$3m
La$_{13}$Hg$_{58}$	87	hP142	P6$_3$/mmc
LaHg$_{-6.5}$	91	o**	Cmcm or C2cm or Cmc2$_1$
(Hg)	100	hR1	R$\overline{3}$m

Hg-Li

From [Hansen]

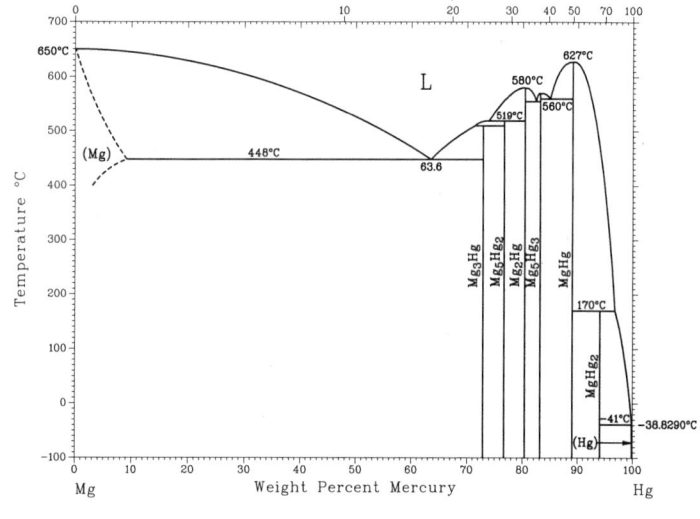

Phase	Composition, wt% Li	Pearson symbol	Space group
(Hg)	0	hR1	$R\bar{3}m$
Hg₃Li	1.1	hP8	$P6_3/mmc$
Hg₂Li	1.7	...	...
HgLi	~2.08 to 5.6	cP2	$Pm\bar{3}m$
HgLi₂	6.5	...	...
HgLi₃	9.4	cF16	$Fm\bar{3}m$
HgLi₆	17.2	...	...
(Li)	? to 100	cI2	$Im\bar{3}m$

Hg-Mg

A.A. Nayeb-Hashemi and J.B. Clark, 1988

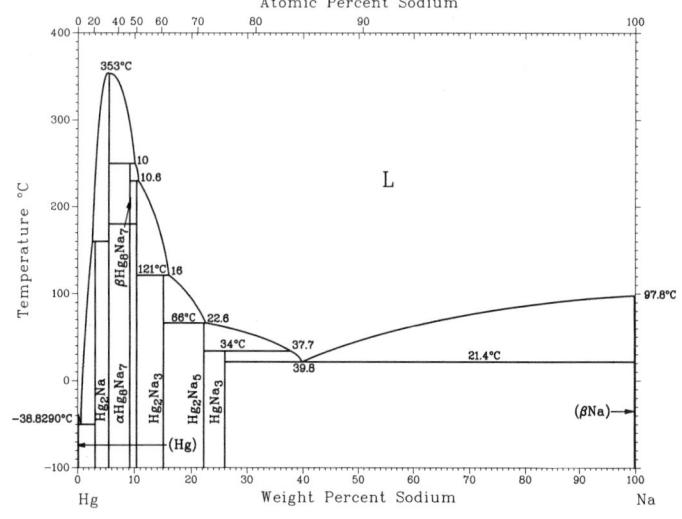

Phase	Composition, wt% Hg	Pearson symbol	Space group
(Mg)	0 to ~9.1	hP2	$P6_3/mmc$
Mg₃Hg	73	hP8	$P6_3/mmc$
Mg₅Hg₂	76.8	...	...
Mg₂Hg	80.5	...	...
Mg₅Hg₃	83.2	hP16	$P6_3/mmc$
MgHg	89.2	cP2	$Pm\bar{3}m$
MgHg₂	94.3	tI6	$I4/mmm$
(Hg)	100	hR1	$R\bar{3}m$

Hg-Na

H. Okamoto, 1990

Phase	Composition, wt% Na	Pearson symbol	Space group
(Hg)	~0	hR1	$R\bar{3}m$
Hg₄Na	3	h**	...
Hg₂Na	5.4	hP3	$P6/mmm$
βHg₈Na₇	~9.1	...	...
αHg₈Na₇	~9.1	...	...
HgNa	10.3	oC16	$Cmcm$
Hg₂Na₃	15	tP20	$P4_2/mnm$
Hg₂Na₅	~22.2	hR*	...
HgNa₃	26	...	...
(βNa)	~100	cI2	$Im\bar{3}m$

Hg-Pb

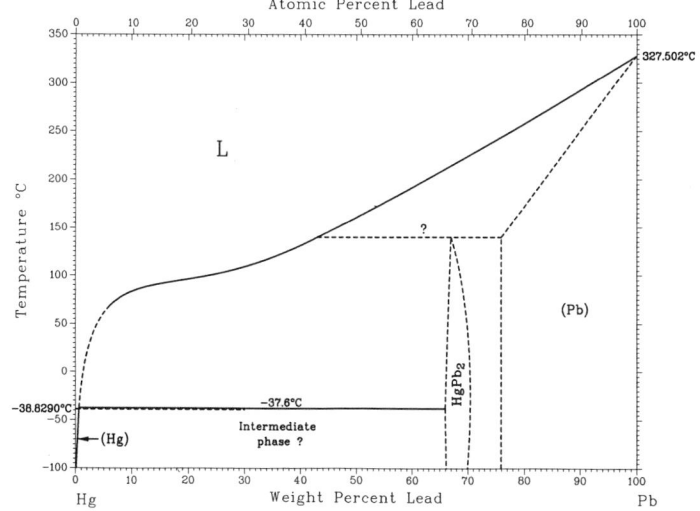

Phase	Composition, wt% Pb	Pearson symbol	Space group
(Hg)	0	$hR1$	$R\overline{3}m$
HgPb$_2$	~66 to ~71	$tP2$	$P4/mmm$
(Pb)	~76 to 100	$cF4$	$Fm\overline{3}m$

Hg-Rb

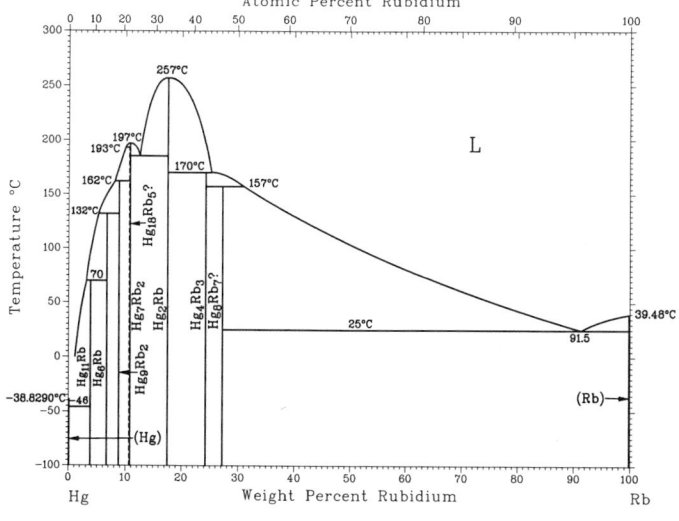

Phase	Composition, wt% Rb	Pearson symbol	Space group
(Hg)	0	$hR1$	$R\overline{3}m$
Hg$_{11}$Rb	3.7	$cP36$	$Pm\overline{3}m$
Hg$_6$Rb	6.6	…	…
Hg$_9$Rb$_2$	8.7	…	…
Hg$_{18}$Rb$_5$	10.6	…	…
Hg$_7$Rb$_2$	10.8	…	…
Hg$_2$Rb	17.5	…	…
Hg$_4$Rb$_3$	24.2	…	…
Hg$_8$Rb$_7$	27.2	…	…
(Rb)	100	$cI2$	$Im\overline{3}m$

Hg-S

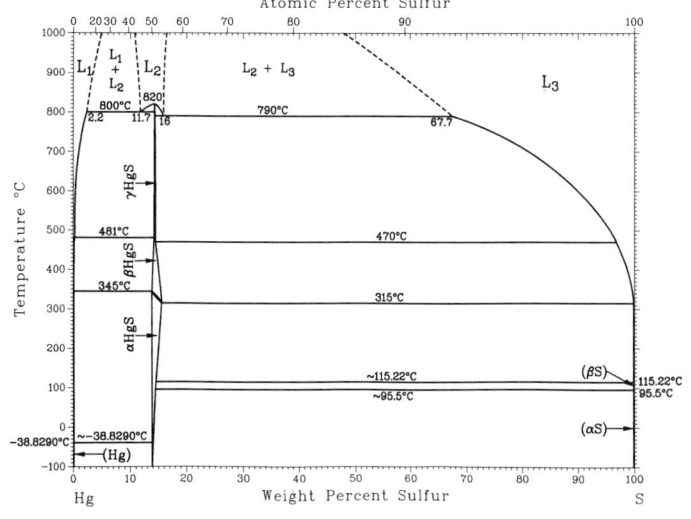

Phase	Composition, wt% S	Pearson symbol	Space group
Hg	0	$hR1$	$R\overline{3}m$
γHgS	14.19 to 14.47	(a)	…
βHgS	13.8 to 15.61	$cF8$	$F\overline{4}3m$
αHgS	13.8 to 15.5	$hP6$	$P3_121$
δHgS(b)	13.8	$cF8$	$Fm\overline{3}m$
(βS)(c)	100	$mP*$	$P2_1/c$
(αS)(d)	100	$oF128$	$Fddd$

(a) Hexagonal. (b) Above 13 GPa. (c) From 95.5 to 115.22 °C. (d) At <95.5 °C

Hg-Se

R.C. Sharma, Y.A. Chang, and C. Guminski, 1992

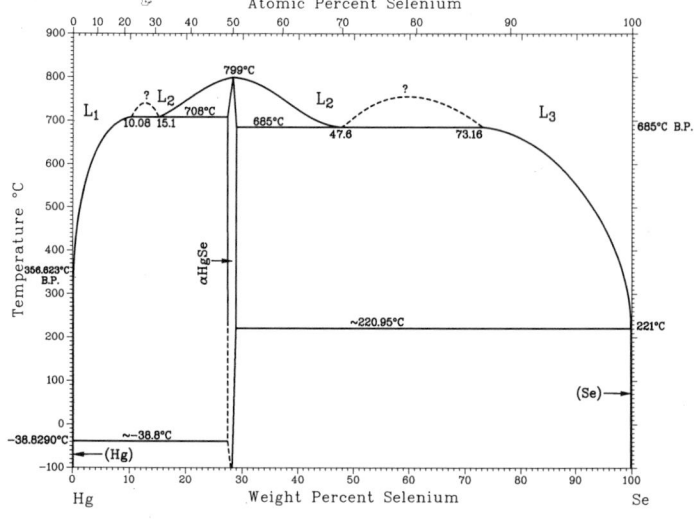

Phase	Composition, wt% Se	Pearson symbol	Space group
(Hg)	0	hR1	$R\bar{3}m$
αHgSe	27.3 to 28.86	cF8	$F\bar{4}3m$
(γSe)	100	hP3	$P3_121$
(βSe)	100	mP64	$P2_1/c$
(αSe)	100	mP32	$P2_1/n$
High-pressure phases			
βHgSe(a)	28.2	hP6	$P3_121$
γHgSe(b)	28.2	cF8	$Fm\bar{3}m$
δHgSe(c)	28.2	tI4	$I4m2$

(a) Between 0.30 and ~7 GPa. (b) Between ~7 GPa and 13.3 GPa. (c) Above 13.3 GPa

Hg-Sn

H. Okamoto, 1990

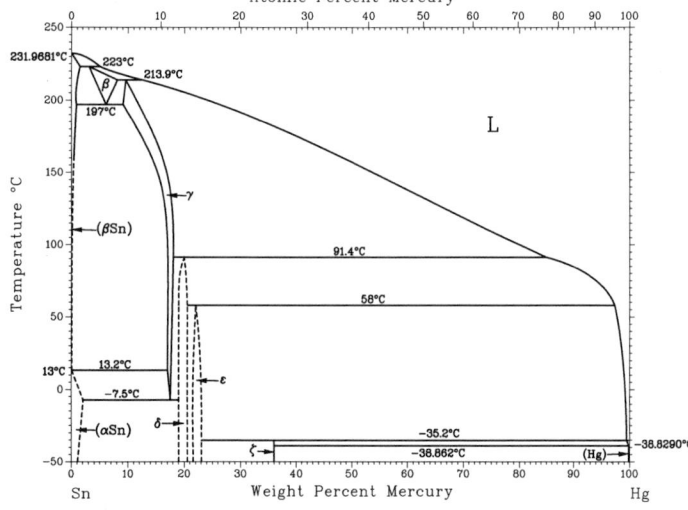

Phase	Composition, wt% Hg	Pearson symbol	Space group
(βSn)	0 to <0.8	tI4	$I4_1/amd$
(αSn)	0 to <2	cF8	$Fd\bar{3}m$
β	~3 to 8	h**	...
γ	~10 to 19	hP2	$P6_3/mmc$
δ	~20	...	...
ε	~22 to 23	...	...
ζ	36	...	...
(Hg)	~100	hR1	$R\bar{3}m$

Hg-Sr

P.R. Subramanian, 1990

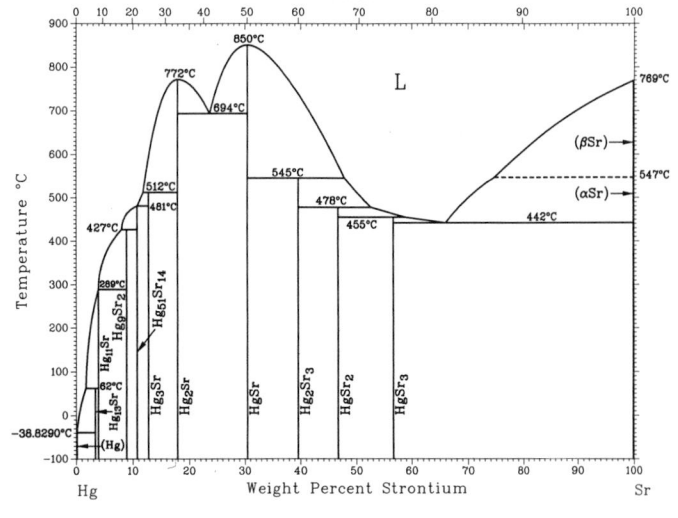

Phase	Composition, wt% Sr	Pearson symbol	Space group
(Hg)	0	hR1	$R\bar{3}m$
Hg₁₃Sr	~3.2	...	...
Hg₁₁Sr	~3.8	cP36	$Pm\bar{3}m$
Hg₉Sr₂	~8.8	hP142	$P6_3mc$
Hg₅₁Sr₁₄	~10.7	hP58	$P6/m$
Hg₃Sr	13	hP8	$P6_3/mmc$
Hg₂Sr	17.9	oI12	$Immc$
		hP3	$P6/mmm$
HgSr	30.4	cP2	$Pm\bar{3}m$
Hg₂Sr₃	40	tP10	$P4/mbm$
HgSr₂	46.7	...	...
HgSr₃	57	oP16	$Pnma$
(βSr)	100	cI2	$Im\bar{3}m$
(αSr)	100	cF4	$Fm\bar{3}m$

Hg-Te

R.C. Sharma and Y.A. Chang, unpublished

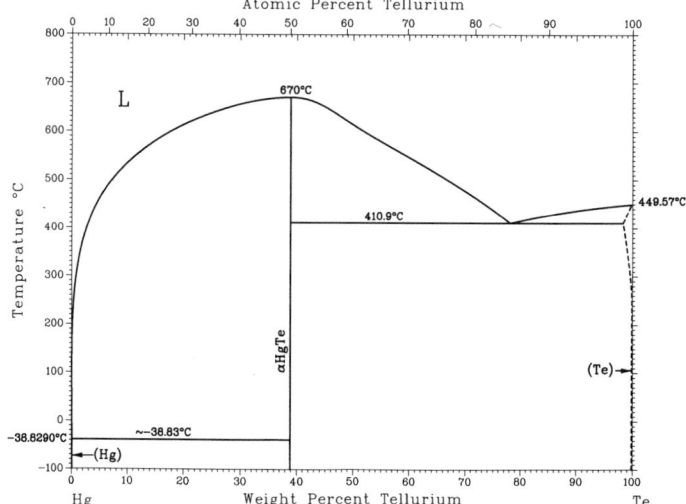

Phase	Composition, wt% Te	Pearson symbol	Space group
(Hg)	0	hR1	$R\bar{3}m$
αHgTe	30.9	cF8	$F\bar{4}3m$
βHgTe(a)	38.9	hP6	$P3_121$
(Te)	~98 to 100	hP3	$P3_121$

(a) High-pressure form

Hg-Tl

C. Guminski, unpublished

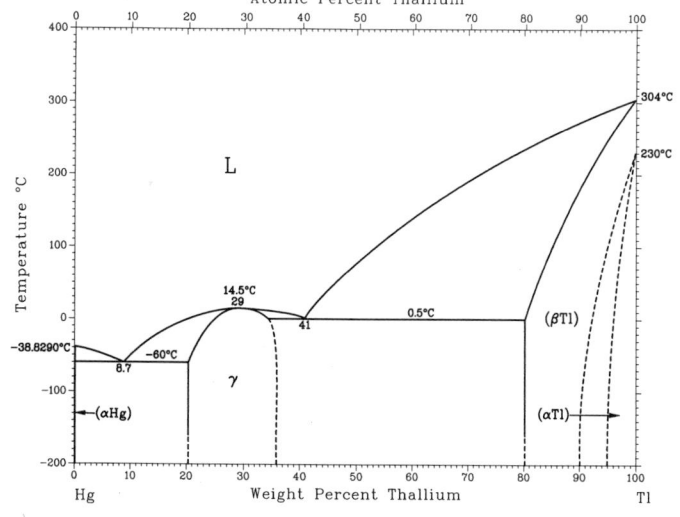

Phase	Composition, wt% Tl	Pearson symbol	Space group
(αHg)	0	hR1	$R\bar{3}m$
γ or Hg$_5$Tl$_2$	~29	cF4	$Fm\bar{3}m$
(βTl)	80 to 100	cI2	$Im\bar{3}m$
(αTl)	? to 100	hP2	$P6_3/mmc$

Hg-Zn

L.A. Zabdyr and C. Guminski, unpublished

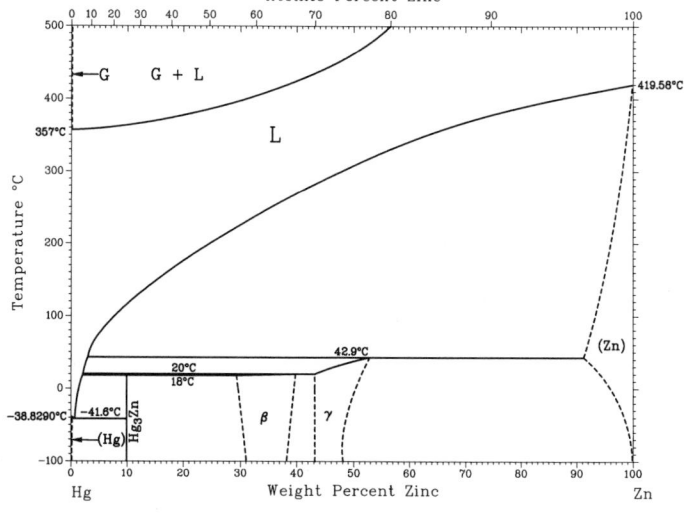

Phase	Composition, wt% Zn	Pearson symbol	Space group
(Hg)	0 to 0.1	hR1	$R\bar{3}m$
Hg$_3$Zn	10	...	...
β	~29 to ~40	...	...
γ(a)	~43 to ~52	oC4	$Cmc2_1$
(Zn)	~95.0 to 100	hP2	$P6_3/mmc$

(a) Possibly a hexagonal structure

Ho-In

H. Okamoto, 1992

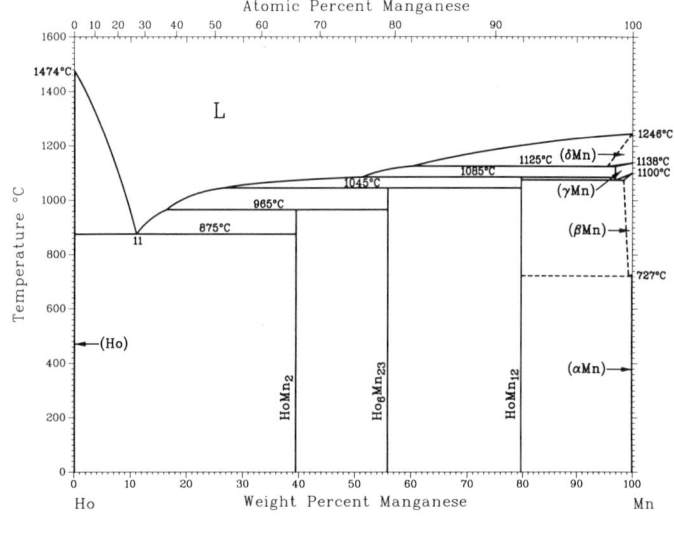

Phase	Composition, wt% In	Pearson symbol	Space group
(αHo)	0 to ~6	hP2	P6₃/mmc
(βHo)	0 to 15	cI2	Im$\overline{3}$m
Ho₂In	25.8	hP6	P6₃/mmc
βHo₅In₃	29.5	hP16	P6₃/mcm
αHo₅In₃	29.5	tI32	I4/mcm
HoIn	33 to 41.0	cP2	Pm$\overline{3}$m
		t**	...
Ho₃In₅	53.7	oC32	Cmcm
HoIn₃	68	cP4	Pm$\overline{3}$m
(In)	100	tI2	I4/mmm

Ho-Mn

H.R. Kirchmayr and W. Lugscheider, 1967

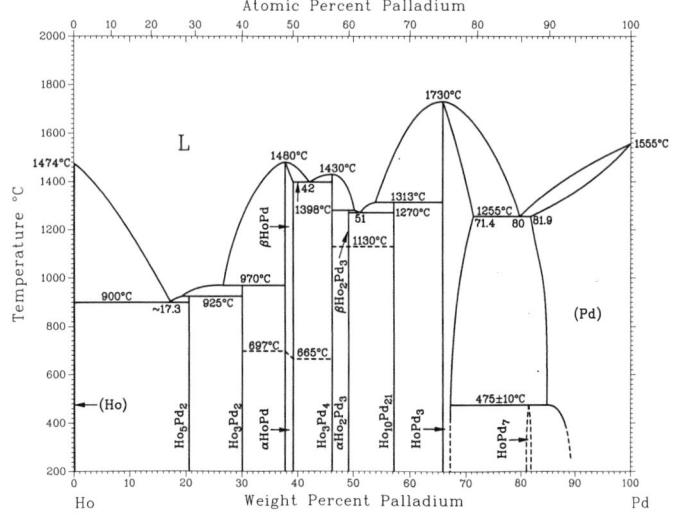

Phase	Composition, wt% Mn	Pearson symbol	Space group
(Ho)	~0	hP2	P6₃/mmc
HoMn₂	40.0	cF24	Fd$\overline{3}$m
Ho₆Mn₂₃	~56.1	cF116	Fm$\overline{3}$m
HoMn₁₂	~80.0	tI26	I4/mmm
(δMn)	~97 to 100	cI2	Im$\overline{3}$m
(γMn)	~97 to 100	cF4	Fm$\overline{3}$m
(βMn)	>97 to 100	cP20	P4₁32
(αMn)	~100	cI58	I$\overline{4}$3m

Ho-Pd

H. Okamoto, 1991

Phase	Composition, wt% Pd	Pearson symbol	Space group
(Ho)	0	hP2	P6₃/mmc
Ho₅Pd₂	20.5	tI49	I4₁a
Ho₃Pd₂	30	tP10	P4/mbm
βHoPd	~39.2	cP2	Pm$\overline{3}$m
αHoPd	~39.2	oP8	Pnma
Ho₃Pd₄	46.2	hR14	R$\overline{3}$
βHo₂Pd₃	49.2	...	...
αHo₂Pd₃	49.2	...	...
Ho₁₀Pd₂₁(a)	57.5	mC124	C2/m
HoPd₃	66 to 71.4	cP4	Pm$\overline{3}$m
HoPd₇	81.9	c**	...
(Pd)	81.9 to 100	cF4	Fm$\overline{3}$m

(a) Similarity to Sm₁₀Pd₂₁ is assumed.

Ho-Sb

H. Okamoto, 1990

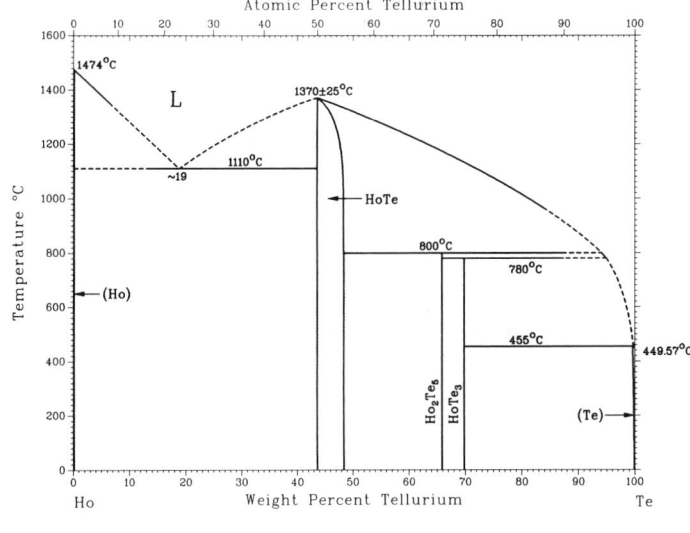

Phase	Composition, wt% Sb	Pearson symbol	Space group
(Ho)	0	hP2	P6₃/mmc
Ho₅Sb₃	30.7	hP16	P6₃/mcm
βHo₄Sb₃	35.7	...	...
αHo₄Sb₃	35.7	cI28	I4̄3d
βHoSb	42.5	...	
αHoSb	42.5	cF8	Fm3̄m
HoSb₂(a)	59.7	oC6	C222
(Sb)	100	hR2	R3̄m

(a) Synthesized under high pressure

Ho-Te

E.I. Yarembash, E.S. Vigileva, A.A. Eliseev, A.V. Zachatskaya, T.G. Aminov, and M.A. Chernitsyna, 1974

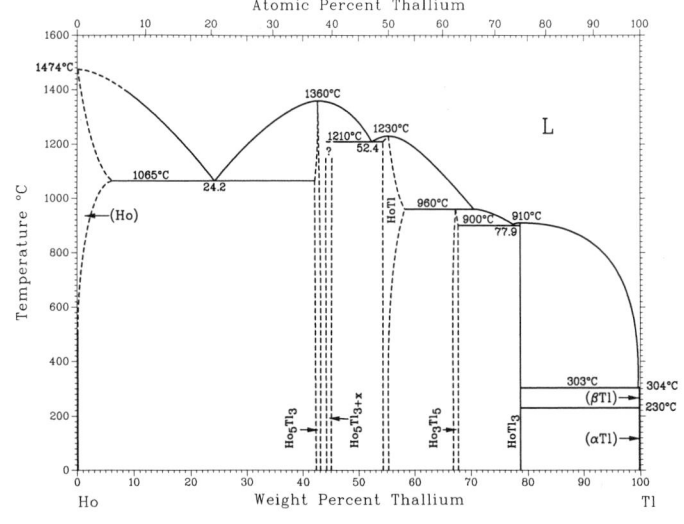

Phase	Composition, wt% Te	Pearson symbol	Space group
(Ho)	~0	hP2	P6₃/mmc
HoTe	44 to <49	cF8	Fm3̄m
Ho₂Te₅	~65.9	oC28	Cmcm
HoTe₃	70	oC16	Cmcm
(Te)	~100	hP3	P3₁21

Ho-Tl

S. Delfino, A. Saccone, A. Palenzona, and R. Ferro, unpublished

Phase	Composition, wt% Tl	Pearson symbol	Space group
(Ho)	0 to ~6	hP2	P6₃/mmc
Ho₅Tl₃	~42 to ~43	hP16	P6₃/mcm
Ho₅Tl₃₊ₓ	?	tI32	I4/mcm
HoTl(a)	~54 to ~58	cP2	Pm3̄m
		(or cI2)	Im3̄m
HoTl(b)	~54 to ~58	tP2	P4/mmm
Ho₃Tl₅	~67 to ~68	oC32	Cmcm
HoTl₃	79	cP4	Pm3̄m
(βTl)	100	cI2	Im3̄m
(αTl)	100	hP2	P6₃/mmc

(a) Cubic structure presumed to be room- and higher-temperature phase. (b) Tetragonal structure presumed to be lower-temperature phase

In-K

H. Okamoto, 1992

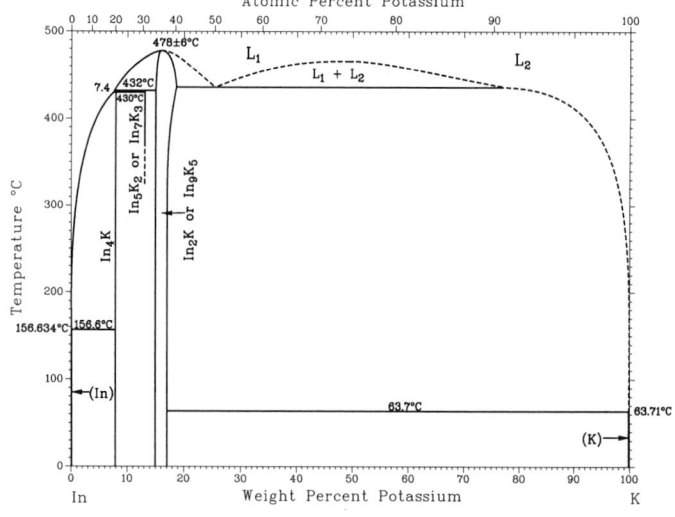

Phase	Composition, wt% K	Pearson symbol	Space group
(In)	0	$tI2$	$I4//mmm$
In_4K	8	$tI10$	$I4/mmm$
In_5K_2(a)	~13	…	…
In_2K(b)	14 to 19	…	…
(K)	100	$cI2$	$Im\overline{3}m$

(a) Or In_7K_3. (b) Or In_9K_5

In-La

A. Palenzona and S. Cirafici, 1992

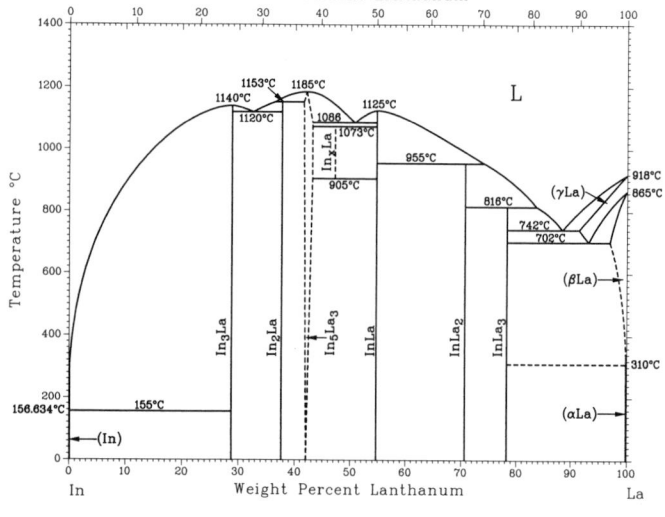

Phase	Composition, wt% La	Pearson symbol	Space group
(In)	0	$tI2$	$I4/mmm$
In_3La	29	$cP4$	$Pm\overline{3}m$
In_2La	37.7	$oI12$	$Imma$
In_5La_3	~42 to ~43.1	$oC32$	$Cmcm$
In_xLa	?	…	…
InLa	~54.7	$cP2$	$Pm\overline{3}m$
		…	$Pmmm$
$InLa_2$	70.8	$hP6$	$P6_3/mmc$
$InLa_3$	78	$cP4$	$Pm\overline{3}m$
(γLa)	91.4 to 100	$cI2$	$Im\overline{3}m$
(βLa)	97.1 to 100	$cF4$	$Fm\overline{3}m$
(αLa)	100	$hP4$	$P6_3/mmc$

In-Li

J. Sangster and A.D. Pelton, 1992

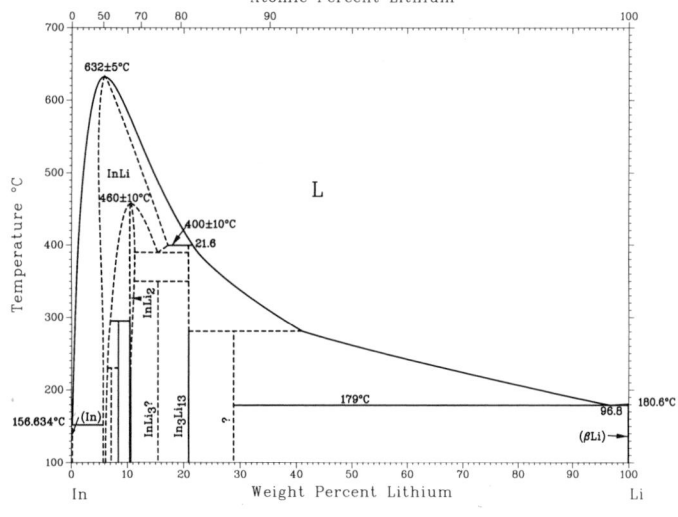

Phase	Composition, wt% Li	Pearson symbol	Space group
(In)	0 to 0.1	$tI2$	$I4/mmm$
InLi	5 to 9(a)	$cF16$	$Fd\overline{3}m$
In_4Li_5	~7.0	$hP9$	$P\overline{3}m1$
In_2Li_3	8	$hR5$	$R\overline{3}m$
$InLi_2$	10.3 to 11.2	$oC12$	$Cmcm$
In_3Li_{13}	20.8	$cF128$	$Fd\overline{3}m$
(αLi)(b)	100	$hP2$	$P6_3/mmc$
(βLi)	100	$cI2$	$Im\overline{3}m$

(a) At 415 °C. (b) Below −193 °C

In-Lu

H. Okamoto, 1992

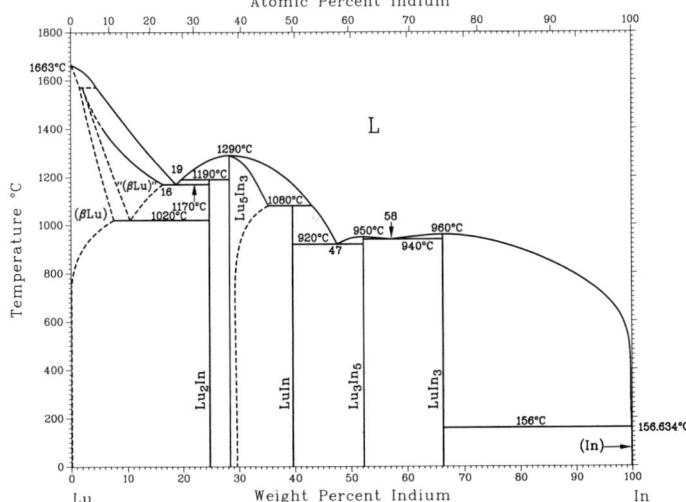

Phase	Composition, wt% In	Pearson symbol	Space group
(Lu)	0 to ?	$cI2$	$Im\bar{3}m$
Lu$_2$In	24.7	$hP6$	$P6_3/mmc$
Lu$_5$In$_5$	28.3 to 35	$hP16$	$P6_3/mcm$
LuIn	39.6	$cP2$	$Pm\bar{3}m$
Lu$_3$In$_5$	52.2	$oC32$	$Cmcm$
LuIn$_3$	66	$cP4$	$Pm\bar{3}m$
(In)	100	$tI2$	$I4/mmm$

In-Mg

A.A. Nayeb-Hashemi and J.B. Clark, 1992

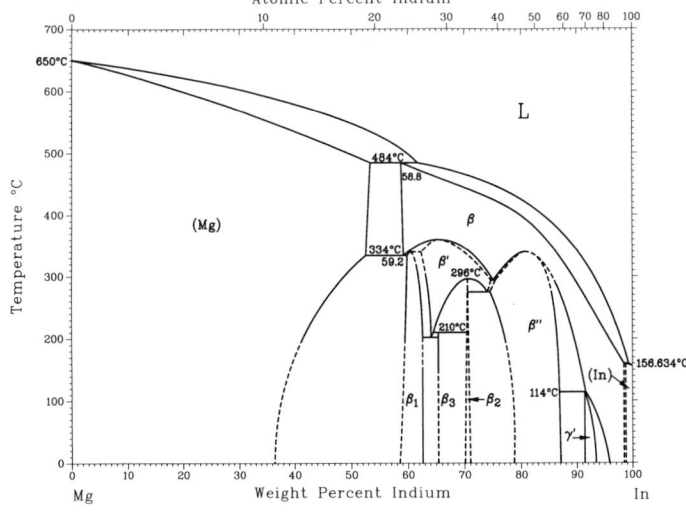

Phase	Composition, wt% In	Pearson symbol	Space group
(Mg)	0 to 53.2	$hP2$	$P6_3/mmc$
β	58.8 to 98.5	$cF4$	$Fm\bar{3}m$
β'	~62 to 74.7	$cP4$	$Pm\bar{3}m$
β$_1$(a)	~58.5 to 62	$hR16$	$R\bar{3}m$
β$_3$	65.4	$oI28$	$Ibam$
β$_2$	~71	$hP9$	$P\bar{6}2m$
β''	~75 to 87	$tP2$	$P4/mmm$
γ	~91.5 to 93.6	$cP4$	$Pm\bar{3}m$
(In)	99 to 100	$tI2$	$I4/mmm$

(a) At high temperatures, β$_1$ has a cubic structure, space group $Pm\bar{3}m$, Pearson symbol $cP4$.

In-Mn

H. Okamoto, 1992

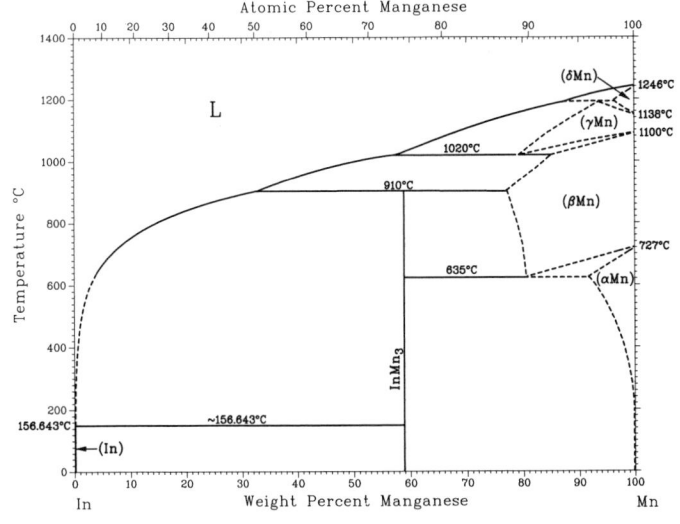

Phase	Composition, wt% Mn	Pearson symbol	Space group
(In)	0	$tI2$	$I4/mmm$
InMn$_3$	59 ± ?	$cP52$	$P\bar{4}3m$
(δMn)	? to 100	$cI2$	$Im\bar{3}m$
(γMn)	? to 100	$cF4$	$Fm\bar{3}m$
(βMn)	? to 100	$cP20$	$P4_132$
(αMn)	? to 100	$cI58$	$I\bar{4}3m$
Metastable phase			
...	75	$cI2$	$Im\bar{3}m$

In-Na

S. Larose and A.D. Pelton, 1992

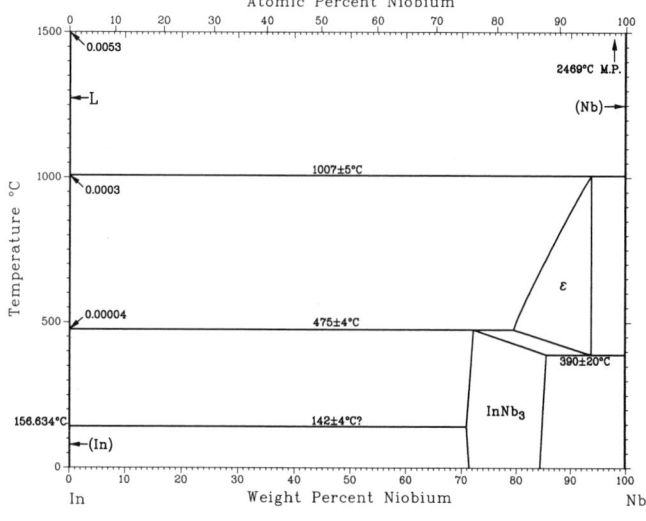

Phase	Composition, wt% Na	Pearson symbol	Space group
(In)	0 to 0.6	tI2	I4/mmm
In₈Na₅(b)	11.1	…	…
InNa	16.7	cF16	Fd$\bar{3}$m
InNa₃(b)	…	…	…
(βNa)	100	cI2	Im$\bar{3}$m
(αNa)	100	hP2	P6₃/mmc

(a) At 160 °C. (b) Stoichiometry uncertain

In-Nb

H. Okamoto, 1992

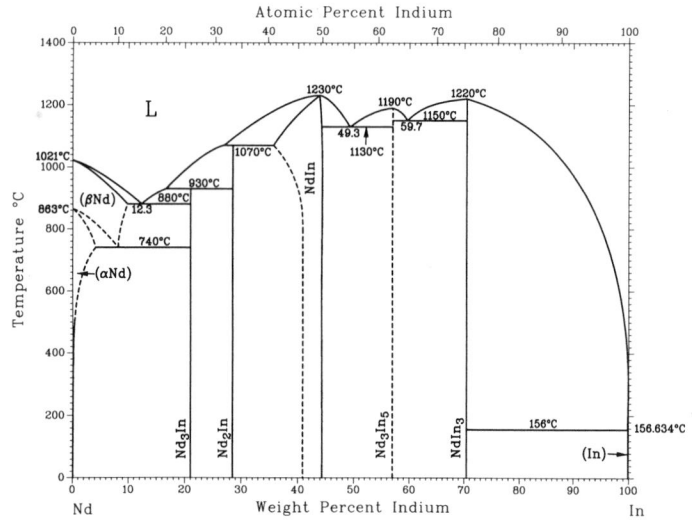

Phase	Composition, wt% Nb	Pearson symbol	Space group
(In)	0	tI2	I4/mmm
InNb₃	71 to 86	cP8	Pm$\bar{3}$n
ε	79 to 94	…	…
(Nb)	100	cI2	Im$\bar{3}$m

In-Nd

H. Okamoto, 1992

Phase	Composition, wt% In	Pearson symbol	Space group
(βNd)	0 to 10	cI2	Im$\bar{3}$m
(αNd)	0 to 4	hP4	P6₃/mmc
Nd₃In	21	cP4	Pm$\bar{3}$m
Nd₂In	28.4	hP6	P6₃/mmc
NdIn	36 to 44.3	cP2	Pm$\bar{3}$m
Nd₃In₅	57.0	oC32	Cmcm
NdIn₃	71	cP4	Pm$\bar{3}$m
(In)	100	tI2	I4/mmm

In-Ni

M.F. Singleton and P. Nash, 1992

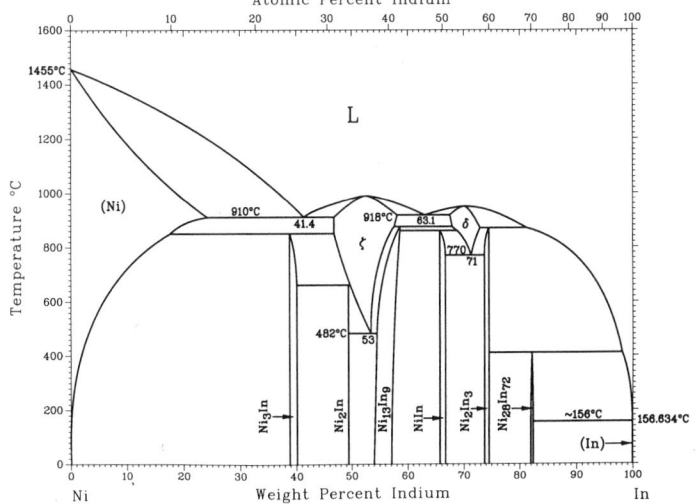

Phase	Composition, wt% In	Pearson symbol	Space group
(Ni)	0 to 24.9	cF4	$Fm\overline{3}m$
Ni₃In	38.8 to 40.1	hP8	$P6_3/mmc$
η(Ni₂In)	49.4	hP6	$P6_3/mmc$
ζ	47 to 58.1	hP4	$P6_3/mmc$
ζ'(Ni₁₃In₉)	55.1 to 58.8	…	…
ε(NiIn)	65.7 to 66.6	hP6	$P6/mmm$
δ(NiIn)	67.5 to 73	cP2	$Pm\overline{3}m$
Ni₂In₃	74 to 75	hP5	$P\overline{3}m1$
Ni₂₈In₇₂	82 to 82.4	…	…
(In)	~100	tI2	$I4/mmm$
Metastable phase			
ε'	27.7 to 40	cI2	$Im\overline{3}m$

In-P

H. Okamoto, 1992

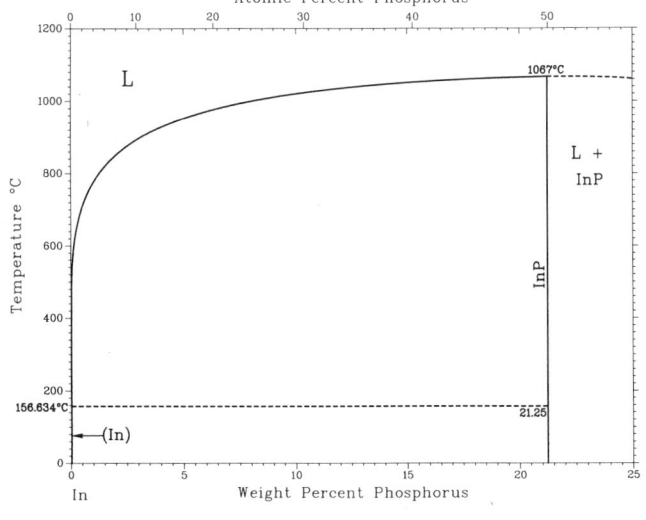

Phase	Composition, wt% P	Pearson symbol	Space group
Stable phases			
(In)	0	tI2	$I4/mmm$
InP	21.2	cF8	$F\overline{4}3m$
High-pressure phases			
InP II(a)	21.2	cF8	$Fm\overline{3}m$
InP III(b)	21.2	tI4	$I4/amd$
InP₃	45	hR8	$R\overline{3}m$

(a) 10.8 to 18.9 GPa. (b) >18.9 GPa

In-Pb

J.P. Nabot and I. Ansara, 1992

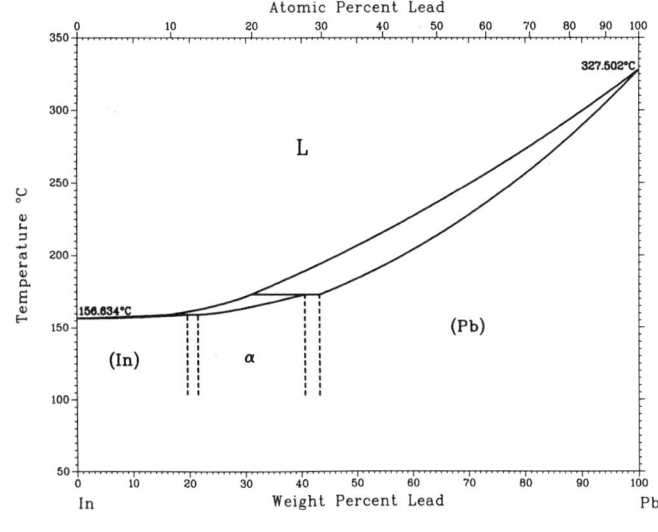

Phase	Composition, wt% Pb	Pearson symbol	Space group
(In)	0 to ?	tI2	$I4/mmm$
α	~24 to ~44	tI2	$I4/mmm$
(Pb)	? to 100	cF4	$Fm\overline{3}m$

In-Pd

H. Okamoto, 1992

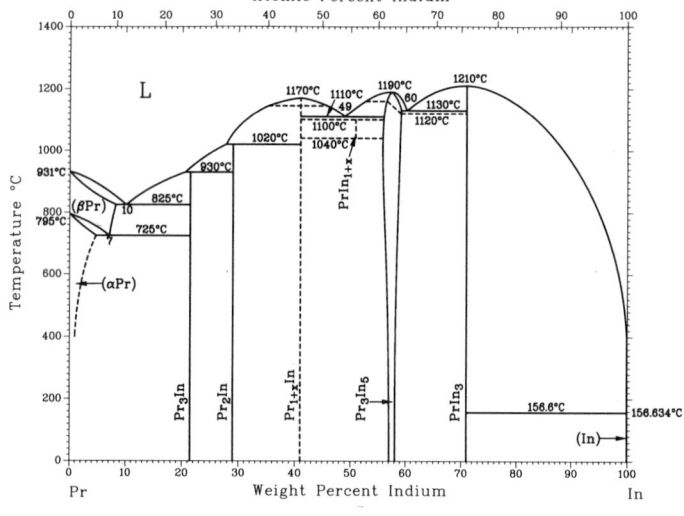

Phase	Composition, wt% Pd	Pearson symbol	Space group
(In)	0	tI2	I4/mmm
In₃Pd	24	c*52	...
In₃Pd₂	37 to 38	hP5	$P\overline{3}m1$
InPd	43 to 59.7	cP2	$Pm\overline{3}m$
In₃Pd₅	60.7	oP16	Pbam
βInPd₂	61.7 to 65.5	...	...
αInPd₂	64 to 65.0	oP12	Pnma
βInPd₃	72 to 74	...	...
αInPd₃	73.0 to 74.1	tI8	I4/mmm
(Pd)	80 to 100	cF4	$Fm\overline{3}m$

In-Pr

H. Okamoto, 1992

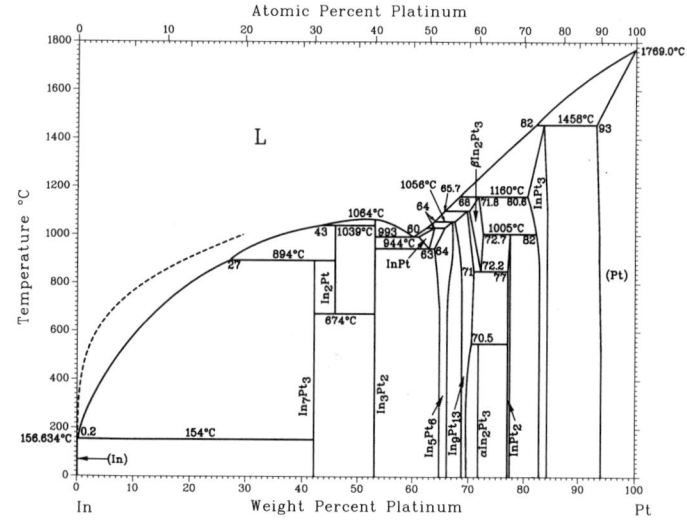

Phase	Composition, wt% In	Pearson symbol	Space group
(βPr)	0 to 8.1	cI2	$Im\overline{3}m$
(αPr)	0 to 4.6	hP4	P6₃/mmc
Pr₃In	21	cP4	$Pm\overline{3}m$
Pr₂In	28.9	hP6	P6₃/mmc
βPr₁₊ₓIn	~41	...	...
αPr₁₊ₓIn	41	...	...
PrIn(a)	44.9	cP2	$Pm\overline{3}m$
PrIn₁₊ₓ	?	...	...
Pr₃In₅	56 to 59	oC32	Cmcm
PrIn₃	70	cP4	$Pm\overline{3}m$
(In)	100	tI2	I4/mmm

(a) Metastable or same as PrIn₁₊ₓ?

In-Pt

H. Okamoto, 1992

Phase	Composition, wt% Pt	Pearson symbol	Space group
(In)	0	tI2	I4/mmm
In₇Pt₃	42	cI40	$Im\overline{3}m$
In₂Pt	45.9	cF12	$Fm\overline{3}m$
In₃Pt₂	53	hP5	$P\overline{3}m1$
InPt	61 to 64	...	...
In₅Pt₆(a)	64 to 67.1	mC20	C2/m
In₉Pt₁₃	68 to 71	mC44	C2/m
βIn₂Pt₃	70 to 72.7	hP4	P6₃/mmc
αIn₂Pt₃	72	hP20	$P\overline{3}1c$
InPt₂	77.1 to 78	oC16	Cmmm
InPt₃	80.6 to 84.3	cP4	$Pm\overline{3}m$
InPt₃	83.6	tP4	P4/mmm
(Pt)	93 to 100	cF4	$Fm\overline{3}m$

(a) InPt₁₊

In-Pu

H. Okamoto, 1992

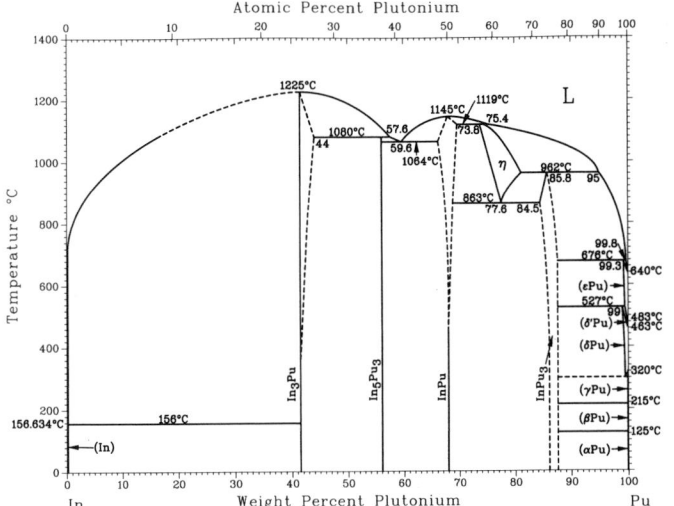

Phase	Composition, wt% Pu	Pearson symbol	Space group
(In)	0	tI2	I4/mmm
In₃Pu	42 to 44	cP4	Pm$\bar{3}$m
In₅Pu₃	56.1	...	(a)
InPu	66 to 70	tP2	P4/mmm
		tI2	I4/mmm
η	73.8 to 81	...	...
InPu₃	84.5 to 88	cP4	Pm$\bar{3}$m
		cF4	Fm$\bar{3}$m
(εPu)	99.3 to 100	cI2	Im$\bar{3}$m
(δ'Pu)	100	tI2	I4/mmm
(δPu)	99 to 100	cF4	Fm$\bar{3}$m
(γPu)	100	oF8	Fddd
(βPu)	100	mC34	C2/m
(αPu)	100	mP16	P2₁/m

(a) Complex

In-Rb

A.D. Pelton and S. Larose, 1992

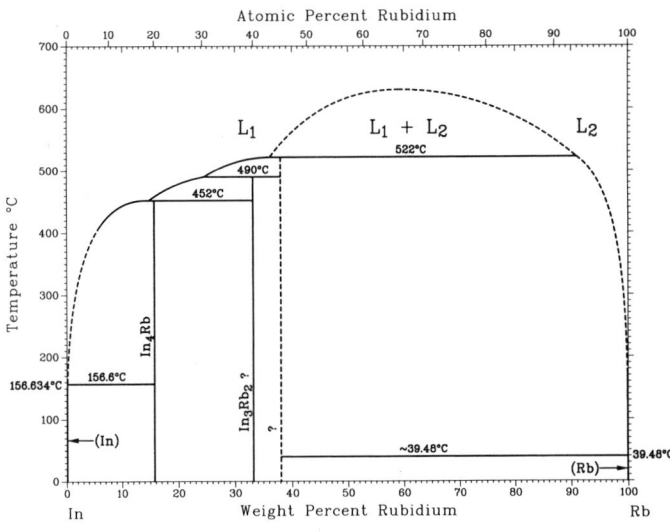

Phase	Composition, wt% Rb	Pearson symbol	Space group
(In)	0	tI2	I4/mmm
In₄Rb	16	tI10	I4/mmm
In₃Rb₂(a)	~33	...	...
In₅Rb₄(b)	...	...	...
(Rb)	100	cI2	Im$\bar{3}$m

(a) Stoichiometry requires verification. (b) Existence and stoichiometry require verification.

In-S

H. Okamoto, 1992

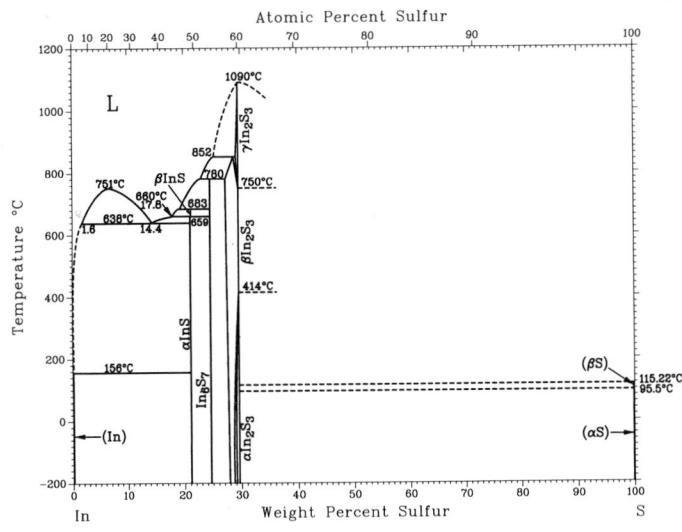

Phase	Composition, wt% S	Pearson symbol	Space group
(In)	0	tI2	I4/mmm
βInS	21.2	...	...
αInS	21.0	oP8	Pnnm
In₆S₇	24.5	mP26	P2₁/m
γIn₂S₃	29.5	hP7	P$\bar{3}$m1
βIn₂S₃	27.8 to 29.5	cF56	Fd$\bar{3}$m
αIn₂S₃	29.5	tI80	I4₁/amd
(βS)	100	mP*	P2₁/c
(αS)	100	oF128	Fddd
Metastable phase			
InS'	21.8	...	...
High-pressure phase			
εIn₂S₃	29.5	hR10	R$\bar{3}$c
Questionable phases			
In₅S₄(a)	18.2	cP72	Pa$\bar{3}$
In₃S₄	27.1	hP*	...
In₂S₃(b)	29.5	cF8	F$\bar{4}$3m
In₂S₃(c)	29.5	o**	...
In₃S₅(c)	31.8	hP*	...

(a) Probably a ternary compound. (b) Low-temperature phase. (c) High-temperature phase. Conflicts with γIn₂S₃

In-Sb

R.C. Sharma, T.L. Ngai, and Y.A. Chang, 1992

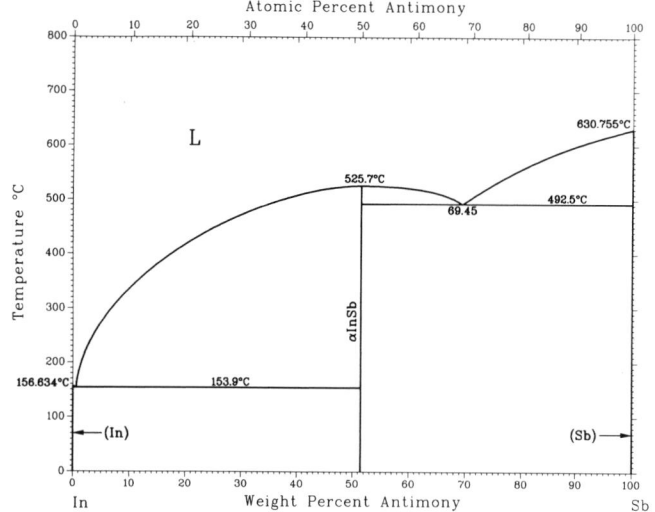

Phase	Composition, wt% Sb	Pearson symbol	Space group
(In)	0	tI2	F4/mmm
ω(a)	~21	hP6	P6₃/mmc
ζ(a)	~41	tI32	I4/mcm
γ₂(a)	49.0	oP4	...
αInSb	51.5	cF8	F4̄3m
βInSb(a)	51.5	tI4	I4₁/amd
γ₁(a)	51.5 to 56	oP2	Pmm2
γInSb(a)	51.5	(b)	...
γInSb(a)	51.5	oP4	Pmmm or Pmmn
δInSb(a)	51.5	oP2	Pmm2
<<βSn>>(a)	56	tI4	I4₁/amd
π'(a)	58.9	hR1	R3̄m
π(c)	61 to 71	cP1	Pm3̄m
InSb (thin films)(c)	...	hP4	P6₃mc
(Sb)	100	hR2	R3̄m

(a) High-pressure phase. (b) Hexagonal. (c) Metastable phase

In-Sc

H. Okamoto, 1992

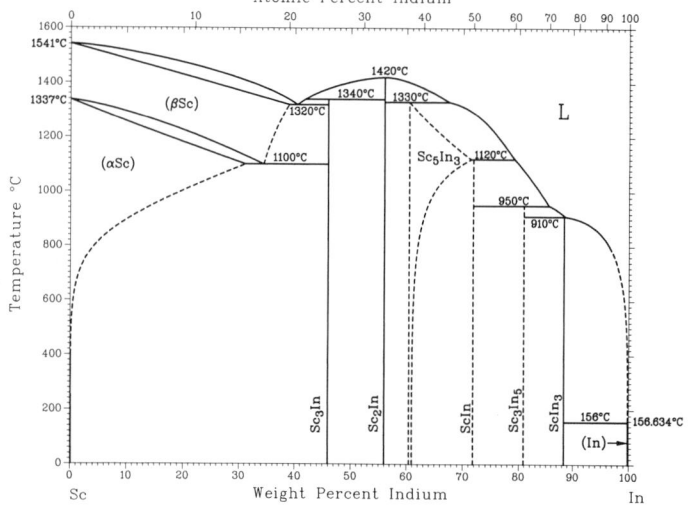

Phase	Composition, wt% In	Pearson symbol	Space group
(βSc)	0 to 39	cI2	Im3̄m
(αSc)	0 to 31	hP2	P6₃/mmc
Sc₃In	46	hP8	P6₃/mmc
Sc₂In	56.0	hP6	P6₃/mmc
Sc₅In₃	60.5 to 71.9	...	...
ScIn	71.9	...	...
Sc₃In₅	81.0	...	...
ScIn₃	89	cP4	Pm3̄m
(In)	100	tI2	I4/mmm

In-Se

H. Okamoto, 1992

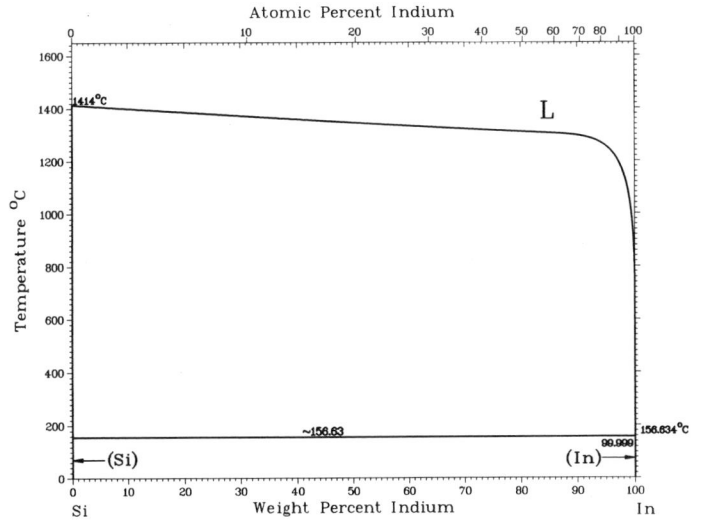

Phase	Composition, wt% Se	Pearson symbol	Space group
(In)	0	tI2	I4/mmm
In₄Se₃	34.1	oP28	Pnnm
αInSe	40.7	hR4	R$\bar{3}$m
βInSe(a)	40.7	hP8	P6₃/mmc
InSe II(b)	40.7	mP8	P2/m
In₆Se₇	44.5	mP26	P2₁/m
In₂Se₃	51	...	...
δ		hP5	P6₁
γ(c)		hP30	P6₁
β(c, d)		hR5	R$\bar{3}$m
α3(d)		hR5	R$\bar{3}$m
α2(d)		hP10	P6₃/mmc
α1(d)		hP5	...
α′(d)		o**	...
In₂Se₃(e)	51	hP160	P6₃
(Se)	100	hP3	P3₁21
Uncertain phases and structures			
In₂Se(f)	25.6	oP24	Pnnm
InSe	40.7	hP*	...
In₆Se₇	44.5	mP26	P2₁/m
In₅Se₆	45.2	...	...
In₃Se₄	47.8	...	...
In₅Se₇	49.0	c**	...
In₂Se₃	51	hP30	P6₅
In₂Se₃(g)	51	m**	...
In₂Se₃(h)	51	c**	...
InSe₄	51	...	...

(a) Probably metastable. (b) High-pressure phase. (c) β ↔ α transition is ambiguous. (d) Metastable. (e) Thin film. (f) Probably In₄Se₃. (g) Probably In₆Se₇. (h) Same as In₅Se₇?

In-Si

R.W. Olesinski, N. Kanani, and G.J. Abbaschian, 1992

Phase	Composition, wt% In	Pearson symbol	Space group
(Si)	~0	cF8	Fd$\bar{3}$m
(In)	~100	tI2	I4/mmm

In-Sm

H. Okamoto, 1992

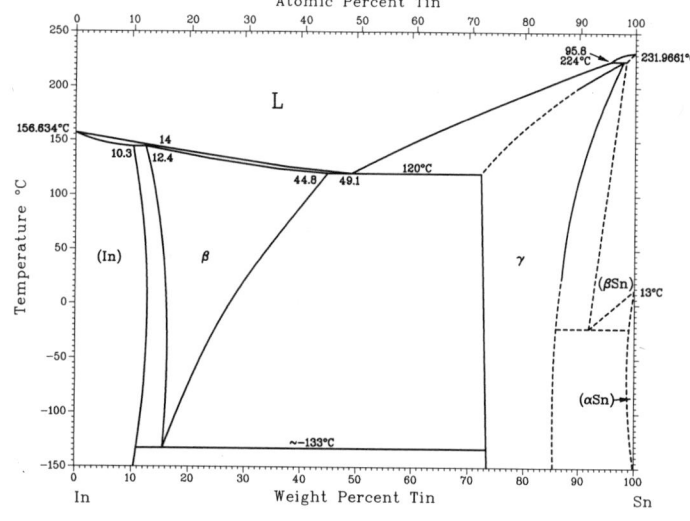

Phase	Composition, wt% In	Pearson symbol	Space group
(γSm)	0 to 11	cI2	$Im\bar{3}m$
(βSm)	0 to 4	hP2	$P6_3/mmc$
(αSm)	0	hR3	$R\bar{3}m$
Sm₃In	20	cP4	$Pm\bar{3}m$
Sm₂In	26 to 28	hP6	$P6_3/mmc$
SmIn	38 to 42	cP2	$Pm\bar{3}m$
Sm₃In₅	56 to 57	oC32	$Cmcm$
SmIn₃	70	cP4	$Pm\bar{3}m$
(In)	100	tI2	$I4/mmm$

In-Sn

H. Okamoto, 1992

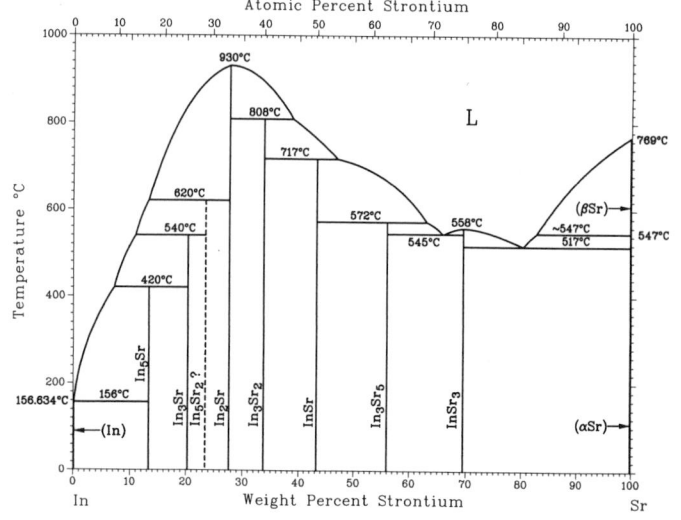

Phase	Composition, wt% Sn	Pearson symbol	Space group
(In)	0 to 12.4	tI2	$I4/mmm$
β	12.4 to 44.8	tI2	$I4/mmm$
γ	73 to ?	hP5	$P6/mmm$
(βSn)	? to 100	tI4	$I4_1/and$
(αSn)	100	cF8	$Fd\bar{3}m$

In-Sr

H. Okamoto, 1992

Phase	Composition, wt% Sr	Pearson symbol	Space group
(In)	0	tI2	$I4/mmm$
In₅Sr	13.3	hP*	...
In₃Sr	20	hP8	$P6_3/mmc$
In₅Sr₂?	23.4	...	...
In₂Sr	27.6	hP6	$P6_3/mmc$
In₃Sr₂	34	...	...
InSr	43.3	o**	...
In₃Sr₅	56.0	tI32	$I4/mcm$
InSr₃	70	cF16	$Fm\bar{3}m$
(βSr)	100	cI2	$Im\bar{3}m$
(αSr)	100	cF4	$Fm\bar{3}m$

In-Tb

H. Okamoto, 1992

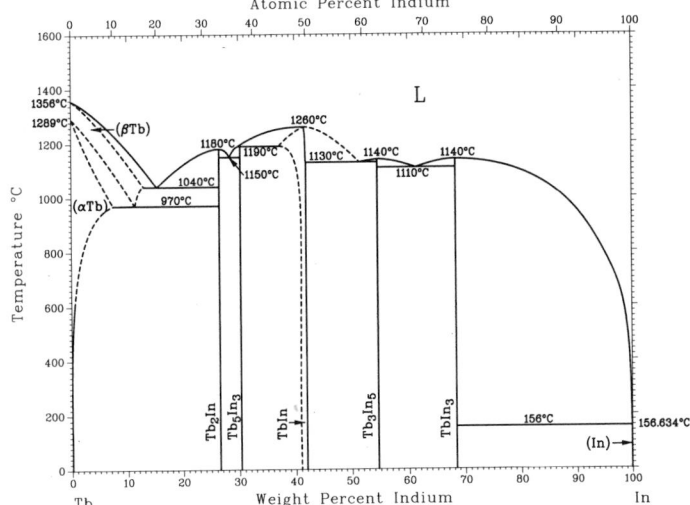

Phase	Composition, wt% In	Pearson symbol	Space group
(βTb)	0 to 13	cI2	Im$\overline{3}$m
(αTb)	0 to 7	hP2	P6₃/mmc
Tb₂In	26.5	hP6	P6₃/mmc
Tb₅In₃	30.2	tI32	I4/mcm
TbIn	37 to 41.9	t**	...
Tb₃In₅	54.6	oC32	Cmcm
TbIn₃	68	cP4	Pm$\overline{3}$m
(In)	100	tI2	I4/mmm

In-Te

H. Okamoto, 1992

Phase	Composition, wt% Te	Pearson symbol	Space group
(In)	0	tI2	I4/mmm
In₄Te₃	45.5	oP28	Pnnm
InTe	53 to 54	tI16	I4/mcm
In₃Te₄	59.7	hR7	R$\overline{3}$m
In₃Te₄	59.7	tI*	I4/mmm
βIn₂Te₃	62.0 to 62.6	cF8	F$\overline{4}$3m
αIn₂Te₃	62.5	cF180	F$\overline{4}$3m
βIn₃Te₅	64.9	hP*	...
αIn₃Te₅	64.9	...	...
In₂Te₅I	73.5	mC28	Cc
(Te)	100	hP3	P3₁21
Metastable or high-pressure phases			
InTeII	52.6	cF8	Fm$\overline{3}$m
InTeIII	52.6	cP2	Pm$\overline{3}$m
InTeII′	52.6	t*8	...
In₂₊ₓTe₃	...	cP*	...
In₂Te₃	62.5	oI*	Imm2
In₂Te₃	62.5	tP*	P4₂mcm or P4₂nm
In₂Te₃(a)	62.5	hP*	...
In₂Te₃II	62.5	hR5	R$\overline{3}$m
In₂Te₅II	73.5	mC84	C2/c

(a) Thin film

In-Th

H. Okamoto, 1992

Phase	Composition, wt% Th	Pearson symbol	Space group
(In)	0	tI2	I4/mmm
In₃Th	40	cP4	Pm$\overline{3}$m
In₅Th₃	54.8	oC32	Cmcm
InTh	66.9	oP24	Pbcm
InTh₂	80.2	tI12	I4/mcm
(βTh)	? to 100	cI2	Im$\overline{3}$m
(αTh)	~95 to 100	cF4	Fm$\overline{3}$m

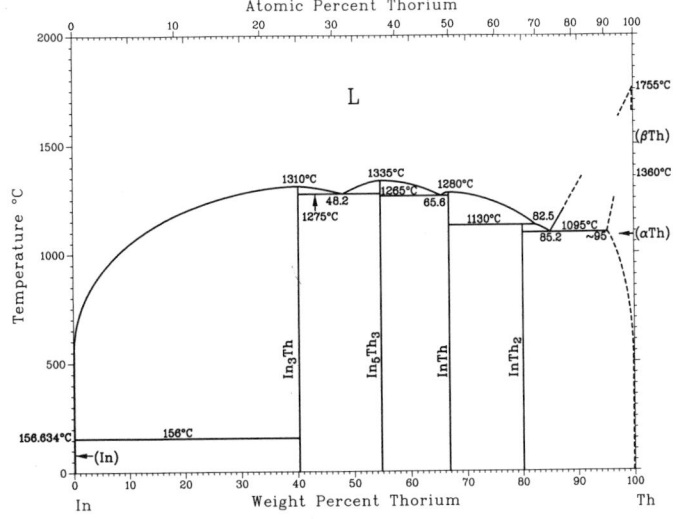

In-Ti

J.L. Murray, 1992

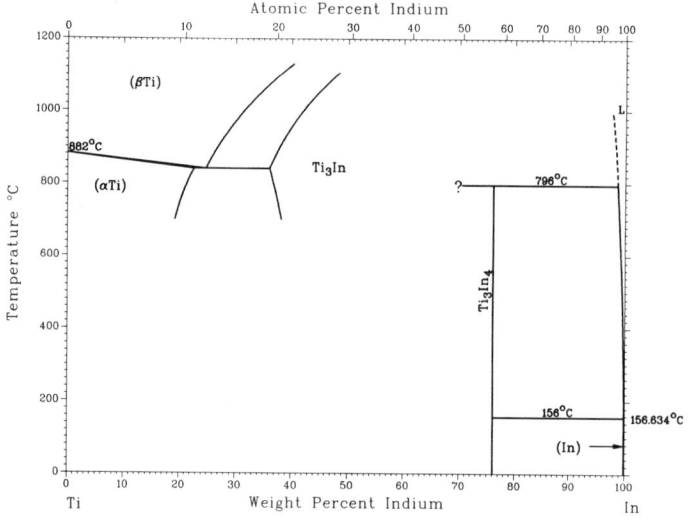

Phase	Composition, wt% In	Pearson symbol	Space group
(βTi)	0 to ?	cI2	$Im\bar{3}m$
(αTi)	0 to ~21	hP2	$P6_3/mmc$
Ti₃In	39 to ?	hP8	$P6_3/mmc$
Ti₃In₂	(a)	tP2	$P4/mmm$
Ti₃In₄	76.1	tP14	$P4/mbm$
(In)	~100	tI2	$I4/mmm$

(a) Unknown

In-Tl

H. Okamoto, 1992

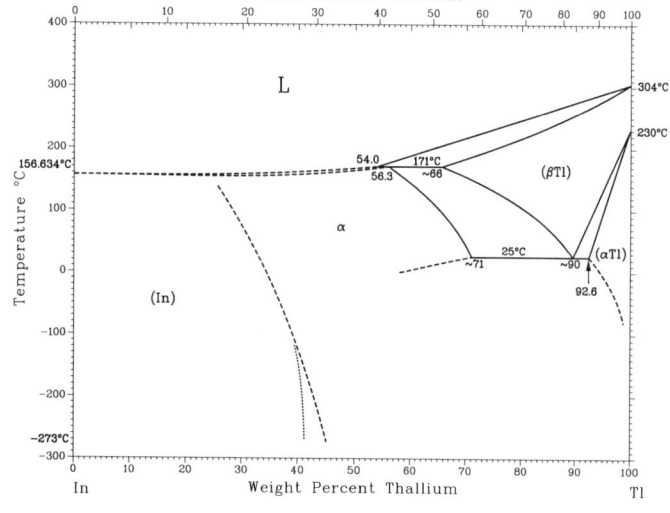

Phase	Composition, wt% Tl	Pearson symbol	Space group
(In)	0 to 44	tI2	$I4/mmm$
α	25 to ~71	cF4	$Fm\bar{3}m$
(βTl)	~66 to 100	cI2	$Im\bar{3}m$
(αTl)	92.6 to 100	hP2	$P6_3/mmc$

In-Tm

H. Okamoto, 1992

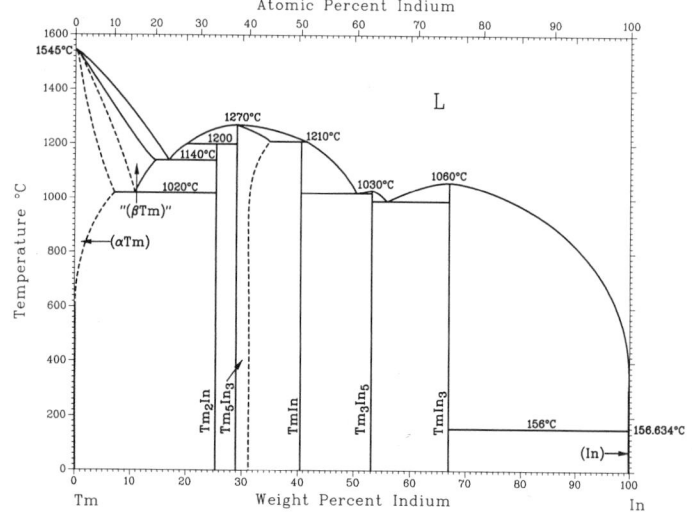

Phase	Composition, wt% In	Pearson symbol	Space group
"(βTm)"	? to 15	cI2	$Im\bar{3}m$
(αTm)	0 to 7	hP2	$P6_3/mmc$
TmIn	25.3	hP6	$P6_3/mmc$
Tm₅In₃	29.0 to 36	hP16	$P6_3/mcm$
TmIn	40.5	cP2	$Pm\bar{3}m$
Tm₃In₅	53.1	oC32	$Cmcm$
TmIn₃	67	cP4	$Pm\bar{3}m$
(In)	100	tI2	$I4/mmm$

In-V

J.F. Smith and K.J. Lee, 1992

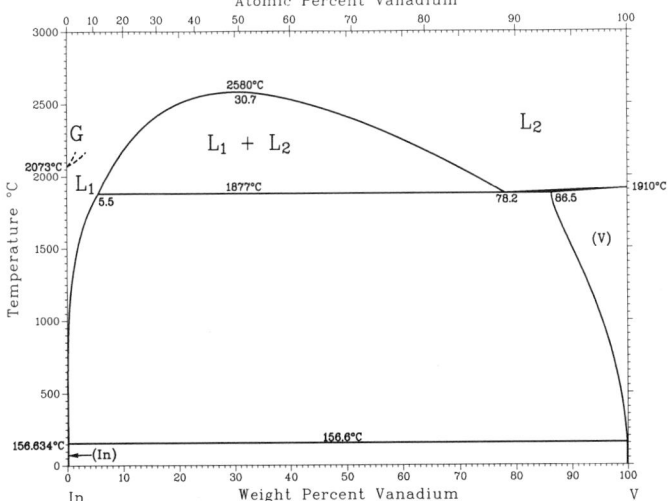

Phase	Composition, wt% V	Pearson symbol	Space group
(In)	0	tI2	I4/mmm
InV₃(a)	57	(b)	...
(V)	100	cI2	Im3̄m

(a) Cr₃Si-type structure reported in impure sample at high pressure. (b) Tetragonal. Pressure-stabilized phase

In-Y

H. Okamoto, 1992

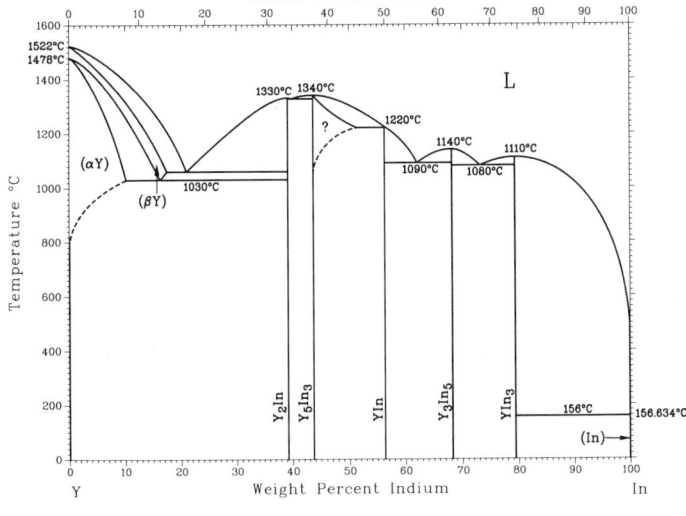

Phase	Composition, wt% In	Pearson symbol	Space group
(βY)	0 to 17.4	cI2	Im3̄m
(αY)	0 to 10.1	hP2	P6₃/mmc
Y₂In	39.2	hP6	P6₃/mmc
Y₅In₃	43.7 to 51.4	hP16	P6₃/mcm
YIn	56.4	cP2	Pm3̄m
Y₃In₅	68.3	oC32	Cmcm
YIn₃	79.5	cP4	Pm3̄m
(In)	100	tI2	I4/mmm

In-Yb

A. Palenzona and S. Cirafici, 1992

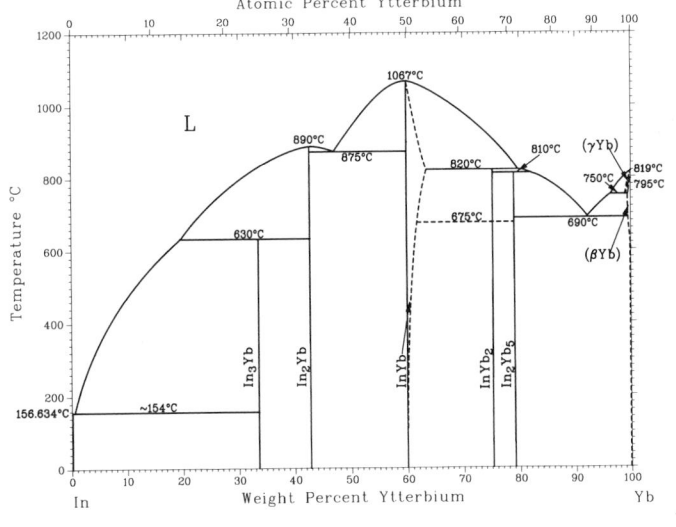

Phase	Composition, wt% Yb	Pearson symbol	Space group
(In)	~0	tI2	I4/mmm
In₃Yb	33.4	cP4	Pm3̄m
In₂Yb	42.9	hP6	P6₃/mmc
InYb	60.1 to ~63	cP2	Pm3̄m
InYb₂	75.1	oP12	Pnma
In₂Yb₅	~79.0	...	R3̄c or R3c
(γYb)	~100	cI2	Im3̄m
(βYb)	~100	cF4	Fm3̄m
(αYb)	~100	hP2	P6₃/mmc

In-Zn

J. Dutkiewicz and W. Zakulski, 1992

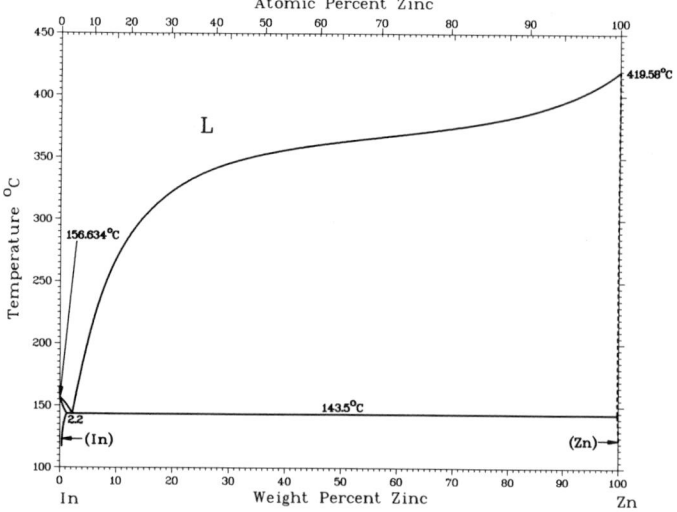

Phase	Composition, wt% Zn	Pearson symbol	Space group
(In)	0 to 1	tI2	I4/mmm
(Zn)	99.8 to 100	hP2	P6₃/mmc

Ir-La

H. Okamoto, 1991

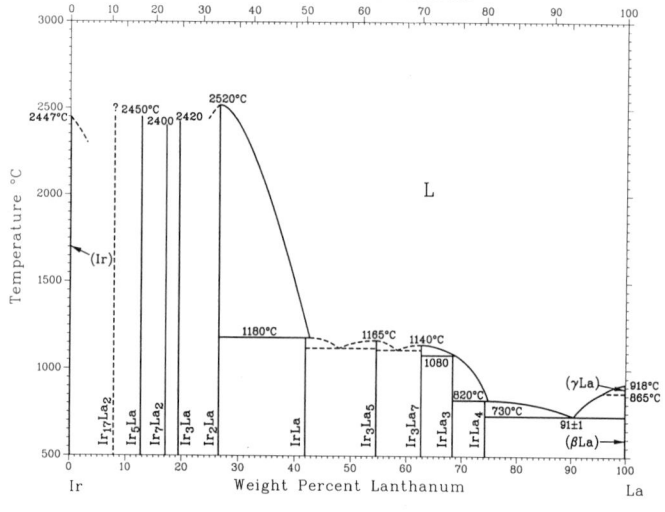

Phase	Composition, wt% La	Pearson symbol	Space group
(Ir)	0	cF4	Fm$\bar{3}$m
Ir₁₇La₂?	7.8	...	...
Ir₅La	12.7	hP6	P6/mmm
Ir₇La₂	17.1	hP36	P6₃/mmc
Ir₃La	19	hR12	R$\bar{3}$m
Ir₂La	26.5	cF24	Fd$\bar{3}$m
IrLa?	41.9	...	...
Ir₃La₅	54.6	tP32	P4/ncc
Ir₃La₇	63	hP20	P6₃mc
IrLa₃	68	oP16	Pnma
IrLa₄	74	...	...
(γLa)	100	cI2	Im$\bar{3}$m
(βLa)	100	cF4	Fm$\bar{3}$m
(αLa)	100	hP4	P6₃/mmc

Ir-Mo

From [Molybdenum]

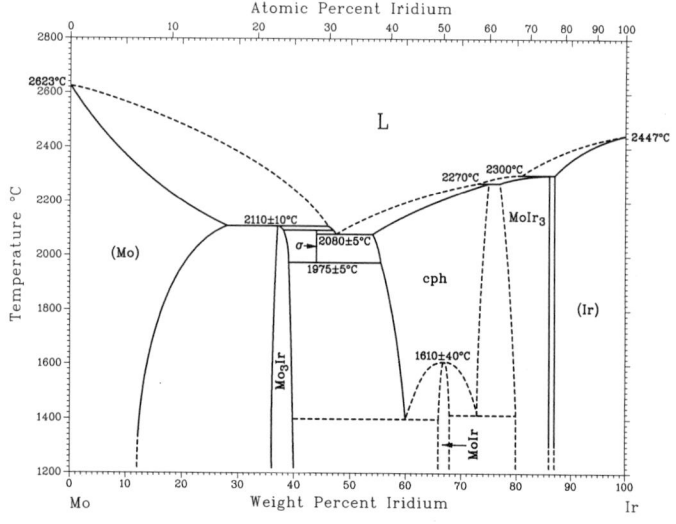

Phase	Composition, wt% Ir	Pearson symbol	Space group
(Mo)	0 to ~28	cI2	Im$\bar{3}$m
Mo₃Ir	<36 to 40	cP8	Pm$\bar{3}$n
σ	~44	tP30	P4₂/mnm
MoIr (cph)	~54 to >75	hP2	P6₃/mmc
MoIr (LT)	~66 to 68	oP4	Pmma
MoIr₃	~77 to 86	hP8	P6₃/mmc
(Ir)	~87 to 100	cF4	Fm$\bar{3}$m

Ir-Nb

H. Okamoto, unpublished

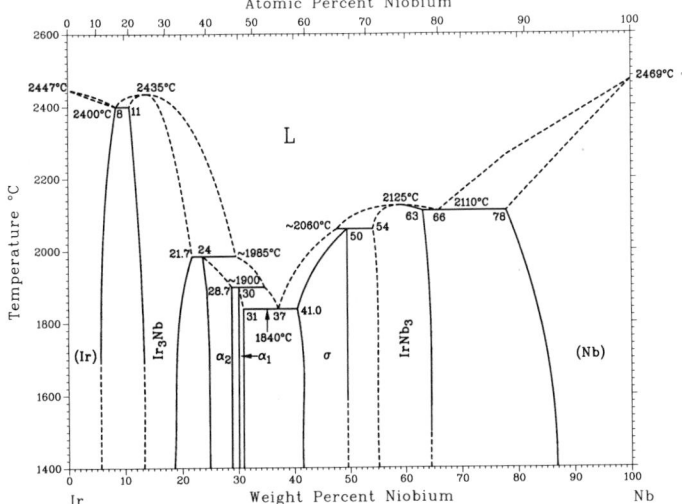

Phase	Composition, wt% Nb	Pearson symbol	Space group
(Ir)	0 to 8	cF4	$Fm\bar{3}m$
Ir₃Nb	11 to 21.7	cP4	$Pm\bar{3}m$
α₂	24 to 28.7	oP12	$Pmma$
α₁	30 to 31	tP2	$P4/mmm$
σ	41.0 to 50	tP30	$P4_2/mnm$
IrNb₃	54 to 63	cP8	$Pm\bar{3}n$
(Nb)	78 to 100	cI2	$Im\bar{3}m$

Ir-Ni

S.C. Yang, N. Chen, and P. Nash, 1991

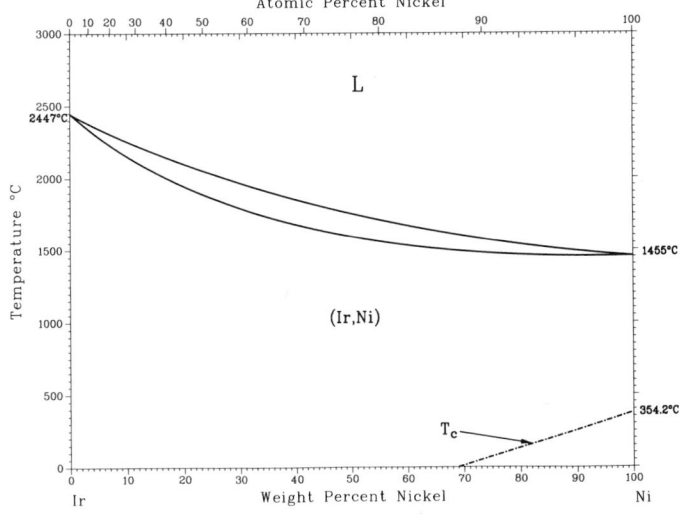

Phase	Composition, wt% Ni	Pearson symbol	Space group
(Ir,Ni)	0 to 100	cF4	$Fm\bar{3}m$

Ir-Pd

S.N. Tripathi, S.R. Bharadwaj, and M.S. Chandrasekharaiah, 1991

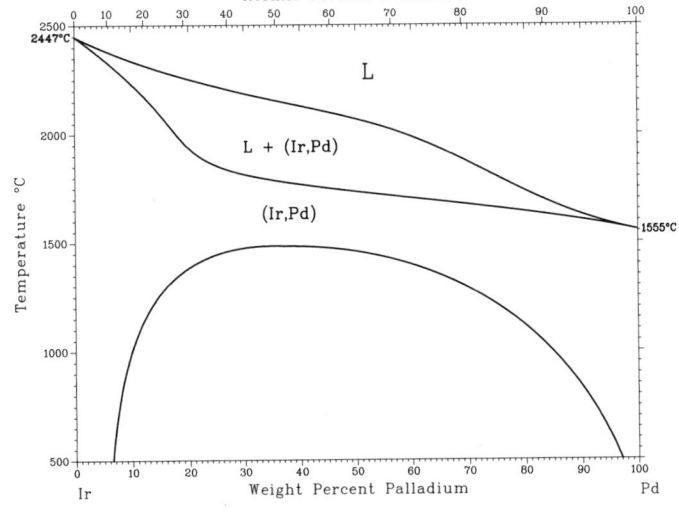

Phase	Composition, wt% Pd	Pearson symbol	Space group
(Ir,Pd)	0 to 100	cF4	$Fm\bar{3}m$

Ir-Pt

L. Muller, 1930; and E. Raub and W. Plate, 1956

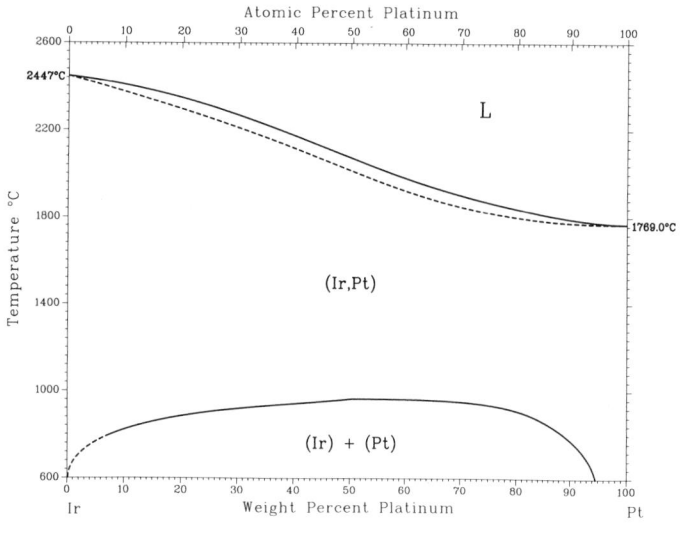

Phase	Composition, wt% Pt	Pearson symbol	Space group
(Ir,Pt)	0 to 100	cF4	Fm$\overline{3}$m

Ir-Rh

S.N. Tripathi, S.R. Bharadwaj, and M.S. Chandrasekharaiah, 1991

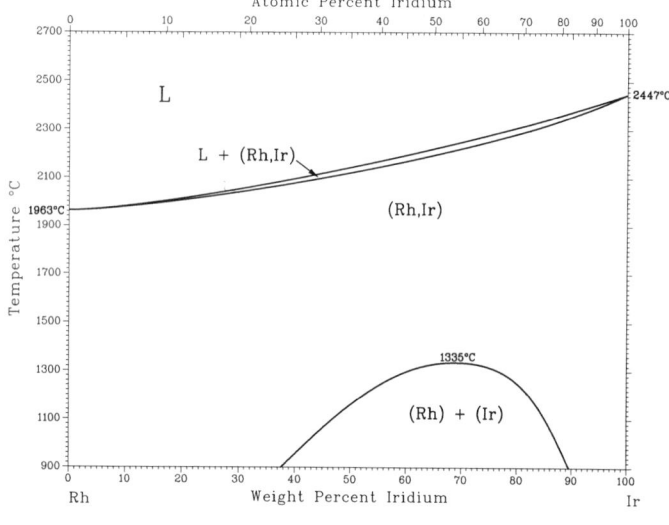

Phase	Composition, wt% Ir	Pearson symbol	Space group
(Ir,Rh)	0 to 100	cF4	Fm$\overline{3}$m

Ir-Ru

H. Okamoto, 1992

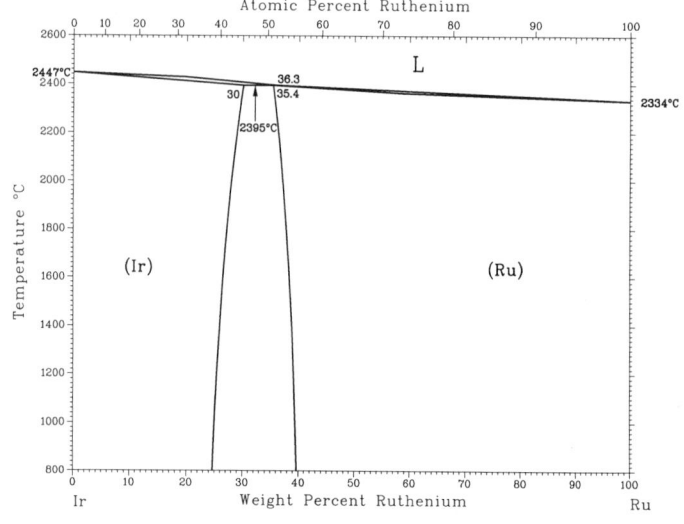

Phase	Composition, wt% Ru	Pearson symbol	Space group
(Ir)	0 to 30	cF4	Fm$\overline{3}$m
(Ru)	35.4 to 100	hP2	P6$_3$/mmc

Ir-Ta

From [Metals]

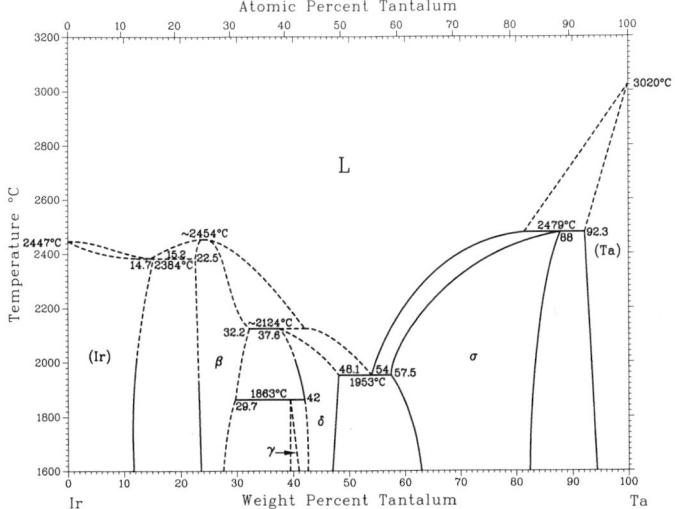

Phase	Composition, wt% Ta	Pearson symbol	Space group
(Ir)	0 to 15.2	cF4	Fm$\overline{3}$m
β	22.5 to 32.2	cP4	Pm$\overline{3}$m
γ	~41	tI2	I4/mmm
δ	37.6 to 48.1	oP12	Pmma
σ	57.5 to 88.0	tP30	P4$_2$/mnm
(Ta)	92.3 to 100	cI2	Im$\overline{3}$m

Ir-Th

H. Okamoto, 1991

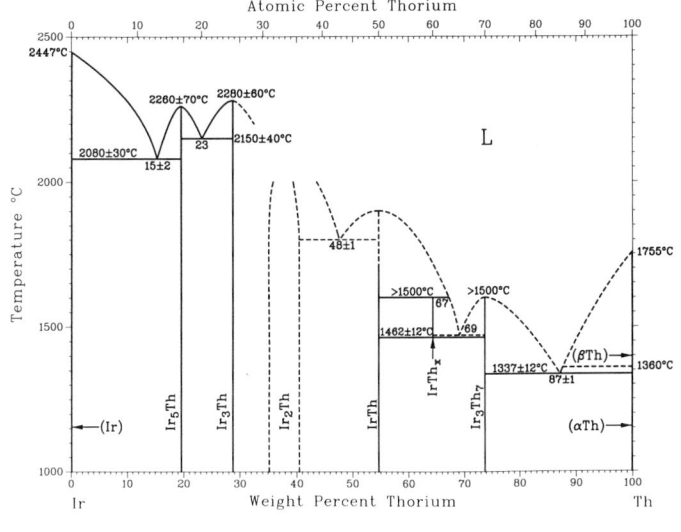

Phase	Composition, wt% Th	Pearson symbol	Space group
(Ir)	0	cF4	Fm$\overline{3}$m
Ir$_5$Th	19.5	hP6	P6/mmm
Ir$_3$Th	29	...	...
Ir$_2$Th	35 to 40	cF24	Fd$\overline{3}$m
IrTh	54.7	oC8	Cmcm
IrTh$_x$	~64	...	...
Ir$_3$Th$_7$	74	hP20	P6$_3$mc
(βTh)	100	cI2	Im$\overline{3}$m
(αTh)	100	cF4	Fm$\overline{3}$m

Ir-Ti

H. Okamoto, 1992

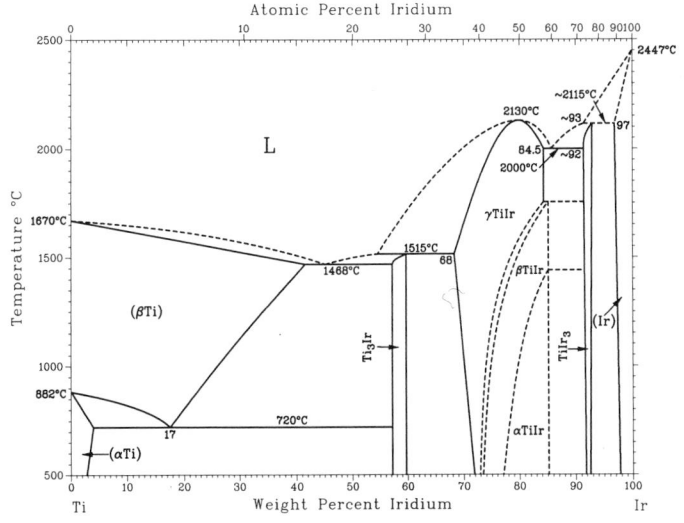

Phase	Composition, wt% Ir	Pearson symbol	Space group
(βTi)	0 to 4	cI2	Im$\overline{3}$m
(αTi)	0 to 40.5	hP2	P6$_3$/mmc
Ti$_3$Ir	57 to 60	cP8	Pm$\overline{3}$n
γTiIr	68 to 84	cP2	Pm$\overline{3}$m
βTiIr	73 to ?	tP2	P4/mmm
αTiIr	77 to ?	c**	...
TiIr$_3$	~92 to ~93	cP4	Pm$\overline{3}$m
(Ir)	~97 to 100	cF4	Fm$\overline{3}$m

Ir-U

H. Okamoto, 1992

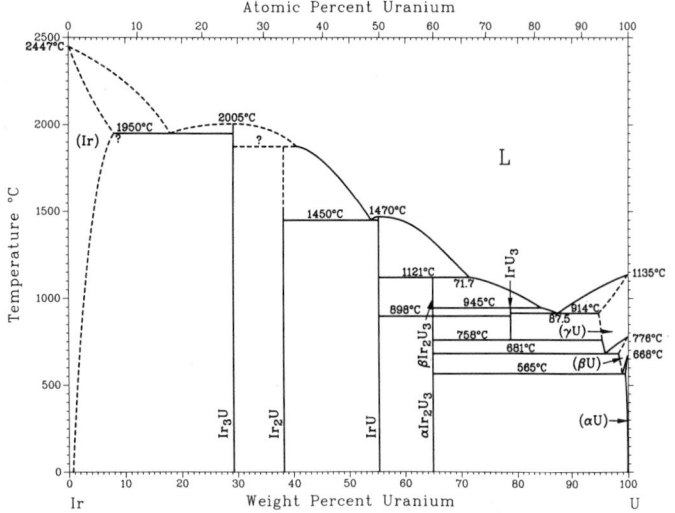

Phase	Composition, wt% U	Pearson symbol	Space group
(Ir)	0 to ?	cF4	$Fm\bar{3}m$
Ir₃U	29	cP4	$Pm\bar{3}m$
Ir₂U	38.2	cF24	$Fd\bar{3}m$
IrU	55.3	...	...
βIr₂U₃	65	...	...
αIr₂U₃	65	...	...
IrU₃	79	...	...
(γU)	? to 100	cI2	$Im\bar{3}m$
(βU)	? to 100	tP30	$P4_2/mnm$
(αU)	? to 100	oC4	$Cmcm$
Possible phase			
IrU₂	71.3	m**	...

Ir-V

J.F. Smith, 1989

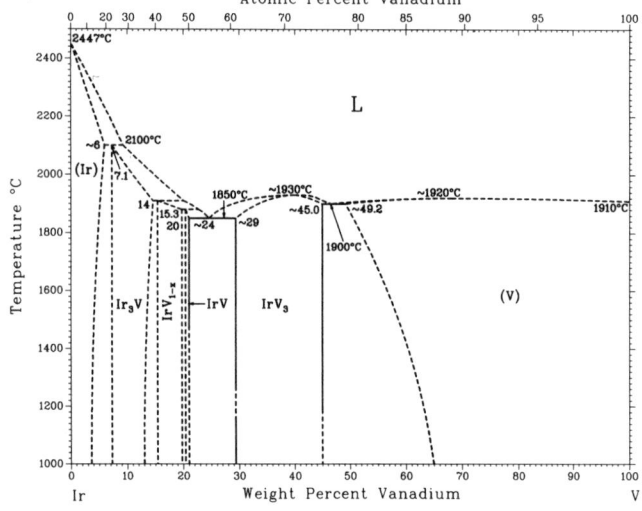

Phase	Composition, wt% V	Pearson symbol	Space group
(Ir)	0 to ~6	cF4	$Fm\bar{3}m$
Ir₃V	7.1 to 14	cP4	$Pm\bar{3}m$
IrV₁₋ₓ	15.3 to 20	tP2	$P4/mmm$
IrV	~20.9	oC8	$Cmmm$
IrV₃	~29 to 45.0	cP8	$Pm\bar{3}n$
(V)	~49.2 to 100	cI2	$Im\bar{3}m$

Ir-W

S.V. Nagender Naidu, A.M. Sriramamurthy, and P. Rama Rao, 1991

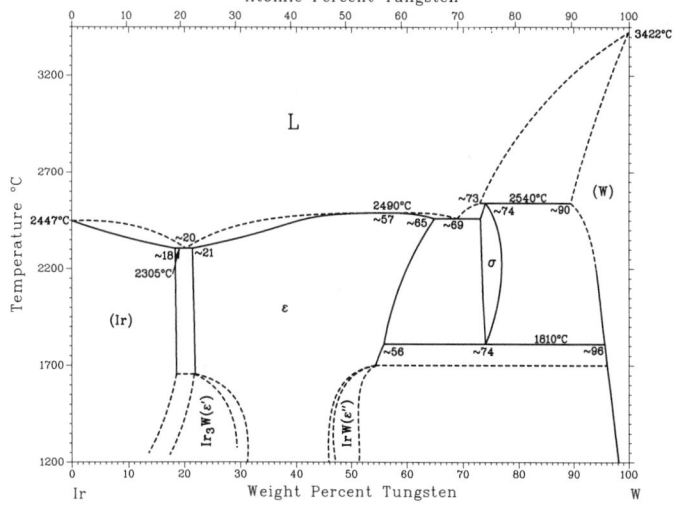

Phase	Composition, wt%W	Pearson symbol	Space group
(Ir)	0 to ~18	cF4	$Fm\bar{3}m$
ε	~21 to ~65	hP2	$P6_3/mmc$
Ir₃W(ε')	~24(a)	hP8	$P6_3/mmc$
IrW(ε'')	48.9(a)	oP4	$Pmma$
σ	74	tP30	$P4_2/mnm$
(W)	~90 to 100	cI2	$Im\bar{3}m$

(a) Ordered structure

Ir-Zr

H. Okamoto, 1992

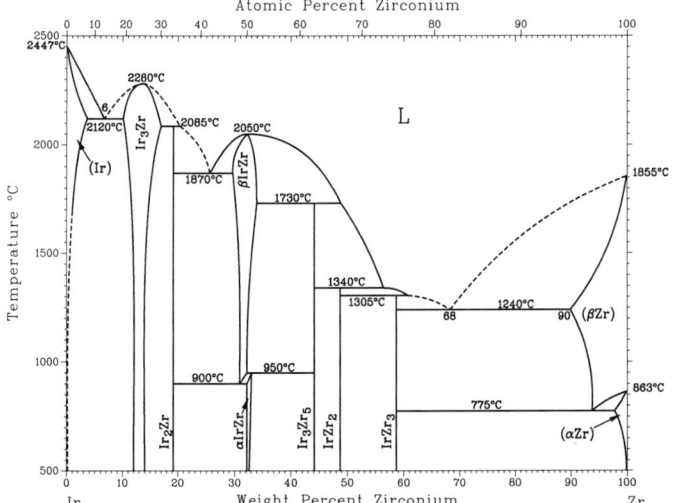

Phase	Composition, wt% Zr	Pearson symbol	Space group
(Ir)	0 to 3	cF4	Fm$\overline{3}$m
Ir$_3$Zr	10 to 17	cP4	Pm$\overline{3}$m
Ir$_2$Zr	19.2	cF24	Fd$\overline{3}$m
βIrZr	30 to 34	cP2	Pm$\overline{3}$m
αIrZr	32.2 to 33	(a)	...
Ir$_3$Zr$_5$	44.2	hP16	P6$_3$/mcm
IrZr$_2$	48.7	tI12	I4/mcm
IrZr$_3$	59	tI32	I$\overline{4}_2$/m
(βZr)	90 to 100	cI2	Im$\overline{3}$m
(αZr)	98 to 100	hP2	P6$_3$/mmc

(a) Complex

K-Na

C.W. Bale, 1982

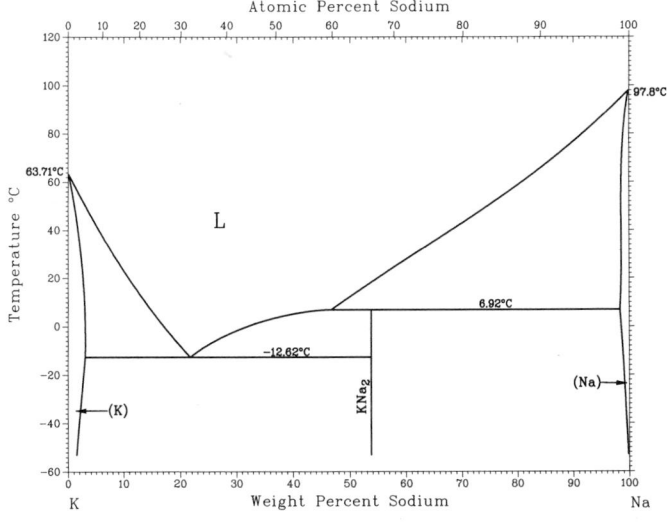

Phase	Composition, wt% Na	Pearson symbol	Space group
(K)	0	cI2	Im$\overline{3}$m
K$_2$Na(a)	22.72	...	...
KNa$_2$	54.05	hP12	P6$_3$/mmc
(Na)	100	cI2	Im$\overline{3}$m

(a) Possible phase (not shown in diagram)

K-Pb

H. Okamoto, 1990

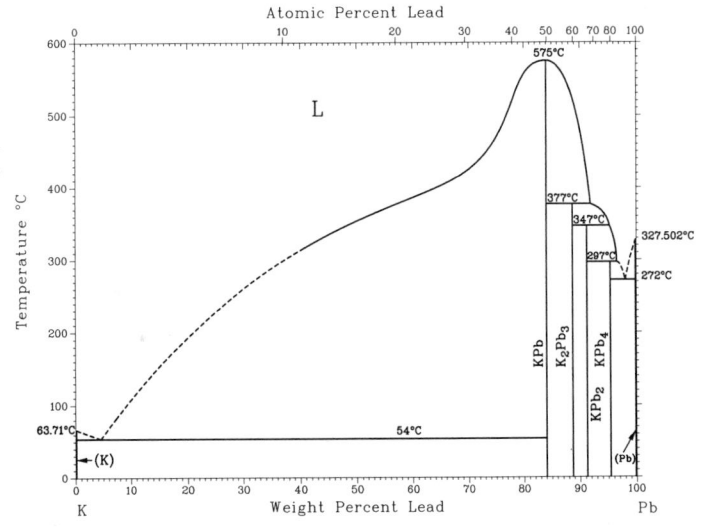

Phase	Composition, wt% Pb	Pearson symbol	Space group
(K)	0	hP2	P6$_3$/mmc
KPb	84.1	tI64	I4$_1$/acd
K$_2$Pb$_3$	89	...	...
KPb$_2$	91.4	hP12	P6$_3$/mmc
KPb$_4$	96	cI*	...
(Pb)	100	cF4	Fm$\overline{3}$m

K-Rb

C.W. Bale and A.D. Pelton, 1983

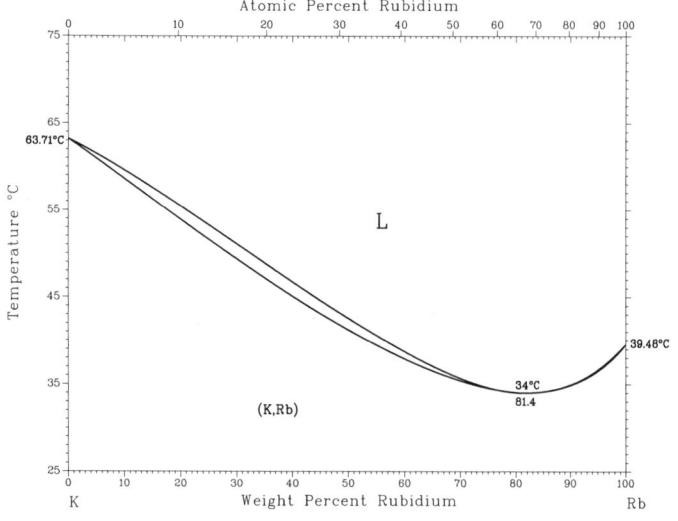

Phase	Composition, wt% Rb	Pearson symbol	Space group
(K,Rb)	0 to 100	$cI2$	$Im\bar{3}m$

K-S

H. Okamoto, 1990

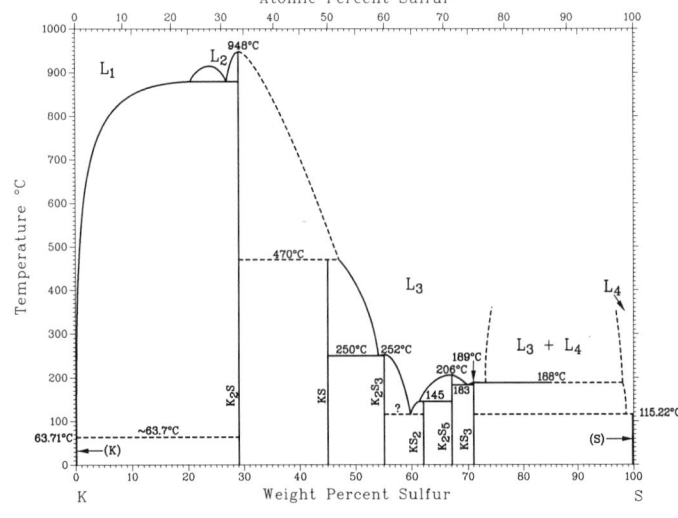

Phase	Composition, wt% S	Pearson symbol	Space group
(K)	0	$hP2$	$P6_3/mmc$
K_2S	29.1	$cF12$	$Fm\bar{3}m$
KS	45.1	$hP12$	$P6_2m$
K_2S_3	55	$oC20$	$Cmc2_1$
KS_2	62.2	$aP42$	...
K_2S_5	67.2	$oP28$	$P2_12_12_1$
KS_3	71	$aP57$	...
(S)	100	$oF128$	$Fddd$

K-Sb

F.W. Dorn and W. Klemm, 1961

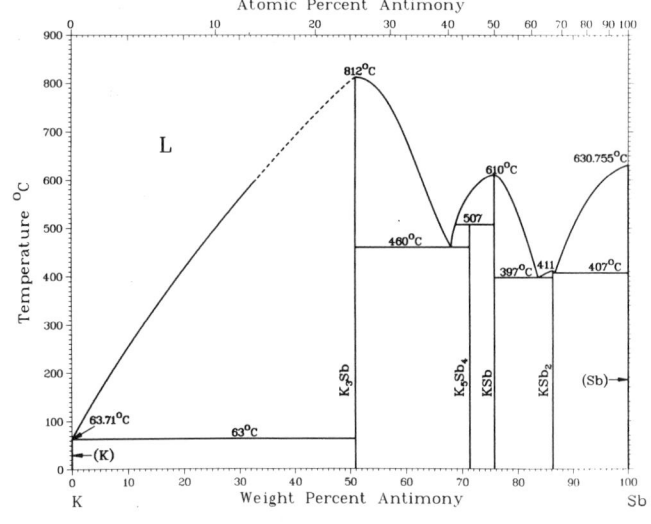

Phase	Composition, wt% Sb	Pearson symbol	Space group
(K)	~0	$cI2$	$Im\bar{3}m$
K_3Sb	51	$hP8$	$P6_3/mmc$
K_5Sb_4	71.3	...	...
KSb	75,7	$mP16$	$P2_1/c$
KSb_2	86.2	...	...
(Sb)	~100	$hR2$	$R\bar{3}m$

K-Se

H. Okamoto, 1990

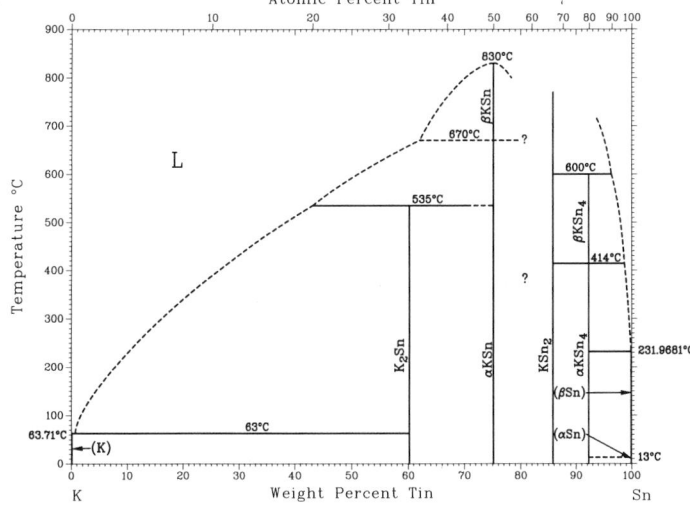

Phase	Composition, wt% Se	Pearson symbol	Space group
(K)	0	hP2	P6₃/mmc
K₂Se	50.2 to 57	cF12	Fm$\overline{3}$m
KSe	66.9	...	...
K₂Se₃	75	oC20	Cmc2₁
KSe₂	80.2	...	...
K₂Se₅	83.4	...	...
(Se)	100	hP3	P3₁21

K-Sn

H. Okamoto, 1990

Phase	Composition, wt% Sn	Pearson symbol	Space group
(K)	~0	cI2	Im$\overline{3}$m
K₂Sn	~60.3	...	...
βKSn	75.2	...	...
αKSn	75.2	tI64	I4₁/acd
KSn₂	85.9	...	...
βKSn₄	92	...	...
αKSn₄	92	...	...
(βSn)	~100	tI4	I4₁/amd
(αSn)	~100	cF8	Fd$\overline{3}$m
Other reported phase			
K₄Sn₂₃	~94.6	cP54	Pm$\overline{3}$n

K-Te

A. Petric and A.D. Pelton, 1990

Phase	Composition, wt% Te	Pearson symbol	Space group
(K)	0	cI2	Im$\overline{3}$m
K₂Te	62.0 to 72.3(a)	cF12	Fm$\overline{3}$m
KTe	76.5	...	...
K₂Te₃	83	oP20	Pnma
(Te)	100	hP3	P3₁2₁

(a) Homogeneity range subject to verification

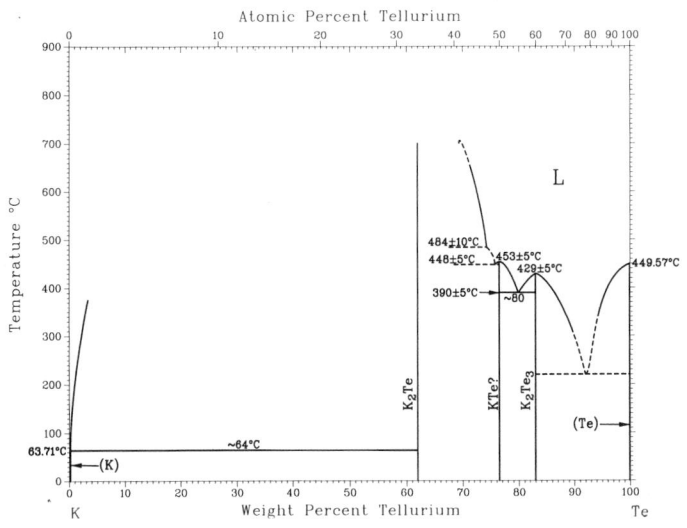

K-Tl

H. Okamoto, 1990

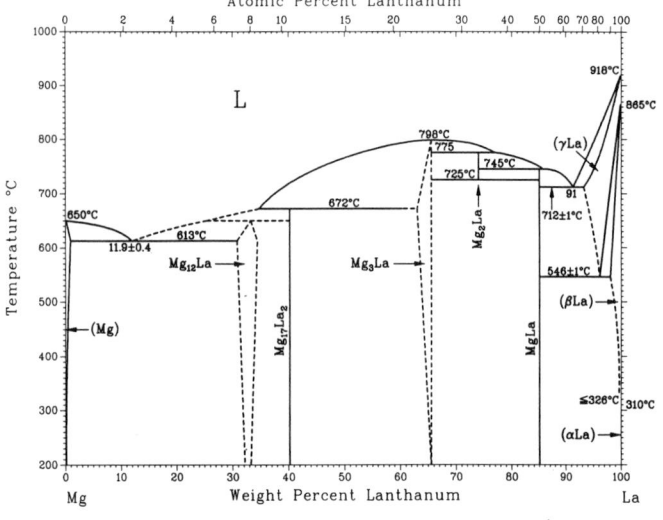

Phase	Composition, wt% Tl	Pearson symbol	Space group
(K)	0	cI2	Im$\overline{3}$m
KTl	83.9	(a)	...
K$_4$Tl$_5$	86.7	...	...
K$_5$Tl$_8$	89.3	...	...
(βTl)	? to 100	cI2	Im$\overline{3}$m
(αTl)	~99.8 to 100	hP2	P6$_3$/mmc

(a) Crystal structure neither the β brass or NaCl type

La-Mg

A.A. Nayeb-Hashemi and J.B. Clark, 1988

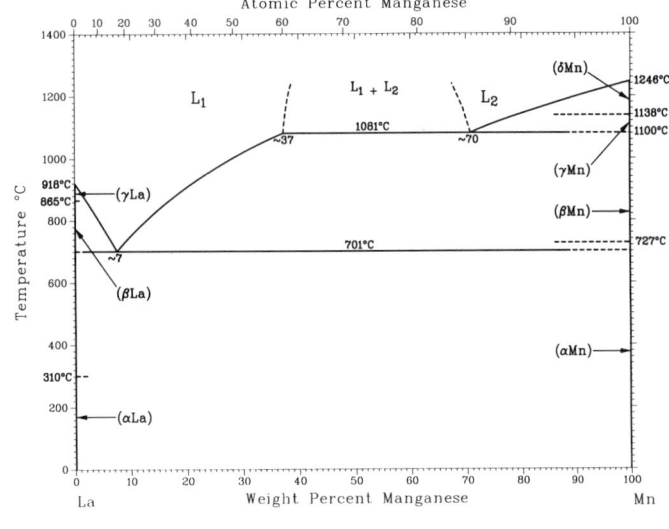

Phase	Composition, wt% La	Pearson symbol	Space group
(Mg)	0 to 0.79	hP2	P6$_3$/mmc
Mg$_{12}$La	30.53 to 34.18(a)	oI338(b)	(Immm)(b)
Mg$_{17}$La$_2$	40.21	hP38	P6$_3$/mmc
Mg$_3$La	? to 66	cF16	Fm$\overline{3}$m
Mg$_2$La	74.07	cF24	Fd$\overline{3}$m
MgLa	85.1	cP2	Pm$\overline{3}$m
(γLa)	~93 to 100	cI2	Im$\overline{3}$m)
(βLa)	~98.2 to 100	cF4	Fm$\overline{3}$m
(αLa)	? to 100	hP4	P6$_3$/mmc

(a) Homogeneity range estimated from lattice parameters. (b) This proposed crystal structure is based on the similarities of the lattice parameters of Mg$_{12}$La with those of Mg$_{12}$Ce(II).

La-Mn

A. Palenzona and S. Cirafici, 1990

Phase	Composition, wt% Mn	Pearson symbol	Space group
(γLa)	0	cI2	Im$\overline{3}$m
(βLa)	0	cF4	Fm$\overline{3}$m
(αLa)	0	hP4	P6$_3$/mmc
(δMn)	~100	cI2	Im$\overline{3}$m
(γMn)	~100	cI4	Im$\overline{3}$m
(βMn)	~100	cP20	P4$_1$32
(αMn)	~100	cI58	I$\overline{4}$3m

La-Ni

H. Okamoto, 1991

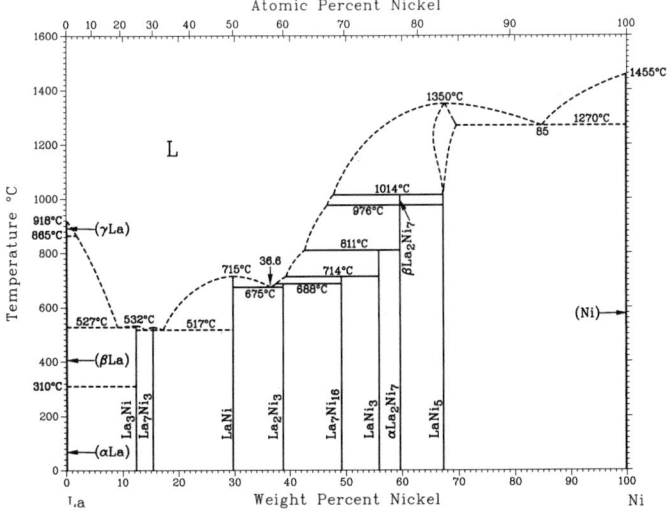

Phase	Composition, wt% Ni	Pearson symbol	Space group
(γLa)	0	cI2	Im$\overline{3}$m
(βLa)	0	cF4	Fm$\overline{3}$m
(αLa)	0	hP4	P6$_3$/mmc
La$_3$Ni	12.3	oP16	Pnma
La$_7$Ni$_3$	15.3	hP20	P6$_3$/mc
LaNi	29.7	oC8	Cmcm
La$_2$Ni$_3$	39.0	oC20	Cmca
La$_7$Ni$_{16}$	49.2	tI46	I$\overline{4}$2m
LaNi$_3$	55.9	hR24	R$\overline{3}$m
βLa$_2$Ni$_7$	59.7	hR18	R$\overline{3}$m
αLa$_2$Ni$_7$	59.7	hP36	P6$_3$/mmc
LaNi$_5$	67.8	hP6	P6/mmm
(Ni)	100	cF4	Fm$\overline{3}$m
Metastable phase			
LaNi$_2$	66.7	cF24	Fd$\overline{3}$m

La-Pb

A. Palenzona and S. Cirafici, 1992

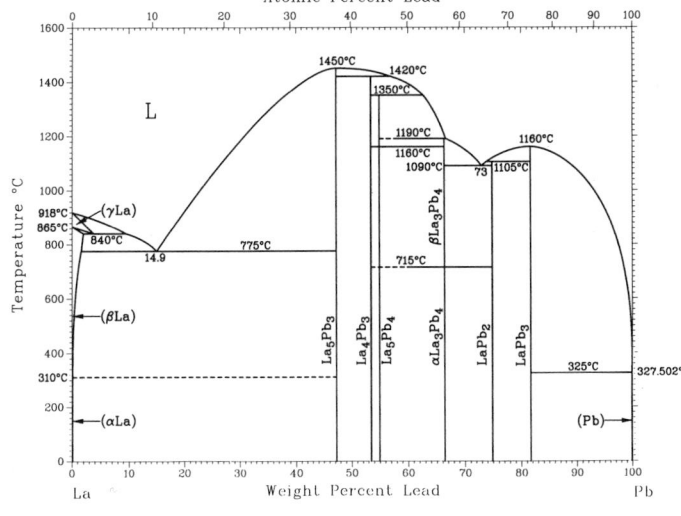

Phase	Composition, wt% Pb	Pearson symbol	Space group
(γLa)	3.7	cI2	Im$\overline{3}$m
(βLa)	0 to <1.5	cF4	Fm$\overline{3}$m
(αLa)	0	hP4	P6$_3$/mmc
La$_5$Pb$_3$	47.2	hP16	P6$_3$/mcm
La$_4$Pb$_3$(a)	52.8	cI28	I$\overline{4}$3d
La$_5$Pb$_4$	54.4	oP36	Pnma
βLa$_3$Pb$_4$	66.5	...	...
αLa$_3$Pb$_4$	66.5	...	...
LaPb$_2$	74.9	...	...
LaPb$_3$	81.7	cP4	Pm$\overline{3}$m
(Pb)	~99.9 to 100	cF4	Fm$\overline{3}$m

(a) Low-temperature modification

La-S

H.F. Franzen, unpublished

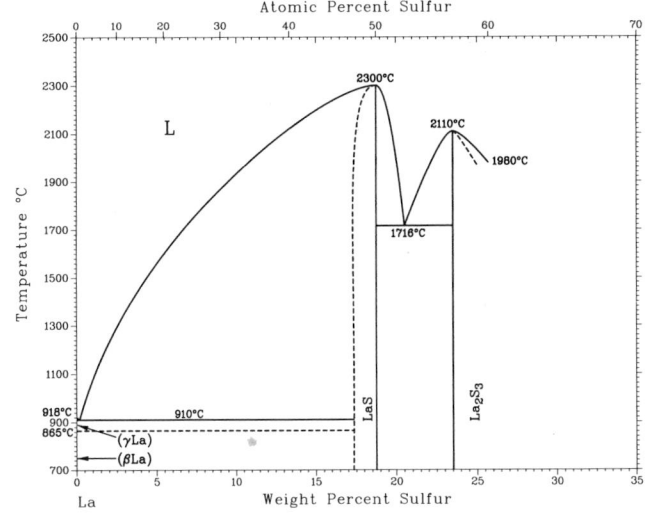

Phase	Composition, wt% S	Pearson symbol	Space group
(γLa)	0	cI2	Im$\overline{3}$m
(βLa)	0	cF4	Fm$\overline{3}$m
LaS	17 to 18.8	cF8	Fm$\overline{3}$m
γLa$_2$S$_3$	23.5 to 26	cI28	I$\overline{4}$3d

La-Sb

R. Vogel and H. Klose, 1954

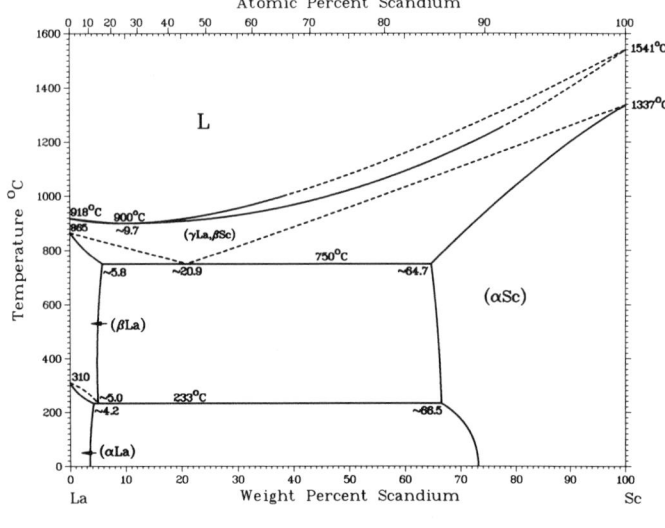

Phase	Composition, wt% Sb	Pearson symbol	Space group
(γLa)	~0	cI2	Im$\overline{3}$m
(βLa)	~0	cF4	Fm$\overline{3}$m
La$_2$Sb	30.4	tI12	I4/mmm
La$_3$Sb$_2$	37	...	...
LaSb	46.7	cF8	Fm$\overline{3}$m
LaSb$_2$	63.7	oC24	Cmca
(Sb)	~100	hR2	R$\overline{3}$m
Other reported phases			
La$_5$Sb$_3$	34.9	hP16	P6$_3$/mcm
La$_4$Sb$_3$	39.7	cI28	I$\overline{4}$3d

La-Sc

K.A. Gschneidner, Jr. and F.W. Calderwood, 1982

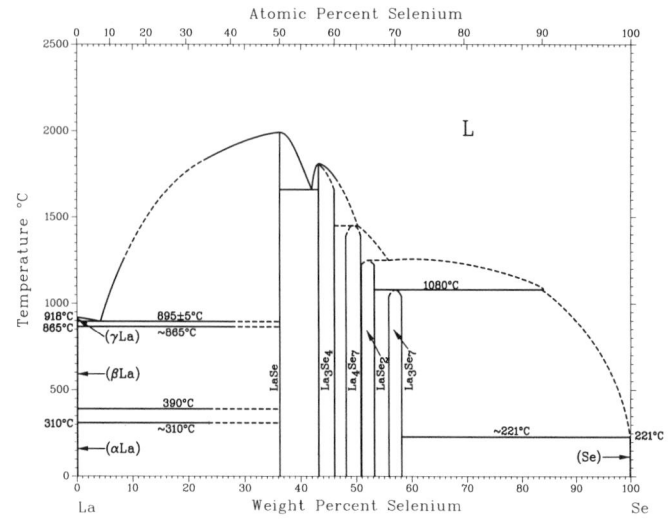

Phase	Composition, wt% Sc	Pearson symbol	Space group
(αLa)	0 to ~4.2	hP4	P6$_3$/mmc
(βLa)	0 to ~5.8	cF4	Fm$\overline{3}$m
(γLa,βSc)	0 to 100	cI2	Im$\overline{3}$m
(αSc)	~64.7 to 100	hP2	P6$_3$/mmc

La-Se

H. Okamoto, 1990

Phase	Composition, wt% Se	Pearson symbol	Space group
(γLa)	0	cI2	Im$\overline{3}$m
(βLa)	0	cF4	Fm$\overline{3}$m
(αLa)	0	hP4	P6$_3$/mmc
LaSe	36.2	cF8	Fm$\overline{3}$m
La$_3$Se$_4$	43.2 to 46	cI28	I$\overline{4}$3d
La$_4$Se$_7$	48 to 50.8	mP6	P2/c
LaSe$_2$	50.9 to 53.2	tP6	P4/nmm
La$_3$Se$_7$	56 to 58	t**	...
(Se)	100	hP3	P3$_1$21

La-Sn

A. Palenzona and S. Cirafici, 1992

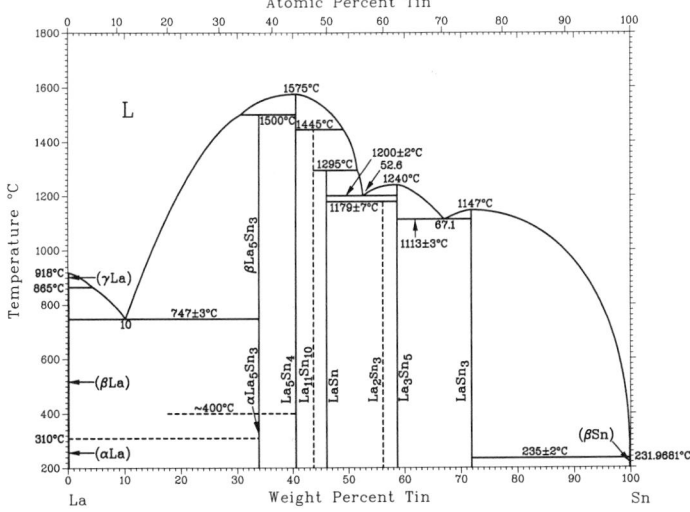

Phase	Composition, wt% Sn	Pearson symbol	Space group
γLa(a)	0	cI2	Im$\overline{3}$m
βLa(b)	0	cF4	Fm$\overline{3}$m
αLa(c)	0	hP4	P6₃/mmc
La₃Sn(d)	22	cP4	Pm$\overline{3}$m
βLa₅Sn₃	33.9	hP16	P6₃/mcm
αLa₅Sn₃	33.9	tI32	I4/mcm
La₅Sn₄	40.6	oP36	Pnma
La₁₁Sn₁₀(e)	43.7	tI84	I4/mmm
LaSn	46.1	oC8	Cmcm
La₂Sn₃	56	...	...
La₃Sn₅	58.8	oC32	Cmcm
LaSn₃	72	cP4	Pm$\overline{3}$m
βSn(f)	100	tI4	I4₁/amd
αSn(g)	100	cF8	Fd$\overline{3}$m

(a) From 918 to 865 °C. (b) From 865 to 310 °C. (c) Up to 310 °C. (d) High-temperature, high-pressure phase   (e) Proposed structure type. (f) From 13 to 231.9681 °C. (g) Up to 13 °C

La-Tl

S. Delfino, A. Saccone, A. Palenzona, and R. Ferro, unpublished

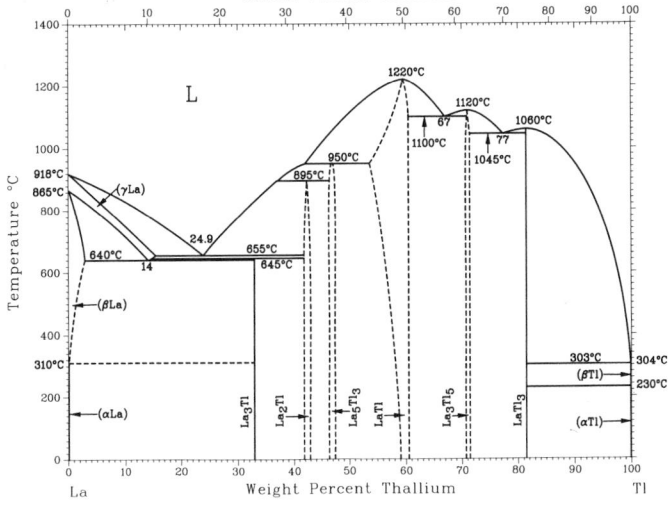

Phase	Composition, wt% Tl	Pearson symbol	Space group
(γLa)	15.4	cI2	Im$\overline{3}$m
(βLa)	0 to 2.8	cF4	Fm$\overline{3}$m
(αLa)	0	hP4	P6₃/mmc
La₃Tl(a)	33	cP4	Pm$\overline{3}$m
	~33	cF4	Fm$\overline{3}$m
La₂Tl	~46.9	...	...
La₅Tl₃	~46 to 47	tI32	I4/mcm
LaTl(b)	~54 to ~61	cP2	Pm$\overline{3}$m
		cI2	Im$\overline{3}$m
LaTl(c)	~54 to ~61	tP2	P4/mmm
La₃Tl₅	~71 to ~72	oC32	Cmcm
LaTl₃	82	cP4	Pm$\overline{3}$m
(βTl)	100	cI2	Im$\overline{3}$m
(αTl)	100	hP2	P6₃/mmc

(a) A cP4-cF4 order-disorder transformation in this phase has been suggested. (b) Cubic structure presumed to be room- and higher-temperature phases. (c) Tetragonal structure presumed to be lower-temperature phase

La-Zn

L. Rolla and A. Iandelli, 1941

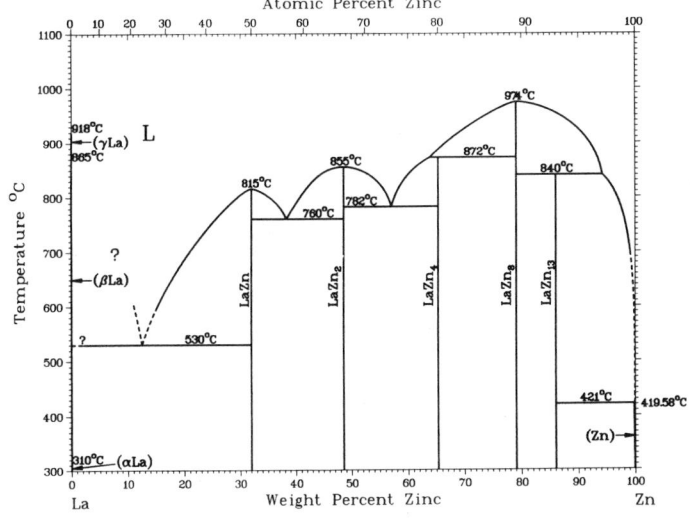

Phase	Composition, wt% Zn	Pearson symbol	Space group
(γLa)	~0	cI2	Im$\overline{3}$m
(βLa)	~0	cF4	Fm$\overline{3}$m
(αLa)	~0	hP4	P6₃/mmc
LaZn	32.0	cP2	Pm$\overline{3}$m
LaZn₂	~48.5	oI12	Imma
LaZn₄	~65	oC20	Cmcm
LaZn₈(La₂Zn₁₇)	~79	hR19	R$\overline{3}$m
LaZn₁₃	~86.0	cF112	Fm$\overline{3}$c
(Zn)	~100	hP2	P6₃/mmc
Other reported phases			
LaZn₅	~70.1	hP6	P6/mmm
La₃Zn₂₂	~78	tI100	I4₁/amd
LaZn₁₁	~83.9	tI48	I4₁/amd

Li-Mg

A.A. Nayeb-Hashemi, A.D. Pelton, and J.B. Clark, 1988

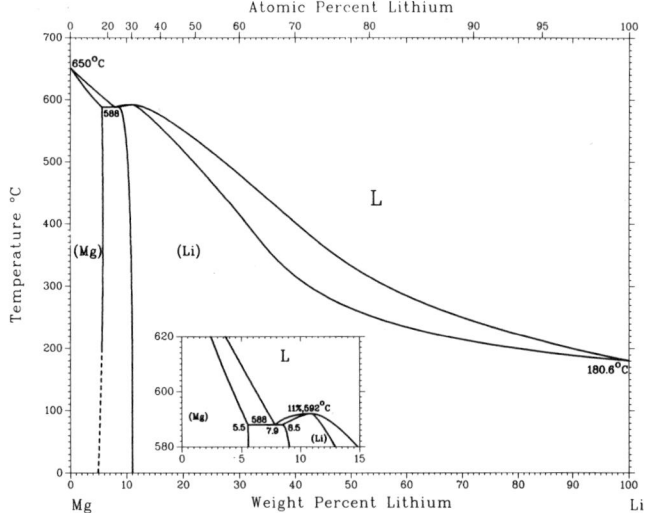

Phase	Composition, wt% Li	Pearson symbol	Space group
(Mg)	0 to 6	hP2	P6₃/mmc
(βLi)	8.5 to 100	cI2	Im3̄m
(αLi)(a)	100	hP2	P6₃/mmc
Cold worked stabilized phase(b)			
(γLi)	100	cF4	Fm3̄m

(a) Below −193 °C. (b) Nonequilibrium

Li-Na

C.W. Bale, 1989

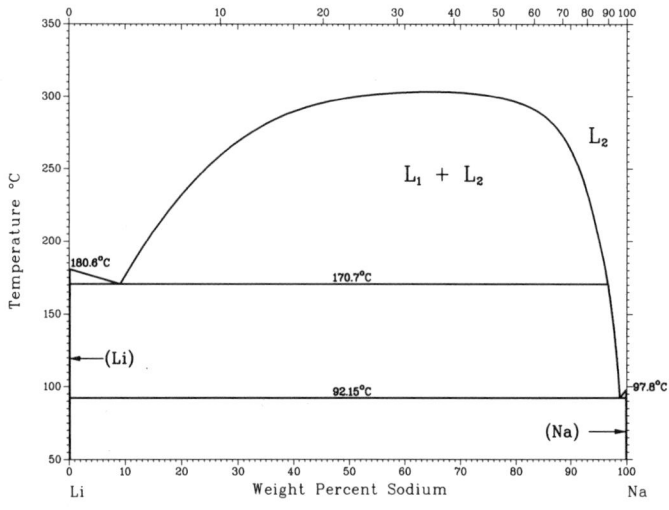

Phase	Composition, wt% Na	Pearson symbol	Space group
(βLi)	0	cI2	Im3̄m
(αLi)(a)	0	hP2	P6₃/mmc
(βNa)	100	cI2	Im3̄m
(αNa)	100	hP2	P6₃/mmc

(a) Below −193 °C

Li-Pb

From [Hansen]

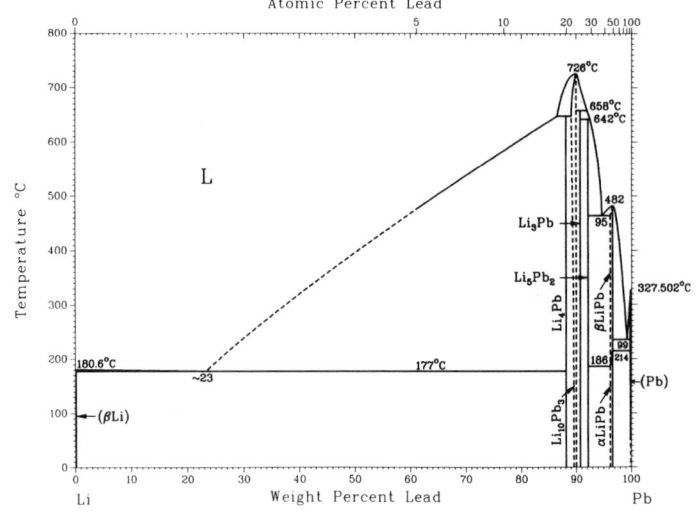

Phase	Composition, wt% Pb	Pearson symbol	Space group
(βLi)	~0	cI2	Im3̄m
(αLi)(a)	0	hP2	P6₃/mmc
Li₄Pb	~88	...	...
Li₁₀Pb₃	~89.7 to ~90.2	cP52	P4̄3m
Li₃Pb	~91	cF16	Fm3̄m
Li₅Pb₂	92.3	...	...
βLiPb	<96 to 96.8	cP2	Pm3̄m
αLiPb	<96 to 96.8	hR2	R3̄m
(Pb)	99.9 to 100	cF4	Fm3̄m
Other reported phases			
Li₂₂Pb₅	~87.1	cF432	F23
Li₇Pb₂	~89.5	hP9	P321
Li₈Pb₃	~91.8	mC22	C2/m

(a) Below −193 °C

Li-Pd

J. Sangster and A.D. Pelton, 1992

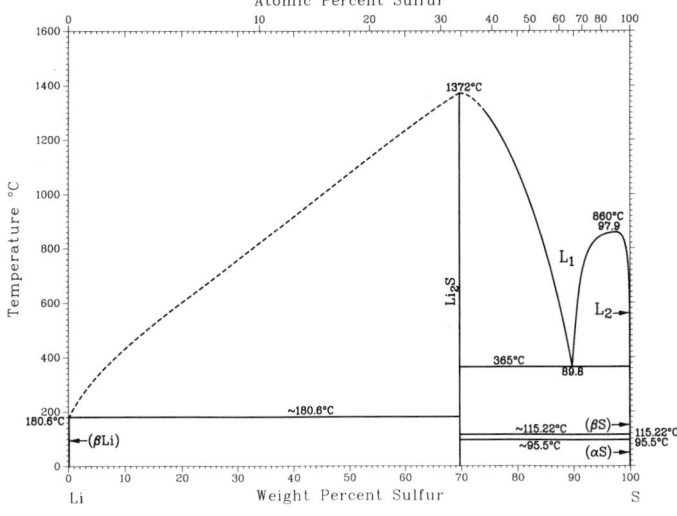

Phase	Composition, wt% Pd	Pearson symbol	Space group
(βLi)	0	cI2	Im$\overline{3}$m
(αLi)(a)	0	hP2	P6$_3$/mmc
Li$_5$Pd(b)	75.5(b)	cF*	...
Li$_{15}$Pd$_4$	80.4	cI76	I$\overline{4}$3d
Li$_3$Pd	84	cF16	Fm$\overline{3}$m
Li$_2$Pd	88.4	hP3	P6/mmm
Li$_3$Pd$_2$	90.9 to 91.5	cP2(?)	Pm$\overline{3}$m
LiPd	92.7 to 94.3	hP2	P$\overline{6}$
LiPd$_2$	~94 to 98	mP8	P2/m
LiPd$_7$	99.1	cF32	Fm$\overline{3}$m
(Pd)	99.7 to 100	cF4	Fm$\overline{3}$m

(a) Below −193 °C. (b) Approximate composition

Li-S

H. Okamoto, unpublished

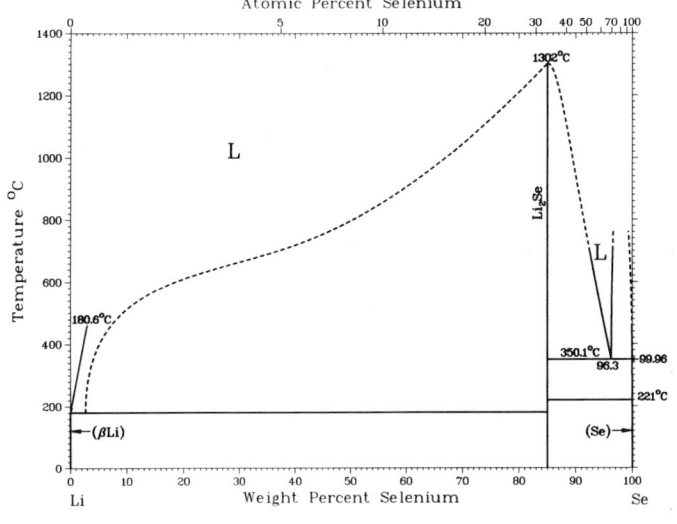

Phase	Composition, wt% S	Pearson symbol	Space group
(βLi)	0	cI2	Im$\overline{3}$m
(αLi)(a)	0	hP2	P6$_3$/mmc
Li$_2$S	69.8	cF12	Fm$\overline{3}$m
(βS)	100	mP48	P2$_1$/a
(αS)	100	oF128	Fddd

(a) Below −193 °C

Li-Se

P.T. Cunningham, S.A. Johnson, and E.J. Cairns, 1971

Phase	Composition, wt% Se	Pearson symbol	Space group
(βLi)	~0	cI2	Im$\overline{3}$m
(αLi)	0	hP2	P6$_3$/mmc
Li$_2$Se	85.0	cF12	Fm$\overline{3}$m
(Se)	~100	hP3	P3$_1$21

(a) Below −193 °C

Li-Si

H. Okamoto, 1990

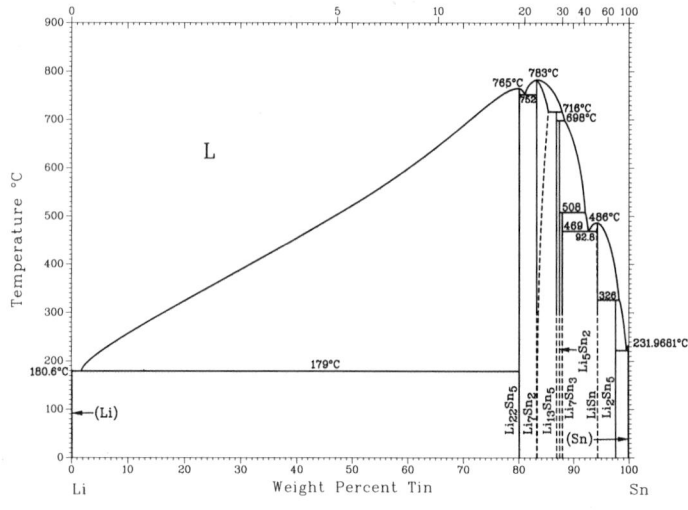

Phase	Composition, wt% Si	Pearson symbol	Space group
(βLi)	0	cI2	Fm$\bar{3}$m
(αLi)(a)	0	hP2	P6$_3$/mmc
Li$_{22}$Si$_5$	47.9	cF432	F23
Li$_{13}$Si$_4$	55.4	oP24	Pbam
Li$_7$Si$_3$	63	hR7	R$\bar{3}$m
Li$_{12}$Si$_7$	70.2	oP152	Pnma
(Si)	100	cF8	Fd$\bar{3}$m
Questionable phases			
Li$_4$Si	50	oP250	?
Li$_7$Si$_2$	53.6	oP36	Pbam
Li$_{10}$Si$_3$	54.9	cF416	?
Li$_2$Si	66.9	mC12	C2/m
Li$_{13}$Si$_7$	69	oP160	Pnma

(a) Below −193 °C

Li-Sn

From [Moffatt]

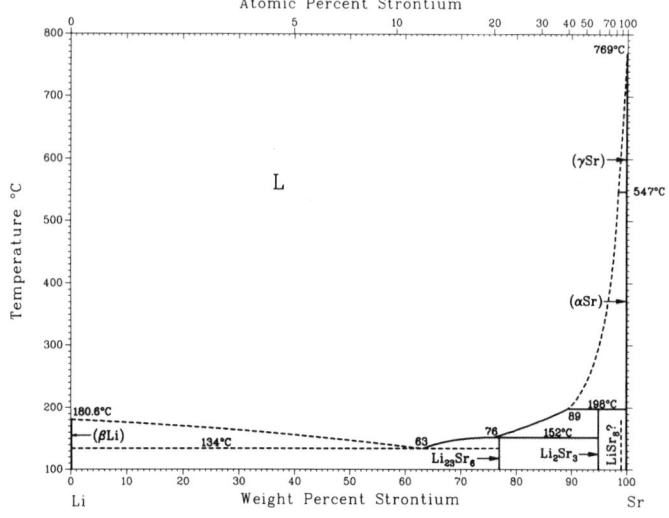

Phase	Composition, wt% Sn	Pearson symbol	Space group
(βLi)	0	cI2	Im$\bar{3}$m
(αLi)(a)	0	hP2	P6$_3$/mmc
Li$_{22}$Sn$_5$	79.5	cF432	F23
Li$_7$Sn$_2$	83.0 to ?	oC36	Cmmm
Li$_{13}$Sn$_5$	86.8	hP18	P$\bar{3}$m1
Li$_5$Sn$_2$	87.3	hR7	R$\bar{3}$m
Li$_7$Sn$_3$	88	mP20	P2$_1$/m
LiSn	94.5	mP6	P2/m
Li$_2$Sn$_5$	97.7	tI14	P4/mbm
(βSn)	100	tI4	I4$_1$/amd
(αSn)	100	cF8	Fd$\bar{3}$m

(a) Below −193 °C

Li-Sr

C.W. Bale and A.D. Pelton, 1989

Phase	Composition, wt% Sr	Pearson symbol	Space group
(βLi)	0	cI2	Im$\bar{3}$m
(αLi)(a)	0	hP2	P6$_3$/mmc
Li$_{23}$Sr$_6$	76.7	cF116	Fm$\bar{3}$m
Li$_2$Sr$_3$	95	tP20	P4$_2$/mnm
LiSr$_7$(?)	98.9	t**	...
LiSr$_8$(?)	99.0	t**	...
		hP*	...
(γSr)	100	cI2	Im$\bar{3}$m
(αSr)	100	cF4	Fm$\bar{3}$m

(a) Below −193 °C

Li-Te

J. Sangster and A.D. Pelton, 1992

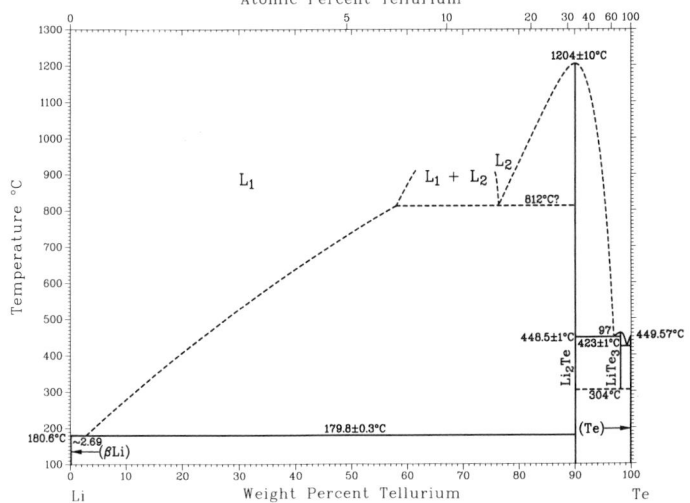

Phase	Composition, wt% Te	Pearson symbol	Space group
(βLi)	0	cI2	$Im\bar{3}m$
(αLi)(a)	0	hP2	$P6_3/mmc$
Li₂Te	90.2	cF12	$Fm\bar{3}m$
LiTe₃	98.2	hP48(b)	$P\bar{3}c1$
(αTe)	100	hP3	$P3_121$

(a) Below –193 °C. (b) Rhombohedrally centered hexagonal supercell, which is imposed on a cubic pseudocell

Li-Tl

G. Grube and G. Schaufler, 1934

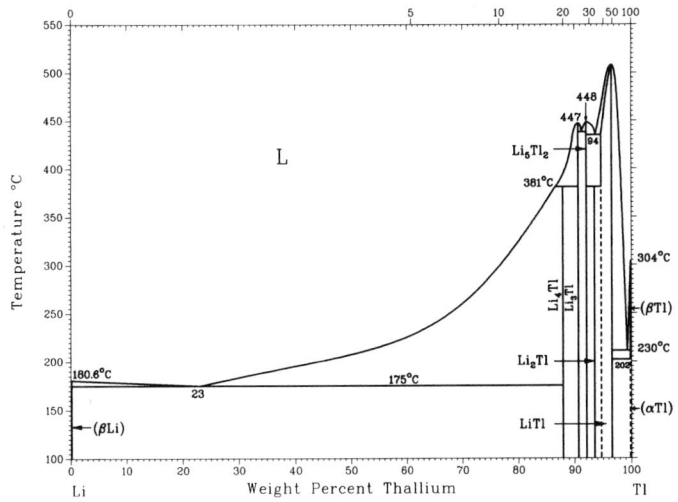

Phase	Composition, wt% Tl	Pearson symbol	Space group
(βLi)	~0	cI2	$Im\bar{3}m$
(αLi)(a)	0	hP2	$P6_3/mmc$
Li₄Tl	88	...	...
Li₃Tl	91	cF16	$Fm\bar{3}m$
Li₅Tl₂	~92.1	hR7	$R\bar{3}m$
Li₂Tl	93.6	oC12	$Cmcm$
LiTl	~94.9 to 96.7	cP2	$Pm\bar{3}m$
(βTl)	>99.9 to 100	cI2	$Im\bar{3}m$
(αTl)	~99.9 to 100	hP2	$P6_3/mmc$
Other reported phase			
Li₂₂Tl₅	87.0	cF432	$F23$

(a) Below –193 °C

Li-Zn

A.D. Pelton, 1991

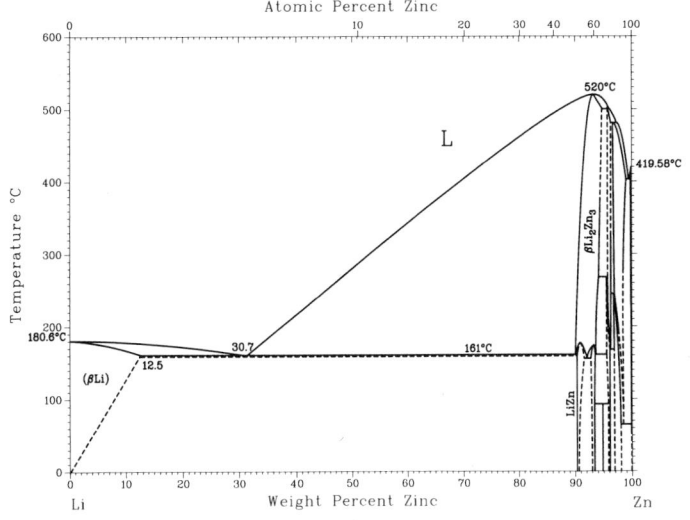

Phase	Composition, wt% Zn	Pearson symbol	Space group
(βLi)	0 to 12.5	cI2	$Im\bar{3}m$
(αLi)(a)	0	hP2	$P6_3/mmc$
LiZn	~90.4 to 92	cF16	$Fd\bar{3}m$
βLi₂Zn₃	~90.4 to 95	...	...
αLi₂Zn₃	~93 to 93	c**?	...
LiZn₂	94.97	...	...
βLi₂Zn₅(b)	95.8 to 99.1	...	...
αLi₂Zn₅(b)	95.6 to 96.2	h**(c)	...
βLiZn₄	~96.6 to 98.8	hP2	$P6_3/mmc$(d)
αLiZn₄	~96.9 to 98.2	h**(e)	...
(Zn)	99.9 to 100	hP2	$P6_3/mmc$

(a) Below –193 °C. (b) Possibly Li₃Zn₈ is a better designation. (c) Pseudocell. (d) Disordered—random distribution. (e) Ordered

Lu-Pb

O.D. McMasters and K.A. Gschneidner, Jr., 1969

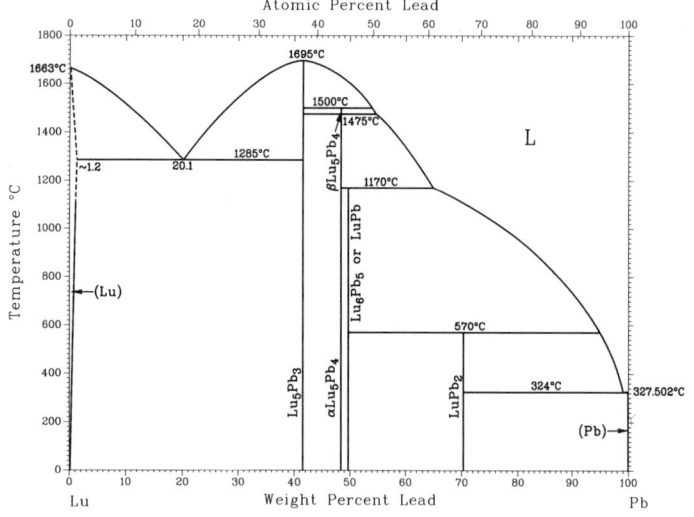

Phase	Composition, wt% Pb	Pearson symbol	Space group
(Lu)	0 to ~1.2	hP2	P6₃/mmc
Lu₅Pb₃	41.5	hP16	P6₃/mcm
βLu₅Pb₄	48.6	…	…
αLu₅Pb₄	48.6	oP*	Pnma
Lu₆Pb₅ or LuPb	49.7	oI*	Ibam
LuPb₂	70.3	tI6	I4/mmm
(Pb)	~100	cF4	Fm3̄m

Lu-Tl

H. Okamoto, 1990

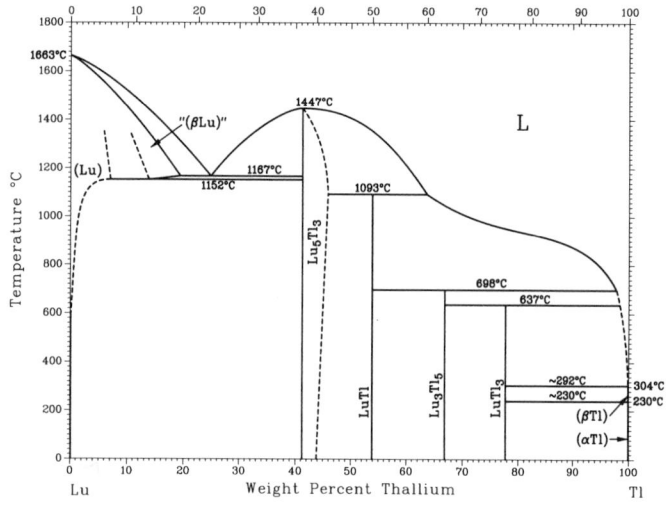

Phase	Composition, wt% Tl	Pearson symbol	Space group
(Lu)	~0	hP6	P6₃/mmc
Lu₅Tl₃	41.2	hP16	P6₃/mcm
LuTl	53.9	…	…
Lu₃Tl₅	66.1	oC32	Cmcm
LuTl₃	78	cP4	Pm3̄m
(βTl)	~100	cI2	Im3̄m
(αTl)	~100	hP2	P6₃/mmc

Mg-Mn

A.A. Nayeb-Hashemi and J.B. Clark, 1988

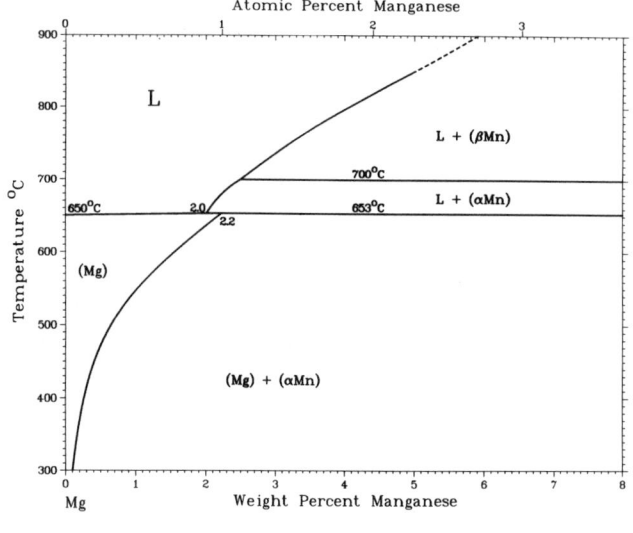

Phase	Composition, wt% Mn	Pearson symbol	Space group
(Mg)	0 to 2.2	hP2	P6₃/mmc
(αMn)	100	cI58	I4̄3m
(βMn)	100	cP20	P4₁32
(γMn)	100	cF4	Fm3̄m
(δMn)	100	cI2	Im3̄m

Mg-Ni

A.A. Nayeb-Hashemi and J.B. Clark, 1991

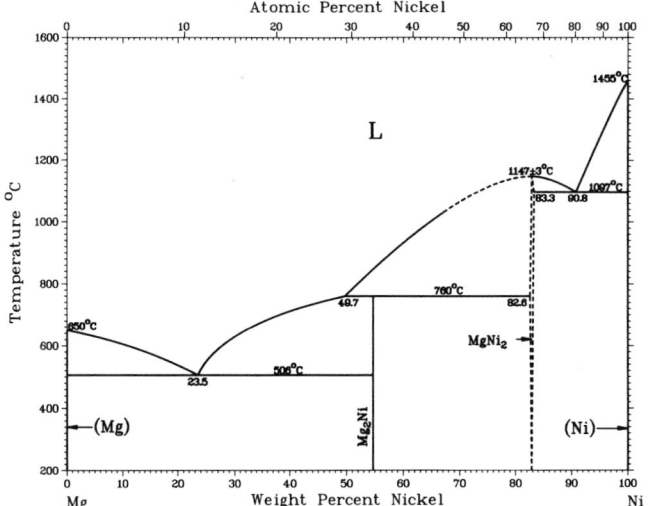

Phase	Composition, wt % Ni	Pearson symbol	Space group
(Mg)	0	$hP2$	$P6_3/mmc$
Mg$_2$Ni	54.7	$hP18$	$P6_222$
MgNi$_2$	82.9	$hP24$	$P6_3/mmc$
(Ni)	100	$cF4$	$Fm\bar{3}m$

Mg-Pb

A.A. Nayeb-Hashemi and J.B. Clark, 1988

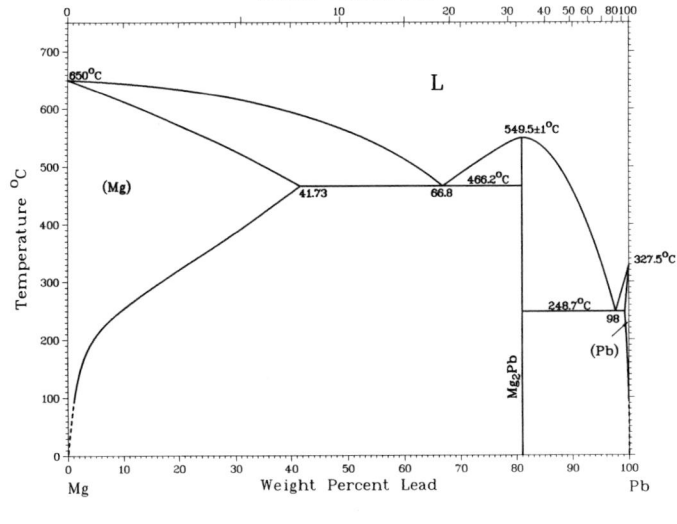

Phase	Composition, wt% Pb	Pearson symbol	Space group
(Mg)	0 to 41.73	$hP2$	$P6_3/mmc$
Mg$_2$Pb	81.00	$cF12$	$Fm\bar{3}m$
(Pb)	~99 to 100	$cF4$	$Fm\bar{3}m$

Mg-Sb

A.A. Nayeb-Hashemi and J.B. Clark, 1988

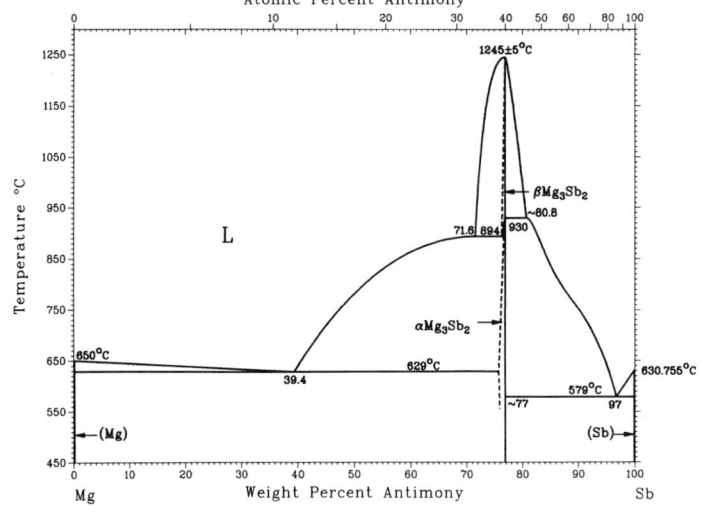

Phase	Composition, wt% Sb	Pearson symbol	Space group
(Mg)	0	$hP2$	$P6_3/mmc$
βMg$_3$Sb$_2$	~77	$cI80$	$Ia\bar{3}$
αMg$_3$Sb$_2$	~77	$hP5$	$P\bar{3}m1$
(Sb)	100	$hR2$	$R\bar{3}m$

Mg-Sc

A.A. Nayeb-Hashemi and J.B. Clark, 1988

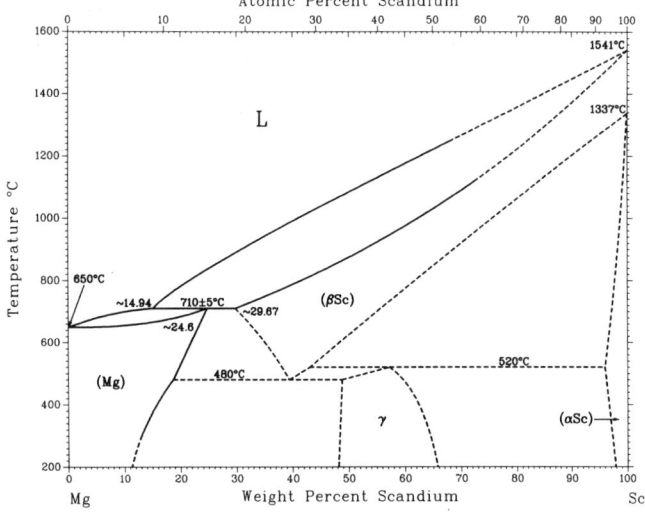

Phase	Composition, wt% Sc	Pearson symbol	Space group
(Mg)	0 to ~24.6	hP2	$P6_3/mmc$
γ	? to ?	cP2	$Pm\overline{3}m$
(βSc)	~29.67 to 100	cI2	$Im\overline{3}m$
(αSc)	? to 100	hP2	$P6_3/mmc$

Mg-Si

A.A. Nayeb-Hashemi and J.B. Clark, 1988

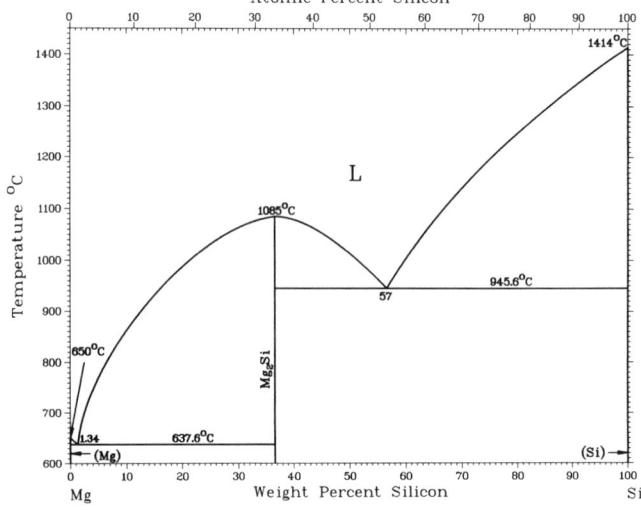

Phase	Composition, wt% Si	Pearson symbol	Space group
(Mg)	~0	hP2	$P6_3/mmc$
Mg_2Si	36.61	cF12	$Fm\overline{3}m$
(Si)	~100	cF8	$Fd\overline{3}m$
High-pressure phases			
Mg_2Si(a)	36.61	...	...
SiII	100	...	...

(a) Above ~2.5 GPa and 900 °C, it forms a hexagonal structure.

Mg-Sm

A.A. Nayeb-Hashemi and J.B. Clark, 1988

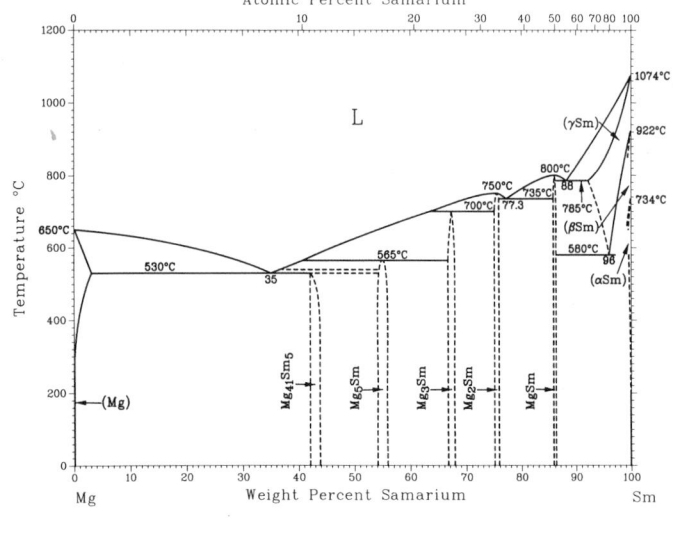

Phase	Composition, wt% Sm	Pearson symbol	Space group
(Mg)	0 to ~6	hP2	$P6_3/mmc$
$Mg_{41}Sm_5$	43.1	tI92	$I4/m$
Mg_5Sm	55.4	cF440-448	$F\overline{4}3m$
Mg_3Sm	67	cF16	$Fm\overline{3}m$
Mg_2Sm	75.57	cF24	$Fd\overline{3}m$
MgSm	86.1	cP2	$Pm\overline{3}m$
(γSm)	~96 to 100	cI2	$Im\overline{3}m$
(βSm)	100	hP2	$P6_3/mmc$
(αSm)	100	hR3	$R\overline{3}m$

Mg-Sn

A.A. Nayeb-Hashemi and J.B. Clark, 1988

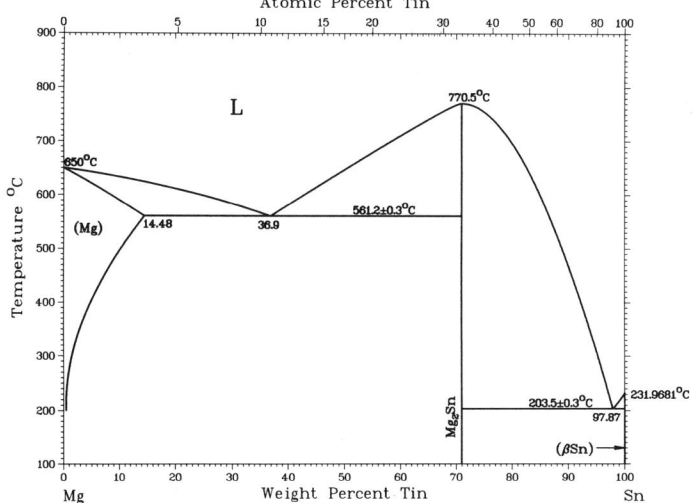

Phase	Composition, wt% Sn	Pearson symbol	Space group
(Mg)	0 to 14.48	hP2	P6₃/mmc
Mg₂Sn	70.9	cF12	Fm$\bar{3}$m
(βSn)	100	tI4	I4₁/amd
(αSn)	100	cF8	Fd$\bar{3}$m
High-pressure phases			
Mg₂Sn	70.9	h**	...
SnII	100	tI2	...
SnIII	100	cI2	...

Mg-Sr

A.A. Nayeb-Hashemi and J.B. Clark, 1988

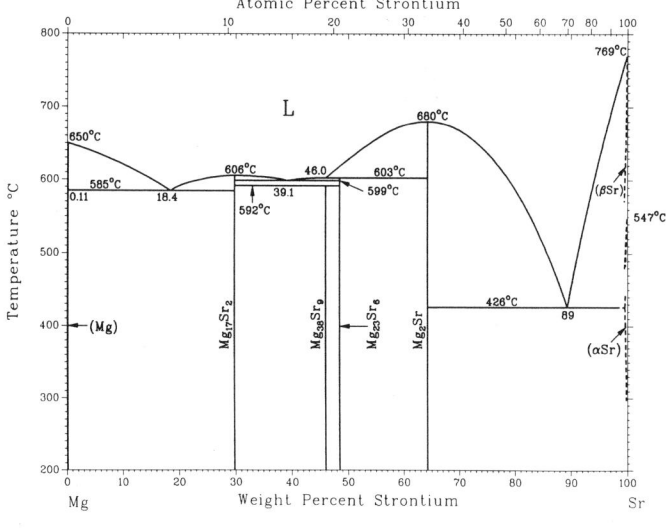

Phase	Composition, wt% Sr	Pearson symbol	Space group
(Mg)	0 to 0.11	hP2	P6₃/mmc
Mg₁₇Sr₂	29.79	hP38	P6₃/mmc
Mg₃₈Sr₉ or Mg₄Sr	46.06	hP94 or hP90	P6₃/mmc
Mg₂₃Sr₆	48.47	cF116	Fm$\bar{3}$m
Mg₂Sr	64.31	hP12	P6₃/mmc
(γSr)	100	cI2	Im$\bar{3}$m
(αSr)	? to 100	cF4	Fm$\bar{3}$m

Mg-Th

A.A. Nayeb-Hashemi and J.B. Clark, 1988

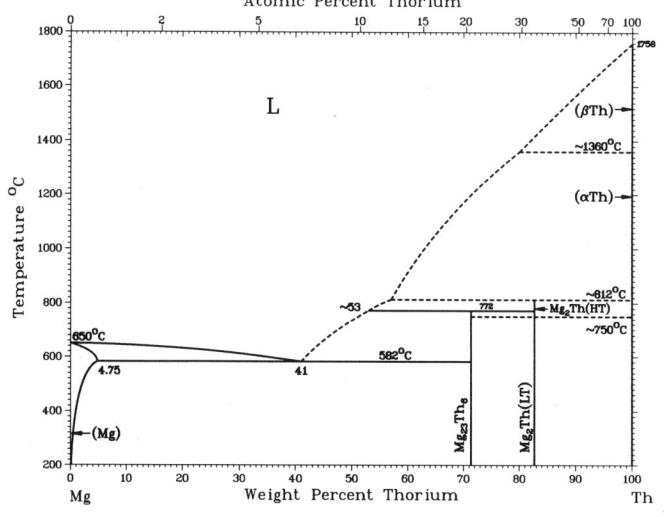

Phase	Composition, wt% Th	Pearson symbol	Space group
(Mg)	0 to 4.75	hP2	P6₃/mmc
Mg₂₃Th₆	71.35	cF116	Fm$\bar{3}$m
Mg₂Th (HT)	82.68	cF4	Fd$\bar{3}$m
Mg₂Th (LT)	82.68	hP4	P6₃/mmc
(βTh)	100	cI2	Im$\bar{3}$m
(αTh)	100	cF4	Fm$\bar{3}$m

Mg-Tl

A.A. Nayeb-Hashemi and J.B. Clark, 1988

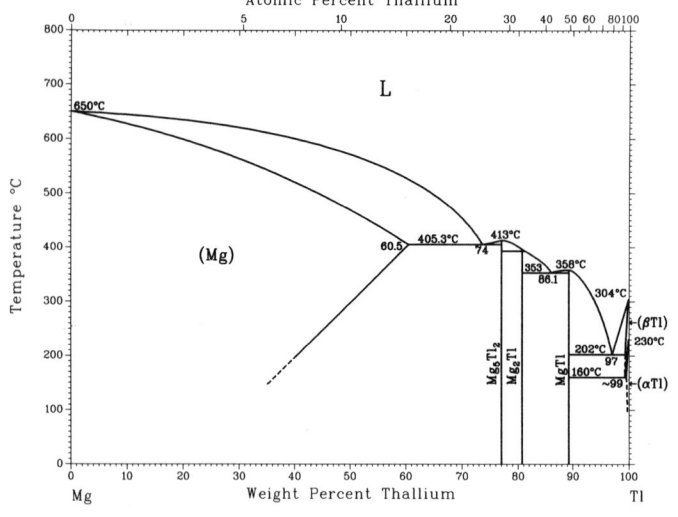

Phase	Composition, wt% Tl	Pearson symbol	Space group
(Mg)	0 to 60.5	hP2	$P6_3/mmc$
Mg_5Tl_2	77.08	oI28	Ibam
Mg_2Tl	80.78	hP9	$P\bar{6}2m$
MgTl	89.4	cP2	$Pm\bar{3}m$
(βTl)	100	cI2	$Im\bar{3}m$
(αTl)	100	hP2	$P6_3/mmc$

Mg-Y

H. Okamoto, 1991

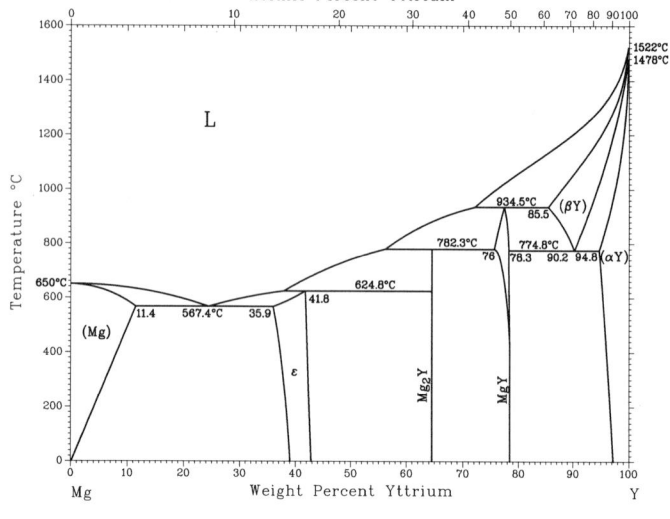

Phase	Composition, wt% Y	Pearson symbol	Space group
(Mg)	0 to 11.4	hP2	$P6_3/mmc$
ε	35.9 to 41.8	cI58	$I\bar{4}3m$
Mg_2Y	64.6	hP12	$P6_3/mmc$
MgY	76 to 78.3	cP2	$Pm\bar{3}m$
(βY)	85.5 to 100	cI2	$Im\bar{3}m$
(αY)	94.8 to 100	hP2	$P6_3/mmc$

Mg-Yb

A.A. Nayeb-Hashemi and J.B. Clark, 1988

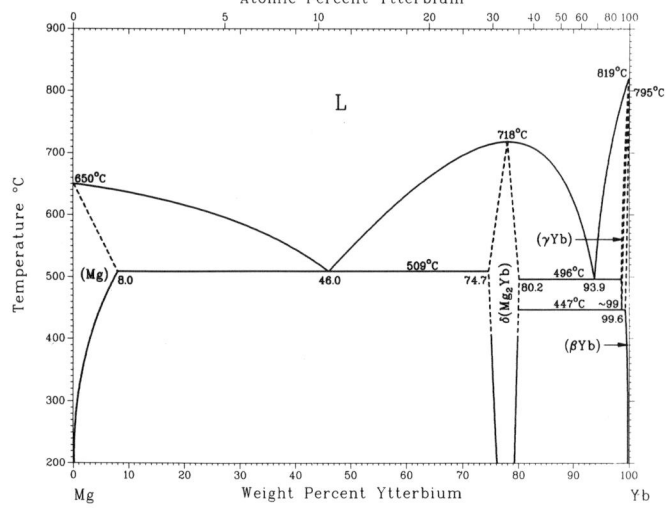

Phase	Composition, wt% Yb	Pearson symbol	Space group
(Mg)	0 to 8.0	hP2	$P6_3/mmc$
(δMg$_2$Yb)	74.7 to 80.2	hP12	$P6_3/mmc$
(γYb)	~99 to 100	cI2	$Im\bar{3}m$
(βYb)	99.6 to 100	cF4	$Fm\bar{3}m$
(αYb)	100	hP2	$P6_3/mmc$

Mg-Zn

J.B. Clark, L. Zabdyr, and Z. Moser, 1988

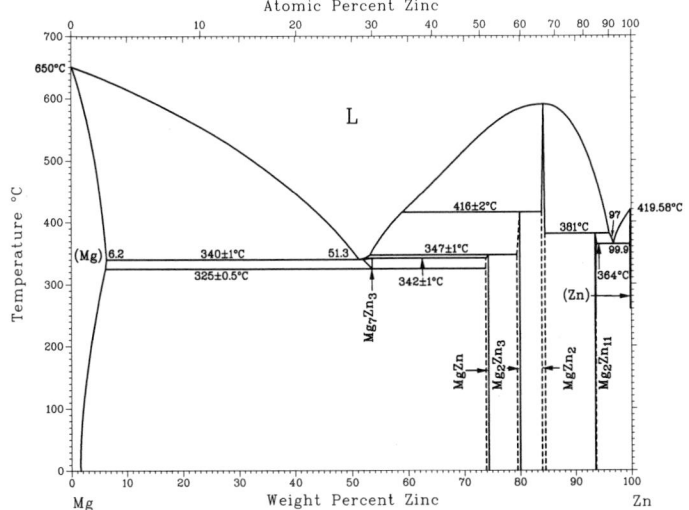

Phase	Composition, wt% Zn	Pearson symbol	Space group
(Mg)	0 to 6.2	hP2	P6₃/mmc
Mg₇Zn₃	53.6	oI142	Immm
MgZn	74.0	...	...
Mg₂Zn₃	80.1	mC110	C2/m
MgZn₂	84 to 84.6	hP12	P6₃/mmc
Mg₂Zn₁₁	93.7	cP39	Pm3̄
(Zn)	99.9 to 100	hP2	P6₃/mmc

Mg-Zr

A.A. Nayeb-Hashemi and J.B. Clark, 1988

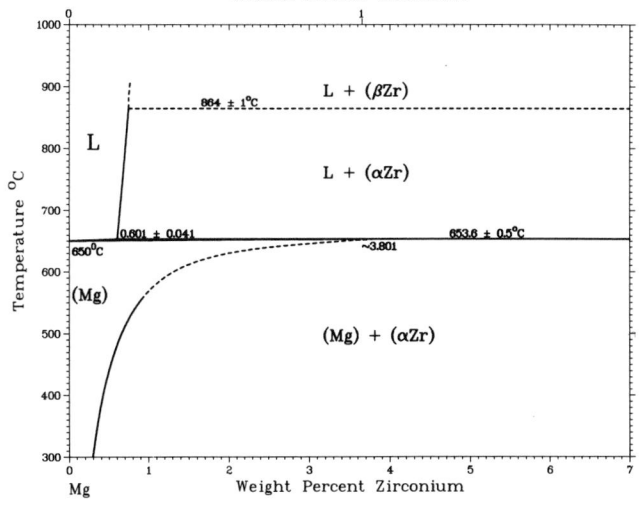

Phase	Composition, wt% Zr	Pearson symbol	Space group
(Mg)	0 to ~3.801	hP2	P6₃/mmc
(αZr)	100	hP2	P6₃/mmc
(βZr)	100	cI2	Im3̄m

Mn-Mo

From [Molybdenum]

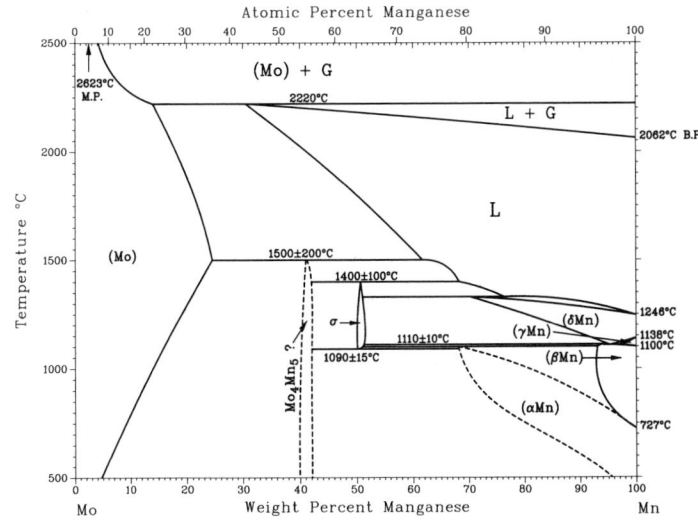

Phase	Composition, wt% Mn	Pearson symbol	Space group
(Mo)	0 to 25	cI2	Im3̄m
Mo₄Mn₅	~42	hR39?	...
σ	~50	tP30	P4₂/mnm
(δMn)	71 to 100	cI2	Im3̄m
(γMn)	97 to 100	cF4	Fm3̄m
(βMn)	78 to 100	cP20	P4₁32
(αMn)	~68 to 100	cI58	I4̄3m

Mn-N

N.A. Gokcen, 1990

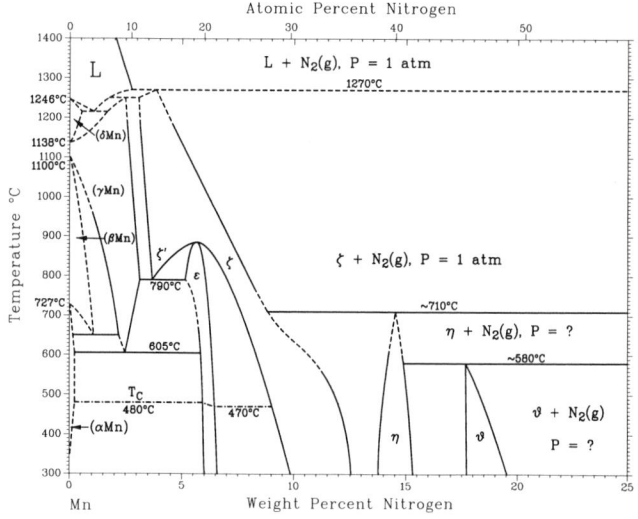

Phase	Composition, wt% N	Pearson symbol	Space group
(δMn)	0 to ~0.5	cI2	$Im\bar{3}m$
(γMn)	0 to 3.2	cF4	$Fm\bar{3}m$
(βMn)	0 to ~1	cP20	$P4_132$
(αMn)	0 to ~0.13	cI58	$I\bar{4}3m$
ε or Mn₄N	5.1 to 6.6	cF5	$Fm\bar{3}m$
ζ′	~13	hP12	$P6_322$
ζ or Mn₁₂N₅	~9	hP12	$P6_322$
ζ or Mn₂N	11.0	hP3	$P6_3/mmc$
ζ or Mn₂N	~11.2	oP12	$Pbcn$
η or Mn₆N₄	~14 to 15	...	$I4/mmm$
θ or Mn₆N₅	~17.7 to 20	...	...

Mn-Nd

H. Okamoto, 1992

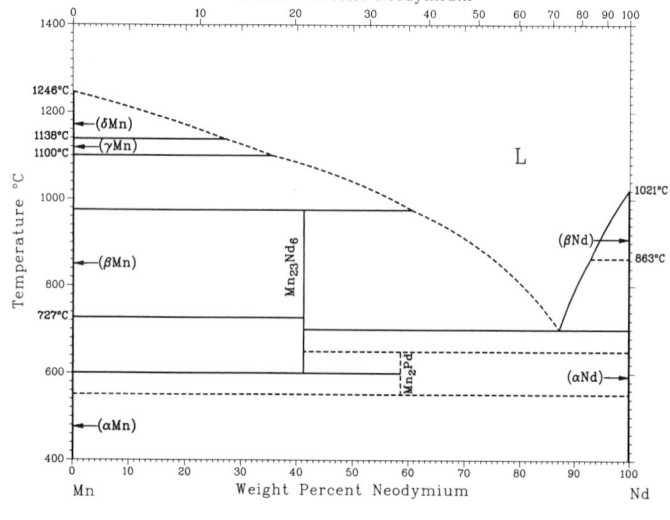

Phase	Composition, wt% Nd	Pearson symbol	Space group
(δMn)	0	cI2	$Im\bar{3}m$
(γMn)	0	cF4	$Fm\bar{3}m$
(βMn)	0	cP20	$P4_132$
(αMn)	0	cI58	$I\bar{4}3m$
Mn₂₃Nd₆	40.7	cF116	$I4/mmm$
βMn₂Nd	56.7	hP12	$P6_3/mmc$
αMn₂Nd	56.7	m**	...
(βNd)	100	cI2	$Im\bar{3}m$
(αNd)	100	hP4	$P6_3/mmc$

Mn-Ni

N.A. Gokcen, 1991

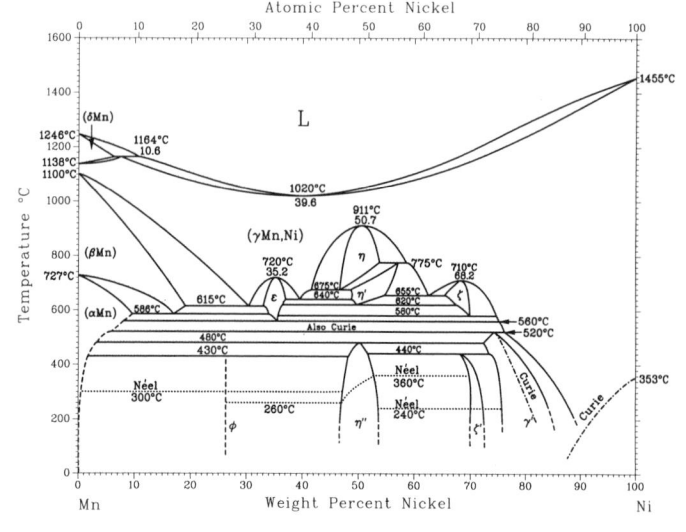

Phase	Composition, wt% Ni	Pearson symbol	Space group
(δMn)	0 to 6	cI2	$Im\bar{3}m$
(γMn, Ni)	0 to 100	cF4	$Fm\bar{3}m$
(βMn)	0 to 19	cP20	$P4_132$
(αMn)	0 to 10	cI58	$I\bar{4}3m$
φ	26	...	...
ε	34 to 38	...	...
η(a)	47 to 54	cP2	$Pm\bar{3}m$
η′	49 to 57.1	tP4	$P4/mmm$
ζ	66 to 70	...	...
ζ′	~71	...	...
γ′	72 to 86	cP4	$Pm\bar{3}m$

(a) At 745 °C; this phase cannot be retained by quenching.

Mn-O

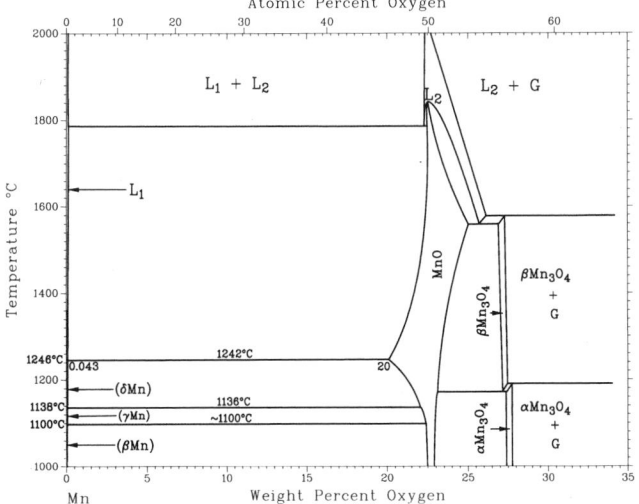

Phase	Composition, wt% O	Pearson symbol	Space group
(δMn)	0	cI2	$Im\bar{3}m$
(γMn)	0	cF4	$Fm\bar{3}m$
(βMn)	0	cP20	$P4_132$
(αMn)	0	cI58	$I\bar{4}3m$
MnO	20 to 25	cF8	$Fm\bar{3}m$
βMn_3O_4	~28	...	...
αMn_3O_4	~28	tI28	$I4_1/amd$

Mn-P

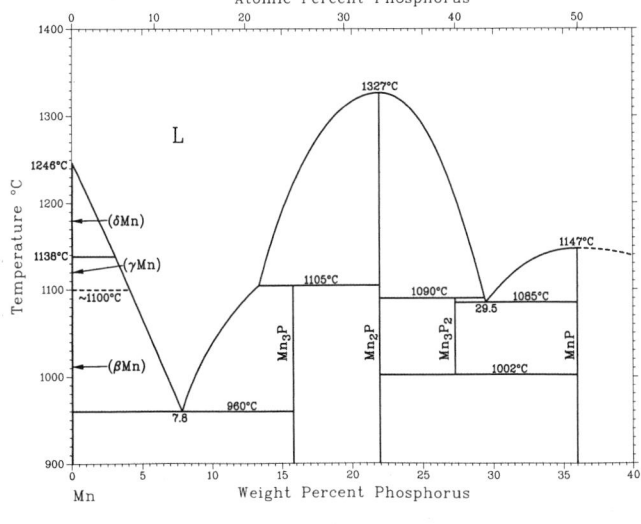

Phase	Composition, wt% P	Pearson symbol	Space group
(δMn)	~0	cI2	$Im\bar{3}m$
(γMn)	~0	cF4	$Fm\bar{3}m$
(βMn)	~0	cP20	$P4_132$
Mn_3P	16	tI32	$I\bar{4}$
Mn_2P	22.0	hP9	$P\bar{6}2m$
Mn_3P_2	27	...	...
MnP	36.1	oP8	Pnma
Other reported phase			
MnP_4	69	aP10	$P\bar{1}$
		aP30	$P\bar{1}$

Mn-Pd

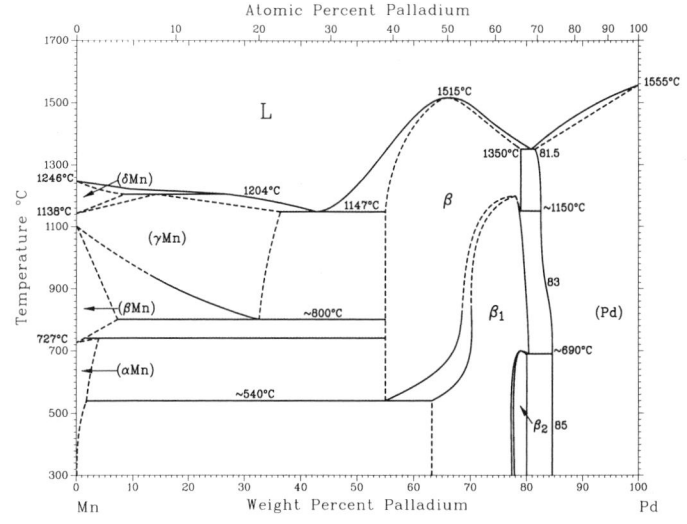

Phase	Composition, wt% Pd	Pearson symbol	Space group
(δMn)	0 to ~9	cI2	$Im\bar{3}m$
(γMn)	0 to ~35	cF4	$Fm\bar{3}m$
(βMn)	0 to ~8	cP20	$P4_132$
(αMn)	0 to ~4	cI58	$I\bar{4}3m$
β(MnPd)	~54 to <79	cP2	$Pm\bar{3}m$
β_1	~63 to <81	...	...
β_2	~77.5 to 80.1	...	...
(Pd)	81.5 to 100	cF4	$Fm\bar{3}m$
Other reported phases			
Mn_2Pd_3(HT)	~74	tP2	P4/mmm
Mn_2Pd_3(LT)	~74	t**	...
Mn_3Pd_5	~76.4	oC16	Cmmm
$Mn_{11}Pd_{21}$	~78.6	tP32	P4/mmm
$MnPd_3$	85	tI16	I4mm

Mn-Pr

H. Okamoto, 1990

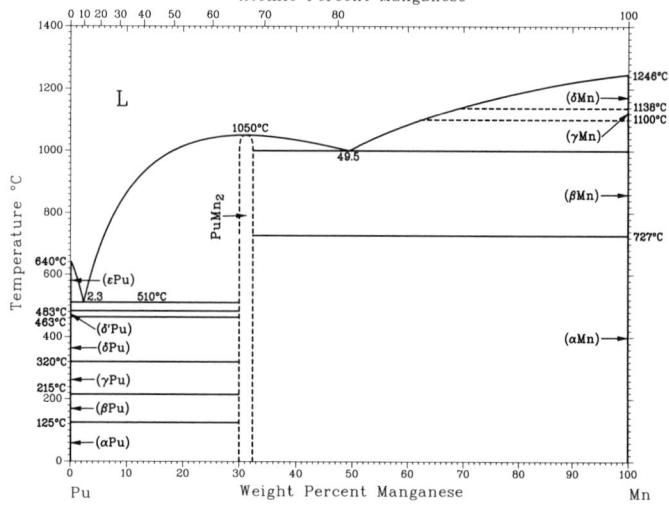

Phase	Composition, wt% Pr	Pearson symbol	Space group
(δMn)	0	cI2	$Im\bar{3}m$
(γMn)	0	cF4	$Fm\bar{3}m$
(βMn)	0	cP20	$P4_132$
(αMn)	0	cI58	$I\bar{4}3m$
Mn$_{23}$Pr$_6$	40.1	cF116	$Fm\bar{3}m$
(βPr)	~96.5 to 100	cI2	$Im\bar{3}m$
(αPr)	? to 100	hP4	$P6_3/mmc$
Metastable phase			
Mn$_2$Pr	56.1	hP12	$P6_3/mmc$

Mn-Pu

S.T. Konobeevsky, 1955

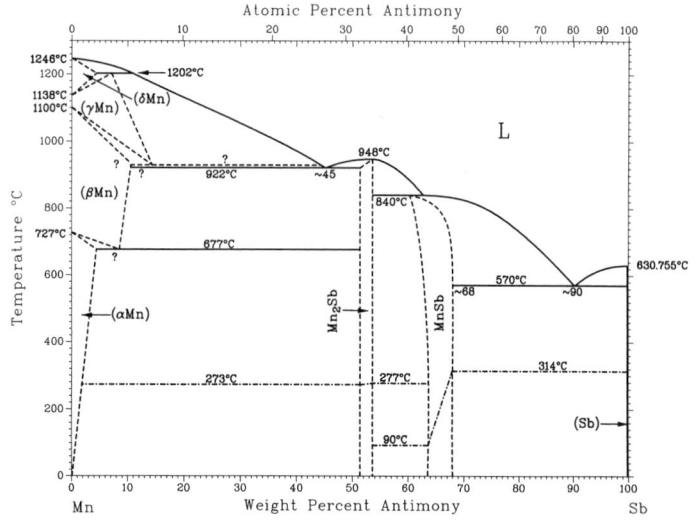

Phase	Composition, wt% Mn	Pearson symbol	Space group
(εPu)	~0	cI2	$Im\bar{3}m$
(δ'Pu)	~0	tI2	$I4/mmm$
(δPu)	~0	cF4	$Fm\bar{3}m$
(γPu)	~0	oF8	$Fddd$
(βPu)	~0	mC34	$C2/m$
(αPu)	~0	mP16	$P2_1/m$
PuMn$_2$	~31.1	cF24	$Fd\bar{3}m$
(δMn)	~100	cI2	$Im\bar{3}m$
(γMn)	~100	cF4	$Fm\bar{3}m$
(βMn)	~100	cP20	$P4_132$
(αMn)	~100	cI58	$I\bar{4}3m$

Mn-Sb

H. Okamoto, 1990

Phase	Composition, wt% Sb	Pearson symbol	Space group
(δMn)	0 to ?	cI2	$Im\bar{3}m$
(γMn)	0 to ?	cF4	$Fm\bar{3}m$
(βMn)	0 to ?	cP20	$P4_132$
(αMn)	0 to ?	cI58	$I\bar{4}3m$
Mn$_2$Sb	~52.5	tP6	$P4/nmm$
MnSb	~61 to ~68	hP4	$P6_3/mmc$
(Sb)	100	hR2	$R\bar{3}m$

Mn-Si

H. Okamoto, 1991

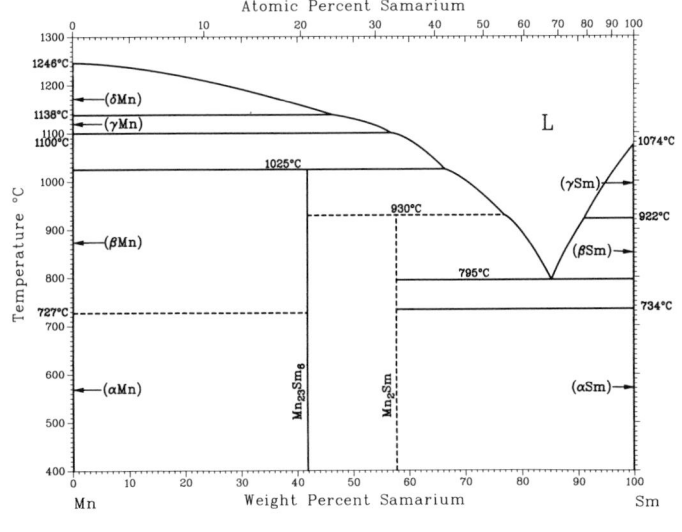

Phase	Composition, wt% Si	Pearson symbol	Space group
(δMn)	0 to ~0.1	cI2	$Im\bar{3}m$
(γMn)	0 to ~1.3	cF4	$Fm\bar{3}m$
(βMn)	0 to ~9.3	cP20	$P4_132$
(αMn)	0 to 3.2	cI58	$I\bar{4}3m$
R	6.5 to 8.72	hR53	$R\bar{3}$
ν	9.0 to 10.55	oI186	$Immm$
βMn₃Si	~14 to 15.0	cF16	$Fm\bar{3}m$
αMn₃Si	14.6 to 15.0		
Mn₅Si₂	17.0	tP56	$P4_12_12$
Mn₅Si₃	23.5	hP16	$P6_3/mcm$
MnSi	33.4 to 34.0	cP8	$P2_13$
Mn₁₁Si₁₉	~46.9	tP120	$P\bar{4}n2$
(Si)	100	cF8	$Fd\bar{3}m$

Mn-Sm

H.R. Kirchmayr and W. Lugscheider, 1970

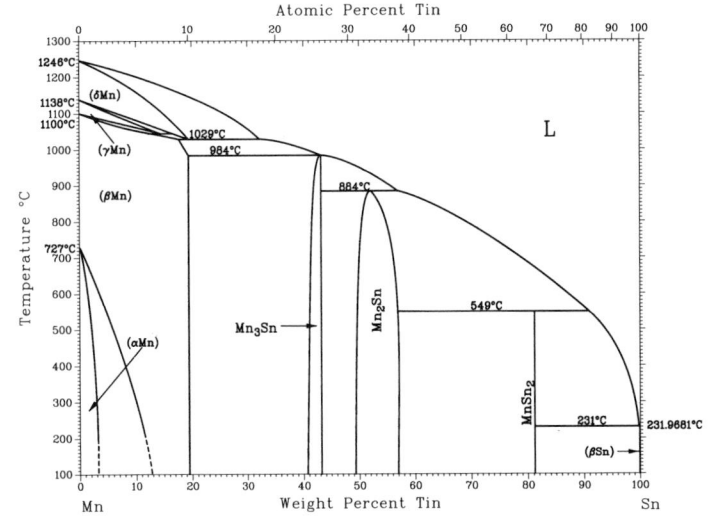

Phase	Composition, wt% Sm	Pearson symbol	Space group
(δMn)	~0	cI2	$Im\bar{3}m$
(γMn)	~0	cF4	$Fm\bar{3}m$
(βMn)	~0	cP20	$P4_132$
(αMn)	~0	cI58	$I\bar{4}3m$
Mn₂₃Sm₆	~41.7	cF116	$Fm\bar{3}m$
Mn₂Sm	57.7	hP12	$P6_3/mmc$
		cF24	$Fd\bar{3}m$
(δSm)	~100	cI2	$Im\bar{3}m$
(βSm)	~100	hP2	$P6_3/mmc$
(αSm)	~100	hR13	$R\bar{3}m$

Mn-Sn

H. Okamoto, 1990

Phase	Composition, wt% Sn	Pearson symbol	Space group
(δMn)	0 to 19	cI2	$Im\bar{3}m$
(γMn)	0 to 14	cF4	$Fm\bar{3}m$
(βMn)	0 to 21	cP20	$P4_132$
(αMn)	0 to 2	cI58	$I\bar{4}3m$
Mn₃Sn	41 to 43	hP8	$P6_3/mmc$
Mn₂Sn	49 to 57	hP6	$P6_3/mmc$
MnSn₂	81.2	tI12	$I4/mcm$
(βSn)	100	tI2	$I4_1/amd$
(αSn)	100	cF8	$Fd\bar{3}m$

Mn-Ti

J.L. Murray, 1987

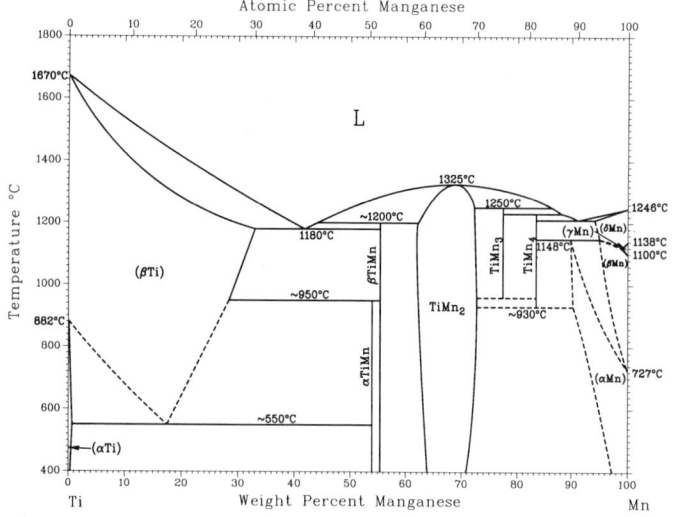

Phase	Composition, wt% Mn	Pearson symbol	Space group
(βTi)	0 to 33	cI2	Im$\bar{3}$m
(αTi)	0 to 0.5	hP2	P6₃/mmc
αTiMn	53.9	t*58	...
βTiMn	55	(a)	...
TiMn₂	63 to 73	hP12	P6₃/mmc
TiMn₃	78	(b)	...
TiMn₄	83.5	hR53	R$\bar{3}$m
(δMn)	92 to 100	cI2	Im$\bar{3}$m
(γMn)	99.5 to 100	cF4	Fm$\bar{3}$m
(βMn)	95 to 100	cP20	P4₁32
(αMn)	89 to 100	cI58	I$\bar{4}$3m
(α'Ti)	(c)	hP2	P6₃/mmc
ω	(c)	hP3	P6/mmm

(a) Undetermined. (b) Orthorhombic. (c) Metastable phase

Mn-U

From [Hansen]

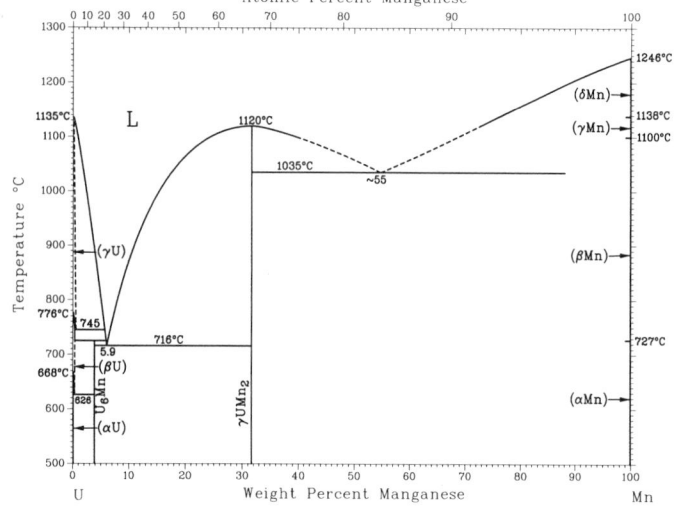

Phase	Composition, wt% Mn	Pearson symbol	Space group
(γU)	0 to ~0.5	cI2	Im$\bar{3}$m
(βU)	0 to ~0.4	tP30	P4₂/mnm
(αU)	~0	oC4	Cmcm
U₆Mn	~3.7	tI28	I4/mcm
γUMn₂	31.6	oI12	Imma
βUMn₂	31.6	cF24	Fd$\bar{3}$m
αUMn₂	31.6	mC24	C2/m
(δMn)	~100	cI2	Im$\bar{3}$m
(γMn)	~100	cF4	Fm$\bar{3}$m
(βMn)	~100	cP20	P4₁32
(αMn)	~100	cI58	I$\bar{4}$3m

Mn-V

H. Okamoto, 1992

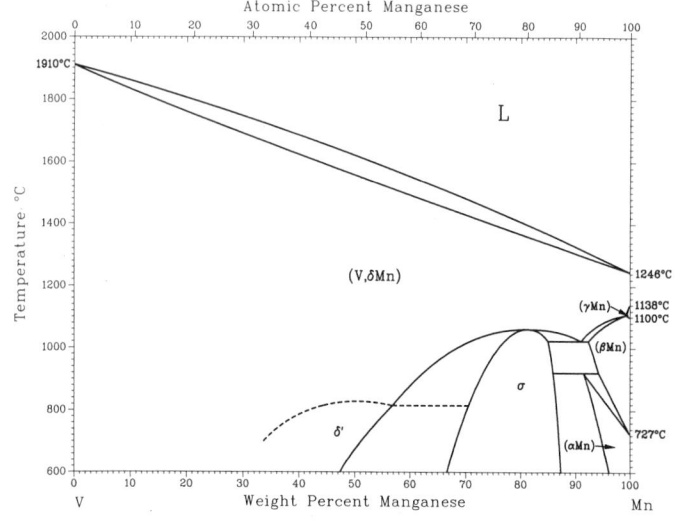

Phase	Composition, wt% Mn	Pearson symbol	Space group
(V, δMn)	0 to 100	cI2	Im$\bar{3}$m
δ'	? to ~57	cP2	Pm$\bar{3}$m
σ	? to ?	tP30	P4₂/mnm
(γMn)	99 to 100	cF4	Fm$\bar{3}$m
(βMn)	93 to 100	cP20	P4₁32
(αMn)	92 to 100	cI58	I$\bar{4}$3m

Mn-Y

A. Palenzona and S. Cirafici, 1991

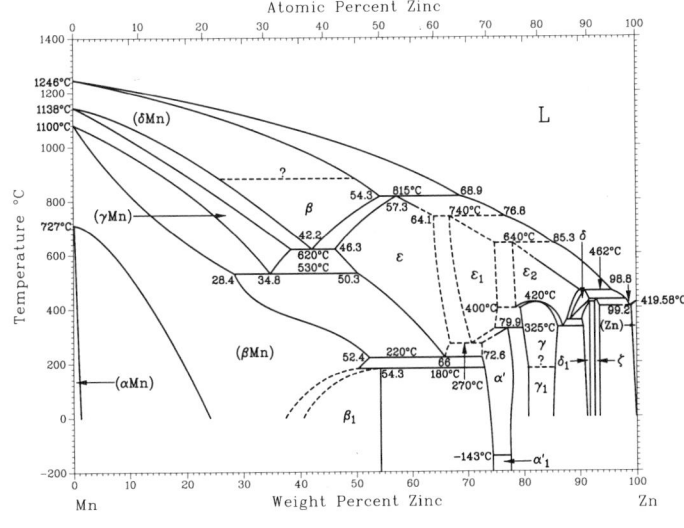

Phase	Composition, wt% Y	Pearson symbol	Space group
(δMn)	0	cI2	Im3̄m
(γMn)	0	cF4	Fm3̄m
(βMn)	0	cP20	P4₁32
(αMn)	0	cI58	I4̄3m
Mn₁₂Y	11.9	tI26	I4/mmm
Mn₂₃Y₆	29.7	cF116	Fm3̄m
Mn₂Y	44.7	cF24	Fd3̄m
(βY)	100	cI2	Im3̄m
(αY)	100	hP2	P6₃/mmc
Other phases			
Mn₂Y(a)	44.7	hP12	P6₃/mmc
Mn₂Y(b)	44.7	...	...

(a) Synthesized under high temperature (1300 °C) and high pressure (40 kbar). (b) Distorted tetragonal Cu₂Mg type obtained below 100 K

Mn-Zn

H. Okamoto and L.E. Tanner, 1990

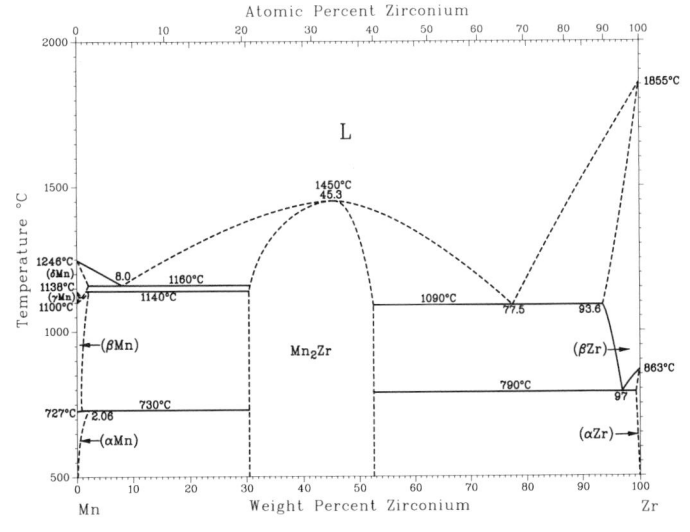

Phase	Composition, wt% Zn	Pearson symbol	Space group
(δMn)	0 to 54.3	cI2	Im3̄m
(γMn)	0 to 38	cF4	Fm3̄m
(γMn₁)	0 to ?	tI2	I4/mmm
(βMn)	0 to 52.4	cP20	P4₁31
(αMn)	0 to 2.0	cI58	I4̄3m
β	? to 54.3	cP2	Pm3̄m
β₁	47.8 to 54.3	cP2	Pm3̄m
ε	46.3 to 67	hP2	P6₃/mmc
ε₁	67 to 75	hP8	P6₃/mmc
ε₂	78 to 90.2	hP*	...
α′	72.6 to 78	cP4	Pm3̄m
α₁′	74 to 78	tP2	P4/mmm
γ	79.9 to 86.6	cI52	...
γ₁	...	cI550 ± 8	...
δ	88.4 to 92.0	hP*	...
δ₁	90.5 to 92.2	hP*	...
ζ(MnZn₁₃)	93.7 to 94.0	mC28	P2/m
(Zn)	99.2 to 100	hP2	P6₃/mmc

Mn-Zr

M. Lasocka, unpublished

Phase	Composition, wt% Zr	Pearson symbol	Space group
(δMn)	0 to 2.06	cI2	Im3̄m
(γMn)	0	cF4	Fm3̄m
(βMn)	0 to ~2	cP20	P4₁32
(αMn)	0 to 2.06	cI58	I4̄3m
Mn₂Zr	30.40 to 53	hP12	P6₃/mmc
(βZr)	93.6 to 100	cI2	Im3̄m
(αZr)	100	hP2	P6₃/mmc

Mo-N

P.R. Subramanian, 1990

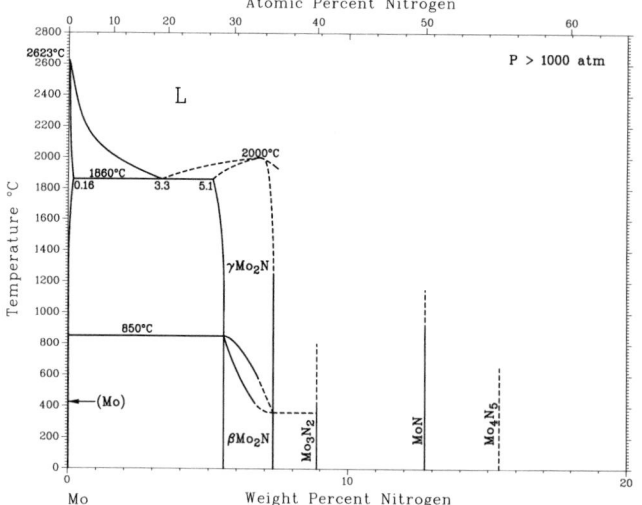

Phase	Composition, wt% N	Pearson symbol	Space group
(Mo)	0 to 0.16	cI2	Im$\bar{3}$m
γMo$_2$N	5.1 to 7	cF8	Fm$\bar{3}$m
βMo$_2$N	5.6 to 7	tI12	I4$_1$/amd
Mo$_3$N$_2$	~9	cP8	Pm$\bar{3}$m
MoN	12.7	hP16	P6$_3$/mmc
Mo$_4$N$_5$	~15.5	hP8	P6$_3$/mmc

Mo-Nb

H. Okamoto, 1991

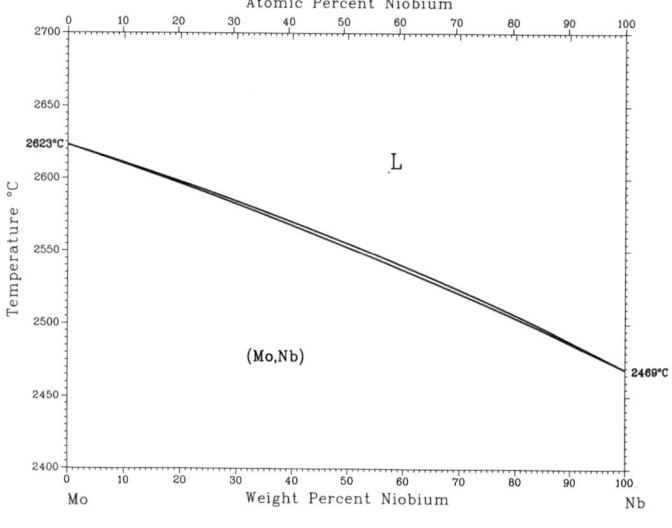

Phase	Composition, wt% Nb	Pearson symbol	Space group
(Mo,Nb)	0 to 100	cI2	Im$\bar{3}$m

Mo-Ni

H. Okamoto, 1991

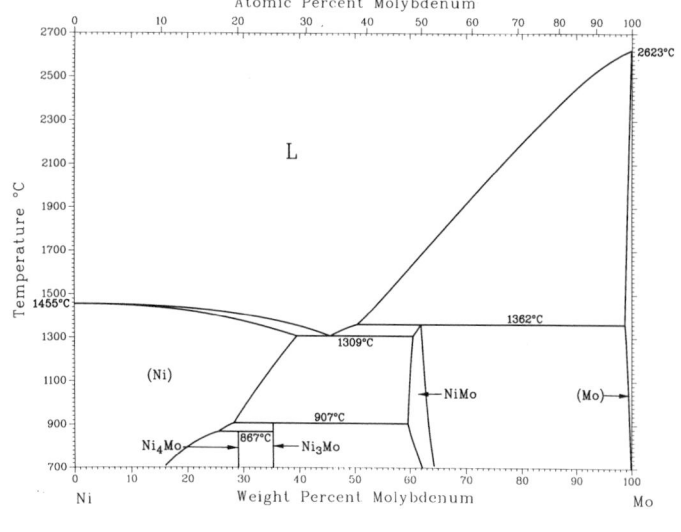

Phase	Composition, wt% Mo	Pearson symbol	Space group
(Ni)	0 to 38(a)	cF4	Fm$\bar{3}$m
Ni$_4$Mo	29.0	tI10	I4/m
Ni$_3$Mo	35.3	oP8	Pmnn
NiMo	63.9 to 65.7	oP112	P2$_1$2$_1$2$_1$
(Mo)	98.9 to 100(b)	cI2	Im$\bar{3}$m
Metastable phases			
Ni$_2$Mo	...	oI6	...
Ni$_3$Mo	...	tI8	I4/mmm
Ni$_4$Mo	...	tI10, cF4	...
Ni$_{17}$Mo$_5$	...	...	...

(a) At 1317 °C. (b) At 1362 °C

Mo-O

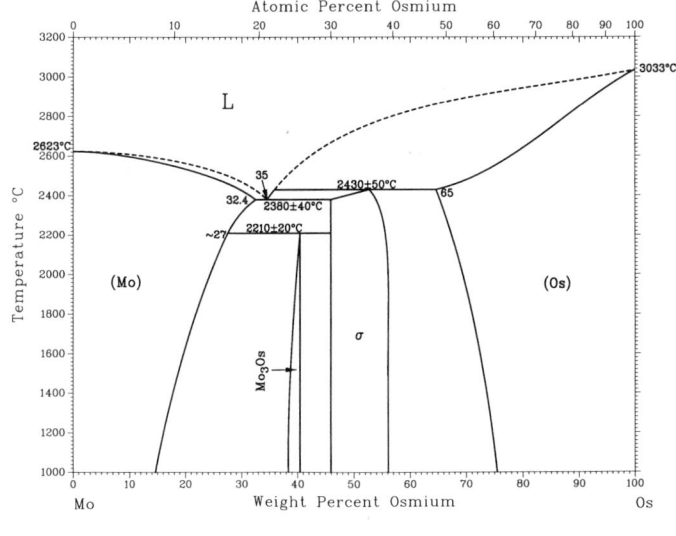

Phase	Composition, wt% O	Pearson symbol	Space group
(Mo)	0	cI2	$Im\bar{3}m$
MoO$_2$	~25.0	mP12	$P2_1/c$(a)
		tP6	$P4_2/mnn$
Mo$_4$O$_{11}$	31.4	oP60	$Pna2_1$
Mo$_8$O$_{23}$	32.4	mP124	Pc
		mP62	$P2/c$
Mo$_9$O$_{24}$(b)	32.5	mC280	$C2/c$
		mP70	$P2/c$
MoO$_3$	33	oP128	$Pba2$

(a) Or $P2_1$. (b) Might be Mo$_9$O$_{26}$

Mo-Os

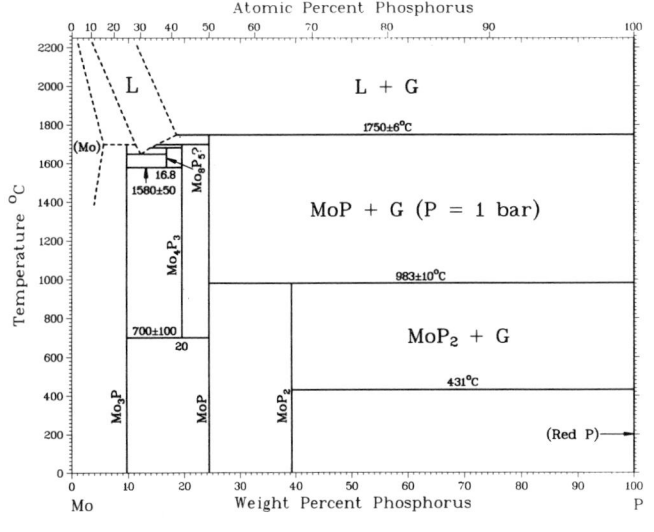

Phase	Composition, wt% Os	Pearson symbol	Space group
(Mo)	0 to 32.4	cI2	$Im\bar{3}m$
Mo$_3$Os	~40	cP8	$Pm\bar{3}n$
σ (Mo$_2$Os)	46 to 56	tP30	$P4_2/mnm$
(Os)	65 to 100	hP2	$P6_3/mmc$

Mo-P

Phase	Composition, wt% P	Pearson symbol	Space group
(Mo)	0 to >5	cI2	$Im\bar{3}m$
Mo$_3$P	10	tI32	$I\bar{4}2m$
Mo$_8$P$_5$	16.8	mP13	Pm
Mo$_4$P$_3$	20	oP56	$Pnma$
MoP	24.4	hP2	$P\bar{6}m2$
MoP$_2$	39.3	oC12	$Cmc2_1$
(P) (red)	~100	...	...
Other reported phase			
Mo$_5$P$_3$	~16.2	h**	...

Mo-Pd

H. Okamoto, 1992

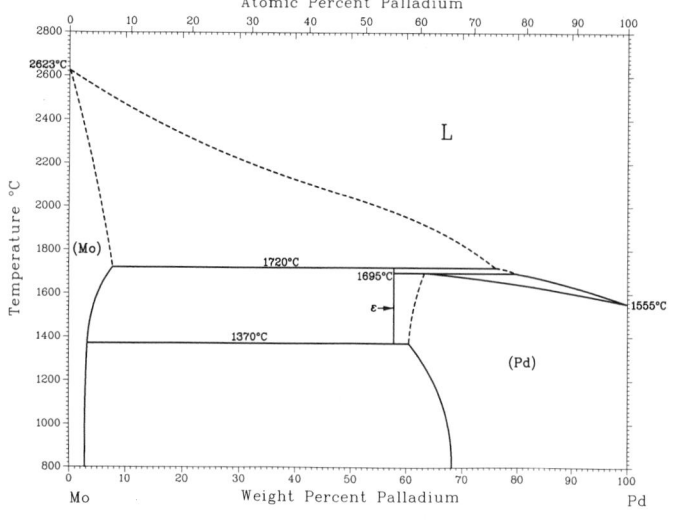

Phase	Composition, wt% Pd	Pearson symbol	Space group
(Mo)	0 to 8	cI2	Im$\bar{3}$m
ε	~58	hP2	P6$_3$/mmc
(Pd)	61 to 100	cF4	Fm$\bar{3}$m

Mo-Pt

L. Brewer and R.H. Lamoreaux, 1980

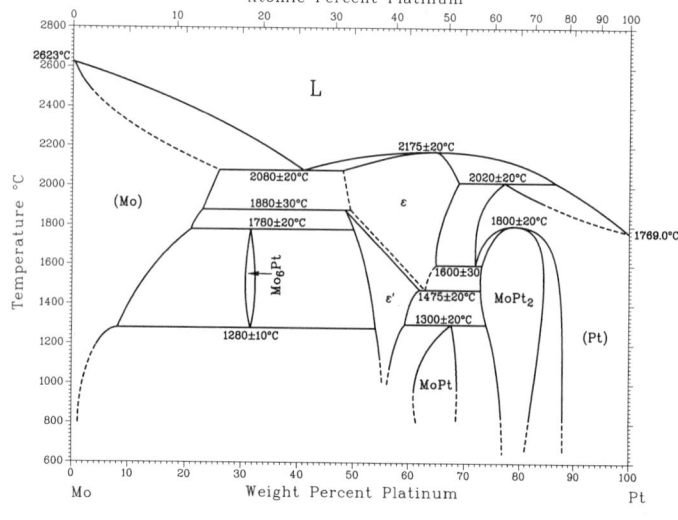

Phase	Composition, wt% Pt	Pearson symbol	Space group
(Mo)	0 to 26 ± 2	cI2	Im$\bar{3}$m
Mo$_6$Pt	31.6 ± 0.7	cP8	Pm$\bar{3}$n
ε?	48 ± 1 to 71 ± 2	hP2	P6$_3$/mmc
ε′	48.4 ± 1 to 62 ± 2	hP8	P6$_3$/mmc
MoPt	61 ± 2 to 70 ± 2	oP4	Pmma
MoPt$_2$	74 ± 2 to 84 ± 1	oI6	Immm
(Pt)	72 ± 2 to 100	cF4	Fm$\bar{3}$m

Mo-Pu

From [Molybdenum]

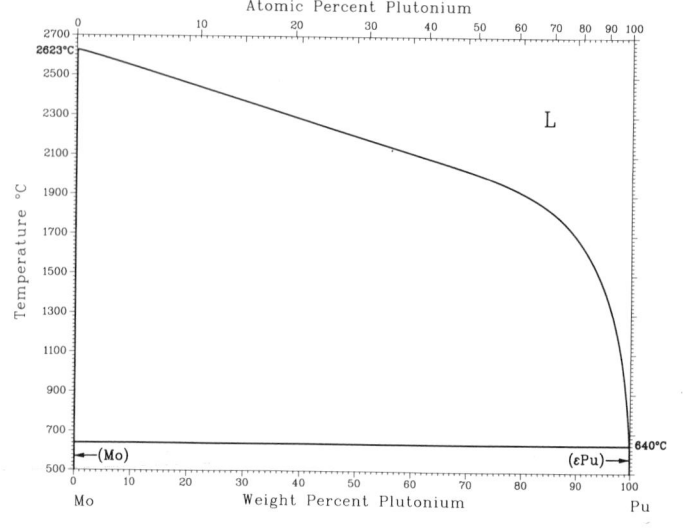

Phase	Composition, wt% Pu	Pearson symbol	Space group
(Mo)	0	cI2	Im$\bar{3}$m
(εPu)	100	cI2	Im$\bar{3}$m

Mo-Rh

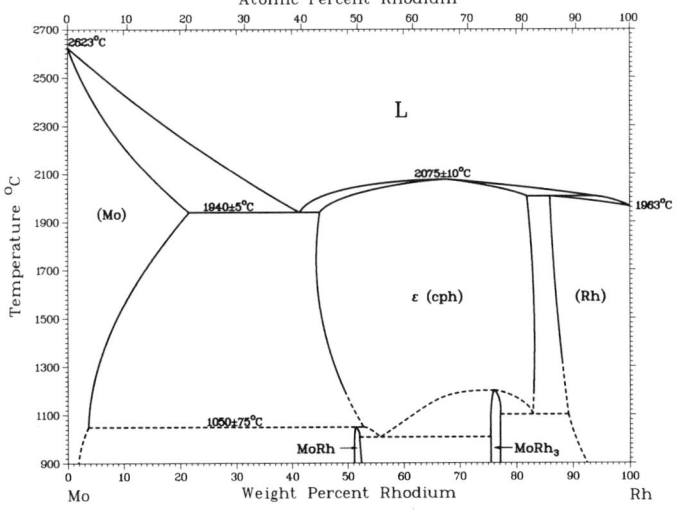

Phase	Composition, wt% Rh	Pearson symbol	Space group
(Mo)	0 to 21	cI2	Im$\bar{3}$m
MoRh	~51.8	oP4	Pmma
ε	~44 to 83	hP2	P6$_3$/mmc
MoRh$_3$	~76	...	...
(Rh)	86 to 100	cF4	Fm$\bar{3}$m

Mo-Ru

H. Okamoto, 1990

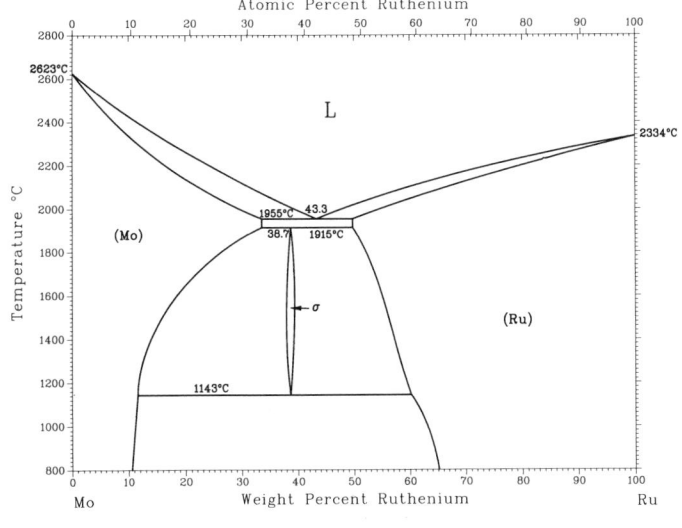

Phase	Composition, wt% Ru	Pearson symbol	Space group
(Mo)	0 to 33.6	cI2	Im$\bar{3}$m
σ	37.9 to 40.7	tP30	P4$_2$/mnm
(Ru)	49.8 to 100	hP2	P6$_3$/mmc

Mo-S

L. Brewer and R.H. Lamoreaux, 1980

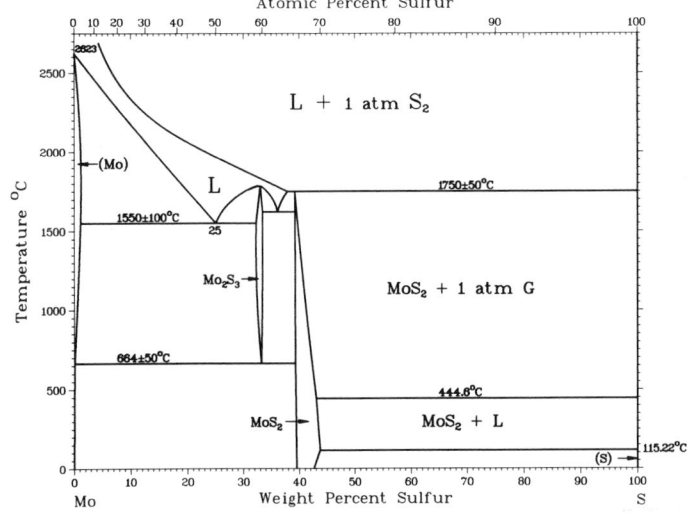

Phase	Composition, wt% S	Pearson symbol	Space group
(Mo)	0 to 1	cI2	Im$\bar{3}$m
Mo$_2$S$_3$	~33	mP10	P2$_1$/m
MoS$_2$	39 to 44	hP6	P6$_3$/mmc
		hR3	R$\bar{3}$m
(βS)	100	mP*	P2$_1$/c
(αS)(a)	100	oF128	Fddd

(a) Below 95.5 °C

Mo-Si

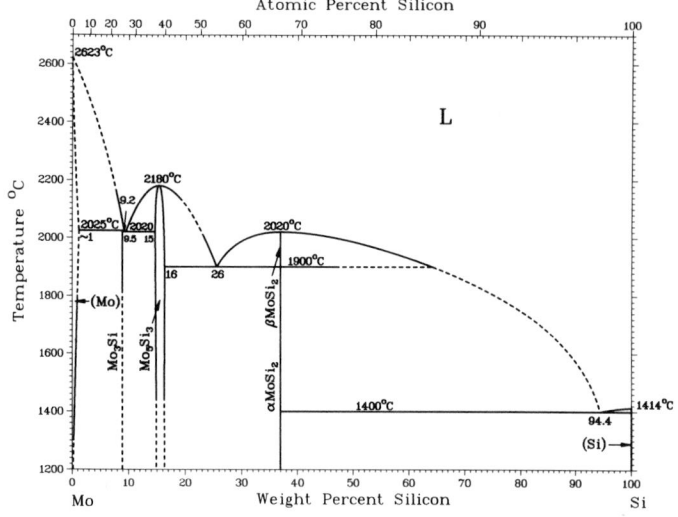

A.B. Gokhale and G.J. Abbaschian, 1991

Phase	Composition, wt% Si	Pearson symbol	Space group
(Mo)	0 to ~1	cI2	Im$\bar{3}$m
Mo₃Si	9	cP8	Pm$\bar{3}$n
Mo₅Si₃	~14.9	tI38	I4/mcm
βMoSi₂	37.0	...	C6₂2
αMoSi₂	37.0	tI6	I4/mmm
(Si)	100	cF8	Fd$\bar{3}$m

Mo-Ta

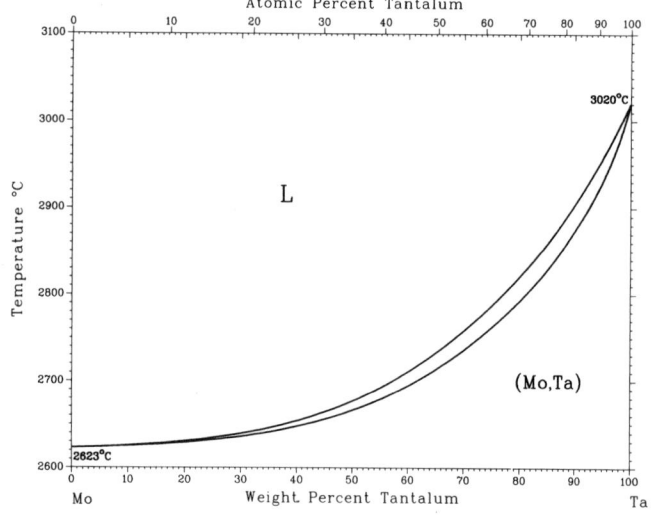

R. Krishnan, S.P. Garg, and N. Krishnamurthy, 1986

Phase	Composition, wt% Ta	Pearson symbol	Space group
(Mo,Ta)	0 to 100	cI2	Im$\bar{3}$m

Mo-Ti

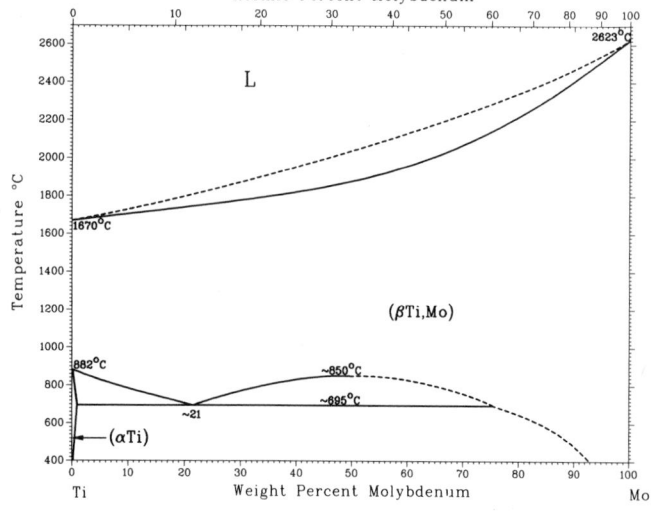

J.L. Murray, 1987

Phase	Composition, wt% Mo	Pearson symbol	Space group
(βTi,Mo)	0 to 100	cI2	Im$\bar{3}$m
(αTi)	0 to 0.8	hP2	P6₃/mmc
α'	(a)	hP2	P6₃/mmc
α''	(a)	oC4	Cmcm
ω	(a)	hP3	P6/mmm

(a) Metastable

Mo-U

H. Okamoto, 1990

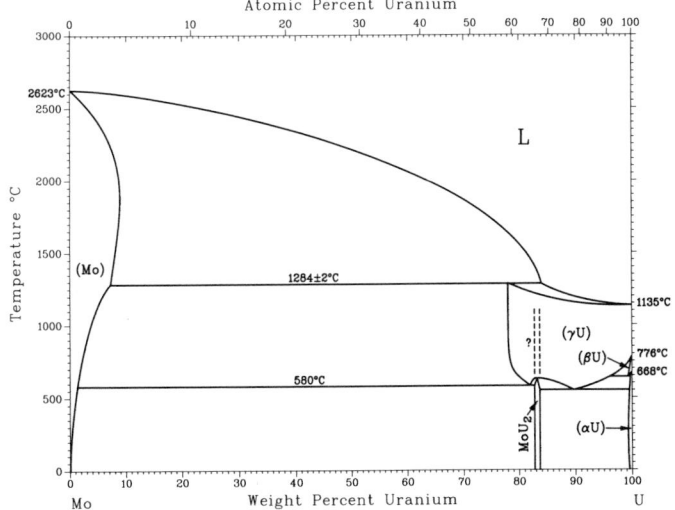

Phase	Composition, wt% U	Pearson symbol	Space group
(Mo)	0 to 9	cI2	$Im\bar{3}m$
MoU$_2$	83.2	tI6	I4/mmm
(γU)	98 to 100	cI2	$Im\bar{3}m$
(βU)	99 to 100	tP30	$P4_2/mnm$
(αU)	99 to 100	oC4	Cmcm

Mo-V

J.F. Smith, 1989

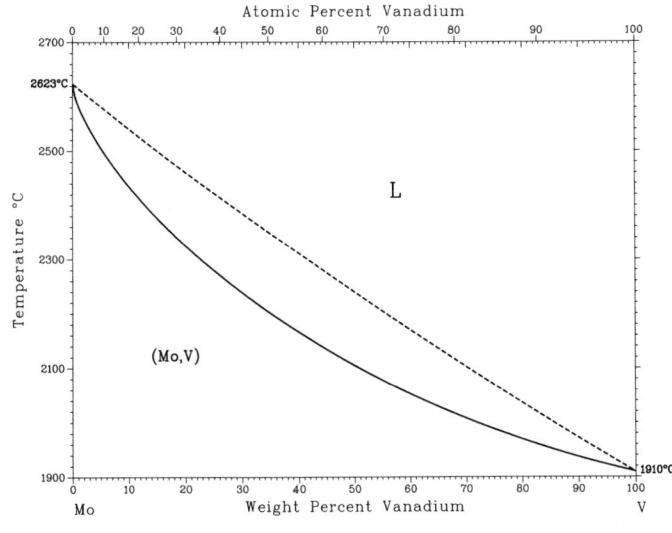

Phase	Composition, wt% V	Pearson symbol	Space group
(Mo,V)	0 to 100	cI2	$Im\bar{3}m$

Mo-W

S.V. Nagender Naidu, A.M. Sriramamurthy, and P. Rama Rao, 1984

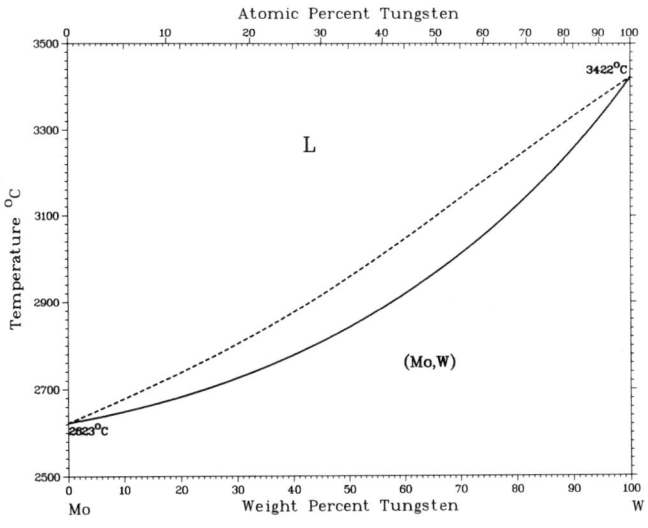

Phase	Composition, wt% W	Pearson symbol	Space group
(Mo,W)	0 to 100	cI2	$Im\bar{3}m$

Mo-Zr

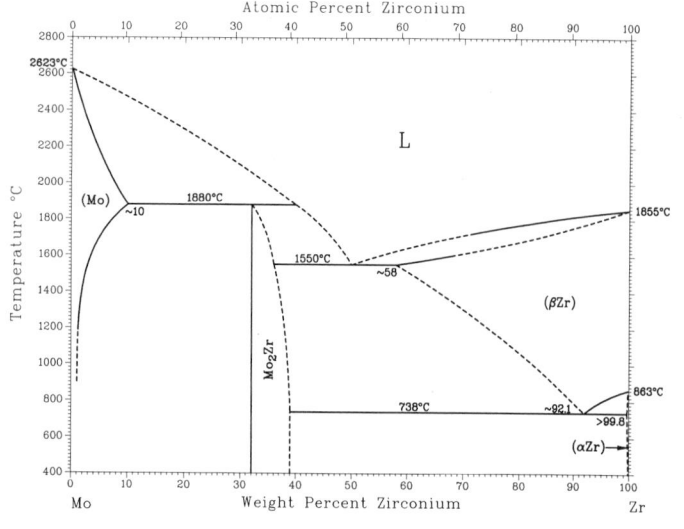

Phase	Composition, wt% Zr	Pearson symbol	Space group
(Mo)	0 to ~10	cI2	$Im\bar{3}m$
Mo$_2$Zr	32 to 39	cF24	$Fd\bar{3}m$
(βZr)	~58 to 100	cI2	$Im\bar{3}m$
(αZr)	~100	hP2	$P6_3/mmc$

N-Nb

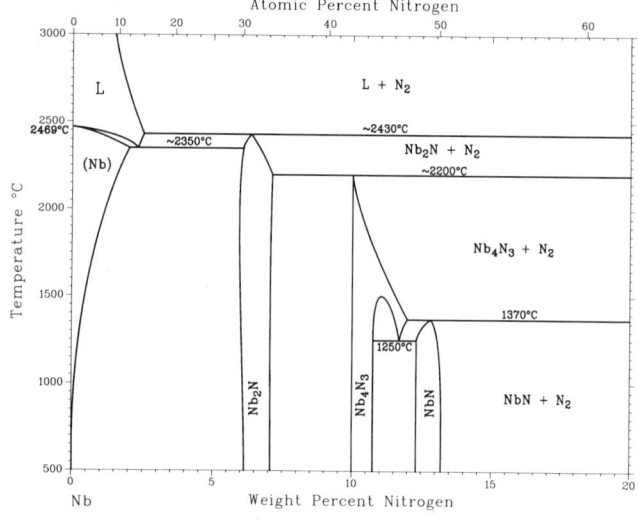

Phase	Composition, wt% N	Pearson symbol	Space group
(Nb)	0 to <3	cI2	$Im\bar{3}m$
Nb$_2$N	~5.9 to 7	hP9	$P\bar{3}1m$
Nb$_4$N$_3$	~10.2	tI14	$I4/mmm$
NbN	~13.1	hP8	$P6_3/mmc$
Other reported phases			
Nb$_3$N	5	tP58	$P4/m$
Nb$_{10}$N$_9$	12.0	hP2	$P\bar{6}m2$
NbN	13.1	hP4	$P6_3/mmc$
Nb$_5$N$_6$	15.3	hP22	$P6_3/mcm$
Nb$_4$N$_5$	15.9	tI18	$I4/m$

N-Ni

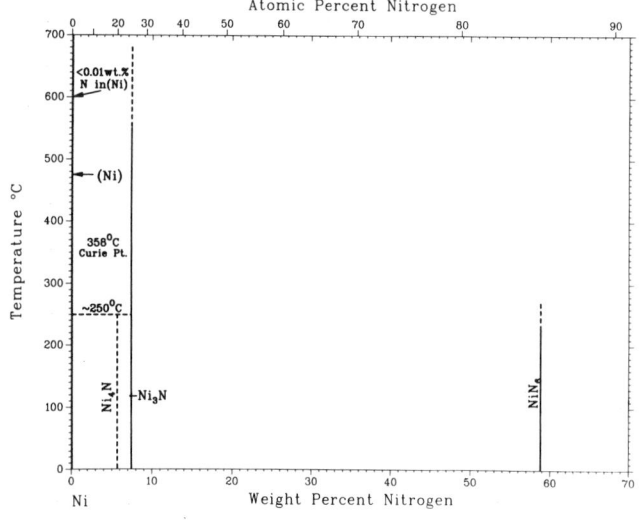

Phase	Composition, wt% N	Pearson symbol	Space group
Stable phases			
(Ni)(a)	~0	cF4	$Fm\bar{3}m$
Ni$_3$N	7	hP*	$P6_322$ or $P312$
Ni(N$_3$)$_2$	58.9	...	...
Other phases			
Ni$_4$N,I	6	c**	...
Ni$_4$N,II	6	t**	...
Ni$_2$N	10.6	t**	...
Ni$_3$N$_2$(b)	14	...	...

(a) At 25 °C. (b) Existence questionable

N-Ta

J. Gatterer, D. Dufek, P. Ettmayer, and R. Kieffer, 1975

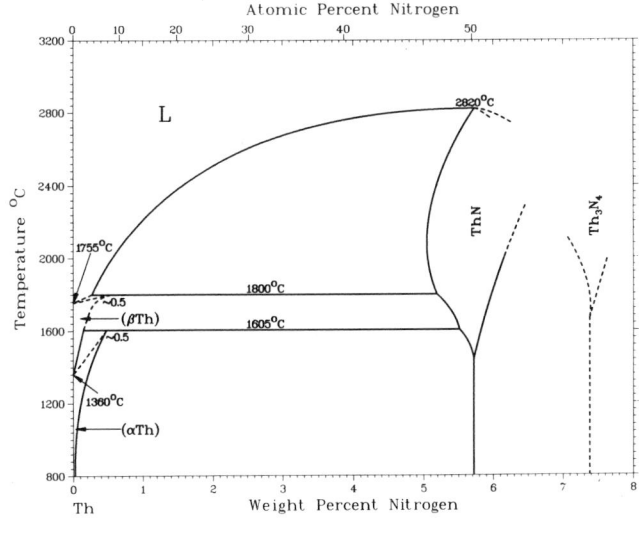

Phase	Composition, wt% N	Pearson symbol	Space group
(Ta)	0 to 1.5	cI2	$Im\bar{3}m$
Ta₂N	2.1 to 4.0	hP3	$P6_3/mmc$
δ	~4.9 to 7.2	cF8	$Fm\bar{3}m$
TaN	7.2	c**	...
Other reported phases			
Ta₉N₂	~1.7	c**	...
Ta₄(HT?)	~1.9	o**	...
Ta₂N	~3.7	hP9	$P\bar{3}1m$
TaN	7.2	hP8	$P6_3/mmc$
Ta₅N₆	~8.5	hP22	$P6_3/mcm$
Ta₄N₅	~8.8	tI18	$I4/m$
Ta₃N₅	~11.4	t**	...
		oC32	$Cmcm$
		mC32	$C2/m$

N-Th

H. Okamoto, 1990

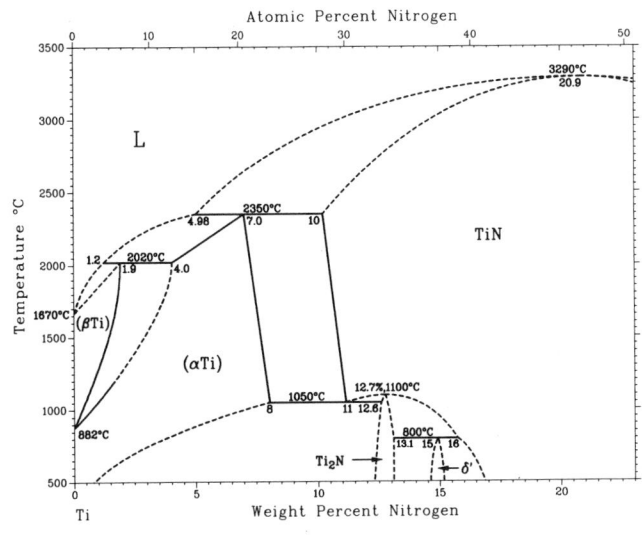

Phase	Composition, wt% N	Pearson symbol	Space group
(βTh)	0	cI2	$Im\bar{3}m$
(αTh)	0	cF4	$Fm\bar{3}m$
ThN	~5.7	cF8	$Fm\bar{3}m$
Th₃N₄	~7.4	mC4	Cm
		o*18	...
		hR7	$R\bar{3}m$
Th₂N₃	8.3	hP5	$P\bar{3}m1$

N-Ti

H.A. Wriedt and J.L. Murray, 1987

Phase	Composition, wt% N	Pearson symbol	Space group
Stable phases			
(αTi)	0 to 8	hP2	$P6_3/mmc$
(βTi)	0 to 1.9	cI2	$Im\bar{3}m$
Ti₂N	~13	tP6	$P4_2/mnm$
TiN	10 to >22.6	cF8	$Fm\bar{3}m$
δ'	~15	tI12	$I4_1/amd$
ω	~0	h**	...
Metastable phase			
α'	...	tP6	$P4_2/mnm$

N-U

From [Metals]

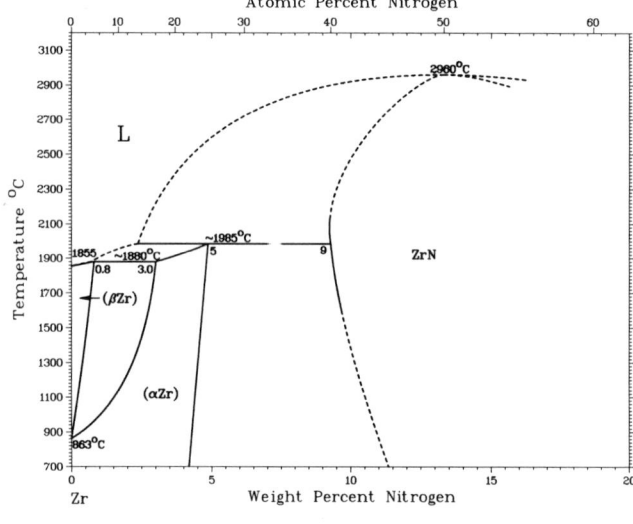

Phase	Composition, wt% N	Pearson symbol	Space group
(γU)	~0	cI2	Im$\bar{3}$m
(βU)	~0	tP30	P4$_2$/mnm
(αU)	~0	oC4	Cmcm
UN	~4.4 to 5.6	cF8	Fm$\bar{3}$m
βU$_2$N$_3$	~7 to 7.5	hP5	P$\bar{3}$m1
αU$_2$N$_3$	~8 to 8.4	cI80	Ia$\bar{3}$
Other reported phases			
U$_4$N$_7$	~9.3	hR*	...
UN$_2$	~10.5	cF12	Fm$\bar{3}$m

N-Zr

From [Zirconium]

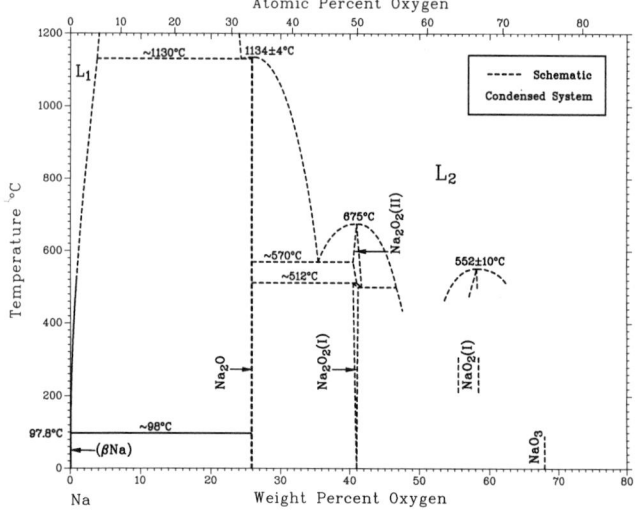

Phase	Composition, wt% N	Pearson symbol	Space group
(βZr)	0 to 0.7	cI2	Im$\bar{3}$m
(αZr)	0 to 5	hP2	P6$_3$/mmc
ZrN	9 to ?	cF8	Fm$\bar{3}$m

Na-O

H.A. Wriedt, 1987

Phase	Composition, wt% O	Pearson symbol	Space group
(βNa)	0	cI2	Im$\bar{3}$m
(αNa)	0	hP2	P6$_3$/mmc
Na$_2$O	25.8	cF12	Fm$\bar{3}$m
Na$_2$O$_2$-II	41.0	...	...
Na$_2$O$_2$-I	41.0	hP9	P$\bar{6}$2m(a)
NaO$_2$ (I)	58.2	cF8	Fm$\bar{3}$m
NaO$_2$ (II)	58.2	cP12	Pa$\bar{3}$
NaO$_2$ (III)	58.2	oP6	Pnnm
NaO$_3$	68	bct	I4/mmm
Other phase			
Na$_2$O$_2$-Q(b)	41.0	...	...

(a) Might be C$\bar{6}$2m. (b) Noncubic

Na-Pb

From [Metals]

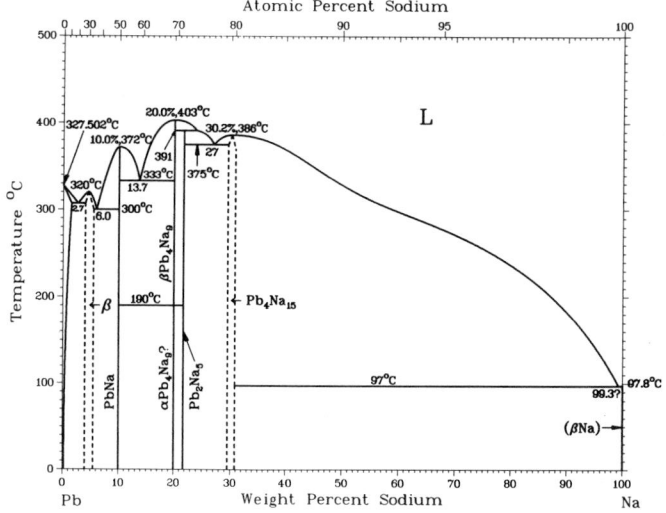

Phase	Composition, wt% Na	Pearson symbol	Space group
(Pb)	0 to 2.7	cF4	$Fm\overline{3}m$
β(Pb₃Na)	>4 to >5	cP4	$Pm\overline{3}m$
PbNa	10.0	tI64	$I4_1/acd$
Pb₄Na₉	~20.0	hP26	$P6_3/mmc$
Pb₂Na₅	~21.7	hR7	$R\overline{3}m$
Pb₄Na₁₅	~29 to 31	cI76	$I\overline{4}3d$
(βNa)	~100	cI2	$Im\overline{3}m$
Other reported phases			
Pb₅Na₁₃	~22.4	hP36	$P6_3/mmc$
PbNa₅	~36.5	hP*	...

Na-Rb

C.W. Bale, 1982

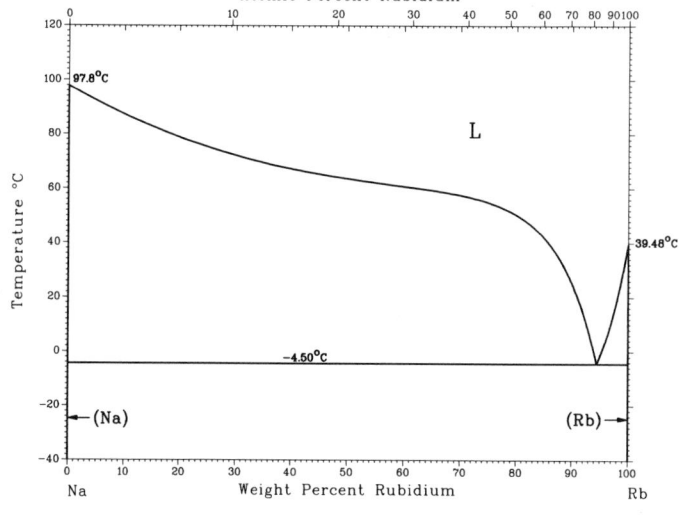

Phase	Composition, wt% Rb	Pearson symbol	Space group
(Na)	0	cI2	$Im\overline{3}m$
(Rb)	100	cI2	$Im\overline{3}m$

Na-S

H. Okamoto, 1990

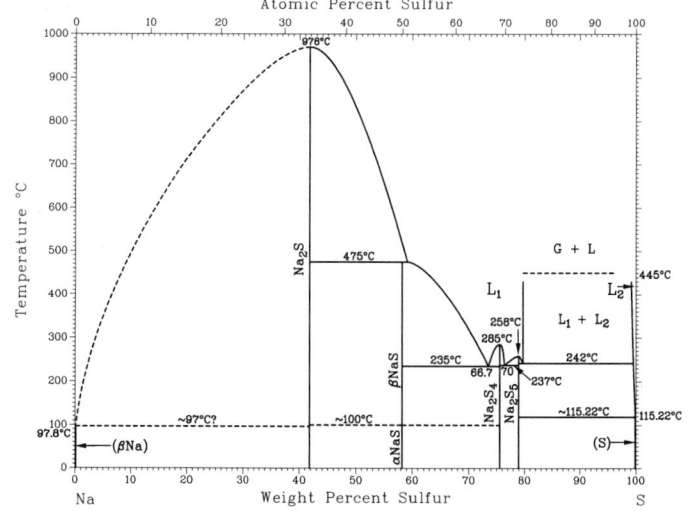

Phase	Composition, wt% S	Pearson symbol	Space group
(βNa)	0	cI2	$Im\overline{3}m$
Na₂S	41.0	cF12	$Fm\overline{3}m$
βNaS	58.2	hP8	$P6_3/mmc$
αNaS	58.2	hP12	$P\overline{6}2m$
Na₂S₄	73.6	tI48	$I\overline{4}2d$
Na₂S₅	~78	oP28	$Pnma$
(S)	0	mP64	$P2_1/c$

Na-Sb

C.H. Mathewson, 1906

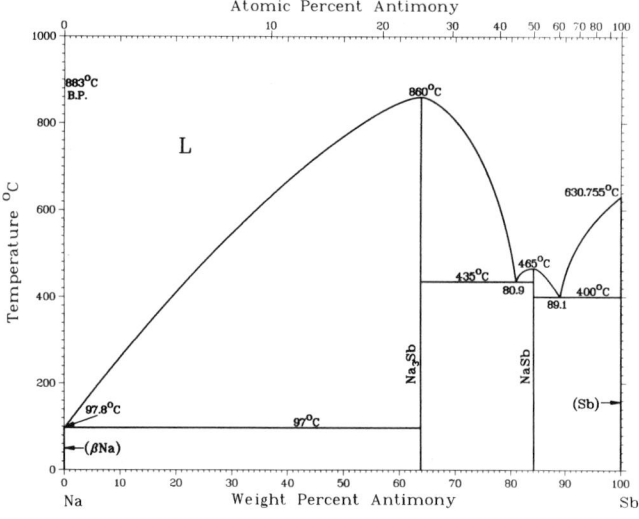

Phase	Composition, wt% Sb	Pearson symbol	Space group
(βNa)	~0	cI2	$Im\bar{3}m$
Na₃Sb	64	hP8	$P6_3/mmc$
NaSb	84.1	mP16	$P2_1/c$
(Sb)	~100	hR2	$R\bar{3}m$

Na-Se

H. Okamoto, 1990

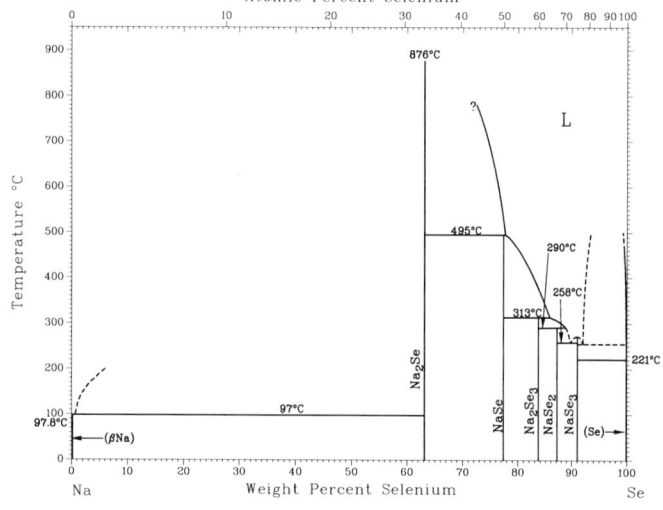

Phase	Composition, wt% Se	Pearson symbol	Space group
(βNa)	0	cI2	$Im\bar{3}m$
Na₂Se	63.2	cF12	$Fm\bar{3}m$
NaSe	77.4	hP8	$P6_3/mmc$
Na₂Se₃	84	...	...
NaSe₂	87.3	...	...
NaSe₃	91	...	...
(Se)	100	hP3	$P3_121$

Na-Sn

H. Okamoto, 1990

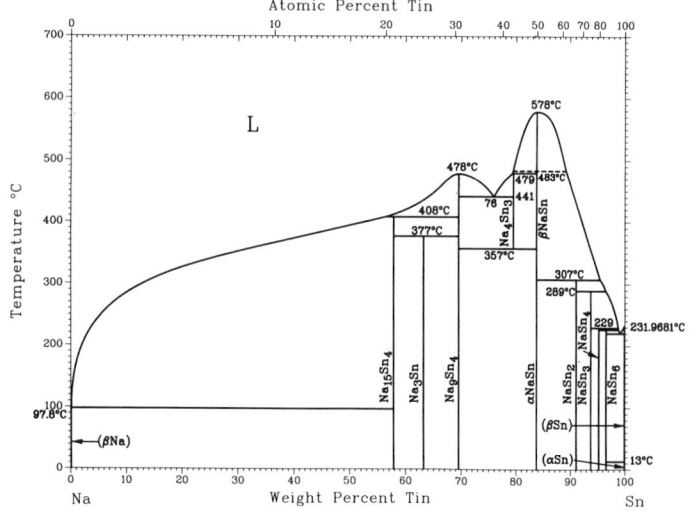

Phase	Composition, wt% Sn	Pearson symbol	Space group
(βNa)	0	cI2	$Im\bar{3}m$
Na₁₅Sn₄	58	cI76	$I\bar{4}3d$
		oP40	$Pnma$
Na₃Sn	63	...	...
Na₉Sn₄	69.7	oC52	$Cmcm$
Na₄Sn₃	79.5	...	...
βNaSn	83.8	...	...
αNaSn	83.8	tI64	$I4_1/acd$
NaSn₂	91.2	...	...
NaSn₃	94	...	...
NaSn₄	95	...	...
NaSn₆	96.9	...	...
(βSn)	100	tI2	$I4_1/amd$
(αSn)	100	cF8	$Fd\bar{3}m$

Na-Sr

A.D. Pelton, 1985

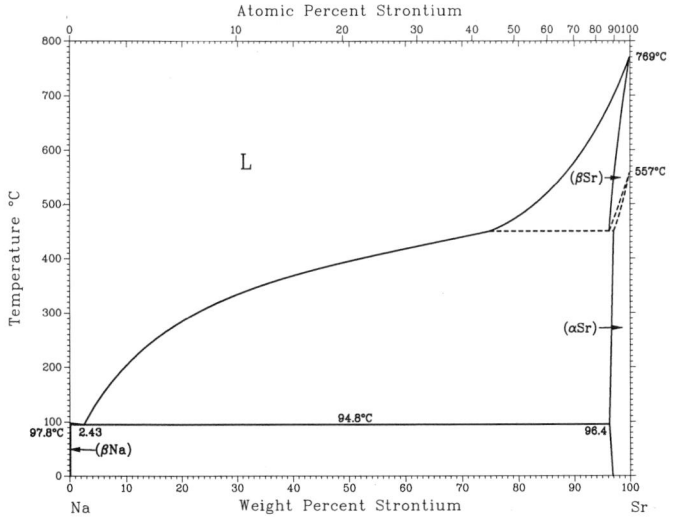

Phase	Composition, wt% Sr	Pearson symbol	Space group
(βNa)	0	cI2	$Im\bar{3}m$
(αNa)	0	hP2	$P6_3/mmc$
(βSr)	97.2 to 100	cI2	$Im\bar{3}m$
(αSr)	96.4 to 100	cF4	$Fm\bar{3}m$

Na-Te

A.D. Pelton and A. Petric, 1990

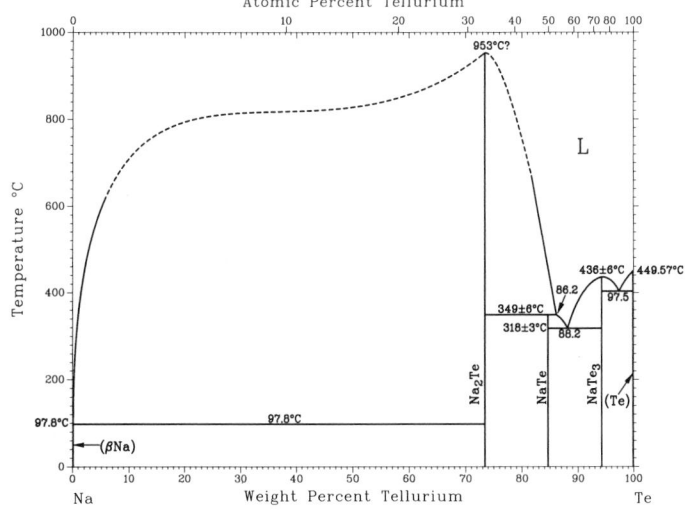

Phase	Composition, wt% Te	Pearson symbol	Space group
(βNa)	0	cI2	$Im\bar{3}m$
(αNa)	0	hP2	$P6_3/mmc$
Na₂Te	73.5	cF12	$Fm\bar{3}m$
NaTe	84.7	...	...
NaTe₃	94	...	...
(Te)	100	hP3	$P3_121$

Na-Tl

G. Grube and A. Schmidt, 1936

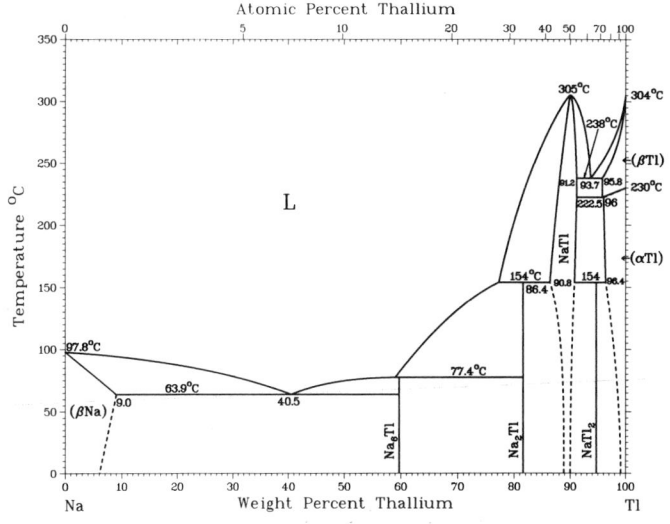

Phase	Composition, wt% Tl	Pearson symbol	Space group
(βNa)	0 to 9.0	cI2	$Im\bar{3}m$
Na₆Tl	~59.7	cF400	$F\bar{4}3m$
Na₂Tl	81.6	oC48	$C222_1$
NaTl	86.4 to 91.2	cF16	$Fd\bar{3}m$
NaTl₂	94.7	...	...
(βTl)	95.8 to 100	cI2	$Im\bar{3}m$
(αTl)	96 to 100	hP2	$P6_3/mmc$

Nb-Ni

H. Okamoto, 1992

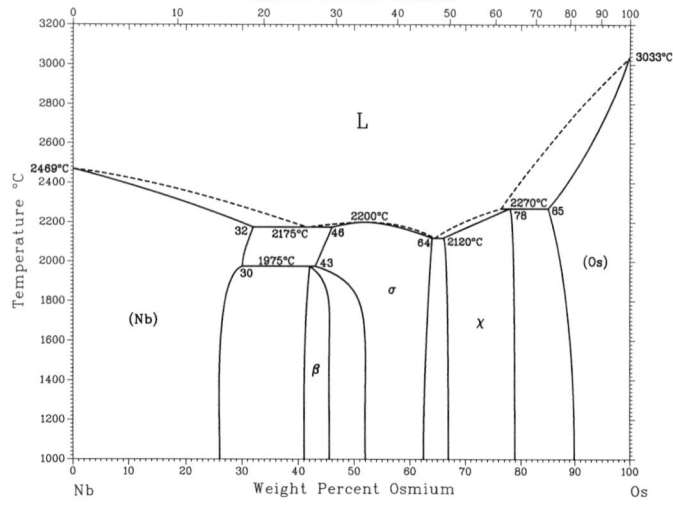

Phase	Composition, wt% Nb	Pearson symbol	Space group
(Ni)	0 to 18.2	$cF4$	$Fm\bar{3}m$
Ni_8Nb	16.5	$tI36$	...
Ni_3Nb	33.1 to 38.0	$oP8$	$Pmmn$
Ni_6Nb_7	60.9 to 65.5	$hR13$	$R\bar{3}m$
(Nb)	97 to 100	$cI2$	$Im\bar{3}m$

Nb-Os

R.M. Waterstrat and R.C. Manuszewski, 1977

Phase	Composition, wt% Os	Pearson symbol	Space group
(Nb)	0 to 32	$cI2$	$Im\bar{3}m$
β	>41 to ~46	$cP8$	$Pm\bar{3}n$
σ	43 to 64	$tP30$	$P4_2/mnm$
χ	66 to 78	$cI58$	$I\bar{4}3m$
(Os)	85 to 100	$hP2$	$P6_3/mmc$

Nb-Pd

M.S. Chandrasekharaiah, 1988

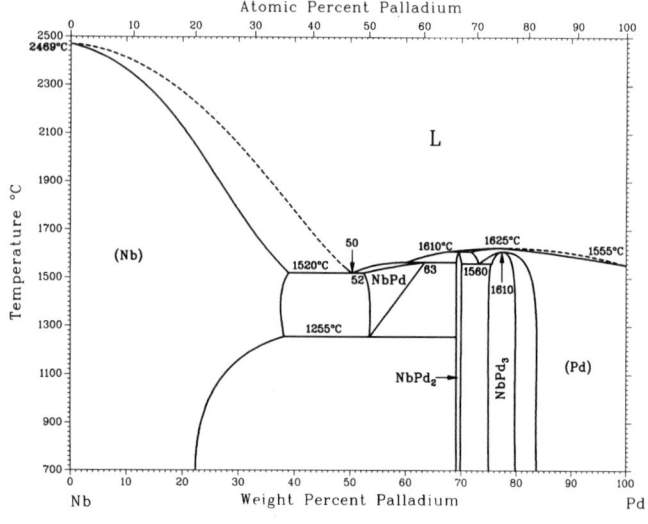

Phase	Composition, wt% Pd	Pearson symbol	Space group
(Nb)	0 to 39	$cI2$	$Im\bar{3}m$
NbPd(a)	52 to 63	$cF4$	$Fm\bar{3}m$
$NbPd_2$	69.2 to 70.1	$oI14$	$Immm$
$\alpha NbPd_3$	78(b)	$tI8$	$I4/mmm$
$\beta NbPd_3$	76 to 78	...	$Pmmn$
(Pd)	73 to 100	$cF4$	$Fm\bar{3}m$

(a) Data from rapidly quenched samples. (b) At 1300 °C

Nb-Pt

H. Okamoto, 1990

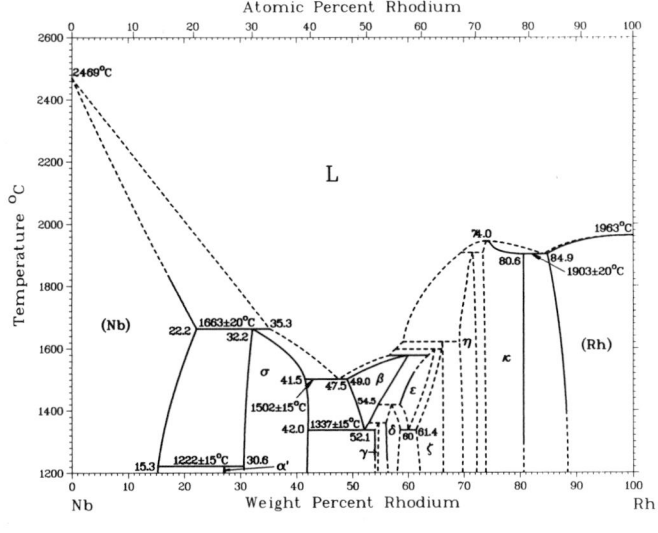

Phase	Composition, wt% Pt	Pearson symbol	Space group
(Nb)	0 to ~22	cI2	Im$\bar{3}$m
Nb$_3$Pt	~33 to ~45	cP8	Pm$\bar{3}$n
Nb$_2$Pt	~49 to ~56	tP30	P4$_2$/mnm
Nb$_{1-x}$Pt$_{1+x}$	69 to 70	oP4	Pmma
α'Pt	~74	...	...
NbPt$_2$	~81	oI6	Immm
βNbPt$_3$	~87	mP48	P2$_1$/m
αNbPt$_3$	~87	oP8	Pmmn
(Pt)	~89 to 100	cF4	Fm$\bar{3}$m

Nb-Rh

D.L. Ritter, B.C. Giessen, and N.J. Grant, 1964

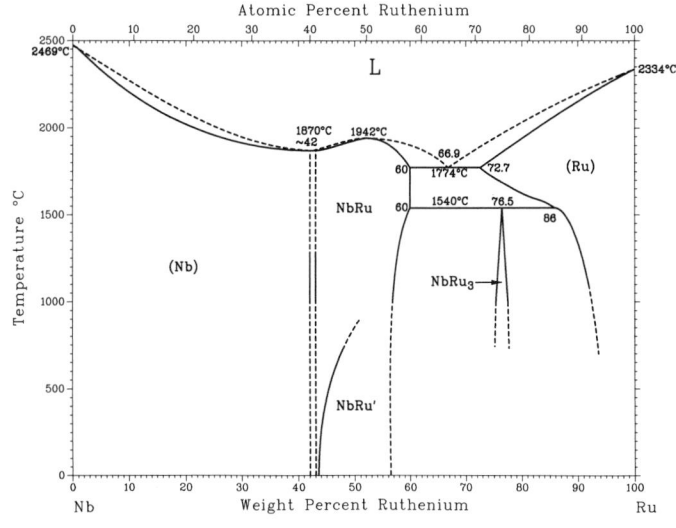

Phase	Composition, wt% Rh	Pearson symbol	Space group
(Nb)	0 to 22.2	cI2	Im$\bar{3}$m
α'(Nb$_3$Rh)	27	cP8	Pm$\bar{3}$n
σ(Nb$_{13}$Rh$_7$)	30.6 to 42.0	tP30	P4$_2$/mnm
β	49.0 to ~61	...	...
γ	~54.0 to 55	tP2	P4/mmm
δ	~56.0 to 59	o**	...
ε(Nb$_2$Rh$_3$)	~59 to 64	oP4	Pmma
ζ(Nb$_2$Rh$_3$)	61 to ~66	mP18	P2/m
η(Nb$_{13}$Rh$_{27}$)	~69 to 72	hP24	P$\bar{6}$m2
κ(NbRh$_3$)	~73 to 80.6	cP4	Pm$\bar{3}$m
(Rh)	84.9 to 100	cF4	Fm$\bar{3}$m
Other reported phases			
NbRh	~52.6	tP2	P4/mmm
		oP4	Pnma
Nb$_9$Rh$_{11}$	58	oP12	Pnma

Nb-Ru

H. Okamoto, 1990

Phase	Composition, wt% Ru	Pearson symbol	Space group
(Nb)	0	cI2	Im$\bar{3}$m
NbRu	43 to 60	cP2	Pm$\bar{3}$m
NbRu'	?	tP2	P4/mmm
NbRu$_3$	76.5	cP4	Pm$\bar{3}$m
(Ru)	72.7 to 100	hP2	P6$_3$/mmc

Nb-Si

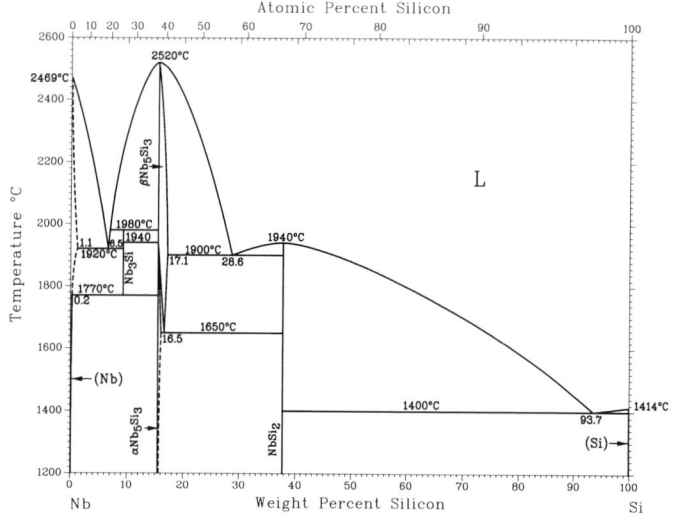

H. Okamoto, A.B. Gokhale, and G.J. Abbaschian, unpublished

Phase	Composition, wt% Si	Pearson symbol	Space group
(Nb)	0 to 1.1	cI2	$Im\bar{3}m$
Nb$_3$Si	9	tP32	$P4_2/n$
βNb$_5$Si$_3$	15.4 to 17.1	tI32	$I4/mcm$
αNb$_5$Si$_3$	15.4 to 15.9	tI32	$I4/mcm$
NbSi$_2$	37.7	hP9	$P6_422$
(Si)	100	cF8	$Fd\bar{3}m$
Metastable phases			
Nb$_7$Si	2.9 to 4.3	c**	...
Nb$_3$Si·m	3.2 to 7.9	cP8	$Pm\bar{3}n$
Nb$_3$Si·m'	3.2 to 10.1	cF4	$Fm\bar{3}m$
Nb$_3$Si·m''	9.2	cF4	$Pm\bar{3}m$
γNb$_5$Si$_3$	15.4	hP16	$P6_3/mcm$
High-pressure phase			
Nb$_3$Si-I	9.2	t**	...

Nb-Ta

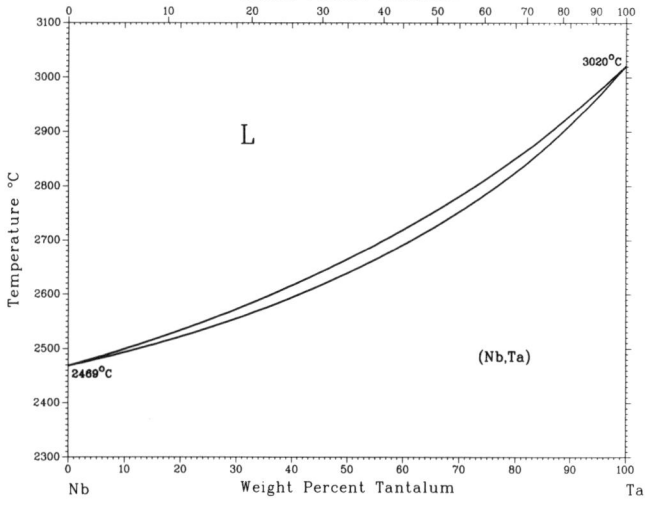

R. Krishnan, S.P. Garg, and N. Krishnamurthy, 1982

Phase	Composition, wt% Ta	Pearson symbol	Space group
(Nb,Ta)	0 to 100	cI2	$Im\bar{3}m$

Nb-Th

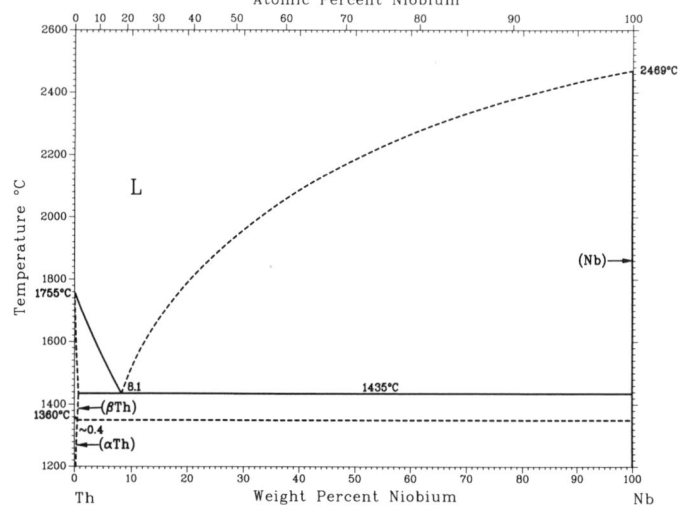

O.N. Carlson, J.M. Dickerson. H.E. Lunt, and H.A. Wilhelm, 1956

Phase	Composition, wt% Nb	Pearson symbol	Space group
(βTh)	0 to ~0.6	cI2	$Im\bar{3}m$
(αTh)	0 to ~0.4	cF4	$Fm\bar{3}m$
(Nb)	100	cI2	$Im\bar{3}m$

Nb-Ti

J.L. Murray, 1987

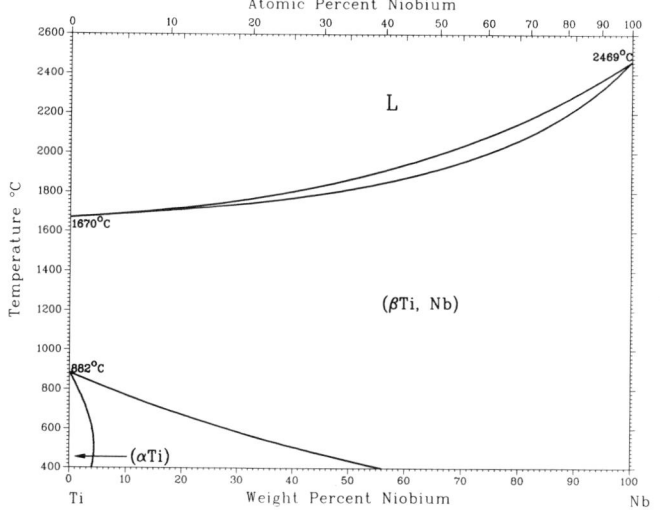

Phase	Composition, wt% Nb	Pearson symbol	Space group
(βTi,Nb)	0 to 100	cI2	Im$\bar{3}$m
(αTi)	0 to 4.7	hP2	P6₃/mmc
Metastable phases			
(α′Ti)	0 to ~9	hP2	P6₃/mmc
(α″Ti)	~14 to 43	oC4	Cmcm
ω	16 to 45	hP3	P6/mmm
τ	26 to 41	(a)	...

(a) bct

Nb-U

H. Okamoto, 1990

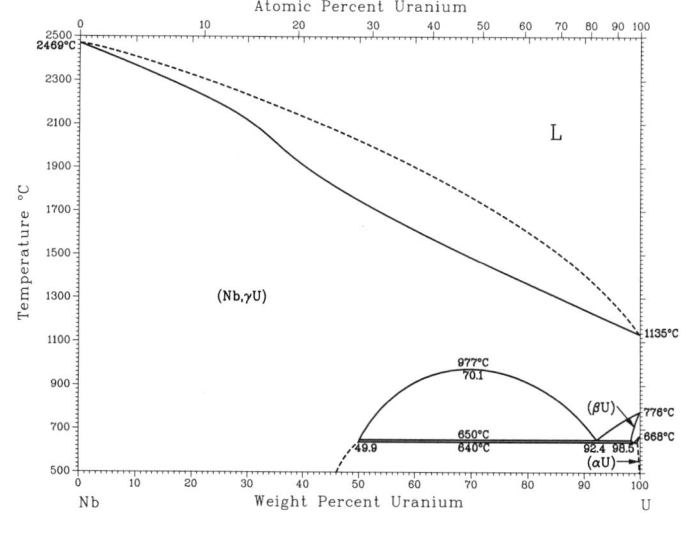

Phase	Composition, wt% U	Pearson symbol	Space group
(Nb,γU)	0 to 100	cI2	Im$\bar{3}$m
(βU)	98.5 to 100	cF4	Fm$\bar{3}$m
(αU)	~100	hP2	P6₃/mmc

Nb-V

J.F. Smith and O.N. Carlson, 1989

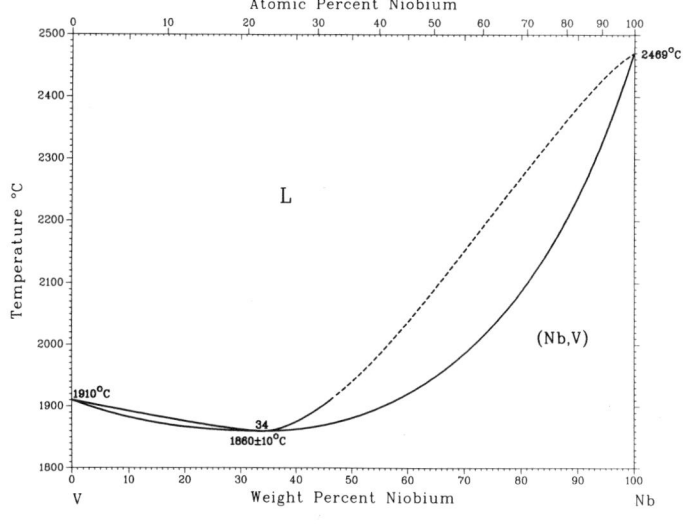

Phase	Composition, wt% Nb	Pearson symbol	Space group
(V,Nb)	0 to 100	cI2	Im$\bar{3}$m

Nb-W

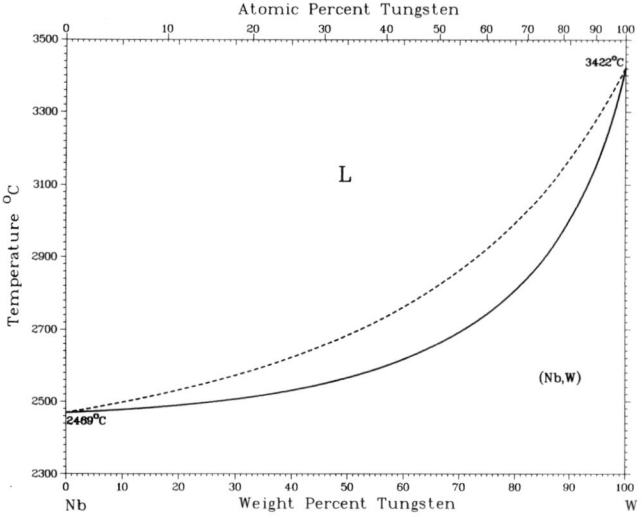

S.V. Nagender Naidu, A.M. Sriramamurthy, and P. Rama Rao, 1988

Phase	Composition, wt% W	Pearson symbol	Space group
(Nb,W)	0 to 100	cI2	Im$\bar{3}$m

Nb-Zr

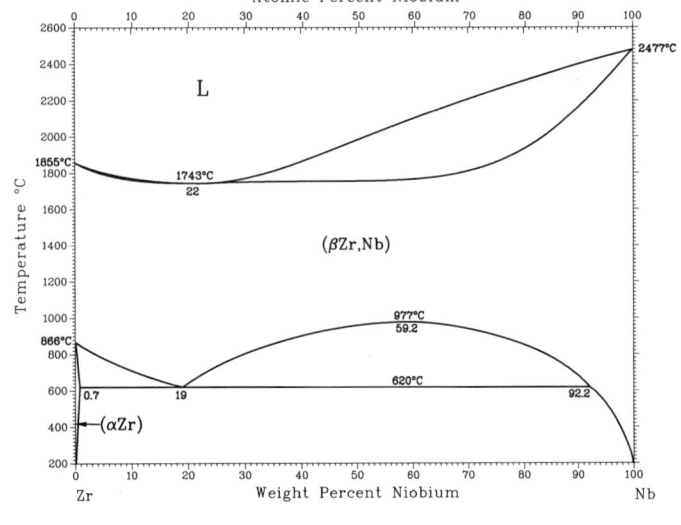

H. Okamoto, 1992

Phase	Composition, wt% Nb	Pearson symbol	Space group
(βZr,Nb)	0 to 100	cI2	Im$\bar{3}$m
(αZr)	0 to 0.7	hP2	P6$_3$/mmc
Metastable phase			
ω	…	hP3	(a)

(a) Changes from P6/mmm to P$\bar{3}$m1 with increasing Nb content

Nd-Ni

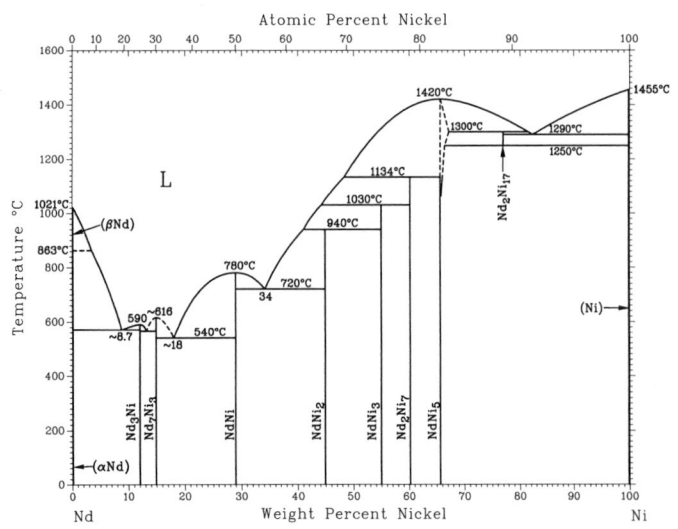

H. Okamoto, 1992

Phase	Composition, wt% Ni	Pearson symbol	Space group
(βNd)	0	cI2	Im$\bar{3}$m
(αNd)	0	hP4	P6$_3$/mmc
Nd$_3$Ni	11.9	oP16	Pnma
Nd$_7$Ni$_3$	14.8	hP20	P6$_3$mc
NdNi	28.9	oC8	Cmcm
NdNi$_2$	44.9	cF24	Fd$\bar{3}$m
NdNi$_3$	55.0	hR12	R$\bar{3}$m
Nd$_2$Ni$_7$	58.8	hP36	P6$_3$/mmc
		hR18	R$\bar{3}$m
NdNi$_5$	67.0	hP6	P6/mmm
Nd$_2$Ni$_{17}$	77.6	hP38	P6$_3$/mmc
(Ni)	100	cF4	Fm$\bar{3}$m

Nd-Pt

H. Okamoto, 1990

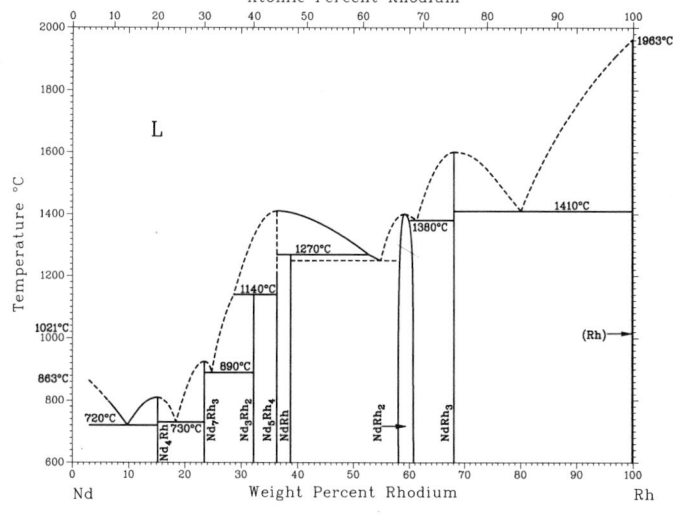

Phase	Composition, wt% Pt	Pearson symbol	Space group
(βNd)	0 to ~4	cI2	Im$\overline{3}$m
(αNd)	0 to ~1.3	hP4	P6$_3$/mmc
Nd$_7$Pt$_3$	37	hP20	P6$_3$mc
Nd$_3$Pt$_2$	47	hR15	R$\overline{3}$
βNdPt	57.5	oC8	Cmcm
αNdPt	57.5	oP8	Pnma
Nd$_3$Pt$_4$	64.3	hR14	R$\overline{3}$
NdPt$_2$	73.0 to 80	cF24	Fd$\overline{3}$m
NdPt$_5$	87.1	hP6	P6/mmm
(Pt)	~100	cF4	Fm$\overline{3}$m

Nd-Rh

H. Okamoto, 1990

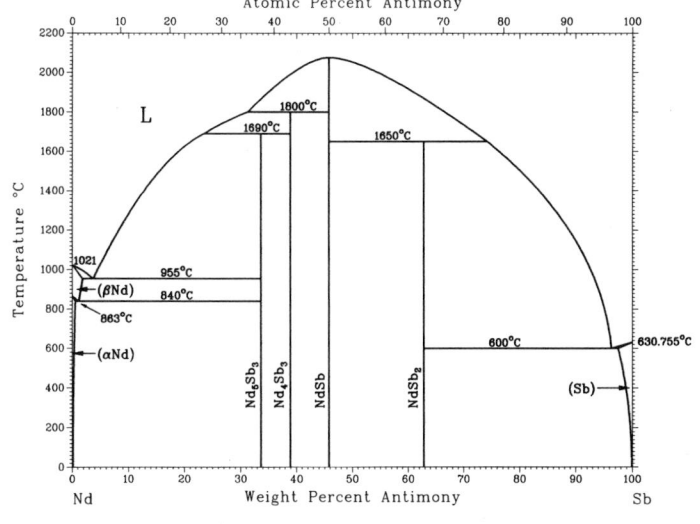

Phase	Composition, wt% Rh	Pearson symbol	Space group
(βNd)	0	cI2	Im$\overline{3}$m
(αNd)	0	hP4	P6$_3$/mmc
Nd$_4$Rh	15	oP16	Pnma
Nd$_7$Rh$_3$	23	hP20	P6$_3$mc
βNd$_3$Rh$_2$	32	...	...
αNd$_3$Rh$_2$	32	hR15	R$\overline{3}$
Nd$_5$Rh$_4$	36.3	oP36	Pnma
NdRh	39	oC8	Cmcm
NdRh$_2$	58 to 60.8	cF24	Fd$\overline{3}$m
NdRh$_3$	68	hP24	P6$_3$/mmc
(Rh)	100	cF4	Fm$\overline{3}$m

Nd-Sb

H. Okamoto, 1990

Phase	Composition, wt% Sb	Pearson symbol	Space group
(βNd)	0 to 1.7	cI2	Im$\overline{3}$m
(αNd)	0 to 0.8	hP4	P6$_3$/mmc
Nd$_5$Sb$_3$	33.6	hP16	P6$_3$/mcm
Nd$_4$Sb$_3$	38.8	cI28	I$\overline{4}$3d
NdSb	45.8	cF8	Fm$\overline{3}$m
NdSb$_2$	62.8	oC24	Cmca
(Sb)	97.6 to 100	hR2	R$\overline{3}$m

Nd-Si

A.B. Gokhale, A. Munitz, and G.J. Abbaschian, 1989

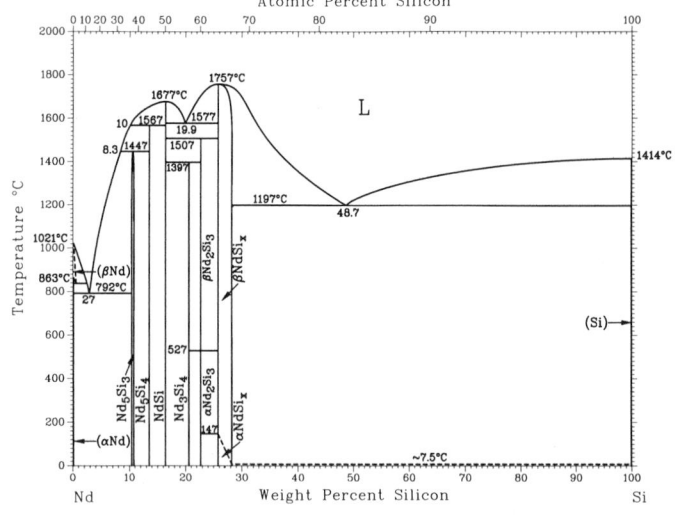

Phase	Composition, wt% Si	Pearson symbol	Space group
(βNd)	0	cI2	Im$\bar{3}$m
(αNd)	0	hP4	P6$_3$/mmc
Nd$_5$Si$_3$	~10.3 to ~10.7	tI32	I4/mcm
Nd$_5$Si$_4$	13.48	...	P4$_1$2$_1$2
NdSi	16.3	oP8	Pnma
Nd$_3$Si$_4$	21	...	...
βNd$_2$Si$_3$	23	...	...
αNd$_2$Si$_3$	22.6	hP3	P6/mmm
βNdSi$_x$	28.14	tI12	I4$_1$/amd
αNdSi$_x$	25.7 to 28.14	...	Imma
(Si)	100	cF8	Fd$\bar{3}$m

Nd-Sn

H. Okamoto, 1990

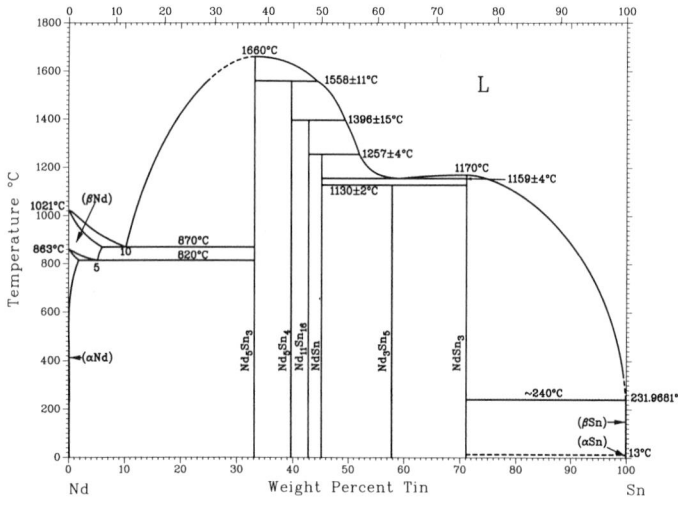

Phase	Composition, wt% Sn	Pearson symbol	Space group
(βNd)	0 to 6	cI2	Im$\bar{3}$m
(αNd)	0 to 2	hP4	P6$_3$/mmc
Nd$_5$Sn$_3$	33.1	hP16	P6$_3$/mcm
Nd$_5$Sn$_4$	39.7	oP36	Pnma
Nd$_{11}$Sn$_{10}$	42.8	tI84	I4/mmm
NdSn	45.1	...	...
Nd$_3$Sn$_5$	57.8	...	...
NdSn$_3$	71	cP4	Pm$\bar{3}$m
(βSn)	100	tI4	I4$_1$/amd
(αSn)	100	cF8	Fd$\bar{3}$m

Nd-Te

H. Okamoto, 1990

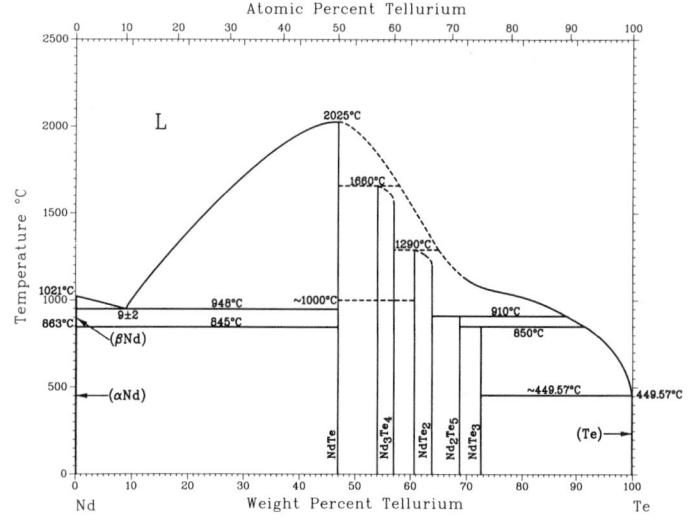

Phase	Composition, wt% Te	Pearson symbol	Space group
(βNd)	0	cI2	Im$\bar{3}$m
(αNd)	0	hP4	P6$_3$/mmc
NdTe	46.9	cF8	Fm$\bar{3}$m
Nd$_3$Te$_4$(a)	54 to 57?	cI28	I$\bar{4}$3d
Nd$_2$Te$_3$(a)	57	oP20	Pnma
NdTe$_2$	60.7 to 63.9	tP6	P4/nmm
Nd$_2$Te$_5$	68.8	oC28	Cmcm
NdTe$_3$	73	oP16	Cmcm
(Te)	100	hP3	P3$_1$21

(a) The phase relationships between Nd$_3$Te$_4$ and Nd$_2$Te$_3$, and the homogeneity range of each, are unknown.

Nd-Ti

J.L. Murray, 1987

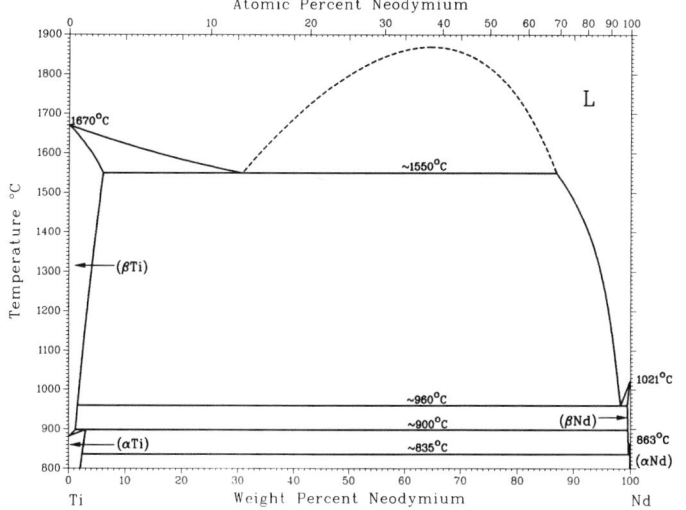

Phase	Composition, wt% Nd	Pearson symbol	Space group
(βTi)	0 to ~9	cI2	$Im\bar{3}m$
(αTi)	0 to ~3	hP2	$P6_3/mmc$
(βNd)	? to 100	cI2	$Im\bar{3}m$
(αNd)	~100	hP2	$P6_3/mmc$

Nd-Tl

S. Delfino, A. Saccone, A. Palenzona, and R. Ferro, unpublished

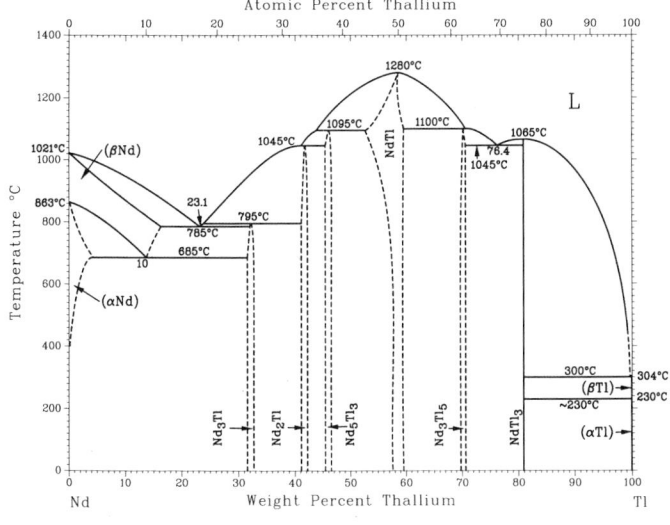

Phase	Composition, wt% Tl	Pearson symbol	Space group
(βNd)	0 to ~16	cI2	$Im\bar{3}m$
(αNd)	0 to ~4	hP4	$P6_3/mmc$
Nd₃Tl(a)	~31.5 to 32.7	cP4	$Pm\bar{3}m$
	~32.1	cF4	$Fm\bar{3}m$
Nd₂Tl	~41 to ~42	hP6	$P6_3/mmc$
Nd₅Tl₃	~45 to ~47	tI32	$I4/mcm$
NdTl(b)	~53 to ~60	cP2	$Pm\bar{3}m$
	(or cI2)		$Im\bar{3}m$
NdTl(c)	~53 to ~60	tP2	$P4/mmm$
Nd₃Tl₅	~70 to ~71	oC32	$Cmcm$
NdTl₃	81	cP4	$Pm\bar{3}m$
(βTl)	100	cI2	$Im\bar{3}m$
(αTl)	100	hP2	$P6_3/mmc$

(a) A cP4-cF4 order-disorder transformation in this phase has been suggested. (b) Cubic structure presumed to be room- and higher-temperature phases. (c) Tetragonal structure presumed to be lower-temperature phase

Nd-Zn

J.T. Mason and P. Chiotti, 1972

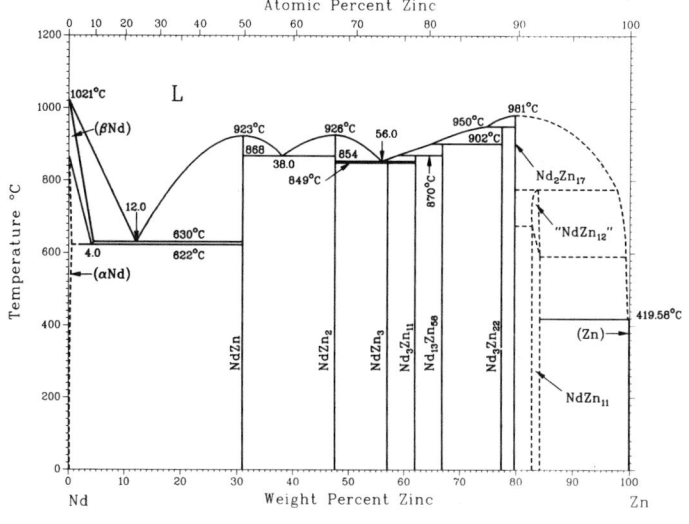

Phase	Composition, wt% Zn	Pearson symbol	Space group
(βNd)	0 to 4.0	cI2	$Im\bar{3}m$
(αNd)	0 to >0.5	hP4	$P6_3/mmc$
NdZn	31.2	cP2	$Pm\bar{3}m$
NdZn₂	47.6	oI12	$Imma$
NdZn₃	~57.0	oP16	$Pnma$
Nd₃Zn₁₁	~62.5	oI28	$Immm$
Nd₁₃Zn₅₈	~66.9	hP142	$P6_3mc$
Nd₃Zn₂₂	>77	tI100	$I4_1/amd$
Nd₂Zn₁₇	~79.4	hP38	$P6_3/mmc$
		hR19	$R\bar{3}m$
NdZn₁₁	~83.4	tI48	$I4_1/amd$
"NdZn₁₂"	~84.5	...	...
(Zn)	~100	hP2	$P6_3/mmc$
Other reported phase			
NdZn₅	~69.3	hP6	$P6/mmm$

Ni-O

J.P. Neumann, T. Zhong, and Y.A. Chang, 1991

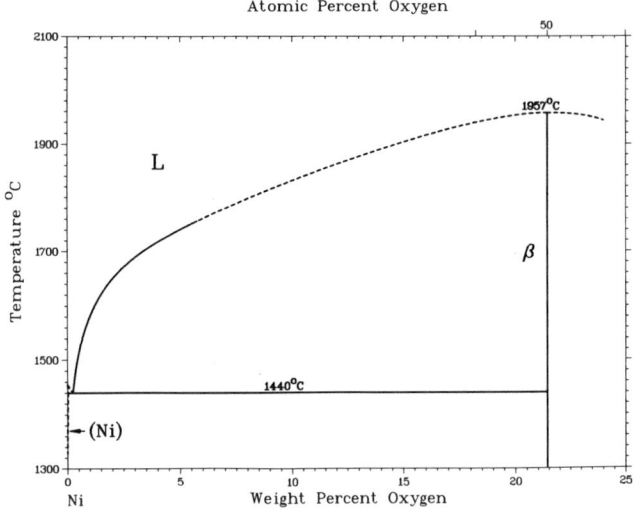

Phase	Composition, wt% O	Pearson symbol	Space group
(Ni)	0 to 0.01	$cF4$	$Fm\bar{3}m$
NiO(HT) or β	21.4	$cF8$	$Fm\bar{3}m$
NiO(LT)	21.4	$rP2$(a)	...
Ni_3O_4	27	...	...
Ni_2O_3	29	...	...
NiO_2	35.3	...	...

(a) The $rP2$ designation for NiO(LT) is an alternative to $hR2$.

Ni-rich region of the Ni-O phase diagram

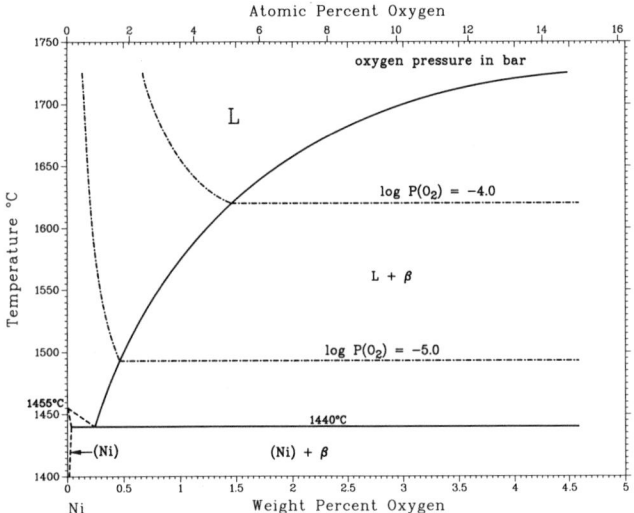

Ni-Os

P. Nash, 1991

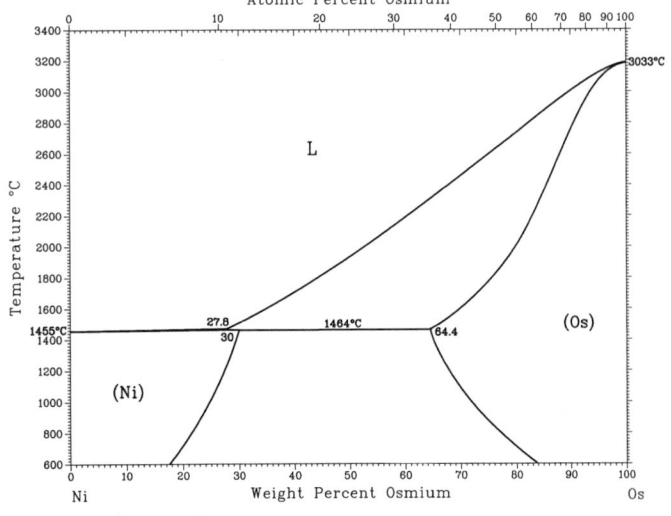

Phase	Composition, wt% Os	Pearson symbol	Space group
(Ni)	0 to 30	$cF4$	$Fm\bar{3}m$
(Os)	64.4 to 100	$hP2$	$P6_3/mmc$

Ni-P

K.J. Lee and P. Nash, 1991

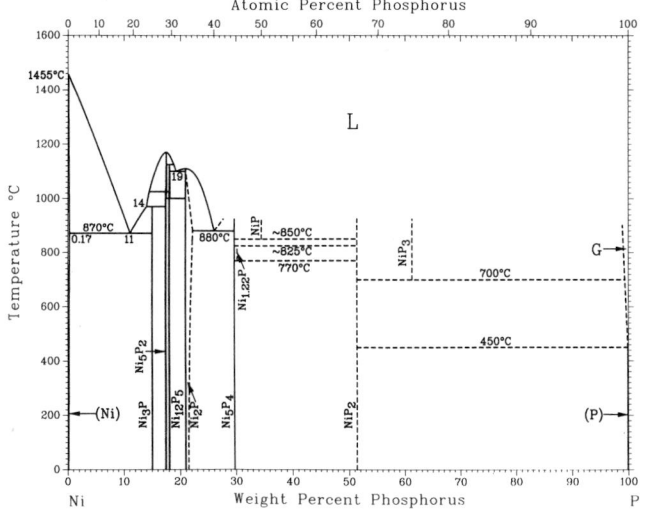

Phase	Composition, wt% P	Pearson symbol	Space group
(Ni)	0 to 0.17	$cF4$	$Fm\overline{3}m$
Ni_3P	15	$tI32$	$I\overline{4}$
βNi_5P_2	17.5	...	...
αNi_5P_2	17.5	$hP168$(a)	$P\overline{3}$
$\delta Ni_{12}P_5$	18.0	...	...
$\gamma Ni_{12}P_5$	18.0	$tI34$	$I4/m$
Ni_2P	20.9 to ?	$hP9$	$P\overline{6}2m$
			$P3_21$
Ni_5P_4	29.6	$hP36$	$P6_3mc$
$Ni_{1.22}P$	30.2	...	...
NiP	34.5	$oP16$	$Pcba$
NiP_2	51.4	$mC12$	$C2/c$
NiP_3	61	$cI32$	$Im\overline{3}$
P (red)	100	...	...
High-pressure phase			
NiP_2	51.4	$cP12$	$Pa\overline{3}$
Metastable phases			
"Ni_5P_2"	11 to 18	$h**$	...
α	8 to 15	$c**$	...
α_1	8 to 15	$h**$	...
α_2	8 to 15	$h**$	...
α_3	8 to 15	$h**$	...
"αNi_3P"	15	$t**$	...
"βNi_3P"	15	$h**$	...
"γNi_3P"	15	$c**$	...
α (amorphous)	~15	(b)	...
β (amorphous)	~15	(c)	...

(a) Might be $hP336$. (b) Liquid-like. (c) Molecular cluster

Ni-Pb

P. Nash, 1991

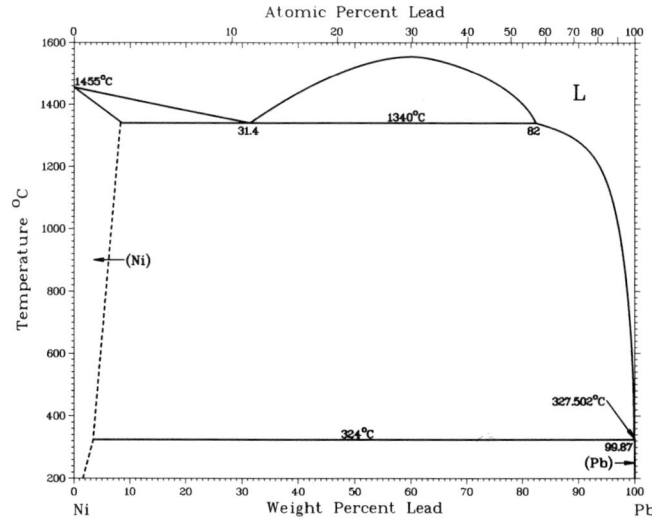

Phase	Composition, wt% Pb	Pearson symbol	Space group
(Ni)	0 to ~4.1	$cF4$	$Fm\overline{3}m$
(Pb)	99.9 to 100	$cF4$	$Fm\overline{3}m$
Metastable phase			
NiPb	77.9	$hP4$	$P6_3/mmc$

Ni-Pd

A. Nash and P. Nash, 1991

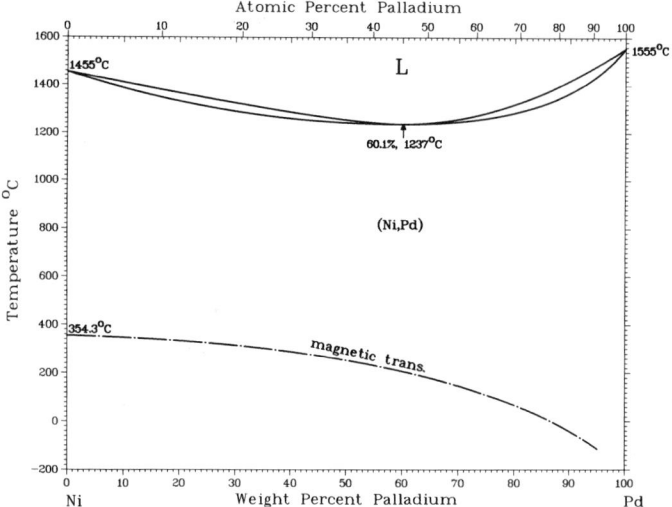

Phase	Composition, wt% Pd	Pearson symbol	Space group
(Ni,Pd)	0 to 100	$cF4$	$Fm\overline{3}m$

Ni-Pr

Y.Y. Pan and P. Nash, 1991

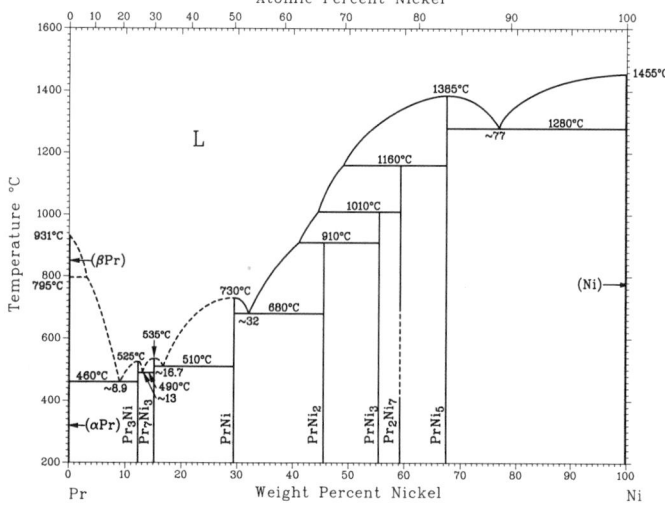

Phase	Composition, wt% Ni	Pearson symbol	Space group
(βPr)	0	$hP4$	$P6_3/mmc$
(αPr)	0	$cI2$	$Im\overline{3}m$
Pr$_3$Ni	12.2	$oP16$	$Pnma$
Pr$_7$Ni$_3$	15.1	$hP20$	$P6_3mc$
PrNi	29.4	$oC8$	$Cmcm$
PrNi$_2$	45.5	$cF24$	$Fd\overline{3}m$
PrNi$_3$	55.5	$hR24$	$R\overline{3}m$
Pr$_2$Ni$_7$	59.3	$hP36$	$P6_3/mmc$
		$hR54$	$R\overline{3}m$
PrNi$_5$	67.5	$hP6$	$P6/mmm$
(Ni)	100	$cF4$	$Fm\overline{3}m$

Ni-Pt

P. Nash and M.F. Singleton, 1991

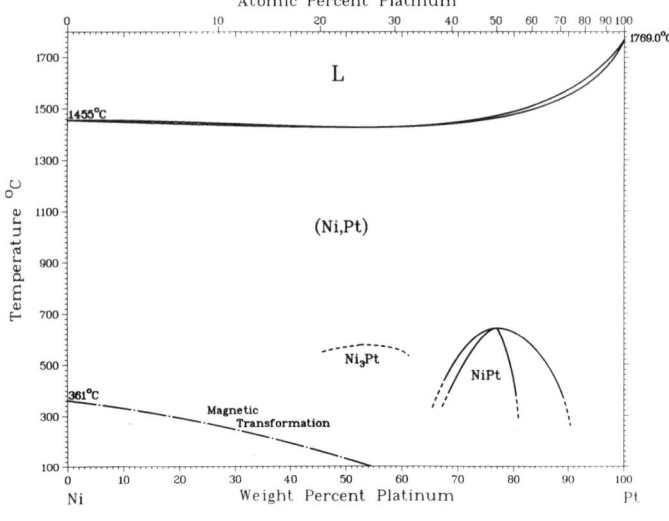

Phase	Composition, wt% Pt	Pearson symbol	Space group
(Ni,Pt)	0 to 100	$cF4$	$Fm\overline{3}m$
Ni$_3$Pt	~53	$cP4$	$Pm\overline{3}m$
NiPt	~76.9	$tP4$	$P4/mmm$

Ni-Pu

D.E. Peterson, 1991

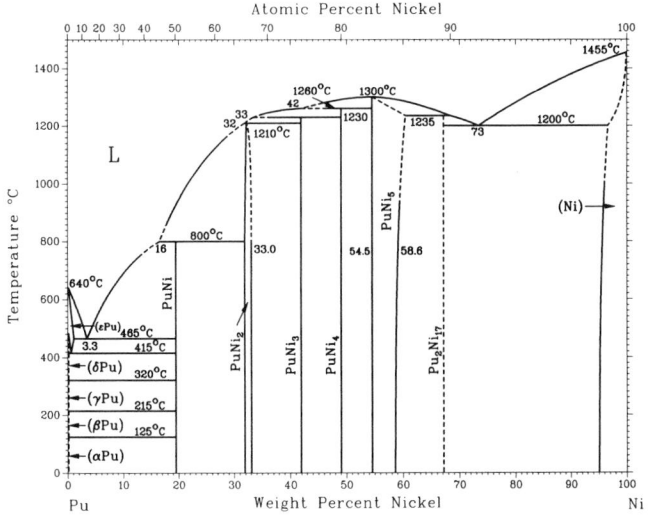

Phase	Composition, wt% Ni	Pearson symbol	Space group
(εPu)	0 to 1.1	cI2	Im$\overline{3}$m
(δ'Pu)	0 to 0.04	tI2	I4/mmm
(δPu)	0 to 0.1	cF4	Fm$\overline{3}$m
(γPu)	0	oF8	Fddd
(βPu)	0	mC34	C2/m
(αPu)	0	mP16	P2$_1$/m
PuNi	19.4	oC8	Cmcm
PuNi$_2$	32.5 to 34	cF24	Fd$\overline{3}$m
PuNi$_3$	42	hR12	R$\overline{3}$m
PuNi$_4$	49	mC30	C2/m
PuNi$_5$	54.5 to 60	hP6	P6/mmm
Pu$_2$Ni$_{17}$	67.2	hP38	P6$_3$/mmc
(Ni)	92.9 to 100	cF4	Fm$\overline{3}$m

Pu-rich region of the Pu-Ni phase diagram

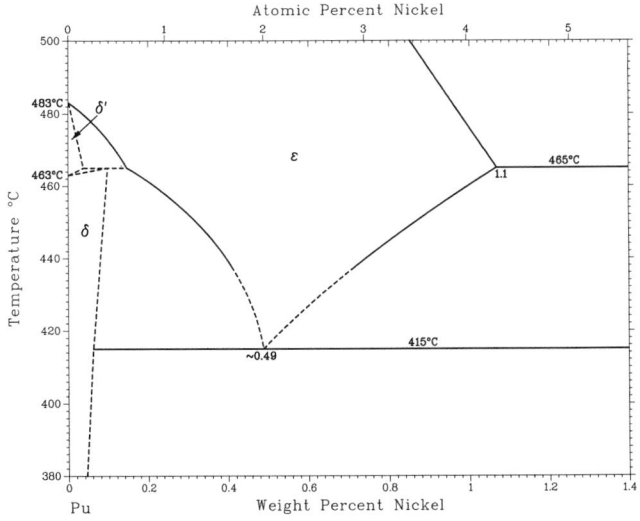

Ni-Re

H. Okamoto, 1992

Phase	Composition, wt% Re	Pearson symbol	Space group
(Ni)	0 to 40.1	cF4	Fm$\overline{3}$m
(Re)	94 to 100	hP2	P6$_3$/mmc

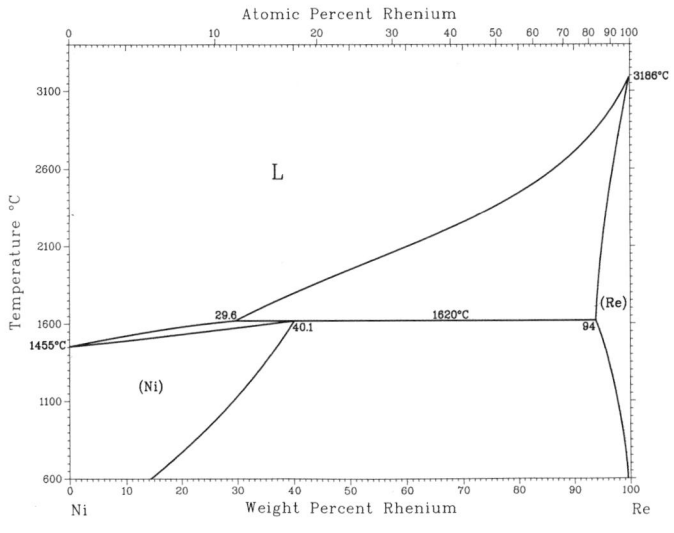

Ni-Rh

A. Nash and P. Nash, 1991

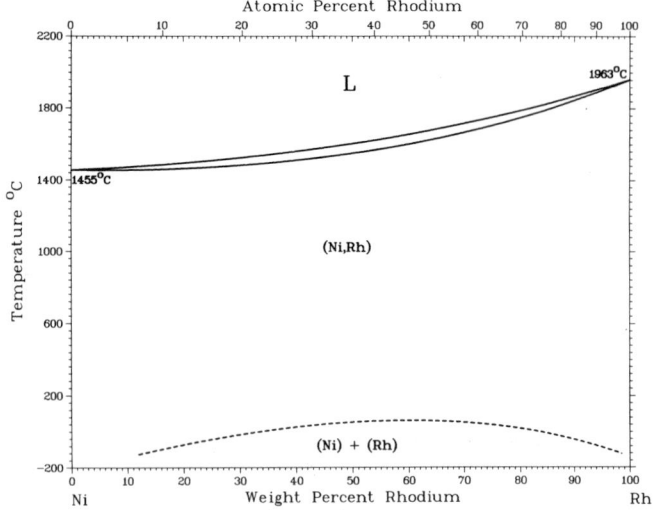

Phase	Composition, wt% Rh	Pearson symbol	Space group
(Ni,Rh)	0 to 100	$cF4$	$Fm\bar{3}m$

Ni-Ru

P. Nash, 1991

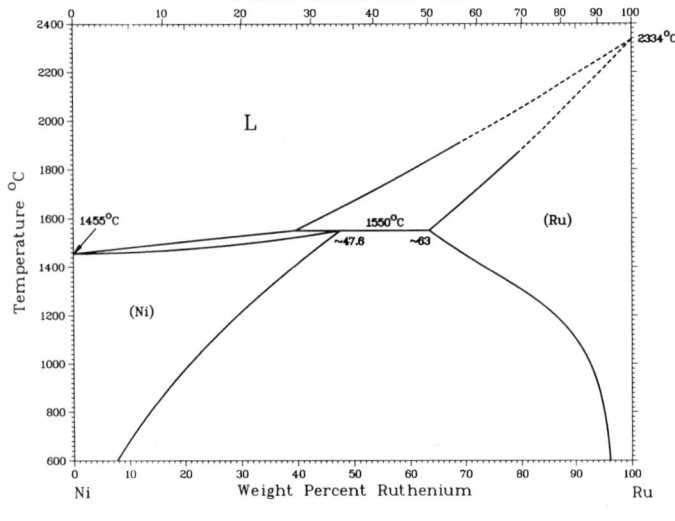

Phase	Composition, wt% Ru	Pearson symbol	Space group
(Ni)	0 to ~47.6	$cF4$	$Fm\bar{3}m$
(Ru)	~63 to 100	$hP2$	$P6_3/mmc$
Metastable phase			
η	?	t^{**}	...

Ni-S

M. Singleton, P. Nash, and K.J. Lee, 1991

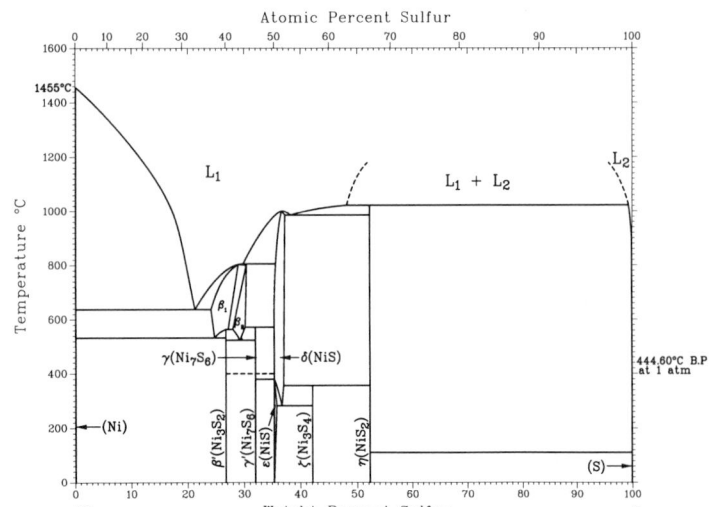

Phase	Composition, wt% S	Pearson symbol	Space group
(Ni)	0	$cF4$	$Fm\bar{3}m$
β′(Ni$_3$S$_2$)	27	$hR5$	$R32$
β$_1$(Ni$_3$S$_2$)	24.1 to ~28	(a)	...
β$_2$(Ni$_4$S$_3$)	28 to 30	...	...
γ(Ni$_7$S$_6$)	31.9	(a)	...
γ′(Ni$_7$S$_6$)	31.9	...	...
ε(NiS)	35.3 to 35.8	$hR6$	$R\bar{3}m$
δ(NiS)	35.1 to 37.7	$hP4$	$P6_3/mmc$
ζ(Ni$_3$S$_4$)	42.1	$cF56$	$Fd\bar{3}m$
η(NiS$_2$)	52.3	$cP12$	$Pa3$
(S)	100	$oF128$	$Fddd$

(a) Hexagonal

Ni-Sb

G.H. Cha, S.Y. Lee, and P. Nash, 1991

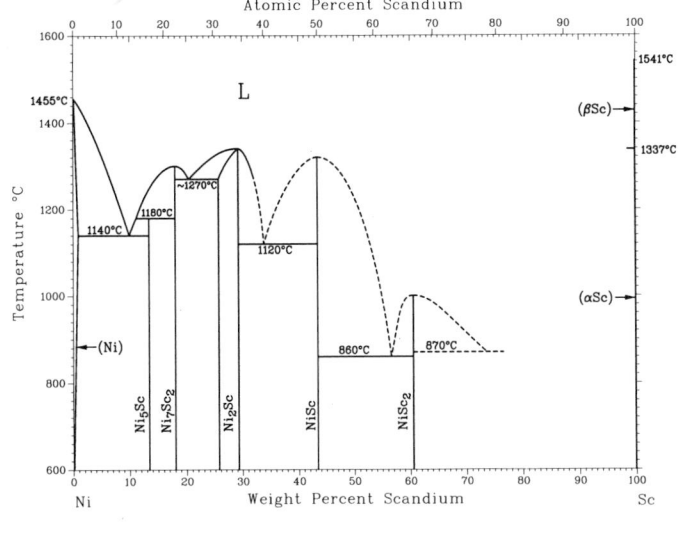

Phase	Composition, wt% Sb	Pearson symbol	Space group
(Ni)	0 to 17.0	cF4	$Fm\overline{3}m$
Ni$_{15}$Sb	12.2	...	...
Ni$_3$Sb	39.2 to 41	oP8	Pmmm
Ni$_5$Sb$_2$	41.1 to 45.6	mC28	...
Ni$_7$Sb$_3$	45	t**	...
NiSb	61.0 to 69.2	hP4	$P6_3/mmc$
NiSb$_2$	80.2 to 80.5	oP6	Pnnm
(Sb)	~100	hR2	$R\overline{3}m$

Ni-Sc

P. Nash and Y.Y. Pan, 1991

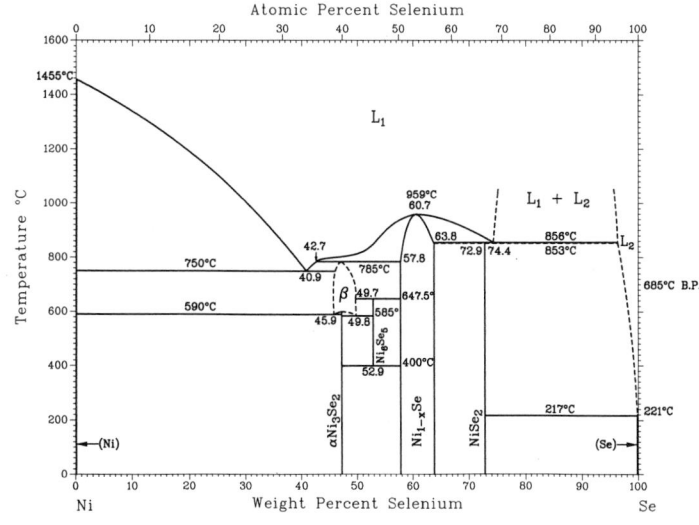

Phase	Composition, wt% Sc	Pearson symbol	Space group
(Ni)	~0	cF4	$Fm\overline{3}m$
Ni$_5$Sc(HT)	13.3	hP6	P6/mmm
Ni$_5$Sc(LT)	13.3	...	...
Ni$_7$Sc$_2$	17.9	hP36	$P6_3/mmc$
Ni$_2$Sc	26 to 29	cF24	$Fd\overline{3}m$
NiSc	43.4	cP2	$Pm\overline{3}m$
NiSc$_2$	60.5	cF96	$Fd\overline{3}m$
(βSc)	100	cI2	$Im\overline{3}m$
(αSc)	100	hP2	$P6_3/mmc$

Ni-Se

S.Y. Lee and P. Nash, 1991

Phase	Composition, wt% Se	Pearson symbol	Space group
(Ni)	~0	cF4	$Fm\overline{3}m$
βNi$_{3\pm x}$Se$_2$	45.9 to 49.8	c**	...
αNi$_3$Se$_2$	47	hR5	R32
Ni$_6$Se$_5$	52.9	oP88	Pca2$_1$
		oC48	Cmcm
Ni$_{1-x}$Se	57.8 to 63.8	hP4	$P6_3/mmc$
NiSe$_2$	72.9	cP12	Pa3
(γSe)	~100	hP2	P3$_1$21
Metastable phase			
α′Ni$_3$Se$_2$	47	tI*	...

Ni-Si

P. Nash and A. Nash, 1991

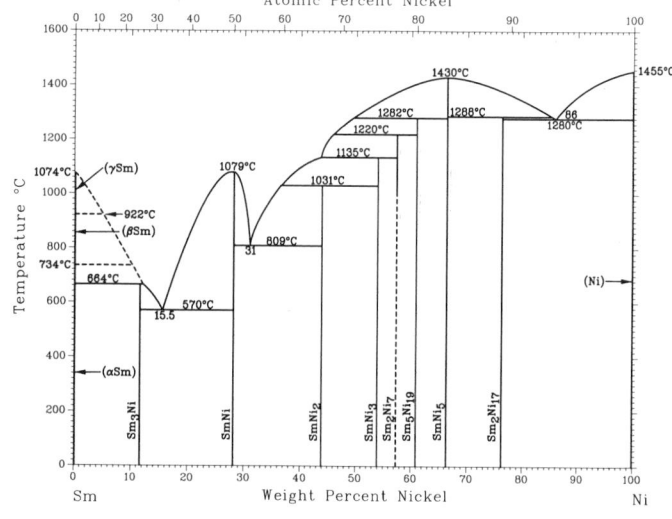

Phase	Composition, wt% Si	Pearson symbol	Space group
(Ni)	0 to 8.2	cF4	$Fm\bar{3}m$
β_1 (Ni$_4$Si)	12.4 to 13.4	cP4	$Pm\bar{3}m$
β_3 (Ni$_3$Si)	~13.4 to 14.1	mC16	...
β_2 (Ni$_3$Si)	~13.4 to 14.1	mC16	...
γ (Ni$_{31}$Si$_{12}$)	15.6	hP14	...
θ (Ni$_2$Si)	19.4 to 25	hP6	...
δ (Ni$_2$Si)	19.3	oP12	...
ε (Ni$_3$Si$_2$)	23 to 25	oP80	...
NiSi	32.4	oP8	Pnma
βNiSi$_2$	48.9	?	...
αNiSi$_2$	48.9	cF12	$Fm\bar{3}m$
(Si)	~100	cF8	$Fd\bar{3}m$

Ni-Sm

Y.Y. Pan and P. Nash, 1991

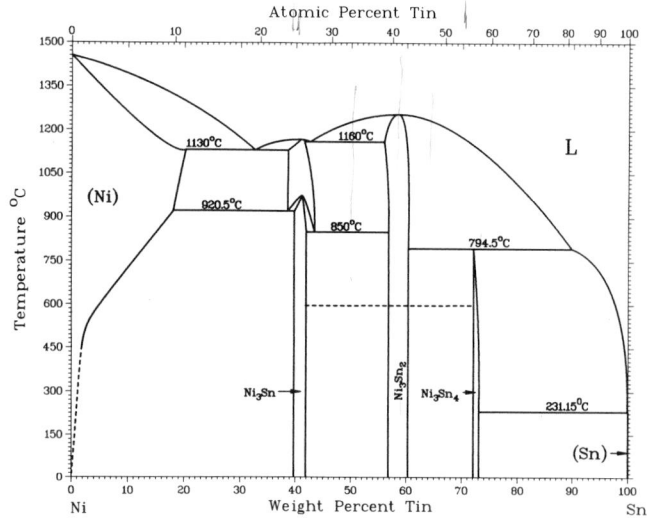

Phase	Composition, wt% Ni	Pearson symbol	Space group
(γSm)	0	cI2	$Im\bar{3}m$
(βSm)	0	hP2	$P6_3/mmc$
(αSm)	0	hR3	$R\bar{3}m$
Sm$_3$Ni	11.5	oP16	Pnma
SmNi	28.1	oC8	Cmcm
SmNi$_2$	43.9	cF24	$Fd\bar{3}m$
SmNi$_3$	53.9	hR24	$R\bar{3}m$
Sm$_2$Ni$_7$	57.8	hP36(a)	$P6_3/mmc$
		hR54(b)	$R\bar{3}m$
Sm$_5$Ni$_{19}$	59.8	(c)	P3m/1
SmNi$_5$	66.1	hP6	P6/mmm
Sm$_2$Ni$_{17}$	76.9	hP38	$P6_3/mmm$
(Ni)	100	cF4	$Fm\bar{3}m$

(a) High-temperature form. (b) Low-temperature form. (c) Trigonal

Ni-Sn

P. Nash and A. Nash, 1991

Phase	Composition, wt% Sn	Pearson symbol	Space group
(Ni)	0 to 19.3	cF4	$Fm\bar{3}m$
Ni$_3$Sn(HT)	37.9 to 43.0	(a)	...
Ni$_3$Sn(LT)	39 to 41.7	hP8	$P6_3/mmc$
Ni$_3$Sn$_2$(HT)	54.8 to 57.9	(a)	...
		(b)	...
Ni$_3$Sn$_2$(LT)	55.9 to 59.9	hP4	$P6_3/mmc$
Ni$_3$Sn$_4$	71.6 to 73	mC14	C2/m
(βSn)	~100	tI4	$I4_1/amd$
Metastable phase			
Ni$_3$Sn	40	oP8	Pmmn

(a) Hexagonal. (b) Orthorhombic

Ni-Ta

A. Nash and P. Nash, 1991

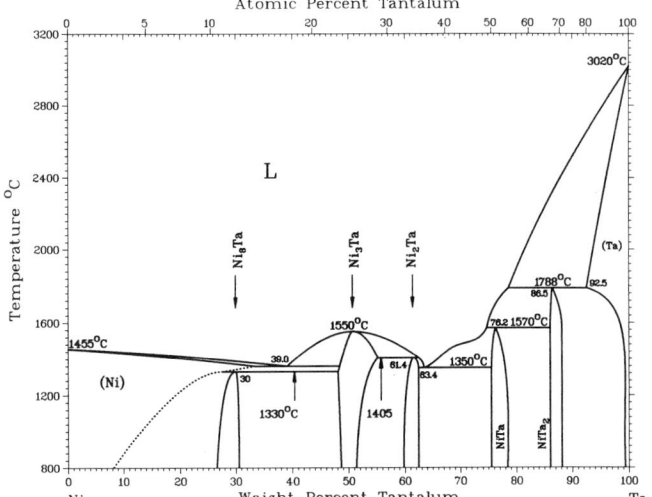

Phase	Composition, wt% Ta	Pearson symbol	Space group
(Ni)	0 to 33	cF4	Fm$\bar{3}$m
Ni$_8$Ta	27.8	tI36	...
Ni$_3$Ta(12)S	47.2 to 55.1	mP48	P2$_1$/m
Ni$_2$Ta	59.7 to 62	tI6	I4/mmm
NiTa	75.5 to 78	hR13	R$\bar{3}$m
NiTa$_2$	86.1 to 88	tI12	I4/mcm
(Ta)	92.5 to 100	cI2	Im$\bar{3}$m
Metastable phases			
ζ	45	...	...
Ni$_3$Ta(2)S	51	mP8	Pmmm
Ni$_3$Ta(3)S	51	tI8	I4/mmm

Note: Number in parentheses indicates stacking period; S identifies the orthogonal layer type.

Ni-Te

S.Y. Lee and P. Nash, 1991

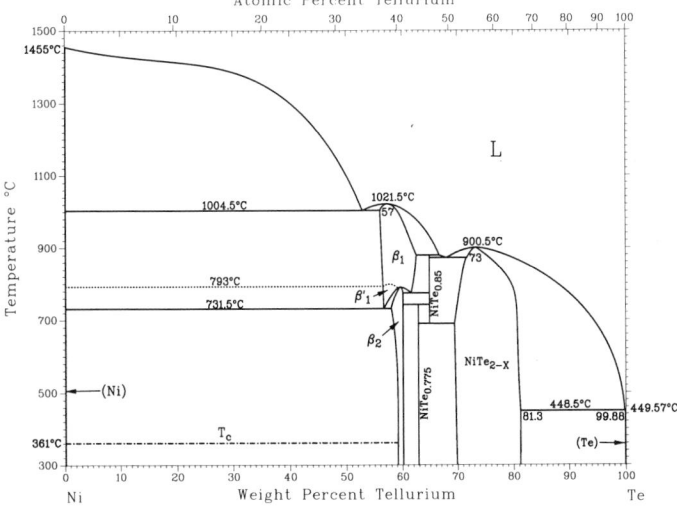

Phase	Composition, wt% Te	Pearson symbol	Space group
(Ni)	~0	cF4	Fm$\bar{3}$m
β$_1$	55.9 to 62.4	cF*	...
β$_2$	58.0 to 59.7	m**	...
	59.7 to 60.1	o**	...
	60.0 to 60.4	t**	...
β′$_1$	56.5 to 58	...	...
NiTe$_{0.775}$	62.7	o**	...
NiTe$_{0.85}$	64.8	...	...
NiTe$_{2-x}$	69.4 to 81.3	hP4	P6$_3$/mmc
		hP3	P$\bar{3}$m1
(Te)	~100	hP3	P3$_1$21

Ni-Ti

J.L. Murray, 1991

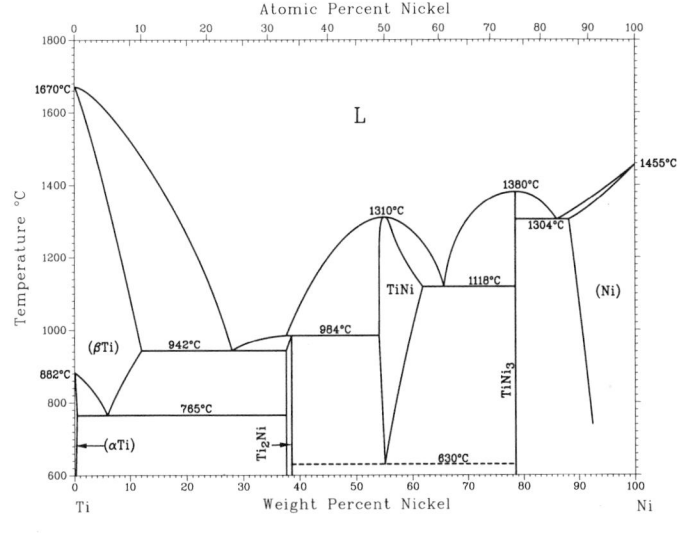

Phase	Composition, wt% Ni	Pearson symbol	Space group
(βTi)	0 to 12	cI2	Im$\bar{3}$m
(αTi)	0 to 0.3	hP2	P6$_3$/mmc
ω(a)	~10	hP3	P6/mmm or P$\bar{3}$m1
Ti$_2$Ni	38.0	cF96	Fd$\bar{3}$m
TiNi′(a)	~54 to 58	mP4	P2$_1$/m
TiNi	54.6 to 62	cP2	Pm$\bar{3}$m
γ″TiNi$_3$(a)	~77	hR21	R$\bar{3}$m
TiNi$_3$	79	hP16	P6$_3$/mmc
γ′TiNi$_3$(a)	~86 to 90	cP4	Pm$\bar{3}$m
(Ni)	88.4 to 100	cF4	Fm$\bar{3}$m

(a) Metastable

Ni-U

D.E. Peterson, 1991

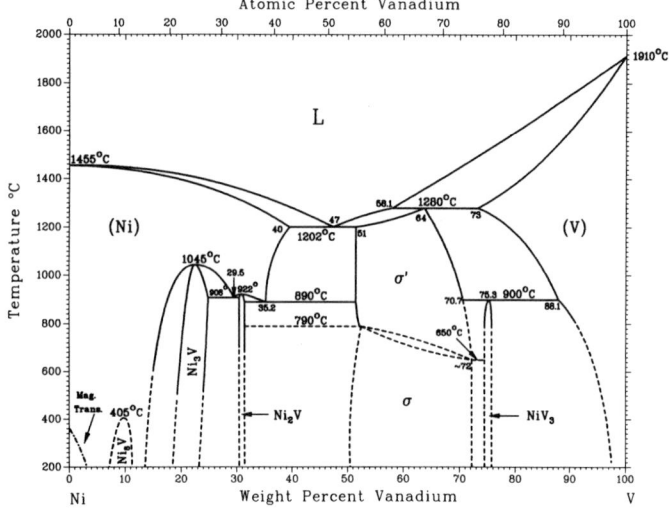

Phase	Composition, wt% Ni	Pearson symbol	Space group
(γU)	0 to 0.5	cI2	Im$\bar{3}$m
(βU)	0 to 0.2	tP30	P4$_2$/mnm
(αU)	0	oC4	Cmcm
U$_6$Ni	4.0	tI28	I4/mcm
U$_7$Ni$_9$	24.0	...	...
U$_5$Ni$_7$	25.6 to 26.6	...	...
UNi$_2$	33.1 to 33.4	hP12	P6$_3$/mmc
δ	45.2	...	...
ε	46.6	...	...
η(a)	52.9	...	...
UNi$_5$	53.8 to 55.2	cF24	F$\bar{4}$3m
(Ni)	93.1 to 100	cF4	Fm$\bar{3}$m

(a) Existence tentative

Ni-V

J.F. Smith, O.N. Carlson, and P. Nash, 1991

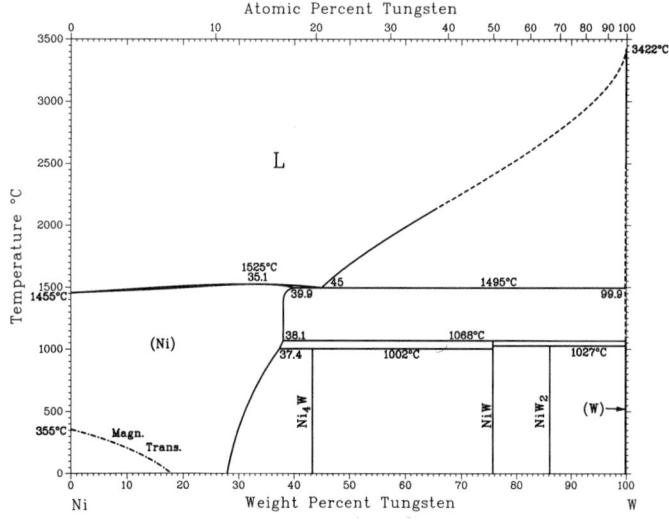

Phase	Composition, wt% V	Pearson symbol	Space group
(Ni)	0 to 40	cF4	Fm$\bar{3}$m
Ni$_8$V	~9.8	tI18	...
Ni$_3$V	~19 to 23.0	tI8	I4/mmm
Ni$_2$V	~30.2	oI6	...
σ′	51 to ~72	...	...
σ	54.0 to ~72	tP30	P4$_2$/mnm
NiV$_3$	74.9 to 76.0	cP8	Pm$\bar{3}$n
(V)	73 to 100	cI2	Im$\bar{3}$m

Ni-W

H. Okamoto, 1991

Phase	Composition, wt% W	Pearson symbol	Space group
(Ni)	0 to 39.9	cF4	Fm$\bar{3}$m
Ni$_4$W	~44	tI10	I4/m
NiW	~75.8	o**	...
NiW$_2$	86.3	tI96	I4
(W)	99.9 to 100	cI2	Im$\bar{3}$m

Ni-Y

P. Nash, 1991

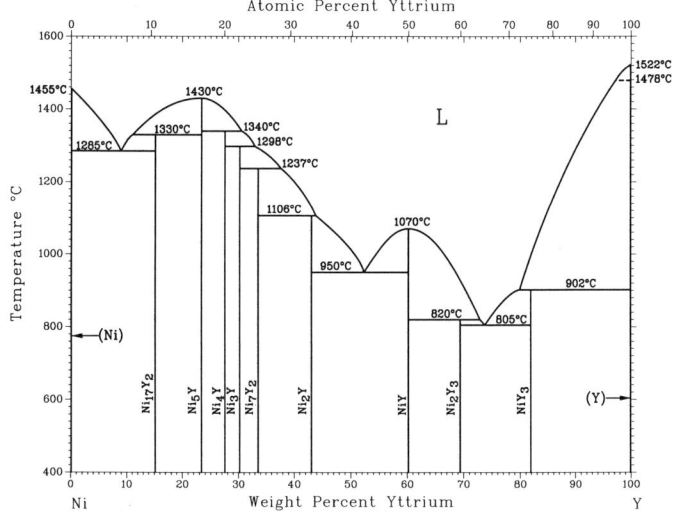

Phase	Composition, wt% Y	Pearson symbol	Space group
(Ni)	0	cF4	$Fm\overline{3}m$
Ni$_{17}$Y$_2$	15.1	hP*	$P6_3/mmc$
Ni$_5$Y	23.3	hP6	$P6/mmm$
Ni$_4$Y	27.5	...	...
Ni$_7$Y$_2$	30.2	hR*	$R\overline{3}m$
Ni$_3$Y	33.6	hP**	$P6_3/mmc$
		hR*	$R\overline{3}m$
Ni$_2$Y	43.1	cF24	$Fd\overline{3}m$
NiY	60.2	oP8	$Pnma$
Ni$_2$Y$_3$	69.4	t**	$P4_12_12$
NiY$_3$	82.0	oP16	$Pnma$
(βY)	100	cF4	$Fm\overline{3}m$
(αY)	100	hP2	$P6_3/mmc$

Ni-Yb

P. Nash, 1991

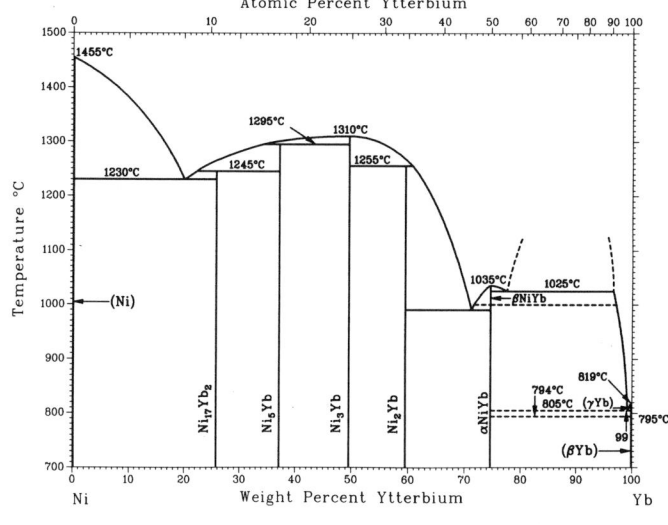

Phase	Composition, wt% Yb	Pearson symbol	Space group
(Ni)	0	cF4	$Fm\overline{3}m$
Ni$_{17}$Yb$_2$	25.7	hP38	$P6_3/mmc$
Ni$_5$Yb	37.2	hP6	$P6/mmm$
Ni$_3$Yb	49.6	hR12	$R\overline{3}m$
Ni$_2$Yb	59.5	cF24	$Fd\overline{3}m$
αNiYb	74.7	oP8	$Pnma$
(γYb)	100	cI2	$Im\overline{3}m$
(βYb)	100	cF4	$Fm\overline{3}m$
(αYb)	100	hP2	$P6_3/mmc$

Ni-Zn

P. Nash and Y.Y. Pan, 1991

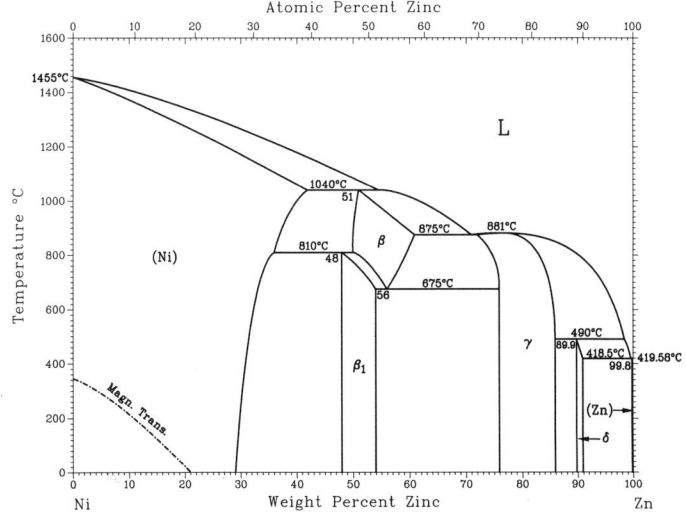

Phase	Composition, wt% Zn	Pearson symbol	Space group
(Ni)	0 to 41.9	cF4	$Fm\overline{3}m$
β	50.0 to 60.9	cP2	$Pm\overline{3}m$
β$_1$	48.0 to 54.5	tP2	$P4/mmm$
γ(a)	72 to 86	cI52	$I\overline{4}3m$
δ	~90	mC6	$C2/m$
(Zn)	100	hP2	$P6_3/mmc$

(a) Might have orthorhombic structure

Ni-Zr

P. Nash and C.S. Jayanth, 1991

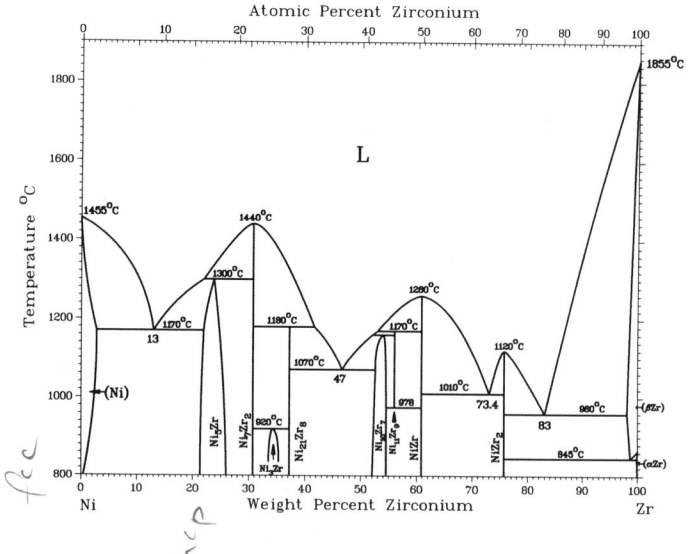

Phase	Composition, wt% Zr	Pearson symbol	Space group
(Ni)	0 to 2.74	cF4	$Fm\bar{3}m$
Ni$_5$Zr	21.32 to 25.95	cF24	$F\bar{4}3m$
Ni$_7$Zr$_2$	30.75	mC36	C2/m
Ni$_3$Zr	33.5 to 35.3	hP8	P6$_3$/mmc
Ni$_{21}$Zr$_8$	37.2	(a)	...
Ni$_{10}$Zr$_7$	52.0 to 54.52	oC68	C2ca(b)
			Pbca(c)
Ni$_{11}$Zr$_9$	56.0	tI40	I4/m
NiZr	60.9	oC8	Cmcm
NiZr$_2$	75.7	tI12	I4/mcm
(βZr)	98.10 to 100	cI2	$Im\bar{3}m$
(αZr)	99.9 to 100	hP2	P6$_3$/mmc

(a) Triclinic. (b) Stoichiometric. (c) Zr-rich

Np-Pu

R.I. Sheldon and D.E. Peterson, 1985

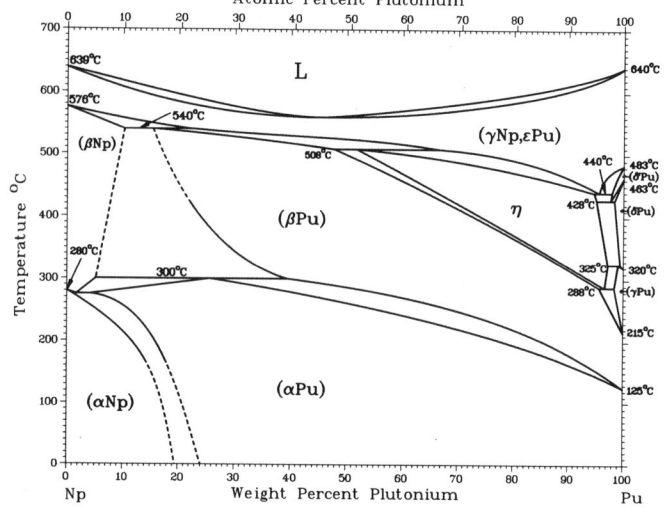

Phase	Composition, wt% Pu	Pearson symbol	Space group
(γNp,εPu)	0 to 100	cI2	$Im\bar{3}m$
(βNp)	0 to 10.3	tP4	P42$_1$2
(αNp)	0 to 19.5	oP8	Pnma
η	52 to 97.1	(a)	...
(δ'Pu)	97.7 to 100	tI2	I4/mmm
(δPu)	98.3 to 100	cF4	$Fm\bar{3}m$
(γPu)	98.3 to 100	oF8	Fddd
(βPu)	15.4 to 100	mC34	C2/m
(αPu)	4.1 to 100	mP16	P2$_1$/m

(a) Orthorhombic (tentative)

Np-U

R.I. Sheldon and D.E. Peterson, 1985

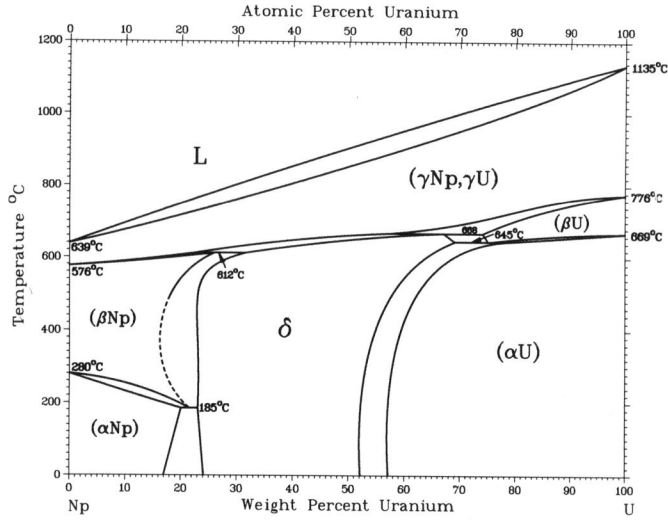

Phase	Composition, wt% U	Pearson symbol	Space group
(γNp,γU)	0 to 100	cI2	$Im\bar{3}m$
(βNp)	0 to 26	tP4	P42$_1$2
(αNp)	0 to 20	oP8	Pnma
δ	23 to 69	cP58(a)	...
(βU)	74 to 100	tP30	P4$_2$/mmm
(αU)	57 to 100	oC4	Cmcm

(a) Tentative

O-Pb (condensed system, 0.1 MPa)

H.A. Wriedt, 1988

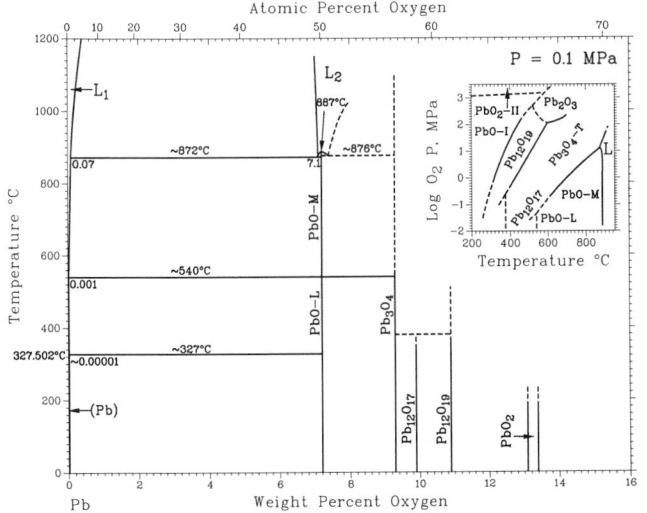

Inset shows equilibrium phase fields under identical hydrostatic and partial O_2 gas pressures

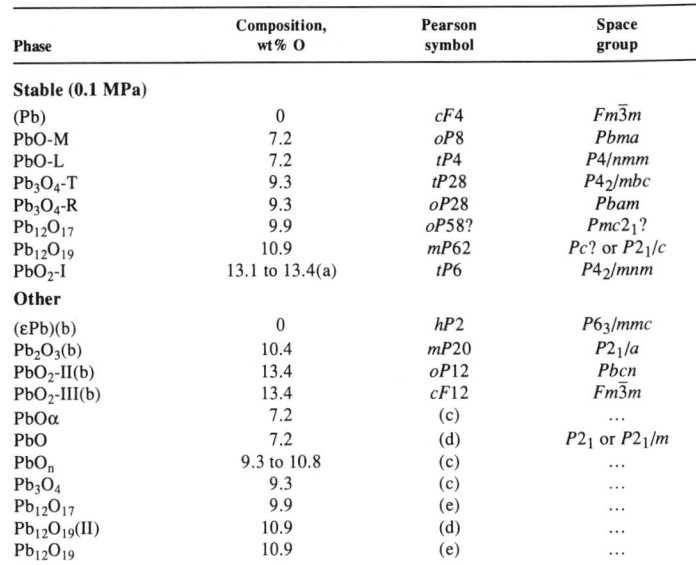

Phase	Composition, wt% O	Pearson symbol	Space group
Stable (0.1 MPa)			
(Pb)	0	cF4	$Fm\bar{3}m$
PbO-M	7.2	oP8	Pbma
PbO-L	7.2	tP4	P4/nmm
Pb_3O_4-T	9.3	tP28	$P4_2/mbc$
Pb_3O_4-R	9.3	oP28	Pbam
$Pb_{12}O_{17}$	9.9	oP58?	$Pmc2_1$?
$Pb_{12}O_{19}$	10.9	mP62	Pc? or $P2_1/c$
PbO_2-I	13.1 to 13.4(a)	tP6	$P4_2/mnm$
Other			
(εPb)(b)	0	hP2	$P6_3/mmc$
Pb_2O_3(b)	10.4	mP20	$P2_1/a$
PbO_2-II(b)	13.4	oP12	Pbcn
PbO_2-III(b)	13.4	cF12	$Fm\bar{3}m$
PbOα	7.2	(c)	...
PbO	7.2	(d)	$P2_1$ or $P2_1/m$
PbO_n	9.3 to 10.8	(c)	...
Pb_3O_4	9.3	(c)	...
$Pb_{12}O_{17}$	9.9	(e)	...
$Pb_{12}O_{19}$(II)	10.9	(d)	...
$Pb_{12}O_{19}$	10.9	(e)	...

(a) Contains a small amount of hydrogen. (b) Stable at hydrostatic pressures elevated from 0.1 MPa. (c) Orthorhombic. (d) Monoclinic. (e) Pseudocubic?

O-Pr

P.R. Subramanian, 1990

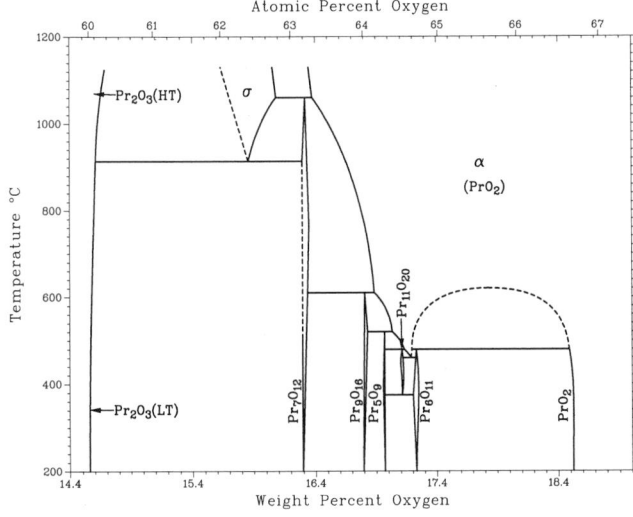

Phase	Composition, wt% O	Pearson symbol	Space group
(βPr)	0	cI2	$Im\bar{3}m$
(αPr)	0	hP4	$P6_3/mmc$
Pr_2O_3(HT)	~15	hP5	$P\bar{3}m1$
Pr_2O_3(LT)	~15	cI80	$Ia\bar{3}$
σ(a)	~16.0	cI80	$Ia\bar{3}$
Pr_7O_{12}	~16.3	hR19	$R\bar{3}$
Pr_9O_{16}	~17	aP*	$P\bar{1}$
Pr_5O_9	~17.0	mP112	$P2_1/c$
$Pr_{11}O_{20}$	~17.1	m**	...
Pr_6O_{11}	~17.2	c**	...
PrO_2(α)	~18.5	cF12	$Fm\bar{3}m$
High-pressure phase			
PrO(b)	10.2	cF*	...

(a) Reported to be a high-temperature phase; stable above ~920 °C. (b) Obtained by reduction of Pr_2O_3 by Pr at 800 °C and 50 kbar

O-Pu (condensed system)

H.A. Wriedt, 1990

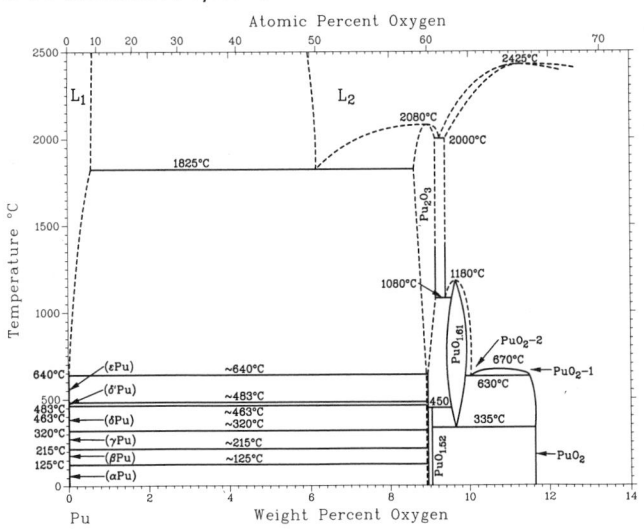

Phase	Composition, wt% O	Pearson symbol	Space group
Stable			
(αPu)	~0	mP16	$P2_1/m$
(βPu)	~0	mC34	C2/m
(γPu)	~0	oF8	Fddd
(δPu)	~0	cF4	$Fm\bar{3}m$
(δ'Pu)	~0	tI2	I4/mmm
(εPu)	~0	cI2	$Im\bar{3}m$
P_2O_3(a)	~8.9 to ~9.0	hP5	$P\bar{3}m1$
$PuO_{1.52}$(b)	~9.1	cI80	$Ia\bar{3}$
$PuO_{1.61}$(b)	~9.6 to ~10.0	cI80	$Ia\bar{3}$
PuO_2	~9 to 11.6(c)	cF12	$Fm\bar{3}m$
Other			
Pu	...	hP2	$P6_3/mmc$

(a) The lower limit at 1100 °C might be 58.8 at.% O. (b) Possibly unconnected ranges of the same phase. (c) At 0.1 MPa O_2 pressure

O-Sn

From [Hansen]

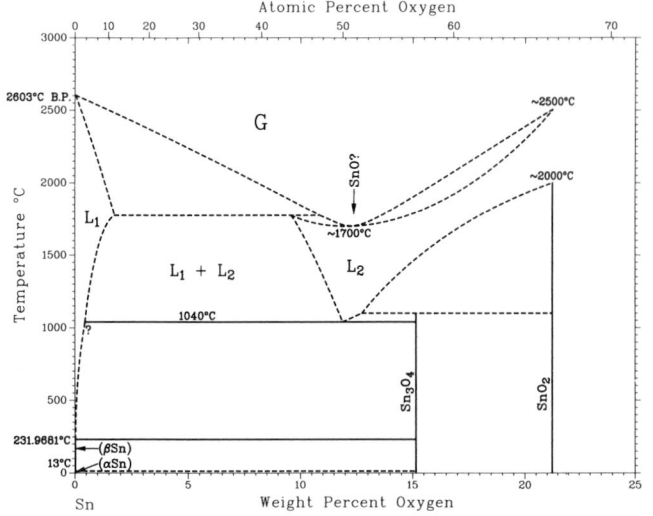

Phase	Composition, wt% O	Pearson symbol	Space group
(βSn)	0	tI4	$I4_1/amd$
(αSn)	0	cF8	$Fd\overline{3}m$
SnO(?)	11.9	tP4	$P4/nmm$
Sn₃O₄	15.2	a**	...
SnO₂	21.3	tP6	$P4_2/mnm$

O-Ti

J.L. Murray and H.A. Wriedt, 1987

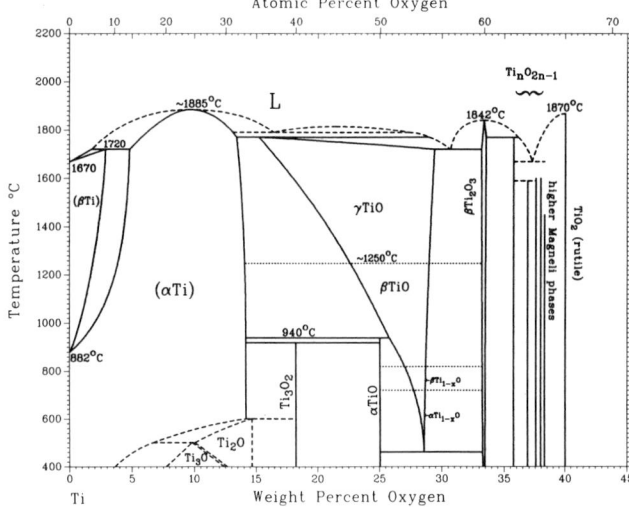

Phase	Composition, wt% O	Pearson symbol	Space group
(βTi)	0 to 3	cI2	$Im\overline{3}m$
(αTi)	0 to 13.5	hP2	$P6_3/mmc$
Ti₃O	~8 to ~13	hP~16	$P\overline{3}1c$
Ti₂O	~10 to 14.4	hP3	$P\overline{3}m1$
γTiO	15.2 to 29.4	cF8	$Fm\overline{3}m$
Ti₃O₂	~18	hP~5	$P6/mmm$
βTiO	~24 to ~29.4	c**	...
αTiO	~25.0	mC16	$A2/m$ or $B*/*$
βTi₁₋ₓO	~29.5	oI12	$I222$
αTi₁₋ₓO	~29.5	tI18	$I4/m$
βTi₂O₃	33.2 to 33.6	hR30	$R\overline{3}c$
αTi₂O₃	33.2 to 33.6	hR30	$R\overline{3}c$
βTi₃O₅	35.8	m**	...
αTi₃O₅	35.8	mC32	$C2/m$
α'Ti₃O₅	35.8	mC32	Cc
γTi₄O₇	36.9	aP44	$P\overline{1}$
βTi₄O₇	36.9	aP44	$P\overline{1}$
αTi₄O₇	36.9	aP44	$P\overline{1}$
γTi₅O₉	37.6	aP28	$P\overline{1}$
βTi₆O₁₁	38.0	aC68	$A\overline{1}$
Ti₇O₁₃	38.3	aP40	$P\overline{1}$
Ti₈O₁₅	38.5	aC92	$A\overline{1}$
Ti₉O₁₇	38.7	aP52	$P\overline{1}$
Rutile	40.1	tP6	$P4_2/mnm$
Metastable phases			
Anatase	...	tI12	$I4_1/amd$
Brookite	...	oP24	$Pbca$
High-pressure phases			
TiO₂-II	...	oP12	$Pbcn$
TiO₂-III	...	hP~48	...

O-V

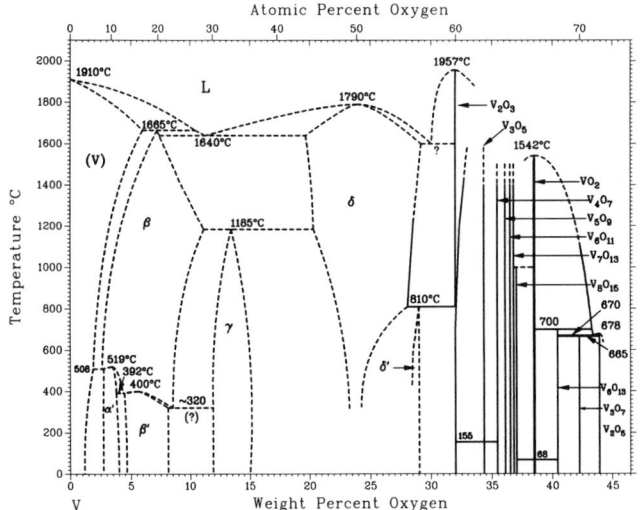

Phase	Composition, wt% O	Pearson symbol	Space group
(V)	0 to 6	cI2	$Im\bar{3}m$
α'	2.7 to 4.0	tI216(a)	...
β	2.6 to 11.1	tI2.5(b)	I4/mmm
β'	4 to 8	tI76(c)	I4/mmm
γ	12 to 15	mC20(d)	C2/m
δ	19 to 29.4	cF8	$Fm\bar{3}m$
δ'	27 to 28.6	tI116	$I4_1/amd$
h-V_2O_3(e)	32.0 to 32.5	hR10	$R\bar{3}c$
l-V_2O_3(f)	32	mI20	I2a
h-V_3O_5(e)	~34.4	mI32	I2/c
l-V_3O_5(f)	34.4 to 34.38	mP32	P2/c
V_4O_7	35.4	aP22	$P\bar{1}$
V_5O_9	36.1	aP28	$P\bar{1}$
V_6O_{11}	36.5	aP34	$P\bar{1}$
V_7O_{13}	36.8	aP40	$P\bar{1}$
V_8O_{15}	37.0	aP46	$P\bar{1}$
βVO_2(e)	38.5 to 38.8	tP6	$P4_2/mnm$
αVO_2(f)	38.6	mP12	$P2_1/c$
h-V_6O_{13}(e)	~40.5	mC38	C2/m
l-V_6O_{13}(f)	~40.5	mP38	$P2_1/a$
V_3O_7	~42	mC120	C2/c
V_2O_5	~43.9	oP14	Pmnm
Other phases			
Martensite-A	2.2 to 2.9	tI*(g)	...
Martensite-B	2.0 to 2.2	tI*(g)	...
ε	8 to 11	mP*	$P2_1/c$
$VO_{1.17}$	~27	...	$I4_1/a$
V_9O_{17}	~37.3	aP52	P1
VO_2-B	~38.6	tI288(?)	...
VO_2-M_2	~38.6	mC24	C2/m
VO_2-T_2	~38.6	tP6	$P4_2/mnm$
VO_2-M_3	38.7 to 39.2	mP6	$P2_1/m$
VO_2-M_4(h)	~38.6	mC24	C2/m
VO_2-D	~38.6	oP12	Pbnm
V_6O_{13}-C	~40.5	cP76(?)	...
V_6O_{13}-D	~40.5	mC38	C2/m
V_4O_9	~41.4	oP52	Pnma
V_4O_9-E	~41.1	oP104(?)	...
V_2O_5	~44.1	...	...

(a) At V_8O. (b) At V_4O. (c) At $V_{16}O_3$. (d) At V_7O_3. (e) Above T_{trs}. (f) Below transformation temperature, T_{trs}. (g) 2 atoms V/unit cell. (h) Also called $VO_2(B)$

O-W (condensed system, 0.1 MPa)

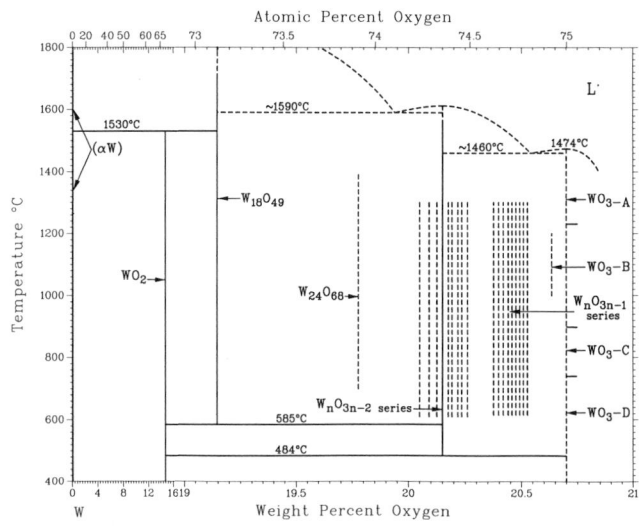

Phase	Composition, wt% O	Pearson symbol	Space group
(αW)	~0	cI2	$Im\bar{3}m$
WO_2	~14.8	mP12	$P2_1/c$
$W_{18}O_{49}$	~19.1	mP67	P2/m
$W_{24}O_{68}$	~19.8	m*92	...
$W_{20}O_{58}$(a)	20.2	mP78	P2/m
$W_{24}O_{70}$(a)	20.3	mP94	...
$W_{25}O_{73}$(a)(b)	20.3	mP98	P2/c
$W_{25}O_{74}$(c)	20.4	mP99	P2/m
WO_3-M	~20.7	...	...
WO_3-J	~20.7	...	...
WO_3-K	~20.7	...	...
WO_3-H	~20.7	...	...
WO_3-G	~20.7	mP16	Pc
WO_3-F	~20.7	aP32	$P\bar{1}$
WO_3-E(d)	~20.7	mP32	$P2_1/n$
WO_3-D(d)	~20.7	oP32	Pmnb
WO_3-C(d)	~20.7	tP8	P4/nmn
WO_3-B	~20.7	tP8	P4/nmm
WO_3-A	~20.7	tP8(?)	P4/nmm(?)
Other			
(βW)	0 to ?	cP8	$Pm\bar{3}n$
$W_{40}O_{118}$(a)	20.4	mP158	P2
WO_3(e)	20.7	hP24	P6/mmm(?)
WO_3(f)	20.7	c*4	...

(a) Member W_nO_{3n-2} series. (b) Identified as $WO_{2.96}(\alpha)$. (c) Probable member W_nO_{3n-1} series, called $WO_{2.96}(\beta)$. (d) Often described as a slightly distorted ReO_3 ($D0_9$). (e) Hexagonal. (f) Cubic

O-Y

O.N. Carlson, 1990

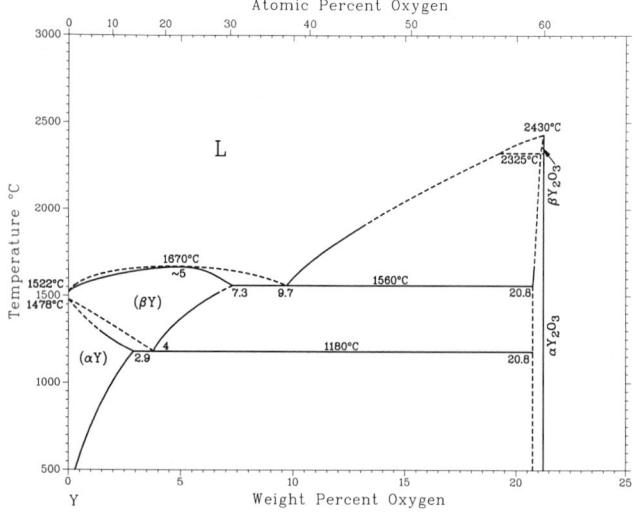

Phase	Composition, wt% O	Pearson symbol	Space group
(βY)	0 to 7.3	cI2	$Im\bar{3}m$
(αY)	0 to 2.9	hP2	$P6_3/mmc$
αY_2O_{3-x}	20.8 to 21	cI80	$Ia\bar{3}$
βY_2O_{3-x}	~21	hP(?)	$P\bar{3}m1$
γY_2O_3(a)	~21	mC(?)	$C2/m$

(a) High-pressure phase

O-Zr

J.P. Abriata, R. Versaci, and J. Garcés, 1986

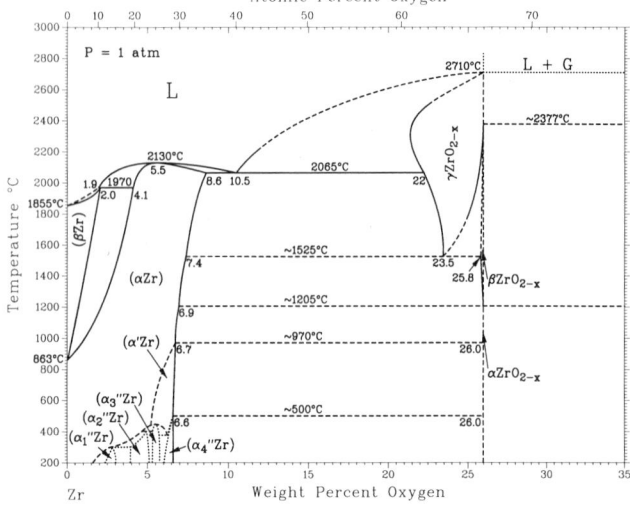

Phase	Composition, wt% O	Pearson symbol	Space group
(αZr)	0 to 8.6	hP2	$P6_3/mmc$
(βZr)	0 to 2.0	cI2	$Im\bar{3}m$
γZrO_{2-x}	22 to 25.9	cF12	$Fm\bar{3}m$
βZrO_{2-x}	25.8 to 25.9	tP6	$P4_2/nmc$
αZrO_{2-x}	25.9	mP12	$P2_1/c$

Os-Pt

H. Okamoto, 1990

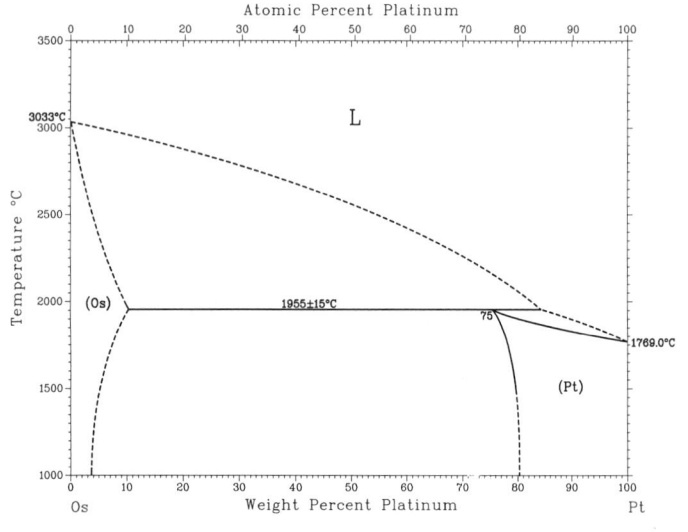

Phase	Composition, wt% Pt	Pearson symbol	Space group
(Os)	0 to ~11	hP2	$P6_3/mmc$
(Pt)	75 to 100	cF2	$Fm\bar{3}m$

Os-Pu

S.T. Konobeevsky, 1955

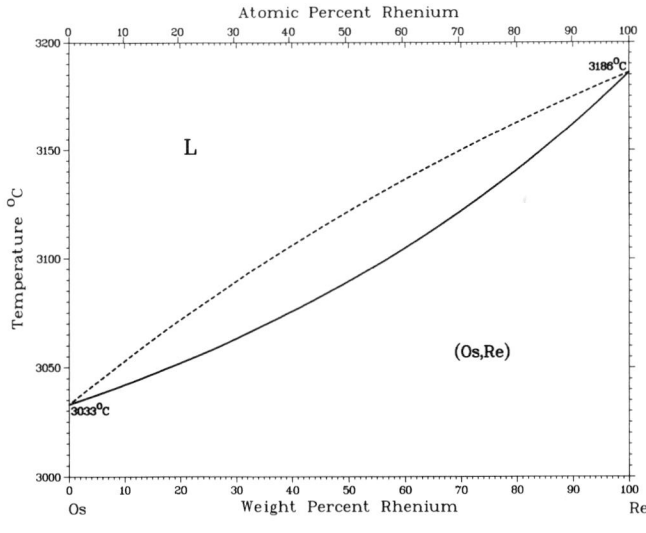

Phase	Composition, wt% Os	Pearson symbol	Space group
(εPu)	0 to ~5	cI2	Im$\bar{3}$m
(δ'Pu)	0 to ~0.4	tI2	I4/mmm
(δPu)	0 to ~0.4	cF4	Fm$\bar{3}$m
(γPu)	~0	oF8	Fddd
(βPu)	~0	mC34	C2/m
(αPu)	~0	mP16	P2$_1$/m
βPu$_{19}$Os	3 to >6	oP52	Pnna
αPu$_{19}$Os	3 to >6	oC40	Cmca
βPu$_3$Os	~21 to <22	...	...
αPu$_3$Os	~17 to >22	...	...
Pu$_5$Os$_3$	~31.9	tI32	I4/mcm
PuOs$_2$	61.0	hP12	P6$_3$/mmc
Other reported phase			
PuOs$_2$	61.0	cF24	Fd$\bar{3}$m

Os-Re

M.A. Tylkina, V.P. Polyakova, and E.M. Savitskii, 1962

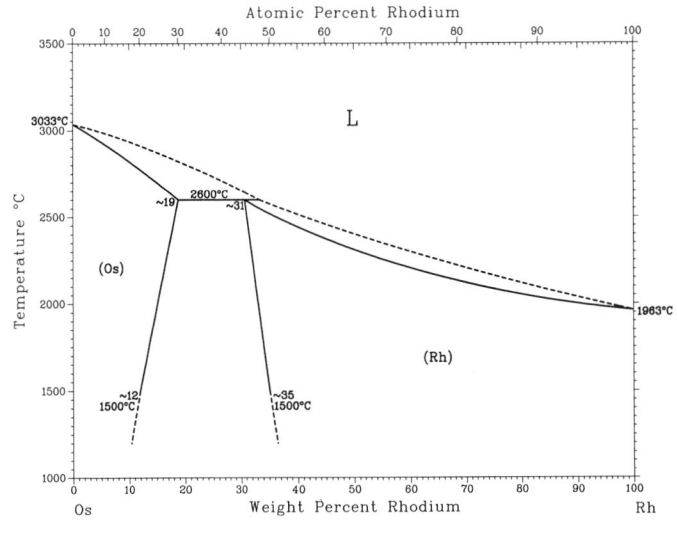

Phase	Composition, wt% Re	Pearson symbol	Space group
(Os,Re)	0 to 100	hP2	P6$_3$/mmc

Os-Rh

H. Okamoto, 1990

Phase	Composition, wt% Rh	Pearson symbol	Space group
(Os)	0 to ~19	hP2	P6$_3$/mmc
(Rh)	~31 to 100	cF2	Fm$\bar{3}$m

Os-Ru

M.A. Tylkina, V.P. Polyakova, and E.M. Savitskii, 1962

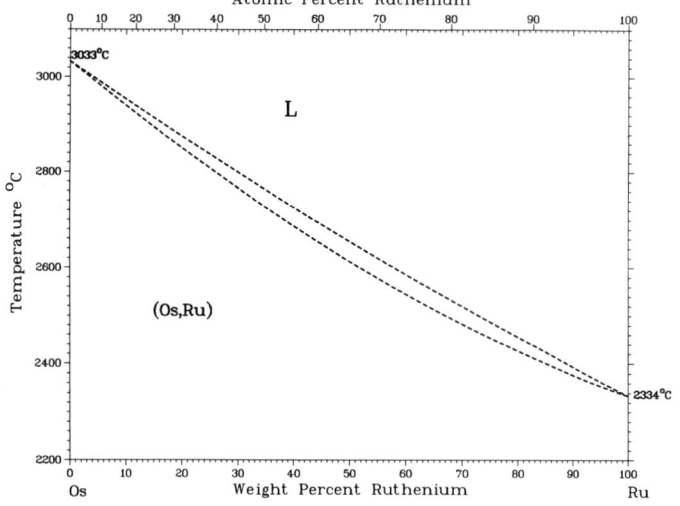

Phase	Composition, wt% Ru	Pearson symbol	Space group
(Os,Ru)	0 to 100	hP2	P6₃/mmc

Os-Si

H. Okamoto, 1990

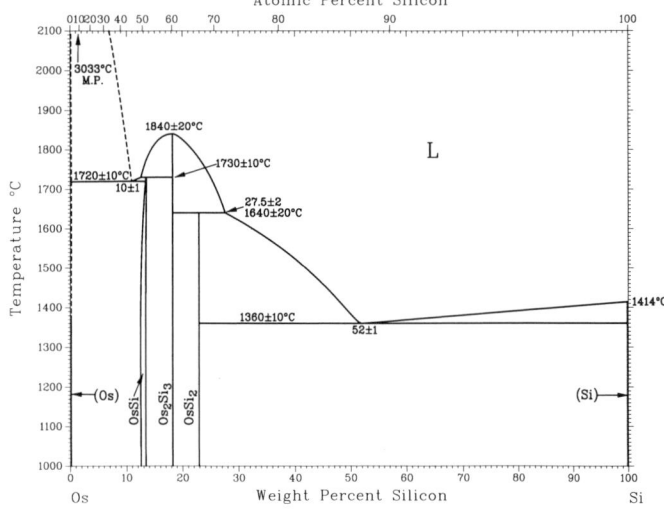

Phase	Composition, wt% Si	Pearson symbol	Space group
(Os)	0	hP2	P6₃/mmc
OsSi	12.9	cP8	P2₁3
Os₂Si₃	18	oP40	Pbcn
OsSi₂	22.8	oC48	Cmca
(Si)	100	cF8	Fd3̄m
Metastable phase			
OsSi₂·m	22.8	mC12	C2/m

Os-Ti

J.L. Murray, 1990

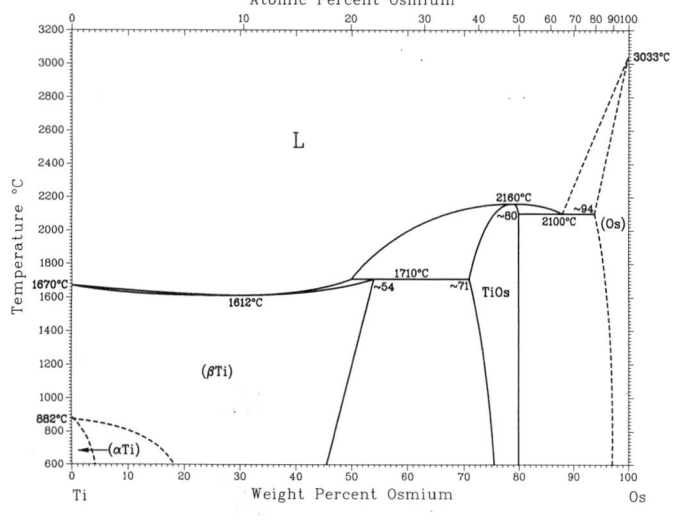

Phase	Composition, wt% Os	Pearson symbol	Space group
(βTi)	0 to 54	cI2	Im3̄m
(αTi)	0 to 4	hP2	P6₃/mmc
TiOs	~71 to ~80	cP2	Pm3̄m
(Os)	~94 to 100	hP2	P6₃/mmc

Os-U

From [Shunk]

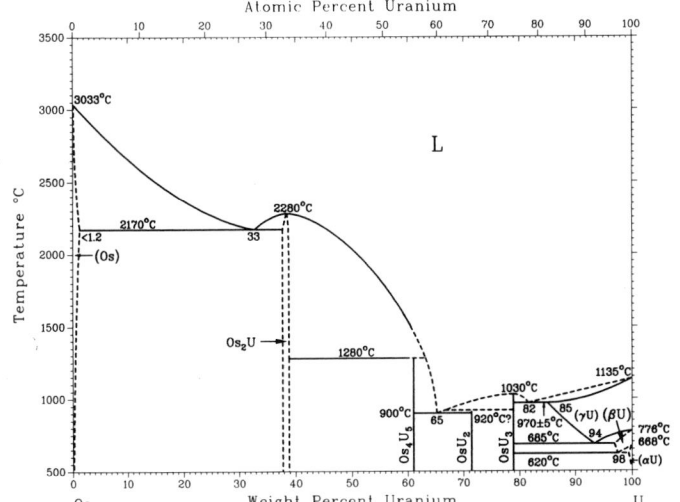

Phase	Composition, wt% U	Pearson symbol	Space group
(Os)	0 to <1.2	hP2	P6₃/mmc
Os₂U	~37.6 to 39	cF24	Fd3̄m
Os₄U₅	~61.0	...	...
OsU₂	~71.5	m*12	...
OsU₃	79	...	...
(γU)	85 to 100	cI2	Im3̄m
(βU)	>97 to 100	tP30	P4₂/mnm
(αU)	>99 to 100	oC4	Cmcm

Os-V

J.F. Smith, 1989

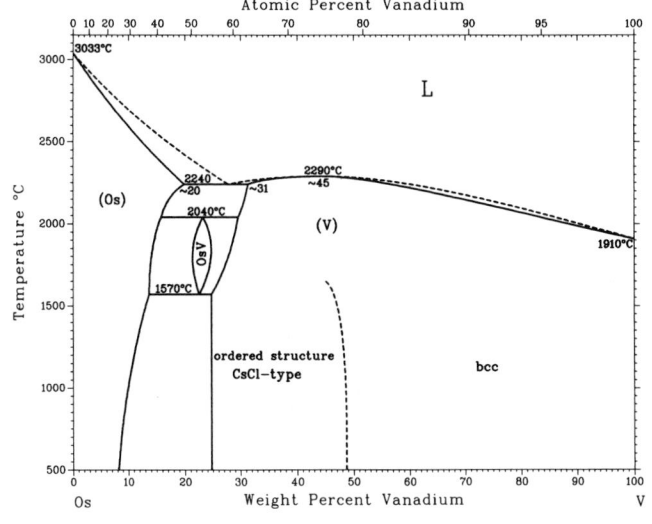

Phase	Composition, wt% V	Pearson symbol	Space group
(Os)	0 to ~20	hP2	P6₃/mmc
OsV	~21.1 to 25	cP8	Pm3̄n
(V)	25 to ?	cP2	Pm3̄m
	? to 100	cI2	Im3̄m

Os-W

S.V. Nagender Naidu and P. Rama Rao, 1991

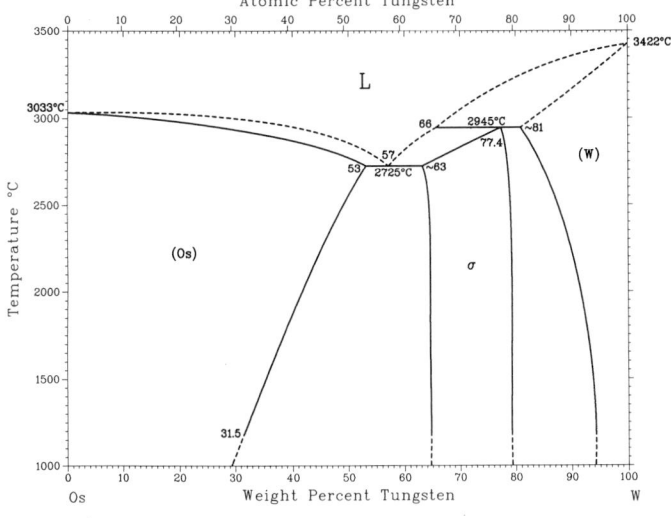

Phase	Composition, wt% W	Pearson symbol	Space group
(Os)	0 to 53	hP2	P6₃/mmc
σ	~63 to ~80	tP30	P4₂/mnm
(W)	~81 to 100	cI2	Im3̄m

Os-Zr

H. Okamoto, 1990

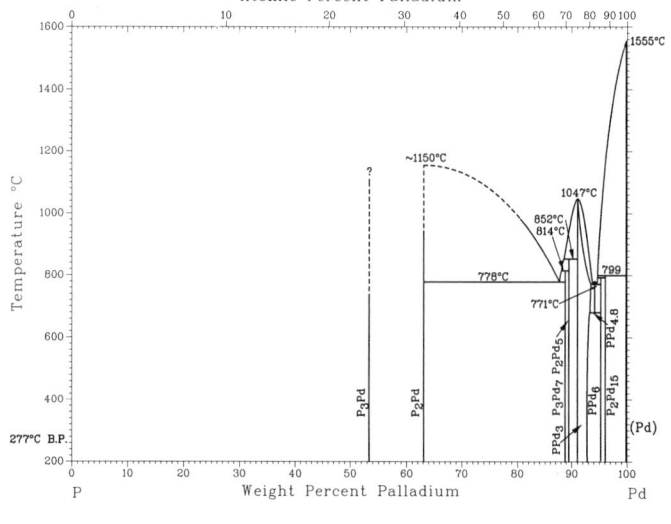

Phase	Composition, wt% Zr	Pearson symbol	Space group
(Os)	0 to 4	hP2	P6₃/mmc
Os₂Zr	~17 to <24	hP12	P6₃/mmc
OsZr	32.4	cP2	Pm$\bar{3}$m
Os₄Zr₁₁	~56.8	cF120	Fm$\bar{3}$m
(βZr)	~75 to 100	cI2	Im$\bar{3}$m
(αZr)	98 to 100	hP2	P6₃/mmc

P-Pd

H. Okamoto, unpublished

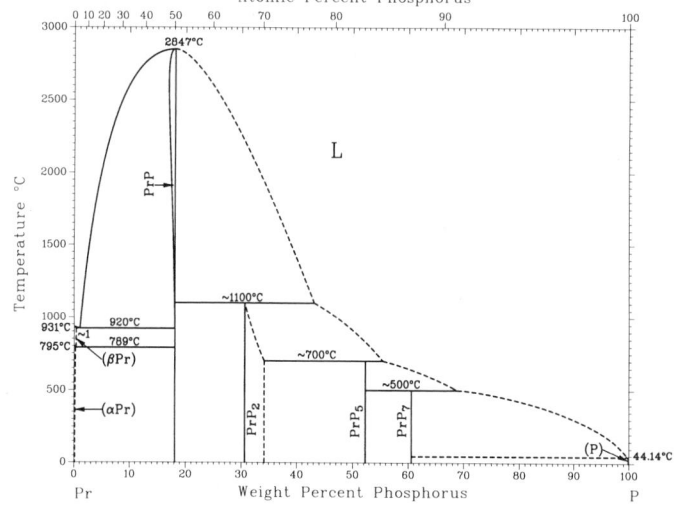

Phase	Composition, wt% Pd	Pearson symbol	Space group
P (white)	0	c**	...
P₃Pd	53	cI32	Im$\bar{3}$
P₂Pd	63.2	mC12	C2/c
P₃Pd₇	88.9	hR20	R$\bar{3}$
P₂Pd₅	89.6	...	...
PPd₃	91 to 93.5	oP16	Pnma
PPd₄.₈	94.3	mP24	P2₁
PPd₆	95.4	mP28	P2₁/c
P₂Pd₁₅	96.3	hR17	R$\bar{3}$
(Pd)	100	cF4	Fm$\bar{3}$m

P-Pr

From [Moffatt]

Phase	Composition, wt% P	Pearson symbol	Space group
(βPr)	0 to 0.2	cI2	Im$\bar{3}$m
(αPr)	0 to ~0.07	hP4	P6₃/mmc
PrP	~17 to 18.0	cF8	Fm$\bar{3}$m
PrP₂	30.6 to ?	mP12	P2₁/c
PrP₅	52.3	mP12	P2₁/m
PrP₇	60.6	...	...
(αP)	100	c**	...

P-Ru

H. Okamoto, 1990

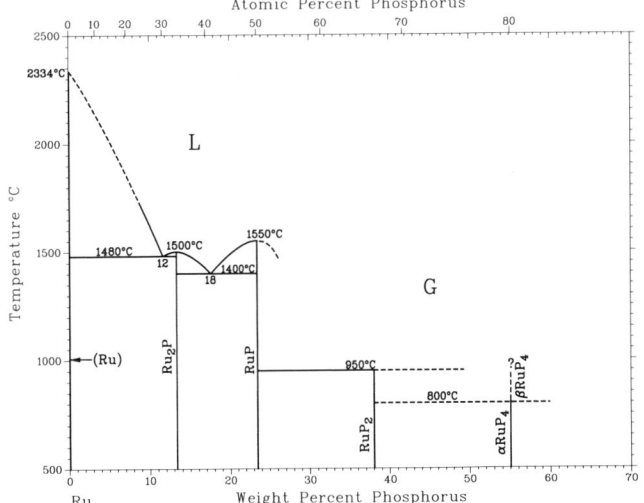

Phase	Composition, wt% P	Pearson symbol	Space group
(Ru)	0	hP2	$P6_3/mmc$
Ru_2P	13.3	oP12	Pnma
RuP	23.5	oP8	Pnma
RuP_2	38.0	oP6	Pnnm
βRuP_4	55	aP15	$P\bar{1}$
αRuP_4	55	mP10	$P2_1/c$

P-Sn

A.C. Vivian, 1920

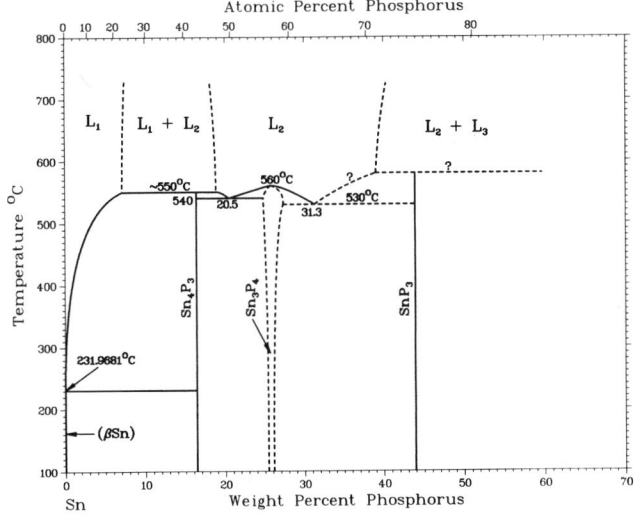

Phase	Composition, wt% P	Pearson symbol	Space group
(βSn)	0	tI4	$I4_1/amd$
(αSn)	0	cF8	$Fd\bar{3}m$
Sn_4P_3	16.4	hR7	$R\bar{3}m$
Sn_3P_4	~25.8	hR7	$R\bar{3}m$
SnP_3	44	hR8	$R\bar{3}m$
Metastable/high-pressure phases			
SnP	20.7	cF8	$Fm\bar{3}m$
	20.7	hP16	P321
	20.7	tI4	I4mm
Sn_7P_{10}	27.1	h**	…

P-Ti

J.L. Murray, 1987

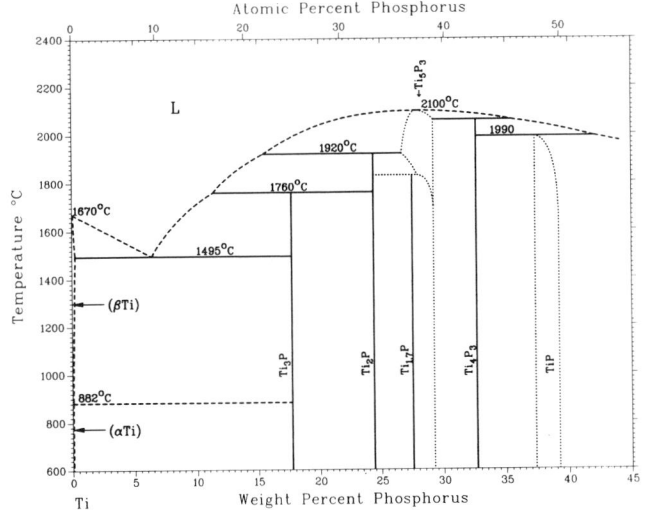

Phase	Composition, wt% P	Pearson symbol	Space group
(βTi)	0 to 0.2	cI2	$Im\bar{3}m$
(αTi)	0 to ?	hP2	$P6_3/mmc$
Ti_3P	18	tP32	$P4_2/n$
Ti_2P	24.4	h**	…
		(a)	…
Ti_5P_3	~27 to ~29	hP16	$P6_3/mcm$
$Ti_{1.7}P$	27.5	oP*	$P2_12_12_1$
Ti_4P_3	32.7	c**	…
Ti_3P_2(b)	28	t**	…
TiP	37 to 39.3	hP8	$P6_3/mmc$
TiP_2	56.4	tI12	I4/mcm

(a) Trigonal. (b) Not shown in diagram

P-Zn

J. Dutkiewicz, 1991

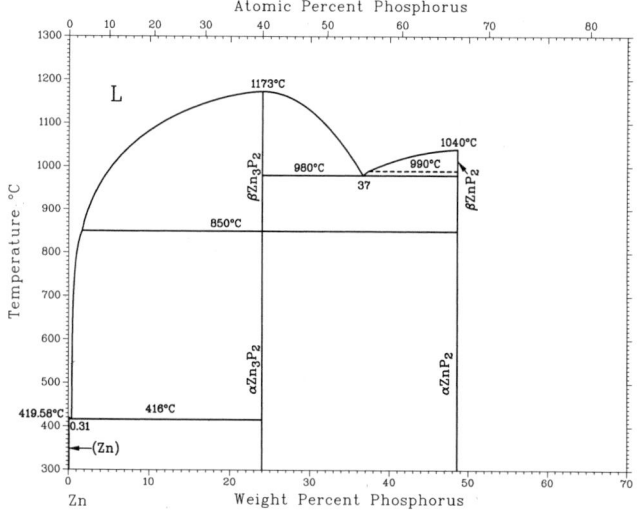

Phase	Composition, wt% P	Pearson symbol	Space group
(Zn)	0	$hP2$	$P6_3/mmc$
βZn_3P_2	24	c^{**}	...
αZn_3P_2	24	$tP40$	$P4_2/nmc$
βZnP_2	48.7	$mP24$	$P2_1/c$
αZnP_2	48.7	$tP24$	$P4_12_12$

Pb-Pd

H. Okamoto, 1990

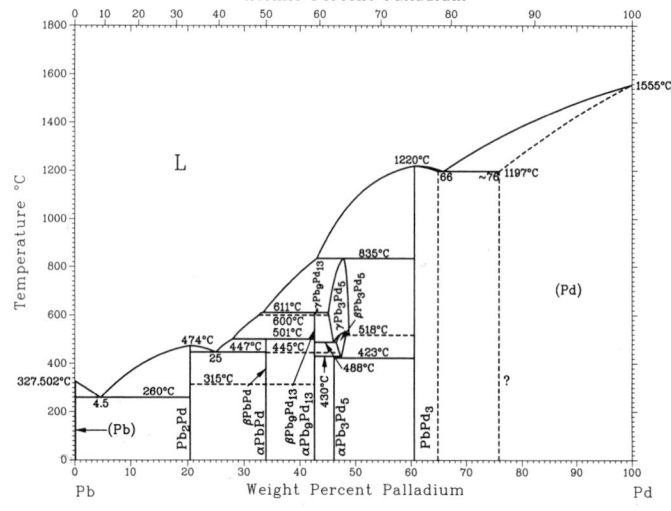

Phase	Composition, wt% Pd	Pearson symbol	Space group
(Pb)	0	$cF4$	$Fm\bar{3}m$
Pb_2Pd	20.4	$tI12$	$I4/mcm$
$PbPd$	33.9	$aP32$	$P\bar{1}$
γPb_9Pd_{13}	42.6	...	...
βPb_9Pd_{13}	42.6	$hP5$	...
αPb_9Pd_{13}	42.6	$mC88$	$C2/c$
γPb_3Pd_5	45 to 48	...	...
βPb_3Pd_5	46 to 47	$hP4$	$P6_3/mmc$
αPb_3Pd_5	46.1	$mC32$	$C2$
$PbPd_3$	61 to 66	$cP4$	$Pm\bar{3}m$
(Pd)	~76 to 100	$cF4$	$Fm\bar{3}m$

Pb-Pr

H. Okamoto, 1990

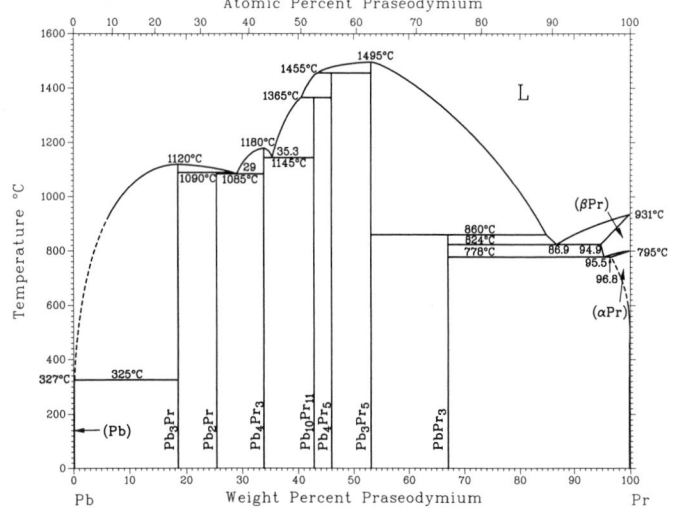

Phase	Composition, wt% Pr	Pearson symbol	Space group
(Pb)	0	$cF4$	$Fm\bar{3}m$
Pb_3Pr	19	$cP4$	$Pm\bar{3}m$
Pb_2Pr	25.3	$tI24$	$I4_1/amd$
Pb_4Pr_3	33.8	...	...
$Pb_{10}Pr_{11}$	42.8	$tI84$	$I4/mmm$
Pb_4Pr_5	46.0	$oP36$	$Pnma$
Pb_3Pr_5	53.1	$hP16$	$P6_3/mcm$
$PbPr_3$	67	$cP4$	$Pm\bar{3}m$
(βPr)	94.9 to 100	$cI2$	$Im\bar{3}m$
(αPr)	96.8 to 100	$hP4$	$P6_3/mmc$

Pb-Pt

From [Hansen]

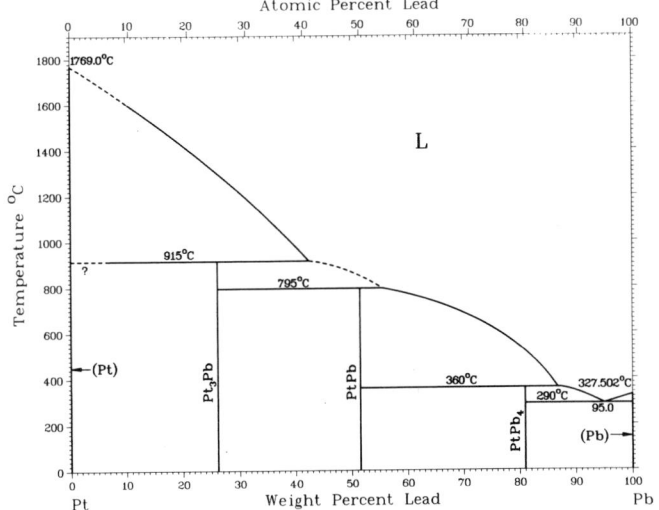

Phase	Composition, wt% Pb	Pearson symbol	Space group
(Pt)	0	cF4	Fm$\overline{3}$m
Pt₃Pb	26	cP4	Pm$\overline{3}$m
PtPb	51.5	hP4	P6₃/mmc
PtPb₄	81	tP10	P4/nbm
(Pb)	100	cF4	Fm$\overline{3}$m

Pb-Pu

E.M. Foltyn and D.E. Peterson, 1988

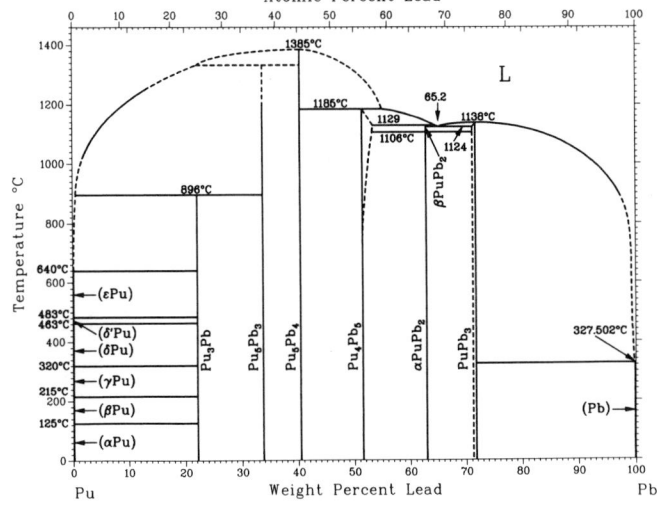

Phase	Composition, wt% Pb	Pearson symbol	Space group
(εPu)	0	cI2	Im$\overline{3}$m
(δ′Pu)	0	tI2	I4/mmm
(δPu)	0	cF4	Fm$\overline{3}$m
(γPu)	0	oF8	Fddd
(βPu)	0	mC34	C2/m
(αPu)	0	mP16	P2₁/m
Pu₃Pb	22	cP4	Pm$\overline{3}$m
Pu₅Pb₃	33.7	tI38	I4/mcm
Pu₅Pb₄	40.4	...	P6₃/mcm
Pu₄Pb₅	51.5 to 53.5	...	P6₃22
βPuPb₂	63.0	...	...
PuPb₂	63.0	tI24	I4₁/amd
PuPb₃	71 to 72	cP4	Pm$\overline{3}$m
(Pb)	100	cF4	Fm$\overline{3}$m

Pb-Rb

A.N. Kuznetsov, K.A. Chuntonov, and S.P. Yatsenko, 1977

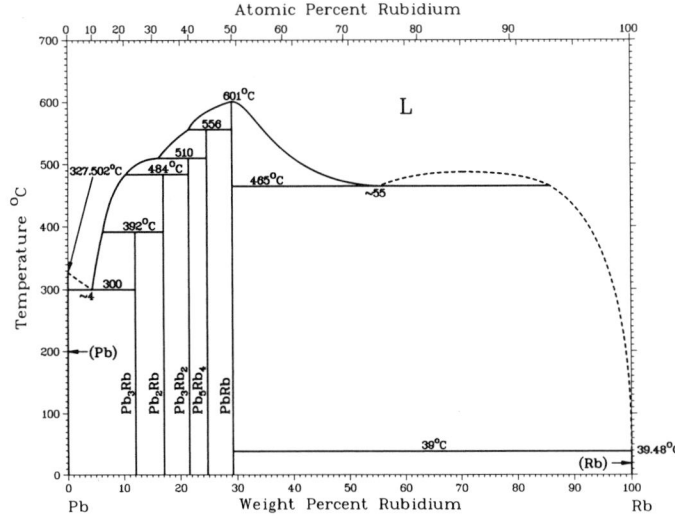

Phase	Composition, wt% Rb	Pearson symbol	Space group
(Pb)	0	cF4	Fm$\overline{3}$m
Pb₃Rb	12	...	...
Pb₂Rb	17.1	...	...
Pb₃Rb₂	22	...	...
Pb₅Rb₄	24.8	...	...
PbRb	29.2	tI64	I4₁/acd
(Rb)	100	cI2	Im$\overline{3}$m

Pb-Rh

H. Okamoto, 1990

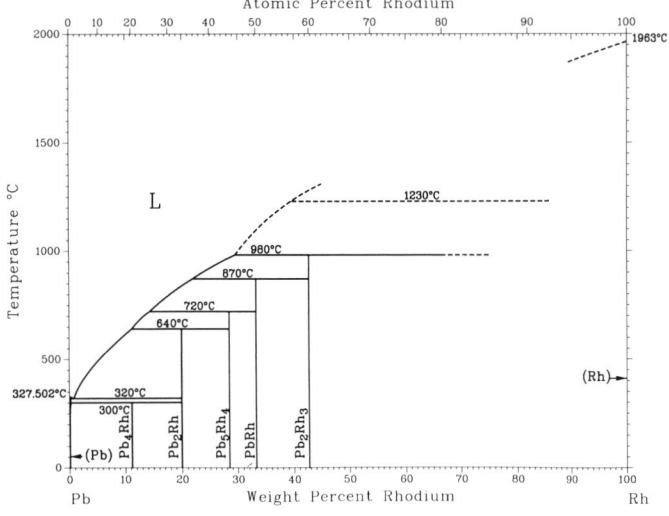

Phase	Composition, wt% Rh	Pearson symbol	Space group
(Pb)	0	cF4	Fm$\bar{3}$m
Pb$_4$Rh	11	...	...
Pb$_2$Rh	19.9	tI12	I4/mcm
Pb$_5$Rh$_4$	28.4	oF72	Fmmm
PbRh	33.2	hP6	P6/mmm
Pb$_2$Rh$_3$	43	hP4	P6$_3$/mmc
(Rh)	100	cF4	Fm$\bar{3}$m

Pb-S

J.-C. Lin, R.C Sharma, and Y.A. Chang, 1986

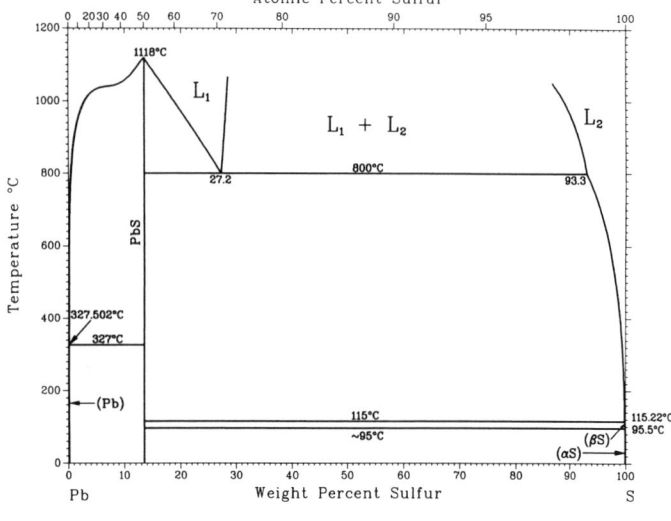

Phase	Composition, wt% S	Pearson symbol	Space group
(Pb)	~0	cF4	Fm$\bar{3}$m
PbS	13.4	cF8	Fm$\bar{3}$m
PbS(a)	13.4	oP8	Pnma
(βS)	100	mP*	P2$_1$/c
(αS)	100	oF128	Fddd

(a) High-pressure phase

Pb-Sb

S. Ashtakala, A.D. Pelton, and C.W. Bale, 1981

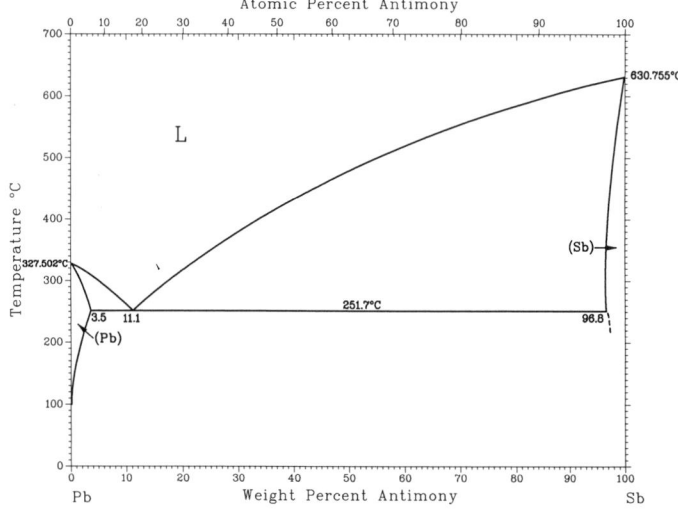

Phase	Composition, wt% Sb	Pearson symbol	Space group
(Pb)	0 to 3.5	cF4	Fm$\bar{3}$m
(Sb)	? to 100	hR2	R$\bar{3}$m

Pb-Se

J.-C. Lin, R.C. Sharma, and Y.A. Chang, unpublished

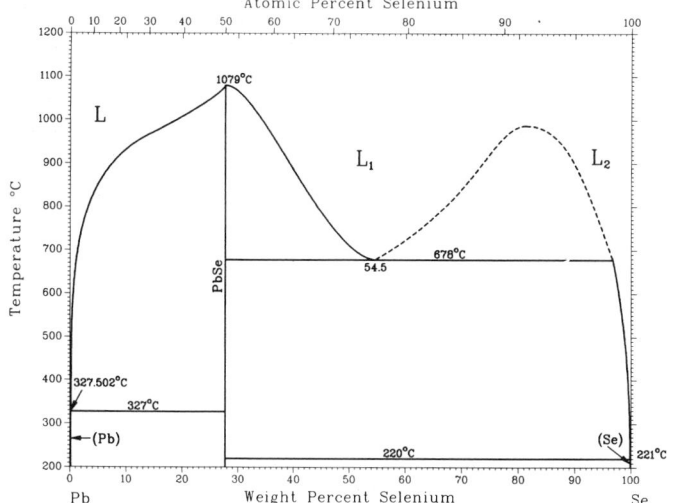

Phase	Composition, wt% Se	Pearson symbol	Space group
(Pb)	~0	cF4	$Fm\overline{3}m$
PbSe	27.6	cF8	$Fm\overline{3}m$
PbSe(HP)	27.6	oP87	$Pnma$
(Se)	~100	hP3	$P3_121$

Pb-Sn

I. Karakaya and W.T. Thompson, 1988

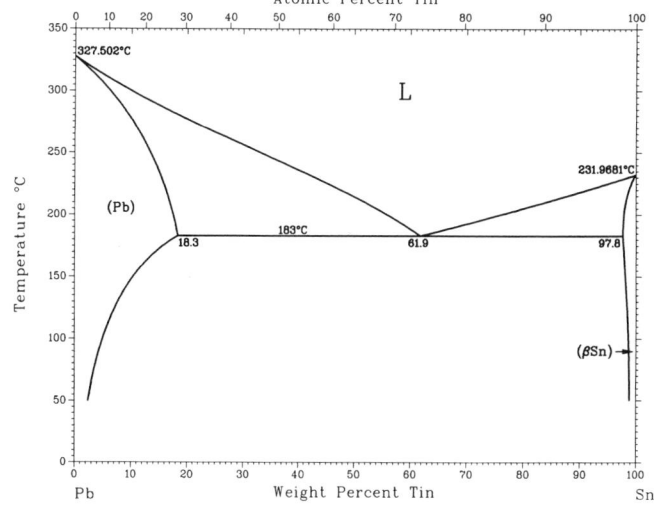

Phase	Composition, wt% Sn	Pearson symbol	Space group
(Pb)	0 to 18.3	cF4	$Fm\overline{3}m$
(βSn)	97.8 to 100	tI4	$I4_1/amd$
(αSn)	100	cF8	$Fd\overline{3}m$
High-pressure phases			
ε(a)	52 to 74	hP1	$P6/mmm$
ε′(b)	52	hP2	$P6_3/mmc$

(a) From phase diagram calculated at 2500 MPa. (b) This phase was claimed for alloys at 350 °C and 5500 MPa.

Pb-Sr

G. Bruzzone, E. Franceschi, and F. Merlo, 1981

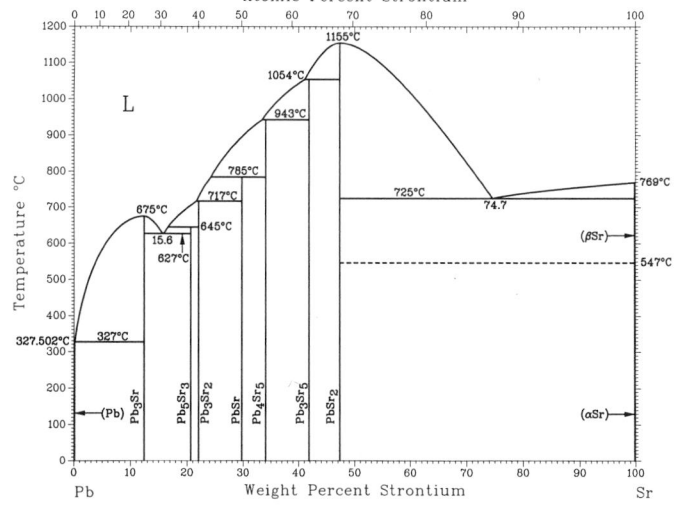

Phase	Composition, wt% Sr	Pearson symbol	Space group
(Pb)	0	cF4	$Fm\overline{3}m$
Pb₃Sr	12	tP4	$P4/mmm$
Pb₅Sr₃	20.2	t**	...
Pb₃Sr₂	22	t**	...
PbSr	29.7	oC8	$Cmcm$
Pb₄Sr₅	34.6	oP36	$Pnma$
Pb₃Sr₅	41.3	tI32	$I4/mcm$
PbSr₂	45.9	oP12	$Pnma$
(βSr)	100	cI2	$Im\overline{3}m$
(αSr)	100	cF4	$Fm\overline{3}m$

Pb-Te

J.-C. Lin, K.C. Hsieh, R.C. Sharma, and Y.A. Chang, 1989

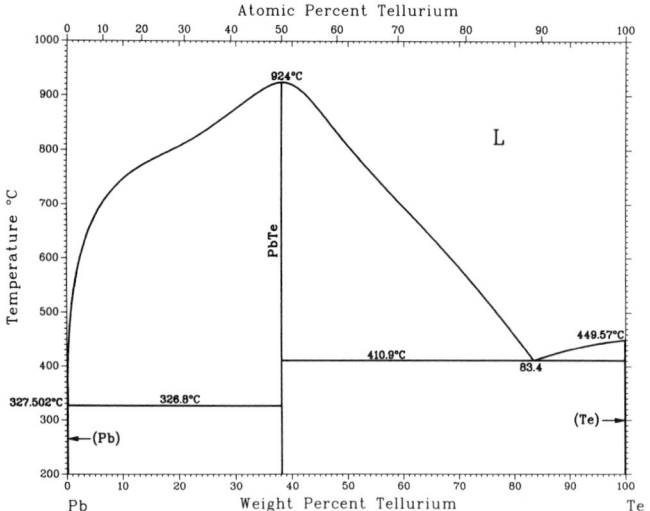

Phase	Composition, wt% Te	Pearson symbol	Space group
(Pb)	0	$cF4$	$Fm\bar{3}m$
PbTe	38.1	$cF8$	$Fm\bar{3}m$
PbTe(HP)	38.1	$oP8$	$Pnma$
(Te)	100	$hP3$	$P3_121$

Pb-Tl

From [Hultgren,B]

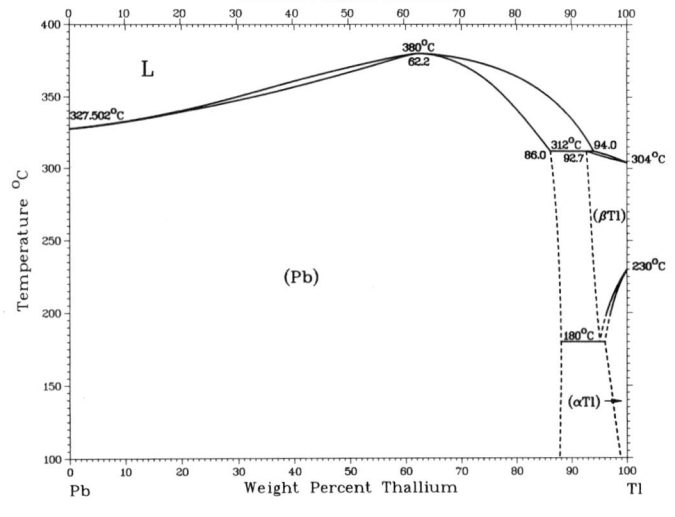

Phase	Composition, wt% Tl	Pearson symbol	Space group
(Pb)	0 to 88	$cF4$	$Fm\bar{3}m$
(βTl)	92.7 to 100	$cI2$	$Im\bar{3}m$
(αTl)	96 to 100	$hP2$	$P6_3/mmc$

Pb-Y

O.N. Carlson, F.A. Schmidt, and D.E. Diesburg, 1967

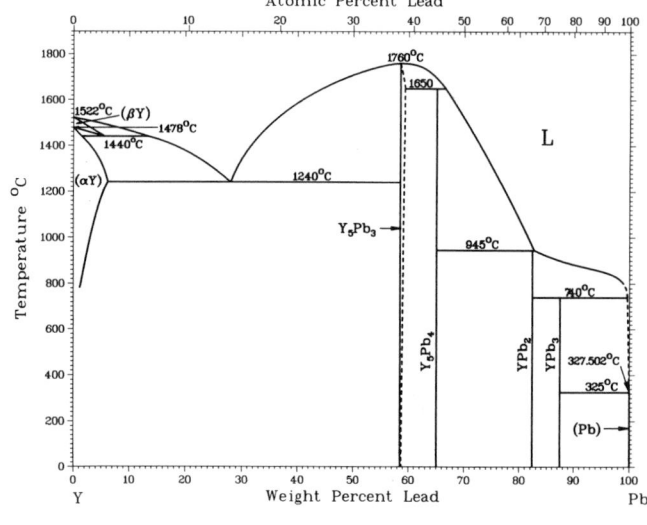

Phase	Composition, wt% Pb	Pearson symbol	Space group
(βY)	0 to 5.6	$cI2$	$Im\bar{3}m$
(αY)	0 to 5.6	$hP2$	$P6_3/mmc$
Y_5Pb_3	~58.3	$hP16$	$P6_3/mcm$
Y_5Pb_4	65.1	$oP6$	$Pnma$
YPb_2	82.3	$oC12$	$Cmcm$
YPb_3	87.5	$cP4$	$Pm\bar{3}m$
(Pb)	100	$cF4$	$Fm\bar{3}m$

Pb-Yb

A. Palenzona and S. Cirafici, 1991

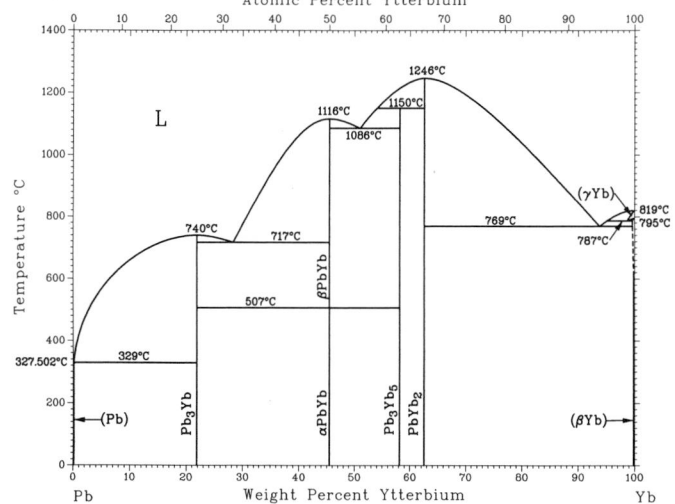

Phase	Composition, wt% Yb	Pearson symbol	Space group
(Pb)	~0	cF4	$Fm\bar{3}m$
Pb₃Yb	22	cP4	$Pm\bar{3}m$
PbYb	45.5	tP4(a)	$P4/mmm$(a)
Pb₃Yb₅	58.2	hP16	$P6_3/mcm$
PbYb₂	62.6	oP12	$Pnma$
(γYb)	~100	cI2	$Im\bar{3}m$
(βYb)	~100	cF4	$Fm\bar{3}m$
(αYb)	~100	hP2	$P6_3/mmc$

(a) Low-temperature modification

Pb-Zn

From [Hansen]

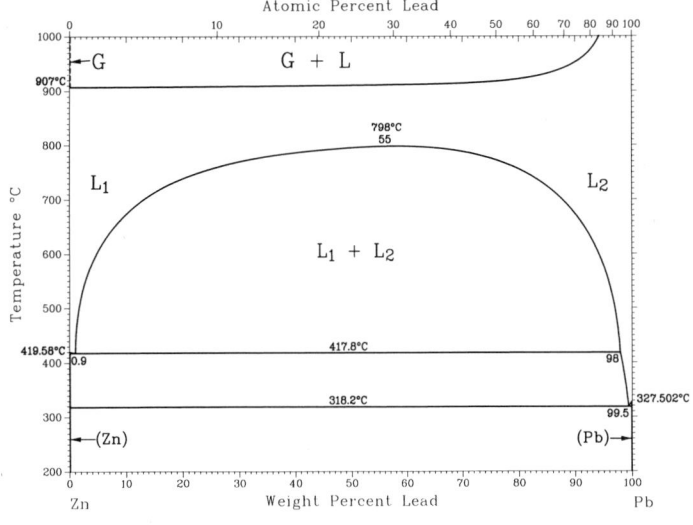

Phase	Composition, wt% Pb	Pearson symbol	Space group
(Zn)	0	hP2	$P6_3/mmc$
(Pb)	100	cF4	$Fm\bar{3}m$

Pd-Pt

H. Okamoto, 1991

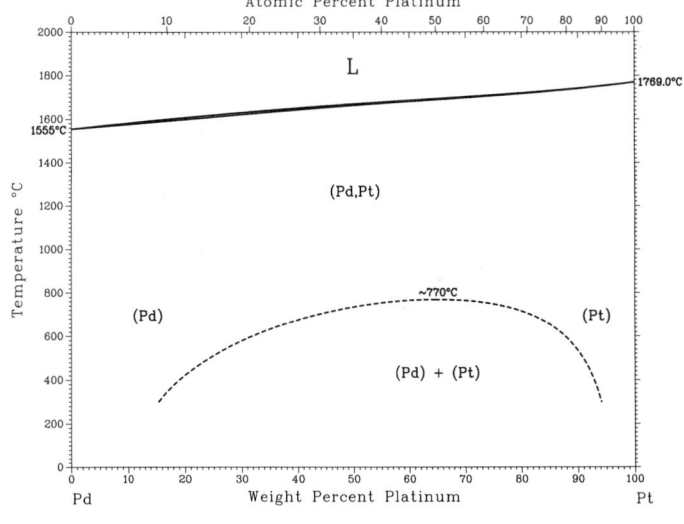

Phase	Composition, wt% Pt	Pearson symbol	Space group
(Pd,Pt)	0 to 100	cF4	$Fm\bar{3}m$

Pd-Pu

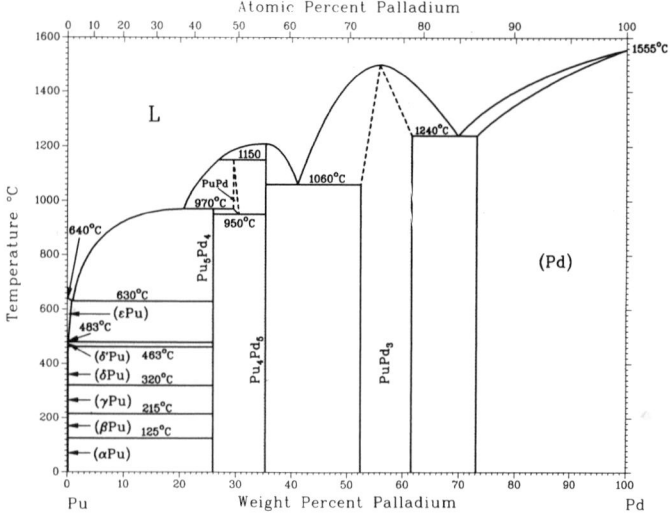

V.I. Kutaitsev, N.T. Chebotarev, I.G. Lebedev, M.A. Andrianov, V.N. Konev, and T.S. Menshikova, 1967

Phase	Composition, wt% Pd	Pearson symbol	Space group
(εPu)	0 to 0.7	cI2	Im$\bar{3}$m
(δ′Pu)	0	tI2	I4/mmm
(δPu)	0	cF4	Fm$\bar{3}$m
(γPu)	0	oF8	Fddd
(βPu)	0	mC34	C2/m
(αPu)	0	mP16	P2$_1$/m
Pu$_5$Pd$_4$	25.8	...	...
PuPd	~30 to 30.4	oP8	Pnma
Pu$_4$Pd$_5$	35.3	...	...
PuPd$_3$	~52.2 to 61.4	cP4	Pm$\bar{3}$m
(Pd)	~73 to 100	cF4	Fm$\bar{3}$m

Pd-Rh

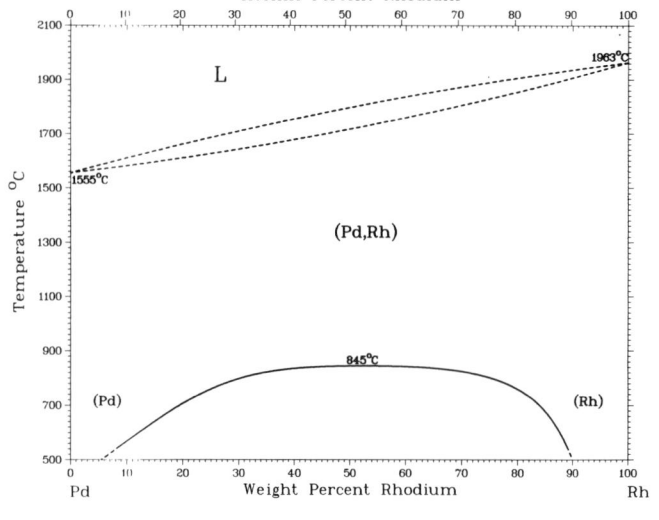

H. Okamoto, 1990

Phase	Composition, wt% Rh	Pearson symbol	Space group
(Pd,Rh)	0 to 100	cF4	Fm$\bar{3}$m

Pd-Ru

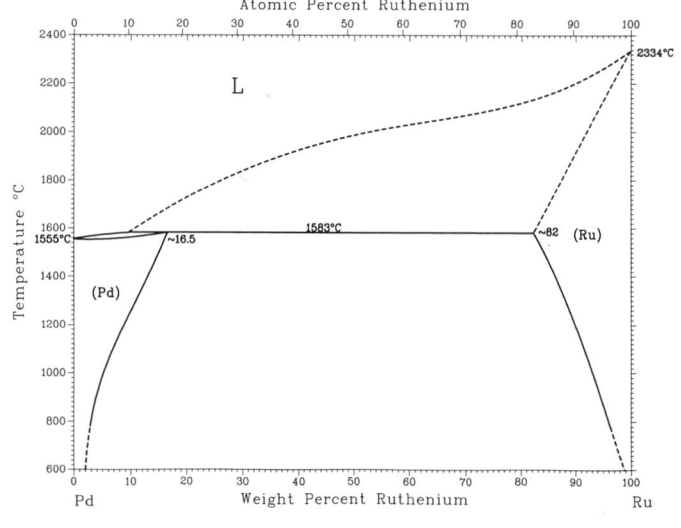

H. Okamoto, 1990

Phase	Composition, wt% Ru	Pearson symbol	Space group
(Pd)	0 to ~16.5	cF4	Fm$\bar{3}$m
(Ru)	~82 to 100	hP2	P6$_3$/mmc

Pd-S

H. Okamoto, 1992

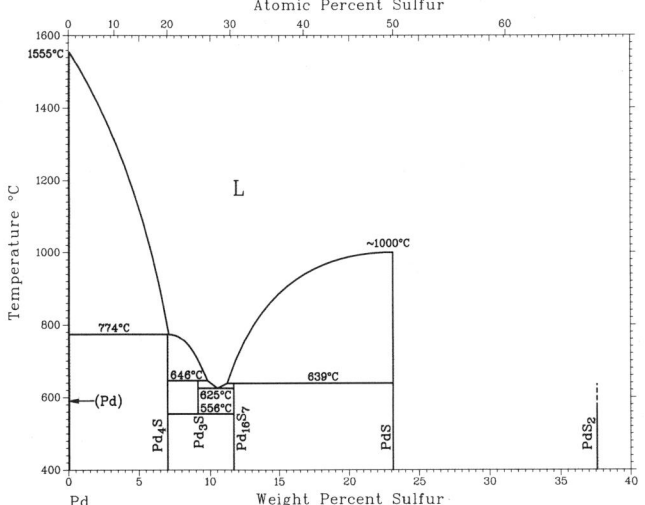

Phase	Composition, wt% S	Pearson symbol	Space group
(Pd)	0	cF4	$Fm\bar{3}m$
Pd_4S	7	tP10	$P\bar{4}2_1c$
Pd_3S	9	oC16	$Ama2$
$Pd_{17}S_7$	11.6	cP64	$Pm\bar{3}m$
PdS	23.2	tP16	$P4_2/m$
PdS_2	37.6	oP12	$Pbca$

Pd-Sb

H. Okamoto, 1992

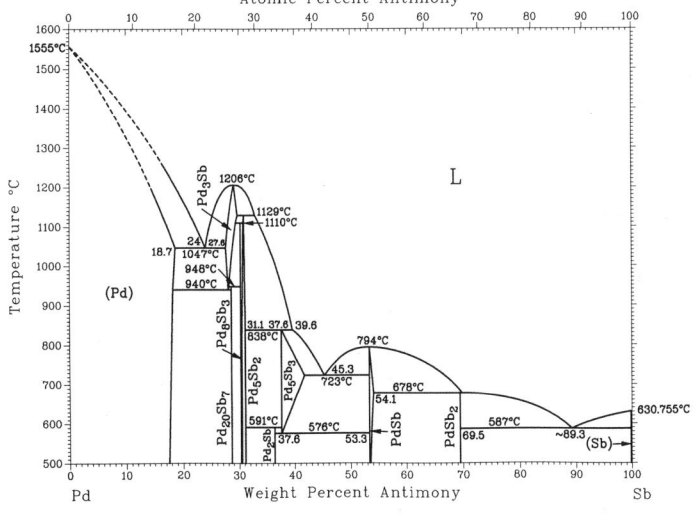

Phase	Composition, wt% Sb	Pearson symbol	Space group
(Pd)	0 to 18.7	cF4	$Fm\bar{3}m$
Pd_3Sb	27.6 to 29.7	cF16	$Fd\bar{3}m$
$Pd_{20}Sb_7$	28.6	hR27	$R\bar{3}$
Pd_8Sb_3	30.3	hR44	$R\bar{3}c$
Pd_5Sb_2	30.5 to 31.1	hP84	$P6_3/mmc$
Pd_2Sb	36.4	oC24	$Cmc2_1$
Pd_5Sb_3	37.4 to 41.7	hP4	$P6_3/mmc$
PdSb	53.4 to 44.2	hP4	$P6_3/mmc$
$PdSb_2$	69.6	cP12	$Pa3$
(Sb)	100	hR2	$R\bar{3}m$

Pd-Se

H. Okamoto, 1992

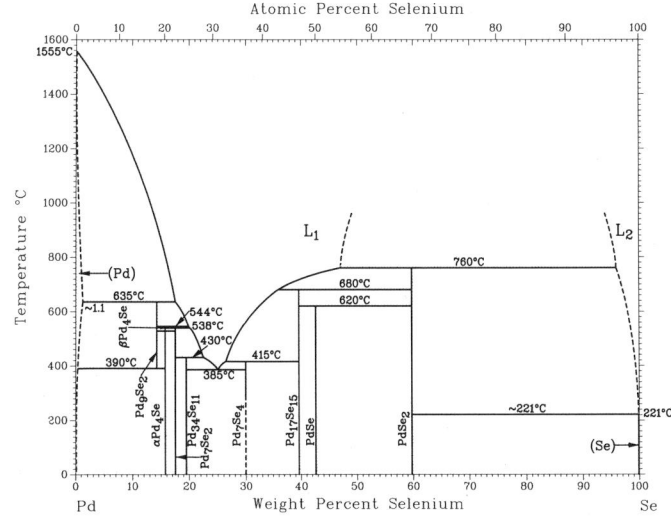

Phase	Composition, wt% Se	Pearson symbol	Space group
(Pd)	0 to ~1.1	cF4	$Fm\bar{3}m$
Pd_9Se_2	14.2	hP*	...
βPd_4Se	16	...	...
αPd_4Se	16	tP10	$P\bar{4}2_1c$
Pd_7Se_2	17.5	m*18	...
$Pd_{34}Se_{11}$	19.3	mP*	$P2_1/n$
Pd_7Se_4	29.8	oP22	$P2_122_1$
$Pd_{17}Se_{15}$	39.6	cP64	$Pm\bar{3}m$
PdSe	42.6	tP16	$P4_2/m$
$PdSe_2$	59.8	oP12	$Pbca$
(γSe)	100	hP3	$P3_121$

Pd-Si

H.C. Baxi and T.B. Massalski, 1991

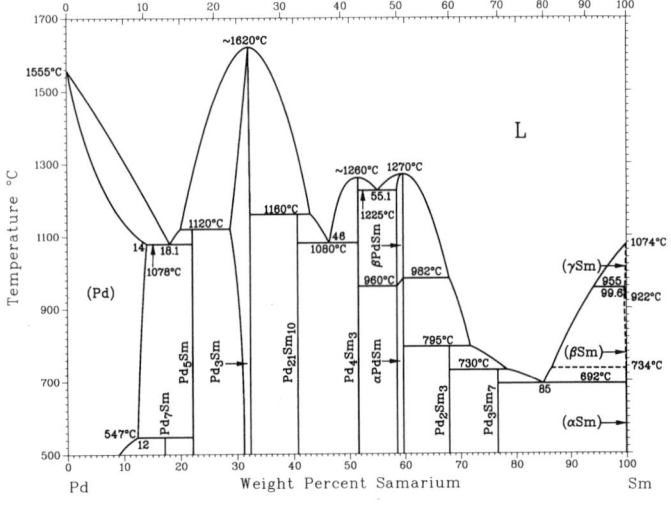

Phase	Composition, wt% Si	Pearson symbol	Space group
(Pd)	0	$cF4$	$Fm\bar{3}m$
Pd₅Si	5.02	$mP24$	$P2_1$
Pd₉Si₂	5.54	$oP44$	$Pnma$
Pd₃Si	8.1	$oP16$	$Pnma$
Pd₂Si	11.5 to 12.1	$hP9$	$P\bar{6}2m$
Pd₂Si'(a)	11.7 to 12.1	(b)	...
PdSi(c)	20.9	$oP8$	$Pnma$
(Si)	100	$cF8$	$Fd\bar{3}m$

(a) Below 1090 °C. (b) Hexagonal superstructure based on the Pd₂Si unit cell. (c) From 972 to 612 °C

Pd-Sm

H. Okamoto, 1990

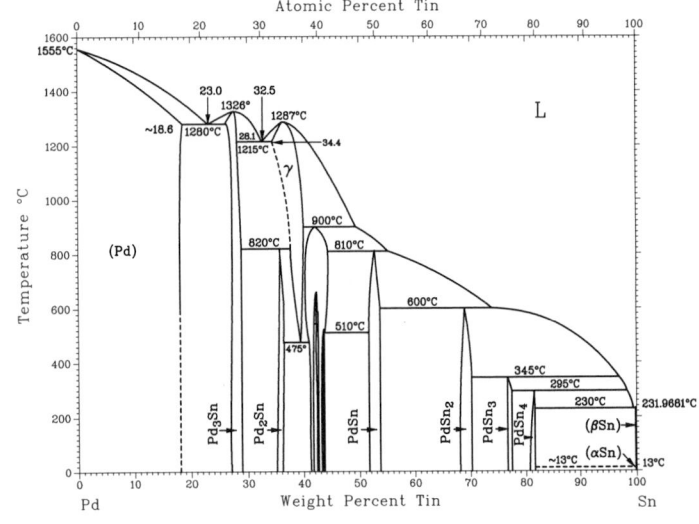

Phase	Composition, wt% Sm	Pearson symbol	Space group
(Pd)	0 to 14	$cF4$	$Fm\bar{3}m$
Pd₇Sm	16.8	$c**$	...
Pd₅Sm	22.1	$o*72$	...
Pd₃Sm	29.1 to 32	$cP4$	$Pm\bar{3}m$
Pd₂₁Sm₁₀	40.3	$mC124$	$C2/m$
Pd₄Sm₃	51.5	$hR14$	$R\bar{3}$
βPdSm	58.6	...	...
αPdSm	58.6	$oC8$	$Cmcm$
Pd₂Sm₃	68	...	...
Pd₂Sm₇	77	$hP20$	$P6_3mc$
(γSm)	100	$cI2$	$Im\bar{3}m$
(βSm)	100	$hP2$	$P6_3/mmc$
(αSm)	100	$hR3$	$R\bar{3}m$

Pd-Sn

H. Okamoto, 1990

Phase	Composition, wt% Sn	Pearson symbol	Space group
(Pd)	0 to ~18.6	$cF4$	$Fm\bar{3}m$
Pd₃Sn	26 to 28.1	$cP4$	$Pm\bar{3}m$
Pd₂Sn	35.8	$oP12$	$Pnma$
γ	34 to 40.1	$hP4$	$P6_3/mmc$
Pd₂₀Sn₁₃	41 to 45	$hP66$	$P3_121$
βPd₃Sn₂	42.4	...	...
αPd₃Sn₂	43	...	...
δ	44	...	...
PdSn	~52.7	$oP8$	$Pnma$
PdSn₂	~69.1	$oC24$	$Aba2$
PdSn₃	~77	$oC32$	$Cmca$
PdSn₄	~82	$oC20$	$Aba2$
(βSn)	100	$tI4$	$I4_1/amd$
(αSn)	100	$cF8$	$Fd\bar{3}m$

(continued)

Pd-Sn phase diagram from 39 to 45 wt% Sn

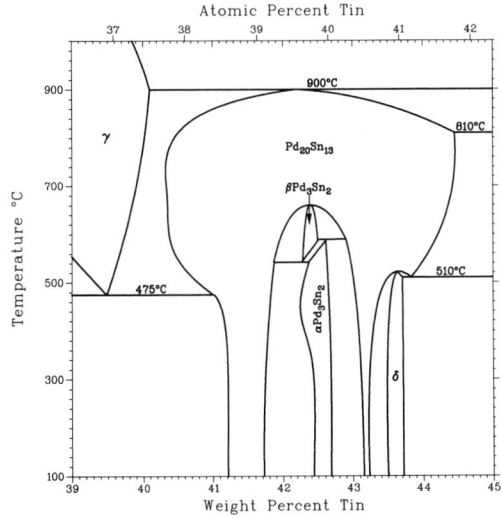

Pd-Te

H. Okamoto, 1992

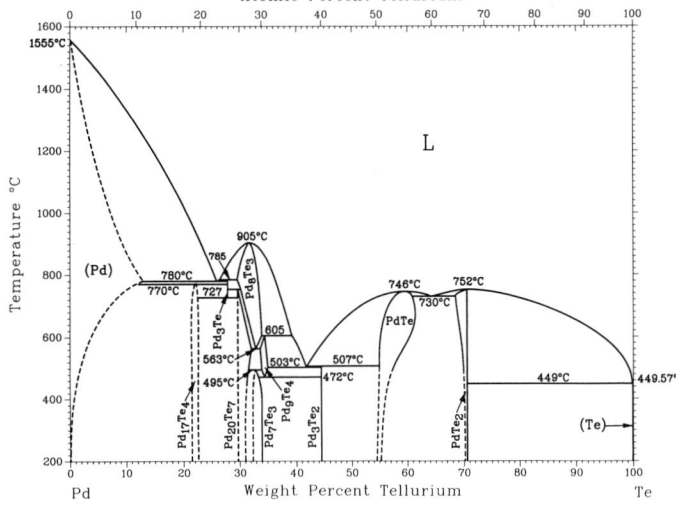

Phase	Composition, wt% Te	Pearson symbol	Space group
(Pd)	0 to 13	cF2	$Fm\overline{3}m$
$Pd_{17}Te_4$	~22	...	...
Pd_3Te	27.8	cI2	$Im\overline{3}m$
$Pd_{20}Te_7$	30 to 34	hR27	R3
Pd_8Te_3	30 to 39	o**	...
Pd_7Te_3	33 to 34	m**	...
Pd_9Te_4	39 to 40	mP52	$P2_1/c$
Pd_3Te_2	44	oC20	Cmcm
PdTe	54.5 to 59	hP4	$P6_3/mmc$
$PdTe_2$	68.5 to 70.6	hP3	$P\overline{3}m1$
(Te)	100	hP3	$P3_121$
Questionable phases			
Pd_4Te	23 to 26	cF104	$F\overline{4}3m$
Pd_3Te_2	44	oP45	$P222_1$

Pd-Ti

J.L. Murray, 1987

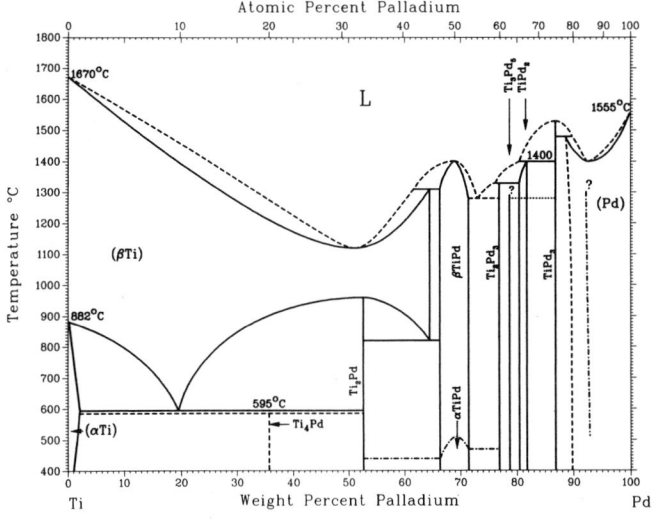

Phase	Composition, wt% Pd	Pearson symbol	Space group
(βTi)	0 to 65	cI2	$Im\overline{3}m$
(αTi)	0 to ~2	hP2	$P6_3/mmc$
Ti_4Pd	36	cP8	$Pm\overline{3}n$
Ti_2Pd	52.6	tI6	I4/mmm
βTiPd	66 to 72	cP2	$Pm\overline{3}m$
αTiPd	66 to 72	oP4	Pmma
Ti_2Pd_3	77	oC20	Cmcm
Ti_3Pd_5	78.7	tP8	P4/mmm
$TiPd_{2-}$	81 to 82	tI6	I4/mmm
$TiPd_2$	81 to 82	(a)	...
$TiPd_3$	87	hP16	$P6_3/mmc$
γ(b)	87 to 92	cP4	P4/mmm
(Pd)	93 to 100	cF4	$Fm\overline{3}m$

(a) Orthorhombic distortion of $MoSi_2$. (b) Possibly an ordered metastable phase. The dot-dash lines show the observed limits of ordering.

Pd-Tl

H. Okamoto, 1990

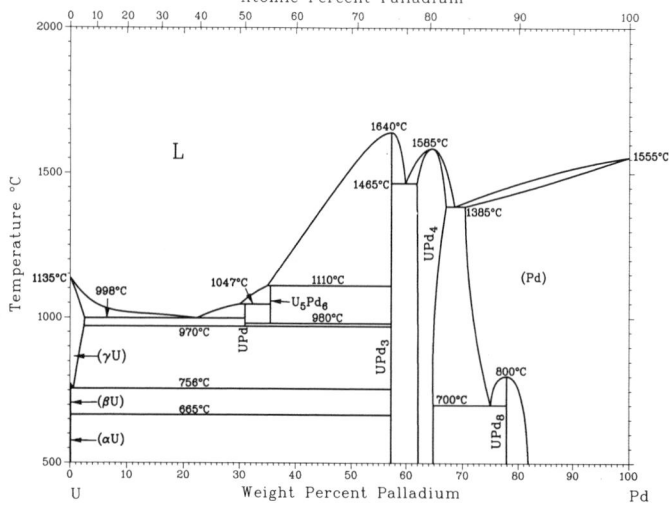

Phase	Composition, wt% Tl	Pearson symbol	Space group
(Pd)	0	cF4	$Fm\overline{3}m$
βPd₃Tl	39	tI8	$I4/mmm$
αPd₃Tl	39	tI16	$I4/mmm$
βPd₂Tl	45 to 56	hP6	$P6_3/mmc$
αPd₂Tl	48.9	oP12	$Pnma$
Pd₁₃Tl₉	56 to 59	hP20	$P\overline{3}1c$
PdTl₂	79.4	tI12	$I4/mcm$
(βTl)	100	cI2	$Im\overline{3}m$
(αTl)	100	hP2	$P6_3/mmc$

Pd-U

H. Okamoto, 1992

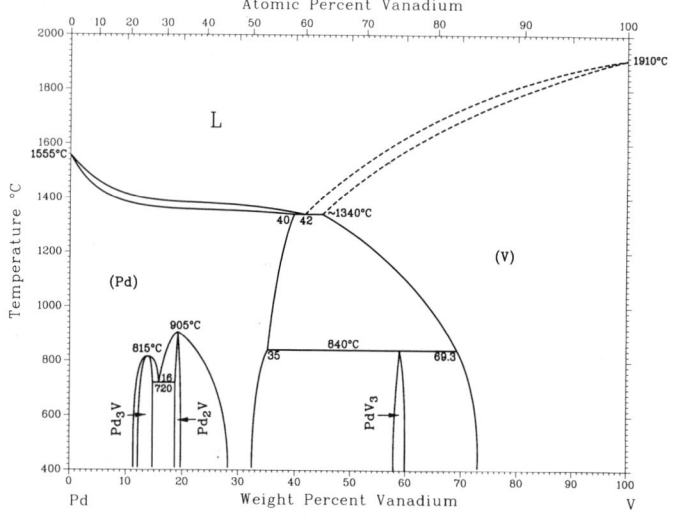

Phase	Composition, wt% Pd	Pearson symbol	Space group
(γU)	0 to 2	cI2	$Im\overline{3}m$
(βU)	0	tP30	$P4_2/mmm$
(αU)	0	oC4	$Cmcm$
UPd	30.9	...	...
U₅Pd₆	34.9	...	...
UPd₃	57	hP16	$P6_3/mmc$
UPd₄	61 to 66	cP4	$Pm\overline{3}m$
UPd₈	78.2	...	...
(Pd)	70 to 100	cF4	$Fm\overline{3}m$

Pd-V

J.F. Smith, 1989

Phase	Composition, wt% V	Pearson symbol	Space group
(Pd)	0 to 40	cF4	$Fm\overline{3}m$
Pd₃V	~14	tI8	$I4/mmm$
Pd₂V	~19.3	oI6	$Immm$
PdV₃	~59	cP8	$Pm\overline{3}n$
(V)	~44.4 to 100	cI2	$Im\overline{3}m$

Pd-W

S.V. Nagender Naidu and P. Rama Rao, 1991

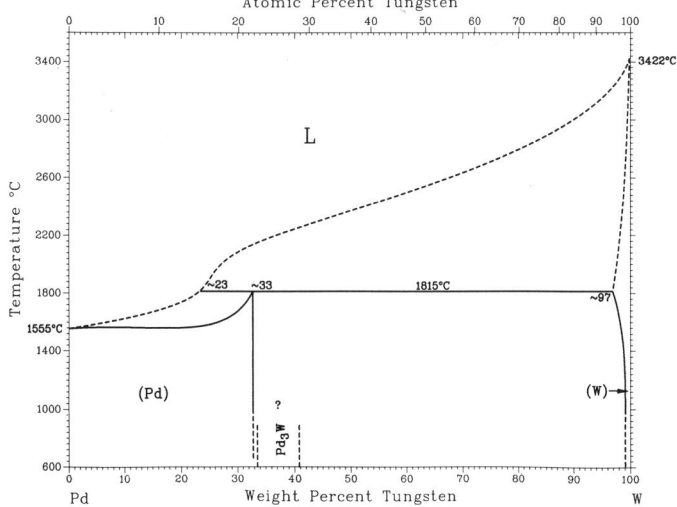

Phase	Composition, wt% W	Pearson symbol	Space group
(Pd)	0 to 33	cF4	$Fm\bar{3}m$
(W)	~97 to 100	cI2	$Im\bar{3}m$

Pd-Y

H. Okamoto, 1990

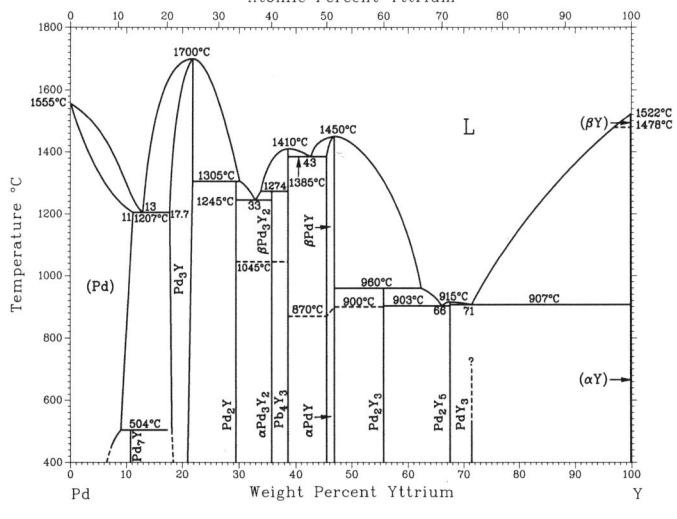

Phase	Composition, wt% Y	Pearson symbol	Space group
(Pd)	0 to 11	cF4	$Fm\bar{3}m$
Pd₇Y	10.7	c**	...
Pd₃Y	17.7 to 22	cP4	$Pm\bar{3}m$
Pd₂Y	29.4	...	...
βPd₃Y₂	36	...	...
αPd₃Y₂	36	...	...
Pd₄Y₃	38.6	hR14	$R\bar{3}$
βPdY	45.5 to ~47	...	...
αPdY	45.5 to ~47	...	...
Pd₂Y₃	56	hR15	$R\bar{3}$
Pd₂Y₅	67.6	cF144	$Fd\bar{3}m$
PdY₃	72	oP16	Pnma
(βY)	100	cI2	$Im\bar{3}m$
(αY)	100	hP2	$P6_3/mmc$

Pd-Yb

A. Iandelli and A. Palenzona, 1973

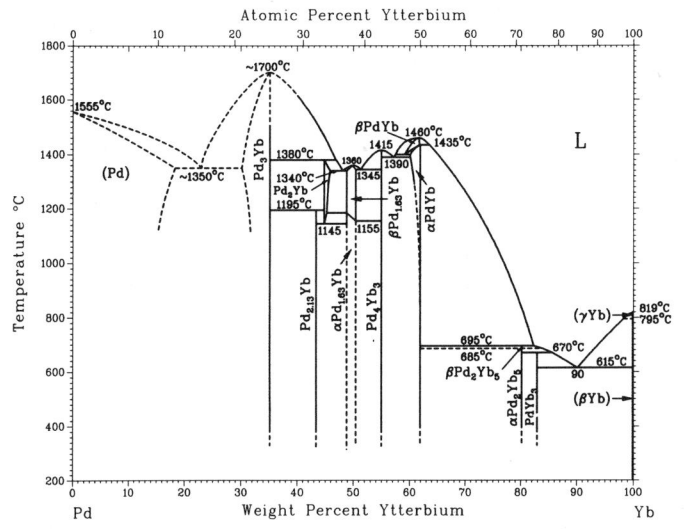

Phase	Composition, wt% Yb	Pearson symbol	Space group
(Pd)	0 to 18	cF4	$Fm\bar{3}m$
Pd₃Yb	30 to 35	cP4	$Pm\bar{3}m$
Pd₂.₁₃Yb	43	...	...
Pd₂Yb	44.8 to 46.1	...	...
βPd₁.₆₃Yb	49 to 50.4	...	...
αPd₁.₆₃Yb	49 to 50.4	...	...
Pd₄Yb₃	55	hR14	$R\bar{3}$
βPdYb	59 to ~61.9	...	...
αPdYb	60 to 61.9	cP2	$Pm\bar{3}m$
βPd₂Yb₅	~80.2	...	...
αPd₂Yb₅	~80.2	...	...
PdYb₃	83	...	...
(γYb)	100	cI2	$Im\bar{3}m$
(βYb)	100	cF4	$Fm\bar{3}m$

Pd-Zn

H. Okamoto, 1990

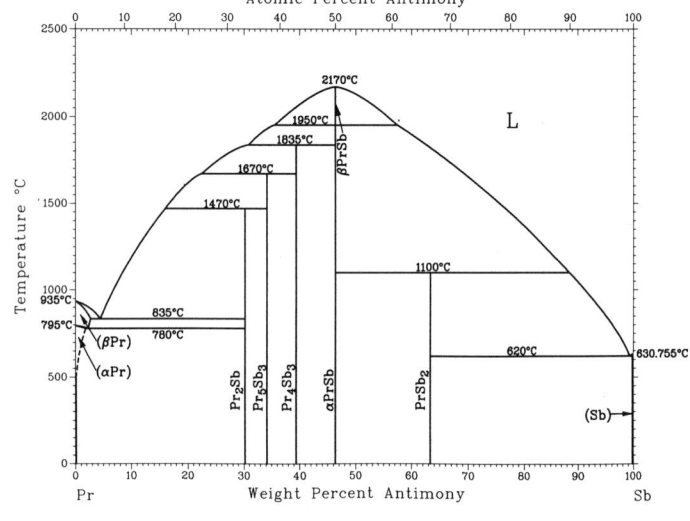

Phase	Composition, wt% Zn	Pearson symbol	Space group
(Pd)	0 to 13	$cF4$	$Fm\bar{3}m$
δ	~23	$oP12$	$Pnma$
β	21 to 53	$cP2$	$Pm\bar{3}m$
$β_1$	27 to 44	$tP4$	$P4/mmm$
$PdZn_2$	55.2	$oC48$	$Cmmm$
γ	66 to 78.4	$cI52$	$I\bar{4}3m$
(Zn)	? to 100	$hP2$	$P6_3/mmc$

Pr-Sb

H. Okamoto, 1990

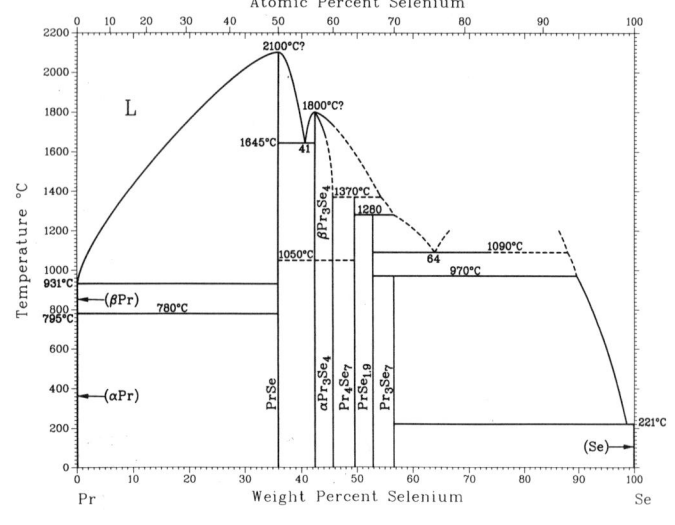

Phase	Composition, wt% Sb	Pearson symbol	Space group
(βPr)	0	$cI2$	$Im\bar{3}m$
(αPr)	0	$hP4$	$P6_3/mmc$
Pr_2Sb	30.1	$tI12$	$I4/mmm$
Pr_5Sb_3	34.1	$hP16$	$P6_3/mcm$
Pr_4Sb_3	39.4	$cI28$	$I\bar{4}3d$
βPrSb	46.4	...	...
αPrSb	46.4	$cF8$	$Fm\bar{3}m$
$PrSb_2$	63.4	$oC24$	$Cmca$
(Sb)	100	$hR2$	$R\bar{3}m$

Pr-Se

E.I. Yarembach, 1970

Phase	Composition, wt% Se	Pearson symbol	Space group
(βPr)	0	$cI2$	$Im\bar{3}m$
(αPr)	0	$hP4$	$P6_3/mmc$
PrSe	35.9	$cF8$	$Fm\bar{3}m$
$βPr_3Se_4$	~42.2 to 46	$cI28$	$I\bar{4}3d$
$αPr_3Se_4$	~42.2 to 46	$tI28$	$I4/mcm$
Pr_4Se_7	49.5	$tP22$	$P4/mmm$
$PrSe_{1.9}$	~52.9	$tP6$	$P4/mmm$
Pr_3Se_7	57	...	...
(Se)	100	$hP3$	$P3_121$

Pr-Si

H. Okamoto, 1990

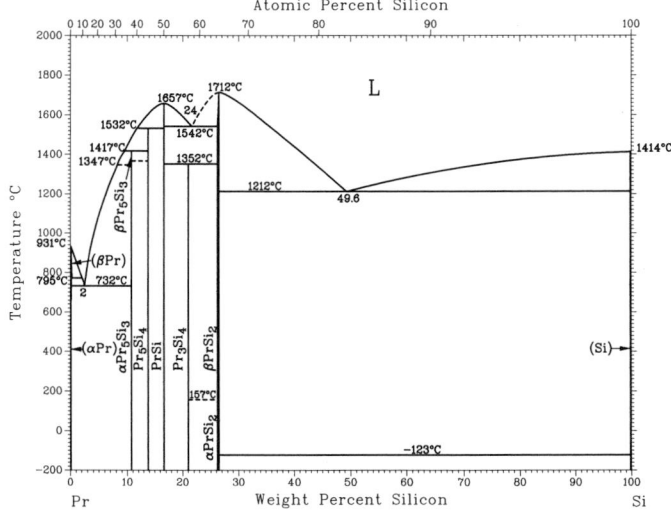

Phase	Composition, wt % Si	Pearson symbol	Space group
(βPr)	0	cI2	Im$\overline{3}$m
(αPr)	0	hP4	P6$_3$/mmc
βPr$_5$Si$_3$	10.7	...	...
αPr$_5$Si$_3$	10.7	tI32	I4/mcm
Pr$_5$Si$_4$	13.7	tP36	P4$_1$2$_1$2
PrSi	16.6	oP8	Pnma
Pr$_3$Si$_4$	21.0	...	...
βPrSi$_2$	26.4	tI12	I4$_1$/amd
αPrSi$_2$	26.4	oI12	Imma
(Si)	100	cF8	Fd$\overline{3}$m

Pr-Sn

H. Okamoto, 1990

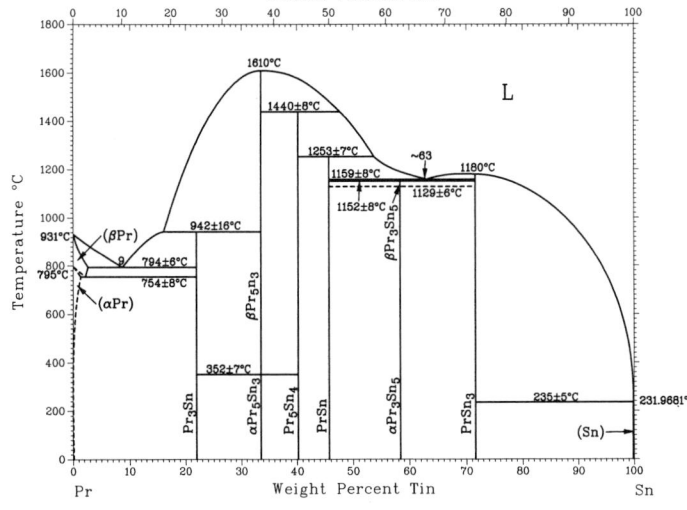

Phase	Composition, wt% Sn	Pearson symbol	Space group
(βPr)	0 to ~3	cI2	Im$\overline{3}$m
(αPr)	0 to ~1.3	hP4	P6$_3$/mmc
Pr$_3$Sn	22	cP4	Pm$\overline{3}$m
βPr$_5$Sn$_3$	33.6	hP16	P6$_3$/mcm
αPr$_5$Sn$_3$	33.6	tI32	I4/mcm
Pr$_5$Sn$_4$	40.2	oP36	Pnma
PrSn	45.7	...	...
βPr$_3$Sn$_5$	58.4	...	...
αPr$_3$Sn$_5$	58.4	...	...
PrSn$_3$	72	cP4	Pm$\overline{3}$m
(βSn)	100	tI4	I4$_1$/amd
(αSn)	100	cF8	Fd$\overline{3}$m

Pr-Te

E.I. Yarembach, 1970

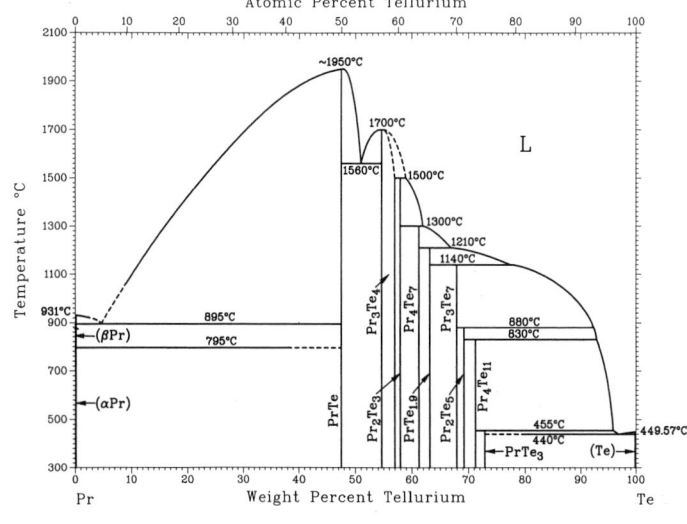

Phase	Composition, wt % Te	Pearson symbol	Space group
(βPr)	0	cI2	Im$\overline{3}$m
(αPr)	0	hP4	P6$_3$/mmc
PrTe	47.5	cF8	Fm$\overline{3}$m
Pr$_3$Te$_4$	54.7 to ~57	cI28	I$\overline{4}$3d
Pr$_2$Te$_3$	58	...	...
Pr$_4$Te$_7$	~61.3	...	...
PrTe$_{1.9}$	~63.2	...	...
Pr$_3$Te$_7$	68	...	...
Pr$_2$Te$_5$	69.3	oC28	Cmcm
Pr$_4$Te$_{11}$	71.3	...	...
PrTe$_3$	73	tP16	P4$_2$/n
(Te)	100	hP3	P3$_1$21

Pr-Tl

S. Delfino, A. Saccone, A. Palenzona, and R. Ferro, unpublished

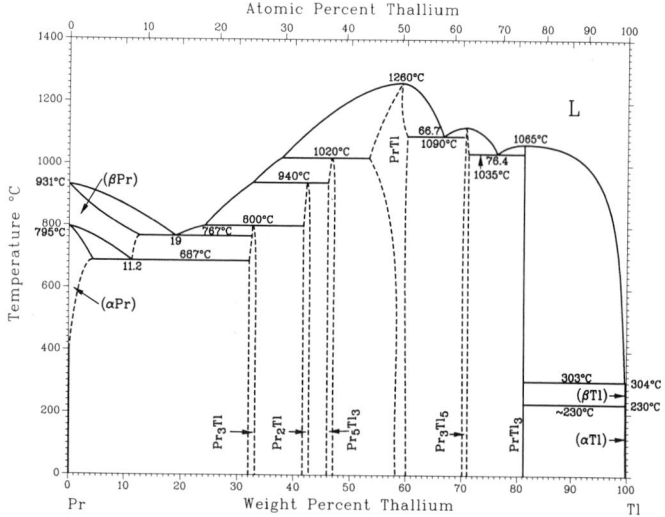

Phase	Composition, wt% Tl	Pearson symbol	Space group
(βPr)	0 to 12.5	cI2	Im$\bar{3}$m
(αPr)	0 to 3.6	hP4	P6₃/mmc
Pr₃Tl(a)	~32.0 to ~33.2	cP4	Pm$\bar{3}$m
	~33	cF4	Fm$\bar{3}$m
Pr₂Tl	~42 to ~43	hP6	P6₃/mmc
Pr₅Tl₃	~46 to ~47	tI32	I4/mcm
PrTl(b)	~53 to ~60	cP2	Pm$\bar{3}$m
		or cI2	Im$\bar{3}$m
PrTl(c)	~53 to ~60	tP2	P4/mmm
Pr₃Tl₅	~70 to ~71	oC32	Cmcm
PrTl₃	81	cP4	Pm$\bar{3}$m
(βTl)	100	cI2	Im$\bar{3}$m
(αTl)	100	hP2	P6₃/mmc

(a) A cP4-cF4 order-disorder transformation in this phase has been suggested. (b) Cubic structure presumed to be room- and higher-temperature phases. (c) Tetragonal structure presumed to be lower-temperature phase

Pr-Zn

J.T. Mason and P. Chiotti, 1970

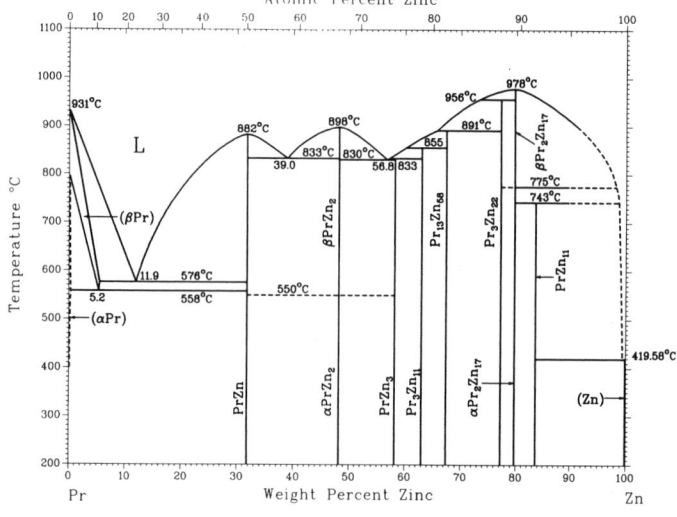

Phase	Composition, wt% Zn	Pearson symbol	Space group
(βPr)	0 to 5	cI2	Im$\bar{3}$m
(αPr)	0 to 0.2	hP4	P6₃/mmc
PrZn	31.7	cP2(a)	Pm$\bar{3}$m
βPrZn₂	48.2	...	...
αPrZn₂	48.2	oI12	Imma
PrZn₃	58	oP16	Pnma
Pr₃Zn₁₁	62.9	oI28	Immm
Pr₁₃Zn₅₈	67.3	hP142	P6₃/mc
Pr₃Zn₂₂	77	tI100	I4₁/amd
βPr₂Zn₁₇	79.6	hR19	R$\bar{3}$m
αPr₂Zn₁₇	79.6	hP38	P6₃/mmc
PrZn₁₁	83.5	tI48	I4₁/amd
(Zn)	100	hP2	P6₃/mmc

(a) t** below 45 K

Pt-Rh

H. Okamoto, 1992

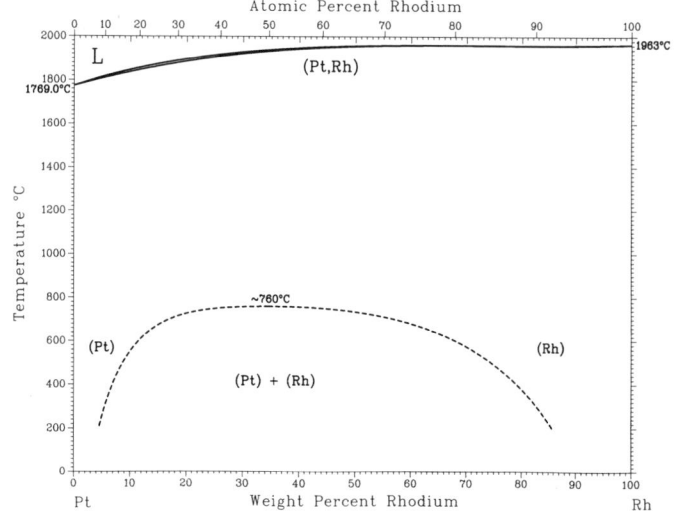

Phase	Composition, wt% Rh	Pearson symbol	Space group
(Pt,Rh)	0 to 100	cF4	Fm$\bar{3}$m

Pt-Si

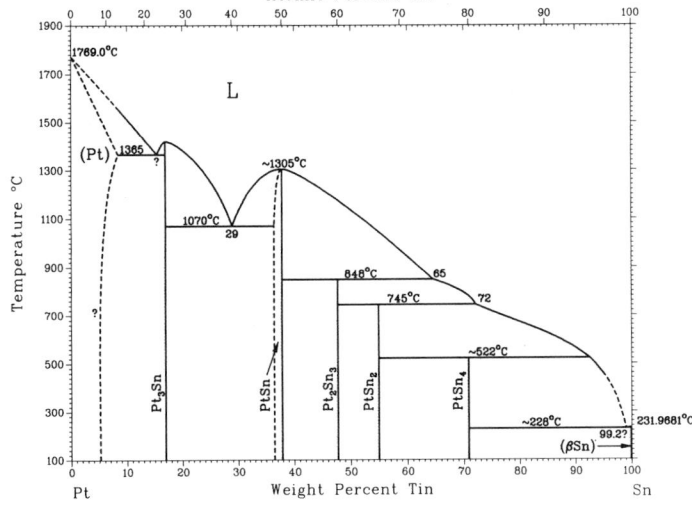

Phase	Composition, wt% Si	Pearson symbol	Space group
(Pt)	0 to 0.2	cF4	Fm$\bar{3}$m
γPt$_3$Si(a)	5	tI16	I4/mcm
βPt$_3$Si	5	oP16	Pnma
αPt$_3$Si	5	mC16	C2/m
βPt$_{12}$Si$_5$	5.7	tI34	I4/m
αPt$_{12}$Si$_5$	5.7	tP68	P4/n
βPt$_2$Si	6.7	hP9	P$\bar{6}$2/m
αPt$_2$Si	6.7	tI6	I4/mmm
Pt$_6$Si$_5$	10.7	mP22	P2$_1$/m
PtSi	12.6	oP8	Pnma
Pt$_2$Si$_3$(b)	18	hP10	P6$_3$/mmc
Pt$_4$Si$_9$(b)	24.4	?	?
(Si)	100	cF8	Fd$\bar{3}$m

(a) Impurity stabilized. (b) Metastable

Pt-Sn

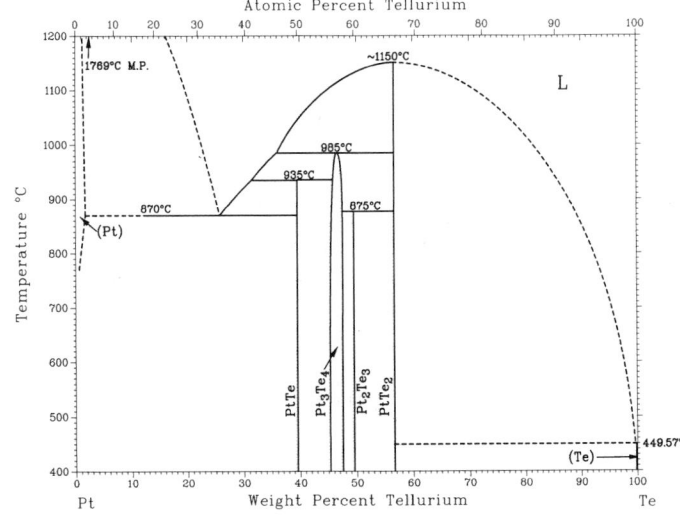

Phase	Composition, wt% Sn	Pearson symbol	Space group
(Pt)	0 to ?	cF4	Fm$\bar{3}$m
Pt$_3$Sn	17	cP4	Pm$\bar{3}$m
PtSn	>36 to 37.8	hP4	P6$_3$/mmc
Pt$_2$Sn$_3$	48	hP10	P6$_3$/mmc
PtSn$_2$	54.9	cF12	Fm$\bar{3}$m
PtSn$_4$	71	oC20	Aba2
(βSn)	100	tI4	I4$_1$/amd
(αSn)	100	cF8	Fd$\bar{3}$m

Pt-Te

Phase	Composition, wt% Te	Pearson symbol	Space group
(Pt)	0 to ?	cF4	Fm$\bar{3}$m
PtTe	39.5	mC8	C2/m
Pt$_3$Te$_4$	~46.5	mC14	C2/m
Pt$_2$Te$_3$	50	mC20	C2/m
PtTe$_2$	56.7	hP3	P$\bar{3}$m1
(Te)	100	hP3	P3$_1$21

Pt-Ti

J.L. Murray, 1987

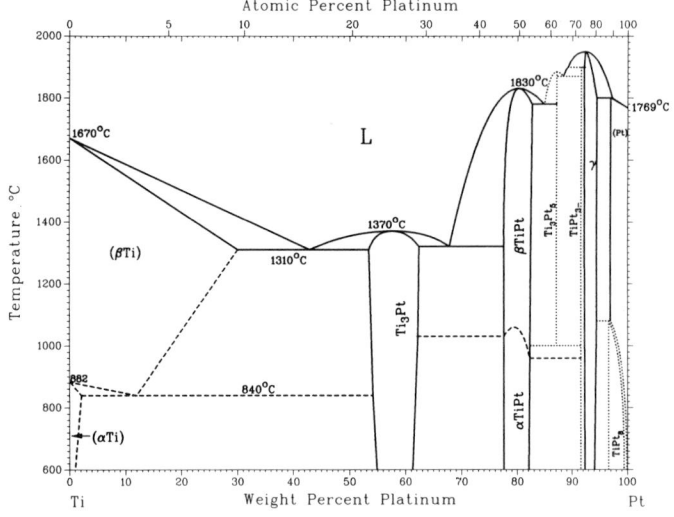

Phase	Composition, wt% Pt	Pearson symbol	Space group
(βTi)	0 to 31	cI2	$Im\bar{3}m$
(αTi)	0 to 2.0	hP2	$P6_3/mmc$
Ti$_3$Pt	54 to 63	cP8	$Pm\bar{3}n$
βTiPt	78 to 83	oP2	Pmma
αTiPt	78 to 83	oP4	Pmma
Ti$_3$Pt$_5$	87.2	oI32	Ibam
TiPt$_{3-}$	<92	hP16	$P6_3/mmc$
γ	92 to 95	tP4	$Pm\bar{3}m$
TiPt$_8$	97 to 99.5	tI18	I4/m
(Pt)	95 to 100	cF4	$Fm\bar{3}m$

Pt-Tl

H. Okamoto, 1990

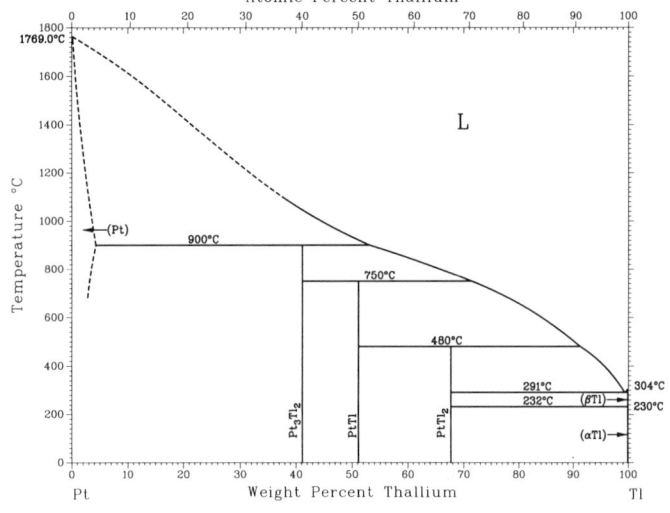

Phase	Composition, wt% Tl	Pearson symbol	Space group
(Pt)	0 to ?	cF4	$Fm\bar{3}m$
Pt$_3$Tl$_2$	41	hP20	$P\bar{3}1c$
PtTl	51.2	hP6	P6/mmm
PtTl$_2$	67.7	tI12	I4/mcm
(βTl)	100	cI2	$Im\bar{3}m$
(αTl)	100	hP2	$P6_3/mmc$

Pt-U

B.A.S. Ross and D.E. Peterson, 1990

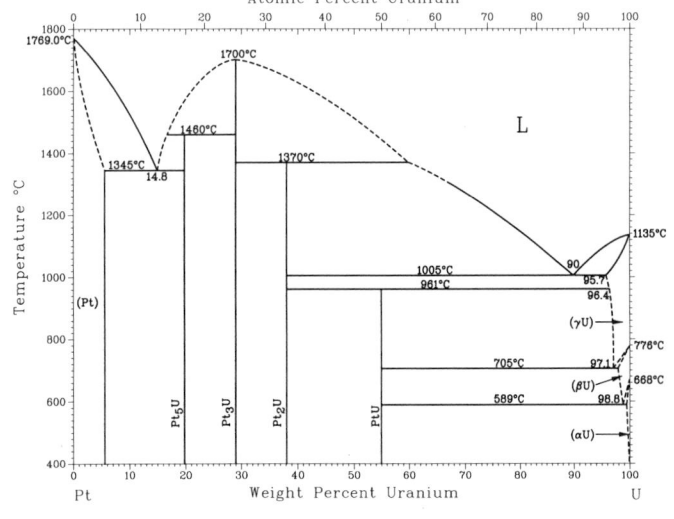

Phase	Composition, wt% U	Pearson symbol	Space group
(Pt)	0 to 5	cF4	$Fm\bar{3}m$
Pt$_5$U	19.7	cF24	$F\bar{4}3m$
Pt$_3$U	29	hP8	$P6_3/mmc$
Pt$_2$U(a)	37.9	oC12	Ama2
PtU	55.0	oC8	Cmcm
(γU)	99.7 to 100	cI2	$Im\bar{3}m$
(βU)	98.1 to 100	tP30	$P4_2/mnm$
(αU)	99.2 to 100	oC4	Cmcm

(a) Distorted structure

Pt-V

J.F. Smith, 1989

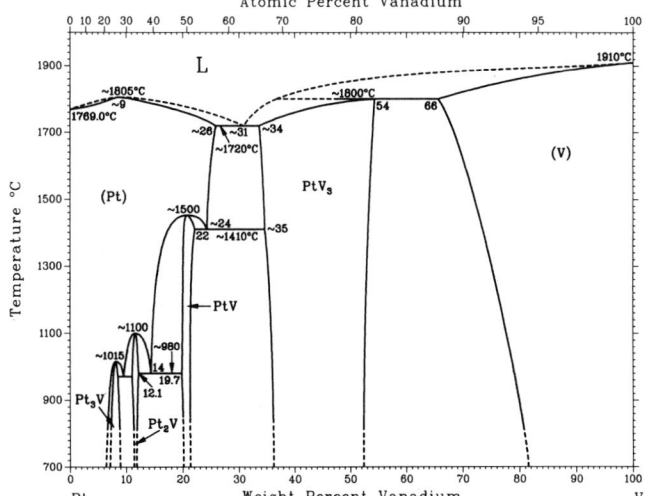

Phase	Composition, wt% V	Pearson symbol	Space group
(Pt)	0 to ~26	cF4	Fm$\bar{3}$m
Pt$_3$V	7 to 8	tI8	I4/mmm
Pt$_2$V	11 to 21.1	oI6	Immm
PtV	19.7 to 22	oP4	Pmma
PtV$_3$	~34 to 54	cP8	Pm$\bar{3}$n
(V)	66 to 100	cI2	Im$\bar{3}$m
Metastable phases			
Pt$_8$V(a)	~3.2	tI18	I4/mmm
Pt$_3$V(b)	6.9 to 7.2	cP4	Pm$\bar{3}$m
PtV	20.7 to 23.5	tP2	P4/mmm
PtV$_3$	~44	cP4	Pm$\bar{3}$m

(a) Possibly misclassified because neither its stability nor metastability is conclusive. (b) Stabilized by oxygen and possibly also by nitrogen and/or carbon

Pt-Zr

H. Okamoto, 1990

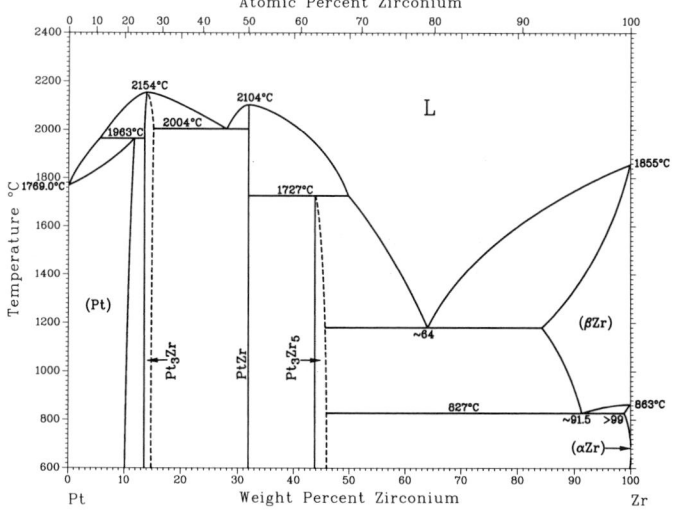

Phase	Composition, wt% Zr	Pearson symbol	Space group
(Pt)	0 to ~12	cF4	Fm$\bar{3}$m
Pt$_3$Zr	14	cP4	Pm$\bar{3}$m
		hP16	P6$_3$/mmc
Pt$_{11}$Zr$_9$(a)	28	tI40	I4/m
βPtZr	31.9	cP2	Pm$\bar{3}$m
αPtZr	31.9	oC8	Cmcm
Pt$_3$Zr$_5$	43.8	hP16	P6$_3$/mcm
(βZr)	~84.3 to 100	cI2	Im$\bar{3}$m
(αZr)	>99 to 100	hP2	P6$_3$/mmc

Note: The polymorphic transformation temperature of PtZr is unknown. (a) Not shown in the diagram

Pu-Sc

H. Okamoto, 1990

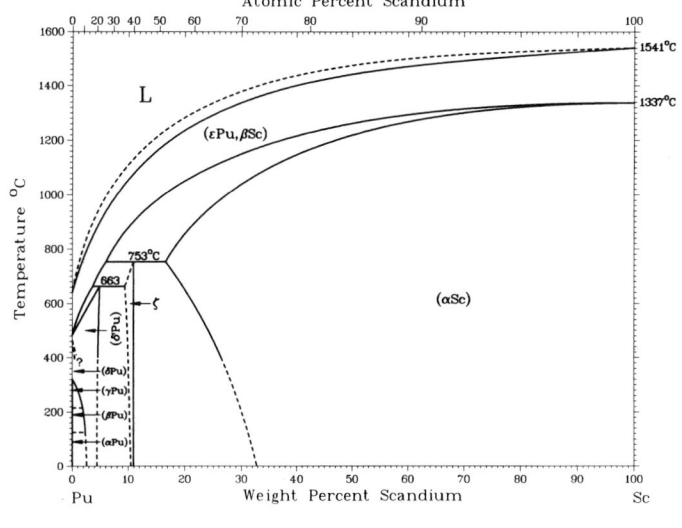

Phase	Composition, wt% Sc	Pearson symbol	Space group
(εPu,βSc)	0 to 100	cI2	Im$\bar{3}$m
(δ'Pu)	0 to ?	tI2	I4/mmm
(δPu)	0 to ?	cF4	Fm$\bar{3}$m
(γPu)	0 to ?	oF8	Fddd
(βPu)	0 to ?	mC34	C2/m
(αPu)	0 to ?	mP16	P2$_1$/m
ζ	? to 11	...	...
(αSc)	17 to 100	hP2	P6$_3$/mmc

Pu-U

H. Okamoto, 1992

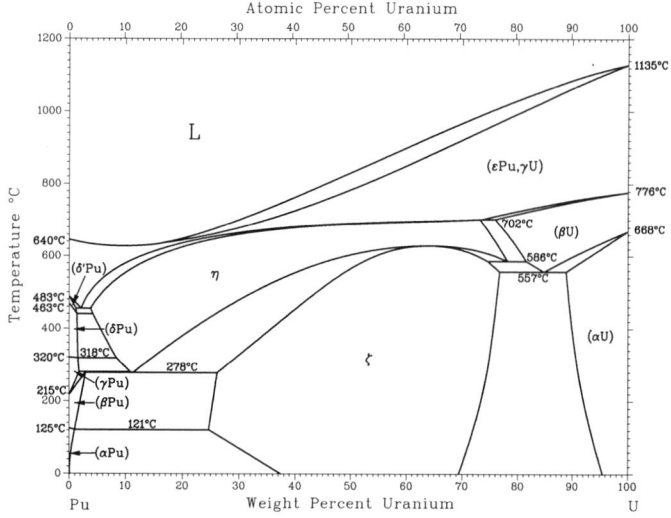

Phase	Composition, wt% U	Pearson symbol	Space group
(εPu,γU)	0 to 100	cI2	Im$\overline{3}$m
(δ'Pu)	0 to 2	tI2	I4/mmm
(δPu)	0 to 2	cF4	Fm$\overline{3}$m
(γPu)	0 to 2	oF8	Fddd
(βPu)	0 to 3	mC34	C2/m
(αPu)	0	mP16	P2$_1$/m
ζ	25 to 77	cP58	...
η	4 to 80	tP52	...
(βU)	77 to 100	tP30	P4$_2$/mnm
(αU)	89 to 100	oC4	Cmcm

Pu-Zn

From [Chiotti]

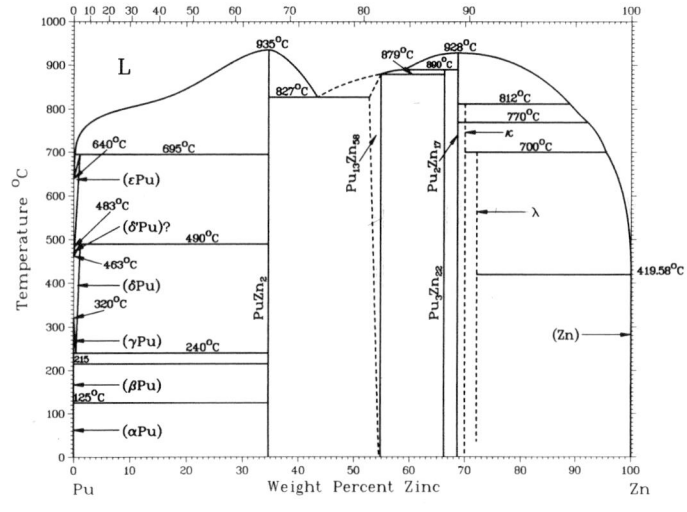

Phase	Composition, wt% Zn	Pearson symbol	Space group
(εPu)	0 to 0.96	cI2	Im$\overline{3}$m
(δ'Pu)	0	tI2	I4/mmm
(δPu)	0 to 1.1	cF4	Fm$\overline{3}$m
(γPu)	0	oF8	Fddd
(βPu)	0	mC34	C2/m
(αPu)	0	mP16	P2$_1$/m
PuZn$_2$	34.9	cF24	Fd$\overline{3}$m
Pu$_3$Zn$_{58}$	~52.5 to 55	hP142	P6$_3$mc
Pu$_3$Zn$_{22}$	66	tI100	I4$_1$/amd
Pu$_2$Zn$_{17}$	69.5	hR*	R$\overline{3}$m
κ (HT)	~71	hP*	P6/mmm
κ (LT)	~71	hP38	P6$_3$/mmc
λ	~71.8	hP*	P6$_3$22
(Zn)	100	hP2	P6$_3$/mmc

Pu-Zr

From [Elliott]

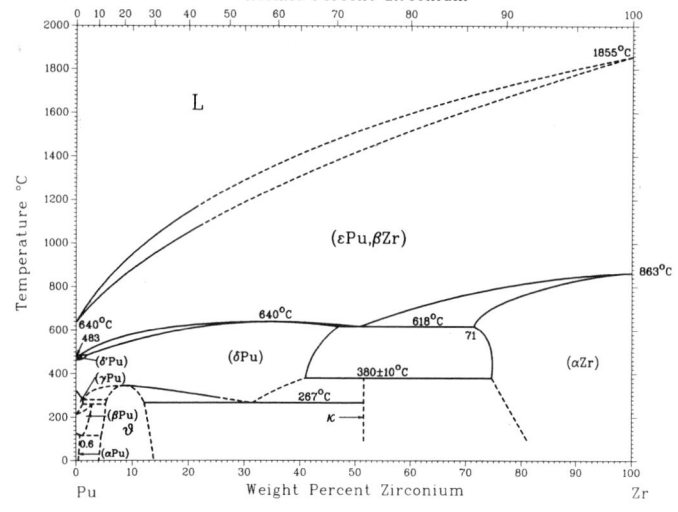

Phase	Composition, wt% Zr	Pearson symbol	Space group
(εPu,βZr)	0 to 100	cI2	Im$\overline{3}$m
(δ'Pu)	0 to 0.76	tI2	I4/mmm
(δPu)	0 to 47	cF4	Fm$\overline{3}$m
(γPu)	0 to 1.1	oF8	Fddd
(βPu)	0 to 2.7	mC34	C2/m
(αPu)	0 to 0.57	mP16	P2$_1$/m
θ (or Pu$_4$Zr)	4 to 14	tP80	P4/ncc
κ (or PuZr$_3$)	52	hP3	P6/mmm
(αZr)	71 to 100	hP2	P6$_3$/mmm

Rb-Sb

F.W. Dorn and W. Klemm, 1961

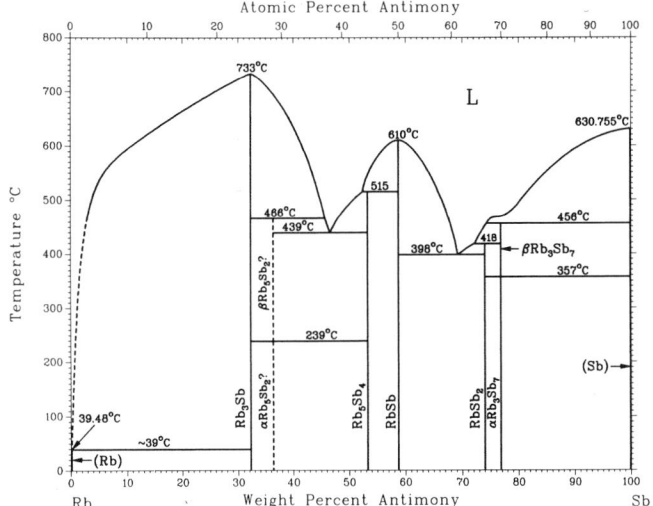

Phase	Composition, wt% Sb	Pearson symbol	Space group
(Rb)	0	cI2	Im$\overline{3}$m
Rb$_3$Sb	30	hP8	P6$_3$/mm
βRb$_5$Sb$_2$	36.3	...	...
αRb$_5$Sb$_2$	36.3	...	...
Rb$_5$Sb$_4$	53.2	...	...
RbSb	58.8	oP16	P2$_1$2$_1$2$_1$
RbSb$_2$	74.0	...	...
βRb$_3$Sb$_7$	77	...	...
αRb$_3$Sb$_7$	77	...	...
(Sb)	100	hR2	R$\overline{3}$m

Rb-Se

H. Okamoto, 1990

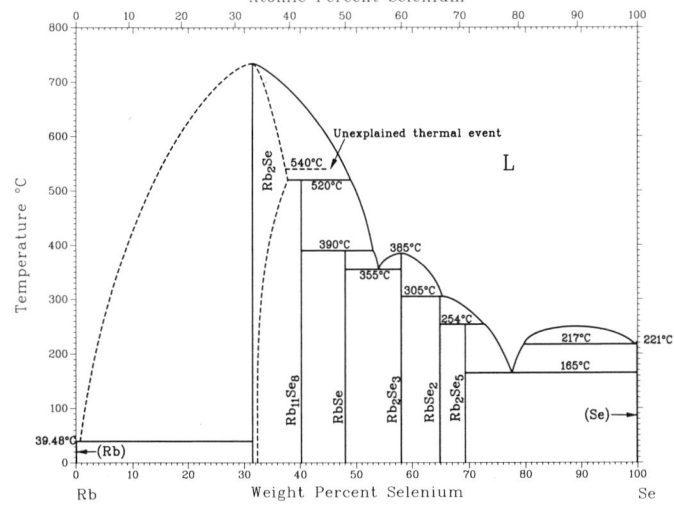

Phase	Composition, wt% Se	Pearson symbol	Space group
(Rb)	0	cI2	Im$\overline{3}$m
Rb$_2$Se	31.6 to ~38	cF12	Fm$\overline{3}$m
Rb$_{11}$Se$_8$	40.2	...	...
RbSe	48.0	...	...
Rb$_2$Se$_3$	58	oC20	Cmc2$_1$
RbSe$_2$	64.9	...	...
Rb$_2$Se$_5$	69.8	oP28	P2$_1$2$_1$2$_1$
(Se)	100	hP3	P3$_1$21

Rb-Tl

R. Thümmel and W. Klemm, 1970

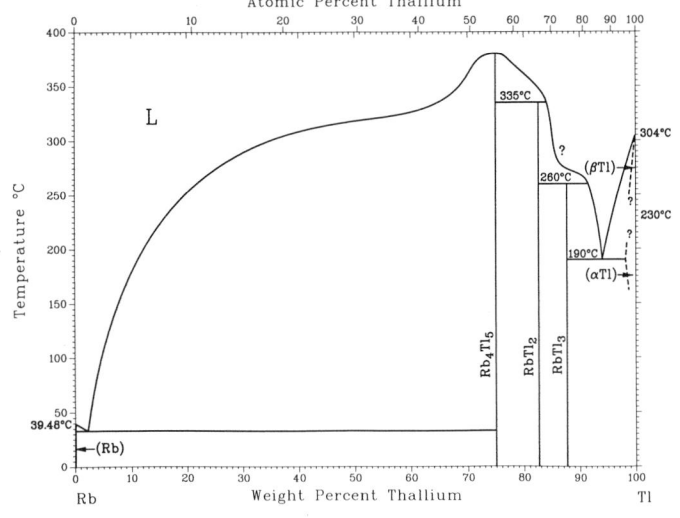

Phase	Composition, wt% Tl	Pearson symbol	Space group
(Rb)	0	cI2	Im$\overline{3}$m
Rb$_4$Tl$_5$	74.9	...	...
RbTl$_2$	82.7	...	...
RbTl$_3$	88	...	...
(βTl)	? to 100	cI2	Im$\overline{3}$m
(αTl)	? to 100	hP2	P6$_3$/mmc

Re-Ru

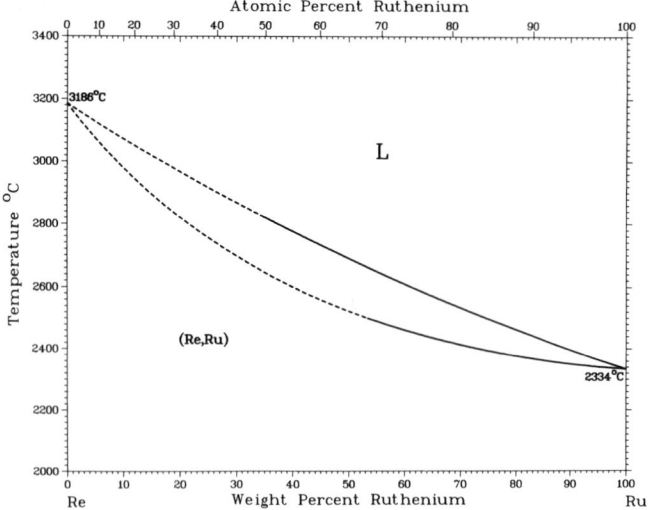

E. Rudy, B. Kietter, and H. Froelich, 1962

Phase	Composition, wt% Ru	Pearson symbol	Space group
(Re,Ru)	0 to 100	hP2	P6₃/mmc

Re-Si

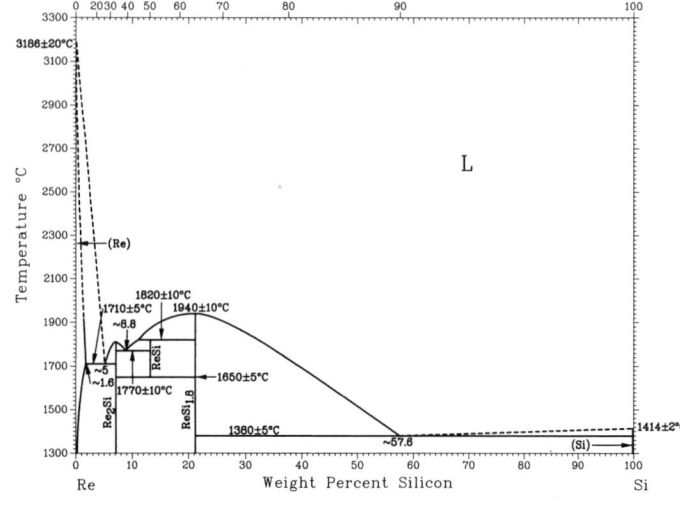

A.B. Gokhale and G.J. Abbaschian, unpublished

Phase	Composition, wt% Si	Pearson symbol	Space group
(Re)	0 to ~1.6	hP2	P6₃/mmc
Re₂Si	7.0	(a)	P2₁/b
ReSi	13.1	cP8	P2₁3
ReSi₁.₈	21.4	tI6	I4/mmm
(Si)	100	cF8	Fd3m

(a) Monoclinic

Re-Te

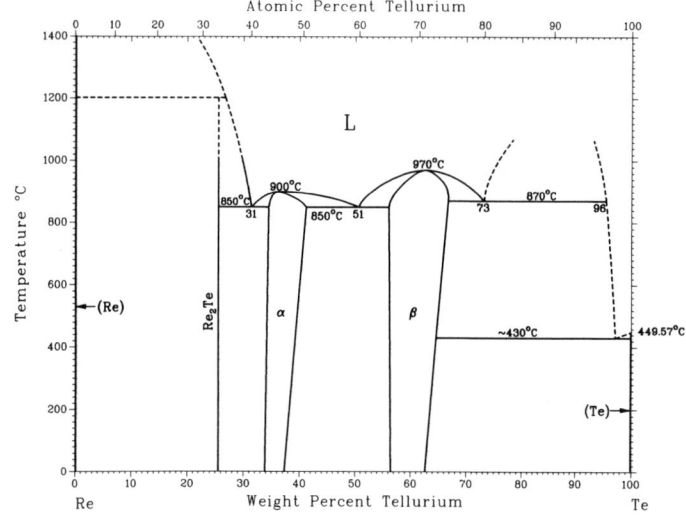

T.Kh. Kurbanov, R.A. Dovlyatshina, I.A. Dzhavodova, and F.A. Akhmenov, 1977

Phase	Composition, wt% Te	Pearson symbol	Space group
(Re)	0	hP2	P6₃/mmc
Re₂Te	25.5	...	...
α	~33.9 to 42.1	...	...
β	~56.5 to 67	oP84	Pbca
(Te)	100	hP3	P3₁21

Re-U

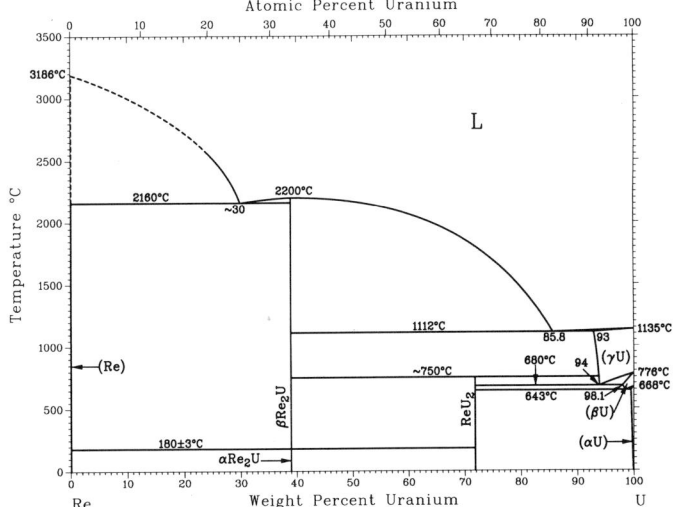

H. Okamoto, 1990

Phase	Composition, wt% U	Pearson symbol	Space group
(Re)	0	hP2	$P6_3/mmc$
Re_2U	39.0	hP12	$P6_3/mmc$
Re_2U	39.0	oC24	Cmcm
ReU_2	71.9	...	...
(γU)	93 to 100	cI2	$Im\bar{3}m$
(βU)	98.1 to 100	tP30	$P4_2/mnm$
(αU)	~100	oC4	Cmcm

Re-V

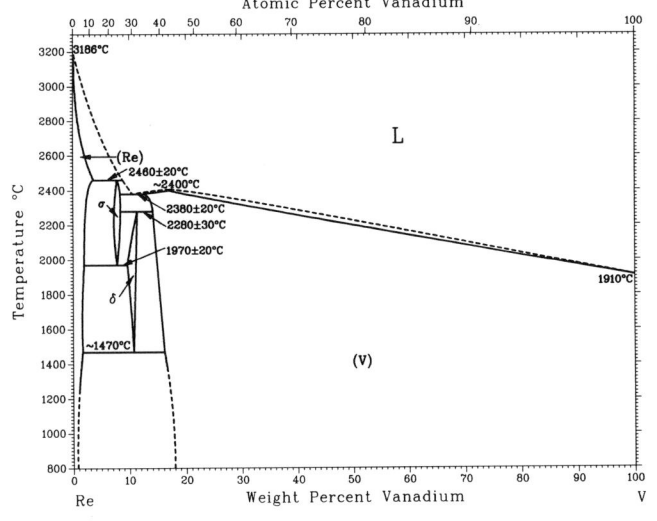

J.F. Smith, 1989

Phase	Composition, wt% V	Pearson symbol	Space group
(Re)	0 to 4	hP2	$P6_3/mmc$
σ	~8.0 to 8.3	tP30	$P4_2/mnm$
δ	9.6 to 11.4	cP8	$Pm\bar{3}n$
(V)	12.8 to 100	cI2	$Im\bar{3}m$

Rh-Se

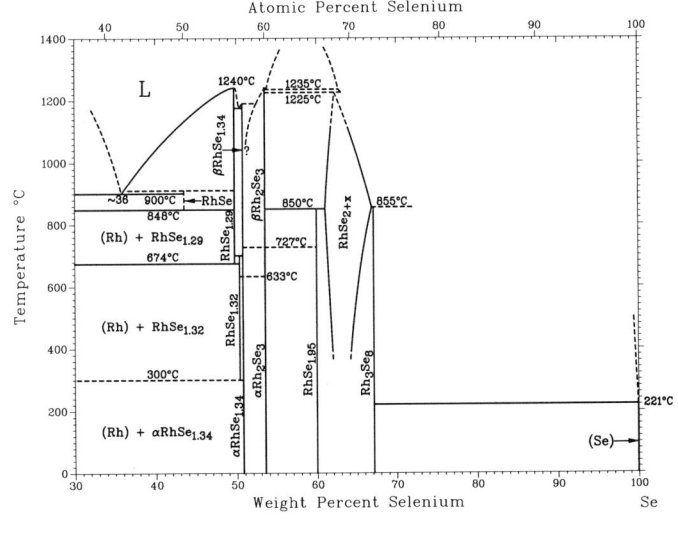

H. Okamoto, 1990

Phase	Composition, wt% Se	Pearson symbol	Space group
(Rh)	0	cF4	$Fm\bar{3}m$
RhSe	43.4	hP4	$P6_3/mmc$
$RhSe_{1.29}$	49.7	...	...
$RhSe_{1.32}$	50.3	o**	...
$\beta RhSe_{1.34}$	50.7	...	...
$\alpha RhSe_{1.34}$	50.7	hP*	...
βRh_2Se_3	54	...	...
αRh_2Se_3	54	oP20	Pbcn
$RhSe_{1.95}$	59.9	oP24	Pnma
$RhSe_{2+x}$	61.0 to 66.9	cP12	$Pa\bar{3}$
Rh_3Se_8	67.1	hR11	$R\bar{3}$
(Se)	100	hP3	$P3_121$

Rh-Ta

B.C. Giessen, H. Ibach, and N.J. Grant, 1964

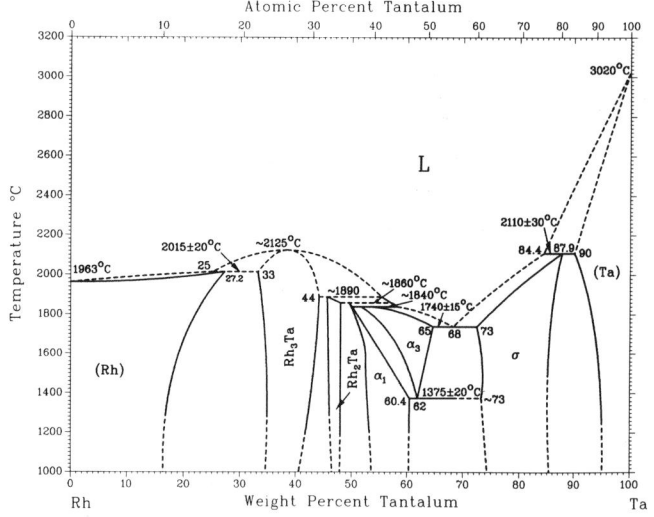

Phase	Composition, wt% Ta	Pearson symbol	Space group
(Rh)	0 to 27.2	cF4	$Fm\bar{3}m$
Rh₃Ta	33 to 44	cP4	$Pm\bar{3}m$
Rh₂Ta	45 to 48	oP12	$Pnma$
α₃	54 to 65	...	...
α₁	51 to 60.4	...	$Pmcm$?
σ	73 to 87.9	tP30	$P4_2/mnm$
(Ta)	90 to 100	cI2	$Im\bar{3}m$

Rh-Ti

J.L. Murray, 1987

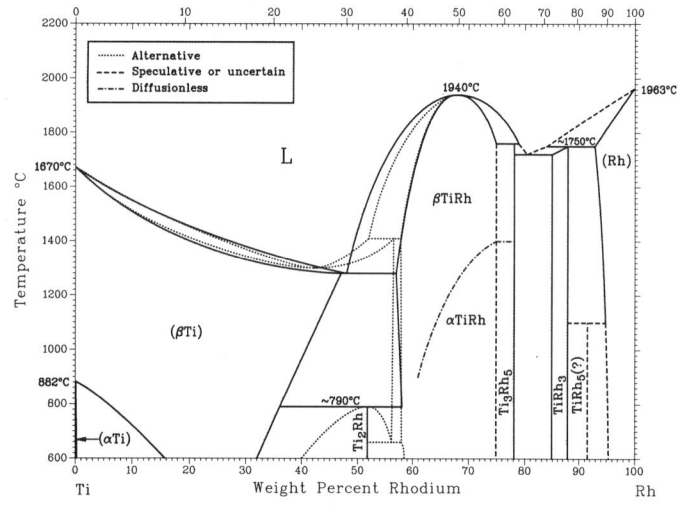

Phase	Composition, wt% Rh	Pearson symbol	Space group
(βTi)	0 to 47	cI2	$Im\bar{3}m$
(αTi)	0 to 0.161	hP2	$P6_3/mmc$
Ti₂Rh	51.8	tI6	$I4/mmm$
βTiRh	~57 to 75	cP2	$Pm\bar{3}m$
αTiRh	~57 to 75	tP2	$Pm\bar{3}m$
Ti₃Rh₅	78.2	oP16	$Pbam$
TiRh₃	85 to 88	cP4	$Pm\bar{3}m$
TiRh₅	~91.7	...	...
(Rh)	93 to 100	cF4	$Fm\bar{3}m$

Rh-U

From [Ivanov]

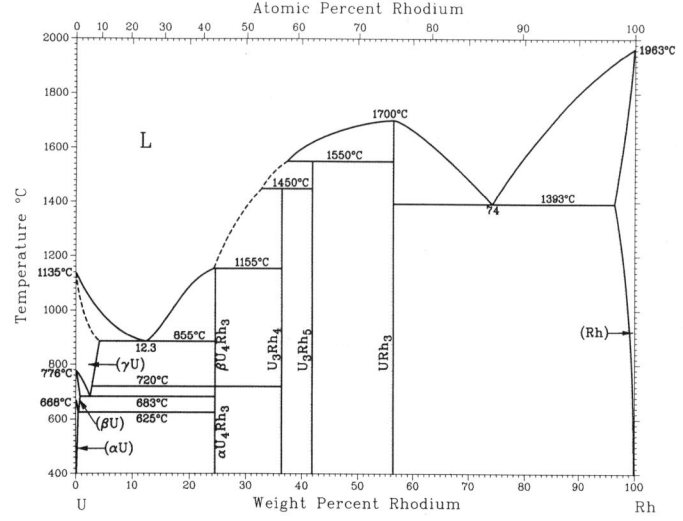

Phase	Composition, wt% Rh	Pearson symbol	Space group
(γU)	0 to 0.41	oC4	$Cmcm$
(βU)	0 to 0.87	tP30	$P\bar{4}n2$
(αU)	0 to 0.43	cI2	$Im\bar{3}m$
βU₄Rh₃	25	...	...
αU₄Rh₃	25	...	...
U₃Rh₄	36	...	...
U₃Rh₅	~42	...	...
URh₃	57	cP4	$Pm\bar{3}m$
(Rh)	96 to 100	cF4	$Fm\bar{3}m$

Rh-V

J.F. Smith, 1989

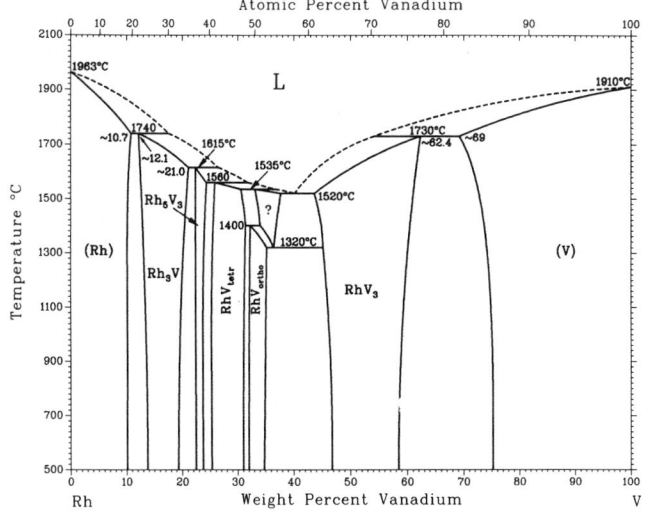

Phase	Composition, wt% V	Pearson symbol	Space group
(Rh)	0 to ~10.7	cF4	$Fm\bar{3}m$
Rh₃V	~12.1 to ~21.0	cP4	$Pm\bar{3}m$
Rh₅V₃	~23 to 24.3	oC16	Cm2m or Cmcm
RhV_tetr	25.2 to 31	tP4	P4/mmm
RhV_ortho	32 to 35.2	oC8	Cmmm
RhV₃	44 to ~62.4	cP8	$Pm\bar{3}n$
(V)	~69 to 100	cI2	$Im\bar{3}m$

Ru-Si

H. Okamoto, 1990

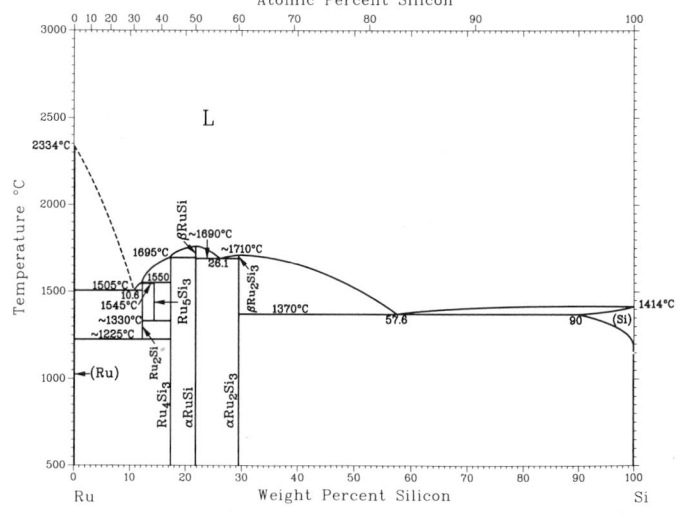

Phase	Composition, wt% Si	Pearson symbol	Space group
(Ru)	0	hP2	P6₃/mmc
Ru₂Si	12.2	oP12	Pnma
Ru₅Si₃	14.3	oP16	Pbam
Ru₄Si₃	17.3	oP28	Pnma
βRuSi	21.7	cP2	$Pm\bar{3}m$
αRuSi	21.7	cP8	P2₁3
βRu₂Si₃	29	tP80	$P\bar{4}c2$
αRu₂Si₃	29	oP40	Pbcn
(Si)	90 to 100	cF8	$Fd\bar{3}m$

Ru-Ta

H. Okamoto, 1991

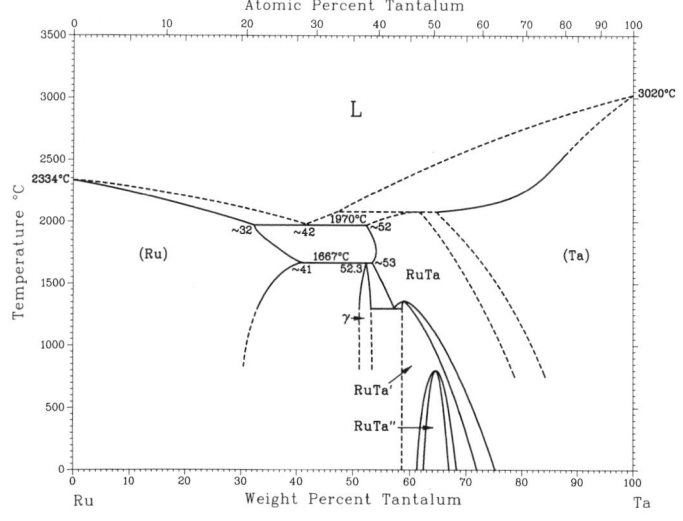

Phase	Composition, wt% Ta	Pearson symbol	Space group
(Ru)	0 to ~41	hP2	P6₃/mmc
γ	~52.3	c**	...
RuTa	~52.3 to ?	cP2	$Pm\bar{3}m$
RuTa′	~58 to 73	tP2	P4/mmm
RuTa″	~62 to 67	oC4	Cmmm
(Ta)	65 to 100	cI2	$Im\bar{3}m$

Ru-Ti

J.L. Murray, 1987

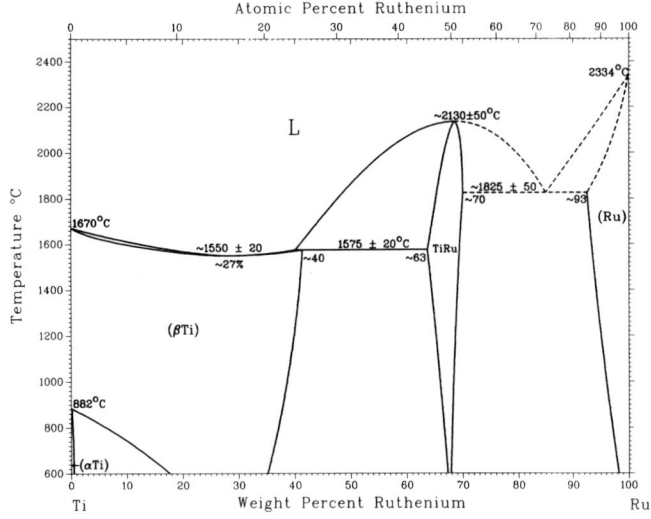

Phase	Composition, wt% Ru	Pearson symbol	Space group
(βTi)	0 to ~40	cI2	Im$\bar{3}$m
(αTi)	0 to >0.2	hP2	P6$_3$/mmc
TiRu	~63 to ~70	cP2	Pm$\bar{3}$m
(Ru)	~93 to 100	hP2	P6$_3$/mmc
Metastable phases			
(α″Ti)	...	hP2	P6$_3$/mmc
(α‴Ti)	...	oC4	Cmcm
ω	...	hP3	P6/mmm

Ru-U

P. Chiotti, V.V. Akhachinskij, I. Ansara, and M.H. Rand, 1982

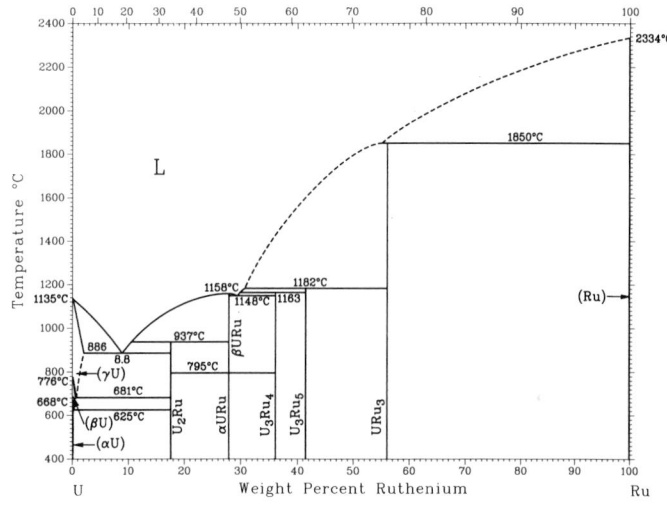

Phase	Composition, wt% Ru	Pearson symbol	Space group
(γU)	0 to 2.0	cI2	Im$\bar{3}$m
(βU)	0 to 0.86	tP30	P4$_2$/mnm
(αU)	~0	oC4	Cmcm
U$_2$Ru	17.5	mP12	P2/m or P2$_1$/m
URu	27.5	...	...
U$_3$Ru$_4$	36	...	...
U$_3$Ru$_5$	41.4	...	...
URu$_3$	56	cP4	Pm$\bar{3}$m
(Ru)	98 to 100	hP2	P6$_3$/mmc

Ru-V

J.F. Smith, 1989

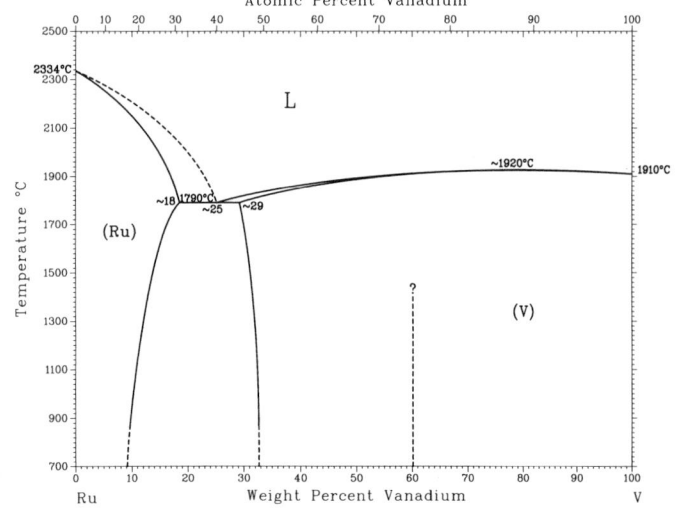

Phase	Composition, wt% V	Pearson symbol	Space group
(Ru)	0 to ~18	hP2	P6$_3$/mmc
RuV	33.5	t**	...
RuV	~29 to 60	cP2	Pm$\bar{3}$m
(V)	60 to 100	cI2	Im$\bar{3}$m

S-Se

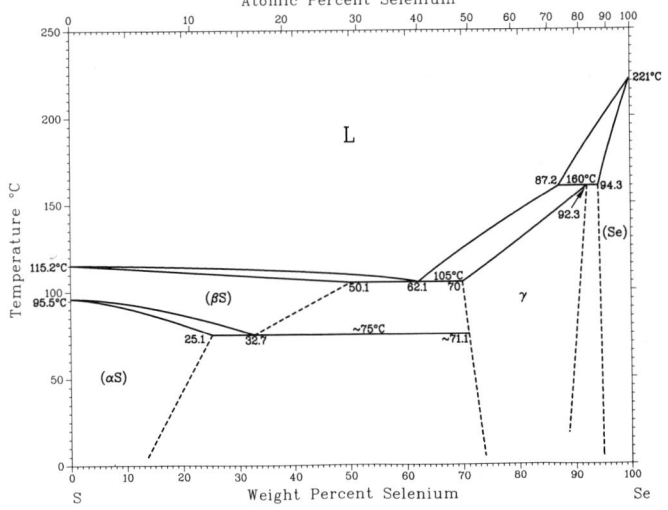

R.C. Sharma and Y.A. Chang, unpublished

Phase	Composition, wt% Se	Pearson symbol	Space group
(βS)	0 to 50.1	mP*	P2₁/c
(αS)	0 to 25.1	oF128	Fddd
γ	70 to 92.3	(a)	...
(Se)	94.3 to 100	hP3	P3₁21
High-pressure phase			
S₀.₅₅₅Se₀.₄₄₅	66.4	(b)	P3₁ or P3₂

(a) Monoclinic. (b) Trigonal

S-Sn

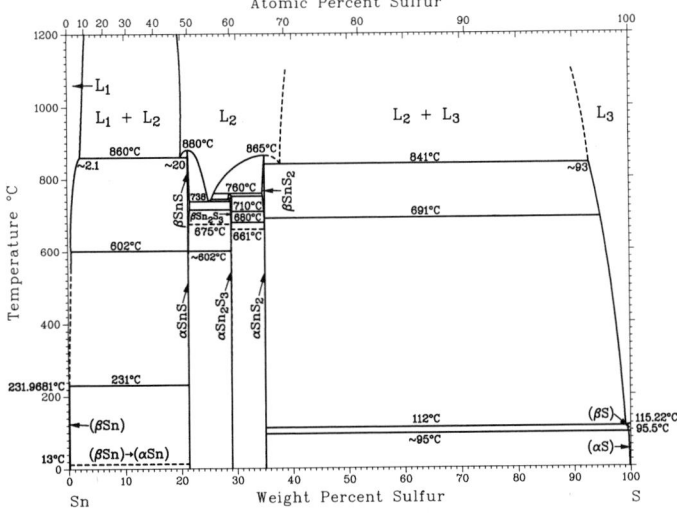

R.C. Sharma and Y.A. Chang, 1986

Phase	Composition, wt% S	Pearson symbol	Space group
(βSn)	0	tI4	I4₁/amd
βSnS	21.3	cC8	Cmcm
αSnS	21.3	oP8	Pnma
δSn₂S₃	29	...	...
γSn₂S₃	29	...	...
βSn₂S₃	29	...	...
αSn₂S₃	29	oP20	Pnma
SnS₂	35.1	hP*	P6₃mc
		hP3	P3̄m1
Metastable phases			
SnS (thin film)	21.3	cF8	Fm3̄m
Sn₄S₅	25.3	...	...
Sn₃S₄	26.4	t**	...

S-Sn phase diagram between 18 and 35 wt%

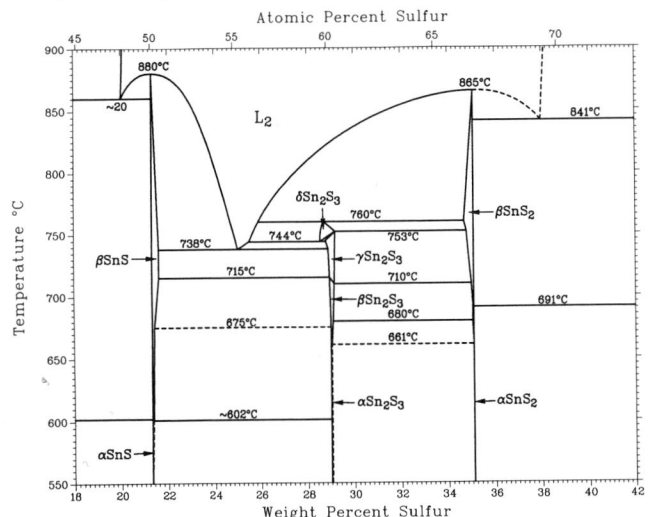

S-Te

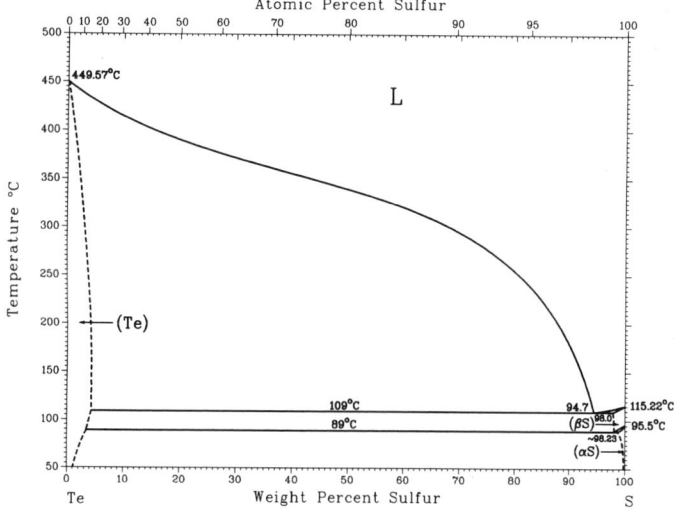

D.T. Li, R.C. Sharma, and Y.A. Chang, 1989

Phase	Composition, wt% S	Pearson symbol	Space group
(Te)	0 to 40.3	hP3	P3$_1$21
Te$_7$S$_{10}$(a)	...	(b)	...
(βS)	98.0 to 100	mP*	P2$_1$/c
(αS)	~98.23 to 100	oF128	Fddd

(a) High-pressure phase. (b) Pseudo-orthorhombic

S-Ti

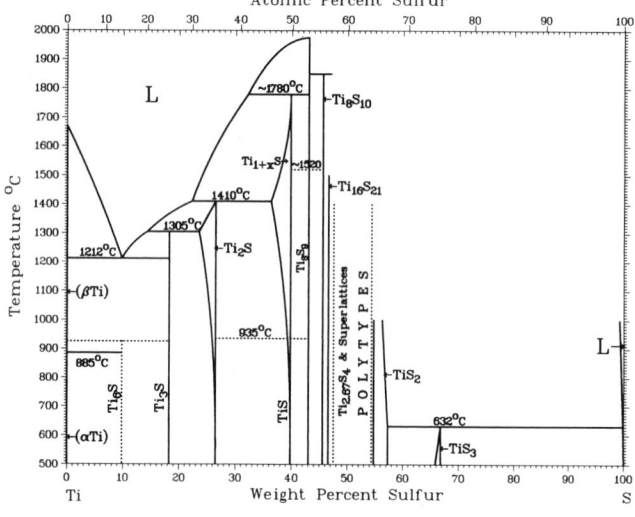

J.L. Murray, 1987

Phase	Composition, wt% S	Pearson symbol	Space group
(βTi)	0 to 0.007	cI2	Im$\bar{3}$m
(αTi)	0 to 0.013	hP2	P6$_3$/mmc
Ti$_6$S	~10	(a)	...
Ti$_3$S	18	t*24	...
Ti$_2$S	23 to 27	(b)	...
Ti$_{1+x}$S	36 to 39.8	hP2	P6m2
TiS	~39.8	hP4	P6$_3$/mmc
Ti$_8$S$_9$	~42.6	hR18	R$\bar{3}$m
Ti$_8$S$_{10}$	~45.6	hP18	P6$_3$/mmc
Ti$_{16}$S$_{21}$	~45.6	hR37.1	R$\bar{3}$m
Ti$_{2.67}$S$_4$	47.9 to 51.6	hP6.8	P6$_3$mc
(4H)$_2$	49.9 to 50.4	mC40.14	Cc
(4H)$_3$	...	mC59.8	Cc
Ti$_7$S$_{12}$	~53.1	hR19.1	R$\bar{3}$m
TiS$_2$	54.8 to 57.3	hP3	P$\bar{3}$m
TiS$_3$	~67	mP8	P2$_1$/m
(S)	100	oF128	Fddd

(a) Hexagonal. (b) Unknown low symmetry

Sb-Se

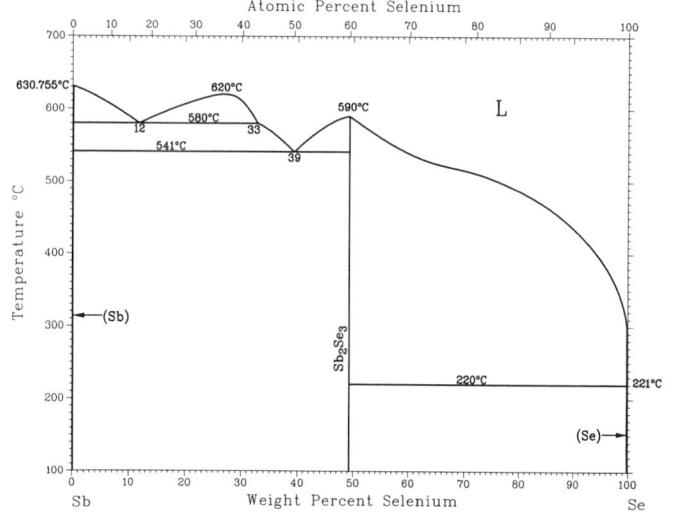

H. Okamoto, 1990

Phase	Composition, wt% Se	Pearson symbol	Space group
(Sb)	0	hR2	R$\bar{3}$m
Sb$_2$Se$_3$	49	oP20	Pnma
(Se)	100	hP3	P3$_1$21

Sb-Si

R.W. Olesinski and G.J. Abbaschian, 1985

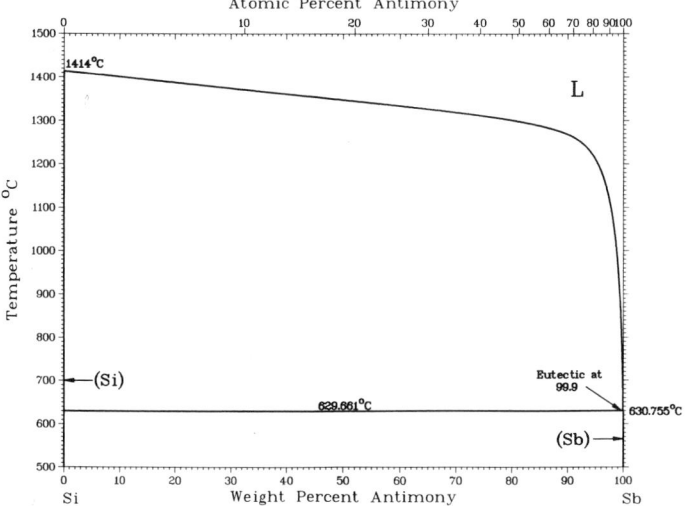

Phase	Composition, wt% Sb	Pearson symbol	Space group
(Si)	0 to 0.09	cF8	Fd$\bar{3}$m
(Sb)	100	hR2	R$\bar{3}$m

Sb-Sm

H. Okamoto, 1990

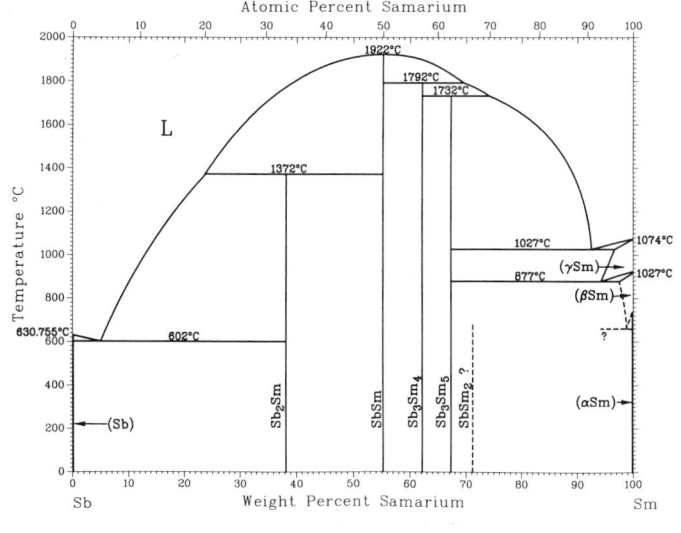

Phase	Composition, wt% Sm	Pearson symbol	Space group
(Sb)	0	hR2	R$\bar{3}$m
Sb$_2$Sm	38.1	oC24	Cmca
SbSm	55.3	cF8	Fm$\bar{3}$m
Sb$_3$Sm$_4$	62.2	cI28	I$\bar{4}$3d
Sb$_3$Sm$_5$	67.3	hP16	P6$_3$/mcm
SbSm$_2$	71.2	tI12	I4/mmm
(γSm)	100	cI2	Im$\bar{3}$m
(βSm)	100	hP2	P6$_3$/mmc
(αSm)	100	hR3	R$\bar{3}$m

Sb-Sn

B. Predel and W. Schwermann, 1971

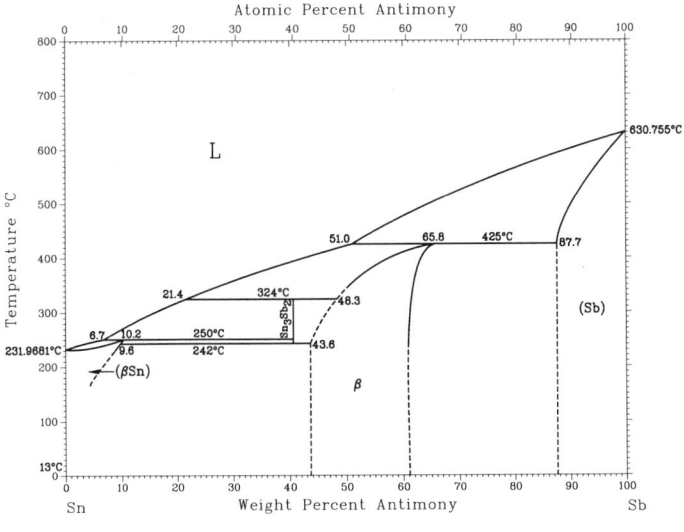

Phase	Composition, wt% Sb	Pearson symbol	Space group
(βSn)	0 to 9.6	tI4	I4$_1$/amd
Sn$_3$Sb$_2$	43.6	...	...
β	43.6 to 65.8	cF8	Fm$\bar{3}$m
(Sb)	87.7 to 100	hR2	R$\bar{3}$m

Sb-Sr

A.V. Vakhobov, Z.V. Niyazova, and B.N. Polev, 1975

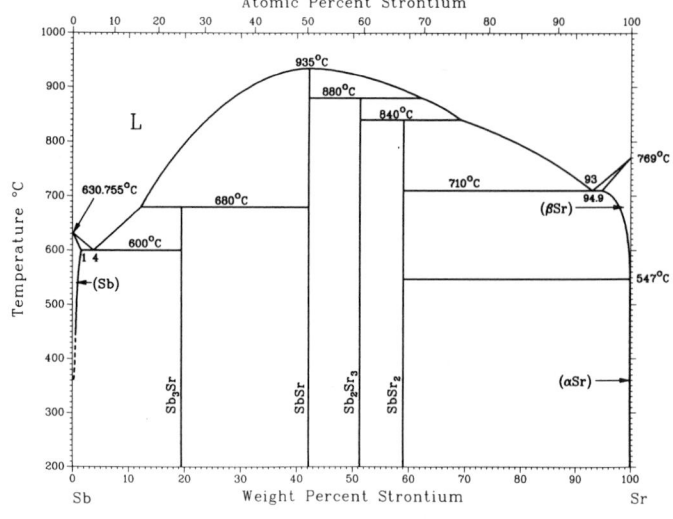

Phase	Composition, wt% Sr	Pearson symbol	Space group
(Sb)	0 to 1	hR2	R3̄m
Sb₃Sr	19	...	...
SbSr	41.8	...	...
Sb₂Sr₃	52	...	...
SbSr₂	59	tI12	I4/mmm
(βSr)	94.9 to 100	cI2	Im3̄m
(αSr)	100	cF4	Fm3̄m

Sb-Tb

H. Okamoto, 1990

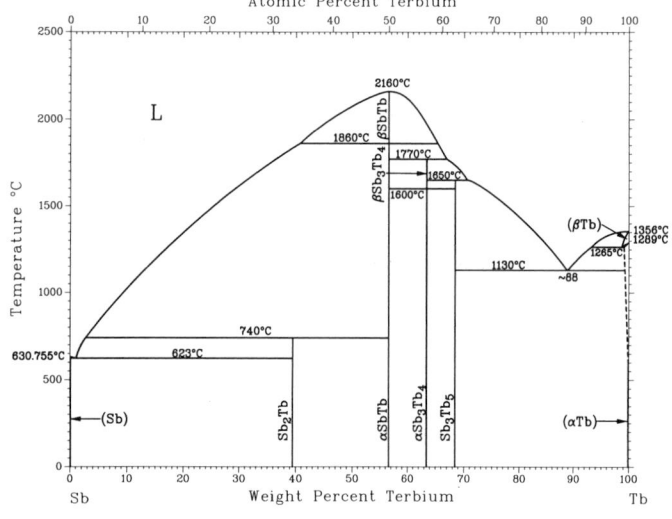

Phase	Composition, wt% Tb	Pearson symbol	Space group
(Sb)	0	hR2	R3̄m
Sb₂Tb	39.5	oC24	Cmca
βSbTb	56.6	...	...
αSbTb	56.6	cF8	Fm3̄m
βSb₃Tb₄	63.5	cI28	I4̄3d
αSb₃Tb₄	63.5	...	...
Sb₃Tb₅	68.5	hP16	P6₃/mcm
(βTb)	100	cI2	Im3̄m
(αTb)	100	hP2	P6₃/mmc

Sb-Te

H. Okamoto, 1990

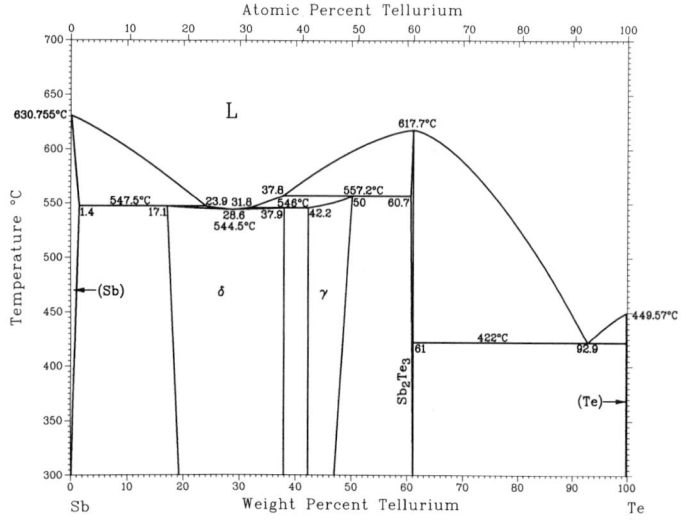

Phase	Composition, wt% Te	Pearson symbol	Space group
(Sb)	0 to 1.4	hR2	R3̄m
δ	17.1 to 37.9	...	...
γ	42.2 to 50	...	...
Sb₂Te₃	60.7 to 61	hR5	R3̄m
(Te)	100	hP3	P3₁21

Sb-Tl

R.C. Sharma and Y.A. Chang, unpublished

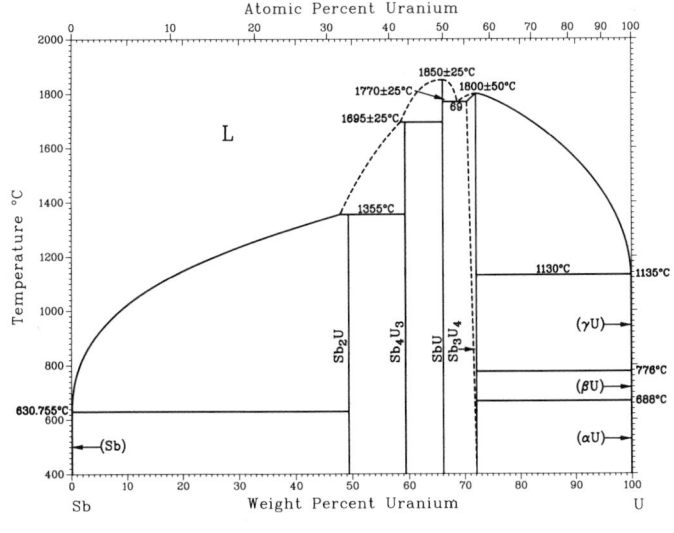

Phase	Composition, wt% Sb	Pearson symbol	Space group
(βTl)	0 to 15.6	cI2	$Im\bar{3}m$
(αTl)	0 to 2	hP2	$P6_3/mmc$
γ	4.0 to 6.0	cF*	...
Tl₇Sb₂	14.7 to 16.9	cI54	$Im\bar{3}m$
TlSb	37.3	...	...
(Sb)	100	hR2	$R\bar{3}m$

Sb-U

P. Chiotti, 1980

Phase	Composition, wt% U	Pearson symbol	Space group
(Sb)	0	hR2	$R\bar{3}m$
Sb₂U	49.4	tP6	$P4/nmm$
Sb₄U₃(a)	59.5	cI28	$I\bar{4}3d$
SbU	66.2	cF8	$Fm\bar{3}m$
Sb₃U₄	72.2	...	$P6_3/mcm$
(γU)	100	cI2	$Im\bar{3}m$
(βU)	100	tP30	$P4_2/mnm$
(αU)	100	oC4	$Cmcm$

(a) Evidence for ferromagnetic ordering of Sb₄U₃ has been presented.

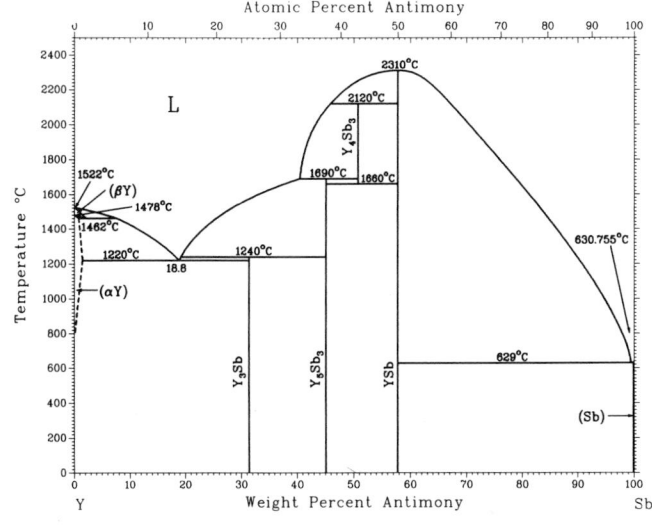

Sb-Y

F.A. Schmidt and O.D. McMasters, 1970

Phase	Composition, wt% Sb	Pearson symbol	Space group
(βY)	0 to 2.7	cI2	$Im\bar{3}m$
(αY)	0 to 1.4	hP2	$P6_3/mmc$
Y₃Sb	31	tP32	$P4_2/n$
Y₅Sb₃	45.1	hP16	$P6_3/mcm$
Y₄Sb₃	50.7	cI28	$I\bar{4}3d$
YSb	57.8	cF8	$Fm\bar{3}m$
(Sb)	100	hR2	$R\bar{3}m$

Sb-Zn

G. Vuillard and J.P. Piton, 1966; and T. Takei, 1927

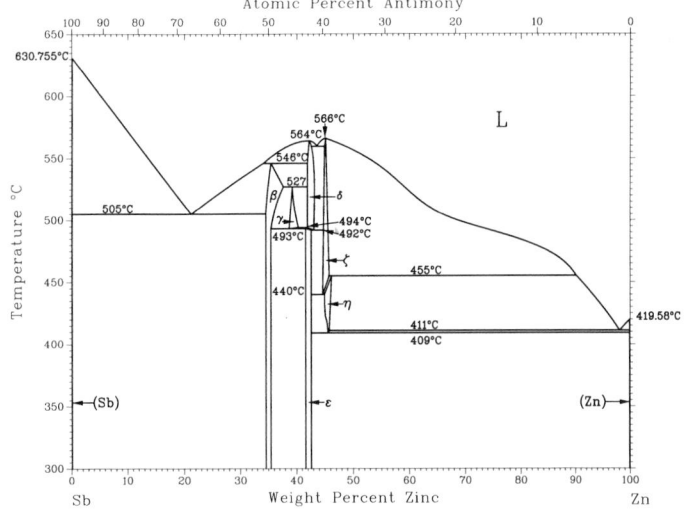

Phase	Composition, wt% Zn	Pearson symbol	Space group
(Sb)	0	hR2	R3̄m
β	~34.9 to ~38	oP16	Pbca
γ	39 to 41	…	…
ε	42 to 43	(a)	…
δ	42 to ~43.1	(a)	…
ζ	45 to 46	oI*	…
η	45 to ~46	oP30	Pmmn
(Zn)	100	hP2	P6₃/mmc

(a) Sb₃Zn₄ (δ,ε?): hR22 or oP28 or mC*?

Sc-Ti

J.L. Murray, 1987

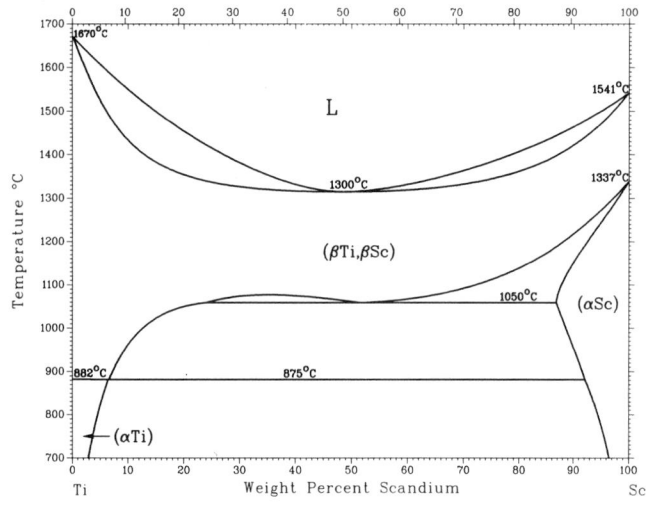

Phase	Composition, wt% Sc	Pearson symbol	Space group
(βTi,βSc)	0 to 100	cI2	Im3̄m
(αTi)	0 to 7.4	hP2	P6₃/mmc
(αSc)	88.2 to 100	hP2	P6₃/mmc

Sc-Y

K.A. Gschneidner, Jr. and F.W. Calderwood, 1983

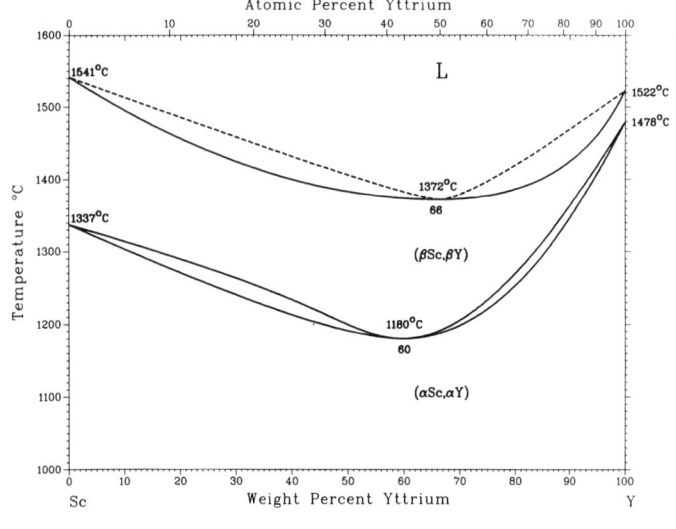

Phase	Composition, wt% Y	Pearson symbol	Space group
(βSc,βY)	0 to 100	cI2	Im3̄m
(αSc,αY)	0 to 100	hP2	P6₃/mmc

Sc-Zr

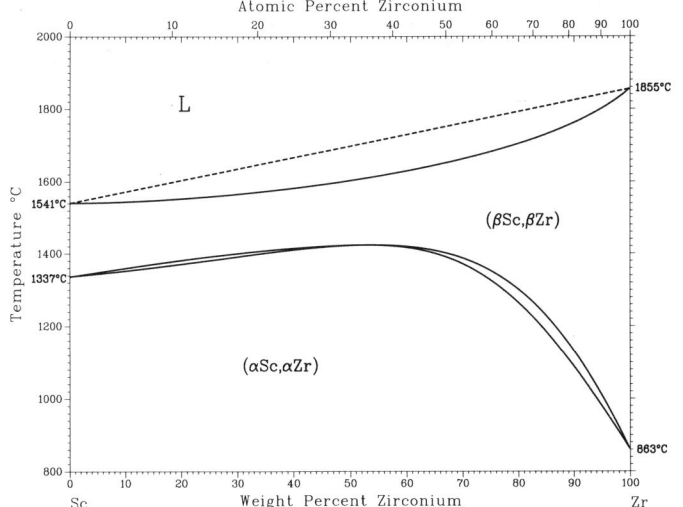

A. Palenzona and S. Cirafici, 1991

Phase	Composition, wt% Zr	Pearson symbol	Space group
(βSc,βZr)	0 to 100	cI2	$Im\bar{3}m$
(αSc,αZr)	0 to 100	hP2	$P6_3/mmc$

Se-Sn

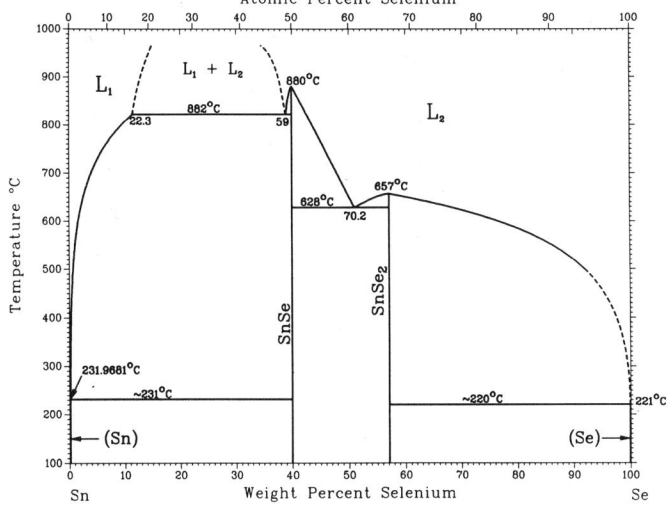

R.C. Sharma and Y.A. Chang, 1986

Phase	Composition, wt% Se	Pearson symbol	Space group
(Sn)	0	tI4	$I4_1/amd$
SnSe	39.9	oP8	Pnma
SnSe₂	57.1	hP3	$P\bar{3}m1$
(Se)	100	hP3	$P3_121$

Se-Sr

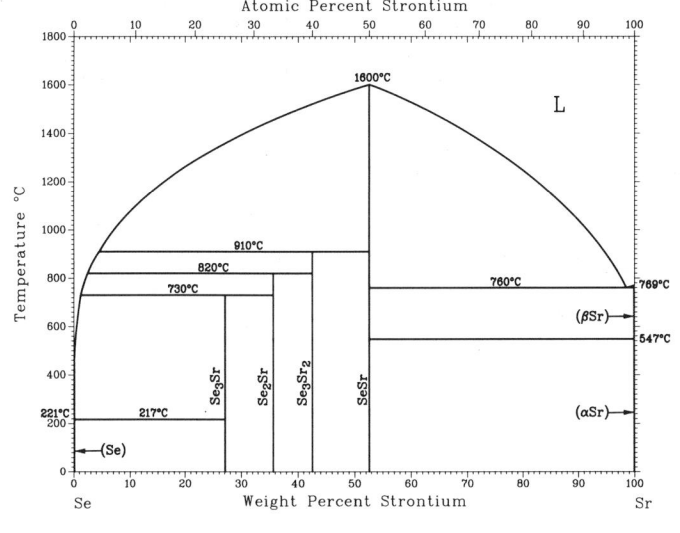

Yu.B. Lyskova and A.V. Vakhobov, 1975

Phase	Composition, wt% Sr	Pearson symbol	Space group
(Se)	0	hP3	$P3_121$
Se₃Sr	27	...	...
Se₂Sr	35.7	...	...
Se₃Sr₂	43	...	...
SeSr	52.6	cF8	$Fm\bar{3}m$
(βSr)	100	cI2	$Im\bar{3}m$
(αSr)	100	cF4	$Fm\bar{3}m$

Se-Te

R.C. Sharma, D.T. Li, and Y.A. Chang, unpublished

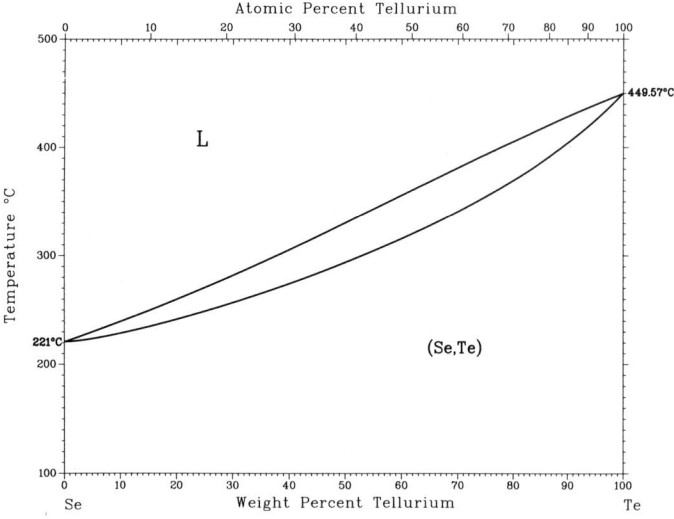

Phase	Composition, wt% Te	Pearson symbol	Space group
(Se,Te)	0 to 100	hP3	P3₁21

Corrected:

Phase	Composition, wt% Te	Pearson symbol	Space group
(Se,Te)	0 to 100	$hP3$	$P3_121$

Se-Tl

G. Morgant, B. Legendre, S. Mareglier-Lacordaire, and C. Souleau, 1981

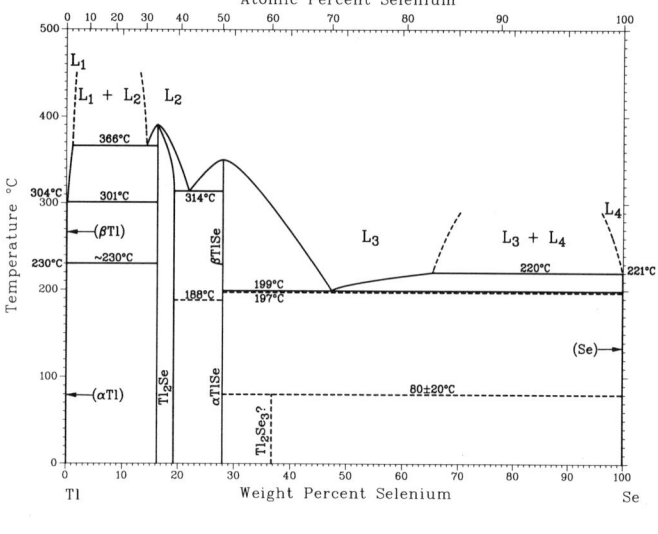

Phase	Composition, wt% Se	Pearson symbol	Space group
(βTl)	0	$cI2$	$Im\bar{3}m$
(αTl)	0	$hP2$	$P6_3/mmc$
Tl₂Se	16.5 to 19	$tP32$	$P4/ncc$
βTlSe	27.9	$tI16$	$I4/mcm$
αTlSe	27.9	...	...
Tl₂Se₃?	37	$hP4$	$P6_3mc$
(Se)	100	$hP3$	$P3_121$

Se-Tm

H. Okamoto, 1990

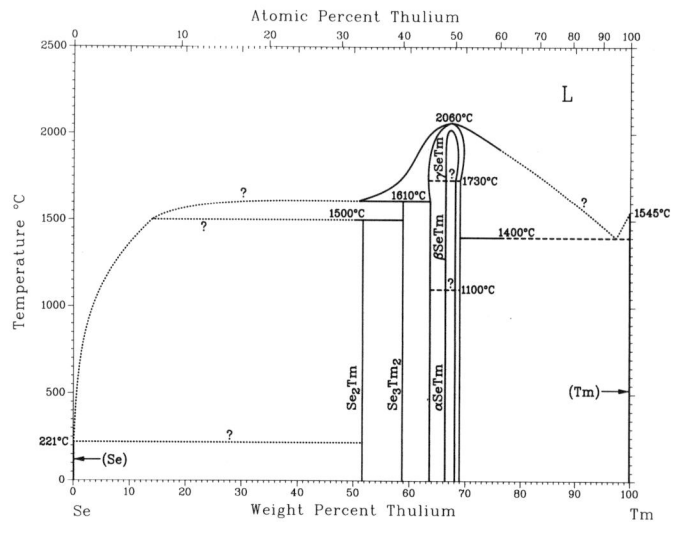

Phase	Composition, wt% Tm	Pearson symbol	Space group
(γSe)	0	$hP3$	$P3_121$
Se₂Tm	51.6	$tP6$	$P4/nmm$
Se₃Tm₂	59	$oF80$	$Fddd$
γSeTm	64 to 69	$cF8$	$Fm\bar{3}m$
βSeTm	65 to 69	...	...
αSeTm	65 to 69	...	...
(Tm)	100	$hP2$	$P6_3/mmc$

Note: "SeTm" is Se₆Tm₅ on the Se-rich side and SeTm₁.₀₅ on the Tm-rich side.

Se-U

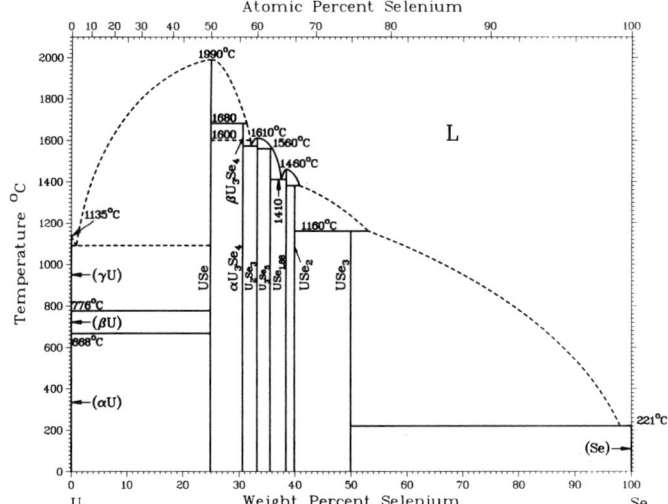

G.V. Ellert, V.G. Sevast'yanov, and V.K. Slovyanskikh, 1975

Phase	Composition, wt% Se	Pearson symbol	Space group
(γU)	0	cI2	$Im\bar{3}m$
(βU)	0	tP30	$P4_2/mnm$
(αU)	0	oC4	Cmcm
USe	24.9	cF8	$Fm\bar{3}m$
βU₃Se₄	31	...	...
αU₃Se₄	31	cI28	$I\bar{4}3d$
U₂Se₃	33	oP20	Pnma
U₃Se₅	35.6	oP32	Pnma
USe₁.₈₈	38.4	hP20	$P6_3/m$
USe₂	39.9	oP12	Pnma
USe₃	49.9	mP8	$P2_1/m$
(Se)	100	hP3	$P3_121$

Si-Sn

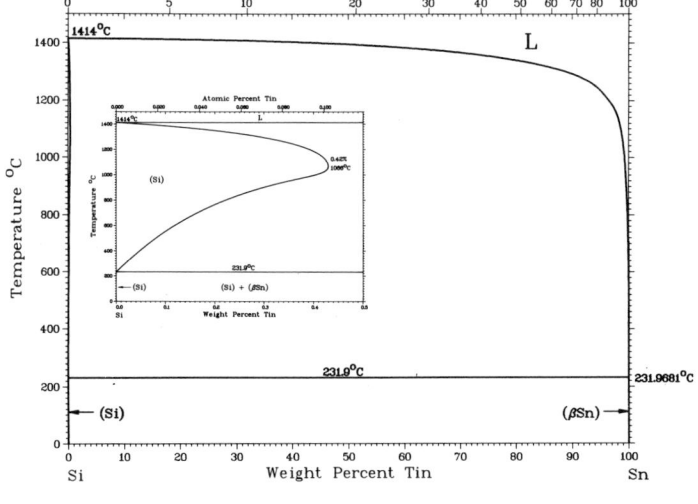

R.W. Olesinski and G.J. Abbaschian, 1984

Phase	Composition, wt% Sn	Pearson symbol	Space group
(Si)	0 to 0.42	cF8	$Fd\bar{3}m$
(βSn)	100	tI4	$I4_1/amd$
(αSn)	100	cF8	$Fd\bar{3}m$

Si-Sr

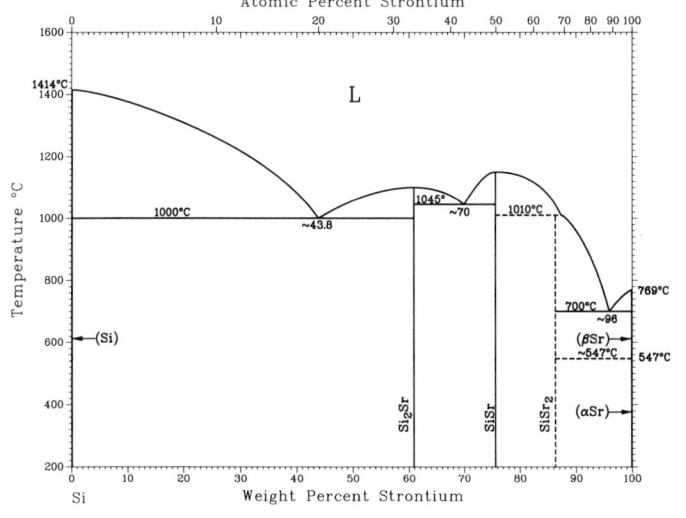

V.P. Itkin and C.B. Alcock, 1989

Phase	Composition, wt% Sr	Pearson symbol	Space group
(Si)	0	cF8	$Fd\bar{3}m$
Si₂Sr	60.9	cP12	$P4_332$
SiSr	75.7	oC8	Cmcm
SiSr₂	86.2	oP12	Pnma
(βSr)	100	cI2	$Im\bar{3}m$
(αSr)	100	cF4	$Fm\bar{3}m$
Other possible phases			
Si₇Sr₄	64.0 to 68(a)	tI12	$I4_1/amd$
αSiSr	75.7	oI40	Immm
Si₃Sr₅	83.9	tI32	I4cm
High-pressure, metastable phase			
Si₂Sr(II)	60.9	tI12	$I4_1/amd$

(a) Possible speculative homogeneity range

Si-Ta

M.E. Schlesinger, unpublished

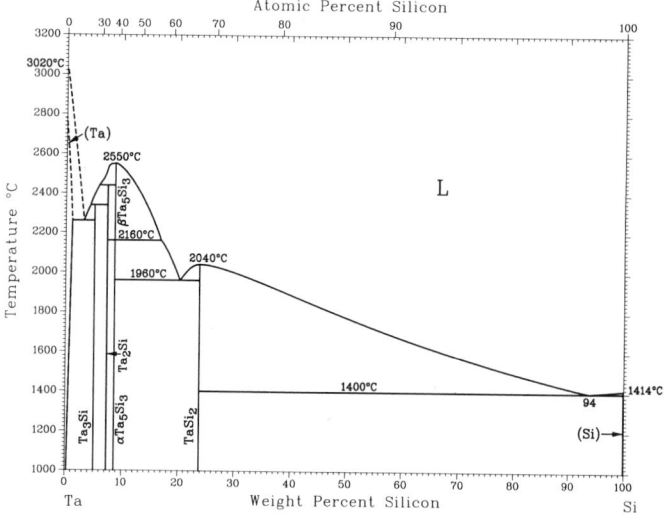

Phase	Composition, wt% Si	Pearson symbol	Space group
(Ta)	0 to ~1	cI2	Im$\overline{3}$m
Ta$_3$Si	5	tP32	P4$_2$/n
Ta$_2$Si	7.2	tI12	I4/m
βTa$_5$Si$_3$	8.5	tI32	I4/mcm
αTa$_5$Si$_3$	8.5	tI32	I4/mcm
TaSi$_2$	23.7	hP9	P6$_2$22
Si	100	cF8	Fd$\overline{3}$m
Metastable phases			
Ta$_{4.5}$Si	3.5	hP8	P6$_3$/mmc
Ta$_5$Si$_3$	8.5	hP16	P6$_3$/mcm

Si-Te

T.G. Davey and E.H. Baker, 1980

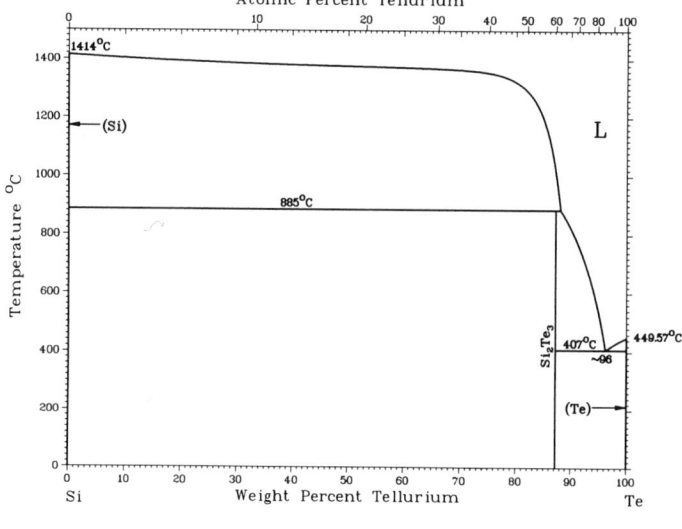

Phase	Composition, wt% Te	Pearson symbol	Space group
(Si)	0	cF8	Fd$\overline{3}$m
Si$_2$Te$_3$	87	hP40	P$\overline{3}$1c
(Te)	100	hP3	P3$_1$21

Si-Th

From [Thorium]

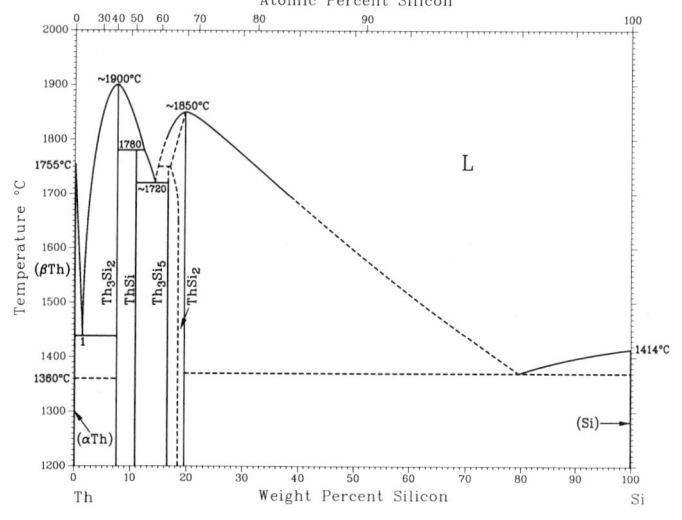

Phase	Composition, wt% Si	Pearson symbol	Space group
(βTh)	0	cI2	Im$\overline{3}$m
(αTh)	0	cF4	Fm$\overline{3}$m
Th$_3$Si$_2$	8	tP10	P4/mbm
ThSi	10.8	oP8	Pnma
Th$_3$Si$_5$	16.8	hP3	P6/mmm
ThSi$_2$	~18 to 19.5	tI12	I4$_1$/amd
(Si)	100	cF8	Fd$\overline{3}$m

Si-Ti

J.L. Murray, 1987

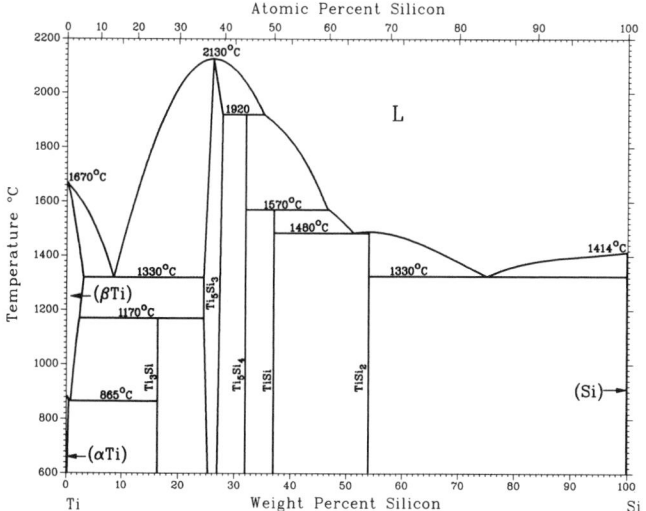

Phase	Composition, wt% Si	Pearson symbol	Space group
(βTi)	0 to 2.1	cI2	Im$\bar{3}$m
(αTi)	0 to 0.3	hP2	P6₃/mmc
Ti₃Si	16	tP32	P4₂/n
Ti₅Si₃	24.4 to 27.7	hP16	P6₃/mcm
Ti₅Si₄	31.9	tP36	P4₁2₁2
Ti₆Si₅(a)	32.9	(b)	…
TiSi	37.0	oP8	Pmm2
		oP8	Pnma
TiSi₂	54.0	oF24	Fddd
(Si)	100	cF8	Fd$\bar{3}$m

(a) Not shown in diagram. (b) Tetragonal, related to σ (D8₈)

Si-U

H. Okamoto, 1990

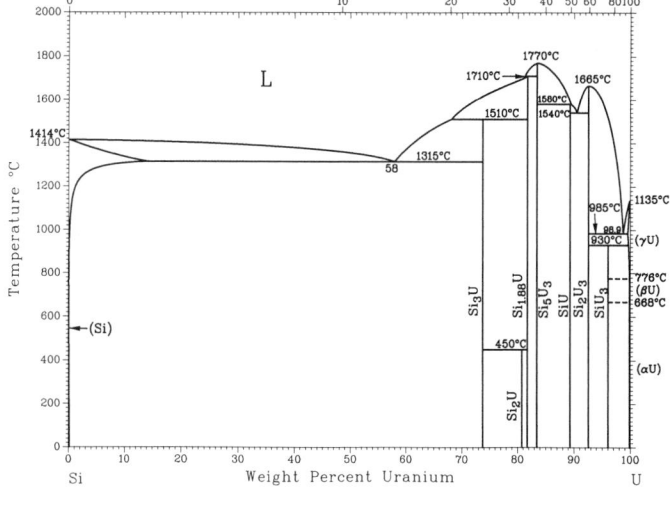

Phase	Composition, wt% U	Pearson symbol	Space group
(Si)	0	cF8	Fd$\bar{3}$m
Si₃U	74	cP4	Pm$\bar{3}$m
Si₂U	80.9	hP3	P6/mmm
Si₁.₈₈U	81.8	tI12	I4₁/amd
Si₅U₃	83.6	hP3	P6/mmm
SiU	89.4	oP8	Pnma
Si₂U₃	93	tP19	P4/mbm
SiU₃	96	cP4	Pm$\bar{3}$m
(γU)	100	cI2	Im$\bar{3}$m
(βU)	100	tP30	P4₂/mnm
(αU)	100	oC4	Cmcm

Si-V

J.F. Smith, 1989

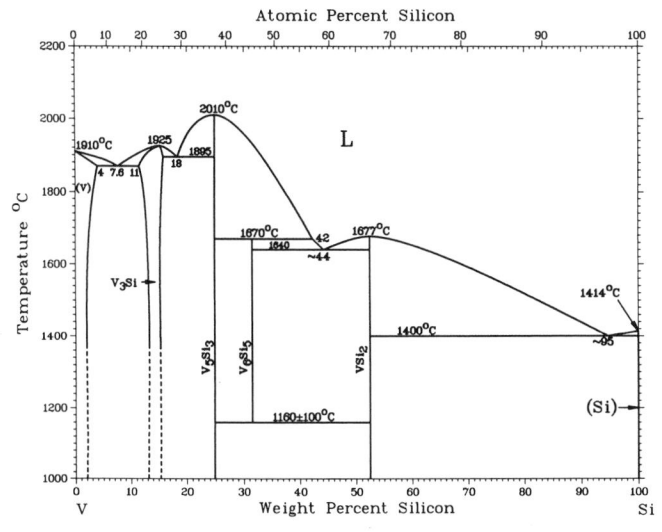

Phase	Composition, wt% Si	Pearson symbol	Space group
(V)	0 to 4	cI2	Im$\bar{3}$m
V₃Si	11 to ~15.9	cP8	Pm$\bar{3}$n
V₅Si₃	24.9	tI32	I4/mcm
V₅Si₃	(a)	hP16	P6₃/mcm
V₆Si₅	~31	oI44	Immm
VSi₂	52.5	hP9	P6₂22
(Si)	100	cF8	Fm$\bar{3}$m

(a) Carbon-stabilized

Si-Zn

R.W. Olesinski and G.J. Abbaschian, 1985

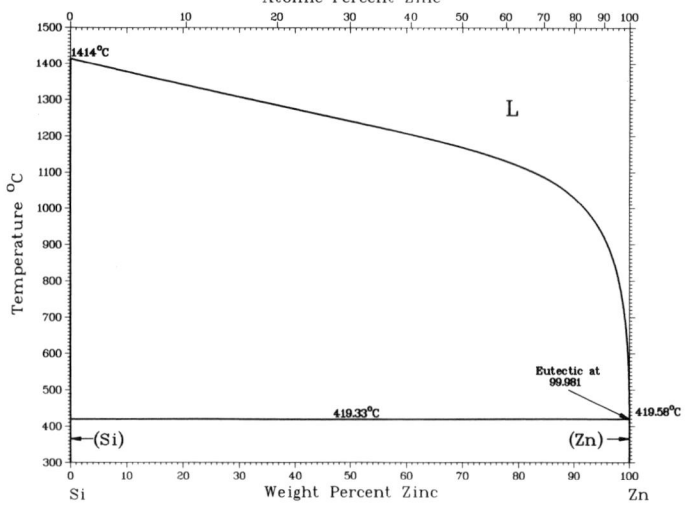

Phase	Composition, wt% Zn	Pearson symbol	Space group
(Si)	0	cF8	Fd$\bar{3}$m
(Zn)	100	hP2	P6$_3$/mmc

Si-Zr

H. Okamoto, 1990

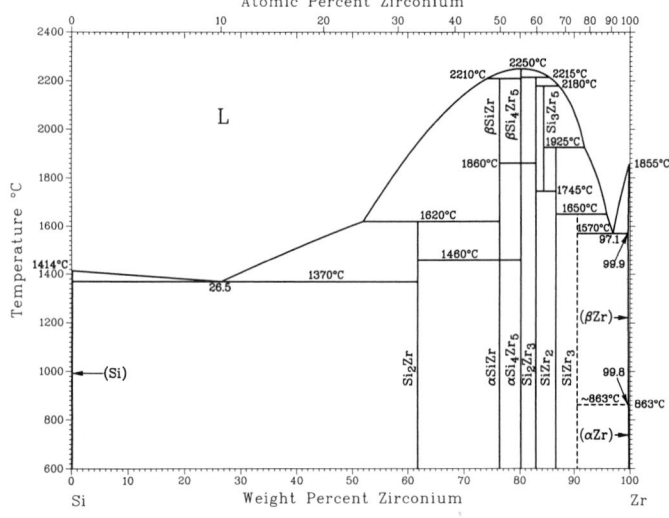

Phase	Composition, wt% Zr	Pearson symbol	Space group
(Si)	0	cF8	Fd$\bar{3}$m
Si$_2$Zr	61.9	oC12	Cmcm
βSiZr	76.5	oC8	Cmcm
αSiZr	76.5	oP8	Pnma
βSr$_4$Zr$_5$	80.3	...	...
αSr$_4$Zr$_5$	80.3	tP36	P4$_1$2$_1$2
Si$_2$Zr$_3$	83	tP10	P4/mbm
Si$_3$Zr$_5$	84.4	hP16	P6$_3$/mcm
SiZr$_2$	86.7	tI12	I4/mcm
SiZr$_3$	~91	tP32	P4$_2$/n
		tI32	I$\bar{4}$
(βZr)	100	cI2	Im$\bar{3}$m
(αZr)	100	hP2	P6$_3$/mmc

Sm-Sn

G. Borzone, A. Borsese, and R. Ferro, 1982

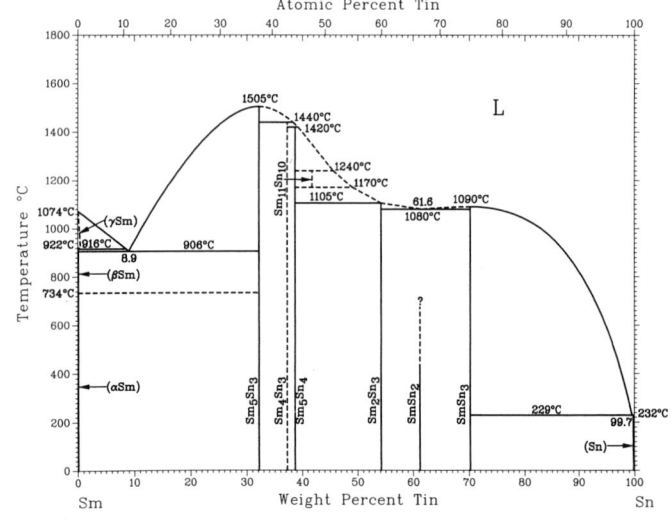

Phase	Composition, wt% Sn	Pearson symbol	Space group
(γSm)	0 to 0.4	cI2	Im$\bar{3}$m
(βSm)	0	hP2	P6$_3$/mmc
(αSm)	0	hR3	R$\bar{3}$m
Sm$_5$Sn$_3$	32.1	hP16	P6$_3$/mcm
Sm$_4$Sn$_3$	37	cI28	I$\bar{4}$3d
Sm$_5$Sn$_4$	38.8	oP36	Pnma
Sm$_{11}$Sn$_{10}$	~42	tI84	I4/mmm
Sm$_2$Sn$_3$	54	t**	...
SmSn$_2$	61.3	...	...
SmSn$_3$	70	cP4	Pm$\bar{3}$m
(βSn)	100	tI4	I4$_1$/amd
(αSn)	100	cF8	Fd$\bar{3}$m

Sm-Tl

S. Delfino, A. Saccone, A. Palenzona, and R. Ferro, unpublished

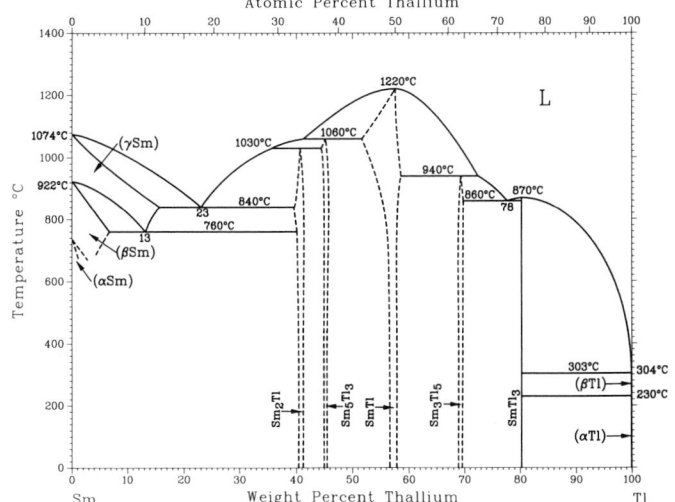

Phase	Composition, wt% Tl	Pearson symbol	Space group
(βSm)	0 to ~3.4	hP2	P6₃/mmc
(αSm)	0 to ?	hR2	R̄3m
(γSm)	0 to ~16	cI2	Im̄3m
Sm₂Tl	~40 to ~41	hP6	P6₃/mmc
Sm₅Tl₃	~44 to ~45	tI32	I4/mcm
SmTl(a)	~52 to ~59	tP2	Pm̄3m
	(or cI2)		Im̄3m
SmTl(b)	~52 to ~59	tP2	P4/mmm
Sm₃Tl₅	~69 to ~70	oC32	Cmcm
SmTl₃	80	cP4	Pm̄3m
(βTl)	100	cI2	Im̄3m
(αTl)	100	hP2	P6₃/mmc

(a) Cubic structure presumed to be room- and high-temperature phases. (b) Tetragonal structure presumed to be low-temperature phase.

Sm-Zn

From [Moffatt]

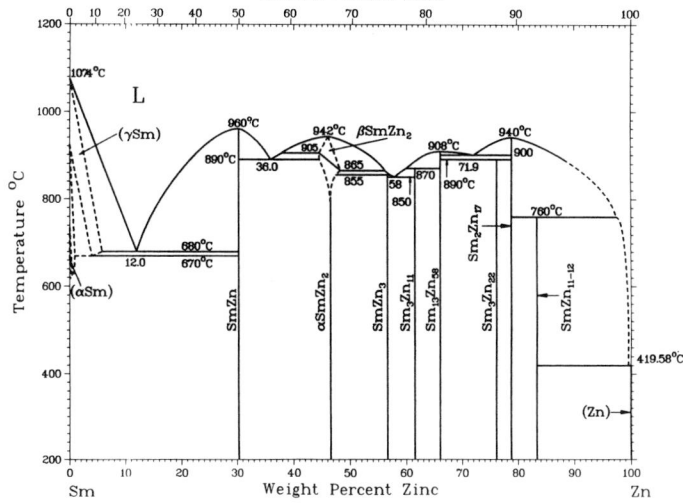

Phase	Composition, wt% Zn	Pearson symbol	Space group
(γSm)	0 to ?	cI2	Im̄3m
(βSm)	0 to ?	hP2	P6₃/mmc
(αSm)	0 to ?	hR3	R̄3m
SmZn	30.3	cP2	Pm̄3m
βSmZn₂	45.2 to 48	oI12	Imma
αSmZn₂	~46.6	oI12	Imma
SmZn₃	57	oP16	Pnma
Sm₃Zn₁₁	~61.5	oI28	Immm
Sm₁₃Zn₅₈	~66.2	hP142	P6₃mc
Sm₃Zn₂₂	76	tI100	I4₁/amd
Sm₂Zn₁₇	~78.8	...	...
SmZn₁₁₋₁₂	83	tI26	I4/mmm
(Zn)	100	hP2	P6₃/mmc

Sn-Sr

P.R. Subramanian, 1990

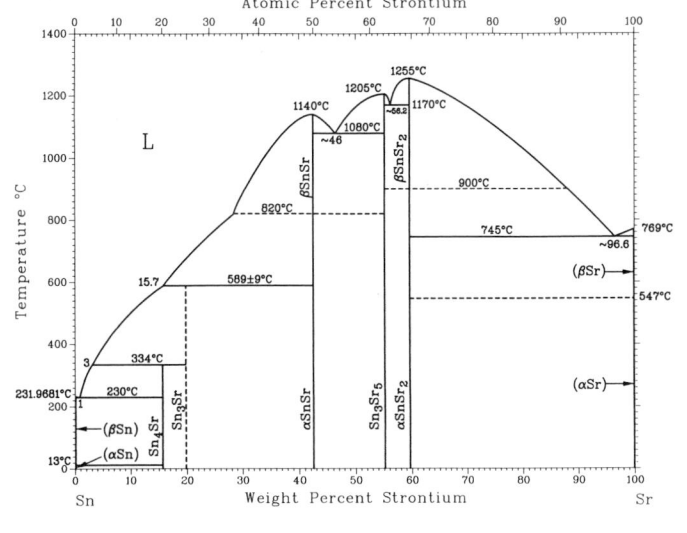

Phase	Composition, wt% Sr	Pearson symbol	Space group
(βSn)	0	tI4	I4₁/amd
(αSn)	0	cF8	Fd̄3m
Sn₄Sr	15.6	...	...
Sn₃Sr	19.7	...	...
βSnSr	42.5	...	...
αSnSr	42.5	oC8	Cmcm
Sn₃Sr₅	55.2	tI32	I4/mcm
βSnSr₂	59.6	...	...
αSnSr₂	59.6	oP12	Pnma
(βSr)	100	cI2	Im̄3m
(αSr)	100	cF4	Fm̄3m

Sn-Te

R.C. Sharma and Y.A. Chang, 1986

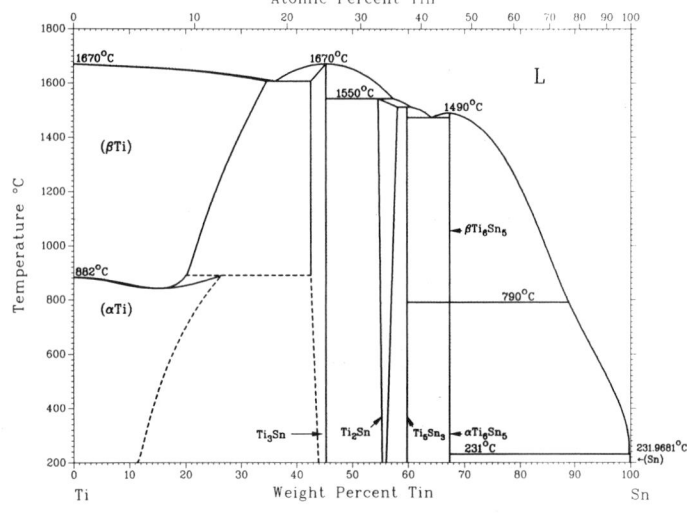

Phase	Composition, wt% Te	Pearson symbol	Space group
(Sn)	~0	tI4	$I4_1/amd$
SnTe	51.8	cF8	$Fm\overline{3}m$
SnTe(HP)	51.8	oP8	$Pnma$
(Te)	100	hP3	$P3_121$

Sn-Ti

J.L. Murray, 1987

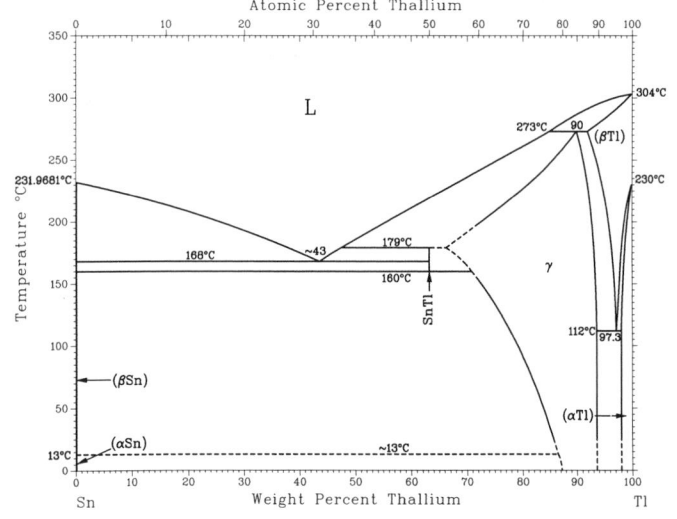

Phase	Composition, wt% Sn	Pearson symbol	Space group
(βTi)	0 to 34	cI2	$Im\overline{3}m$
(αTi)	0 to >16.7	hP2	$P6_3/mmc$
Ti₃Sn	43 to 45	hP8	$P6_3/mmc$
Ti₂Sn	54.6 to 58.1	hP6	$P6_3/mmc$
Ti₅Sn₃	59.8	hP16	$P6_3/mcm$
βTi₆Sn₅	67.4	hP22	$P6_3/mmc$
			$P\overline{3}1c$
αTi₆Sn₅	67.4	oI44	$Immm$
(Sn)	99.99 to 100	tI4	$I4_1/amd$

Sn-Tl

H. Okamoto, 1990

Phase	Composition, wt% Tl	Pearson symbol	Space group
(βSn)	0	tI4	$I4_1/amd$
(αSn)	0	cF8	$Fd\overline{3}m$
SnTl	63.3	tP2	$P4/mmm$
γ	68 to 94	cF4	$Fm\overline{3}m$
(βTl)	92 to 100	cI2	$Im\overline{3}m$
(αTl)	98 to 100	hP2	$P6_3/mmc$

Sn-U

R.I. Sheldon, E.M. Foltyn, and D.E. Peterson, 1987

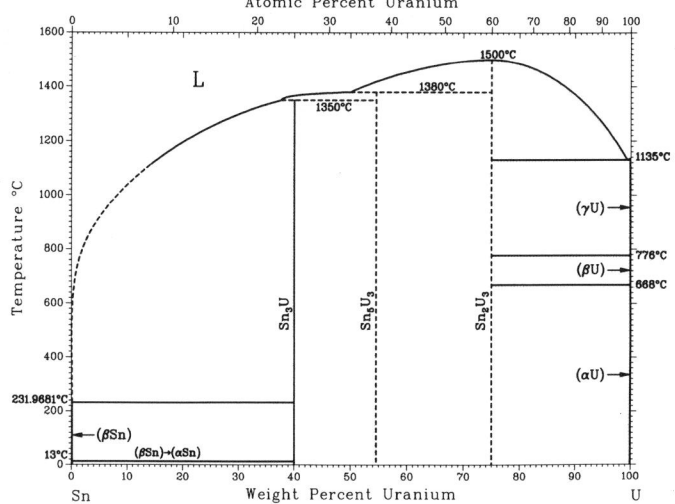

Phase	Composition, wt% U	Pearson symbol	Space group
(βSn)	0	tI4	I4₁/amd
(αSn)	0	cF8	Fd$\overline{3}$m
Sn₃U(a)	40.1	cP4	Pm$\overline{3}$m
Sn₅U₃	54.6	...	...
Sn₂U₃	75.0	...	...
(γU)	100	cI2	Im$\overline{3}$m
(βU)	100	tP30	P4₂/mnm
(αU)	100	oC4	Cmcm

(a) No tendency to disorder was observed.

Sn-Y

H. Okamoto, 1990

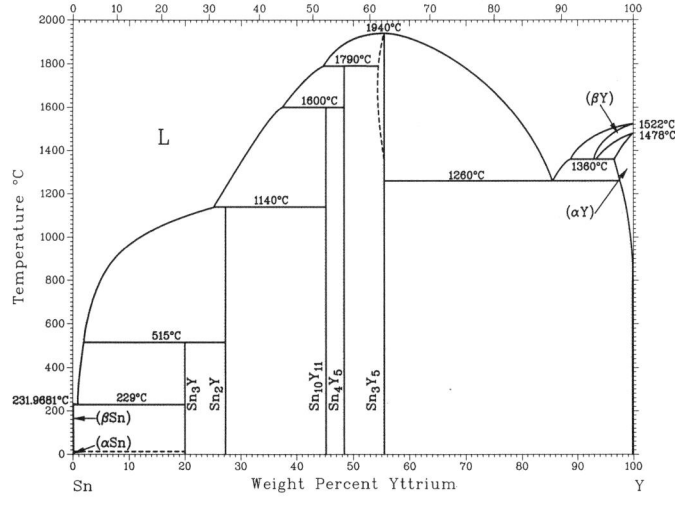

Phase	Composition, wt% Y	Pearson symbol	Space group
(βSn)	0	tI4	I4₁/amd
(αSn)	0	cF8	Fd$\overline{3}$m
Sn₃Y	20	cP4	Pm$\overline{3}$m
Sn₂Y	27.2	oC12	Cmcm
Sn₁₀Y₁₁	45.2	tI84	I4/mmm
Sn₄Y₅	48.4	oP36	Pnma
Sn₃Y₅	55.5	hP16	P6₃/mcm
(βY)	100	cI2	Im$\overline{3}$m
(αY)	100	hP2	P6₃/mmc

Sn-Yb

A. Palenzona and S. Cirafici, 1991

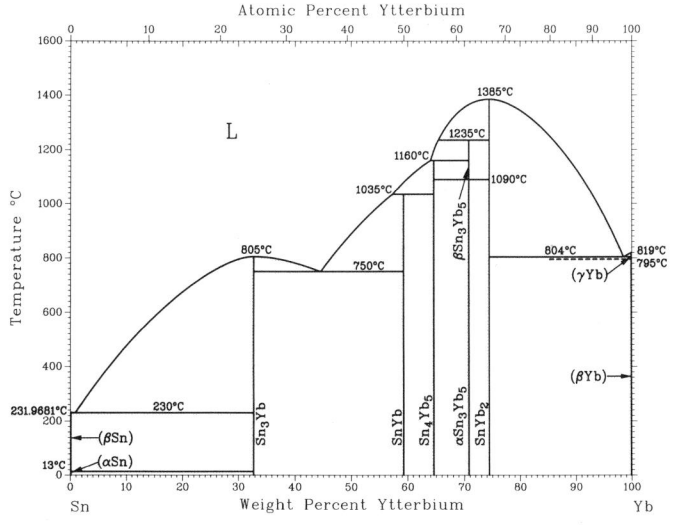

Phase	Composition, wt% Yb	Pearson symbol	Space group
(βSn)	0	tI4	I4₁/amd
(αSn)	0	cF8	Fd$\overline{3}$m
Sn₃Yb	32.7	cP4	Pm$\overline{3}$m
SnYb	59.3	tP2	P4/mmm
Sn₄Yb₅	64.6	oP36	Pnma
βSn₃Yb₅	70.8	tI32	I4/mcm
αSn₃Yb₅	70.8	hP16	P6₃/mcm
SnYb₂	74.5	hP6	P6₃/mmc
(γYb)	100	cI2	Im$\overline{3}$m
(βYb)	100	cF4	Fm$\overline{3}$m
(αYb)	100	hP2	P6₃/mmc

Sn-Zn

Z. Moser, J. Dutkiewicz, W. Gasior, and J. Salawa, 1985

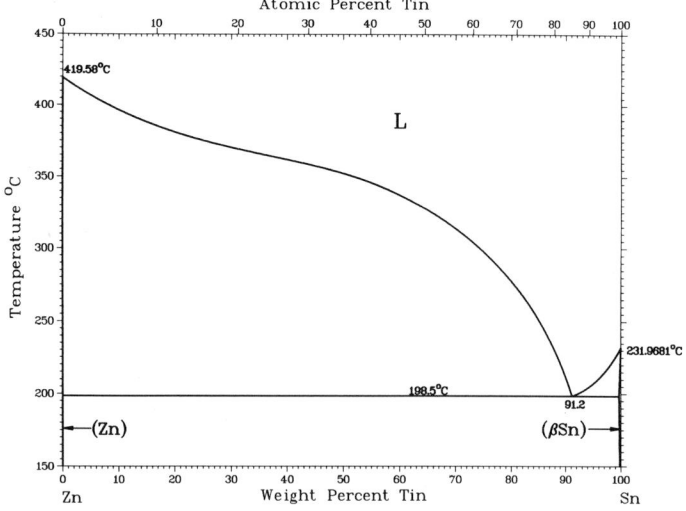

Phase	Composition, wt% Sn	Pearson symbol	Space group
(Zn)	0	hP2	$P6_3/mmc$
(βSn)	~100	tI4	$I4_1/amd$

Sn-Zr

J.P. Abriata, J.C. Bolcich, and D. Arias, 1983

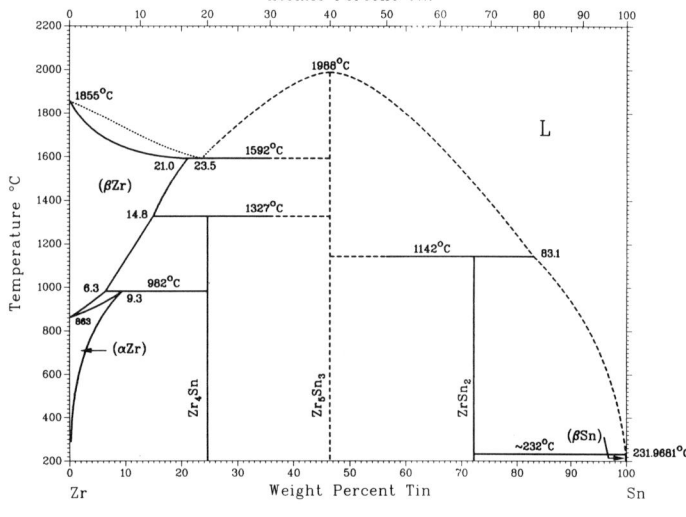

Phase	Composition, wt% Sn	Pearson symbol	Space group
(βZr)	0 to 21.0	cI2	$Im\bar{3}m$
(αZr)	0 to 9.3	hP2	$P6_3/mmc$
Zr_4Sn	~25	cP8	$Pm\bar{3}n$
Zr_5Sn_3	40 to ~47	hP16	$P6_3/mcm$
$ZrSn_2$	72.8	oF24	Fddd
(βSn)	100	tI4	$I4_1/amd$
(αSn)	100	cF8	$Fd\bar{3}m$
Possible additional phase			
Zr_5Sn_4	~52	hP18	...

Sr-Te

Yu.B. Lyskova and A.V. Vakhobov, 1975

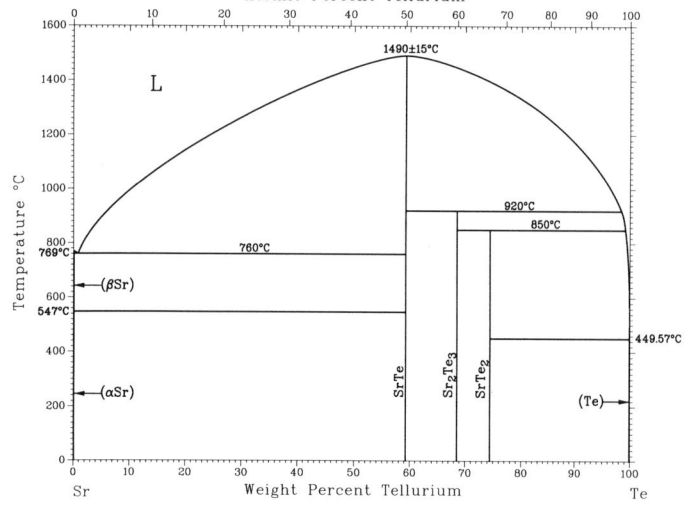

Phase	Composition, wt% Te	Pearson symbol	Space group
(βSr)	0	cI2	$Im\bar{3}m$
(αSr)	0	cF4	$Fm\bar{3}m$
SrTe	59.3	cF8	$Fm\bar{3}m$
Sr_2Te_3	69	...	...
$SrTe_2$	74.5	...	...
(Te)	100	hP3	$P3_121$

Sr-Tl

H. Okamoto, 1990

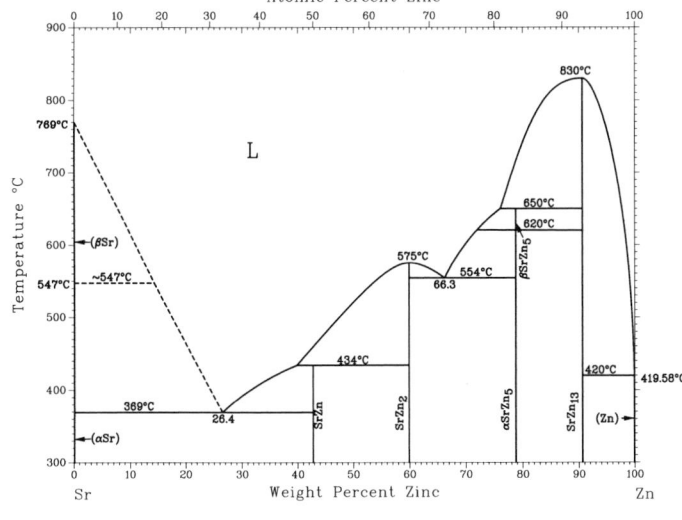

Phase	Composition, wt% Tl	Pearson symbol	Space group
(βSr)	0	cI2	Im$\bar{3}$m
(αSr)	0	cF4	Fm$\bar{3}$m
Sr$_3$Tl	44	...	...
Sr$_5$Tl$_3$	58.3	tI32	I4/mcm
SrTl	70.0	cP2	Pm$\bar{3}$m
Sr$_2$Tl$_3$	78	...	...
SrTl$_2$	82.4	hP6	P6$_3$/mmc
SrTl$_3$	88	...	...
(βTl)	? to 100	cI2	Im$\bar{3}$m
(αTl)	98.0 to 100	hP2	P6$_3$/mmc

Sr-Zn

P.R. Subramanian, 1990

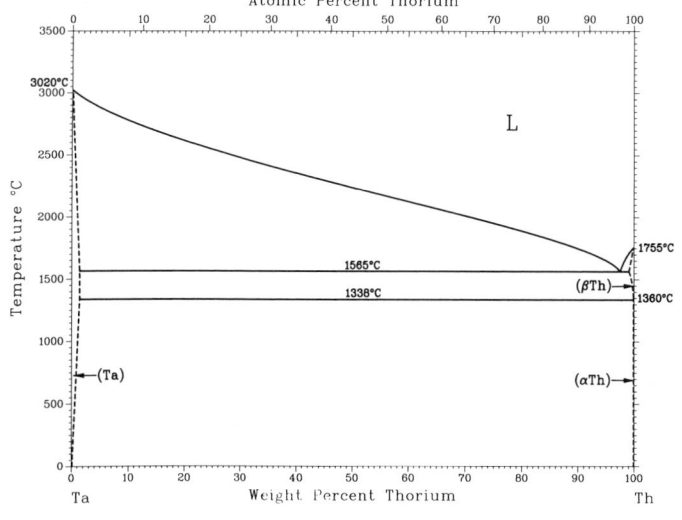

Phase	Composition, wt% Zn	Pearson symbol	Space group
(βSr)	0	cI2	Im$\bar{3}$m
(αSr)	0	cF4	Fm$\bar{3}$m
SrZn	42.7	oP8	Pnma
SrZn$_2$	59.9	oI12	Imma
SrZn$_5$(HT)	78.8	hP6	P6/mmm
SrZn$_5$(LT)	78.8	oP24	Pnma
SrZn$_{13}$	~90.7	cF112	Fm$\bar{3}$c
(Zn)	100	hP2	P6$_3$/mmc

Ta-Th

R. Krishnan, S.P. Garg, and N. Krishnamurthy, 1989

Phase	Composition, wt% Th	Pearson symbol	Space group
(Ta)	0 to <1	cI2	Im$\bar{3}$m
(βTh)	99.85 to 100	cI2	Im$\bar{3}$m
(αTh)	>99.9 to 100	cF4	Fm$\bar{3}$m

Ta-Ti

J.L. Murray, 1987

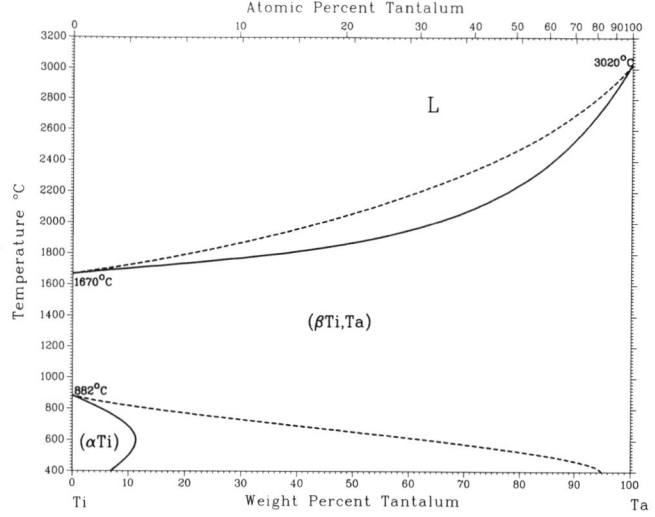

Phase	Composition, wt% Ta	Pearson symbol	Space group
(βTi,Ta)	0 to 100	cI2	$Im\bar{3}m$
(αTi)	0 to 12.4	hP2	$P6_3/mmc$
Metastable phases			
(α')	...	hP2	$P6_3/mmc$
(α'')	...	oC4	Cmcm
ω	...	hP3	$P6/mmm$ or $P\bar{3}m1$

Ta-U

R. Krishnan, S.P. Garg, and N. Krishnamurthy, 1988

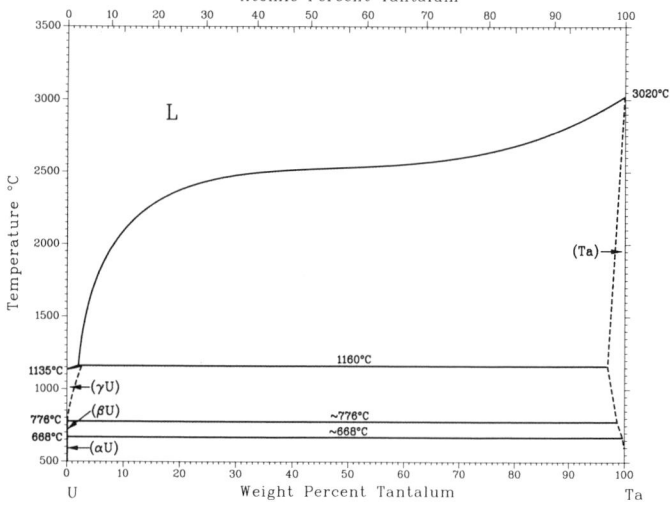

Phase	Composition, wt% Ta	Pearson symbol	Space group
(γU)	0 to ~2	cI2	$Im\bar{3}m$
(βU)	0	tP30	$P4_2/mnm$
(αU)	0	oC4	Cmcm
(Ta)	? to 100	cI2	$Im\bar{3}m$

Ta-V

J.F. Smith and O.N. Carlson, 1989

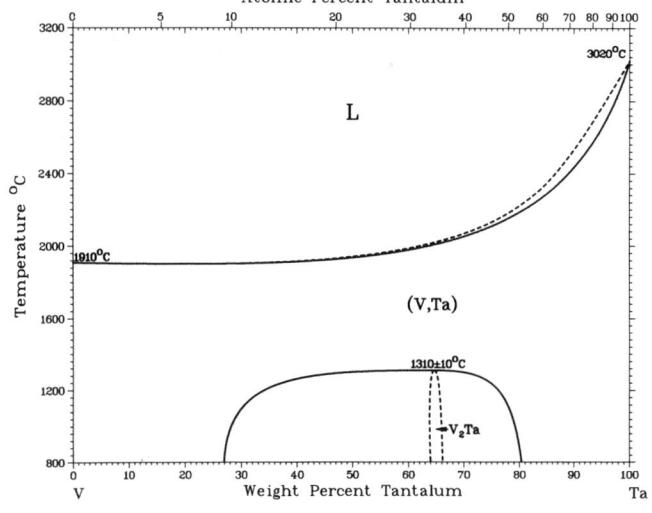

Phase	Composition, wt% Ta	Pearson symbol	Space group
(V,Ta)	0 to 100	cI2	$Im\bar{3}m$
V₂Ta(a)	~64 to ~67	cF24	$Fd\bar{3}m$

(a) A high-temperature polymorph of V_2Ta has been reported to be a hexagonal $MgZn_2$-type structure, with $hP12$ and $P6_3/mmc$.

Ta-W

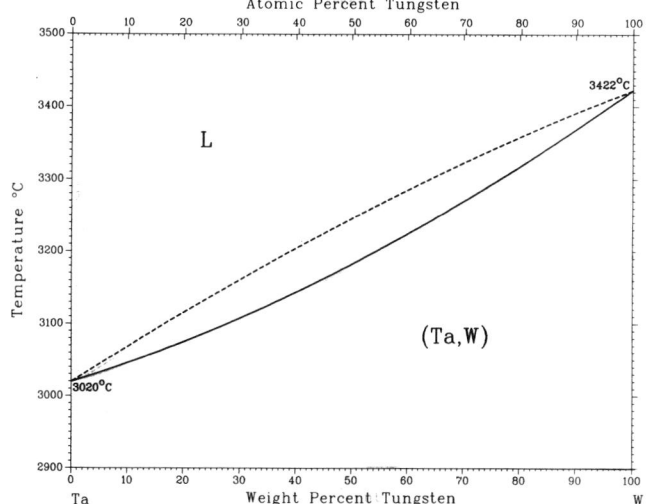

R. Krishnan, S.P. Garg, and N. Krishnamurthy, 1985

Phase	Composition, wt% W	Pearson symbol	Space group
(Ta,W)	0 to 100	cI2	Im$\bar{3}$m

Ta-Zr

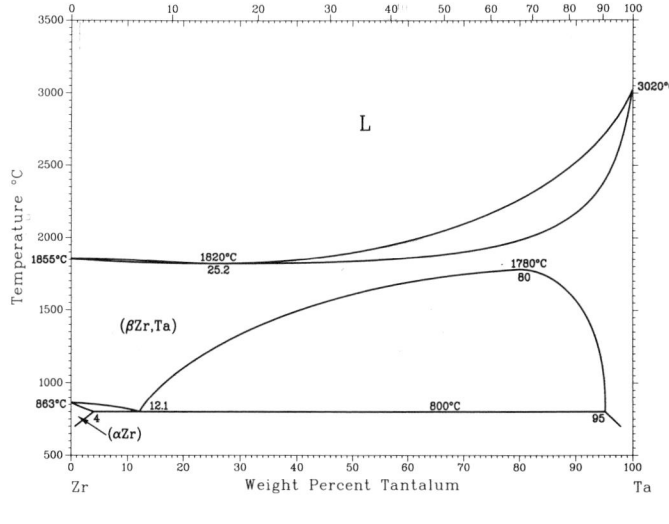

R. Krishnan, S.P. Garg, S. Banerjee, and N. Krishnamurthy, 1989

Phase	Composition, wt% Ta	Pearson symbol	Space group
(βZr,Ta)	0 to 100	cI2	Im$\bar{3}$m
(αZr)	0 to 4	hP2	P6$_3$/mmc

Tb-Tl

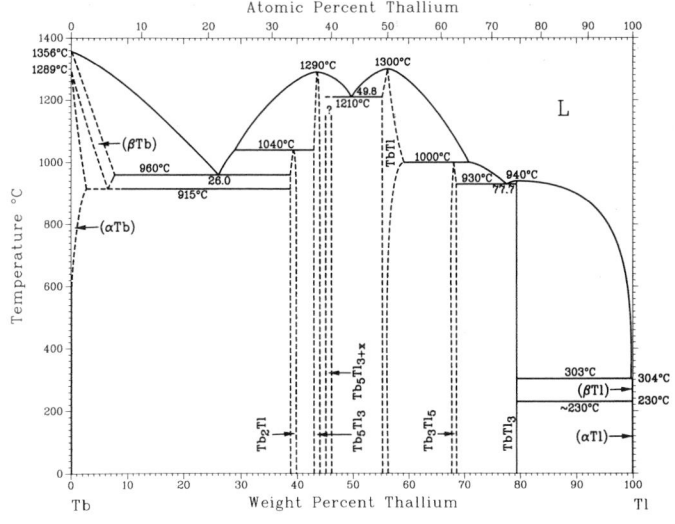

S. Delfino, A. Saccone, A. Palenzona, and R. Ferro, unpublished

Phase	Composition, wt% Tl	Pearson symbol	Space group
(βTb)	0 to ~6	cI2	Im$\bar{3}$m
(αTb)	0 to ?	hP2	P6$_3$/mmc
Tb$_2$Tl	~39 to ~40	hP6	P6$_3$/mmc
βTb$_5$Tl$_3$	~43 to ~44	hP16	P6$_3$/mcm
αTb$_5$Tl$_3$	~43 to ~44	tI32	I4/mcm
Tb$_5$Tl$_{3+x}$	?	tI32	I4/mcm
TbTl	~55 to ~59	cP2(a)	Pm$\bar{3}$m
		(or cI2)	Im$\bar{3}$m
	~55 to ~59	tP2(b)	P4/mmm
Tb$_3$Tl$_5$	~68 to ~69	oC32	Cmcm
TbTl$_3$	79	cP4	Pm$\bar{3}$m
(βTl)	100	cI2	Im$\bar{3}$m
(αTl)	100	hP2	P6$_3$/mmc

(a) High-temperature phase (>250 K). (b) Low-temperature phase

Te-Tl

H. Okamoto, 1991

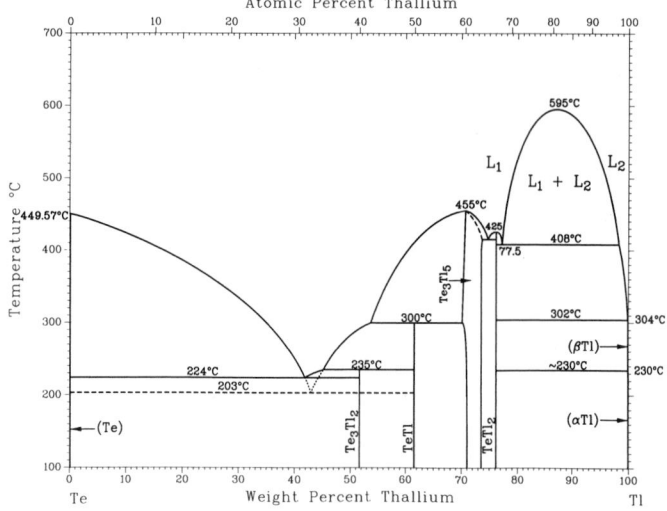

Phase	Composition, wt% Tl	Pearson symbol	Space group
(Te)	0	hP3	P3₁21
Te₃Tl₂	52	mC20	Cc
TeTl	61.6	tI32	I4/mcm
Te₃Tl₅	72.7	tI32	I4/m
TeTl₂	76.2	...	...
(βTl)	100	cI2	Im3̄m
(αTl)	100	hP2	P6₃/mmc

Te-U

From [Moffatt]

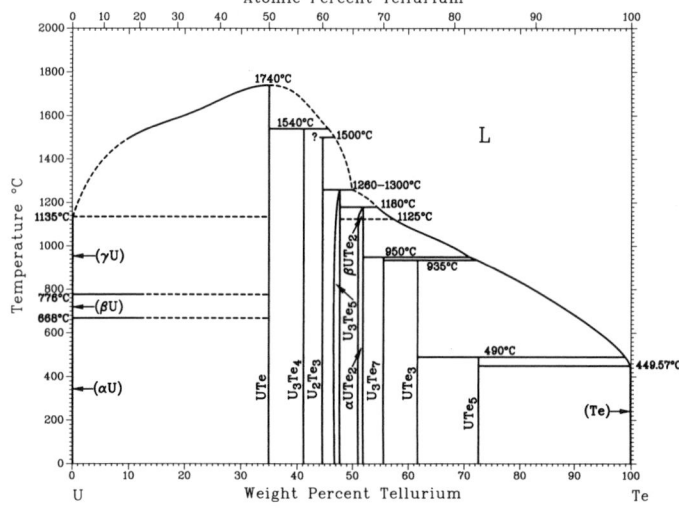

Phase	Composition, wt% Te	Pearson symbol	Space group
(γU)	0	cI2	Im3̄m
(βU)	0	tP30	P4₂mnm
(αU)	0	oC4	Cmcm
UTe	34.9	cF8	Fm3̄m
U₃Te₄	~41.0	cI28	Ī43d
U₂Te₃	45	hP16	P6₃/mcm
U₃Te₅	47 to 48	oP32	Pnma
β/αUTe₂	51 to 52	oI12	Immm
		tP6	P4/nmm
U₃Te₇	56	...	...
UTe₃	62	t**	...
UTe₅	73.1	...	...
(Te)	100	hP3	P3₁21

Te-Yb

H. Okamoto, 1990

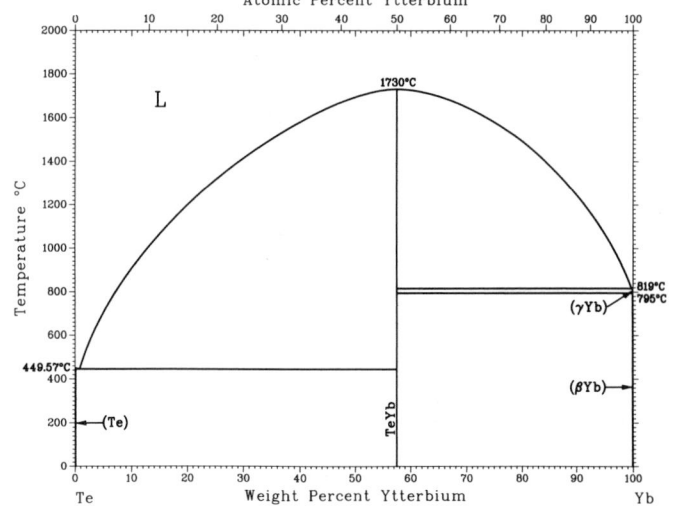

Phase	Composition, wt% Yb	Pearson symbol	Space group
(Te)	0	hP3	P3₁21
TeYb	57.6	cF8	Fm3̄m
(γYb)	100	cI2	Im3̄m
(βYb)	100	cF4	Fm3̄m
(αYb)	100	hP2	P6₃/mmc

Te-Zn

R.C. Sharma and Y.A. Chang, 1987

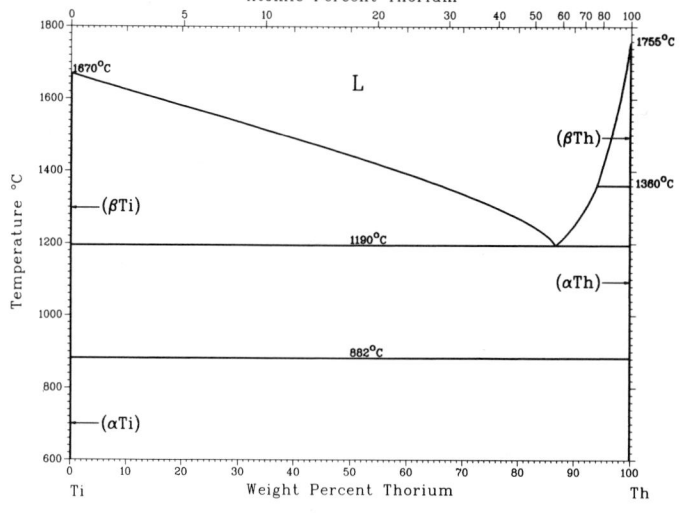

Phase	Composition, wt% Te	Pearson symbol	Space group
(Zn)	0	hP2	$P6_3/mmc$
αZnTe	66.1	cF8	$F\overline{4}3m$
(Te)	100	hP3	$P3_121$

Th-Ti

J.L. Murray, 1987

Phase	Composition, wt% Th	Pearson symbol	Space group
(βTi)	0	cI2	$Im\overline{3}m$
(αTi)	0	hP2	$P6_3/mmc$
(βTh)	100	cI2	$Im\overline{3}m$
(αTh)	100	cF4	$Fm\overline{3}m$

Th-Tl

H. Okamoto, 1990

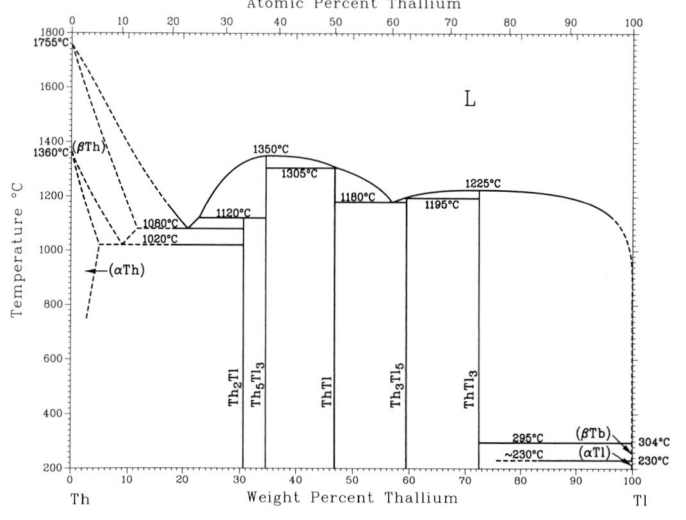

Phase	Composition, wt% Tl	Pearson symbol	Space group
(βTh)	0 to ?	cI2	$Im\overline{3}m$
(αTh)	0 to ?	cF4	$Fm\overline{3}m$
Th₂Tl	30.5	tI12	$I4/mcm$
Th₅Tl₃	34.6	hP16	$P6_3/mcm$
ThTl	46.8	oP24	$Pbcm$
Th₃Tl₅	59.5	oC32	$Cmcm$
ThTl₃	73	cP4	$Pm\overline{3}m$
(βTl)	100	cI2	$Im\overline{3}m$
(αTl)	100	hP2	$P6_3/mmc$

Th-Zn

P. Chiotti and K.J. Gill, 1961

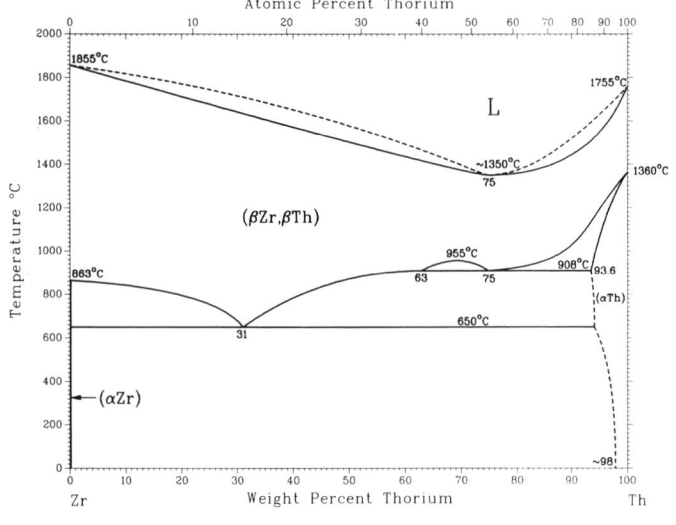

Phase	Composition, wt% Zn	Pearson symbol	Space group
(βTh)	0	cI2	$Im\overline{3}m$
(αTh)	0	cF2	$Fm\overline{3}m$
Th₂Zn	12.3	tI12	I4/mcm
ThZn₂	36.1	hP3	P6/mmm
ThZn₄	53	tI10	I4/mmm
βTh₂Zn₁₇	70.6	hR19	$R\overline{3}m$
αTh₂Zn₁₇	70.6	hP38	P6₃/mmc
(Zn)	100	hP2	P6₃/mmc

Th-Zr

E.D. Gibson, B.A. Loomis, and O.N. Carlson, 1958;
R.H. Johnson and R.W.K. Honeycombe, 1961

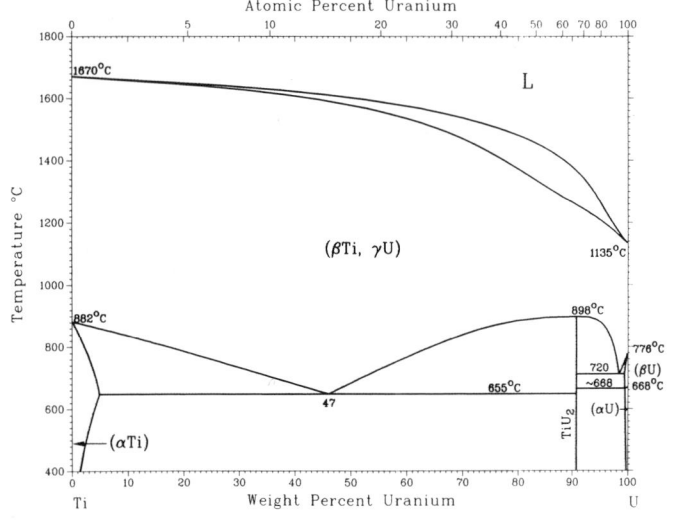

Phase	Composition, wt% Th	Pearson symbol	Space group
(βZr,βTh)	0 to 100	cI2	$Im\overline{3}m$
(αZr)	0	hP2	P6₃/mmc
(αTh)	93.6 to 100	cF4	$Fm\overline{3}m$

Ti-U

J.L. Murray, 1987

Phase	Composition, wt% U	Pearson symbol	Space group
(βTi,γU)	0 to 100	cI2	$Im\overline{3}m$
(αTi)	0 to ~5	hP2	P6₃/mmc
TiU₂	90.9	hP3	P6/mmm
(βU)	~99.6 to 100	tP30	P4₂nm
(αU)	~99.6 to 100	oC4	Cmcm
α_b″(a)	38	(b)	...

(a) Metastable. (b) Monoclinic

Ti-V

J.L. Murray, 1989

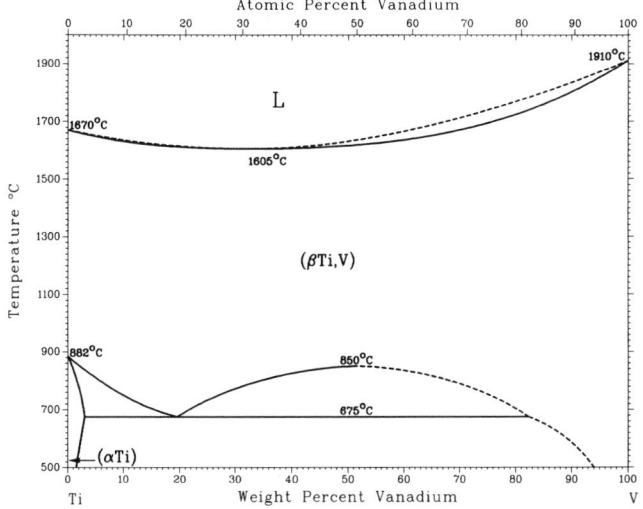

Phase	Composition, wt% V	Pearson symbol	Space group
(βTi,V)	0 to 100	cI2	Im$\bar{3}$m
(αTi)	0 to ~3	hP2	P6₃/mmc
Metastable phases			
α′	0 to 5	hP2	P6₃/mmc
α″	5 to 16	oC4	Cmcm
ω	12 to ~51.5	hP3	P6/mmm or P$\bar{3}$m1

Ti-W

J.L. Murray, 1987

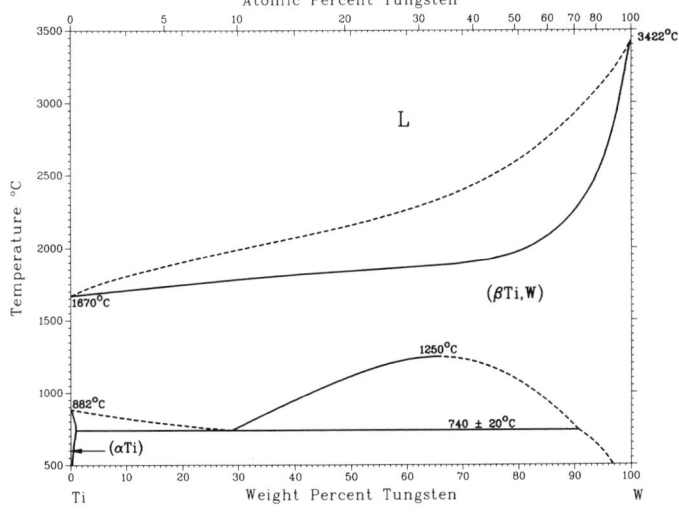

Phase	Composition, wt% W	Pearson symbol	Space group
(βTi,W)	0 to 100	cI2	Im$\bar{3}$m
(αTi)	0 to 0.8	hP2	P6₃/mmc
α′(a)	0 to 7	hP2	P6₃/mmc
α″(a)	7 to 18.3	oC4	Cmcm
ω(a)	20 to 30	hP3	P6/mmm

(a) Metastable

Ti-Y

J.L. Murray, 1987

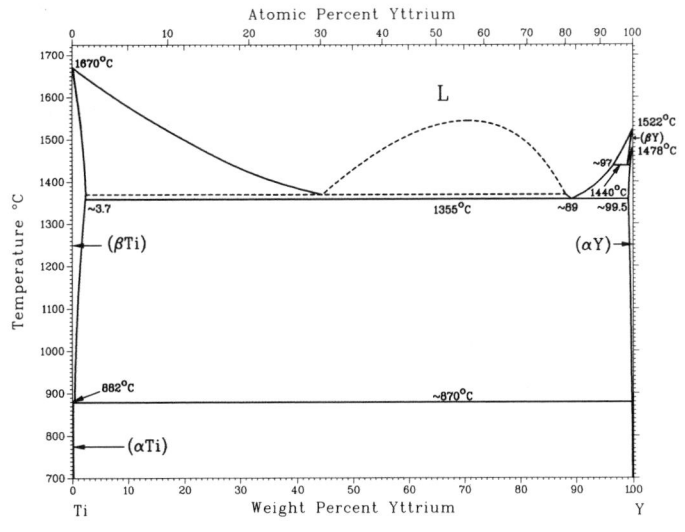

Phase	Composition, wt% Y	Pearson symbol	Space group
(βTi)	0 to ~3.7	cI2	Im$\bar{3}$m
(αTi)	0 to ~0.02	hP2	P6₃/mmc
(βY)	~99.5 to 100	cI2	Im$\bar{3}$m
(αY)	~99.5 to 100	hP2	P6₃/mmc

Ti-Zr

J.L. Murray, 1987

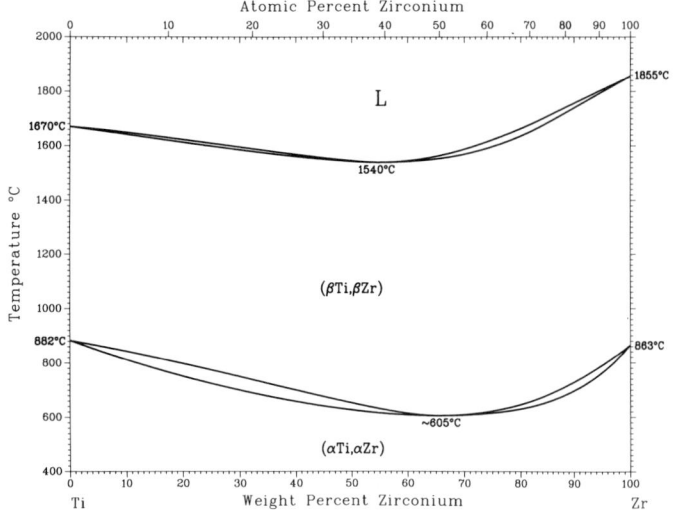

Phase	Composition, wt% Zr	Pearson symbol	Space group
(βTi,βZr)	0 to 100	cI2	Im$\overline{3}$m
(αTi,αZr)	0 to 100	hP2	P6₃/mmc
Metastable phases			
α′	...	hP2	P6₃/mmc
ω	...	hP3	P6/mmm or P$\overline{3}$m1

Tl-Yb

S. Delfino, A. Saccone, A. Palenzona, and R. Ferro, unpublished

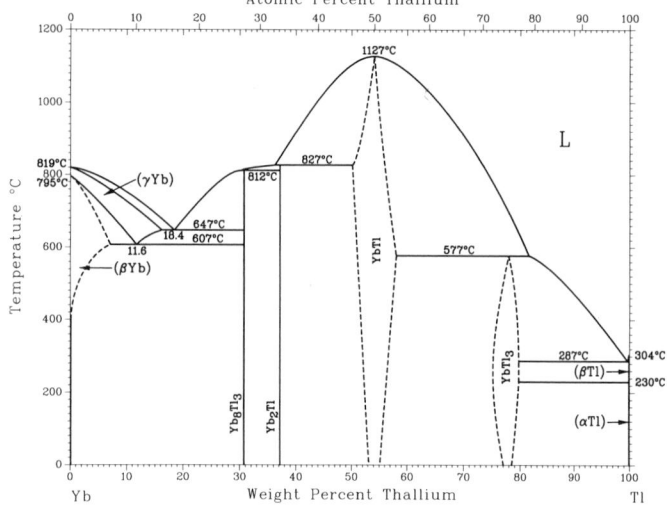

Phase	Composition, wt% Tl	Pearson symbol	Space group
(βYb)	0 to ~7	cF2	Fm$\overline{3}$m
(γYb)	0 to ~16	cI2	Im$\overline{3}$m
Yb₈Tl₃	30.69	aP22	P$\overline{1}$
Yb₂Tl	37.13	oP12	Pnma
YbTl	~50 to ~58	cP2	Pm$\overline{3}$m
		(or cI2)	Im$\overline{3}$m
YbTl₃	~75 to ~80	cP4	Pm$\overline{3}$m
(βTl)	100	cI2	Im$\overline{3}$m
(αTl)	100	hP2	P6₃/mmc

Tl-Zn

A.V. Vegesack, 1907; and W. Seith, H. Johnson, and J. Wagner, 1952

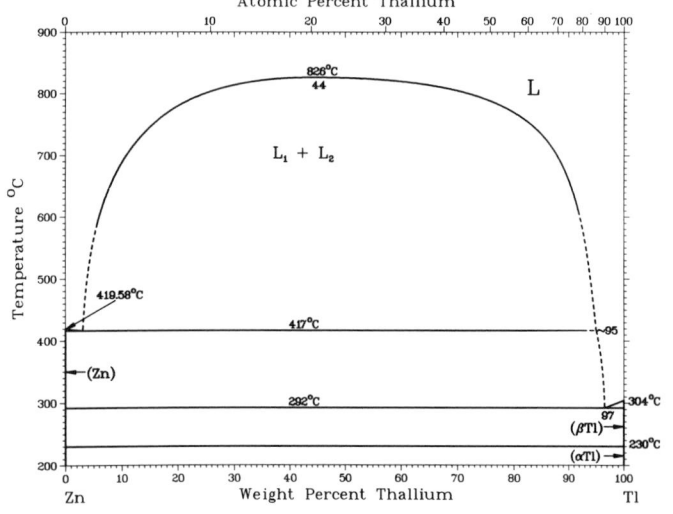

Phase	Composition, wt% Tl	Pearson symbol	Space group
(Zn)	0	hP2	P6₃/mmc
(βTl)	100	cI2	Im$\overline{3}$m
(αTl)	100	hP2	P6₃/mmc

U-Zr

H. Okamoto, 1992

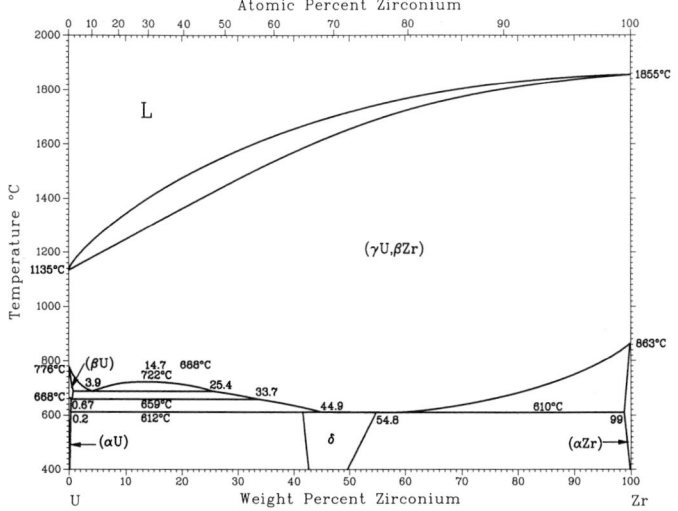

Phase	Composition, wt% Zr	Pearson symbol	Space group
(γU,βZr)	0 to 100	cI2	Im$\overline{3}$m
(βU)	0 to 0.4	tP30	P4$_2$/mnm
(αU)	0 to 0.2	oC4	Cmcm
δ	42 to 55	hP3	P6/mmm
(αZr)	99 to 100	hP2	P6$_3$/mmc

V-W

S.V. Nagender Naidu, A.M. Sriramamurthy, M. Vijayakumar, and
P. Rama Rao, 1989

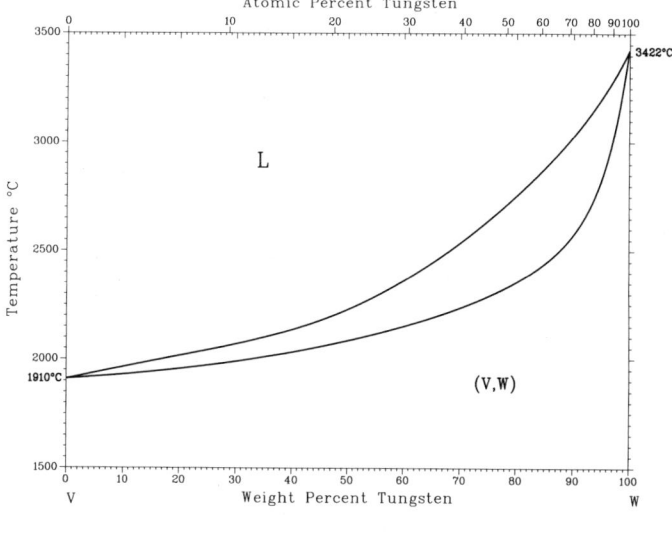

Phase	Composition, wt% W	Pearson symbol	Space group
(V,W)	0 to 100	cI2	Im$\overline{3}$m

V-Zr

J.F. Smith, 1989

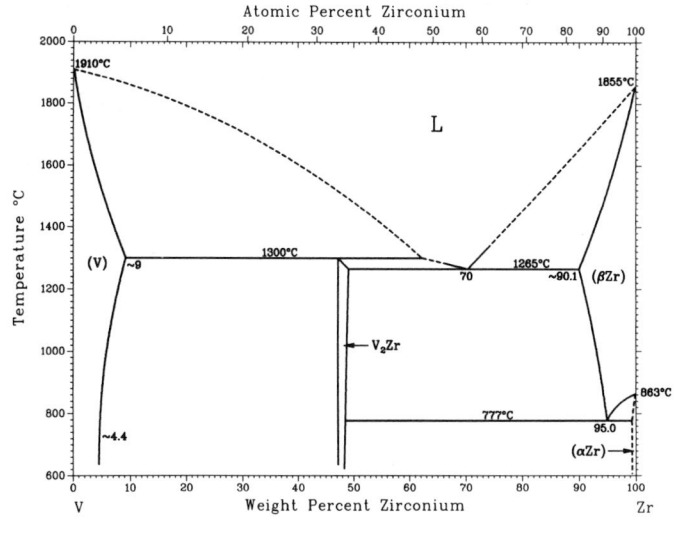

Phase	Composition, wt% Zr	Pearson symbol	Space group
(V)	0 to ~9	cI2	Im$\overline{3}$m
V$_2$Zr	~47.2	cF24	Fd$\overline{3}$m
(βZr)	~90.1 to 100	cI2	Im$\overline{3}$m
(αZr)	~100	hP2	P6$_3$/mmc

W-Zr

S.V. Nagender Naidu and P. Rama Rao, 1991

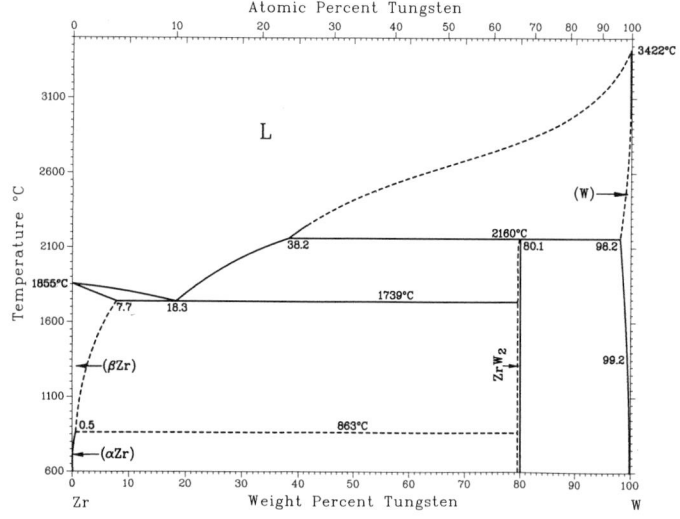

Phase	Composition, wt% W	Pearson symbol	Space group
(βZr)	0 to 7.7	cI2	$Im\bar{3}m$
(αZr)	0 to 0.50	hP2	$P6_3/mmc$
ZrW₂	~80.1	cF24	$Fd\bar{3}m$
(W)	98.2 to 100	cI2	$Im\bar{3}m$

Y-Zn

H. Okamoto, 1990

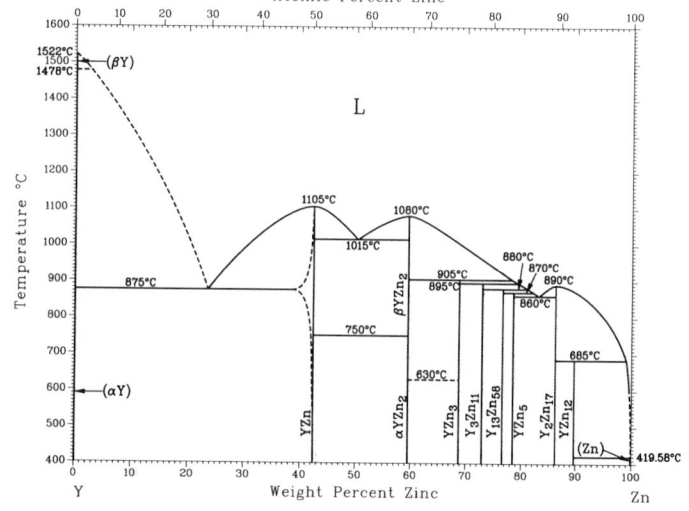

Phase	Composition, wt% Zn	Pearson symbol	Space group
(βY)	0	cI2	$Im\bar{3}m$
(αY)	0	hP2	$P6_3/mmc$
YZn	? to 42.4	cP2	$Pm\bar{3}m$
βYZn₂	59.6	...	...
αYZn₂	59.6	oI12	Imma
YZn₃	69	oP16	Pnma
Y₃Zn₁₁	73.0	oI28	Immm
Y₁₃Zn₅₈	76.7	hP142	$P6_3mc$
YZn₅	76.6	hP36	$P6_3/mmc$
Y₂Zn₁₇	86.2	hP38	$P6_3/mmc$
YZn₁₂	89.8	tI26	I4/mmm
(Zn)	100	hP2	$P6_3/mmc$

Y-Zr

A. Palenzona and S. Cirafici, 1991

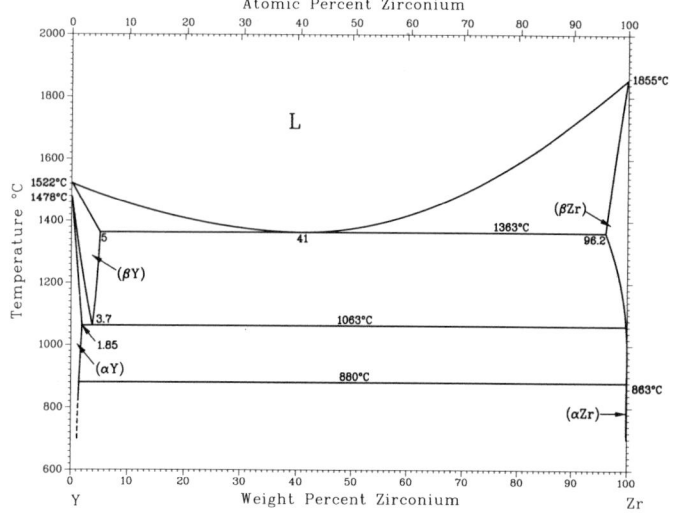

Phase	Composition, wt% Zr	Pearson symbol	Space group
(βY)	0 to 5	cI2	$Im\bar{3}m$
(αY)	0 to 1.85	hP2	$P6_3/mmc$
(βZr)	96.2 to 100	cI2	$Im\bar{3}m$
(αZr)	100	hP2	$P6_3/mmc$

Yb-Zn

J.T. Mason and P. Chiotti, 1968

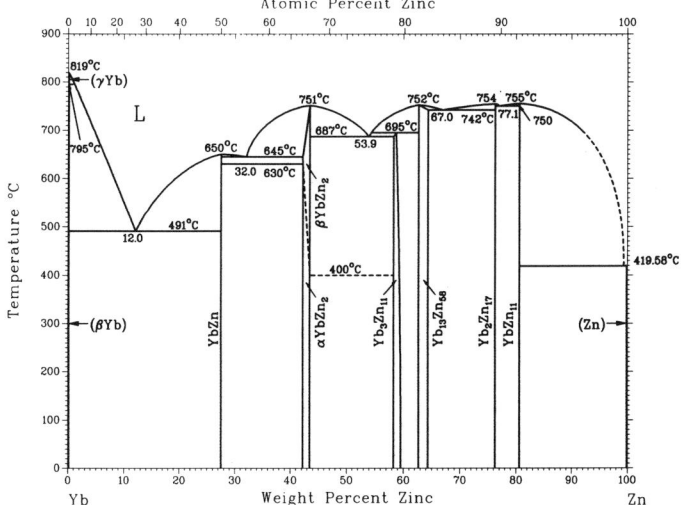

Phase	Composition, wt% Zn	Pearson symbol	Space group
(γYb)	0	cI2	Im$\overline{3}$m
(βYb)	0	cF4	Fm$\overline{3}$m
YbZn	27.4	cP2	Pm$\overline{3}$m
βYbZn$_2$	~42 to 43	...	...
αYbZn$_2$	~42 to 43	oI12	Imma
Yb$_3$Zn$_{11}$	~58.0 to 59.4	oI28	Immm
Yb$_{13}$Zn$_{58}$	~62.5 to 64.0	hP142	P6$_3$mc
Yb$_2$Zn$_{17}$	76.3	...	...
YbZn$_{11}$	80.3	tI48	I4$_1$/amd
(Zn)	100	hP2	P6$_3$/mmc
Other reported phases			
Yb$_3$Zn$_{17}$	68	cI160	Im$\overline{3}$
YbZn$_{13}$	83.2	cF112	Fm$\overline{3}$c

Section 3
Ternary Alloy Phase Diagrams

List of Systems Included:

Introduction to Ternary Alloy Phase Diagrams

THE 80 TERNARY SYSTEMS covered in this Section were selected for their commercial importance from the thousands of systems scheduled for inclusion in the *Handbook of Ternary Alloy Phase Diagrams*, to be published by ASM in 1994. The 313 diagrams shown here were chosen from the more than 12,000 assembled for that project. Wherever a recent compilation of diagrams assessed under the International Programme covered one of these systems, priority was given to those evaluated diagrams in preference to older, unassessed work. The remaining diagrams, although not yet assessed, were selected as the best available.

When a single source covered a system, a set of compatible diagrams was selected from it. For some systems, however, diagrams from more than one source were needed. Except for occasional conversion of composition scale from atomic to weight percent or change in orientation or labeling, each author's diagram has been redrawn, but shown as originally presented. *Therefore, the diagrams do not, in all instances, agree with one another and with the binary diagrams published in this Volume.* The reference source for each diagram is identified by a code consisting of two numbers (indicating the year of publication) followed by the first three letters of the first author's (or editor's) surname. The complete citation for each source code is listed at the end of this Section.

Because this Handbook is designed to be used primarily by engineers to solve industrial problems, the composition scale is plotted in weight percent. Conversions between weight and atomic composition can be made using the standard atomic weights listed in the Appendix. For the sake of clarity, grid lines are not superimposed on the phase diagrams. However, tick marks are provided along the composition scales as well as the temperature scale, which is shown in degrees Celsius. Celsius temperatures can be easily converted to degrees Fahrenheit using the table in the Appendix. When an arrowhead appears on a temperature trough line in a liquidus projection, it indicates the direction of decreasing temperature in the trough. Dashed lines are used to denote uncertain or speculative boundaries. Dotted lines indicate the limit of the investigated region.

The diagrams presented in this Section are for stable equilibrium conditions, with the exception of metastable conditions for some diagrams involving carbon and iron. These latter ternary diagrams can be identified by the presence of Fe_3C on the Fe-C binary portion of the diagram. In some ternary diagrams involving carbon and iron, the symbol M is used to represent both iron and the other metallic element when the two metals substitute for each other in a carbide phase—for example, M_3C.

Ag-Au-Cu liquidus projection [90Pri]

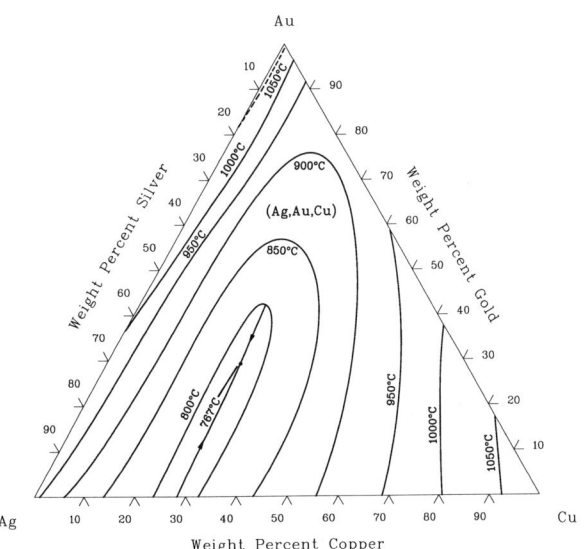

Ag-Au-Cu isothermal section at 775 °C [90Pri]

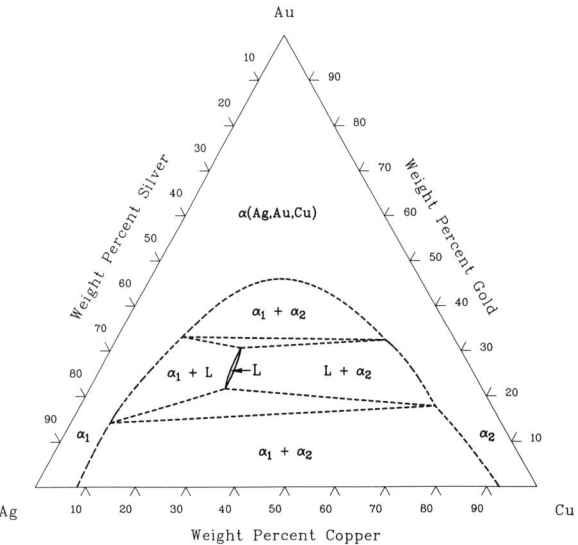

Ag-Au-Cu isothermal section at 950 °C [90Pri]

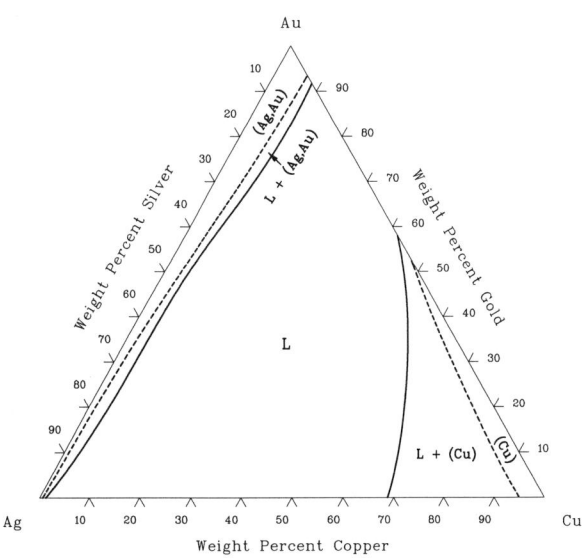

Ag-Au-Cu isothermal section at 300 °C [90Pri]

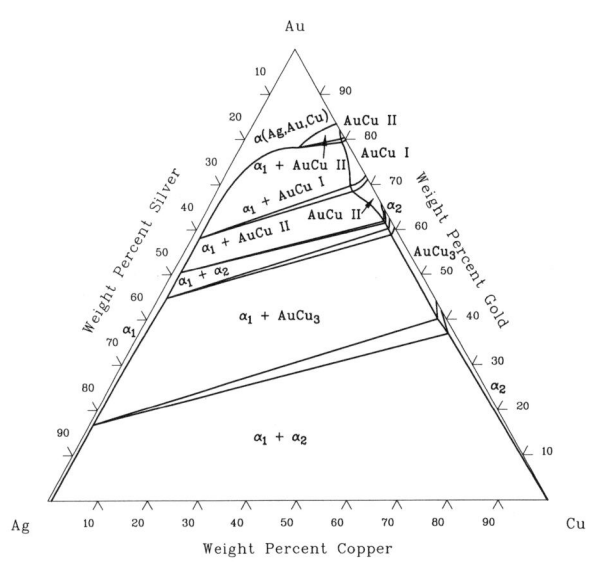

Ag-Au-Cu isothermal section at 850 °C [90Pri]

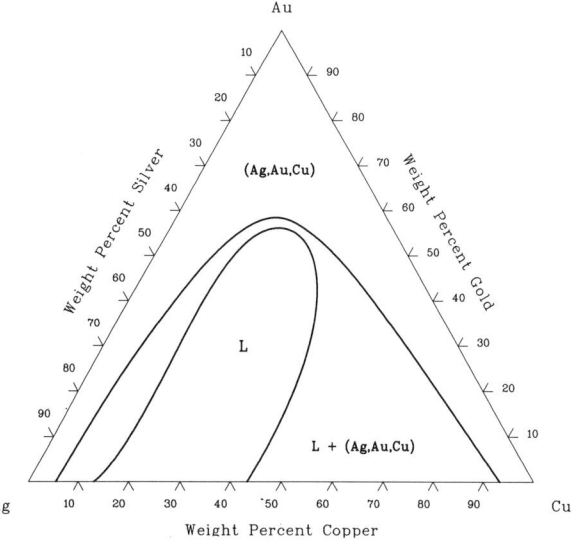

Ag-Cd-Cu liquidus projection [88Pet]

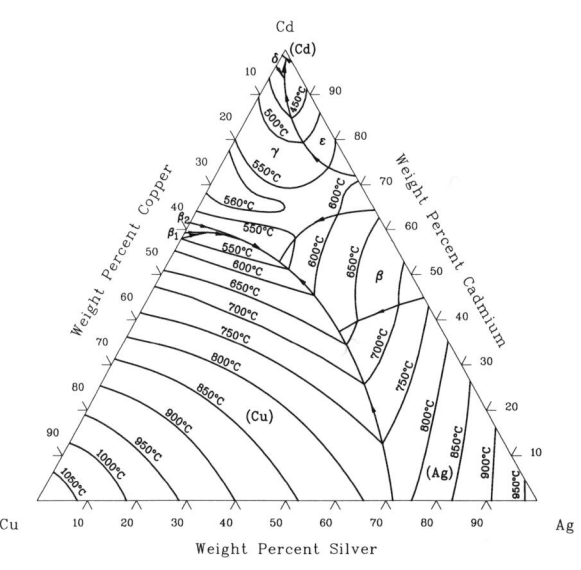

3•6/Ternary Alloy Phase Diagrams

Ag-Cd-Cu isothermal section at 600 °C [88Pet]

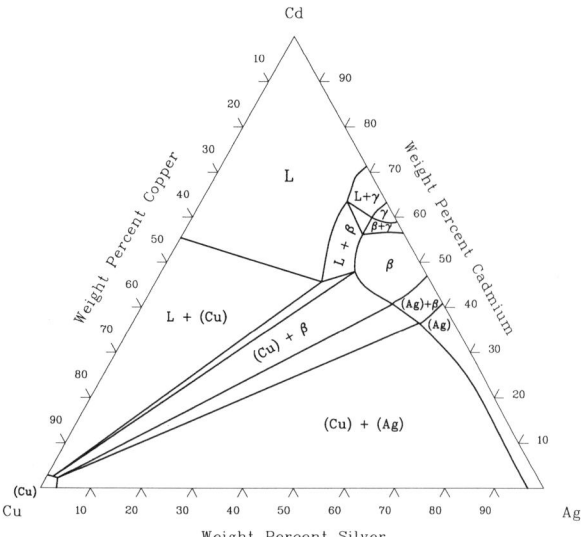

Ag-Cd-Zn liquidus projection with regions of primary crystallization [88Pet]

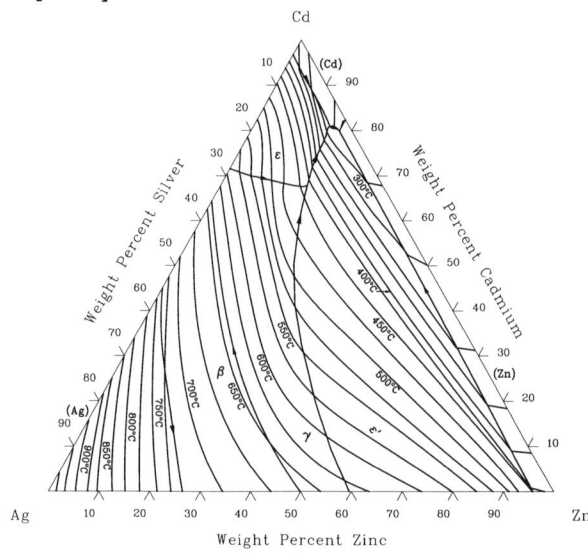

Ag-Cd-Cu isothermal section at 500 °C [88Pet]

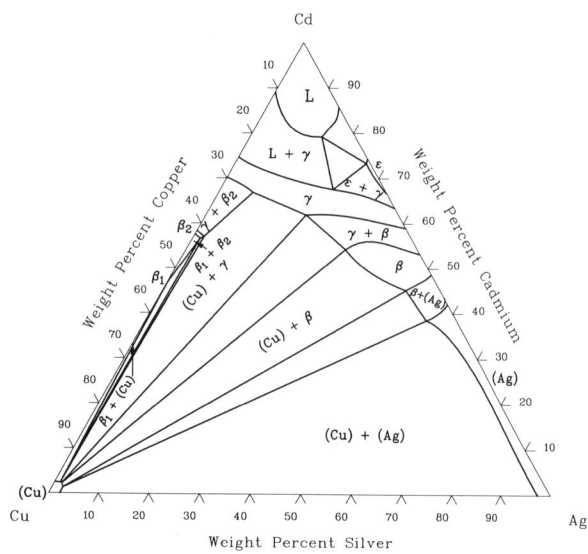

Ag-Cd-Zn isothermal section at 600 °C [88Pet]

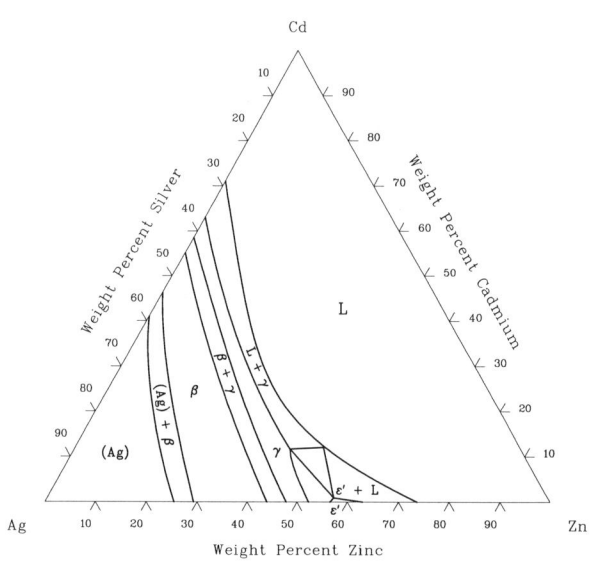

Ag-Cd-Cu isothermal section at 300 °C [88Pet]

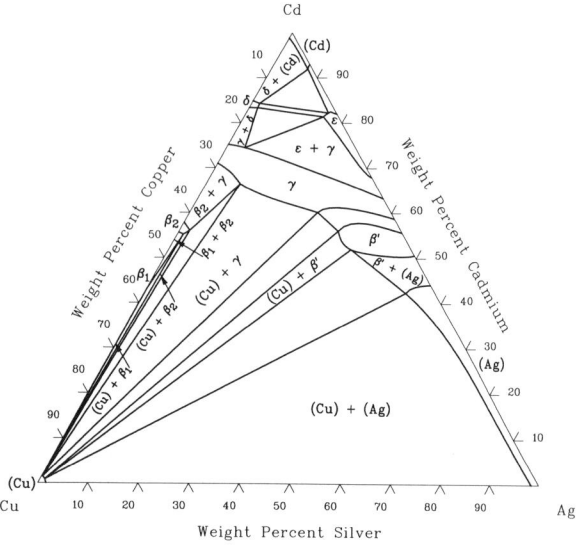

Ag-Cd-Zn isothermal section at 400 °C [88Pet]

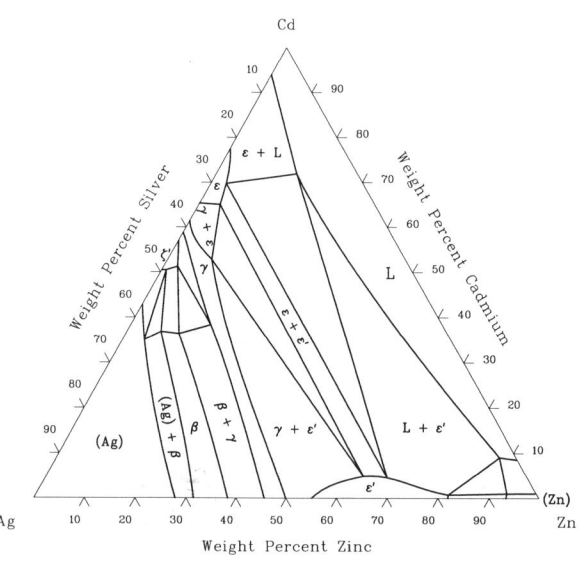

Ag-Cd-Zn isothermal section at 200 °C [88Pet]

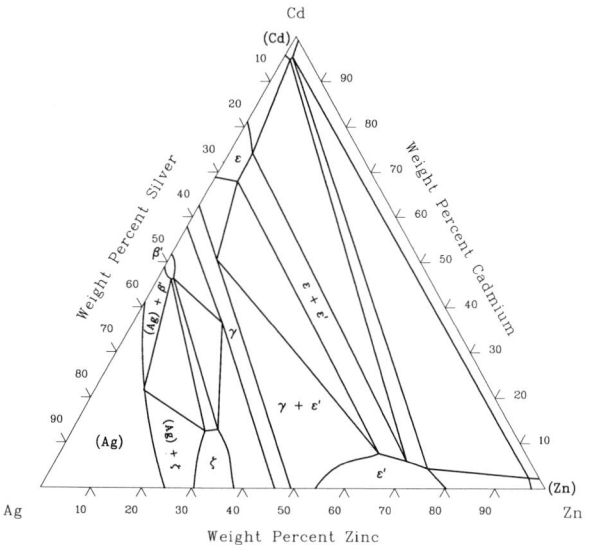

Ag-Cu-Zn isothermal section at 350 °C [88Pet]

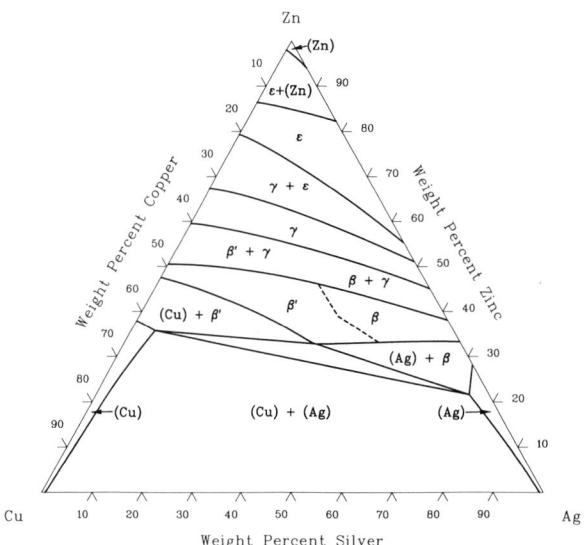

Ag-Cu-Zn liquidus projection [88Pet]

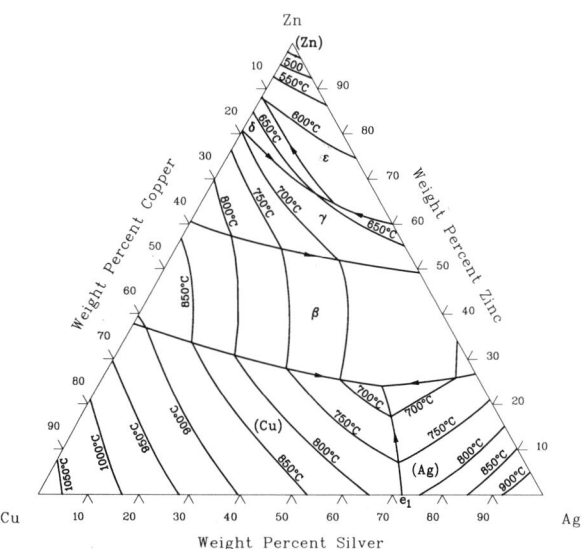

Ag-Pb-Sn [11Par]

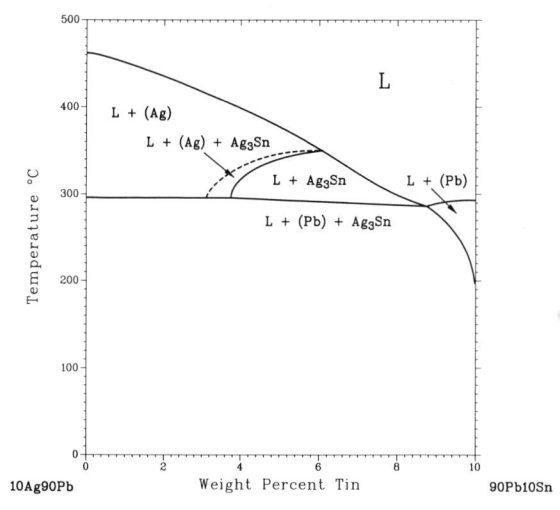

Ag-Cu-Zn isothermal section at 600 °C [88Pet]

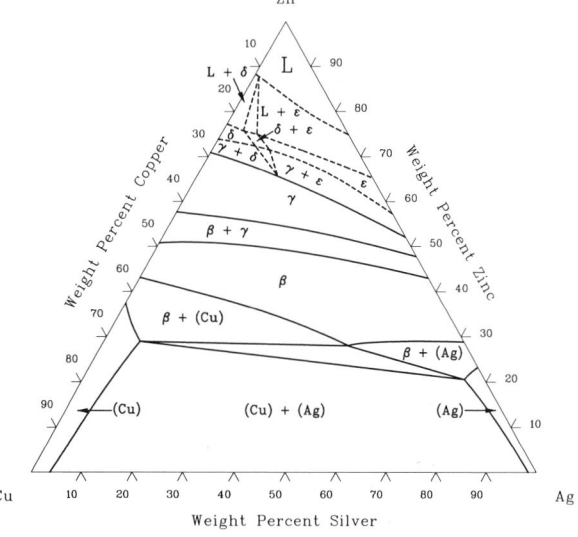

Ag-Pb-Sn [11Par]

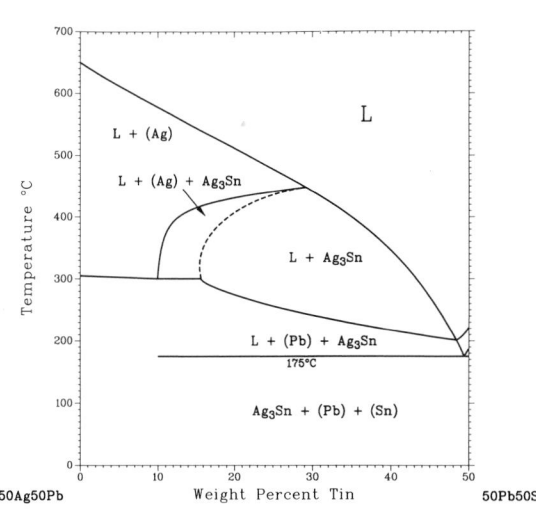

Ag-Pb-Sn [11Par]

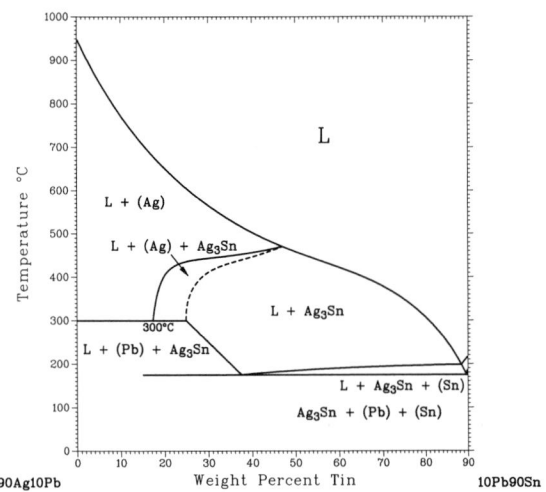

Al-Cr-Fe isothermal section at 750 °C [88Ray]

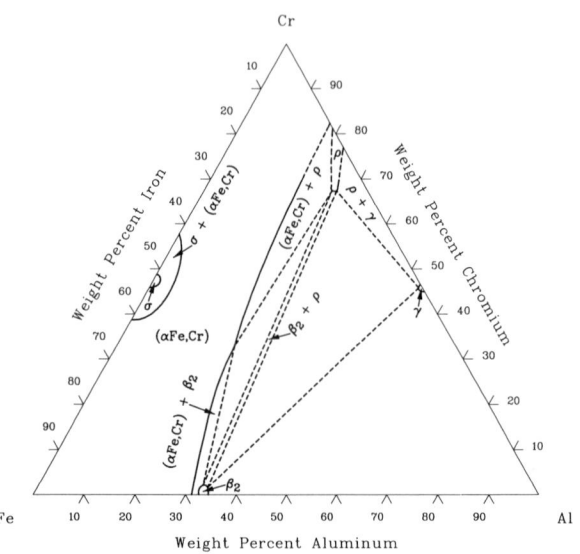

Al-Cr-Fe liquidus projection [88Ray]

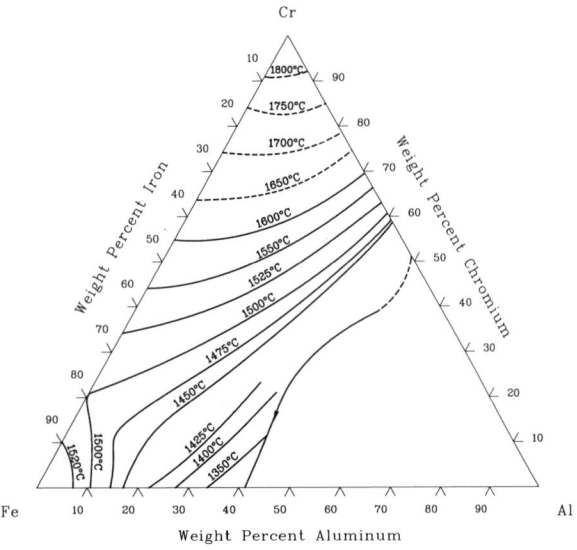

Al-Cr-Fe isothermal section at 600 °C [88Ray]

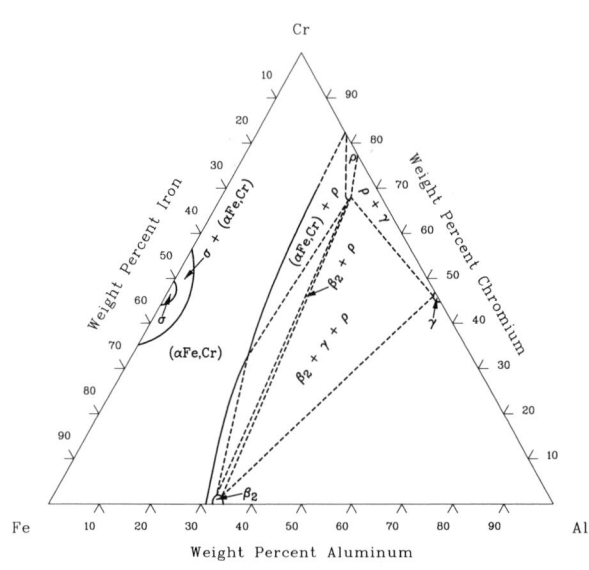

Al-Cr-Fe isothermal section at 900 °C [88Ray]

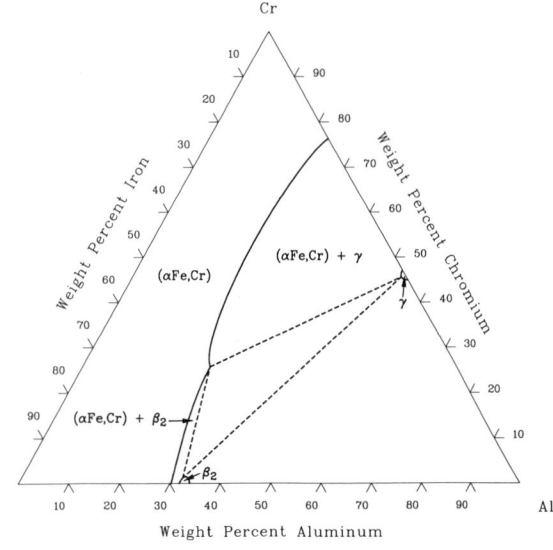

Al-Cr-Mn isothermal section at 690 °C [73Wil]

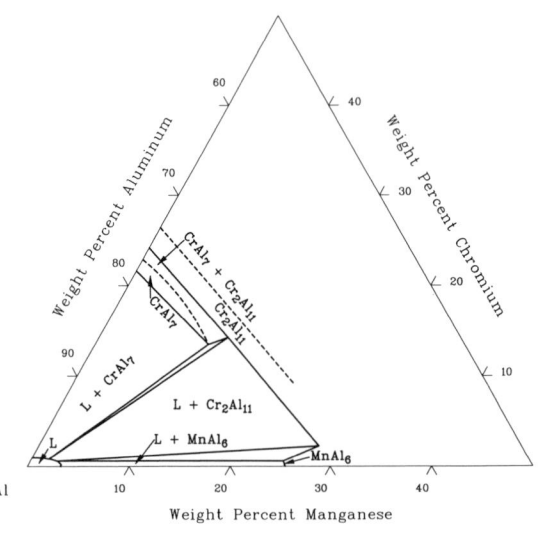

Al-Cr-Mn isothermal section at 600 °C [73Wil]

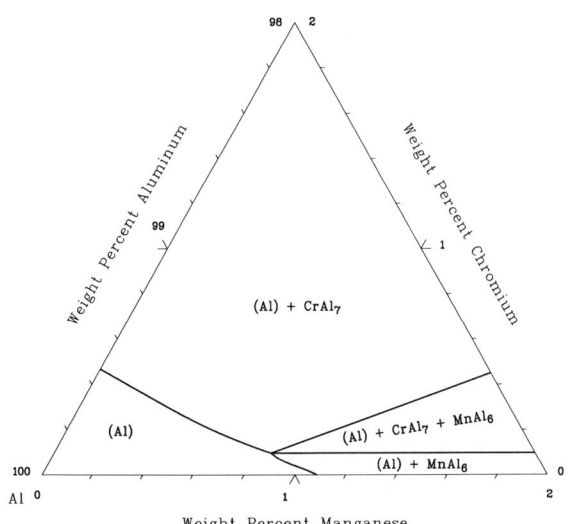

Al-Cr-Ni isothermal section at 1150 °C [87Ofo]

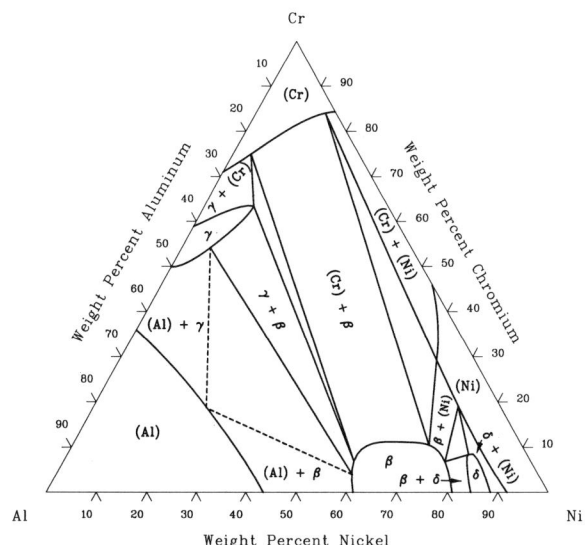

Al-Cr-Mn isothermal section at 550 °C [73Wil]

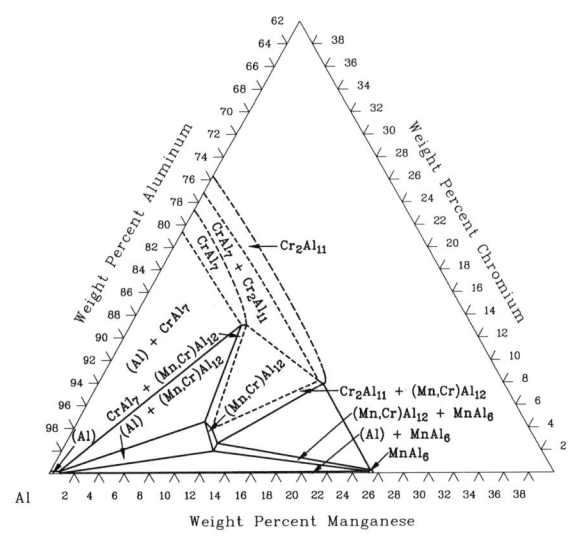

Al-Cr-Ti isothermal section at 760 °C [56Zwi]

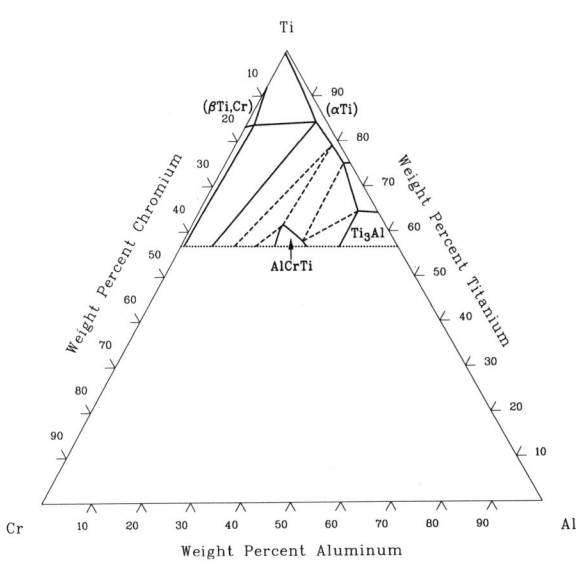

Al-Cr-Mn (Al) isothermal section at 550 °C [73Wil]

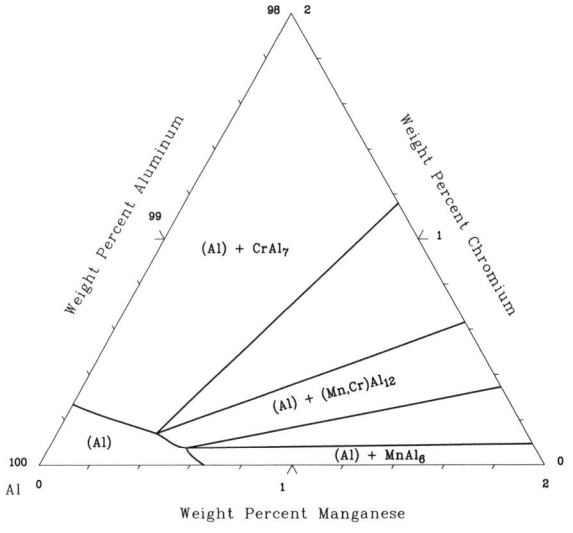

Al-Cu-Fe liquidus projection [73Wil]

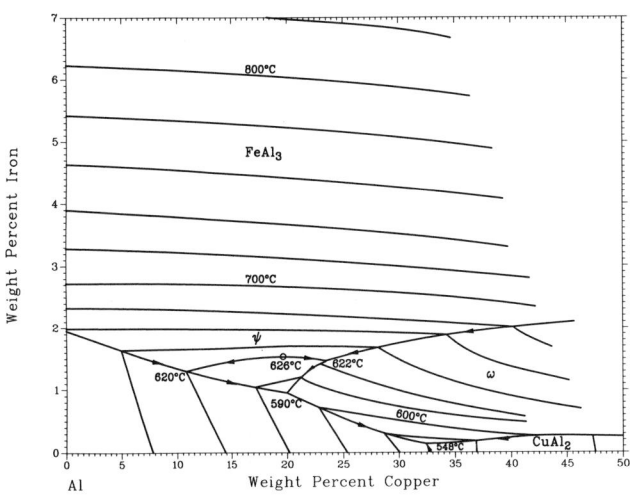

3•10/Ternary Alloy Phase Diagrams

Al-Cu-Fe solidus projection [73Wil]

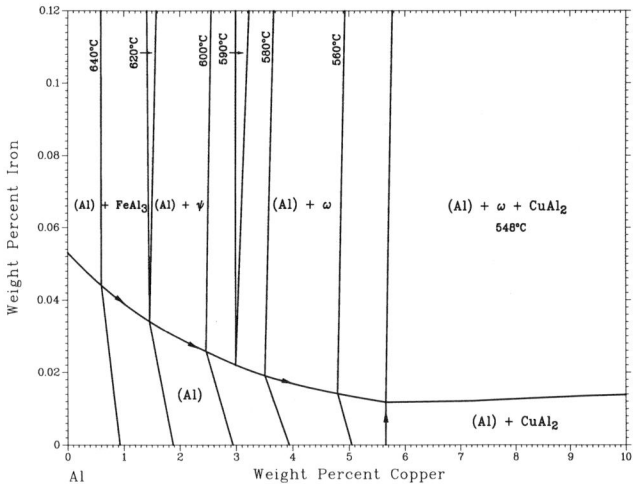

Al-Cu-Mn liquidus projection [73Wil]

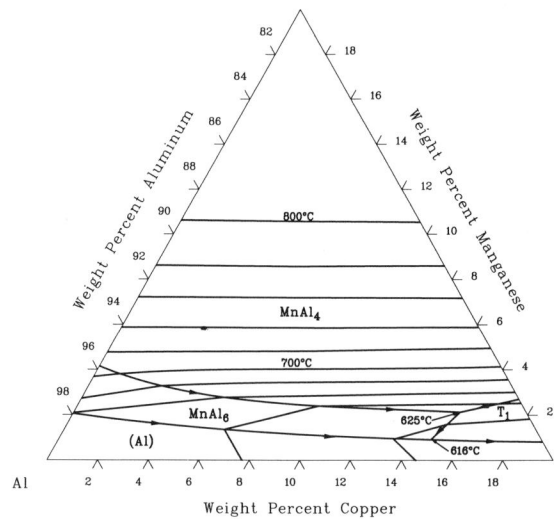

Al-Cu-Fe solvus projection [73Wil]

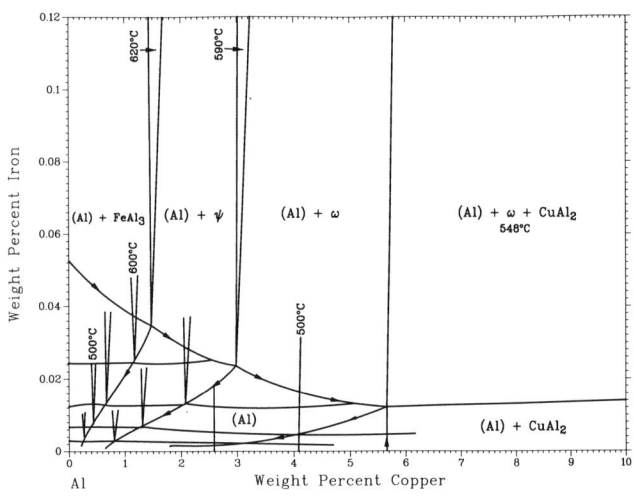

Al-Cu-Mn solidus projection [73Wil]

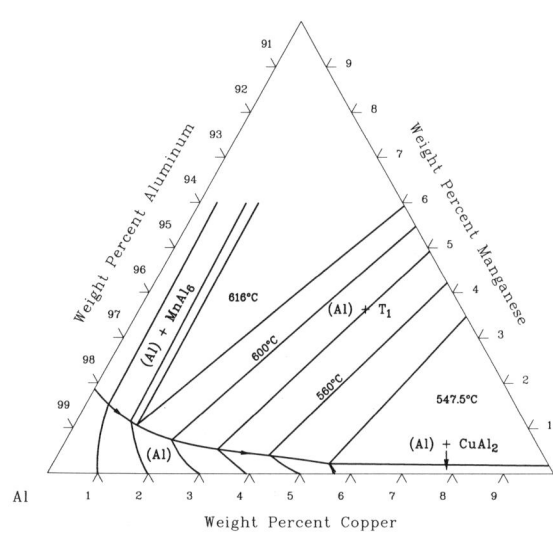

Al-Cu-Fe isothermal section at 600 °C [71Pre]

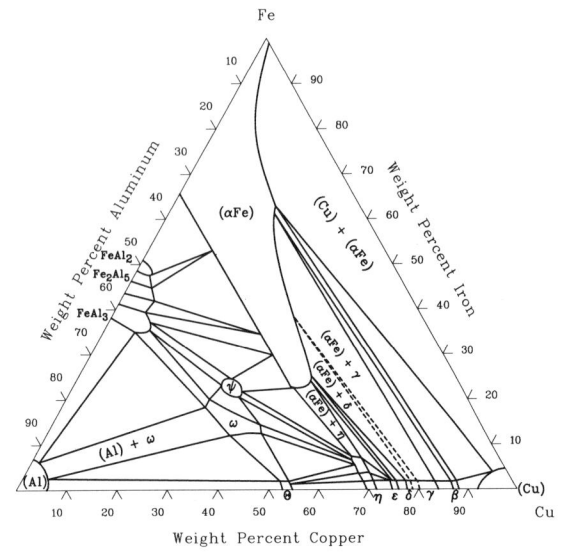

Al-Cu-Mn solvus projection [73Wil]

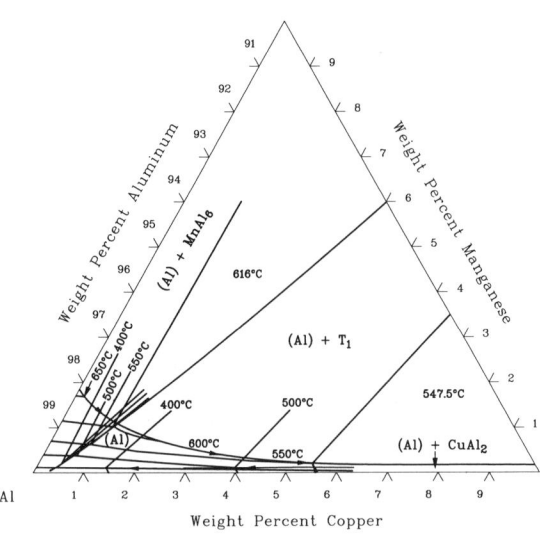

Al-Cu-Mn isothermal section at 950 °C [66Kos]

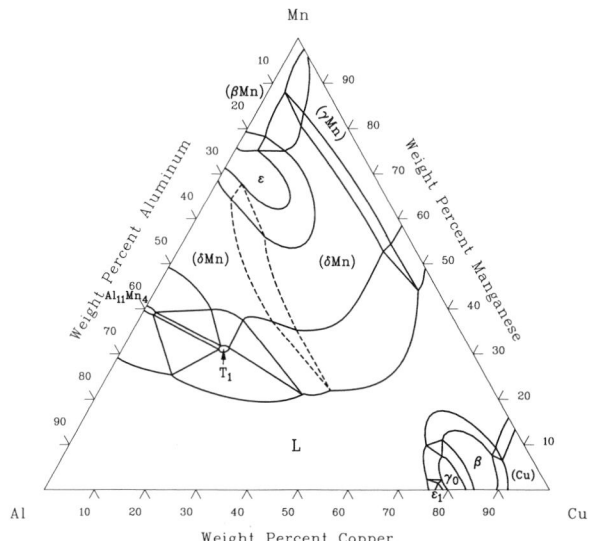

Al-Cu-Ni liquidus projection [73Wil]

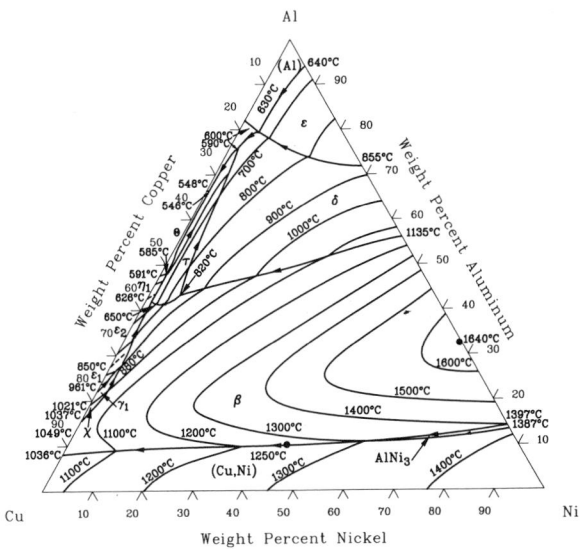

Al-Cu-Mn isothermal section at 700 °C [66Kos]

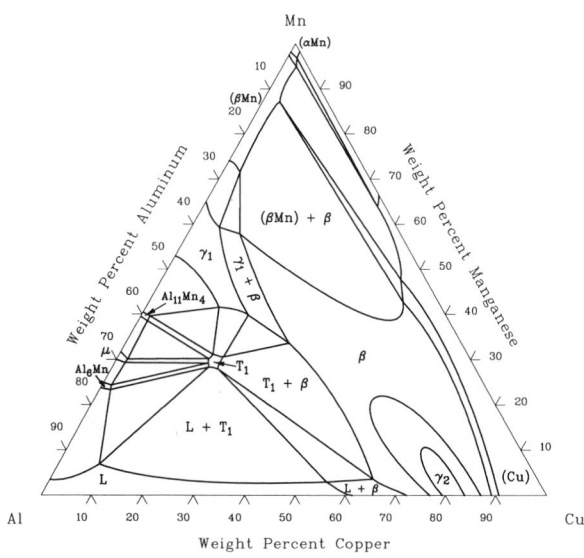

Al-Cu-Ni isothermal section at 900 °C [48Kos]

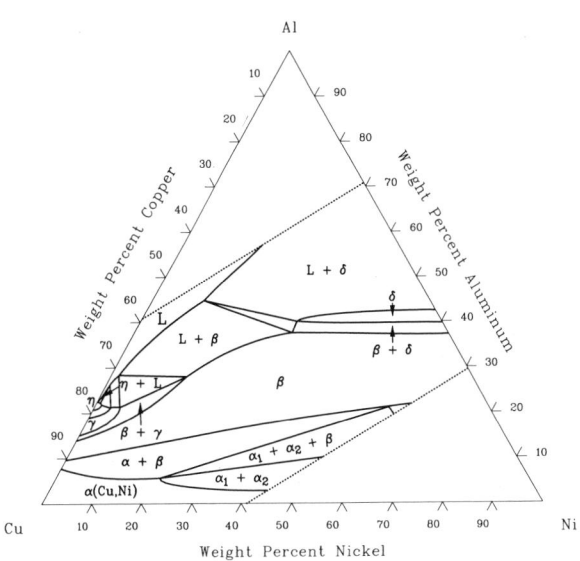

Al-Cu-Mn isothermal section at 25 °C [66Kos]

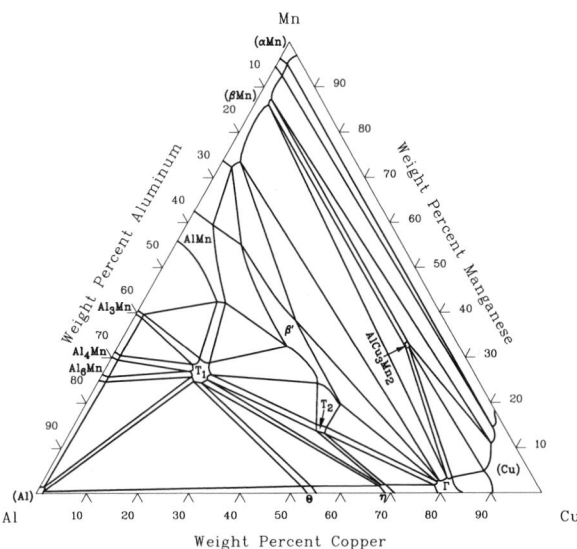

Al-Cu-Ni isothermal section at 700 °C [48Kos]

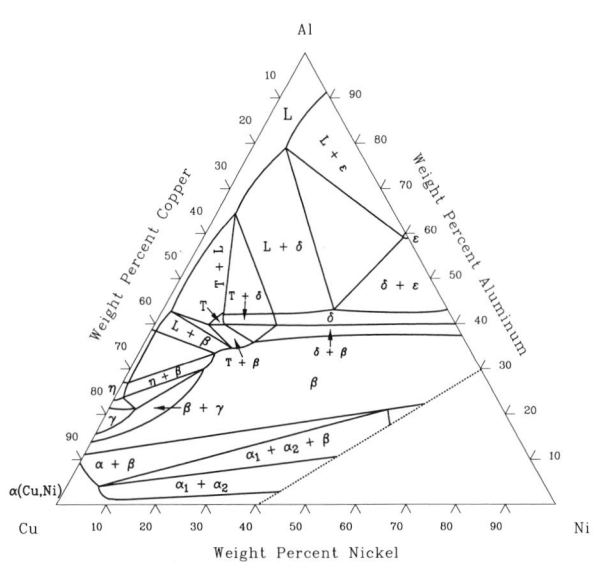

3•12/Ternary Alloy Phase Diagrams

Al-Cu-Ni isothermal section at 500 °C [73Wil]

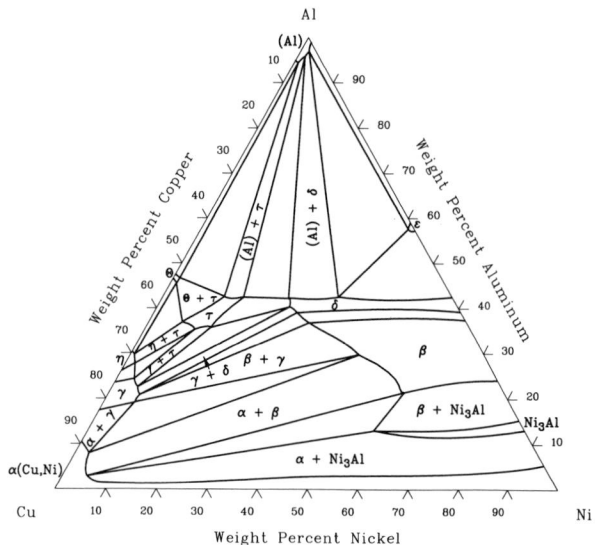

Al-Cu-Si liquidus projection [79Cha]

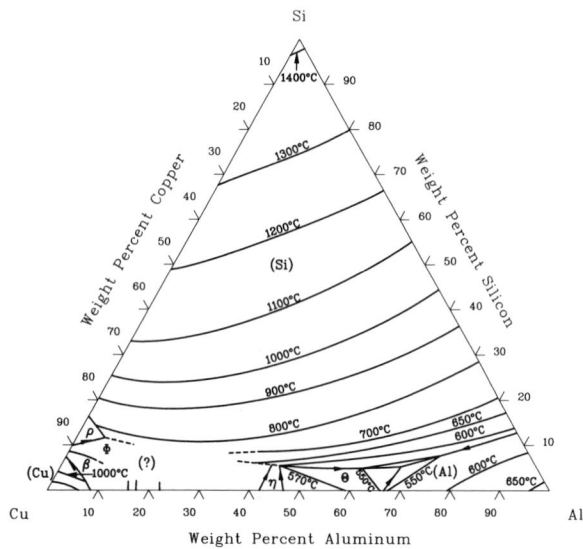

Al-Cu-Si isothermal section at 955 °C [48Wil]

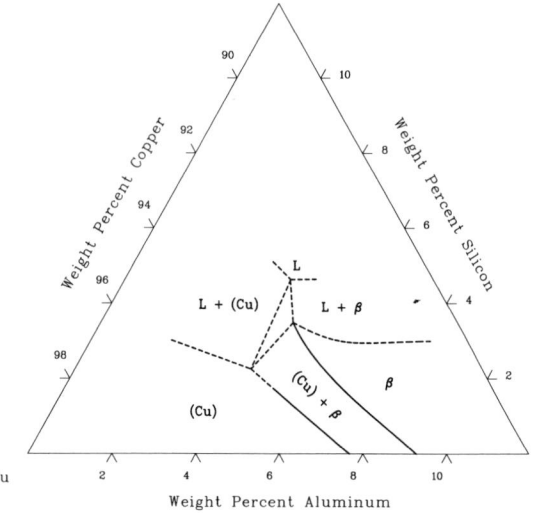

Al-Cu-Si isothermal section at 750 °C [48Wil]

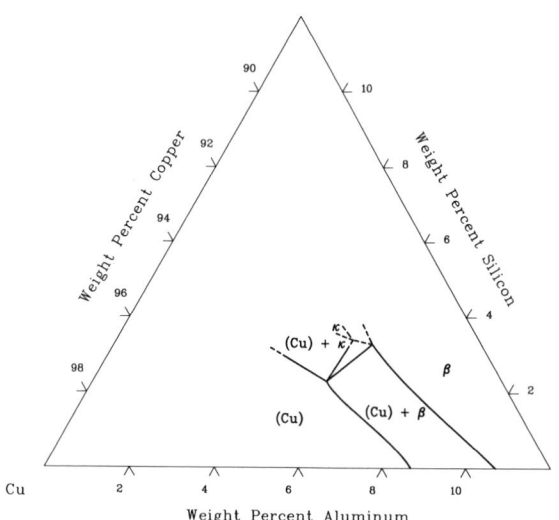

Al-Cu-Si isothermal section at 400 °C [48Wil]

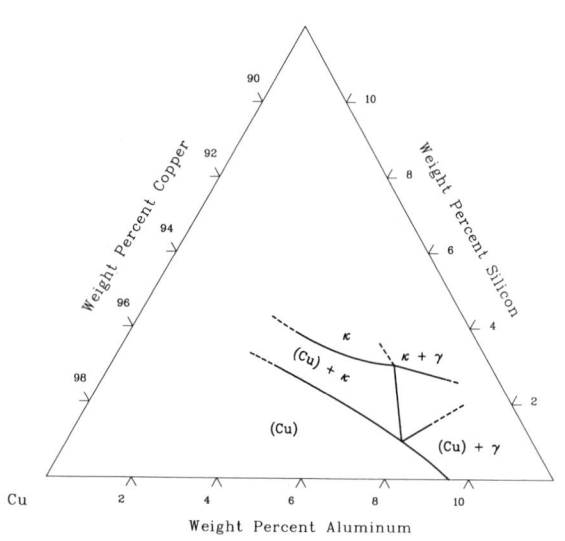

Al-Cu-Zn liquidus projection [73Wil]

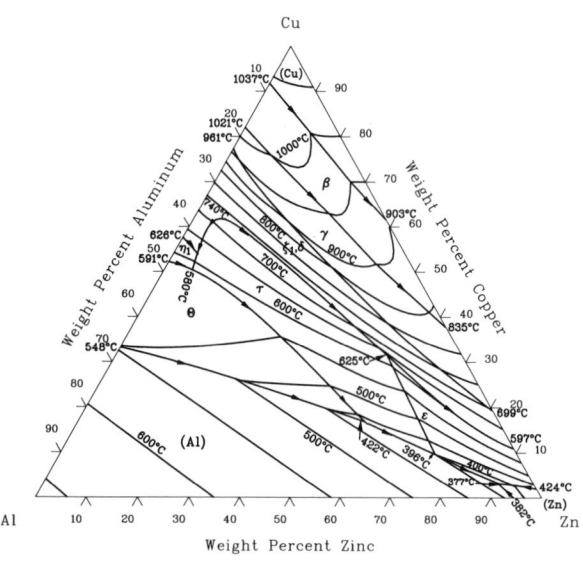

Al-Cu-Zn isothermal section at 700 °C [73Wil]

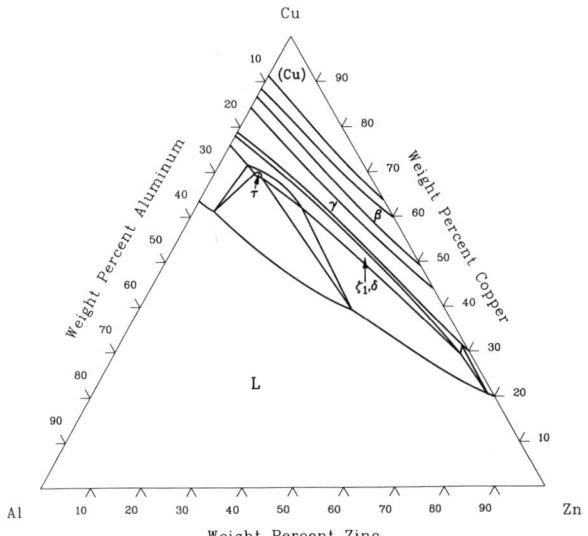

Al-Cu-Zn isothermal section at 200 °C [73Wil]

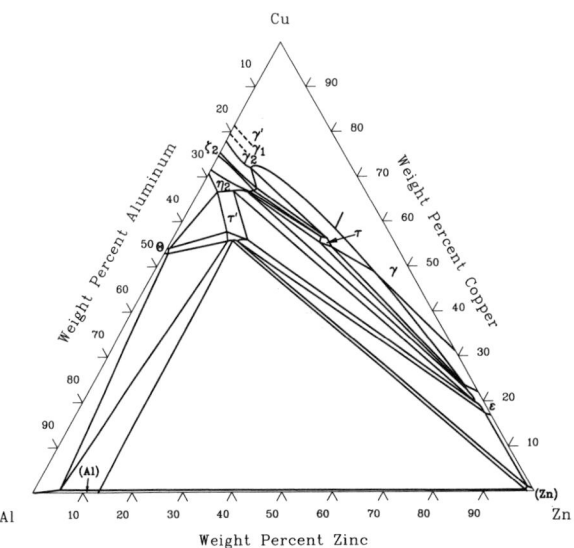

Al-Cu-Zn isothermal section at 550 °C [73Wil]

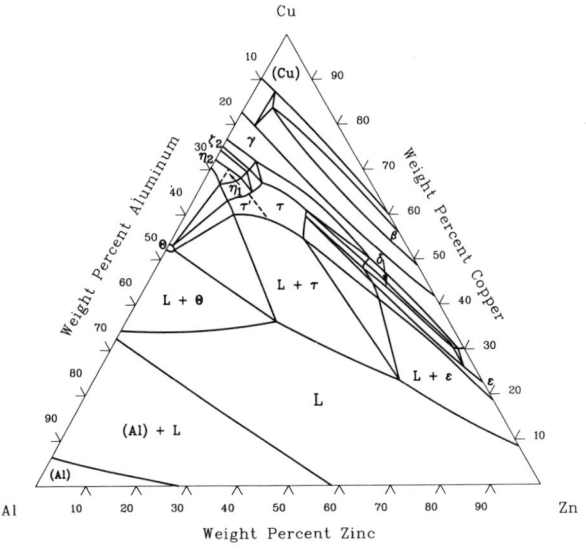

Al-Fe-Mn (Al) liquidus projection [88Ray]

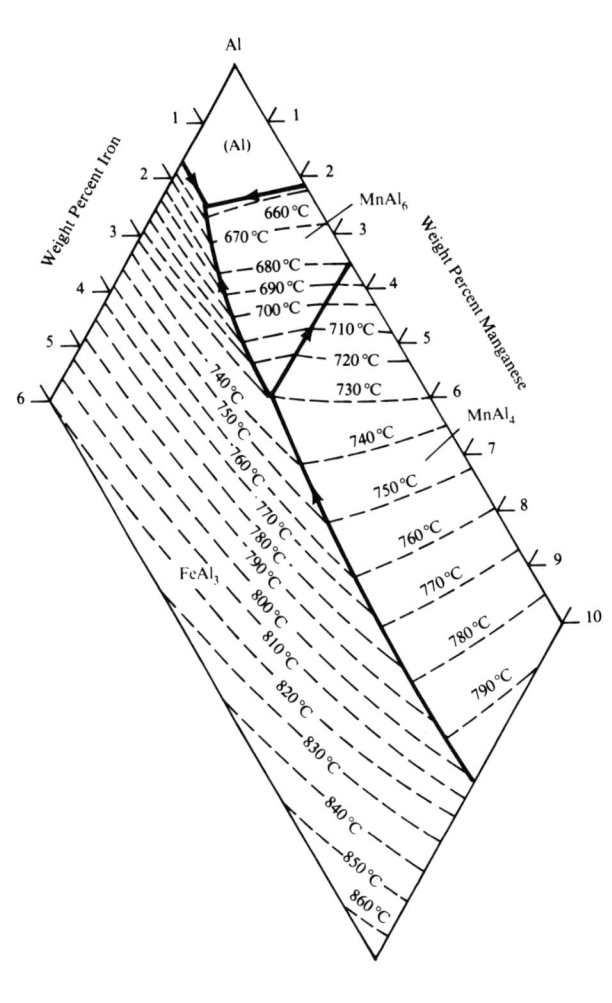

Al-Cu-Zn isothermal section at 350 °C [73Wil]

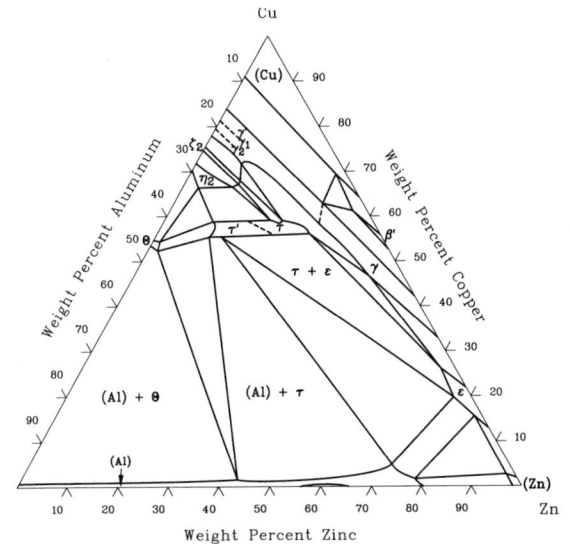

3•14/Ternary Alloy Phase Diagrams

Al-Fe-Mn liquidus projection [88Ray]

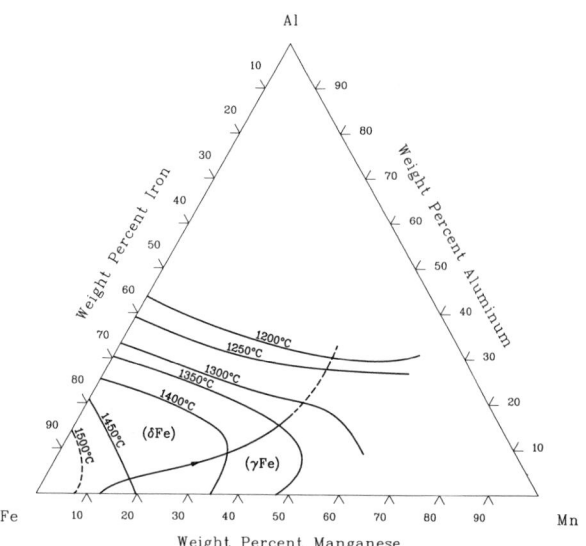

Al-Fe-Ni liquidus projection [88Ray]

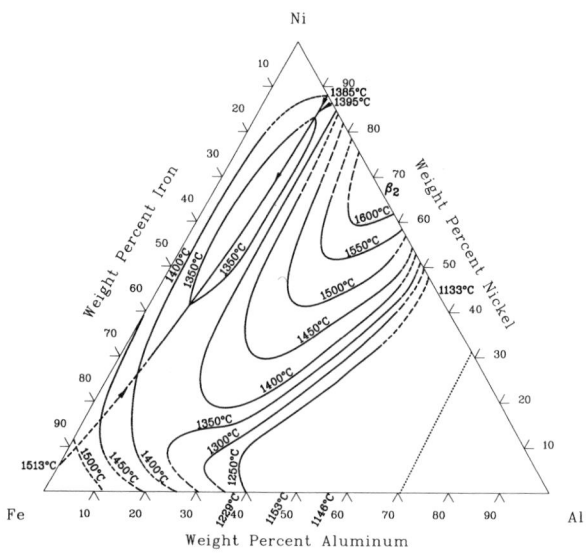

Al-Fe-Mn isothermal section at 1000 °C [88Ray]

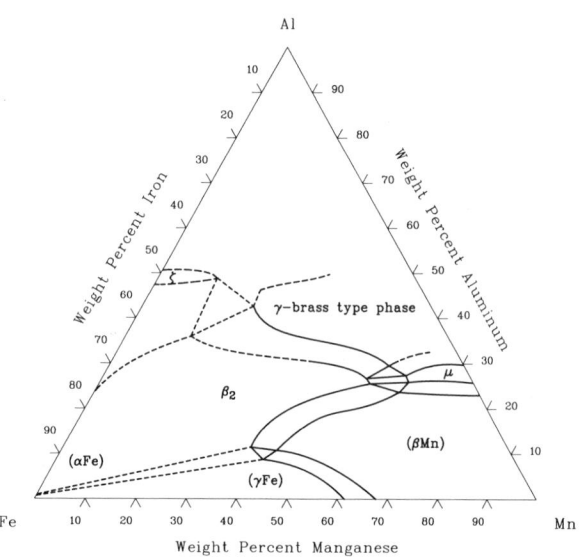

Al-Fe-Ni (Al) liquidus projection [88Ray]

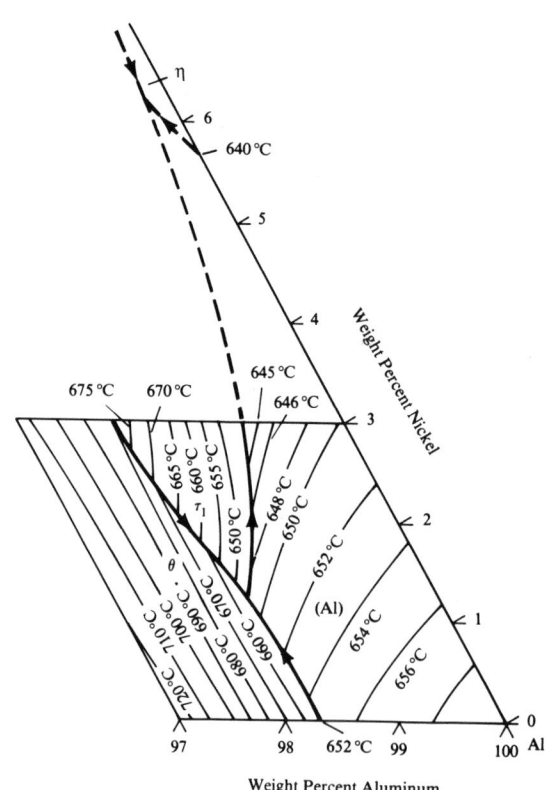

Al-Fe-Mn isothermal section at 600 °C [88Ray]

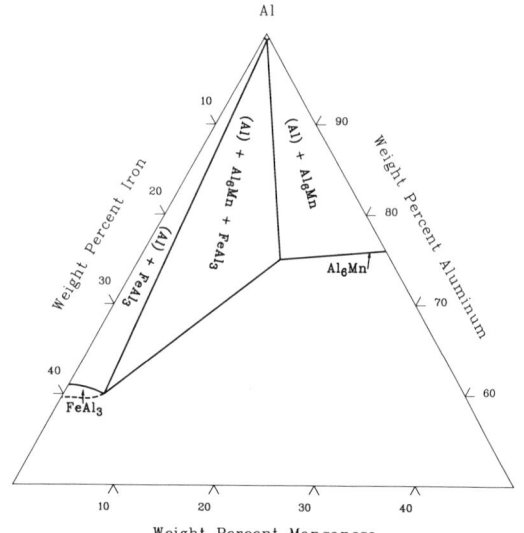

Al-Fe-Ni isothermal section at 1250 °C [88Ray]

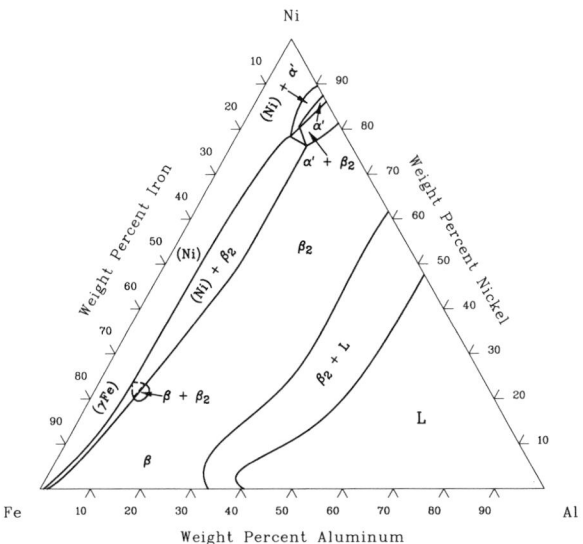

Al-Fe-Ni isothermal section at 600 °C [88Ray]

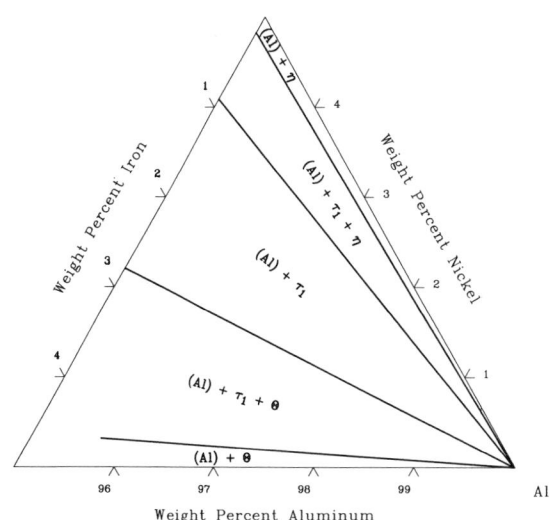

Al-Fe-Ni isothermal section at 950 °C [88Ray]

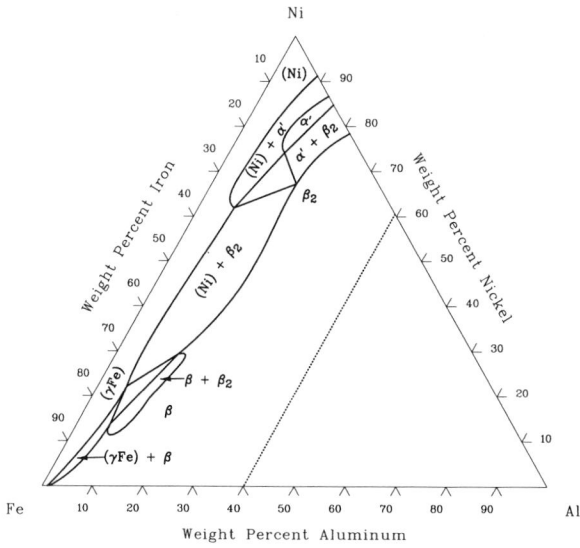

Al-Fe-Si liquidus projection [88Ray]

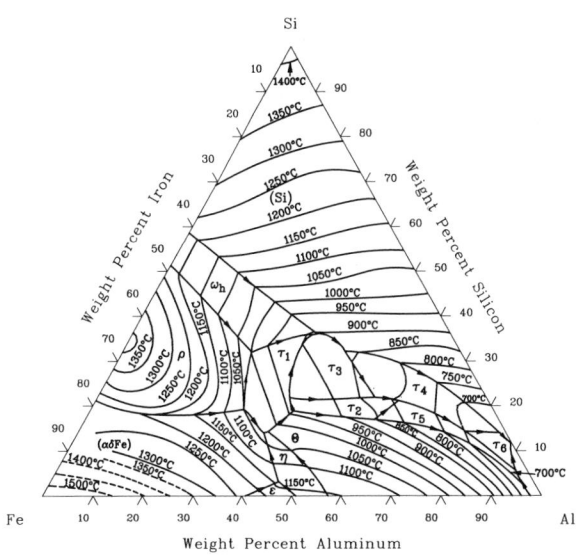

Al-Fe-Ni isothermal section at 750 °C [88Ray]

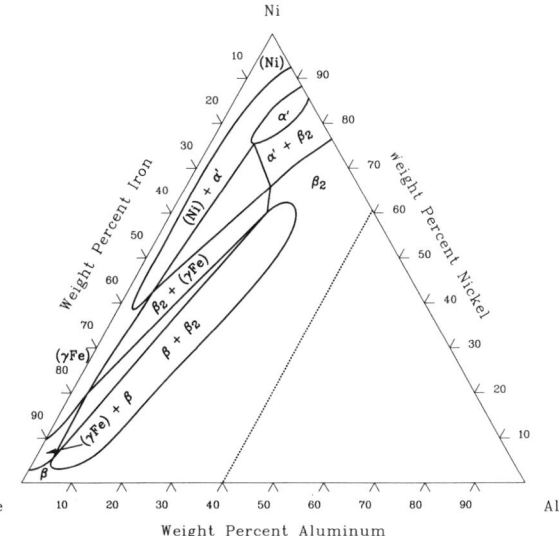

Al-Fe-Si isothermal section at 1000 °C [88Ray]

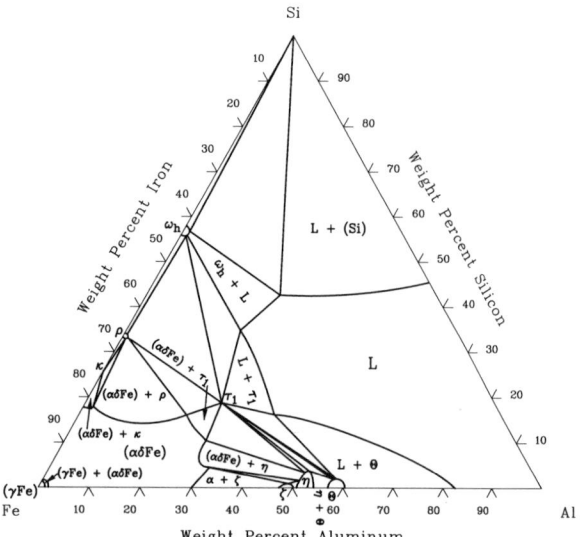

Al-Fe-Zn isothermal section at 700 °C [70Kos]

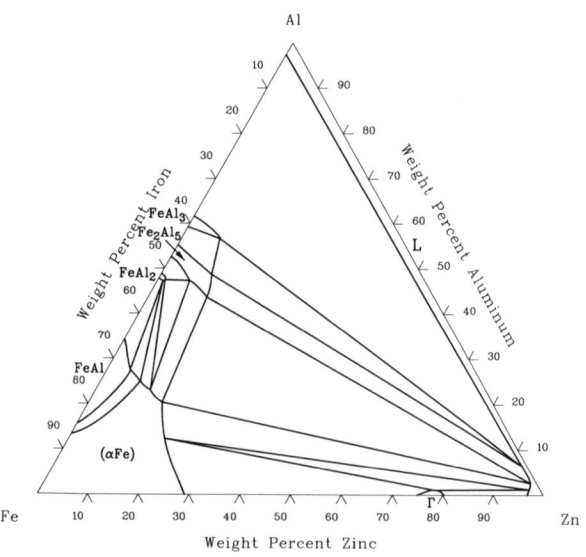

Al-Fe-Si isothermal section at 550 °C [88Ray]

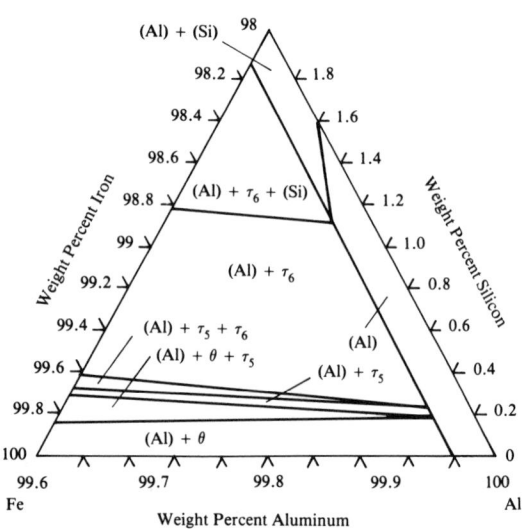

Al-Fe-Zn isothermal section at 500 °C [70Kos]

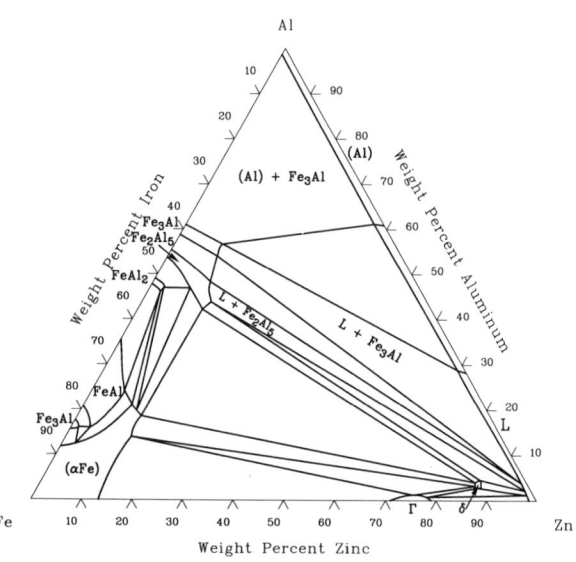

Al-Fe-Si isothermal section at 450 °C [88Ray]

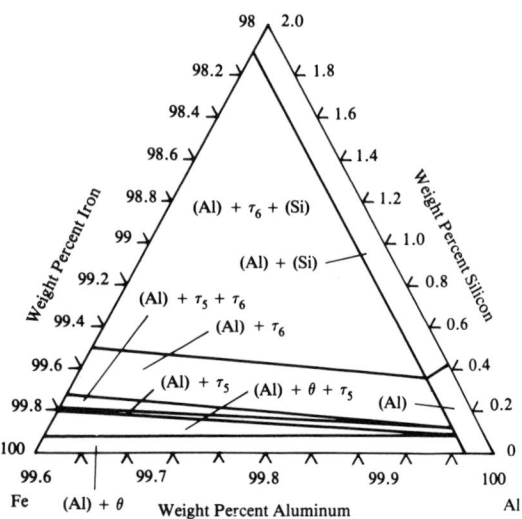

Al-Fe-Zn isothermal section at 330 °C [70Kos]

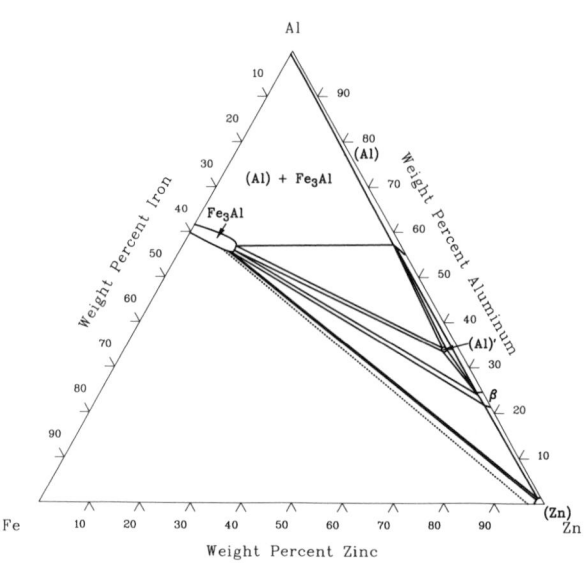

Al-Mg-Mn liquidus projection [73Wil]

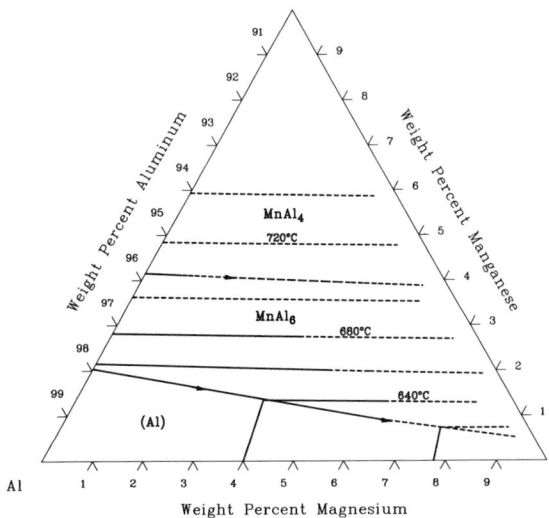

Al-Mg-Mn isothermal section at 400 °C [73Wil]

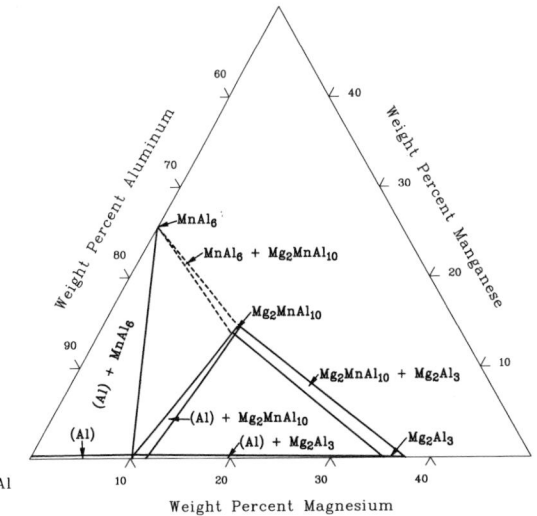

Al-Mg-Mn isothermal section at 750 °C [88Sim]

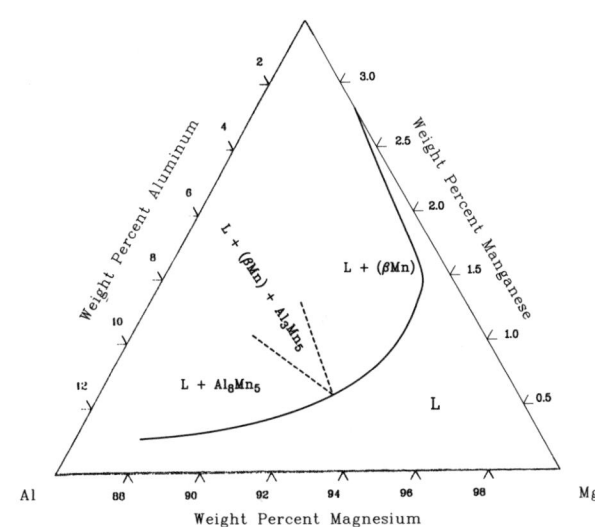

Al-Mg-Si liquidus projection [73Wil]

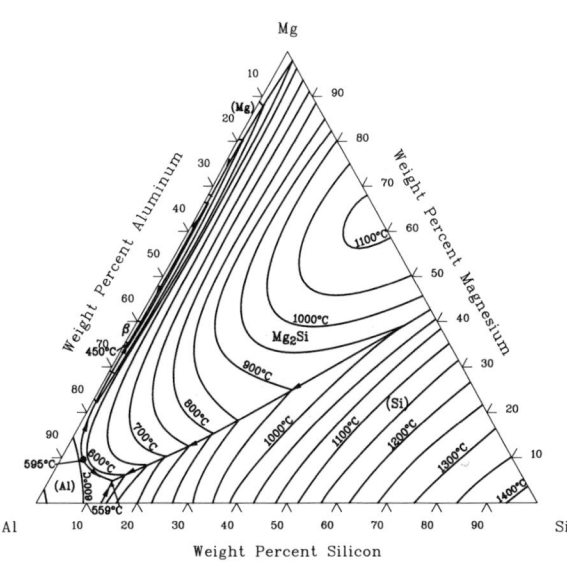

Al-Mg-Mn isothermal section at 670 °C [88Sim]

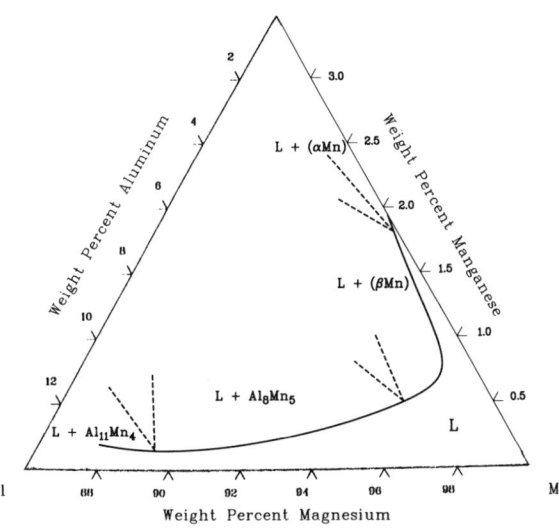

Al-Mg-Si solidus projection [73Wil]

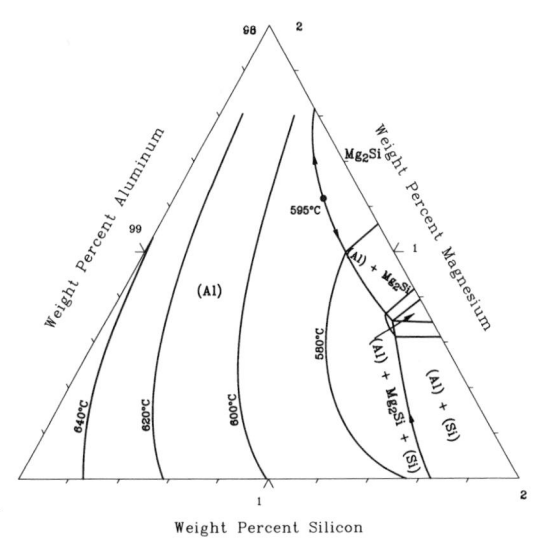

3•18/Ternary Alloy Phase Diagrams

Al-Mg-Si solvus projection [73Wil]

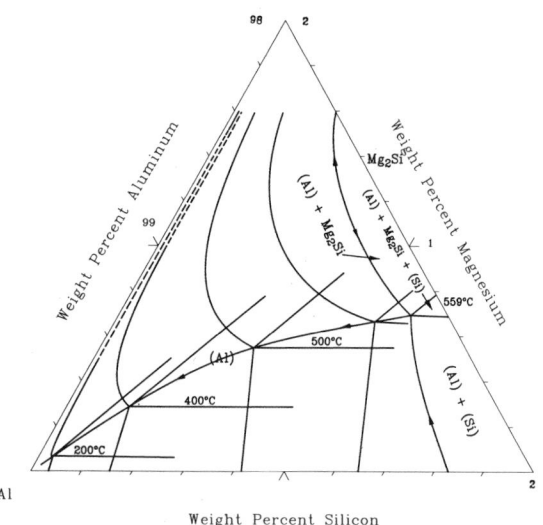

Al-Mg-Zn liquidus projection [73Wil]

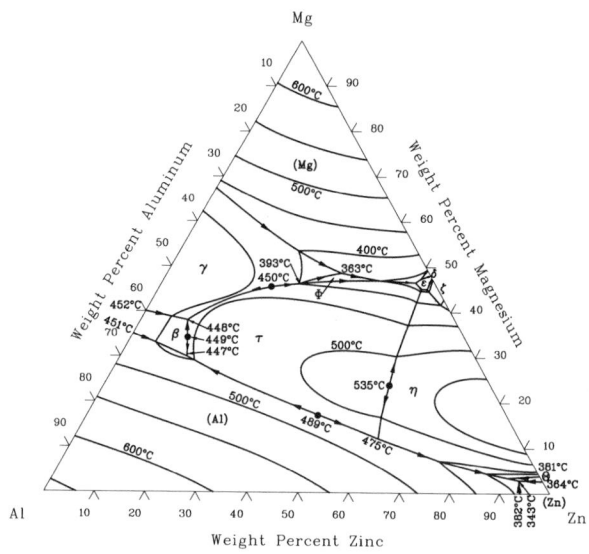

Al-Mg-Si isothermal section at 800 °C [88Rok]

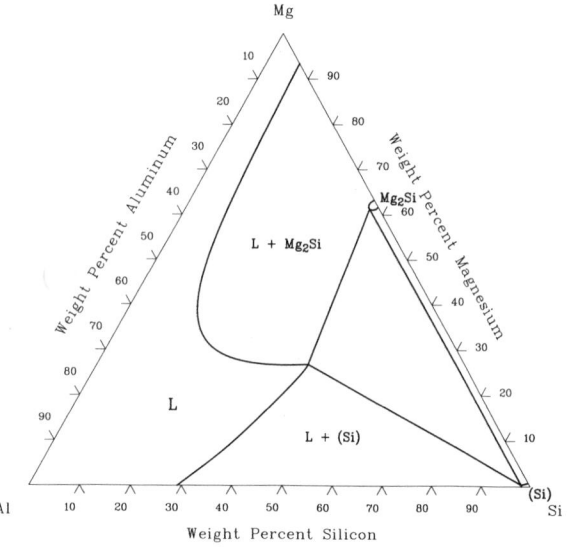

Al-Mg-Zn solvus projection [73Wil]

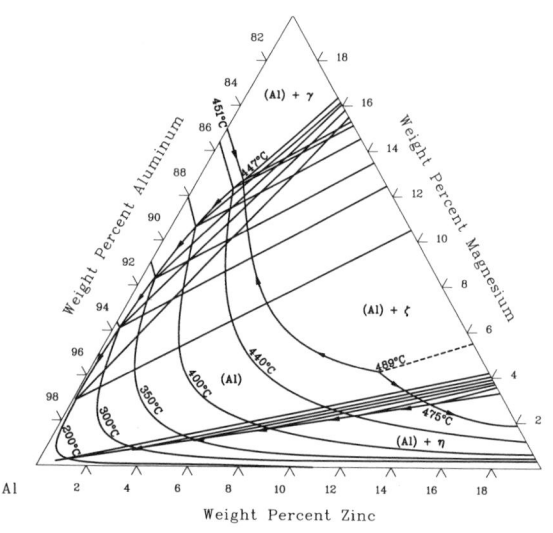

Al-Mg-Si isothermal section at 430 °C [88Rok]

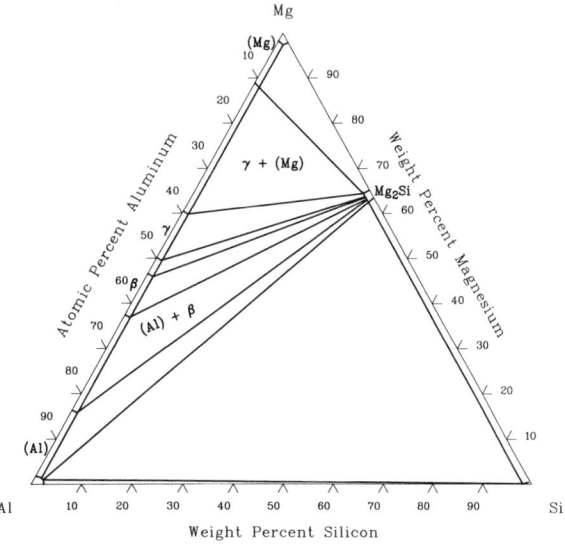

Al-Mg-Zn solidus projection [73Wil]

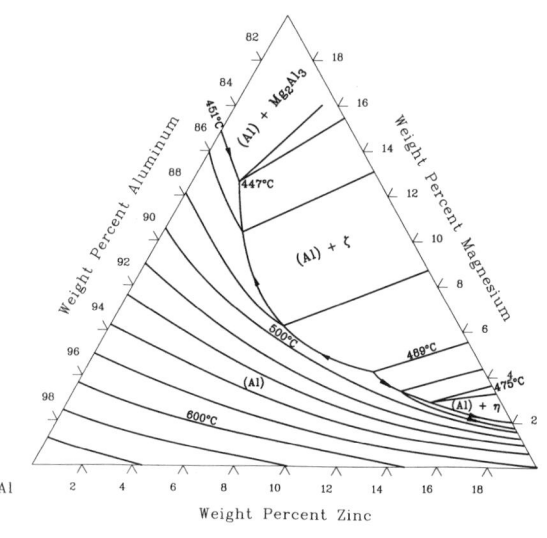

Al-Mg-Zn isothermal section at 335 °C [73Wil]

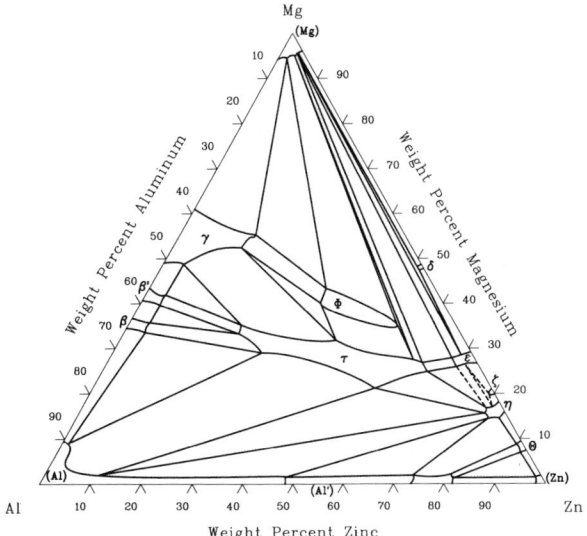

Al-Mg-Zn isothermal section at 20 °C [36Kos]

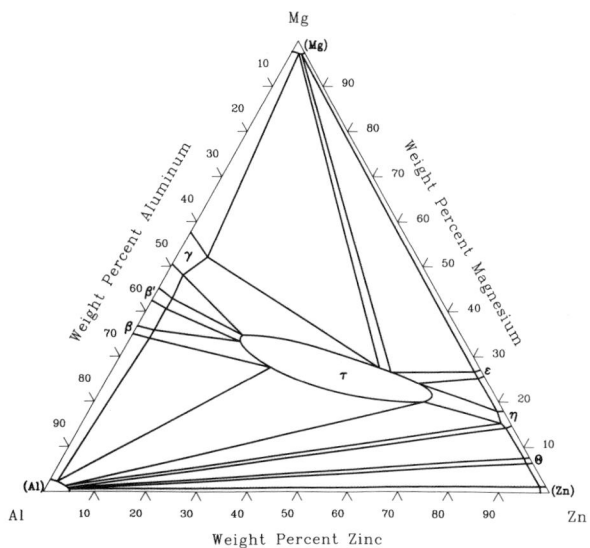

Al-Mn-Si liquidus projection [73Wil]

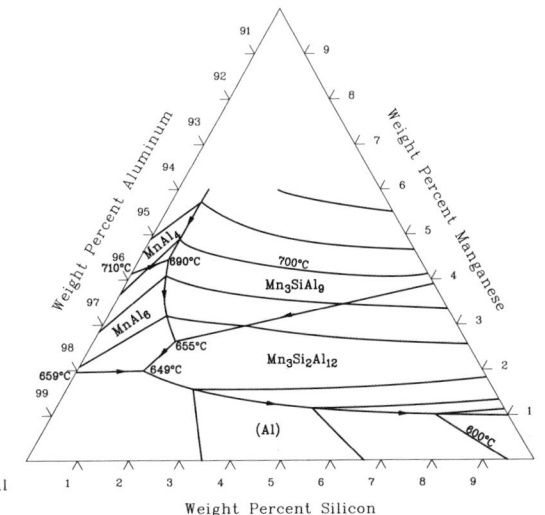

Al-Mn-Si solidus projection {73Will}

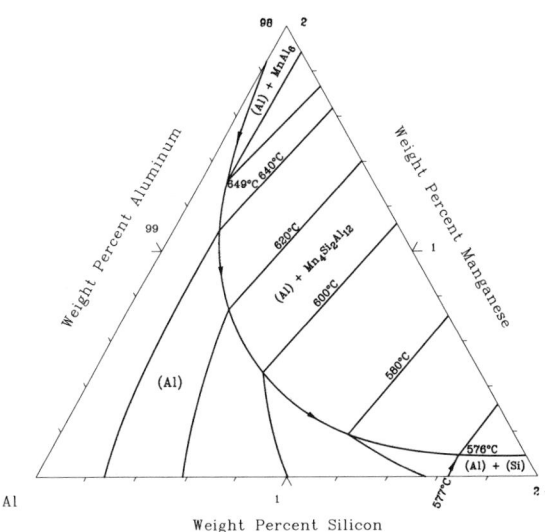

Al-Mn-Si isothermal section at 800 °C [64Kus]

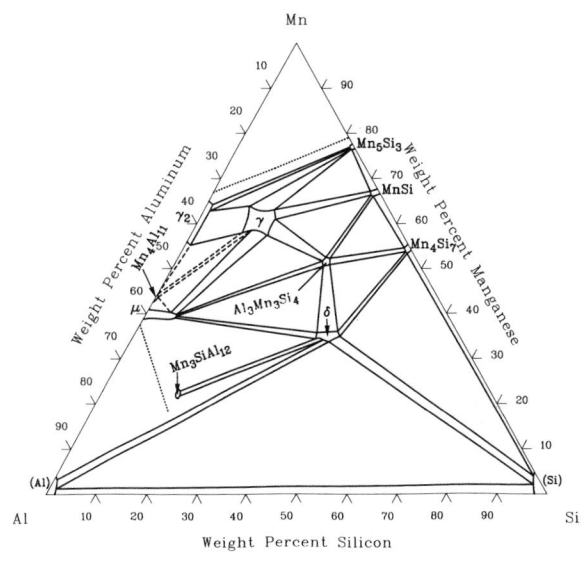

Al-Mn-Si isothermal section at 460 °C [73Wil]

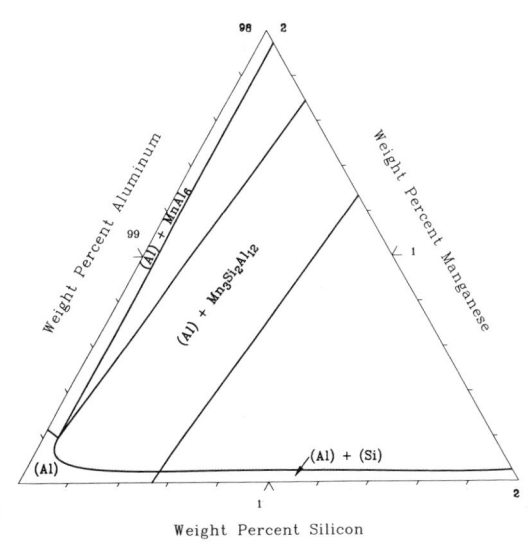

3•20/Ternary Alloy Phase Diagrams

Al-Mo-Ni isothermal section at 1260 °C [84Mir]

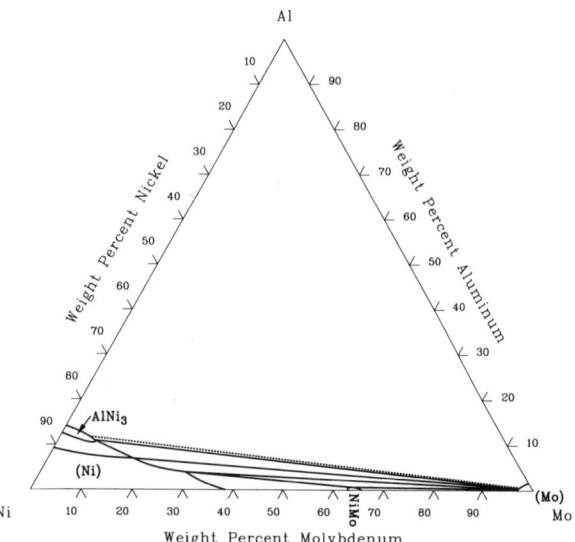

Al-Mo-Ni isothermal section at 1093 °C [84Mir]

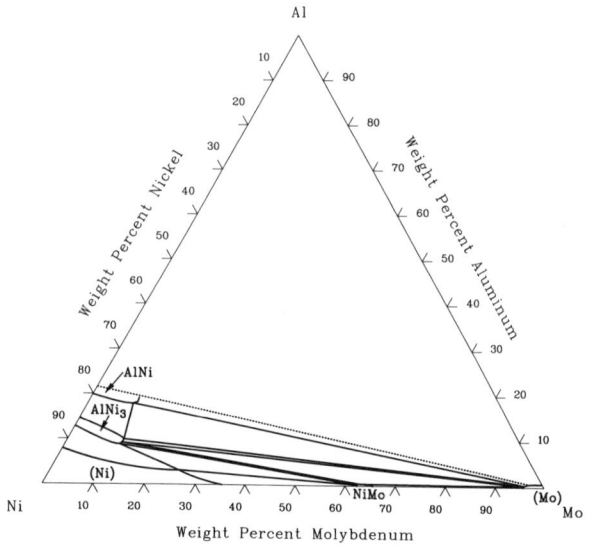

Al-Mo-Ni isothermal section at 927 °C [84Mir]

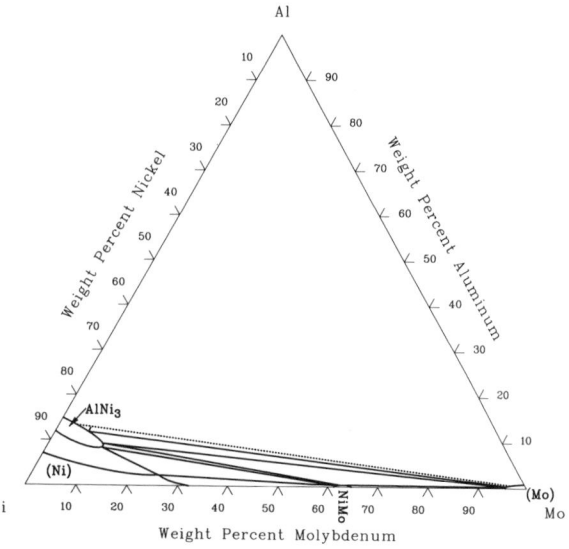

Al-Mo-Ti isothermal section at 925 °C [70Han]

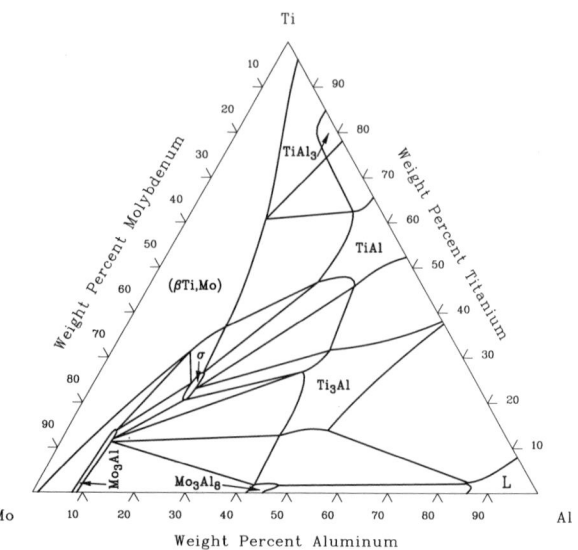

Al-Ni-Ti liquidus projection [85Nas]

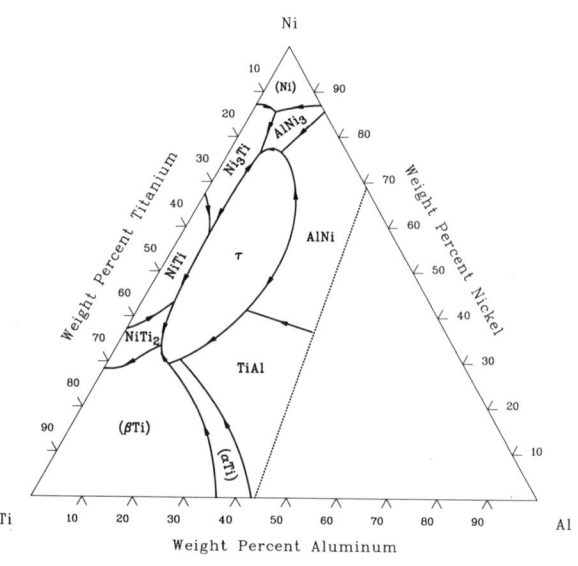

Al-Ni-Ti isothermal section at 900 °C [85Nas]

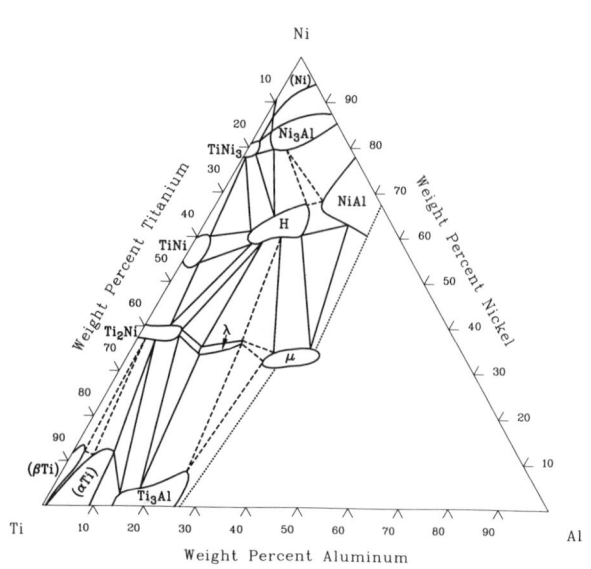

Al-Ni-Ti isothermal section at 800 °C [73Mar]

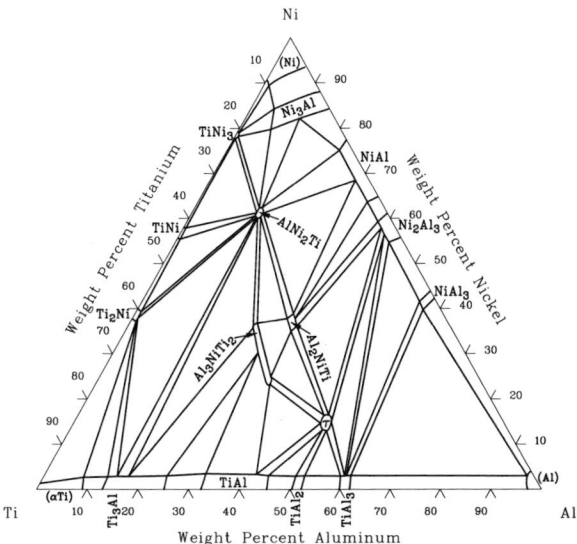

Al-Si-Zn schematic liquidus projection

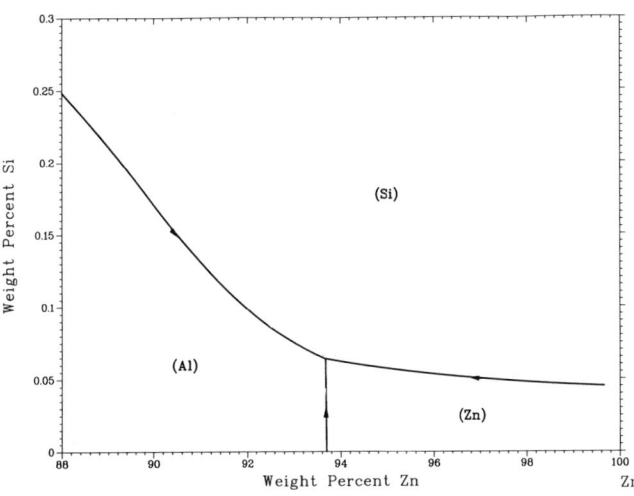

Al-Ni-Ti isothermal section at 600 °C [85Oma]

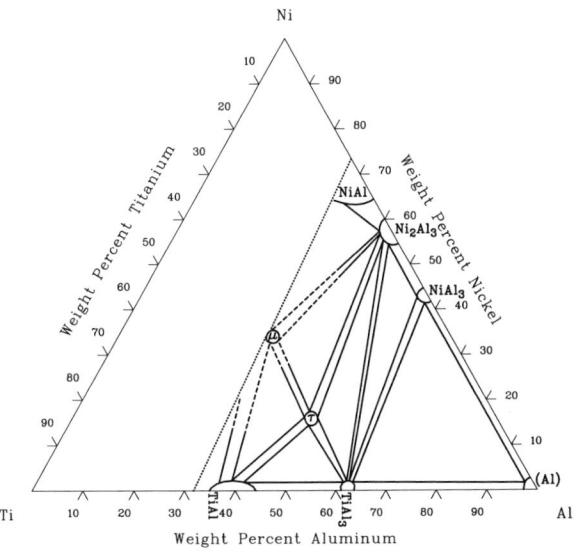

Al-Si-Zn isothermal section at 527 °C [86Mey]

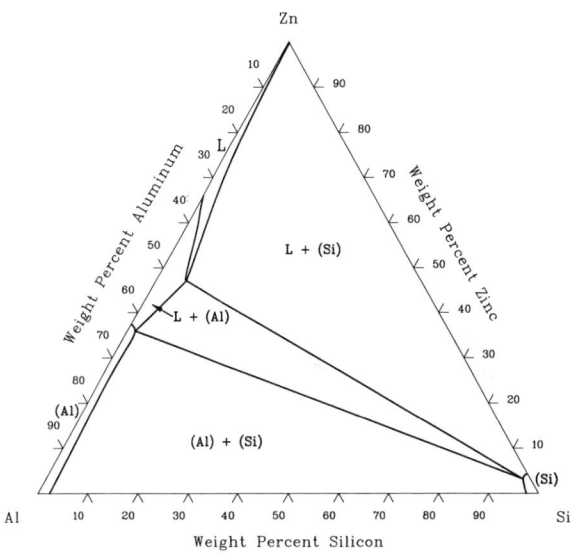

Al-Si-Zn liquidus projection [86Mey]

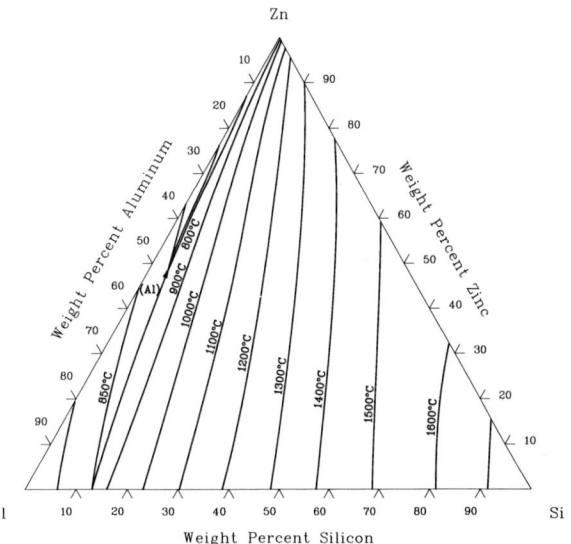

Al-Si-Zn isothermal section at 357 °C [86Mey]

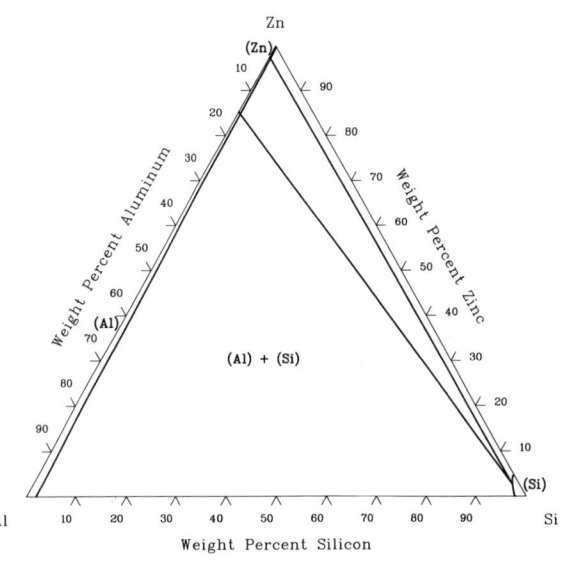

3•22/Ternary Alloy Phase Diagrams

Al-Si-Zn isothermal section at 307 °C [86Mey]

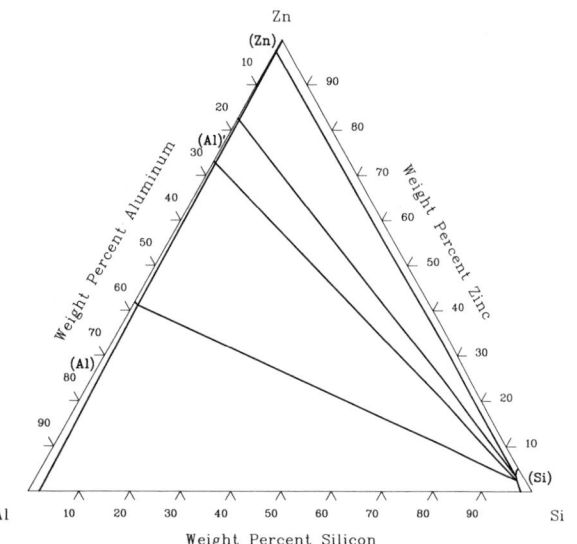

Al-Ti-V isothermal section at 980 °C [56Zwi]

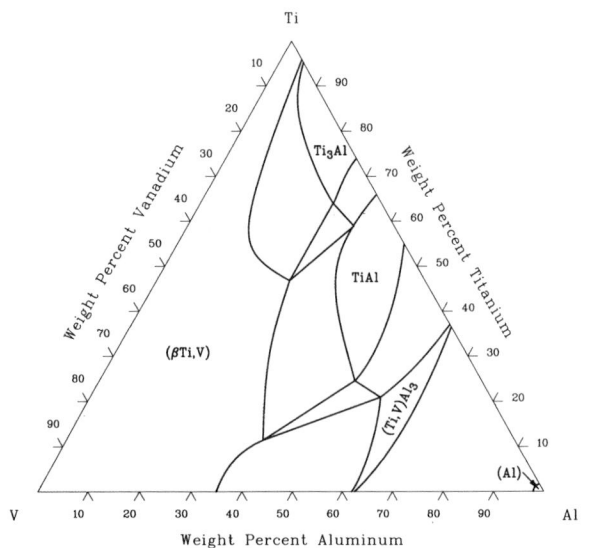

Al-Ti-V isothermal section at 1400 °C [61Far]

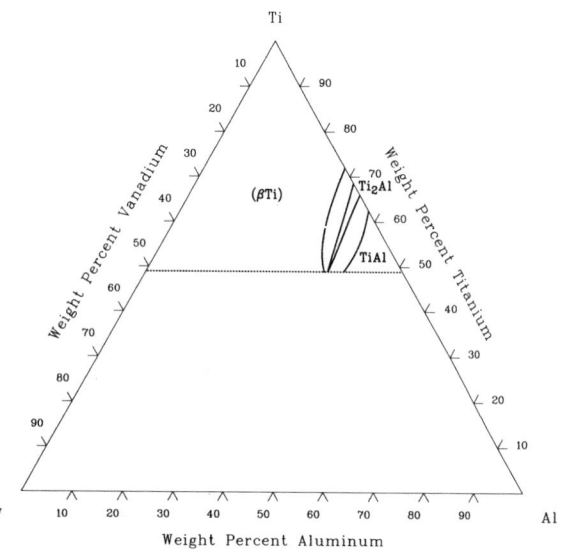

Al-Ti-V isothermal section at 900 °C [61Far]

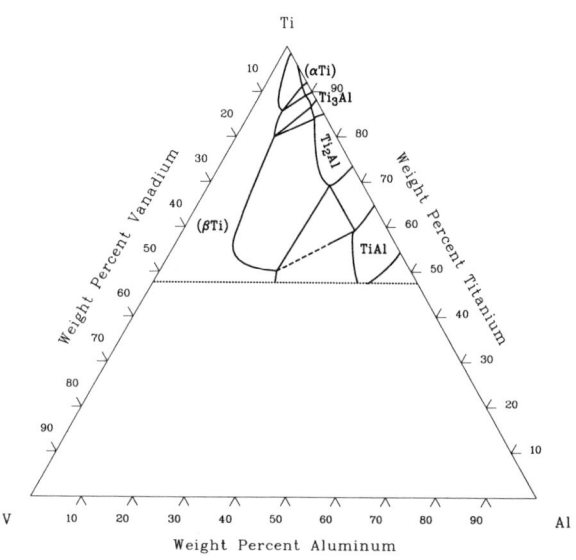

Al-Ti-V isothermal section at 1200 °C [61Far]

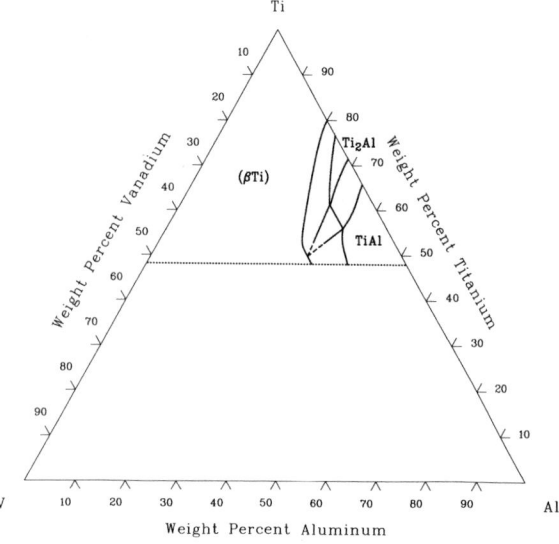

Au-Cu-Ni liquidus projection [90Pri]

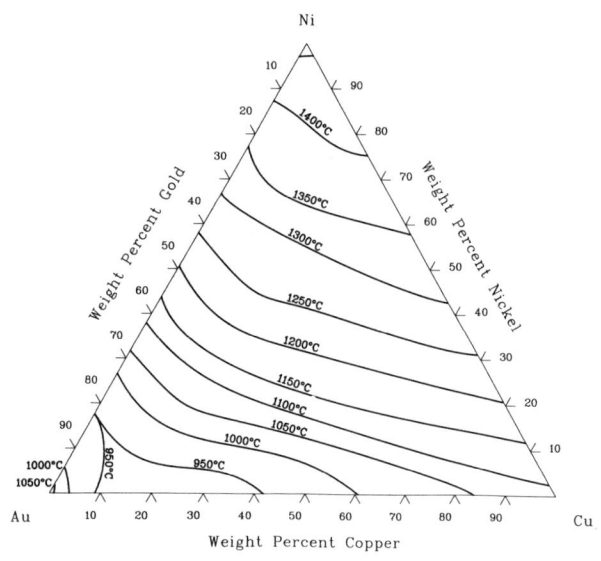

Au-Cu-Ni boundaries of solid-state miscibility gap [90Pri]

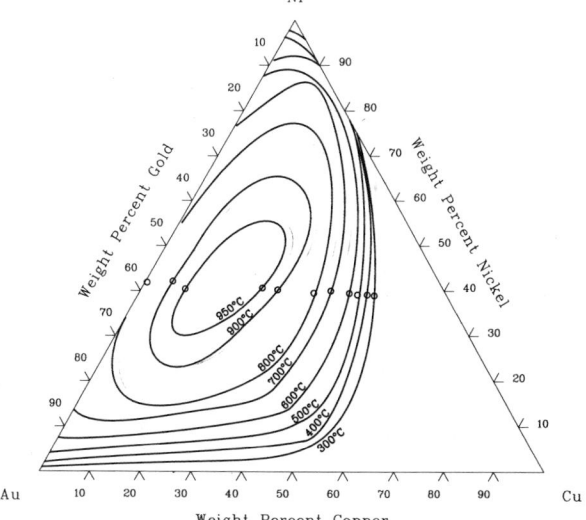

The open circles represent the compositions at which the gap closes.

Au-Cu-Ni boundary of miscibility gap at 400 °C, with tie lines [90Pri]

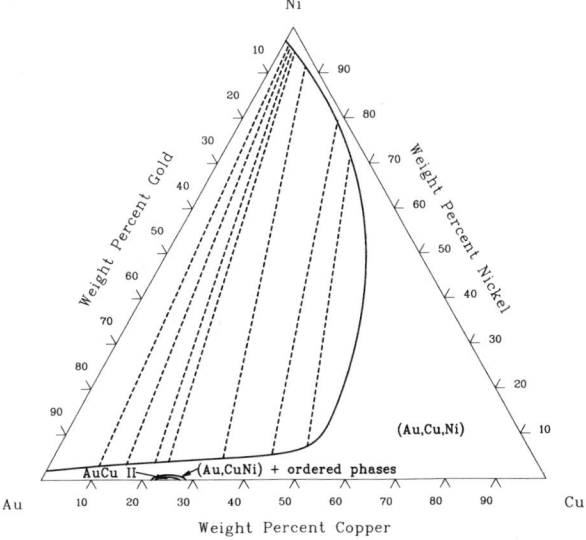

Au-Cu-Ni boundary of miscibility gap at 700 °C, with tie lines [90Pri]

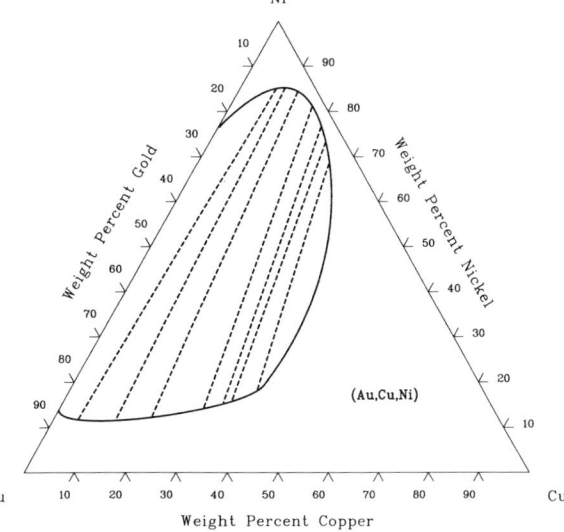

B-C-Fe liquidus projection [63Sta]

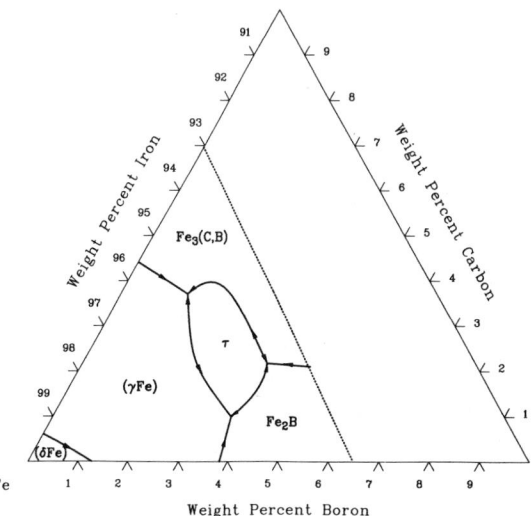

B-C-Fe isothermal section at 1000 °C [73Bre]

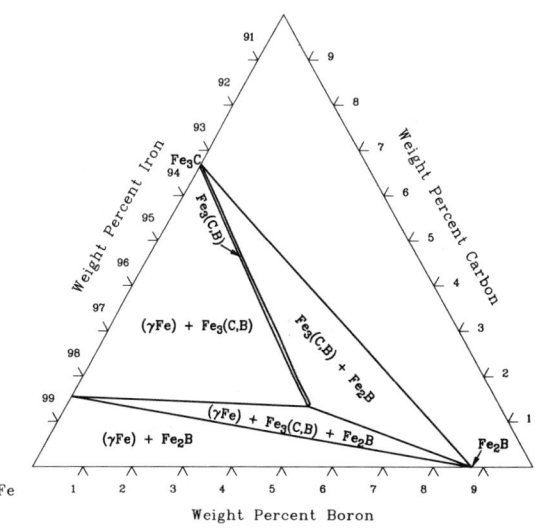

B-C-Fe isothermal section at 900 °C [73Bre]

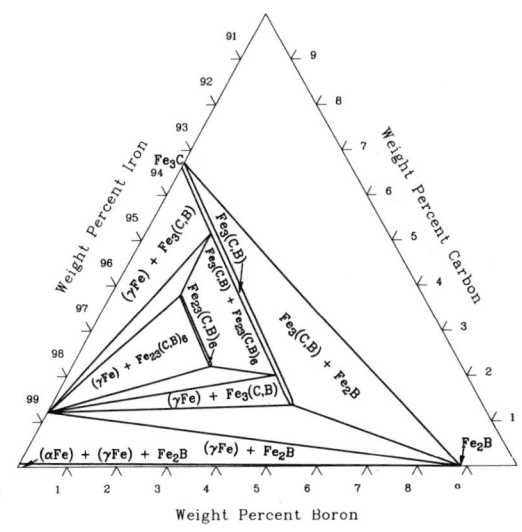

B-C-Fe isothermal section at 800 °C [73Bre]

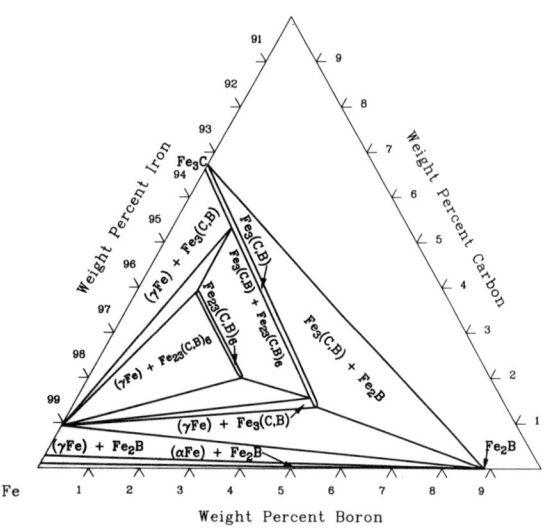

C-Cr-Fe isothermal section at 1000 °C [88Ray]

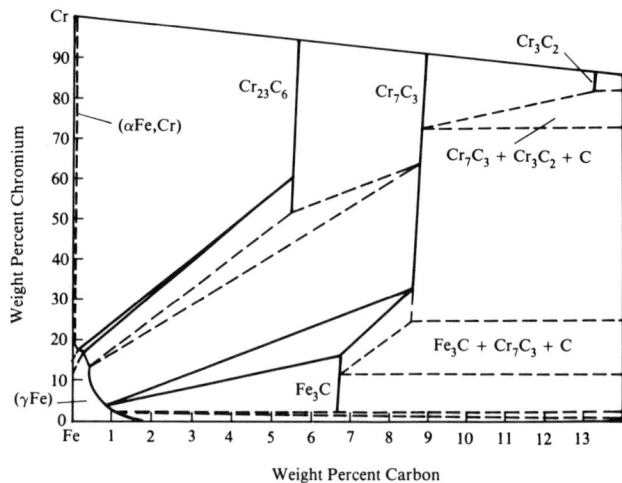

B-C-Fe isothermal section at 700 °C [73Bre]

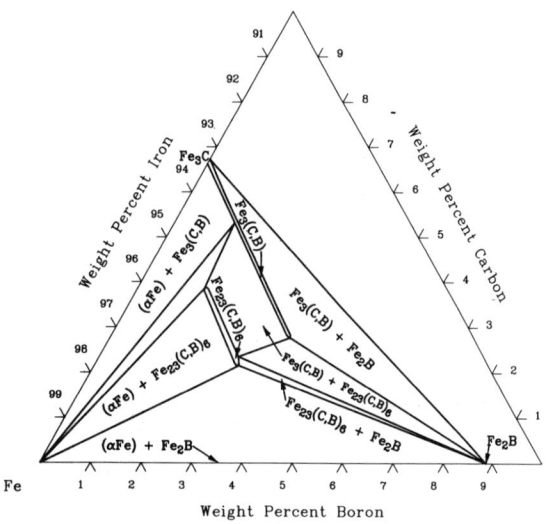

C-Cr-Fe liquidus projection [88Ray]

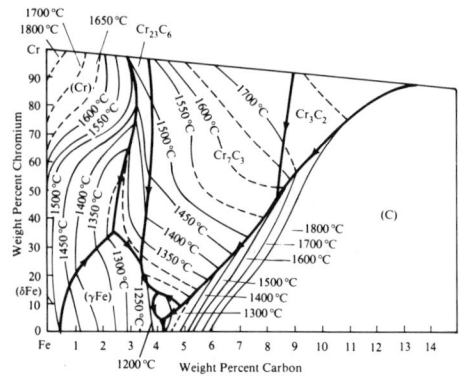

C-Cr-Fe isothermal section at 870 °C [88Ray]

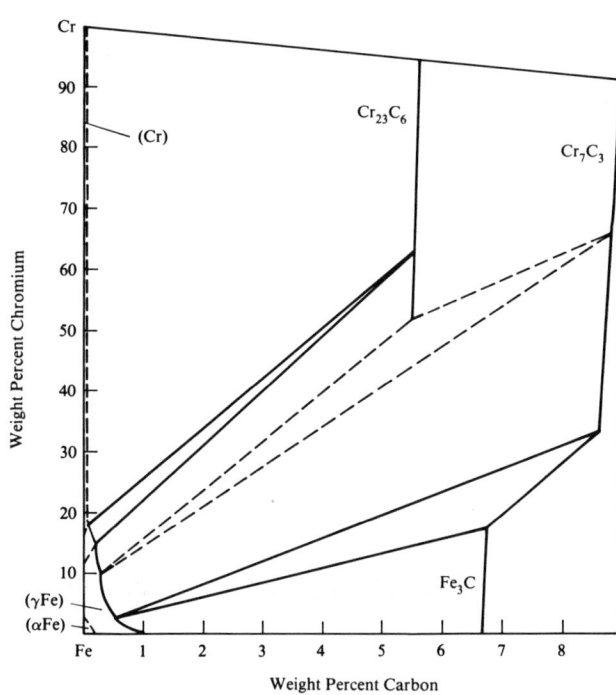

C-Cr-Fe isothermal section at 700 °C [88Ray]

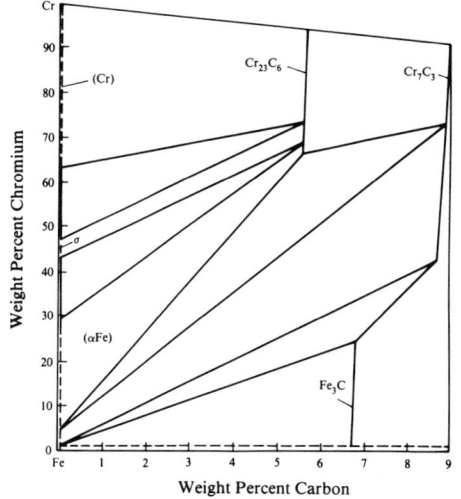

C-Cr-Fe isothermal section at 900 °C [88Ray]

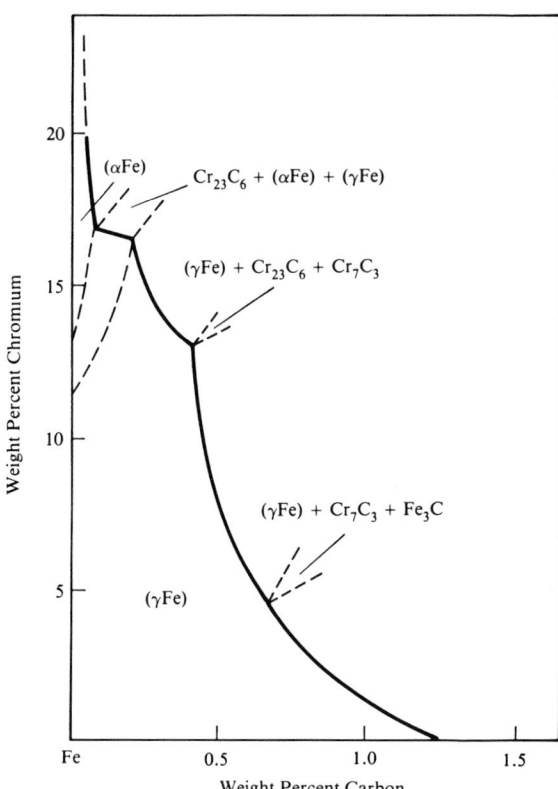

C-Cr-Fe (Fe) isothermal section at 1100 °C [88Ray]

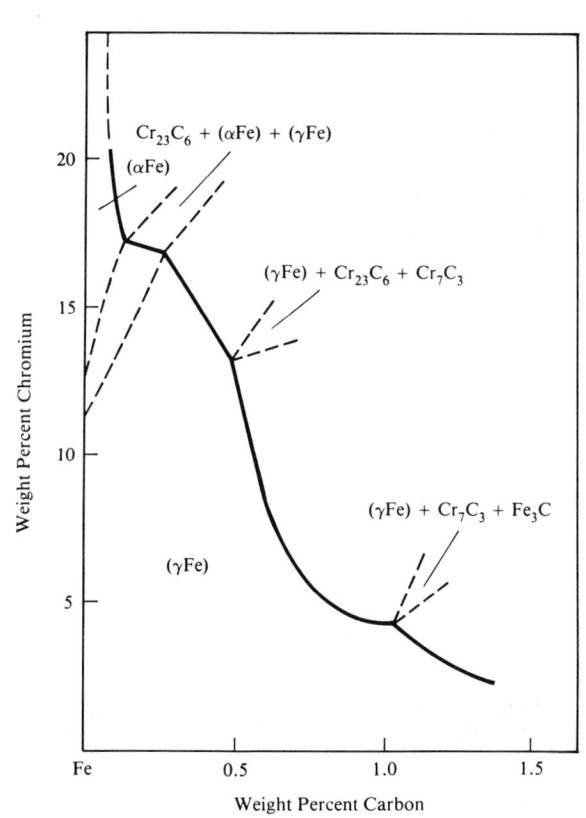

C-Cr-Mo liquidus projection [87Ere]

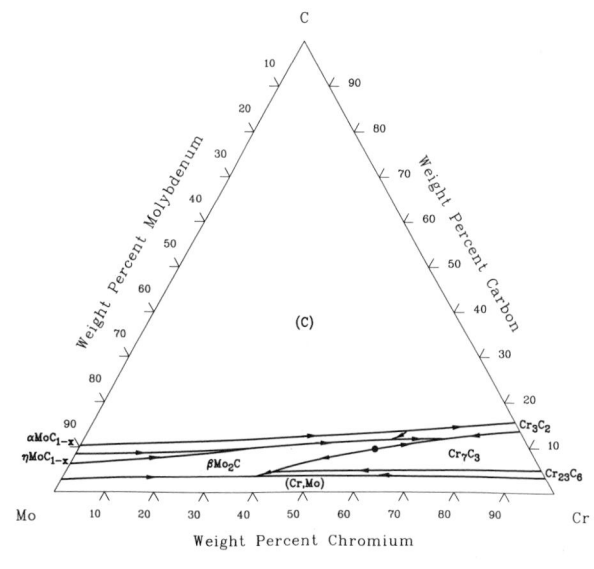

C-Cr-Mo isothermal section at 1350 °C [65Kuz]

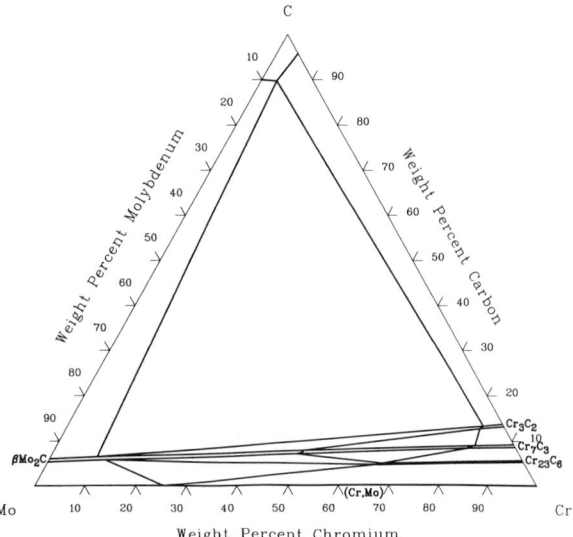

C-Cr-N isothermal section at 1400 °C [73Bre]

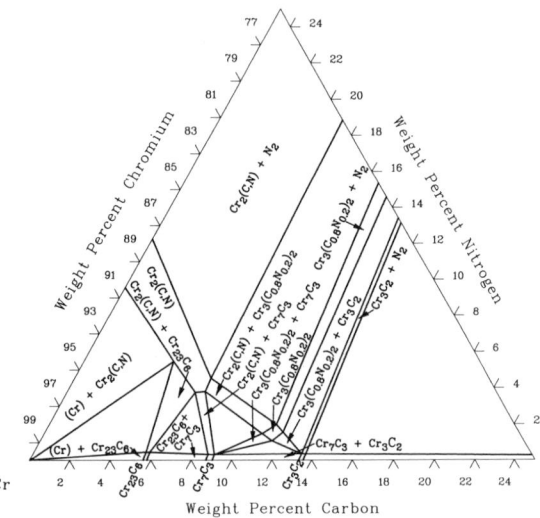

Nitrogen pressure: ~3 MPa.

C-Cr-N isothermal section at 1400 °C [73Bre]

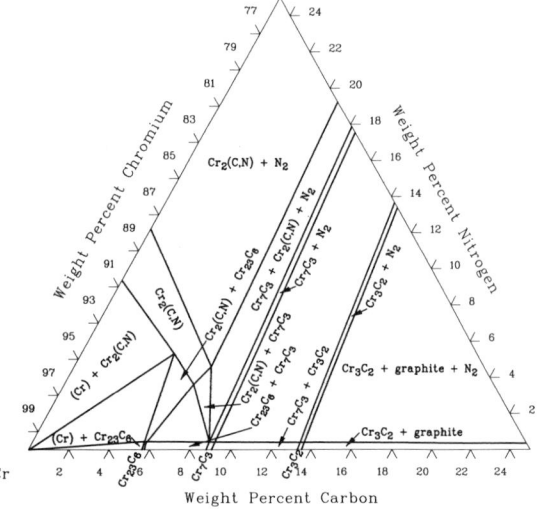

Nitrogen pressure: ≤0.1 MPa.

C-Cr-N isothermal section at 1100 °C [73Bre]

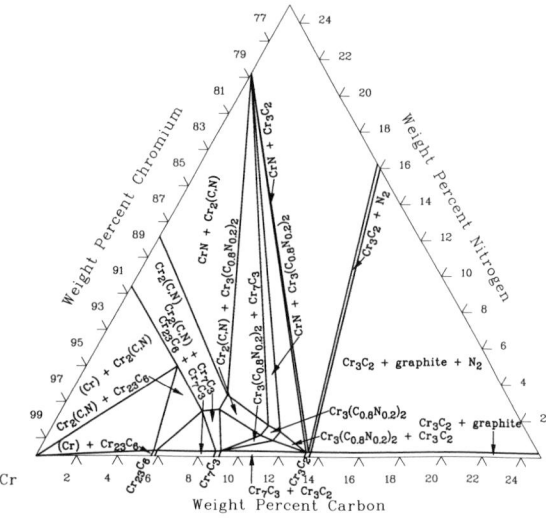

Nitrogen pressure: 0.2 to 3 MPa.

C-Cr-N isothermal section at 1100 °C [73Bre]

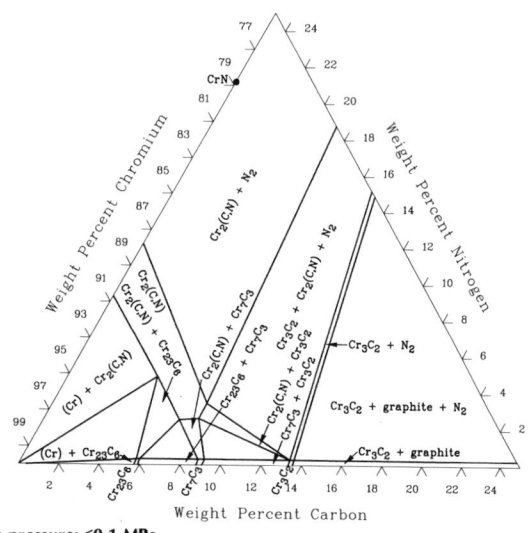

Nitrogen pressure: ≤0.1 MPa.

C-Cr-V liquidus projection [66Kie]

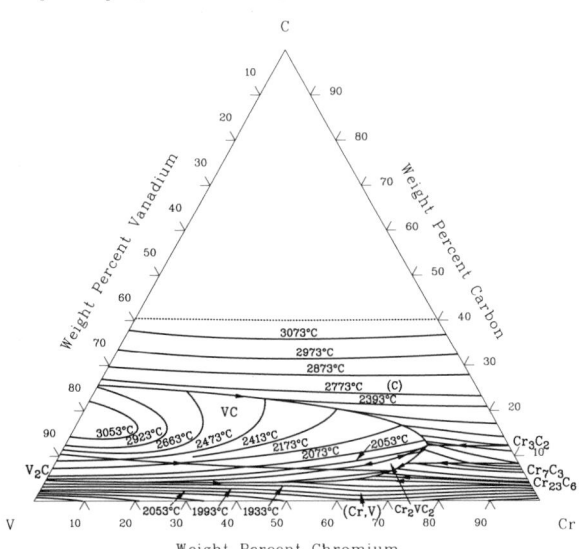

C-Cr-V isothermal section at 1350 °C [66Kie]

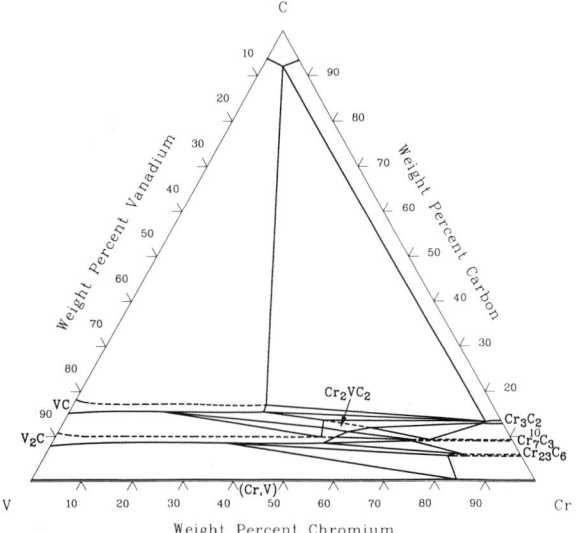

C-Cr-W isothermal section at 1600 °C [86Ere]

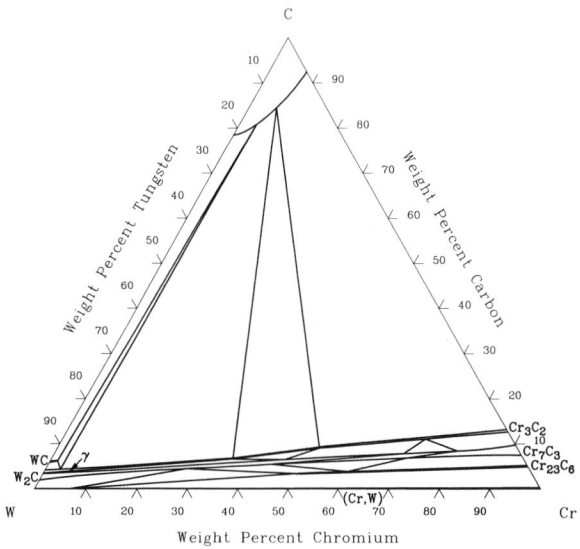

C-Cr-W isothermal section at 1350 °C [64Ste]

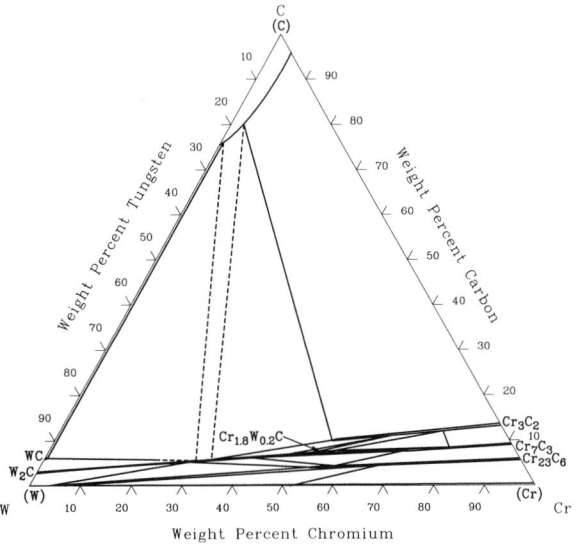

C-Cu-Fe liquidus projection [88Ray]

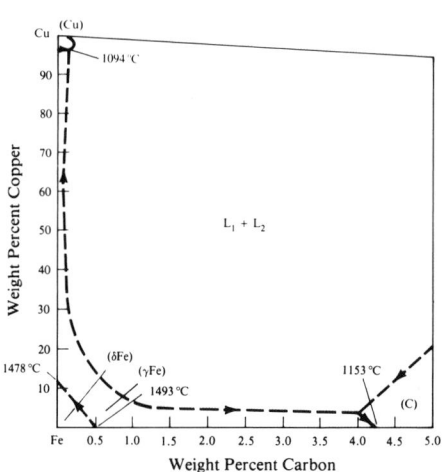

C-Cu-Fe isothermal section at 1172 °C [88Ray]

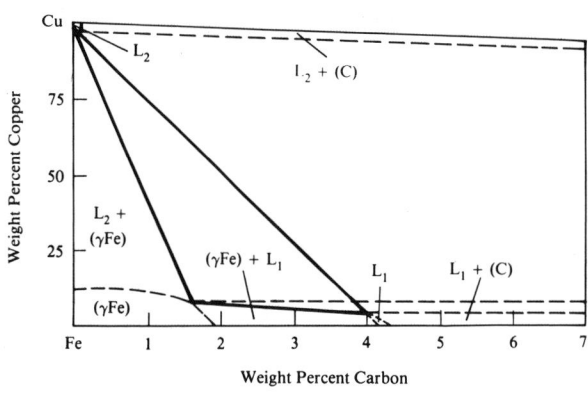

C-Cu-Fe isothermal section at 1050 °C [88Ray]

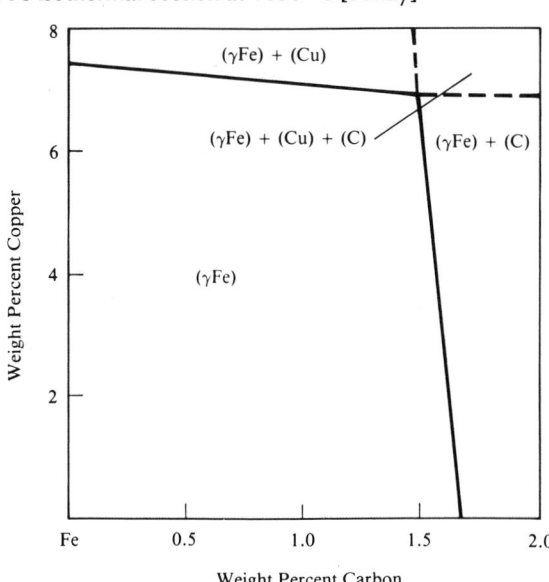

3•28/Ternary Alloy Phase Diagrams

C-Cu-Fe isothermal section at 925 °C [88Ray]

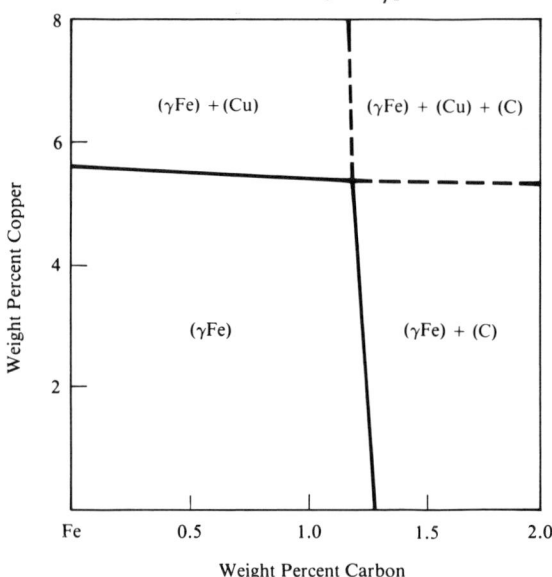

C-Cu-Fe isothermal section at 850 °C [88Ray]

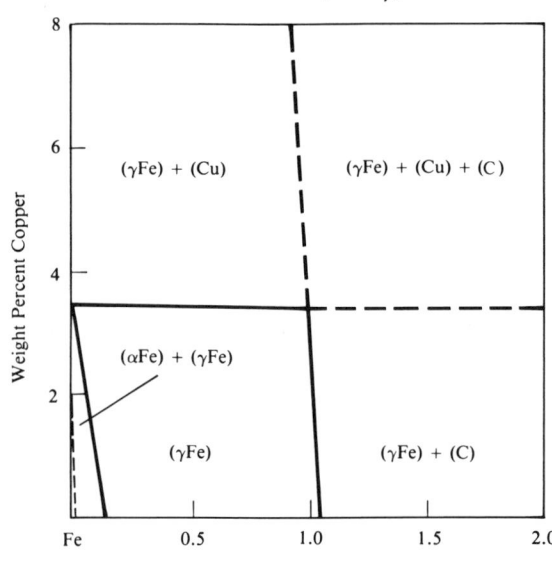

C-Cu-Fe schematic isothermal section at 850 °C [88Ray]

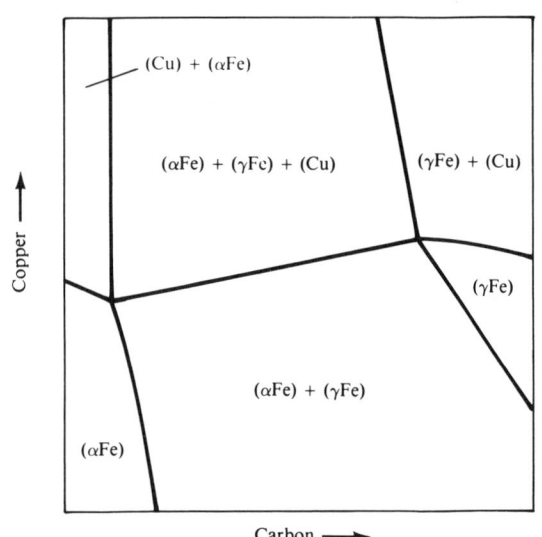

C-Fe-Mn liquidus projection [88Ray]

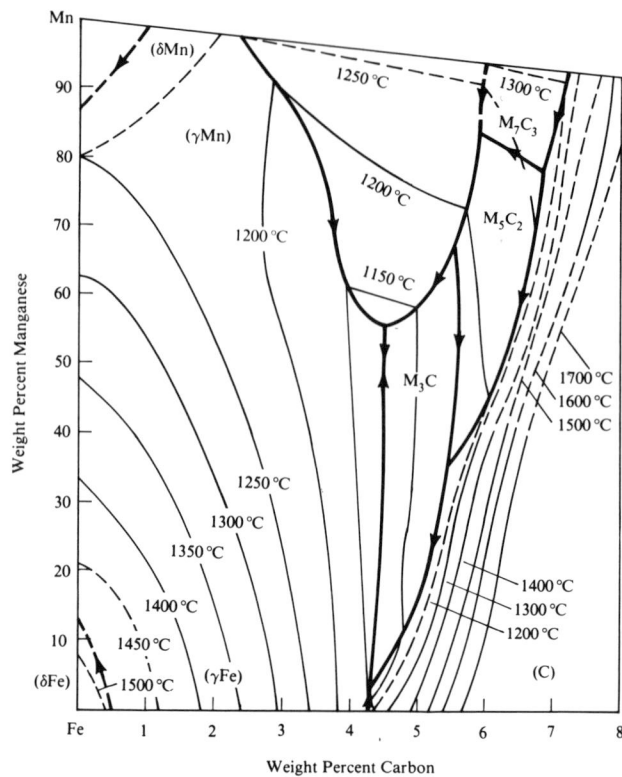

C-Fe-Mn isothermal section at 1100 °C [73Ben]

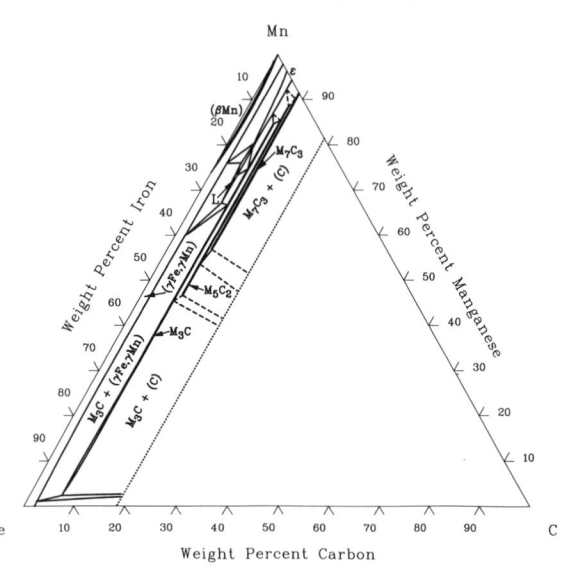

C-Fe-Mn isothermal section at 1000 °C [73Ben]

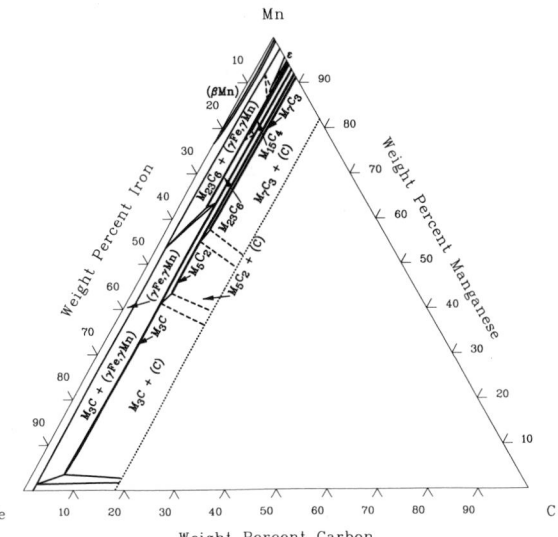

C-Fe-Mn isothermal section at 600 °C [73Ben]

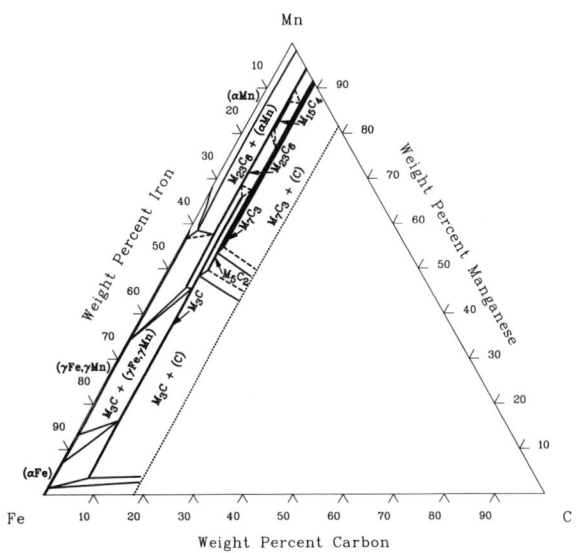

C-Fe-Mn isothermal section at 900 °C [73Ben]

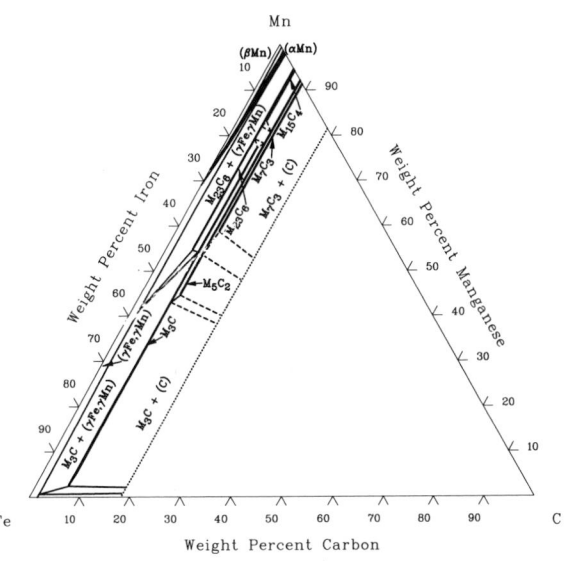

C-Fe-Mn [73Bre]

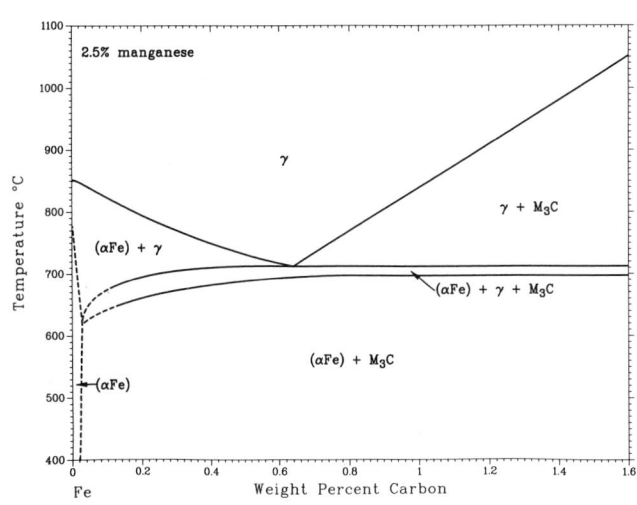

C-Fe-Mn isothermal section at 800 °C [73Ben]

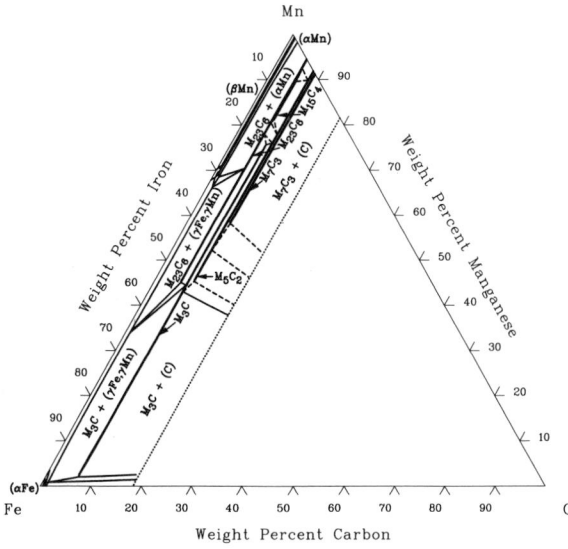

C-Fe-Mn [73Bre]

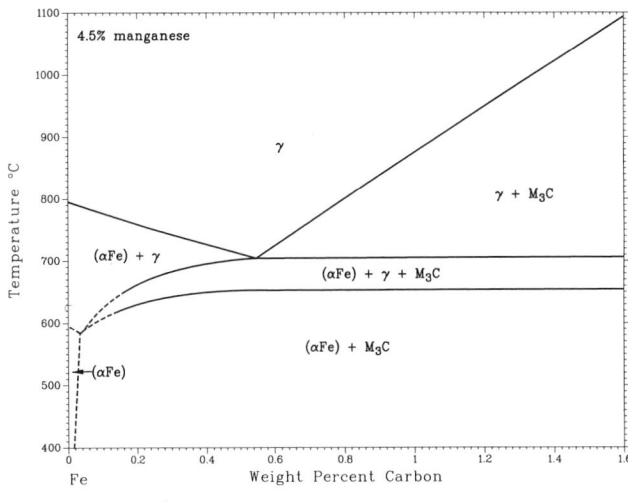

C-Fe-Mn [73Bre]

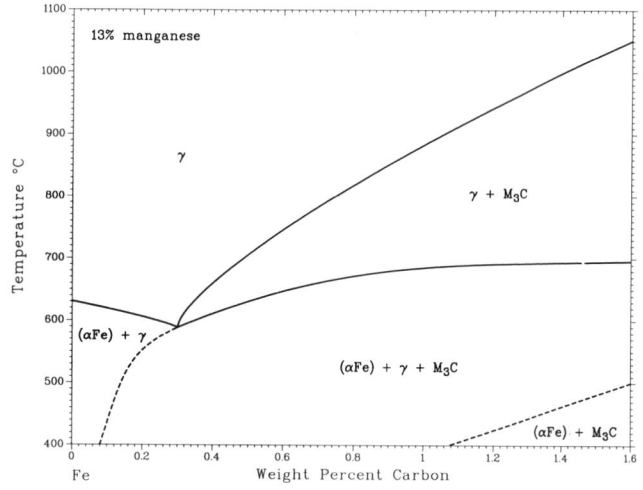

C-Fe-Mo isothermal section at 1000 °C [88Ray]

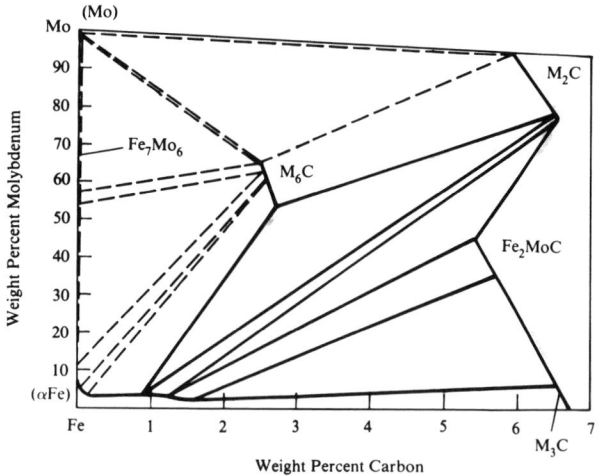

C-Fe-Mo (Fe) isothermal section at 1000 °C [88Ray]

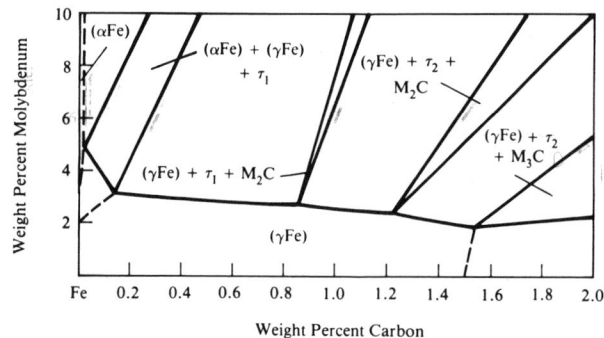

C-Fe-Mo liquidus projection [88Ray]

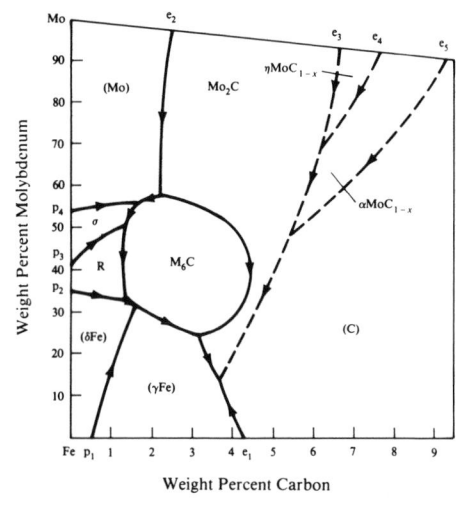

C-Fe-Mo isothermal section at 700 °C (calculated) [88Ray]

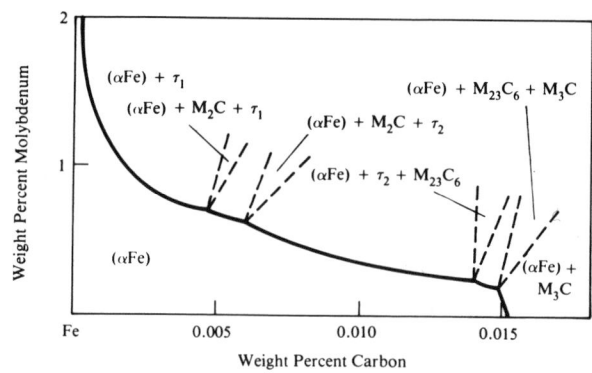

C-Fe-Mo (Fe) isothermal section at 700 °C [Ray]

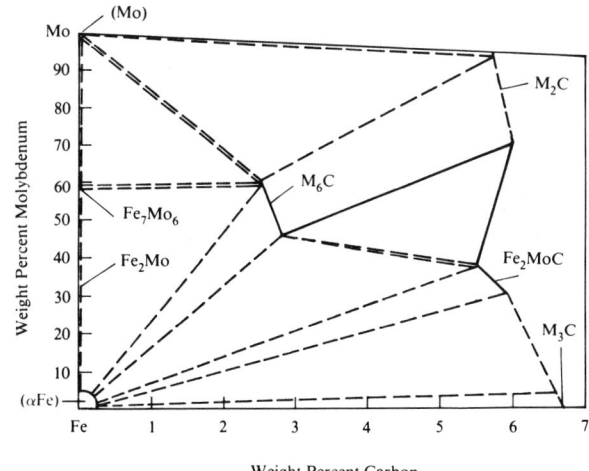

C-Fe-Mo [88Ray]

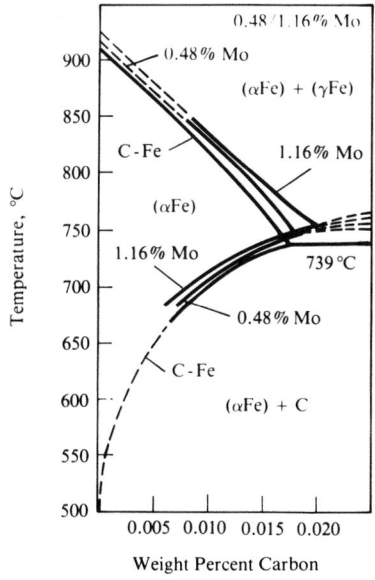

C-Fe-N isothermal section at 700 °C [87Rag]

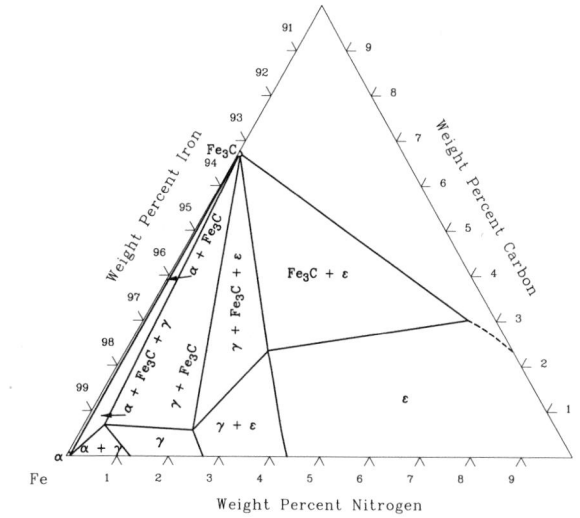

C-Fe-N isothermal section at 600 °C [87Rag]

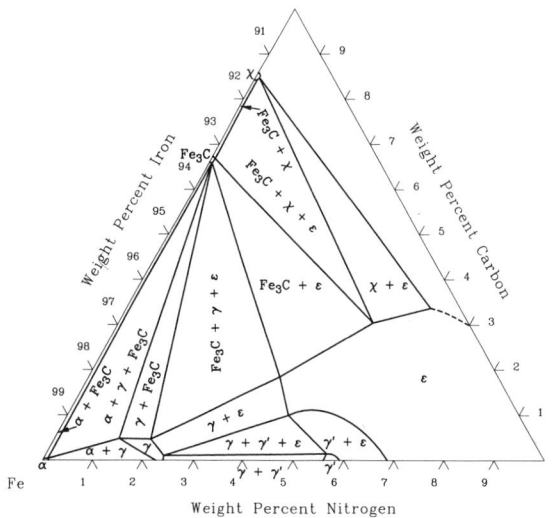

C-Fe-N isothermal section at 575 °C [87Rag]

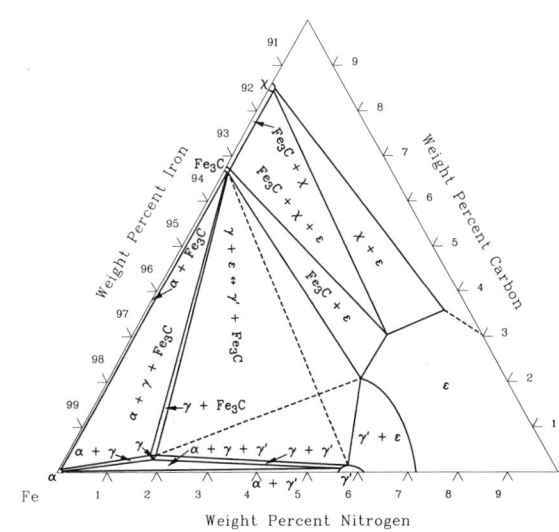

C-Fe-N isothermal section at 565 °C [87Rag]

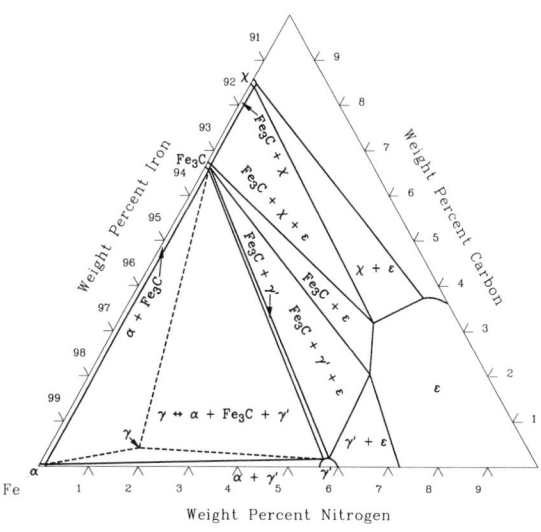

C-Fe-N isothermal section at 500 °C [87Rag]

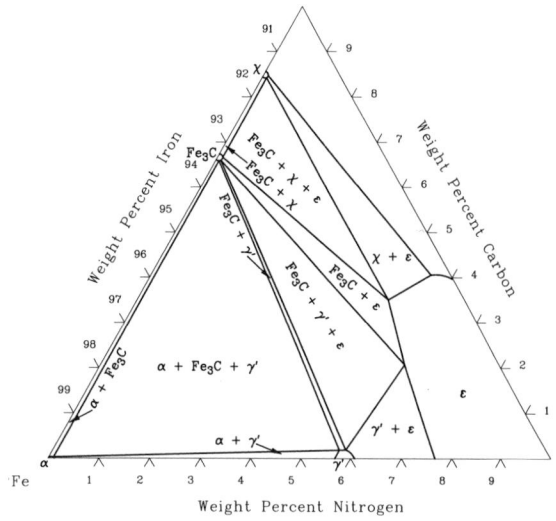

C-Fe-Ni solidus projection [88Ray]

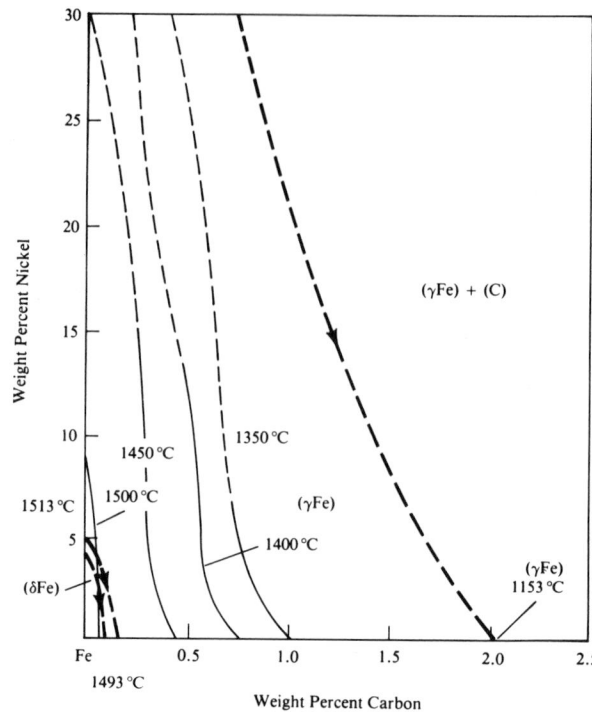

C-Fe-Ni liquidus projection [88Ray]

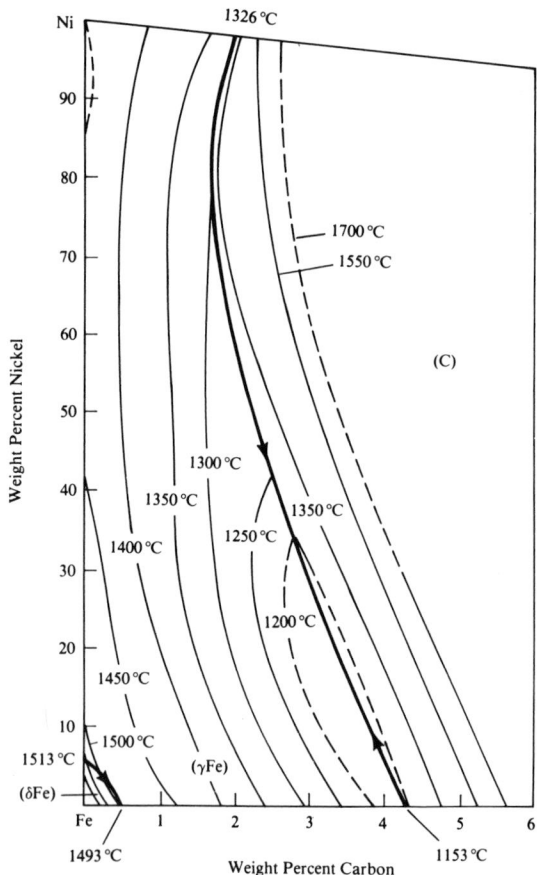

C-Fe-Ni γFe/ (γFe + C) boundary at 800 and 1000 °C [88Ray]

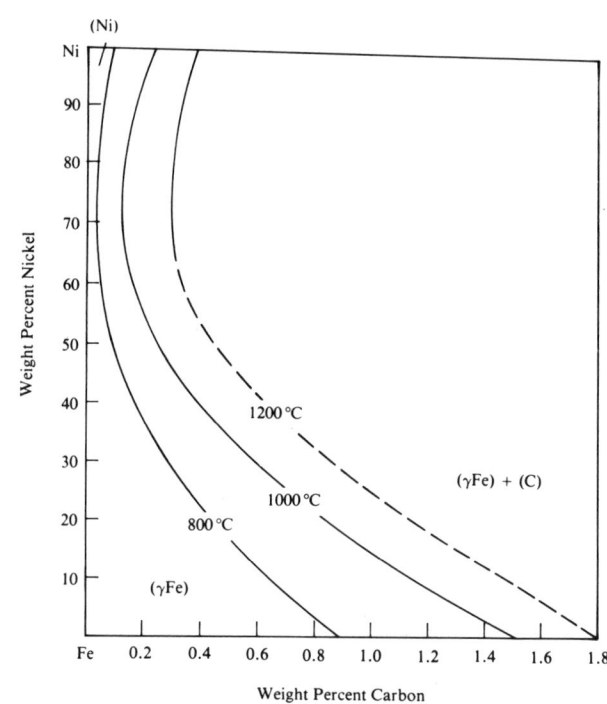

Note that at 800 °C the (αFe) phase will also appear at low Ni contents.

C-Fe-Si liquidus projection (stable equilibrium) [86Rag]

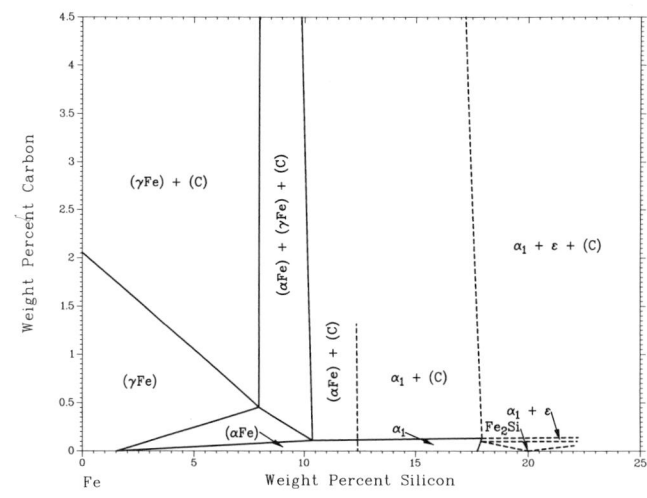

C-Fe-Si isothermal section at 1150 °C (stable equilibrium) [86Rag]

C-Fe-Si liquidus projection (metastable equilibrium) [86Rag]

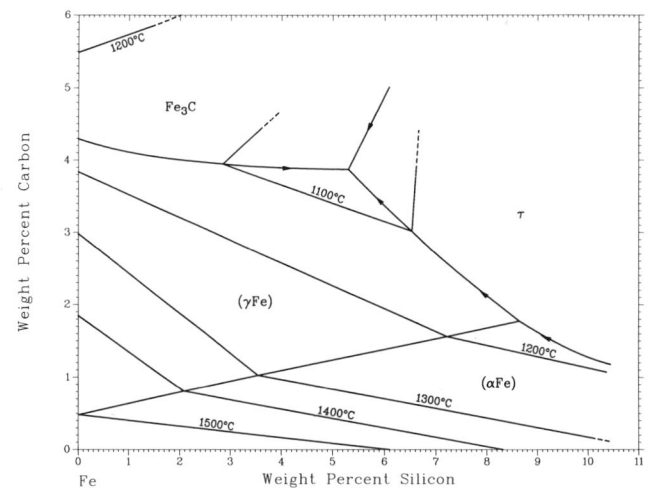

C-Fe-Si isothermal section at 1100 °C (metastable equilibrium) [86Rag]

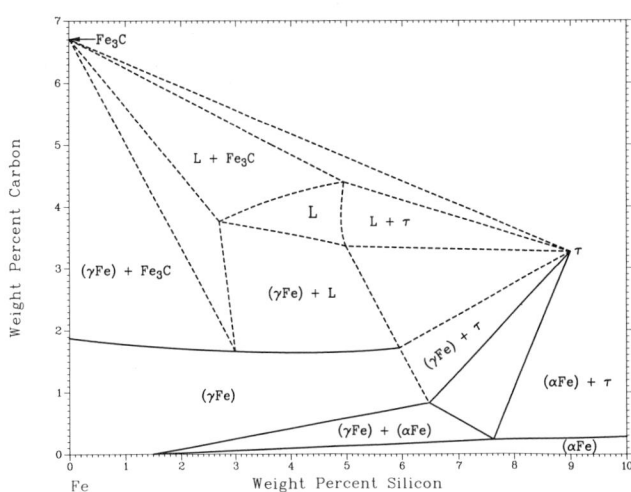

C-Fe-Si isothermal section at 1300 °C (stable equilibrium) [86Rag]

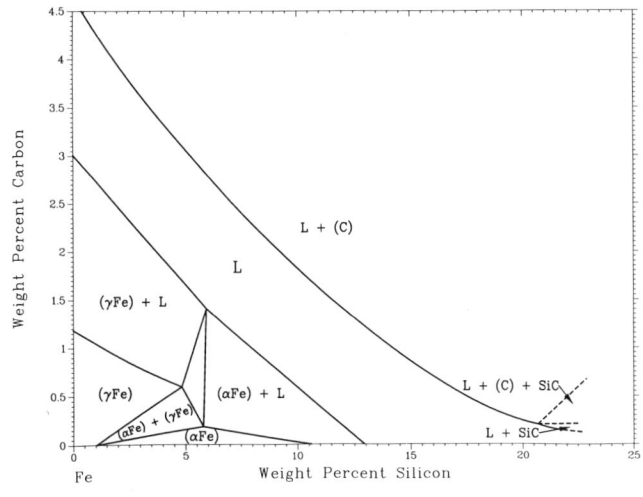

C-Fe-Si isothermal section at 1000 °C (stable equilibrium) [86Rag]

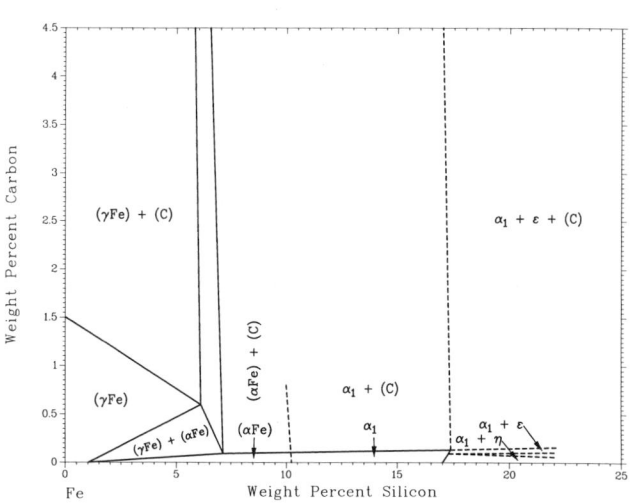

C-Fe-Si isothermal section at 900 °C (metastable equilibrium) [86Rag]

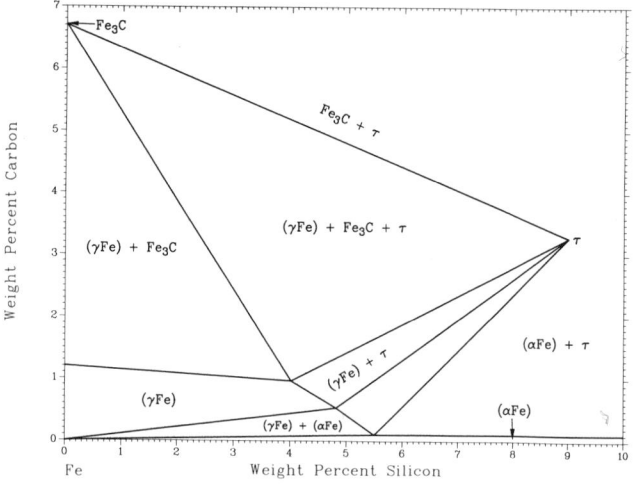

C-Fe-V isothermal section at 1100 °C [87Rag]

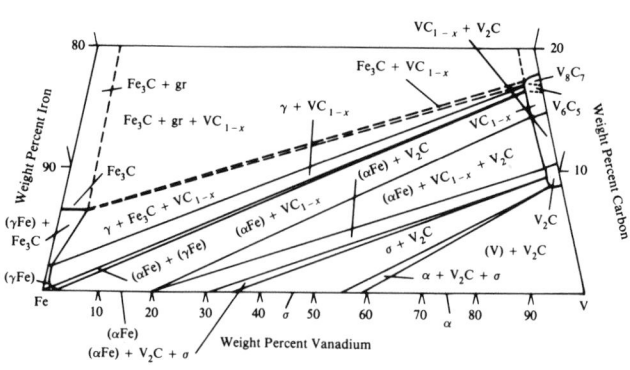

C-Fe-Si isothermal section at 800 °C (metastable equilibrium) [86Rag]

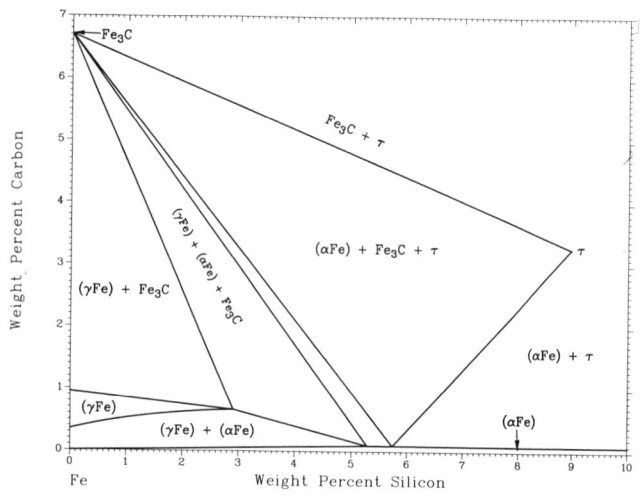

C-Fe-V isothermal section at 1000 °C [87Rag]

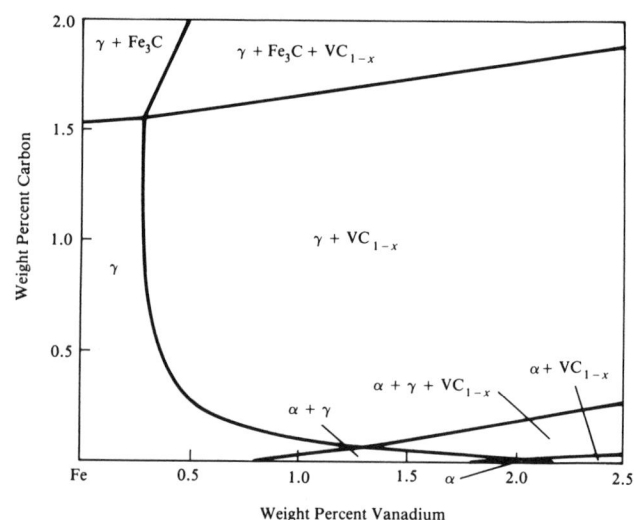

C-Fe-V liquidus projection [87Rag]

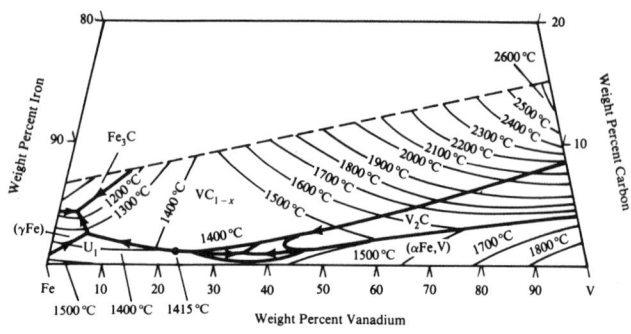

C-Fe-V isothermal section at 500 °C [87Rag]

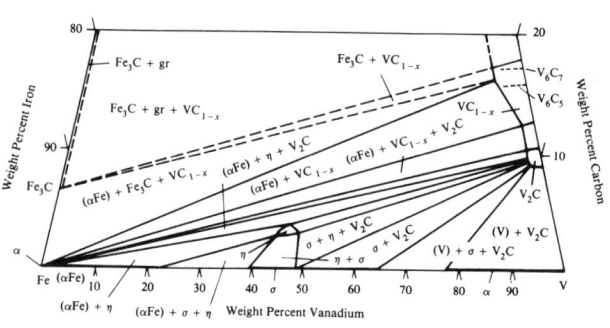

C-Fe-W liquidus projection [88Ray]

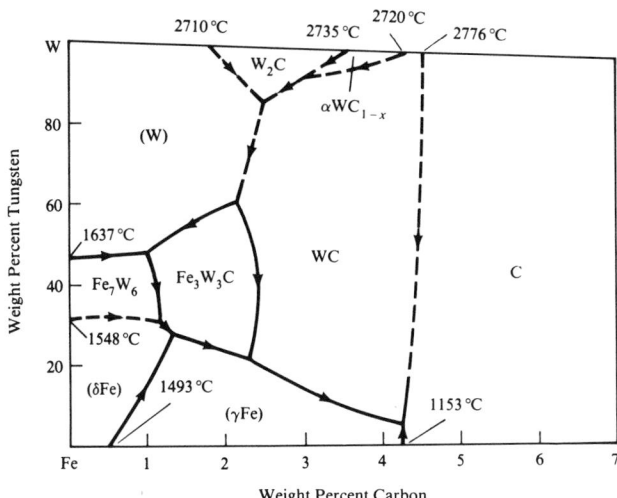

C-Fe-W isothermal section at 1000 °C [88Ray]

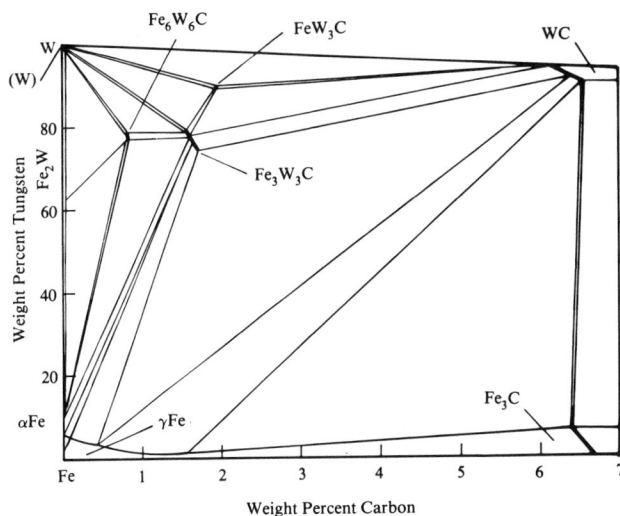

C-Fe-W isothermal section at 1250 °C [88Ray]

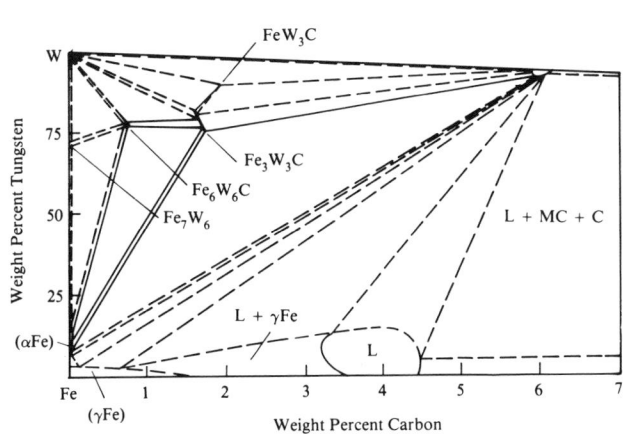

C-Fe-W (Fe) isothermal section at 1000 °C [88Ray]

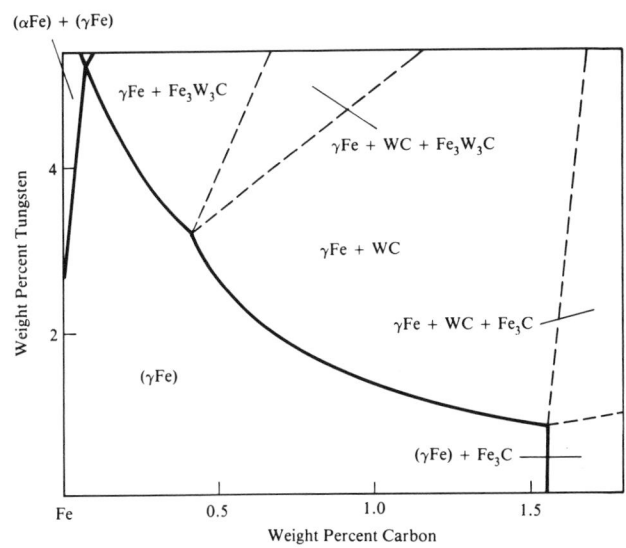

C-Fe-W (Fe) isothermal section at 1250 °C [88Ray]

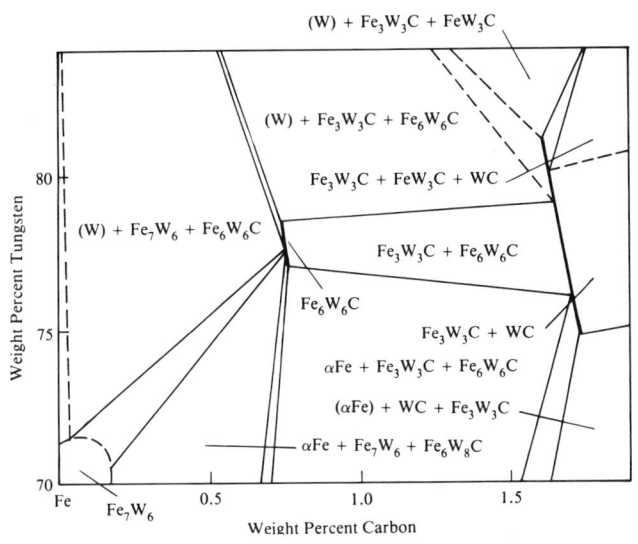

Cd-Sb-Sn liquidus projection [73Pel]

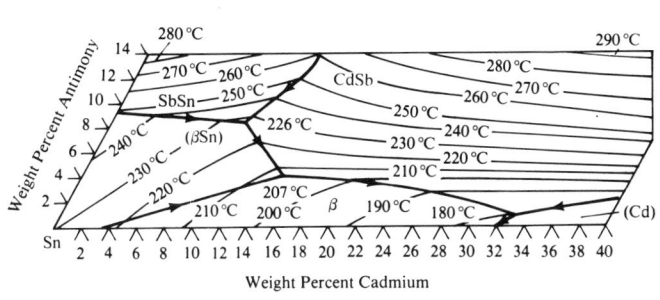

Cd-Sb-Sn isothermal section at 212 °C [73Pel]

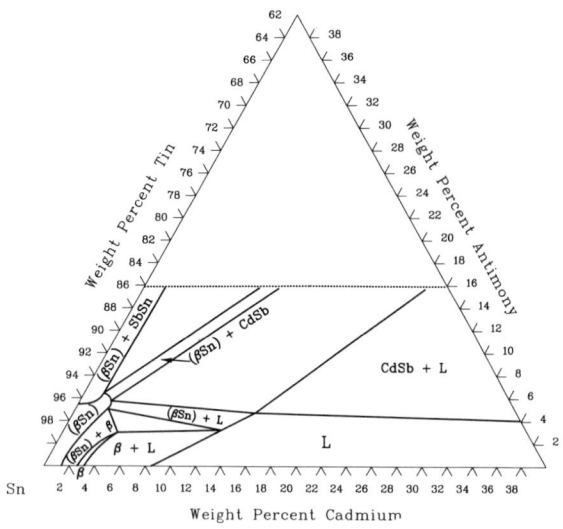

Cd-Sb-Sn isothermal section at 175 °C [73Pel]

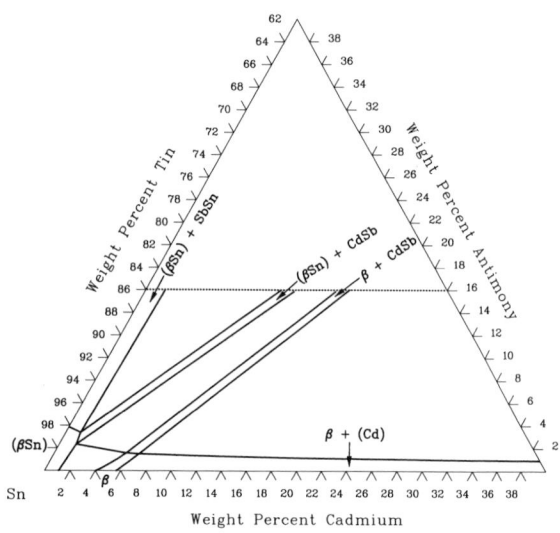

Cd-Sb-Sn isothermal section at 20 °C [73Pel]

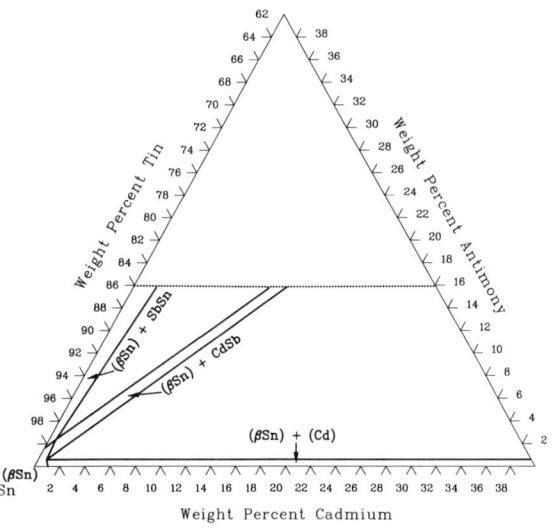

Cd-Sb-Sn [73Pel]

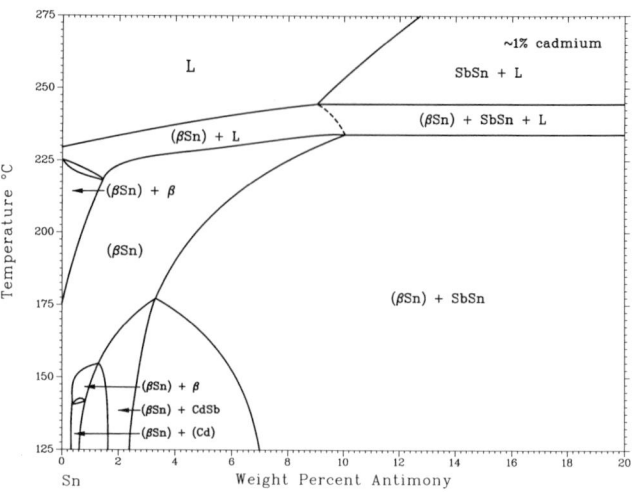

Cd-Sb-Sn [73Pel]

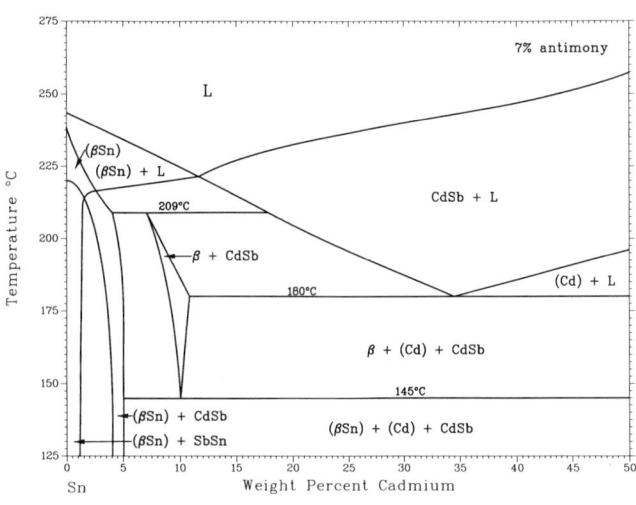

Co-Cr-Fe liquidus projection [88Ray]

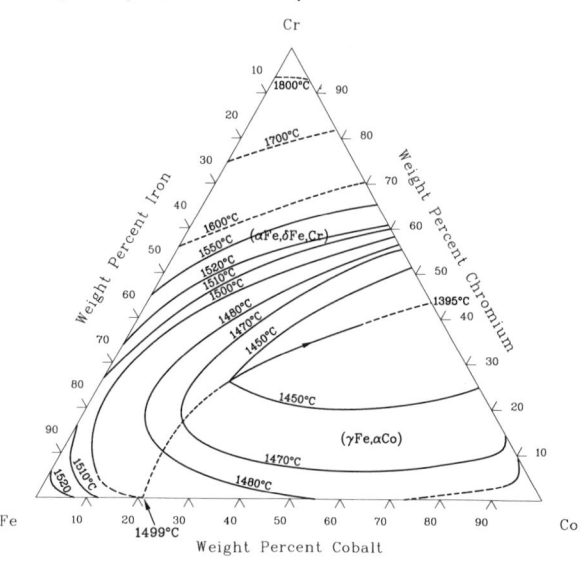

Co-Cr-Fe solidus projection [88Ray]

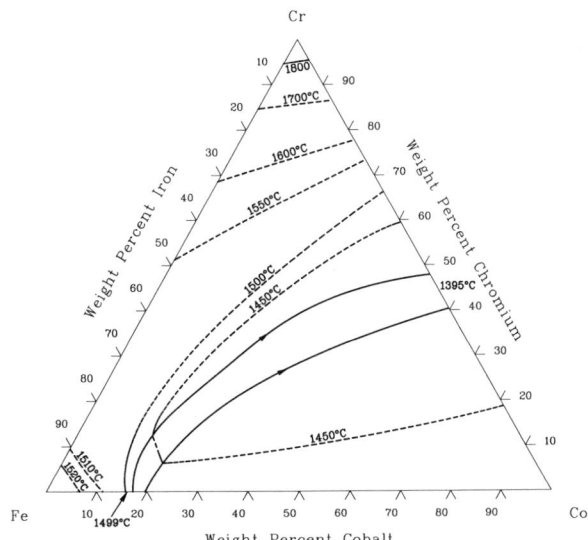

Co-Cr-Fe isothermal section at 800 °C [88Ray]

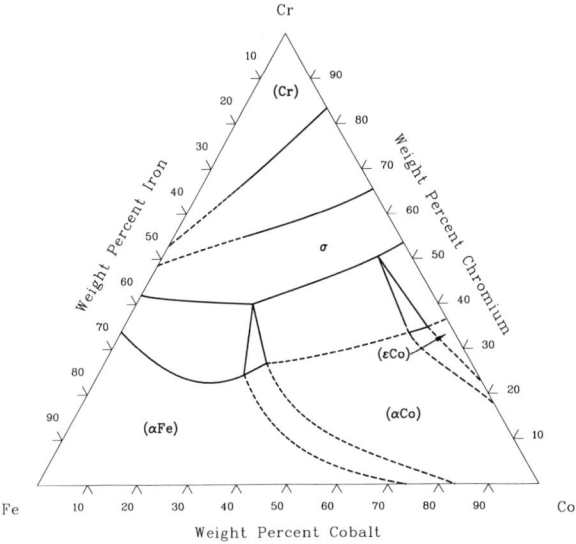

Co-Cr-Fe isothermal section at 1200 °C [88Ray]

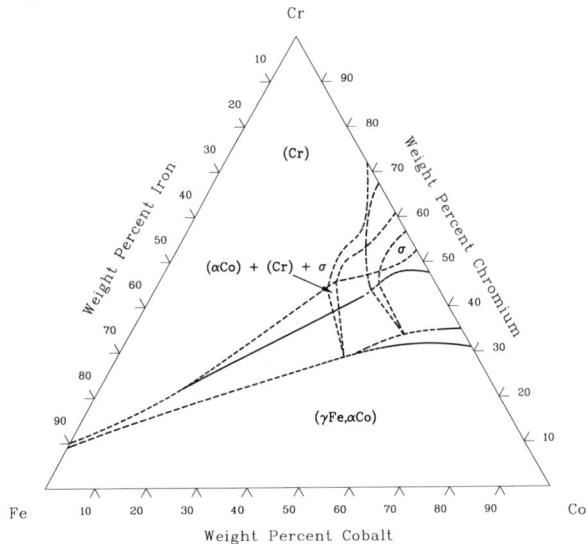

Co-Cr-Fe isothermal section at 600 °C [88Ray]

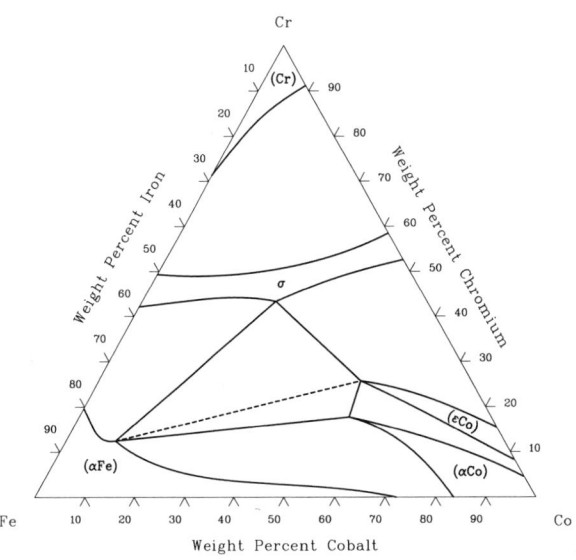

Co-Cr-Fe isothermal section at 1000 °C [88Ray]

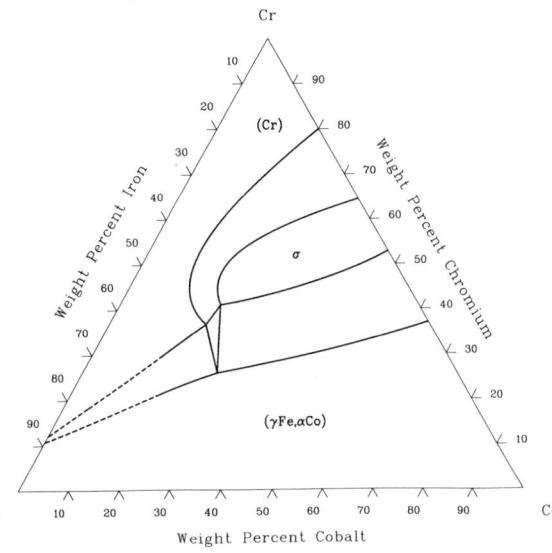

Co-Cr-Ni isothermal section at 1200 °C [81Zha]

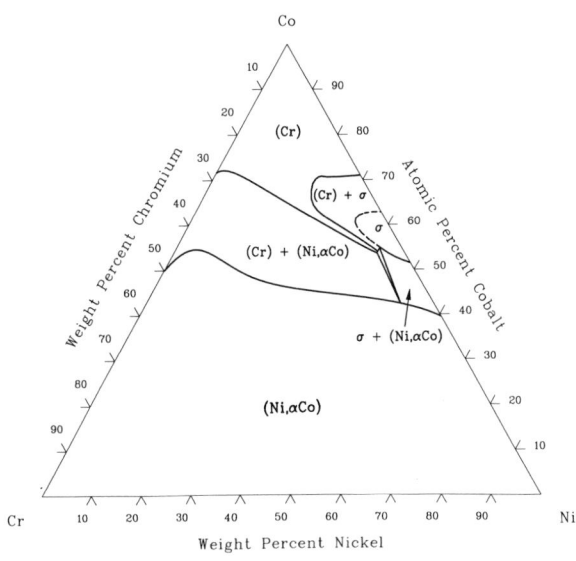

Co-Cr-Ti liquidus projection [62Zak]

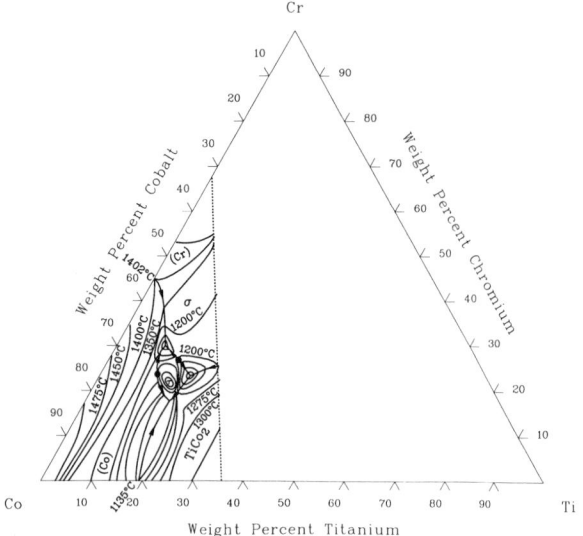

Co-Cr-W isothermal section at 1350 °C [73Dra]

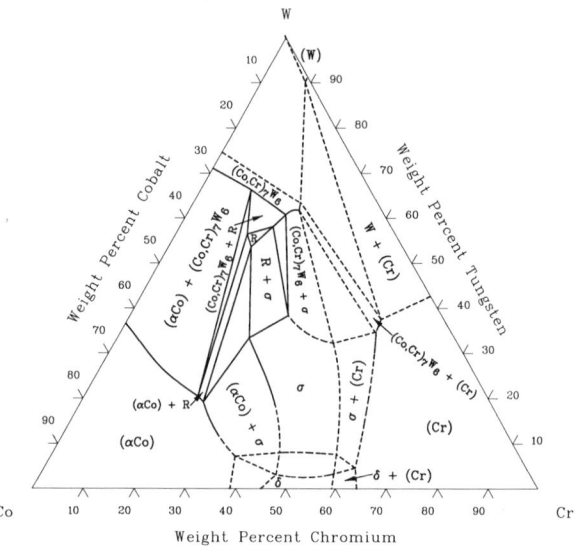

Co-Cr-Ti solidus projection [62Zak]

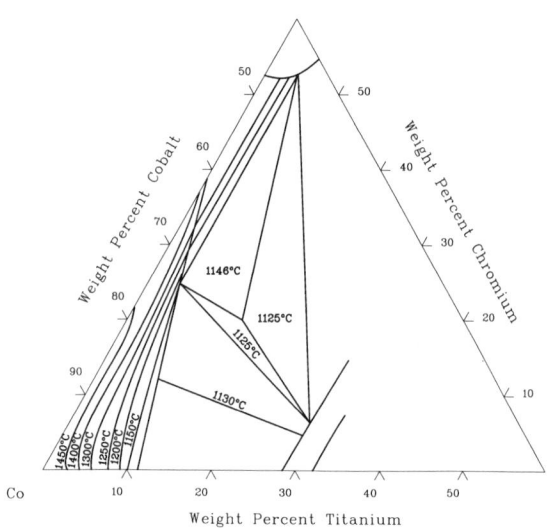

Co-Cr-W isothermal section at 700 °C [73Dra]

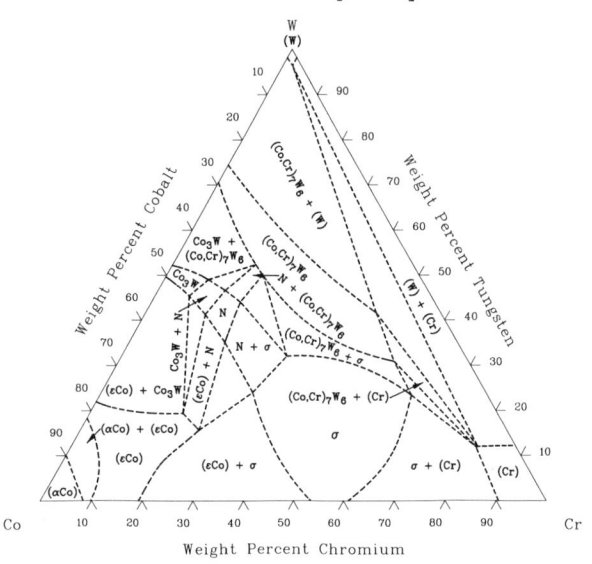

Co-Cr-Ti isothermal section at 1050 °C [58Liv]

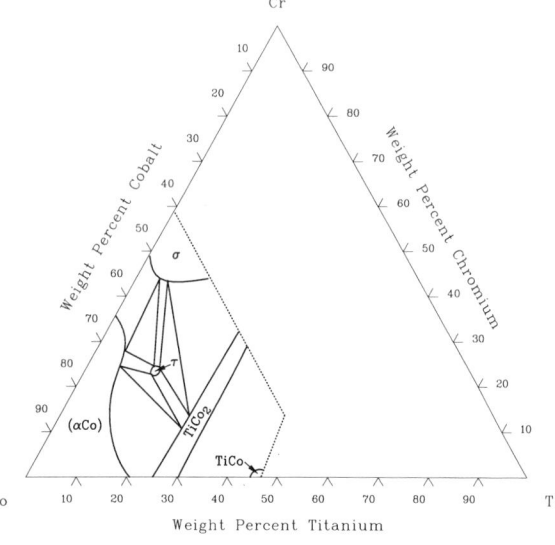

Co-Fe-Mo liquidus projection [88Ray]

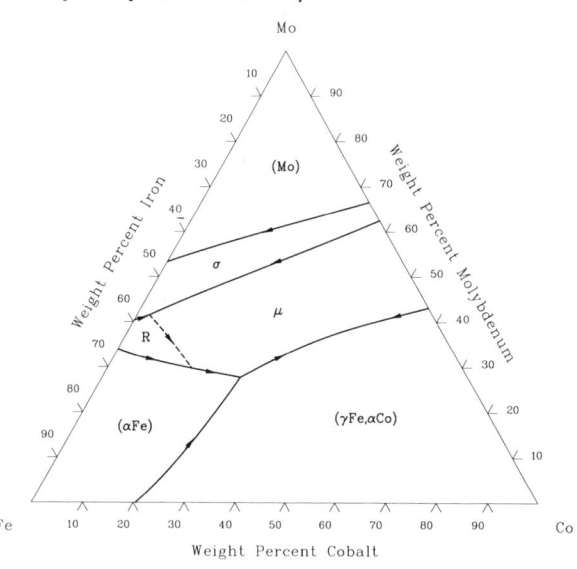

Co-Fe-Mo isothermal section at 1300 °C [88Ray]

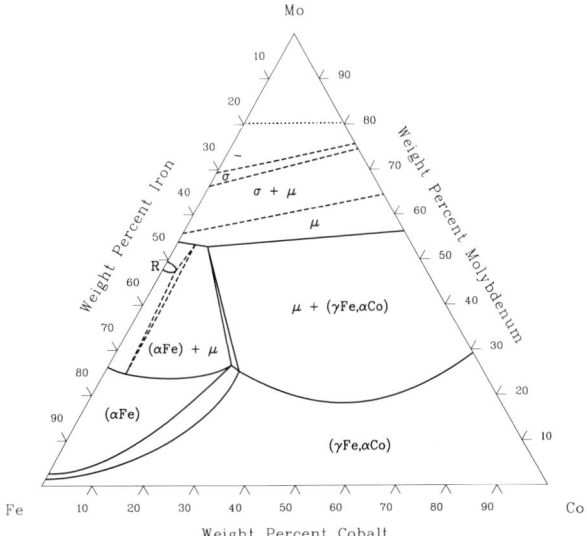

Co-Fe-Mo isothermal section at 800 °C [88Ray]

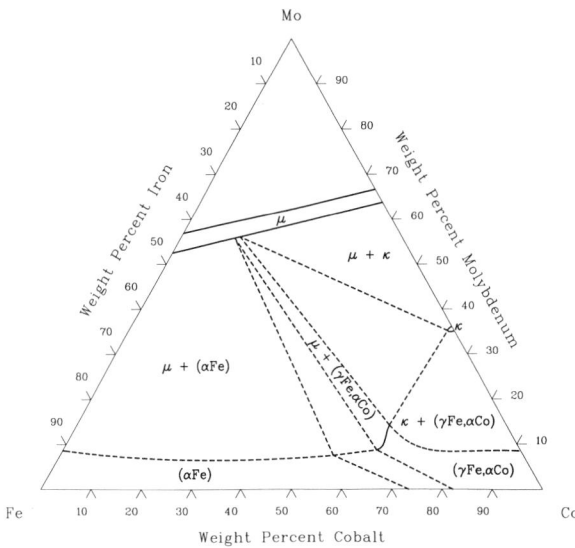

Co-Fe-Mo isothermal section at 1093 °C [88Ray]

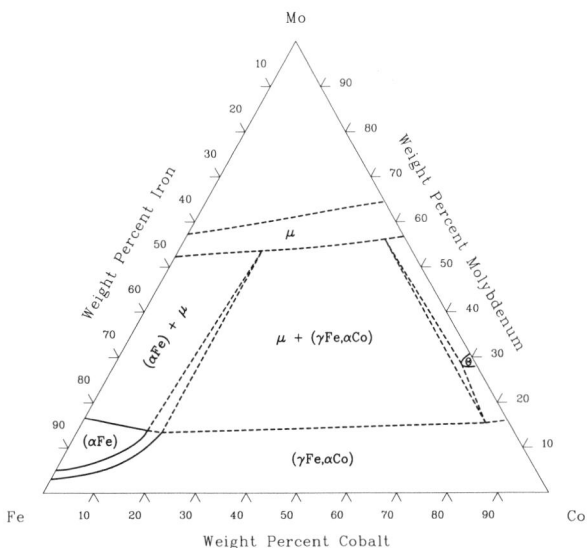

Co-Fe-Mo isothermal section at 20 °C [88Ray]

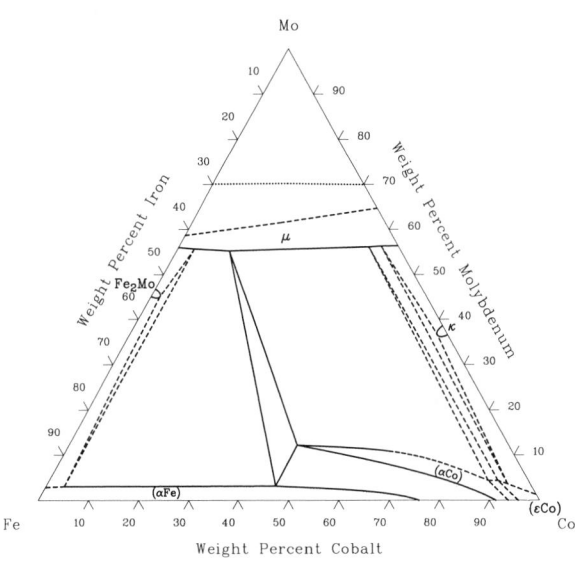

Co-Fe-Mo isothermal section at 982 °C [88Ray]

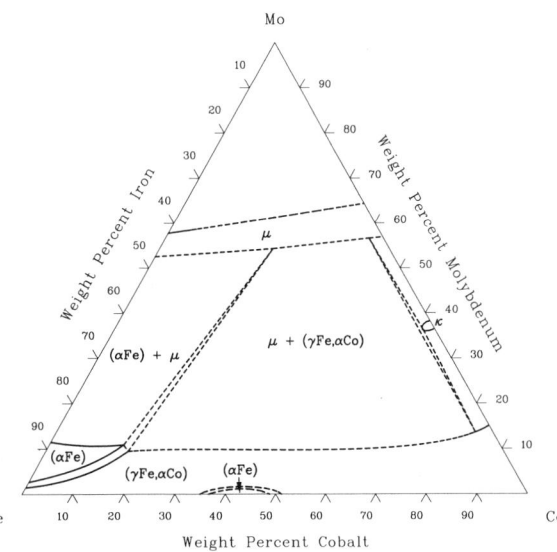

Co-Fe-Ni liquidus projection [88Ray]

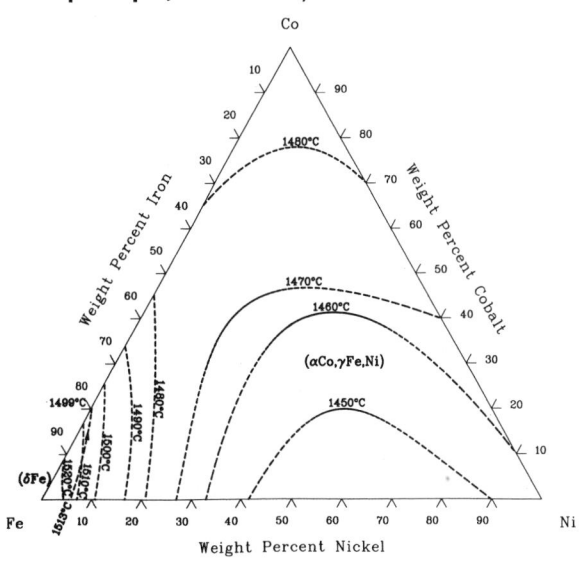

3•40/Ternary Alloy Phase Diagrams

Co-Fe-Ni solidus projection [88Ray]

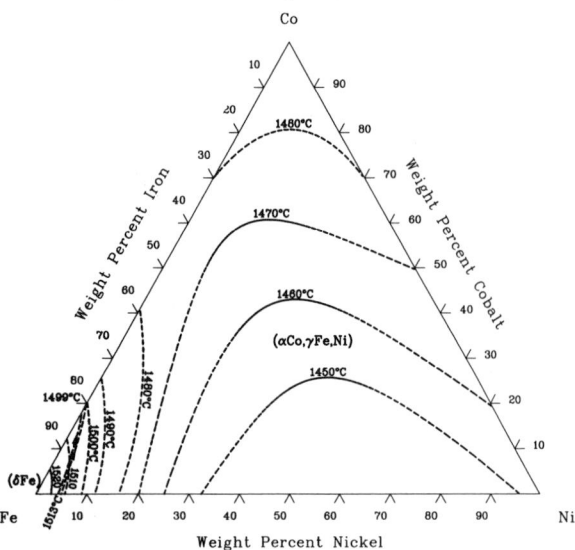

Co-Fe-W liquidus and solidus projections [88Ray]

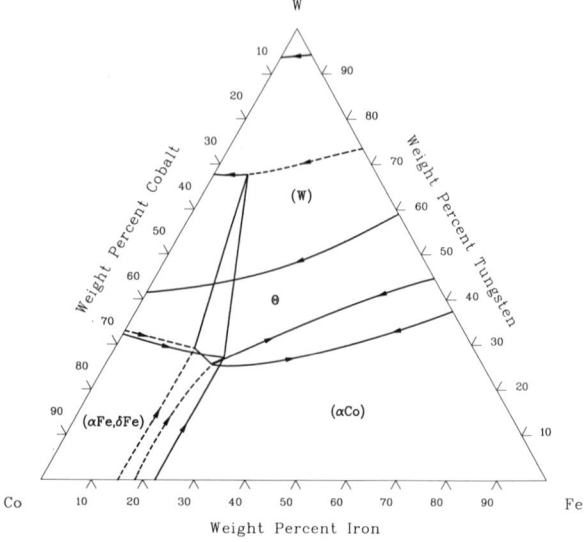

Co-Fe-Ni isothermal section at 800 °C [88Ray]

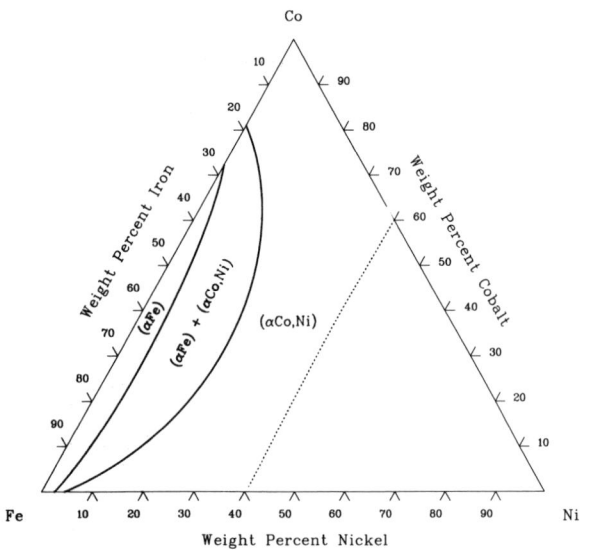

Co-Fe-W isothermal section at 1200 °C [88Ray]

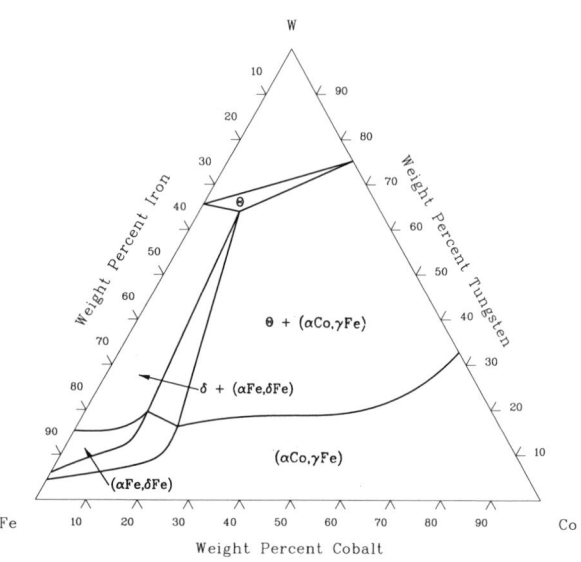

Co-Fe-Ni isothermal section at 600 °C [88Ray]

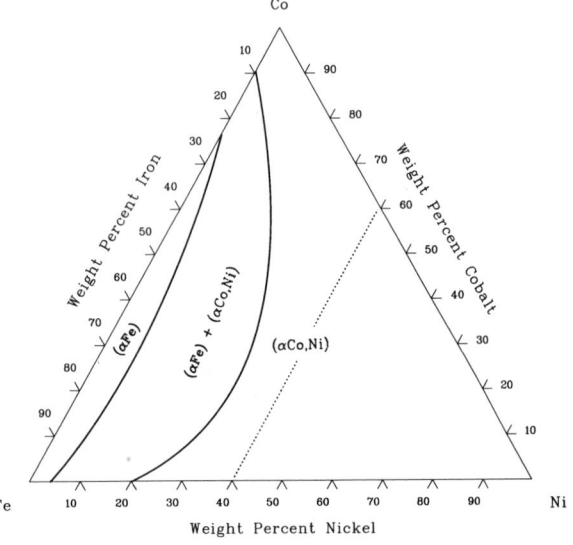

Co-Fe-W isothermal section at 1000 °C [88Ray]

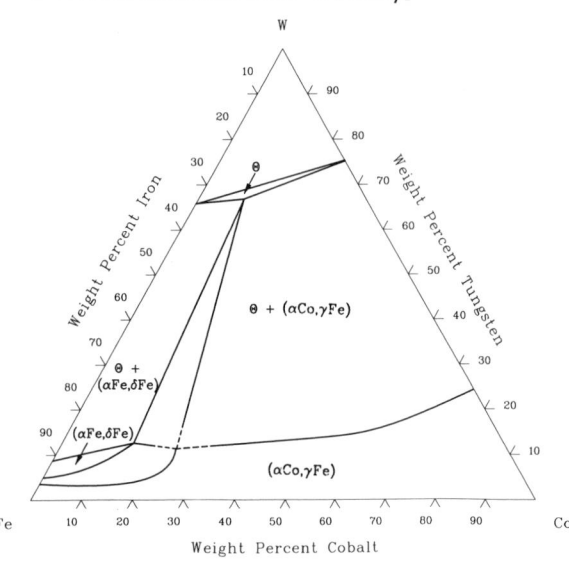

Co-Fe-W isothermal section at 800 °C [88Ray]

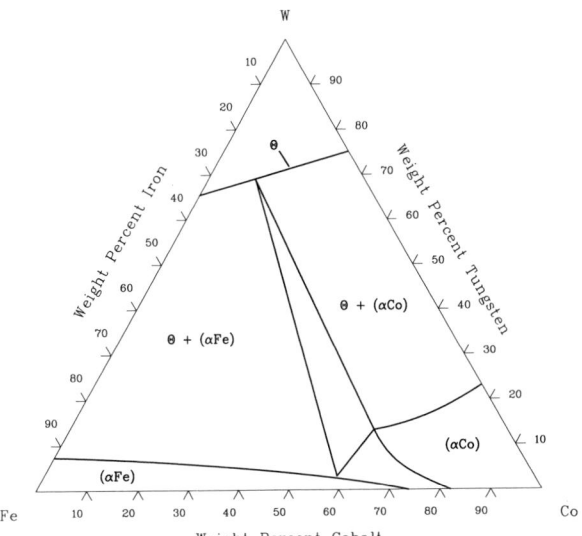

Co-Mo-Ni isothermal section at 1100 °C [80Loo]

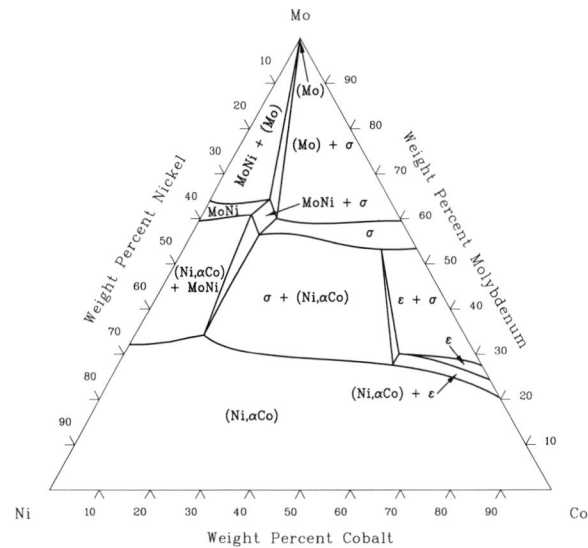

Co-Mo-Ni liquidus projection [84Gup]

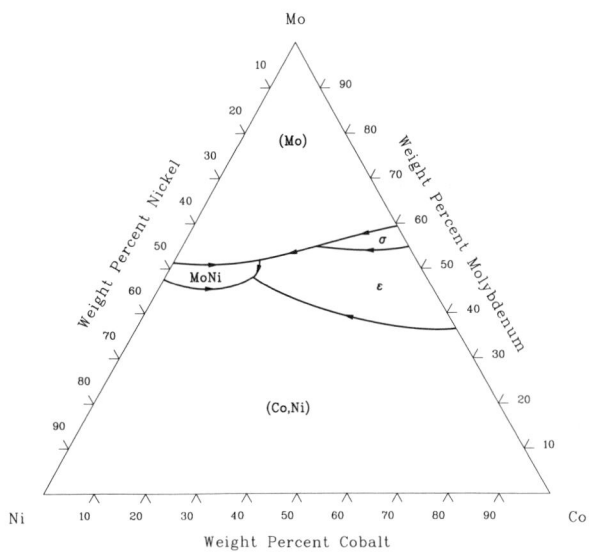

Co-Ni-Ti isothermal section at 1000 °C [83Gry]

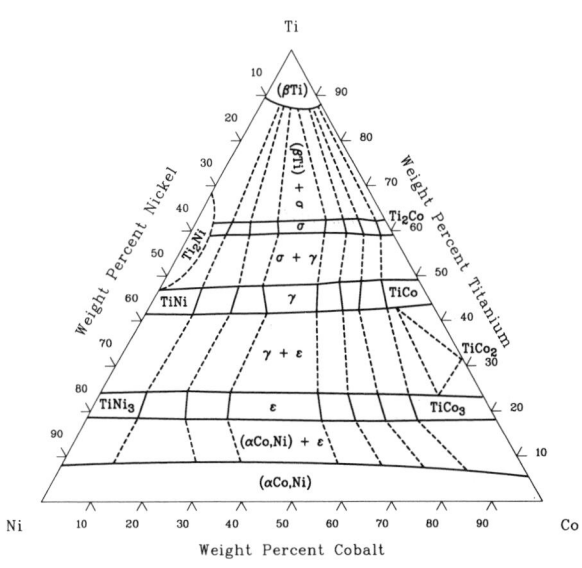

Co-Mo-Ni isothermal section at 1200 °C [52Das]

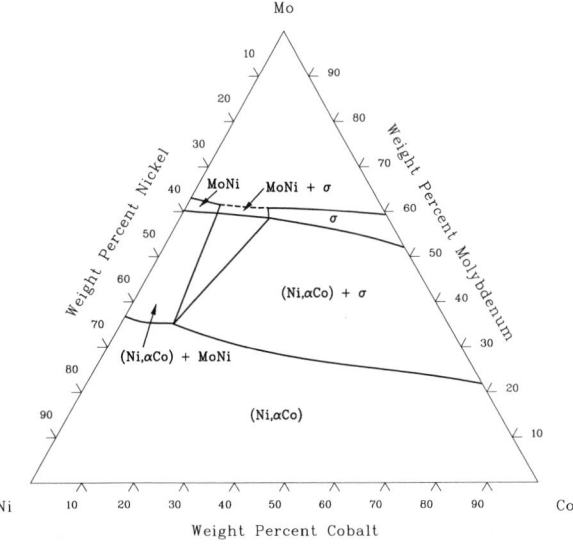

Co-Ni-Ti isothermal section at 800 °C [80Gry]

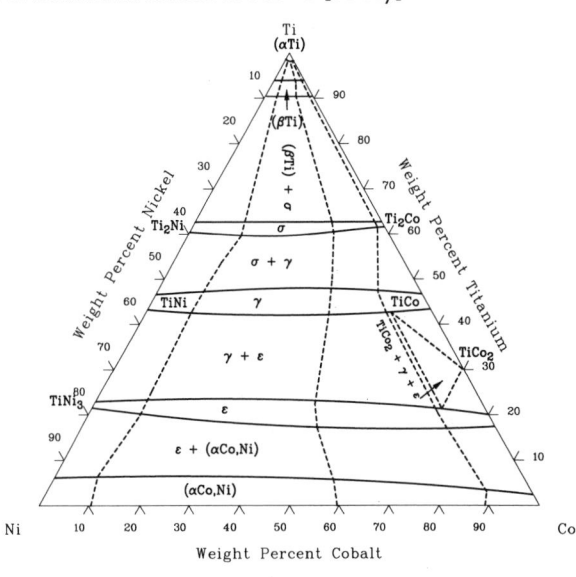

Cr-Fe-Mo liquidus projection [88Ray]

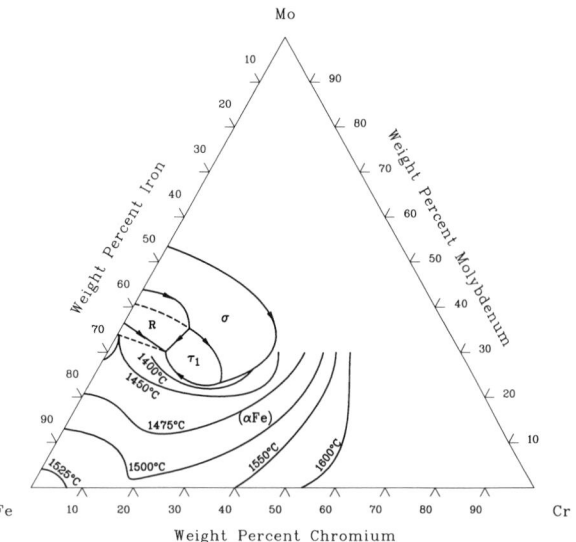

Cr-Fe-Mo isothermal section at 815 °C [88Ray]

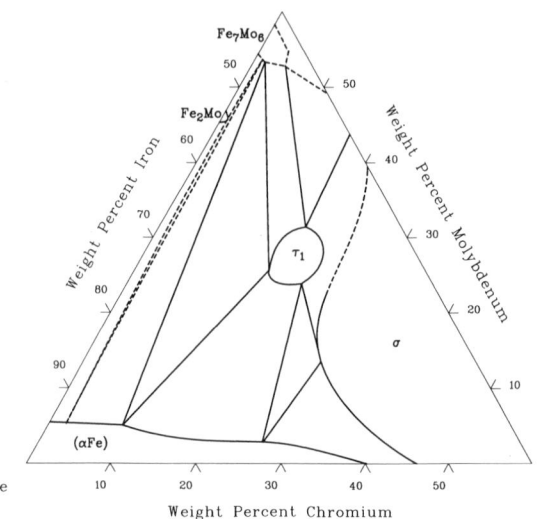

Cr-Fe-Mo isothermal section at 1250 °C [88Ray]

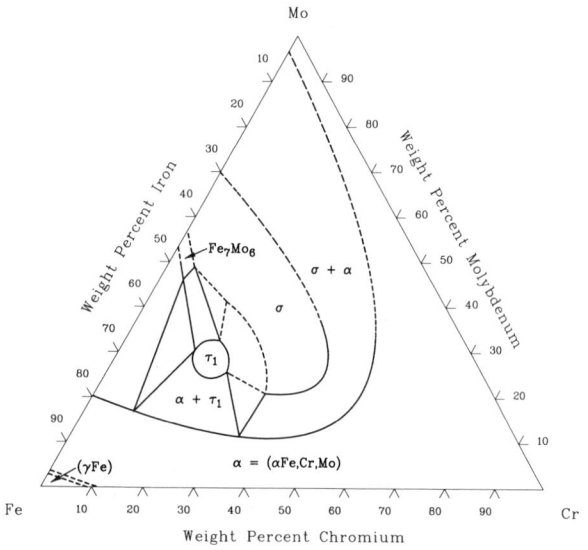

Cr-Fe-Mo [88Ray]

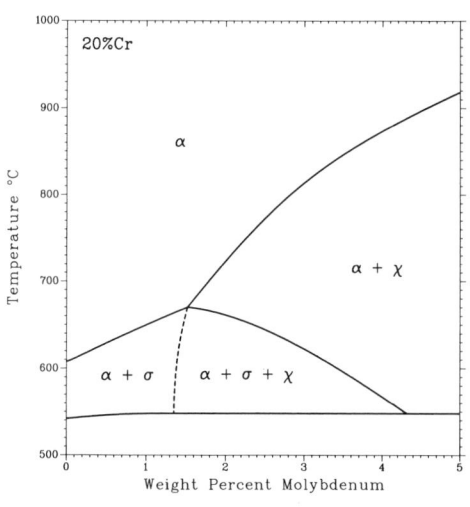

Cr-Fe-Mo isothermal section at 1100 °C [88Ray]

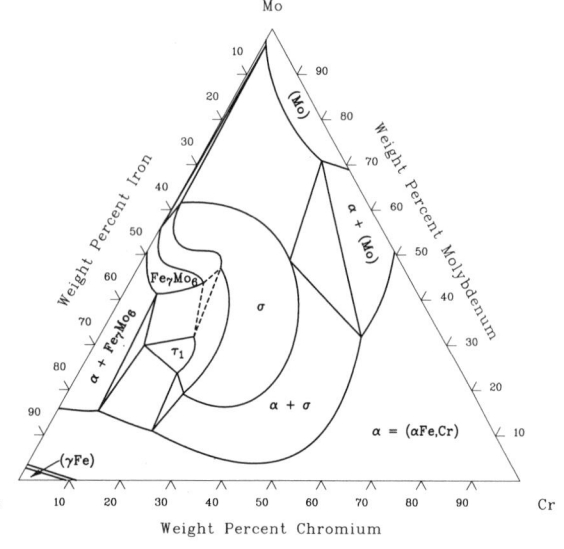

Cr-Fe-Mo [88Ray]

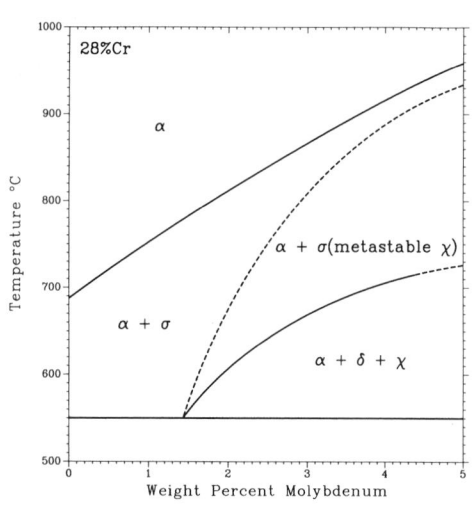

Cr-Fe-N liquidus projection [87Rag]

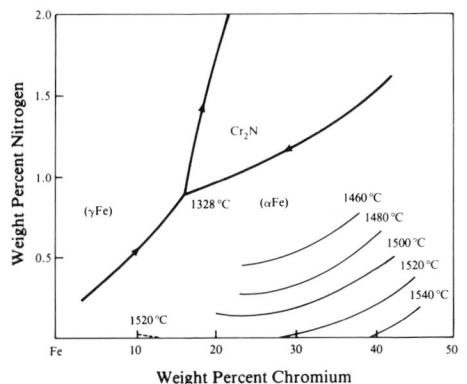

Cr-Fe-N isothermal section at 700 °C [87Rag]

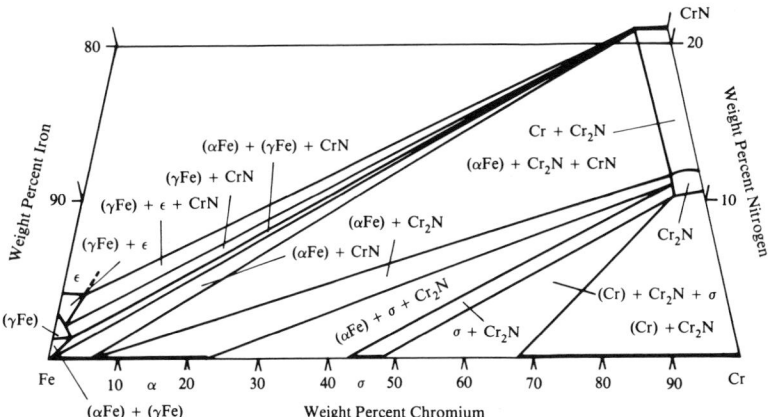

Cr-Fe-N isothermal section at 1200 °C [87Rag]

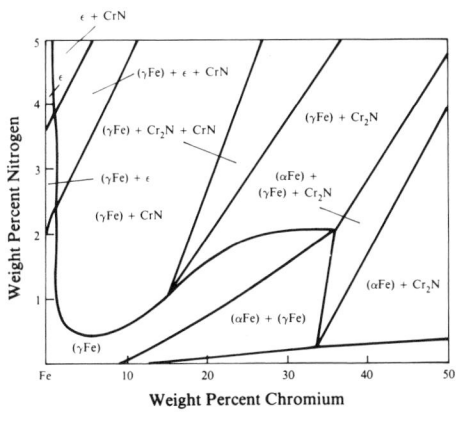

Cr-Fe-N isothermal section at 567 °C [87Rag]

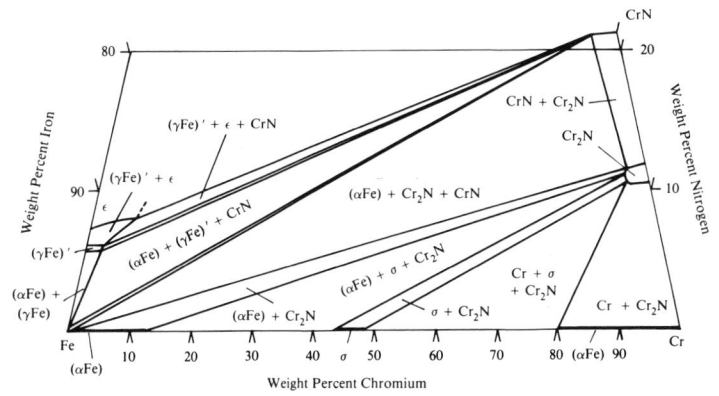

Cr-Fe-N isothermal section at 1000 °C [87Rag]

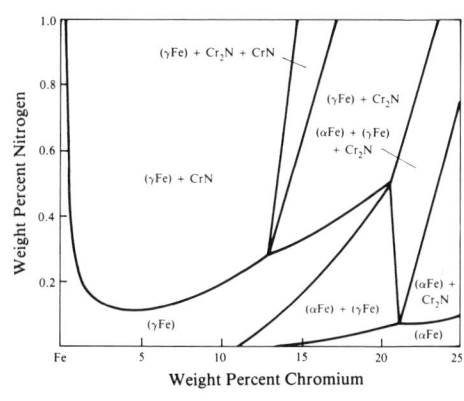

Cr-Fe-Ni liquidus projection [88Ray]

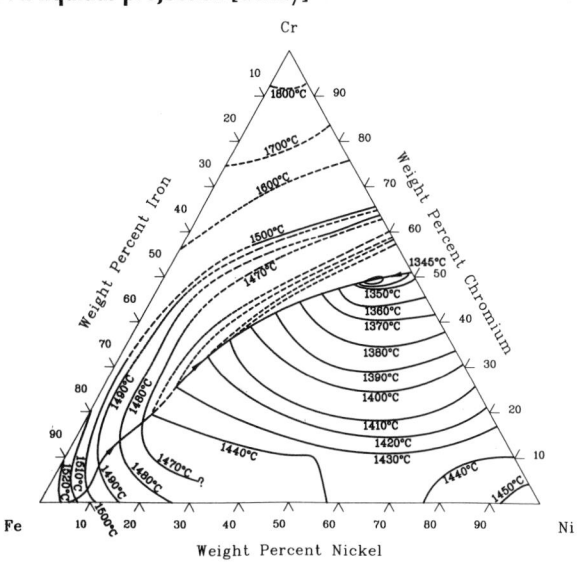

3•44/Ternary Alloy Phase Diagrams

Cr-Fe-Ni solidus projection [88Ray]

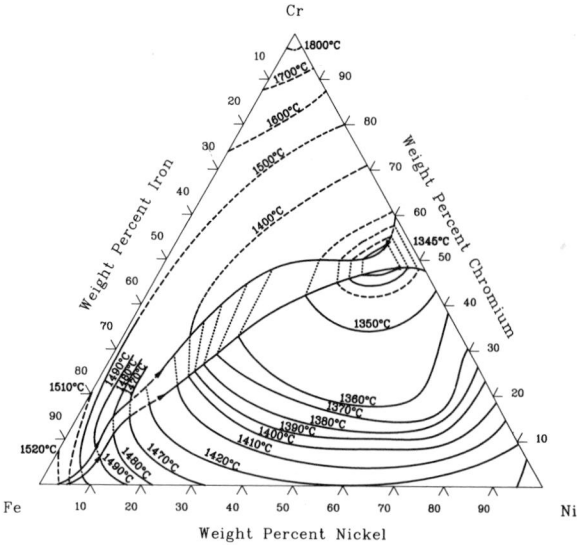

Cr-Fe-Ni isothermal section at 900 °C [88Ray]

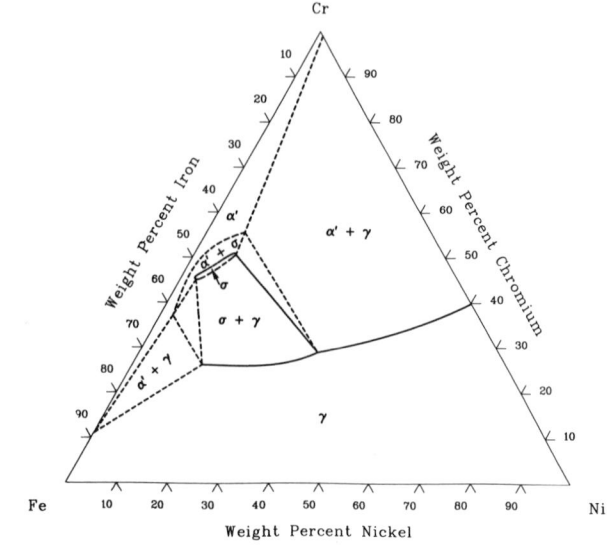

Note: α = (αFe,Cr); γ = (γFe,Ni)

Cr-Fe-Ni isothermal section at 1300 °C [88Ray]

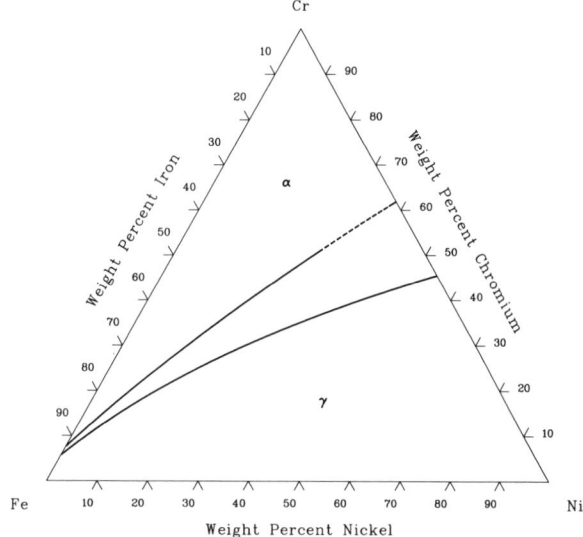

Note: α = (αFe,Cr); γ = (γFe,Ni)

Cr-Fe-Ni isothermal section at 800 °C [88Ray]

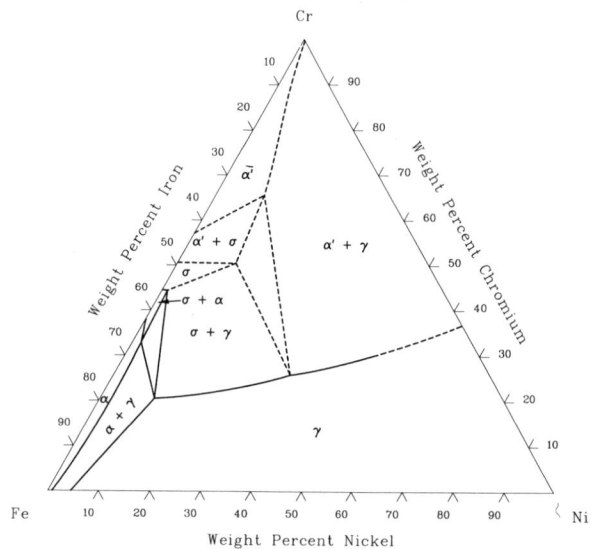

Note: α = (αFe,Cr); γ = (γFe,Ni)

Cr-Fe-Ni isothermal section at 1000 °C [88Ray]

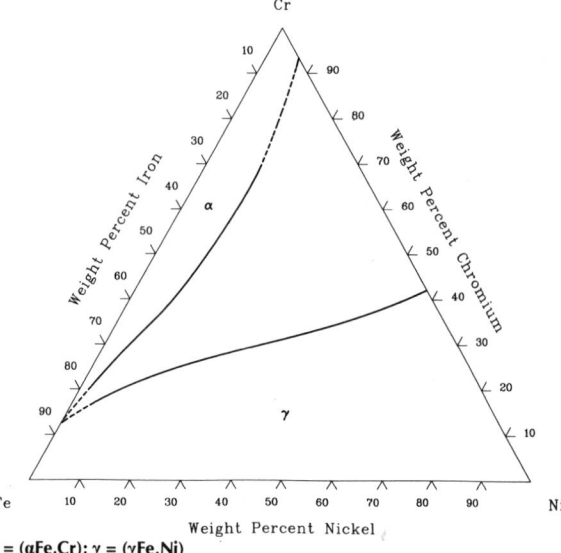

Note: α = (αFe,Cr); γ = (γFe,Ni)

Cr-Fe-Ni isothermal section at 650 °C [88Ray]

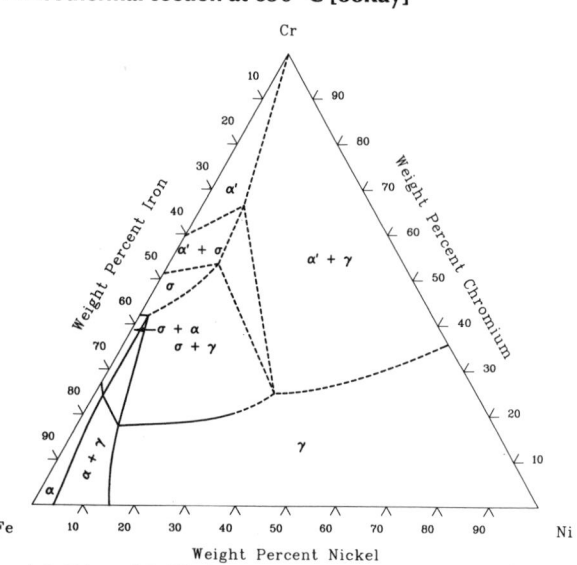

Note: α = (αFe,Cr); γ = (γFe,Ni)

Cr-Fe-W isothermal section at 1200 °C [88Ray]

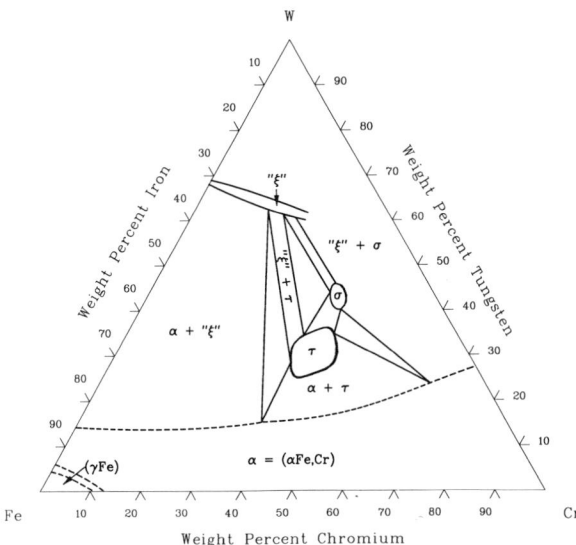

Cr-Fe-W isothermal section at 600 °C [88Ray]

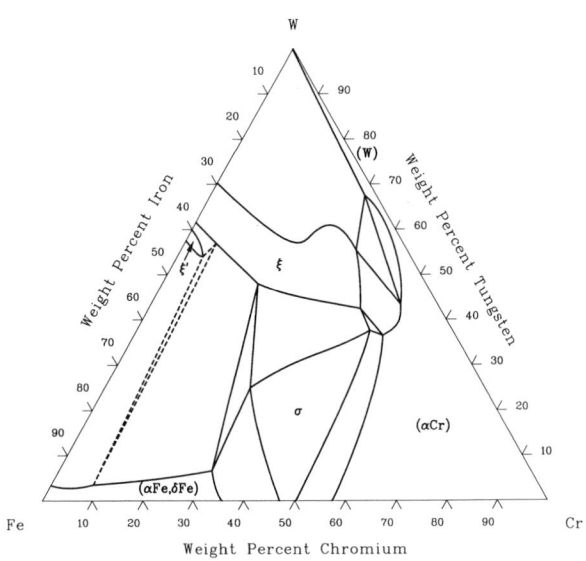

Cr-Mo-Ni liquidus projection [90Gup]

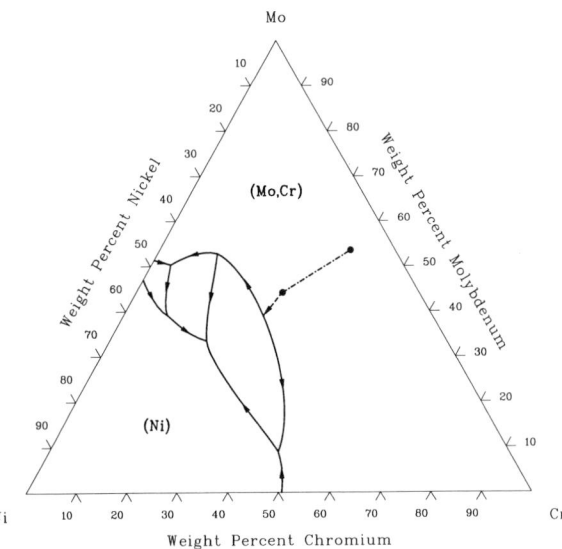

Cr-Mo-Ni isothermal section at 1250 °C [90Gup]

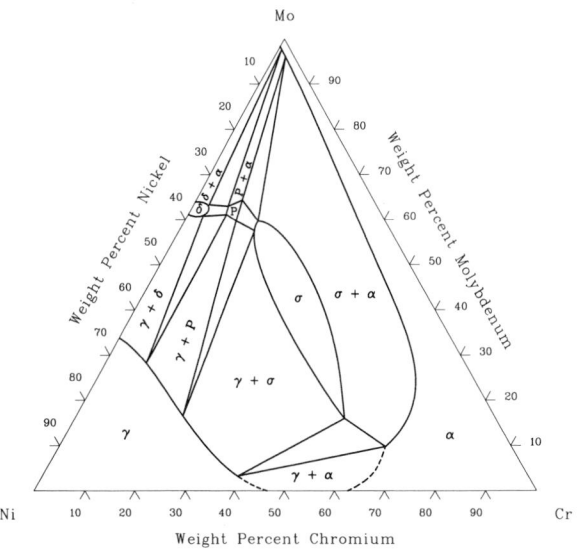

Cr-Mo-Ni isothermal section at 1200 °C [90Gup]

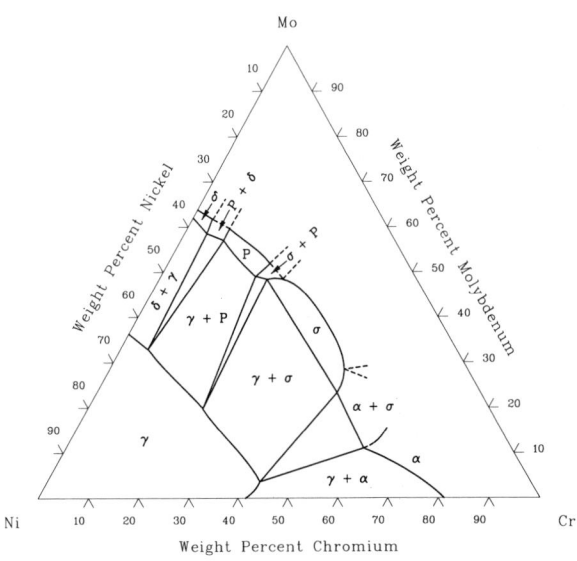

Cr-Mo-Ni isothermal section at 600 °C [90Gup]

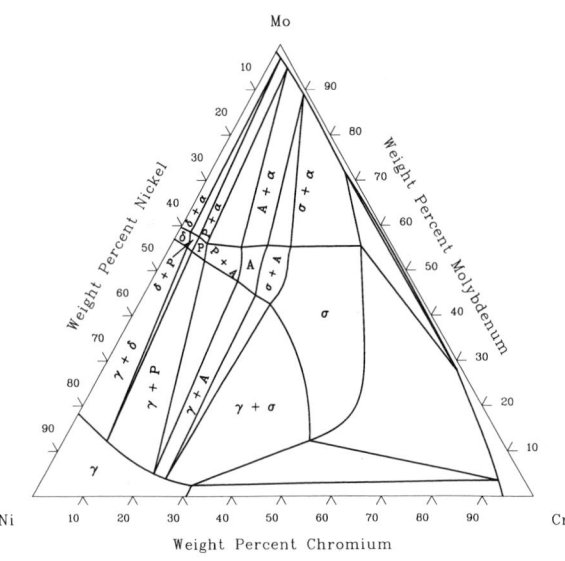

Cr-Mo-W isothermal section at 2227 °C [75Kau]

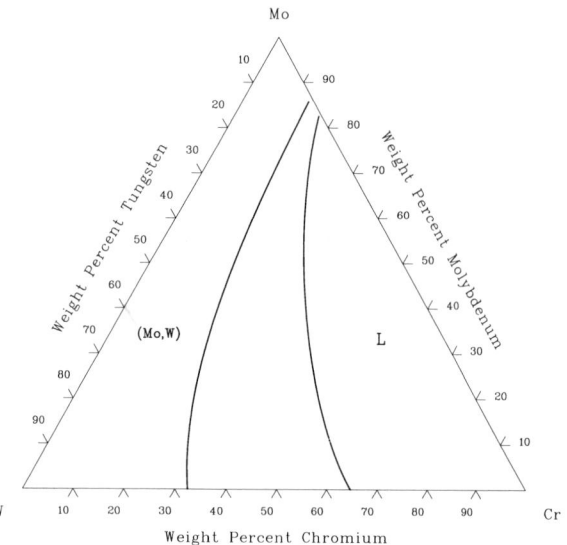

Cr-Nb-Ni liquidus projection [90Gup]

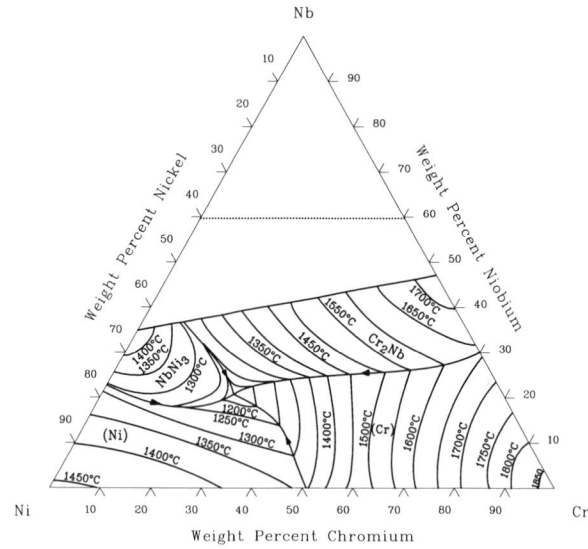

Cr-Mo-W isothermal section at 1300 °C [75Kau]

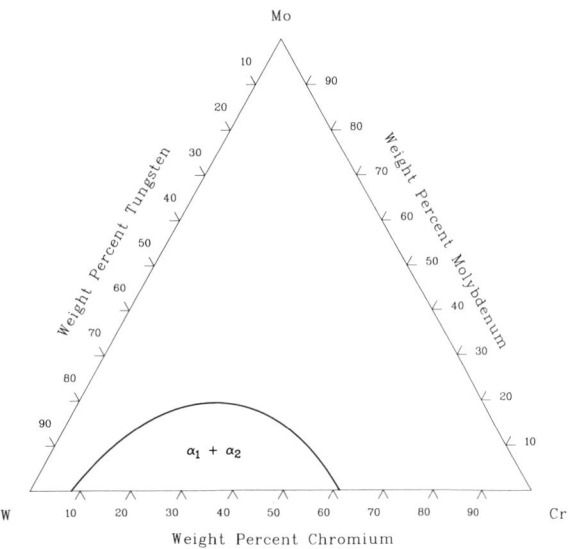

Cr-Nb-Ni isothermal section at 1200 °C [90Gup]

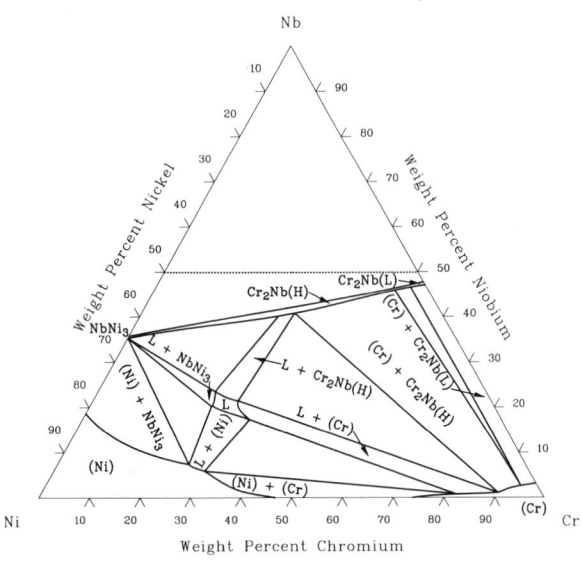

Cr-Mo-W isothermal section at 1000 °C [75Kau]

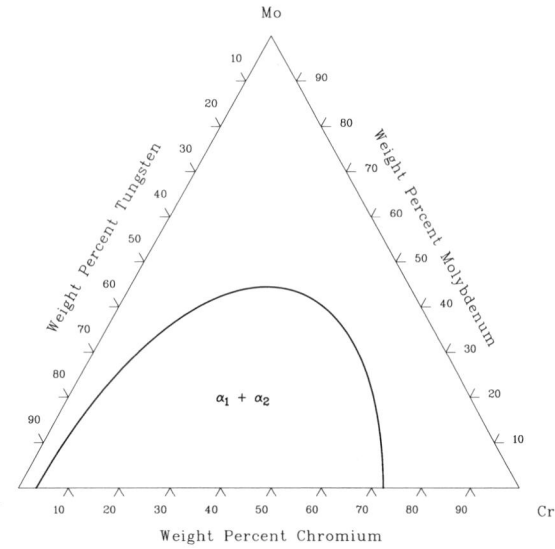

Cr-Nb-Ni isothermal section at 1175 °C [90Gup]

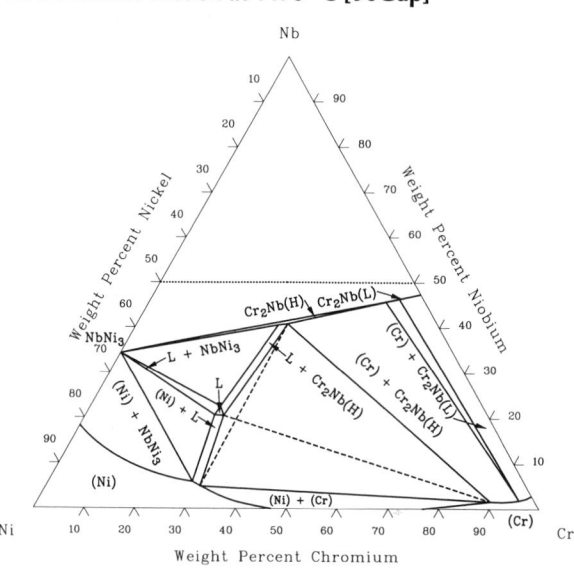

Cr-Nb-Ni isothermal section at 1100 °C [90Gup]

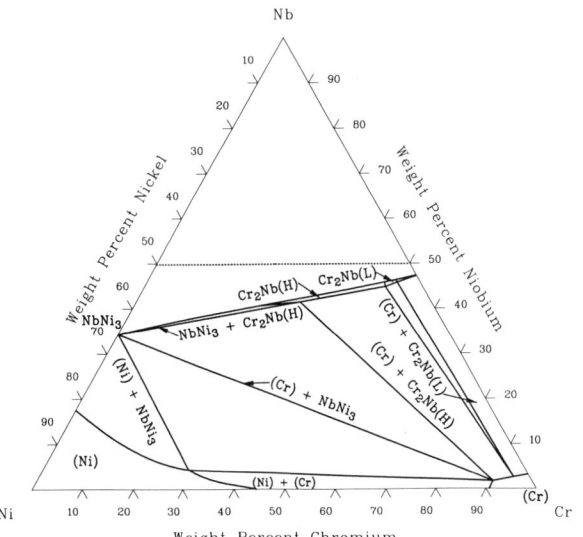

Cr-Ni-Ti liquidus projection [90Gup]

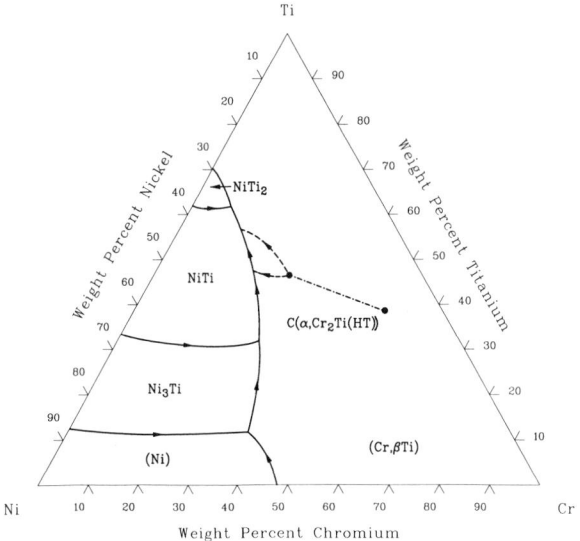

Cr-Nb-W isothermal section at 1500 °C [61Eng]

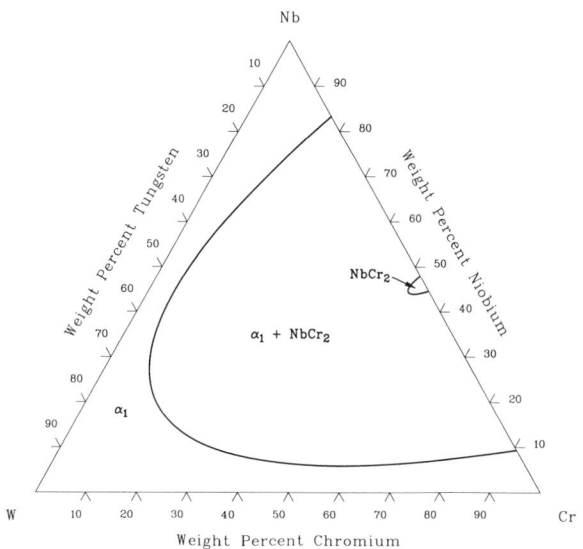

Cr-Ni-Ti isothermal section at 1352 °C [74Kau]

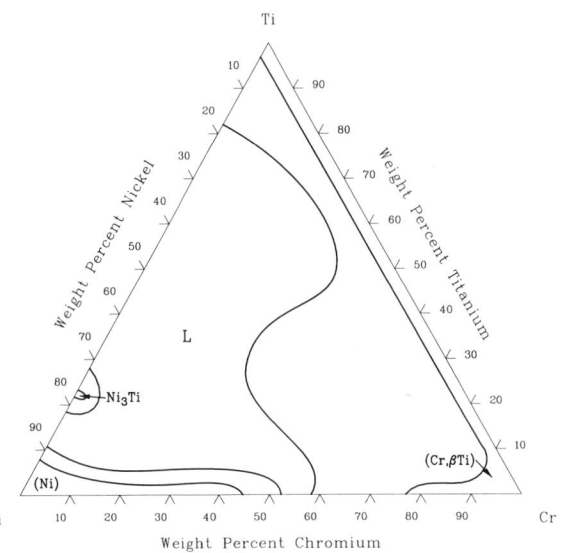

Cr-Nb-W isothermal section at 1000 °C [61Eng]

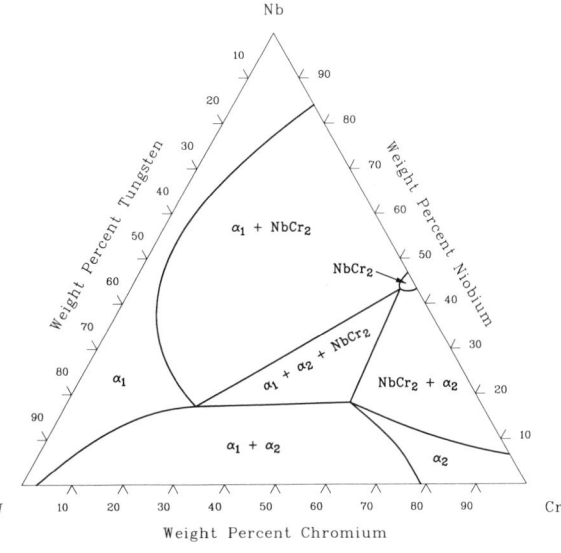

Cr-Ni-Ti isothermal section at 1277 °C [74Kau]

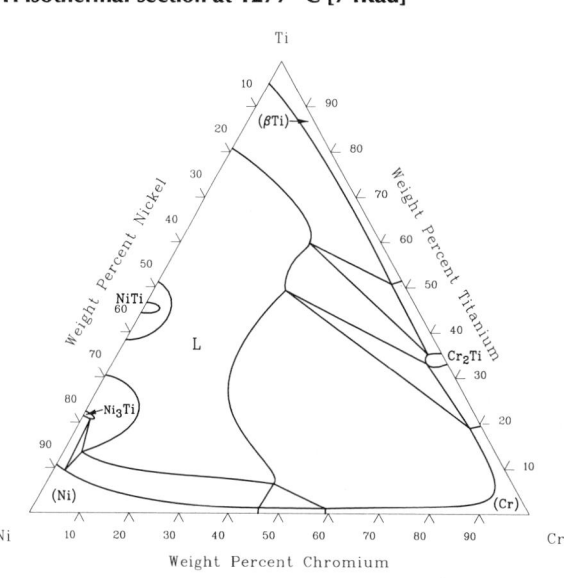

Cr-Ni-Ti isothermal section at 1027 °C [74Kau]

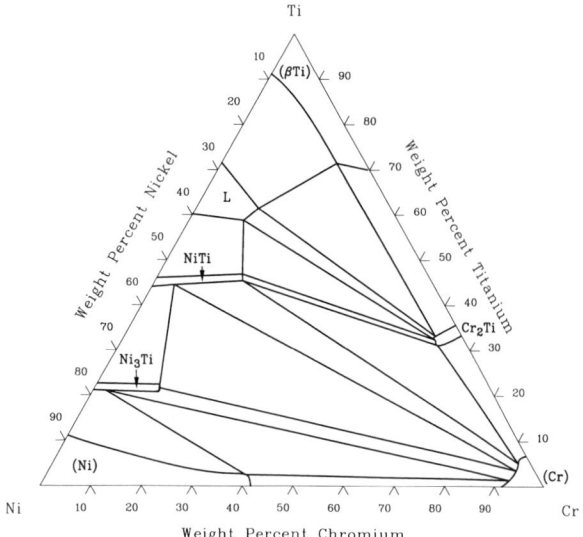

Cr-Ni-W isothermal section at 1000 °C [90Gup]

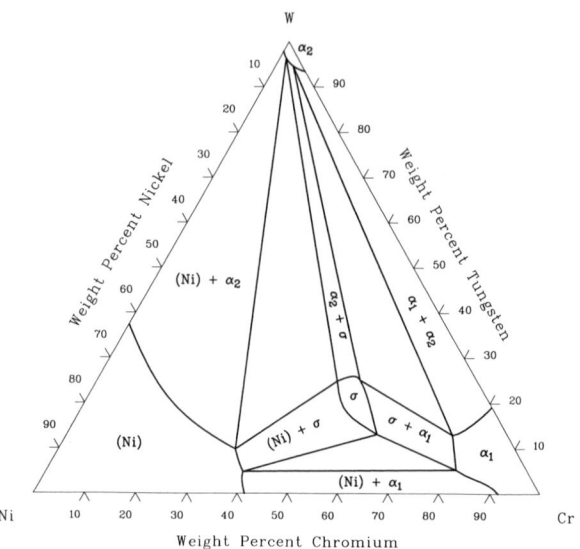

Cr-Ni-W liquidus projection [90Gup]

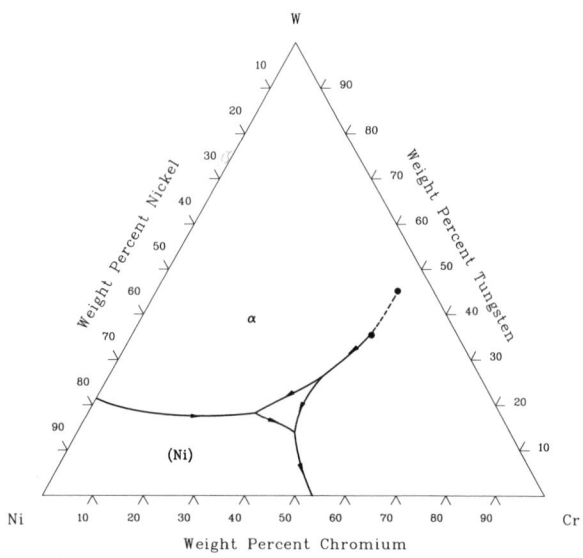

Cr-Ni-W isothermal section at 900 °C [90Gup]

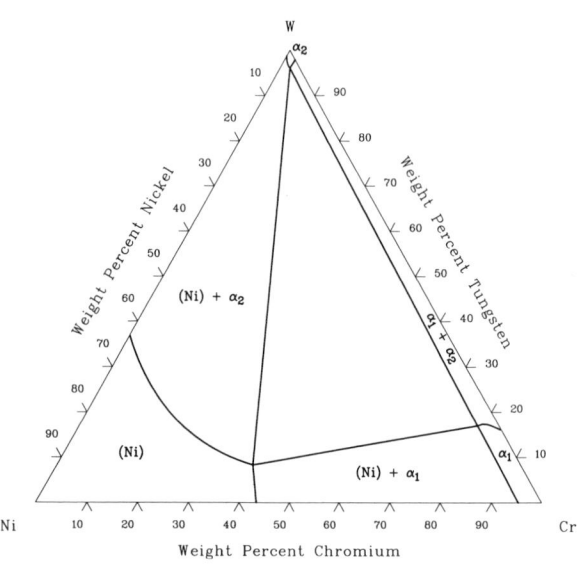

Cr-Ni-W isothermal section at 1250 °C [90Gup]

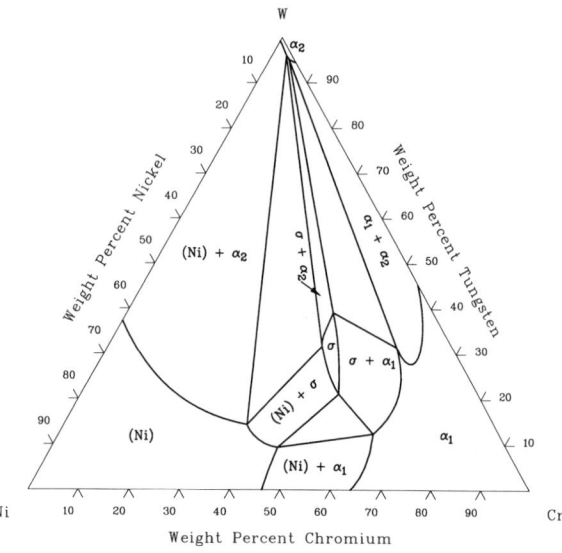

Cr-Ni-W isothermal section at 800 °C [990Gup]

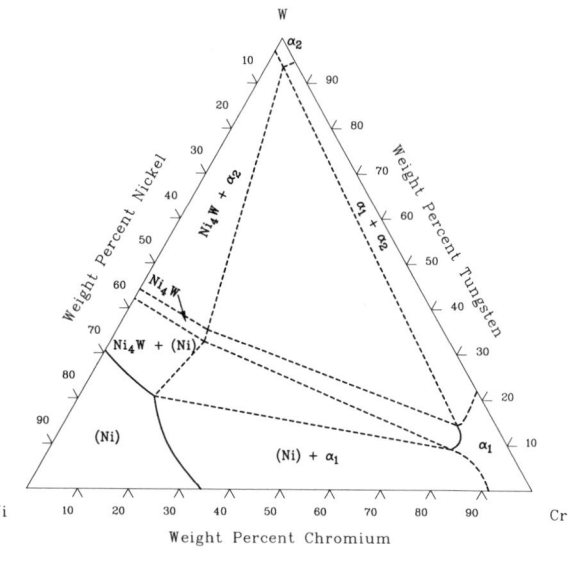

Cr-Ti-W isothermal section at 800 °C [58Bag]

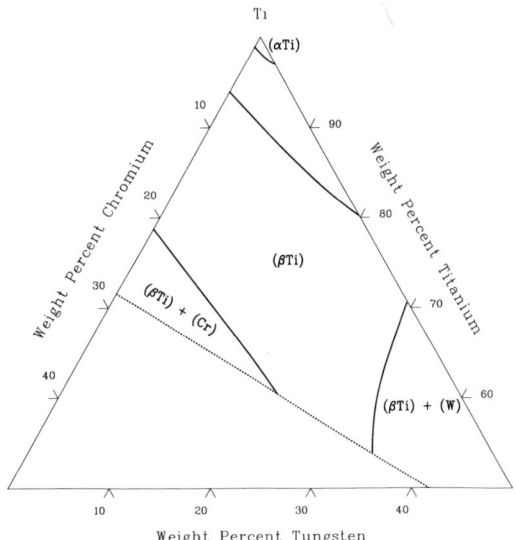

Cu-Fe-Ni liquidus projection [90Gup]

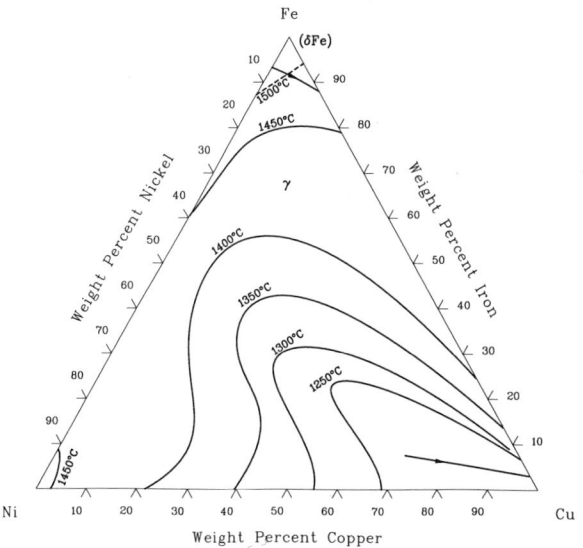

Cr-Ti-W isothermal section at 750 °C [58Bag]

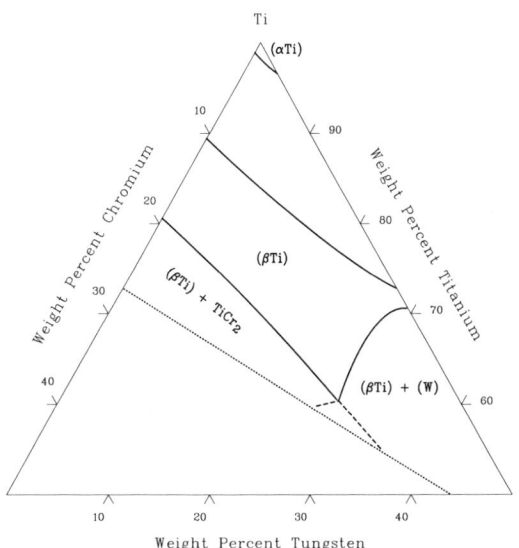

Cu-Fe-Ni miscibility gap [90Gup]

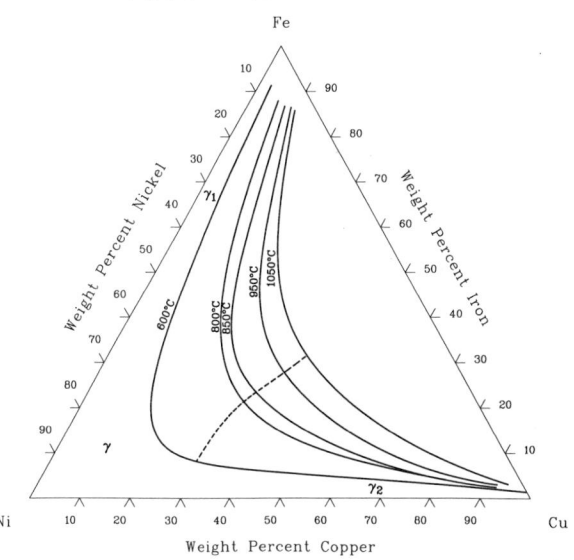

Cr-Ti-W isothermal section at 600 °C [58Bag]

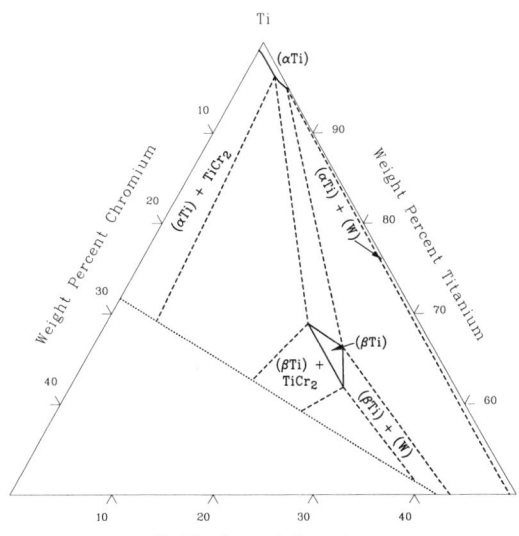

Cu-Fe-Ni isothermal section at 400 °C [90Gup]

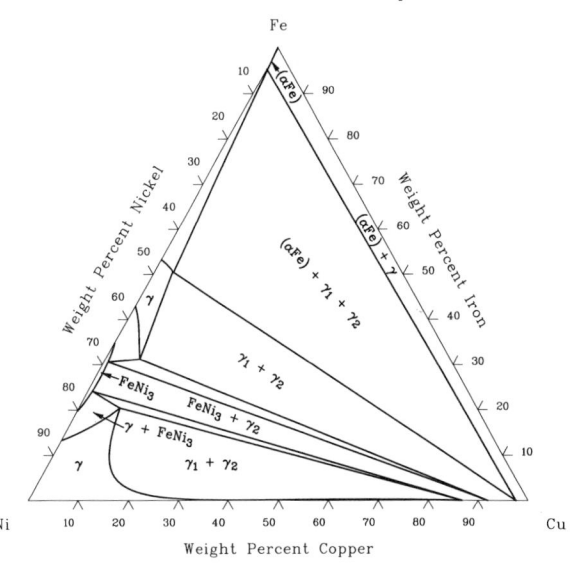

Cu-Fe-Ni isothermal section at 20 °C [90Gup]

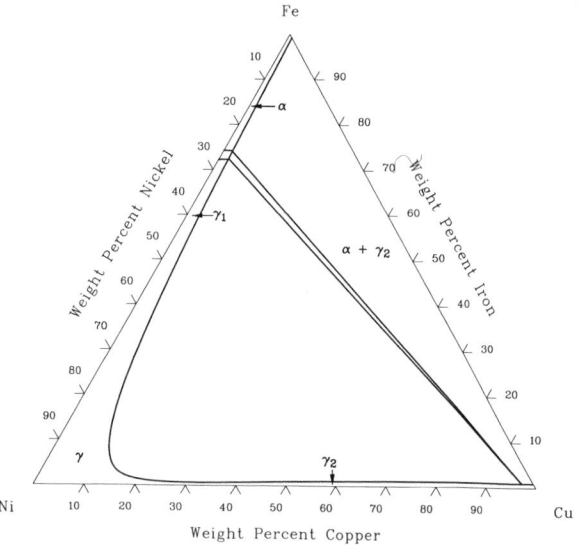

Cu-Ni-Sn solidus projection [90Gup]

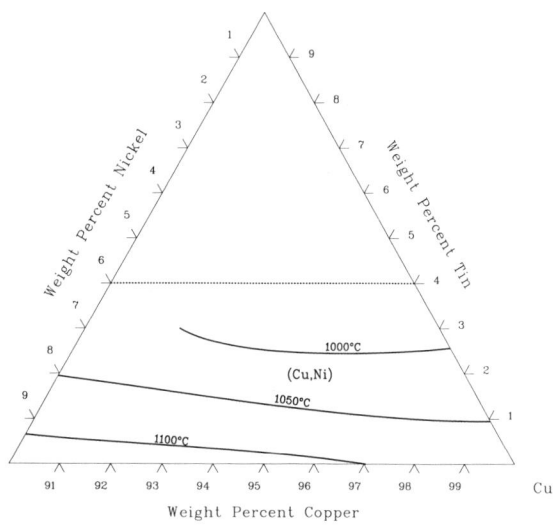

Cu-Fe-Ni [90Gup]

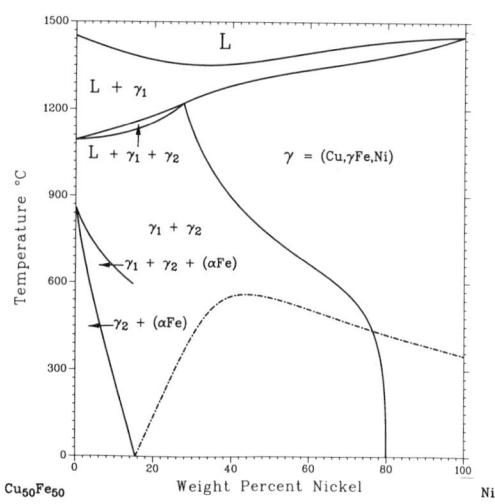

Cu-Ni-Sn isothermal section at 700 °C [90Gup]

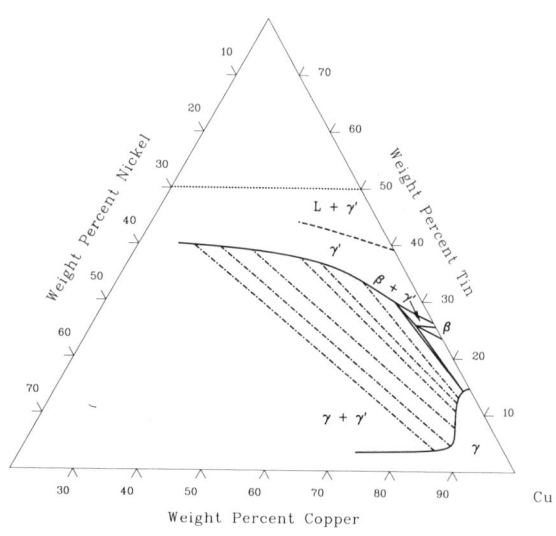

Cu-Ni-Sn liquidus projection [90Gup]

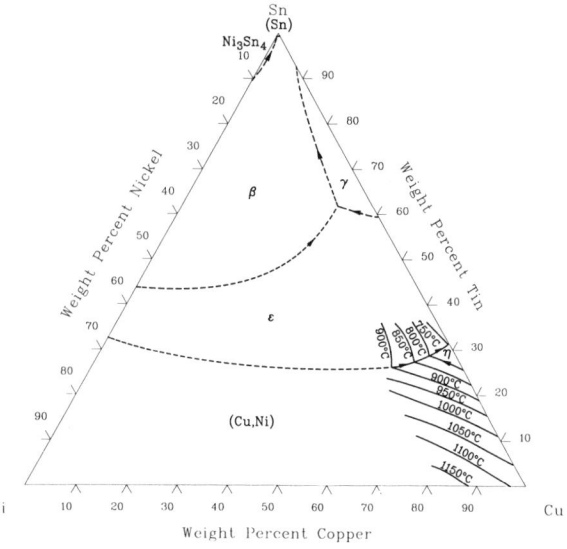

Cu-Ni-Sn isothermal section at 550 °C [90Gup]

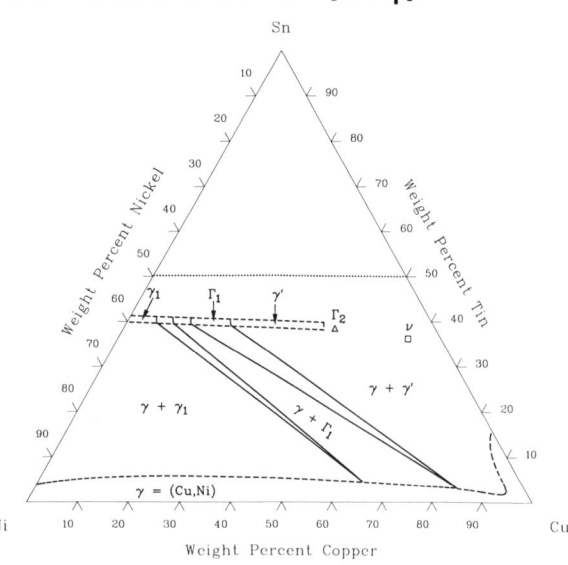

Cu-Ni-Zn liquidus projection [79Cha]

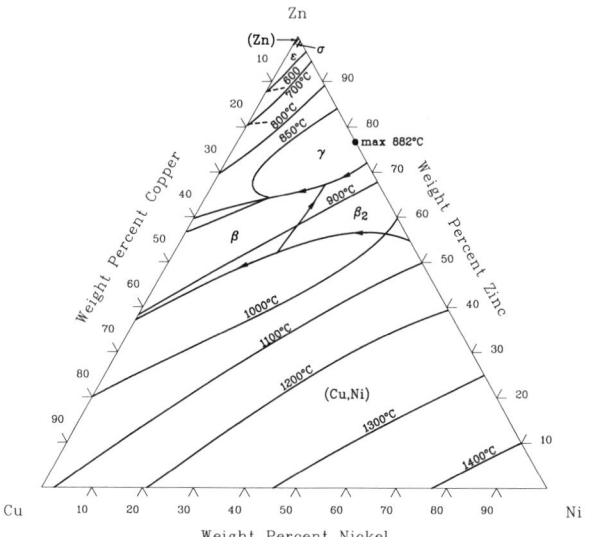

Cu-Ni-Zn isothermal section at 20 °C [73Lev]

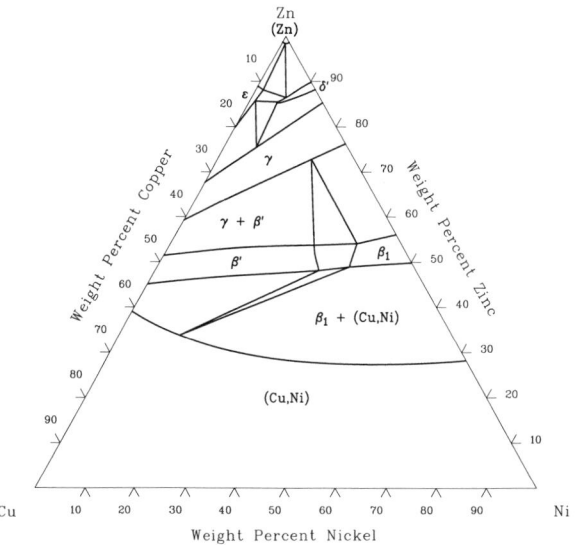

Cu-Ni-Zn isothermal section at 775 °C [79Cha]

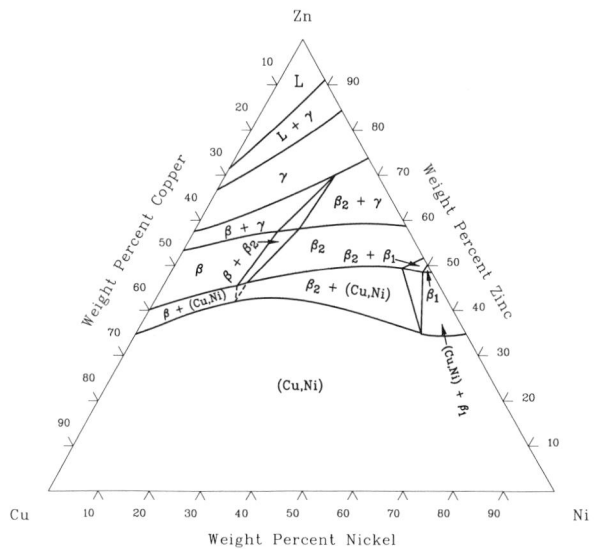

Cu-Pb-Zn liquidus projection [79Cha]

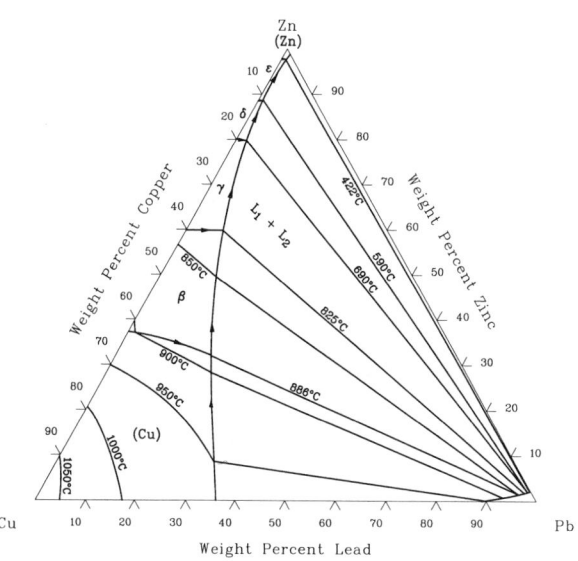

Cu-Ni-Zn isothermal section at 650 °C [73Lev]

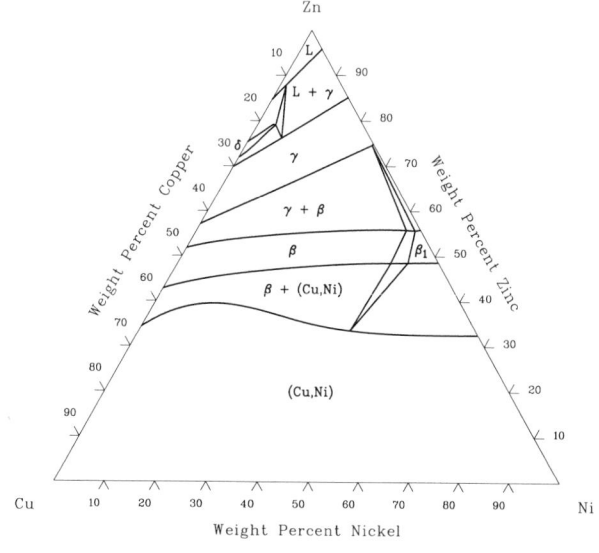

Cu-Pb-Zn (Pb) liquidus projection [79Cha]

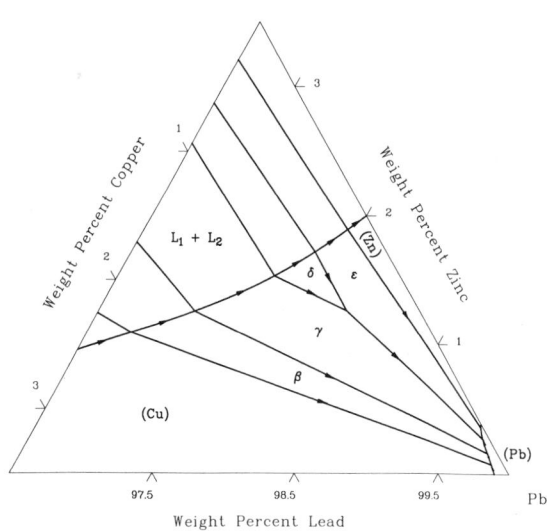

Cu-Pb-Zn isothermal section at 25 °C [29Bau]

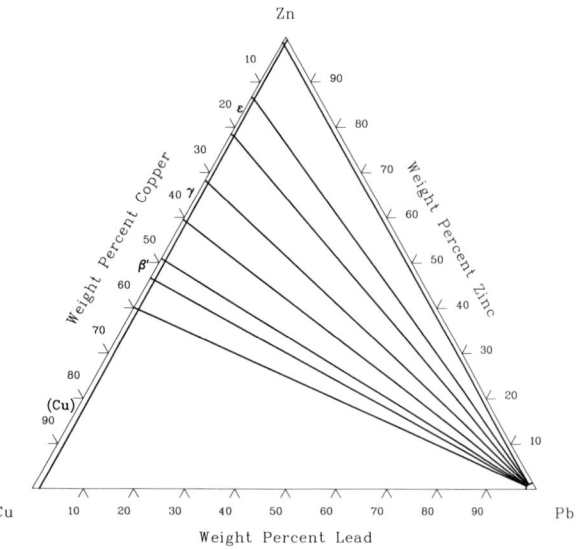

Cu-Sb-Sn phases present at temperatures below the reactions in the solid state [73Bla]

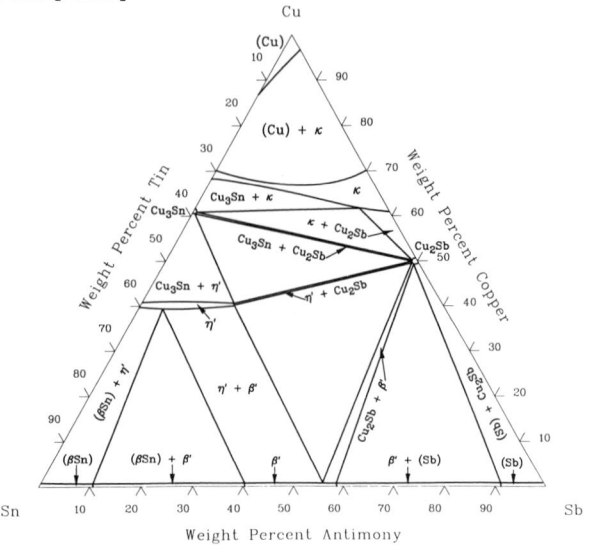

Cu-Sb-Sn liquidus projection [73Bla]

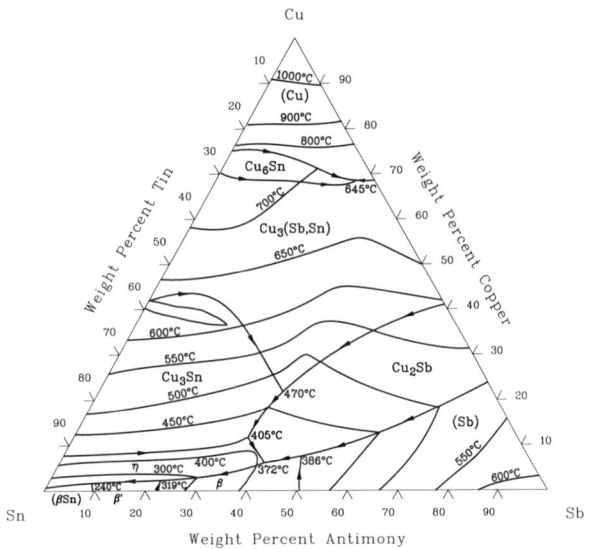

Cu-Sn-Zn liquidus projection [73Smi]

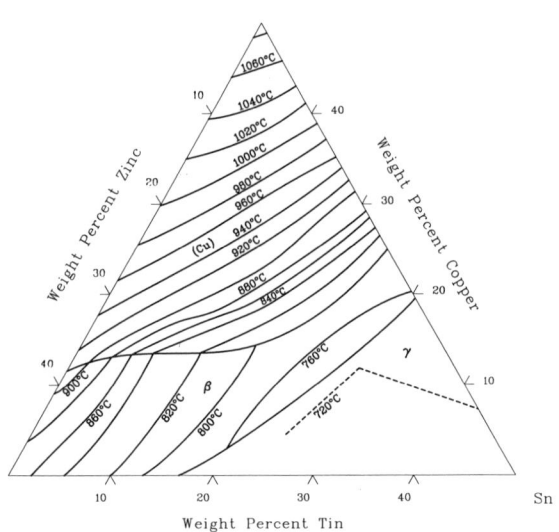

Cu-Sb-Sn (Sn) liquidus projection [73Bla]

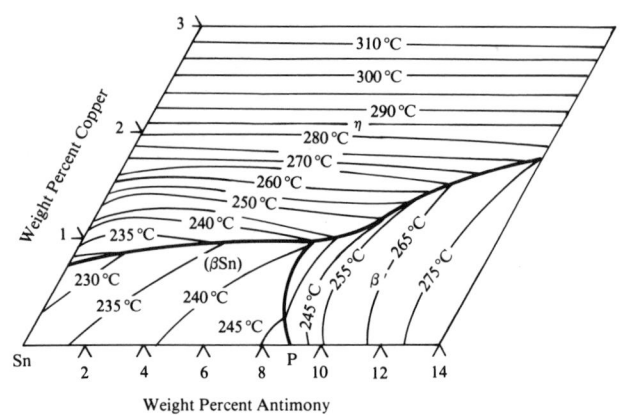

Cu-Sn-Zn isothermal section at 500 °C [73Smi]

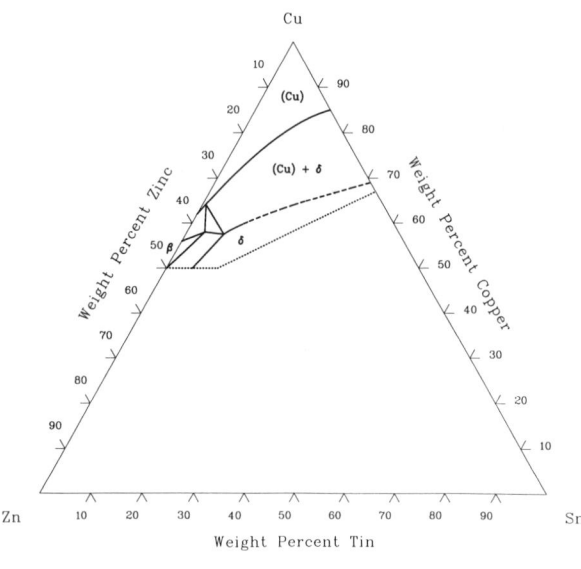

Fe-Mn-Ni liquidus projection [88Ray]

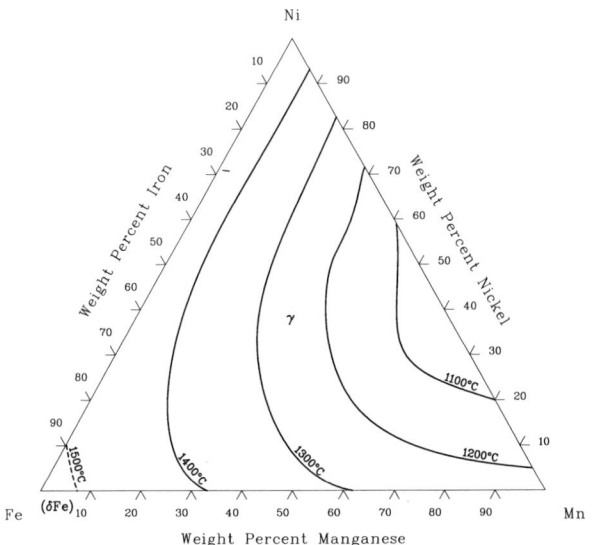

Fe-Mn-Ni isothermal section at 650 °C [89Har]

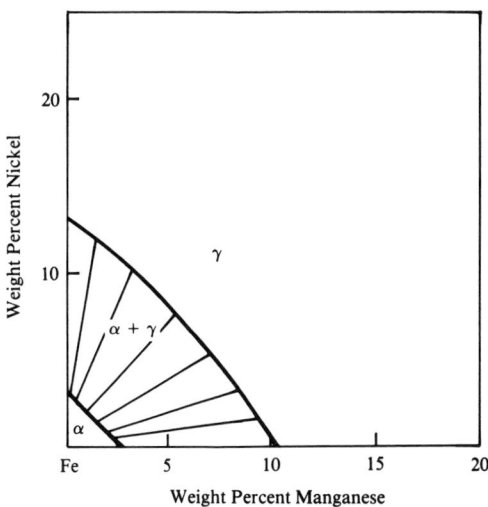

Fe-Mn-Ni isothermal section at 850 °C [89Har]

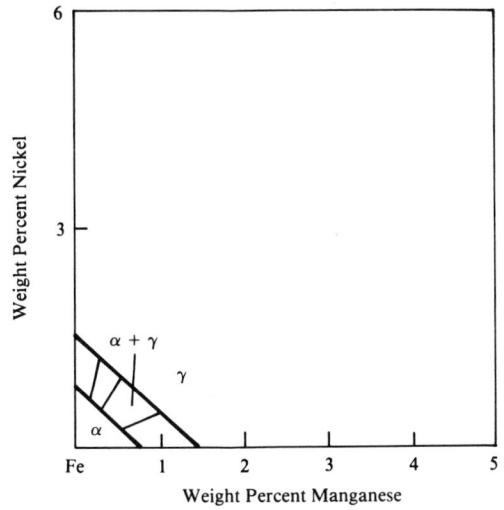

Fe-Mn-Ni isothermal section at 550 °C [89Har]

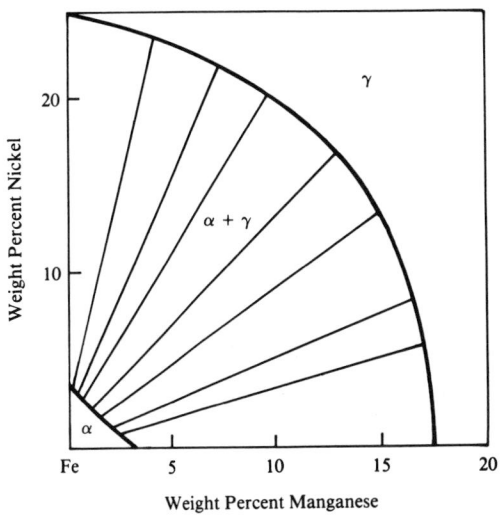

Fe-Mn-Ni isothermal section at 750 °C [89Har]

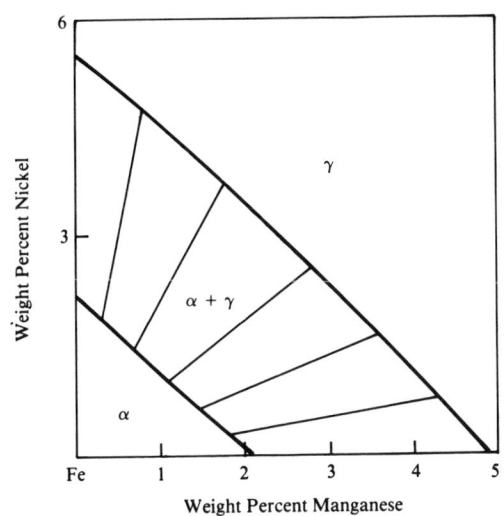

Fe-Mo-Nb isothermal section at 1250 °C [89Har]

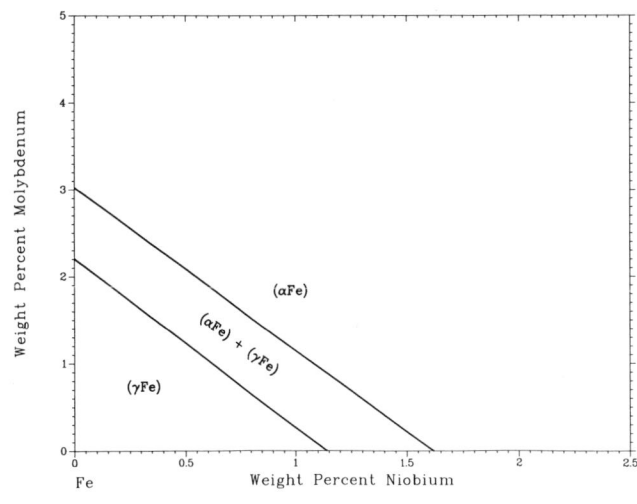

Fe-Mo-Nb isothermal section at 1150 °C [89Har]

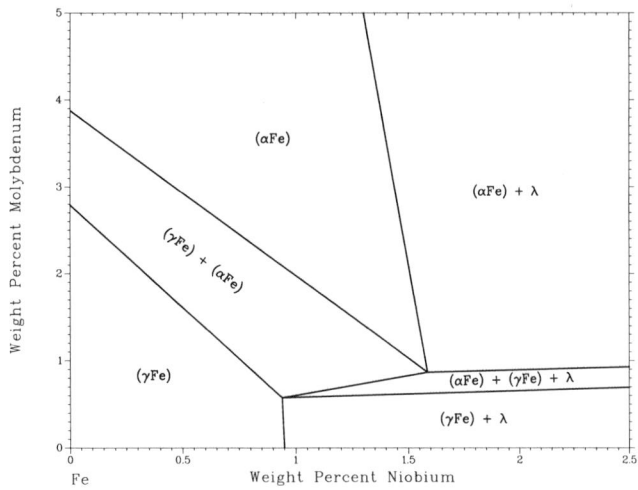

Fe-Mo-Nb isothermal section at 1050 °C [89Har]

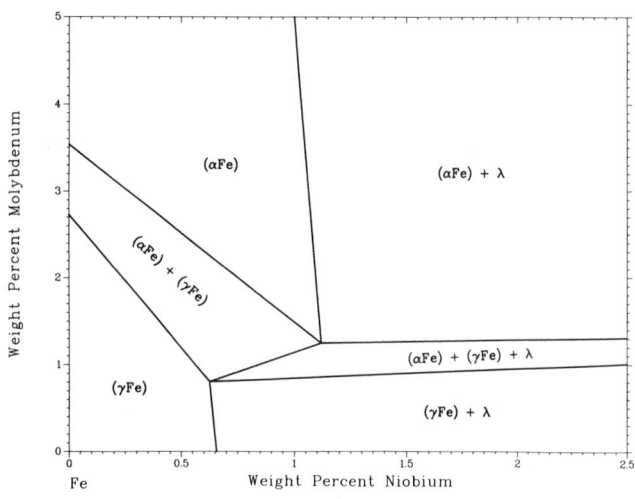

Fe-Mo-Nb isothermal section at 950 °C [89Har]

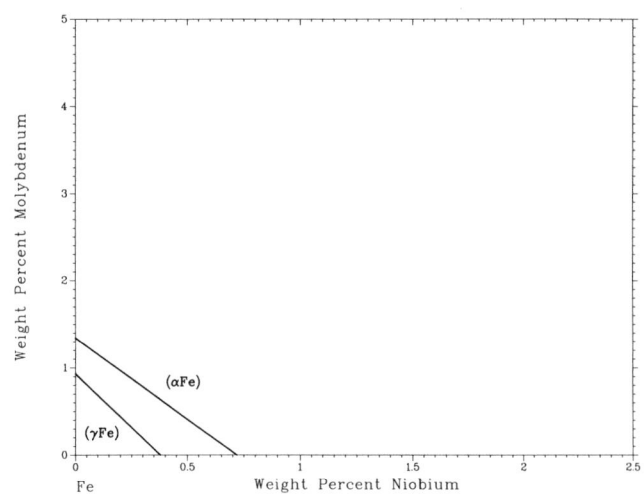

Fe-Mo-Nb isothermal section at 900 °C [87Smi]

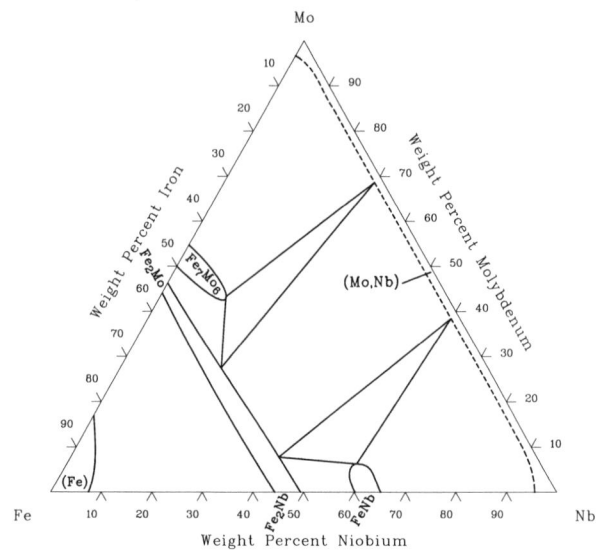

Fe-Mo-Ni liquidus projection [34Kos]

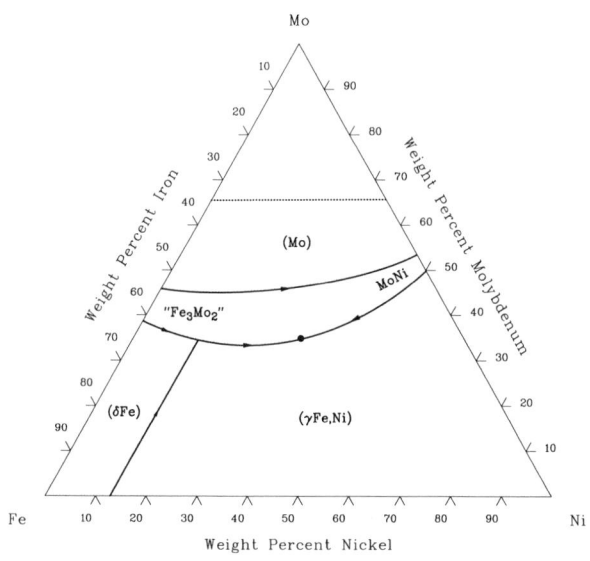

Fe-Mo-Ni isothermal section at 1200 °C [52Das]

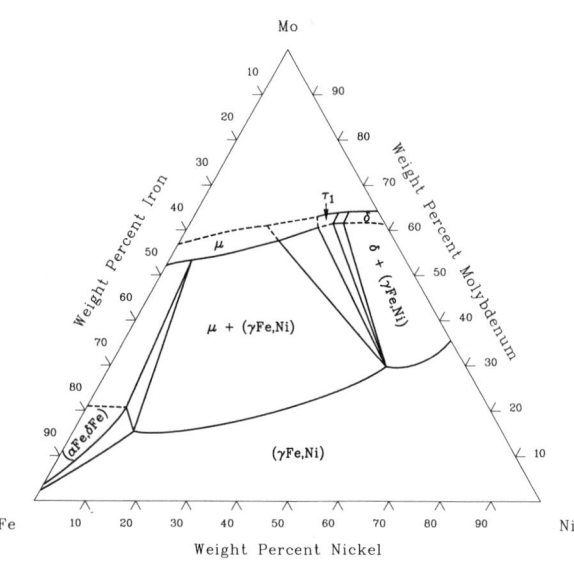

Fe-Mo-Ni isothermal section at 1100 °C [88Ray]

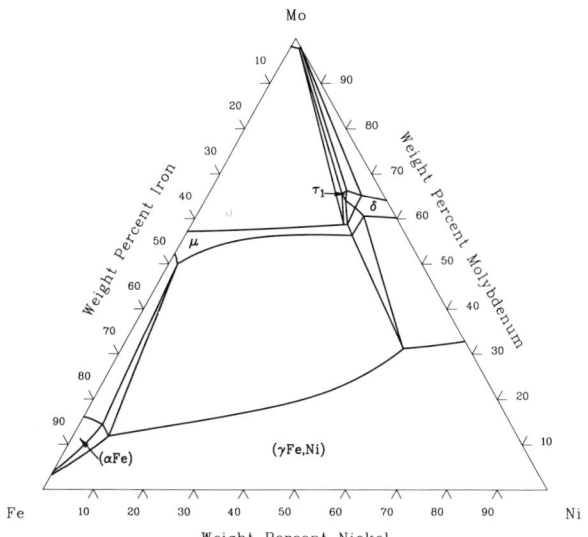

Fe-Ni-W isothermal section at 1465 °C [88Ray]

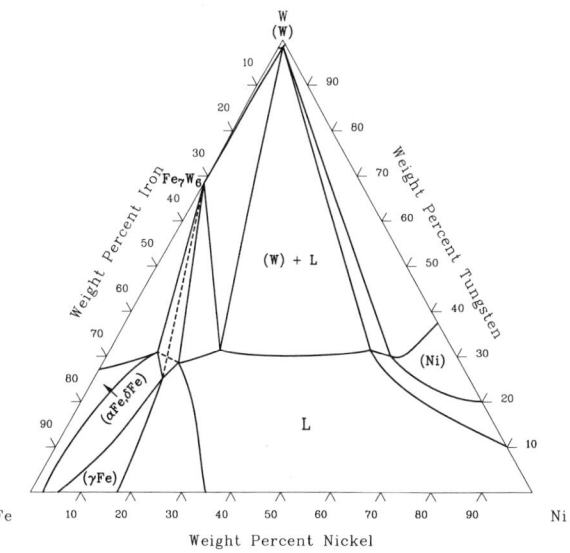

Fe-Ni-W liquidus and solidus projections [88Ray]

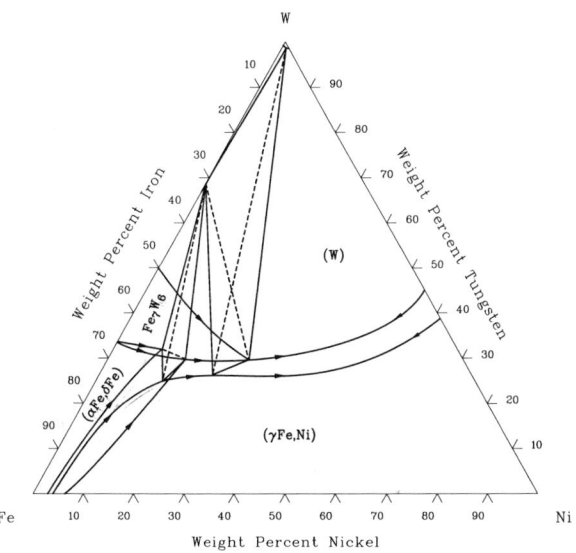

Fe-Ni-W isothermal section at 1455 °C [88Ray]

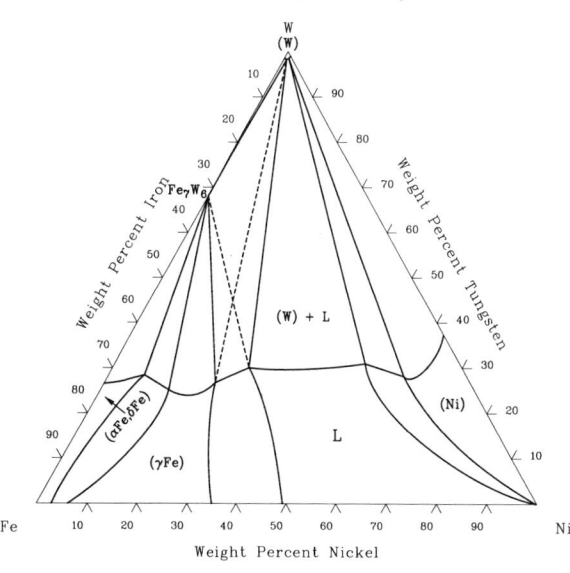

Fe-Ni-W isothermal section at 1500 °C [88Ray]

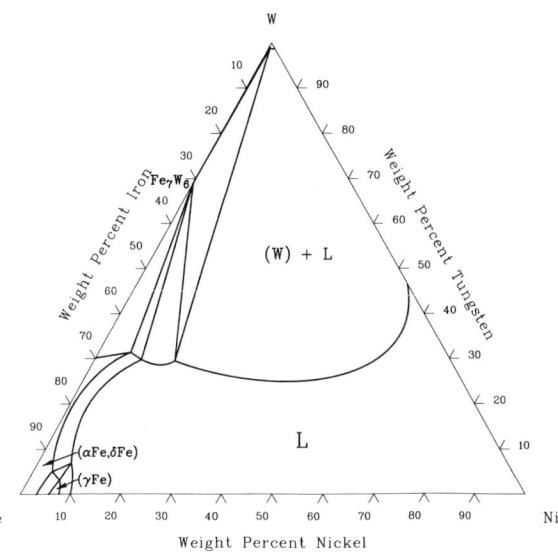

Fe-Ni-W isothermal section at 1400 °C [88Ray]

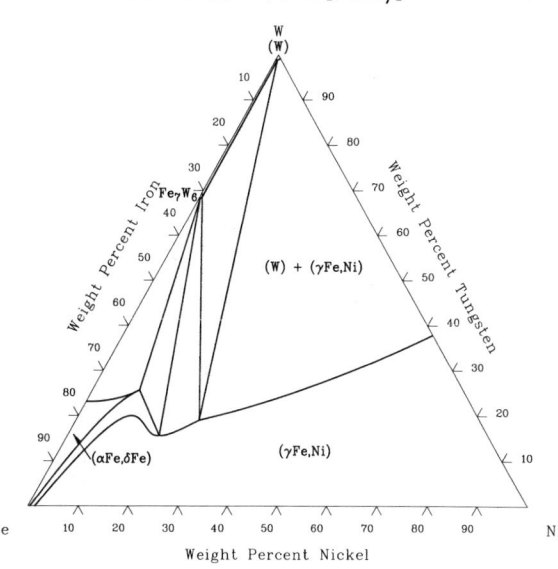

Mo-Nb-Ti isothermal section at 1100 °C [58Kor]

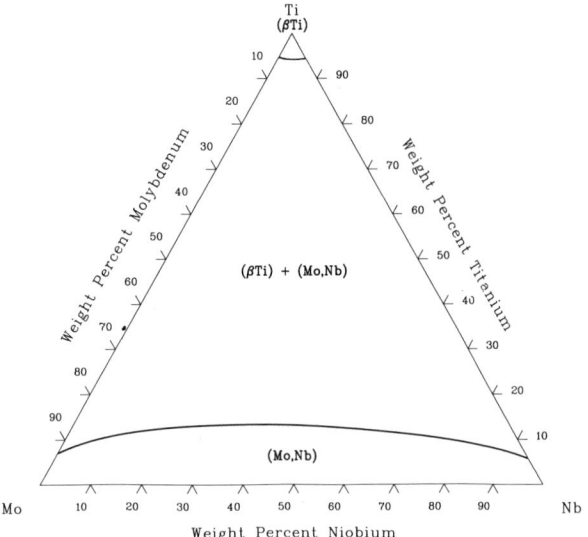

Mo-Ni-Ti isothermal section at 900 °C [84Ere]

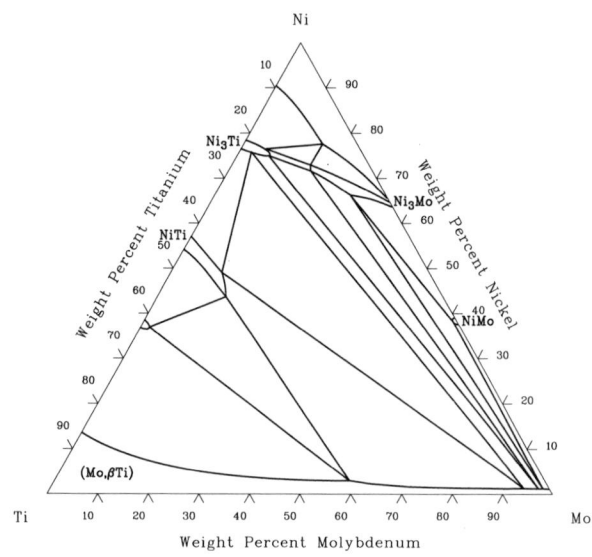

Mo-Nb-Ti isothermal section at 600 °C [58Kor]

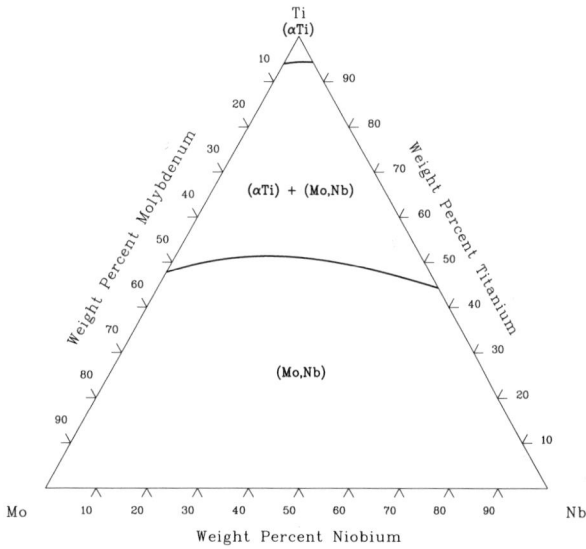

Mo-Ni-W isothermal section at 1000 °C [80Mas]

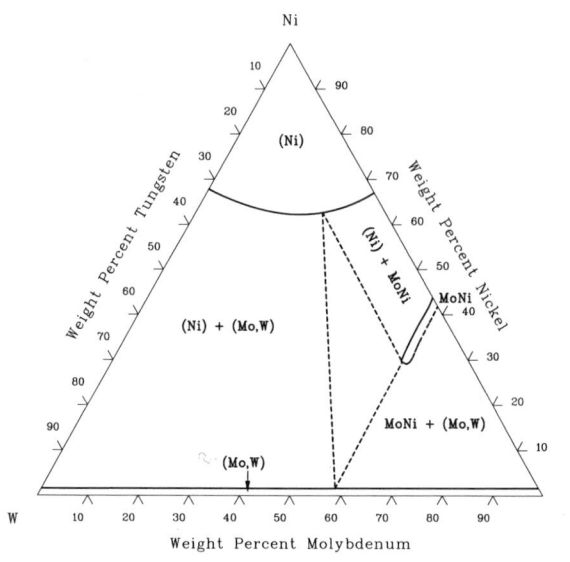

Mo-Ni-Ti isothermal section at 1200 °C [86Pri]

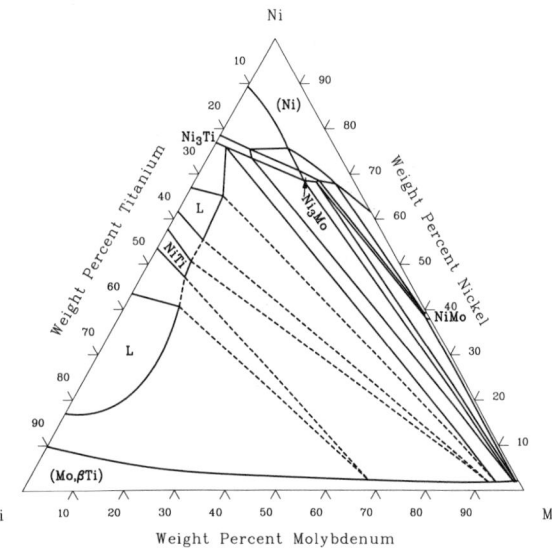

Mo-Ni-W isothermal section at 700 °C [85Mes]

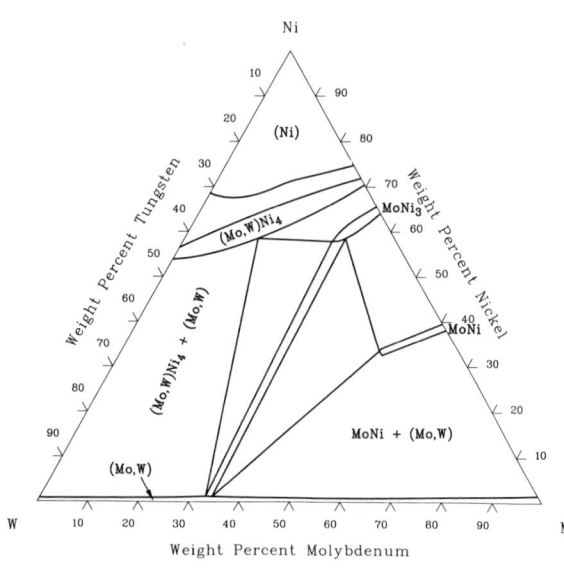

Mo-Ti-W isothermal section at 2227 °C [75Kau]

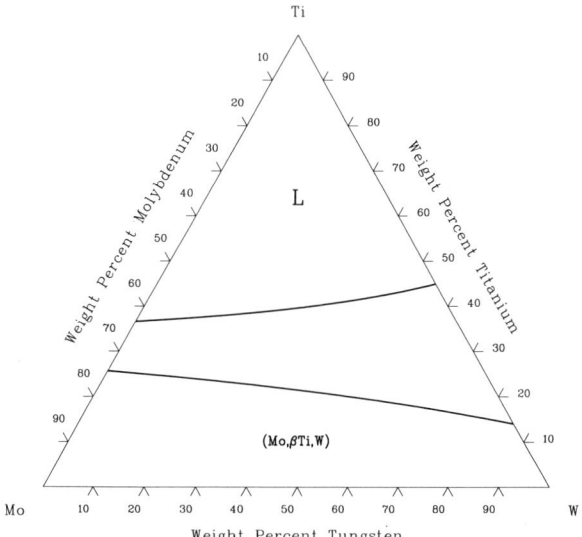

Nb-Ti-W isothermal section at 600 °C [77Lev]

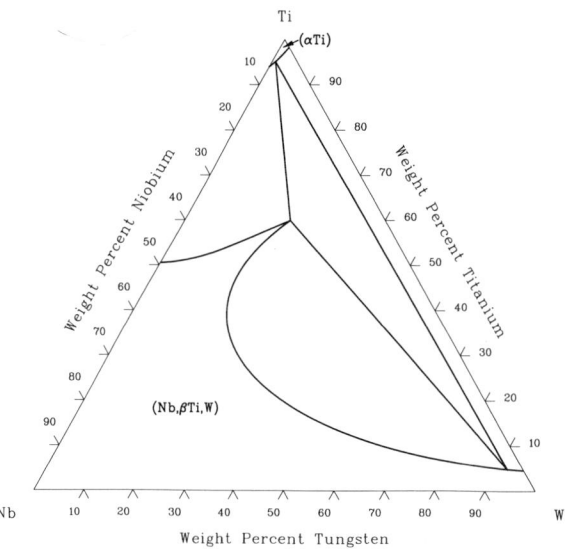

Mo-Ti-W isothermal section at 1000 °C [75Kau]

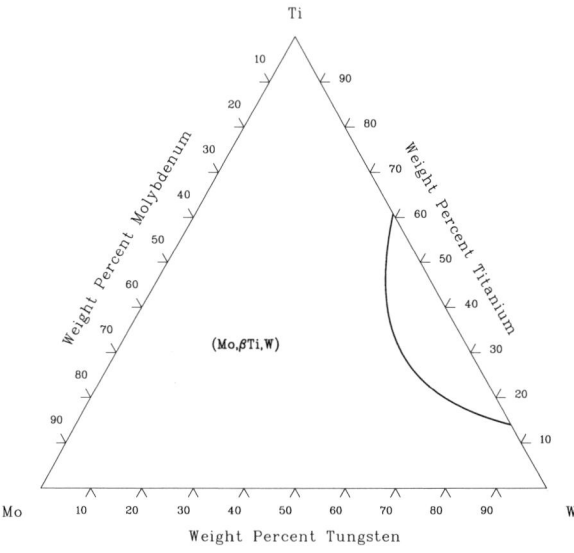

Pb-Sb-Sn liquidus projection [73Bre]

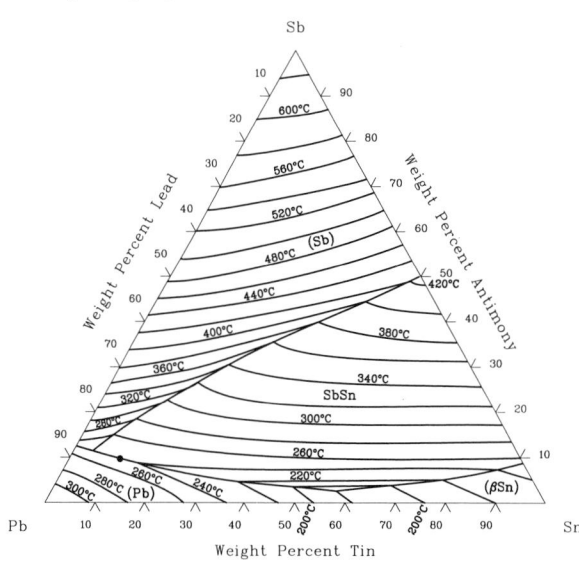

Nb-Ti-W isothermal section at 1000 °C [75Kau]

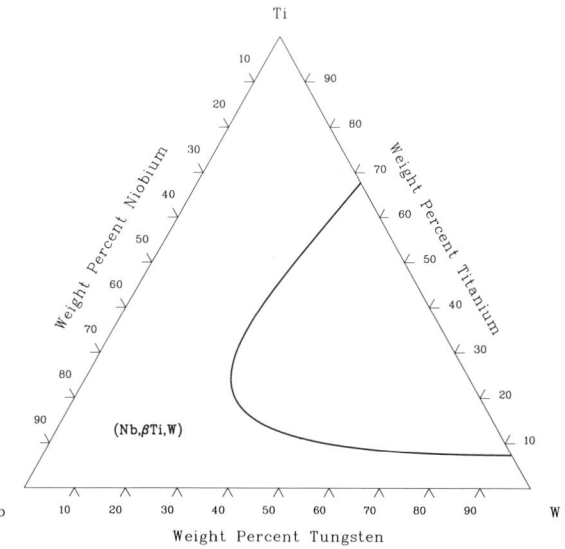

Pb-Sb-Sn isothermal section at 240 °C [85Osa]

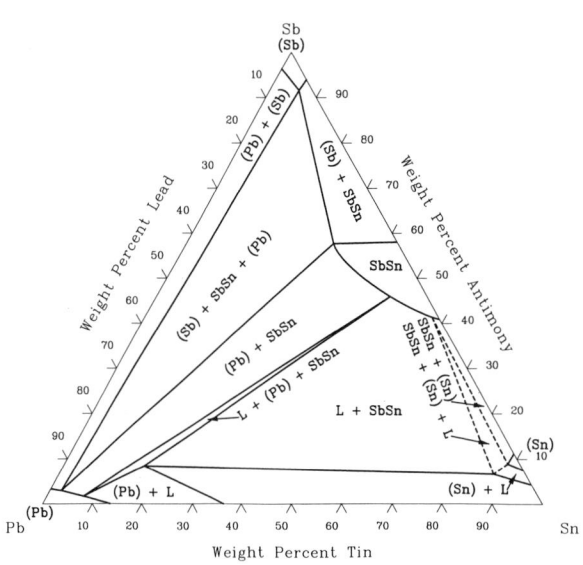

3•58/Ternary Alloy Phase Diagrams

Pb-Sb-Sn (Pb) liquidus projection [73Bre]

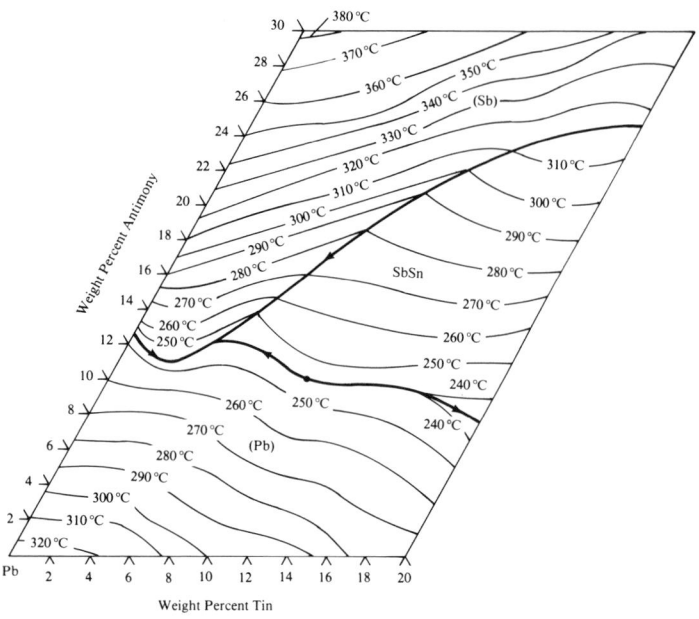

Pb-Sb-Sn [85Osa]

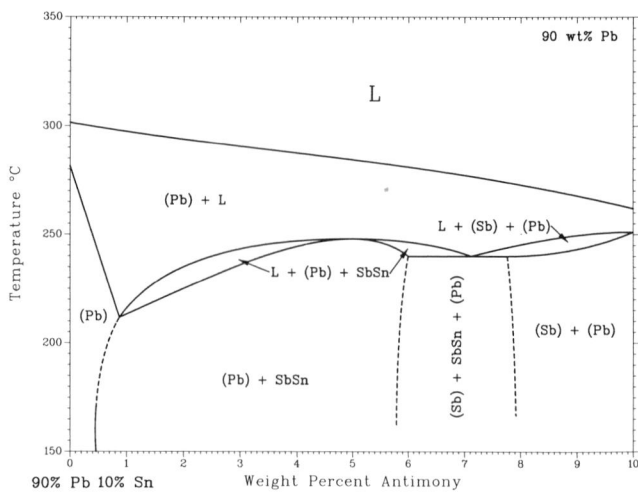

Pb-Sn-Zn liquidus projection [51Lin]

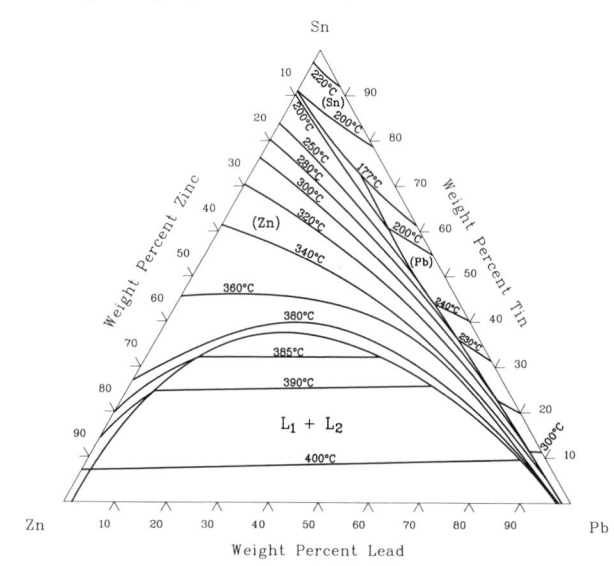

Pb-Sb-Sn isothermal section at 189 °C [85Osa]

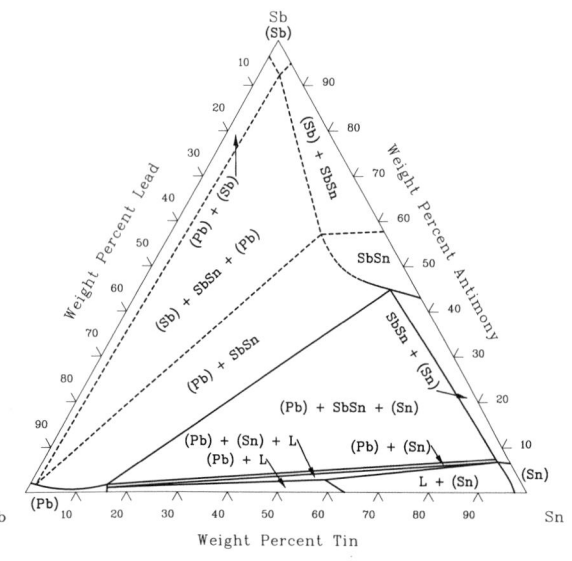

Pb-Sn-Zn isothermal section at 532 °C [67Pta]

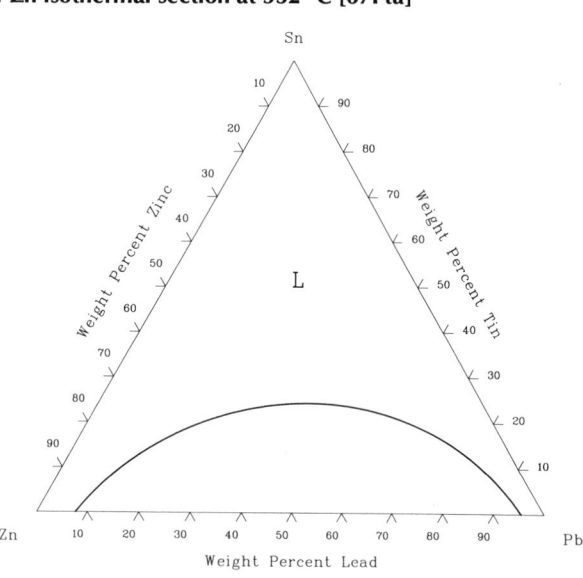

Ternary System References

11Par: N. Parravano, "Das Ternäre System Silber-Zinn-Blei," *Z. Metallkd.*, Vol 1, 1911, p 89-108

29Bau: O. Bauer and M. Hansen, "Der Einfluss von dritten Metallen auf die Konstitution der Messingle ierungen. I. Der Einfluss von Blei," *Z. Metallkd.*, Vol 21, 1929, p 190-196

36Kos: W. Köster and W. Dullenkopf, "Das Dreistoffsystem Aluminium-Magnesium-Zink. III. Der Teilbereich Mg-Al$_3$Mg$_4$-Al$_2$Mg$_3$Zn$_3$ -MgZn$_2$-Mg," *Z. Metallkd.*, Vol 28, 1936, p 363-367

48Kos: W. Köster, U. Zwicker, and K. Moeller, "Mikroskopische und röntgenographische Untersuchungen zur Kenntnis des Systems Kupfer-Nickel-Aluminium," *Z. Metallkd.*, Vol 39, 1948, p 225-231

48Wil: F.H. Wilson, "The Copper-Rich Corner of the Copper-Aluminum-Silicon Diagram," *Trans. AIME*, Vol 175, 1948, p 262-273

51Lin: E. Linder, "Eine Methode zur Erforschung von Vierstoffsystemen Dargestellt am System Blei-Zink-Kadmium-Zinn," *Z. Metallkd.*, Vol 43, 1951, p 377-387

52Das: D.K. Das, S.P. Rideout, and P.A. Beck, "Intermediate Phases in the Mo-Fe-Co, Mo-Fe-Ni, and Mo-Ni-Co Ternary Systems," *Trans. AIME*, Vol 194, 1952, p 1071-1075

56Zwi: U. Zwicker, "Die Systeme Titan-Aluminium-Chrom und Titan-Aluminium-Vanadin und die technishcen Titanlegierungen mit 5% Cr und 3% Al sowie mit 6% Al und 4% V," *Z. Metallkd.*, Vol 47, 1956, p 535-548

58Bag: Yu.A. Bagaryatskiy, G.I. Nosova, and T.V. Tagunova, "Study of the Phase Diagrams of the Alloys Titanium-Chromium, Titanium-Tungsten, and Titanium-Chromium-Tungsten, Prepared by the Method of Powder Metallurgy, *Russ. J. Inorganic Chem.*; TR: *Zh. Neorg. Khim.*, Vol 3 (No. 3), 1958, p 330-341

58Kor: I.I. Kornilov and R.S. Polyakov, Phase Diagram of the Ternary Sytem Titanium-Niobium-Molybdenum, *Russ. J. Inorganic Chem.*, Tr. *Zh. Neorg. Khim.*, Vol 3 (No. 4), 1958, p 62-74

58Liv: B. G. Livshits and Ya.D. Khorin, "Study of Equilibrium Phase Diagram of the System Co-Cr-Ti," *Russ. J. Inorganic Chem.*; TR: *Zh. Neorg. Khim.*, Vol 3 (No. 3), 1958, p 193-205

59Cla: J.W.H. Clare, "The Constitution of Aluminium-Rich Alloys of the Aluminium-Chromium-Manganese System," *Trans. AIME*, Vol 215, 1959, p 429-433

61Eng: J.J. English, "Binary and Ternary Phase Diagrams of Niobium, Molybdenum and Tungsten (1961)," Available as NTIS Document AD 257,739

61Far: P. Farrar and H. Margolin, "The Titanium Rich Region of the Titanium-Aluminium-Vanadium System," *Trans. AIME*, Vol 221, 1961, p 1214-1221

62Zak: E.K. Zakharov and B.G. Livshits, "Phase Composition Diagram of the Cobalt-Chromium-Titanium Ternary System," *Russ. Metall. Fuels*, (No. 5), 1962, p 88-97

63Sta: H.H. Stadelmaier and R.A. Gregg, "Die Ternäre Phase Fe$_{23}$C$_3$B$_3$ im Dreistoffsystem EisenKohlenstoff-Bor," *Metall. Berlin*, Vol 17, 1963, p 412-414

64Kus: J.B. Kusma and H. Nowotny, "Untersuchungen im Dreistoff: Mn-Al-Si," *Monatsh. Chem.*, Vol 95, 1964, p 1266-1271

64Ste: P. Stecher, F. Benesovsky, and H. Nowotny, "Untersuchungen im System Chrom-Wolfram-Kohlenstoff," Vol. 12, 1964, p 89-95

65Kuz: Yu.B. Kuz'ma and T.F. Fedorov, "Phase Equilibria in the System Molybdenum-Chromium-Carbon," *Sov. Powder Metall. Met. Ceram.*; TR: *Poroshk. Metall. Kiev*, Vol 4, 1965, p 920-922

66Kie: R. Kieffer and H. Rassaerts, "Über das System Vanadium-Chrom-Kohlenstoff und über den Einsatz von Vanadin- und Chromcarbiden in Hartmetallen, Teil I," *Metall, Berlin*, Vol 20, 1966, p 691-695

66Kos: W. Köster and T. Gödecke, "Das Dreistoffsystem Kupfer-Mangan-Aluminium," *Z. Metallkd.*, Vol 57, 1966, p 889-901

67Pta: W. Ptak and Z. Moser, "The Range of Occurrence of Two Liquid Phases in Zn-Sn-Cd-Pb Alloys," *Bull. Acad. Pol. Sci. Ser. Sci. Tech.*, Vol 15 (No. 9), 1967, p 809-815

70Han: R.C. Hansen and A. Raman, "Alloy Chemistry of sigma (beta-U)-Related Phases. III. sigma-Phases with Non-Transition Elements," *Z. Metallkd.*, Vol 61, 1970, p 115-120

70Kos: W. Köster and T. Gödecke, "Das Dreistoffsystem Eisen-Aluminum-Zink," *Z. Metallkd.*, Vol 61, 1970, p 649-658

71Pre: A.P. Prevarskiy, "Investigation of Fe-Cu-Al Alloys," *Russ. Metall.*; TR: *Izv. Akad. Nauk SSSR, Metall.*, (No. 4), 1971, p 154-156

73Ben: R. Benz, J.F. Elliott, and J. Chipman, "Thermodynamics of the Solid Phases in the System Fe-Mn-C," *Metall. Trans.*, Vol 4, 1973, p 1975-1986

73Bla: J.M. Blalock, Jr., J.V. Harding, and W.T. Pell-Walpole, *Metallography, Structures and Phase Diagrams*, Vol 8, *Metals Handbook*, 8th ed., American Society for Metals, Metals, Park, OH, 1973

73Bre: L. Brewer and S.-G. Chang, *Metallography, Structures and Phase Diagrams*, Vol 8, *Metals Handbook*, 8th ed., American Society for Metals, Metals Park, OH, 1973

73Dra: J.M. Drapier and D. Coutsouradis, *Metallography, Structures and Phase Diagrams*, Vol 8, *Metals Handbook*, 8th ed., American Society for Metals, Metals Park, OH 1973

73Lev: E.D. Levine, *Metallography, Structures and Phase Diagrams*, Vol 8, *Metals Handbook*, 8th ed., American Society for Metals, Metals Park, OH 1973

73Mar: V.Ya. Markiv, V.V. Burnashova, and V.R. Ryabov, "The Systems Titanium-Iron-Aluminium, Titanium-Nickel-Aluminium, and Titanium-Copper-Aluminium," *Met. Allofizika, Kiev (Akad. Nauk Ukr. SSSR, Metallofiz.*, Vol 46, 1973, p 103-109

73Pel: W.T. Pell-Walpole and C.T. Thwaites, *Metallography, Structures and Phase Diagrams*, Vol 8, *Metals Handbook*, 8th ed., American Society for Metals, Metals Park, OH 1973

73Smi: C.S. Smith and E.D. Levine, *Metallography, Structures and Phase Diagrams*, Vol 8, *Metals Handbook*, 8th ed., American Society for Metals, Metals Park, OH 1973

73Wil: L.A. Willey, *Metallography, Structures and Phase Diagrams*, Vol 8, *Metals Handbook*, 8th ed., American Society for Metals, Metals Park, OH 1973

74Kau: L. Kaufman and H. Nesor, "Calculation of Superalloy Phase Diagrams: Part I, " *Metall. Trans.*, Vol 5, 1974, p 1617-1621

75Kau: L. Kaufman and H. Nesor, "Calculation of Superalloy Phase Diagrams: Part IV," *Metall. Trans. A*, Vol 6, 1975, p 2123-2131

77Lev: V.I. Levanov, V.S. Mikheyev, and A.I. Chernitysn, "Investigation of the Ti-Nb-W System (Nb + W up to 50 wt.%)," *Russ. Metall.*; TR: *Izv. Akad. Nauk SSSR, Met.*, (No. 1), 1977, p 186-191

79Cha: Y.A. Chang, J.P. Neumann, A. Mikula, and D. Goldberg, *Phase Diagrams and Thermodynamic Properties of Ternary Copper-Metal Systems*, INCRA Monograph VI, International Copper Research Association, 1979

80Gry: V.I. Gryzunov and A.S. Sagyndykov, "Mutual Diffusion in the System Ti-Ni-Co," *Phys. Met. Metallogr.*, Tr: *Fiz. Met. Metalloved.*, Vol 49 (No. 5), 1980, p 178-182

80Loo: F.J.J. van Loo, G.F. Bastin, J.W.G.A. Vrolijk, and J.J.M. Hendriks, "Phase Relations in the Systems Fe-Ni-Mo, Fe-Co-Mo and Ni-Co-Mo at 1100 °C," *J. Less-Common Met.*, Vol 72, 1980, p 225-230

80Mas: S.B. Maslenkov and E.A. Nikandrova, "Examination of the Ni-Mo-W Phase Diagram," *Russ. Metall.*, Tr: *Izv. Akad. Nauk SSSR, Met.*, (No. 2), 1980, p 184-187

81Zha: Jin Zhanpeng, "A Study of the Range of Stability of sigma Phase in Some Ternary Systems," *Scand. J. Metall.*, Vol 10, 1981, p 279-287

83Gry: V.I. Gryzunov, G.V. Shcherbedinskiy, Ye.M. Sokolovskaya, B.K. Aytbayev, and A.S. Sagyndykov, "Kinetics of Phase Growth During Mutual Diffusion in Ternary Multiphase Metallic Systems," *Phys. Met. Metallogr.;* TR: *Fiz. Met. Metalloved.*, Vol 56 (No. 1), 1983, p 183-186

84Ere: V.N. Eremenko, L.A. Tret'yachenko, S.B. Prima, and E.L. Semenova, "Constitution Diagrams of Titanium-Nickel-Groups IV-VIII Transition Metal Systems," *Sov. Powder Metall. Met. Ceram.*; TR: *Poroshk. Metall. Kiev*, Vol 23 (No. 8), 1984, p 613-621

84Gup: K.P. Gupta, S.B. Rajendraprasad, A.K. Jena, and R.C. Sharma, "The Co-Mo-Ni System," *Trans. Indian Inst. Met.*, Vol 37 (No. 6), 1984, p 691-697

84Mir: D.B. Miracle, K.A. Lark, V. Srinivasan, and H.A. Lipsitt, "Nickel-Aluminium-Molybdenum Phase Equilibria," *Metall. Trans. A*, Vol 15, 1984, p 481-486

85Mes: L.L. Meshkov, S.N. Nesterenko, and T.V. Ishchenko, "Structural Features of Phase Diagrams Formed by Molybdenum and Tungsten with Iron-Group Metals," *Russ. Metall.;* TR: *Izv. Akad Nauk SSSR, Met.*, (No. 2), 1985, p 204-207

85Nas: P. Nash and W.W. Liang, Phase Equilibria in the Ni-Al-Ti System at 1173 K," *Metall. Trans. A*, Vol 16, 1985, p 319-322

85Oma: A.K. Omarov, S.V. Sejtzhanov, and A.I. Idirisov, "Isothermal Sections of the Ternary System Al-Ni-Ti for the Temperature Range 1150-600 °C," *Izv. Akad. Nauk Kazakh. SSSR, Khim*, (No. 1), 1985, p 36-42

85Osa: K. Osamura, "The Pb-Sb-Sn (Lead-Antimony-Tin) System," *Bull. Alloy Phase Diagrams*, Vol 6 (No. 4), 1985, p 372-379

86Ere: V.N. Eremenko, T.Ja. Velikanova, and A.A. Bondar, "The Ternary Phase Diagram Cr-W-C System," *Dop. Akad. Nauk Ukr. RSR, A, Fiz.- Mat. Tekh.*, Vol 48 (No. 11), 1986, p 74-78

86Mey: S.a. Mey and K. Hack, "A Thermochemical Evaluation of the Silicon-Zinc, Aluminum-Silicon, and Aluminum-Silicon-Zinc Systems," *Z. Metallkd.*, Vol 77 (No. 7), 1986, p 454-459

86Pri: S.B. Prima, L.A. Tret'yachenko, and V.N. Eremenko, "Investigation of Phase Equilibria in the Ti-Ni-Mo System at 1200 °C," *Russ. Metall.;* TR: *Izv. Akad. Nauk SSSR, Met.*, (No. 2), 1986, p 205-210

86Rag: V. Raghavan, "The Carbon-Iron-Silicon System," *J. Alloy Phase Diagrams, India*, Vol 2 (No. 2), 1986, p 97-107

87Ere: V.N. Eremenko, T.Ya. Velikanova, and A.A. Bondar, "The Phase Diagram of the Cr-Mo-C System, II. Phase Equilibria in the Partial System Mo_2C-Cr_7C_3-C," *Sov. Powder Metall. Met. Ceram.*, TR: *Poroshk. Metall. Kiev*, Vol 26 (No. 6), 1987, p 506-511

87Ofo: N.C. Oforka and C.W. Haworth, "Phase Equilibria of Aluminum-Chromium-Nickel

System at 1423 K," *Scand. J. Metall.*, Vol 16, 1987, p 184-188

87Rag: V. Raghavan, *Phase Diagrams of Ternary Iron Alloys*, The Indian Institute of Metals, Calcutta, India, (No. 1), 1987

87Smi: S.V. Smirnova, L.L. Meshkov, and O.N. Kosolapova, "Physicochemical Interaction and Magnetic Properties on the Phases in the Iron-Molybdenum-Niobium System," *Moscow Univ. Chem. Bull.*, Tr: *Vest. Mosk. Univ. Khim.*, Vol 42 (No. 1), 1987, p 84-87

88Pet: G. Petzow and G. Effenberg, *Ternary Alloys*, VCH Verlagsgesellschaft, Weinheim, Germany, Vol 1, 1988

88Ray: G.V. Raynor and V.G. Rivlin, *Phase Equilibria in Iron Ternary Alloys*, The Institute of Metals, London, (No. 4), 1988

88Rok: L.L. Rokhlin and A.G. Pepelyan, "Phase Equilibria in the Mg-Rich Region of the Mg-Al-Si System," *Russ. Metall.*, Tr: *Izv. Akad. Nauk SSSR, Met.*, (No. 6), 1988, p 172-174

88Sim: C.J. Simensen, B.C. Oberländer, J. Svalestuen, and A. Thornvaldsen, "The Phase Diagram for Magnesium-Aluminum-Manganese Above 650 °C," *Z. Metallkd.*, Vol 79 (No. 11), 1988, p 696-699

89Har: K.C. Harikumar and V. Raghavan, "BCC-FCC Equilibrium in Ternary Iron Alloys II, *"J. Alloy Phase Diagrams*, India, Vol 5 (No. 2), 1989, p 77-96

90Gup: K.P. Gupta, *Phase Diagrams of Ternary Nickel Alloys*, Indian Institute of Metals, Calcutta, (No. 1), 1990

90Pri: A. Prince, G.V. Raynor, and D.S. Evans, *Phase Diagrams of Ternary Gold Alloys*, The Institute of Metals, London, 1990

Section 4
Appendix

Symbols for the Chemical Elements

Actinium .... Ac	Gold .... Au	Praseodymium .... Pr
Aluminum .... Al	Hafnium .... Hf	Promethium .... Pm
Americium .... Am	Helium .... He	Protactinium .... Pa
Antimony .... Sb	Holmium .... Ho	Radium .... Ra
Argon .... Ar	Hydrogen .... H	Radon .... Rn
Arsenic .... As	Indium .... In	Rhenium .... Re
Astatine .... At	Iodine I	Rhodium .... Rh
Barium .... Ba	Iridium .... Ir	Rubidium .... Rb
Berkelium .... Bk	Iron .... Fe	Ruthenium .... Ru
Beryllium .... Be	Krypton .... Kr	Samarium .... Sm
Bismuth .... Bi	Lanthanum .... La	Scandium .... Sc
Boron .... B	Lawrencium .... Lr	Selenium .... Se
Bromine .... Br	Lead .... Pb	Silicon .... Si
Cadmium .... Cd	Lithium Li	Silver .... Ag
Calcium .... Ca	Lutetium .... Lu	Sodium .... Na
Californium .... Cf	Magnesium .... Mg	Strontium .... Sr
Carbon C	Manganese .... Mn	Sulfur .... S
Cerium .... Ce	Mendelevium .... Md	Tantalum .... Ta
Cesium .... Cs	Mercury .... Hg	Technetium .... Tc
Chlorine Cl	Molybdenum .... Mo	Tellurium .... Te
Chromium .... Cr	Neodymium .... Nd	Terbium .... Tb
Cobalt .... Co	Neon .... Ne	Thallium .... Tl
Columbium (Niobium) .... Nb	Neptunium .... Np	Thorium .... Th
Copper .... Cu	Nickel .... Ni	Thulium .... Tm
Curium .... Cm	Niobium .... Nb	Tin .... Sn
Dysprosium .... Dy	Nitrogen .... N	Titanium .... Ti
Einsteinium .... Es	Nobelium .... No	Tungsten .... W
Erbium .... Er	Osmium .... Os	Uranium .... U
Europium .... Eu	Oxygen .... O	Vanadium V
Fermium .... Fm	Palladium .... Pd	Xenon .... Xe
Fluorine .... F	Phosphorus .... P	Ytterbium .... Yb
Francium .... Fr	Platinum .... Pt	Yttrium .... Y
Gadolinium .... Gd	Plutonium .... Pu	Zinc .... Zn
Gallium .... Ga	Polonium .... Po	Zirconium .... Zr
Germanium .... Ge	Potassium .... K	

Standard Atomic Weights of the Elements

Key

Chemical symbol → Cd 48 ← Atomic number
112.41 ← Atomic weight

IA	IIA		IIIA	IVA	VA	VIA	VIIA	VIIIA	IXA	XA	IB	IIB	IIIB	IVB	VB	VIB	VIIB	Inert Gases
H 1 1.00794																		He 2 4.002602
Li 3 6.941	Be 4 9.01218												B 5 10.811	C 6 12.011	N 7 14.0067	O 8 15.9994	F 9 18.998403	Ne 10 20.179
Na 11 22.98977	Mg 12 24.305												Al 13 26.98154	Si 14 28.0855	P 15 30.97376	S 16 32.066	Cl 17 35.453	Ar 18 39.948
K 19 39.0983	Ca 20 40.078		Sc 21 44.95591	Ti 22 47.88	V 23 50.9415	Cr 24 51.9961	Mn 25 54.9380	Fe 26 55.847	Co 27 58.9332	Ni 28 58.69	Cu 29 63.546	Zn 30 65.39	Ga 31 69.723	Ge 32 72.59	As 33 74.9216	Se 34 78.96	Br 35 79.904	Kr 36 83.80
Rb 37 85.4678	Sr 38 87.62		Y 39 88.9059	Zr 40 91.224	Nb 41 92.9064	Mo 42 95.94	Tc 43 (98)	Ru 44 101.07	Rh 45 102.9055	Pd 46 106.42	Ag 47 107.8682	Cd 48 112.41	In 49 114.82	Sn 50 118.710	Sb 51 121.75	Te 52 127.60	I 53 126.9045	Xe 54 131.29
Cs 55 132.9054	Ba 56 137.33	La·Lu	Hf 72 178.49	Ta 73 180.9479	W 74 183.85	Re 75 186.207	Os 76 190.2	Ir 77 192.22	Pt 78 195.08	Au 79 196.9665	Hg 80 200.59	Tl 81 204.383	Pb 82 207.2	Bi 83 208.9804	Po 84 (209)	At 85 (210)	Rn 86 (222)
Fr 87 (223)	Ra 88 226.0254	Ac·Lr																

Alkali Metals

Alkaline Earth Metals

Transition Metals

Pnictide Elements (VB)

Chalcogens (VIB)

Halogens (VIIB)

Lanthanide Metals

La 57 138.9055	Ce 58 140.12	Pr 59 140.9077	Nd 60 144.24	Pm 61 (145)	Sm 62 150.36	Eu 63 151.96	Gd 64 157.25	Tb 65 158.9254	Dy 66 162.50	Ho 67 164.9304	Er 68 167.26	Tm 69 168.9342	Yb 70 173.04	Lu 71 174.967

Actinide Metals

Ac 89 227.0278	Th 90 232.0381	Pa 91 231.0359	U 92 238.0289	Np 93 237.0482	Pu 94 (244)	Am 95 (243)	Cm 96 (247)	Bk 97 (247)	Cf 98 (251)	Es 99 (252)	Fm 100 (257)	Md 101 (258)	No 102 (259)	Lr 103 (260)

Melting and Boiling Points of the Elements at Atmospheric Pressure

Symbol	Melting point		Error limits	Boiling point	
	°C	K		°C	K
Ac	1051	1324	±50	3200	3473(a)
Ag	961.93	1235.08	...	2163	2436
Al	660.452	933.602	...	2520	2793
Am	1176	1449	...	...	...
Ar	−189.352(T.P.)	83.798(T.P.)	...	−185.9	87.3
As	614(S.P.)	887(S.P.)	...	...	...
At	(302)	(575)	...	...	...
Au	1064.43	1337.58	...	2857	3130
B	2092	2365	...	4002	4275
Ba	727	1000	±2	1898	2171
Be	1289	1562	±5	2472	2745
Bi	271.442	544.592	...	1564	1837
Bk	1050	1323	...	...	...
Br	−7.25(T.P.)	265.90(T.P.)	...	59.10	332.25
C	3827(S.P.)	4100(S.P.)	±50	...	...
Ca	842	1115	±2	1484	1757
Cd	321.108	594.258	...	767	1040
Ce	798	1071	±3	3426	3699
Cf	900	1173	...	...	...
Cl	−100.97(T.P.)	172.18(T.P.)	...	−34.05	239.10
Cm	1345	1618	...	...	...
Co	1495	1768	...	2928	3201
Cr	1863	2136	±20	2672	2945
Cs	28.39	301.54	±0.05	671	944
Cu	1084.87	1358.02	±0.04	2563	2836
Dy	1412	1685	...	2562	2835
Er	1529	1802	...	2863	3136
Es	860	1133	...	...	...
Eu	822	1095	...	1597	1870
F	−219.67(T.P.)	53.48(T.P.)	...	−188.20	84.95
Fe	1538	1811	...	2862	3135
Fm	(1527)	(1800)	...	...	...
Fr	(27)	(300)	...	...	...
Ga	29.7741(T.P.)	302.9241(T.P.)	±0.001	2205	2478
Gd	1313	1586	...	3266	3539
Ge	938.3	1211.5	...	2834	3107
H	−259.34(T.P.)	13.81(T.P.)	...	−252.882	20.268
He	−271.69(T.P.)	1.46(T.P.)	(b)	−268.935	4.215
Hf	2231	2504	±20	4603	4876
Hg	−38.836	234.210	...	356.623	629.773
Ho	1474	1747	...	2695	2968
I	113.6	386.8	...	185.25	458.40
In	156.634	429.784	...	2073	2346
Ir	2447	2720	...	4428	4701
K	63.71	336.86	±0.5	759	1032
Kr	−157.385	115.765	±0.001	−153.35	119.80
La	918	1191	...	3457	3730
Li	180.6	453.8	±0.5	1342	1615
Lr	(1627)	(1900)	...	...	...
Lu	1663	1936	...	3395	3668
Md	(827)	(1100)	...	...	...
Mg	650	923	±0.5	1090	1363
Mn	1246	1519	±5	2062	2335
Mo	2623	2896	...	4639	4912
N	−210.0042(T.P.)	63.1458(T.P.)	±0.0002	−195.80	77.35
Na	97.8	371.0	±0.1	883	1156
Nb	2469	2742	...	4744	5017

(continued)

Symbol	Melting point			Boiling point	
	°C	K	Error limits	°C	K
Nd	1021	1294	...	3068	3341
Ne	−248.587(T.P.)	24.563(T.P.)	±0.002	−246.054	27.096
Ni	1455	1728	...	2914	3187
No	(827)	(1100)	...	...	...
Np	639	912	±2	...	...
O	−218.789(T.P.)	54.361(T.P.)	...	−182.97	90.18
Os	3033	3306	±20	5012	5285
P(white)	44.14	317.29	±0.1	277	550
P(red)	589.6(T.P.)	862.8(T.P.)	(c)	431	704
Pa	1572	1845	...	...	...
Pb	327.502	600.652	...	1750	2023
Pd	1555	1828	±0.4	2964	3237
Pm	1042	1315	...	...	...
Po	254	527	...	...	...
Pr	931	1204	...	3512	3785
Pt	1769.0	2042.2	...	3827	4100
Pu	640	913	±1	3230	3503
Ra	700	973	...	...	...
Rb	39.48	312.63	±0.5	688	961
Re	3186	3459	±20	5596	5869
Rh	1963	2236	...	3697	3970
Rn	−71	202	...	−62	211
Ru	2334	2607	±10	4150	4423
S	115.22	388.37	...	444.60	717.75
Sb	630.755	903.905	...	1587	1860
Sc	1541	1814	...	2831	3104
Se	221	494	...	685	958
Si	1414	1687	±2	3267	3540
Sm	1074	1347	...	1791	2064
Sn	231.9681	505.1181	...	2603	2876
Sr	769	1042	...	1382	1655
Ta	3020	3293	...	5458	5731
Tb	1356	1629	...	3223	3496
Tc	2155	2428	±50	4265	4538
Te	449.57	722.72	±0.3	988	1261
Th	1755	2028	±10	4788	5061
Ti	1670	1943	±6	3289	3562
Tl	304	577	±2	1473	1746
Tm	1545	1818	...	1947	2220
U	1135	1408	...	4134	4407
V	1910	2183	±6	3409	3682
W	3422	3695	...	5555	5828
Xe	−111.7582(T.P.)	161.3918(T.P.)	±0.0002	−108.12	165.03
Y	1522	1795	...	3338	3611
Yb	819	1092	...	1194	1467
Zn	419.58	692.73	...	907	1180
Zr	1855	2128	±5	4409	4682

Note: T.P. = triple point; S.P. = sublimation point at atmospheric pressure. Measurements in parentheses are approximate. (a) ±300. (b) There are various triple points. (c) Red P sublimes without melting at atmospheric pressure.

Allotropic Transformations of the Elements at Atmospheric Pressure

Allotropic transformation of the chemical elements is discussed in the Introduction to Alloy Phase Diagrams, page 1•1 of this Handbook.

Element	Atomic number	Transfor-mation	Temperature, °C	Element	Atomic number	Transfor-mation	Temperature, °C
Ag	47	L ↔ S	0961.93	Ir	77	L ↔ S	2447
Al	13	L ↔ S	660.452	K	19	L ↔ S	63.71
Am	95	L ↔ γ	1176	Kr	36	L ↔ S	115.65 K
		γ ↔ β	1077	La	57	L ↔ γ	918
β ↔ α	769					γ ↔ β	865
Ar	18	L ↔ S	83.798 K			β ↔ α	310
Au	79	L ↔ S	1064.43	Li	3	L ↔ β	180.6
B	5	L ↔ β	2092			β ↔ α	−193
Ba	56	L ↔ S	727	Lu	71	L ↔ S	1663
Be	4	L ↔ β	1289	Mg	12	L ↔ S	650
		β ↔ α	1270	Mb	25	L ↔ δ	1246
Bi	83	L ↔ S	271.442			δ ↔ γ	1138
Bk	97	L ↔ S	1050			γ ↔ β	1100
Br	35	L ↔ S	265.9 K			β ↔ α	727
Ca	20	L ↔ β	842	Mo	42	L ↔ S	2623
		β ↔ α	443	N	7	L ↔ β	63.146 K
Cd	48	L ↔ S	321.108			β ↔ α	35.61 K
Ce	58	L ↔ δ	798	Na	11	L ↔ β	97.8
		δ ↔ γ	726			β ↔ α	−233
		γ ↔ β	61	Nb	41	L ↔ S	2469
		β ↔ α	...	Nd	60	L ↔ β	1021
Cf	98	L ↔ β	900			β ↔ α	863
		β ↔ α	590	Ne	10	L ↔ S	24.563 K (T.P.)
Cl	17	L ↔ S	172.16 K	Ni	28	L ↔ S	1455
Cm	96	L ↔ β	1345	Np	93	L ↔ γ	639
		β ↔ γ	1277			γ ↔ β	576
Co	27	L ↔ α	1495			β ↔ α	280
		α ↔ ε	422	O	8	L ↔ γ	54.361 K
Cr	24	L ↔ S	1863			γ ↔ β	43.801 K
Cs	55	L ↔ S	28.39			β ↔ α	23.867 K
Cu	29	L ↔ S	1084.87	Os	76	L ↔ S	3033
Dy	66	L ↔ β	1412	P (white α)	15	L ↔ α	44.14
		β ↔ α	1381	Pa	91	L ↔ β	1572
		α ↔ α′	−187			β ↔ α	1170
Er	68	L ↔ S	1529	Pb	82	L ↔ S	327.502
Es	99	L ↔ S	860	Pd	46	L ↔ S	1555
Eu	63	L ↔ S	822	Pm	61	L ↔ β	1042
F	9	L ↔ β	53.48 K			β ↔ α	890
		β ↔ α	45.55 K	Po	84	L ↔ β	254
Fe	26	L ↔ δ	1538			β ↔ α	54
		δ ↔ γ	1394	Pr	59	L ↔ β	931
		γ ↔ α	912			β ↔ α	795
Ga	31	L ↔ S	29.7741	Pt	78	L ↔ S	1769.0
Gd	64	L ↔ β	1313	Pu	94	L ↔ ε	640
		β ↔ α	1235			ε ↔ δ′	483
Ge	32	L ↔ S	938.3			δ′ ↔ δ	463
H	1	L ↔ S	13.81 K			δ ↔ γ	320
Hf	72	L ↔ β	2231			γ ↔ β	215
		β ↔ α	1743			β ↔ α	125
Hg	80	L ↔ α	−38.290	Rb	37	L ↔ S	39.48
Ho	67	L ↔ S	1474	Re	75	L ↔ S	3186
I	53	L ↔ S	113.6	Rh	45	L ↔ S	1963
In	49	L ↔ S	156.634	Rn	86	L ↔ S	−71

(continued)

Element	Atomic number	Transfor-mation	Temperature, °C	Element	Atomic number	Transfor-mation	Temperature, °C
Ru	44	L ↔ S	2334			β ↔ α	1360
S	16	L ↔ β	115.22	Ti	22	L ↔ β	1670
		β ↔ α	95.5			β ↔ α	882
Sb	51	L ↔ S	630.755	Tl	81	L ↔ β	304
Sc	21	L ↔ β	1541			β ↔ α	230
		β ↔ α	1337	T	69	L ↔ S	1545
Se	34	L ↔ S	221	U	92	L ↔ γ	1135
Si	14	L ↔ S	1414			γ ↔ β	776
Sm	62	L ↔ γ	1074			β ↔ α	668
		γ ↔ β	922	V	23	L ↔ S	1910
		β ↔ α	734	W	74	L ↔ S	3422
Sn	50	L ↔ β	231.9681	Xe	54	L ↔ S	161.918 (T.P.)
		β ↔ α	13	Y	39	L ↔ β	1522
Sr	38	L ↔ β	769			β ↔ α	1478
		β ↔ α	547	Yb	70	L ↔ γ	819
Ta	73	L ↔ S	3020			γ ↔ β	795
Tb	65	L ↔ β	1356			β ↔ α	−3
		β ↔ α	1289	Zn	30	L ↔ S	419.58
		α ↔ α′	−53	Zr	40	L ↔ β	1855
Te	52	L ↔ S	449.57			β ↔ α	863
Th	90	L ↔ β	1755				

Note: T.P. = triple point.

Magnetic Phase Transition Temperatures of the Elements

Magnetic phase transition, and other higher-order transitions of the chemical elements, is discussed in the Introduction to Alloy Phase Diagrams, page 1•1 of this Handbook.

Chemical symbol	Atomic number	Allotrope	Phase transition temperature (T_c), K	Type of magnetic ordering(a)	Phase transition temperature (T_{c2}), K	Type of magnetic ordering(a)	Phase transition temperature (T_{c3}), K	Type of magnetic ordering(a)	Saturation magnetic moment, µB
Ce(b)............	58	β-dcph	13.7	AC?	12.5	AC?	...	...	2.61
		γ-fcc	14.4	AC?	...	...	...	...	
Cm	96	α-dcph	52	AC	...	...	...	...	...
Co	27	fcc	1388(1115 °C)	FM	...	...	...	...	1.715
Cr................	24	bcc	312.7(39.5 °C)	AI	...	...	...	...	0.45
Dy	66	α-cph	179.0	AI	89.0	FM	...	...	10.33
Er	68	cph	85.0	AI	53	AC	20.0	CF	9.1
Eu	63	bcc	90.4	AC	...	...	...	...	5.9
Fe(c)............	26	α-bcc	1044(771 °C)	FM	...	...	...	...	2.216
		γ-fcc	67	AC	...	...	...	...	0.75
Gd	64	α-cph	293.4(20.2 °C)	FM	...	...	...	...	0.75
Ho	67	cph	132.0	AI	20.0	CF	...	...	10.34
Mn	25	α-bcc	100	AC	...	...	...	...	(d)
Nd	60	α-dcph	19.9	AI	7.5	AC	...	...	1.84
Ni................	28	fcc	627.4(354.2 °C)	FM	...	...	...	...	0.616
Pm	61	α-dcph	98	FM?	...	...	...	...	0.24
Pr	59	α-dcph	0.06	AC	...	...	...	...	0.36
Sm	62	α-rhomb	106	h, A(e)	13.8	c, A(e)	...	...	0.1
Tb	65	α-cph	230.0	AI	219.5	FM	...	...	9.34
Tm	69	cph	58.0	AI	40 to 32	FI	...	...	7.14

(a) Type of magnetic ordering indicated by symbols in the table below and the chart on the reverse side: FM = transition from paramagnetic to ferromagnetic state, AC = transition to periodic (antiferromagnetic) state that is commensurate with the lattice periodicity (*e.g.*, spins on three atom layers directed up followed by three layers down, *etc.*), AI = transition to periodic (antiferromagnetic) state that is generally not commensurate with the lattice periodicity (*e.g.*, helical spin ordering), CF = transition to conical ferromagnetic state (combination of planar helical antiferromagnetic plus ferromagnetic component), and FI = transition to ferromagnetic periodic structure (unequal number of up and down spin layers). (b) Ce exists in five crystal structures, two of which are magnetic (γ–fcc; and β–dcph). γCe is estimated to be antiferromagnetic below 14.4 K by extrapolation from fcc Ce-La alloys. (αCe does not exist in pure form below ~100 K.) βCe is thought to exhibit antiferromagnetism on the hexagonal lattice sites below 13.7 K and on the cubic sites below 12.5 K. (c) Magnetic measurements quoted in table for γFe are for fcc Fe precipitated in copper. (d) The magnetic moment assignments of Mn are complex. (e) h, A; c, A = indicate that sites of hexagonal and cubic point symmetry order antiferromagnetically, but at different temperatures.
Source: J.J. Rhyne, *Bull. Alloy Phase Diagrams*, 3(3), 402 (1982).

Crystal Structures and Lattice Parameters of Allotropes of the Metallic Elements

The crystal structure of the allotropic forms of the metallic elements are presented here in terms of the Pearson symbol, space group, and prototype of the structure. The temperatures of the phase transformations are listed in degrees Celsius and the pressures are in GPa. A consistent nomenclature is used, whereby all allotropes are labeled by Greek letters. The lattice parameters of the unit cells are given in nanometers (nm) and are considered to be accurate ± 2 in the last reported digit. Both crystal structure and lattice parameters are discussed in the Introduction to Alloy Phase Diagrams, page 1•1 of this Handbook.

This compilation is restricted to changes in crystal structure that occur as a result of a change in temperature or pressure. Low-temperature structures are included for the diatomic and rare gases, which show many similarities with respect to the metallic elements.

Note that there may be differences between values quoted below and similar values given in another table in this Handbook that has been reproduced from another source. For example, the allotropic transformation temperatures of Mn may differ by as much as 23 °C, etc.

Element	Temperature, °C	Pressure, GPa	Pearson symbol	Space group	Proto-type	a	Lattice parameters, nm b	c	Comment, c/a, or α or β
Ac	25	atm	cF4	Fm$\bar{3}$m	Cu	0.5311	...	...	...
Ag	25	atm	cF4	Fm$\bar{3}$m	Cu	0.40857	...	...	...
αAl	25	atm	cF4	Fm$\bar{3}$m	Cu	0.40496	...	...	...
βAl	25	>20.5	hP2	P6$_3$/mmc	Mg	0.2693	...	0.4398	1.6331
αAm	25	atm	hP4	P6$_3$/mmc	αLa	0.34681	...	1.1241	2 × 1.621
βAm	>769	atm	cF4	Fm$\bar{3}$m	Cu	0.4894	...	...	...
γAm	>1077	atm	cI2	Im$\bar{3}$m	W	?	...	...	...
δAm	25	>15	oC4	Cmcm	αU	0.3063	0.5968	0.5169	...
αAr	<−189.2	atm	cF4	Fm$\bar{3}$m	Cu	0.53109	...	...	...
As	25	atm	hR2	R$\bar{3}$m	αAs	0.41319	...	...	α = 54.12°
Au	25	atm	cF4	Fm$\bar{3}$m	Cu	0.40782	...	...	...
βB	25	atm	hR105	R$\bar{3}$m	βB	1.017	...	...	α = 65.12°
αBa	25	atm	cI2	Im$\bar{3}$m	W	0.50227	...	...	...
βBa	25	>5.33	hP2	P6$_3$/mmc	Mg	0.3901	...	0.6154	1.5775
γBa	25	>23	?	?	...	...	...	...	...
αBe	25	atm	hP2	P6$_3$/mmc	Mg	0.22859	...	0.35845	1.5681
βBe	>1270	atm	cI2	Im$\bar{3}$m	W	0.25515	...	...	...
BeII	25	>28.3	hP*	...	...	0.4328	...	0.3416	0.7893
αBi	25	atm	hR2	R$\bar{3}$m	αAs	0.47460	...	...	α = 57.23°
βBi	25	>2.6	mC4	C2/m	βBi	0.6674	0.6117	0.3304	β = 110.33°
γBi	25	>3.0	mP4	P2$_1$/m	...	0.665	0.420	0.465	β = 85.33°
δBi	25	>4.3	?	?	...	...	...	...	...
ζBi	25	>9.0	cI2	Im$\bar{3}$m	W	0.3800	...	...	...
αBk	25	atm	hP4	P6$_3$/mmc	αLa	0.3416	...	1.1069	2 × 1.620
βBk	>977	atm	cF4	Fm$\bar{3}$m	Cu	0.4997	...	...	...
Br	<−7.25	atm	oC8	Cmca	I$_2$	0.668	0.449	0.874	...
C(graphite)	25	atm	hP4	P6$_3$/mmc	C(graphite)	0.24612	...	0.6709	2.7258
C(diamond)	25	>60	cF8	Fd$\bar{3}$m	C(diamond)	0.35669	...	...	...
αCa	25	atm	cF4	Fm$\bar{3}$m	Cu	0.55884	...	...	...
βCa	>443	atm	cI2	Im$\bar{3}$m	W	0.4480	...	...	...
γCa	25	>1.5	?	...	...	...	...	...	...
Cd	25	atm	hP2	P6$_3$/mmc	Mg	0.29793	...	0.56196	1.8862
αCe	<−177	atm	cF4	Fm$\bar{3}$m	Cu	0.485	...	...	...
βCe	25	atm	hP4	P6$_3$/mmc	αLa	0.36810	...	1.1857	2 × 1.611
γCe	>61	atm	cF4	Fm$\bar{3}$m	Cu	0.51610	...	...	...
δCe	>726	atm	cI2	Im$\bar{3}$m	W	0.412	...	...	...
α′Ce	25	>5.4	oC4	Cmcm	αU	0.3049	0.5998	0.5215	...
αCf	25	atm	hP4	P6$_3$/mmc	αLa	0.339	...	1.1015	2 × 1.625
βCf	>590	atm	cF4	Fm$\bar{3}$m	Cu	?	...	...	...
Cl	<−100.97	atm	oC8	Cmca	I$_2$	0.624	0.448	0.826	...
αCm	25	atm	hP4	P6$_3$/mmc	αLa	0.3496	...	1.1331	2 × 1.621
βCm	>1277	atm	cF4	Fm$\bar{3}$m	Cu	0.4382	...	...	...
εCo	25	atm	hP2	P6$_3$/mmc	Mg	0.25071	...	0.40686	1.6228
αCo	>422	atm	cF4	Fm$\bar{3}$m	Cu	0.35447	...	...	...
αCr	25	atm	cI2	Im$\bar{3}$m	W	0.28848	...	...	...
α′Cr	25	HP	tI2	I4/mmm	α′Cr	0.2882	...	0.2887	1.002
αCs	25	atm	cI2	Im$\bar{3}$m	W	0.6141	...	...	...
βCs	25	>2.37	cF4	Fm$\bar{3}$m	Cu	0.6465	...	...	...

(continued)

Element	Temperature, °C	Pressure, GPa	Pearson symbol	Space group	Proto-type	a	Lattice parameters, nm b	c	Comment, c/a, or α or β
β′Cs	25	>4.22	cF4	Fm3̄m	Cu	0.5800	...	...	...
γCs	25	>4.27	?	...	...	...	...	...	...
Cu	25	atm	cF4	Fm3̄m	Cu	0.36146	...	...	...
α′Dy	<−187	atm	oC4	Cmcm	α′Dy	0.3595	0.6184	0.5678	...
αDy	25	atm	hP2	P6₃/mmc	Mg	0.35915	...	0.56501	1.5732
βDy	>1381	atm	cI2	Im3̄m	W	(0.398)	...	...	...
γDy	25	>7.5	hR3	R3̄m	CdCl₂	0.3436	...	2.483	4.5 × 1.606
Er	25	atm	hP2	P6₃/mmc	Mg	0.35592	...	0.55850	1.5692
αEs	25	atm	hP4	P6₃/mmc	αLa	?	...	...	...
βEs	?	atm	cF4	Fm3̄m	Cu	?	...	...	...
Eu	25	atm	cI2	Im3̄m	W	0.45827	...	...	...
αF	<−227.60	atm	mC8	C2/c	αF	0.550	0.338	0.728	β = 102.17°
βF	<−219.67	atm	cP16	Pm3̄n	γO	0.667	...	...	...
αFe	25	atm	cI2	Im3̄m	W	0.28665	...	...	...
γFe	>912	atm	cF4	Fm3̄m	Cu	0.36467	...	...	...
δFe	>1394	atm	cI2	Im3̄m	W	0.29315	...	...	...
εFe	25	>13	hP2	P6₃/mmc	Mg	0.2468	...	0.396	1.603
αGa	25	atm	oC8	Cmca	αGa	0.45186	0.76570	0.45258	...
βGa	25	>1.2	tI2	I4/mmm	In	0.2808	...	0.4458	1.588
γGa	−53	>3.0	oC40	Cmcm	γGa	1.0593	1.3523	0.5203	...
αGd	25	atm	hP2	P6₃/mmc	Mg	0.36336	...	0.57810	1.5910
βGd	>1235	atm	cI2	Im3̄m	W	0.406	...	...	...
γGd	25	>3.0	hR3	R3̄m	αSm	0.361	...	2.603	4.5 × 1.60
αGe	25	atm	cF8	Fd3̄m	C(diamond)	0.56574	...	...	...
βGe	25	>12	tI4	I4₁/amd	βSn	0.4884	...	0.2692	0.551
γGe	25	>12 → atm	tP12	P4₁2₁2	γGe	0.593	...	0.698	1.18
δGe	LT	>12	cI16	Im3̄m	γSi	0.692	...	...	...
αH	<−271.9	atm	cF4	Fm3̄m	Cu	0.5338	...	...	...
βH₃	<−259.34	atm	hP2	P6₃/mmc	Mg	0.3776	...	0.6162	1.632
He₄	−269.67	0.163	hP2	P6₃/mmc	Mg	0.3501	...	0.5721	1.634
He	−269.2	0.129	hP2	P6₃/mmc	Mg	0.3470	...	0.5540	1.597
αHf	25	atm	hP2	P6₃/mmc	Mg	0.31946	...	0.50510	1.5811
βHf	>1743	atm	cI2	Im3̄m	W	0.3610	...	...	...
αHg	<−38.836	atm	hR1	R3̄m	αHg	0.3005	...	...	α = 70.53°
βHg	<−194	HP	tI2	I4/mmm	βHg	0.3995	...	0.2825	0.707
αHo	25	atm	hP2	P6₃/mmc	Mg	0.35778	...	0.56178	1.5702
βHo	25	>7.5	hR3	R3̄m	αSm	0.334	...	2.45	4.5 × 1.63
I	25	atm	oC8	Cmca	I₂	0.72697	0.47903	0.97942	...
In	25	atm	tI2	I4/mmm	In	0.3253	...	0.49470	1.5210
Ir	25	atm	cF4	Fm3̄m	Cu	0.38392	...	...	...
K	25	atm	cI2	Im3̄m	W	0.5321	...	...	...
Kr	<−157.385	atm	cF4	Fm3̄m	Cu	0.5810	...	...	...
αLa	25	atm	hP4	P6₃/mmc	αLa	0.37740	...	1.2171	2 × 1.6125
βLa	>310	atm	cF4	Fm3̄m	Cu	0.5303	...	...	...
γLa	>865	atm	cI2	Im3̄m	W	0.426	...	...	...
β′La	25	>2.0	cF4	Fm3̄m	Cu	0.517	...	...	...
αLi	<−193	atm	hP2	P6₃/mmc	Mg	0.3111	...	0.5093	1.637
βLi	25	atm	cI2	Im3̄m	W	0.35093	...	...	...
Lu	25	atm	hP2	P6₃/mmc	Mg	0.35052	...	0.55494	1.5832
Mg	25	atm	hP2	P6₃/mmc	Mg	0.32094	...	0.52107	1.6236
αMn	25	atm	cI58	I4̄3m	αMn	0.89126	...	...	...
βMn	>727	atm	cP20	P4₁32	βMn	0.63152	...	...	...
γMn	>1100	atm	cF4	Fm3̄m	Cu	0.3860	...	...	...
δMn	>1138	atm	cI2	Im3̄m	W	0.3080	...	...	...
Mo	25	atm	cI2	Im3̄m	W	0.31470	...	...	...
αN	<−237.54	atm	cP8	Pa3	αN	0.5661	...	...	...
βN	<−210.004	atm	hP4	P6₃/mmc	βN	0.4050	...	0.6604	1.631
γN	<−253	>3.3	tP4	P4₂/mnm	γN	0.3957	...	0.5109	1.291
αNa	<−233	atm	hP2	P6₃/mmc	Mg	0.3767	...	0.6154	1.634
βNa	25	atm	cI2	Im3̄m	W	0.42906	...	...	...
Nb	25	atm	cI2	Im3̄m	W	0.33004	...	...	...
αNd	25	atm	hP4	P6₃/mmc	αLa	0.36582	...	1.17966	2 × 1.6124
βNd	>863	atm	cI2	Im3̄m	W	0.413	...	...	...
γNd	25	>5.0	cF4	Fm3̄m	Cu	0.480	...	...	...
Ne	<−248.587	atm	cF4	Fm3̄m	Cu	0.4462	...	...	...
Ni	25	atm	cF4	Fm3̄m	Cu	0.35240	...	...	...
αNp	25	atm	oP8	Pnma	αNp	0.6663	0.4723	0.4887	...
βNp	>280	atm	tP4	P42₁2	βNp	0.4883	...	0.3389	0.694
γNp	>576	atm	cI2	Im3̄m	W	0.352	...	...	...
αO	<−249.283	atm	mC4	C2m	αO	0.5403	0.3429	0.5086	β = 132.53°
βO	<−229.349	atm	hR2	R3̄m	βO	0.4210	...	...	α = 46.27°
γO	<−218.789	atm	cP16	Pm3̄n	γO	0.683	...	...	...
Os	25	atm	hP2	P6₃/mmc	Mg	0.27341	...	0.43198	1.5800
αP(white)	25	atm	c**	...	P(white)	0.718	...	...	...
P(black)	25	atm	oC8	Cmca	P(black)	0.33136	1.0478	0.43763	...
αPa	25	atm	tI2	I4/mmm	αPa	0.3921	...	0.3235	0.825
βPa	>1170	atm	cI2	Im3̄m	W	0.381	...	...	...
αPb	25	atm	cF4	Fm3̄m	Cu	0.49502	...	...	...
βPb	25	>10.3	hP2	P6₃/mmc	Mg	0.3265	...	0.5387	1.650
Pd	25	atm	cF4	Fm3̄m	Cu	0.38903	...	...	...
αPm	25	atm	hP4	P6₃/mmc	αLa	0.365	...	1.165	2 × 1.60

(continued)

Element	Temperature, °C	Pressure, GPa	Pearson symbol	Space group	Proto-type	a	Lattice parameters, nm b	c	Comment, c/a, or α or β
βPm	>890	atm	cI2	Im$\bar{3}$m	W	?	...	...	...
αPo	25	atm	cP1	Pm$\bar{3}$m	αPo	0.3366	...	...	...
βPo	>54	atm	hR1	R$\bar{3}$m	βPo	0.3373	...	...	α = 98.08°
αPr	25	atm	hP4	P6$_3$/mmc	αLa	0.36721	...	1.18326	2 × 1.6111
βPr	>795	atm	cI2	Im$\bar{3}$m	W	0.413	...	...	...
γPr	25	>4.0	cF4	Fm$\bar{3}$m	Cu	0.488	...	...	...
Pt	25	atm	cF4	Fm$\bar{3}$m	Cu	0.39236	...	...	...
αPu	25	atm	mP16	P2$_1$/m	αPu	0.6183	0.4822	1.0963	β = 101.97°
βPu	>125	atm	mC34	C2/m	βPu	0.9284	1.0463	0.7859	β = 92.13°
γPu	>215	atm	oF8	Fddd	γPu	0.31587	0.57682	1.0162	...
δPu	>320	atm	cF4	Fm$\bar{3}$m	Cu	0.46371	...	...	...
δ'Pu	>463	atm	tI2	I4/mmm	In	0.33261	...	0.44630	1.3418
εPu	>483	atm	cI2	Im$\bar{3}$m	W	0.36343	...	...	...
Ra	25	atm	cI2	Im$\bar{3}$m	W	0.5148	...	...	...
αRb	25	atm	cI2	Im$\bar{3}$m	W	0.5705	...	...	...
βRb	25	>1.08	?	...	...	...	...	...	...
γRb	25	>2.05	?	...	...	...	...	...	...
Re	25	atm	hP2	P6$_3$/mmc	Mg	0.27609	...	0.4458	1.6145
Rh	25	atm	cF4	Fm$\bar{3}$m	Cu	0.38032	...	...	...
Ru	25	atm	hP2	P6$_3$/mmc	Mg	0.27058	...	0.42816	1.5824
αS	25	atm	oF128	Fddd	αS	1.0464	1.28660	2.44860	...
βS	>95.5	atm	mP64	P2$_1$/c	βS	1.102	1.096	1.090	β = 96.7°
αSb	25	atm	hR2	R$\bar{3}$m	αAs	0.45067	...	...	α = 57.11°
βSb	25	>5.0	cP1	Pm$\bar{3}$m	αPo	0.2992	...	...	...
γSb	25	>7.5	hP2	P6$_3$/mmc	Mg	0.3376	...	0.5341	1.582
δSb	25	>14.0	mP3	?	...	0.556	0.404	0.422	β = 86.0°
αSc	25	atm	hP2	P6$_3$/mmc	Mg	0.33088	...	0.52680	1.5921
βSc	>1337	atm	cI2	Im$\bar{3}$m	W	0.373	...	...	...
γSe	25	atm	hP3	P3$_1$21	γSe	0.43659	...	0.49537	1.1346
αSi	25	atm	cF8	Fd$\bar{3}$m	C(diamond)	0.54306	...	...	...
βSi	25	>9.5	tI4	I4$_1$/amd	βSn	0.4686	...	0.2585	0.552
γSi	25	>16.0	cI16	Im$\bar{3}$m	γSi	0.6636	...	...	...
δSi	25	>16 → atm	hP4	P6$_3$/mmc	αLa	0.380	...	0.628	1.653
αSm	25	atm	hR3	R$\bar{3}$m	αSm	0.36290	...	2.6207	4.5 × 1.6048
βSm	>734	atm	hP2	P6$_3$/mmc	Mg	0.36630	...	0.58448	1.5956
γSm	>922	atm	cI2	Im$\bar{3}$m	W	?	...	...	...
δSm	25	>4.0	hP4	P6$_3$/mmc	αLa	0.3618	...	1.166	2 × 1.611
αSn	<13	atm	cF8	Fd$\bar{3}$m	C(diamond)	0.64892	...	...	...
βSn	25	atm	tI4	I4$_1$/amd	βSn	0.58318	...	0.31818	0.5456
γSn	25	>9.0	tI2	?	γSn	0.370	...	0.337	0.91
αSr	25	atm	cF4	Fm$\bar{3}$m	Cu	0.6084	...	...	...
βSr	>547	atm	cI2	Im$\bar{3}$m	W	0.487	...	...	...
β'Sr	25	>3.5	cI2	Im$\bar{3}$m	W	0.4437	...	...	...
Ta	25	atm	cI2	Im$\bar{3}$m	W	0.33030	...	...	...
αTb	<−53	atm	oC4	Cmcm	α'Dy	0.3605	0.6244	0.5706	...
α'Tb	25	atm	hP2	P6$_3$/mmc	Mg	0.36055	...	0.56966	1.5800
βTb	>1289	atm	cI2	Im$\bar{3}$m	W	(0.402)	...	...	...
γTb	25	>6.0	hR3	R$\bar{3}$m	αSm	0.341	...	2.45	4.5 × 1.60
Tc	25	atm	hP2	P6$_3$/mmc	Mg	0.2738	...	0.4393	1.604
αTe	25	atm	hP3	P3$_1$21	γSe	0.44566	...	0.59264	1.3298
βTe	25	>2.0	hR2	R$\bar{3}$m	αAs	0.469	...	...	α = 53.30°
γTe	25	>7.0	hR1	R$\bar{3}$m	βPo	0.3002	...	...	α = 103.3°
αTh	25	atm	cF4	Fm$\bar{3}$m	Cu	0.50842	...	...	...
βTh	>1360	atm	cI2	Im$\bar{3}$m	W	0.411	...	...	...
αTi	25	atm	hP2	P6$_3$/mmc	Mg	0.29506	...	0.46835	1.5873
βTi	>882	atm	cI2	Im$\bar{3}$m	W	0.33065	...	...	...
ωTi	25	HP → atm	hP3	P6/mmm	ωTi	0.4625	...	0.2813	0.6082
αTl	25	atm	hP2	P6$_3$/mmc	Mg	0.34566	...	0.55248	1.5983
βTl	>230	atm	cI2	Im$\bar{3}$m	W	0.3879	...	...	...
γTl	25	HP	cF4	Fm$\bar{3}$m	Cu	?	...	...	...
Tm	25	atm	hP2	P6$_3$/mmc	Mg	0.35375	...	0.55540	1.5700
αU	25	atm	oC4	Cmcm	αU	0.28537	0.58695	0.49548	...
βU	>668	atm	tP30	P4$_2$/mnm	βU	1.0759	...	0.5656	0.526
γU	>776	atm	cI2	Im$\bar{3}$m	W	0.3524	...	...	...
V	25	atm	cI2	Im$\bar{3}$m	W	0.30240	...	...	...
W	25	atm	cI2	Im$\bar{3}$m	W	0.31652	...	...	...
Xe	<−111.758	atm	cF4	Fm$\bar{3}$m	Cu	0.6350	...	...	...
αY	25	atm	hP2	P6$_3$/mmc	Mg	0.36482	...	0.57318	1.5711
βY	>1478	atm	cI2	Im$\bar{3}$m	W	(0.407)	...	...	...
αYb	<−3	atm	hP2	P6$_3$/mmc	Mg	0.38799	...	0.63859	1.6459
βYb	25	atm	cF4	Fm$\bar{3}$m	Cu	0.54848	...	...	...
γYb	>795	atm	cI2	Im$\bar{3}$m	W	0.444	...	...	...
Zn	25	atm	hP2	P6$_3$/mmc	Mg	0.26650	...	0.49470	1.8563
αZr	25	atm	hP2	P6$_3$/mmc	Mg	0.32316	...	0.51475	1.5929
βZr	>863	atm	cI2	Im$\bar{3}$m	W	0.36090	...	...	...
ωZr	25	HP → atm	hP2	P6/mmm	ωTi	0.5036	...	0.3109	0.617

Note: Values in parentheses are estimated.

Crystal Structure Nomenclature

The various designation systems for describing crystal structure are discussed in the Introduction to Alloy Phase Diagrams, page 1•1 of this Handbook.

Arranged Alphabetically by Pearson-Symbol Designation

Pearson symbol	Prototype	Strukturbericht designation	Space group
$cF4$	Cu	$A1$	$Fm\bar{3}m$
$cF8$	C(diamond)	$A4$	$Fd\bar{3}m$
	NaCl	$B1$	$Fm\bar{3}m$
	ZnS(sphalerite)	$B3$	$F\bar{4}3m$
$cF12$	CaF_2	$C1$	$Fm\bar{3}m$
	MgAgAs	$C1_b$	$F\bar{4}3m$
$cF16$	$AlCu_2Mn$	$L2_1$	$Fm\bar{3}m$
	BiF_3	$D0_3$	$Fm\bar{3}m$
	NaTl	$B32$	$Fd\bar{3}m$
$cF24$	$AuBe_5$	$C15_b$	$F\bar{4}3m$
	SiO_2(β cristobalite)	$C9$	$Fd\bar{3}m$
	Cu_2Mg	$C15$	$Fd\bar{3}m$
$cF32$	$CuPt_3$	$L1_a$	$Fm\bar{3}c$
$cF52$	UB_{12}	$D2_f$	$Fm\bar{3}m$
$cF56$	Al_2MgO_4	$H1_1$	$Fd\bar{3}m$
	Co_3S_4	$D7_2$	$Fd\bar{3}m$
$cF68$	Co_9S_8	$D8_9$	$Fm\bar{3}m$
$cF80$	Sb_2O_3(senarmontite)	$D5_4$	$Fd\bar{3}m$
$cF112$	Fe_3W_3C	$E9_3$	$Fd\bar{3}m$
	$NaZn_{13}$	$D2_3$	$Fm\bar{3}c$
$cF116$	$Cr_{23}C_6$	$D8_4$	$Fm\bar{3}m$
	$Mn_{23}Th_6$	$D8_a$	$Fm\bar{3}m$
$cI2$	W	$A2$	$Im\bar{3}m$
$cI16$	CoU	B_a	$I2_13$
$cI28$	Th_3P_4	$D7_3$	$I\bar{4}3d$
$cI32$	$CoAs_3$	$D0_2$	$Im\bar{3}$
$cI40$	Ge_7Ir_3	$D8_f$	$Im\bar{3}m$
	Pu_2C_3	$D5_c$	$I\bar{4}3d$
$cI52$	Cu_5Zn_8	$D8_2$	$I\bar{4}3m$
	Fe_3Zn_{10}	$D8_1$	$Im\bar{3}m$
$cI54$	Sb_2Tl_7	$L2_2$	$Im\bar{3}m$
$cI58$	αMn	$A12$	$I\bar{4}3m$
$cI76$	$Cu_{15}Si_4$	$D8_6$	$I\bar{4}3d$
$cI80$	Mn_2O_3	$D5_3$	$Ia\bar{3}$
$cI96$	$AlLi_3N_2$	$E9_d$	$Ia\bar{3}$
$cI162$	$Mg_{32}(Al,Zn)_{49}$	$D8_e$	$Im\bar{3}$
$cP1$	αPo	A_h	$Pm\bar{3}m$
$cP2$	CsCl	$B2$	$Pm\bar{3}m$
$cP4$	$AuCu_3$	$L1_2$	$Pm\bar{3}m$
	ReO_3	$D0_9$	$Pm\bar{3}m$
$cP5$	$AlFe_3C$	$L'1_2$	$Pm\bar{3}m$
	$CaTiO_3$	$E2_1$	$Pm\bar{3}m$
	Fe_4N	$L'1$	$P\bar{4}3m$
$cP6$	Ag_2O	$C3$	$Pn\bar{3}m$
$cP7$	CaB_6	$D2_1$	$Pm\bar{3}m$
$cP8$	Cr_3Si	$A15$	$Pm\bar{3}n$
	FeSi	$B20$	$P2_13$
	Cu_3VS_4	$H2_4$	$P\bar{4}3m$
$cP12$	FeS_2(pyrite)	$C2$	$Pa3$
	NiSbS	$F0_1$	$P2_13$
$cP20$	βMn	$A13$	$P4_132$
$cP36$	$BaHg_{11}$	$D2_e$	$Pm\bar{3}m$
$cP39$	Mg_2Zn_{11}	$D8_c$	$Pm\bar{3}$
$cP52$	Cu_9Al_4	$D8_3$	$P\bar{4}3m$
$hP1$	$HgSn_{6-10}$	A_f	$P6/mmm$
$hP2$	Mg	$A3$	$P6_3/mmc$

Pearson symbol	Prototype	Strukturbericht designation	Space group
$hP2$(continued)	WC	B_h	$P\bar{6}m2$
$hP3$	AlB_2	$C32$	$P6/mmm$
	CdI_2	$C6$	$P\bar{3}m1$
	Fe_2N	$L'3$	$P6_3/mmc$
	$LiZn_2$	C_k	$P6_3/mmc$
	γSe	$A8$	$P3_121$
$hP4$	αLa	$A3'$	$P6_3/mmc$
	BN	B_k	$P6_3/mmc$
	C(graphite)	$A9$	$P6_3/mmc$
	NiAs	$B8_1$	$P6_3/mmc$
	ZnS(wurtzite)	$B4$	$P6_3mc$
$hP5$	La_2O_3	$D5_2$	$P\bar{3}m1$
	Ni_2Al_3	$D5_{13}$	$P\bar{3}m1$
$hP6$	$CaCu_5$	$D2_d$	$P6/mmm$
	CoSn	$B35$	$P6/mmm$
	Cu_2Te	C_h	$P6/mmm$
	HgS	$B9$	$P3_121$
	MoS_2	$C7$	$P6_3/mmc$
	Ni_2In	$B8_2$	$P6_3/mmc$
$hP8$	Na_3As	$D0_{18}$	$P6_3/mmc$
	Ni_3Sn	$D0_{19}$	$P6_3/mmc$
	TiAs	B_i	$P6_3/mmc$
$hP9$	$CrSi_2$	$C40$	$P6_222$
	Fe_2P	$C22$	$P\bar{6}2m$
	ζAgZn	B_b	$P\bar{3}$
	SiO_2(high quartz)	$C8$	$P6_222$
$hP10$	Pt_2Sn_3	$D5_b$	$P6_3/mmc$
$hP12$	CuS	$B18$	$P6_3/mmc$
	$MgZn_2$	$C14$	$P6_3/mmc$
	SiO_2(βtridymite)	$C10$	$P6_3/mmc$
$hP14$	W_2B_5	$D8_h$	$P6_3/mmc$
$hP16$	Mn_5Si_3	$D8_8$	$P6_3/mcm$
	Ni_3Ti	$D0_{24}$	$P6_3/mmc$
$hP18$	Al_4C_4Si	$E9_4$	$P6_3mc$
	$Al_8FeMg_3Si_6$	$E9_b$	$P\bar{6}2m$
	Mg_2Ni	C_a	$P6_222$
$hP20$	Fe_3Th_7	$D10_2$	$P6_3mc$
	Th_7S_{12}	$D8_k$	$P6_3/m$
$hP24$	Cu_3P	$D0_{21}$	$P6_3cm$
	$MgNi_2$	$C36$	$P6_3/mmc$
$hP28$	Co_2Al_5	$D8_{11}$	$P6_3/mmc$
$hR1$	αHg	$A10$	$R\bar{3}m$
	βPo	A_i	$R\bar{3}m$
$hR2$	αAs	$A7$	$R\bar{3}m$
$hR3$	αSm	$C19$	$R\bar{3}m$
$hR4$	$NaCrS_2$	$F5_1$	$R\bar{3}m$
$hR5$	Bi_2Te_3	$C33$	$R\bar{3}m$
	Ni_3S_2	$D5_e$	$R32$
$hR6$	$CaSi_2$	$C12$	$R\bar{3}m$
	NiS	$B13$	$R\bar{3}m$
$hR7$	Al_4C_3	$D7_1$	$R\bar{3}m$
	Mo_2B_5	$D8_i$	$R\bar{3}m$
$hR10$	αAl_2O_3	$D5_1$	$R\bar{3}c$
$hR13$	Fe_7W_6	$D8_5$	$R\bar{3}m$

(continued)

Arranged Alphabetically by Pearson-Symbol Designation (continued)

Pearson symbol	Prototype	Strukturbericht designation	Space group
$hR15$	B_4C	$D1_g$	$R\bar{3}m$
$hR26$	Cr_5Al_8	$D8_{10}$	$R\bar{3}m$
$hR32$	$CuPt$	$L1_1$	$R\bar{3}m$
$mC6$	$AuTe_2$(calaverite)	$C34$	$C2/m$
$mC8$	CuO	$B26$	$C2/c$
$mC12$	ThC_2	C_g	$C2/c$
$mC14$	δNi_3Sn_4	$D7_a$	$C2/m$
$mC16$	$FeKS_2$	$F5_a$	$C2/c$
$mP12$	$AgAuTe_4$	$E1_b$	$P2/c$
	ZrO_2	$C43$	$P2_1/c$
$mP20$	As_2S_3	$D5_f$	$P2_1/c$
$mP22$	Co_2Al_9	$D8_d$	$P2_1/c$
$mP24$	$FeAsS$	$E0_7$	$P2_1/c$
$mP32$	AsS	B_l	$P2_1/c$
	βSe	A_l	$P2_1/c$
$mP64$	αSe	A_k	$P2_1/c$
$oC4$	αU	$A20$	$Cmcm$
$oC8$	$CaSi$	B_c	$Cmmc$
	αGa	$A11$	$Cmca$
	CrB	B_f	$Cmcm$
	I_2	$A14$	$Cmca$
	P(black)	$A17$	$Cmca$
$oC12$	$ZrSi_2$	$C49$	$Cmcm$
$oC16$	BRe_3	$E1_a$	$Cmcm$
$oC20$	$PdSn_4$	$D1_c$	$Aba2$
$oC24$	$PdSn_2$	C_e	$Aba2$
$oC28$	Al_6Mn	$D2_h$	$Cmcm$
$oF24$	$TiSi_2$	$C54$	$Fddd$
$oF40$	Mn_4B	$D1_f$	$Fddd$
$oF48$	$CuMg_2$	C_b	$Fddd$
$oF72$	GeS_2	$C44$	$Fdd2$
$oF128$	αS	$A16$	$Fddd$
$oI12$	SiS_2	$C42$	$Ibam$
$oI14$	Ta_3B_4	$D7_b$	$Immm$
$oI20$	Al_4U	$D1_b$	$Imma$
$oI28$	Ga_2Mg_5	$D8_g$	$Ibam$
$oP4$	$AuCd$	$B19$	$Pmma$
$oP6$	FeS_2(marcasite)	$C18$	$Pnnm$
	$CaCl_2$	$C35$	$Pnnm$
$oP8$	αNp	A_c	$Pnma$
	$\eta NiSi$	B_d	$Pbnm$
	βCu_3Ti	$D0_a$	$Pmmn$
	FeB	$B27$	$Pnma$
	GeS	$B16$	$Pnma$
	SnS	$B29$	$Pmcn$
	MnP	$B31$	$Pnma$
	TiB	B_m	$Pnma$
$oP12$	Co_2Si	$C23$	$Pnma$
	Co_2Si	$C37$	$Pbnm$
	$HgCl_2$	$C28$	$Pmnb$
$oP16$	Al_3Ni	$D0_{20}$	$Pnma$
	$AsMn_3$	$D0_d$	$Pmmn$
	BaS_3	$D0_{17}$	$P42_1m$
	$CdSb$	B_e	$Pbca$
$oP16$ (continued)	CuS_2Sb	$F5_6$	$Pnma$
	Fe_3C	$D0_{11}$	$Pnma$
$oP20$	Cr_3C_2	$D5_{10}$	$Pnma$
	Sb_2S_3	$D5_8$	$Pnma$
$oP24$	$AuTe_2$(krennerite)	$C46$	$Pma2$
	$CuFe_2S_3$	$E9_e$	$Pnma$
	TiO_2(brookite)	$C21$	$Pbca$
$oP20$	Sb_2O_3(valentinite)	$D5_{11}$	$Pccn$
$oP40$	Cr_7C_3	$D10_1$	$Pnma$
$tI2$	αPa	A_a	$I4/mmm$
	In	$A6$	$I4/mmm$
$tI4$	βSn	$A5$	$I4_1/amd$
$tI6$	CaC_2	$C11_a$	$I4/mmm$
	$FeCu_2SnS_4$	$H2_6$	$I\bar{4}2m$
	$MoSi_2$	$C11_b$	$I4/mmm$
	ThH_2	$L'2_b$	$I4/mmm$
$tI8$	Al_3Ti	$D0_{22}$	$I4/mmm$
$tI10$	Al_4Ba	$D1_3$	$I4/mmm$
	$MoNi_4$	$D1_a$	$I4/m$
$tI12$	Al_2Cu	$C16$	$I4/mcm$
	$ThSi_2$	C_c	$I4_1/amd$
$tI14$	Al_2CdS_4	$E3$	$I\bar{4}$
$tI16$	Al_3Zr	$D0_{23}$	$I4/mmm$
	$CuFeS_2$	$E1_1$	$I\bar{4}2d$
	Ir_3Si	$D0_c'$	$I4/mcm$
	MoB	B_g	$I4_1/amd$
	SiU_3	$D0_c$	$I4/mcm$
	$TlSe$	$B37$	$I4/mcm$
$tI18$	Fe_8N	$D2_g$	$I4/mmm$
$tI26$	$Mn_{12}Th$	$D2_b$	$I4/mmm$
$tI28$	MnU_6	$D2_c$	$I4/mcm$
$tI32$	Cr_5B_3	$D8_l$	$I4/mcm$
	Ni_3P	$D0_e$	$I\bar{4}$
	W_5Si_3	$D8_m$	$I4/mcm$
$tP2$	$AuCu$	$L1_0$	$P4/mmm$
	$\delta CuTi$	$L2_a$	$P4/mmm$
$tP4$	βNp	A_d	$P42_12$
	$CuTi_3$	$L6_0$	$P4/mmm$
	$\gamma CuTi$	$B11$	$P4/nmm$
	PbO	$B10$	$P4/nmm$
	PtS	$B17$	$P4_2/mmc$
$tP6$	Cu_2Sb	$C38$	$P4/nmm$
	$PbFCl$	$E0_1$	$P4/nmm$
	TiO_2(rutile)	$C4$	$P4_2/mnm$
$tP10$	Pb_4Pt	$D1_d$	$P4/nbm$
	Si_2U_3	$D5_a$	$P4/mbm$
$tP16$	PdS	$B34$	$P4_2/m$
$tP20$	B_4Th	$D1_e$	$P4/mbm$
$tP30$	βU	A_b	$P4_2/mnm$
	$\sigma CrFe$	$D8_b$	$P4_2/mnm$
$tP40$	Al_7Cu_2Fe	$E9_a$	$P4/mnc$
	Zn_3P_2	$D5_9$	$P4_2/nmc$
$tP50$	γB	A_g	$P4_2/nnm$

Arranged Alphabetically by Strukturbericht Designation

Strukturbericht designation	Prototype	Pearson symbol	Space group
A_a	αPa	$tI2$	$I4/mmm$
A_b	βU	$tP30$	$P4_2/mnm$
A_c	αNp	$oP8$	$Pnma$
A_d	βNp	$tP4$	$P42_12$
A_f	$HgSn_{6-10}$	$hP1$	$P6/mmm$
A_g	γB	$tP50$	$P4_2/nnm$
A_h	αPo	$cP1$	$Pm\bar{3}m$
A_i	βPo	$hR1$	$R\bar{3}m$
A_k	αSe	$mP64$	$P2_1/c$
A_l	βSe	$mP32$	$P2_1/c$
$A1$	Cu	$cF4$	$Fm\bar{3}m$
$A2$	W	$cI2$	$Im\bar{3}m$
$A3$	Mg	$hP2$	$P6_3/mmc$
$A3'$	αLa	$hP4$	$P6_3/mmc$
$A4$	C(diamond)	$cF8$	$Fd\bar{3}m$

(continued)

Arranged Alphabetically by Strukturbericht Designation (continued)

Strukturbericht designation	Prototype	Pearson symbol	Space group	Strukturbericht designation	Prototype	Pearson symbol	Space group
A5	βSn	$tI4$	$I4_1/amd$	C36	MgNi$_2$	$hP24$	$P6_3/mmc$
A6	In	$tI2$	$I4/mmm$	C37	Co$_2$Si	$oP12$	$Pbnm$
A7	αAs	$hR2$	$R\bar{3}m$	C38	Cu$_2$Sb	$tP6$	$P4/nmm$
A8	γSe	$hP3$	$P3_121$	C40	CrSi$_2$	$hP9$	$P6_222$
A9	C(graphite)	$hP4$	$P6_3/mmc$	C42	SiS$_2$	$oI12$	$Ibam$
A10	αHg	$hR1$	$R\bar{3}m$	C43	ZrO$_2$	$mP12$	$P2_1/c$
A11	αGa	$oC8$	$Cmca$	C44	GeS$_2$	$oF72$	$Fdd2$
A12	αMn	$cI58$	$I\bar{4}3m$	C46	AuTe$_2$(krennerite)	$oP24$	$Pma2$
A13	βMn	$cP20$	$P4_132$	C49	ZrSi$_2$	$oC12$	$Cmcm$
A14	I$_2$	$oC8$	$Cmca$	C54	TiSi$_2$	$oF24$	$Fddd$
A15	Cr$_3$Si	$cP8$	$Pm\bar{3}n$	D0$_a$	βCu$_3$Ti	$oP8$	$Pmmn$
A16	αS	$oF128$	$Fddd$	D0$_c$	SiU$_3$	$tI16$	$I4/mcm$
A17	P(black)	$oC8$	$Cmca$	D0$'_c$	Ir$_3$Si	$tI16$	$I4/mcm$
A20	αU	$oC4$	$Cmcm$	D0$_d$	AsMn$_3$	$oP16$	$Pmmn$
B$_a$	CoU	$cI16$	$I2_13$	D0$_e$	Ni$_3$P	$tI32$	$I\bar{4}$
B$_b$	ζAgZn	$hP9$	$P\bar{3}$	D0$_2$	CoAs$_3$	$cI32$	$Im\bar{3}$
B$_c$	CaSi	$oC8$	$Cmmc$	D0$_3$	BiF$_3$	$cF16$	$Fm\bar{3}m$
B$_d$	ηNiSi	$oP8$	$Pbnm$	D0$_9$	ReO$_3$	$cP4$	$Pm\bar{3}m$
B$_e$	CdSb	$oP16$	$Pbca$	D0$_{11}$	Fe$_3$C	$oP16$	$Pnma$
B$_f$	CrB	$oC8$	$Cmcm$	D0$_{17}$	BaS$_3$	$oP16$	$P42_1m$
B$_g$	MoB	$tI16$	$I4_1/amd$	D0$_{18}$	Na$_3$As	$hP8$	$P6_3/mmc$
B$_h$	WC	$hP2$	$P\bar{6}m2$	D0$_{19}$	Ni$_3$Sn	$hP8$	$P6_3/mmc$
B$_i$	TiAs	$hP8$	$P6_3/mmc$	D0$_{20}$	Al$_3$Ni	$oP16$	$Pnma$
B$_k$	BN	$hP4$	$P6_3/mmc$	D0$_{21}$	Cu$_3$P	$hP24$	$P6_3cm$
B$_l$	AsS	$mP32$	$P2_1/c$	D0$_{22}$	Al$_3$Ti	$tI8$	$I4/mmm$
B$_m$	TiB	$oP8$	$Pnma$	D0$_{23}$	Al$_3$Zr	$tI16$	$I4/mmm$
B1	NaCl	$cF8$	$Fm\bar{3}m$	D0$_{24}$	Ni$_3$Ti	$hP16$	$P6_3/mmc$
B2	CsCl	$cP2$	$Pm\bar{3}m$	D1$_a$	MoNi$_4$	$tI10$	$I4/m$
B3	ZnS(sphalerite)	$cF8$	$F\bar{4}3m$	D1$_b$	Al$_4$U	$oI20$	$Imma$
B4	ZnS(wurtzite)	$hP4$	$P6_3mc$	D1$_c$	PdSn$_4$	$oC20$	$Aba2$
B8$_1$	NiAs	$hP4$	$P6_3/mmc$	D1$_d$	Pb$_4$Pt	$tP10$	$P4/nbm$
B8$_2$	Ni$_2$In	$hP6$	$P6_3/mmc$	D1$_e$	B$_4$Th	$tP10$	$P4/mbm$
B9	HgS	$hP6$	$P3_121$	D1$_f$	Mn$_4$B	$oF40$	$Fddd$
B10	PbO	$tP4$	$P4/nmm$	D1$_g$	B$_4$C	$hR15$	$R\bar{3}m$
B11	γCuTi	$tP4$	$P4/nmm$	D1$_3$	Al$_4$Ba	$tI10$	$I4/mmm$
B13	NiS	$hR6$	$R\bar{3}m$	D2$_b$	Mn$_{12}$Th	$tI26$	$I4/mmm$
B16	GeS	$oP8$	$Pnma$	D2$_c$	MnU$_6$	$tI28$	$I4/mcm$
B17	PtS	$tP4$	$P4_2/mmc$	D2$_d$	CaCu$_5$	$hP6$	$P6/mmm$
B18	CuS	$hP12$	$P6_3/mmc$	D2$_e$	BaHg$_{11}$	$cP36$	$Pm\bar{3}m$
B19	AuCd	$oP4$	$Pmma$	D2$_f$	UB$_{12}$	$cF52$	$Fm\bar{3}m$
B20	FeSi	$cP8$	$P2_13$	D2$_g$	Fe$_8$N	$tI18$	$I4/mmm$
B26	CuO	$mC8$	$C2/c$	D2$_h$	Al$_6$Mn	$oC28$	$Cmcm$
B27	FeB	$oP8$	$Pnma$	D2$_1$	CaB$_6$	$cP7$	$Pm\bar{3}m$
B29	SnS	$oP8$	$Pmcn$	D2$_3$	NaZn$_{13}$	$cF112$	$Fm\bar{3}c$
B31	MnP	$oP8$	$Pnma$	D5$_a$	Si$_2$U$_3$	$tP10$	$P4/mbm$
B32	NaTl	$cF16$	$Fd\bar{3}m$	D5$_b$	Pt$_2$Sn$_3$	$hP10$	$P6_3/mmc$
B34	PdS	$tP16$	$P4_2/m$	D5$_c$	Pu$_2$C$_3$	$cI40$	$I\bar{4}3d$
B35	CoSn	$hP6$	$P6/mmm$	D5$_e$	Ni$_3$S$_2$	$hR5$	$R32$
B37	TlSe	$tI16$	$I4/mcm$	D5$_f$	As$_2$S$_3$	$mP20$	$P2_1/c$
C$_a$	Mg$_2$Ni	$hP18$	$P6_222$	D5$_1$	αAl$_2$O$_3$	$hR10$	$R\bar{3}c$
C$_b$	CuMg$_2$	$oF48$	$Fddd$	D5$_2$	La$_2$O$_3$	$hP5$	$P\bar{3}m1$
C$_c$	ThSi$_2$	$tI12$	$I4_1/amd$	D5$_3$	Mn$_2$O$_3$	$cI80$	$Ia\bar{3}$
C$_e$	PdSn$_2$	$oC24$	$Aba2$	D5$_4$	Sb$_2$O$_3$(senarmontite)	$cF80$	$Fd\bar{3}m$
C$_g$	ThC$_2$	$mC12$	$C2/c$	D5$_8$	Sb$_2$S$_3$	$oP20$	$Pnma$
C$_h$	Cu$_2$Te	$hP6$	$P6/mmm$	D5$_9$	Zn$_3$P$_2$	$tP40$	$P4_2/nmc$
C$_k$	LiZn$_2$	$hP3$	$P6_3/mmc$	D5$_{10}$	Cr$_3$C$_2$	$oP20$	$Pnma$
C1	CaF$_2$	$cF12$	$Fm\bar{3}m$	D5$_{11}$	Sb$_2$O$_3$(valentinite)	$oP20$	$Pccn$
C1$_b$	MgAgAs	$cF12$	$F\bar{4}3m$	D5$_{13}$	Ni$_2$Al$_3$	$hP5$	$P\bar{3}m1$
C2	FeS$_2$(pyrite)	$cP12$	$Pa3$	D7$_a$	δNi$_3$Sn$_4$	$mC14$	$C2/m$
C3	Ag$_2$O	$cP6$	$Pn\bar{3}m$	D7$_b$	Ta$_3$B$_4$	$oI14$	$Immm$
C4	TiO$_2$(rutile)	$tP6$	$P4_2/mnm$	D7$_1$	Al$_4$C$_3$	$hR7$	$R\bar{3}m$
C6	CdI$_2$	$hP3$	$P\bar{3}m1$	D7$_2$	Co$_3$S$_4$	$cF56$	$Fd\bar{3}m$
C7	MoS$_2$	$hP6$	$P6_3/mmc$	D7$_3$	Th$_3$P$_4$	$cI28$	$I\bar{4}3d$
C8	SiO$_2$(high quartz)	$hP9$	$P6_222$	D8$_a$	Mn$_{23}$Th$_6$	$cF116$	$Fm\bar{3}m$
C9	SiO$_2$(β cristobalite)	$cF24$	$Fd\bar{3}m$	D8$_b$	σCrFe	$tP30$	$P4_2/mnm$
C10	SiO$_2$(β tridymite)	$hP12$	$P6_3/mmc$	D8$_c$	Mg$_2$Zn$_{11}$	$cP39$	$Pm\bar{3}$
C11$_a$	CaC$_2$	$tI6$	$I4/mmm$	D8$_d$	Co$_2$Al$_9$	$mP22$	$P2_1/c$
C11$_b$	MoSi$_2$	$tI6$	$I4/mmm$	D8$_e$	Mg$_{32}$(Al,Zn)$_{49}$	$cI162$	$Im\bar{3}$
C12	CaSi$_2$	$hR6$	$R\bar{3}m$	D8$_f$	Ge$_7$Ir$_3$	$cI40$	$Im\bar{3}m$
C14	MgZn$_2$	$hP12$	$P6_3/mmc$	D8$_g$	Ga$_2$Mg$_5$	$oI28$	$Ibam$
C15	Cu$_2$Mg	$cF24$	$Fd\bar{3}m$	D8$_h$	W$_2$B$_5$	$hP14$	$P6_3/mmc$
C15$_b$	AuBe$_5$	$cF24$	$F\bar{4}3m$	D8$_i$	Mo$_2$B$_5$	$hR7$	$R\bar{3}m$
C16	Al$_2$Cu	$tI12$	$I4/mcm$	D8$_k$	Th$_7$S$_{12}$	$hP20$	$P6_3/m$
C18	FeS$_2$(marcasite)	$oP6$	$Pnnm$	D8$_l$	Cr$_5$B$_3$	$tI32$	$I4/mcm$
C19	αSm	$hR3$	$R\bar{3}m$	D8$_m$	W$_5$Si$_3$	$tI32$	$I4/mcm$
C21	TiO$_2$(brookite)	$oP24$	$Pbca$	D8$_1$	Fe$_3$Zn$_{10}$	$cI52$	$Im\bar{3}m$
C22	Fe$_2$P	$hP9$	$P\bar{6}2m$	D8$_2$	Cu$_5$Zn$_8$	$cI52$	$I\bar{4}3m$
C23	Co$_2$Si	$oP12$	$Pnma$	D8$_3$	Cu$_9$Al$_4$	$cP52$	$P\bar{4}3m$
C28	HgCl$_2$	$oP12$	$Pmnb$	D8$_4$	Cr$_{23}$C$_6$	$cF116$	$Fm\bar{3}m$
C32	AlB$_2$	$hP3$	$P6/mmm$	D8$_5$	Fe$_7$W$_6$	$hR13$	$R\bar{3}m$
C33	Bi$_2$Te$_3$	$hR5$	$R\bar{3}m$	D8$_6$	Cu$_{15}$Si$_4$	$cI76$	$I\bar{4}3d$
C34	AuTe$_2$(calaverite)	$mC6$	$C2/m$				
C35	CaCl$_2$	$oP6$	$Pnnm$				

(continued)

Arranged Alphabetically by Strukturbericht Designation (continued)

Struktur-bericht designation	Prototype	Pearson symbol	Space group	Struktur-bericht designation	Prototype	Pearson symbol	Space group
$D8_8$	Mn_5Si_3	$hP16$	$P6_3/mcm$	$E9_4$	Al_4C_4Si	$hP18$	$P6_3mc$
$D8_9$	Co_9S_8	$cF68$	$Fm\overline{3}m$	$F5_a$	$FeKS_2$	$mC16$	$C2/c$
$D8_{10}$	Cr_5Al_8	$hR26$	$R\overline{3}m$	$F0_1$	$NiSbS$	$cP12$	$P2_13$
$D8_{11}$	Co_2Al_5	$hP28$	$P6_3/mmc$	$F5_1$	$NaCrS_2$	$hR4$	$R\overline{3}m$
$D10_1$	Cr_7C_3	$oP40$	$Pnma$	$F5_6$	CuS_2Sb	$oP16$	$Pnma$
$D10_2$	Fe_3Th_7	$hP20$	$P6_3mc$	$H1_1$	Al_2MgO_4	$cF56$	$Fd\overline{3}m$
$D8_{11}$	Co_2Al_5	$hP28$	$P6_3/mmc$	$H2_4$	Cu_3VS_4	$cP8$	$P\overline{4}3m$
$E0_1$	$PbFCl$	$tP6$	$P4/nmm$	$H2_6$	$FeCu_2SnS_4$	$tI16$	$I\overline{4}2m$
$E0_7$	$FeAsS$	$mP24$	$P2_1/c$	$L'1$	Fe_4N	$cF5$	$Fm\overline{3}m$
$E1_a$	$MgCuAl_2$	$oC16$	$Cmcm$	$L'1_2$	$AlFe_3C$	$cP5$	$Pm\overline{3}m$
$E1_b$	$AgAuTe_4$	$mP12$	$P2/c$	$L'2$	ThH_2	$tI6$	$I4/mmm$
$E1_1$	$CuFeS_2$	$tI16$	$I\overline{4}2d$	$L'3$	Fe_2N	$hP3$	$P6_3/mmc$
$E2_1$	$CaTiO_3$	$cP5$	$Pm\overline{3}m$	$L1_a$	$CuPt_3$	$cF32$	$Fm\overline{3}c$
$E3$	Al_2CdS_4	$tI14$	$I\overline{4}$	$L1_0$	$AuCu$	$tP2$	$P4/mmm$
$E9_a$	Al_7Cu_2Fe	$tP40$	$P4/mnc$	$L1_1$	$CuPt$	$hR32$	$R\overline{3}m$
$E9_b$	$Al_8FeMg_3Si_6$	$hP18$	$P\overline{6}2m$	$L1_2$	$AuCu_3$	$cP4$	$Pm\overline{3}m$
$E9_d$	$AlLi_3N_2$	$cI96$	$Ia\overline{3}$	$L2_a$	$\delta CuTi$	$tP2$	$P4/mmm$
$E9_e$	$CuFe_2S_3$	$oP24$	$Pnma$	$L2_1$	$AlCu_2Mn$	$cF16$	$Fm\overline{3}m$
$E9_c$	Mn_3Al_9Si	$hP26$	$P6_3/mmc$	$L2_2$	Sb_2Tl_7	$cI54$	$Im\overline{3}m$
$E9_3$	Fe_3W_3C	$cF112$	$Fd\overline{3}m$	$L6_0$	$CuTi_3$	$tP4$	$P4/mmm$

Temperature Conversions

The general arrangement of this table was devised by Sauveur and Boylston more than 40 years ago. The middle column of figures (in **boldface** type) contains the readings (°F or °C) to be converted. If converting from degrees Fahrenheit to degrees Celsius, read the Celsius equivalent in the column headed "C". If converting from Celsius to Fahrenheit, read the Fahrenheit equivalent in the column headed "F".

F		C	F		C	F		C	F		C	F		C
.....	**−458**	−272.22	.....	**−308**	−188.89	−252.4	**−158**	−105.56	+17.6	**−8**	−22.22	287.6	**142**	61.11
.....	**−456**	−271.11	.....	**−306**	−187.78	−248.8	**−156**	−104.44	+21.2	**−6**	−21.11	291.2	**144**	62.22
.....	**−454**	−270.00	.....	**−304**	−186.67	−245.2	**−154**	−103.33	+24.8	**−4**	−20.00	294.8	**146**	63.33
.....	**−452**	−268.89	.....	**−302**	−185.56	−241.6	**−152**	−102.22	+28.4	**−2**	−18.89	298.4	**148**	64.44
.....	**−450**	−267.78	.....	**−300**	−184.44	−238.0	**−150**	−101.11	+32.0	**±0**	−17.78	302.0	**150**	65.56
.....	**−448**	−266.67	.....	**−298**	−183.33	−234.4	**−148**	−100.00	+35.6	**+2**	−16.67	305.6	**152**	66.67
.....	**−446**	−265.56	.....	**−296**	−182.22	−230.8	**−146**	−98.89	+39.2	**+4**	−15.56	309.2	**154**	67.78
.....	**−444**	−264.44	.....	**−294**	−181.11	−227.2	**−144**	−97.78	+42.8	**+6**	−14.44	312.8	**156**	68.89
.....	**−442**	−263.33	.....	**−292**	−180.00	−223.6	**−142**	−96.67	+46.4	**+8**	−13.33	316.4	**158**	70.00
.....	**−440**	−262.22	.....	**−290**	−178.89	−220.0	**−140**	−95.56	+50.0	**+10**	−12.22	320.0	**160**	71.11
.....	**−438**	−261.11	.....	**−288**	−177.78	−216.4	**−138**	−94.44	+53.6	**+12**	−11.11	323.6	**162**	72.22
.....	**−436**	−260.00	.....	**−286**	−176.67	−212.8	**−136**	−93.33	+57.2	**+14**	−10.00	327.2	**164**	73.33
.....	**−434**	−258.89	.....	**−284**	−175.56	−209.2	**−134**	−92.22	+60.8	**+16**	−8.89	330.8	**166**	74.44
.....	**−432**	−257.78	.....	**−282**	−174.44	−205.6	**−132**	−91.11	+64.4	**+18**	−7.78	334.4	**168**	75.56
.....	**−430**	−256.67	.....	**−280**	−173.33	−202.0	**−130**	−90.00	+68.0	**+20**	−6.67	338.0	**170**	76.67
.....	**−428**	−255.56	.....	**−278**	−172.22	−198.4	**−128**	−88.89	+71.6	**+22**	−5.56	341.6	**172**	77.78
.....	**−426**	−254.44	.....	**−276**	−171.11	−194.8	**−126**	−87.78	+75.2	**+24**	−4.44	345.2	**174**	78.89
.....	**−424**	−253.33	.....	**−274**	−170.00	−191.2	**−124**	−86.67	+78.8	**+26**	−3.33	348.8	**176**	80.00
.....	**−422**	−252.22	−457.6	**−272**	−168.89	−187.6	**−122**	−85.56	+82.4	**+28**	−2.22	352.4	**178**	81.11
.....	**−420**	−251.11	−454.0	**−270**	−167.78	−184.0	**−120**	−84.44	+86.0	**+30**	−1.11	356.0	**180**	82.22
.....	**−418**	−250.00	−450.4	**−268**	−166.67	−180.4	**−118**	−83.33	+89.6	**+32**	±0.00	359.6	**182**	83.33
.....	**−416**	−248.89	−446.8	**−266**	−165.56	−176.8	**−116**	−82.22	+93.2	**+34**	+1.11	363.2	**184**	84.44
.....	**−414**	−247.78	−443.2	**−264**	−164.44	−173.2	**−114**	−81.11	+96.8	**+36**	+2.22	366.8	**186**	85.56
.....	**−412**	−246.67	−439.6	**−262**	−163.33	−169.6	**−112**	−80.00	+100.4	**+38**	+3.33	370.4	**188**	86.67
.....	**−410**	−245.56	−436.0	**−260**	−162.22	−166.0	**−110**	−78.89	+104.0	**+40**	+4.44	374.0	**190**	87.78
.....	**−408**	−244.44	−432.4	**−258**	−161.11	−162.4	**−108**	−77.78	107.6	**42**	5.56	377.6	**192**	88.89
.....	**−406**	−243.33	−428.8	**−256**	−160.00	−158.8	**−106**	−76.67	111.2	**44**	6.67	381.2	**194**	90.00
.....	**−404**	−242.22	−425.2	**−254**	−158.89	−155.2	**−104**	−75.56	114.8	**46**	7.78	384.8	**196**	91.11
.....	**−402**	−241.11	−421.6	**−252**	−157.78	−151.6	**−102**	−74.44	118.4	**48**	8.89	388.4	**198**	92.22
.....	**−400**	−240.00	−418.0	**−250**	−156.67	−148.0	**−100**	−73.33	122.0	**50**	10.00	392.0	**200**	93.33
.....	**−398**	−238.89	−414.4	**−248**	−155.56	−144.4	**−98**	−72.22	125.6	**52**	11.11	395.6	**202**	94.44
.....	**−396**	−237.78	−410.8	**−246**	−154.44	−140.8	**−96**	−71.11	129.2	**54**	12.22	399.2	**204**	95.56
.....	**−394**	−236.67	−407.2	**−244**	−153.33	−137.2	**−94**	−70.00	132.8	**56**	13.33	402.8	**206**	96.67
.....	**−392**	−235.56	−403.6	**−242**	−152.22	−133.6	**−92**	−68.89	136.4	**58**	14.44	406.4	**208**	97.78
.....	**−390**	−234.44	−400.0	**−240**	−151.11	−130.0	**−90**	−67.78	140.0	**60**	15.56	410.0	**210**	98.89
.....	**−388**	−233.33	−396.4	**−238**	−150.00	−126.4	**−88**	−66.67	143.6	**62**	16.67	413.6	**212**	100.00
.....	**−386**	−232.22	−392.8	**−236**	−148.89	−122.8	**−86**	−65.56	147.2	**64**	17.78	417.2	**214**	101.11
.....	**−384**	−231.11	−389.2	**−234**	−147.78	−119.2	**−84**	−64.44	150.8	**66**	18.89	420.8	**216**	102.22
.....	**−382**	−230.00	−385.6	**−232**	−146.67	−115.6	**−82**	−63.33	154.4	**68**	20.00	424.4	**218**	103.33
.....	**−380**	−228.89	−382.0	**−230**	−145.56	−112.0	**−80**	−62.22	158.0	**70**	21.11	428.0	**220**	104.44
.....	**−378**	−227.78	−378.4	**−228**	−144.44	−108.4	**−78**	−61.11	161.6	**72**	22.22	431.6	**222**	105.56
.....	**−376**	−226.67	−374.8	**−226**	−143.33	−104.8	**−76**	−60.00	165.2	**74**	23.33	435.2	**224**	106.67
.....	**−374**	−225.56	−371.2	**−224**	−142.22	−101.2	**−74**	−58.89	168.8	**76**	24.44	438.8	**226**	107.78
.....	**−372**	−224.44	−367.6	**−222**	−141.11	−97.6	**−72**	−57.78	172.4	**78**	25.56	442.4	**228**	108.89
.....	**−370**	−223.33	−364.0	**−220**	−140.00	−94.0	**−70**	−56.67	176.0	**80**	26.67	446.0	**230**	110.00
.....	**−368**	−222.22	−360.4	**−218**	−138.89	−90.4	**−68**	−55.56	179.6	**82**	27.78	449.6	**232**	111.11
.....	**−366**	−221.11	−356.8	**−216**	−137.78	−86.8	**−66**	−54.44	183.2	**84**	28.89	453.2	**234**	112.22
.....	**−364**	−220.00	−353.2	**−214**	−136.67	−83.2	**−64**	−53.33	186.8	**86**	30.00	456.8	**236**	113.33
.....	**−362**	−218.89	−349.6	**−212**	−135.56	−79.6	**−62**	−52.22	190.4	**88**	31.11	460.4	**238**	114.44
.....	**−360**	−217.78	−346.0	**−210**	−134.44	−76.0	**−60**	−51.11	194.0	**90**	32.22	464.0	**240**	115.56
.....	**−358**	−216.67	−342.4	**−208**	−133.33	−72.4	**−58**	−50.00	197.6	**92**	33.33	467.6	**242**	116.67
.....	**−356**	−215.56	−338.8	**−206**	−132.22	−68.8	**−56**	−48.89	201.2	**94**	34.44	471.2	**244**	117.78
.....	**−354**	−214.44	−335.2	**−204**	−131.11	−65.2	**−54**	−47.78	204.8	**96**	35.56	474.8	**246**	118.89
.....	**−352**	−213.33	−331.6	**−202**	−130.00	−61.6	**−52**	−46.67	208.4	**98**	36.67	478.4	**248**	120.00
.....	**−350**	−212.22	−328.0	**−200**	−128.89	−58.0	**−50**	−45.56	212.0	**100**	37.78	482.0	**250**	121.11
.....	**−348**	−211.11	−324.4	**−198**	−127.78	−54.4	**−48**	−44.44	215.6	**102**	38.89	485.6	**252**	122.22
.....	**−346**	−210.00	−320.8	**−196**	−126.67	−50.8	**−46**	−43.33	219.2	**104**	40.00	489.2	**254**	123.33
.....	**−344**	−208.89	−317.2	**−194**	−125.56	−47.2	**−44**	−42.22	222.8	**106**	41.11	492.8	**256**	124.44
.....	**−342**	−207.78	−313.6	**−192**	−124.44	−43.6	**−42**	−41.11	226.4	**108**	42.22	496.4	**258**	125.56
.....	**−340**	−206.67	−310.0	**−190**	−123.33	−40.0	**−40**	−40.00	230.0	**110**	43.33	500.0	**260**	126.67
.....	**−338**	−205.56	−306.4	**−188**	−122.22	−36.4	**−38**	−38.89	233.6	**112**	44.44	503.6	**262**	127.78
.....	**−336**	−204.44	−302.8	**−186**	−121.11	−32.8	**−36**	−37.78	237.2	**114**	45.56	507.2	**264**	128.89
.....	**−334**	−203.33	−299.2	**−184**	−120.00	−29.2	**−34**	−36.67	240.8	**116**	46.67	510.8	**266**	130.00
.....	**−332**	−202.22	−295.6	**−182**	−118.89	−25.6	**−32**	−35.56	244.4	**118**	47.78	514.4	**268**	131.11
.....	**−330**	−201.11	−292.0	**−180**	−117.78	−22.0	**−30**	−34.44	248.0	**120**	48.89	518.0	**270**	132.22
.....	**−328**	−200.00	−288.4	**−178**	−116.67	−18.4	**−28**	−33.33	251.6	**122**	50.00	521.6	**272**	133.33
.....	**−326**	−198.89	−284.8	**−176**	−115.56	−14.8	**−26**	−32.22	255.2	**124**	51.11	525.2	**274**	134.44
.....	**−324**	−197.78	−281.2	**−174**	−114.44	−11.2	**−24**	−31.11	258.8	**126**	52.22	528.8	**276**	135.56
.....	**−322**	−196.67	−277.6	**−172**	−113.33	−7.6	**−22**	−30.00	262.4	**128**	53.33	532.4	**278**	136.67
.....	**−320**	−195.56	−274.0	**−170**	−112.22	−4.0	**−20**	−28.89	266.0	**130**	54.44	536.0	**280**	137.78
.....	**−318**	−194.44	−270.4	**−168**	−111.11	−0.4	**−18**	−27.78	269.6	**132**	55.56	539.6	**282**	138.89
.....	**−316**	−193.33	−266.8	**−166**	−110.00	+3.2	**−16**	−26.67	273.2	**134**	56.67	543.2	**284**	140.00
.....	**−314**	−192.22	−263.2	**−164**	−108.89	+6.8	**−14**	−25.56	276.8	**136**	57.78	546.8	**286**	141.11
.....	**−312**	−191.11	−259.6	**−162**	−107.78	+10.4	**−12**	−24.44	280.4	**138**	58.89	550.4	**288**	142.22
.....	**−310**	−190.00	−256.0	**−160**	−106.67	+14.0	**−10**	−23.33	284.0	**140**	60.00	554.0	**290**	143.33

Temperature Conversions (continued)

F		C	F		C	F		C	F		C	F		C
557.6	292	144.44	870.8	466	241.11	1832.0	1000	537.78	3398.0	1870	1021.1	4964.0	2740	1504.4
561.2	294	145.56	874.4	468	242.22	1850.0	1010	543.33	3416.0	1880	1026.7	4982.0	2750	1510.0
564.8	296	146.67	878.0	470	243.33	1868.0	1020	548.89	3434.0	1890	1032.2	5000.0	2760	1515.6
568.4	298	147.78	881.6	472	244.44	1886.0	1030	554.44	3452.0	1900	1037.8	5018.0	2770	1521.1
572.0	300	148.89	885.2	474	245.56	1904.0	1040	560.00	3470.0	1910	1043.3	5036.0	2780	1526.7
575.6	302	150.00	888.8	476	246.67	1922.0	1050	565.56	3488.0	1920	1048.9	5054.0	2790	1532.2
579.2	304	151.11	892.4	478	247.78	1940.0	1060	571.11	3506.0	1930	1054.4	5072.0	2800	1537.8
582.8	306	152.22	896.0	480	248.89	1958.0	1070	576.67	3524.0	1940	1060.0	5090.0	2810	1543.3
586.4	308	153.33	899.6	482	250.00	1976.0	1080	582.22	3542.0	1950	1065.6	5108.0	2820	1548.9
590.0	310	154.44	903.2	484	251.11	1994.0	1090	587.78	3560.0	1960	1071.1	5126.0	2830	1554.4
593.6	312	155.56	906.8	486	252.22	2012.0	1100	593.33	3578.0	1970	1076.7	5144.0	2840	1560.0
597.2	314	156.67	910.4	488	253.33	2030.0	1110	598.89	3596.0	1980	1082.2	5162.0	2850	1565.6
600.8	316	157.78	914.0	490	254.44	2048.0	1120	604.44	3614.0	1990	1087.8	5180.0	2860	1571.1
604.4	318	158.89	917.6	492	255.56	2066.0	1130	610.00	3632.0	2000	1093.3	5198.0	2870	1576.7
608.0	320	160.00	921.2	494	256.67	2084.0	1140	615.56	3650.0	2010	1098.9	5216.0	2880	1582.2
611.6	322	161.11	924.8	496	257.78	2102.0	1150	621.11	3668.0	2020	1104.4	5234.0	2890	1587.8
615.2	324	162.22	928.4	498	258.89	2120.0	1160	626.67	3686.0	2030	1110.0	5252.0	2900	1593.3
618.8	326	163.33	932.0	500	260.00	2138.0	1170	632.22	3704.0	2040	1115.6	5270.0	2910	1598.9
622.4	328	164.44	935.6	502	261.11	2156.0	1180	637.78	3722.0	2050	1121.1	5288.0	2920	1604.4
626.0	330	165.56	939.2	504	262.22	2174.0	1190	643.33	3740.0	2060	1126.7	5306.0	2930	1610.0
629.6	332	166.67	942.8	506	263.33	2192.0	1200	648.89	3758.0	2070	1132.2	5324.0	2940	1615.6
633.2	334	167.78	946.4	508	264.44	2210.0	1210	654.44	3776.0	2080	1137.8	5342.0	2950	1621.1
636.8	336	168.89	950.0	510	265.56	2228.0	1220	660.00	3794.0	2090	1143.3	5360.0	2960	1626.7
640.4	338	170.00	953.6	512	266.67	2246.0	1230	665.56	3812.0	2100	1148.9	5378.0	2970	1632.2
644.0	340	171.11	957.2	514	267.78	2264.0	1240	671.11	3830.0	2110	1154.4	5396.0	2980	1637.8
647.6	342	172.22	960.8	516	268.89	2282.0	1250	676.67	3848.0	2120	1160.0	5414.0	2990	1643.3
651.2	344	173.33	964.4	518	270.00	2300.0	1260	682.22	3866.0	2130	1165.6	5432.0	3000	1648.9
654.8	346	174.44	968.0	520	271.11	2318.0	1270	687.78	3884.0	2140	1171.1	5450.0	3010	1654.4
658.4	348	175.56	971.6	522	272.22	2336.0	1280	693.33	3902.0	2150	1176.7	5468.0	3020	1660.0
662.0	350	176.67	975.2	524	273.33	2354.0	1290	698.89	3920.0	2160	1182.2	5486.0	3030	1665.6
665.6	352	177.78	978.8	526	274.44	2372.0	1300	704.44	3938.0	2170	1187.8	5504.0	3040	1671.1
669.2	354	178.89	982.4	528	275.56	2390.0	1310	710.00	3956.0	2180	1193.3	5522.0	3050	1676.7
672.8	356	180.00	986.0	530	276.67	2408.0	1320	715.56	3974.0	2190	1198.9	5540.0	3060	1682.2
676.4	358	181.11	989.6	532	277.78	2426.0	1330	721.11	3992.0	2200	1204.4	5558.0	3070	1687.8
680.0	360	182.22	993.2	534	278.89	2444.0	1340	726.67	4010.0	2210	1210.0	5576.0	3080	1693.3
683.6	362	183.33	996.8	536	280.00	2462.0	1350	732.22	4028.0	2220	1215.6	5594.0	3090	1698.9
687.2	364	184.44	1000.4	538	281.11	2480.0	1360	737.78	4046.0	2230	1221.1	5612.0	3100	1704.4
690.8	366	185.56	1004.0	540	282.22	2498.0	1370	743.33	4064.0	2240	1226.7	5702.0	3150	1732.2
694.4	368	186.67	1007.6	542	283.33	2516.0	1380	748.89	4082.0	2250	1232.2	5792.0	3200	1760.0
698.0	370	187.78	1011.2	544	284.44	2534.0	1390	754.44	4100.0	2260	1237.8	5882.0	3250	1787.7
701.6	372	188.89	1014.8	546	285.56	2552.0	1400	760.00	4118.0	2270	1243.3	5972.0	3300	1815.5
705.2	374	190.00	1018.4	548	286.67	2570.0	1410	765.56	4136.0	2280	1248.9	6062.0	3350	1843.3
708.8	376	191.11	1022.0	550	287.78	2588.0	1420	771.11	4154.0	2290	1254.4	6152.0	3400	1871.1
712.4	378	192.22	1040.0	560	293.33	2606.0	1430	776.67	4172.0	2300	1260.0	6242.0	3450	1898.8
716.0	380	193.33	1058.0	570	298.89	2624.0	1440	782.22	4190.0	2310	1265.6	6332.0	3500	1926.6
719.6	382	194.44	1076.0	580	304.44	2642.0	1450	787.78	4208.0	2320	1271.1	6422.0	3550	1954.4
723.2	384	195.56	1094.0	590	310.00	2660.0	1460	793.33	4226.0	2330	1276.7	6512.0	3600	1982.2
726.8	386	196.67	1112.0	600	315.56	2678.0	1470	798.89	4244.0	2340	1282.2	6602.0	3650	2010.0
730.4	388	197.78	1130.0	610	321.11	2696.0	1480	804.44	4262.0	2350	1287.8	6692.0	3700	2037.7
734.0	390	198.89	1148.0	620	326.67	2714.0	1490	810.00	4280.0	2360	1293.3	6782.0	3750	2065.5
737.6	392	200.00	1166.0	630	332.22	2732.0	1500	815.56	4298.0	2370	1298.9	6872.0	3800	2093.3
741.2	394	201.11	1184.0	640	337.78	2750.0	1510	821.11	4316.0	2380	1304.4	6962.0	3850	2121.1
744.8	396	202.22	1202.0	650	343.33	2768.0	1520	826.67	4334.0	2390	1310.0	7052.0	3900	2148.8
748.4	398	203.33	1220.0	660	348.89	2786.0	1530	832.22	4352.0	2400	1315.6	7142.0	3950	2176.6
752.0	400	204.44	1238.0	670	354.44	2804.0	1540	837.78	4370.0	2410	1321.1	7232.0	4000	2204.4
755.6	402	205.56	1256.0	680	360.00	2822.0	1550	843.33	4388.0	2420	1326.7	7322.0	4050	2232.2
759.2	404	206.67	1274.0	690	365.56	2840.0	1560	848.89	4406.0	2430	1332.2	7412.0	4100	2260.0
762.8	406	207.78	1292.0	700	371.11	2858.0	1570	854.44	4424.0	2440	1337.8	7502.0	4150	2287.7
766.4	408	208.89	1310.0	710	376.67	2876.0	1580	860.00	4442.0	2450	1343.3	7592.0	4200	2315.5
770.0	410	210.00	1328.0	720	382.22	2894.0	1590	865.56	4460.0	2460	1348.9	7682.0	4250	2343.3
773.6	412	211.11	1346.0	730	387.78	2912.0	1600	871.11	4478.0	2470	1354.4	7772.0	4300	2371.1
777.2	414	212.22	1364.0	740	393.33	2930.0	1610	876.67	4496.0	2480	1360.0	7862.0	4350	2398.8
780.8	416	213.33	1382.0	750	398.89	2948.0	1620	882.22	4514.0	2490	1365.6	7952.0	4400	2426.6
784.4	418	214.44	1400.0	760	404.44	2966.0	1630	887.78	4532.0	2500	1371.1	8042.0	4450	2454.4
788.0	420	215.56	1418.0	770	410.00	2984.0	1640	893.33	4550.0	2510	1376.7	8132.0	4500	2482.2
791.6	422	216.67	1436.0	780	415.56	3002.0	1650	898.89	4568.0	2520	1382.2	8222.0	4550	2510.0
795.2	424	217.78	1454.0	790	421.11	3020.0	1660	904.44	4586.0	2530	1387.8	8312.0	4600	2537.7
798.8	426	218.89	1472.0	800	426.67	3038.0	1670	910.00	4604.0	2540	1393.3	8402.0	4650	2565.5
802.4	428	220.00	1490.0	810	432.22	3056.0	1680	915.56	4622.0	2550	1398.9	8492.0	4700	2593.3
806.0	430	221.11	1508.0	820	437.78	3074.0	1690	921.11	4640.0	2560	1404.4	8582.0	4750	2621.1
809.6	432	222.22	1526.0	830	443.33	3092.0	1700	926.67	4658.0	2570	1410.0	8672.0	4800	2648.8
813.2	434	223.33	1544.0	840	448.89	3110.0	1710	932.22	4676.0	2580	1415.6	8762.0	4850	2676.6
816.8	436	224.44	1562.0	850	454.44	3128.0	1720	937.78	4694.0	2590	1421.1	8852.0	4900	2704.4
820.4	438	225.56	1580.0	860	460.00	3146.0	1730	943.33	4712.0	2600	1426.7	8942.0	4950	2732.2
824.0	440	226.67	1598.0	870	465.56	3164.0	1740	948.89	4730.0	2610	1432.2	9032.0	5000	2760.0
827.6	442	227.78	1616.0	880	471.11	3182.0	1750	954.44	4748.0	2620	1437.8	9122.0	5050	2787.7
831.2	444	228.89	1634.0	890	476.67	3200.0	1760	960.00	4766.0	2630	1443.3	9212.0	5100	2815.5
834.8	446	230.00	1652.0	900	482.22	3218.0	1770	965.56	4784.0	2640	1448.9	9302.0	5150	2843.3
838.4	448	231.11	1670.0	910	487.78	3236.0	1780	971.11	4802.0	2650	1454.4	9392.0	5200	2871.1
842.0	450	232.22	1688.0	920	493.33	3254.0	1790	976.67	4820.0	2660	1460.0	9482.0	5250	2898.8
845.6	452	233.33	1706.0	930	498.89	3272.0	1800	982.22	4838.0	2670	1465.6	9572.0	5300	2926.6
849.2	454	234.44	1724.0	940	504.44	3290.0	1810	987.78	4856.0	2680	1471.1	9662.0	5350	2954.4
852.8	456	235.56	1742.0	950	510.00	3308.0	1820	993.33	4874.0	2690	1476.7	9752.0	5400	2982.2
856.4	458	236.67	1760.0	960	515.56	3326.0	1830	998.89	4892.0	2700	1482.2	9842.0	5450	3010.0
860.0	460	237.78	1778.0	970	521.11	3344.0	1840	1004.4	4910.0	2710	1487.8	9932.0	5500	3037.7
863.6	462	238.89	1796.0	980	526.67	3362.0	1850	1010.0	4928.0	2720	1493.3	10022.0	5550	3065.5
867.2	464	240.00	1814.0	990	532.22	3380.0	1860	1015.6	4946.0	2730	1498.9	10112.0	5600	3093.3

Abbreviations

antiphase structure	APS	gas	G	megapascal	MPa
atomic percent	at.%	Gibbs energy	G	melting point	M.P.
body-centered cubic	bcc	gigapascal	GPa	metallic element	M
body-centered tetragonal	bct	greater than	>	nanometer	nm
boiling point	B.P.	heat capacity	C	percent	%
Celsius	°C	heat energy	Q	pressure	P
close-packed hexagonal	cph	high temperature	HT	room temperature	RT
components	c	increment (finite)	δ	solid	S
composition	X	increment (infinitesimally		stable phases	p
Curie temperature	T_C	small)	Δ	sublimation point	S.P.
degree (Angular)	°	interaxial angle	A, B, Γ	temperature	T
degrees of freedom	f	internal energy	E	transformation temperature	A
differential	d	Kelvin	K	triple point	T.P.
edge length	a,b,c	kilobar	kbar	unknown	…
enthalpy	H	kilopascal	kPa	volume	V
entropy	S	less than	<	weight percent	wt.%
face-centered cubic	fcc	liquid	L	work energy	W
Fahrenheit	°F	low temperature	LT		

Greek Alphabet

Greek letter	Name	English equivalent	Greek letter	Name	English equivalent	Greek letter	Name	English equivalent
A, α	Alpha	A, a	I, ι	Iota	I, i	P, ρ	Rho	R, r
B, β	Beta	B, b	K, κ	Kappa	K, k	Σ, σ	Sigma	S, s
Γ, γ	Gamma	G, g	Λ, λ	Lambda	L, l	T, τ	Tau	T, t
Δ, δ	Delta	D, d	M, μ	Mu	M, m	Υ, υ	Upsilon	U, u
E, ε	Epsilon	E, e	N, ν	Nu	N, n	Φ, φ	Phi	Ph
Z, ζ	Zeta	Z, ζ	Ξ, ξ	Xi	X, x	X, χ	Chi	Ch
H, η	Eta	E, e	O, o	Omicron	O, o	Ψ, ψ	Psi	Ps
Θ, θ	Theta	Th	Π, π	Pi	P, p	Ω, ω	Omega	O, o

Alloys Index

5•4/Index

Subject Index